# ILLUSTRATED HANDBOOK

OF

# PHYSICAL-CHEMICAL PROPERTIES

AND

# ENVIRONMENTAL FATE

FOR

# ORGANIC CHEMICALS

**Volume III**

Volatile Organic Chemicals

Donald Mackay
Wan Ying Shiu
Kuo Ching Ma

Library of Congress Cataloging-in-Publication Data
(Revised for v. 3)

Mackay, Donald, Ph.D.
Illustrated handbook of physical-chemical properties and environmental fate for organic chemicals.

Includes bibliographical references.
Contents: v. 1. Monoaromatic hydrocarbons, chlorobenzenes, and PCBs -- v. 2. Polynuclear aromatic hydrocarbons, polychlorinated dioxins, and dibenzofurans -- v. 3. Volatile organic chemicals.
1. Organic compounds -- Environmental aspects -- Handbooks, manuals, etc. 2. Environmental chemistry -- Handbooks, manuals, etc. I. Shiu, Wan Ying. II. Ma, Kuo Ching. III. Title.
~~TD196.073M32 1992~~ 628.5'2 91-33888



ISBN 0-87371-513-6 (v. 1)
ISBN 0-83731-583-7 (v. 2)
ISBN 0-83731-973-5 (v. 3)



LEWIS PUBLISHERS
121 South Main Street, Chelsea, Michigan 48118

Printed in the United States of America 1 2 3 4 5 6 7 8 9 0
Printed on acid-free paper

# PREFACE

This series of Handbooks brings together physical-chemical data for similarly structured groups of chemical substances which influence their fate in the multimedia environment of air, water, soils, sediments and their resident biota. The task of assessing chemical fate locally, regionally and globally is complicated by the large (and increasing) number of chemicals of potential concern, by uncertainties in their physical-chemical properties, and by lack of knowledge of prevailing environmental conditions such as temperature, pH and deposition rates of solid matter from the atmosphere to water, or from water to bottom sediments. Further, reported values of properties such as solubility are often in conflict. Some are measured accurately, some approximately and some are estimated by various correlation schemes from molecular structure. In some cases, units or chemical identity are wrongly reported. The user of such data thus has the difficult task of selecting the "best" or "right" values. There is justifiable concern that the resulting deductions of environmental fate may be in substantial error. For example, the potential for evaporation may be greatly underestimated if an erroneously low vapor pressure is selected.

To assist the environmental scientist and engineer in such assessments, this Handbook contains compilations of physical-chemical property data for series of chemicals such as, in this case, the volatile organic chemicals or VOCs. It has long been recognized that within such series, properties vary systematically with molecular size, thus providing guidance about the properties of one substance from those of its homologs. Plots of these systematic property variations are provided to check the reported data and provide an opportunity for interpolation and even modest extrapolation to estimate unreported properties of other homologs. Most handbooks treat chemicals only on an individual basis, and do not contain this feature of chemical-to-chemical comparison which can be valuable for identifying errors and estimating properties.

The data are taken a stage further and used to estimate likely environmental partitioning tendencies, i.e., how the chemical is likely to become distributed between the various media which comprise our biosphere. The results are presented numerically and pictorially to provide a visual impression of likely environmental behavior. This will be of interest to those assessing environmental fate by confirming the general fate characteristics or behavior profile. It is, of course, only possible here to assess fate in a "typical" or "generic" or "evaluative" environment; thus, no claim is made that a chemical will behave in this manner in all situations, but this assessment should reveal the broad characteristics of behavior. These evaluative fate assessments are generated using simple fugacity models which flow naturally from the compilations of data on physical-chemical properties of relevant chemicals.

It is hoped that this series of Handbooks will be of value to environmental scientists and engineers and to students and teachers of environmental science. Its aim is to contribute to better assessments of chemical fate in our multimedia environment by serving as a reference source for environmentally relevant physical-chemical property data of classes of chemicals and by illustrating the likely behavior of these chemicals as they migrate throughout our biosphere.

Donald Mackay, born and educated in Scotland, received his degrees in Chemical Engineering from The University of Glasgow. After a period of time in the petrochemical industry he joined The University of Toronto, where he is now a Professor in the Department of Chemical Engineering and Applied Chemistry, and in the Institute for Environmental Studies. Professor Mackay's primary research is the study of organic environmental contaminants, their sources, fates, effects, and control, and particularly in understanding and modeling their behavior with the aid of the fugacity concept. His work has focused especially on the Great Lakes Basin and on cold northern climates.

Wan Ying Shiu is a Research Associate in the Department of Chemical Engineering and Applied Chemistry, and the Institute for Environmental Studies, University of Toronto. She received her Ph.D. in Physical Chemistry from the Department of Chemistry, University of Toronto, M.Sc. in Physical Chemistry from St. Francis Xavier University and B.Sc. in Chemistry from Hong Kong Baptist College. Her research interest is in the area of physical-chemical properties and thermodynamics for organic chemicals of environmental concern.

Kuo Ching Ma obtained his Ph.D. from The Florida State University, M.Sc. from The University of Saskatchewan and B.Sc. from The National Taiwan University; all in Physical Chemistry. After working many years in the Aerospace, Battery Research, Fine Chemicals and Metal Finishing industries in Canada as Research Scientist, Technical Supervisor/Director, he is now dedicating his time and interests to environmental research.

## TABLE OF CONTENTS

**1. Introduction** . . . 1
1.1 The Incentive . . . 1
1.2 Physical-Chemical Properties . . . 3
1.2.1 The Key Physical-Chemical Properties . . . 3
1.2.2 Experimental Methods . . . 6
1.3 Quantitative Structure-Property Relationships (QSPRs) . . . 11
1.3.1 Objectives . . . 11
1.3.2 Examples . . . 13
1.4 Fate Models . . . 18
1.4.1 Evaluative Environmental Calculations . . . 18
1.4.2 Level I Fugacity Calculation . . . 18
1.4.3 Level II Fugacity Calculation . . . 22
1.4.4 Level III Fugacity Calculation . . . 23
1.5 Data Sources and Presentations . . . 33
1.5.1 Data Sources . . . 33
1.5.2 Data Format . . . 33
1.5.3 Explanation of Data Presentations . . . 34
1.5.4 Evaluative Calculations . . . 37
1.6 References . . . 52
**2. Hydrocarbons** . . . 67
2.1 List of Chemicals and Data Compilations . . . 70
2.2 Summary Tables and QSPR Plots . . . 276
2.3 Illustrative Fugacity Calculations: Levels I, II and III . . . 299
2.4 Commentary on the Physical-Chemical Properties and Environmental Fate . . 368
2.5 References . . . 372
**3. Halogenated hydrocarbons** . . . 392
3.1 List of Chemicals and Data Compilations . . . 394
3.2 Summary Tables and QSPR Plots . . . 616
3.3 Illustrative Fugacity Calculations: Levels I, II and III . . . 633
3.4 Commentary on the Physical-Chemical Properties and Environmental Fate . . 714
3.5 References . . . 717
**4. Ethers** . . . 747
4.1 List of Chemicals and Data Compilations . . . 749
4.2 Summary Tables and QSPR Plots . . . 820
4.3 Illustrative Fugacity Calculations: Levels I, II and III . . . 829
4.4 Commentary on the Physical-Chemical Properties and Environmental Fate . . 874
4.5 References . . . 876
**List of Symbols and Abbreviations** . . . 886
**Appendices** . . . 891
A1 Listing of BASIC Program for Fugacity Calculations . . . 892
A2 Listing of Spreadsheet Program for Lotus 123® Calculations . . . 910

# ILLUSTRATED HANDBOOK

OF

# PHYSICAL-CHEMICAL PROPERTIES

AND

# ENVIRONMENTAL FATE

FOR

# ORGANIC CHEMICALS

**Volume III**

Volatile Organic Chemicals

# 1. INTRODUCTION

## 1.1 THE INCENTIVE

It is alleged that there are some 60,000 chemicals in current commercial production, with approximately 1000 being added each year. Most are organic chemicals. Of these, perhaps 500 are of environmental concern because of their presence in detectable quantities in various components of the environment, their toxicity, their tendency to bioaccumulate, or their persistence. A view is emerging that some of these chemicals are of such extreme environmental concern that all production and use should be ceased, i.e., as a global society we should elect not to synthesize or use these chemicals. They should be "sunsetted". PCBs, "dioxins" and freons are examples. A second group consists of chemicals which are of concern because they are used or discharged in large quantities, or they are toxic or persistent. They are, however, of sufficient value to society that their continued use is justified, but only under conditions in which we fully understand their sources, fate and effects. This understanding is essential if society is to be assured that there are no adverse ecological or human health effects. Other groups of increasingly benign chemicals can presumably be treated with less rigor.

A key feature of this "cradle-to-grave" approach is that society must improve its skills in assessing chemical fate in the environment. We must better understand where chemicals originate, how they migrate in, and between, the various media of air, water, soils, sediments and their biota which comprise our biosphere. We must understand how these chemicals are transformed by chemical and biochemical processes and thus how long they will persist in the environment. We must seek a fuller understanding of the effects which they will have on the multitude of interacting organisms which occupy these media, including ourselves.

It is now clear that the fate of chemicals in the environment is controlled by a combination of two groups of factors. First are the prevailing environmental conditions such as temperatures, flows and accumulations of air, water and solid matter and the composition of these media. Second are the properties of the chemicals which influence partitioning and reaction tendencies, i.e., whether the chemical evaporates or associates with sediments, and how the chemical is eventually destroyed by conversion to other chemical species.

In recent decades there has emerged a discipline within environmental science concerned with increasing our understanding of how chemicals behave in our multimedia environment. It has been termed "chemodynamics". Practitioners of this discipline include scientists and engineers, students and teachers who attempt to measure, assess and predict how this large number of chemicals will behave in laboratory, local, regional and global environments. These individuals need data on physical-chemical and reactivity properties, as well as information on how these properties translate into environmental fate. This Handbook provides a compilation of such data and uses them to estimate the broad features of environmental fate. It does so for classes or groups of chemicals, instead of the usual approach of treating chemicals on an individual basis. This has the advantage that systematic variations in properties with molecular size can be revealed and used to check reported values, interpolate and even extrapolate to other chemicals of similar structure.

With the advent of inexpensive and rapid computation there has been a remarkable growth in interest in this general area of Quantitative Structure-Property Relationships (QSPRs). The ultimate goal is to use information about chemical structure to deduce physical-chemical properties, environmental partitioning and reaction tendencies, and even uptake and effects on biota. The goal is far from being realized, but considerable progress has been made, as is briefly reviewed in a following section. In this Handbook we adopt a simple, and well tried, approach of using molecular structure to deduce a molar volume, which in turn is related to physical-chemical properties. Undoubtedly, other molecular descriptors such as surface area or topological indices have the potential to give more accurate correlations and will be increasingly used in the future, but at this stage we believe that the improvements in accuracy obtained by using these more complex descriptors do not justify the computational effort of generating them, at least for the purposes of routine, general assessments. In some cases, the fundamental causes of the relationships remain obscure.

A major benefit of this simple QSPR analysis is that it reveals likely errors in reported data. Regrettably, the scientific literature contains a great deal of conflicting data with reported values often varying over several orders of magnitude. There are some good, but more not-so-good reasons for this lack of accuracy. Many of these properties are difficult to measure because they involve analyzing very low concentrations of 1 part in $10^9$ or $10^{12}$. For many purposes an approximate value, for example, that a solubility is less than 1 mg/L, is adequate. There has been a mistaken impression that if a vapor pressure is low, as is the case with DDT, it is not important. DDT evaporates appreciably from solution in water, despite its low vapor pressure, because of its low solubility in water. In some cases the units are reported incorrectly or there are uncertainties about temperature or pH. In other cases the chemical is wrongly identified. One aim of this Handbook is to assist the user to identify such problems and provide guidance when selecting appropriate values.

The final aspect of chemical fate treated in this Handbook is the depiction or illustration of likely chemical fate. This is done using a series of multimedia "fugacity" models as is described in a later section. The authors' aim is to convey an impression of likely environmental partitioning and transformation characteristics, i.e., we seek to generate a "behavior profile". A fascinating feature of chemodynamics is that chemicals differ so greatly in their behavior. Some, such as chloroform, evaporate rapidly and are dissipated in the atmosphere. Others, such as DDT, partition into the organic matter of soils and sediments and the lipids of fish, birds and mammals. Phenols tend to remain in water subject to fairly rapid transformation processes such as biodegradation and photolysis. By entering the physical-chemical data into a model of chemical fate in a generic or evaluative environment, it is possible to estimate the likely general features of the chemical's behavior and fate. The output of these calculations is presented numerically and pictorially.

In total, the aim of this series of Handbooks is to provide a useful reference work for those concerned with the assessment of the fate of existing and new chemicals in the environment.

## 1.2 PHYSICAL-CHEMICAL PROPERTIES

### 1.2.1 The key physical-chemical properties

The major differences between behavior profiles of organic chemicals in the environment are attributable to physical-chemical properties. The key properties are believed to be solubility in water, vapor pressure, octanol-water partition coefficient, dissociation constant in water (when relevant) and susceptibility to degrading or transformation reactions. Other essential molecular descriptors are molecular mass and molar volume, with properties such as critical temperature and pressure and molecular area being occasionally useful for specific purposes.

Chemical identity may appear to present a trivial problem, but many chemicals have several names, and subtle differences between isomers (e.g., cis and trans) may be ignored. The most commonly accepted identifiers are the IUPAC name and the Chemical Abstracts System (CAS) number. More recently, methods have been sought of expressing the structure in line notation form so that computer entry of a series of symbols can be used to define a three-dimensional structure. The Wiswesser Line Notation is quite widely used, but it appears that for environmental purposes it will be superceded by the SMILES (Simplified Molecular Identification and Line Entry System, Anderson et al. 1987).

Molecular mass is readily obtained from structure. Also of interest are molecular volume and area, which may be estimated by a variety of methods.

Solubility in water and vapor pressure are both "saturation" properties, i.e., they are measurements of the maximum capacity which a phase has for dissolved chemical. Vapor pressure P (Pa) can be viewed as a "solubility in air", the corresponding concentration C ($mol/m^3$) being P/RT where R is the ideal gas constant (8.314 J/mol • K) and T is absolute temperature (K). Although most chemicals are present in the environment at concentrations well below saturation, these concentrations are useful for estimating air-water partition coefficients as ratios of saturation values. It is usually assumed that the same partition coefficient applies at lower sub-saturation concentrations. Vapor pressure and solubility thus provide estimates of air-water partition coefficients $K_{AW}$ or Henry's law constants H ($Pa \cdot m^3/mol$), and thus the relative air-water partitioning tendency.

The octanol-water partition coefficient $K_{OW}$ provides a direct estimate of hydrophobicity or of partitioning tendency from water to organic media such as lipids, waxes and natural organic matter such as humin or humic acid. It is invaluable as a method of estimating $K_{OC}$, the organic carbon-water partition coefficient, the usual correlation invoked being that of Karickhoff (1981)

$$K_{OC} = 0.41\ K_{OW}$$

It is also used to estimate fish-water bioconcentration factors $K_B$ or BCF using a correlation similar to that of Mackay (1982)

$$K_B = 0.05\ K_{OW}$$

where the term 0.05 corresponds to a 5% lipid content of the fish.

For ionizing chemicals it is essential to quantify the extent of ionization as a function of pH using the dissociation constant pKa. The parent and ionic forms behave and partition quite differently, thus pH and the presence of other ions may profoundly affect chemical fate.

Characterization of chemical reactivity presents a severe problem in Handbooks. Whereas radioisotopes have fixed half-lives, the half-life of a chemical in the environment depends not only on the intrinsic properties of the chemical, but also on the nature of the surrounding environment. Factors such as sunlight intensity, hydroxyl radical concentration and the nature of the microbial community, as well as temperature, affect the chemical's half-life so it is impossible (and misleading) to document a single reliable half-life. The compilation by Howard et al. (1991) provides an excellent review of the existing literature for a large number of chemicals. It is widely used as a source document in this work. The best that can be done is to suggest a semi-quantitative classification of half-lives into groups, assuming average environmental conditions to apply. Obviously, a different class will generally apply in air and bottom sediment. In this compilation we use the following class ranges for chemical reactivity in a single medium such as water.

| Class | Mean half-life (hours) | Range (hours) |
|---|---|---|
| 1 | 5 | < 10 |
| 2 | 17 (~ 1 day) | 10-30 |
| 3 | 55 (~ 2 days) | 30-100 |
| 4 | 170 (~ 1 week) | 100-300 |
| 5 | 550 (~ 3 weeks) | 300-1,000 |
| 6 | 1700 (~ 2 months) | 1,000-3,000 |
| 7 | 5500 (~ 8 months) | 3,000-10,000 |
| 8 | 17000 (~ 2 years) | 10,000-30,000 |
| 9 | 55000 (~ 6 years) | > 30,000 |

These times are divided logarithmically with a factor of approximately 3 between adjacent classes. With the present state of knowledge it is probably misleading to divide the classes into finer groupings; indeed, a single chemical may experience half-lives ranging over three classes, depending on season.

When compiling suggested reactivity classes the authors have examined the available information as reaction rates of the chemical in each medium, by all relevant processes. These were expressed as an overall half-life for transformation. For example, a chemical may be subject to biodegradation with a half-life of 20 days (rate constant 0.0014 $h^{-1}$) and a photolysis with a rate constant of 0.0011 $h^{-1}$ (half-life 630 hours). The overall rate constant is thus 0.0025 $h^{-1}$ and the half-life 277 h or 12 days. Data for homologous chemicals were also compiled, and insights into the reactivity of various functional groups considered. In most cases a single reaction class was assigned to the series; in this case, class 4 with a mean half-life of 170 hours was chosen. It must be appreciated that this chemical could fall into class 3 in summer and class 5 in winter. These half-lives must be used with caution and it is wise to test the implications of selecting longer and shorter half-lives.

The volatile organic chemicals considered in this volume tend to have high vapor pressures. Most are sparingly soluble in water; however, some unsaturated compounds are fairly soluble and some ethers are very soluble, in some cases being miscible with water. These compounds tend to have relatively high Henry's law constants, i.e., $10^3$ - $10^5$ Pa·m$^3$/mol for hydrocarbons, 10 - $10^5$ Pa·m$^3$/mol for halogenated hydrocarbons and 0.1 - 100 Pa·m$^3$/mol for ethers. Although at first sight it might appear appropriate to classify volatile organic chemicals (VOCs) on the basis of vapor pressure or boiling point, it transpires that Henry's law constant is a better descriptor. H, of course, is also affected by solubility in water. When VOCs are discharged into the environment, they are likely to partition or evaporate fairly rapidly into the atmosphere. The major degradation and removal processes thus occur in the air phase; hence the atmospheric photochemistry or photodegradation of this type of volatile chemical is very important.

Extensive research has been conducted into the atmospheric chemistry of the VOCs because of air quality concerns. Recently, Atkinson and coworkers (1984, 1985, 1987, 1990, 1991), Altshuller (1980, 1991) and Sabljic and Güsten (1990) have reviewed the photochemistry of many organic chemicals of environmental interest for their gas phase reactions with hydroxyl radicals (OH), ozone ($O_3$) and nitrate radicals ($NO_3$) and have provided detailed information on reaction rate constants and experimental conditions, which allowed the estimation of atmospheric lifetimes. Klöpffer (1991) has estimated the atmospheric lifetimes for the reaction with OH radicals to range from 1 hour to 130 years, based on these reaction rate constants and an assumed constant concentration of OH radicals in air. As Atkinson (1985) has pointed out, the gas-phase reactions with OH radicals are the major tropospheric loss process for the alkanes, haloalkanes, the lower alkenes, the aromatic hydrocarbons, and a majority of the oxygen-containing organics. In addition, photooxidation reactions with $O_3$ and $NO_3$ radicals can result in transformation of these compounds. The night-time reaction with $NO_3$ radicals may also be important (Atkinson and Carter 1984, Sabljic and Güsten 1990).

There are fewer studies on direct or indirect photochemical degradation in the water phase; however, Klöpffer (1991) had pointed out that the rate constant or lifetimes derived from these studies "is valid only for the top layer or surface waters". Mill (1982, 1988, 1993) and Mill and Mabey (1985) have estimated half-lives of various VOCs in aqueous solutions from their reaction rate constants with singlet oxygen, as well as photooxidation with OH and peroxy radicals. Buxton et al. (1988) gave a critical review of rate constants for reactions with hydrated electrons, hydrogen atoms and hydroxyl radicals in aqueous solutions. Mabey and Mill (1978) also reviewed the hydrolysis of organic chemicals in water under environmental conditions.

Recently, Ellington and coworkers (1987a,b, 1988, 1989) also reported the hydrolysis rate constants in aqueous solutions for a variety of organic chemicals.

Other processes, such as biodegradation, may be important, especially in the subsurface environment where there is widespread concern about groundwater contamination by VOCs.

In most cases a review of the literature suggested that reaction rates in water by chemical processes are 1 to 2 orders of magnitude slower than in air, but with biodegradation often being significant, especially for hydrocarbons. Generally, the water half-life class is three more than that in air, i.e., a factor of about 30 slower. Chemicals in soils tend to be shielded from photolytic processes, and they are less bioavailable, thus the authors have frequently assigned a reactivity class to soil of one more than that for water. Bottom sediments are assigned an additional class to that of soils largely on the basis that there is little or no photolysis, there may be lack of oxygen, and the intimate sorption to sediments renders the chemicals less bioavailable.

The chemical reactivity of VOCs is a topic which continues to be the subject of extensive research; thus there is often detailed, more recent information about the fate of chemical species which is of particular relevance to air or water quality. The reader is thus urged to consult the original and recent references because it is impossible in a volume such as this, which considers the entire multimedia picture, to treat this subject in the detail it deserves.

In summary, when selecting physical-chemical properties or reactivity classes the authors have been guided by:

(1) the age of the data and acknowledgment of previous conflicting or supporting values,
(2) the method of determination,
(3) the perception of the objectives of the authors, not necessarily as an indication of competence, but often as an indication of the need of the authors for accurate values, and
(4) the reported values for structurally similar, or homologous compounds.

In this Handbook we have used these considerations as well as information derived from the QSPR analyses.

It is appropriate, therefore, to review briefly the experimental methods which are commonly used for property determinations and comment on their accuracy.

### 1.2.2. Experimental methods

**Solubility in water**

The conventional method of preparing saturated solutions for the determination of solubility is batch equilibration. An excess amount of solute chemical is added to water and equilibrium is achieved by shaking gently (generally referred as the "shake flask method") or slow stirring with a magnetic stirrer. The aim is to prevent formation of emulsions or suspensions and thus avoid extra experimental procedures such as filtration or centrifuging which may be required to ensure that a true solution is obtained. Experimental difficulties can still occur because of the formation of emulsion or microcrystal suspensions with the sparingly soluble chemicals such as higher normal alkanes and polycyclic aromatic hydrocarbons (PAHs). An alternative approach is to coat a thin layer of the chemical on the surface of the equilibration flask before water is added. An accurate "generator column" method has also been developed (Weil et al. 1974, May et al. 1978a,b) in which a column is packed with an inert solid support, such as glass beads or Chromosorb, and then coated with the solute chemical. Water is pumped through the column at a controlled, known flow rate to achieve saturation.

The method of concentration measurement of the saturated solution depends on the solute solubility and its chemical properties. Some common methods used for solubility measurement are listed below.

1. Gravimetric or volumetric methods (Booth and Everson 1948)
   An excess amount of solid compound is added to a flask containing water to achieve saturation solution by shaking, stirring, centrifuging until the water is saturated with solute and undissolved solid or liquid residue appears, often as a cloudy phase. For liquids, successive known amounts of solute may be added to water and allowed to reach equilibrium, and the volume of excess undissolved solute is measured.
2. Instrumental methods
   a. UV spectrometry (Andrews and Keffer 1950, Bohon and Claussen 1951, Yalkowsky et al. 1976);
   b. Gas chromatographic analysis with FID, ECD or other detectors (McAuliffe 1968, Mackay et al. 1975, Chiou et al. 1982);
   c. Fluorescence spectrophotometry (Mackay and Shiu 1977);
   d. Interferometry (Gross and Saylor 1931);
   e. High-pressure liquid chromatography (HPLC) with R.I., UV or fluorescence detection (May et al. 1978a,b, Wasik et al. 1983, Shiu et al. 1988, Doucette and Andren 1988a);
   f. Liquid phase elution chromatography (Schwarz 1980, Schwarz and Miller 1980);
   g. Nephelometric methods (Davis and Parke 1942, Davis et al. 1942, Hollifield 1979);
   h. Radiotracer or liquid scintillation counting (LSC) method (Banerjee et al. 1980, Lo et al. 1986).

For most organic chemicals the solubility is reported at a defined temperature in distilled water. For substances which ionize (e.g. phenols, carboxylic acids and amines) it is essential to report the pH of the determination because the extent of ionization affects the solubility. It is common to maintain the desired pH by buffering with an appropriate electrolyte mixure. This raises the complication that the presence of electrolytes modifies the water structure and changes the solubility. The effect is usually "salting-out". For example, many hydrocarbons have solubilities in seawater about 75% of their solubilities in distilled water. Care must thus be taken to interpret and use reported data properly when electrolytes are present.

The most common problem encountered with reported data is inaccuracy associated with very low solubilities, i.e., those less than 1.0 mg/L. Such solutions are difficult to prepare, handle and analyze, and reported data are often contain appreciable errors.

**Octanol-water partition coefficient $K_{OW}$**

The experimental approaches are similar to those for solubility, i.e., employing shake flask or generator-column techniques. Concentrations in both the water and octanol phases may be determined after equilibration. Both phases can then be analyzed by the instrumental methods discussed above and the partition coefficient is calculated from the concentration ratio $C_O/C_W$. This is actually the ratio of solute concentration in octanol saturated with water to that in water saturated with octanol.

As with solubility, $K_{OW}$ is a function of the presence of electrolytes and for dissociating chemicals it is a function of pH. Accurate values can generally be measured up to about $10^6$, but accurate measurement beyond this requires meticulous technique. A common problem is that the presence of small quantities of emulsified octanol in the water phase could create a high concentration of chemical in that emulsion which would cause an erroneously high apparent water phase concentration.

Considerable success has been achieved by calculating $K_{OW}$ from molecular structure; thus, there has been a tendency to calculate $K_{OW}$ rather than measure it, especially for "difficult" hydrophobic chemicals. These calculations are, in some cases, extrapolations and can be in serious error. Any calculated log $K_{OW}$ value above 7 should be regarded as suspect, and any experimental or calculated value above 8 should be treated with extreme caution.

Details of experimental methods are described by Fujita et al. (1964), Leo et al. (1971); Hansch and Leo (1979), Rekker (1977), Chiou et al. (1977), Miller et al. (1984), Bowman and Sans (1983), Woodburn et al. (1984), Doucette and Andren (1987), and De Bruijn et al. (1989).

**Vapor pressure**

In principle, the determination of vapor pressure involves the measurement of the saturation concentration or pressure of the solute in a gas phase. The most reliable methods involve direct determination of these concentrations, but convenient indirect methods are also available based on evaporation rate measurement or chromatographic retention times. Some methods and approaches are listed below.

a. Direct measurement by use of pressure gauges: diaphragm gauge (Ambrose et al. 1975), Rodebush gauge (Sears and Hopke 1947), inclined-piston gauge (Osborn & Douslin 1975);
b. Comparative ebulliometry (Ambrose 1981);
c. Effusion methods, torsion and weight-loss (Balson 1947, Bradley and Cleasby 1953, Hamaker and Kerlinger 1969, De Kruif 1980);
d. Gas saturation or transpiration methods (Spencer and Cliath 1970, 1972, Sinke 1974, Macknick and Prausnitz 1979, Westcott et al. 1981, Rodorf 1985a,b, 1986);
e. Dynamic coupled-column liquid chromatographic method - a gas saturation method (Sonnefeld et al. 1983);
f. Calculation from evaporation rates and vapor pressures of reference compound (Gückel et al. 1974, 1982, Dobbs and Grant 1980, Dobbs and Cull 1982);
g. Calculation from GC retention time data (Hamilton 1980, Westcott and Bidleman 1982, Bidleman 1984, Kim et al. 1984, Foreman and Bidleman 1985, Burkhard et al. 1985a, Hinckley et al. 1990).

The greatest difficulty and uncertainty arises when determining the vapor pressure of chemicals of low volatility, i.e., those with vapor pressures below 1.0 Pa. Vapor pressures are strongly dependent on temperature, thus accurate temperature control is essential. Data are often regressed against temperature and reported as Antoine or Clapeyron constants. Care must be taken when using the Antoine or other equations to extrapolate data beyond the temperature range specified. It must be clear if the data apply to the solid or liquid phase of the chemical.

**Henry's law constant**

The Henry's law constant is essentially an air-water partition coefficient which can be determined by measurement of solute concentrations in both phases. This raises the difficulty of accurate analytical determination in two very different media which require different techniques. Accordingly, some effort has been devoted to devising techniques in which concentrations are measured in only one phase and the other concentration is deduced by a mass balance. These methods are generally more accurate. The principal difficulty arises with hydrophobic, low-volatility chemicals which can establish only very small concentrations in both phases.

Henry's law constant can be regarded as a ratio of vapor pressure to solubility, thus it is subject to the same effects which electrolytes have on solubility and temperature has on both properties. Some methods are as follows:

a. Multiple equilibration method (McAuliffe 1971, Munz & Roberts 1987);
b. Equilibrium batch stripping (Mackay et al. 1979, Dunnivant et al. 1988);
c. GC-determined distribution coefficients (Leighton & Calo 1981);
d. GC analysis of both air/water phases (Vejrosta et al. 1982, Jonsson et al. 1982);
e. EPICS (Equilibrium Partioning In Closed Systems) method (Lincoff and Gossett 1984; Gossett 1987, Ashworth et al. 1988);
f. Wetted-wall column (Fendinger and Glotfelty 1988, 1990);
g. Headspace analyses (Hussam and Carr 1985);

h. Calculation from vapor pressure and solubility (Mackay and Shiu 1981);
i. GC retention volume/time determined activity coefficient at infinite dilution $\gamma^\infty$ (Karger et al. 1971, Sugiyama et al. 1975, Tse et al. 1992).

When using vapor pressure and solubility data, it is essential to ensure that both properties apply to the same chemical phase, i.e., both are of the liquid, or of the solid. Occasionally, a solubility is of a solid while a vapor pressure is extrapolated from higher temperature liquid phase data.

## 1.3 QUANTITATIVE-STRUCTURE-PROPERTY RELATIONSHIPS (QSPRs)

### 1.3.1 Objectives

Because of the large number of chemicals of actual and potential concern, the difficulties and cost of experimental determinations, and scientific interest in elucidating the fundamental molecular determinants of physical-chemical properties, a considerable effort has been devoted to generating quantitative structure-activity relationships (QSARs). This concept of structure-property relationships or structure-activity relationships is based on observations of linear free-energy relationships, and usually takes the form of a plot or regression of the property of interest as a function of an appropriate molecular descriptor which can be obtained from merely a knowledge of molecular structure.

Such relationships have been applied to solubility, vapor pressure, $K_{OW}$, Henry's law constant, reactivities, bioconcentration data and several other environmentally relevant partition coefficients. Of particular value are relationships involving various manifestations of toxicity, but these are beyond the scope of this handbook. These relationships are valuable because they permit values to be checked for "reasonableness" and (with some caution) interpolation is possible to estimate undetermined values. They may be used (with extreme caution!) for extrapolation.

A large number of descriptors have been, and are being, proposed and tested. Dearden (1990) and the compilation by Karcher and Devillers (1990) and Hermens and Opperhuizen (1991) give comprehensive accounts of descriptors and their applications.

Among the most commonly used molecular descriptors are molecular weight and volume, the number of specific atoms (e.g., carbon or chlorine), surface areas (which may be defined in various ways), refractivity, parachor, steric parameters, connectivities and various topological parameters. Several quantum chemical parameters can be calculated from molecular orbital calculations including charge, electron density and superdelocalizability.

It is likely that existing and new descriptors will be continued to be tested, and that eventually a generally preferred set of readily accessible parameters will be adopted of routine use for correlating purposes. From the viewpoint of developing quantitative correlations it is very desirable to seek a linear relationship between descriptor and property, but a nonlinear or curvilinear relationship is quite adequate for illustrating relationships and interpolating purposes. In this handbook we have elected to use the simple descriptor of molar volume at the normal boiling point as estimated by the LeBas method (Reid et al. 1987). This parameter is very easily calculated and proves to be adequate for the present purposes of plotting property versus relationship without seeking linearity.

The LeBas method is based on a summation of atomic volumes with adjustment for the volume decrease arising from ring formation. The full method is described by Reid et al. (1987), but for the purposes of this compilation, the volumes and rules as listed in Table 1.1 are used.

Table 1.1 LeBas Molar Volume

| | increment, $cm^3$ /mol |
|---|---|
| carbon | 14.8 |
| hydrogen | 3.7 |
| oxygen | 7.4 |
| in methyl esters and ethers | 9.1 |
| in ethyl esters and ethers | 9.9 |
| join to S, P, or N | 8.3 |
| nitrogen | |
| doubly bonded | 15.6 |
| in primary amines | 10.5 |
| in secondary amines | 12.0 |
| bromine | 27.0 |
| chlorine | 24.6 |
| fluorine | 8.7 |
| iodine | 37.0 |
| sulfur | 25.6 |
| ring | |
| three-membered | -6.0 |
| four-membered | -8.5 |
| five-membered | -11.5 |
| six-membered | -15.0 |
| naphthalene | -30.0 |
| anthracene | -47.5 |

Example: The experimental molar volume of chlorobenzene 115 $cm^3$/mol (Reid et al. 1987). From the above rules, the LeBas molar volume for chlorobenzene ($C_6H_5Cl$) is:

$$V = 6 \times 14.8 + 5 \times 3.7 + 24.6 - 15 = 117 \text{ cm}^3/\text{mol}$$

Accordingly, plots are presented at the end of each chapter for solubility, vapor pressure, $K_{OW}$, and Henry's law constant versus LeBas molar volume.

A complication arises in that two of these properties (solubility and vapor pressure) are dependent on whether the solute is in the liquid or solid state. Solid solutes have lower solubilities and vapor pressures than they would have if they had been liquids. The ratio of the (actual) solid to the (hypothetical subcooled) liquid solubility or vapor pressure is termed the fugacity ratio and can be estimated from the melting point and the entropy of fusion $\Delta S_{fus}$, as discussed by Mackay and Shiu (1981). For solid solutes, the correct property to plot is the calculated or extrapolated subcooled liquid solubility. This is calculated in this handbook using the relationship suggested by Yalkowsky (1979) which implies an entropy of fusion of 56 J/mol K or 13.5 cal/mol • K

$$C^S_S/C^S_L = P^S_S/P^S_L = \exp\{6.79(1 - T_M/T)\}$$

where $C^S$ is solubility, $P^S$ is vapor pressure, subcripts S and L refer to solid and liquid phases, $T_M$ is melting point and T is the system temperature, both in absolute (K) units. The fugacity ratio is given in the data tables at 25 °C, the usual temperature at which physical-chemical property data are reported. For liquids, the fugacity ratio is 1.0.

### 1.3.2 Examples

Recently, there have been efforts to extend the long-established concept of Quantitative Structure-Activity Relationships (QSARs) to Quantitative Structure-Property Relationships (QSPRs) to compute all relevant environmental physical-chemical properties (such as aqueous solubility, vapor pressure, octanol-water partition coefficient, Henry's law constant, bioconcentration factor (BCF)), sorption coefficient and envrionmental reaction rate constants from molecular structure. Examples are Burkhard (1984) and Burkhard et al. (1985a) who calculated solubility, vapor pressure, Henry's law constant, $K_{OW}$ and $K_{OC}$ for all PCB congeners. Hawker and Connell (1988) also calculated log $K_{OW}$; Abramowitz and Yalkowsky (1990) calculated melting point and solubility for all PCB congeners based on the correlation with total surface area (planar TSAs). Doucette and Andren (1988b) used six molecular descriptors to compute the $K_{OW}$ of some chlorobenzenes, PCBs and PCDDs. Mailhot and Peters (1988) employed seven molecular descriptors to compute physical-chemical properties of some 300 compounds. Isnard and Lambert (1988, 1989) correlated solubility, $K_{OW}$ and BCF for a large number of organic chemicals. Nirmalakhandan and Speece (1988a,b, 1989) used molecular connectivity indices to predict aqueous solubility and Henry's law constants for 300 compounds over 12 logarithmic units in solubility. Kamlet and coworkers (1987, 1988) have developed the solvatochromic parameters with the intrinsic molar volume to predict solubility, log $K_{OW}$ and toxicity of organic chemicals. Warne et al. (1990) correlated solubility and $K_{OW}$ for lipophilic organic compound with 39 molecular descriptors and physical-chemical properties. Atkinson (1987, 1988) has used the structure-activity relationship (SAR) to estimated gas-phase reaction rate constants of hydroxyl radicals for organic chemicals. Mabey et al. (1984) have reviewed the estimation methods from SAR correlation for reaction rate constants and physical-chemical properties in environmental fate assessment. Other correlations are reviewed by Lyman et al. (1982) and Yalkowsky and Banerjee (1992). As Dearden (1990) has pointed out, "new parameters are continually being devised and tested, although the necessity of that may be questioned, given the vast number already available". It must be emphasized, however, that regardless of how accurate these predicted or estimated properties are claimed to be, utimately they have to be confirmed or verified by experimental measurement.

A fundamental problem encountered in these correlations is that the molecular descriptors can be calculated with relatively high precision, usually within a few percent. The accuracy may not always be high, but for empirical correlation purposes precision is more important than accuracy. The precision and accuracy of the experimental data are often poor, frequently ranging over a factor of two or more. Certain isomers may yield identical descriptors, but have different properties. There is thus an inherent limit to the applicability of QSPRs imposed by the quality of the experimental data, and further efforts to improve descriptors, while interesting and potentially useful, are unlikely to yield demonstrably improved QSPRs.

For correlation of **solubility**, the correct thermodynamic quantities for correlation are the activity coefficient $\gamma$, or the excess Gibbs free energy $\Delta G$, as discussed by Pierotti et al. (1959) and Tsonopoulos and Prausnitz (1971). Examples of such correlations are given below.

1. Carbon number or carbon plus chlorine number (Tsonopoulos and Prausnitz 1971, Mackay and Shiu 1977);
2. Molar volume $cm^3/mol$
   a. Liquid molar volume - from density (McAuliffe 1966, Lande and Banerjee 1981, Chiou et al. 1982, Abernethy et al. 1988, Wang et al. 1992);
   b. Molar volume by additive group contribution method, e.g., LeBas method, Schroeder method (Reid et al. 1987, Miller et al. 1985);
   c. Intrinsic molar volume, $V_I$, $cm^3/mol$ - from van der Waals radius with solvatochromic parameters $\alpha$ and $\beta$ (Leahy 1986, Kamlet et al. 1987, 1988);
   d. Characteristic molecular volume, $m^3/mol$ (McGowan and Mellors 1986);
3. Group contribution method (Irmann 1965, Koremnan et al. 1971, Polak and Lu 1973);
4. Molecular volume - $Å^3/mol$ (cubic Angstrom per mole)
   a. van der Waals volume (Bondi 1964);
   b. Total Molecular Volume (TMV) (Pearlman et al. 1984, Pearlman 1986);
5. Total Surface Area (TSA) - $Å^2/mol$ (Hermann 1971, Amidon et al. 1975, Yalkowsky and Valvani 1976, Yalkowsky et al. 1979, Pearlman 1986, Andren et al. 1987, Hawker and Connell 1988, Dunnivant et al. 1992);
6. Molecular Connectivity indices, $\chi$ (Kier and Hall 1976, Andren et al. 1987, Nirmalakhandan and Speece 1988b, 1989);
7. Boiling point (Almgren et al. 1979);
8. Melting point (Amidon and Williams 1982);
9. Melting point and TSA (Abramowitz and Yalkowsky 1990);
10. High Pressure Liquid Chromatography (HPLC) - retention data (Locke 1974, Whitehouse & Cooke 1982, Brodsky and Ballschmiter 1988);
11. Adsorbability index (AI) (Okouchi et al. 1992);
12. Fragment solubility constants (Wakita et al. 1986).

Several workers have explored the linear relationship between octanol-water partition coefficient and solubility as means of estimating solubility.

Hansch et al. (1968) established the linear free-energy relationship between aqueous and octanol-water partition of organic liquid. Others, such as Tulp and Hutzinger (1978), Yalkowsky et al. (1979), Mackay et al. (1980), Banerjee et al. (1980), Chiou et al. (1982), Bowman and Sans (1983), Miller et al. (1985), Andren et al. (1987) and Andren and Doucette (1988b) have all presented similar but modified relationships.

The UNIFAC (UNIQUAC Functional Group Activity Coefficient) group contribution (Fredenslund et al. 1975, Kikic et al. 1980, Magnussen et al. 1980, Gmehling et al. 1982 and Hansen et al. 1991) is widely used for predicting the activity coefficient in nonelectrolyte liquid mixtures by using group-interaction parameters. This method has been used by Kabadi and Danner (1979), Banerjee (1985), Arbuckle (1983, 1986), Banerjee and Howard (1988) and Al-Sahhaf (1989) for predicting solubility (as a function of the infinite dilution acitivity coefficient, $\gamma^{\infty}$) in aqueous systems. Its performance is reviewed by Yalkowsky and Banerjee (1992).

HPLC retention time data have been used as a psuedo-molecular descriptor by Whitehouse and Cooke (1982), Hafkenscheid and Tomlinson (1981), Tomlinson and Hafkenscheid (1986) and Swann et al. (1983).

The **octanol-water partition coefficient** $K_{OW}$ is widely used as a descriptor of hydrophobicity. Variation in $K_{OW}$ is primarily attributable to variation in activity coefficient in the aqueous phase (Miller et al. 1985); thus, the same correlations used for solubility in water are applicable to $K_{OW}$. Most widely used is the Hansch-Leo compilation of data (Leo et al. 1971, Hansch and Leo 1979) and related predictive methods. Examples of $K_{OW}$ correlations are

I. Molecular descriptors
  1. Molar volumes: LeBas method; from density; intrinsic molar volume; characteristic molecular volume (Abernethy et al. 1988, Chiou 1985, Kamlet et al. 1988, McGowan and Mellors 1986);
  2. TMV (De Bruijn and Hermens 1990);
  3. TSA (Yalkowsky et al. 1979, Yalkowsky et al. 1983, Hawker and Connell 1988);
  4. Molecular connectivity indices (Doucette and Andren 1988b);
  5. Molecular weight (Doucette and Andren 1988b).

II. Group contribution methods
  1. $\pi$-constant or hydrophobic substituent method (Hansch et al. 1968, Hansch & Leo 1979, Doucette and Andren 1988b);
  2. Fragmental constants or f-constant (Rekker 1977, Yalkowsky et al. 1983);
  3. Hansch & Leo's f-constant (Hansch & Leo 1979; Doucette and Andren 1988b).

III. From solubility - $K_{OW}$ relationship

IV. HPLC retention data
  1. HPLC-k' capacity factor (Könemann et al. 1979, McDuffie 1981);
  2. HPLC-RT retention time (Veith et al. 1979, Rappaport and Eisenreich 1984, Doucette and Andren 1988b);
  3. HPLC-RV retention volume (Garst 1984);
  4. HPLC-RT/MS HPLC retention time with mass spectrometry (Burkhard et al. 1985c).

V. Reversed-phase thin-layer chromatography (TLC) (Bruggeman et al. 1982).

VI. Molar refractivity (Yoshida et al. 1983).

As with solubility and octanol-water partition coefficient, **vapor pressure** can be estimated with a variety of correlations as discussed in detail by Burkhard et al. (1985a) and summarized as follows:

1. Interpolation or extrapolation from equation for correlating temperature relationships, e.g., the Clausius-Clapeyron, Antoine equations (Burkhard et al. 1985a);
2. Carbon or chlorine numbers (Mackay et al. 1980, Shiu and Mackay 1986);
3. LeBas molar volume (Shiu et al. 1987, 1988);
4. Boiling point $T_B$ and heat of vaporization $\Delta H_v$ (Mackay et al. 1982);
5. Group contribution method (Macknick and Prausnitz 1979);
6. UNIFAC group contribution method (Jenson et al. 1981, Yair and Fredenslund 1983, Burkhard et al. 1985a, Banerjee et al.1990);
7. Molecular weight and Gibbs' free energy of vaporization $\Delta G_v$ (Burkhard et al. 1985a);
8. TSA and $\Delta G_v$ (Amidon and Anik 1981, Burkhard et al. 1985a, Hawker 1989);
9. Molecular connectivity indices (Kier and Hall 1976, 1986, Burkhard et al. 1985a);
10. Melting point $T_M$ and GC retention index (Bidleman 1984, Burkhard et al. 1985a);
11. Solvatochromic parameters and intrinsic molar volume (Banerjee et al. 1990).

As described earlier, **Henry's law constants** can be calculated from the ratio of vapor pressure and aqueous solubility. Henry's law constants do not show a simple linear pattern as solubility, $K_{OW}$ or vapor pressure when plotted against simple molecular descriptors, such as numbers of chlorine or LeBas molar volume, e.g. PCBs (Burkhard et al. 1985b), pesticides (Suntio et al. 1988), and chlorinated dioxins (Shiu et al. 1988). Henry's law constants can be estimated from:

1. UNIFAC-derived infinite dilution activity coefficients (Arbuckle 1983);
2. Group contribution and bond contribution method (Hine and Mookerjee 1975, Meylan and Howard 1991);
3. Molecular connectivity indices (Nirmalakhandan and Speece 1988b, Sabljic and Güsten 1989, Dunnivant et al. 1992);
4. Total surface area - planar TSA (Hawker 1989);

**Bioconcentration factors:**

1. Correlation with $K_{OW}$ (Neely et al. 1974, Könemann and van Leeuwen 1980, Veith et al. 1980, Chiou et al. 1977, Mackay 1982, Briggs 1981, Garten and Trabalka 1983, Davies and Dobbs 1984, Oliver and Niimi 1988, Isnard and Lambert 1988);
2. Correlation with solubility (Kenaga 1980, Kenaga and Goring 1980, Briggs 1981, Garten and Trabalka 1983, Davies and Dobbs 1984, Isnard and Lambert 1988);
3. Correlation with $K_{OC}$ (Kenaga 1980, Kenaga and Goring 1980, Briggs 1981);
4. Calculation with HPLC retention data (Swann et al. 1983);
5. Calculation with solvatochromic parameters (Hawker 1989a, 1990).

**Sorption coefficients:**

1. Correlation with $K_{ow}$ (Karickhoff et al. 1979, Schwarzenbach and Westall 1981, Mackay 1982, Oliver 1984);
2. Correlation with solubility (Karickhoff et al. 1979);
3. Molecular connectivity indices (Sabljic 1984, 1987, Sabljic et al. 1989, Meylan et al. 1992);
4. Estimation from molecular connectivity index/fragment contribution method (Meylan et al. 1992);
5. From HPLC retention data (Swann et al. 1983, Szabo et al. 1990).

## 1.4 FATE MODELS

### 1.4.1 Evaluative Environmental Calculations

The nature of these calculations has been described in a series of papers, notably Mackay (1979), Paterson and Mackay (1985), Mackay and Paterson (1990, 1991), and a recent text (Mackay 1991). Only the salient features are presented here. Three calculations are completed for each chemical, namely the Level I, II and III fugacity calculations.

### 1.4.2 Level I Fugacity Calculation

The Level I calculation describes how a given amount of chemical partitions at equilibrium between six media: air, water, soil, bottom sediment, suspended sediment and fish. No account is taken of reactivity. Whereas most early evaluative environments have treated a one square kilometer region with about 70% water surface (simulating the global proportion of ocean surface), it has become apparent that a more useful approach is to treat a larger, principally terrestrial area similar to a jurisdictional region such as a U.S. state. The area selected is 100,000 $km^2$ or $10^{11}$ $m^2$, which is about the area of Ohio, Greece or England.

The atmospheric height is selected as a fairly arbitrary 1000 m reflecting that region of the troposphere which is most affected by local air emissions. A water surface area of 10% or 10,000 $km^2$ is used, with a water depth of 20 m. The water volume is thus $2x10^{11}$ $m^3$. The soil is viewed as being well mixed to a depth of 10 cm and is considered to be 2% organic carbon. It has a volume of $9x10^9$ $m^3$. The bottom sediment has the same area as the water, a depth of 1 cm and an organic carbon content of 4%. It thus has a volume of $10^8$ $m^3$.

For the Level I calculation both the soil and sediment are treated as simple solid phases with the above volumes, i.e., the presence of air or water in the pores of these phases is ignored.

Two other phases are included for interest. Suspended matter in water is often an important medium when compared in sorbing capacity to that of water. It is treated as having 20% organic carbon and being present at a volume fraction in the water of $5x10^{-6}$, i.e., it is about 5 mg/L. Fishes are also included at an entirely arbitrary volume fraction of $10^{-6}$ and are assumed to contain 5% lipid, equivalent in sorbing capacity to octanol. These two phases are small in volume and rarely contain an appreciable fraction of the chemical present, but it is in these phases that the highest concentration of chemical often exists.

Another phase which is introduced later in the Level III model is aerosol particles with a volume fraction in air of $2x10^{-11}$, i.e., approximately 30 $\mu g/m^3$. Although negligible in volume, an appreciable fraction of the chemical present in the air phase may be associated with aerosols. Aerosols are not treated in Level I or II calculations because their capacity for chemical is usually negligible when compared with soil.

These dimensions and properties are summarized in Table 1.2. The user is encouraged to modify these dimensions to reflect conditions in a specific area of interest.

Table 1.2a Compartment Dimensions and Properties for Level I and II Calculations

| Compartment | Air | Water | Soil | Sediment | Suspended Sediment | Fish |
|---|---|---|---|---|---|---|
| Volume, V ($m^3$) | $10^{14}$ | $2x10^{11}$ | $9x10^{9}$ | $10^{8}$ | $10^{6}$ | $2x10^{5}$ |
| Depth, h (m) | 1000 | 20 | 0.1 | 0.01 | - | - |
| Area, A ($m^2$) | $100x10^{9}$ | $10x10^{9}$ | $90x10^{9}$ | $10x10^{9}$ | - | - |
| Org. Fraction ($\phi_{OC}$) | - | - | 0.02 | 0.04 | 0.2 | - |
| Density, $\rho$ ($kg/m^3$) | 1.2 | 1000 | 2400 | 2400 | 1500 | 1000 |
| Adv. Residence Time, t (hours) | 100 | 1000 | - | 50,000 | - | - |
| Adv. flow, G ($m^3/h$) | $10^{12}$ | $2x10^{8}$ | - | 2000 | - | - |

Table 1.2b Bulk Compartment Dimensions and Volume Fraction (v) for Level III Calculations

| | | |
|---|---|---|
| Air | Total volume | $10^{14}$ $m^3$ (as above) |
| | Air phase | $10^{14}$ $m^3$ |
| | Aerosol phase | 2000 $m^3$ (v = $2x10^{-11}$) |
| Water | Total volume | $2x10^{11}$ $m^3$ |
| | Water phase | $2x10^{11}$ $m^3$ (as above) |
| | Suspended sediment phase | $10^{6}$ $m^3$ (v = $5x10^{-6}$) |
| | Fish phase | $2x10^{5}$ $m^3$ (v = $1x10^{-6}$) |
| Soil | Total volume | $18x10^{9}$ $m^3$ |
| | Air phase | $3.6x10^{9}$ $m^3$ (v = 0.2) |
| | Water phase | $5.4x10^{9}$ $m^3$ (v = 0.3) |
| | Solid phase | $9.0x10^{9}$ $m^3$ (v = 0.5) (as above) |
| Sediment | Total volume | $500x10^{6}$ $m^3$ |
| | Water phase | $400x10^{6}$ $m^3$ (v = 0.8) |
| | Solid phase | $100x10^{6}$ $m^3$ (v = 0.2) (as above) |

The amount of chemical introduced in the Level I calculation is an arbitrary 100,000 kg or 100 tonnes. If dispersed entirely in the air, this amount yields a concentration of 1 $\mu g/m^3$ which is not unusual for ubiquitous contaminants such as hydrocarbons. If dispersed entirely in the water, the concentration is a higher 500 $\mu g/m^3$ or 500 ng/L, which again is reasonable for a well-used chemical of commerce. The corresponding value in soil is about 0.0046 $\mu g/g$. It is believed that this amount is a reasonable common value for evaluative purposes. Clearly for restricted chemicals such as PCBs, this amount is too large, but it is preferable to adopt a common evaluative amount for all substances. No significance should, of course, be attached to the absolute values of the concentrations which are deduced from this arbitrary amount. Only the relative values have significance.

The Level I calculation proceeds by deducing the fugacity capacities, Z values for each medium (see Table 1.3), following the procedures described by Mackay (1991). These working equations show the necessity of having data on molecular mass, water solubility, vapor pressure, and octanol-water partition coefficient. The fugacity f (Pa) common to all media is deduced as

$$f = M/\Sigma\ V_i Z_i$$

where M is the total amount of chemical (mol), $V_i$ is the medium volume ($m^3$) and $Z_i$ is the corresponding fugacity capacity for the chemical in each medium.

The molar concentration C ($mol/m^3$) can then be deduced as Zf $mol/m^3$ or as WZf $g/m^3$ or 1000 $WZf/\rho$ $\mu g/g$ where $\rho$ is the phase density ($kg/m^3$) and W is the molecular mass (g/mol). The amount $m_i$ in each medium is $C_i V_i$ mol, and the total in all media is M mol. The **BASIC** computer program for undertaking this calculation is appended. For those who prefer a spreadsheet format, an identical Lotus 123® program is also provided.

The information obtained from this calculation includes the concentrations, amounts and distribution. In the figures, a pie chart illustrates the distribution between the four primary compartments of air, water, soil and sediment, the amount in fish and suspended sediment being ignored. This information is useful as an indication of the relative concentrations.

Note that this simple treatment assumes that the soil and sediment phases are entirely solid, i.e., there are no air or water phases present to "dilute" the solids. Later in the Level III calculation these phases and aerosols are included.

Table 1.3a Equations for Phase Z values used in Levels I and II and the Bulk Phase values used in Level III

| | |
|---|---|
| Air | $Z_1 = 1/RT$ |
| Water | $Z_2 = 1/H = C^S/P^S$ |
| Soil | $Z_3 = Z_2 \cdot \rho_3 \cdot \phi_3 \cdot K_{OC}/1000$ |
| Sediment | $Z_4 = Z_2 \cdot \rho_4 \cdot \phi_4 \cdot K_{OC}/1000$ |
| Suspended Sediment | $Z_5 = Z_2 \cdot \rho_5 \cdot \phi_5 \cdot K_{OC}/1000$ |
| Fish | $Z_6 = Z_2 \cdot \rho_6 \cdot L \cdot K_{OW}/1000$ |
| Aerosol | $Z_7 = Z_1 \cdot 6x10^6/P^S_L$ |
| where | $R$ = gas constant (8.314 J/mol K) |
| | $T$ = absolute temperature (K) |
| | $C^S$ = solubility in water ($mol/m^3$) |
| | $P^S$ = vapor pressure (Pa) |
| | $H$ = Henry's law constant ($Pa \cdot m^3/mol$) |
| | $P^S_L$ = liquid vapor pressure (Pa) |
| | $K_{OW}$ = octanol-water partition coefficient |
| | $K_{OC}$ = organic-carbon partition coefficient (= 0.41 $K_{OW}$) |
| | $\rho_i$ = density of phase i ($kg/m^3$) |
| | $\phi_i$ = mass fraction organic-carbon in phase i (g/g) |
| | $L$ = lipid content of fish |

Note for solids $P^S_L = P^S_S/\exp\{6.79(1 - T_M/T)\}$ where $T_M$ is melting point (K) of the solute.

Table 1.3b Bulk Phase Z values, $Z_{Bi}$ deduced as $\Sigma\, v_i Z_i$, in which the coefficients, e.g., $2x10^{-11}$, are the volume fractions $v_i$ of each pure phase as specified in Table 1.2b

| | | |
|---|---|---|
| Air | $Z_{B1} = Z_1 + 2x10^{-11}\, Z_7$ | (approximately 30 $\mu g/m^3$ aerosols) |
| Water | $Z_{B2} = Z_2 + 5x10^{-6}\, Z_5 + 1x10^{-6}\, Z_6$ | (5 ppm solids, 1 ppm fish by volume) |
| Soil | $Z_{B3} = 0.2\, Z_1 + 0.3\, Z_2 + 0.5\, Z_3$ | (20% air, 30% water, 50% solids) |
| Sediment | $Z_{B4} = 0.8\, Z_2 + 0.2\, Z_4$ | (80% water, 20% solids) |

### 1.4.3 Level II Fugacity Calculation

The Level II calculation simulates a situation in which chemical is continuously discharged into the multimedia environment and achieves a steady-state equilibrium condition at which input and output rates are equal. The task is to deduce the rates of loss by reaction and advection.

The reaction rate data developed for each chemical in the tables are used to select a reactivity class as described earlier, and hence a first-order rate constant for each medium. Often these rates are in considerable doubt, thus the quantities selected should be used with extreme caution because they may not be widely applicable. The rate constants $k_i$ $h^{-1}$ are used to calculate reaction D values for each medium $D_{Ri}$ as $V_iZ_ik_i$. The rate of reactive loss is then $D_{Ri}f$ mol/h.

For advection, it is necessary to select flow rates. This is conveniently done in the form of advective residence times, t h, thus the advection rate $G_i$ is $V_i/t$ $m^3/h$ for each medium. For air, a residence time of 100 hours is used (approximately 4 days), which is probably too long for the geographic area considered, but shorter residence times tend to cause air advective loss to be a dominant mechanism. For water, a figure of 1000 hours (42 days) is used, reflecting a mixture of rivers and lakes. For sediment burial (which is treated as an advective loss), a time of 50,000 hours or 5.7 years is used. Only for very persistent, hydrophobic chemicals is this process important. No advective loss from soil is included. The D value for loss by advection $D_{Ai}$ is $G_iZ_i$ and the rates are $D_{Ai}f$ mol/h. These rates are listed in Table 1.2.

There may thus be losses caused by both reaction and advection D values for the four primary media. These loss processes are not included for fish or suspended matter. At steady state, equilibrium conditions the input rate E mol/h can be equated to the sum of the output rates, from which the common fugacity can be calculated as follows

$$E = f \cdot \Sigma D_{Ai} + f \cdot \Sigma D_{Ri}$$

thus,

$$f = E/(\Sigma D_{Ai} + \Sigma D_{Ri})$$

The common assumed emission rate is 1000 kg/h or 1 tonne/h. To achieve an amount equivalent to the 100 tonnes in the Level I calculation requires an overall residence time of 100 hours. Again, the concentrations and amounts $m_i$ and $\Sigma$ $m_i$ or M can be deduced, as well as the reaction and advection rates. These rates obviously total to give the input rate E. Of particular interest are the relative rates of these loss processes, and the overall persistence or residence time which is calculated as

$$t_O = M/E$$

where M is the total amount present. It is also useful to calculate a reaction and an advection persistence $t_R$ and $t_A$ as

$$t_R = M/\Sigma D_{Ri}f \qquad t_A = M/\Sigma D_{Ai}f$$

Obviously, $1/t_O = 1/t_R + 1/t_A$

These persistences indicate the likelihood of the chemical being lost by reaction as distinct from advection. The percentage distribution of chemical between phases is identical to that in Level I. A pie chart depicting the distribution of losses is presented.

### 1.4.4 Level III Fugacity Calculation

Whereas the Level I and II calculations assume equilibrium to prevail between all media, this is recognized as being excessively simplistic and even misleading. In the interests of algebraic simplicity only the four primary media are treated for this level. The task is to develop expressions for intermedia transport rates by the various diffusive and nondiffusive processes as described by Mackay (1991). This is done by selecting values for 12 intermedia transport velocity parameters which have dimensions of velocity (m/h or m/year), are designated as $U_i$ m/h and are applied to all chemicals. These parameters are used to calculate seven intermedia transport D values.

It is desirable to calculate new "bulk phase" Z values for the four primary media which include the contribution of dispersed phases within each medium as described by Mackay and Paterson (1991) and as listed in Tables 1.2 and 1.3. The air is now treated as an air-aerosol mixture, water as water plus suspended particles and fish, soil as solids, air and water, and sediment as solids and porewater. The Z values thus differ from the Level I and Level II "pure phase" values. The necessity for introducing this complication arises from the fact that much of the intermedia transport of the chemicals occurs in association with the movement of chemical in these dispersed phases. To accommodate this change the same volumes of the soil solids and sediment solids are retained, but the total phase volumes are increasd. These Level III volumes are also given in Table 1.2. The reaction and advection D values employ the generally smaller bulk phase Z values but the same residence times, thus the G values are increased and the D values are generally larger.

#### Intermedia D values

The justification for each intermedia D value follows. It is noteworthy that, for example, air-to-water and water-to-air values differ because of the presence of one-way nondiffusive processes. A fuller description of the background to these calculations is given by Mackay (1991).

#### 1. Air to water ($D_{12}$)

Four processes are considered: diffusion (absorption), dissolution in rain of gaseous chemical, and wet and dry deposition of particle-associated chemical.

For diffusion, the conventional two-film approach is taken with water-side ($k_W$) and air-side ($k_A$) mass transfer coefficients (m/h) being defined. Values of 0.05 for $k_W$ and 5 m/h for $k_A$ are used. The absorption D value is then

$$D_{VW} = 1/(1/(k_A A_W Z_1) + 1/(k_W A_W Z_2))$$

where $A_W$ is the air-water area ($m^2$) and $Z_1$ and $Z_2$ are the pure air and water Z values. The velocities $k_A$ and $k_W$ are designated as $U_1$ and $U_2$.

For rain dissolution, a rainfall rate of 0.876 m/year is used, i.e., $U_R$ or $U_3$ is $10^{-4}$ m/h. The D value for dissolution $D_{RW}$ is then

$$D_{RW} = U_R A_W Z_2 = U_3 A_W Z_2$$

For wet deposition, it is assumed that the rain scavenges Q (scavenging ratio) or about 200,000 times its volume of air. Using a particle concentration (volume fraction) $v_Q$ of $2 \times 10^{-11}$, this corresponds to the removal of $Qv_Q$ or $4 \times 10^{-6}$ volumes of aerosol per volume of rain. The total rate of particle removal by wet deposition is then $Qv_Q U_R A_W$ $m^3/h$, thus the wet "transport velocity" $Qv_Q U_R$ is $4 \times 10^{-10}$ m/h.

For dry deposition, a typical deposition velocity $U_Q$ of 10 m/h is selected yielding a rate of particle removal of $U_Q v_Q A_W$ or $2 \times 10^{-10} A_W$ $m^3/h$ corresponding to a transport velocity of $2 \times 10^{-10}$ m/h. Thus,

$$U_4 = Qv_Q U_R + U_Q v_Q = v_Q(QU_R + U_Q)$$

The total particle transport velocity $U_4$ for wet and dry deposition is thus $6 \times 10^{-10}$ m/h and the total D value $D_{QW}$ is

$$D_{QW} = U_4 A_W Z_7$$

where $Z_7$ is the aerosol Z value.

The overall D value is given by

$$D_{12} = D_{VW} + D_{RW} + D_{QW}$$

**2. Water to air ($D_{21}$)**

Evaporation is treated as the reverse of absorption; thus $D_{21}$ is simply $D_{VW}$ as before.

**3. Air to soil ($D_{13}$)**

A similar approach is adopted as for air to water transfer. Four processes are considered with rain dissolution ($D_{RS}$) and wet and dry deposition ($D_{QS}$) being treated identically except that the area term is now the air-soil area $A_S$.

For diffusion, the approach of Jury et al. (1983, 1984) is used as described by Mackay and Stiver (1991) and Mackay (1991) in which three diffusive processes are treated. The air boundary layer is characterized by a mass transfer coefficient $k_S$ or $U_7$ of 5 m/h, equal to that of the air-water MTC coefficient $k_A$ used in $D_{12}$.

For diffusion in the soil air-pores, a molecular diffusivity of 0.02 $m^2/h$ is reduced to an effective diffusivity using a Millington-Quirk type of relationship by a factor of about 20 to $10^{-3}$ $m^2/h$. Combining this with a path length of 0.05 m gives an effective air to soil mass transfer coefficient $k_{SA}$ of 0.02 m/h which is designated as $U_5$.

Similarly, for diffusion in water a molecular diffusivity of $2x10^{-6}$ $m^2/h$ is reduced by a factor of 20 to an effective diffusivity of $10^{-7}$ $m^2/h$, which is combined with a path length of 0.05 m to give an effective soil to water mass transfer coefficient of $k_{SW}$ $2x10^{-6}$ m/h.

It is probable that capillary flow of water contributes to transport in the soil. For example, a rate of 7 cm/year would yield an equivalent water velocity of $8x10^{-6}$ m/h which exceeds the water diffusion rate by a factor of four. For illustrative purposes we thus select a water transport velocity or coefficient $U_6$ in the soil of $10x10^{-6}$ m/h, recognizing that this may be in error by a substantial amount, and will vary with rainfall characteristics and soil type.

The soil processes are in parallel with boundary layer diffusion in series, so the final equation is

$$D_{VS} = 1/[1/D_S + 1/(D_{SW} + D_{SA})]$$

where

$$D_S = U_7 A_S Z_1 \qquad (U_7 = 5 \text{ m/h})$$
$$D_{SW} = U_6 A_S Z_2 \qquad (U_6 = 10x10^{-6} \text{ m/h})$$
$$D_{SA} = U_5 A_S Z_1 \qquad (U_5 = 0.02 \text{ m/h})$$

where $A_S$ is the soil horizontal area.

Air-soil diffusion thus appears to be much slower than air-water diffusion because of the slow migration in the soil matrix. In practice, the result will be a nonuniform composition in the soil with the surface soil (which is much more accessible to the air than the deeper soil) being closer in fugacity to the atmosphere.

The overall D value is given as

$$D_{13} = D_{VS} + D_{QS} + D_{RS}$$

**4. Soil to air ($D_{31}$)**

Evaporation is treated as the reverse of absorption, thus the D value is simply $D_{VS}$.

**5. Water to sediment ($D_{24}$)**

Two processes are treated, diffusion and deposition.

Diffusion is characterized by a mass transfer coefficient $U_8$ of $10^{-4}$ m/h which can be regarded as a molecular diffusivity of $2x10^{-6}$ $m^2/h$ divided by a path length of 0.02 m. In practice, bioturbation may contribute substantially to this exchange process, and in shallow water current-induced turbulence may also increase the rate of transport. Diffusion in association with organic colloids is not included. The D value is thus given as $U_8A_WZ_2$.

Deposition is assumed to occur at a rate of 5000 $m^3/h$ which corresponds to addition of a depth of solids of 0.438 cm/year; thus 43.8% of the solids resident in the accessible bottom sediment is added each year. This rate is about 12 $cm^3/m^2 \cdot$ day which is high compared to values observed in large lakes. The velocity $U_9$, corresponding to the addition of 5000 $m^3/h$ over the area of $10^{10}$ $m^2$, is thus $5 \times 10^{-7}$ m/h.

It is assumed that of this 5000 $m^3/h$ deposited, 2000 $m^3/h$ or 40% is buried (yielding the advective flow rate in Table 1.2), 2000 $m^3/h$ or 40% is resuspended (as discussed later) and the remaining 20% is mineralized organic matter. The organic carbon balance is thus only approximate.

The transport velocities are thus:

| | |
|---|---|
| deposition $U_9$ | $5.0 \times 10^{-7}$ m/h or 0.438 cm/years |
| resuspension $U_{10}$ | $2.0 \times 10^{-7}$ m/h or 0.175 cm/year |
| burial $U_B$ | $2.0 \times 10^{-7}$ m/h or 0.175 cm/year<br>(included as an advective residence time of 50,000 h) |

The water to sediment D value is thus

$$D_{24} = U_8A_WZ_2 + U_9A_WZ_5$$

where $Z_5$ is the Z value of the particles in the water column.

**6. Sediment to water ($D_{42}$)**

This is treated similarly to $D_{24}$ giving:

$$D_{42} = U_8A_WZ_2 + U_{10}A_WZ_4$$

where $U_{10}$ is the sediment resuspension velocity of $2.0 \times 10^{-7}$ m/h and $Z_4$ is the Z value of the sediment solids.

**7. Sediment advection ($D_{A4}$)**

This D value is $U_BA_WZ_4$ where $U_B$, the sediment burial rate, is $2.0 \times 10^{-7}$ m/h. It can be viewed as $G_BZ_{B4}$ where $G_B$ is the total burial rate specified as $V_S/t_B$ where $t_B$ (residence time) is 50,000 h, and $V_S$ (the sediment volume) is the product of sediment depth (0.01 cm) and area $A_W$. $Z_4$, $Z_{B4}$ are the Z values of the sediment solids and of the bulk sediment respectively. Since there are 20% solids, $Z_{B4}$ is about 0.2 $Z_4$. There is a slight difference between these approaches because in the advection approach (which is used here) there is burial of water as well as solids.

**8. Soil to water ($D_{32}$)**

It is assumed that there is run-off of water at a rate of 50% of the rain rate, i.e., the D value is

$$D = 0.5\ U_3 A_S Z_2 = U_{11} A_S Z_2$$

thus the transport velocity term $U_{11}$ is $0.5U_3$ or $5x10^{-5}$ m/h.

For solids run-off it is assumed that this run-off water contains 200 parts per million by volume of solids; thus the corresponding velocity term $U_{12}$ is $200x10^{-6}U_{11}$, i.e., $10^{-8}$ m/h. This corresponds to the loss of soil at a rate of about 0.1 mm per year. If these solids were completely deposited in the aquatic environment (which is about 1/10th the soil area), they would accumulate at about 0.1 cm per year, which is about a factor of four less than the deposition rate to sediments. The implication is that most of this deposition is of naturally generated organic carbon and from sources such as bank erosion.

**Summary**

The twelve intermedia transport parameters are listed in Table 1.4 and the equations are summarized in Table 1.5.

Table 1.4 Intermedia Transport Parameters

| U | | m/h | m/year |
|---|---|---|---|
| 1 | Air side, air-water MTC*, $k_A$ | 5 | 43800 |
| 2 | Water side, air-water MTC, $k_W$ | 0.05 | 438 |
| 3 | Rain rate, $U_R$ | $10^{-4}$ | 0.876 |
| 4 | Aerosol deposition | $6x10^{-10}$ | $5.256x10^{-6}$ |
| 5 | Soil-air phase diffusion MTC, $k_{SA}$ | 0.02 | 175.2 |
| 6 | Soil-water phase diffusion MTC | $10x10^{-6}$ | 0.0876 |
| 7 | Soil-air boundary layer MTC, $k_S$ | 5 | 43800 |
| 8 | Sediment-water MTC | $10^{-4}$ | 0.876 |
| 9 | Sediment deposition | $5.0x10^{-7}$ | 0.00438 |
| 10 | Sediment resuspension | $2.0x10^{-7}$ | 0.00175 |
| 11 | Soil-water run-off | $5x10^{-5}$ | 0.438 |
| 12 | Soil-solids run-off | $10^{-8}$ | 0.0000876 |

* Mass transfer coefficient
with,
Scavenging ratio Q = $2x10^5$
Dry deposition velocity $U_Q$ = 10 m/h
Sediment burial rate $U_B$ = $2.0x10^{-7}$ m/h

Table 1.5 Intermedia Transport D Value Equations

Air-Water

$$D_{12} = D_{VW} + D_{RW} + D_{QW}$$
$$D_{VW} = A_W/(1/U_1Z_1 + 1/U_2Z_2)$$
$$D_{RW} = U_3A_WZ_2$$
$$D_{QW} = U_4A_WZ_7$$

Water-Air

$$D_{21} = D_{VW}$$

Air-Soil

$$D_{13} = D_{VS} + D_{RS} + D_{QS}$$
$$D_{VS} = 1/(1/D_S + 1/(D_W + D_A))$$
$$D_S = U_7A_SZ_1$$
$$D_{SA} = U_5A_SZ_1$$
$$D_{SW} = U_6A_SZ_2$$
$$D_{RS} = U_3A_SZ_2$$
$$D_{QS} = U_4A_SZ_7$$

Soil-Air

$$D_{31} = D_{VS}$$

Water-Sediment

$$D_{24} = U_8A_WZ_2 + U_9A_WZ_5$$

Sediment-Water

$$D_{42} = U_8A_WZ_2 + U_{10}A_WZ_4$$

Soil-Water

$$D_{32} = U_{11}A_SZ_2 + U_{12}A_SZ_3$$

**Algebraic solution**

Four mass balance equations can be written, one for each medium, resulting in a total of four unknown fugacities, enabling simple algebraic solution as shown in Table 1.6. From the four fugacities, the concentration, amounts and rates of all transport and transformation processes can be deduced, yielding a complete mass balance.

The new information from the Level III calculations are the intermedia transport data, i.e., the extent to which chemical discharged into one medium tends to migrate into another. This migration pattern depends strongly on the proportions of the chemical discharged into each medium; indeed, the relative amounts in each medium are largely a reflection of the locations of discharge. It is difficult to interpret these mass balance diagrams because, for example, chemical depositing from air to water may have been discharged to air, or to soil from which it evaporated, or even to water from which it is cycling to and from air.

To simplify this interpretation, it is best to conduct three separate Level III calculations in which unit amounts (1000 kg/h) are introduced individually into air, soil and water. Direct discharges to sediment are unlikely and are not considered here. These calculations show clearly the extent to which intermedia transport occurs. If, for example, the intermedia D values are small compared to the reaction and advection values, the discharged chemical will tend to remain in the discharge or "source" medium with only a small proportion migrating to other media. Conversely, if the intermedia D values are relatively large the chemical becomes very susceptible to intermedia transport. This behavior is observed for persistent substances such as PCBs, which have very low rates of reaction.

A direct assessment of multimedia behavior is thus possible by examining the proportions of chemical found at steady state in the "source" medium and in other media. For example, when discharged to water, an appreciable fraction of the benzene is found in air, whereas for atrazine, only a negligible fraction of atrazine reaches air.

Table 1.6 Level III Solutions to Mass Balance Equations

---

Mass balance equations:

Air $E_1 + f_2D_{21} + f_3D_{31} = f_1D_{T1}$

Water $E_2 + f_1D_{12} + f_3D_{32} + f_4D_{42} = f_2 D_{T2}$

Soil $E_3 + f_1D_{13} = f_3D_{T3}$

Sediment $E_4 + f_2D_{24} = f_4D_{T4}$

where $E_i$ is discharge rate, $E_4$ usually being zero.

$$D_{T1} = D_{R1} + D_{A1} + D_{12} + D_{13}$$

$$D_{T2} = D_{R2} + D_{A2} + D_{21} + D_{23} + D_{24}, \quad (D_{23} = 0)$$

$$D_{T3} = D_{R3} + D_{A3} + D_{31} + D_{32}, \quad (D_{A3} = 0)$$

$$D_{T4} = D_{R4} + D_{A4} + D_{42}$$

Solution:

$$f_2 = [E_2 + J_1J_4/J_3 + E_3D_{32}/D_{T3} + E_4D_{42}/D_{T4}]/(D_{T2} - J_2J_4/J_3 - D_{24} \bullet D_{42}/D_{T4})$$

$$f_1 = (J_1 + f_2J_2)/J_3$$

$$f_3 = (E_3 + f_1D_{13})/D_{T3}$$

$$f_4 = (E_4 + f_2D_{24})/D_{T4}$$

where

$$J_1 = E_1/D_{T1} + E_3D_{31}/(D_{T3} \bullet D_{T1})$$

$$J_2 = D_{21}/D_{T1}$$

$$J_3 = 1 - D_{31} \bullet D_{13}/(D_{T1} \bullet D_{T3})$$

$$J_4 = D_{12} + D_{32} \bullet D_{13}/D_{T3}$$

**Linear additivity**

Because these equations are entirely linear, the solutions can be scaled linearly. The concentrations resulting from a discharge of 2000 kg/h are simply twice those of 1000 kg/h. Further, if discharge of 1000 kg/h to air causes 500 kg in water and discharge of 1000 kg/h to soil causes 100 kg in water, then if both discharges occur simultaneously, there will be 600 kg in water. If the discharge to soil is increased to 3000 kg/h, the total amount in the water will rise to (500 + 300) or 800 kg. It is thus possible to deduce the amount in any medium arising from any combination of discharge rates by scaling and adding the responses from the unit inputs. This "linear additivity principle" is more fully discussed by Stiver and Mackay (1989).

In the diagrams presented later, these three-unit (1000 kg/h) responses are given. Also, an illustrative "three discharge" mass balance is given in which a total of 1000 kg/h is discharged, but in proportions judged to be typical of chemical use and discharge to the environment. For example, benzene is believed to be mostly discharged to air with minor amounts to soil and water.

Also given in the tables are the rates of reaction, advection and intermedia transport for each case.

The reader can deduce the fate of any desired discharge pattern by appropriate scaling and addition. It is important to re-emphasize that because the values of transport velocity parameters are only illustrative, actual environmental conditions may be quite different; thus, simulation of conditions in a specific region requires determination of appropriate parameter values as well as the site specific dimensions, reaction rate constants and the physical-chemical properties which prevail at the desired temperature.

In total, the aim is to convey an impression of the likely environmental behavior of the chemical in a readily assimilable form.

## 1.5. DATA SOURCES AND PRESENTATIONS

### 1.5.1 Data sources

Most physical properties such as molecular weight (MW, g/mol), melting point (M.P., °C), boiling point (B.P., °C), and density have been obtained from commonly used handbooks such as the CRC Handbook of Physics and Chemistry (Weast 1972, 1984), Lange's Handbook of Chemistry (Dean 1979, 1985), Dreisbach's Physical Properties of Chemical Compounds, Vol. I, II and III (1955, 1959, 1961), Organic Solvents, Physical Properties and Methods of Purification (Riddick et al. 1986) and the Merck Index (1983, 1987). Other physical-chemical properties such as aqueous solubility, vapor pressure, octanol-water partition coefficient, Henry's law constant, bioconcentration factor and sorption coefficient have been obtained from scientific journals or other environmental handbooks, notably Verschueren's Handbook of Environmental Data on Organic Chemicals (1977, 1983) and Howard et al.'s Handbook of Environmental Fate and Exposure Data, Vol. I and II (1989, 1990). Other important sources of vapor pressure are the CRC Handbook of Physics and Chemistry (Weast 1972), Lange's Handbook of Chemistry (Dean 1985), the Handbook of Vapor Pressures and Heats of Vaporization of Hydrocarbons and Related Compounds (Zwolinski and Wilhoit 1971), the Vapor Pressure of Pure Substances (Boublik et al. 1973, 1984), the Handbook of the Thermodynamics of Organic Compounds (Stephenson and Malanowski 1987). For aqueous solubilities, valuable sources include the IUPAC Solubility Data Series (1985, 1989a,b) and Horvath's Halogenated Hydrocarbons, Solubility-Miscibility with Water (Horvath 1982). Octanol-water partition coefficients are conveniently obtained from the compilation by Leo et al. (1971) and Hansch and Leo (1979), or can be calculated from molecular structure by the methods of Hansch and Leo (1979) or Rekker (1977). Lyman et al. (1982) also outline methods of estimating solubility, $K_{OW}$, vapor pressure, and bioconcentration factor for organic chemicals. The recent Handbook of Environmental Degradation Rates by Howard et al. (1991) is a valuable source of rate constants and half-lives for inclusion in subsequent fugacity calculations.

The most reliable sources of data are the original citations in the reviewed scientific literature. Particularly reliable are those papers which contain a critical review of data from a number of sources as well as independent experimental determinations. Calculated or correlated values are reported in the tables but are viewed as being less reliable. A recurring problem is that a value is frequently quoted, then requoted and the original paper may not be cited. The aim in this work has been to gather and list the citations, interpret them and select a "best" or "most likely" value. To assist in this process, plots are prepared of properties as a function of molar volume as the molecular descriptors. These are discussed at the end of each chapter.

### 1.5.2 Data Format

Each data sheet lists the following properties, although not all quantities are included for all chemicals. In all cases citations are provided.

Common Name:
Synonym:
Chemical Name:

CAS Registry No:
Molecular Formula:
Molecular Weight (g/mol):
Melting Point (°C):
Boiling Point (°C):
Density ($g/cm^3$ at 20 °C):
Molar Volume ($cm^3/mol$):
Molecular Volume, TMV ($Å^3$):
Total Surface Area, TSA ($Å^2$):
Heat of Fusion, $\Delta H_{fus}$, kcal/mol:
Entropy of Fusion, $\Delta S_{fus}$, cal/mol K (or e.u.):
Fugacity Ratio at 25 °C:
Water Solubility ($g/m^3$ or mg/L at 25°C):
Vapor Pressure (Pa at 25°C):
Henry's Law Constant (Pa $m^3$/mol) or Air-Water Partition Coefficient:
Octanol-Water Partition Coefficient $K_{OW}$ or log $K_{OW}$:
Bioconcentration Factor $K_B$ or BCF (or log $K_B$):
Sorption Partition Coefficient to Organic Carbon $K_{OC}$ or to Organic Matter $K_{OM}$:
Half-lives in the Environment:

Air:
Surface water:
Groundwater:
Soil:
Sediment:
Biota:

Environmental Fate Rate Constants or Half-Lives:

Volatilization/Evaporation:
Photolysis:
Oxidation or Photooxidation:
Hydrolysis:
Biotransformation/Biodegradation:
Bioconcentration, Uptake ($k_1$) and Elimination ($k_2$) Rate Constants:

### 1.5.3 Explanation of Data Presentations

**Example: Trichloroethylene or trichloroethene** (data sheets presented in Chapter 3)

**1. Chemical Properties.**

The names, formula, melting and boiling point and density data are self-explanatory.

The molar volumes are in some cases at the stated temperature and in others at the normal boiling point. Certain calculated molecular volumes are also used; thus the reader is cautioned to ensure that when using a molar volume in any correlation, it is correctly selected.

In the case of polynuclear aromatic hydrocarbons, the LeBas molar volume is regarded as suspect because of the compact nature of the multi-ring compounds. It should thus be regarded as merely an indication of relative volume, not absolute volume.

The total surface areas (TSAs) are calculated in various ways and may contain the hydration shell, thus giving a much larger area. Again, the reader is cautioned to ensure that values are consistent.

Heats of fusion, $\Delta H_{fus}$, are generally expressed in kcal/mol or kJ/mol and entropies of fusion, $\Delta S_{fus}$ in cal/mol·K (e.u. or entropy unit) or J/mol·K. In the case of liquids such as benzene, it is 1.0. For solids it is a fraction representing the ratio of solid to liquid solubility or vapor pressure. It is generally assumed that for a rigid organic molecule, the entropy of fusion is 13.5 e.u. or 56 J/mol·K, which is an average value of a number of organic compounds (Yalkowsky 1979, Miller et al. 1984). The fugacity ratio, F, given is calculated using $\Delta S_{fus}$ = 56 J/mol·K in the following expression

$$F = \exp\{(\Delta H_{fus}/RT)(1 - T_M/T)\} = \exp\{(\Delta S_{fus}/R)(1 - T_M/T)\}$$

where R is the ideal gas constant (8.314 J/mol·K or 1.987 cal/mol·K) and $T_M$ is the melting point and T is the system temperature (K).

As is apparent, a wide variety of solubilities (in units of g/m$^3$ or the equivalent mg/L) have been reported. Experimental data have the method of determinations indicated. In other compilations of data the reported value has merely been quoted from another secondary source. In some cases the value has been calculated. The abbreviations are generally self-explanatory and usually include two entries, the method of equilibration followed by the method of determination. From these values a single value is selected for inclusion in the summary data table. In the case of trichloroethylene (TCE), the selected solubility value is 1100 g/m$^3$ at 25°C. From an examination of the data it is judged that the true value almost certainly lies between 1000 and 1500 g/m$^3$.

The vapor pressure data are treated similarly with a value of 9900 Pa being selected. The true value is judged to lie between 9100 and 10000 Pa. Vapor pressures are, of course very temperature-dependent.

The Henry's law constant data are measured and reported to be in the range from 770 to 1050 by various methods, such as multiequilibration, batch stripping, EPICS (see Chapter for detailed listing and references), but in other cases are calculated from the ratio of vapor pressure and solubility (in units of mol/m$^3$). In this case a value of 1180 Pa • m$^3$/mol is selected based on the selected vapor pressure and solubility, the actual value probably lying between 700 and 1200 Pa • m$^3$/mol as determined experimentally. Care must be exercised when water is appreciably soluble in the chemical because the assumption that the Henry's law constant is the ratio of vapor pressure to solubility may be invalid.

The octanol-water partition coefficient data are similarly a combination of calculated and experimental values. A value of 2.53 is selected as being the most likely value of log $K_{OW}$, i.e.,

$K_{OW}$ is 339. A lower value of about 2.3 to 2.4 is frequently used, but the authors believe that the actual value is somewhat higher. It is likely that the true value lies between 2.3 to 2.6.

A number of (log) bioconcentration factors are listed. Most lie in the range from 1.03 to 2.0, which corresponds to BCFs of 11 to 100. This range could be interpreted as lipid contents ranging from 3 to 29%, but it is possible that some metabolism occurs, giving lower-than-equilibrium BCFs. If no metabolism occurrs, it is expected that a 5% lipid fish would have a BCF of about 17, i.e., log BCF is 1.23. This agrees well with reported values from uptake experiments.

The (log) organic-carbon partition coefficients listed range mostly from 1.50 to 2.50. It is expected that $K_{OC}$ usually lies in the range of 20 to 80% of $K_{OW}$, i.e., $K_{OC}$ will range from 68 to 270. This is clustering of experimental values around 160, i.e., log $K_{OC}$ is 2.2. Organic matter partition coefficients are also reported. Since organic carbon accounts for some 50 to 60% of the content of organic matter, $K_{OM}$ is expected to be about half $K_{OC}$.

The reader is advised to consult the original reference when using these values of BCF, $K_{OC}$ and $K_{OM}$, to ensure that conditions are as close as possible to those of specific interest.

The "Half-life in the Environment" data reflect observations of the rate of disappearance of the chemical from a medium, without necessarily identifying the cause of mechanism of loss. For example, loss from water may be a combination of evaporation, biodegradation and photolysis. Clearly these times are highly variable and depend on factors such as temperature, meteorology and the nature of the media. Again, the reader is urged to consult the original reference.

The "Environmental Fate Rate Constants" refer to specific degradation processes rather than media. As far as possible the original numerical quantities are given and thus there is a variety of time units with some expressions being rate constants and others half-lives.

The conversion is

$$k = 0.693/t_{1/2}$$

where k is the first-order rate constant ($h^{-1}$) and $t_{1/2}$ is the half-life (h).

From these data a set of medium-specific degradation reaction half-lives was selected for use in Level II and III calculations. Emphasis was based on the fastest and the most plausible degradation process for each of the environmental compartments considered. Instead of assuming an equal half-life for both the water and soil compartment, as suggested by Howard et al. (1991), a slower active class (in the reactivity table described earlier) was assigned for soil and sediment compared to that of the water compartment. This is in part because the major degradation processes are often photolysis (or photooxidation) and biodegradation. There is an element of judgement in this selection and it may be desirable to explore the implications of selecting other values. The selected values of the volatile halogenated hydrocarbons are given in Table 3.3 at the end of Chapter 3.

In summary, the physical-chemical and environmental fate data listed result in the selection of values of solubility, vapor pressure, $K_{OW}$ and reaction half-lives which are used in the evaluative environmental calculations.

The physical-chemical data of the halogenated VOCs are also plotted in the appropriate QSPR plots on Figures 1.1 to 1.6 (which are also as Figures 3.1 to 3.6 later). These plots show that the chlorinated alkanes data are relatively "well-behaved" and are consistent with data obtained for homologous chemicals. In the case of trichloroethylene this QSPR plot is of little value because this is a well-studied chemical, but for other less-studied chemicals the plots are invaluable as a means of checking the reasonableness of data. The plots can also be used, with appropriate caution, to estimate data for untested chemicals.

Figure 1.1 shows the linear relationships among various molecular descriptors. Figures 1.2 to 1.5 show the dependence of the physical-chemical properties on LeBas molar volume. Figure 1.2 shows that solubilities of the chlorinated VOCs decrease steadily with increasing molar volume. The vapor pressure data are similar. $K_{OW}$ increases with increasing molar volume also in a linear fashion. The plot between Henry's law constant and molar volume (Figure 1.5) is more scattered. Figure 1.6 shows the often-reported inverse relationship between octanol-water partition coefficient and the subcooled liquid solubility. Plots such as these are discussed in more detail at the conclusion of each chapter.

The net result is a marked increase in hydrophobicity with molecular weight and also an increased tendency to partition from air to water. This latter behavior contrasts with chlorinated benzenes and biphenyls in which the Henry's law constant is relatively constant for the series.

### 1.5.4. Evaluative Calculations

The illustrative evaluative environmental calculations discussed here and presented later for a number of chemicals have been modified from those in Volume I of this series to give more data in a more spacious layout. Level I and II diagrams are assigned to separate pages and the physical-chemical properties are included in the Level I diagram. The Level III diagram is identical to that in Volume I, but additional pie-chart diagrams are included to show how the mass of chemical is distributed between compartments, and how loss processes are distributed as a function of the compartment of discharge. Generally, if discharge is into a compartment such as water, most chemical will be found in that compartment, and will react there but a quantity does migrate to other compartments and is lost from these media. Three pie-charts corresponding to discharges of 1000 kg/h into air, water and soil are included. A fourth pie-chart with discharges to all three compartments is also given. This latter chart is in principle the linear sum of the first three, but since the overall residence times differ, the diagram with the longer residence time, and greater resident mass, tends to dominate.

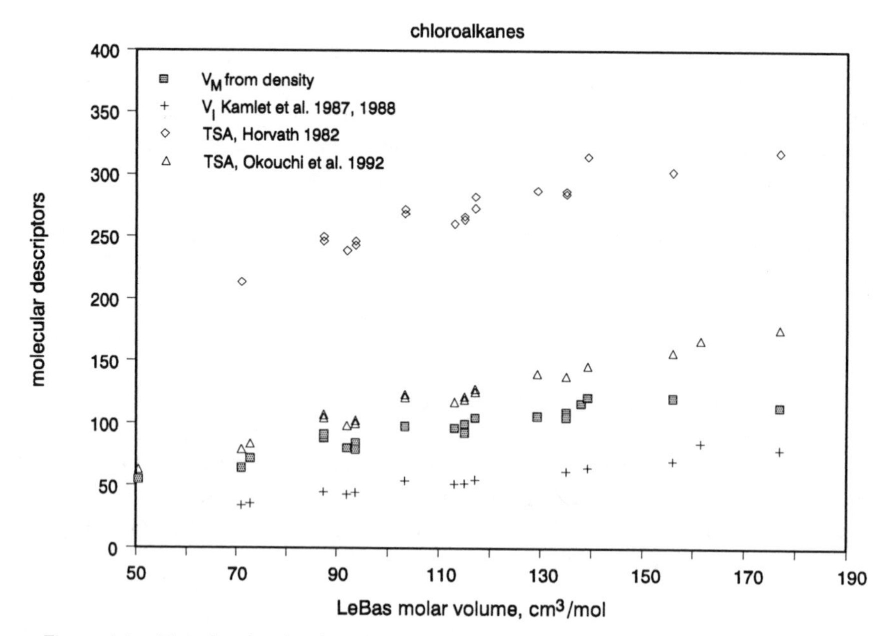


Figure 1.1a Plot of molecular descriptors versus LeBas molar volume.

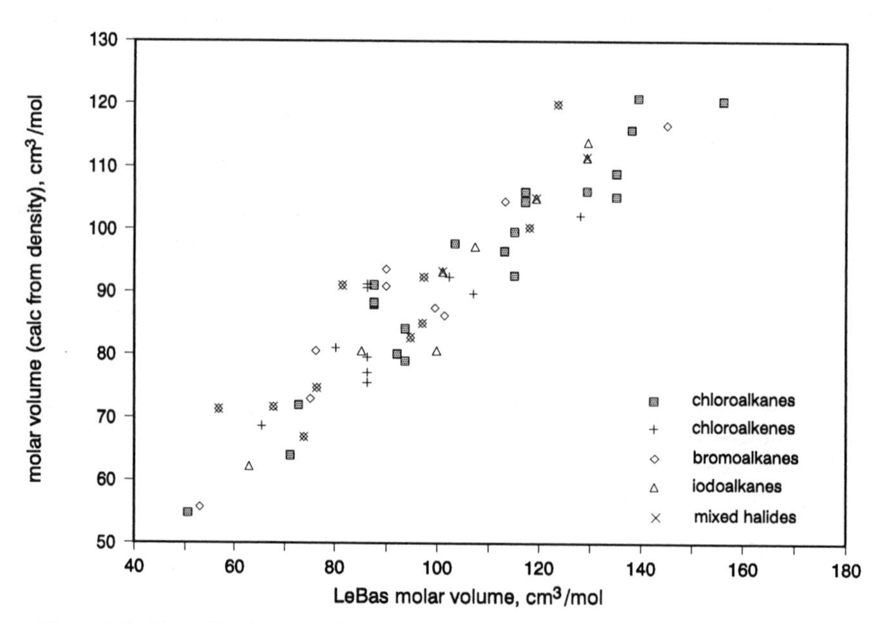


Figure 1.1b Plot of liquid molar volume versus LeBas molar volume.

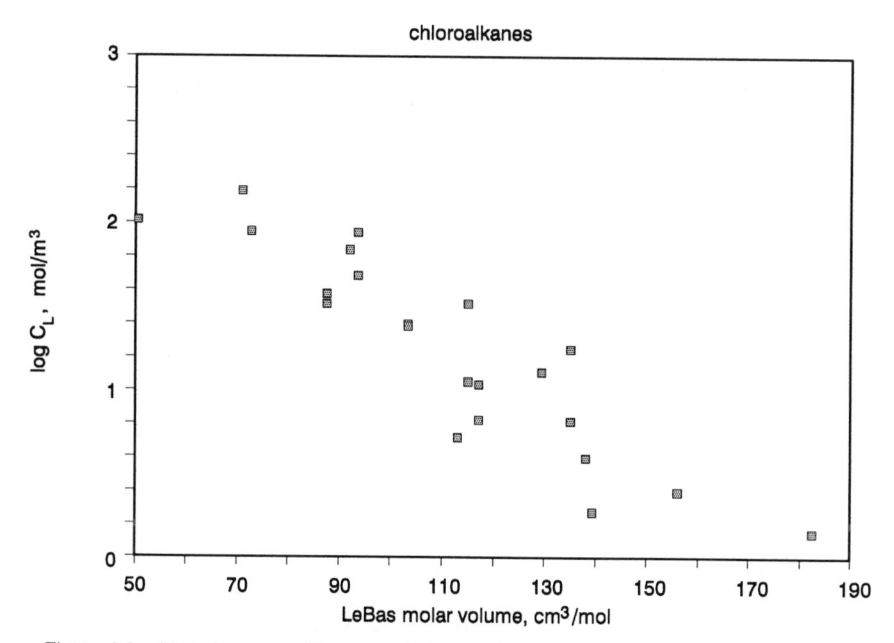


Figure 1.2a Plot of log solubility versus LeBas molar volume.

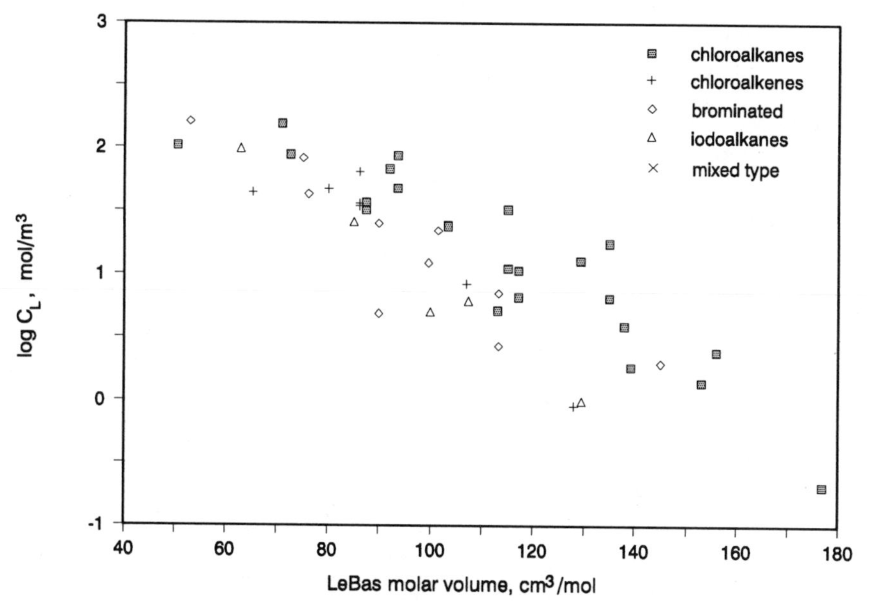


Figure 1.2b Plot of log solubility versus LeBas molar volume.

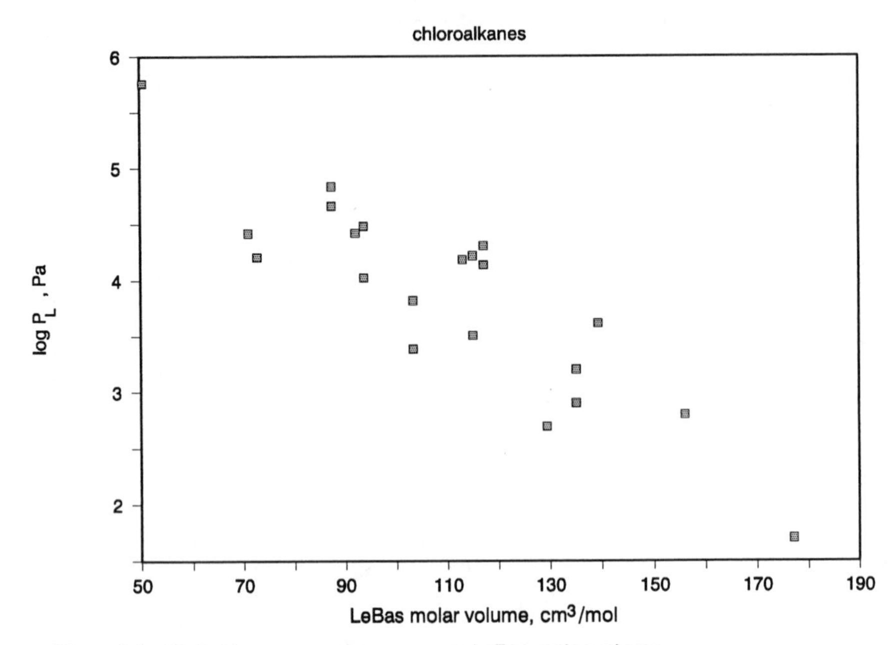


Figure 1.3a Plot of log vapor pressure versus LeBas molar volume.

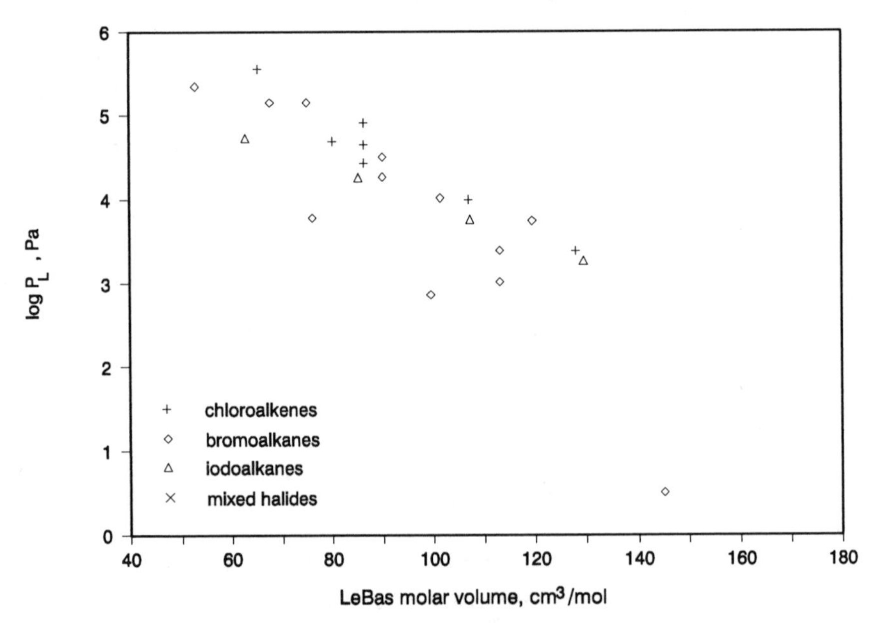


Figure 1.3b Plot of log vapor pressure versus LeBas molar volume.

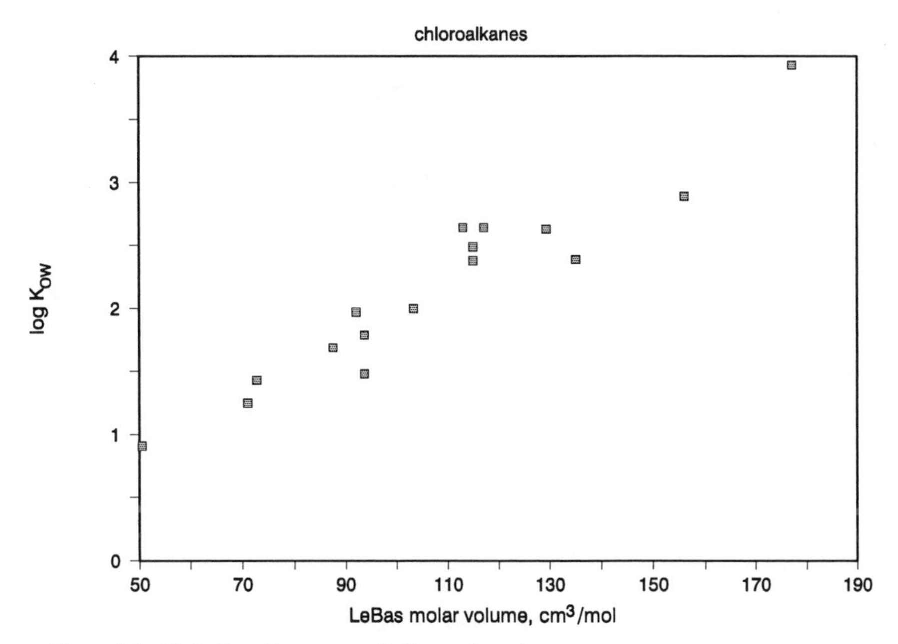


Figure 1.4a Plot of log $K_{OW}$ versus LeBas molar volume.

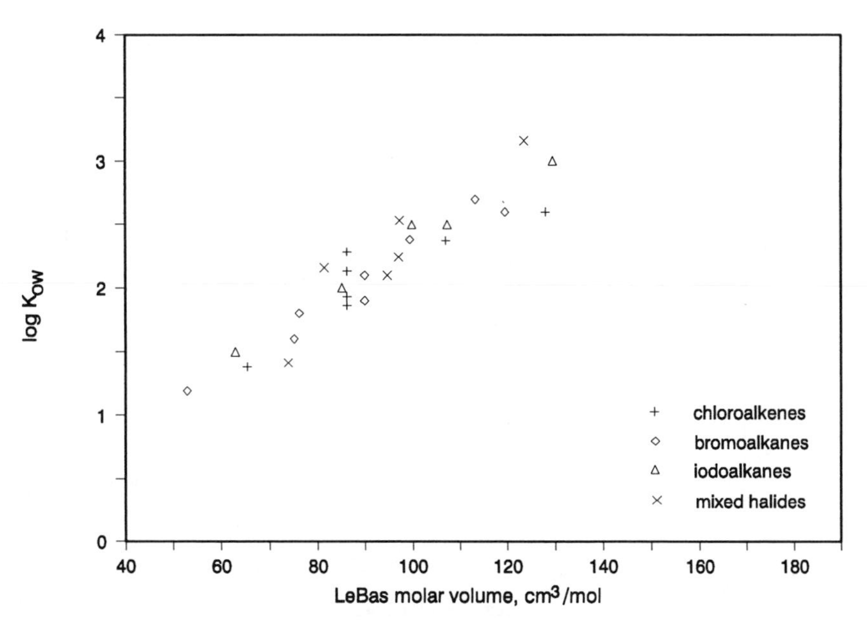


Figure 1.4b Plot of log $K_{OW}$ versus LeBas molar volume.

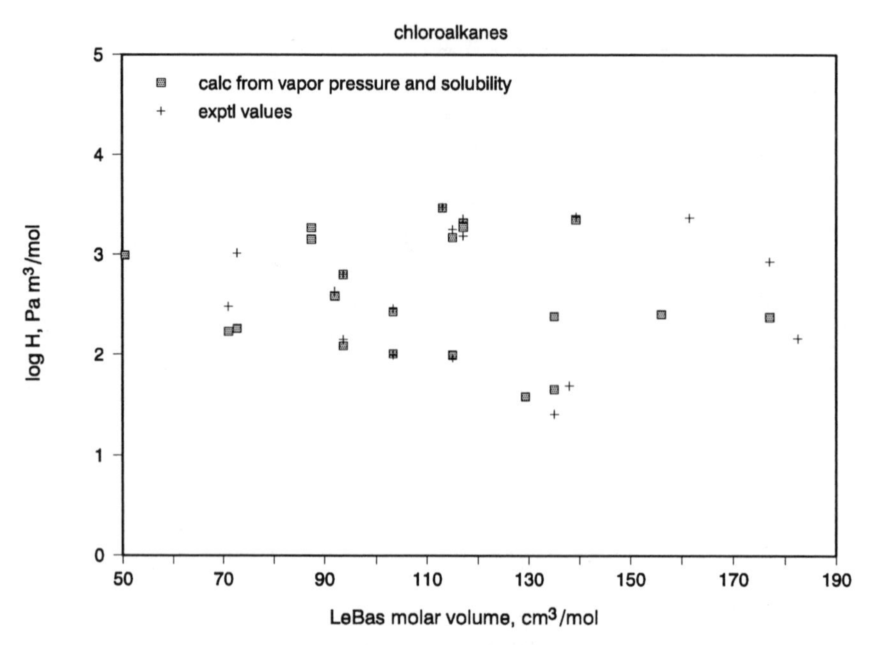


Figure 1.5a Plot of log Henry's law constant versus LeBas molar volume.

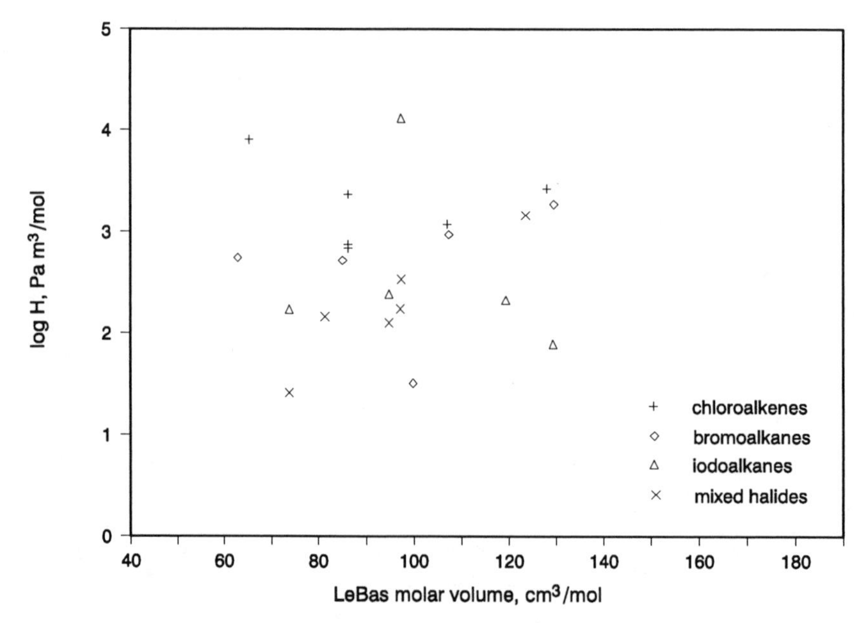


Figure 1.5b Plot of log Henry's law constant versus LeBas molar volume.

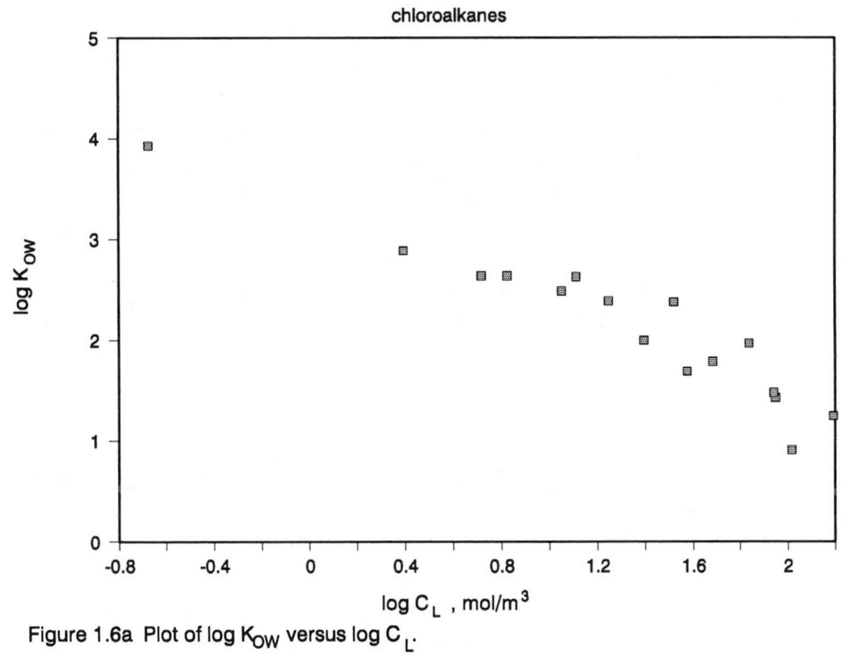

Figure 1.6a Plot of log $K_{OW}$ versus log $C_L$.

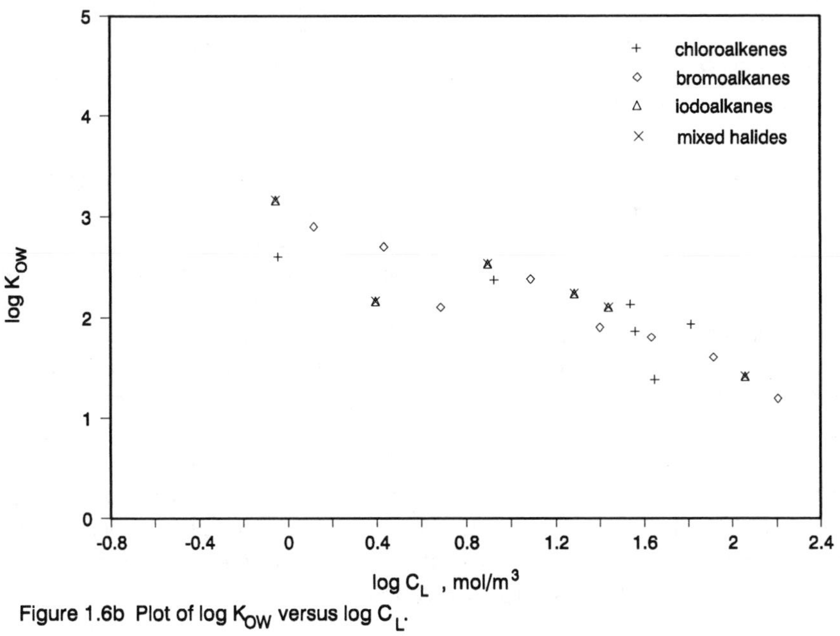

Figure 1.6b Plot of log $K_{OW}$ versus log $C_L$.

**Level I**

The Level I calculation illustrated in Figure 1.7 suggests that if 100,000 kg (100 tonnes) of trichloroethylene (TCE) are introduced into the 100,000 $km^2$ environment, 99.5% will partition into air at a concentration of $9.95 \times 10^{-7}$ $g/m^3$ or about 1.0 $\mu g/m^3$. The water will contain 0.42% at a low concentration of 2.1 $\mu g/m^3$ or equivalently 2.1 ng/L. Soils would contain 0.125% of the TCE at $5.8 \times 10^{-6}$ $\mu g/g$ and sediments about 0.003% at $11.6 \times 10^{-6}$ $\mu g/g$. These soil and sediment values would barely be detectable as a result of the moderate tendency of TCE to sorb to organic matter in these media. The fugacity is calculated to be $1.88 \times 10^{-5}$ Pa. The dimensionless soil-water and sediment-water partition coefficients or ratios of Z values are 6.7 and 13.3 as a result of a $K_{OC}$ of about 139 and a few percent organic carbon in these media. There is little evidence of bioconcentration with a very low fish concentration of $3.53 \times 10^{-5}$ $\mu g/g$. The pie chart in Figure 1.7 (which is the same as the Level I diagram for TCE in Chapter 3) clearly shows that air is the primary medium of accumulation. The more hydrophobic and less volatile VOCs tend to partition less into air and more into soil and sediment. Note that only four media (air, water, soil and bottom sediment) are depicted in the pie chart, therefore the sum of the percent distribution figures is slightly less than 100%. In almost all cases the VOCs partition nearly entirely into the air phase because of its large volume compared to the other media. The air-water partition coefficient is about 0.48.

**Level II**

The Level II calculation illustrated in Figure 1.8 includes the reaction half-lives of 170 h in air, 550 h in water, 1700 h in soil and 5500 h in sediment. No reaction is included for suspended sediment or fish. The input of 1000 kg/h results in an overall fugacity of $1.34 \times 10^{-5}$ Pa which is about 71% of the Level I value. The concentrations and amounts in each medium are thus about 71% of the Level I values. The relative mass distribution is identical to Level I. The primary loss mechanism is advection in air which accounts for 710 kg/h or 71% of the input. Most of the remainder is lost by reaction in air. The water, soil and sediment loss processes are unimportant largely because so little of the TCE is present in these media, and because of the slower reaction and advection rates. The overall residence time is 71.4 h; thus, there is an inventory of TCE in the system of 71.4 x 1000 or 71,400 kg. The pie chart in Figure 1.8 shows the dominance of air reaction and advection.

The primary loss mechanisms of atmospheric advection and reaction have D values of $4.0 \times 10^8$ and $1.6 \times 10^8$ mol/Pa·h; thus, for any other process to compete with these would require a D value of at least $10^7$ mol/Pa·h. The next largest D value is only $1.7 \times 10^5$ for advection in water, which is a factor of about 1000 smaller. Only if the water advection or reaction rates are increased by about this factor will these processes become significant. This seems inconceivably large, corresponding to very short half-lives, thus the actual values of the rate constants in media other than air are relatively unimportant in this context. They need not be known accurately for Level II calculations. The most sensitive quantities are clearly the atmospheric advection and reaction rates.

The amounts in the compartments can be calculated easily from the total amount and the percentages of mass distribution in Level I. For example, the amount in water is 0.417% of 71400 kg or 298 kg. Although TCE has a relatively short residence time, it is not degraded, it is merely relocated, so long distance transport is indicated.

**Level III**

The Level III calculation includes an estimation of intermedia transport. Examination of the magnitude of the intermedia D values given in the fate diagrams (Figure 1.9 and Figure 1.10, which are the same as the Level III calculation and distribution graphs for TCE in Chapter 3) suggests that air-water and air-soil exchange are most important with water-sediment and soil-water transport being slower by about a factor or 100 in potential transfer rate. The magnitude of these larger intermedia transport D values (approximately 5 x $10^5$ mol/Pa·h) compared to the atmospheric reaction and advection values of $10^8$ suggests that reaction and advection in air will be very fast relative to transport from air to water and soil.

The bulk Z values are similar for air and water to the values for the "pure" phases in Level I and II, but they are lower for soil and sediment because of the "dilution" of the solid soil and sediment phases with air or water.

The first row of figures at the tables at the foot of Figure 1.9 describes the condition if 1000 kg/h is emitted into the air. The result is similar to the Level II calculation with 71010 kg in air, 157 kg in water, 97.4 kg in soil and only 1.28 kg in sediment. It can be concluded that TCE discharged to the atmosphere has very little potential to enter other media. The rate of transfer from air to water ($T_{12}$) is only about 0.73 kg/h and that from air to soil ($T_{13}$) 1.3 kg/h. Even if the transfer coefficients were increased by a factor of 10, the rates would remain negligible. The reason for this is the value of the mass transfer coefficients which control this transport process. The overall residence time is 71 hours, similar to Level II.

If 1000 kg/h of TCE is discharged to water, as in the second row, there is as expected a much higher concentration in water (by a factor of over 1000). There is reaction of 268 kg/h in water, advective outflow of 212 kg/h and transfer to air ($T_{21}$) of 520 kg/h with negligible loss of 1.28 kg/h to sediment. The amount in the water is 212,400 kg, thus the residence time in the water is 212 hours and the overall environmental residence time is a longer 251 hours. The key processes are thus reaction in water (half-life 550 h), evaporation (half-life 283 h) and advective outflow (residence time 1000 h). The evaporation half-life can be calculated as (0.693 x mass in water)/rate of transfer, i.e., (0.693 x 212,400)/520 = 283 h. Clearly competition between advection, reaction and evaporation in the water determines the overall fate. Eighty-five percent of the TCE discharged is now found in the water and the concentration is fairly high, namely 1.06 x $10^{-3}$ g/m$^3$ or 1.06 $\mu$g/L.

The third row shows the fate if TCE is discharged into soil. The amount in soil is 75700 kg, with 68650 kg in air. The overall residence time is 146 hours, which is largely controlled by the evaporation rate from soil. The rate of reaction in soil is only 30.9 kg/h, there is no advection, thus the other loss mechanism is transfer to air ($T_{31}$) at a rate of 965 kg/h, with a relatively minor 5.08 kg/h to water by run-off. The soil concentration of 0.0042 g/m$^3$ is controlled almost entirely by the rate at which the TCE evaporates.

Chemical name: Trichloroethylene

Level I calculation: (six-compartment model)

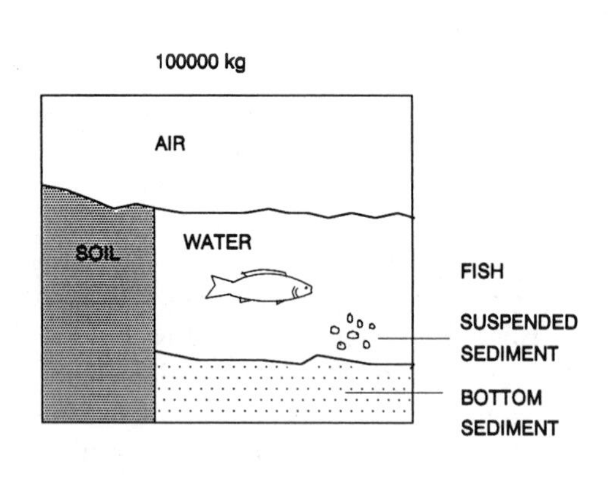


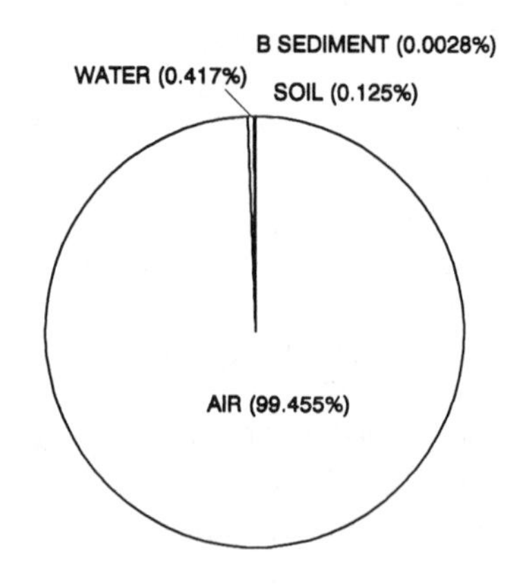


Distribution of mass

physical-chemical properties:

MW: 131.39 g/mol

M.P.: - 73 °C

Fugacity ratio: 1.0

vapor pressure: 9900 Pa

solubility: 1100 g/m$^3$

log $K_{OW}$: 2.53

| Compartment | Z | Concentration | | | Amount | Amount |
|---|---|---|---|---|---|---|
| | mol/m3 Pa | mol/m3 | mg/L (or g/m3) | ug/g | kg | % |
| Air | 4.034E-04 | 7.569E-09 | 9.946E-07 | 8.390E-04 | 99455 | 99.455 |
| Water | 8.457E-04 | 1.587E-08 | 2.085E-06 | 2.085E-06 | 416.96 | 0.4170 |
| Soil | 5.639E-03 | 1.058E-07 | 1.390E-05 | 5.793E-06 | 125.12 | 0.1251 |
| Biota (fish) | 1.433E-02 | 2.688E-07 | 3.532E-05 | 3.532E-05 | 7.06E-03 | 7.06E-06 |
| Suspended sediment | 3.525E-02 | 6.613E-07 | 8.689E-05 | 5.793E-05 | 8.69E-02 | 8.69E-05 |
| Bottom sediment | 1.128E-02 | 2.116E-07 | 2.780E-05 | 1.159E-05 | 2.7805 | 2.78E-03 |
| | Total | | | | 100000 | 100 |
| | | f = | 1.876E-05 | Pa | | |

Figure 1.7 Fugacity Level I calculation for trichloroethylene in a generic environment (dimensions defined in Table 1.2).

Chemical name: Trichloroethylene
Level II calculation: (six-compartment model)

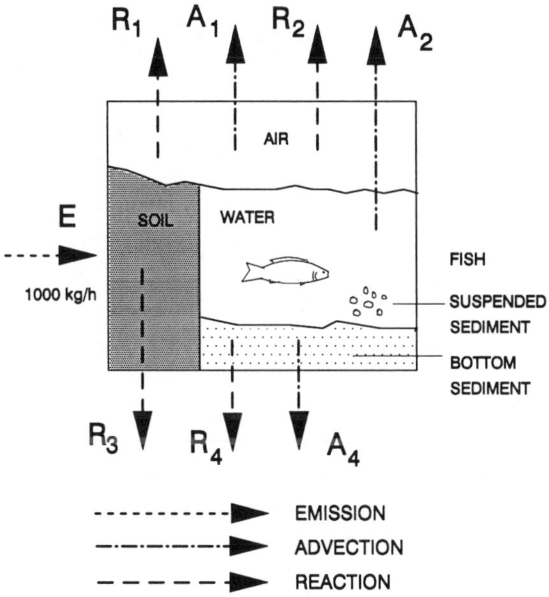

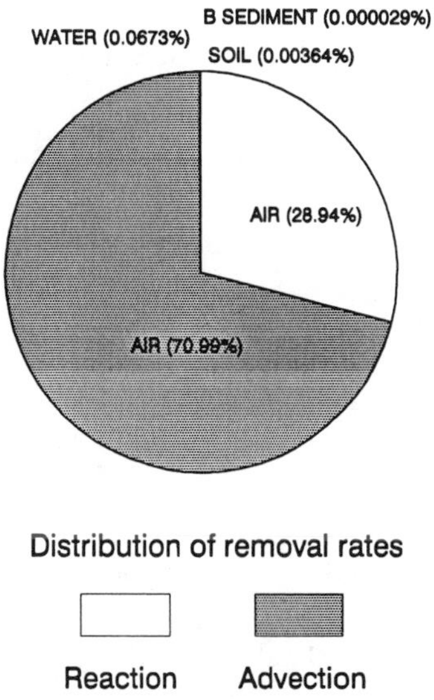

| Compartment | Half-life h | D Values Reaction mol/Pa h | Advection mol/Pa h | Conc'n mol/m3 | Loss Reaction kg/h | Loss Advection kg/h | Removal % |
|---|---|---|---|---|---|---|---|
| Air | 170 | 1.64E+08 | 4.03E+08 | 5.40E-09 | 289.389 | 709.901 | 99.929 |
| Water | 550 | 2.13E+05 | 1.69E+05 | 1.13E-08 | 0.3750 | 0.2976 | 0.0673 |
| Soil | 1700 | 2.07E+04 | | 7.55E-08 | 0.0364 | | 3.64E-03 |
| Biota (fish) | | | | 1.92E-07 | | | |
| Suspended sediment | | | | 4.72E-07 | | | |
| Bottom sediment | 5500 | 1.42E+02 | 2.26E+01 | 1.51E-07 | 2.50E-04 | 3.97E-05 | 2.90E-05 |
| | Total | 1.65E+08 | 4.04E+08 | | 289.80 | 710.20 | 100 |
| | R + A | | 5.68E+08 | | | 1000 | |

f = 1.339E-05 Pa
Total Amt 71379 kg

Overall residence time = 71.38 h
Reaction time = 246.30 h
Advection time = 100.51 h

Figure 1.8 Fugacity Level II calculation for trichloroethylene in a generic environment (dimensions defined in Table 1.2).

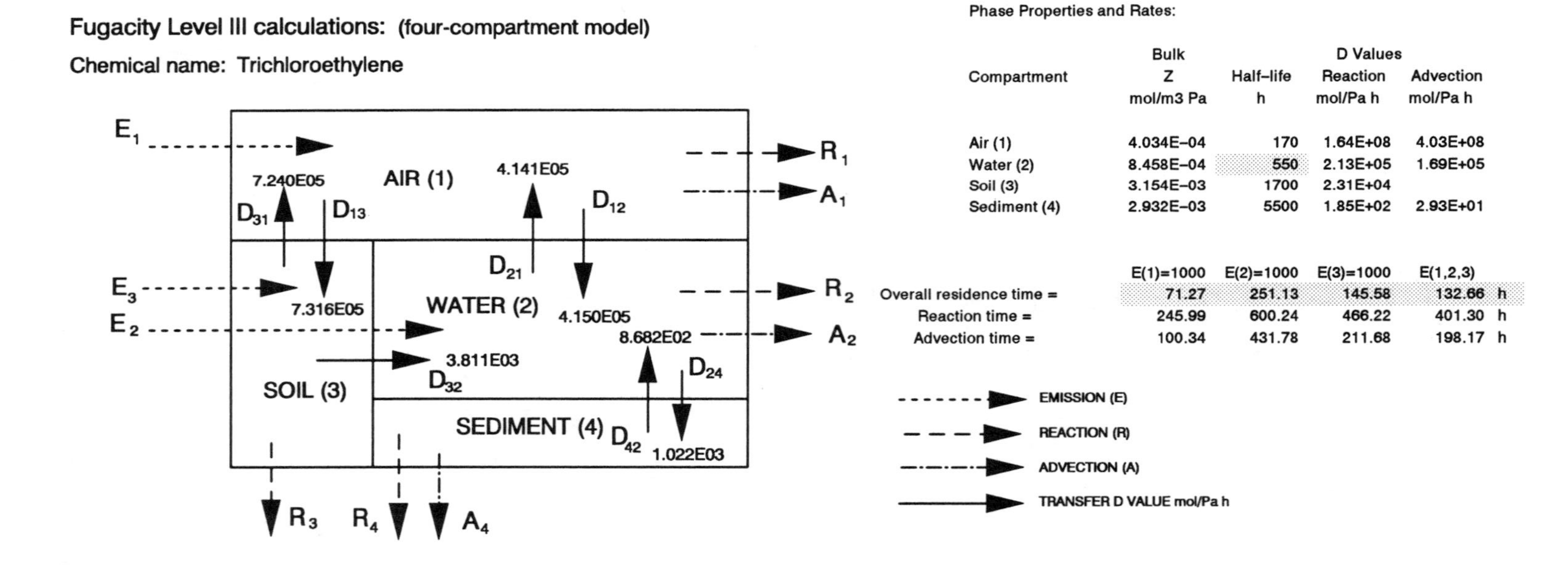


Phase Properties and Rates:

| Compartment | Bulk Z mol/m3 Pa | Half-life h | D Values Reaction mol/Pa h | D Values Advection mol/Pa h |
|---|---|---|---|---|
| Air (1) | 4.034E−04 | 170 | 1.64E+08 | 4.03E+08 |
| Water (2) | 8.458E−04 | 550 | 2.13E+05 | 1.69E+05 |
| Soil (3) | 3.154E−03 | 1700 | 2.31E+04 | |
| Sediment (4) | 2.932E−03 | 5500 | 1.85E+02 | 2.93E+01 |

| | E(1)=1000 | E(2)=1000 | E(3)=1000 | E(1,2,3) | |
|---|---|---|---|---|---|
| Overall residence time = | 71.27 | 251.13 | 145.58 | 132.66 | h |
| Reaction time = | 245.99 | 600.24 | 466.22 | 401.30 | h |
| Advection time = | 100.34 | 431.78 | 211.68 | 198.17 | h |



Phase Properties, Compositions, Transport and Transformation Rates:

| Emission, kg/h | | | Fugacity, Pa | | | | Concentration, g/m3 | | | | Amounts, kg | | | | Total Amount, kg |
|---|---|---|---|---|---|---|---|---|---|---|---|---|---|---|---|
| E(1) | E(2) | E(3) | f(1) | f(2) | f(3) | f(4) | C(1) | C(2) | C(3) | C(4) | m(1) | m(2) | m(3) | m(4) | |
| 1000 | 0 | 0 | 1.340E−05 | 7.041E−06 | 1.305E−05 | 6.648E−06 | 7.101E−07 | 7.825E−07 | 5.409E−06 | 2.561E−06 | 7.101E+04 | 1.565E+02 | 9.736E+01 | 1.281E+00 | 7.127E+04 |
| 0 | 1000 | 0 | 6.965E−06 | 9.557E−03 | 6.785E−06 | 9.024E−03 | 3.692E−07 | 1.062E−03 | 2.812E−06 | 3.477E−03 | 3.692E+04 | 2.124E+05 | 5.061E+01 | 1.738E+03 | 2.511E+05 |
| 0 | 0 | 1000 | 1.295E−05 | 5.528E−05 | 1.015E−02 | 5.220E−05 | 6.865E−07 | 6.144E−06 | 4.205E−03 | 2.011E−05 | 6.865E+04 | 1.229E+03 | 7.569E+04 | 1.006E+01 | 1.456E+05 |
| 600 | 300 | 100 | 1.142E−05 | 2.877E−03 | 1.025E−03 | 2.716E−03 | 6.055E−07 | 3.197E−04 | 4.246E−04 | 1.047E−03 | 6.055E+04 | 6.394E+04 | 7.643E+03 | 5.233E+02 | 1.327E+05 |

| Emission, kg/h | | | Loss, Reaction, kg/h | | | | Loss, Advection, kg/h | | | Intermedia Rate of Transport, kg/h T12 | T21 | T13 | T31 | T32 | T24 | T42 |
|---|---|---|---|---|---|---|---|---|---|---|---|---|---|---|---|---|
| E(1) | E(2) | E(3) | R(1) | R(2) | R(3) | R(4) | A(1) | A(2) | A(4) | air−water | water−air | air−soil | soil−air | soil−water | water−sed | sed−water |
| 1000 | 0 | 0 | 2.895E+02 | 1.972E−01 | 3.97E−02 | 1.614E−04 | 7.101E+02 | 1.565E−01 | 2.561E−05 | 7.305E−01 | 3.831E−01 | 1.288E+00 | 1.242E+00 | 6.535E−03 | 9.454E−04 | 7.584E−04 |
| 0 | 1000 | 0 | 1.505E+02 | 2.677E+02 | 2.06E−02 | 2.190E−01 | 3.692E+02 | 2.124E+02 | 3.477E−02 | 3.798E−01 | 5.200E+02 | 6.695E−01 | 6.455E−01 | 3.397E−03 | 1.283E+00 | 1.029E+00 |
| 0 | 0 | 1000 | 2.799E+02 | 1.548E+00 | 3.09E+01 | 1.267E−03 | 6.865E+02 | 1.229E+00 | 2.011E−04 | 7.062E−01 | 3.008E+00 | 1.245E+00 | 9.653E+02 | 5.080E+00 | 7.423E−03 | 5.955E−03 |
| 600 | 300 | 100 | 2.468E+02 | 8.057E+01 | 3.12E+00 | 6.593E−02 | 6.055E+02 | 6.394E+01 | 1.047E−02 | 6.228E−01 | 1.565E+02 | 1.098E+00 | 9.747E+01 | 5.130E−01 | 3.863E−01 | 3.099E−01 |

Figure 1.9 Fugacity Level III calculation for trichloroethylene in a generic environment (dimensions defined in Table 1.2).

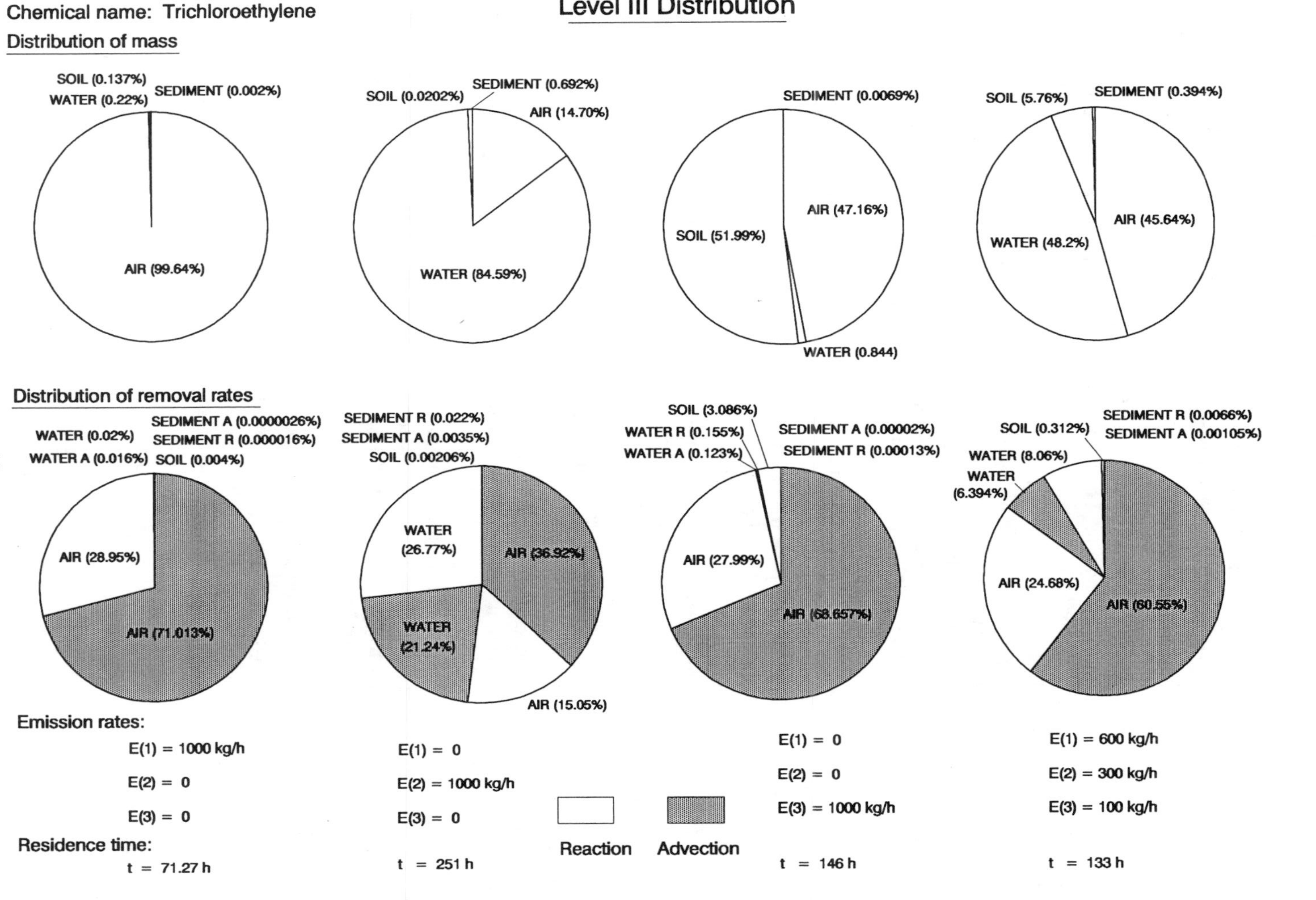


Figure 1.10 Fugacity Level III distributions of trichloroethylene for four emission scenarios in a generic environment.

The net result is that TCE behaves entirely differently when discharged to the three media. If discharged to air, it advects rapidly and reacts with a residence time of 71 h or about 3 days with little transport to soil or water. If discharged to water, it reacts and evaporates to air with a residence time of 251 h or 11 days. If discharged to soil, it mostly evaporates with an overall residence time of about 146 h or 6 days.

The final scenario is a combination of discharges, 600 kg/h to air, 300 kg/h to water, and 100 kg/h to soil. The concentrations, amounts and transport and transformation rates are merely linearly combined versions of the three initial scenarios. For example, the rate of reaction in air is now 247 kg/h. This is 0.6 of the first (air emission) rate of 290 kg/h, i.e., 174 kg/h, plus 0.3 of the second (water emission) rate of 151 kg/h, i.e., 45 kg/h and 0.1 of the third (soil emission) rate of 280 kg/h, i.e., 28 kg yielding a total of (174 + 45 + 28) or 247 kg/h. It is also apparent that the amount in the air of 60550 kg causing a concentration of 0.606 $\mu g/m^3$ is attributable to emissions to air (0.6 x 0.71 or 0.426 $\mu g/m^3$), emissions to water (0.3 x 0.369 or 0.111 $\mu g/m^3$) and emissions to soil (0.1 x 0.687 or 0.0687 $\mu g/m^3$). The concentration in water of 3.20 x $10^{-4}$ $g/m^3$ or 320 $\mu g/m^3$ or ng/L is largely attributable to the discharges to water which alone cause 0.3 x 1.06 x $10^{-3}$ $g/m^3$ or 318 $\mu g/m^3$. Although more is emitted to air it contributes only about 0.5 $\mu g/m^3$ to the water with soil emissions accounting for about 0.60 $\mu g/m^3$. Similarly, the prevailing soil concentration is controlled by the rate of discharge to the soil.

In this multimedia discharge scenario the overall residence time is 133 hours, which can be viewed as 60% of the air residence time of 71 h, 30% of the water residence time of 251 h and 10% of the overall soil residence time of 146 h. The overall amount in the environment of 132700 kg is thus largely controlled by the discharges to water and air.

Figure 1.10 shows the distributions of mass and removal process rates for these four scenarios. Generally, when TCE is discharged into a specific medium, most of the chemical is found in that medium.

Finally, it is interesting to note that the fugacities in this final case (in units of $\mu$Pa) are for the four media 11.4 (air), 2877 (water), 1025 (soil) and 2716 (sediment). The soil, sediment and water are fairly close to equilibrium, i.e., within a factor of about 3, with the air notably "undersaturated" by a factor of about 200. This is the result of the rapid loss processes from air.

The volatile organic chemicals treated in this volume tend to have similar environmental fates of TCE. Partitioning into air with subsequent loss by advection and reaction are the primary processes. The key parameters are thus half-life in air and the mass transfer coefficients governing evaporation from soil and water.

It is believed that these three behavior profiles, when combined in the fourth, give a comprehensive illustration and explanation of the environmental fate characteristics of this and other chemicals. They show which intermedia transport processes are important and how levels in various media arise from discharges into other media. The same broad fate characteristics as described in the generic environment are believed to be generally applicable to other environments. Certainly this evaluation should, in most cases, identify the key physical-chemical properties, reactions and intermedia transport parameters. With a knowledge of the key parameters, more effort can be devoted to obtaining more accurate site-specific values, and sensitivity analyses can be conducted. In essence, the evaluation translates physical-chemical data into environmental fate information.

## 1.6 REFERENCES


Abernethy, S., Mackay, D., McCarty, L.S. (1988) "Volume fraction" correlation for narcosis in aquatic organisms: The key role or partitioning. *Environ. Toxicol. Chem.* 7, 469-481.

Abramowitz, R., Yalkowsky, S.H. (1990) Estimation of aqueous solubility and melting point of PCB congeners. *Chemosphere* 21, 1221-1229.

Almgren, M., Grieser, F., Powell, J.R., Thomas, J.K. (1979) A correlation between the solubility of aromatic hydrocarbons in water and micellar solutions, with their normal boiling points. *J. Chem. Eng. Data* 24, 285-287.

Al-Sahhaf, T.A. (1989) Prediction of the solubility of hydrocarbons in water using UNIFAC. *J. Environ. Sci. Health* A24, 49-56.

Altshuller, A.P. (1980) Lifetimes of organic molecules in the troposphere and lower stratosphere. *Adv. Environ. Sci. Technol.* 10, 181-219.

Altshuller, A.P. (1991) Chemical reactions and transport of alkanes and their products in the troposphere. *J. Atmos. Chem.* 12, 19-61.

Ambrose, D. (1981) Reference value of vapor pressure. The vapor pressures of benzene and hexafluorobenzene. *J. Chem. Thermodyn.* 13, 1161-1167.

Ambrose, D., Lawrenson, L.J., Sprake, C.H.S. (1975) The vapour pressure of naphthalene. *J. Chem. Thermodyn.* 7, 1173-1176.

Amidon, G.L., Yalkowsky, S.H., Anik, S.T., Leung, S. (1975) Solubility of nonelectrolytes in polar solvents. V. Estimation of the solubility of aliphatic monofunctional compounds in water using a molecular surface area approach. *J. Phys. Chem.* 9, 2239-2245.

Amidon, G.L., Anik, S.T. (1981) Application of the surface area approach to the correlation and estimation of aqueous solubility and vapor pressure. Alkyl aromatic hydrocarbons. *J. Chem. Eng. Data* 26, 28-33.

Amidon, G.L., Williams, N.A. (1982) A solubility equation for non-electrolytes in water. *Intl. J. Pharm.* 11, 249-156.

Anderson, E., Veith, G.D., Weininger, D. (1987) *SMILES: A Line Notation and Computerized Interpreter for Chemical Structures.* US EPA Environmental Research Brief, EPA/600/M-87/021.

Andren, A.W., Doucette, W.J., Dickhut, R.M. (1987) Methods for estimating solubilities of hydrophobic organic compounds: Environmental modeling efforts. In: *Sources and Fates of Aquatic Pollutants.* Hites, R.A., Eisenreich, S.J., Eds., pp. 3-26, Advances in Chemistry Series 216, American Chemical Society, Washington, D.C.

Andrews, L.J., Keffer, R.M. (1950a) Cation complexes of compounds containing carbon-carbon double bonds. IV. The argentation of aromatic hydrocarbons. *J. Am. Chem. Soc.* 72, 3644-3647.

Andrews, L.J., Keefer, R.M. (1950b) Cation complexes of compounds containing carbon-carbon double bonds. VII. Further studies on the argentation of substituted benzenes. *J. Am. Chem. Soc.* 72, 5034-5037.

Arbuckle, W.B. (1983) Estimating activity coefficients for use in calculating environmental parameters. *Environ. Sci. Technol.* 17, 537-542.

Arbuckle, W.B. (1986) Using UNIFAC to calculate aqueous solubilities. *Environ. Sci. Technol.* 20, 1060-1064.

Ashworth, R.A., Howe, G.B., Mullins, M.E., Roger, T.N. (1988) Air-water partitioning coefficients of organics in dilute aqueous solutions. *J. Hazard. Materials* 18, 25-36.

Atkinson, R. (1985) Kinetics and mechanisms of the gas phase reaction of hydroxyl radicals with organic compounds under atmospheric conditions. *Chem. Rev.* 85, 69-201.

Atkinson, R. (1987) A structure-activity relationship for the estimation of the rate constants for the gas phase reactions of OH radicals with organic compounds. *Int. J. Chem. Kinetics* 19, 790-828.

Atkinson, R. (1988) Estimation of gas-phase hydroxyl radical rate constants for organic chemicals. *Environ. Toxicol. Chem.* 7, 435-442.

Atkinson, R. (1990) Gas-phase tropospheric chemistry of organic compounds, a review. *Atmos. Environ.* 24A, 1-41.

Atkinson, R. (1991) Kinetics and mechanisms of the gas-phase reactions of the $NO_3$ radicals with organic compounds. *J. Phys. Chem. Data* 20, 450-507.

Atkinson, R., Carter, W.L. (1984) Kinetics and mechanisms of the gas-phase reactions of ozone with organic compounds under atmospheric conditions. *Chem. Rev.* 84, 437-470.

Balson, E.W. (1947) Studies in vapour pressure measurement. Part III. An effusion manometer sensitive to $5x10^{-6}$ millimetres of mercury: vapour pressure of D.D.T. and other slightly volatile substances. *Trans. Farad. Soc.* 43, 54-60.

Banerjee, S. (1985) Calculation of water solubility of organic compounds with UNIFAC-derived parameters. *Environ. Sci. Technol.* 19, 369-370.

Banerjee, S., Howard, P.H. (1988) Improved estimation of solubility and partitioning through correction of UNIFAC-derived activity coefficients. *Environ. Sci. Technol.* 22, 839-841.

Banerjee, S., Howard, P.H., Lande, S.S. (1990) General structure-vapor pressure relationships for organics. *Chemosphere* 21, 1173-1180.

Banerjee, S., Yalkowsky, S.H., Valvani, S.C. (1980) Water solubiltiy and octanol/water partition coefficients of organics. Limitations of the solubility-partition coefficient correlation. *Environ. Sci. Technol.* 14, 1227-1229.

Bidleman, T.F. (1984) Estimation of vapor pressures for nonpolar organic compounds by capillary gas chromatography. *Anal. Chem.* 56, 2490-2496.

Bohon, R.L., Claussen, W.F. (1951) The solubility of aromatic hydrocarbons in water. *J. Am. Chem. Soc.* 73, 1571-1576.

Bondi, A. (1964) van der Waals volumes and radii. *J. Phys. Chem.* 68, 441-451.

Booth, H.S., Everson, H.E. (1948) Hydrotropic solublities: solublities in 40 percent sodium xylenesulfonate. *Ind. Eng. Chem.* 40, 1491-1493.

Boublik, T., Fried, V., Hala, E. (1973) *The Vapor Pressure of Pure Substances*, Elsevier, Amsterdam.

Boublik, T., Fried, V., Hala, E. (1984) *The Vapor Pressure of Pure Substances*, 2nd revised ed., Elsevier, Amsterdam.

Bowman, B.T., Sans, W.W. (1983) Determination of octanol-water partitioning coefficient ($K_{OW}$) of 61 organophosphorus and carbamate insecticides and their relationship to respective water solubility (S) values. *J. Environ. Sci. Health* B18, 667-683.

Bradley, R.S., Cleasby, T.G. (1953) The vapour pressure and lattice energy of some aromatic ring compounds. *J. Chem. Soc.* 1953, 1690-1692.

Briggs, G.G. (1981) Theoretical and experimental relationships between soil adsorption, octanol-water partition coefficients, water solubilities, bioconcentration factors, and the Parachor. *J. Agric. Food Chem.* 29, 1050-1059.

Brodsky, J., Ballschmiter, K. (1988) Reversed phase liquid chromatography of PCBs as a basis for the calculation of water solubility and log $K_{OW}$ for polychlorobiphenyls.

Bruggeman, W.A., van der Steen, J., Hutzinger, O. (1982) Reversed-phase thin-layer chromatography of polynuclear aromatic hydrocarbons and chlorinated biphenyls. Relationship with hydrophobicity as measured by aqueous solubility and octanol-water partition coefficient. *J. Chromatogr.* 238, 335-346.

Budavari, S., Ed. (1989) *The Merck Index. An Encyclopedia of Chemicals, Drugs and Biologicals.* 11th ed., Merck & Co. Inc., Rahway, New Jersey.

Burkhard, L.P. (1984) *Physical-Chemical Properties of the Polychlorinated Biphenyls: Measurement, Estimation, and Application to Environmental Systems.* Ph.D. Thesis, University of Wisconsin-Madison, Wisconsin.

Burkhard, L.P., Andren, A.W., Armstrong, D.E. (1985a) Estimation of vapor pressures for polychlorinated biphenyls: A comparison of eleven predictive methods. *Environ. Sci. Technol.* 19, 500-507.

Burkhard, L.P., Armstrong, D.E., Andren, A.W. (1985b) Henry's law constants for polychlorinated biphenyls. *Environ. Sci. Technol.* 590-595.

Burkhard, L.P., Kuehl, D.W., Veith G.D. (1985c) Evaluation of reversed phase liquid chromatograph/mass spectrometry for estimation of n-octanol/water partition coefficients of organic chemicals. *Chemosphere* 14, 1551-1560.

Buxton, G.V., Greenstock, G.L., Helman, W.P., Ross, A.B. (1988) Critical review of rate constants for reactions of hydrated electrons, hydrogen atoms and hydroxyl radicals in aqueous solutions. *J. Phys. Chem. Data* 17, 513-886.

Chiou, C.T. (1981) Partition coefficient and water solubility in environmental chemistry. In: *Hazard Assessment of Chemicals Current Developments*. Vol. 1, pp. 117-153. Academic Press, New York.

Chiou, C.T. (1985) Partition coefficients of organic compounds in lipid-water systems and correlations with fish bioconcentration factors. *Environ. Sci. Technol.* 19, 57-62.

Chiou, C.T., Freed, V.H., Schmedding, D.W. (1977) Partition coefficient and bioaccumulation of selected organic chemicals. *Environ. Sci. Technol.* 11, 475-478.

Chiou, C.T., Schmedding, D.W., Manes, M. (1982) Partitioning of organic compounds in octanol-water system. *Environ. Sci. Technol.* 16, 4-10.

Coates, M., Connell, D.W., Barron, D.M. (1985) Aqueous solubility and octan-1-ol to water partition coefficients of aliphatic hydrocarbons. *Environ. Sci. Technol.* 19, 628-632.

Davies, R.P., Dobbs, A.J. (1984) The prediction of bioconcentration in fish. *Water Res.* 18, 1253-1262.

Davis, W.W., Krahl, M.E., Clowes, G.H. (1942) Solubility of carcinogenic and related hydrocarbons in water *J. Am. Chem. Soc.* 64, 108-110.

Davis, W.W., Parke, Jr, T.V. (1942) A nephelometric method for determination of solubilities of extremely low order. *J. Am. Chem. Soc.* 64, 101-107.

Dean, J.D., Ed. (1979) *Lange's Handbook of Chemistry.* 12th ed., McGraw-Hill, New York, N.Y.

Dean, J.D., Ed. (1985) *Lange's Handbook of Chemistry.* 13th ed., McGraw-Hill, New York, N.Y.

Dearden, J.C. (1990) Physico-chemical descriptors. In: *Practical Applications of Quantitative Structure-Activity Relationships (QSAR) in Environmental Chemistry and Toxicology.* Karcher, W. and Devillers, J., Eds., pp. 25-60. Kluwer Academic Publisher, Dordrecht, Netherlands.

De Bruijn, J., Busser, G., Seinen, W., Hermens, J. (1989) Determination of octanol/water partition coefficient for hydrophobic organic chemicals with the "slow-stirring" method. *Environ. Toxicol. Chem.* 8, 499-512.

De Bruijn, J., Hermens, J. (1990) Relationships between octanol/water partition coefficients and total molecular surface area and total molecular volume of hydrophobic organic chemicals. *Quant. Struct.-Act. Relat.* 9, 11-21.

Dobbs, A.J., Grant, C. (1980) Pesticide volatilization rate: a new measurement of the vapor pressure of pentachlorophenol at room temperature. *Pestic. Sci.* 11, 29-32.

Dobbs, A.J., Cull, M.R. (1982) Volatilization of chemical relative loss rates and the estimation of vapor pressures. *Environ. Pollut. (Ser. B)* 3, 289-298.

Doucette, W.J., Andren, A.W. (1987) Correlation of octanol/water partition coefficients and total molecular surface area for highly hydrophobic aromatic compounds. *Environ. Sci. Technol.* 21, 521-524.

Doucette, W.J., Andren, A.W. (1988a) Aqueous solubility of selected biphenyl, furan, and dioxin congeners. *Chemosphere* 17, 243-252.

Doucette, W.J., Andren, A.W. (1988b) Estimation of octanol/water partition coefficients: Evaluation of six methods for highly hydrophobic aromatic hydrocarbons. *Chemosphere* 17, 345-359.

Dreisbach, R.R. (1955) *Physical Properties of Chemical Compounds.* No. 15 of the Adv. in Chemistry Series, American Chemical Society, Washington, D.C.

Dreisbach, R.R. (1959) *Physical Properties of Chemical Compounds, II.* No. 22, Adv. in Chemistry Series, American Chemical Society, Washington, D.C.

Dreisbach, R.R. (1961) *Physical Properties of Chemical Compounds, III.* No. 25, Adv. in Chemistry Series, American Chemical Society, Washington, D.C.

Dunnivant, F.M., Coate, J.T., Elzerman, A.W. (1988) Experimentally determined Henry's law constants for 17 polychlorobiphenyl congeners. *Environ. Sci. Technol.* 22, 448-453.

Dunnivant, F.M., Elzerman, A.W., Jurs, P.C., Hansen, M.N. (1992) Quantitative structure-property relationships for aqueous solubilities and Henry's law constants of polychlorinated biphenyls. *Environ. Sci. Technol.* 26, 1567-1573.

Ellington, J.J. (1989) *Hydrolysis Rate Constants for Enhancing Property-Reactivity Relationships*. USEPA, EPA/600/3-89/063, Athens, GA.

Ellington, J.J., Stancil, Jr., F.E., Payne, W.D. (1987a) *Measurements of Hydrolysis Rate Constant for Evaluation of Hazardous Land Disposal: Volume I. Data on 32 Chemicals*. USEPA, EPA/600/3-86/043, Athens, GA.

Ellington, J.J., Stancil, Jr., F.E., Payne, W.D., Trusty, C.D. (1987b) *Measurements of Hydrolysis Rate Constant for Evaluation of Hazardous Land Disposal: Volume II. Data on 54 Chemicals*. USEPA, EPA/600/3-88/028, Athens, GA.

Ellington, J.J., Stancil, Jr., F.E., Payne, W.D., Trusty, C.D. (1988) *Interim Protocol for Measurement Hydrolysis Rate Constants in Aqueous Solutions*. USEPA, EPA/600/3-88/014, Athens, GA.

Fendinger, N.J., Glotfelty, D.E. (1988) A laboratory method for the experimental determination of air-water Henry's law constants for several pesticides. *Environ. Sci. Technol.* 22, 1289-1293.

Fendinger, N.J., Glotfelty, D.E. (1990) Henry's law constants for selected pesticides, PAHs and PCBs. *Environ. Toxicol. Chem.* 9, 731-735.

Foreman, W.T., Bidleman, T.F. (1985) Vapor pressure estimates of individual polychlorinated biphenyls and commercial fluids using gas chromatographic retention data. *J. Chromatogr.* 330, 203-216.

Fredenslund, A., Jones, R.L., Prausnitz, J.M. (1975) Group-contribution estimation of activity coefficients in nonideal liquid mixtures. *AIChE J.* 21, 1086-1099.

Fujita, T., Iwasa, J., Hansch, C. (1964) A new substituent constant, "pi" derived from partition coefficients. *J. Am. Chem. Soc.* 86, 5175-5180.

Garst, J.E. (1984) Accurate, wide-range, automated, high-performance chromatographic method for the estimation of octanol/water partition coefficients. II: Equilibrium in partition coefficient measurements, additivity of substituent constants, and correlation of biological data. *J. Pharm. Sci.* 73, 1623-1629.

Garten, C.T., Trabalka, J.R. (1983) Evaluation of models for predicting terrestrial food chain behavior of xenobiotics. *Environ. Sci. Technol.* 17, 590-595.

Gmeheling, J., Rasmussen, P., Fredenslund, A. (1982) Vapor-liquid equilibria by UNIFAC group contribution. Revision and extension. 2 *Ind. Eng. Chem. Process Des. Dev.* 21, 118-127.

Gossett, R. (1987) Measurement of Henry's law constants for $C_1$ and $C_2$ chlorinated hydrocarbons. *Environ. Sci. Technol.* 21, 202-208.

Gross, P.M., Saylor, J.H. (1931) The solubilities of certain slightly soluble organic compounds in water. *J. Am. Chem. Soc.* 1931, 1744-1751.

Gückel, W., Rittig, R., Synnatschke, G. (1974) A method for determining the volatility of active ingredients used in plant protection. II. Application to formulated products. *Pestic. Sci.* 5, 393-400.

Gückel, W. Kästel, R., Lawerenz, J., Synnatschke, G. (1982) A method for determining the volatility of active ingredients used in plant protection. Part III: The temperature relationship between vapour pressure and evaporation rate. *Pestic. Sci.* 13, 161-168.

Hafkenscheid, T.L., Tomlinson, E. (1981) Estimation of aqueous solubilities of organic non-electrolytes using liquid chromatographic retention data. *J. Chromatogr.* 218, 409-425.

Hamaker, J.W. Kerlinger, H.O. (1969) Vapor pressures of pesticides. *Adv. Chem. Ser.* 86, 39-54.

Hamliton, D.J. (1980) Gas chromatographic measurement of volatility of herbicide esters. *J. Chromatogr.* 195, 75-83.

Hansch, C., Leo, A. (1979) *Substituent Constants for Correlation Analysis in Chemistry and Biology*. Wiley-Interscience, New York, N.Y.

Hansch, C., Quinlan, J.E., Lawrence, G.L. (1968) The linear-free energy relationship between partition coefficient and aqueous solubility of organic liquids. *J. Org. Chem.* 33, 347-350.

Hansen, H.K., Schiller, M., Gmehling, J. (1991) Vapor-liquid equlibria by UNIFAC group contribution. 5. Revision and extension. *Ind. Eng. Chem. Res.* 30, 2362-2356.

Hawker, D.W. (1989) The relationship between octan-1-ol/water partition coefficient and aqueous solubility in terms of solvatochromic parameters. *Chemosphere* 19, 1586-1593.

Hawker, D.W. (1990a) Vapor pressures and Henry's law constants of polychlorinated biphenyls. *Environ. Sci. Technol.* 23, 1250-1253.

Hawker, D.W. (1990b) Description of fish bioconcentration factors in terms of solvatochromic parameters. *Chemosphere* 20, 267-477.

Hawker, D.W., Connell, D.W. (1988) Octanol-water partition coefficients of polychlorinated biphenyl congeners. *Environ. Sci. Technol.* 22, 382-387.

Hermann, R.B. (1971) Theory of hydrophobic bonding. II. The correlation of hydrocarbon solubility in water with solvent cavity surface area. *J. Phys. Chem.* 76, 2754-2758.

Hermens, J.L.M., Opperhuizen, A., Eds. (1991) *QSAR in Environmental Toxicology IV.* Elsevier, Amsterdam, The Netherlands. Also published as Volumes 109/110 in *Science for the Total Environment* 1991.

Hinckley, D.A., Bidleman, T.F., Foreman, W.T. (1990) Determination of vapor pressures for nonpolar and semipolar organic compounds from gas chromatographic retention data. *J. Chem. Eng. Data* 35, 232-237.

Hine, J., Mookerjee, P.K. (1975) The intrinsic hydrophilic character of organic compounds. Correlations in terms of structural contributions. *J. Org. Chem.* 40, 292-298.

Hollifield, H.C. (1979) Rapid nephelometric estimate of water solubility of highly insoluble organic chemicals of environmental interest. *Bull. Environ. Contam. Toxicol.* 23, 579-586.

Horvath, A.L. (1982) *Halogenated Hydrocarbons, Solubility - Miscibility with Water.* Marcel Dekker, Inc., New York, N.Y.

Howard, P.H., Ed. (1989) *Handbook of Fate and Exposure Data for Organic Chemicals. Vol. I. Large Production and Priority Pollutants.* Lewis Publishers, Chelsea, Michigan.

Howard, P.H., Ed. (1990) *Handbook of Fate and Exposure Data for Organic Chemicals. Vol. - II - Solvents.* Lewis Publishers, Inc., Chelsea, Michigan.

Howard, P.H., Boethling, R.S., Jarvis, W.F., Meylan, W.M., Michalenko, E.M. (1991) *Handbook of Environmental Degradation Rates.* Lewis Publishers, Inc., Chelsea, Michigan.

Hussam, A., Carr, P.W. (1985) A study of a rapid and precise methodology for the measurement of vapor liquid equilibria by headspace gas chromatography. *Anal. Chem.* 57, 793-801.

Irmann, F. (1965) Eine einfache korrelation zwishen wasserlöslichkeit und strucktur von kohlenwasserstoffen und halogenkohlenwasserstoffen. *Chemie-Ing.-Techn.* 37, 789-798.

Isnard, P., Lambert, S. (1988) Estimating bioconcentration factors for octanol-water partition coefficient and aqueous solubility. *Chemosphere* 17, 21-34.

Isnard, P., Lambert, S. (1989) Aqueous solubility/n-octanol-water partition coefficient correlations. *Chemosphere* 18, 1837-1853.

*IUPAC Solubility Data Series* (1985) *Vol. 20: Halogenated Benzenes, Toluenes and Phenols with Water.* Horvath, A.L., Getzen, F.W., Eds., Pergamon Press, Oxford, England.

*IUPAC Solubility Data Series* (1989a) *Vol. 37: Hydrocarbons ($C_5$ - $C_7$) with Water and Seawater.* Shaw, D.G., Ed., Pergamon Press, Oxford, England.

*IUPAC Solubility Data Series* (1989b) *Vol. 38: Hydrocarbons ($C_8$ -$C_{36}$) with Water and Seawater.* Shaw, D.G., Ed., Pergamon Press, Oxford, England.

Jensen, T., Fredenslund, A., Rasmussen, P. (1981) Pure-compound vapor pressures using UNIFAC group contribution. *Ind. Eng. Chem. Fundam.* 20, 239-246.

Jönsson, J.A., Vejrosta, J., Novak, J. (1982) Air/water partition coefficients for normal alkanes (*n*-pentane to *n*-nonane). *Fluid Phase Equil.* 9, 279-286.

Jury, W.A., Spencer, W.F., Farmer, W.J. (1983) Behavior assessment model for trace organics in soil: I. Model description. *J. Environ. Qual.* 12, 558-566.

Jury, W.A., Farmer, W.J., Spencer, W.F. (1984a) Behavior assessment model for trace organics in soil: II. Chemical classification and parameter sensitivity. *J. Environ. Qual.* 13, 567-572.

Jury, W.A., Farmer, W.J., Spencer, W.F. (1984b) Behavior assessment model for trace organics in soil: III. Application of screening model. *J. Environ. Qual.* 13, 573-579.

Jury, W.A., Spencer, W.F., Farmer, W.J. (1984) Behavior assessment model for trace organics in soil: IV. Review of experimental evidence. *J. Environ. Qual.* 13, 580-587.

Kabadi, V.N., Danner, R.P. (1979) Nomograph solves for solubilities of hydrocarbons in water. *Hydrocarbon Processing* 68, 245-246.

Kamlet, M.J., Doherty, R.M., Abraham, M.H., Carr, P.W., Doherty, R.F., Raft, R.W. (1987) Linear solvation energy relationships. Important differences between aqueous solublity relationships for aliphatic and aromatic solutes. *J. Phy. Chem.* 91, 1996.

Kamlet, M.J., Doherty, R.M., Carr, P.W., Mackay, D., Abraham, M.H., Taft, R.W. (1988) Linear solvation energy relationships. 44. Parameter estimation rules that allow accurate prediction of octanol/water partition coefficients and other solubility and toxicity properties of polychlorinated biphenyls and polycyclic aromatic hydrocarbons. *Environ. Sci. Technol.* 22, 503-509.

Karcher, W., Devillers, J., Eds., (1990) *Practical Applications of Quantitative-Structure-Activity Relationships (QSAR) in Environmental Chemistry and Toxicology.* Kluwer Academic Publisher, Dordrecht, Netherlands.

Karger, B.L., Sewell, P.A., Castells, R.C., Hartkopf, A. (1976) Gas chromatographic study of the adsorption of insoluble vapors on water. *J. Colloid & Interface Sci.* 35(2), 328-339.

Karickhoff, S.W. (1981) Semiempirical estimation of sorption of hydrophobic pollutants on natural sediments and soil. *Chemosphere* 10, 833-846.

Karickhoff, S.W., Brown, D.S., Scott, T.A. (1979) Sorption of hydrophobic pollutants on natural water sediments. *Water Res.* 13, 241-248.

Kenaga, E.E. (1980) Predicted bioconcentration factors and soil sorption coefficients of pesticides and other chemicals. *Ecotox. Environ. Saf.* 4, 26-38.

Kenaga, E.E., Goring, C.A.I. (1980) Relationship between water solubility, soil sorption, octanol-water partitioning, and concentration of chemicals in biota. In: *Aquatic Toxicology.* Eaton, J.G., Parrish, P.R., Hendrick, A.C., Eds., pp. 78-115, Am. Soc. for Testing and Materials, STP 707, Philadelphia.

Kier, L.B., Hall, L.H. (1976) Molar properties and molecular connectivity. In: *Molecular Connectivity in Chemistry and Drug Design.* Medicinal Chem. Vol. 14, pp. 123-167, Academic Press, New York.

Kier, L.B., Hall, L.H. (1986) *Molecular Connectivity in Structure-Activity Analysis.* Wiley, New York.

Kikic, I., Alesse, P., Rasmussen, P., Fredenslunds, A. (1980) On the combinatorial part of the UNIFAC and UNIQUAC models. *Can. J. Chem. Eng.* 58, 253-258.

Kim, Y.-H., Woodrow, J.E., Seiber, J.N. (1984) Evaluation of a gas chromatographic method for calculating vapor pressures with organophosphorus pesticides. *J. Chromotagr.* 314, 37-53.

Klöpffer, W. (1991) Photochemistry in environmental research: Its role in abiotic degradation and exposure analysis. *EPA Newsletter* 41, 24-39.

Könemann, H., van Leeuewen, K. (1980) Toxicokinetics in fish: accumulation of six chlorobenzenes by guppies. *Chemosphere* 9, 3-19.

Könemann, H., Zelle, R., Busser, F. (1979) Determination of log $P_{oct}$ values of chloro-substituted benzenes, toluenes and anilines by high-performance liquid chromatography on ODS-silica. *J. Chromatogr.* 178, 559-565.

Korenman, I.M., Gur'ev, I.A., Gur'eva, Z.M. (1971) Solubility of liquid aliphatic compounds in water. *Russ. J. Phys. Chem.* 45, 1065-1066.

Lande, S.S., Banerjee, S. (1981) Predicting aqueous solubility of organic nonelectrolytes from molar volume. *Chemosphere* 10, 751-759.

Leahy, D.E. (1986) Intrinsic molecular volume as a measure of the cavity term in linear solvation energy relationships: octanol-water partition coefficients and aqueous solubilities. *J. Pharm. Sci.* 75, 629-636.

Leighton, D.T.,Jr., Calo, J.M. (1981) Distribution coefficients of chlorinated hydrocarbons in dilute air-water systems for groundwater contamination applications. *J. Chem. Eng. Data* 26, 382-385.

Leo, A., Hansch, C., Elkins, D. (1971) Partition coefficients and their uses. *Chem. Rev.* 71, 525-616.

Lincoff, A.H., Gossett, J.M. (1984) The determination of Henry's law constants for volatile organics by equilibrium partitioning in closed systems. In: *Gas Transfer at Water Surfaces*. Brutsaert, W., Jirka, G.H., Eds., pp. 17-26, D. Reidel Publishing Co., Dordrecht, Holland.

Lo, J.M., Tseng, C.L., Yang, J.Y. (1986) Radiometric method for determining solubility of organic solvents in water. *Anal. Chem.* 58, 1596-1597.

Locke, D. (1974) Selectivity in reversed-phase liquid chromatography using chemically bonded stationary phases. *J. Chromatogr. Sci.* 12, 433-437.

Lyman, W.J., Reehl, W.F., Rosenblatt, D.H. (1982) *Handbook of Chemical Property Estimation Methods*. McGraw-Hill, New York.

Mabey, W., Mill, T. (1978) Critical review of hydrolysis of organic compounds in water under environmental conditions. *J. Phys. Chem. Ref. Data* 7, 383-414.

Mabey, W.J., Mill, T., Podoll, R.T. (1984) *Estimation Methods for Process Constants and Properties used in Fate Assessment.* USEPA, EPA-600/3-84-035, Athens, GA.

Mackay, D. (1979) Finding fugacity feasible. *Environ. Sci. Technol.* 13, 1218-1223.

Mackay, D. (1982) Correlation of bioconcentration factors. *Environ. Sci. Technol.* 16, 274-278.

Mackay, D. (1991) *Multimedia Environmental Models. The Fugacity Approach.* Lewis Publishers, Inc., Chelsea, Michigan.

Mackay, D., Bobra, A.M., Shiu. W.Y., Yalkowsky, S.H. (1980) Relationships between aqueous solubility and octanol-water partition coefficient. *Chemosphere* 9, 701-711.

Mackay, D., Bobra, A.M., Chan, D.W., Shiu, W.Y. (1982) Vapor pressure correlation for low-volatility environmental chemicals. *Environ. Sci. Technol.* 16, 645-649.

Mackay, D., Paterson, S. (1990) Fugacity models. In: *Practical Applications of Quantitative Structure-Activity Relationships (QSAR) in Environmental Chemistry and Toxicology.* Karcher, W., Devillers, J., Eds., pp. 433-460, Kluwer Academic Publishers, Dordrecht, Holland.

Mackay, D., Paterson, S. (1991) Evaluating the multimedia fate of organic chemicals: A Level III fugacity model. *Environ. Sci. Technol.* 25, 427-436.

Mackay, D., Shiu, W.Y. (1977) Aqueous solubility of polynuclear aromatic hydrocarbons. *J. Chem. Eng. Data* 22, 339-402.

Mackay, D., Shiu, W.Y. (1981) A critical review of Henry's law constants for chemicals of environmental interest. *J. Phys. Chem. Ref. Data* 11, 1175-1199.

Mackay, D., Shiu, W.Y., Sutherland, R.P. (1979) Determination of air-water Henry's law constants for hydrophobic pollutants. *Envrion. Sci. Technol.* 13, 333-337.

Mackay, D., Shiu, W.Y., Wolkoff, A.W. (1975) Gas chromatographic determination of low concentration of hydrocarbons in water by vapor phase extraction. *ASTM STP 573,* pp. 251-258, American Society for Testing and Materials, Philadelphia, Pa.

Mackay, D., Stiver, W.H. (1991) Predictability and environmental chemistry. In: *Environmental Chemistry of Herbicides*. Vol. II. Grover, R., Lessna, A.J., Eds., pp. 281-297. CRC Press, Boca Raton, FL.

Macknick, A.B., Prausnitz, J.M. (1979) Vapor pressure of high-molecular weight hydrocarbons. *J. Chem. Eng. Data* 24, 175-178.

Magnussen, T., Rasmussen, P., Fredenslund, A. (1981) UNIFAC parameter table for prediction of liquid-liquid equilibria. *Ind. Eng. Chem. Process Des. Dev.* 20, 331-339.

Mailhot, H., Peters, R.H. (1988) Empirical relationships between the 1-octanol/water partition coefficient and nine physicochemical properties. *Environ. Sci. Technol.* 22, 1479-1488.

May, W.E., Wasik, S.P., Freeman, D.H. (1978a) Determination of the aqueous solubility of polynuclear aromatic hydrocarbons by a coupled-column liquid chromatographic technique. *Anal. Chem.* 50, 175-179.

May, W.E., Wasik, S.P., Freeman, D.H. (1978b) Determination of the solubility behavior of some polycyclic aromatic hydrocarbons in water. *Anal. Chem.* 50, 997-1000.

McAuliffe, C. (1966) Solubility in water of paraffin, cycloparaffin, olefin, acetylene, cycloolefin and aromatic hydrocarbons. *J. Phys. Chem.* 76, 1267-1275.

McAuliffe, C. (1971) GC determination of solutes by multiple phase equilibration. *Chem. Tech.* 1, 46-51.

McDuffie, B. (1981) Estimation of octanol/water partition coefficient for organic pollutants using reversed phase HPLC. *Chemosphere* 10, 73-83.

McGowan, J.C., Mellors, A. (1986) *Molecular Volumes in Chemistry and Biology-Applications including Partitioning and Toxicity.* Ellis Horwood Limited, Chichester, England.

The Merck Index (1989) *An Encyclopedia of Chemicals, Drugs and Biologicals.* 11th ed., Budavari, S., Ed., Merck and Co., Inc., Rahway, N.J.

Meylan, W.M., Howard, P.H. (1991) Bond contribution method for estimating Henry's law constants. *Environ. Toxicol. Chem.* 10, 1283-1293.

Meylan, W.M., Howard, P.H., Boethling, R.S. (1992) Molecular topology/fragment contribution for predicting soil sorption coefficient. *Environ. Sci. Technol.* 26, 1560-1567.

Mill, T. (1982) Hydrolysis and oxidation processes in the environment. *Environ. Toxicol. Chem.* 1, 135-141.

Mill, T. (1989) Structure-activity relationships for photooxidation processes in the environment. *Environ. Toxicol. Chem.* 8, 31-45.

Mill, T. (1993) Environmental chemistry. In: *Ecological Risk Assessment*. Suter, II, G.W., Ed., pp. 91-127.

Mill, T., Mabey, W. (1985) Photodegradation in water. In: *Environmental Exposure from Chemicals.* Vol. 1. Neely, W.B., Blau, G.E., Eds., pp. 175-216, CRC Press, Boca Raton, FL.

Miller, M.M., Ghodbane, S., Wasik, S.P., Tewari, Y.B., Martire, D.E. (1984) Aqueous solubilities, octanol/water partition coefficients and entropies of melting of chlorinated benzenes and biphenyls. *J. Chem. Eng. Data* 29, 184-190.

Miller, M.M., Wasik, S.P., Huang, G.L., Shiu, W.Y., Mackay, D. (1985) Relationships between octanol-water partition coefficient and aqueous solublity. *Environ. Sci. Technol.* 19, 522-529.

Munz, C., Roberts, P.V. (1987) Air-water phase equilibria of volatile organic solutes. *J. Am. Wat. Works Assoc.* 79, 62-69.

Neely, W.B., Branson, D.R., Blau, G.E. (1974) Partition coefficient to measure bioconcentration potential of organic chemicals in fish. *Environ. Sci. Technol.* 8, 1113-1115.

Nirmalakhandan, N.N., Speece, R.E. (1988a) Prediction of aqueous solubility of organic chemicals based on molecular structure. *Environ. Sci. Technol.* 22, 328-338.

Nirmalakhandan, N.N., Speece, R.E. (1988b) QSAR model for predicting Henry's law constant. *Environ. Sci. Technol.* 22, 1349-1357.

Nirmalakhandan, N.N., Speece, R.E. (1989) Prediction of aqueous solubility of organic chemicals based on molecular structure. 2. Application to PNAs, PCBs, PCDDs, etc. *Environ. Sci. Technol.* 23, 708-713.

Okouchi, H., Saegusa, H., Nojima, O. (1992) Prediction of environmental parameters by adsorbability index: water solubilities of hydrophobic organic pollutants. *Environ. Intl.* 18, 249-261.

Oliver, B.G. (1984) The relationship between bioconcentration factor in rainbow trout and physical-chemical properties for some halogenated compounds. In: *QSAR in Environmental Toxicology.* Kaiser, K.L.E., Ed., pp. 300-317, D. Reidel Publishing Co., Dordrecht, Holland.

Oliver, B.G., Niimi, A.J. (1988) Trophodynamic analysis of polychlorinated biphenyl congeners and other chlorinated hydrocarbons in the Lake Ontario ecosystem. *Environ. Sci. Technol.* 22, 388-397.

Osborn, A.G., Douslin, D.R. (1975) Vapor pressures and derived enthalpies of vaporization of some condensed-ring hydrocarbons. *J. Chem. Eng. Data* 20, 229-231.

Paterson, S., Mackay, D. (1985) The fugacity concept in environmental modelling. In: *The Handbook of Environmental Chemistry.* Vol. 2/Part C, Hutzinger, O., Ed., pp. 121-140. Springer-Verlag, Heidelberg, Germany.

Pearlman, R.S. (1980) Molecular surface areas and volumes and their use in structure/activity relationships. In: *Physical Chemical Properties of Drugs.* Yalkowsky, S.H., Sinkula, A.A., Valvani, S.C., Eds., Medicinal Research Series, Vol. 10., pp. 321-317, Marcel Dekker, Inc., New York.

Pearlman, R.S. (1986) Molecular surface area and volume: Their calculation and use in predicting solubilities and free energies of desolvation. In: *Partition coefficient, Determination and Estimation.* Dunn III, W.J., Block, J.H., Pearlman R.S., Eds., pp. 3-20, Pergamon Press, New York.

Pearlman, R.S., Yalkowsky, S.H., Banerjee, S. (1984) Water solubilities of polynuclear aromatic and heteroaromatic compounds. *J. Phys. Chem. Ref. Data* 13, 555-562.

Pierotti, C., Deal, C., Derr, E. (1959) Activity coefficient and molecular structure. *Ind. Eng. Chem. Fundam.* 51, 95-101.

Polak, J., Lu, B.C.Y. (1973) Mutual solubilities of hydrocarbons and water at 0° and 25°C. *Can. J. Chem.* 51, 4018-4023.

Rapaport, R.A., Eisenreich, S.J. (1984) Chromatographic determination of octanol-water partition coefficients ($K_{OW}$'s) for 58 polychlorinated biphenyl congeners. *Environ. Sci. Technol.* 18, 163-170.

Reid, R.C., Prausnitz, J.M., Polling, B.E. (1987) *The Properties of Gases and Liquids.* 4th ed., McGraw-Hill, New York, N.Y.

Rekker, R.F. (1977) *The Hydrophobic Fragmental Constant.* Elsevier, Amsterdam/New York N.Y.

Riddick, J.A., Bunger, W.B., Sakano, T.K. (1986) *Organic Solvents, Physical Properties and Methods of Purification.* 4th ed., Wiley-Science Publication, John Wiley & Sons, New York, N.Y.

Rordorf, B.F. (1985a) Thermodynamic and thermal properties of polychlorinated compounds: the vapor pressures and flow tube kinetic of ten dibenzo-para-dioxins. *Chemosphere* 14, 885-892.

Rordorf, B.F. (1985b) Thermodynamic properties of polychlorinated compounds: the vapor pressures and enthalpies of sublimation of ten dibenzo-p-dioxins. *Thermochimica Acta*, 85, 435-438.

Rordorf, B.F. (1986) Thermal properties of dioxins, furans and related compounds, *Chemosphere* 15, 1325-1332.

Sabljic, A. (1984) Predictions of the nature and strength of soil sorption of organic pollutants by molecular topology. *J. Agric. Food Chem.* 32, 243-246.

Sabljic, A. (1987) On the prediction of soil sorption coefficients of organic pollutants from molecular structure: Application of molecular topology model. *Environ. Sci. Technol.* 21, 358-366.

Sabljic, A., Lara, R., Ernst, W. (1989) Modelling association of highly chlorinated biphenyls with marine humic substances. *Chemosphere* 19, 1665-1676.

Sabljic, A., Güsten, H. (1989) Predicting Henry's law constants for polychlorinated biphenyls, *Chemosphere* 19, 1503-1511.

Sabljic, A., Güsten, H. (1990) Predicting the night-time $NO_3$ radical reactivity in the troposphere. *Atmos. Environ.* 24A, 73-78.

Schwarz, F.P. (1980) Measurement of the solubilities of slightly soluble organic liquids in water by elution chromatography. *Anal. Chem.* 52, 10-15.

Schwarz, F.P., Miller, J. (1980) Determination of the aqueous solubilities of organic liquids at 10.0, 20.0, 30.0 °C by elution chromatography. *Anal. Chem.* 52, 2162-2164.

Schwarzenbach, R.P., Westall, J. (1981) Transport of nonpolar compounds from surface water to groundwater. Laboratory sorption studies. *Environ. Sci. Technol.* 11, 1360-1367.

Sears, G.W., Hopke, E.R. (1947) Vapor pressures of naphthalene, anthracene and hexachlorobenzene in a low pressure region. *J. Am. Chem. Soc.* 71, 1632-1634.

Shiu, W.Y., Mackay, D. (1986) A critical review of aqueous solubilities, vapor pressures, Henry's law constants, and octanol-water partition coefficients of the polychlorinated biphenyls. *J. Phys. Chem. Ref. Data* 15, 911-929.

Shiu, W.Y., Gobas, F.A.P.C., Mackay, D. (1987) Physical-chemical properties of three congeneric series of chlorinated aromatic hydrocarbons. In: *QSAR in Environmental Toxicology -II.* Kaiser, K.L.E., Ed., pp. 347-362. D. Reidel Publishing Co., Dordrecht, Holland.

Shiu, W.Y., Doucette, W., Gobas, F.A.P.C., Mackay, D., Andren, A.W. (1988) Physical-chemical properties of chlorinated dibenzo-p-dioxins. *Environ. Sci. Technol.* 22, 651-658.

Sinke, G.C. (1974) A method for measurement of vapor pressures of organic compounds below 0.1 torr. Naphthalene as reference substance. *J. Chem. Thermodyn.* 6, 311-316.

Sonnefeld, W.J., Zoller, W.H., May, W.E. (1983) Dynamic coupled-column liquid chromatographic determination of ambient temperature vapor pressures of polynuclear aromatic hydrocarbons. *Anal. Chem.* 55, 275-280.

Spencer, W.F., Cliath, M.M. (1969) Vapor density of dieldrin. *Environ. Sci. Technol.* 3, 670-674.

Spencer, W.F., Cliath, M.M. (1970) Vapor density and apparent vapor pressure of lindane (γ-BHC). *J. Agric. Food Chem.* 18, 529-530.

Spencer, W.F., Cliath, M.M. (1972) Volatility of DDT and related compounds. *J. Agric. Food Chem.* 20, 645-649.

Stephenson, R.M., Malanowski, A. (1987) *Handbook of the Thermodynamics of Organic Compounds.* Elsevier, New York.

Stiver, W., Mackay, D. (1989) The linear additivity principle in environmental modelling: Application to chemical behaviour in soil. *Chemosphere* 19, 1187-1198.

Sugiyama, T., Takeuchi, T., Suzuki, Y. (1975) Thermodynamic properties of solute molecules at infinite dilution determined by gas-liquid chromatography. I. Intermolecular energies of *n*-alkane solutes in $C_{28}$ - $C_{36}$ *n*-alkane solvents. *J. Chromatogr.* 105, 265-272.

Suntio, L.R., Shiu, W.Y., Mackay, D. (1988) Critical review of Henry's law constants for pesticides. *Rev. Environ. Contam. Toxicol.* 103, 1-59.

Swann, R.L., Laskowski, D.A., McCall, P.J., Vander Kuy, K., Dishburger, H.J. (1983) A rapid method for the estimation of the environmental parameters octanol/water partition coefficient, soil sorption constant, water to air ratio, and water solubility. *Residue Rev.* 85, 17-28.

Szabo, G., Prosser, S., Bulman, R.A. (1990) Determination of the adsorption coefficient ($K_{OC}$) of some aromatics for soil by RP-HPLC on two immobilized humic acid phases. *Chemosphere* 21, 777-788.

Tomlinson, E., Hafkenscheid, T.L. (1986) Aqueous solution and partition coefficient estimation from HPLC data. In: *Partition Coefficient, Determination and Estimation.* Dunn III, W.J., Block, J.H., Pearlman, R.S., Eds., pp. 101-141, Pergamon Press, New York.

Tse, G., Orbey, H., Sandler, S.I. (1992) Infinite dilution activity coefficients and Henry's law coefficients for some priority water pollutants determined by a relative gas chromatographic method. *Environ. Sci. Technol.* 26, 2017-2022.

Tsonopoulos, C., Prausnitz, J.M. (1971) Activity coefficients of aromatic solutes in dilute aqueous solutions. *Ind. Eng. Chem. Fundam.* 10, 593-600.

Tulp, M.T.M., Hutzinger, O. (1978) Some thoughts on the aqueous solubilities and partition coefficients of PCB, and the mathematical correlation between bioaccumulation and physico-chemical properties. *Chemosphere* 7, 849-760.

Veith, G.D., Austin, N.M., Morris, R.T. (1979) A rapid method for estimating log P for organic chemicals. *Water Res.* 13, 43-47.

Veith, G.D., Macek, K.J., Petrocelli, S.R., Caroll, J. (1980) An evaluation of using partition coefficients and water solubilities to estimate bioconcentration factors for organic chemicals in fish. In: *Aquatic Toxicology.* Eaton, J.G., Parrish, P.R., Hendrick, A.C., Eds, pp. 116-129, ASTM ATP 707, Am. Soc. for Testing and Materials, Philadelphia, Pa.

Vejrosta, J., Novak, J., Jönsson, J. (1982) A method for measuring infinite-dilution partition coefficients of volatile compounds between the gas and liquid phases of aqueous systems. *Fluid Phase Equil.* 8, 25-35.

Verschueren, K. (1977) *Handbook of Environmental Data on Organic Chemicals.* Van Nostrand Reinhold, New York, N.Y.

Verschueren, K. (1983) *Handbook of Environmental Data on Organic Chemicals.* Van Nostrand Reinhold, New York, N.Y.

Wakita, K., Yoshimoto, M., Miyamoto, S., Watsnabe, H. (1986) A method for calculations of the aqueous solubility of organic compounds by using new fragment solubility constants. *Chem. Pharm. Bull.* 34, 4663-4681.

Wang, L., Zhao, Y., Hong, G. (1992) Predicting aqueous solubility and octanol/water partition coefficients of organic chemicals from molar volume. *Environ. Chem.* 11, 55-70.

Warne, M., St. J., Connell, D.W., Hawker, D.W. (1990) Prediction of aqueous solubility and the octanol-water partition coefficient for lipophilic organic compounds using molecular descriptors and physicochemical properties. *Chemosphere* 16, 109-116.

Wasik, S.P., Miller, M.M., Tewari, Y.B., May, W.E., Sonnefeld, W.J., DeVoe, H., Zoller, W.H. (1983) Determination of the vapor pressure, aqueous solubility, and octanol/water partition coefficient of hydrophobic substances by coupled generator column/liquid chromatographic methods. *Res. Rev.* 85, 29-42.

Weast, R. (1972-73) *Handbook of Chemistry and Physics.* 53th ed., CRC Press, Cleveland, OH.

Weast, R. (1984) *Handbook of Chemistry and Physics.* 64th ed., CRC Press, Boca Raton, FL.

Weil, L., Dure, G., Quentin, K.L. (1974) Solubility in water of insecticide, chlorinated hydrocarbons and polychlorinated biphenyls in view of water pollution. *Z. Wasser Abwasser Forsch.* 7, 169-175.

Westcott, J.W., Bidleman, T.F. (1982) Determination of polychlorinated biphenyl vapor pressures by capillary gas chromatography. *J. Chromatogr.* 210, 331-336.

Westcott, J.W., Simon, J.J., Bidleman, T.F. (1981) Determination of polychlorinated biphenyl vapor pressures by a semimicro gas saturation method. *Environ. Sci. Technol.* 15, 1375-1378.

Whitehouse, B.G., Cooke, R.C. (1982) Estimating the aqueous solubility of aromatic hydrocarbons by high performance liquid chromatography. *Chemosphere* 11, 689-699.

Windholz, M. Ed. (1983) *The Merck Index, An Encyclopedia of Chemicals, Drugs and Biologicals.* 10th ed., Merck & Co. Inc. Rahway, New Jersey.

Woodburn, K.B., Doucette, W.J., Andren, A.W. (1984) Generator column determination of octanol/water partition coefficients for selected polychlorinated biphenyl congeners. *Environ. Sci. Technol.* 18, 457-459.

Yair, O.B., Fredenslund, A. (1983) Extension of the UNIFAC group-contribution method for the prediction of pure-component vapor pressure. *Ind. Eng. Chem. Fundam. Des. Dev.* 22, 433-436.

Yalkowsky, S.H. (1979) Estimation of entropies of fusion of organic compounds. *Ind. Eng. Chem. Fundam.* 18, 108-111.

Yalkowsky, S.H., Banerjee, S. (1992) *Aqueous Solubility, Methods of Estimation for Organic Compounds.* Marcel Dekker, Inc., New York, N.Y.

Yalkowsky, S.H., Valvani, S.C. (1976) Partition coefficients and surface areas of some alkylbenzenes. *J. Med. Chem.* 19, 727-728.

Yalkowsky, S.H., Valvani, S.C. (1979) Solubility and partitioning. I: Solubility of nonelectrolytes in water. *J. Pharm. Sci.* 69, 912-922.

Yalkowsky, S.H., Orr, R.J., Valvani, S.C. (1979) Solubility and partitioning. 3. The solubility of halobenzenes in water. *I&EC Fundam.* 18, 351-353.

Yalkowsky, S.H., Valvani, S.S., Mackay, D. (1983) Estimation of the aqueous solubility of some aromatic compounds. *Res. Rev.* 85, 43-55.

Yoshida, K., Shigeoka, T., Yamauchi, F. (1983) Relationship between molar refraction and n-octanol/water partition coefficient. *Ecotox. Environ. Saf.* 7, 558-565.

Zwolinski, B.J. Wilhoit, R.C. (1971) *Handbook of Vapor Pressures and Heats of Vaporization of Hydrocarbons and Related Compounds.* API-44, TRC Publication No. 101, Texas A&M University, College Station, TX.

# 2. Hydrocarbons

2.1 List of Chemicals and Data Compilations:

2.1.1 Saturated Hydrocarbons

Alkanes


Isobutane (2-methylpropane) . . . . . 71
2,2-Dimethylpropane (neopentane) . . . . . 74
*n*-Butane . . . . . 77
2-Methylbutane (isopentane) . . . . . 81
2,2-Dimethylbutane . . . . . 84
2,3-Dimethylbutane . . . . . 87
2,2,3-Trimethylbutane . . . . . 90
*n*-Pentane . . . . . 92
2-Methylpentane (isohexane) . . . . . 97
3-Methylpentane . . . . . 100
2,2-Dimethylpentane . . . . . 103
2,4-Dimethylpentane . . . . . 105
3,3-Dimethylpentane . . . . . 108
2,2,4-Trimethylpentane (isooctane) . . . . . 110
2,3,4-Trimethylpentane . . . . . 113
*n*-Hexane . . . . . 115
2-Methylhexane (isoheptane) . . . . . 120
3-Methylhexane . . . . . 123
2,2,5-Trimethylhexane . . . . . 126
*n*-Heptane . . . . . 129
2-Methylheptane . . . . . 134
3-Methylheptane . . . . . 136
*n*-Octane . . . . . 138
4-Methyloctane . . . . . 143
*n*-Nonane . . . . . 145
*n*-Decane . . . . . 148
*n*-Undecane . . . . . 151
*n*-Dodecane . . . . . 153


Cycloalkanes


Cyclopentane . . . . . 155
Methylcyclopentane . . . . . 159
1,1,3-Trimethylcyclopentane . . . . . 162
Propylcyclopentane . . . . . 164
Pentylcyclopentane . . . . . 166
Cyclohexane . . . . . 168
Methylcyclohexane . . . . . 173
1,2-*cis*-Dimethylcyclohexane . . . . . 176

1,4-*trans*-Dimethylcyclohexane . . . . . . . . . . 179
1,1,3-Trimethylcyclohexane . . . . . . . . . . 181
Cycloheptane . . . . . . . . . . 183
Cyclooctane . . . . . . . . . . 185
Decalin . . . . . . . . . . 187
2.1.2 Unsaturated Hydrocarbons
Alkenes
2-Methylpropene . . . . . . . . . . 189
1-Butene . . . . . . . . . . 192
2-Methyl-1-butene . . . . . . . . . . 195
3-Methyl-1-butene . . . . . . . . . . 197
2-Methyl-2-butene . . . . . . . . . . 199
1-Pentene . . . . . . . . . . 201
*cis*-2-Pentene . . . . . . . . . . 204
2-Methyl-1-pentene . . . . . . . . . . 206
4-Methyl-1-pentene . . . . . . . . . . 208
1-Hexene . . . . . . . . . . 210
1-Heptene . . . . . . . . . . 213
1-Octene . . . . . . . . . . 216
1-Nonene . . . . . . . . . . 219
1-Decene . . . . . . . . . . 221
Dienes
1,3-Butadiene . . . . . . . . . . 223
2-Methyl-1,3-butadiene (isoprene) . . . . . . . . . . 227
2,3-Dimethyl-1,3-butadiene . . . . . . . . . . 230
1,4-Pentadiene . . . . . . . . . . 232
1,5-Hexadiene . . . . . . . . . . 235
1,6-Heptadiene . . . . . . . . . . 238
Alkynes
1-Butyne . . . . . . . . . . 240
1-Pentyne . . . . . . . . . . 242
1-Hexyne . . . . . . . . . . 244
1-Heptyne . . . . . . . . . . 246
1-Octyne . . . . . . . . . . 248
1-Nonyne . . . . . . . . . . 250
Cycloalkenes
Cyclopentene . . . . . . . . . . 252
Cyclohexene . . . . . . . . . . 255
1-Methylcyclohexene . . . . . . . . . . 258
Cycloheptene . . . . . . . . . . 260
1,4-Cyclohexadiene . . . . . . . . . . 262
Cycloheptatriene . . . . . . . . . . 265

*dextro*-Limonene [(R)-(+)-limonene] . . . . . 267
2.1.3 Aromatic Hydrocarbons
Styrene . . . . . 269
Methylstyrene . . . . . 272
Tetralin . . . . . 274
Others (see Volume I and II)
2.2 Summary Tables and QSPR Plots . . . . . 276
2.3 Illustrative Fugacity Calculations: Levels I, II and III . . . . . 299
*n*-Pentane . . . . . 300
*n*-Hexane . . . . . 304
*n*-Octane . . . . . 308
*n*-Decane . . . . . 312
*n*-Dodecane . . . . . 316
Cyclopentane . . . . . 320
Cyclohexane . . . . . 324
Methylcyclohexane . . . . . 328
Cyclooctane . . . . . 332
Cyclohexene . . . . . 336
1-Pentene . . . . . 340
1-Octene . . . . . 344
1-Hexyne . . . . . 348
1,3-Butadiene . . . . . 352
1,4-Pentadiene . . . . . 356
1,4-Cyclohexadiene . . . . . 360
Styrene . . . . . 364
2.4 Commentary on the Physical-Chemical Properties and Environmental Fate . . . . . 368
2.5 References . . . . . 372

## 2.1 List of Chemicals and Data Compilations

Common Name: Isobutane
Synonym:
Chemical Name: 2-methylpropane
CAS Registry No: 75-28-5
Molecular Formula: $C_4H_{10}$
Molecular Weight: 58.123
Melting Point (°C):
-159.4 (Weast 1984)
-159.6 (Riddick et al. 1986)
Boiling Point (°C):
-11.7 (Dreisbach 1959, Weast 1984)
-11.73 (Riddick et al. 1986)
Density (g/cm³ at 20°C):
0.549 (Weast 1984)
0.5572 (20°C, Dreisbach 1959)
0.5510 (25°C, Dreisbach 1959)
0.55711 (20°C, Riddick et al. 1986)
0.55092 (25°C, Riddick et al. 1986)
Molar Volume (cm³/mol):
104.3 (20°C, calculated from density, McAuliffe 1968)
96.2 (calculated-LeBas method)
Molecular Volume, TMV (Å³):
75.82 (Pearlman 1986)
Total Surface Area, TSA (Å²):
249.1 (Amidon et al. 1974, 1975)
249.1, 234.0 (Kier et al. 1975)
251.7, 247.2, 115.9 (Valvani et al. 1976)
270.1 (Amidon & Anik 1976; Kier et al. 1975; quoted, Mailhot & Peters 1988)
197.6 (Iwase et al. 1985)
115.64 (Pearlman 1986)
111.1 (Okouchi et al. 1992)
Heat of Fusion, $\Delta H_{fus}$, (kcal/mol):
1.085 (Riddick et al. 1986)
Entropy of Fusion, $\Delta S_{fus}$ (cal/mol K or e.u.):
Fugacity Ratio at 25°C, F: 1.0

Water Solubility (g/m³ or mg/L at 25°C):
48.9 (shake flask-GC, McAuliffe 1963,1966; quoted, Mackay & Shiu 1975,1981; Brookman et al. 1985)
200, 49; 178 (quoted values; calculated-group contribution, Irmann 1965)
165, 108 (quoted, calculated-TSA, Amidon et al. 1974)
49.5 (quoted, Hine & Mookerjee 1975)

49.5 (quoted, Amidon et al. 1975)
164, 159 (quoted, calculated-$\chi$ , Hall et al. 1975)
164, 159 (quoted, calculated-$\chi$ , Kier & Hall 1976)
48.9 (quoted, Riddick et al. 1986)
48.9 (calculated-AI, Okouchi et al. 1992)
39.3, 35.0 (quoted, calculated-$V_M$, Wang et al. 1992)

Vapor Pressure (Pa at 25°C):
348105 (Antoine eqn., Dreisbach 1959)
357000 (Antoine eqn., Zwolinski & Wilhoit 1971)
356600 (Antoine eqn., Dean 1985)
348100 (quoted, Riddick et al. 1986)
351130 (Antoine eqn., Stephenson & Malanowski 1987)

Henry's Law Constant (Pa $m^3$/mol):
118644 (calculated as $1/K_{AW}$, $C_W/C_A$ reported as exptl., Hine & Mookerjee 1975)
100982, 22092 (calculated-group contribution, bond contribution, Hine & Mookerjee 1975)
120000 (calculated-P/C, Mackay & Shiu 1981)
116710 (calculated-P/C, Yaws et al. 1991)

Octanol/Water Partition Coefficient, log $K_{OW}$:
2.76 (shake flask-GC, Leo et al. 1975)
2.76, 2.69 (quoted, calculated-hydrophobicity const., Iwase et al. 1985)
3.11 (misquoted from 2,2-dimethylpropane, Mailhot & Peters 1988)
2.76, 2.75 (quoted, calculated-MO, Bodor et al. 1989)
2.76, 2.68 (quoted, calculated-$\pi$, Bodor et al. 1989)
2.76, 2.53 (misquoted from 1-butane, calculated-$V_M$, Wang et al. 1992)

Bioconcentration Factor, log BCF:

Sorption Partition Coefficient, log $K_{OC}$:

Half-Lives in the Environment:
Air: atmospheric lifetime was estimated to be 59 hours, based on a photooxidation rate constant of $2.34 \times 10^{-12}$ $cm^3$ $molecule^{-1}$ $sec^{-1}$ in summer daylight with OH radicals (Altshuller 1991).

Surface water:
Ground water:
Sediment:
Soil:
Biota:

Environmental Fate Rate Constant and Half-Lives:
Volatilization:
Photolysis:
Oxidation: rate constant of $1.28 \times 10^9$ liter $mol^{-1}$ $sec^{-1}$ for the reaction with hydroxyl radicals at 300 K (Greiner 1967; quoted, Altshuller & Bufalini 1971); rate constant of $6.03 \times 10^{12}$ $cm^3$ $mol^{-1}$ $sec^{-1}$ (observed) and $5.07 \times 10^{12}$ $cm^3$ $mol^{-1}$ $sec^{-1}$ (correlated) for the reaction with hydroxyl radicals in the air at 296 K (Greiner 1970); rate constant of $2.52 \times 10^{-12}$ $cm^3$ $molecule^{-1}$ $sec^{-1}$ for the reaction with OH radicals at atmospheric pressure and 300 K (Darnall et al. 1978); experimentally determined rate constant of $(2.34 \pm 0.33) \times 10^{-12}$ $cm^3$ $molecule^{-1}$ $sec^{-1}$ with OH radicals at $(24.6 \pm 0.4)$°C (Edney et al. 1986); rate constant of $2.34 \times 10^{-12}$ $cm^3$ $molecule^{-1}$ $sec^{-1}$ with OH radicals and $9.70 \times 10^{-17}$ $cm^3$ $molecule^{-1}$ $sec^{-1}$ with $NO_3$ radicals in air (Atkinson 1990, 1991); reaction rate constant of $2.34 \times 10^{-12}$ $cm^3$ $molecule^{-1}$ $sec^{-1}$ and an estimated lifetime was 59 hours (Altshuller 1991); reaction rate constant of $7.38 \times 10^{-13}$ $cm^3$ $molecule^{-1}$ $sec^{-1}$ with OH radicals and $6.50 \times 10^{-17}$ $cm^3$ $molecule^{-1}$ $sec^{-1}$ with $NO_3$ radicals (Sabljic & Güsten 1990).
Hydrolysis:
Biodegradation:
Biotransformation:
Bioconcentration, Uptake ($k_1$) and Elimination ($k_2$) Rate Constants or Half-Lives:

Common Name: 2,2-Dimethylpropane
Synonym: neopentane, tetramethylmethane
Chemical Name: 2,2-dimethylpropane
CAS Registry No: 75-83-2
Molecular Formula: $C_5H_{12}$
Molecular Weight: 72.15
Melting Point (°C): -16.55
Boiling Point (°C): 9.503
Density (g/cm$^3$ at 20°C):
    0.591 (20°C, Dreisbach 1959; Riddick et al. 1986)
    0.5852 (25°C, Dreisbach 1959; Riddick et al. 1986)
Molar Volume (cm$^3$/mol):
    122.1 (20°C, calculated-density, McAuliffe 1966; Wang et al. 1992)
    123.29 (calculated-density)
    116.4 (calculated-LeBas method)
Molecular Volume, TMV (Å$^3$):
    107.29 (Pearlman 1986)
Total Surface Area, TSA (Å$^2$):
    270.14 (Hermann 1972; quoted, Amidon et al. 1974)
    270.1, 249.7 (Kier et al. 1975)
    265.1, 266.0, 120.3 (Valvani et al. 1976)
    152.29 (Pearlman 1986)
    134.6 (Okouchi et al. 1992)
Heat of Fusion, $\Delta H_{fus}$, (kcal/mol):
    0.778 (Dreisbach 1959)
Entropy of Fusion, $\Delta S_{fus}$ (cal/mol K or e.u.):
Fugacity Ratio at 25°C, F: 1.0

Water Solubility (g/m$^3$ or mg/L at 25°C):
    33.20 (shake flask-GC, McAuliffe 1966; quoted, Mackay & Shiu 1981; Brookman et al. 1985; quoted, IUPAC 1989)
    54.0, 54.3 (quoted, calculated-TSA, Amidon et al. 1974)
    33.0 (quoted, Hine & Mookerjee 1975)
    54.0, 101 (quoted, calculated-$\chi$, Hall et al. 1975)
    54.0, 101 (quoted, calculated-$\chi$, Kier & Hall 1976)
    32.2, 99.6 (quoted, calculated-$K_{OW}$, Yalkowsky & Morozowich 1980)
    33.0 (quoted, Riddick et al. 1986)
    33.2 (calculated-AI, Okouchi et al. 1992)
    33.0, 19.4 (quoted, calculated-$V_M$, Wang et al. 1992)

Vapor Pressure (Pa at 25°C):

177932 (Antoine eqn. regression, Stull 1947)
171346 (calculated from determined data, Dreisbach 1959)
172000 (Antoine eqn., Zwolinski & Wilhoit 1971)
171300 (quoted, Riddick et al. 1986)
171518, 171449 (Antoine eqn., Stephenson & Malanowski 1987)

Henry's Law Constant (Pa $m^3$/mol):

220925 (calculated as $1/K_{AW}$, $C_W/C_A$, reported as exptl., Hine & Mookerjee 1975)
125640 (calculated-group contribution, Hine & Mookerjee 1975)
33438 (calculated-bond contribution, Hine & Mookerjee 1975)
373000 (calculated-P/C, Mackay & Shiu 1981)
213253 (calculated-P/C, Yaws et al. 1991)

Octanol/Water Partition Coefficient, log $K_{OW}$:

3.11 (Leo et al. 1971; Hansch & Leo 1979)
3.11 (shake flask-GC, Leo et al. 1975)
3.11; 2.95, 3.41, 3.22 (quoted; calculated-f const., Rekker 1977)
3.11 (quoted, Yalkowsky & Morozowich 1980)
3.11, 3.30 (quoted, calculated-MO, Bodor et al. 1989)
3.11, 3.08 (quoted, calculated- $\pi$, Bodor et al. 1989)
3.11 (recommended, Sangster 1989)
3.11, 2.98 (quoted, calculated-$V_M$, Wang et al. 1992)

Bioconcentration Factor, log BCF:

Sorption Partition Coefficient, log $K_{OC}$:

Half-Lives in the Environment:

Air:
Surface water:
Ground water:
Sediment:
Soil:
Biota:

Environmental Fate Rate Constant and Half-Lives:

Volatilization:

Photolysis:

Oxidation: photooxidation rate constant of $6.50 \times 10^{11}$ $cm^3$ $mol^{-1}$ $sec^{-1}$ (observed) and $5.37 \times 10^{11}$ $cm^3$ $mol^{-1}$ $sec^{-1}$ (correlated) for the reaction with OH radicals in air at 298 K (Greiner 1970); rate constant of $5.50 \times 10^{-15}$ $cm^3$ $molecule^{-1}$ $sec^{-1}$ for the reaction with $O(^3P)$ (Herron & Huie 1973; quoted, Gaffney & Levine 1979) and $9.30 \times 10^{-13}$ $cm^3$ $molecule^{-1}$ $sec^{-1}$ for the reaction with OH radicals in air both at room temperature (Atkinson et al. 1979; quoted, Gaffney & Levine 1979); photooxidation reaction rate constant of $9.0 \times 10^{-13}$ $cm^3$ $molecule^{-1}$ $sec^{-1}$ for the reaction with hydroxyl radicals in air (Winer et al. 1979); photooxidation rate constant of $8.49 \times 10^{-13}$ $cm^3$ $molecule^{-1}$ $sec^{-1}$ for the reaction with OH radicals in air (Atkinson 1990, 1991).

Hydrolysis:

Biodegradation:

Biotransformation:

Bioconcentration, Uptake ($k_1$) and Elimination ($k_2$) Rate Constants or Half-Lives:

Common Name: n-Butane
Synonym: 1-butane
Chemical Name: n-butane
CAS Registry No: 106-97-8
Molecular Formula: $C_4H_{10}$
Molecular Weight: 58.123
Melting Point (°C):
-135.0 (Stull 1947)
-138.55 (Weast 1973)
-138.35 (Stephenson & Malanowski 1987)
Boiling Point (°C):
-0.5 (Stephenson & Malanowski 1987)
Density (g/cm$^3$ at 20 °C):
0.57861 (20°C, Riddick et al. 1986)
0.57287 (25°C, Riddick et al. 1986)
Molar Volume (cm$^3$/mol):
100.46 (20°C, calculated-density)
101.47 (25°C, calculated-density)
100.4 (20°C, calculated-density, McAuliffe 1966)
96.2 (calculated-LeBas method)
98.0 (Valsaraj 1988)
Molecular Volume, TMV (Å$^3$):
79.39 (Cramer 1977)
75.87 (Pearlman 1986)
Total Surface Area, TSA (Å$^2$):
255.2 (Hermann 1972; quoted, Amidon et al. 1974,1975; Amidon & Anik 1976; Kier et al. 1975; Kier & Hall 1976; Mailhot & Peters 1988)
225.0, 247.6, 115.9 (Valvani et al. 1976)
255.0 (Amidon & Anik 1981)
255.2, 254.058 (Horvath 1982)
202.1 (Iwase et al. 1985)
116.1 (Pearlman 1986)
223.4 (Mailhot & Peters 1988)
115.0 (Valsaraj 1988)
108.6 (Okouchi et al. 1992)
Heat of Fusion, $\Delta H_{fus}$, (kcal/mol):
1.050 (Parks & Huffman 1931)
1.114 (Riddick et al. 1986)
Entropy of Fusion, $\Delta S_{fus}$ (cal/mol K or e.u.):
Fugacity Ratio at 25°C, F: 1.0

Water Solubility ($g/m^3$ or mg/L at 25°C):

65.6 (UV, Morrison & Billett 1952)
67.0 (UV, Claussen & Polglase 1952)
72.7 (shake flask-GC, Franks et al. 1966)
180, 68, 61; 178 (quoted values; calculated-group contribution, Irmann 1965)
61.4 (shake flask-GC, McAuliffe 1963. 1966; quoted, Mackay & Shiu 1975, 1981)
136, 83.1 (quoted, calculated-TSA, Amidon et al. 1974)
62.2 (quoted, Hine & Mookerjee 1975)
136, 101 (quoted, calculated-$\chi$ , Hall et al. 1975)
136, 101 (quoted, calculated-$\chi$ , Kier & Hall 1976)
133, 146 (quoted, calculated-$K_{OW}$, Yalkowsky & Morozowich 1980)
61.4 (quoted, Mackay et al. 1979; Mackay 1981; Mackay & Shiu 1981)
61.66 (shake flask-GC, Coates et al. 1985)
75.3, 61.4 (quoted, Riddick et al. 1986)
61.4 (calculated-AI, Okouchi et al. 1992)
60.9, 41.2 (quoted, calculated-$V_M$, Wang et al. 1992)

Vapor Pressure (Pa at 25°C):

288200 (Antoine eqn. regression, Stull 1947)
243000 (Antoine eqn., Zwolinski & Wilhoit 1971)
243000 (quoted, Mackay & Shiu 1975,1981)
242835 (Antoine eqn., Dean 1985)
243000 (quoted, Riddick et al. 1986)
242814 (Antoine eqn., Stephenson & Malanowski 1987)

Henry's Law Constant (Pa $m^3$/mol):

94242 (calculated-$1/K_{AW}$, $C_W/C_A$, reported as exptl., Hine & Mookerjee 1975; quoted, Nirmalakhandan & Speece 1988)
82080 (calculated-group contribution, Hine & Mookerjee 1975)
22095 (calculated-bond contribution, Hine & Mookerjee 1975)
95900 (calculated-P/C, Mackay & Shiu 1975; Mackay 1981; Mackay & Shiu 1981)
93955 (selected, Mills et al. 1982)
94242, 80213 (quoted, calculated-$\chi$ , Nirmalakhandan & Speece 1988)
92910 (calculated-P/C, Yaws et al. 1991)

Octanol/Water Partition Coefficient, log $K_{OW}$:

2.89 (shake flask-GC, Leo et al. 1975)
2.89 (concn. ratio, Cramer 1977)
2.89; 2.46, 2.84, 2.96 (quoted; calculated-f const., Rekker 1977)
2.89 (quoted, Yalkowsky & Morozowich 1980)

2.89, 2.76 (Hansch & Leo 1979; quoted, Iwase et al. 1985)
2.89, 2.79 (quoted, calculated-hydrophobicity const., Iwase et al. 1985)
2.89 (quoted, Mailhot & Peters 1988)
2.86, 2.43 (quoted, calculated-$V_M$, Wang et al. 1992)

Bioconcentration Factor, log BCF:

Sorption Partition Coefficient, log $K_{OC}$:

Half-Lives in the Environment:

Air: half-life of 6.5 hours in ambient air based on reaction with OH radicals at 300 K (Doyle et al. 1975); photolysis half-life of 2.4 to 24 hours (Darnall et al. 1976); atmospheric lifetimes were calculated to be $4x10^7$ hours for reaction with $O_3$, 107 hours with OH radicals and 32150 hours with $NO_3$ radicals based on reaction rate constants and environmental concentraions of OH, $NO_3$ radicals and $O_3$ in the gas phase (Atkinson & Carter 1984); atmospheric lifetimes were calculated to be 222 hours for the reaction with OH radicals, $4x10^7$ hours with $O_3$ and 32150 hours with $NO_3$ radicals based on the rate constants and environmental concentrations of OH, $NO_3$ radicals and $O_3$ in the gas phase (Atkinson 1985); atmospheric lifetime was estimated to be 54 hours based on a photoooxidation reaction rate constant of $2.54x10^{-12}$ $cm^3$ $molecule^{-1}$ $sec^{-1}$ with OH radicals during summer daylight hours (Altshuller 1991).

Surface water:

Ground water:

Sediment:

Soil:

Biota:

Environmental Fate Rate Constant and Half-Lives:

Volatilization:

Photolysis:

Oxidation: room temperature rate constants of $3.1x10^{-14}$ $cm^3$ $molecule^{-1}$ $sec^{-1}$ for the reaction with $O(^3P)$ (Herron & Huie 1973; quoted, Gaffney & Levine 1979) and $3.0x10^{-12}$ $cm^3$ $molecule^{-1}$ $sec^{-1}$ for the reaction with OH radicals (Atkinson et al. 1979; quoted, Gaffney & Levine 1979); photooxidation rate constant of $2.33x10^{11}$ $cm^3$ $mol^{-1}$ $sec^{-1}$ (observed value) and $2.05x10^{11}$ $cm^3$ $mol^{-1}$ $sec^{-1}$ (correlated value) with hydroxyl radicals in air at 298 K (Greiner 1970); rate constant of $1.8x10^9$ L $mol^{-1}$ $sec^{-1}$ with half-life of 6.5 hours in ambient air, based on reaction with OH radicals at 300 K (Doyle et al. 1975); photooxidation rate constant of 1.8 x $10^9$ $M^{-1}$ $sec^{-1}$ with hydroxyl radicals in polluted atmosphere at 305 K (Lloyd et al. 1976; Darnall et al. 1976) and atmospheric half-life of 2.4 to 24 hours (Darnall et al. 1976); photooxidation reaction rate constant of $3.0x10^{-12}$ $cm^3$ $molecule^{-1}$ $sec^{-1}$ for the reaction with OH radicals with an average OH concentration calculated to be $1.2x10^6$ molecules/$cm^3$ and with a loss rate at 0.11 per day with OH radicals (Zafonte & Bonamassa 1977); rate constant for the reaction with OH radicals of $(2.35\text{-}4.22)x10^{12}$ $cm^3$ $molecule^{-1}$ $sec^{-1}$ (exptl.) and $2.71x10^{-12}$ $cm^3$ $molecule^{-1}$ $sec^{-1}$ (calculated) in air at 298 K (Darnall et al. 1978); rate constant of $< 10^{-23}$ $cm^3$ $molecule^{-1}$ $sec^{-1}$ with a loss rate of $< 6x10^{-7}$ per day with ozone and $3.6x10^{-17}$ $cm^3$ $molecule^{-1}$ $sec^{-1}$ with a loss rate of 0.0007 per day with $NO_3$ radicals in air (Atkinson 1985); rate constant of $2.52x10^{-12}$ $cm^3$ $molecule^{-1}$ $sec^{-1}$ for the reaction of OH radicals at 297 K (Atkinson 1986; quoted, Edney et al. 1986); rate constant of $2.54x10^{-12}$ $cm^3$ $molecule^{-1}$ $sec^{-1}$ for the reaction of OH radicals at 298 K and rate constant of $6.50x10^{-17}$ $cm^3$ $molecule^{-1}$ $sec^{-1}$ for the reaction with $NO_3$ radicals (Atkinson 1990, 1991; Altshuller 1991) and the atmospheric lifetime was estimated to be 54 hours, based on the photooxidation reaction rate constant with OH radicals during summer daylight hours (Altshuller 1991); rate constant of 6.5 $x10^{-17}$ $cm^3$ $molecule^{-1}$ $sec^{-1}$ for the reaction with $NO_3$ radicals at 296 K (Atkinson 1990); 6.6 $x10^{-17}$ $cm^3$ $molecule^{-1}$ $sec^{-1}$ for the reaction with the $NO_3$ radicals in the gas phase (Atkinson 1991).

Hydrolysis:

Biodegradation:

Biotransformation:

Bioconcentration, Uptake ($k_1$) and Elimination ($k_2$) Rate Constants or Half-Lives:

Common Name: Isopentane
Synonym:
Chemical Name: 2-methylbutane
CAS Registry No: 78-78-4
Molecular Formula: $C_5H_{12}$
Molecular Weight: 72.15
Melting Point (°C): -159.9
Boiling Point (°C): 27.875
Density (g/cm³ at 20 °C):
 0.61967 (20°C, Dreisbach 1959)
 0.61462 (25°C, Dreisbach 1959)
 0.6193 (20°C, Riddick et al. 1986)
 0.6142 (25°C, Riddick et al. 1986)
Molar Volume (cm³/mol):
 116.5 (20°C, calculated-density)
 117.47 (25°C, calculated-density)
 116.4 (20°C, calculated-density, McAuliffe 1966)
 118.4 (calculated-LeBas method)
 116.0 (calculated-density, Lande & Banerjee 1981; Wang et al. 1992)
 116.4 (20°C, calculated-density, Stephenson & Malanowski 1987)
 0.533 (intrinsic volume: $V_I/100$, Kamlet et al. 1987)
Molecular Volume, TMV (Å³):
 91.12 (Pearlman 1986)
Total Surface Area, TSA (Å²):
 273.58, 268.72 (Herman 1972)
 274.6 (Amidon et al. 1974, 1975)
 274.6, 265.6 (Kier et al. 1975)
 274.4, 269.0,121.1 (Valvani et al. 1976)
 130.5 (Pearlman 1986)
 374.6 (Mailhot & Peters 1988)
 132.0 (Okouchi et al. 1992)
Heat of Fusion, $\Delta H_{fus}$, (kcal/mol):
 1.231 (Dreisbach 1959, Riddick et al. 1986)
Entropy of Fusion, $\Delta S_{fus}$ (cal/mol K or e.u.):
Fugacity Ratio at 25°C, F: 1.0

Water Solubility (g/m³ or mg/L at 25°C):
 47.8 (shake flask-GC, McAuliffe 1963,1966; quoted, Hermann 1972)
 48.0, 70.8 (quoted, calculated-group contribution, Irmann 1965)
 47.7, 58.4 (quoted, calculated-$K_{OW}$, Hansch et al. 1968)
 49.6, 55.2 (shake flask-GC, calculated-group contribution, Polak & Lu 1973)
 47.7, 44.8 (quoted, calculated-TSA, Amidon et al. 1974)

38.75 (quoted, Hine & Mookerjee 1975)
47.7, 50.9 (quoted, calculated- $\chi$ , Hall et al. 1975)
47.7, 50.9 (quoted, calculated- $\chi$ , Kier & Hall 1976)
48.0 (shake flask-GC, Price 1976)
38.5 (quoted, Mackay et al. 1979; Mackay 1981; Mackay & Shiu 1981)
47.67 (quoted, Lande & Banerjee 1981)
47.8, 49.6, 48.0 (quoted, Brookman et al. 1985)
48.0 (quoted, Riddick et al. 1986)
47.7, 46.6 (quoted, calculated-fragment solubility constants, Wakita et al. 1986)
47.8, 48.6, 49.6, 48.0; 48.5 (quoted exptl values; recommended best value, IUPAC 1989)
47.7, 25.6 (quoted, calculated-$V_M$, Wang et al. 1992)
47.8 (calculated-AI, Okouchi et al. 1992)
48.2 (quoted literature average, Myrdal et al. 1992)

Vapor Pressure (Pa at 25°C):
82786 (22.04 °C, Schumann et al. 1942)
83722 (22.44 °C, Willingham et al. 1945)
99546 (Antoine eqn. regression, Stull 1947)
91740 (Antoine eqn., Dreisbach 1959)
92600 (Antoine eqn, Zwolinski & Wilhoit 1971)
92600 (quoted, Mackay & Shiu 1981)
91730, 92096 (interpolated, Antoine equations, Boublik et al. 1984)
91656 (interpolated, Antoine eqn., Dean 1985)
91700 (quoted, Riddick et al. 1986)
91640 (interpolated, Antoine eqn., Stephenson & Malanowski 1987)

Henry's Law Constant (Pa $m^3$/mol):
140000 (calculated-P/C, Mackay et al. 1979; Mackay 1981)
138000; 140000, 139000, 134700 (recommended; calculated-P/C, Mackay & Shiu 1981)
138207 (selected, Mills et al. 1982)
138285 (calculated-P/C, Yaws et al. 1991)

Octanol/Water Partition Coefficient, log $K_{OW}$:
2.30 (calculated-$\pi$ constant, Hansch et al. 1968; quoted, Mailhot & Peters 1988)
2.30, 2.41 (quoted, calculated- $\chi$ , Murray et al. 1975)
2.30, 2.83 (quoted, calculated-$V_M$, Wang et al. 1992)

Bioconcentration Factor, log BCF:

Sorption Partition Coefficient, log $K_{OC}$:

Half-Lives in the Environment:

Air: photooxidation reaction rate constant of $2.0 \times 10^{-9}$ $cm^3$ $mol^{-1}$ $sec^{-1}$ for the reaction with hydroxyl radicals in air (Darnall et al. 1976; Lloyd et al. 1976) with atmospheric half-life of 2.4-24 hours (Darnall et al. 1976); atmospheric lifetime of 36 hours, based on rate constant of $3.90 \times 10^{-12}$ $cm^3$ $molecule^{-1}$ $sec^{-1}$ for the reaction with OH radicals during summer daylight (Altshuller 1991).

Surface water:

Ground water:

Sediment:

Soil:

Biota:

Environmental Fate Rate Constant and Half-Lives:

Volatilization:

Photolysis:

Oxidation: photooxidation rate constant of $2.0 \times 10^{-9}$ $cm^3$ $mol^{-1}$ $sec^{-1}$ for the reaction with hydroxyl radicals in air with atmospheric half-life of 2.4-24 hours (Darnall et al. 1976); rate constant of $3.78 \times 10^{-12}$ $cm^3$ $molecule^{-1}$ $sec^{-1}$ for the reaction with OH radicals at atmospheric pressure and 300 K (Darnall et al. 1978); rate constant of $3.9 \times 10^{-12}$ $cm^3$ $molecule^{-1}$ $sec^{-1}$ for the reaction with OH radicals in air (Atkinson 1990, 1991; Altshuller 1991) with an estimated atmospheric lifetime of 36 hours (Altshuller 1991).

Hydrolysis:

Biodegradation:

Biotransformation:

Bioconcentration, Uptake ($k_1$) and Elimination ($k_2$) Rate Constants or Half-Lives:

Common Name: 2,2-Dimethylbutane
Synonym: neohexane, dimethylpropylmethane
Chemical Name: 2,2-dimethylbutane
CAS Registry No: 75-83-2
Molecular Formula: $C_6H_{14}$
Molecular Weight: 86.177
Melting Point (°C): -99.865
Boiling Point (°C): 49.741
Density (g/cm³ at 20 °C):
    0.64916 (20°C, Dreisbach 1959; Riddick et al. 1986)
    0.64446 (25°C, Dreisbach 1959; Riddick et al. 1986)
Molar Volume (cm³/mol):
    132.74, 133.72 (calculated-density)
    130.0 (calculated-density, Lende & Banerjee 1981; Wang et al. 1992)
    140.6 (calculated-LeBas method, Eastcott et al. 1988)
    133.7 (20°C, calculated-density, Stephenson & Malanowski 1987)
Molecular Volume, TMV (Å³):
    107.29 (Pearlman 1986)
Total Surface Area, TSA (Å²):
    290.76 (Hermann 1972)
    290.8 (Amidon et al. 1974)
    289.2, 287.4, 135.7 (Valvani et al. 1975)
    290.8, 282.6 (Kier et al. 1975)
    152.29 (Pearlman 1986)
    290.8 (Mailhot & Peters 1988)
    155.5 (Okouchi et al. 1992)
Heat of Fusion, $\Delta H_{fus}$, (kcal/mol):
    1.384 (Riddick et al. 1986)
Entropy of Fusion, $\Delta S_{fus}$ (cal/mol K or e.u.):
Fugacity Ratio at 25°C, F: 1.0

Water Solubility (g/m³ or mg/L at 25°C):
    18.4 (shake flask-GC, McAuliffe 1963,1966; quoted, Hermann 1972; Price 1976; Lande & Banerjee 1981)
    18.0, 28.0 (quoted, calculated-group contribution, Irmann 1965)
    23.8, 23.0 (shake flask-GC, Polak & Lu 1973)
    18.4, 26.7 (quoted, calculated-TSA, Amidon et al. 1974)
    18.4 (quoted, Hine & Mookerjee 1975)
    18.4, 29.1 (quoted, calculated- $\chi$ , Hall et al. 1975)
    18.4, 29.1 (quoted, calculated- $\chi$ , Kier & Hall 1976)
    21.2 (shake flask-GC, Price 1976)
    18.4, 21.1, 23.81 (quoted, Mackay & Shiu 1981)

18.4, 21.2, 23.8 (quoted, Brookman et al. 1985)
18.0 (quoted, Riddick et al. 1986)
18.4, 17.2 (quoted, calculated-fragment solubility constants, Wakita et al. 1986)
18.4, 24.4 (quoted, calculated-UNIFAC, Al-Sahhaf 1989)
18.4, 23.8, 21.1; 21.0 (quoted exptl values; recommended best value, IUPAC 1989)
18.4, 13.66 (quoted, calculated-$V_M$, Wang et al. 1992)
21.45 (quoted literature average, Myrdal et al. 1992)
18.4 (calculated-AI, Okouchi et al. 1992)

Vapor Pressure (Pa at 25°C):
43320 (24.47°C, Willingham et al. 1945)
43478 (Antoine eqn. regression, Stull 1947)
42570 (25°C, Nicolini & Laffitte 1949)
42540 (Antoine eqn., Dreisbach 1959)
42600 (Antoine eqn., Zwolinski & Wilhoit 1971; quoted, Mackay & Shiu 1981)
42540 (quoted, Hine & Mookerjee 1975)
42585, 42550 (interpolated, Antoine equations, Boublik et al. 1984)
42540 (Antoine eqn., Dean 1985)
42700 (quoted, Riddick et al. 1986)
42560 (interpolated, Antoine eqn., Stephenson & Malanowski 1987)

Henry's Law Constant (Pa $m^3$/mol):
173000; 199000, 173000, 154000 (recommended; calculated-P/C, Mackay & Shiu 1981)
196800 (calculated as $1/K_{AW}$, $C_W/C_A$, reported as exptl., Hine & Mookerjee 1975; quoted, Nirmalakhandan & Speece 1988)
196800 (calculated-group contribution, Hine & Mookerjee 1975)
49430 (calculated-bond contribution, Hine & Mookerjee 1975)
173182 (calculated-P/C, Eastcott et al. 1988)
196800, 188038 (quoted, calculated-$\chi$, Nirmalakhandan & Speece 1988)
153887 (calculated-P/C, Yaws et al. 1991)

Octanol/Water Partition Coefficient, log $K_{OW}$:
3.82 (calculated-f const., Valvani et al. 1981)
3.82 (quoted, Eastcott et al. 1988)
3.85 (quoted selected value from Pomona College Med. Chem. project, Sangster 1989)
3.85 (recommended, Sangster 1989)
3.82, 3.25 (quoted, calculated-$V_M$, Wang et al. 1992)

Bioconcentration Factor, log BCF:

Sorption Partition Coefficient, log $K_{OC}$:

Half-Lives in the Environment:

Air:

Surface water:

Ground water:

Sediment:

Soil:

Biota:

Environmental Fate Rate Constant and Half-Lives:

Volatilization:

Photolysis:

Oxidation: photooxidation reaction rate constant of $2.59 \times 10^{-12}$ $cm^3$ $molecule^{-1}$ $sec^{-1}$ for the reaction with OH radicals at 297 K and $6.16 \times 10^{-12}$ $cm^3$ $molecule^{-1}$ $sec^{-1}$ at 299 K (Atkinson 1985); gas-phase rate constant of $2.32 \times 10^{-12}$ $cm^3$ $molecule^{-1}$ $sec^{-1}$ for the reaction with OH radicals at 298 K (Atkinson 1990).

Hydrolysis:

Biodegradation:

Biotransformation:

Bioconcentration, Uptake ($k_1$) and Elimination ($k_2$) Rate Constants or Half-Lives:

Common Name: 2,3-Dimethylbutane
Synonym: diisopropyl
Chemical Name: 2,3-dimethylbutane
CAS Registry No: 79-29-8
Molecular Formula: $C_6H_{14}$
Molecular Weight: 86.177
Melting Point (°C):
-128.5 (Weast 1984)
-128.54 (Dreisbach 1959; Riddick et al. 1986)
Boiling Point (°C):
57.99 (Dreisbach 1959; Riddick et al. 1986)
58.0 (Weast 1984)
Density (g/cm$^3$ at 20°C):
0.6616 (Weast 1984)
0.66164 (20°C, Dreisbach 1959; Riddick et al. 1986)
0.65702 (25°C, Dreisbach 1959; Riddick et al. 1986)
Molar Volume (cm$^3$/mol):
130.24 (20°C, calculated-density)
131.15 (25°C, calculated-density)
140.6 (calculated-LeBas method)
130.3 (20°C, calculated-density, Stephenson & Malanowski 1987)
Molecular Volume, TMV (Å$^3$):
107.29 (Pearlman 1986)
Total Surface Area, TSA (Å$^2$):
253.76 (Pearlman 1986)
155.4 (Okouchi et al. 1992)
Heat of Fusion, $\Delta H_{fus}$, (kcal/mol):
0.191 (Riddick et al. 1986)
Entropy of Fusion, $\Delta S_{fus}$ (cal/mol K or e.u.):
Fugacity Ratio at 25°C, F: 1.0

Water Solubility (g/m$^3$ or mg/L at 25°C):
22.5, 22.0 (shake flask-GC, calculated-group contribution, Polak & Lu 1973)
19.1 (shake flask-GC, Price 1976)
22.5, 28.2 (quoted, calculated-group contribution method of Irmann 1865, Horvath 1982)
22.5, 19.1 (quoted, Mackay & Shiu 1981)
22.5, 19.1 (quoted, Brookman et al. 1985)
11.0 (quoted, Riddick et al. 1986)
22.5, 19.1; 21.0 (quoted exptl values; recommended tentative value, IUPAC 1989)
116, 47.4 (calculated-$K_{OW}$, calculated-$V_M$, Wang et al. 1992)
20.2 (quoted literature average, Myrdal et al. 1992)
19.1 (calculated-AI, Okouchi et al. 1992)

Vapor Pressure (Pa at 25°C):

28955 (23.10°C, Willingham et al. 1945)
31204 (Antoine eqn. regression, Stull 1947)
31280 (Antoine eqn., Dreisbach 1959)
31300 (Antoine eqn., Zwolinski & Wilhoit 1971; quoted, Mackay & Shiu 1981)
31280 (interpolated, Antoine equations, Boublik et al. 1984)
32010 (Antoine eqn., Dean 1985)
31300 (quoted, Riddick et al. 1986)
31287 (interpolated, Antoine eqn., Stephenson & Malanowski 1987)

Henry's Law Constant (Pa $m^3$/mol):

130000 (recommended, Mackay & Shiu 1981)
141000, 120000 (calculated-P/C, Mackay & Shiu 1981)
131194 (calculated-P/C, Yaws et al. 1991)

Octanol/Water Partition Coefficient, log $K_{OW}$:

3.85 (calculated-f const., Valvani et al. 1981)
3.85 (quoted from Pomona College Med. Chem. Project, Sangster 1989)
3.85 (recommended, Sangster 1989)
2.42 (calculated-S, Wang et al. 1992)
2.63 (calculated-$V_M$, Wang et al. 1992)

Bioconcentration Factor, log BCF:

Sorption Partition Coefficient, log $K_{OC}$:

Half-Lives in the Environment:

Air: photooxidation reaction rate constant of $6.30 \times 10^{-12}$ $cm^3$ $molecule^{-1}$ $sec^{-1}$ with hydroxyl radicals and an estimated atmospheric lifetime of 22 hours during summer daylight (Altshuller 1991).

Surface water:

Ground water:

Sediment:

Soil:

Biota:

Environmental Fate Rate Constant and Half-Lives:

Volatilization:

Photolysis:

Oxidation: photooxidation reaction rate constant of $5.16 \times 10^{12}$ $cm^3$ $mol^{-1}$ $sec^{-1}$ (observed) and $4.49 \times 10^{12}$ $cm^3$ $mol^{-1}$ $sec^{-1}$ (correlated) with hydroxyl radicals in air (Greiner 1970); rate constant of $2.0 \times 10^{-13}$ $cm^3$ $molecule^{-1}$ $sec^{-1}$ for the reaction with $O(^3P)$ (Herron & Huie 1973; quoted, Gaffney & Levine 1979) and $5.50 \times 10^{-12}$ $cm^3$ $molecule^{-1}$ $sec^{-1}$ for the reaction with OH radicals both at room temperature (Atkinson et al. 1979; quoted, Gaffney & Levine 1979); rate constant of $5.67 \times 10^{-12}$ $cm^3$ $molecule^{-1}$ $sec^{-1}$ for the reaction with hydroxyl radicals at atmospheric pressure and 300 K (Darnall et al. 1978); rate constant of $6.30 \times 10^{-12}$ $cm^3$ $molecule^{-1}$ $sec^{-1}$ for the reaction with hydroxyl radicals in air at 298 K (Atkinson 1990, 1991); rate constant of $4.06 \times 10^{-16}$ $cm^3$ $molecule^{-1}$ $sec^{-1}$ for gas-phase reaction with $NO_3$ radicals (Atkinson 1990); an estimated atmospheric lifetime of 22 hours, based on photooxidation reaction rate constant of $6.3 \times 10^{-12}$ $cm^3$ $molecule^{-1}$ $sec^{-1}$ with OH radicals in air during summer daylight (Altshuller 1991); rate constants of $6.19 \times 10^{-12}$ $cm^{-1}$ $molecule^{-1}$ $sec^{-1}$ for the reaction with OH radicals and $4.06 \times 10^{-16}$ $cm^3$ $molecule^{-1}$ $sec^{-1}$ for the reaction with $NO_3$ radicals in the gas phase at 298 K (Sabljic & Güsten 1990).

Hydrolysis:

Biodegradation:

Biotransformation:

Bioconcentration, Uptake ($k_1$) and Elimination ($k_2$) Rate Constants or Half-Lives:

Common Name: 2,2,3-Trimethylbutane
Synonym: triptene
Chemical Name: 2,2,3-trimethylbutane
CAS Registry No: 464-06-2
Molecular Formula: $C_7H_{16}$
Molecular Weight: 100.203
Melting Point (°C):
-24.912 (Dreisbach 1959)
-24.20 (Stephenson & Malanowski 1987)
Boiling Point (°C):
80.882 (Dreisbach 1959)
80.90 (Stephenson & Malanowski 1987)
Density (g/cm³ at 20°C):
0.69011 (20°C, Dreisbach 1959)
0.68588 (25°C, Dreisbach 1959)
0.6901 (Weast 1984)
Molar Volume (cm³/mol):
145.2 (20°C, calculated-density)
146.09 (25°C, calculated-density)
145.2 (20°C, calculated-density, Stephenson & Malanowski 1987)
162.8 (calculated-LeBas method)
Molecular Volume (A³):
122.54 (Pearlman 1986)
Total Surface Area, TSA (A²):
166.7 (Pearlman 1986)
Heat of Fusion, $\Delta H_{fus}$, (kcal/mol):
0.54 (Dreisbach 1959)
Entropy of Fusion, $\Delta S_{fus}$ (cal/mol K or e.u.):
Fugacity Ratio at 25°C, F: 1.0

Water Solubility (g/m³ or mg/L at 25°C):
4.38 (estimated-nomograph of Kabadi & Danner 1979; Brookman et al. 1985)

Vapor Pressure (Pa at 25°C):
13840 (25.3°C, Forziati et al. 1950)
13648, 13662 (interpolated, Antoine equations, Boublik et al. 1984)
13652 (interpolated, Antoine eqn., Stephenson & Malanowski 1987)

Henry's Law Constant (Pa m³/mol):
241012 (calculated-P/C, Yaws et al. 1991)

Octanol/Water Partition Coefficient, log $K_{OW}$:

Bioconcentration Factor, log BCF:

Sorption Partition Coefficient, log $K_{OC}$:

Half-Lives in the Environment:

Air:
Surface water:
Ground water:
Sediment:
Soil:
Biota:

Environmental Fate Rate Constant and Half-Lives:

Volatilization:
Photolysis:
Oxidation: photooxidation rate constant $3.84 \times 10^{12}$ $cm^3$ $mol^{-1}$ $sec^{-1}$ (observed) and $3.15 \times 10^{12}$ $cm^3$ $mol^{-1}$ $sec^{-1}$ (correlated) for the reaction with hydroxyl radicals in air at 296 K (Greiner 1970); rate constant of $5.23 \times 10^{-12}$ $cm^3$ $molecule^{-1}$ $sec^{-1}$ at 296 K, and $4.09 \times 10^{-12}$ $cm^3$ $molecule^{-1}$ $sec^{-1}$ at 297 K for the gas-phase reaction with OH radicals (Atkinson 1985).
Hydrolysis:
Biodegradation:
Biotransformation:
Bioconcentration, Uptake ($k_1$) and Elimination ($k_2$) Rate Constants or Half-Lives:

Common Name: n-Pentane
Synonym: pentane
Chemical Name: n-pentane
CAS Registry No: 109-66-0
Molecular Formula: $C_5H_{12}$
Molecular Weight: 72.15
Melting Point (°C):
-129.7 (Stull 1947; Dreisbach 1959)
-129.13 (Riddick et al. 1986)
-129.73 (Stephenson & Malanowski 1987)
Boiling Point (°C):
36.056 (Riddick et al. 1986)
36.07 (Dreisbach 1959; Stephenson & Malanowski 1987)
Density (g/cm³ at 20°C):
0.62624 (20°C, Dreisbach 1959; Riddick et al. 1986)
0.62139 (25°C, Dreisbach 1959; Riddick et al. 1986)
Molar Volume (cm³/mol):
115.3 (20°C, calculated-density)
116.11 (25°C, calculated-density)
115.2 (20°C, calculated-density, McAuliffe 1966)
118.0 (calculated-LeBas method)
115 (calculated-density, Lande & Banerjee 1981)
115.2 (calculated-density, Taft et al. 1985; Kamlet et al. 1986)
0.553 (intrinsic volume: $V_I/100$, Leahy 1986; Kamlet 1987,1988)
115.3 (20°C, calculated-density, Stephenson & Malanowski 1987)
115 (20°C, calculated-density, Wang et al. 1992)
Molecular Volume (A³):
87.4 (Moriguchi et al. 1976)
96.36 (van der Waals based values, Cramer 1977)
92.15 (Pearlman 1986)
Total Surface Area, TSA (A²):
286.97, 280.93, 273,18 (Herman 1972)
287.0 (quoted, Amidon et al. 1974,1975; Amidon & Anik 1981; Mailhot & Peters 1988)
124.3 (Moriguchi et al. 1976)
255.0 (solvent radius 1.5 Å, Amidon et al. 1979; Amidon & Anik 1981)
289.6, 277.6, 124.0 (Valvani et al. 1976)
287, 286.55 (Horvath 1982)
138.8 (Pearlman 1986)
134.8 (molecular surface area without solvent, Warne et al. 1990)
129.5 (Okouchi et al. 1992)
Heat of Fusion, $\Delta H_{fus}$, (kcal/mol):
2.006 (Riddick et al. 1986)

Entropy of Fusion, $\Delta S_{fus}$ (cal/mol K or e.u.):
Fugacity Ratio at 25°C, F: 1.0

Water Solubility ($g/m^3$ or mg/L at 25°C):

360 (16°C, cloud point, Fühner 1924)
120 (radiotracer, Black et al. 1948)
360 (quoted, Deno & Berkheimer 1960)
38.5 (shake flask-GC, McAuliffe 1963)
38.5 (shake flask-GC, McAuliffe 1963, 1966; quoted, Mackay & Shiu 1975; Price 1976; Mackay et al. 1979; Bobra et al. 1979; Mackay 1981; Bobra et al. 1984; Abernethy et al. 1986; Eastcott et al. 1988)
39, 137, 360, 48 (quoted values, Irmann 1965)
28.7, 56.2 (calculated-group contribution, Irmann 1965)
49.7 (vapor saturation-GC, Barone et al. 1966)
40.0 (Baker 1967; quoted, Price 1976)
40.3 (shake flask-GC, Nelson & De Ligny 1968)
38.5, 33.0 (quoted, calculated-$K_{OW}$, Hansch et al. 1968)
11.8 (shake flask-GC, Pierotti & Liabastre 1972)
47.6, 44.6 (shake flask-GC, calculated-group contribution, Polak & Lu 1973)
38.7, 26.3 (quoted, calculated-TSA, Amidon et al. 1974)
38.75 (quoted, Hine & Mookerjee 1975)
38.9, 35.4 (quoted, calculated- $\chi$ , Hall et al. 1975)
38.9, 35.4 (quoted, calculated- $\chi$ , Kier & Hall 1976)
39.5 (shake flask-GC, Price 1976)
39.0 (shake flask-GC, Kryzanowska & Szeliga 1978)
40.0 (partition coefficient, Rudakov & Lutsyk 1979)
38.5, 39.5, 40, 40.4, 47.6 (quoted, Mackay 1981; Mackay & Shiu 1981)
38.5, 64.3 (quoted, calculated-$K_{OW}$, Yalkowsky & Morozowich 1980)
38.75 (quoted, Lande & Banerjee 1981)
40.75 (gen. col.-GC, Tewari et al. 1982a; quoted, Wasik et al. 1982)
40.75, 36.9 (quoted exptl., calculated- $\gamma$ and $K_{OW}$, Tewari 1982b)
40.6 (vapor saturation-GC, Jönsson et al. 1982)
38.5, 38.9 (quoted, shake flask-GC, Coates et al. 1985)
38.5, 47.6, 39.6, 39.5 (quoted, Brookman et al. 1985)
38.0 (quoted, Riddick et al. 1986)
38.7, 37.0 (quoted, calculated-fragment solubility constants, Wakita et al. 1986)
38.7, 34.5 (quoted, calculated-$V_M$ and solvatochromic p., Kamlet et al. 1986)
38.7, 38.7 (calculated-$K_{AW}$, calculated-$V_I$ & solvatochromic p., Kamlet et al. 1987)
38.8 (quoted, Isnard & Lambert 1989)
49.7, 38.5, 40.5, 47.6, 39.5, 40.6; 42.0 (quoted exptl values; recommended best value, IUPAC 1989)
40.0 (quoted literature average, Warne et al. 1990)

32.4 (quoted, literature average, Myrdal et al. 1992)
38.5 (calculated-AI, Okouchi et al. 1992)
38.7, 26.2 (quoted, calculated-$V_M$, Wang et al. 1992)

Vapor Pressure (Pa at 25°C):
66756 (24.374°C, Willingham et al. 1945)
71050 (Antoine eqn. regression, Stull 1947)
68330 (calculated from determined data, Dreisbach 1959)
57820 (20.57°C, Osborn & Douslin 1974)
68400 (Antoine eqn., Zwolinski & Wilhoit 1971; quoted, Bobra et al. 1979)
68328 (quoted, Hine & Mookerjee 1975)
68400 (quoted, Mackay & Shiu 1975,1981; Eastcott et al. 1988)
68330 (interpolated, Antoine eqn, Boublik et al. 1984)
70915 (Antoine eqn., Dean 1985)
69810, 68880, 68330 (headspace-GC, correlated, Antoine eqn., Hussam & Carr 1985)
68330 (quoted, Riddick et al. 1986)
68350 (quoted, Kamlet et al. 1986)
68355 (interpolated, Antoine eqn., Stephenson & Malanowski 1987)

Henry's Law Constant (Pa $m^3$/mol):
128000 (calculated-P/C, Mackay & Shiu 1975,1990; Bobra et al. 1979; Mackay et al. 1979; Mackay 1981)
125000 (recommended, Mackay & Shiu 1981)
128000, 125000, 123000, 122200, 10370 (calculated-P/C, Mackay & Shiu 1981)
121450 (concentration ratio-GC, Jönsson et al. 1982)
127050 (calculated as $1/K_{AW}$, $C_W/C_A$, reported as exptl., Hine & Mookerjee 1975; quoted, Nirmalakhandan & Speece 1988)
115885 (calculated-group contribution, Hine & Mookerjee 1975)
33433 (calculated-bond contribution, Hine & Mookerjee 1975)
127670 (selected, Mills et al. 1982)
120966 (calculated-P/C, Eastcott et al. 1988)
127050, 100982 (quoted, calculated-$\chi$, Nirmalakhandan & Speece 1988)
128053 (calculated-P/C, Yaws et al. 1991)

Octanol/Water Partition Coefficient, log $K_{OW}$:
2.50 (shake flask-GC, Hansch et al. 1968; quoted, Moriguchi et al. 1976)
3.39 (shake flask-GC, Leo et al. 1975)
2.50, 2.54 (quoted, calculated-$\chi$, Murray et al. 1975)
2.50, 2.49 (quoted, calculated-hydrophobicity const., Moriguchi et al. 1976)
3.39 (quoted, Cramer 1977)

3.39; 2.99, 3.42, 3.48 (quoted; calculated-f const., Rekker 1977)
3.23 (Hansch & Leo 1979)
3.39, 2.17 (quoted observed & calculated values, Chou & Jurs 1979)
2.50, 3.40 (calculated-$\pi$, f consts., Chou & Jurs 1979)
3.30 (quoted, Yalkowsky & Morozowich 1980)
3.62, 3.64 (quoted explt., calculated- $\gamma$ , Wasik et al. 1981,1982)
3.62 (gen. col.-GC, Tewari et al. 1982a,b)
2.37 (estimated-HPLC-k', Coates et al. 1985)
3.39, 3.49 (quoted, calculated-molar volume & solvatochromic p., Taft et al. 1985)
3.39, 3.40 (quoted, calculated-molar volume & solvatochromic p., Leahy 1986)
3.39, 3.49 (quoted, calculated-molar volume & solvatochromic p., Kamlet et al. 1986)
3.39, 3.33 (quoted, calculated-molar volume & solvatochromic p., Kamlet et al. 1988)
3.62, 3.60 (gen. col.-GC, calculated- $\gamma$ , Schantz & Martire 1987)
3.31, 3.28 (quoted, calculated-MO, Bodor et al. 1989)
3.39, 3.34 (calculated-CLOGP, $\pi$, Bodor et al. 1989)
3.45 (recommended, Sangster 1989)
2.50 (quoted, Isnard & Lambert 1989)
3.21 (quoted literature average, Warne et al. 1990)
3.50, 2.81 (quoted, calculated-$V_M$, Wang et al. 1992)

Bioconcentration Factor, log BCF:

Sorption Partition Coefficient, log $K_{OC}$:

Half-Lives in the Environment:

Air: photooxidation reaction rate constant of $3.94 \times 10^{-12}$ $cm^3$ $molecule^{-1}$ $sec^{-1}$ with hydroxyl radicals and an estimated atmospheric lifetime of 35 hours (Altshuller 1990).

Surface water:

Ground water:

Sediment:

Soil:

Biota:

Environmental Fate Rate Constant and Half-Lives:

Volatilization:

Photolysis:

Oxidation: photooxidation reacton rate constant of $3.74 \times 10^{-12}$ $cm^3$ $molecule^{-1}$ $sec^{-1}$ for the reaction with OH radicals at atmospheric pressure and 300 K (Darnall et al.

1978); room temperature rate constants of $5.8 \times 10^{-14}$ $cm^3$ $molecule^{-1}$ $sec^{-1}$ for the reaction with $O(^3P)$ (Herron & Huie 1973; quoted, Gaffney & Levine 1979) and $5.0 \times 10^{-12}$ $cm^3$ $molecule^{-1}$ $sec^{-1}$ for the reaction with OH radicals (Atkinson et al. 1979; quoted, Gaffney & Levine 1979); rate constant of $4.06 \times 10^{-12}$ $cm^3$ $molecule^{-1}$ $sec^{-1}$ for the reaction with OH radicals at 297 K (Atkinson 1986; quoted, Edney et al. 1986); rate constant of $3.94 \times 10^{-12}$ $cm^3$ $molecule^{-1}$ $sec^{-1}$ for the reaction with OH radicals at 298 K (Atkinson 1990, 1991; Altshuller 1991) and summer daylight atmospheric lifetime of 35 hours, based on reaction rate constant with OH radicals in air (Altshuller 1991); rate constant of $4.06 \times 10^{-12}$ $cm^3$ $molecule^{-1}$ $sec^{-1}$ and $8.0 \times 10^{-17}$ $cm^3$ $molecule^{-1}$ $sec^{-1}$ for the reaction with OH and $NO_3$ radicals in air respectively (Sabljic & Güsten 1990); rate constant of $8.1 \times 10^{-17}$ $cm^3$ $molecule^{-1}$ $sec^{-1}$ for the gas-phase reaction with $NO_3$ radical (Altshuller 1991).

Hydrolysis:

Biodegradation:

Biotransformation:

Bioconcentration, Uptake ($k_1$) and Elimination ($k_2$) Rate Constants or Half-Lives:

Common Name: 2-Methylpentane
Synonym: isohexane
Chemical Name: 2-methylpentane
CAS Registry No: 107-83-5
Molecular Formula: $C_6H_{14}$
Molecular Weight: 86.177
Melting Point (°C):
-154.0 (Stull 1947)
-153.66 (Dreisbach 1959; Stephenson & Malanowski 1987)
Boiling Point (°C):
60.127 (Riddick et al. 1986)
60.27 (Dreisbach 1959; Stephenson & Malanowski 1987)
Density ($g/cm^3$ at 20°C):
0.63215 (20°C, Dreisbach 1959; Riddick et al. 1986)
0.64852 (25°C, Dreisbach 1959; Riddick et al. 1986)
Molar Volume ($cm^3/mol$):
131.92 (20°C, calculated-density)
132.88 (25°C, calculated-density)
131.9 (20°C, calculated-density, McAuliffe 1966)
140.6 (calculated-LeBas method)
132 (calculated-density, Lande & Banerjee 1981)
131.9 (20°C, calculated-density, Stephenson & Malanowski 1987)
132 (20°C, calculated-density, Wang et al. 1992)
Molecular Volume, TMV ($Å^3$):
107.4 (Pearlman 1986)
Total Surface Area, TSA ($Å^2$):
306.41, 301.54, 298.94 (Hermann 1972)
274.4, 266.5 (Kier et al. 1975)
153.2 (Pearlman 1986)
306.4 (Mailhot & Peters 1988)
152.9 (Okouchi et al. 1992)
Heat of Fusion, $\Delta H_{fus}$, (kcal/mol):
1.498 (Riddick et al. 1986)
Entropy of Fusion, $\Delta S_{fus}$ (cal/mol K or e.u.):
Fugacity Ratio at 25°C, F: 1.0

Water Solubility ($g/m^3$ or mg/L at 25°C):
13.8 (shake flask-GC, McAuliffe 1963,1966; quoted, Hermann 1972; Price 1976)
14.0, 22.4 (quoted, calculated-group contribution, Irmann 1965)
16.21 (vapor saturation-GC, Barone et al. 1966)
14.0, 16.8 (quoted, calculated-$K_{OW}$, Hansch et al. 1968)
15.7, 15.6 (shake flask-GC, calculated-group contribution, Polak & Lu 1973)

14.2 (shake flask-GC, Leinonen & Mackay 1973)
14.0 (quoted, Hine & Mookerjee 1975)
13.0 (shake flask-GC, Price 1976)
13.8, 13, 15.7 (quoted, Mackay & Shiu 1981)
13.7 (quoted, Lande & Banerjee 1981)
13.8, 13, 15.7 (quoted, Brookman et al. 1985)
14.0 (quoted, Riddick et al. 1986)
14.0, 13.7 (quoted, calculated-fragment solubility constants, Wakita et al. 1986)
13.8, 14.2, 13.0; 13.7 (quoted exptl. values; recommended best value, IUPAC 1989)
12.8 (calculated-AI, Okouchi et al. 1992)
13.9 (quoted literature average, Myrdal et al. 1992)
13.7, 14.6 (quoted, calculated-$V_M$, Wang et al. 1992)
16.2 (quoted, Müller & Klein 1992)

Vapor Pressure (Pa at 25°C):
29037 (25.64°C, Willingham et al. 1945)
27818 (calculated-Antoine eqn. regression, Stull 1947)
28240 (Antoine eqn., Dreisbach 1959)
28200 (Antoine eqn., Zwolinski & Wilhoit 1971)
27780 (Antoine eqn., Weast 1972-73)
28238 (quoted, Hine & Mookerjee 1975)
28240 (interpolated, Antoine eqn., Boublik et al. 1984)
28230 (Antoine eqn., Dean 1985)
28300 (quoted, Riddick et al. 1986)
28249 (calculated-Antoine eqn., Stephenson & Malanowski 1987)
33638 (interpolated, Antoine eqn., Stephenson & Malanowski 1987)

Henry's Law Constant (Pa m$^3$/mol):
170000; 175000, 188000, 154000 (recommended; calculated-P/C, Mackay & Shiu 1981)
175490 (calculated as $1/K_{AW}$, $C_W/C_A$, reported as exptl., Hine & Mookerjee 1975; quoted, Nirmalakhandan & Speece 1988)
196800 (calculated-group contribution, Hine & Mookerjee 1975)
49434 (calculated-bond contribution, Hine & Mookerjee 1975)
175292 (selected, Mills et al. 1982)
176155 (calculated-P/C, Eastcott et al. 1988)
175490, 149363 (quoted, calculated- $\chi$ , Nirmalakhandan & Speece 1988)
83593 (EPICS-GC, Ashworth et al. 1988)
176276 (calculated-P/C, Yaws et al. 1991)

Octanol/Water Partition Coefficient, log $K_{OW}$:

2.80 (calculated-$\pi$ constant, Hansch et al. 1968)
2.80, 2.85 (quoted, calculated-$\chi$ , Murray et al. 1975)
2.80 (quoted, Mailhot & Peters 1988)
2.80, 3.23 (quoted, calculated-$V_M$m Wang et al. 1992)
3.74 (calculated-f constant, Müller & Klein 1992)

Bioconcentration Factor, log BCF:

Sorption Partition Coefficient, log $K_{OC}$:

Half-Lives in the Environment:

Air: atmospheric lifetime was estimated to be 25 hours during summer daylight, based on photooxidation rate constant of $5.6 \times 10^{-12}$ $cm^3$ $molecule^{-1}$ $sec^{-1}$ for the reaction with hydroxyl radicals in air (Altshuller 1991).
Surface water:
Ground water:
Sediment:
Soil:
Biota:

Environmental Fate Rate Constant and Half-Lives:

Volatilization:
Photolysis:
Oxidation: photooxidation rate constant of $5.6 \times 10^{-12}$ $cm^3$ $molecule^{-1}$ $sec^{-1}$ for the reaction with OH radicals (Atkinson 1990); atmospheric lifetime was estimated to be 25 hours during summer daylight hours, based on photooxidation rate constant of $5.6 \times 10^{-12}$ $cm^3$ $molecule^{-1}$ $sec^{-1}$ for the reaction with hydroxyl radicals in air (Altshuller 1991).
Hydrolysis:
Biodegradation:
Biotransformation:
Bioconcentration, Uptake ($k_1$) and Elimination ($k_2$) Rate Constants or Half-Lives:

Common Name: 3-Methylpentane
Synonym: diethylmethylmethane
Chemical Name: 3-methylpentane
CAS Registry No: 96-14-1
Molecular Formula: $C_6H_{14}$
Molecular Weight: 86.177
Melting Point (°C): -118.0
Boiling Point (°C): 63.282
Density ($g/cm^3$ at 25 °C):
0.66431, 0.65976 (20°C, 25°C, Dreisbach 1959; Riddick et al. 1986)
Molar Volume ($cm^3/mol$):
129.7 (20°C, calculated-density)
130.62 (25°C, calculated-density)
129.7 (20°C, calculated-density, McAuliffe 1966)
140.6 (calculated-LeBas method)
130 (calculated-density, Lande & Banerjee 1981)
131.8 (calculated-density, Leahy 1986)
0.648 (intrinsic volume: $V_I/100$, Leahy 1986)
129.7 (20°C, calculated-density, Stephenson & Malanowski 1987)
130.0 (20°C, calculated-density, Wang et al. 1992)
Molecular Volume, TMV ($Å^3$):
107.4 (Pearlman 1986)
Total Surface Area, TSA ($Å^2$):
300.08, 295.2, 294.07 (Hermann 1972)
300.1 (quoted, Amidon et al. 1974)
300.8, 294.1, 137.5 (Valvani et al. 1975)
300.1, 297.8 (quoted, correlated, Kier et al. 1975)
153.2 (Pearlman 1986)
152.9 (Okouchi et al. 1992)
Heat of Fusion, $\Delta H_{fus}$, (kcal/mol):
1.2675 (Riddick et al. 1986)
Entropy of Fusion, $\Delta S_{fus}$ (cal/mol K or e.u.):
Fugacity Ratio at 25°C, F: 1.0

Water Solubility ($g/m^3$ or mg/L at 25°C):
12.8 (shake flask-GC, McAuliffe 1966; quoted, Hermann 1972; Price 1976)
12.8, 16.8 (quoted, calculated-$K_{OW}$, Hansch et al. 1968)
17.9, 17.2 (shake flask-GC, calculated-group contribution, Polak & Lu 1973)
12.8, 17.92 (quoted, calculated-TSA, Amidon et al. 1974)
12.7 (quoted, Hine & Mookerjee 1975)
12.8, 13.1 (quoted, calculated- $\chi$ , Hall et al. 1975)
12.8, 13.1 (quoted, calculated- $\chi$ , Kier & Hall 1976)

13.1 (shake flask-GC, Price 1976)
12.9 (partition coefficient-GC, Rudakov & Lutsyk 1979)
12.8, 13.1 (quoted, Mackay & Shiu 1981)
12.7 (quoted, Lande & Banerjee 1981)
12.8 (quoted, Bobra et al. 1984)
12.8, 13.1, 17.9 (quoted, Brookman et al. 1985)
13.0 (quoted, Riddick et al. 1986)
12.8, 13.7 (quoted, calculated-fragment solubility constants, Wakita et al. 1986)
12.8, 13.1, 12.9; 12.9 (quoted exptl. values; recommended best value, IUPAC 1989)
14.1 (quoted literature average, Myrdal et al. 1992)
12.8 (calculated-AI, Okouchi et al. 1992)
12.7, 16.04 (quoted, calculated-$V_M$, Wang et al. 1992)
12.7 (quoted, Müller & Klein 1992)

Vapor Pressure (Pa at 25°C):
29037 (23.2°C, Willingham et al. 1945)
24971 (Antoine eqn. regression, Stull 1947)
25300 (Antoine eqn., Zwolinski & Wilhoit 1971)
24935 (Antoine eqn., Weast 1972-73)
25305 (quoted, Hine & Mookerjee 1975)
25310 (interpolated, Antoine eqn, Boublik et al. 1984)
25307 (Antoine eqn., Dean 1985)
25300 (quoted, Riddick et al. 1986)
25320 (interpolated, Antoine eqn., Stephenson & Malanowski 1987)

Henry's Law Constant (Pa m$^3$/mol):
172000; 172000, 186000, 154000 (recommended; calculated-P/C, Mackay & Shiu 1981)
171492 (calculated as $1/K_{AW}$, $C_W/C_A$, reported as exptl., Hine & Mookerjee 1975; quoted, Nirmalakhandan & Speece 1988)
196800 (calculated-contribution, Hine & Mookerjee 1975)
49434 (calculated-bond contribution, Hine & Mookerjee 1975)
170210 (calculated-P/C, Eastcott et al. 1988)
171492, 139394 (quoted, calculated- $\chi$ , Nirmalakhandan & Speece 1988)
113668 (calculated-P/C, Yaws et al. 1991)

Octanol/Water Partition Coefficient, log $K_{OW}$:
2.80 (calculated-$\pi$, Hansch et al. 1968)
2.80 (Hansch & Leo 1979)
2.80, 2.88 (quoted, calculated- $\chi$ , Murray et al. 1975)
3.81 (calculated-$V_I$ and solvatochromic p., Leahy 1986)

3.60 (recommended, Sangster 1989)
2.80, 3.18 (quoted, calculated-$V_M$, Wang et al. 1992)
3.74 (calculated-f constant, Müller & Klein 1992)

Bioconcentration Factor, log BCF:

Sorption Partition Coefficient, log $K_{OC}$:

Half-Lives in the Environment:

Air: photooxidation reaction rate constant of $4.30 \times 10^{-9}$ cm$^3$ mol$^{-1}$ sec$^{-1}$ with hydroxyl radicals with half-life of 2.4-24 hours (Darnall et al. 1976); rate constant of $5.7 \times 10^{-12}$ cm$^3$ molecule$^{-1}$ sec$^{-1}$ for the reaction with OH radicals with an estimated atmospheric lifetime of 25 hours during summer daylight (Altshuller 1991).

Surface water:

Ground water:

Sediment:

Soil:

Biota:

Environmental Fate Rate Constant and Half-Lives:

Volatilization:

Photolysis:

Oxidation: photooxidation reaction rate constant of $4.30 \times 10^{-9}$ cm$^3$ mol$^{-1}$ sec$^{-1}$ with hydroxyl radicals in air (Lloyd et al. 1976, Darnall et al. 1976); atmospheric half-life of 2.4-24 hours (Darnall et al. 1976); rate constant of $6.8 \times 10^{-12}$ cm$^3$ molecule$^{-1}$ sec$^{-1}$ for the reaction with OH radicals at 305 K (Darnall et al. 1978); rate constant of $5.7 \times 10^{-12}$ cm$^3$ molecule$^{-1}$ sec$^{-1}$ for the reaction with OH radicals (Atkinson 1990, 1991; Altshuller 1991); atmospheric lifetime was estimated to be 25 hours, based on the reaction rate constant with OH radicals during summer daylight (Altshuller 1991).

Hydrolysis:

Biodegradation:

Biotransformation:

Bioconcentration, Uptake ($k_1$) and Elimination ($k_2$) Rate Constants or Half-Lives:

Common Name: 2,2-Dimethylpentane
Synonym:
Chemical Name: 2,2-dimethylpentane
CAS Registry No: 590-35-2
Molecular Formula: $C_6H_{14}$
Molecular Weight: 100.21
Melting Point (°C): -123.8
Boiling Point (°C): 79.2
Density ($g/cm^3$ at 20°C):
0.6739 (Weast 1984)
0.67385 (20°C, Dreisbach 1959)
0.66953 (25°C, Dreisbach 1959)
Molar Volume ($cm^3/mol$):
148.7 (calculated from density)
162.8 (calculated-LeBas method)
Molecular Volume, TMV ($Å^3$):
123.57 (Pearlman 1986)
Total Surface Area, TSA ($Å^2$):
174.99 (Pearlman 1986)
176.4 (Okouchi et al. 1992)
Heat of Fusion, $\Delta H_{fus}$, (kcal/mol):
1.389 (Dreisbach 1959)
Entropy of Fusion, $\Delta S_{fus}$ (cal/mol K or e.u.):
Fugacity Ratio at 25°C, F: 1.0

Water Solubility ($g/m^3$ or mg/L at 25°C):
4.40 (shake flask-GC, Price 1976; quoted, Mackay & Shiu 1981; Brookman et al. 1985)
21.4, 8.30 (quoted, calculated-$K_{OW}$, Hansch et al. 1968)
21.4, 4.91 (quoted, calculated-fragment solubility constants, Wakita et al. 1986)
4.40 (quoted literature average, Myrdal et al. 1992)
21.4, 8.14 (quoted, calculated-$V_M$, Wang et al. 1992)
4.40 (calculated-AI, Okouchi et al. 1992)
21.4 (quoted, Müller & Klein 1992)

Vapor Pressure (Pa at 25°C):
14030 (Antoine eqn., Dreisbach 1959)
14000 (Antoine eqn, Zwolinski & Wilhoit 1971; quoted, Mackay & Shiu 1981)
13500 (Antoine eqn., Weast 1972-72)

Henry's Law Constant (Pa $m^3$/mol):

318000 (calculated-P/C, Mackay & Shiu 1981)
319174 (selected, Mills et al. 1982)
319424 (calculated-P/C, Yaws et al. 1991)

Octanol/Water Partition Coefficient, log $K_{OW}$:

3.10 (calculated-$\pi$ substituent constant, Hansch et al. 1968)
3.10, 3.62 (quoted, calculated-$V_M$, Wang et al. 1992)
4.14 (calculated-f constant, Müller & Klein 1992)

Bioconcentration Factor, log BCF:

Sorption Partition Coefficient, log $K_{OC}$:

Half-Lives in the Environment:

Air:
Surface water:
Ground water:
Sediment:
Soil:
Biota:

Environmental Fate Rate Constant and Half-Lives:

Volatilization:
Photolysis:
Oxidation: photooxidation rate constant of $3.4 \times 10^{-12}$ $cm^3$ $molecule^{-1}$ $sec^{-1}$ for the reaction with hydroxyl radicals in air at 298 K (Atkinson 1990).
Hydrolysis:
Biodegradation:
Biotransformation:
Bioconcentration, Uptake ($k_1$) and Elimination ($k_2$) Rate Constants or Half-Lives:

Common Name: 2,4-Dimethylpentane
Synonym: diisopropylmethane
Chemical Name: 2,4-dimethylpentane
CAS Registry No: 108-08-7
Molecular Formula: $C_7H_{16}$
Molecular Weight: 100.203
Melting Point (°C): -119.52
Boiling Point (°C): 80.5
Density (g/cm$^3$ at 20°C):
0.6727 (20°C, Dreisbach 1959; Riddick et al. 1986)
0.66832 (25°C, Dreisbach 1959; Riddick et al. 1986)
Molar Volume (cm$^3$/mol):
148.97 (20°C, calculated-density)
149.93 (25°C, calculated-density)
148.9 (20°C, calculated-density, McAuliffe 1966)
162.8 (calculated-LeBas method)
149 (calculated-density, Lande & Banerjee 1981)
148.7 (20°C, calculated-density, Stephenson & Malanowski 1987)
149 (20°C, calculated-density, Wang et al. 1992)
Molecular Volume, TMV (Å$^3$):
123.63 (Pearlman 1986)
Total Surface Area, TSA (Å$^2$):
324.71 (Hermann 1972)
324.7 (Amidon et al. 1974)
323.4, 319.0, 154.0 (Valvani et al. 1975)
171.53 (Pearlman 1986)
176.3 (Okouchi et al. 1992)
Heat of Fusion, $\Delta H_{fus}$, (kcal/mol):
1.636 (Riddick et al. 1986)
Entropy of Fusion, $\Delta S_{fus}$ (cal/mol K or e.u.):
Fugacity Ratio at 25°C, F: 1.0

Water Solubility (g/m$^3$ or mg/L at 25°C):
3.62 (shake flask-GC, McAuliffe 1963)
4.06 (shake flask-GC, McAuliffe 1966; quoted, Hermann 1972; Price 1976)
3.60, 8.91 (quoted, calculated-group contribution, Irmann 1965)
4.08, 8.30 (quoted, calculated-$K_{OW}$, Hansch et al. 1968)
5.50, 5.26 (shake flask-GC, calculated-group contribution, Polak & Lu 1973)
4.08, 7.21 (quoted, calculated-TSA, Amidon et al. 1974)
4.08, 8.15 (quoted, calculated- $\chi$ , Hall et al 1975)
4.08, 8.15 (quoted, calculated- $\chi$ , Kier & Hall 1976)
4.08 (quoted, Hine & Mookerjee 1975)

4.41 (shake flask-GC, Price 1976)
4.06, 4.41, 5.50 (quoted, Mackay & Shiu 1981)
4.08 (quoted, Lande & Banerjee 1981)
4.06, 4.41, 5.5 (quoted, Brookman et al. 1985)
4.10 (quoted, Riddick et al. 1986)
4.08, 4.90 (quoted, calculated-fragment solubility constants, Wakita 1986)
4.06, 4.40; 4.20 (quoted exptl. values; recommended best value IUPAC 1989)
4.37 (quoted literature average, Myrdal et al. 1992)
4.06 (calculated-AI, Okouchi et al. 1992)
4.08, 7.78 (quoted, calculated-$V_M$, Wang 1992)
4.08 (quoted, Müller & Klein 1992)

Vapor Pressure (Pa at 25°C):
12635 (Antoine eqn. regression, Stull 1947)
11730 (22.54°C, Forziati et al. 1950)
13120 (Antoine eqn., Dreisbach 1959)
13100 (Antoine eqn, Zwolinski & Wilhoit 1971; quoted, Mackay & Shiu 1981)
12620 (Antoine eqn., Weast 1972-73)
13120 (quoted, Hine & Mookerjee 1975)
13120 (interpolated, Antoine eqn, Boublik et al. 1984)
15454 (Antoine eqn., Dean 1985)
13000 (quoted, Riddick et al. 1986)
13125 (interpolated, Antoine eqn., Stephenson & Malanowski 1987)

Henry's Law Constant (Pa m$^3$/mol):
300000; 323000, 298000, 239000 (recommended; calculated-P/C, Mackay & Shiu 1981)
319334 (calculated-1/$K_{AW}$, $C_W/C_A$, reported as exptl., Hine & Mookerjee 1975; quoted, Nirmalakhandan & Speece 1988)
326608 (calculated-group contribution, Hine & Mookerjee 1975)
73118 (calculated-bond contribution, Hine & Mookerjee 1975)
297310 (calculated-P/C, Eastcott et al. 1988)
319334, 160046 (quoted, calculated- $\chi$ , Nirmalakhandan & Speece 1988)
298048 (calculated-P/C, Yaws et al. 1991)

Octanol/Water Partition Coefficient, log $K_{OW}$:
3.10 (calculated-$\pi$ constant, Hansch et al. 1968)
3.10, 3.17 (quoted, calculated-$\chi$ , Murray et al. 1975)
3.10, 3.66 (quoted, calculated-$V_M$, Wang et al. 1992)
4.14 (calculated-f constant, Müller & Klein 1992)

Bioconcentration Factor, log BCF:

Sorption Partition Coefficient, log $K_{OC}$:

Half-Lives in the Environment:

Air: photooxidation reaction rate constant of $5.10 \times 10^{-12}$ $cm^3$ $molecule^{-1}$ $sec^{-1}$ with hydroxyl radicals and an estimated lifetime of 27 hours during summer daylight (Altshuller 1991).

Surface water:

Ground water:

Sediment:

Soil:

Biota:

Environmental Fate Rate Constant and Half-Lives:

Volatilization:

Photolysis:

Oxidation: photooxidation reaction rate constant of $5.10 \times 10^{-12}$ $cm^3$ $molecule^{-1}$ $sec^{-1}$ with hydroxyl radicals in air (Atkinson 1990, 1991; Altshuller 1991); atmospheric lifetime was estimated to be 27 hours during summer daylight (Altshuller 1991).

Hydrolysis:

Biodegradation:

Biotransformation:

Bioconcentration, Uptake ($k_1$) and Elimination ($k_2$) Rate Constants or Half-Lives:

Common Name: 3,3-Dimethylpentane
Synonym:
Chemical Name: 3,3-dimethylpentane
CAS Registry No: 562-49-2
Molecular Formula: $C_7H_{16}$
Molecular Weight: 100.203
Melting Point (°C): -134.5
Boiling Point (°C): 86.06
Density (g/cm³ at 20°C):
 0.6936 (Weast 1984)
 0.69327 (20°C, Dreisbach 1959)
 0.68908 (25°C, Dreisbach 1959)
Molar Volume (cm³/mol):
 144.5 (calculated-density)
 162.8 (calculated-LeBas method)
Molecular Volume, TMV (Å³):
 122.6 (Pearlman 1986)
Total Surface Area, TSA (Å²):
 167.15 (Pearlman 1986)
 176.4 (Okouchi et al. 1992)
Heat of Fusion, $\Delta H_{fus}$, (kcal/mol):
 1.69 (Dreisbach 1959)
Entropy of Fusion, $\Delta S_{fus}$ (cal/mol K or e.u.):
Fugacity Ratio at 25°C, F: 1.0

Water Solubility (g/m³ or mg/L at 25°C):
 5.94 (shake flask-GC, Price 1976; quoted, Mackay & Shiu 1981; Brookman et al. 1985)
 5.93 (quoted literature average, Myrdal et al. 1992)
 5.94 (calculated-AI, Okouchi et al. 1992)

Vapor Pressure (Pa at 25°C):
 10601 (Antoine eqn. regression, Stull 1947)
 11044 (Antoine eqn., Dreisbach 1959)
 11000 (Antoine eqn., Zwolinski & Wilhoit 1971; quoted, Mackay & Shiu 1981)
 10594 (Anotine eqn., Weast 1972-73)

Henry's Law Constant (Pa m³/mol):
 186000 (calculated-P/C, Mackay & Shiu 1981)
 186305 (calculated-P/C, Yaws et al. 1991)

Octanol/Water Partition Coefficient, log $K_{OW}$:

Bioconcentration Factor, log BCF:

Sorption Partition Coefficient, log $K_{OC}$:

Half-Lives in the Environment:

Air:
Surface water:
Ground water:
Sediment:
Soil:
Biota:

Environmental Fate Rate Constant and Half-Lives:

Volatilization:
Photolysis:
Oxidation:
Hydrolysis:
Biodegradation:
Biotransformation:
Bioconcentration, Uptake ($k_1$) and Elimination ($k_2$) Rate Constants or Half-Lives:

Common Name: 2,2,4-Trimethylpentane
Synonym: isooctane, isobutyltrimethylmethane
Chemical Name: 2,2,4-trimethylpentane
CAS Registry No: 504-84-1
Molecular Formula: $C_8H_{18}$
Molecular Weight: 114.23
Melting Point (°C):
-107.30 (Stull 1947)
-107.388 (Weast 1972-73)
-107.40 (Stephenson & Malanowski 1987)
Boiling Point (°C): 99.238
Density (g/cm$^3$ at 20°C):
0.69192 (20°C, Dreisbach 1959)
0.68777 (25°C, Dreisbach 1959)
0.69193 (20°C, Riddick et al. 1986)
0.68781 (25°C, Riddick et al. 1986)
Molar Volume (cm$^3$/mol):
166.08 (calculated-density)
165.1 (20°C, calculated-density, McAuliffe 1966; Stephenson & Malanowski 1987)
165 (calculated-density, Lande & Banerjee 1981)
185 (calculated-LeBas method)
165 (20°C, calculated-density, Wang et al. 1992)
Molecular Volume, TMV (Å$^3$):
Total Surface Area, TSA (Å$^2$):
338.93 (Hermann 1972; quoted, Valvani et al. 1974)
338.9, 332.8 (Kier et al. 1975)
332.4, 329.2, 163.1 (Valvani et al. 1975)
199.8 (Okouchi et al. 1992)
Heat of Fusion, $\Delta H_{fus}$, (kcal/mol):
2.198 (Riddick et al. 1986)
Entropy of Fusion, $\Delta S_{fus}$ (cal/mol K or e.u.):
Fugacity Ratio at 25°C, F: 1.0

Water Solubility (g/m$^3$ or mg/L at 25°C):
2.44 (shake flask-GC, McAuliffe 1963,1966; quoted, Hermann et al. 1972; Mackay & Wolkoff 1973; Mackay & Leinonen 1975; Lande & Banerjee 1981; Lyman et al. 1982)
2.40, 3.55 (quoted, calculated-group contribution, Irmann 1965)
2.05, 2.11 (shake flask-GC, calculated-group contribution, Polak & Lu 1973)
8.54, 4.47 (quoted, calculated-TSA, Amidon et al. 1974)
8.54, 4.47 (quoted, calculated- $\chi$ , Hall et al. 1975)
1.14 (shake flask-GC, Price 1976; quoted, Hine & Mooerjee 1975)

2.44, 2.05 (quoted, Mackay & Shiu 1981)
2.44, 2.05, 1.14 (quoted, Brookman et al. 1985)
2.40 (quoted, Riddick 1986)
2.44, 1.83 (quoted, calculated-fragment solubility constants, Wakita et al. 1986)
2,44, 2.05, 1.14; 2.20 (quoted; recommended, IUPAC 1989)
2.44 (calculated-AI, Okoucki et al. 1992)
2.44, 4.34 (quoted, calculated-$V_M$, Wang et al. 1992)

Vapor Pressure (Pa at 25°C):
6371 (24.4°C, Willingham et al. 1945)
6578 (Antoine eqn., Dreisbach 1959)
6250 (calculated-Antoine eqn. regression, Stull 1947)
6560 (Antoine eqn., Zwolinski & Wilhoit 1971; quoted, Mackay & Wolkoff 1973; Mackay & Shiu 1981; Eastcott et al. 1988)
6573 (interpolated, Antoine eqn., Boublik et al. 1973, 1984)
6242 (Antoine eqn., Weast 1972-73)
6578 (quoted, Hine & Mookerjee 1975)
6573 (quoted, Mackay & Leinonen 1975; Lyman et al. 1982)
6580 (interpolated, Antoine eqn., Dean 1985)
6500 (quoted, Riddick et al. 1986)
6577 (interpolated, Antoine eqn., Stephenson & Malanowski 1987)

Henry's Law Constant (Pa $m^3$/mol):
308028 (calculated-P/C, Mackay & Leinonen 1975)
330000; 308000, 365000 (recommended; calculated-P/C, Mackay & Shiu 1981)
304961 (calculated as $1/K_{AW}$, $C_W/C_A$, reported as exptl., Hine & Mookerjee 1975; quoted, Nirmalakhandan & Speece 1988)
472092 (calculated-group contribution, Hine & Mookerjee 1975)
110669 (calculated-bond contribution, Hine & Mookerjee 1975)
308028 (selected, Mills et al. 1982)
314108 (calculated-P/C, Lyman et al. 1982)
304961, 350143 (quoted, calculated- $\chi$ , Nirmalakhandan & Speece 1988)
327204 (calculated-P/C, Eastcott et al. 1988)
338267 (calculated-P/C, Yaws et al. 1991)

Octanol/Water Partition Coefficient, log $K_{OW}$:
5.83 (estimated-HPLC/MS, Burkhard et al. 1985)
4.54 (calculated-f const., Burkhard et al. 1985)
5.02 (calculated-regression eqn. from Lyman et al. 1982, Wang et al. 1992)
4.06 (calculated-$V_M$, Wang et al. 1992)

Bioconcentration Factor, log BCF:

Sorption Partition Coefficient, log $K_{OC}$:

Half-Lives in the Environment:

Air: atmospheric lifetime was estimated to be 16 hours, based on the photooxidation reaction rate constant of $3.68 \times 10^{-12}$ $cm^3$ $molecule^{-1}$ $sec^{-1}$ with OH radicals in air during summer daylight (Altshuller 1991).

Surface water: volatilization half-life of 5.5 hours from a water column 1 $m^2$ in cross section of depth 1 m (Mackay & Leinonen 1975); estimated half-life of 3.1 hours at 20°C in a river 1 meter deep flowing at 1 m/s with a wind velocity of 3 m/s (Lyman et al. 1982).

Ground water:

Sediment:

Soil:

Biota:

Environmental Fate Rate Constant and Half-Lives:

Volatilization: volatilization half-life of 5.5 hours from a water column 1 $m^2$ in cross section of depth 1 m (Mackay & Leinonen 1975); estimated half-life of 3.1 hours at 20°C in a river 1 meter deep flowing at 1 m/s with a wind velocity of 3 m/s (Lyman et al. 1982).

Photolysis:

Oxidation: photooxidation reaction rate constants of $2.83 \times 10^{-12}$ $cm^3$ $molecule^{-1}$ $sec^{-1}$ (observed) and $2.35 \times 10^{-12}$ $cm^3$ $molecule^{-1}$ $sec^{-1}$ (correlated) with OH radicals in air at 298 K (Greiner 1970); rate constant of $3.73 \times 10^{-12}$ $cm^3$ $molecule^{-1}$ $sec^{-1}$ for the reaction with OH radicals in air at 298-305 K (Darnall et al. 1978); room temperature rate constants of $9.10 \times 10^{-14}$ $cm^3$ $molecule^{-1}$ $sec^{-1}$ for the reaction with $O(^3P)$ (Herron & Huie 1973; quoted, Gaffney & Levine 1979) and $3.7 \times 10^{-12}$ $cm^3$ $molecule^{-1}$ $sec^{-1}$ for the reaction with hydroxyl radicals in air (Atkinson et al. 1979; quoted, Gaffney & Levine 1979); rate constant of $3.90 \times 10^{-12}$ $cm^3$ $molecule^{-1}$ $sec^{-1}$at 298 K and $3.56 \times 10^{-12}$ $cm^3$ $molecule^{-1}$ $sec^{-1}$ at 297 K for reaction with OH radicals in air (Atkinson 1985); rate constants of $7.0 \times 10^{-12}$ $cm^3$ $molecule^{-1}$ $sec^{-1}$ for the reaction with hydroxyl radicals in air at 298 K (Atkinson 1990); atmospheric lifetime was estimated to be 16 hours, based on the photooxidation reaction rate constant of $3.68 \times 10^{-12}$ $cm^3$ $molecule^{-1}$ $sec^{-1}$ with OH radicals in air during summer daylight (Altshuller 1991).

Hydrolysis:

Biodegradation:

Biotransformation:

Bioconcentration, Uptake ($k_1$) and Elimination ($k_2$) Rate Constants or Half-Lives:

Common Name: 2,3,4-Trimethylpentane
Synonym:
Chemical Name: 2,3,4-trimethylpentane
CAS Registry No: 565-75-3
Molecular Formula: $C_8H_{18}$
Molecular Weight: 114.23
Melting Point (°C): -109.21
Boiling Point (°C): 113.6
Density (g/cm$^3$ at 20°C):
0.7191 (Weast 1984)
0.71906 (20°C, Dreisbach 1959)
0.71503 (25°C, Dreisbach 1959)
Molar Volume (cm$^3$/mol):
Molecular Volume, TMV (Å$^3$):
158.9 (calculated-density)
185 (calculated-LeBas method)
Total Surface Area, TSA (Å$^2$):
199.7 (Okouchi et al. 1992)
Heat of Fusion, $\Delta H_{fus}$, (kcal/mol):
Entropy of Fusion, $\Delta S_{fus}$ (cal/mol K or e.u.):
Fugacity Ratio at 25°C, F: 1.0

Water Solubility (g/m$^3$ or mg/L at 25°C):
2.30, 2.38 (shake flask-GC, calculated-group contribution, Polak & Lu 1973)
1.36 (shake flask-GC, Price 1976)
2.30, 23.55 (quoted, calculated-group contribution method per Irmann 1965, Horvath 1982)
2.30, 1.36 (quoted, Mackay & Shiu 1981; Brookman et al. 1985)
2.30, 1.36 (quoted, IUPAC 1989)
1.63 (quoted literature average, Myrdal et al. 1992)
2.56 (calculated-AI, Okouchi et al. 1992)

Vapor Pressure (Pa at 25°C):
3431 (Antoine eqn. regression, Stull 1947)
3600 (Antoine eqn., Dreisbach 1959)
3600 (Antoine eqn., Zwolinski & Wilhoit 1971)
3430 (Antoine eqn., Weast 1972-73)
3600, 3232 (quoted, calculated-B.P., Mackay et al. 1982)
3602 (Antoine eqn., Dean 1985)

Henry's Law Constant (Pa $m^3$/mol):

190000; 302000, 179000 (recommended; calculated-P/C, Mackay & Shiu 1981)

178707 (calculated-P/C, Yaws et al. 1991)

Octanol/Water Partition Coefficient, log $K_{OW}$:

Bioconcentration Factor, log BCF:

Sorption Partition Coefficient, log $K_{OC}$:

Half-Lives in the Environment:

Air: atmospheric lifetime was estimated to be 20 hours, based on the photooxidation reaction rate constant of $7.0 \times 10^{-12}$ $cm^3$ $molecule^{-1}$ $sec^{-1}$ with OH radicals in air during summer daylight (Altshuller 1991).

Surface water:

Ground water:

Sediment:

Soil:

Biota:

Environmental Fate Rate Constant and Half-Lives:

Volatilization:

Photolysis:

Oxidation: photooxidation rate constant of $7.0 \times 10^{-12}$ $cm^3$ $molecule^{-1}$ $sec^{-1}$ for the reaction with hydroxyl radicals in the gas phase at 298 K (Atkinson 1990); and the atmospheric lifetime was estimated to be 20 hours, based on the reaction rate constant of $7.0 \times 10^{-12}$ $cm^3$ $molecule^{-1}$ $sec^{-1}$ with OH radicals in air during summer daylight (Altshuller 1991).

Hydrolysis:

Biodegradation:

Biotransformation:

Bioconcentration, Uptake ($k_1$) and Elimination ($k_2$) Rate Constants or Half-Lives:

Common Name: n-Hexane
Synonym: hexane
Chemical Name: n-hexane
CAS Registry No: 110-54-3
Molecular Formula: $C_6H_{14}$
Molecular Weight: 86.17
Melting Point (°C):
-95.30 (Stull 1947)
-95.32 (Stephenson & Malanowski 1987)
Boiling Point (°C):
68.736 (Riddick et al. 1986)
68.74 (Dreisbach 1959; Stephenson & Malanowski 1987)
Density ($g/cm^3$ at 20 °C):
0.65933 (20°C, Riddick et al. 1986)
0.65484 (25°C, Riddick et al. 1986)
Molar Volume ($cm^3/mol$):
130.69 (20°C, calculated-density)
131.59 (25°C, calculated-density)
140.6 (calculated-LeBas method)
130.7 (20°C, calculated-density, McAuliffe 1966)
130.0 (calculated-density, Lande & Banerjee 1981)
126.7 (calculated-density, Miller et al. 1985; Eastcott et al. 1988)
130.5 (calculated-density, Taft et al. 1985; Kamlet et al. 1986)
131.6 (calculated-density, Stephenson & Malanowski 1987)
131 (Abernethy et al. 1988; Mackay & Shiu 1990)
0.648 (intrinsic volume: $V_I/100$, Leahy 1986; Kamlet et al. 1987,1988)
0.65 (intrinsic volume: $V_I/100$, Abernethy et al. 1988)
140.6 (Warne et al. 1991)
130 (20°C, calculated-density, Wang et al. 1992)
Molecular Volume, TMV ($Å^3$):
108.44 (Pearlman 1986)
Total Surface Area, TSA ($Å^2$):
319.0 (Hermann 1972, quoted, Amidon et al. 1974,1975; Amidon & Anik 1981)
319.0, 303.3 (Kier et al. 1975)
318.7, 317,5, 142.1 (Valani et al. 1976)
287.0 (solvent radius 1.5 Å, Amidon et al. 1979)
142.1, 147.7; 319.0, 305.2 (calculated without hydration cell; calculated with hydration cell of r=1.5 Å, Lande et al. 1985)
254.5 (Iwase et al. 1985)
161.51 (Pearlman 1986)
156.3 (Warne et al. 1990)
150.4 (Okouchi et al. 1992)

Heat of Fusion, $\Delta H_{fus}$, (kcal/mol):
3.126 (Riddick et al. 1986)
Entropy of Fusion, $\Delta S_{fus}$ (cal/mol K or e.u.):
Fugacity Ratio at 25°C, F: 1.0

Water Solubility ($g/m^3$ or mg/L at 25°C):
140 (15.5°C, cloud point, Fühner 1924; quoted, Deno & Berkheimer 1960)
< 262 (residue volume, Booth & Everson 1948)
36.0 (cloud point, Durand 1948)
120 (cloud point, McBain & Lissant 1951)
9.50 (shake flask-GC, McAuliffe 1963,1966; quoted, Mackay & Shiu 1975; Price 1976; Bobra et al. 1979; Mackay 1981; Bobra et al. 1984; Abernethy et al. 1986; Mackay & Shiu 1990,1991)
9.5, 34, 36, 140; 18.8 (quoted values; calculated-group contribution, Irmann 1965)
16.21 (vapor saturation-GC, Barone et al. 1966)
9.52 (quoted, Baker 1967; Price 1976)
18.3 (shake flask-GC, Nelson & De Ligny 1968)
9.50, 9.49 (quoted, calculated-$K_{OW}$, Hansch et al. 1968)
12.3 (shake flask-GC, Leinonen & Mackay 1973; quoted, Bobra et al. 1979; Eastcott et al. 1988)
12.4, 12.6 (shake flask-GC, calculated-group contribution, Polak & Lu 1973)
13.0 (shake flask-GC, Krasnoshchekova & Gubertrits 1973)
9.56, 10.6 (quoted, calculated-TSA, Amidon et al. 1974)
16.2 (shake flask-GC, Mackay et al. 1975)
9.44 (quoted, Hine & Mookerjee 1975)
9.56, 11.98 (quoted, calculated- $\chi$ , Hall et al. 1975)
9.56, 11.98 (quoted, calculated- $\chi$ , Kier & Hall 1976)
9.47 (shake flask-GC, Price 1976)
12.3 (shake flask-GC, Aquan-Yuen et al. 1979)
13.7, 46.3 (quoted, calculated-$K_{OW}$, Yalkowsky & Morozowich 1980)
12.24 (gen. col.-GC, Tewari et al. 1982a; quoted, Wasik et al. 1982; Miller et al. 1985)
12.24, 14.1 (quoted exptl., calculated- $\gamma$ & $K_{OW}$, Tewari et al. 1982b)
10.09 (vapor saturation-partition coefficient-GC, Jonsson et al. 1982)
9.50, 9.55 (quoted, shake flask-GC, Coates et al. 1985)
9.50, 8.44 (quoted, calculated-TSA, Lande et al. 1985)
9.60, 12.4, 9.47 (quoted, Brookman et al. 1985)
12.3 (quoted, Riddick et al. 1986)
9.50, 10.85 (quoted, calculated-fragment solubility constants, Wakita et al. 1986)
9.67, 13.04 (quoted, calculated-molar volume & solvatochromic p., Kamlet et al. 1986)
9.67, 23.2 (quoted, calculated-molar volume & solvatochromic p., Leahy 1986)
9.23, 14.3 (calculated-$K_{AW}$, calculated-$V_I$ & solvatochromic p., Kamlet et al. 1988)

10.0, 14.0 (quoted, shake flask-purge and trap-GC, Coutant & Keigley 1988)
9.50, 23.4 (quoted, calculated-UNIFAC, Al-Sahhaf 1989)
12.2 (quoted, Isnard & Lambert 1989)
9.95, 9.0, 1.07; 9.8 (quoted exptl values; recommended best value, IUPAC 1989)
11.8 (quoted literature average, Warne 1990)
14.8 (quoted literature average, Myrdal et al. 1992)
9.49 (calculated-AI, Okouchi et al. 1992)
10.8, 16.05 (quoted, calculated-$V_M$, Wang et al. 1992)

Vapor Pressure (Pa at 25°C):
19704 (calculated-Antoine eqn. regression, Stull 1947)
20172 (Antoine eqn., Dreisbach 1959)
20200 (Antoine eqn., Zwolinski & Wilhoit 1971; quoted, Bobra et al. 1979; Mackay 1981; Mackay & Shiu 1981; Eastcott et al. 1988; Mackay & Shiu 1975,1990)
19920 (24.7°C, Willingham et al. 1945)
20132 (Campbell et al. 1968)
20153 (Harris & Dunlop 1970)
20124, 20141 (differential pressure gauge, Bissell & Williamson 1975)
22091 (27.1°C, Letcher & Marrisicano 1974)
20160 (interpolated, Antoine eqn., Boublik et al. 1984)
20700, 20184, 20164 (headspace-GC, correlated, Antoine eqn., Hussam & Carr 1985)
20170 (quoted, Riddick et al. 1986)
20170 (quoted, Kamlet et al. 1986)
20180, 20302, 20165 (interpolated, Antoine equations, Stephenson & Malanowski 1987)
19865, 10854 (quoted, calculated-UNIFAC, Banerjee et al. 1990)

Henry's Law Constant (Pa m$^3$/mol):
183690 (calculated-$1/K_{AW}$, $C_W/C_A$, reported as exptl., Hine & Mookerjee 1975)
159965 (calculated-group contribution, Hine & Mookerjee 1975)
50586 (calculated-bond contribution, Hine & Mookerjee 1975)
183400 (calculated-P/C, Mackay 1981)
190000 (calculated-P/C, Mackay & Shiu 1975,1981; Bobra et al. 1979)
172504 (concentration ratio-GC, Jönsson et al. 1982)
187450 (selected, Mills et al. 1982)
77818 (EPICS-GC, Ashworth et al. 1988; quoted, Mackay & Shiu 1990)
130789 (calculated-P/C, Yaws et al. 1991)

Octanol/Water Partition Coefficient, log $K_{OW}$:
3.00 (calculated-$\pi$ constant, Hansch et al. 1968; Hansch & Leo 1979)
3.00, 2.98 (quoted, calculated-$\chi$, Murray et al. 1975)

3.52 (calculated-f const, Yalkowsky & Morozowich 1980)
3.90 (concn. ratio-GC, Platford 1979; quoted, Platford 1983)
4.11, 4.20 (quoted exptl., calculated- $\gamma$, Wasik et al. 1981, 1982)
4.11 (gen. col.-GC, Tewari et al. 1982a,b; quoted, Miller et al. 1985; Eastcott et al. 1988; Abernethy et al. 1988; Mackay & Shiu 1990)
2.90 (estimated-HPLC-k', Coates et al. 1985)
3.82, 3.78 (quoted, calculated-hydrophobicity const., Iwase et al. 1985)
4 25 (Berti et al. 1986)
3.90, 3.83 (quoted, calculated-$V_I$ & solvatochromic p., Leahy 1986)
4.11, 4.16 (quoted, gen. col.-GC, Schantz & Martire 1987)
3.90, 3.82 (quoted, calculated-molar volume & solvatochromic p., Taft et al. 1985)
3.90, 3.80 (quoted, calculated-$V_I$ & solvatochromic p., Kamlet et al. 1988)
4.11 (quoted, Isnard & Lambert 1989)
4.00 (recommended, Sangster 1989)
3.53 (quoted literature average, Warne 1991)
3.53 (Warne et al. 1990, quoted, Thoms & Lion 1992)
3.0, 3.18 (quoted, calculated-$V_M$, Wang et al. 1992)

Bioconcentration Factor, log BCF:

Sorption Partition Coefficient, log $K_{OC}$:

Half-Lives in the Environment:

Air: photooxidation reaction half-life of 2.4 to 24 hours in air, based on reaction rate constant of $3.8 \times 10^9$ L $mol^{-1}$ $sec^{-1}$ for the reaction with hydroxyl radicals (Darnall et al. 1976); atmospheric lifetime was estimated to be 25 hours, based on a rate constant of $5.61 \times 10^{-12}$ $cm^3$ $molecule^{-1}$ $sec^{-1}$ for the reaction with OH radicals in summer daylight (Altshuller 1991).

Surface water:

Ground water:

Sediment:

Soil:

Biota:

Environmental Fate Rate Constant and Half-Lives:

Volatilization:

Photolysis:

Oxidation: room temperature rate constants of $9.30 \times 10^{-14}$ $cm^3$ $molecule^{-1}$ $sec^{-1}$ and $5.90 \times 10^{-12}$ $cm^3$ $molecule^{-1}$ $sec^{-1}$ for the gas-phase reaction with $O(^3P)$ (Herron

& Huie 1973; quoted, Gaffney & Levine 1979) and OH radicals in air (Atkinson et al. 1979; quoted, Gaffney & Levine 1979); photooxidation reaction rate constant of $3.8 \times 10^{9}$ $cm^3$ $mol^{-1}$ $sec^{-1}$ for reaction with OH radicals in environmental chamber studies (Lloyd et al. 1976; Darnall et al. 1976) with a disappearance half-life of 2.4-24 hours (Darnall et al. 1976); rate constant of $(6.1\text{-}6.8) \times 10^{-12}$ $cm^3$ $molecule^{-1}$ $sec^{-1}$ for the reaction with OH radicals between 292-303 K and calculated value of $6.96 \times 10^{-12}$ $cm^3$ $molecule^{-1}$ $sec^{-1}$ at 300 K (Darnall et al. 1978); rate constant of $5.61 \times 10^{-12}$ $cm^3$ $molecule^{-1}$ $sec^{-1}$ for reaction with OH radicals in air (Atkinson 1990, 1991; Altshuller 1991) with an estimated atmospheric lifetime of 25 hours in summer daylight (Altshuller 1991); photooxidation reaction rate constants of $5.58 \times 10^{-12}$ $cm^3$ $molecule^{-1}$ $sec^{-1}$ and $1.05 \times 10^{-16}$ $cm^3$ $molecule^{-1}$ $sec^{-1}$ for the reaction with OH and $NO_3$ radicals respectively in air (Sabljic & Güsten 1990).

Hydrolysis:

Biodegradation:

Biotransformation:

Bioconcentration, Uptake ($k_1$) and Elimination ($k_2$) Rate Constants or Half-Lives:

Common Name: 2-Methylhexane
Synonym: isoheptane, ethylisobutylmethane
Chemical Name: 2-methylhexane
CAS Registry No: 591-76-4
Molecular Formula: $C_7H_{16}$
Molecular Weight: 100.203
Melting Point (°C):
-118.20 (Stull 1947)
-118.276 (Dreisbach 1959; Riddick et al. 1986)
-119.10 (Stephenson & Malanowski 1987)
Boiling Point (°C):
90.05 (Dreisbach 1959)
90.0 (Stephenson & Malanowski 1987)
Density (g/cm$^3$ at 20°C):
0.67859 (20°C, Dreisbach 1959)
0.67439 (25°C, Dreisbach 1959)
0.67856 (20°C, Riddick et al. 1986)
0.65484 (25°C, Riddick et al. 1986)
Molar Volume (cm$^3$/mol):
147.68 (20°C, calculated-density)
148.58 (25°C, calculated-density)
162.80 (calculated-LeBas method)
0.745 (intrinsic volume: $V_I/100$, Kamlet et al. 1987)
147.6 (20°C, calculated-density, Stephenson & Malanowski 1987)
Molecular Volume (Å$^3$):
123.68 (Pearlman 1986)
Total Surface Area, TSA (Å$^2$):
175.9 (Pearlman 1986)
173.8 (Okouchi et al. 1992)
Heat of Fusion, $\Delta H_{fus}$, (kcal/mol):
2.195 (Riddick et al. 1986)
Entropy of Fusion, $\Delta S_{fus}$ (cal/mol K or e.u.):
Fugacity Ratio at 25°C, F: 1.0

Water Solubility (g/m$^3$ or mg/L at 25°C):
2.54 (shake flask-GC, Price 1976; quoted, Mackay & Shiu 1981; Brookman et al. 1985; Eastcott et al. 1988)
10.3 (quoted, Riddick et al. 1986)
2.54 (quoted, IUPAC 1989)
2.54 (quoted literature average, Myrdal et al. 1992)
2.54 (calculated-AI, Okouchi et al. 1992)

Vapor Pressure (Pa at 25°C):

8383 (calculated-Antoine eqn. regression, Stull 1947)
8783 (Antoine eqn., Dreisbach 1959)
8780 (Antoine eqn., Zwolinski & Wilhoit 1971; quoted, Mackay & Shiu 1981; Eastcott et al. 1988)
8372 (Antoine eqn., Weast 1972-73)
8786 (interpolated, Antoine eqn., Dean 1985)
8800 (quoted, Riddick et al. 1986)
8790 (Antoine eqn., Stephenson & Malanowski 1987)

Henry's Law Constant (Pa $m^3$/mol):

346000 (calculated-P/C, Mackay & Shiu 1981)
346532 (selected, Mills et al. 1982)
347000 (calculated-P/C, Eastcott et al. 1988)
346473 (calculated-P/C, Yaws et al. 1991)

Octanol/Water Partition Coefficient, log $K_{OW}$:

Bioconcentration Factor, log BCF:

Sorption Partition Coefficient, log $K_{OC}$:

Half-Lives in the Environment:

Air: photooxidation reaction rate constant of $6.80 \times 10^{-12}$ $cm^3$ $molecule^{-1}$ $sec^{-1}$ with hydroxyl radicals and an estimated lifetime of 25 hours in summer daylight (Altshuller 1990).
Surface water:
Ground water:
Sediment:
Soil:
Biota:

Environmental Fate Rate Constant and Half-Lives:

Volatilization:
Photolysis:

Oxidation: photooxidation reaction rate constant of $6.80 \times 10^{-12}$ $cm^3$ $molecule^{-1}$ $sec^{-1}$ with hydroxyl radicals in air (Atkinson 1990,1991; Altshuller 1990); atmospheric lifetime was estimated to be 25 hours, based on photooxidation reaction with OH radicals in summer daylight (Altshuller 1990).

Hydrolysis:

Biodegradation:

Biotransformation:

Bioconcentration, Uptake ($k_1$) and Elimination ($k_2$) Rate Constants or Half-Lives:

Common Name: 3-Methylhexane
Synonym: ethylmethylpropylmethane
Chemical Name: 3-methylhexane
CAS Registry No: 589-34-4
Molecular Formula: $C_7H_{16}$
Molecular Weight: 100.203
Melting Point (°C):
-119.4 (Riddick et al. 1986)
Boiling Point (°C):
91.85 (Stephenson & Malanowski 1987)
Density (g/cm$^3$ at 20°C):
0.6860 (Weast 1984)
0.68713 (20°C, Riddick et al. 1986)
0.68295 (25°C, Riddick et al. 1986)
Molar Volume (cm$^3$/mol):
145.84, 146,73 (calculated-density)
145.9 (20°C, calculated-density, Stephenson & Malanowski 1987)
162.8 (calculated-LeBas method)
Molecular Volume, TMV (Å$^3$):
123.68 (Pearlman 1986)
Total Surface Area, TSA (Å$^2$):
175.9 (Pearlman 1986)
173.8 (Okouchi et al. 1992)
Heat of Fusion, $\Delta H_{fus}$, (kcal/mol):
Entropy of Fusion, $\Delta S_{fus}$ (cal/mol K or e.u.):
Fugacity Ratio at 25°C, F: 1.0

Water Solubility (g/m$^3$ or mg/L at 25°C):
4.95, 4.62 (shake flask-GC, calculated-group contribution, Polak & Lu 1973)
2.64 (shake flask-GC, Price 1976; quoted, Mackay & Shiu 1981; Brookman et al. 1985; Eastcott et al. 1988)
4.95, 7.08 (quoted, calculated-group contribution method per Irmann 1965, Horvath 1982)
2.64, 4.95 (quoted, Brookman et al. 1985)
3.80 (suggested tentative value, IUPAC 1989)
3.28 (quoted literature average, Myrdal et al. 1992)
3.01 (calculated-AI, Okouchi 1992)

Vapor Pressure (Pa at 25°C):
7782 (Antoine eqn. regression, Stull 1947)
8210 (Antoine eqn., Dreisbach 1959)

8210 (Antoine eqn., Zwolinski & Wilhoit 1971; quoted, Mackay & Shiu 1981; Eastcott et al. 1988)
7772 (interpolated, Antoine eqn., Weast 1972-73)
8210 (Antoine eqn., Dean 1985)
8300 (quoted, Riddick et al. 1986)
8215 (Antoine eqn., Stephenson & Malanowski 1987)

Henry's Law Constant (Pa $m^3$/mol):
172000 (recommended, Mackay & Shiu 1981)
172000, 171000 (calculated-P/C, Mackay & Shiu 1981)
312170 (calculated-P/C, Eastcott et al. 1988)
311623 (calculated-P/C, Yaws et al. 1991)

Octanol/Water Partition Coefficient, log $K_{OW}$:

Bioconcentration Factor, log BCF:

Sorption Partition Coefficient, log $K_{OC}$:

Half-Lives in the Environment:
Air: atmospheric half-life was estimated to be 2.4-24 hours for the reaction with hydroxyl radicals, based on the EPA Reactivity Classification of Organics (Darnall et al. 1976); photooxidation reaction rate constant of $7.20 \times 10^{-12}$ $cm^3$ $molecule^{-1}$ $sec^{-1}$ with hydroxyl radicals and an estimated lifetime of 20 hours during summer daylight (Altshuller 1991).
Surface water:
Ground water:
Sediment:
Soil:
Biota:

Environmental Fate Rate Constant and Half-Lives:
Volatilization:
Photolysis:

Oxidation: photooxidation reaction rate constant of $7.20 \times 10^{-12}$ $cm^3$ $molecule^{-1}$ $sec^{-1}$ with hydroxyl radicals in air at 298 K (Atkinson 1990,1991; Altshuller 1991); atmospheric lifetime was estimated to be 20 hours, based on the photooxidation reaction with OH radicals in air during summer daylight (Altshuller 1991).

Hydrolysis:
Biodegradation:
Biotransformation:
Bioconcentration, Uptake ($k_1$) and Elimination ($k_2$) Rate Constants or Half-Lives:

Common Name: 2,2,5-Trimethylhexane
Synonym:
Chemical Name: 2,2,5-trimethylhexane
CAS Registry No: 3522-94-9
Molecular Formula: $C_9H_{20}$
Molecular Weight: 128.257
Melting Point (°C): -105.78
Boiling Point (°C): 124.09
Density (g/cm$^3$ at 25°C):
    0.7072 (Weast 1984)
    0.70721 (20°C, Dreisbach 1959; Riddick et al. 1986)
    0.70322 (25°C, Dreisbach 1959; Riddick et al. 1986)
Molar Volume (cm$^3$/mol):
    181.36 (20°C, calculated-density)
    182.39 (25°C, calculated-density)
    181.3 (20°C, calculated-density, McAuliffe 1966)
    181.0 (calculated-density, Lande & Banerjee 1981)
    181.4 (20°C, calculated-density, Stephenson & Malanowski 1987)
    207.2 (calculated-LeBas method)
    181.0 (20°C, calculated-density, Wang et al. 1992)
Molecular Volume, TMV (Å$^3$):
Total Surface Area, TSA (Å$^2$):
    373.01, 367.33 (Hermann 1972)
    373, 362.7 (quoted, correlated, Kier et al. 1975)
    370.2, 367.9, 186.6 (Valvani et al. 1976)
    220.7 (Okouchi et al. 1992)
Heat of Fusion, $\Delta H_{fus}$, (kcal/mol):
    1.48 (Riddick et al. 1986)
Entropy of Fusion, $\Delta S_{fus}$ (cal/mol K or e.u.):
Fugacity Ratio at 25°C, F: 1.0

Water Solubility (g/m$^3$ or mg/L at 25°C):
    1.15 (shake flask-GC, McAuliffe 1966; quoted, Hermann 1972)
    0.54, 0.54 (shake flask-GC, calculated-group contribution, Polak & Lu 1973)
    0.54, 1.12 (quoted, calculated-group contribution method of Irmann 1965, Horvath 1982)
    1.15, 0.54 (quoted, Mackay & Shiu 1981; Brookman et al. 1985)
    1.15, 1.16 (quoted, calculated-TSA, Amidon et al. 1974)
    1.15, 1.41 (quoted, calculated- $\chi$ , Hall et al. 1975)
    1.15, 1.41 (quoted, calculated- $\chi$ , Kier & Hall 1976)
    1.28 (misquoted from 2,2,5-trimethylpentane, Lande & Banerjee 1981)
    1.14, 0.476 (quoted, calculated-fragment solubility constants, Wakita et al. 1986)
    1.15, 1.51 (quoted, calculated-UNIFAC, Al-Sahhaf 1989)

1.15, 0.54; 0.80 (quoted; recommended best value, IUPAC 1989)
0.79 (quoted literature average, Myrdal et al. 1992)
1.11 (calculated-AI, Okouchi et al. 1992)

Vapor Pressure (Pa at 25°C):
2212 (Antoine eqn., Dreisbach 1959)
2210 (Antoine eqn., Zwolinski & Wilhoit 1971; quoted, Mackay & Shiu 1981)
2208, 2200 (quoted, calculated-B.P., Mackay et al. 1982)
2207 (extrapolated, Antoine eqn., Dean 1985)
2216 (quoted, Riddick et al. 1986)
2218 (interpolated, Antoine eqn., Stephenson & Malanowski 1987)

Henry's Law Constant (Pa $m^3$/mol):
21900, 46700 (calculated-P/C, Mackay & Shiu 1981)
350000 (recommended, Mackay & Shiu 1981)
523762 (calculated-P/C, Yaws et al. 1991)

Octanol/Water Partition Coefficient, log $K_{OW}$:
4.63 (calculated-regression eqn. from Lyman et al. 1982, Wang et al. 1992)
4.46 (calculated-$V_M$, Wang et al. 1992)

Bioconcentration Factor, log BCF:

Sorption Partition Coefficient, log $K_{OC}$:

Half-Lives in the Environment:
Air:
Surface water:
Ground water:
Sediment:
Soil:
Biota:

Environmental Fate Rate Constant and Half-Lives:
Volatilization:
Photolysis:
Oxidation:

Hydrolysis:
Biodegradation:
Biotransformation:
Bioconcentration, Uptake ($k_1$) and Elimination ($k_2$) Rate Constants or Half-Lives:

Common Name: n-Heptane
Synonym: heptane
Chemical Name: n-heptane
CAS Registry No: 142-82-5
Molecular Formula: $C_7H_{16}$
Molecular Weight: 100.203
Melting Point (°C):
-90.58 (Stull 1947)
-90.6 (Dreisbach 1959; Weast 1972-73,1984)
-91.0 (Stephenson & Malanowski 1987)
Boiling Point (°C):
98.4 (Dreisbach 1959, Weast 1972-73)
98.424 (Riddick et al. 1986)
98.30 (Stephenson & Malanowski 1987)
Density (g/cm$^3$ at 20 °C):
0.6837 (Weast 1972-73)
0.68372 (20°C, Dreisbach 1959)
0.67951 (25°C, Dreisbach 1959)
0.6837 (20°C, Riddick et al. 1986)
0.67946 (25°C, Riddick et al. 1986)
Molar Volume (cm$^3$/mol):
146.57 (20°C, calculated-density)
147.47 (25°C, calculated-density)
146.5 (20°C, calculated-density, McAuliffe 1966)
147.0 (calculated-density, Lande & Banerjee 1981)
146.5 (calculated-density, Taft et al. 1985ab; Kamlet et al. 1986,1987)
146.6 (20°C, calculated-density, Stephenson & Malanowski 1987)
146.5 (calculated-density, Miller et al. 1985; Eastcott et al. 1988)
0.745 (intrinsic volume: $V_I$/100, Leahy 1986; Kamlet et al. 1987,1988; quoted, Thoms & Lion 1992)
162.8 (calculated-LeBas method, Mackay & Shiu 1990)
147 (20°C, calculated-density, Wang et al. 1992)
Molecular Volume (A$^3$):
124.72 (Pearlman 1986)
Total Surface Area, TSA (A$^2$):
351.0 (Hermann 1972, quoted, Amidon et al. 1974)
350.5, 337.4, 160.3 (Valvani et al. 1975)
351, 332.6 (Kier et al. 1975)
351.0 (solvent radius 1.5 Å, Amidon et al. 1979; Amidon & Anik 1981)
351, 351.58 (Horvath 1982)
160.3, 158.6; 351.0, 320.9 (calculated without hydration cell; with hydration cell of r=1.5 Å, Lande et al. 1985)
184.21 (Pearlman 1986)

178.1 (Warne et al. 1990)
171.3 (Okouchi et al. 1992)

Heat of Fusion, $\Delta H_{fus}$, (kcal/mol):
3.355 (Riddick et al. 1986)

Entropy of Fusion, $\Delta S_{fus}$ (cal/mol K or e.u.):

Fugacity Ratio at 25°C, F: 1.0

Water Solubility (g/m³ or mg/L at 25°C):
50.0 (cloud point, Fühner 1924; quoted, Price 1976)
150 (radiotracer, Black et al. 1948)
15.0 (16°C, cloud point, Durand 1948)
70.9 (quoted, Deno & Berkheimer 1960)
2.93 (shake flask-GC, McAuliffe 1963,1966; quoted, Mackay & Shiu 1981; Eastcott et al. 1988; Mackay & Shiu 1990,1991; Thoms & Lion 1992)
2.9, 100, 50; 5.62 (quoted values; calculated-group contribution, Irmann 1965)
11.0 (cloud point, Connolly 1966)
2.19 (quoted, Baker 1967; Price 1976)
2.66 (shake flask-GC, Nelson & De Ligny 1968)
2.96, 2.65 (quoted, calculated-$K_{OW}$, Hansch et al. 1968)
3.37, 3.41 (shake flask-GC, calculated-group contribution, Polak & Lu 1973)
2.57 (shake flask-GC, Krasnoshchekova & Gubergrits 1973)
2.93, 2.33 (quoted, calculated-TSA, Amidon et al. 1974)
2.93, 3.96 (quoted, calculated- $\chi$ , Hall et al. 1975; Kier & Hall 1976)
2.24 (shake flask-GC, Price 1976)
3.70 (shake flask-GC, Bittrich et al. 1979)
2.90 (partition coefficient-GC, Rudakov & Lutsyk 1979)
2.95, 14.5 (quoted, calculated-$K_{OW}$, Yalkowsky & Morozowich 1980)
2.95 (quoted, Lande & Banerjee 1981)
3.58 (gen. col.-GC, Tewari et al. 1982a; quoted, Wasik et al. 1982)
3.58, 4.62 (quoted exptl., calculated- $\gamma$ & $K_{OW}$, Tewari et al. 1982b)
2.51 (vapor saturation-partition coefficient-GC, Jonsson et al. 1982)
2.93, 2.92 (quoted, calculated-TSA, Lande et al. 1985)
2.95 (shake flask-GC, Coates et al. 1985)
2.96, 3.90 (quoted, calculated-molar volume & solvatochromic p., Taft et al. 1985b)
2.93, 3.37, 2.24 (quoted, Brookman et al. 1985)
3.57 (quoted, Riddick et al. 1986)
2.95, 3.10 (quoted, calculated-fragment solubility constants, Wakita et al. 1992)
2.95, 7.76 (quoted, calculated-molar volume & solvatochromic p., Leahy 1986)
2.95, 4.57 (quoted, calculated-molar volume & solvatochromic p., Kamlet et al. 1986)
2.95, 4.69 (calculated-$K_{AW}$, calculated-$V_I$ and solvatochromic p., Kamlet et al. 1987)
2.90 (shake flask-purge and trap-GC, Coutant & Keigley 1988)
3.10 (quoted, Isnard & Lambert 1989)

2.93, 1.66, 2,80, 3.37, 2.24, 2.90, 2.51; 2.40 (quoted exptl values; recommended best value, IUPAC 1989)
3.12 (quoted literature average, Warne et al. 1990)
2.96, 8.73 (quoted, calculated-$V_M$, Wang et al. 1992)
2.93 (calculated-AI, Okouchi et al. 1992)
2.79 (quoted literature average, Myrdal et al. 1992)

Vapor Pressure (Pa at 25°C):
5795 (calculated-Antoine eqn. regression, Stull 1947)
6108 (Antoine eqn., Dreisbach 1959)
6110 (Antoine eqn., Zwolinski & Wilhoit 1971; quoted, Mackay 1981; Mackay & Shiu 1981; Eastcott et al. 1988; Mackay & Shiu 1990)
6107 (quoted, Hine & Mookerjee 1975)
6370 (25.9°C, Willingham et al. 1945)
6105 (Harris & Dunlop 1970)
5080 (22.5°C, Carruth & Kobayashi 1973)
6037, 6057 (differential pressure gauge, Bissell & Williamson 1975)
6090 (interpolated, Antoine eqn., Boublik et al. 1984)
6122 (Antoine eqn., Dean 1985)
6110, 5958, 6090 (headspace-GC, correlated, Antoine eqn., Hussam & Carr 1985)
6093 (interpolated, Antoine eqn., Dean 1985)
6090 (quoted, Riddick et al. 1986)
6105 (quoted, Kamlet et al. 1986)
6114 (interpolated, Antoine eqn., Stephenson & Malanowski 1987)

Henry's Law Constant (Pa $m^3$/mol):
206179 (calculated-$1/K_{AW}$, $C_W/C_A$, reported as exptl., Hine & Mookerjee 1975)
225957 (calculated-group contribution, Hine & Mookerjee 1975)
73118 (calculated-bond contribution, Hine & Mookerjee 1975)
209000, 230000 (calculated-P/C, recommended, Mackay & Shiu 1981)
242899 (concentration ratio-GC, Jönsson et al. 1982)
209743 (selected, Mills et al. 1982)
208900 (calculated-P/C, quoted, Mackay & Shiu 1990)
196900, 188040 (quoted, calculated- $\chi$ , Nirmalakhandan & Speece 1988)
273430 (calculated-P/C, Yaws et al. 1991)
205740 (quoted, Thoms & Lion 1992)

Octanol/Water Partition Coefficient, log $K_{OW}$:
3.50 (calculated-$\pi$ constant, Hansch et al. 1968; quoted, Hansch & Leo 1979)
3.50, 3.42 (quoted, calculated- $\chi$ , Murray et al. 1975)

4.09 (calculated-f const., Yalkowsky & Morozowich 1980)
4.66, 4.76 (quoted exptl., calculated- $\gamma$ , Wasik et al. 1981,1982)
4.66 (gen. col.-GC, Tewari et al. 1982a,b; quoted, Miller et al. 1985; Eastcott et al. 1988; Mackay & Shiu 1990)
3.44 (estimated-HPLC-k', Coates et al. 1985)
4.48 (Berti et al. 1986)
4.66, 4.72 (gen. col.-GC, calculated- $\gamma$ , Schantz & Martire 1987)
4.66, 4.30 (quoted, calculated-$V_I$ & solvatochromic p., Kamlet et al. 1988)
4.66 (quoted, Isnard & Lambert 1989)
4.50 (recommended, Sangster 1989)
4.065 (quoted literature average, Warne et al. 1990)
4.07 (calculated- $\chi$ , Warne et al. 1990; quoted, Thoms & Lion 1992)
3.50, 3.61 (quoted, calculated-$V_M$, Wang et al. 1992)

Bioconcentration Factor, log BCF:

Sorption Partition Coefficient, log $K_{OC}$:

Half-Lives in the Environment:

Air: photooxidation reaction rate constant of $7.15x10^{-12}$ $cm^3$ $molecule^{-1}$ $sec^{-1}$ with hydroxyl radicals with an estimated lifetime of 19 hours in summer daylight (Altshuller 1991).

Surface water:

Ground water:

Sediment:

Soil:

Biota:

Environmental Fate Rate Constant and Half-Lives:

Volatilization:

Photolysis:

Oxidation: photooxidation reaction rate constant of $7.15x10^{-12}$ $cm^3$ $molecule^{-1}$ $sec^{-1}$ with hydroxyl radicals in air (Atkinson 1990,1991; Altshuller 1991) and $1.36x10^{-16}$ $cm^3$ $molecule^{-1}$ $sec^{-1}$ with $NO_3$ radicals in the gas-phase at 296 K (Atkinson 1988,1990); atmospheric lifetime was estimated to be 19 hours, based on the photooxidation reaction with OH radicals (Altshuller 1991); rate constant of $7.19x10^{-12}$ $cm^3$ $molecule^{-1}$ $sec^{-1}$ for the reaction with hydroxyl radicals and $1.36x10^{-16}$ $cm^3$ $molecule^{-1}$ $sec^{-1}$ for the reaction with $NO_3$ radicals in air (Sabljic & Güsten 1990).

Hydrolysis:
Biodegradation:
Biotransformation:
Bioconcentration, Uptake ($k_1$) and Elimination ($k_2$) Rate Constants or Half-Lives:

Common Name: 2-Methylheptane
Synonym:
Chemical Name: 2-methylheptane
CAS Registry No: 592-27-8
Molecular Formula: $C_8H_{18}$
Molecular Weight: 114.23
Melting Point (°C):
    -109.50 (Stull 1947)
    -111.30 (Stephenson & Malanowski 1987)
Boiling Point (°C): 117.6
Density ($g/cm^3$ at 20°C): 0.698
Molar Volume ($cm^3/mol$):
    163.7 (20°C, calculated-density, Stephenson & Malanowski 1987)
    185.0 (calculated-LeBas method)
Molecular Volume, TMV ($Å^3$):
Total Surface Area, TSA ($Å^2$):
Heat of Fusion, $\Delta H_{fus}$, (kcal/mol):
Entropy of Fusion, $\Delta S_{fus}$ (cal/mol K or e.u.):
Fugacity Ratio at 25°C, F: 1.0

Water Solubility ($g/m^3$ or mg/L at 25°C):
    0.95, 4.55 (quoted, calculated-$V_M$, Wang et al. 1992)

Vapor Pressure (Pa at 25°C):
    2623 (calculated-Antoine eqn. regression, Stull 1947)
    6386 (23.4°C, Nicilini & Laffitte 1949)
    6850 (interpolated, Antoine eqn., Boublik et al. 1973, 1984)
    2620 (interpolated, Antoine eqn., Weast 1972-73)
    2747 (extrapolated, Antoine eqn., Dean 1985)
    2750 (interpolated, Antoine eqn., Stephenson & Malanowski 1987)

Henry's Law Constant (Pa $m^3$/mol):
    369876 (calculated-P/C, Yaws et al. 1991)

Octanol/Water Partition Coefficient, log $K_{OW}$:
    3.91 (calculated-regression eqn. from Lyman et al. 1982, Wang et al. 1992)
    4.04 (calculated-$V_M$, Wang et al. 1992)

Bioconcentration Factor, log BCF:

Sorption Partition Coefficient, log $K_{OC}$:

Half-Lives in the Environment:

Air: photooxidation reaction rate constant of $8.20 \times 10^{-12}$ $cm^3$ $molecule^{-1}$ $sec^{-1}$ with hydroxyl radicals and an estimated lifetime of 17 hours during summer daylight (Altshuller 1991).

Surface water:

Ground water:

Sediment:

Soil:

Biota:

Environmental Fate Rate Constant and Half-Lives:

Volatilization:

Photolysis:

Oxidation: QSAR calculated photooxidation reaction rate constant of $8.20 \times 10^{-12}$ $cm^3$ $molecule^{-1}$ $sec^{-1}$ with hydroxyl radicals in air (Atkinson 1987; Altshuller 1991); atmospheric lifetime was estimated to be 17 hours, based on photooxidation reaction with OH radicals during summer daylight (Altshuller 1991).

Hydrolysis:

Biodegradation:

Biotransformation:

Bioconcentration, Uptake ($k_1$) and Elimination ($k_2$) Rate Constants or Half-Lives:

Common Name: 3-Methylheptane
Synonym:
Chemical Name: 3-methylheptane
CAS Registry No: 589-81-1
Molecular Formula: $C_7H_{18}$
Molecular Weight: 114.23
Melting Point (°C): -120.5
Boiling Point (°C): 115
Density (g/cm$^3$ at 20°C):
0.7075 (Weast 1984)
0.70582 (20°C, Dreisbach 1959)
0.70175 (25°C, Dreisbach 1959)
Molar Volume (cm$^3$/mol):
161.8 (20°C, calculated-density, Stephenson & Malanowski 1987)
185.0 (calculated-LeBas method)
Molecular Volume, TMV (Å$^3$):
Total Surface Area, TSA (Å$^2$):
194.7 (Okouchi et al. 1992)
Heat of Fusion, $\Delta H_{fus}$, (kcal/mol):
Entropy of Fusion, $\Delta S_{fus}$ (cal/mol K or e.u.):
Fugacity Ratio at 25°C, F: 1.0

Water Solubility (g/m$^3$ or mg/L at 25°C):
0.792 (shake flask-GC, Price 1976; quoted, Mackay & Shiu 1981)
0.85 (estimated-nomograph, Brookman et al. 1985)
0.792 (quoted, IUPAC 1989)
0.792 (quoted literature average, Myrdal et al. 1992)
0.792 (calculated-AI, Okouchi et al. 1992)

Vapor Pressure (Pa at 25°C):
2610 (Antoine eqn., Dreisbach 1959)
2600 (Antoine eqn., Zwolinski & Wilhoit 1971; quoted, Mackay & Shiu 1981)
2466 (Antoine eqn., Weast 1972-73)
2600, 3232 (quoted, calculated-B.P., Mackay et al. 1982)
2600 (extrapolated, Antoine eqn., Boublik et al. 1984)
2605 (extrapolated, Antoine eqn., Dean 1985)
2630 (interpolated, Antoine eqn., Stephenson & Malanowski 1987)

Henry's Law Constant (Pa $m^3$/mol):

376000 (calculated-P/C, Mackay & Shiu 1981)
375916 (selected, Mills et al. 1982)
375751 (calculated-P/C, Yaws et al. 1991)

Octanol/Water Partition Coefficient, log $K_{OW}$:

Bioconcentration Factor, log BCF:

Sorption Partition Coefficient, log $K_{OC}$:

Half-Lives in the Environment:

Air: photooxidation reaction rate constant of $8.90 \times 10^{-12}$ $cm^3$ $molecule^{-1}$ $sec^{-1}$ with hydroxyl radicals with an estimated lifetime of 16 hours during summer daylight (Altshuller 1991).
Surface water:
Ground water:
Sediment:
Soil:
Biota:

Environmental Fate Rate Constant and Half-Lives:

Volatilization:
Photolysis:
Oxidation: QSAR calculated photooxidation reaction rate constant of $8.90 \times 10^{-12}$ $cm^3$ $molecule^{-1}$ $sec^{-1}$ with hydroxyl radicals in air (Atkinson 1987; Altshuller 1991); atmospheric lifetime was estimated to be 16 hours, based on photooxidation reaction with OH radicals during summer daylight (Altshuller 1991).
Hydrolysis:
Biodegradation:
Biotransformation:
Bioconcentration, Uptake ($k_1$) and Elimination ($k_2$) Rate Constants or Half-Lives:

Common Name: n-Octane
Synonym: octane
Chemical Name: n-octane
CAS Registry No: 111-65-9
Molecular Formula: $C_8H_{18}$
Molecular Weight: 114.23
Melting Point (°C):
-56.8 (Stull 1947; Dreisbach 1959; Stephenson & Malanowski 1987)
-56.2 (Weast 1972-73)
Boiling Point (°C):
125.67 (Dreisbach 1959, Weast 1972-73)
125.66 (Stephenson & Malanowski 1987)
Density ($g/cm^3$ at 20°C):
0.70252 (20°C, Dreisbach 1959)
0.69849 (25°C, Dreisbach 1959)
0.70267 (20°C, Riddick et al. 1986)
0.68862 (25°C, Riddick et al. 1986)
Molar Volume ($cm^3/mol$):
162.57 (20°C, calculated-density)
163.51 (25°C, calculated-density)
162.6 (20°C, calculated-density, McAuliffe 1966)
162.6 (20 °C, calculated-density, Stephenson & Malanowski 1987)
163.0 (calculated-density, Lande & Banerjee 1981)
162.5 (calculated-density, Miller et al. 1985; Eastcott et al. 1988)
162.6 (calculated-density, Leahy 1986; Kamlet et al. 1987)
0.842 (intrinsic volume: $V_I/100$, Leahy 1986; Kamlet et al. 1988; quoted, Thoms & Lion 1992)
185.0 (calculated-LeBas method)
182.0 (Valsaraj 1988)
163.0 (20°C, calculated-density, Wang et al. 1992)
Molecular Volume, TMV ($Å^3$):
141.0 (Pearlman 1986)
Total Surface Area, TSA ($Å^2$):
383.0 (Hermann 1972; quoted, Amidon et al. 1974)
382.3, 367.3, 178.4 (Valvani et al. 1975)
383.0 (solvent radius 1,5 Å, Amidon et al. 1979; Amidon & Anik 1981)
383.0, 384.07 (Horvath 1982)
178.4, 171.3; 383.0, 338.3 (calculated without hydration cell; calculated with hydration cell radius 1.5 Å, Lande et al. 1985)
206.9 (Pearlman 1986)
178.4 (Valsaraj 1988)
199.5 (Warne et al. 1990)
192.2 (Okouchi et al. 1992)

Heat of Fusion, $\Delta H_{fus}$, (kcal/mol):
4.957 (Riddick et al. 1986)
Entropy of Fusion, $\Delta S_{fus}$ (cal/mol K or e.u.):
Fugacity Ratio at 25°C, F: 1.0

Water Solubility (g/m$^3$ or mg/L at 25°C):
14.0 (cloud point, Fühner 1924, quoted, Deno & Berkheimer 1960)
0.66 (shake flask-GC, McAuliffe 1963, 1966; quoted, Mackay & Wolkoff 1973; Mackay & Leinonen 1975; Price 1976; Bobra et al. 1979; Mackay 1981; Mackay & Shiu 1981,1990,1991; Bobra et al. 1984; Abernethy et al. 1986; Eastcott et al. 1988; Thoms & Lion 1992)
0.66, 14.0; 8.91 (quoted values; calculated-group contribution, Irmann 1965)
0.493 (radiotracer, Baker 1967; quoted, Price 1976)
0.88 (shake flask-GC, Nelson & De Ligny 1968)
0.66, 0.73 (quoted, calculated-$K_{OW}$, Hansch et al. 1968)
0.70 (shake flask-GC, Krzsnoshchekova & Gubergrits 1973)
0.85, 0.92 (shake flask-GC, calculated-group contribution, Polak & Lu 1973)
0.66, 0.67 (quoted, calculated-TSA, Amidon et al. 1974)
0.66, 1.28 (quoted, calculated- $\chi$ , Hall et al. 1975)
0.431 (shake flask-GC, Price 1976)
1.103 (gen. col.-GC, Tewari et al. 1982a; quoted, Wasik et al. 1982)
1.10, 1.56 (quoted exptl., calculated- $\gamma$ & $K_{OW}$, Tewari et al. 1982b)
0.615 (vapor saturation-partition coefficient-GC, Jönsson et al. 1982)
0.66 (quoted, Lyman 1982)
0.66 (shake flask-GC, Coates et al. 1985)
0.66, 0.85. 0.43 (quoted, Brookman et al. 1985)
0.660, 0.810 (quoted, calculated-TSA, Lande et al. 1985)
0.884, 0.949 (20°C, shake flask-GC, Burris & MacIntyre 1986)
0.66 (quoted, Riddick et al. 1986)
0.66, 0.87 (quoted, calculated-fragment solubility constants, Wakita et al. 1986)
0.66, 2.62 (quoted, calculated-molar volume and solvatochromic p., Leahy 1986)
0.66, 1.58 (quoted, calculated-molar volume and solvatochromic p., Kamlet et al. 1986)
0.66, 1.58 (calculated-$K_{AW}$, calculated-$V_I$ & solvatochromic p., Kamlet et al. 1987)
1.25 (shake flask-purge and trap-GC, Coutant & Keigley 1988)
0.66 (quoted, Valsaraj 1988)
0.66, 1.62 (quoted, calculated-UNIFAC, Al-Sahhaf 1989)
0.66, 0.70, 0.85, 0.62; 0.71 (quoted; recommended, IUPAC 1989)
0.673 (quoted, Isnard & Lambert 1989)
0.756 (quoted literature average, Warne et al. 1990)
0.66, 4.55 (quoted, calculated-$V_M$, Wang et al. 1992)
0.89 (quoted literature average, Myrdal et al. 1992)
0.66 (calculated-AI, Okouchi et al. 1992)

Vapor Pressure (Pa at 25°C):

1777 (calculated-Antoine eqn. regression, Stull 1947)
1871 (Antoine eqn., Dreisbach 1959)
1880 (extrapolated, Antoine eqn., Zwolinski & Wilhoit 1971; quoted, Mackay & Wolkoff 1973; Bobra et al. 1979; Mackay 1981; Mackay & Shiu 1981; Lyman et al. 1982; Eastcott et al. 1988; Mackay & Shiu 1990)
1825 (Antoine eqn., Weast 1972-73)
1872 (quoted, Hine & Mookerjee 1975)
1885, 2060 (quoted, calculated-B.P., Mackay et al. 1982)
1860 (extrapolated, Antoine eqn., Boublik et al. 1984)
1860 (interpolated, Antoine eqn., Dean 1985)
1854, 1814, 1854 (headspace-GC, correlated, Antoine eqn., Hussam & Carr 1985)
1870 (quoted, Riddick et al. 1986)
1870 (quoted, Kamlet et al. 1986)
1862 (interpolated, Antoine eqn., Stephenson & Malanowski 1987)
1885 (quoted, Valrsaraj 1988)

Henry's Law Constant (Pa $m^3$/mol):

326772 (calculated-1/$K_{AW}$, $C_W/C_A$, reported as exptl., Hine & Mookerjee 1975; quoted, Nirmalakhandan & Speece 1988)
311908 (calculated-group contribution, Hine & Mookerjee 1975)
110669 (calculated-bond contribution, Hine & Mookerjee 1975)
325320 (calculated-P/C, Mackay & Leinonen 1975)
323230 (calculated-P/C, Bobra et al. 1979; Mackay et al. 1979)
332346 (calculated-P/C, Mackay 1981)
300000; 325000, 499000, 438000, 253000, 244000 (recommended; calculated-P/C, Mackay 1981)
325000 (calculated-P/C, Mackay & Shiu 1981)
450468 (concentration ratio-GC, Jönsson et al. 1982)
326267 (selected, Mills et al. 1982)
324240 (calculated-P/C, Lyman et al. 1982)
314652 (calculated-P/C, Eastcott et al. 1988)
326165 (Valsaraj 1988)
326772, 201486 (quoted, calculated- $\chi$ , Nirmalakhandan & Speece 1988)
325300 (calculated-P/C, Mackay & Shiu 1990)
499448 (calculated-P/C, Yaws et al. 1991)
327204 (quoted, Thoms & Lion 1992)

Octanol/Water Partition Coefficient, log $K_{OW}$:

4.00 (calculated-$\pi$ substituent const., Hansch et al. 1968; Hansch & Leo 1979)
4.00, 3.89 (quoted, calculated- $\chi$ , Murray et al. 1975)

5.18, 5.29 (quoted, exptl., calculated- $\gamma$ , Wasik et al. 1981,1982)
5.18 (gen. col.-GC, Tewari et al. 1982a,b; quoted, Miller et al. 1985; Eastcott et al. 1988; Mackay & Shiu 1990)
4.00 (calculated-f const., Lyman 1982)
4.00 (estimated-HPLC-k', Coates et al. 1985)
5.18, 4.77 (quoted, calculated-$V_I$ & solvatochromic p., Leahy 1986)
5.18, 5.24 (quoted, gen. col.-GC, Schantz & Martire 1987)
4.66, 4.30 (quoted, calculated-$V_I$ & solvatochromic p., Kamlet et al. 1988)
5.28 (quoted, Valsaraj 1988)
5.18 (quoted, Isnard & Lambert 1989)
5.15 (recommended, Sangster 1989)
4.59 (quoted literature average, Warne et al. 1990; Thoms & Lion 1992)
4.59 (quoted, Thoms & Lion 1992)
4.0, 4.01 (quoted, calculated-$V_M$, Wang et al. 1992)

Bioconcentration Factor, log BCF:

Sorption Partition Coefficient, log $K_{OC}$:

Half-Lives in the Environment:

Air: atmospheric half-life was estimated to be 2.4-24 hours for $C_4H_{10}$ and higher paraffins, based on the EPA Reactivity Classification of Organics (Darnall et al. 1976); photooxidation reaction rate constant of $8.68 \times 10^{-12}$ $cm^3$ $molecule^{-1}$ $sec^{-1}$ with OH radicals with an estimated lifetime of 16 hours in air during summer daylight (Altshuller 1991).

Surface water: volatilization half-life of 5.55 hours for a water column of 1 $m^2$ minimum cross section of depth 1 m (Mackay & Leinonen 1975); estimated volatilization half-life of 3.1 hours at 20°C in a river 1 meter deep flowing at 1 m/s and with a wind velocity of 3 m/s (Lyman et al. 1982).

Ground water:

Sediment:

Soil:

Biota:

Environmental Fate Rate Constant and Half-Lives:

Volatilization: half-life of 5.55 hours for a water column of 1 $m^2$ minimum cross section of depth 1 m (Mackay & Leinonen 1975); estimated half-life of 3.1 hours at 20°C in a river 1 meter deep flowing at 1 m/s and with a wind velocity of 3 m/s (Lyman et al. 1982).

Photolysis:

Oxidation: photooxidation reaction rate constant of $1.46x10^{-12}$ $cm^3$ $molecule^{-1}$ $sec^{-1}$ (observed) and $1.34x10^{-12}$ $cm^3$ $molecule^{-1}$ $sec^{-1}$ (correlated) with hydroxyl radicals at 296 K in air (Greiner 1970); room temperature rate constants of $1.70x10^{-13}$ $cm^3$ $molecule^{-1}$ $sec^{-1}$ for the reaction with $O(^3P)$ (Herron & Huie 1973; quoted, Gaffney & Levine 1979) and $8.40x10^{-12}$ $cm^3$ $molecule^{-1}$ $sec^{-1}$ for the reaction with OH radicals (Atkinson et al. 1979; quoted, Gaffney & Levine 1979); rate constant of $8.42x10^{-12}$ $cm^3$ $molecule^{-1}$ $sec^{-1}$ (reported exptl. value at 295 K) and $7.35x10^{-12}$ $cm^3$ $molecule^{-1}$ $sec^{-1}$ (calculated value at 300 K) for the reaction with OH radicals in gas phase (Darnall et al. 1978); rate constant of $8.68x10^{-12}$ $cm^3$ $molecule^{-1}$ $sec^{-1}$ for reaction with OH radicals (Atkinson 1990, 1991; Altshuller 1991); rate constant of $1.82x10^{-16}$ $cm^3$ $molecule^{-1}$ $sec^{-1}$ for the gas-phase reaction with $NO_3$ radicals at 296 K (Atkinson 1990); estimated atmospheric lifetime of 16 hours, based on photooxidation reaction with OH radicals (Altshuller 1991); rate constant of $8.71x10^{-12}$ $cm^3$ $molecule^{-1}$ $sec^{-1}$ for the reaction with OH radicals and $1.81x10^{-18}$ $cm^3$ $molecule^{-1}$ $sec^{-1}$ for the reaction with $NO_3$ radicals in air (Sabljic & Güsten 1990); rate constant of $1.82x10^{-16}$ $cm^3$ $molecule^{-1}$ $sec^{-1}$ for reaction with $NO_3$ radicals in the gas phase at 296 K (Atkinson 1991); photooxidation rate constant $8.68x10^{-12}$ $cm^3$ $molecule^{-1}$ $sec^{-1}$ for reaction with OH radicals in air (Paulson & Seifeld 1992).

Hydrolysis:

Biodegradation:

Biotransformation:

Bioconcentration, Uptake ($k_1$) and Elimination ($k_2$) Rate Constants or Half-Lives:

Common Name: 4-Methyloctane
Synonym:
Chemical Name: 4-methyloctane
CAS Registry No: 2216-34-4
Molecular Formula: $C_9H_{20}$
Molecular Weight: 128.26
Melting Point (°C): -113.2
Boiling Point (°C): 142.4
Density (g/cm³ at 20°C):
  0.7199 (20°C, Dreisbach 1959)
  0.7169 (25°C, Dreisbach 1959)
Molar Volume (cm³/mol):
  178.16 (calculated-density)
  207.2 (calculated-LeBas method)
Molecular Volume, TMV (Å³):
Total Surface Area, TSA (Å²):
  215.6 (Okouchi et al. 1992)
Heat of Fusion, $\Delta H_{fus}$, (kcal/mol):
Entropy of Fusion, $\Delta S_{fus}$ (cal/mol K or e.u.):
Fugacity Ratio at 25°C, F: 1.0

Water Solubility (g/m³ or mg/L at 25°C):
  0.115 (shake flask-GC, Price 1976; quoted, Mackay & Shiu 1981; Brookman 1985; IUPAC 1989)
  0.115 (quoted, Krzyzanowska & Szeliga 1978)
  0.115 (calculated-AI, Okouchi et al. 1992)
  0.115 (quoted literature average, Myrdal et al. 1992)

Vapor Pressure (Pa at 25°C):
  901 (Antoine eqn., Dreisbach 1959)
  903 (Antoine eqn., Zwolinski & Wilhoit 1971; quoted, Mackay & Shiu 1981)
  903, 992 (quoted, calculated-B.P., Mackay et al. 1982)

Henry's Law Constant (Pa m³/mol):
  1010000 (calculated-P/C, Mackay & Shiu 1981)
  1000000 (recommended, Mackay & Shiu 1981)
  1006765 (selected, Mills et al. 1982)
  1007103 (calculated-P/C, Yaws et al. 1991)

Octanol/Water Partition Coefficient, log $K_{OW}$:

Bioconcentration Factor, log BCF:

Sorption Partition Coefficient, log $K_{OC}$:

Half-Lives in the Environment:

Air:
Surface water:
Ground water:
Sediment:
Soil:
Biota:

Environmental Fate Rate Constant and Half-Lives:

Volatilization:
Photolysis:
Oxidation:
Hydrolysis:
Biodegradation:
Biotransformation:
Bioconcentration, Uptake ($k_1$) and Elimination ($k_2$) Rate Constants or Half-Lives:

Common Name: n-Nonane
Synonym: nonane
Chemical Name: n-nonane
CAS Registry No: 111-84-2
Molecular Formula: $C_9H_{20}$
Molecular Weight: 128.257
Melting Point (°C):
-53.70 (Stull 1947)
-53.49 (Weast 1972-73)
-51.0 (Stephenson & Malanowski 1987)
Boiling Point (°C):
150.81 (Weast 1972-73)
150.80 (Stephenson & Malanowski 1987)
Density (g/cm$^3$ at 20 °C):
0.71763 (20°C, Dreisbach 1959)
0.71381 (25°C, Dreisbach 1959)
0.71772 (20°C, Riddick et al. 1986)
0.71375 (25°C, Riddick et al. 1986)
Molar Volume (cm$^3$/mol):
178.7 (20°C, calculated-density)
179.69 (25°C, calculated-density)
178.7 (20°C, calculated-density, Stephenson & Malanowski 1987)
207.2 (calculated-LeBas method, Eastcott et al. 1988)
179 (20°C, calculated-density, Wang et al. 1992)
Molecular Volume, TMV (Å$^3$):
157.28 (Pearlman 1986)
Total Surface Area, TSA (Å$^2$):
418.8 (Horvath 1982)
187.7, 361.7 (calculated without hydration cell, with hydration cell radius 1.5 Å; Lande et al. 1985)
229.61 (Pearlman 1986)
213.1 (Okouchi et al. 1992)
Heat of Fusion, $\Delta H_{fus}$, (kcal/mol):
3.697 (Riddick et al. 1986)
Entropy of Fusion, $\Delta S_{fus}$ (cal/mol K or e.u.):
Fugacity Ratio at 25°C, F: 1.0

Water Solubility (g/m$^3$ or mg/L at 25°C):
0.22 (shake flask-GC, McAuliffe 1969; quoted, Price 1976; Mackay 1981; Mackay & Shiu 1981)
0.098 (quoted, Baker 1967; Price 1976)
0.071 (shake flask-GC, Krasnoshchekova & Gubergrits 1973)

0.122 (shake flask-GC, Price 1976; quoted, Eastcott et al. 1988; Mackay & Shiu 1991)
0.272 (vapor saturation-partition coefficient-GC, Jönsson et al. 1982)
0.122 (quoted, Brookman et al. 1985)
0.219 (shake flask-GC, Coates et al. 1985)
0.22, 0.15 (quoted, calculated-TSA, Lande et al. 1985)
0.123 (quoted, Riddick et al. 1986)
0.22, 0.071, 0.122, 0.272; 1.70 (quoted; recommended best value, IUPAC 1989)
0.20, 2.62 (quoted, calculated-$V_M$, Wang et al. 1992)
0.122 (calculated-AI, Okouchi et al. 1992)
0.136 (quoted literature average, Myrdal et al. 1992)

Vapor Pressure (Pa at 25°C):
623 (calculated-Antoine eqn. regression, Stull 1947)
580 (Antoine eqn., Dreisbach 1959)
571 (Antoine eqn., Zwolinski & Wilhoit 1971; quoted, Mackay 1981; Mackay & Shiu 1981; Eastcott et al. 1988; Mackay & Shiu 1991)
571, 684 (quoted, calculated-B.P., Mackay et al. 1982)
570, 713 (extrapolated, Antoine eqn., Boublik et al. 1984)
571 (extrapolated, Antoine eqn., Dean 1985)
570 (quoted, Riddick et al. 1986)
517 (calculated-Antoine eqn., Stephenson & Malanowski 1987)
571 (interpolated, Antoine eqn., Stephenson & Malanowski 1987)

Henry's Law Constant (Pa $m^3$/mol):
601000, 748000, 333000 (calculated-P/C, Mackay & Shiu 1981)
500000 (recommended, Mackay & Shiu 1981)
235170 (concn ratio-GC, Jönsson et al. 1982)
333360 (selected, Mills et al. 1982)
41950 (EPICS-GC, Ashworth et al. 1988)
599572 (calculated-P/C, Eastcott et al. 1988)
600959 (calculated-P/C, Yaws et al. 1991)

Octanol/Water Partition Coefficient, log $K_{OW}$:
4.51 (estimated-HPLC-k', Coates et al. 1985)
5.65 (recommended, Sangster 1989; quoted, Mackay & Shiu 1991)
4.50 (calculated-regression eqn. from Lyman et al. 1982, Wang et al. 1992)
4.42 (calculated-$V_M$, Wang et al. 1992)

Bioconcentration Factor, log BCF:

Sorption Partition Coefficient, log $K_{OC}$:

Half-Lives in the Environment:

Air: atmospheric half-life was estimated to be 2.4-24 hours for $C_4H_{10}$ and higher paraffins for the reaction with hydroxyl radicals, based on the EPA Reactivity Classification of Organics (Darnall et al. 1976); photooxidation reaction rate constant of $1.02 \times 10^{-11}$ $cm^3$ $molecule^{-1}$ $sec^{-1}$ with OH radicals with an estimated lifetime of 14 hours during summer daylight (Altshuller 1991).

Surface water:

Ground water:

Sediment:

Soil:

Biota:

Environmental Fate Rate Constant and Half-Lives:

Volatilization:

Photolysis:

Oxidation: photooxidation reaction rate constant of $1.02 \times 10^{-11}$ $cm^3$ $molecule^{-1}$ $sec^{-1}$ with OH radicals in air (Atkinson 1990, 1991; Altshuller 1991); estimated atmospheric lifetime of 14 hours, based on reaction with OH radicals in air (Altshuller 1991); photooxidation rate constant of $1.0 \times 10^{-11}$ $cm^3$ $molecule^{-1}$ $sec^{-1}$ with hydroxyl radicals and $2.30 \times 10^{-16}$ $cm^3$ $molecule^{-1}$ $sec^{-1}$ with $NO_3$ radicals in air (Sabljic & Güsten 1990); rate constant of $2.41 \times 10^{-16}$ $cm^3$ $molecule^{-1}$ $sec^{-1}$ for reaction with $NO_3$ in the gas phase (Atkinson 1991).

Hydrolysis:

Biodegradation:

Biotransformation:

Bioconcentration, Uptake ($k_1$) and Elimination ($k_2$) Rate Constants or Half-Lives:

Common Name: n-Decane
Synonym: decane
Chemical Name: n-decane
CAS Registry No: 124-18-5
Molecular Formula: $C_{10}H_{22}$
Molecular Weight: 142.284
Melting Point (°C):
-29.70 (Stull 1947; Dreisbach 1959; Weast 1984)
-29.58 (Weast 1972-73)
-30.0 (Stephenson & Malanowski 1987)
Boiling Point (°C):
174.123 (Dreisbach 1959; Weast 1984)
98.64 (Weast 1972-73)
174.0 (Stephenson & Malanowski 1987)
Density (g/cm³ at 20 °C):
0.7300 (Weast 1984)
0.73005 (20°C, Dreisbach 1959)
0.72725 (25°C, Dreisbach 1959)
0.73012 (20°C, Riddick et al. 1986)
0.72635 (25°C, Riddick et al. 1986)
Molar Volume (cm³/mol):
194.89 (20°C, calculated-density)
195.89 (25°C, calculated-density)
194.9 (20°C, calculated-density, Stephenson & Malanowski 1987)
1.036 (intrinsic volume: $V_I$/100, Kamlet et al. 1988; quoted, Thoms & Lion 1992)
229.4 (calculated-LeBas method)
195 (20°C, calculated-density, Wang et al. 1992)
Molecular Volume, TMV (Å³):
Total Surface Area, TSA (Å²):
446.7 (Horvath 1982)
198.8, 377.0 (calculated without hydration cell, with hydration radius 1.5Å; Lande et al. 1985)
234 (Okouchi et al. 1992)
Heat of Fusion, $\Delta H_{fus}$, (kcal/mol):
6.864 (Riddick et al. 1986)
Entropy of Fusion, $\Delta S_{fus}$ (cal/mol K or e.u.):
Fugacity Ratio at 25°C, F: 1.0

Water Solubility (g/m³ or mg/L at 25°C):
0.0016 (radiotracer, Baker 1958,1959)
0.0198 (shake flask-GC, Franks 1966)
0.022 (Baker 1967)

0.052 (shake flask-GC, McAuliffe 1969; quoted, Bobra et al. 1979; Mackay et al. 1979; Mackay 1981; Mackay & Shiu 1981; Eastcott et al. 1988; Mackay & Shiu 1990,1991; Thoms & Lion 1992)
0.00087 (shake flask-GC, Krasnoshchekova & Gubergrits 1973)
0.052; 0.182, 1.22 (quoted; shake flask-headspace-GC, Mackay et al. 1975)
0.00292 (refractometer, Becke & Quitzsch 1977)
0.050 (quoted, Bobra et al. 1984; Abernethy et al. 1986)
0.022, 0.020 (quoted, Brookman et al. 1985)
0.046, 0.0466 (quoted, calculated-TSA, Lande et al. 1985)
0.052, 0.0524 (quoted, shake flask-GC, Coates et al. 1985)
0.052 (quoted, Riddick et al. 1986)
0.0016, 0.00198, 0.00087; 0.0015 (quoted; recommended best value, IUPAC 1989)
0.914, 1.450 (quoted, calculated-$V_M$, Wang et al. 1992)
0.027 (quoted literature average, Myrdal et al. 1992)
0.052 (calculated-AI, Okouchi et al. 1992)

Vapor Pressure (Pa at 25°C):
238 (calculated-Antoine eqn. regression, Stull 1947)
182 (Antoine eqn., Dreisbach 1959)
175 (Antoine eqn., Zwolinski & Wilhoit 1971; quoted, Bobra et al. 1979; Mackay et al. 1979; Mackay 1981; Mackay & Shiu 1975,1981; Eastcott et al. 1988; Mackay & Shiu 1990)
170 (quoted, Hine & Mookerjee 1975)
175, 238 (quoted, calculated-B.P., Mackay et al. 1982)
174 (Antoine eqn., Boublik et al. 1984)
173 (extrapolated, Antoine eqn., Dean 1985)
170 (quoted, Riddick et al. 1986)
171 (extrapolated, Antoine eqn., Stephenson & Malanowski 1987)

Henry's Law Constant (Pa $m^3$/mol):
326270 (calculated-P/C, Mackay & Shiu 1975)
499530 (calculated-P/C, Bobra et al. 1979; Mackay et al. 1979)
489400 (calculated-P/C, Mackay 1981)
700000; 500000, 108000 (recommended; calculated-P/C, Mackay & Shiu 1981)
499530 (selected, Mills et al. 1982)
431100 (calculated-P/C, Eastcott et al. 1988)
477870 (calculated-P/C, Yaws et al. 1991)
699030 (quoted, Thoms & Lion 1992)

Octanol/Water Partition Coefficient, log $K_{OW}$:

5.67 (estimated-f const., Lyman 1982; quoted, Thoms & Lion 1992)
6.69, 5.98 (estimated-HPLC/MS, calculated-f const., Burkhard et al. 1985)
5.01 (estimated, Coates et al. 1985)
6.25 (recommended, Sangster 1989; quoted, Mackay & Shiu 1991)
5.00 (calculated-regression eqn. from Lyman et al. 1982, Wang et al. 1992)
4.82 (calculated-$V_M$, Wang et al. 1992)

Bioconcentration Factor, log BCF:

Sorption Partition Coefficient, log $K_{OC}$:

Half-Lives in the Environment:

Air: atmospheric half-life was estimated to be 2.4-24 hours for $C_4H_{10}$ and higher paraffins for the reaction with hydroxyl radicals, based on the EPA Reactivity Classification of Organics (Darnall et al. 1976); photooxidation reaction rate constant of $1.16 \times 10^{-11}$ $cm^3$ $molecule^{-1}$ $sec^{-1}$ with hydroxyl radicals with an estimated lifetime of 12 hours during summer daylight (Altshuller 1991).
Surface water:
Ground water:
Sediment:
Soil:
Biota:

Environmental Fate Rate Constant and Half-Lives:

Volatilization:
Photolysis:
Oxidation: photooxidation reaction rate constant of $1.16 \times 10^{-11}$ $cm^3$ $molecule^{-1}$ $sec^{-1}$ with hydroxyl radicals in air (Atkinson 1990, 1991; Altshuller 1991); atmospheric lifetime of 12 hours was estimated based on photooxidation reaction with OH radicals during summer daylight (Altshuller 1991).
Hydrolysis:
Biodegradation:
Biotransformation:
Bioconcentration, Uptake ($k_1$) and Elimination ($k_2$) Rate Constants or Half-Lives:

Common Name: *n*-Undecane
Synonym: undecane
Chemical Name: *n*-undecane
CAS Registry No: 1120-21-4
Molecular Formula: $C_{11}H_{24}$
Molecular Weight: 156.32
Melting Point (°C): -25.59
Boiling Point (°C): 195.9
Density (g/cm$^3$ at 20°C):
0.74017 (20°C, Dreisbach 1959)
0.73655 (25°C, Dreisbach 1959)
Molar Volume (cm$^3$/mol):
211.2 (calculated-density)
251.6 (calculated-LeBas method)
Molecular Volume, TMV (Å$^3$):
Total Surface Area, TSA (Å$^2$):
Heat of Fusion, $\Delta H_{fus}$, (kcal/mol):
Entropy of Fusion, $\Delta S_{fus}$ (cal/mol K or e.u.):
Fugacity Ratio at 25°C, F: 1.0

Water Solubility (g/m$^3$ or mg/L at 25°C):
0.0044 (shake-GC, McAuliffe 1969; quoted, Mackay & Shiu 1981)
0.0036 (shake-GC, Krashoshhchekova & Gubergrits 1973)
0.0044, 0.0036; 0.0040 (quoted values; recommended, IUPAC 1989)
0.0042 (quoted lit. average, Myrdal et al. 1992)

Vapor Pressure (Pa at 25°C):
57.18 (Antoine eqn., Dreisbach 1959)
52.2 (Antoine eqn., Zwolinski & Wilhoit 1971; quoted, Mackay & Shiu 1981)
52.18, 86.23 (quoted, calculated-B.P., Mackay et al. 1982)

Henry's Law Constant (Pa m$^3$/mol):
185000 (calculated-P/C, Mackay & Shiu 1981)
185394 (calculated-P/C, Yaws et al. 1991)

Octanol/Water Partition Coefficient, log $K_{OW}$:
6.94 (estimated-HPLC/MS, Burkhard et al. 1985)
6.51 (calculated-f const., Burkhard et al. 1985)

Bioconcentration Factor, log BCF:

Sorption Partition Coefficient, log $K_{OC}$:

Half-Lives in the Environment:

Air:
Surface water:
Ground water:
Sediment:
Soil:
Biota:

Environmental Fate Rate Constant and Half-Lives:

Volatilization:
Photolysis:
Oxidation:
Hydrolysis:
Biodegradation:
Biotransformation:
Bioconcentration, Uptake ($k_1$) and Elimination ($k_2$) Rate Constants or Half-Lives:

Common Name: *n*-Dodecane
Synonym: dodecane
Chemical Name: *n*-dodecane
CAS Registry No: 112-40-3
Molecular Formula: $C_{12}H_{26}$
Molecular Weight: 170.33
Melting Point (°C): -9.6
Boiling Point (°C): 216.3
Density (g/cm$^3$ at 20°C):

0.74869 (20°C, Dreisbach 1959; Riddick et al. 1986)
0.74516 (25°C, Dreisbach 1959; Riddick et al. 1986)

Molar Volume (cm$^3$/mol):

227.5 (calculated-density)
273.8 (calculated-LeBas method)

Molecular Volume, TMV (Å$^3$):
Total Surface Area, TSA (Å$^2$):
Heat of Fusion, $\Delta H_{fus}$, (kcal/mol):

8.57 (Riddick et al. 1986)

Entropy of Fusion, $\Delta S_{fus}$ (cal/mol K or e.u.):
Fugacity Ratio at 25°C, F: 1.0

Water Solubility (g/m$^3$ or mg/L at 25°C):

0.0084 (shake flask-GC, Franks 1966)
0.0034 (shake flask-GC, McAuliffe 1969; quoted, Mackay & Shiu 1981; Brookman et al. 1985)
0.0037 (shake-GC, Sutton & Calder 1974; quoted, Bobra et al. 1979; Riddick et al. 1986)
0.0038 (quoted, Bobra et al. 1984)
0.0084, 0.0037 (quoted values, IUPAC 1989)
0.0037 (recommended, IUPAC 1989)
0.0053 (quoted lit. average, Myrdal et al. 1992)

Vapor Pressure (Pa at 25°C):

17.60 (Antoine eqn., Dreisbach 1959)
15.70 (Antoine eqn., Zwolinski & Wilhoit 1971; quoted, Bobra et al. 1979; Mackay & Shiu 1981)
15.40 (Antoine eqn., Boublik et al. 1973)
15.7, 32.53 (quoted, calculated-B.P., Mackay et al. 1982)
16.0 (quoted, Riddick et al. 1986)

Henry's Law Constant (Pa $m^3$/mol):

722953 (calculated-P/C, Bobra et al. 1979)
723000, 786000, 317000 (calculated-P/C, Mackay & Shiu 1981)
750000 (recommended, Mackay & Shiu 1981)
721434 (selected, Mills et al. 1982)
726885 (calculated-P/C, Yaws et al. 1991)

Octanol/Water Partition Coefficient, log $K_{OW}$:

7.24 (estimated-HPLC/MS, Burkhard et al. 1985)
7.04 (calculated-f const., Burkhard et al. 1985)
6.10 (Coates et al. 1985)
5.64, 6.10 (quoted, Sangster 1989)
5.80 (recommended, Sangster 1989)

Bioconcentration Factor, log BCF:

Sorption Partition Coefficient, log $K_{OC}$:

Half-Lives in the Environment:

Air:
Surface water:
Ground water:
Sediment:
Soil:
Biota:

Environmental Fate Rate Constant and Half-Lives:

Volatilization:
Photolysis:
Oxidation:
Hydrolysis:
Biodegradation:
Biotransformation:
Bioconcentration, Uptake ($k_1$) and Elimination ($k_2$) Rate Constants or Half-Lives:

Common Name: Cyclopentane
Synonym: pentamethylene
Chemical Name: cyclopentane
CAS Registry No: 287-92-37
Molecular Formula: $C_5H_{10}$
Molecular Weight: 70.134
Melting Point (°C):
-93.7 (Stull 1947)
-92.9 (Weast 1984)
-93.87 (Dreisbach 1959; Stephenson & Malanowski 1987)
Boiling Point (°C):
49.2 (Weast 1984)
49.26 (Stephenson & Malanowski 1987)
Density (g/cm$^3$ at 20°C):
0.7457 (Weast 1984)
0.74538 (20°C, Dreisbach 1959; Riddick et al. 1986)
0.74394 (25°C, Dreisbach 1959; Riddick et al. 1986)
Molar Volume (cm$^3$/mol):
94.1 (20°C, calculated-density)
94.72 (25°C, calculated-density)
94.1 (20°C, calculated-density, McAuliffe 1966)
93.4 (20°C, calculated-density, Stephenson & Malanowski 1987)
0.500 (intrinsic volume: $V_I$/100, Leahy 1986; Kamlet et al. 1988)
100 (calculated-LeBas method, Abernathy et al. 1988; Mackay & Shiu 1990)
94.0 (calculated-density, Wang et al. 1992)
Molecular Volume, TMV (Å$^3$):
Total Surface Area, TSA (Å$^2$):
255.4 (Hermann 1972; Amidon et al. 1974)
201.9 (Iwase et al. 1985)
104.5 (Okouchi et al. 1992)
Heat of Fusion, $\Delta H_{fus}$, (kcal/mol):
0.1455 (Riddick et al. 1986)
Entropy of Fusion, $\Delta S_{fus}$ (cal/mol K or e.u.):
Fugacity Ratio at 25°C, F: 1.0

Water Solubility (g/m$^3$ or mg/L at 25°C):
156 (shake flask-GC, McAuliffe 1963, 1966; quoted, Irmann 1965; Price 1976)
157, 116 (quoted, calculated-$K_{OW}$, Hansch et al. 1968)
342 (shake flask-GC, Pierotti & Liabastre 1972)
160 (shake flask-GC, Price 1976)
157 (quoted, Hine & Mookerjee 1975)
160 (shake flask-GC, Krzyzanowska & Szeliga 1978)

156, 160 (quoted, Mackay & Shiu 1981; Brookman et al. 1985)
157 (quoted, Lande & Banerjee 1981)
156 (quoted, Bobra et al. 1984; Abernethy et al. 1986)
159 (quoted, Riddick et al. 1986)
157, 157 (quoted, calculated-fragment solubility constants, Wakita et al. 1986)
157, 150 (quoted, calculated-$V_I$ & solvatochromic p., Leahy 1986)
164 (shake flask-GC, Groves 1988)
30, 40.1 (misquoted from cycloheptane, calculated-UNIFAC, Al-Sahhaf 1989)
156, 342, 160 (quoted, IUPAC 1989)
157, 153 (quoted, calculated-$V_M$, Wang et al. 1992)
176 (quoted literature average, Myrdal et al. 1992)
156 (calculated-AI, Okouchi et al. 1992)
157 (quoted, Müller & Klein 1992)

Vapor Pressure (Pa at 25°C):
34892 (20.2°C, Willingham et al. 1945)
43154 (Antoine eqn. regression, Stull 1947)
42330 (Antoine eqn., Dreisbach 1959)
42400 (Antoine eqn., Zwolinski & Wilhoit 1971; quoted, Mackay & Shiu 1981)
42330 (quoted, Hine & Mookerjee 1975)
42320, 42567 (interpolated, Antoine equations, Boublik et al. 1984)
42322 (interpolated, Antoine eqn., Dean 1985)
42400 (quoted, Riddick et al. 1986)
42340 (interpolated, Antoine eqn., Stephenson & Malanowski 1987)

Henry's Law Constant (Pa $m^3$/mol):
18804 (calculated as $1/K_{AW}$, $C_W/C_A$, reported as exptl., Hine & Mookerjee 1975; quoted, Nirmalakhandan & Speece 1988)
13312 (calculated-group contribution, Hine & Mookerjee 1975)
18376 (calculated-bond contribution, Hine & Mookerjee 1975)
18500; 19100, 18600 (recommended; calculated-P/C, Mackay & Shiu 1981)
18948 (selected, Mills et al. 1982)
18804, 17550 (quoted, calculated- $\chi$ , Nirmalakhandan & Speece 1988)
19026 (calculated-P/C, Yaws et al. 1991)

Octanol/Water Partition Coefficient, log $K_{OW}$:
2.05 (calculated-$\pi$ substituent constant, Hansch et al. 1968)
3.00 (shake flask-GC, Leo et al. 1975)
2.05, 2.17 (quoted, calculated- $\chi$ , Murray et al. 1975)
3.00 (Hansch & Leo 1979)

3.00, 2.78 (quoted, calculated-hydrophobicity const., Iwase et al. 1985)
3.00, 3.02 (quoted, calculated-molar volume & solvatochromic p., Leahy 1986)
3.00, 2.98 (quoted, calculated-$V_I$ & solvatochromic p., Kamlet et al. 1988)
3.00 (quoted, Abernethy et al. 1988)
3.00 (recommended, Sangster 1989)
2.05, 2.06 (quoted, calculated-$V_M$, Wang et al. 1992)
2.80 (calculated-f const., Müller & Klein 1992)

Bioconcentration Factor, log BCF:

Sorption Partition Coefficient, log $K_{OC}$:

Half-Lives in the Environment:

Air: atmospheric half-life was estimated to be 2.4-24 hours for cycloparaffins, based on the EPA Reactivity Classification of Organics (Darnall et al. 1976); photooxidation reaction rate constant of $5.16x10^{-12}$ $cm^3$ $molecule^{-1}$ $sec^{-1}$ with hydroxyl radicals in air at 298 K (Atkinson 1990; Altshuller 1991) with an estimated lifetime of 27 hours, based on reaction rate with OH radicals in summer daylight (Altshuller 1991).

Surface water:

Ground water:

Sediment:

Soil:

Biota:

Environmental Fate Rate Constant and Half-Lives:

Volatilization:

Photolysis:

Oxidation: room temperature rate constants of $1.30x10^{-13}$ $cm^3$ $molecule^{-1}$ $sec^{-1}$ for the gas-phase reaction with $O(^3P)$ (Herron & Huie 1973; quoted, Gaffney & Levine 1979) and $5.40x10^{-12}$ $cm^3$ $molecule^{-1}$ $sec^{-1}$ for the gas-phase reaction with hydroxyl radicals (Atkinson et al. 1979; quoted, Gaffney & Levine 1979); photooxidation rate constant of $6.1x10^{-12}$ to 4.72 $x10^{-12}$ $cm^3$ $molecule^{-1}$ $sec^{-1}$ between 298-300 K with calculated rate constant of $5.80x10^{-12}$ $cm^3$ $molecule^{-1}$ $sec^{-1}$ at 300 K for the reaction with hydroxyl radicals in the gas phase (Danrall et al. 1978); rate constants for the gas-phase reaction with hydroxyl radicals: $6.20x10^{-12}$ $cm^3$ $molecule^{-1}$ $sec^{-1}$ at 298 K and $5.18x10^{-12}$ $cm^3$ $molecule^{-1}$ $sec^{-1}$ at 298 K and $5.24x10^{-12}$ $cm^3$ $molecule^{-1}$ $sec^{-1}$ at 299 K and $4.43x10^{12}$ $cm^3$ $molecule^{-1}$ $sec^{-1}$ at 300 K(Atkinson 1985); photooxidation reaction rate constant of $5.16 \times 10^{-12}$ $cm^3$ $molecule^{-1}$ $sec^{-1}$ with hydroxyl radicals in air at 298 K (Atkinson 1990; Altshuller 1991) with an estimated lifetime of 27 hours, based on the gas-phase reaction with OH radicals in summer daylight (Altshuller 1991).

Hydrolysis:

Biodegradation:

Biotransformation:

Bioconcentration, Uptake ($k_1$) and Elimination ($k_2$) Rate Constants or Half-Lives:

Common Name: Methylcyclopentane
Synonym:
Chemical Name: methylcyclopentane
CAS Registry No: 96-37-3
Molecular Formula: $C_6H_{12}$
Molecular Weight: 84.161
Melting Point (°C): -142.434
Boiling Point (°C): 71.812
Density (g/cm$^3$ at 25°C):
 0.7496 (Weast 1984)
 0.74864 (20°C, Dreisbach 1955; Riddick et al. 1986)
 0.74394 (25°C, Dreisbach 1955; Riddick et al. 1986)
Molar Volume (cm$^3$/mol):
 121.70 (calculated-LeBas method)
 112.42 (20°C, calculated from density)
 113.13 (25°C, calculated from density)
 112.4 (20°C, calculated-density, McAuliffe 1966)
 112.0 (calculated-density, Lande & Banerjee 1981; Wang et al. 1992)
 112.4 (20°C, calculated-density, Stephenson & Malanowski 1987)
Molecular Volume, TMV (Å$^3$):
Total Surface Area, TSA (Å$^2$):
 127.9 (Okouchi et al. 1992)
Heat of Fusion, $\Delta H_{fus}$, (kcal/mol):
 1.656 (Riddick et al. 1986)
Entropy of Fusion, $\Delta S_{fus}$ (cal/mol K or e.u.):
Fugacity Ratio at 25°C, F: 1.0

Water Solubility (g/m$^3$ or mg/L at 25°C):
 42,6 (shake flask-GC, McAuliffe 1963; quoted, Irmann 1965)
 42.0 (shake flask-GC, McAuliffe 1966; quoted, Price 1976)
 36.0, 50.4 (quoted, calculated-$K_{OW}$, Hansch et al. 1968)
 55.6, 33.2 (quoted, calculated-$\chi$ , Hall et al. 1975)
 42.4 (quoted, Hine & Mookerjee 1975)
 41.8 (shake flask-GC, Price 1976)
 45.0 (partition coefficient-GC, Rudakov & Lutsyk 1979)
 42.2 (quoted, Lande & Banerjee 1981)
 42.7 (quoted, Lyman 1982)
 42.0, 41.8 (quoted, Mackay & Shiu 1981; Brookman et al. 1985)
 42.0 (quoted, Bobra et al. 1984)
 42.6 (quoted, Riddick et al. 1986)
 42.4, 58.2 (quoted, calculated-fragment solubility constants, Wakita et al. 1986)
 42.0, 41.8, 45.0; 43.0 (quoted values; recommended, IUPAC 1989)

42.0, 43.0 (quoted, calculated-UNIFAC, Al-Sahhaf 1989)
42.2, 43.2 (quoted, calculated-$V_M$, Wang et al. 1992)
42.0 (quoted literature average, Myrdal et al. 1992)
42.0 (calculated-AI, Okouchi et al. 1992)
42.2 (quoted, Müller & Klein 1992)

Vapor Pressure (Pa at 25°C):
16680 (22.75°C, Willingham et al. 1945)
17868 (Antoine eqn. regression, Stull 1947)
18332 (Antoine eqn., Dreisbach 1955)
18300 (Antoine eqn., Zwolinski & Wilhoit 1971; quoted, Mackay & Shiu 1981; Eastcott et al. 1988)
17845 (Antoine eqn., Weast 1972-73)
18332 (quoted, Hine & Mookerjee 1975)
18332 (interpolated, Antoine eqn., Boublik et al. 1984)
18330 (interpolated, Antoine eqn., Dean 1985)
18400 (quoted, Riddick et al. 1986)
18338 (interpolated, Antoine eqn., Stephenson & Malanowski 1987)
18400, 17066 (quoted, calculated-UNIFAC, Banerjee et al. 1990)

Henry's Law Constant (Pa $m^3$/mol):
36664 (calculated as $1/K_{AW}$, $C_W/C_A$ reported as exptl., Hine & Mookerjee 1975; quoted, Nirmalakhandan & Speece 1988)
22092 (calculated-group contribution, Hine & Mookerjee 1975)
27810 (calculated-bond contribution, Hine & Mookerjee 1975)
36700; 36700, 36800 (recommended, calculated-P/C, Mackay & Shiu 1981)
36680 (selected, Mills et al. 1982)
36664, 25366 (quoted, calculated-$\chi$, Nirmalakhandan & Speece 1988)
36934 (calculated-P/C, Eastcott et al. 1988)
36177 (calculated-P/C, Yaws et al. 1991)

Octanol/Water Partition Coefficient, log $K_{OW}$:
2.35 (calculated-$\pi$ substituent constant, Hansch et al. 1968)
2.35, 2.52 (quoted, calculated-$\chi$, Murray et al. 1975)
2.35 (calculated-f const., Lyman et al. 1982)
3.37 (recommended, Sangster 1989)
2.35, 2.50 (quoted, calculated-$V_M$, Wang et al. 1992)
3.31 (calculated-f const., Müller & Klein 1992)

Bioconcentration Factor, log BCF:

Sorption Partition Coefficient, log $K_{OC}$:

Half-Lives in the Environment:

Air: atmospheric half-life was estimated to be 2.4-24 hours for cylcoparaffins, based on the EPA Reactivity Classification of Organics (Darnall et al. 1976); rate constant of $7.10 \times 10^{-12}$ $cm^3$ $molecule^{-1}$ $sec^{-1}$ for the reaction with hydroxyl radicals in air (Atkinson 1990, 1991, Altshuller 1991); and an estimated reaction lifetime of 20 hours in summer daylight (Altshuller 1991).

Surface water:

Ground water:

Sediment:

Soil:

Biota:

Environmental Fate Rate Constant and Half-Lives:

Volatilization:

Photolysis:

Oxidation: rate constant of $7.10 \times 10^{-12}$ $cm^3$ $molecule^{-1}$ $sec^{-1}$ for the reaction with hydroxyl radicals in air (Atkinson 1990, 1991, Altshuller 1991); and an estimated reaction lifetime of 20 hours in summer daylight (Altshuller 1991).

Hydrolysis:

Biodegradation:

Biotransformation:

Bioconcentration, Uptake ($k_1$) and Elimination ($k_2$) Rate Constants or Half-Lives:

Common Name: 1,1,3-Trimethylcyclopentane
Synonym:
Chemical Name: 1,1,3-trimethylcyclopentane
CAS Registry No: 4516-69-2
Molecular Formula: $C_8H_{16}$
Molecular Weight: 112.21
Melting Point (°C): -14.2
Boiling Point (°C): 104.9
Density (g/cm$^3$ at 20°C):
0.7703 (Weast 1984)
0.74825 (20°C, Dreisbach 1955)
0.74392 (25°C, Dreisbach 1955)
Molar Volume (cm$^3$/mol):
146.0 (20°C, calculated-density, Wang et al. 1992)
145.67 (calculated-density)
166.1 (calculated-LeBas method)
Molecular Volume, TMV (Å$^3$):
Total Surface Area, TSA (Å$^2$):
174.8 (Okouchi et al. 1992)
Heat of Fusion, $\Delta H_{fus}$, (kcal/mol):
Entropy of Fusion, $\Delta S_{fus}$ (cal/mol K or e.u.):
Fugacity Ratio at 25°C, F: 1.0

Water Solubility (g/m$^3$ or mg/L at 25°C):
3.73 (shake flask-GC, Price 1976; quoted, Mackay & Shiu 1981; Brookman et al. 1985)
3.73 (quoted literature average, Myrdal et al. 1992)
5.50, 3.72 (calculated-$K_{OW}$, calculated-$V_M$, Wang et al. 1992)
3.73 (calculated-AI, Okouchi et al. 1992)

Vapor Pressure (Pa at 25°C):
5297 (Anotine eqn., Dresibach 1955)
5300 (Antoine eqn., Zwolinski & Wilhoit 1971; quoted, Mackay & Shiu 1981)
5071 (Antoine eqn., Dean 1985)
5300, 4924 (quoted, calculated-B.P., Mackay et al. 1982)

Henry's Law Constant (Pa m$^3$/mol):
159000 (calculated-P/C, Mackay & Shiu 1981)

Octanol/Water Partition Coefficient, log $K_{OW}$:

3.28 (calculated-regression eqn. from Lyman et al. 1982, Wang et al. 1992)

3.34 (calculated-$V_M$, Wang et al. 1992)

Bioconcentration Factor, log BCF:

Sorption Partition Coefficient, log $K_{OC}$:

Half-Lives in the Environment:

Air: atmospheric half-life was estimated to be 2.4-24 hours for cycloparaffins, based on the EPA Reactivity Classification of Organics (Darnall et a. 1976).

Surface water:

Ground water:

Sediment:

Soil:

Biota:

Environmental Fate Rate Constant and Half-Lives:

Volatilization:

Photolysis:

Oxidation:

Hydrolysis:

Biodegradation:

Biotransformation:

Bioconcentration, Uptake ($k_1$) and Elimination ($k_2$) Rate Constants or Half-Lives:

Common Name: *n*-Propylcyclopentane
Synonym:
Chemical Name: *n*-propylcyclopentane
CAS Registry No: 2040-96-2
Molecular Formula: $C_8H_{16}$
Molecular Weight: 112.21
Melting Point (°C): -117.3
Boiling Point (°C): 130.95
Density ($g/cm^3$ at 20°C):
    0.77633 (20°C, Dreisbach 1955)
    0.77229 (25°C, Dreisbach 1955)
    0.7763 (Weast 1984)
Molar Volume ($cm^3$/mol):
    144.5 (calculated-density)
    166.1 (calculated-LeBas method)
    159.0 (20°C, calculated-density, Wang et al. 1992)
Molecular Volume, TMV ($Å^3$):
Total Surface Area, TSA ($Å^2$):
    167.9 (Okouchi et al. 1992)
Heat of Fusion, $\Delta H_{fus}$, (kcal/mol):
Entropy of Fusion, $\Delta S_{fus}$ (cal/mol K or e.u.):
Fugacity Ratio at 25°C, F: 1.0

Water Solubility ($g/m^3$ or mg/L at 25°C):
    2.04 (shake flask-GC, Price 1976; quoted, Mackay & Shiu 1981; Brookman et al. 1985)
    1.77 (shake flask-GC, Krzyzanowska & Szeliga 1978)
    2.92 (calculated-QSAR, Passino & Smith 1987)
    2.04 (quoted literature average, Myrdal et al. 1992)
    2.04 (calculated-AI, Okouchi et al. 1992)
    8.32, 1.30 (calculated-$K_{OW}$, $V_M$, Wang et al. 1992)
    2.92 (quoted, Müller & Klein 1992)

Vapor Pressure (Pa at 25°C):
    1650 (Antoine eqn., Dreisbach 1955)
    1640 (Antoine eqn., Zwolinski & Wilhoit 1971; quoted, Mackay & Shiu 1981)
    1644 (Antoine eqn., Dean 1985)
    1640, 5320 (quoted, calculated-B.P., Mackay et al. 1982)

Henry's Law Constant (Pa $m^3$/mol):

90200 (calculated-P/C, Mackay & Shiu 1981)

90428 (calculated-P/C, Yaws et al. 1991)

Octanol/Water Partition Coefficient, log $K_{OW}$:

3.95 (calculated-regression eqn. of Lyman et al. 1982, Wang et al. 1992)

2.65 (calculated-$V_M$, Wang et al. 1992)

4.37 (calculated-f const., Müller & Klein 1992)

Bioconcentration Factor, log BCF:

Sorption Partition Coefficient, log $K_{OC}$:

Half-Lives in the Environment:

Air:

Surface water:

Ground water:

Sediment:

Soil:

Biota:

Environmental Fate Rate Constant and Half-Lives:

Volatilization:

Photolysis:

Oxidation:

Hydrolysis:

Biodegradation:

Biotransformation:

Bioconcentration, Uptake ($k_1$) and Elimination ($k_2$) Rate Constants or Half-Lives:

Common Name: Pentylcyclopentane
Synonym: 1-cyclopentylpentane
Chemical Name: pentylcyclopentane
CAS Registry No: 3741-00-2
Molecular Formula: $C_{10}H_{20}$
Molecular Weight: 140.26
Melting Point (°C): -83
Boiling Point (°C): 180
Density ($g/cm^3$ at 20°C):
0.7912 (20°C, Dreisbach 1959)
0.7874 (25°C, Dreisbach 1959)
Molar Volume ($cm^3/mol$):
177.3 (20°C, calculated-density)
178.1 (25°C, calculated-density)
210.5 (calculated-LeBas method)
Molecular Volume ($A^3$):
Total Surface Area, TSA ($A^2$):
211.5 (Okouchi et al. 1992)
Heat of Fusion, $\Delta H_{fus}$, (kcal/mol):
Entropy of Fusion, $\Delta S_{fus}$ (cal/mol K or e.u.):
Fugacity Ratio at 25°C, F: 1.0

Water Solubility ($g/m^3$ or mg/L at 25°C):
0.115 (shake flask-GC, Price 1976; quoted, Mackay & Shiu 1981; Brookman et al. 1985)
0.115 (quoted literature average, Myrdal et al. 1992)

Vapor Pressure (Pa at 25°C):
159 (Anotine eqn., Dreisbach 1959)
152 (Antoine eqn., Zwolinski & Wilhoit 1971; quoted, Mackay & Shiu 1981)

Henry's Law Constant (Pa $m^3/mol$):
18500 (calculated-P/C, Mackay & Shiu 1981)
18596 (calculated-P/C, Yaws et al. 1991)

Octanol/Water Partition Coefficient, log $K_{OW}$:

Bioconcentration Factor, log BCF:

Sorption Partition Coefficient, log $K_{OC}$:

Half-Lives in the Environment:

Air:
Surface water:
Ground water:
Sediment:
Soil:
Biota:

Environmental Fate Rate Constant and Half-Lives:

Volatilization:
Photolysis:
Oxidation:
Hydrolysis:
Biodegradation:
Biotransformation:
Bioconcentration, Uptake ($k_1$) and Elimination ($k_2$) Rate Constants or Half-Lives:

Common Name: Cyclohexane
Synonym: hexahydrobenzene, hexamethylene
Chemical Name: cyclohexane
CAS Registry No: 110-82-7
Molecular Formula: $C_6H_{12}$
Molecular Weight: 84.161
Melting Point (°C):
6.60 (Stull 1947)
6.55 (Stephenson & Malanowski 1987)
Boiling Point (°C):
80.74 (Stephenson & Malanowski 1987)
Density (g/cm$^3$ at 20°C):
0.77855 (20°C, Dreisbach 1955; Riddick et al. 1986)
0.77389 (25°C, Dreisbach 1955; Riddick et al. 1986)
Molar Volume (cm$^3$/mol):
108.1 (20°C, calculated-density)
108.75 (25°C, calculated-density)
108.1 (20°C, calculated-density, McAuliffe 1966)
108 (calculated-density, Lande & Banerjee 1981)
101.4 (20°C, calculated-density, Stephenson & Malanowski 1987)
118.0 (calculated-density, Leahy 1986; Taft et al. 1985; Kamlet et al. 1987,1988; Warne et al. 1991)
0.598 (intrinsic volume: $V_I$/100, Leahy 1986; Kamlet et al. 1988; Thoms & Lion 1992)
118 (calculated-LeBas method, Abernathy et al. 1988; Mackay & Shiu 1990)
118 (calculated-density, Wang et al. 1992)
Molecular Volume, TMV (Å$^3$):
93.45 (Pearlman 1986)
96.36 (van der Waals based value, Cramer 1977)
Total Surface Area, TSA (Å$^2$):
279.1 (Hermann 1972, quoted, Amidon et al. 1974)
279.5, 266.6, 120.8 (Valvani et al. 1976)
128.82 (Pearlman 1986)
125.4 (Okouchi et al. 1992)
Heat of Fusion, $\Delta H_{fus}$, (kcal/mol):
0.6399 (Riddick et al. 1986)
Entropy of Fusion, $\Delta S_{fus}$ (cal/mol K or e.u.):
Fugacity Ratio at 25°C, F: 1.0

Water Solubility (g/m$^3$ or mg/L at 25°C):
55.0 (shake flask-GC, McAuliffe 1963,1966; quoted, Hermann 1972; Mackay & Shiu 1975,1981,1990; Price 1976; Thoms & Lion 1992)
55.0, 80.0, 64.0 (quoted values, Irmann 1965)

55.6, 43.2 (quoted, calculated-$K_{OW}$, Hansch et al. 1968)
56.7 (shake-GC, Lionenon & Mackay 1973)
55.6, 43.0 (quoted, calculated-TSA, Amidon et al. 1974)
55.6, 33.2 (quoted, calculated- $\chi$ , Hall et al. 1975)
55.8, 50.2, 61.7 (shake flask-GC, Mackay et al. 1975)
55.6 (quoted, Hine & Mookerjee 1975)
66.5 (shake flask-GC, Price 1976)
55.6 (quoted, Verschueren 1977; quoted, Könemann 1981)
55.0, 52.0 (quoted, elution chromatography, Schwarz 1980)
68.4, 71.6 (quoted, calculated-$K_{OW}$, Yalkowsky & Morozowich 1980)
55.6 (quoted, Lyman 1982)
71.6, 86.12 (quoted, calculated-$K_{OW}$, Amidon & Williams 1982)
55.0, 57.5, 66.5 (quoted, Mackay & Shiu 1981; Brookman et al. 1985)
72.4 (calculated-HPLC-k', converted from reported $\gamma_W$, Hafkenscheid & Tomlinson 1983)
54.9 (quoted, Bobra et al. 1984; Abernethy et al. 1986)
100 (quoted, Riddick et al. 1986)
55.6, 46.2 (quoted, calculated-fragment solubility constants, Wakita et al. 1986)
55, 48.2 (quoted, calculated-$V_I$ and solvatochromic parameters, Leahy 1986)
55, 35.1 (quoted, calculated-$V_I$ and solvatochromic parameters, Kamlet et al. 1986)
55, 29.9 (quoted, calculated-$V_I$ and solvatochromic parameters, Kamlet et al. 1987)
58.4 (shake flask-GC, Groves 1988)
55, 56.7, 57.5, 6.65; 58 (quoted values; recommended best value, IUPAC 1989)
55.0, 47.8 (quoted, calculated-UNIFAC, Al-Sahhaf 1989)
54.3 (quoted, Isnard & Lambert 1989)
54.8 (quoted, Howard 1990)
66.85, 59.6 (quoted, calculated-$V_M$, Wang et al. 1992)
61.2 (quoted literature average, Myrdal et al. 1992)
55.0 (calculated-AI, Okouchi et al. 1992)
55.6 (quoted, Müller & Klein 1992)

Vapor Pressure (Pa at 25°C):
11695 (20.96°C, Willingham et al. 1945)
12281 (calculated-Antoine eqn. regression, Stull 1947)
13010 (Antoine eqn., Dreisbach 1955)
12700 (Antoine eqn., Zwolinski & Wilhoit 1971; quoted, Mackay & Shiu 1981; Eastcott et al 1988)
11172 (Antoine eqn., Weast 1972-73)
13010 (quoted, Hine & Mookerjee 1975)
13012, 13040 (interpolated, Antoine equations, Boublik et al. 1984)
13016 (interpolated, Antoine eqn., Dean 1985)
13040 (quoted, Riddick et al. 1986)

13023 (quoted, Kamlet et al. 1986)
11850 (interpolated, Antoine eqn., Stephenson & Malanowski 1987)
13014 (calculated-Antoine eqn., Stephenson & Malanowski 1987)
12920, 1733 (quoted, calculated-UNIFAC, Banerjee et al. 1990)
13012 (quoted, Howard 1990)

Henry's Law Constant (Pa $m^3$/mol):
19860 (calculated-P/C, Mackay & Shiu 1975)
19690 (calculated as $1/K_{AW}$, $C_W/C_A$, reported as exptl., Hine & Mookerjee 1975; quoted, Nirmalakhandan & Speece 1988)
18376 (calculated-group contribution, Hine & Mookerjee 1975)
27813 (calculated-bond contribution, Hine & Mookerjee 1975)
18000; 19400, 18600, 16100 (recommended; calculated-P/C, Mackay & Shiu 1981)
19860 (selected, Mills et al. 1982)
19690, 22092 (quoted, calculated- $\chi$ , Nirmalakhandan & Speece 1988)
17935 (EPICS-GC, Ashworth et al. 1988)
5532, 1450 (concn. ratio, calculated-UNIFAC, Ashworth et al. 1988)
19556 (quoted, Howard 1990)
19978 (calculated-P/C, Yaws et al. 1991)

Octanol/Water Partition Coefficient, log $K_{OW}$:
2.46 (calculated-$\pi$ substituent constant, Hansch et al. 1968)
3.44 (shake flask-GC, Leo et al. 1975; Hansch & Leo 1979; quoted, Yalkowsky & Morozowich 1980; Hafkenscheid & Tomlinson 1983; Bodor et al. 1989)
3.40 (quoted, Cramer 1977)
3.44; 3.18, 3.48, 3.48 (quoted; calculated-f const., Rekker 1977)
3.18 (calculated-f const., Könemann 1981)
2.46 (calculated-f const., Lyman 1982)
3.44 (Valvani et al. 1981; quoted, Amidon & Williams 1982)
3.44, 3.69 (quoted, HPLC-k', Hafkenscheid & Tomlinson 1983)
3.70 (calculated-$\gamma$ , Berti et al. 1986; quoted, Sangster 1989)
3.44, 3.49 (quoted, calculated-molar volume and solvatochromic p., Taft et al. 1985)
3.44, 3.53 (quoted, calculated-molar volume and solvatochromic p., Leahy 1986)
3.44, 3.50 (quoted, calculated-$V_I$ and solvatochromic parameters, Kamlet et al. 1988)
3.44 (quoted, Abernethy et al. 1988; Isnard & Lambert 1989)
3.44, 2.64 (quoted, calculated-MO, Bodor et al. 1989)
3.44, 3.35 (quoted, calculated- $\pi$, Bodor et al. 1989)
3.44 (recommended, Sangster 1989)
3.44 (quoted, Howard 1990; Thoms & Lion 1992)
2.46, 2.40 (quoted, calculated-$V_M$, Wang et al. 1992)
3.35 (calculated-f const., Müller & Klein 1992)

Bioconcentration Factor, log BCF:

2.38 (estimated, Howard 1990)

Sorption Partition Coefficient, log $K_{OC}$:

2.68 (estimated-S, Howard 1990)

Half-Lives in the Environment:

Air: atmospheric half-life was estimated to be 2.4-24 hours for cycloparaffins, based on the EPA Reactivity Classification of Organics (Darnall et al. 1976); an atmospheric lifetime was estimated to be 19 hours in summer daylight, based on the photooxidation reaction rate constant of $7.49 \times 10^{-12}$ $cm^3$ $molecule^{-1}$ $sec^{-1}$ for the reaction with OH radicals in air during summer daylight (Altshuller 1991); will degrade photochemically by hydroxyl radicals with a half-life of 52 hours and much faster under photochemical smog conditions with half-life as low as 6 hours (Howard 1990); 8.7 to 87 hours, based on photooxidation half-life in air (Howard et al. 1991).

Surface water: volatilization half-life of 2 hours in a model river (Howard 1990); 672 to 4320 hours, based on estimated unacclimated aqueous aerobic biodegradation half-life (Howard et al. 1991).

Ground water: 1344-8640 hours, based on estimated unacclimated aqueous aerobic biodegradation half-life (Howard et al. 1991).

Sediment:

Soil: 672-4320 hours, based on unacclimated grab sample of aerobic soil and aerobic aqueous screening test data (Howard et al. 1991).

Biota:

Environmental Fate Rate Constant and Half-Lives:

Volatilization: 2.8 hours from a model river 1 m deep with a 1 m/s current and a 3 m/s wind (Lyman et al. 1982, quoted, Howard 1990).

Photolysis:

Oxidation: photooxidation reaction rate constant of $5.38 \times 10^{12}$ $cm^3$ $mol^{-1}$ $sec^{-1}$ (observed) and $4.79 \times 10^{12}$ $cm^3$ $mol^{-1}$ $sec^{-1}$ (correlated) for the reaction with hydroxyl radicals in air at 295 K (Greiner 1970); exptl. rate constant of $6.7 \times 10^{-12}$ $cm^3$ $molecule^{-1}$ $sec^{-1}$ at 298 K and the calculated rate constant of $6.96 \times 10^{-12}$ $cm^3$ $molecule^{-1}$ $sec^{-1}$ at 300 K for the reaction with OH radicals in the gas phase (Darnall et al. 1978); gas-phase rate constant of $1.40 \times 10^{-13}$ $cm^3$ $molecule^{-1}$ $sec^{-1}$ for the reaction with $O(^3P)$ (Herron & Huie 1973; quoted, Gaffney & Levine 1979) and $7.0 \times 10^{-12}$ $cm^3$ $molecule^{-1}$ $sec^{-1}$ with OH radicals both at room temperature (Atkinson et al. 1979; quoted, Gaffney & Levine 1979); rate constant of $(6.20 \pm 0.44)x10^{-12}$ $cm^3$ $molecule^{-1}$ $sec^{-1}$ with reference to *n*-butane for the reaction with OH radicals at $(24.4 \pm 0.4)$°C with an atmospheric lifetime of 1.9 day for an average OH radical concentration of $1.0x10^6$ $molecules/cm^3$ (Edney et al. 1986); rate constant of $7.34x10^{-12}$ $cm^3$ $molecule^{-1}$ $sec^{-1}$ for the reaction with hydroxyl radicals in the gas phase at 297 K (Atkinson 1986; quoted, Edney et al. 1986); gas-phase rate constant of $7.38x10^{-12}$ $cm^3$ $molecule^{-1}$ $sec^{-1}$ for the reaction with hydroxyl radicals (Dilling et al. 1988); reaction rate constant of $7.49x10^{-12}$ $cm^3$ $molecule^{-1}$ $sec^{-1}$ for the reaction with OH radicals at 298 K and $1.34x10^{-16}$ $cm^3$ $molecule^{-1}$ $sec^{-1}$ for reaction with $NO_3$ radicals at 296 K in air (Atkinson 1990); will react with photochemically produced hydroxyl radicals with a half-life of 52 hours, based on a rate constant of 7.38 x $10^{-10}$ $cm^3$ $molecule^{-1}$ $sec^{-1}$ with OH radical concentration of 5 $x10^5$ $cm^3$ $sec^{-1}$ (Howard 1990); gas-phase reaction rate constant of $1.35x10^{-16}$ $cm^3$ $molecule^{-1}$ $sec^{-1}$ for the reaction with $NO_3$ radicals at 296 K (Atkinson 1991); an atmospheric lifetime was estimated to be 19 hours in summer daylight, based on the photooxidation reaction rate of $7.49x10^{-12}$ $cm^3$ $molecule^{-1}$ $sec^{-1}$ with OH radicals in air (Altshuller 1991); photooxidation half-life of 1.4 $x10^9$ to 6.9 x $10^{10}$ hours (16000 to 780000 years), based on measured rate data for alkylperoxyl radicals in aqueous solution (Howard et al. 1991); photooxidation half-life of 8.7 to 87 hours, based on measured rate data for the vapor phase reaction with OH radicals in air (Howard et al. 1991).

Hydrolysis:

Biodegradation: highly resistant to biodegradation (Howard 1990); aqueous aerobic half-life of 672 to 4032 hours, based on unacclimated grab sample of aerobic soil and aerobic aqueous screening test data (Howard et al. 1991); aqueous anaerobic half-life of 2688 to 16280 hours, based on estimated unacclimated aqueous aerobic biodegradation half-life (Howard et al. 1991).

Biotransformation:

Bioconcentration, Uptake ($k_1$) and Elimination ($k_2$) Rate Constants or Half-Lives:

Common Name: Methylcyclohexane
Synonym: hexahydrotoluene, cyclohexylmethane
Chemical Name: methylcyclohexane
CAS Registry No: 108-87-2
Molecular Formula: $C_7H_{14}$
Molecular Weight: 98.188
Melting Point (°C):
-126.40 (Stull 1947)
-126.60 (Weast 1984; Stephenson & Malanowski 1987)
-126.593 (Dreisbach 1959; Riddick et al. 1986)
Boiling Point (°C):
100.934 (Riddick et al. 1986)
100.40 (Stephenson & Malanowski 1987)
Density (g/cm$^3$ at 20°C):
0.7694 (Weast 1984)
0.7694 (20°C, Dreisbach 1955; Riddick et al. 1986)
0.76506 (25°C, Dreisbach 1955; Riddick et al. 1986)
Molar Volume (cm$^3$/mol):
127.62 (20°C, calculated-density)
128.34 (25°C, calculated-density)
127.6 (20°C, calculated-density, McAuliffe 1966)
128.0 (calculated-density, Lande & Banerjee 1981; Wang et al. 1992)
127.6 (20°C, calculated-density, Stephenson & Malanowski 1987)
118.0 (calculated-density, Leahy 1986; Taft et al. 1985; Kamlet et al. 1987,1988; Warne et al. 1991)
0.70 (intrinsic volume: $V_I/100$, Abernethy et al. 1988)
0.695 (intrinsic volume: $V_I/100$, Kamlet et al. 1988; Thoms & Lion 1992))
140.4 (calculated-LeBas method, Mackay & Shiu 1990)
Molecular Volume, TMV (Å$^3$):
Total Surface Area, TSA (Å$^2$):
304.9 (Amidon et al. 1974)
304.9, 302.1 (Kier et al. 1975)
305.2, 292.7, 137.7 (Valvani et al. 1976)
137.7 (Valvani et al. 1981)
148.8 (Okouchi et al. 1992)
Heat of Fusion, $\Delta H_{fus}$, (kcal/mol):
1.615 (Riddick et al. 1986)
Entropy of Fusion, $\Delta S_{fus}$ (cal/mol K or e.u.):
Fugacity Ratio at 25°C, F: 1.0

Water Solubility ($g/m^3$ or mg/L at 25°C):

14.0 (shake flask-GC, McAuliffe 1963,1966; quoted, Irmann 1965; Price 1976)
13.9, 21.9 (quoted, calculated-$K_{OW}$, Hansch et al. 1968)
13.8, 14.3 (quoted, calculated- $\chi$ , Hall et al. 1975)
13.8, 14.3 (quoted, calculated- $\chi$ , Kier & Hall 1976)
14.2 (quoted, Hine & Mookerjee 1975)
16.0 (shake flask-GC, Price 1976; quoted, Eastcott et al. 1988)
15.3 (partition coefficient-GC, Rudakov & Lutsyk 1979)
14, 16 (quoted, Mackay & Shiu 1981; Brookman et al. 1985; Thoms & Lion 1992)
15.2 (20°C, shake flask-GC, Burris & MacIntyre 1986)
14.0 (quoted, Bobra et al. 1984; Abernethy et al. 1986; Riddick et al. 1986)
14.0, 16.7 (quoted, calculated-fragment solubility constants, Wakita et al. 1986)
16.7 (shake flask-GC, Groves 1988)
14, 16, 15.3; 15.1 (quoted values; recommended, IUPAC 1989)
14.0, 14.0 (quoted, calculated-UNIFAC, AL-Sahhaf 1989)
13.8, 13.8 (quoted, calculated-$V_M$, Wang et al. 1992)
15.2 (quoted, Thoms & Lion 1992)
14.0 (calculated-AI, Okouchi et al. 1992)
13.9 (quoted, Müller & Klein 1992)

Vapor Pressure (Pa at 25°C):

6354 (25.59°C, Willingham et al. 1945)
5887 (calculated-Antoine eqn. regression, Stull 1947)
6180 (Antoine eqn., Dreisbach 1955)
6180 (Antoine eqn., Zwolinski & Wilhoit 1971; quoted, Mackay & Shiu 1981; Eastcott et al. 1988)
6180, 5806 (quoted, calculated-B.P., Mackay et al. 1982)
6179 (quoted, Hine & Mookerjee 1975)
5364, 6111, 6177 (Antoine equations, Boublik et al. 1984)
6180 (interpolated, Antoine eqn., Dean 1985)
6100 (quoted, Riddick et al. 1986)
6163 (interpolated, Antoine eqn., Stephenson & Malanowski 1987)

Henry's Law Constant (Pa $m^3$/mol):

44080 (calculated as $1/K_{AW}$, $C_W/C_A$, reported as exptl., Hine & Mookerjee 1975; quoted, Nirmalakhandan & Speece 1988)
31026 (calculated-group contribution, Hine & Mookerjee 1975)
41338 (calculated-bond contribution, Hine & Mookerjee 1975)
40000; 42800, 38000 (recommended; calculated-P/C, Mackay & Shiu 1981)
43367 (selected, Mills et al. 1982)
44080, 31934 (quoted, calculated- $\chi$ , Nirmalakhandan & Speece 1988)

37926 (calculated-P/C, Eastcott et al. 1988)
43269 (calculated-P/C, Yaws et al. 1991)

Octanol/Water Partition Coefficient, log $K_{OW}$:

2.76 (calculated-$\pi$ substituent constant, Hansch et al. 1968)
2.76, 2.96 (quoted, calculated- $\chi$ , Murray et al. 1975)
2.82 (Hansch & Leo 1979)
2.76 (Hutchinson et al. 1980, quoted, Sangster 1989)
3.44 (quoted, Abernethy et al. 1988)
3.88 (recommended, Sangster 1989)
4.10 (calculated-f const. per Lyman 1982, Thoms & Lion 1992)
2.76, 2.89 (quoted, calculated-$V_M$, Wang et al. 1992)
3.87 (calculated-f const., Müller & Klein 1992)

Bioconcentration Factor, log BCF:

Sorption Partition Coefficient, log $K_{OC}$:

Half-Lives in the Environment:

Air: an atmospheric lifetime was estimated to be 13 hours in summer daylight, based on the photooxidation rate constant of $1.04 \times 10^{-11}$ $cm^3$ $molecule^{-1}$ $sec^{-1}$ with hydroxyl radicals in air (Altshuller 1991).
Surface water:
Ground water:
Sediment:
Soil:
Biota:

Environmental Fate Rate Constant and Half-Lives:

Volatilization:
Photolysis:
Oxidation: photooxidation reaction rate constant of $1.04 \times 10^{-11}$ $cm^3$ $molecule^{-1}$ $sec^{-1}$ with hydroxyl radicals in air at 298 K (Atkinson 1990; Altshuller 1991) with an estimated atmospheric lifetime of 13 hours, based on rate constant for the reaction with OH radicals in air (Altshuller 1991).
Hydrolysis:
Biotransformation:
Biodegradation:
Bioconcentration, Uptake ($k_1$) and Elimination ($k_2$) Rate Constants or Half-Lives:

Common Name: 1,2-*cis*-Dimethylcyclohexane
Synonym:
Chemical Name: 1,2-*cis*-dimethylcyclohexane
CAS Registry No: 2207-01-4
Molecular Formula: $C_8H_{16}$
Molecular Weight: 112.214
Melting Point (°C): -50.023
Boiling Point (°C): 129.728
Density (g/cm³ at 20°C):
    0.7963 (Weast 1984)
    0.79627 (20°C, Dreisbach 1955; Riddick et al. 1986)
    0.79222 (25°C, Dreisbach 1955; Riddick et al. 1986)
Molar Volume (cm³/mol):
    140.9 (20°C, calculated-density)
    141.64 (25°C, calculated-density)
    140.9 (20°C, calculated-density, McAuliffe 1966)
    162.6 (calculated-LeBas method)
Molecular Volume, TMV (Å³):
Total Surface Area, TSA (Å²):
    315.5 (Hermann 1972, Amidon et al. 1974)
    315.5, 326.6 (Kier et al. 1975)
    319.1, 309.2, 150.2 (Valvani et al. 1976)
    172.2 (Okouchi et al. 1992)
Heat of Fusion, $\Delta H_{fus}$, (kcal/mol):
    0.3932 (Dreisbach 1955; Riddick et al. 1986)
Entropy of Fusion, $\Delta S_{fus}$ (cal/mol K or e.u.):
Fugacity Ratio at 25°C, F: 1.0

Water Solubility (g/m³ or mg/L at 25°C):
    6.0 (shake flask-GC, McAuliffe 1966; quoted, Hermann 1972; Mackay & Shiu 1981; IUPAC 1898)
    6.0, 10.4 (quoted, calculated-$K_{OW}$, Hansch et al. 1968)
    6.0, 12.0 (quoted, calculated-TSA, Amidon et al. 1974)
    6.0 (quoted, Hine & Mookerjee 1975)
    5.80, 5.28 (quoted, calculated- $\chi$ , Hall et al. 1975)
    5.80, 5.28 (quoted, calculated- $\chi$ , Kier & Hall 1976)
    2.76 (calculated-QSAR, Passino & Smith 1987)
    6.0, 5.90 (quoted, calculated-fragment solubility constants, Wakita et al. 1986)
    10.4, 5.59 (quoted, calculated-$V_M$, Wang et al. 1992)
    5.20 (quoted literature average, Myrdal et al. 1992)
    6.0 (calculated-AI, Okouchi et al. 1992)

Vapor Pressure (Pa at 25°C):

1929 (Antoine eqn., Dreisbach 1959)
1930 (Antoine eqn., Zwolinski & Wilhoit 1971; quoted, Mackay & Shiu 1981)
1929 (quoted, Hine & Mookerjee 1975)
1925, 1723 (quoted, calculated-B.P., Mackay et al. 1982)

Henry's Law Constant (Pa $m^3$/mol):

36000 (calculated-P/C, Mackay & Shiu 1981)
35830 (calculated-1/$K_{AW}$, $C_W/C_A$, reported as exptl., Hine & Mookerjee 1975; quoted, Nirmalakhandan & Speece 1988)
52996 (calculated-group contribution, Hine & Mookerjee 1975)
62265 (calculated-bond contribution, Hine & Mookerjee 1975)
35830, 44080 (quoted, calculated- $\chi$ , Nirmalakhandan & Speece 1988)
36045 (calculated-P/C, Yaws et al. 1991)

Octanol/Water Partition Coefficient, log $K_{OW}$:

3.06 (calculated-$\pi$ substituent constant, Hansch et al. 1968)
3.06, 3.33 (quoted, calculated- $\chi$ , Murray et al. 1975)
3.06, 3.21 (quoted, calculated-$V_M$, Wang et al. 1992)

Bioconcentration Factor, log BCF:

Sorption Partition Coefficient, log $K_{OC}$:

Half-Lives in the Environment:

Air: atmospheric half-life was estimated to be 2.4-24 hours for cycloparaffins, based on the EPA Reactivity Classification of Organics (Darnall et al. 1976).
Surface water:
Ground water:
Sediment:
Soil:
Biota:

Environmental Fate Rate Constant and Half-Lives:

Volatilization:
Photolysis:
Oxidation:

Hydrolysis:
Biodegradation:
Biotransformation:
Bioconcentration, Uptake ($k_1$) and Elimination ($k_2$) Rate Constants or Half-Lives:

Common Name: 1,4-*trans*-Dimethylcyclohexane
Synonym:
Chemical Name: 1,4-*trans*-dimethylcyclohexane
CAS Registry No: 2207-04-7
Molecular Formula: $C_8H_{16}$
Molecular Weight: 112.21
Melting Point (°C): -37
Boiling Point (°C): 119.4
Density ($g/cm^3$ at 20°C):
0.76255 (20°C, Dreisbach 1955)
0.75835 (25°C, Dreisbach 1955)
Molar Volume ($cm^3/mol$):
162.6 (calculated-LeBas method)
147.15 (20°C, calculated-density)
148.0 (25°C, calculated-density)
Molecular Volume, TMV ($Å^3$):
Total Surface Area, TSA ($Å^2$):
172.2 (Okouchi et al. 1992)
Heat of Fusion, $\Delta H_{fus}$, (kcal/mol):
2.73 (Dreisbach 1955)
Entropy of Fusion, $\Delta S_{fus}$ (cal/mol K or e.u.):
Fugacity Ratio at 25°C, F: 1.0

Water Solubility ($g/m^3$ or mg/L at 25°C):
3.84 (shake flask-GC, Price 1976; quoted, Mackay & Shiu 1981; IUPAC 1989)
3.84 (quoted, Krzyzanowska & Szeliga 1978; quoted IUPAC 1989)
3.84 (calculated-AI, Okouchi et al. 1992)
3.84 (quoted literature average, Myrdal et al. 1992)

Vapor Pressure (Pa at 25°C):
3025 (Antoine eqn., Dreisbach 1955)
3020 (Antoine eqn., Zwolinski & Wilhoit 1971; quoted, Mackay & Shiu 1981)
3020, 2685 (quoted, calculated-B.P., Mackay et al. 1982)

Henry's Law Constant (Pa $m^3$/mol):
88200 (calculated-P/C, Mackay & Shiu 1981)
88360 (calculated-P/C, Yaws et al. 1991)

Octanol/Water Partition Coefficient, log $K_{OW}$:

Bioconcentration Factor, log BCF:

Sorption Partition Coefficient, log $K_{OC}$:

Half-Lives in the Environment:

Air: atmospheric half-life was estimated to be 2.4-24 hours for cycloparaffins, based on the EPA Reactivity Classification of Organics (Darnall et al. 1976).

Surface water:

Ground water:

Sediment:

Soil:

Biota:

Environmental Fate Rate Constant and Half-Lives:

Volatilization:

Photolysis:

Oxidation:

Hydrolysis:

Biodegradation:

Biotransformation:

Bioconcentration, Uptake ($k_1$) and Elimination ($k_2$) Rate Constants or Half-Lives:

Common Name: 1,1,3-Trimethylcyclohexane
Synonym:
Chemical Name: 1,1,3-trimethylcyclohexane
CAS Registry No: 3073-66-3
Molecular Formula: $C_9H_{18}$
Molecular Weight: 126.24
Melting Point (°C): -65.7
Boiling Point (°C): 138.94
Density (g/cm$^3$ at 20°C): 0.7664
Molar Volume (cm$^3$/mol):
164.72 (calculated-density)
184.8 (calculated-LeBas method)
Molecular Volume, TMV (Å$^3$):
Total Surface Area, TSA (Å$^2$):
Heat of Fusion, $\Delta H_{fus}$, (kcal/mol):
Entropy of Fusion, $\Delta S_{fus}$ (cal/mol K or e.u.):
Fugacity Ratio at 25°C, F: 1.0

Water Solubility (g/m$^3$ or mg/L at 25°C):
1.77 (shake flask-GC, Price 1976; quoted, Mackay & Shiu 1981; Brookman 1985)
1.77 (quoted literature average, Myrdal et al. 1992)

Vapor Pressure (Pa at 25°C):
1480 (Antoine eqn., Dean 1985)

Henry's Law Constant (Pa m$^3$/mol):
105557 (calculated-P/C from selected data)

Octanol/Water Partition Coefficient, log $K_{OW}$:

Bioconcentration Factor, log BCF:

Sorption Partition Coefficient, log $K_{OC}$:

Half-Lives in the Environment:

Air:

Surface water:

Ground water:

Sediment:

Soil:

Biota:

Environmental Fate Rate Constant and Half-Lives:

Volatilization:

Photolysis:

Oxidation:

Hydrolysis:

Biodegradation:

Biotransformation:

Bioconcentration, Uptake ($k_1$) and Elimination ($k_2$) Rate Constants or Half-Lives:

Common Name: Cycloheptane
Synonym: suberane
Chemical Name: cycloheptane
CAS Registry No: 291-64-5
Molecular Formula: $C_7H_{14}$
Molecular Weight: 98.188
Melting Point (°C): -12
Boiling Point (°C): 118.5
Density ($g/cm^3$ at 20°C):

0.8098 (Weast 1984)
0.8011 (20°C, Riddick et al. 1986)
0.811 (25°C, Riddick et al. 1986)

Molar Volume ($cm^3/mol$):

136.4 (calculated-LeBas method)
121.0 (20°C, calculated-density, McAuliffe 1966)
121.3 (20°C, calculated-density, Stephenson & Malanowski 1987)
121.0 (calculated-density, Wang et al. 1992)

Molecular Volume, TMV ($Å^3$):
Total Surface Area, TSA ($Å^2$):

301.94 (Hermann, 1972)
301.9 (Amidon et al. 1974)
301.9. 308.3 (Kier et al. 1975)

Heat of Fusion, $\Delta H_{fus}$, (kcal/mol):
Entropy of Fusion, $\Delta S_{fus}$ (cal/mol K or e.u.):
Fugacity Ratio at 25°C, F: 1.0

Water Solubility ($g/m^3$ or mg/L at 25°C):

30.0 (shake flask-GC, McAuliffe 1966; quoted, Brookman et al. 1985)
30, 15.7 (quoted, calculated-$K_{OW}$, Hansch et al. 1968)
30, 18.85 (quoted, calculated-TSA, Amidon et al. 1974)
30, 11 (quoted, calculated- $\chi$ , Hall et al. 1975)
30, 11 (quoted, calculated- $\chi$ , Kier & Hall 1976)
29.7, 30.3 (quoted, calculated-$K_{OW}$, Yalkowsky & Morozowich 1980)
30.3, 13.2 (quoted, calculated-fragment solubility constants, Wakita 1986)
27.1 (30°C, Groves 1988)
30, 40.1 (misquoted from cyclopentane, calculated-UNIFAC, Al-Sahhaf 1989)
31.1, 24.7 (quoted, calculated-$V_M$, Wang et al. 1992)
24.44 (quoted literature average, Myrdal et al. 1992)
30.3 (quoted, Müller & Klein 1992)

Vapor Pressure (Pa at 25°C):

2895 (extrapolated, Antoine eqn., Dean 1985)
2895, 2898 (extrapolated, Antoine eqn., Boublik et al. 1986)
2930 (interpolated, Antoine eqn., Stephenson & Malanowski 1987)

Henry's Law Constant (Pa $m^3$/mol):

9492 (calculated-P/C from selected data)
9977 (calculated-P/C, Yaws et al. 1991)

Octanol/Water Partition Coefficient, log $K_{OW}$:

2.87 (calculated-$\pi$ substituent constant, Hansch et al. 1968)
2.87, 3.06 (quoted, calculated- $\chi$ , Murray et al. 1975)
3.76 (calculated-f const., Yalkowsky & Morozowich 1980)
2.87, 2.72 (quoted, calculated-$V_M$, Wang et al. 1992)
3.91 (calculated-f const., Müller & Klein 1992)

Bioconcentration Factor, log BCF:

Sorption Partition Coefficient, log $K_{OC}$:

Half-Lives in the Environment:

Air: atmospheric half-life was estimated to be 2.4-24 hours for cycloparaffins, based on the EPA Reactivity Classification of Organics (Darnall et al. 1976).
Surface water:
Ground water:
Sediment:
Soil:
Biota:

Environmental Fate Rate Constant and Half-Lives:

Volatilization:
Photolysis:
Oxidation: photooxidation reaction rate constant of $1.31 \times 10^{-12}$ $cm^3$ $molecule^{-1}$ $sec^{-1}$ for the gas-phase reaction with hydroxyl radicals at 298 K (Atkinson 1985); rate constant of $1.25 \times 10^{-11}$ $cm^3$ $molecule^{-1}$ $sec^{-1}$ for the reaction with OH radicals in air at 298 K (Atkinson 1990).
Hydrolysis:
Biodegradation:
Biotransformation:
Bioconcentration, Uptake ($k_1$) and Elimination ($k_2$) Rate Constants or Half-Lives:

Common Name: Cyclooctane
Synonym:
Chemical Name: cyclooctane
CAS Registry No: 292-64-8
Molecular Formula: $C_8H_{16}$
Molecular Weight: 112.214
Melting Point (°C): 10-13
Boiling Point (°C): 151
Density ($g/cm^3$ at 20°C):
 0.834 (Weast 1984)
 0.8349 (20°C, Riddick et al. 1986)
 0.834 (25°C, Riddick et al. 1986)
Molar Volume ($cm^3/mol$):
 135.0 (20°C, calculated-density, McAuliffe 1966; Lande & Banerjee 1981)
 134.4 (20°C, calculated-density, Stephenson & Malanowski 1987)
 134.0 (calculated-density, Wang et al. 1992)
 154.1 (calculated-LeBas method)
Molecular Volume, TMV ($Å^3$):
Total Surface Area, TSA ($Å^2$):
 322.38 (Hermann 1972)
 383 (Amidon et al. 1974)
 322.6, 337.6 (Kier et al. 1975)
 148.8, 305.7, 315.8 (Valvani et al. 1976)
Heat of Fusion, $\Delta H_{fus}$, (kcal/mol):
Entropy of Fusion, $\Delta S_{fus}$ (cal/mol K or e.u.):
Fugacity Ratio at 25°C, F: 1.0

Water Solubility ($g/m^3$ or mg/L at 25°C):
 7.90 (shake flask-GC, McAuliffe 1966; quoted, Hermann 1972)
 7.90, 5.57 (quoted, calculated-$K_{OW}$, Hansch et al. 1968)
 7.91, 8.85 (quoted, calculated-TSA, Amidon et al. 1974)
 8.62, 3.56 (quoted, calculated- $\chi$ , Hall et al. 1975)
 8.62, 3.56 (quoted, calculated- $\chi$ , Kier & Hall 1976)
 7.94 (quoted, Lande & Banerjee 1981)
 7.94 (quoted, Lyman 1982)
 7.90 (quoted, Bobra et al. 1984)
 7.94, 3.72 (quoted, calculated-fragment solubility constants, Wakita et al. 1986)
 7.90, 5.10 (quoted, calculated-UNIFAC, Al-Sahhaf 1989)
 7.90, 9.77 (quoted, calculated-$V_M$, Wang et al. 1992)
 7.90 (quoted literature average, Myrdal et al. 1992)
 7.94 (quoted, Müller & Klein 1992)

Vapor Pressure (Pa at 25°C):

767, 740 (Antoine eqns., extrapolated, Boublik et al. 1984)
740 (Antoine eqn., Dean 1985)
753 (Antoine eqn., Stephenson & Malanowski 1987)

Henry's Law Constant (Pa $m^3$/mol):

10520 (calculated-P/C from selected data)
10485 (calculated-P/C, Yaws et al. 1991)

Octanol/Water Partition Coefficient, log $K_{OW}$:

3.28 (calculated-$\pi$ substituent constant, Hansch et al. 1968)
3.28, 3.50 (quoted, calculated- $\chi$ , Murray et al. 1975)
3.28 (quoted, Hutchinson et al. 1980; Sangster 1989)
3.28 (calculated-f const., Lyman 1982)
4.45 (recommended, Sangster 1989)
3.28, 3.04 (quoted, calculated-$V_M$, Wang et al. 1992)
4.47 (calculated-f const., Müller & Klein 1992)

Bioconcentration Factor, log BCF:

Sorption Partition Coefficient, log $K_{OC}$:

Half-Lives in the Environment:

Air: atmospheric half-life was estimated to be 2.4-24 hours for cycloparaffins, based on the EPA Reactivity Classification of Organics (Darnall et al. 1976).
Surface water:
Ground water:
Sediment:
Soil:
Biota:

Environmental Fate Rate Constant and Half-Lives:

Volatilization:
Photolysis:
Oxidation:
Hydrolysis:
Biodegradation:
Biotransformation:
Bioconcentration, Uptake ($k_1$) and Elimination ($k_2$) Rate Constants or Half-Lives:

Common Name: Decalin
Synonym: bicyclo[4.4.0]decane, naphthalane, naphthane
Chemical Name: decahydronaphthalene (mixed isomer)
CAS Registry No: 91-17-8
Molecular Formula: $C_{10}H_{18}$
Molecular Weight: 138.252
Melting Point (°C): -124
Boiling Point (°C): 191.7
Density (g/cm$^3$ at 20°C):
    0.8865 (20°C, Riddick et al. 1986)
    0.8789 (25°C, Riddick et al. 1986)
Molar Volume (cm$^3$/mol):
    156.0 (20°C, calculated-density)
    157.3 (25°C, calculated-density)
    184.6 (calculated-LeBas method)
Molecular Volume, TMV (Å$^3$):
Total Surface Area, TSA (Å$^2$):
Heat of Fusion, $\Delta H_{fus}$, (kcal/mol):
    2.268 (*cis*-form, Riddick et al. 1986)
    3.445 (*trans*-form, Riddick et al. 1986)
Entropy of Fusion, $\Delta S_{fus}$ (cal/mol K or e.u.):
Fugacity Ratio at 25°C, F: 1.0

Water Solubility (g/m$^3$ or mg/L at 25°C):
    < 175.7 (residue volume, reported as < 0.020 mL/100 mL, Booth & Everson 1948)
    0.889 (shake flask-GC, Price 1976; quoted as more reliable value, IUPAC 1989)
    6.21 (shake flask-GC, Shiu & Mackay 1979)
    6.21 (quoted, Hutchinson et al. 1980)
    < 200 (quoted, Riddick et al. 1986)
    1.99 (calculated-QSAR, Passino & Smith 1987, quoted, Müller & Klein 1992)

Vapor Pressure (Pa at 25°C):
    104 (*cis*-decalin, Antoine eqn., Zwolinski & Wilhoit 1971)
    164 (*trans*-decalin, Antoine eqn., Zwolinski & Wilhoit 1971)
    133.3 (23°C, quoted, Verschueren 1983)
    100 (quoted, Riddick et al. 1986)
    100, 164 (quoted, *cis-*, *trans*-decalin, Riddick et al. 1986)
    105 (*cis*-decalin, Antoine eqn., extrapolated, Stephenson & Malanowski 1987)
    168 (*trans*-decalin, Antoine eqn., extrapolated, Stephenson & Malanowski 1987)
    104, 168 (*cis-*, *trans*-decalin, Antoine eqn., Dean 1985)

Henry's Law Constant (Pa $m^3$/mol):

Octanol/Water Partition Coefficient, log $K_{OW}$:

4.79 (calculated-f const., Müller & Klein 1992)

Bioconcentration Factor, log BCF:

Sorption Partition Coefficient, log $K_{OC}$:

Half-Lives in the Environment:

Air:

Surface water:

Ground water:

Sediment:

Soil:

Biota:

Environmental Fate Rate Constant and Half-Lives:

Volatilization:

Photolysis:

Oxidation: photooxidation reaction rate constants: $1.96 \times 10^{-11}$ $cm^3$ $molecule^{-1}$ $sec^{-1}$ of *cis*-decalin and $2.02 \times 10^{-11}$ $cm^3$ $molecule^{-1}$ $sec^{-1}$ of *trans*-decalin for the reaction with hydroxyl radicals in air at 299 K (Atkinson 1985); rate constant for the mixed isomers of $2.00 \times 10^{-11}$ $cm^3$ $molecule^{-1}$ $sec^{-1}$ for the reaction with hydroxyl radicals in the gas phase at 298 K (Atkinson 1990).

Hydrolysis:

Biodegradation:

Biotransformation:

Bioconcentration, Uptake ($k_1$) and Elimination ($k_2$) Rate Constants or Half-Lives:

Common Name: 2-Methylpropene
Synonym: isobutene
Chemical Name: 2-methylpropene
CAS Registry No: 115-11-7
Molecular Formula: $C_4H_8$
Molecular Weight: 56.11
Melting Point (°C): -140.35
Boiling Point (°C): -6.9
Density (g/cm³ at 20°C):
 0.5942 (20°C, Dreisbach 1959)
 0.5879 (25°C, Dreisbach 1959)
Molar Volume (cm³/mol):
 94.4 (20°C, calculated-density, McAuliffe 1966)
 94.0 (calculated-density, Wang et al. 1992)
 94.4 (20°C, calculated-density)
 95.4 (25°C, calculated-density)
 88.8 (calculated-LeBas method)
Molecular Volume, TMV (Å³):
Total Surface Area, TSA (Å²):
 103.3 (Okouchi et al. 1992)
Heat of Fusion, $\Delta H_{fus}$, (kcal/mol):
Entropy of Fusion, $\Delta S_{fus}$ (cal/mol K or e.u.):
Fugacity Ratio at 25°C, F: 1.0

Water Solubility (g/m³ or mg/L at 25°C):
 263 (shake flask-GC, McAuliffe 1966; quoted, Mackay & Shiu 1981; Brookman et al. 1985)
 263 (quoted, Hine & Mookerjee 1975)
 209, 330 (quoted, calculated-$V_M$, Wang et al. 1992)
 260 (calculated-AI, Okouchi et al. 1992)

Vapor Pressure (Pa at 25°C):
 303710 (Antoine eqn., Dreisbach 1959)
 304000 (Antoine eqn., Zwolinski & Wilhoit 1971; quoted, Mackay & Shiu 1981)

Henry's Law Constant (Pa m³/mol):
 21590 (calculated-$1/K_{AW}$, $C_W/C_A$, reported as exptl., Hine & Mookerjee 1975; quoted, Nirmalakhandan & Speece 1988)
 23134 (calculated-group contribution, Hine & Mookerjee 1975)
 10820 (calculated-bond contribution, Hine & Mookerjee 1975)

14800 (calculated-P/C, Mackay & Shiu 1981)
21590, 35800 (quoted, calculated- $\chi$ , Nirmalakhandan & Speece 1988)

Octanol/Water Partition Coefficient, log $K_{OW}$:
0.64, 1.32 (quoted, calculated-$V_M$, Wang et al. 1992)

Bioconcentration Factor, log BCF:

Sorption Partition Coefficient, log $K_{OC}$:

Half-Lives in the Environment:
Air: atmospheric lifetime was estimated to be 5.3 hours, based on photooxidation rate constant of $5.14 \times 10^{-11}$ $cm^3$ $molecule^{-1}$ $sec^{-1}$ with OH radicals in air during summer daylight (Altshuller 1991).
Surface water: half-life was estimated to be 320 hours and $9 \times 10^4$ days for oxidation by OH and $RO_2$ radicals for olefins and 8.0 days for substituted olefins, based on rate constant of $1 \times 10^6$ $M^{-1}$ $sec^{-1}$ for oxidation by singlet oxygen in aquatic system (Mill & Mabey 1985).
Ground water:
Sediment:
Soil:
Biota:

Environmental Fate Rate Constant and Half-Lives:
Volatilization:
Photolysis:
Oxidation: room temperature rate constants: $6.20 \times 10^{-18}$ $cm^3$ $molecule^{-1}$ $sec^{-1}$ (Hanst et al. 1958; quoted, Adeniji et al. 1981), $15 \times 10^{-18}$ $cm^3$ $molecule^{-1}$ $sec^{-1}$ (Schuck et al. 1960; quoted, Adeniji et al. 1981), $23 \times 10^{-18}$ $cm^3$ $molecule^{-1}$ $sec^{-1}$ (Bufalini & Altshuller 1965; quoted, Adeniji et al. 1981), $13.6 \times 10^{-18}$ $cm^3$ $molecule^{-1}$ $sec^{-1}$ (Japar et al. 1974; quoted, Adeniji et al. 1981), $11.7 \times 10^{-18}$ $cm^3$ $molecule^{-1}$ $sec^{-1}$ (Huie & Herron 1975; quoted, Adeniji et al. 1981) and $11.7 \times 10^{-18}$ $cm^3$ $molecule^{-1}$ $sec^{-1}$ at (294 $\pm$ 2) K (Adeniji et al. 1981) all reacted with $O_3$ in the gas phase; photooxidation rate constant of $1.40 \times 10^4$ L $mole^{-1}$ $sec^{-1}$ for reaction with ozone at 30°C (Bufalini & Altshuller 1965); room temperature rate constants of $1.60 \times 10^{-11}$ $cm^3$ $molecule^{-1}$ $sec^{-1}$ for the reaction with $O(^3P)$ (Singleton & Cvetanovic 1976; Atkinson & Pitts Jr. 1977; quoted, Gaffney & Levine 1979) and $5.40 \times 10^{-11}$ $cm^3$ $molecule^{-1}$ $sec^{-1}$ for the reaction

with OH radicals (Atkinson et al. 1979; quoted, Gaffney & Levine 1979); rate constant of $1.45 \times 10^{-17}$ to $2.32 \times 10^{-17}$ $cm^3$ $molecule^{-1}$ $sec^{-1}$ for the reaction with ozone between 295 to 303 K in the gas phase (Atkinson & Carter 1984); rate constants of $5.14 \times 10^{-11}$ $cm^3$ $molecule^{-1}$ $sec^{-1}$ for the reaction with hydroxyl radicals and $1.21 \times 10^{-17}$ $cm^3$ $molecule^{-1}$ $sec^{-1}$ with $O_3$ in the gas phase at 298 K (Atkinson 1990); gas phase reaction rate constants of $5.14 \times 10^{-11}$ $cm^3$ $molecule^{-1}$ $sec^{-1}$ with OH radicals and $31.3 \times 10^{-12}$ $cm^3$ $molecule^{-1}$ $sec^{-1}$ with $NO_3$ radicals at 298 K (Sabljic & Güsten 1990).

Hydrolysis:

Biodegradation:

Biotransformation:

Bioconcentration, Uptake ($k_1$) and Elimination ($k_2$) Rate Constants or Half-Lives:

Common Name: 1-Butene
Synonym: butylene
Chemical Name: 1-butene
CAS Registry No: 106-98-9
Molecular Formula: $C_4H_8$
Molecular Weight: 56.11
Melting Point (°C): -185.5
Boiling Point (°C): -6.26
Density (g/cm$^3$ at 20°C):
0.5951 (20°C, at saturation pressure, Dreisbach 1959)
0.5888 (25°C, at saturation pressure, Dreisbach 1959)
Molar Volume (cm$^3$/mol):
94.3 (20°C, calculated-density, McAuliffe 1966)
94.3 (20°C, calculated-density)
95.3 (25°C, calculated-density)
89.8 (calculated-LeBas method)
Molecular Volume, TMV (Å$^3$):
Total Surface Area, TSA (Å$^2$):
102.7 (Okouchi et al. 1992)
Heat of Fusion, $\Delta H_{fus}$, (kcal/mol):
Entropy of Fusion, $\Delta S_{fus}$ (cal/mol K or e.u.):
Fugacity Ratio at 25°C, F: 1.0

Water Solubility (g/m$^3$ or mg/L at 25°C):
222 (shake flask-GC, McAuliffe 1966; quoted, Mackay & Shiu 1981)
223 (quoted, Hine & Mookerjee 1975)
177, 330 (quoted, calculated-$V_M$, Wang et al. 1992)
222 (calculated-AI, Okouchi et al. 1992)

Vapor Pressure (Pa at 25°C):
361100 (Antoine eqn. regression, Stull 1947)
296000 (Antoine eqn., Dreisbach 1959)
297000 (Antoine eqn., Zwolinski & Wilhoit 1971; quoted, Mackay & Shiu 1981)

Henry's Law Constant (Pa m$^3$/mol):
25610 (calculated-P/C, Mackay & Shiu 1981)
25370 (calculated-1/$K_{AW}$, $C_W/C_A$, reported as exptl., Hine & Mookerjee 1975; quoted, Nirmalakhandan & Speece 1988)
26560 (calculated-group contribution, Hine & Mookerjee 1975)
15284 (calculated-bond contribution, Hine & Mookerjee 1975)

25370, 29800 (quoted, calculated- $\chi$ , Nirmalakhandan & Speece 1988)
24800 (calculated-P/C, Yaws et al. 1991)

Octanol/Water Partition Coefficient, log $K_{OW}$:
2.40; 2.17, 2.26, 2.43 (quoted; calculated-f const., Rekker 1977)
1.59, 1.32 (quoted, calculated-$V_M$, Wang et al. 1992)

Bioconcentration Factor, log BCF:

Sorption Partition Coefficient, log $K_{OC}$:

Half-Lives in the Environment:
Air: atmospheric lifetime was estimated to be 5.5 hours, based on the reaction rate constant of $3.14 \times 10^{-11}$ $cm^3$ $molecule^{-1}$ $sec^{-1}$ with OH radicals during summer daylight in the gas phase (Altshuller 1991).
Surface water: half-lives were reported to be 320 hours and $9 \times 10^4$ days for reaction with OH and $RO_2$ radicals of olefins in aquatic system, and 7.3 days, based on oxidation reaction rate constant of $3 \times 10^5$ $M^{-1}$ $sec^{-1}$ with singlet $O_2$ for unsubstituted olefins in aquatic system (Mill & Mabey 1985).
Ground water:
Sediment:
Soil:
Biota:

Environmental Fate Rate Constant and Half-Lives:
Volatilization:
Photolysis:
Oxidation: photooxidation reaction rate constants: $1.0 \times 10^{-17}$ $cm^3$ $molecule^{-1}$ $sec^{-1}$ (Bufalini & Altshuller 1965; quoted, Adeniji et al. 1981), $1.23 \times 10^{-17}$ $cm^3$ $molecule^{-1}$ $sec^{-1}$ (Japar et al. 1974; quoted, Adeniji et al. 1981) and $1.03 \times 10^{-17}$ $cm^3$ $molecule^{-1}$ $sec^{-1}$ (Huie & Herron 1975; quoted, Adeniji et al. 1981) all for reaction with ozone in gas phase at room temperatue; while $1.26 \times 10^{-17}$ $cm^3$ $molecule^{-1}$ $sec^{-1}$ was at (294 $\pm$ 2) K (Adeniji et al. 1981); room temperature rate constants of $4.20 \times 10^{-12}$ $cm^3$ $molecule^{-1}$ $sec^{-1}$ for the reaction with $O(^3P)$ (Singleton & Cvetanovic 1976; Atkinson & Pitts Jr. 1977; quoted, Gaffney & Levine 1979) and $2.70 \times 10^{-11}$ $cm^3$ $molecule^{-1}$ $sec^{-1}$ for the reaction with OH radicals (Atkinson et al. 1979; quoted, Gaffney & Levine 1979); rate constant of $3.14 \times 10^{-11}$ $cm^3$ $molecule^{-1}$ $sec^{-1}$ for the reaction with hydroxyl radicals and

$1.10 \times 10^{-17}$ $cm^3$ $molecule^{-1}$ $sec^{-1}$ for the reaction with ozone in the gas phase at 298 K (Atkinson 1990); rate constants of $3.14 \times 10^{-11}$ $cm^3$ $molecule^{-1}$ $sec^{-1}$ for the reaction with OH radicals and $1.23 \times 10^{-14}$ $cm^3$ $molecule^{-1}$ $sec^{-1}$ for the reaction with $NO_3$ radicals in air at 298 K (Sabljic & Güsten 1990); atmospheric lifetime was estimated to be 5.5 hours, based on the reaction rate constant of $3.14 \times 10^{-11}$ $cm^3$ $molecule^{-1}$ $sec^{-1}$ with OH radicals during summer daylight in the gas phase (Altshuller 1991).

Hydrolysis:

Biodegradation:

Biotransformation:

Bioconcentration, Uptake ($k_1$) and Elimination ($k_2$) Rate Constants or Half-Lives:

Common Name: 2-Methyl-1-butene
Synonym:
Chemical Name: 2-methyl-1-butene
CAS Registry No:
Molecular Formula: $C_5H_{10}$, $CH_3CH_2C(CH_3)=CH_2$
Molecular Weight: 70.14
Melting Point (°C): -137.56
Boiling Point (°C): 31.163
Density (g/cm$^3$ at 20°C):

0.6504 (20°C, at saturation pressure, Dreisbach 1959; Dean 1985)
0.6451 (25°C, at saturation pressure, Dreisbach 1959)

Molar Volume (cm$^3$/mol):

107.8 (20°C, calculated-density)
108.7 (25°C, calculated-density)
111.0 (calculated-LeBas method)
108 (calculated-density, Wang et al. 1992)

Molecular Volume, TMV (Å$^3$):
Total Surface Area, TSA (Å$^2$):
Heat of Fusion, $\Delta H_{fus}$, (kcal/mol):
Entropy of Fusion, $\Delta S_{fus}$ (cal/mol K or e.u.):
Fugacity Ratio at 25°C, F: 1.0

Water Solubility (g/m$^3$ or mg/L at 25°C):

155 (estimated-nomograph of Kabadi & Danner 1979; Brookman et al. 1985)
130 (misquoted from 3-methyl-1-butene, Wakita et al. 1986)
168 (calculated-fragment solubility constants, Wakita et al. 1986)
260 (calculated-regression eqn. of Lyman et al. 1982, Wang et al. 1992)
137 (calculated-$V_M$, Wang et al. 1992)

Vapor Pressure (Pa at 25°C):

135539 (Antoine eqn. regression, Stull 1947)
81320 (Antoine eqn., Dreisbach 1959)
81330 (Antoine eqn., Zwolinski & Wilhoit 1971)

Henry's Law Constant (Pa m$^3$/mol):

54230 (quoted, Nirmalakhandan & Speece 1988)
43080 (calculated- $\chi$ , Nirmalakhandan & Speece 1988)

Octanol/Water Partition Coefficient, log $K_{OW}$:

2.07 (calculated-regression of Lyman et al. 1982, Wang et al. 1992)

1.89 (calculated-$V_M$, Wang et al. 1992)

Bioconcentration Factor, log BCF:

Sorption Partition Coefficient, log $K_{OC}$:

Half-Lives in the Environment:

Air: atmospheric lifetime was estimated to be 2.3 hours, based on the photooxidation rate constant of $6.10 \times 10^{-11}$ $cm^3$ $molecule^{-1}$ $sec^{-1}$ with hydroxyl radicals in air during summer daylight (Altshuller 1991).

Surface water: half-lives were reported to be 320 hours and $9 \times 10^4$ days for reaction with OH and $RO_2$ radicals for olefins in aquatic system, and 7.3 days, based on oxidation reaction rate constant of $3 \times 10^5$ $M^{-1}$ $sec^{-1}$ with singlet oxygen for unsubstituted olefins in aquatic system (Mill & Mabey 1985).

Ground water:

Sediment:

Soil:

Biota:

Environmental Fate Rate Constant and Half-Lives:

Volatilization:

Photolysis:

Oxidation: photooxidation reaction rate constant of $9.01 \times 10^{-11}$ $cm^3$ $molecule^{-1}$ $sec^{-1}$ to $6.07 \times 10^{-11}$ $cm^3$ $molecule^{-1}$ $sec^{-1}$ with hydroxyl radicals in gas phase at 298 K (Atkinson 1985); atmospheric lifetime was estimated to be 2.3 hours, based on photooxidation rate constant of $6.10 \times 10^{-11}$ $cm^3$ $molecule^{-1}$ $sec^{-1}$ for the reaction with hydroxyl radicals in air during summer daylight (Altshuller 1991).

Hydrolysis:

Biodegradation:

Biotransformation:

Bioconcentration, Uptake ($k_1$) and Elimination ($k_2$) Rate Constants or Half-Lives:

Common Name: 3-Methyl-1-butene
Synonym:
Chemical Name: 3-methyl-1-butene
CAS Registry No: 563-45-1
Molecular Formula: $C_5H_{10}$, $(CH_3)_2CCHCH=CH_2$
Molecular Weight: 70.14
Melting Point (°C): -168.5
Boiling Point (°C): 20.0
Density ($g/cm^3$ at 20°C):
0.6272 (20°C, at saturation pressure, Dreisbach 1959)
0.6219 (25°C, at saturation pressure, Dreisbach 1959)
Molar Volume ($cm^3/mol$):
111.0 (calculated-LeBas method)
111.8 (20°C, calculated-density, McAuliffe 1966)
109.0 (calculated from density, Wang et al. 1992)
Molecular Volume ($A^3$):
Total Surface Area, TSA ($A^2$):
126.1 (Okouchi et al. 1992)
Heat of Fusion, $\Delta H_{fus}$, (kcal/mol):
Entropy of Fusion, $\Delta S_{fus}$ (cal/mol K or e.u.):
Fugacity Ratio at 25°C, F: 1.0

Water Solubility ($g/m^3$ or mg/L at 25°C):
130 (shake flask-GC, McAuliffe 1966; quoted, Mackay & Shiu 1981; Brookman et al. 1985)
130 (quoted, Hine & Mookerjee 1975)
130, 161 (quoted, calculated-fragment solubility constants, Wakita et al. 1986)
130 (calculated-AI, Okouchi et al. 1992)
130, 101 (quoted, calculated-$V_M$, Wang et al. 1992)

Vapor Pressure (Pa at 25°C):
120260 (Antoine eqn., Dreisbach 1959)
120000 (Antoine eqn., Zwolinski & Wilhoit 1971; quoted, Mackay & Shiu 1981)

Henry's Law Constant (Pa $m^3$/mol):
54230 (calculated-$1/K_{AW}$, $C_W/C_A$, reported as exptl., Hine & Mookerjee 1975; quoted, Nirmalakhandan & Speece 1988)
63715 (calculated-group contribution, Hine & Mookerjee 1975)
22607 (calculated-bond contribution, Hine & Mookerjee 1975)
54700 (calculated-P/C, Mackay & Shiu 1981)

54230, 43080 (quoted, calculated- $\chi$ , Nirmalakhandan & Speece 1988)
52944 (calculated-P/C, Yaws et al. 1991)

Octanol/Water Partition Coefficient, log $K_{OW}$:
2.07 (calculated-regression of Lyman et al. 1982, Wang et al. 1992)
2.05 (calculated-$V_M$, Wang et al. 1992)

Bioconcentration Factor, log BCF:

Sorption Partition Coefficient, log $K_{OC}$:

Half-Lives in the Environment:
Air:
Surface water: half-lives were reported to be 320 hours and $9\times10^4$ days for reaction with OH and $RO_2$ radicals for olefins in aquatic system, and 7.3 days, based on oxidation reaction rate constant of $3\times10^5$ $M^{-1}$ $sec^{-1}$ with singlet oxygen for unsubstituted olefins in aquatic system (Mill & Mabey 1985).
Ground water:
Sediment:
Soil:
Biota:

Environmental Fate Rate Constant and Half-Lives:
Volatilization:
Photolysis:
Oxidation: room temperature rate constants of $4.30\times10^{-12}$ $cm^3$ $molecule^{-1}$ $sec^{-1}$ for the reaction with $O(^3P)$ (Singleton & Cvetanovic 1976; quoted, Gaffney & Levine 1979) and $3.10\times10^{-11}$ $cm^3$ $molecule^{-1}$ $sec^{-1}$ for reaction with OH radicals in air (Atkinson et al. 1979; quoted, Gaffney & Levine 1979); photooxidation reaction rate constants of $9.01\times10^{-11}$ $cm^3$ $molecule^{-1}$ $sec^{-1}$ and $6.07\times10^{-11}$ $cm^3$ $molecule^{-1}$ $sec^{-1}$ with hydroxyl radicals in gas phase at 298 K (Atkinson 1985); rate constant of $3.18\times10^{-11}$ $cm^3$ $molecule^{-1}$ $sec^{-1}$ for reaction with OH radicals in air at 298 K (Atkinson 1990).
Hydrolysis:
Biodegradation:
Biotransformation:
Bioconcentration, Uptake ($k_1$) and Elimination ($k_2$) Rate Constants or Half-Lives:

Common Name: 2-Methyl-2-butene
Synonym:
Chemical Name: 2-methyl-2-butene
CAS Registry No:
Molecular Formula: $C_5H_{10}$, $CH_3CH=C(CH_3)CH_3$
Molecular Weight: 80.14
Melting Point (°C): -133.8
Boiling Point (°C): 38.57
Density (g/cm$^3$ at 20°C):
    0.6623 (20°C, Dreisbach 1959)
    0.6570 (25°C, Dreisbach 1959)
Molar Volume (cm$^3$/mol):
    105,.9 (20°C, calculated-density)
    106.8 (25°C, calculated-density)
    111.0 (calculated-LeBas method)
Molecular Volume, TMV (Å$^3$):
Total Surface Area, TSA (Å$^2$):
Heat of Fusion, $\Delta H_{fus}$, (kcal/mol):
Entropy of Fusion, $\Delta S_{fus}$ (cal/mol K or e.u.):
Fugacity Ratio at 25°C, F: 1.0

Water Solubility (g/m$^3$ or mg/L at 25°C):
    195, 282 (quoted, calculated-group contribution, Irmann 1965)
    221, 192 (quoted, calculated-fragment solubility constants, Wakita et al. 1986)

Vapor Pressure (Pa at 25°C):
    62140 (Antoine eqn., Dreisbach 1959)
    62143 (Antoine eqn., Zwolinski & Wilhoit 1971)

Henry's Law Constant (Pa m$^3$/mol):
    24653 (calculated-P/C from selected data)

Octanol/Water Partition Coefficient, log $K_{OW}$:

Bioconcentration Factor, log BCF:

Sorption Partition Coefficient, log $K_{OC}$:

Half-Lives in the Environment:

Air: atmospheric half-life was estimated < 0.24 hour, based on the photooxidation rate constant of $4.8x10^9$ L $mol^{-1}$ $sec^{-1}$ with hydroxyl radicals in air (Darnall et al. 1976); atmospheric lifetimes were calculated to be 0.95 hour for the reaction with $O_3$, 3.2 hours with OH radicals and 0.12 hour with $NO_3$ radicals, based on the rate constants and environmental concentrations of OH, $O_3$ and $NO_3$ in the gas phase (Atkinson & Carter 1984); atmospheric lifetimes were calculated to be 6.38 hours for the reaction with OH radicals, 0.92 hour with $O_3$ and 0.12 hour with $NO_3$ radicals in the gas phase (Atkinson 1985).

Surface water: half-lives were estimated to be 320 hours and $9x10^4$ days for reaction with OH and $RO_2$ radicals respectively in aquatic system, and 8.0 days, based on rate constant of $10^6$ $M^{-1}$ $sec^{-1}$ for the reaction with singlet oxygen in aquatic system (Mill & Mabey 1985).

Ground water:

Sediment:

Soil:

Biota:

Environmental Fate Rate Constant and Half-Lives:

Volatilization:

Photolysis:

Oxidation: room temperature rate constants of $5.40x10^{-11}$ $cm^3$ $molecule^{-1}$ $sec^{-1}$ for the reaction with $O(^3P)$ (Herron & Huie 1973; Furuyama et al. 1974; Atkinson & Pitts Jr. 1978; quoted, Gaffney & Levine 1979) and $8.40x10^{-11}$ $cm^3$ $molecule^{-1}$ $sec^{-1}$ for the reaction with OH radicals (Atkinson et al. 1979; quoted, Gaffney & Levine 1979); photooxidation reaction rate constant of $4.93x10^{-16}$ $cm^3$ $molecule^{-1}$ $sec^{-1}$ with ozone at 299 K (Japar et al. 1974); rate constant of $4.8x10^9$ L $mol^{-1}$ $sec^{-1}$ for the reaction with OH radicals in air with atmospheric half-life of < 0.24 hour (Darnall et al. 1976); rate constants of $(6.79\text{-}7.97)x10^{-16}$ $cm^3$ $molecule^{-1}$ $sec^{-1}$ for the reaction with $O_3$ between 296-299 K in the gas phase (Atkinson & Carter 1984); rate constants of $7.7x10^{-11}$ to $1.19x10^{-10}$ $cm^3$ $molecule^{-1}$ $sec^{-1}$ for the reaction with OH radicals between 297.7-299.5 K (Atkinson 1985); rate constants of $6.89x10^{-11}$ $cm^3$ $molecule^{-1}$ $sec^{-1}$ for the reaction with OH radicals and $4.23x10^{-16}$ $cm^3$ $molecule^{-1}$ $sec^{-1}$ for the reaction with $O_3$ in the gas phase at 198 K (Atkinson 1990); rate constants of $8.69x10^{-11}$ $cm^3$ $molecule^{-1}$ $sec^{-1}$ for the reaction with OH radicals and $9.33x10^{-12}$ $cm^3$ $molecule^{-1}$ $sec^{-1}$ for the reaction with $NO_3$ radicals in the gas phase at 298 K (Sabljic & Güsten 1990).

Hydrolysis:

Biodegradation:

Biotransformation:

Bioconcentration, Uptake ($k_1$) and Elimination ($k_2$) Rate Constants or Half-Lives:

Common Name: 1-Pentene
Synonym: amylene, $\alpha$-*n*-amylene, propylethylene
Chemical Name: 1-pentene
CAS Registry No: 109-67-1
Molecular Formula: $C_5H_{10}$
Molecular Weight: 70.134
Melting Point (°C):

-165.219 (Dreisbach 1959; Riddick et al. 1986)
-165.00 (Stephenson & Malanowski 1987)

Boiling Point (°C):

29.962 (Riddick et al. 1986)
29.95 (Stephenson & Malanowski 1987)

Density ($g/cm^3$ at 20°C):

0.64050 (Dreisbach 1959)
0.63533 (Riddick et al. 1986)

Molar Volume ($cm^3/mol$):

110.39 (calculated-density)
109.5 (20°C, calculated-density, McAuliffe 1966)
109.4 (20°C, calculated-density, Stephenson & Malanowski 1987)
111.0 (calculated-LeBas method)

Molecular Volume, TMV ($Å^3$):
Total Surface Area, TSA ($Å^2$):

123.6 (Okouchi et al. 1992)

Heat of Fusion, $\Delta H_{fus}$, (kcal/mol):

1.338 (Riddick et al. 1986)

Entropy of Fusion, $\Delta S_{fus}$ (cal/mol K or e.u.):
Fugacity Ratio at 25°C, F: 1.0

Water Solubility ($g/m^3$ or mg/L at 25°C):

148 (shake flask-GC, McAuliffe 1966; quoted, Mackay & Shiu 1975,1981; Eastcott et al. 1988; Brookman et al. 1985)
147 (quoted, Hine & Mookerjee 1975)
150, 1062 (quoted, calculated-$K_{OW}$, Hansch et al. 1968; quoted, Müller & Klein 1992)
147, 255 (quoted, calculated-$K_{OW}$. Yalkowsky & Morozowich 1980)
148 (quoted, Riddick et al. 1986)
150, 134 (quoted, calculated-fragment solubility constants, Wakita et al. 1986)
150 (quoted, Müller & Klein 1992)
148 (calculated-AI, Okouchi et al. 1992)

Vapor Pressure (Pa at 25°C):

91418 (Antoine eqn. regression, Stull 1947)
83754 (24.6°C, Forziati et al. 1950)
86500 (Antoine eqn., Dreisbach 1959)
85000 (Antoine eqn., Zwolinski & Wilhoit 1971; quoted, Mackay & Shiu 1975,1981; Eastcott et al. 1988)
86500 (quoted, Hine & Mookerjee 1975)
85020 (interpolated, Antoine eqn., Boublik et al. 1984)
85200 (Antoine eqn., Dean 1985)
85100 (quoted, Riddick et al. 1986)
85040 (interpolated, Antoine eqn., Stephenson & Malanowski 1987)

Henry's Law Constant (Pa $m^3$/mol):

40327 (calculated-P/C, Mackay & Shiu 1975)
41138 (calculated as $1/K_{AW}$, $C_W/C_A$, reported as exptl., Hine & Mookerjee 1975; quoted, Nirmalakhandan & Speece 1988)
37518 (calculated-group contribution, Hine & Mookerjee 1975)
22607 (calculated-bond contribution, Hine & Mookerjee 1975)
40300 (calculated-P/C, Mackay & Shiu 1981)
40327 (selected, Mills et al. 1982)
41138, 37518 (quoted, calculated- $\chi$ , Nirmalakhandan & Speece 1988)
40405 (calculated-P/C, Eastcott et al. 1988)
40280 (calculated-P/C, Yaws et al. 1991)

Octanol/Water Partition Coefficient, log $K_{OW}$:

2.20 (calculated-$\pi$ substituent constant, Hansch et al. 1968)
2.69 (calculated-f const., Yalkowsky & Morozowich 1980)
2.20, 2.20 (quoted, calculated- $\chi$ , Murray et al. 1975, quoted, Mailhot & Peters 1988)
2.80 (selected, Müller & Klein 1992)

Bioconcentration Factor, log BCF:

Sorption Partition Coefficient, log $K_{OC}$:

Half-Lives in the Environment:

Air:

Surface water: half-lives for olefins in aquatic system by oxidation with OH and $RO_2$ radicals were estimated to be 320 hours and $9x10^4$ days; while 7.3 days based on rate constant of $3x10^5$ $M^{-1}$ $sec^{-1}$ for the oxidation of unsubstituted olefins with singlet oxygen in aquatic system (Mill & Mabey 1985).

Ground water:

Sediment:

Soil:

Biota:

Environmental Fate Rate Constant and Half-Lives:

Volatilization:

Photolysis:

Oxidation: photooxidation rate constant of $1.07x10^{-17}$ $cm^3$ $molecule^{-1}$ $sec^{-1}$ for the reation with ozone in air (Japar et al. 1974); rate constants of $5.3x10^{18}$, $7.4x10^{-18}$, and $1.07x10^{-17}$ $cm^3$ $molecule^{-1}$ $sec^{-1}$ for the reaction with ozone under atmospheric conditions (Atkinson & Carter 1984); rate constant of $3.14x10^{-11}$ $cm^3$ $molecule^{-1}$ $sec^{-1}$ for the reaction with hydroxyl radicals and $1.10x10^{-17}$ $cm^3$ $molecule^{-1}$ $sec^{-1}$ for the reaction with ozone in air phase (Atkinson 1990).

Hydrolysis:

Biodegradation:

Biotransformation:

Bioconcentration, Uptake ($k_1$) and Elimination ($k_2$) Rate Constants or Half-Lives:

Common Name: *cis*-2-Pentene
Synonym: *(Z)*-2-pentene
Chemical Name: *cis*-2-pentene
CAS Registry No: 627-20-3
Molecular Formula: $C_5H_{10}$
Molecular Weight: 70.134
Melting Point (°C): -151.4
Boiling Point (°C): 36.9
Density (g/cm$^3$ at 20°C):
    0.6556 (20°C, Dreisbach 1959)
    0.6504 (25°C, Dreisbach 1959)
Molar Volume (cm$^3$/mol):
    107.0-108.2 (20°C, calculated-density, for *cis-trans* forms, McAuliffe 1966)
    107.1 (20°C, Stephenson & Malanowski 1987)
    107.0 (20°C, calculated-density)
    107.8 (25$^0$C, calculated-density)
    111.0 (calculated-LeBas method)
Molecular Volume, TMV (Å$^3$):
Total Surface Area, TSA (Å$^2$):
    123.3 (Okouchi et al. 1992)
Heat of Fusion, $\Delta H_{fus}$, (kcal/mol):
    1.6998 (Riddick et al. 1986)
Entropy of Fusion, $\Delta S_{fus}$ (cal/mol K or e.u.):
Fugacity Ratio at 25°C, F: 1.0

Water Solubility (g/m$^3$ or mg/L at 25°C):
    203 (shake flask-GC, *cis-trans* form not specified, McAuliffe 1966)
    202, 1062 (quoted, calculated-$K_{OW}$, *cis-trans* form not specified, Hansch et al. 1968)
    203 (quoted, Mackay & Shiu 1981; Brookman et al. 1985)
    203 (calculated-AI, Okouchi et al. 1992)

Vapor Pressure (Pa at 25°C):
    65940 (Antoine eqn., Dreisbach 1959)
    66000 (Antoine eqn., Zwolinski & Wilhoit 1971; quoted, Mackay & Shiu 1981)
    65948 (Antoine eqn., Boublik et al. 1984)
    65970 (Antoine eqn., Stephenson & Malanowski 1987)

Henry's Law Constant (Pa m$^3$/mol):
    22800 (calculated-P/C, Mackay & Shiu 1981)
    22774 (calculated-P/C, Yaws et al. 1991)

Octanol/Water Partition Coefficient, log $K_{OW}$:

2.20 (calculated-$\pi$ substituent constant, Hansch et al. 1968)

2.20, 2.20 (quoted, calculated- $\chi$ , *cis-trans* form not specified, Murray et al. 1975)

2.20 (quoted, Mailhot & Peters 1988)

Bioconcentration Factor, log BCF:

Sorption Partition Coefficient, log $K_{OC}$:

Half-Lives in the Environment:

Air: photooxidation reaction rate constant of $6.50 \times 10^{-11}$ $cm^3$ molucule$^{-1}$ sec$^{-1}$ with hydroxyl radicals in air (Atkinson 1990, Altshuller 1991) with an estimated atmospheric lifetime of 2.29 hours in summer daylight (Altshuller 1991).

Surface water: half-lives for oxidation by OH and $RO_2$ radicals were estimated to be 320 hours and $9 \times 10^4$ days for olefins in aquatic system, and 7.3 days, based on rate constant of $3 \times 10^5$ $M^{-1}$ sec$^{-1}$ for oxidation of unsubstituted olefins by singlet oxygen in aquatic system (Mill & Mabey 1985).

Ground water:

Sediment:

Soil:

Biota:

Environmental Fate Rate Constant and Half-Lives:

Volatilization:

Photolysis:

Oxidation: room temperature rate constants of $1.80 \times 10^{-11}$ $cm^3$ molecule$^{-1}$ sec$^{-1}$ for the reaction with $O(^3P)$ (Herron & Huie 1973; quoted, Gaffney & Levine 1979) and $6.20 \times 10^{-11}$ $cm^3$ molecule$^{-1}$ sec$^{-1}$ for the reaction with hydroxyl radicals in air (Atkinson et al. 1979; quoted, Gaffney & Levine 1979); photooxidation reaction rate constant of $6.50 \times 10^{-11}$ $cm^3$ molecule$^{-1}$ sec$^{-1}$ with hydroxyl radicals in air (Atkinson 1990, Altshuller 1991) with an estimated atmospheric lifetime of 2.29 hours in summer daylight (Altshuller 1991).

Hydrolysis:

Biodegradation:

Biotransformation:

Bioconcentration, Uptake ($k_1$) and Elimination ($k_2$) Rate Constants or Half-Lives:

Common Name: 2-Methyl-1-pentene
Synonym:
Chemical Name: 2-methyl-1-pentene
CAS Registry No: 76-20-3
Molecular Formula: $C_6H_{12}$
Molecular Weight: 84.16
Melting Point (°C): -135.7
Boiling Point (°C): 60.7
Density ($g/cm^3$ at 20°C):
  0.6799 (20°C, Dreisbach 1959)
  0.6751 (25°C, Dreisbach 1959)
Molar Volume ($cm^3/mol$):
  123.4 (20°C, calculated-density, McAuliffe 1966)
  123.8 (20°C, calculated-density)
  124.7 (25°C, calculated-density)
  133.2 (calculated-LeBas method)
Molecular Volume, TMV ($Å^3$):
Total Surface Area, TSA ($Å^2$):
  157.0 (Okouchi et al. 1992)
Heat of Fusion, $\Delta H_{fus}$, (kcal/mol):
Entropy of Fusion, $\Delta S_{fus}$ (cal/mol K or e.u.):
Fugacity Ratio at 25°C, F: 1.0

Water Solubility ($g/m^3$ or mg/L at 25°C):
  78.0 (shake flask-GC, McAuliffe 1966; quoted, Mackay & Shiu 1981; Brookman et al. 1985)
  78.0, 73.2 (quoted, calculated-UNIFAC, Al-Sahhaf 1989)
  78.5, 41.3 (quoted, calculated-fragment solubility constants, Wakita et al. 1986)
  78.0 (calculated-AI, Okouchi et al. 1992)

Vapor Pressure (Pa at 25°C):
  27464 (Antoine eqn., Dreisbach 1959)
  26000 (Antoine eqn., Zwolinski & Wilhoit 1971; quoted, Mackay & Shiu 1981)
  26060 (Antoine eqn., Stephenson & Malanowski 1987)

Henry's Law Constant (Pa $m^3$/mol):
  28100 (calculated-P/C, Mackay & Shiu 1981)
  28093 (calculated-P/C, Yaws et al. 1991)

Octanol/Water Partition Coefficient, log $K_{OW}$:

Bioconcentration Factor, log BCF:

Sorption Partition Coefficient, log $K_{OC}$:

Half-Lives in the Environment:

Air:

Surface water: half-lives for oxidation by OH and $RO_2$ radicals were estimated to be 320 hours and $9x10^4$ days for olefins in aquatic system, and 8.0 days, based on rate constant of $1.0x10^6$ $M^{-1}$ $sec^{-1}$ for oxidation of substituted olefins with singlet oxygen in aquatic system (Mill & Mabey 1985).

Ground water:

Sediment:

Soil:

Biota:

Environmental Fate Rate Constant and Half-Lives:

Volatilization:

Photolysis:

Oxidation: photooxidation reaction rate constant of $1.05x10^{-17}$ $cm^3$ $molecule^{-1}$ $sec^{-1}$ with ozone under atmospheric conditions (Atkinson & Carter 1984); photooxidation reaction rate constant of $62.6x10^{-12}$ $cm^3$ $molecule^{-1}$ $sec^{-1}$ with hydroxyl radicals in air at 298 K (Atkinson 1985).

Hydrolysis:

Biodegradation:

Biotransformation:

Bioconcentration, Uptake ($k_1$) and Elimination ($k_2$) Rate Constants or Half-Lives:

Common Name: 4-Methyl-1-pentene
Synonym:
Chemical Name: 4-methyl-1-pentene
CAS Registry No: 691-37-2
Molecular Formula: $C_6H_{12}$
Molecular Weight: 84.16
Melting Point (°C): -153.6
Boiling Point (°C): 53.9
Density (g/cm$^3$ at 20°C):

0.6642 (20°C, Dreisbach 1059)
0.6594 (25°C, Dreisbach 1059)

Molar Volume (cm$^3$/mol):

126.7 (20°C, calculated-density, McAuliffe 1966)
126.7 (20°C, calculated-density)
127.6 (25°C, calculated-density)
133.2 (calculated-LeBas method)

Molecular Volume, TMV (Å$^3$):
Total Surface Area, TSA (Å$^2$):

147.0 (Okouchi et al. 1992)

Heat of Fusion, $\Delta H_{fus}$, (kcal/mol):
Entropy of Fusion, $\Delta S_{fus}$ (cal/mol K or e.u.):
Fugacity Ratio at 25°C, F: 1.0

Water Solubility (g/m$^3$ or mg/L at 25°C):

48.0 (shake flask-GC, McAuliffe 1966; quoted, Mackay & Shiu 1981; Brookman et al. 1985)
48.4, 550 (quoted, calculated-$K_{OW}$, Hansch et al. 1968)
48.4 (quoted, Hine & Mookerjee 1975)
48.4, 49.6 (quoted, calculated-fragment solubility constants, Wakita et al. 1986)
42.9 (calculated-UNIFAC, Al-Sahhaf 1989)
48.0 (calculated-AI, Okouchi et al. 1992)

Vapor Pressure (Pa at 25°C):

35600 (Antoine eqn., Dreisbach 1959)
36100 (Antoine eqn., Zwolinski & Wilhoit 1971)
35730 (quoted, Hine & Mookerjee 1975)
36110 (Antoine eqn., Stephenson & Malanowski 1987)

Henry's Law Constant (Pa $m^3$/mol):

62265 (calculated as $1/K_{AW}$, $C_W/C_A$, reported as exptl., Hine & Mookerjee 1975)
65200 (calculated-group contribution, Hine & Mookerjee 1975)
34217 (calculated-bond contribution, Hine & Mookerjee 1975)
63200 (calculated-P/C, Mackay & Shiu 1981)
63267 (calculated-P/C, Yaws et al. 1991)

Octanol/Water Partition Coefficient, log $K_{OW}$:

2.50 (calculated-$\pi$ constant, Hansch et al. 1968)
2.50, 2.51 (quoted, calculated- $\chi$ , Murray et al. 1975)

Bioconcentration Factor, log BCF:

Sorption Partition Coefficient, log $K_{OC}$:

Half-Lives in the Environment:

Air:
Surface water: half-lives for oxidation by OH and $RO_2$ radicals were estimated to be 320 hours and $9 \times 10^4$ days for olefins in aquatic system, and 8.0 days, based on rate constant of $1.0 \times 10^6$ $M^{-1}$ $sec^{-1}$ for oxidation of substituted olefins with singlet oxygen in aquatic system (Mill & Mabey 1985).
Ground water:
Sediment:
Soil:
Biota:

Environmental Fate Rate Constant and Half-Lives:

Volatilization:
Photolysis:
Oxidation: photooxidation rate constant of $1.06 \times 10^{-17}$ $cm^3$ $molecule^{-1}$ $sec^{-1}$ for the gas-phase reaction with $O_3$ in air (Atkinson & Carter 1984).
Hydrolysis:
Biodegradation:
Biotransformation:
Bioconcentration, Uptake ($k_1$) and Elimination ($k_2$) Rate Constants or Half-Lives:

Common Name: 1-Hexene
Synonym: α-hexene
Chemical Name: 1-hexene
CAS Registry No: 646-04-8
Molecular Formula: $C_6H_{12}$
Molecular Weight: 84.161
Melting Point (°C): -139.813
Boiling Point (°C): 63.478
Density ($g/cm^3$ at 20°C):
0.67317 (20°C, Dreisbach 1959; Riddick et al. 1986)
0.66848 (25°C, Dreisbach 1959; Riddick et al. 1986)
Molar Volume ($cm^3/mol$):
125.0 (20°C, calculated-density)
125.9 (25°C, calculated-density)
125.0 (20°C, calculated-density, McAuliffe 1966)
133.2 (calculated-LeBas method, Eastcott et al. 1988)
125.0 (calculated-density, Wang et al. 1992)
Molecular Volume, TMV ($Å^3$):
Total Surface Area, TSA ($Å^2$):
147.3 (Warne et al. 1990)
144.5 (Okouchi et al. 1992)
Heat of Fusion, $\Delta H_{fus}$, (kcal/mol):
2.2341 (Riddick et al. 1986)
Entropy of Fusion, $\Delta S_{fus}$ (cal/mol K or e.u.):
Fugacity Ratio at 25°C, F: 1.0

Water Solubility ($g/m^3$ or mg/L at 25°C):
50.0 (shake flask-GC, McAuliffe 1966; quoted, Mackay & Shiu 1975,1981; Eastcott et al. 1988; Brookman 1985)
49.6, 31.4 (quoted, calculated-$K_{OW}$, Hansch et al. 1968; quoted, Müller & Klein 1992)
55.0, 55.4 (quoted, shake flask-GC, Leinonen & Mackay 1973)
49.5 (quoted, Hine & Mookerjee 1975)
46.24, 88.12 (quoted, calculated-$K_{OW}$, Yalkowsky & Morozowich 1980)
69.7 (gen. col.-GC, Tewari et al. 1982a; quoted, Wasik et al. 1982)
69.2, 84.8 (quoted exptl., calculated- $\gamma$ & $K_{OW}$, Tewari et al. 1982b)
69.7 (quoted, Riddick et al. 1986)
49.6, 39.4 (quoted, calculated-fragment solubility constants, Wakita 1986)
50.0, 43.0 (quoted, calculated- $\chi$ , Al-Sahhaf 1989)
70.0 (quoted, Isnard & Lambert 1989)
60.1 (quoted literature average, Warne et al. 1990)
49.6 (quoted, Müller & Klein 1992)
49.6, 43.2 (quoted, calculated-$V_M$, Wang et al. 1992)

50.0 (calculated-AI, Okouchi et al. 1992)
49.6 (quoted, Müller & Klein 1992)

Vapor Pressure (Pa at 25°C):
22317 (Antoine eqn. regression, Stull 1947)
23485 (23.7°C, Forziati et al. 1950)
24960 (Antoine eqn., Dreisbach 1959)
24800 (Antoine eqn., Zwolinski & Wilhoit 1971; quoted, Mackay & Shiu 1975,1981; Eastcott et al. 1988)
24958 (quoted, Hine & Mookerjee 1975)
24795 (interpolated, Boublik et al. 1984)
24805 (Antoine eqn., Dean 1985)
24800 (quoted, Riddick et al. 1986)

Henry's Law Constant (Pa $m^3$/mol):
41746 (calculated-P/C, Mackay & Shiu 1975)
44080 (calculated-1/$K_{AW}$, $C_W/C_A$, reported as exptl., Hine & Mookerjee 1975; quoted, Nirmalakhandan & Speece 1988)
51790 (calculated-group contribution, Hine & Mookerjee 1975)
34217 (calculated-bond contribution, Hine & Mookerjee 1975)
41800 (calculated-P/C, Mackay & Shiu 1981)
44080, 47233 (quoted, calculated- $\chi$ , Nirmalakhandan & Speece 1988)
41644 (calculated-P/C, Eastcott et al. 1988)
29937 (calculated-P/C, Yaws et al. 1991)

Octanol/Water Partition Coefficient, log $K_{OW}$:
2.70 (calculated-$\pi$ substituent constant, Hansch et al. 1968)
2.70, 2.64 (quoted, calculated- $\chi$ , Murray et al. 1975)
2.70 (Hansch & Leo 1979)
3.33 (calculated-f const., Yalkowsky & Morozowich 1980)
3.39 (gen. col.-GC, Tewari et al. 1982a,b; quoted, Eastcott et al. 1988)
3.47, 3.47 (quoted exptl., calculated- $\gamma$ , Wasik et al. 1981)
3.39, 3.48 (quoted exptl., calculated- $\gamma$ , Wasik et al. 1982)
3.39, 3.40 (gen. col.-GC, calculated- $\gamma$ , Schantz & Martire 1987)
3.40 (recommended, Sangster 1989)
3.39 (quoted, Isnard & Lambert 1989)
2.78 (quoted literature average, Warne et al. 1990)
3.32 (selected, Müller & Klein 1992)
2.70, 2.57 (quoted, calculated-$V_M$, Wang et al. 1992)
3.32 (calculated-f const., Müller & Klein 1992)

Bioconcentration Factor, log BCF:

Sorption Partition Coefficient, log $K_{OC}$:

Half-Lives in the Environment:

Air:

Surface water: half-lives for oxidation by OH and $RO_2$ radicals were estimated to be 320 hours and $9x10^4$ days for olefins in aquatic system, and 7.3 days, based on rate constant of $3x10^5$ $M^{-1}$ $sec^{-1}$ for oxidation of unsubstituted olefins with singlet oxygen in aquatic system (Mill & Mabey 1985).

Ground water:

Sediment:

Soil:

Biota:

Environmental Fate Rate Constant and Half-Lives:

Volatilization:

Photolysis:

Oxidation: room temperature rate constants: $0.90x10^{-17}$ $cm^3$ $molecule^{-1}$ $sec^{-1}$ (Cadle & Schadt 1952; quoted, Adeniji et al. 1981), $1.00x10^{-17}$ $cm^3$ $molecule^{-1}$ $sec^{-1}$ (Hanst et al. 1958; quoted, Adeniji et al. 1981), $0.92x10^{-17}$ $cm^3$ $molecule^{-1}$ $sec^{-1}$ (Saltzmann & Gilbert 1959; quoted, Adeniji et al. 1981), $1.10x10^{-17}$ $cm^3$ $molecule^{-1}$ $sec^{-1}$ (Bufalini & Altshuller 1965; quoted, Adeniji et al. 1981), $1.40x10^{-17}$ $cm^3$ $molecule^{-1}$ $sec^{-1}$ (Cox & Penkett 1972; quoted, Adeniji et al. 1981), $1.10x10^{-17}$ $cm^3$ $molecule^{-1}$ $sec^{-1}$ (Stedman et al. 1973; quoted, Adeniji et al. 1981), $1.11x10^{-17}$ $cm^3$ $molecule^{-1}$ $sec^{-1}$ (Japar et al. 1974; quoted, Adeniji et al. 1981) and $1.08x10^{-17}$ $cm^3$ $molecule^{-1}$ $sec^{-1}$ at $(294 \pm 2)$ K (Adeniji et al. 1981) were all for the gas-phase reaction with ozone in air; rate constants of $9.1x10^{-18}$ to $1.36x10^{-17}$ $cm^3$ $molecule^{-1}$ $sec^{-1}$ for the reaction with $O_3$ in the gas phase at 294-303 K (Atkinson & Carter 1984); photooxidation reaction rate constant of $3.75x10^{-11}$ $cm^3$ $molecule^{-1}$ $sec^{-1}$ with OH radicals in air at 295 K (Atkinson & Aschmann 1984); rate constants of $(3.75\text{-}3.25)x10^{-11}$ $cm^3$ $molecule^{-1}$ $sec^{-1}$ for the reaction with hydroxyl radicals between 295 and 303 K (Atkinson 1985); rate constants of $3.18x10^{11}$ $cm^3$ $molecule^{-1}$ $sec^{-1}$ for the reaction with OH radicals and $1.17x10^{-17}$ $cm^3$ $molecule^{-1}$ $sec^{-1}$ for the reaction with $O_3$ in the gas phase at 298 K (Atkinson 1990).

Hydrolysis:

Biodegradation:

Biotransformation:

Bioconcentration, Uptake ($k_1$) and Elimination ($k_2$) Rate Constants or Half-Lives:

Common Name: 1-Heptene
Synonym: 1-heptylene, $\alpha$-heptene
Chemical Name: 1-heptene
CAS Registry No: 592-76-7
Molecular Formula: $C_7H_{14}$
Molecular Weight: 98.188
Melting Point (°C): -118.856
Boiling Point (°C): 93.639
Density (g/cm$^3$ at 20°C):

0.69698 (20°C, Dreisbach 1959)
0.69267 (25°C, Dreisbach 1959)

Molar Volume (cm$^3$/mol):

140.90 (20°C, calculated-density)
141.75 (25°C, calculated-density)
140.9 (20°C, calculated-density, Stephenson & Malanowski 1987)
155.4 (calculated-LeBas method)
139.0 (calculated-density, Wang et al. 1992)

Molecular Volume, TMV (Å$^3$):
Total Surface Area, TSA (Å$^2$):
Heat of Fusion, $\Delta H_{fus}$, (kcal/mol):

2.964 (Riddick et al. 1986)

Entropy of.Fusion, $\Delta S_{fus}$ (cal/mol K or e.u.):
Fugacity Ratio at 25°C, F: 1.0

Water Solubility (g/m$^3$ or mg/L at 25°C):

29.7, 31.0 (quoted, calculated-$K_{OW}$, 1- or 2- and *cis-trans* form not specified, Yalkowsky & Morozowich 1980)
18.16 (gen. col.-GC, Tewari et al. 1982a; quoted, Wasik et al. 1982)
18.18, 23.6 (quoted exptl., calculated- $\gamma$ & $K_{OW}$, Tewari et al. 1982b)
14.1 (estimated-nomograph, Brookman et al. 1985)
18.30 (quoted, Riddick et al. 1986; Isnard & Lambert 1989)
12.65, 15.6 (quoted, calculated-$V_M$, Wang et al. 1992)
18.0 (quoted, Müller & Klein 1992)

Vapor Pressure (Pa at 25°C):

7690 (25.5°C, Forziati et al. 1950)
7510 (Antoine eqn., Dreisbach 1959)
7510 (Antoine eqn., Zwolinski & Wilhoit 1971)
7517, 7533 (Antoine eqns., Boublik et al. 1984)
7515 (Antoine eqn., Dean 1985)

7500 (quoted, Riddick et al. 1986)
7500 (extrapolated, Antoine eqn., Stephenson & Malanowski 1987)

Henry's Law Constant (Pa $m^3$/mol):
40584 (calculated-P/C, Yaws et al. 1991)

Octanol/Water Partition Coefficient, log $K_{OW}$:
3.76 (calculated-f const., 1- or 2- and *cis-trans* form not specified, Yalkowsky & Morozowich 1980)
3.99 (gen. col.-concn. ratio-GC, Tewari et al. 1982a,b)
3.99, 4.09 (quoted exptl., calculated- $\gamma$ , Wasik et al. 1982)
3.99, 4.06 (quoted, gen. col.-GC, Schantz & Martire 1987)
3.99 (quoted, Isnard & Lambert 1989)
3.99 (recommended, Sangster 1989)
3.20 (calculated-regression eqn. from Lyman et al. 1982, Wang et al. 1992)
3.18 (calculated-$V_M$, Wang et al. 1992)
3.85 (calculated-f const., Müller & Klein 1992)

Bioconcentration Factor, log BCF:

Sorption Partition Coefficient, log $K_{OC}$:

Half-Lives in the Environment:
Air:
Surface water: half-lives for oxidation by OH and $RO_2$ radicals were estimated to be 320 hours and $9x10^4$ days for olefins in aquatic system, and 7.3 days, based on rate constant of $3x10^5$ $M^{-1}$ $sec^{-1}$ for oxidation of unsubstituted olefins with singlet oxygen in aquatic system (Mill & Mabey 1985).
Ground water:
Sediment:
Soil:
Biota:

Environmental Fate Rate Constant and Half-Lives:
Volatilization:
Photolysis:

Oxidation: photooxidation reaction rate constant of $4.05 \times 10^{-11}$ $cm^3$ $molecule^{-1}$ $sec^{-1}$ with hydroxyl radicals in air at 295 K (Atkinson & Aschmann 1984); photooxidation reaction rate constant of $4.0 \times 10^{-11}$ $cm^3$ $molecule^{-1}$ $sec^{-1}$ for the reaction with hydroxyl radicals and $1.73 \times 10^{-17}$ $cm^3$ $molecule^{-1}$ $sec^{-1}$ for the reaction with ozone in gas phase at 298 K (Atkinson 1990).

Hydrolysis:

Biodegradation:

Biotransformation:

Bioconcentration, Uptake ($k_1$) and Elimination ($k_2$) Rate Constants or Half-Lives:

Common Name: 1-Octene
Synonym: α-octene, caprylene, α-octylene
Chemical Name: 1-octene
CAS Registry No: 111-66-0
Molecular Formula: $C_8H_{16}$
Molecular Weight: 112.214
Melting Point (°C): -101.856
Boiling Point (°C): 121.286
Density (g/cm³ at 20°C):
  0.71492 (20°C, Dreisbach 1959; Riddick et al. 1986)
  0.71085 (25°C, Dreisbach 1959; Riddick et al. 1986)
Molar Volume (cm³/mol):
  157.0 (20°C, calculated-density)
  157.86 (25°c, calculated-density)
  157.0 (20°C, calculated-density, McAuliffe 1966)
  154.9 (20°C, calculated-density, Stephenson & Malanowski 1987)
  157 (calculated-density, Lande & Banerjee 1981; Wang et al. 1992)
  177.6 (calculated-LeBas method)
Molecular Volume, TMV (Å³):
Total Surface Area, TSA (Å²):
  191.5 (without consideration of solvent, Warne et al. 1990)
  186.3 (Okouchi et al. 1992)
Heat of Fusion, $\Delta H_{fus}$, (kcal/mol):
  3.721 (Riddick et al. 1986)
Entropy of Fusion, $\Delta S_{fus}$ (cal/mol K or e.u.):
Fugacity Ratio at 25°C, F: 1.0

Water Solubility (g/m³ or mg/L at 25°C):
  2.70 (shake flask-GC, McAuliffe 1966; quoted, Mackay & Shiu 1975,1981; Lande & Banerjee; Brookman et al. 1985)
  2.70, 25.6 (quoted, calculated-$K_{OW}$, Hansch et al. 1968)
  2.22 (shake flask-titration with bromine, Natarajan & Venkatachalam 1972)
  2.70 (quoted, Hine & Mookerjee 1975)
  4.10 (gen. col.-GC, Tewari et al. 1982a; quoted, Wasik et al. 1982)
  4.10, 6.82 (quoted exptl., calculated- $\gamma$ & $K_{OW}$, Tewari et al. 1982b)
  4.10 (quoted, Riddick et al. 1986)
  2.69, 3.16 (quoted, calculated-fragment solubility constants, Wakita et al. 1986)
  2.7, 3.60 (quoted, calculated-UNIFAC, Al-Sahhaf 1989)
  2.70, 2.22 (quoted, IUPAC 1989)
  4.07 (quoted, Isnard & Lambert 1989)
  3.07 (quoted literature average, Warne et al. 1990)
  2.70 (quoted, Müller & Klein 1992)

2.70 (calculated-AI, Okouchi et al. 1992)
2.7, 4.68 (quoted, calculated-$V_M$, Wang et al. 1992)
2.70 (quoted, Müller & Klein 1992)

Vapor Pressure (Pa at 25°C):
2317 (Antoine eqn., Driesbach 1959)
2320 (Antoine eqn., Zwolinski & Wilhoit 1971; quoted, Mackay & Shiu 1975,1981; Eastcott et al. 1988)
2317 (quoted, Hine & Mookerjee 1975)
2322 (interpolated, Antoine eqn., Dean 1985)
2300 (quoted, Riddick et al. 1986)
2317 (interpolated, Antoine eqn., Stephenson & Malanowski 1987)

Henry's Law Constant (Pa $m^3$/mol):
91700 (calculated-P/C, Mackay & Shiu 1975)
96437 (calculated-1/$K_{AW}$, $C_W/C_A$, reported as exptl., Hine & Mookerjee 1975; quoted, Nirmalakhandan & Speece 1988 )
100982 (calculated-group contribution, Hine & Mookerjee 1975)
74860 (calculated-bond contribution, Hine & Mookerjee 1975)
96400 (calculated-P/C, Mackay et al. 1981)
91699 (selected, Mills et al. 1982)
96426 (calculated-P/C, Eastcott et al. 1988)
96437, 74860 (quoted, calculated- $\chi$ , Nirmalakhandan & Speece 1988)
63480 (calculated-P/C, Yaws et al. 1991)

Octanol/Water Partition Coefficient, log $K_{OW}$:
3.70 (calculated-$\pi$ substituent constant, Hansch et al. 1968)
3.70 (Hansch & Leo 1979)
3.70, 3.52 (quoted, calculated-$\chi$ , Murray et al. 1975)
4.57 (gen. col.-GC, Tewari et al. 1982a; quoted, Eastcott et al. 1988)
4.88, 4.76 (quoted exptl., calculated- $\gamma$ , Wasik et al. 1981)
4.57, 4.76 (quoted exptl., calculated- $\gamma$ , Wasik et al. 1982)
4.56, 4.72 (gen. col.-GC, calculated- $\gamma$ , Schantz & Martire 1987)
4.57 (quoted, Isnard & Lambert 1989)
4.57 (recommended, Sangster 1989)
4.57 (quoted literature average, Warne et al. 1990)
4.38 (selected, Müller & Klein 1992)
3.70, 3.85 (quoted, calculated-$V_M$, Wang et al. 1992)
4.39 (calculated-f const., Müller & Klein 1992)

Bioconcentration Factor, log BCF:

Sorption Partition Coefficient, log $K_{OC}$:

Half-Lives in the Environment:

Air:

Surface water: half-lives for oxidation by OH and $RO_2$ radicals were estimated to be 320 hours and $9x10^4$ days for olefins in aquatic system, and 7.3 days, based on rate constant of $3x10^5$ $M^{-1}$ $sec^{-1}$ for oxidation of unsubstituted olefins with singlet oxygen in aquatic system (Mill & Mabey 1985).

Ground water:

Sediment:

Soil:

Biota:

Environmental Fate Rate Constant and Half-Lives:

Volatilization:

Photolysis:

Oxidation: photooxidation reaction rate constant of $8.1x10^{-18}$ $cm^3$ $molecule^{-1}$ $sec^{-1}$ for the reaction with ozone in air (Atkinson & Carter 1984); atmospheric rate constants of $4.0x10^{-11}$ $cm^3$ $molecule^{-1}$ $sec^{-1}$ for the reaction with hydroxyl radicals, $1.70x10^{-17}$ $cm^3$ $molecule^{-1}$ $sec^{-1}$ for the reaction with ozone and $1.10x10^{-11}$ $cm^3$ $molecule^{-1}$ $sec^{-1}$ for the reaction with $O(^3P)$ in gas phase (Paulson & Seinfeld 1992).

Hydrolysis:

Biodegradation:

Biotransformation:

Bioconcentration, Uptake ($k_1$) and Elimination ($k_2$) Rate Constants or Half-Lives:

Common Name: 1-Nonene
Synonym: $\alpha$-nonene, *n*-heptylethylene, 1-nonylene
Chemical Name: 1-nonene
CAS Registry No: 124-11-8
Molecular Formula: $C_9H_{18}$
Molecular Weight: 126.241
Melting Point (°C): -81.344
Boiling Point (°C): 146.883
Density ($g/cm^3$ at 20°C):
    0.72922 (20°C, Dreisbach 1959; Riddick et al. 1986)
    0.72351 (25°C, Dreisbach 1959; Riddick et al. 1986)
Molar Volume ($cm^3/mol$):
    173.11 (20°C, calculated-density)
    174.05 (25°C, calculated-density)
    174.1 (20°C, calculated-density, Stephenson & Malanowski 1987)
    170 (calculated-density, Wang et al. 1992)
    199.8 (calculated-LeBas method)
Molecular Volume, TMV ($Å^3$):
Total Surface Area, TSA ($Å^2$):
Heat of Fusion, $\Delta H_{fus}$, (kcal/mol):
    4.32 (Riddick et al. 1986)
Entropy of Fusion, $\Delta S_{fus}$ (cal/mol K or e.u.):
Fugacity Ratio at 25°C, F: 1.0

Water Solubility ($g/m^3$ or mg/L at 25°C):
    0.63 (estimated-nomograph, Brookman et al. 1986)
    1.12 (gen. col.-GC, Tewari et al. 1982a; quoted, Wasik et al. 1982)
    1.12, 2.09 (quoted exptl., calculated- $\gamma$ & $K_{OW}$, Tewari et al. 1982b)
    1.12 (quoted, Isnard & Lambert 1989)
    0.74, 1.87 (quoted exptl., calculated-$V_M$, Wang et al. 1992)
    1.12 (quoted, Müller & Klein 1992)

Vapor Pressure (Pa at 25°C):
    712 (Antoine eqn., Dreisbach 1959)
    712 (Antoine eqn., Zwolinski & Wilhoit 1971)
    710 (quoted, Riddick et al. 1986)
    712 (extrapolated, Antoine eqn., Stephenson & Malanowski 1987)

Henry's Law Constant (Pa $m^3/mol$):

80449 (calculated-P/C, Yaws et al. 1991)

Octanol/Water Partition Coefficient, log $K_{OW}$:

5.15 (gen. col.-GC, Tewari et al. 1982a,b)
5.35, 5.34 (quoted exptl., calculated- $\gamma$ , Wasik et al. 1981)
5.15, 5.34 (quoted exptl., calculated- $\gamma$ , Wasik et al. 1982)
5.15, 5.31 (quoted exptl., calculated- $\gamma$ , Schantz & Martire 1987)
5.15 (quoted, Isnard & Lambert 1989)
5.15 (recommended, Sangster 1989)
4.24 (calculated-regression eqn. from Lyman et al. 1982, Wang et al. 1992)
4.38 (calculated-$V_M$, Wang et al. 1992)
4.91 (calculated-f const., Müller & Klein 1992)

Bioconcentration Factor, log BCF:

Sorption Partition Coefficient, log $K_{OC}$:

Half-Lives in the Environment:

Air:
Surface water: half-lives for oxidation by OH and $RO_2$ radicals were estimated to be 320 hours and $9x10^4$ days for olefins in aquatic system, and 7.3 days, based on rate constant of $3x10^5$ $M^{-1}$ $sec^{-1}$ for oxidation of unsubstituted olefins with singlet oxygen in aquatic system (Mill & Mabey 1985).
Ground water:
Sediment:
Soil:
Biota:

Environmental Fate Rate Constant and Half-Lives:

Volatilization:
Photolysis:
Oxidation:
Hydrolysis:
Biodegradation:
Biotransformation:
Bioconcentration, Uptake ($k_1$) and Elimination ($k_2$) Rate Constants or Half-Lives:

Common Name: 1-Decene
Synonym: $\alpha$-decene
Chemical Name: 1-decene
CAS Registry No: 872-05-9
Molecular Formula: $C_{10}H_{20}$
Molecular Weight: 140.26
Melting Point (°C): -66.31
Boiling Point (°C): 170.57
Density ($g/cm^3$ at 20°C):
    0.74081 (20°C, Dreisbach 1959)
    0.73693 (25°C, Dreisbach 1959)
Molar Volume ($cm^3/mol$):
    189.33 (20°C, calculated-density)
    190.33 (25°C, calculated-density)
    189.30 (20°C, calculated-density, Stephenson & Malanowski 1987)
    190.0 (calculated-density, Wang et al. 1992)
    222.0 (calculated-LeBas method)
Molecular Volume, TMV ($Å^3$):
Total Surface Area, TSA ($Å^2$):
Heat of Fusion, $\Delta H_{fus}$, (kcal/mol):
Entropy of Fusion, $\Delta S_{fus}$ (cal/mol K or e.u.):
Fugacity Ratio at 25°C, F: 1.0

Water Solubility ($g/m^3$ or mg/L at 25°C):
    5.70 (shake flask-titration with bromine, Natarajan & Venkatachalam 1972; quoted, IUPAC 1989)
    0.161 (calculated-$K_{OW}$, Wang et al. 1992)
    0.433 (calculated-$V_M$, Wang et al. 1992)

Vapor Pressure (Pa at 25°C):
    218 (Antoine eqn., Dreisbach 1959)
    218 (Antoine eqn., Zwolinski & Wilhoit 1971)
    215 (Antoine eqn., Dean 1985)
    223 (Antoine eqn., extrapolated, Stephenson & Malanowski 1987)

Henry's Law Constant (Pa $m^3$/mol):

Octanol/Water Partition Coefficient, log $K_{OW}$:

4.78 (calculated-regression eqn. of Lyman et al. 1982, Wang et al. 1992)

5.18 (calculated-$V_M$, Wang et al. 1992)

Bioconcentration Factor, log BCF:

Sorption Partition Coefficient, log $K_{OC}$:

Half-Lives in the Environment:

Air:

Surface water: half-lives for oxidation by OH and $RO_2$ radicals were estimated to be 320 hours and $9x10^4$ days for olefins in aquatic system, and 7.3 days, based on rate constant of $3x10^5$ $M^{-1}$ $sec^{-1}$ for the oxidation of unsubstituted olefins with singlet oxygen in aquatic system (Mill & Mabey 1985).

Ground water:

Sediment:

Soil:

Biota:

Environmental Fate Rate Constant and Half-Lives:

Volatilization:

Photolysis:

Oxidation: photooxidation reaction rate constant of $1.08x10^{-17}$ $cm^3$ $molecule^{-1}$ $sec^{-1}$ for the reaction with ozone in the gas phase (Atkinson & Carter 1984).

Hydrolysis:

Biodegradation:

Biotransformation:

Bioconcentration, Uptake ($k_1$) and Elimination ($k_2$) Rate Constants or Half-Lives:

Common Name: 1,3-Butadiene
Synonym: $\alpha,\gamma$-butadiene, bivinyl, divinyl, erythrene, vinylethylene, biethylene, pyrrolylene
Chemical Name: 1,3-butadiene
CAS Registry No: 106-99-0
Molecular Formula: $C_4H_6$
Molecular Weight: 54.09
Melting Point (°C):
-108.9 (Weast 1984; Stephenson & Malanowski 1987; Ma et al. 1990)
Boiling Point (°C):
-4.41 (Dreisbach 1959; Stephenson & Malanowski 1987; Ma et al. 1990)
-4.40 (Weast 1984)
-4.50 (Howard 1989)
Density (g/cm$^3$ at 20°C):
0.6211 (20°C, at saturation pressure, Dreisbach 1959)
0.6149 (25°C, at saturation pressure, Dreisbach 1959)
0.6211 (Weast 1984)
Molar Volume (cm$^3$/mol):
87.07 (20°C, calculated-density)
88.0 (25°C, calculated-density)
81.4 (calculated-LeBas method)
87.1 (20°C, calculated-density, McAuliffe 1966)
87.0 (calculated-density, Wang et al. 1992)
Molecular Volume (A$^3$):
Total Surface Area, TSA (A$^2$):
99.3 (Okouchi et al. 1992)
Heat of Fusion, $\Delta H_{fus}$, (kcal/mol):
Entropy of Fusion, $\Delta S_{fus}$ (cal/mol K or e.u.):
Fugacity Ratio at 25°C, F: 1.0

Water Solubility (g/m$^3$ or mg/L at 25°C):
735 (shake flask-GC, McAuliffe 1966; quoted, Mackay & Shiu 1981; Verschueren 1983; Howard 1989; Müller & Klein 1992)
730 (quoted, Hine & Mookerjee 1975)
764, 984 (quoted, calculated-$K_{OW}$, Yalkowsky & Morozowich 1980)
730 (Lande & Banerjee 1981; quoted, Ma et al. 1990)
434 (calculated-UNIFAC, Banerjee & Howard 1988)
735, 728 (quoted, calculated-UNIFAC, Al-Sahhaf 1989)
730, 1239 (quoted, calculated-$V_M$, Wang et al. 1992)
735 (calculated-AI, Okouchi et al. 1992)
735 (quoted, Müller & Klein 1992)

Vapor Pressure (Pa at 25°C):

336200 (calculated-Antoine eqn. regression, Stull 1947)
280590 (Antoine eqn., Dreisbach 1959)
281000 (Antoine eqn., Zwolinski & Wilhoit 1971; quoted, Mackay & Shiu 1981; Ma et al. 1990)
247680 (Antoine eqn., Weast 1972-73)
281231, 310387 (Antoine equations, Boublik et al. 1984)
280988 (calculated-Antoine eqn., Stephenson & Malanowski 1987)

Henry's Law Constant (Pa $m^3$/mol):

7460 (calculated-P/C, Mackay & Shiu 1981)
6372 (calculated-$1/K_{AW}$, $C_W/C_A$, reported as exptl., Hine & Mookerjee 1975; quoted, Nirmalakhandan & Speece 1988)
7149 (calculated-group contribution, Hine & Mookerjee 1975)
6227 (calculated-bond contribution, Hine & Mookerjee 1975)
6372, 10820 (quoted, calculated- $\chi$ , Nirmalakhandan & Speece 1988)
6370 (quoted, Howard 1989)
7720 (calculated-P/C, Yaws et al. 1991)

Octanol/Water Partition Coefficient, log $K_{OW}$:

1.99 (shake flask-GC, Leo et al. 1975)
1.99; 1.87, 1.68, 1.90 (quoted; calculated-f const., Rekker 1977)
1.99 (Hansch & Leo 1979)
1.99 (quoted, Yalkowsky & Morozowich 1980)
1.99, 2.22 (quoted, calculated-UNIFAC, Banerjee & Howard 1988)
1.99 (quoted, Howard 1989)
1.99 (recommended, Sangster 1989)
1.90 (selected, Müller & Klein 1992)
1.63, 1.56 (quoted, calculated-$V_M$, Wang et al. 1992)
1.90 (calculated-f const., Müller & Klein 1992)

Bioconcentration Factor, log BCF:

1.28 (calculated-$K_{OW}$, Lyman et al. 1982; quoted, Howard 1989)

Sorption Partition Coefficient, log $K_{OC}$:

1.86-2.36 (soils and sediments, calculated-$K_{OW}$ & S, Lyman et al. 1982; quoted, Howard 1989)

Half-Lives in the Environment:

Air: photooxidation rate constant of $4.64 \times 10^{-8}$ $cm^3$ $mol^{-1}$ $sec^{-1}$ (Darnall et al. 1976; Lloyd et al. 1976) with an estimated half-life of 0.24-24 hours (Darnall et al 1976) for the reaction with hydroxyl radicals; photooxidation with OH radicals with an estimated half-life of 3.1 hours (Lyman et al. 1982; quoted, Howard 1989); completely degraded within 6 hours in a smog chamber irradiated by sunlight (Kopcynski et al 1972; quoted, Howard 1989); half-life in air for the reaction with nitrate radicals is 15 hours (Atkinson et al. 1984; quoted, Howard 1989); 0.76-7.8 hours, based on measured photooxidation rate constants in air (Howard et al. 1991).

Surface water: volatilizes rapidly with a half-life estimated to be several hours (Howard 1989); estimated half-life to be 3.8 hours for evaporation from a model river 1 m deep with a 1 m/s current and a 3 m/s wind (Lyman et al. 1982; quoted, Howard 1989); half-lives for oxidation by OH and $RO_2$ radicals were estimated to be 320 hours and $9 \times 10^4$ days for olefins in aquatic system, and 19 hours, based on rate constant of $1.0 \times 10^7$ $M^{-1}$ $sec^{-1}$ for oxidation of dienes with singlet oxygen in aquatic system (Mill & Mabey 1985); 168-672 hours, based on estimated aqueous aerobic biodegradation half-lives (Howard et al. 1991).

Ground water: 336-1344 hours, based on estimated aqueous aerobic biodegradation half-lives (Howard et al. 1991).

Sediment:

Soil: 168-672 hours, based on estimated aqueous aerobic biodegradation half-lives (Howard et al. 1991).

Biota:

Environmental Fate Rate Constant and Half-Lives:

Volatilization: volatilizes rapidly from water and land (Howard 1989).

Photolysis:

Oxidation: photooxidation rate constant of $8.4 \times 10^{-18}$ $cm^3$ $molecule^{-1}$ $sec^{-1}$ for the gas phase reaction with ozone at 298 K (Japar et al. 1974); rate constant of $4.64 \times 10^{-8}$ $cm^3$ $mol^{-1}$ $sec^{-1}$ for the reaction with OH radicals with half-life of 0.24 to 24 hours (Darnall et al. 1976); room temp. rate constants of $1.9 \times 10^{-11}$ $cm^3$ $molecule^{-1}$ $sec^{-1}$ for the gas phase reaction with $O(^3P)$ (Atkinson & Pitts 1977; quoted, Gaffney & Levine 1979) and $6.9 \times 10^{-11}$ $cm^3$ $molecule^{-1}$ $sec^{-1}$ for the gas phase reaction with OH radicals in air (Atkinson et al. 1979; quoted, Gaffney & Levine 1979); rate constants: $8.1 \times 10^{-18}$, $8.4 \times 10^{-18}$, $6.7 \times 10^{-18}$ $cm^3$ $molecule^{-1}$ $sec^{-1}$ for the photooxidation with $O_3$ (Atkinson & Carter 1984); rate constant of $7.7 \times 10^{-9}$ $cm^3$ $molecule^{-1}$ $sec^{-1}$ for the reaction with hydroxyl radicals with a half-life of 3.1 hours in air (Howard 1989); half-life of 1200 to 48000 hours, based on measured photooxidation rate constants with OH radicals in water (Güsten et al. 1981; quoted, Howard et al. 1991); rate constant of $2.1 \times 10^{-13}$ $cm^3$ $molecule^{-1}$ $sec^{-1}$ for the reaction with $NO_3$ radicals (Benter &

Schindler 1988); photooxidation rate constants of $6.7x10^{11}$ $cm^3$ $molecule^{-1}$ $sec^{-1}$ for the reaction with OH radicals and $9.80x10^{-14}$ $cm^3$ $molecule^{-1}$ $sec^{-1}$ for the reaction with $NO_3$ radicals (Sabljic & Gusten 1990); photooxdation rate constants of $6.66x10^{-12}$ $cm^3$ $molecule^{-1}$ $sec^{-1}$ for the reaction with OH radicals and $7.50x10^{-18}$ $cm^3$ $molecule^{-1}$ $sec^{-1}$ for the reaction with $NO_3$ radicals (Atkinson 1990).

Hydrolysis: will hydrolyze appreciably (Howard 1989).

Biodegradation:

Biotransformation:

Bioconcentration, Uptake ($k_1$) and Elimination ($k_2$) Rate Constants or Half-Lives:

Common Name: 2-Methyl-1,3-butadiene
Synonym: isoprene
Chemical Name: 2-methyl-1,3-butadiene
CAS Registry No: 78-79-5
Molecular Formula: $C_5H_8$
Molecular Weight: 68.13
Melting Point (°C): -146
Boiling Point (°C): 34
Density (g/cm$^3$ at 20°C):

0.68092 (20°C, Dreisbach 1959)
0.67587 (25°C, Dreisbach 1959)

Molar Volume (cm$^3$/mol):

100.05 (20°C, calculated-density)
100.8 (25°C, calculated-density)
100.0 (20°C, calculated-density, McAuliffe 1966)
100.0 (calculated-density, Wang et al. 1992)
103.6 (calculated-LeBas method)

Molecular Volume, TMV (Å$^3$):
Total Surface Area, TSA (Å$^2$):

118.3 (Okouchi et al. 1986)

Heat of Fusion, $\Delta H_{fus}$, (kcal/mol):
Entropy of Fusion, $\Delta S_{fus}$ (cal/mol K or e.u.):
Fugacity Ratio at 25°C, F: 1.0

Water Solubility (g/m$^3$ or mg/L at 25°C):

642 (shake flask-GC, McAuliffe 1966; quoted, Mackay & Shiu 1981; Brookman et al. 1985)
545 (20°C, shake flask-GC, Pavlova et al. 1966)
636 (quoted, Hine & Mookerjee 1975)
636, 607 (quoted, calculated-fragment solubility constants, Wakita et al. 1986)
610 (recommended best value, IUPAC 1989)
642, 680 (quoted, calculated-UNIFAC, Al-Sahhaf 1989)
636, 517 (quoted, calculated-$V_M$, Wang et al. 1992)
642 (calculated-AI, Okouchi et al. 1992)

Vapor Pressure (Pa at 25°C):

73330 (Antoine eqn., Dreisbach 1959)
73300 (Antoine eqn., Zwolinski & Wilhoit 1971, quoted, Mackay & Shiu 1981)
73330 (quoted, Hine & Mookerjee 1975)
78383, 72925 (interpolated, Antoine eqn., Boublik et al. 1984)
73327, 10772 (quoted, calculated-UNIFAC, Banerjee et al. 1990)

Henry's Law Constant (Pa $m^3$/mol):

7840 (calculated-1/$K_{AW}$, $C_W/C_A$, reported as exptl., Hine & Mookerjee 1975; quoted Nirmalakhandan & Speece 1988)

6227 (calculated-group contribution, Hine & Mookerjee 1975)

6520 (calculated-bond contribution, Hine & Mookerjee 1975)

7780 (calculated-P/C, Mackay & Shiu 1981)

7840, 14936 (quoted, calculated- $\chi$ , Nirmalakhandan & Speece 1988)

7777 (calculated-P/C, Yaws et al. 1991)

Octanol/Water Partition Coefficient, log $K_{OW}$:

2.05 (calculated-regression eqn. from Lyman et al. 1982, Wang et al. 1992)

1.91 (calculated-$V_M$, Wang et al. 1992)

Bioconcentration Factor, log BCF:

Sorption Partition Coefficient, log $K_{OC}$:

Half-Lives in the Environment:

Air: atmospheric lifetime of 1.4-3.6 hours, based on photooxidation rate constant of $9.60x10^{-11}$ $cm^3$ $molecule^{-1}$ $sec^{-1}$ with OH radicals, and lifetime of 10-32 hours, based on photooxidation rate constant of $1.20x10^{-17}$ $cm^3$ $molecule^{-1}$ $sec^{-1}$ with ozone and lifetime of 22 minutes to 3.6 hours, based on photooxidation reaction rate constant of $3.23x10^{-13}$ $cm^3$ $molecule^{-1}$ $sec^{-1}$ with $NO_3$ radicals in the gas phase (Atkinson et al. 1984); atmospheric lifetimes are calculated to be 28.3 hours for the reaction with $O_3$, 2.9 hours with OH radicals and 0.083 hour with $NO_3$ radicals, all based on the reaction rate constants with $O_3$, OH and $NO_3$ radicals in the gas phase (Atkinson & Carter 1984).

Surface water: half-lives for oxidation by OH and $RO_2$ radicals were estimated to be 320 hours and $9x10^4$ days for olefins in aquatic system, and 19 hours, based on rate constant of $1.0x10^7$ $M^{-1}$ $sec^{-1}$ for oxidation of dienes with singlet oxygen in aquatic system (Mill & Mabey 1985).

Ground water:

Sediment:

Soil:

Biota:

Environmental Fate Rate Constant and Half-Lives:

Volatilization:

Photolysis:

Oxidation: rate constant of $16.5 \times 10^{-18}$ $cm^3$ $molecule^{-1}$ $sec^{-1}$ for the reaction with ozone at (294 ± 2) K in air under atmospheric conditions (Adeniji et al. 1981); photooxidation rate constant of $5.80 \times 10^{-18}$ to $1.25 \times 10^{-17}$ $cm^3$ $molecule^{-1}$ $sec^{-1}$ for the reaction with ozone in air between 260-296 K (Atkinson & Carter 1984); rate constants of $9.98 \times 10^{-11}$ $cm^3$ $molecule^{-1}$ $sec^{-1}$, $9.26 \times 10^{-11}$ $cm^3$ $molecule^{-1}$ $sec^{-1}$ for the reaction with OH radicals in gas phase at 299 K (Atkinson 1985); rate constant of $(101 \pm 2) \times 10^{-12}$ $cm^3$ $molecule^{-1}$ $sec^{-1}$ with reference to propylene for the reaction with OH radicals at (23.7 ± 0.5)°C (Edney et al. 1986); rate constant of $1.3 \times 10^{-12}$ $cm^3$ $molecule^{-1}$ $sec^{-1}$ for the reaction with $NO_3$ radicals in air (Benter & Schindler 1988); rate constants of $5.91 \times 10^{-13}$ $cm^3$ $molecule^{-1}$ $sec^{-1}$ for the reaction with OH radicals and $1.01 \times 10^{-10}$ $cm^3$ $molecule^{-1}$ $sec^{-1}$ for the reaction with $NO_3$ radicals in the gas phase at 298 K (Sabljic & Güsten 1990); rate constants of $1.10 \times 10^{-10}$ $cm^3$ $molecule^{-1}$ $sec^{-1}$ for the reaction with OH radicals and $1.43 \times 10^{-17}$ $cm^3$ $molecule^{-1}$ $sec^{-1}$ for the reaction with $O_3$ in air at 298 K (Atkinson 1990); rate constant of $5.94 \times 10^{-13}$ $cm^3$ $molecule^{-1}$ $sec^{-1}$ for the reaction with OH radicals at 298 K, and $6.52 \times 10^{-13}$ $cm^3$ $molecule^{-1}$ $sec^{-1}$ for the reaction with $NO_3$ radicals at 297 K (Atkinson 1991).

Hydrolysis:

Biodegradation:

Biotransformation:

Bioconcentration, Uptake ($k_1$) and Elimination ($k_2$) Rate Constants or Half-Lives:

Common Name: 2,3-Dimethyl-1,3-butadiene
Synonym:
Chemical Name: 2,3-dimethyl-1,3-butadiene
CAS Registry No: 513-82-5
Molecular Formula: $C_6H_{10}$
Molecular Weight: 82.15
Melting Point (°C): -76
Boiling Point (°C): 69
Density (g/cm$^3$ at 20°C):
  0.7267 (20°C, Dreisbach 1959)
  0.7222 (25°C, Dreisbach 1959)
Molar Volume (cm$^3$/mol):
  113.5 (20°C, calculated-density)
  113.75 (25°C, calculated-density)
  113.7 (calculated-density, Stephenson & Malanowski 1987)
  125.8 (calculated-LeBas method)
Molecular Volume (A$^3$):
Total Surface Area, TSA (A$^2$):
Heat of Fusion, $\Delta H_{fus}$, (kcal/mol):
Entropy of Fusion, $\Delta S_{fus}$ (cal/mol K or e.u.):
Fugacity Ratio at 25°C, F: 1.0

Water Solubility (g/m$^3$ or mg/L at 25°C):
  327 (quoted, Hine & Mookerjee 1975)
  327, 226 (quoted, calculated-fragment solubility constants, Wakita et al. 1986)

Vapor Pressure (Pa at 25°C):
  19177 (Antoine eqn., Dreisbach 1959)
  19172 (quoted, Hine & Mookerjee 1975)
  20147 (interpolated, Stephenson & Malanowski 1987)

Henry's Law Constant (Pa m$^3$/mol):
  4833 (calculated-1/$K_{AW}$, $C_W/C_A$, reported as exptl., Hine & Mookerjee 1975)
  5423 (calculated-group contribution, Hine & Mookerjee 1975)
  6986 (calculated-bond contribution, Hine & Mookerjee 1975)
  4833, 21098 (quoted, calculated- $\chi$ , Nirmalakhandan & Speece 1988)

Octanol/Water Partition Coefficient, log $K_{OW}$:

Bioconcentration Factor, log BCF:

Sorption Partition Coefficient, log $K_{OC}$:

Half-Lives in the Environment:

Air:

Surface water: half-lives for oxidation by OH and $RO_2$ radicals were estimated to be 320 hours and $9x10^4$ days for olefins in aquatic system, and 19 hours, based on rate constant of $1.0x10^7$ $M^{-1}$ $sec^{-1}$ for oxidation of dienes by singlet oxygen in aquatic system (Mill & Mabey 1985).

Ground water:

Sediment:

Soil:

Biota:

Environmental Fate Rate Constant and Half-Lives:

Volatilization:

Photolysis:

Oxidation: photooxidation rate constant of $1.22x10^{-10}$ $cm^3$ $molecule^{-1}$ $sec^{-1}$ for the reaction in gas phase with hydroxyl radicals at 297 K (Atkinson 1985); rate constant of $2.3x10^{-12}$ $cm^3$ $molecule^{-1}$ $sec^{-1}$ for the reaction with $NO_3$ radicals in air (Benter & Shindler 1988); rate constants of $1.22x10^{-10}$ $cm^3$ $molecule^{-1}$ $sec^{-1}$ for the reaction with OH radicals and $1.052x10^{-12}$ $cm^3$ $molecule^{-1}$ $sec^{-1}$ for the reaction with $NO_3$ radicals in the gas phase at 298 K (Sabljic & Güsten 1990); rate constants were reported to be $2.3x10^{-12}$ $cm^3$ $molecule^{-1}$ $sec^{-1}$ and $1.96x10^{-12}$ $cm^3$ $molecule^{-1}$ $sec^{-1}$ for the reaction in the gas phase with $NO_3$ radicals at 298 K (Atkinson 1991).

Hydrolysis:

Biodegradation:

Biotransformation:

Bioconcentration, Uptake ($k_1$) and Elimination ($k_2$) Rate Constants or Half-Lives:

Common Name: 1,4-Pentadiene
Synonym:
Chemical Name: 1,4-pentadiene
CAS Registry No: 591-93-5
Molecular Formula: $C_5H_8$
Molecular Weight: 68.13
Melting Point (°C): -148.3
Boiling Point (°C): 26
Density (g/cm³ at 20°C):
0.66076 (20°C, Dreisbach 1959)
0.65571 (25°C, Dreisbach 1959)
Molar Volume (cm³/mol):
102.01 (20°C, calculated-density)
103.9 (25°C, calculated-density)
103.6 (calculated-LeBas method)
103.1 (20°C, calculated-density, McAuliffe 1966)
98.0 (20°C, calculated-density, Stephenson & Malanowski 1987)
103.6 (calculated-density, Wang et al. 1992)
Molecular Volume (A³):
Total Surface Area, TSA (A²):
117.7 (Okouchi et al. 1992)
Heat of Fusion, $\Delta H_{fus}$, (kcal/mol):
Entropy of Fusion, $\Delta S_{fus}$ (cal/mol K or e.u.):
Fugacity Ratio at 25°C, F: 1.0

Water Solubility (g/m³ or mg/L at 25°C):
558 (shake flask-GC, McAuliffe 1966; quoted, Mackay & Shiu 1981; Brookman et al. 1985; IUPAC 1989)
567, 2384 (quoted, calculated-$K_{OW}$, Hansch et al. 1968; quoted Müller & Klein 1992)
554 (quoted, Hine & Mookerjee 1975)
1525, 4012 (quoted, calculated-$K_{OW}$, Yalkowsky & Morozowich 1980)
554, 6972 (quoted, calculated-$K_{OW}$, Valvani et al. 1981)
567, 482 (quoted, calculated-fragment solubility constants, Wakita et al. 1986)
558, 524 (quoted, calculated-UNIFAC, Al-Sahhaf 1989)
553.8 (quoted, Isnard & Lambert 1989)
554, 410.5 (quoted, calculated-$V_M$, Wang et al. 1992)
558 (calculated-AI, Okouchi et al. 1992)
567 (quoted, Müller & Klein 1992)

Vapor Pressure (Pa at 25°C):
105142 (Antoine eqn. regression, Stull 1947)
97860 (Antoine eqn., Dreisbach 1959)

98000 (Antoine eqn., Zwolinski & Wilhoit 1971; quoted, Mackay & Shiu 1981)
97900 (quoted, Hine & Mookerjee 1975)
99384, 98286 (Antoine equtations, Boublik et al. 1984)
97934 (interpolated, Antoine eqn., Stephenson & Malanowski 1987)

Henry's Law Constant (Pa $m^3$/mol):
12140 (calculated1/$K_{AW}$, $C_W/C_A$, reported as exptl., Hine & Mookerjee 1975; quoted, Nirmalakhandan & Speece 1988)
15640 (calculated-bond contribution, Hine & Mookerjee 1975)
12000 (calculated-P/C, Mackay & Shiu 1981)
12140, 13622 (quoted, calculated- $\chi$ , Nirmalakhandan & Speece 1988)
11944 (calculated-P/C, Yaws et al. 1991)

Octanol/Water Partition Coefficient, log $K_{OW}$:
1.90 (calculated-$\pi$ substituent constant, Hansch et al. 1968)
1.90, 1.85 (quoted, calculated- $\chi$ , Murray et al. 1975)
1.48 (quoted, Isnard & Lambert 1989)
2.48 (recommended, Sangster 1989)
2.25 (selected, Müller & Klein 1992)
1.90, 1.99 (quoted, calculated-$V_M$, Wang et al. 1992)
2.25 (calculated-f const., Müller & Klein 1992)

Bioconcentration Factor, log BCF:

Sorption Partition Coefficient, log $K_{OC}$:

Half-Lives in the Environment:
Air:
Surface water: half-lives for oxidation by OH and $RO_2$ radicals were estimated to be 320 hours and $9x10^4$ days for olefins in aquatic system, and 19 hours, based on rate constant of $1.0x10^7$ $M^{-1}$ $sec^{-1}$ for oxidation of dienes with singlet oxygen in aquatic system (Mill & Mabey 1985).
Ground water:
Sediment:
Soil:
Biota:

Environmental Fate Rate Constant and Half-Lives:

Volatilization:

Photolysis:

Oxidation: photooxidation rate constant of $5.33 \times 10^{-11}$ $cm^3$ $molecule^{-1}$ $sec^{-1}$ for the gas phase reaction of hydroxyl radicals at 297 K (Atkinson 1985).

Hydrolysis:

Biodegradation:

Biotransformation:

Bioconcentration, Uptake ($k_1$) and Elimination ($k_2$) Rate Constants or Half-Lives:

Common Name: 1,5-Hexadiene
Synonym:
Chemical Name: 1,5-Hexadiene
CAS Registry No: 592-42-7
Molecular Formula: $C_6H_{10}$
Molecular Weight: 82.15
Melting Point (°C): -141
Boiling Point (°C): 60
Density (g/cm$^3$ at 20°C):
0.692 (Weast 1984)
0.6923 (20°C, Dreisbach 1959)
0.6878 (25°C, Dreisbach 1959)
Molar Volume (cm$^3$/mol):
118.7 (20°C, calculated-density)
119.4 (25°C, calculated-density)
118.7 (20°C, calculated-density, McAuliffe 1966)
119.4 (20°C, calculated-density, Stephenson & Malanowski 1987)
119.0 (calculated-density, Wang et al. 1992)
125.8 (calculated-LeBas method)
Molecular Volume (A$^3$):
Total Surface Area, TSA (A$^2$):
249.6 (Iwase et al. 1985)
138.6 (Okouchi et al. 1992)
Heat of Fusion, $\Delta H_{fus}$, (kcal/mol):
Entropy of Fusion, $\Delta S_{fus}$ (cal/mol K or e.u.):
Fugacity Ratio at 25°C, F: 1.0

Water Solubility (g/m$^3$ or mg/L at 25°C):
169 (shake flask-GC, McAuliffe 1966)
168, 711 (quoted, calculated-$K_{OW}$, Hansch et al. 1968; quoted, Müller & Klein 1992)
168 (quoted, Hine & Mookerjee 1975)
168, 51.8 (quoted, calculated-$K_{OW}$, Yalkowsky & Morozowich 1980)
168, 157 (quoted, calculated-fragment solubility constants, Wakita et al. 1986)
169, 138 (quoted, calculated-UNIFAC, Al-Sahhaf 1989)
168, 150 (quoted, calculated-$V_M$, Wang et al. 1992)
164 (calculated-AI, Okouchi et al. 1992)
168 (quoted, Müller & Klein 1992)

Vapor Pressure (Pa at 25°C):
27732 (Antoine eqn., Dreisbach 1959)
27732 (quoted, Hine & Mookerjee 1975)

29670 (interpolated, Antoine eqn., Stephenson & Malanowski 1987)
29686 (extrapolated, Antoine eqn., Stephenson & Malanowski 1987)

Henry's Law Constant (Pa $m^3$/mol):
13622 (calculated-1/$K_{AW}$, $C_W/C_A$, reported as exptl., Hine & Mookerjee 1975; quoted, Nirmalakhandan & Speece 1988)
17550 (calculated-group contribution, Hine & Mookerjee 1975)
23134 (calculated-bond contribution, Hine & Mookerjee 1975)
13622, 17150 (quoted, calculated-$\chi$, Nirmalakhandan & Speece 1988)

Octanol/Water Partition Coefficient, log $K_{OW}$:
2.40 (calculated-$\pi$ substituent constant, Hansch et al. 1968)
2.40, 2.29 (quoted, calculated-$\chi$, Murray et al. 1975)
2.45 (calculated-f const., Yalkowsky & Morozowich 1980)
2.45, 2.68 (quoted, calculated-hydrophobicity const., Iwase et al. 1985)
2.80 (recommended, Sangster 1989)
2.78 (calculated-f const., Müller & Klein 1992)
2.40, 2.43 (quoted, calculated-$V_M$, Wang et al. 1992)

Bioconcentration Factor, log BCF:

Sorption Partition Coefficient, log $K_{OC}$:

Half-Lives in the Environment:
Air:
Surface water: half-lives for oxidation by OH and $RO_2$ radicals were estimated to be 320 hours and $9 \times 10^4$ days for olefins in aquatic system, and 19 hours, based on rate constant of $1.0 \times 10^7$ $M^{-1}$ $sec^{-1}$ for oxidation of dienes by singlet oxygen in aquatic system (Mill & Mabey 1985).
Ground water:
Sediment:
Soil:
Biota:

Environmental Fate Rate Constant and Half-Lives:
Volatilization:
Photolysis:

Oxidation: photooxidation reaction rate constants of $6.25 \times 10^{-11}$ $cm^3$ $molecule^{-1}$ $sec^{-1}$ and $6.17 \times 10^{-11}$ $cm^3$ $molecule^{-1}$ $sec^{-1}$ both for the reaction with hydroxyl radicals in gas phase at 297 K (Atknson 1985).

Hydrolysis:

Biodegradation:

Biotransformation:

Bioconcentration, Uptake ($k_1$) and Elimination ($k_2$) Rate Constants or Half-Lives:

Common Name: 1,6-Heptadiene
Synonym:
Chemical Name: 1,6-heptadiene
CAS Registry No: 3070-53-9
Molecular Formula: $C_7C_{12}$
Molecular Weight: 96.17
Melting Point (°C): -129
Boiling Point (°C): 89-90
Density (g/cm³ at 20°C): 0.714
Molar Volume (cm³/mol):

134.0 (20°C, calculated-density, McAuliffe 1966)
134.0 (calculated-density, Wang et al. 1992)
148.0 (calculated-LeBas method)

Molecular Volume (A³):
Total Surface Area, TSA (A²):
Heat of Fusion, $\Delta H_{fus}$, (kcal/mol):
Entropy of Fusion, $\Delta S_{fus}$ (cal/mol K or e.u.):
Fugacity Ratio at 25°C, F: 1.0

Water Solubility (g/m³ or mg/L at 25°C):

44.0 (shake flask-GC, McAuliffe 1966)
44, 206 (quoted, calculated-$K_{OW}$, Hansch et al. 1968; quoted, Müller & Klein 1992)
44, 41 (quoted, calculated-fragment solubility constants, Wakita et al. 1986)
44, 59 (quoted, calculated-UNIFAC, Al-Sahhaf 1989)
44.0 (quoted, Müller & Klein 1992)
44, 56.6 (quoted, calculated-$V_M$, Wang et al. 1992)

Vapor Pressure (Pa at 25°C):

Henry's Law Constant (Pa m³/mol):

Octanol/Water Partition Coefficient, log $K_{OW}$:

2.90 (calculated-$\pi$ substituent constants, Hansch et al. 1968)
2.90, 2.73 (quoted, calculated-$\chi$, Murray et al. 1975)
3.31 (calculated-f const., Müller & Klein 1992)
2.90, 2.85 (quoted exptl., calculated-$V_M$, Wang et al. 1992)

Bioconcentration Factor, log BCF:

Sorption Partition Coefficient, log $K_{OC}$:

Half-Lives in the Environment:

Air:

Surface water: half-lives for oxidation by OH and $RO_2$ radicals were estimated to be 320 hours and $9x10^4$ days for olefins in aquatic system, and 19 hours, based on rate constant of $1.0x10^7$ $M^{-1}$ $sec^{-1}$ for oxidation of dienes by singlet oxygen in aquatic system (Mill & Mabey 1985).

Ground water:

Sediment:

Soil:

Biota:

Environmental Fate Rate Constant and Half-Lives:

Volatilization:

Photolysis:

Oxidation:

Hydrolysis:

Biodegradation:

Biotransformation:

Bioconcentration, Uptake ($k_1$) and Elimination ($k_2$) Rate Constants or Half-Lives:

Common Name: 1-Butyne
Synonym: ethyl acetylene
Chemical Name: 1-butyne
CAS Registry No: 107-00-6
Molecular Formula: $C_4H_6$, $CH_3CH_2C{\equiv}CH$
Molecular Weight: 54.09
Melting Point (°C): -125.7
Boiling Point (°C): 8.1
Density (g/cm³ at 20°C):
0.650 (20°C, at saturation pressure, Dreisbach 1959)
0.650 (25°C, at saturation pressure, Dreisbach 1959)
Molar Volume (cm³/mol):
83.0 (20°C, calculated-density, McAuliffe 1966; Wang et al. 1992)
83.21 (calculated-density)
81.4 (calculated-LeBas method)
Molecular Volume, TMV (Å³):
Total Surface Area, TSA (Å²):
96.5 (Okouchi et al. 1992)
Heat of Fusion, $\Delta H_{fus}$, (kcal/mol):
Entropy of Fusion, $\Delta S_{fus}$ (cal/mol K or e.u.):
Fugacity Ratio at 25°C, F: 1.0

Water Solubility (g/m³ or mg/L at 25°C):
2870 (shake flask-GC, McAuliffe 1966)
2900 (quoted, Hine & Mookerjee 1975)
2630, 4166 (quoted, calculated-$V_M$, Wang et al. 1992)
2870 (calculated-AI, Okouchi et al. 1992)

Vapor Pressure (Pa at 25°C):
188000 (Antoine eqn., Zwolinski & Wilhoit 1971; quoted, Mackay & Shiu 1981)
188220 (Antoine eqn., extrapolated, Stephenson & Malanowski 1987)

Henry's Law Constant (Pa m³/mol):
1880 (calculated-$1/K_{AW}$, $C_W/C_A$, reported as exptl., Hine & Mookerjee 1975; quoted, Nirmalakhandan & Speece 1988)
2210 (calculated-group contribution, Hine & Mookerjee 1975)
1800 (calculated-bond contribution, Hine & Mookerjee 1975)
1910 (calculated-P/C, Mackay & Shiu 1981)
1880, 2820 (quoted, calculated- $\chi$ , Nirmalakhandan & Speece 1988)

Octanol/Water Partition Coefficient, log $K_{OW}$:

1.44, 1.48 (quoted, calculated-$V_M$, Wang et al. 1992)

Bioconcentration Factor, log BCF:

Sorption Partition Coefficient, log $K_{OC}$:

Half-Lives in the Environment:

Air:

Surface water:

Ground water:

Sediment:

Soil:

Biota:

Environmental Fate Rate Constant and Half-Lives:

Volatilization:

Photolysis:

Oxidation: rate constants for the reaction with $O_3$ in the gas phase: $1790 \times 10^{-21}$ $cm^3$ $molecule^{-1}$ $sec^{-1}$ at 298 K (Dillemuth et al. 1963; quoted, Atkinson & Aschmann 1984), $(33 \pm 5) \times 10^{-21}$ $cm^3$ $molecule^{-1}$ $sec^{-1}$ at $(294 \pm 1)$ K (DeMore 1971; quoted, Atkinson & Aschmann 1984) and $(19.7 \pm 2.6) \times 10^{-21}$ $cm^3$ $molecule^{-1}$ $sec^{-1}$ at $(294 \pm 2)$ K (Atkinson & Aschmann 1984); rate constants of $(8.25 \pm 0.23) \times 10^{-12}$ $cm^3$ $molecule^{-1}$ $sec^{-1}$ relative to cyclohexane for the reaction with hydroxyl radicals in the gas-phase at $(294 \pm 2)$ K and atmospheric pressure (Atkinson & Aschmann 1984); photooxidation rate constants of $8.0 \times 10^{-12}$ $cm^3$ $molecule^{-1}$ $sec^{-1}$ for the reaction with hydroxyl radicals, and $2.0 \times 10^{-20}$ $cm^3$ $molecule^{-1}$ $sec^{-1}$ for the reaction with $O_3$ in the gas phase at 298 K (Atkinson 1990); rate constants of $1.79 \times 10^{-18}$ $cm^3$ $molecule^{-1}$ $sec^{-1}$ (at 298 K) and $1.79 \times 10^{-20}$ $cm^3$ $molecule^{-1}$ $sec^{-1}$, $4.0 \times 10^{-21}$ $cm^3$ $molecule^{-1}$ $sec^{-1}$ (at 294 K) for the reaction with ozone in the gas phase (Atkinson 1991).

Hydrolysis:

Biodegradation:

Biotransformation:

Bioconcentration, Uptake ($k_1$) and Elimination ($k_2$) Rate Constants or Half-Lives:

Common Name: 1-Pentyne
Synonym:
Chemical Name: 1-pentyne
CAS Registry No: 627-19-0
Molecular Formula: $C_5H_8$
Molecular Weight: 68.13
Melting Point (°C): -106
Boiling Point (°C): 48
Density (g/cm$^3$ at 20°C):
    0.6901 (20°C, Dreisbach 1959)
    0.6849 (25°C, Dreisbach 1959)
Molar Volume (cm$^3$/mol):
    98.75 (20°C, calculated-density)
    99.47 (25°C, calculated-density)
    98.7 (20°C, calculated-density, McAuliffe 1966; Stephenson & Malanowski 1987)
    99.0 (calculated-density, Wang et al. 1992)
    103.6 (calculated-LeBas method)
Molecular Volume, TMV (Å$^3$):
Total Surface Area, TSA (Å$^2$):
    225.6 (Iwase et al. 1985)
Heat of Fusion, $\Delta H_{fus}$, (kcal/mol):
Entropy of Fusion, $\Delta S_{fus}$ (cal/mol K or e.u.):
Fugacity Ratio at 25°C, F: 1.0

Water Solubility (g/m$^3$ or mg/L at 25°C):
    1570 (shake flask-GC, McAuliffe 1966)
    1560, 1031 (quoted, calculated-$K_{OW}$, Hansch et al. 1968)
    1560 (quoted, Hine & Mookerjee 1975)
    1457, 1270 (quoted, calculated-$K_{OW}$, isomer not specified, Yalkowsky & Morozowich 1980)
    1049 (gen. col.-GC, Tewari et al. 1982a,b)
    1560, 898 (quoted, calculated-fragment solubility constants, Wakita et al. 1986)
    801 (quoted, Isnard & Lambert 1989)
    1560, 1457 (quoted, calculated-$V_M$, Wang et al. 1992)
    1560 (quoted, Müller & Klein 1992)

Vapor Pressure (Pa at 25°C):
    57515 (Antoine eqn, Dreisbach 1959)
    57600 (Antoine eqn., Zwolinski & Wilhoit 1971; quoted, Mackay & Shiu 1981)
    57515 (quoted, Hine & Mookerjee 1975)
    57536 (interpolated, Antoine eqn., Dean 1985)
    57536 (interpolated, Antoine eqn., Stephenson & Malanowski 1987)

Henry's Law Constant (Pa $m^3$/mol):

2536 (calculated-1/$K_{AW}$, $C_W/C_A$, reported as exptl., Hine & Mookerjee 1975; quoted Nirmalakhandan & Speece 1988)
2980 (calculated-group contribution, Hine & Mookerjee 1975)
2656 (calculated-bond contribution, Hine & Mookerjee 1975)
2500 (calculated-P/C, Mackay & Shiu 1981)
2536, 3422 (quoted, calculated- $\chi$ , Nirmalakhandan & Speece 1988)
4983 (calculated-P/C, Yaws et al. 1991)

Octanol/Water Partition Coefficient, log $K_{OW}$:

1.98 (shake flask-UV, Hansch et al. 1968, Hansch & Anderson 1967)
1.98, 2.04 (quoted, calculated- $\chi$ , Murray et al. 1975)
1.98, 1.96 (quoted observed & calculated values, Chou & Jurs 1979)
1.98, 1.98 (calculated-$\pi$, f consts., Chou & Jurs 1979)
1.98 (quoted, Yalkowsky & Morozowich 1980)
2.12 (gen. col.-GC, Tewari et al. 1982a,b)
1.98, 1.97 (quoted, calculated-hydrophobicity const., Iwase et al. 1985)
2.12 (quoted, Isnard & Lambert 1989)
1.98 (recommended, Sangster 1989)
1.98, 1.97 (quoted, calculated-$V_M$, Wang et al. 1992)
1.93 (calculated-f const., Müller & Klein 1992)

Bioconcentration Factor, log BCF:

Sorption Partition Coefficient, log $K_{OC}$:

Half-Lives in the Environment:

Air:
Surface water:
Ground water:
Sediment:
Soil:
Biota:

Environmental Fate Rate Constant and Half-Lives:

Volatilization:
Photolysis:
Oxidation: photooxidation reaction rate constant of $7.54 \times 10^{-16}$ $cm^3$ $molecule^{-1}$ $sec^{-1}$ for the reaction with $NO_3$ radicals in the gas phase at 295 K (Atkinson 1991).
Hydrolysis:
Biodegradation:
Biotransformation:
Bioconcentration, Uptake ($k_1$) and Elimination ($k_2$) Rate Constants or Half-Lives:

Common Name: 1-Hexyne
Synonym:
Chemical Name: 1-hexyne
CAS Registry No: 693-02-7
Molecular Formula: $C_6H_{10}$
Molecular Weight: 82.15
Melting Point (°C): -132.06
Boiling Point (°C): 73.15
Density ($g/cm^3$ at 20°C):
0.7155 (20°C, Dreisbach 1959)
0.7106 (25°C, Dreisbach 1959)
Molar Volume ($cm^3/mol$):
114.8 (20°C, calculated-density, McAuliffe 1966; Stephenson & Malanowski 1987)
115.0 (calculated-density, Lande & Banerjee 1981; Wang et al. 1992)
125.8 (calculated-LeBas method)
Molecular Volume, TMV ($Å^3$):
Total Surface Area, TSA ($Å^2$):
Heat of Fusion, $\Delta H_{fus}$, (kcal/mol):
Entropy of Fusion, $\Delta S_{fus}$ (cal/mol K or e.u.):
Fugacity Ratio at 25°C, F: 1.0

Water Solubility ($g/m^3$ or mg/L at 25°C):
360 (shake flask-GC, McAuliffe 1966; quoted, Lande & Banerjee 1981)
360, 568 (quoted, calculated-$K_{OW}$, Hansch et al. 1968)
367 (quoted, Hine & Mookerjee 1975)
335, 451 (quoted, calculated-$K_{OW}$, Yalkowsky & Morozowich 1980)
686 (gen. col.-GC, Tewari et al. 1982a,b)
688 (gen. col.-GC, Miller et al. 1985; quoted, Müller & Klein 1992)
360, 266 (quoted, calculated-fragment solubility constants, Wakita et al. 1986)
360 (quoted, IUPAC 1989)
699 (quoted, Isnard & Lambert 1989)
360, 462 (quoted, calculated-$V_M$, Wang et al. 1992)

Vapor Pressure (Pa at 25°C):
18140 (Antoine eqn., Dreisbach 1959)
18132 (quoted, Hine & Mookerjee 1975)
18145 (interpolated, Antoine eqn., Stephenson & Malanowski 1987)

Henry's Law Constant (Pa $m^3$/mol):

4020 (calculated-$1/K_{AW}$, $C_W/C_A$, reported as exptl., Hine & Mookerjee 1975; quoted, Nirmalakhandan & Speece 1988)

4210 (calculated-group contribution, Hine & Mookerjee 1975)

4020 (calculated-bond contribution, Hine & Mookerjee 1975)

4020, 4308 (quoted, calculated- $\chi$ , Nirmalakhandan & Speece 1988)

2166 (calculated-P/C, Yaws et al. 1991)

Octanol/Water Partition Coefficient, log $K_{OW}$:

2.48 (calculated-$\pi$ substituent constants, Hansch et al. 1968)

2.48, 2.48 (quoted, calculated- $\chi$ , Murray et al. 1975)

2.51 (calculated-f const., Yalkowsky & Morozowich 1980)

2.73 (gen. col.-GC, Tewari et al. 1982a,b)

2.73 (quoted, Isnard & Lambert 1989)

2.73 (recommended, Sangster 1989)

2.46 (selected, Müller & Klein 1992)

2.36, 2.47 (quoted, calculated-$V_M$, Wang et al. 1992)

Bioconcentration Factor, log BCF:

Sorption Partition Coefficient, log $K_{OC}$:

Half-Lives in the Environment:

Air:

Surface water:

Ground water:

Sediment:

Soil:

Biota:

Environmental Fate Rate Constant and Half-Lives:

Volatilization:

Photolysis:

Oxidation: photooxidation reaction rate constant of $1.60 \times 10^{-15}$ $cm^3$ $molecule^{-1}$ $sec^{-1}$ for the reaction with $NO_3$ radicals in the gas phase at 295 K (Atkinson 1991).

Hydrolysis:

Biodegradation:

Biotransformation:

Bioconcentration, Uptake ($k_1$) and Elimination ($k_2$) Rate Constants or Half-Lives:

Common Name: 1-Heptyne
Synonym:
Chemical Name: 1-heptyne
CAS Registry No: 628-71-7
Molecular Formula: $C_7H_{14}$
Molecular Weight: 96.17
Melting Point (°C): -81
Boiling Point (°C): 99
Density (g/cm³ at 20°C):
0.733 (Weast 1984)
0.7328 (20°C, Dreisbach 1959)
0.7283 (25°C, Dreisbach 1959)
Molar Volume (cm³/mol):
131.2 (20°C, calculated-density, McAuliffe 1966; Stephenson & Malanowski 1987)
131.0 (calc-density, Wang et al. 1992)
155.4 (calculated-LeBas method)
Molecular Volume (A³):
Total Surface Area, TSA (A²):
Heat of Fusion, $\Delta H_{fus}$, (kcal/mol):
Entropy of Fusion, $\Delta S_{fus}$ (cal/mol K or e.u.):
Fugacity Ratio at 25°C, F: 1.0

Water Solubility (g/m³ or mg/L at 25°C):
94.0 (shake flask-GC, McAuliffe 1966, quoted, IUPAC 1989)
94, 164 (quoted, calculated-$K_{OW}$, Hansch et al. 1968)
98.4 (quoted, Hine & Mookerjee 1976)
94, 76 (quoted, calculated-fragment solubility constants, Wakita et al. 1986)
94, 139 (quoted, calculated-$V_M$, Wang et al. 1992)
94.0 (quoted, Müller & Klein 1992)

Vapor Pressure (Pa at 25°C):
7000 (Antoine eqn., Dreisbach 1959)
7000 (quoted, Hine & Mookerjee 1975)
7500 (extrapolated, Antoine eqn., Stephenson & Malanowski 1987)

Henry's Law Constant (Pa m³/mol):
6827 (calculated-1/$K_{AW}$, $C_W/C_A$, reported as exptl., Hine & Mookerjee 1975; quoted, Nirmalakhandan & Speece 1988)
5946 (calculated-group contribution, Hine & Mookerjee 1975)
6085 (calculated-bond contribution, Hine & Mookerjee 1975)

6827, 5423 (quoted, calculated- $\chi$ , Nirmalakhandan & Speece 1988)
7160 (calculated-P/C, Yaws et al. 1991)

Octanol/Water Partition Coefficient, log $K_{OW}$:
2.98 (calculated-$\pi$ substituent constants, Hansch et al. 1968)
2.98, 2.93 (quoted, calculated- $\chi$ , Murray et al. 1975)
2.98, 2.98 (quoted, calculated-$V_M$, Wang et al. 1992)
2.99 (calculated-f const., Müller & Klein 1992)

Bioconcentration Factor, log BCF:

Sorption Partition Coefficient, log $K_{OC}$:

Half-Lives in the Environment:
Air:
Surface water:
Ground water:
Sediment:
Soil:
Biota:

Environmental Fate Rate Constant and Half-Lives:
Volatilization:
Photolysis:
Oxidation:
Hydrolysis:
Biodegradation:
Biotransformation:
Bioconcentration, Uptake ($k_1$) and Elimination ($k_2$) Rate Constants or Half-Lives:

Common Name: 1-Octyne
Synonym:
Chemical Name: 1-octyne
CAS Registry No: 629-05-0
Molecular Formula: $C_8H_{14}$
Molecular Weight: 110.2
Melting Point (°C): -80
Boiling Point (°C): 127-128
Density ($g/cm^3$ at 20°C):
0.747 (Weast 1984)
0.7461 (20°C, Dreisbach 1959)
0.7419 (25°C, Dreisbach 1959)
Molar Volume ($cm^3/mol$):
148.5 (calculated-density)
147.7 (20°C, calculated-density, McAuliffe 1966; Stephenson & Malanowski 1987)
170.2 (calculated-LeBas method)
148.0 (calculated-density, Wang et al. 1992)
Molecular Volume ($A^3$):
Total Surface Area, TSA ($A^2$):
Heat of Fusion, $\Delta H_{fus}$, (kcal/mol):
Entropy of Fusion, $\Delta S_{fus}$ (cal/mol K or e.u.):
Fugacity Ratio at 25°C, F: 1.0

Water Solubility ($g/m^3$ or mg/L at 25°C):
24.0 (shake flask-GC, McAuliffe 1966; quoted, IUPAC 1989)
24.1, 46.6 (quoted, calculated-$K_{OW}$, Hansch et al. 1968)
24.1 (quoted, Hine & Mookerjee 1975)
24.1, 21.5 (quoted, calculated-fragment solubility constants, Wakita et al. 1986)
24.1, 38.2 (quoted, calculated-$V_M$, Wang et al. 1992)
24.1 (quoted, Müller & Klein 1992)

Vapor Pressure (Pa at 25°C):
1813 (Antoine eqn., Dreisbach 1959)
1813 (quoted, Hine & Mookerjee 1975)
1723 (extrapolated, Antoine eqn., Boublik et al. 1984)
1715 (extrapolated, Antoine eqn., Stephenson & Malanowski 1987)

Henry's Law Constant (Pa $m^3$/mol):
8208 (calculated-$1/K_{AW}$, $C_W/C_A$, reported as exptl., Hine & Mookerjee 1975; quoted, Nirmalakhandan & Speece 1988)

8208 (calculated-group contribution, Hine & Mookerjee 1975)
9000 (calculated-bond contribution, Hine & Mookerjee 1975)
8208, 6827 (quoted, calculated- $\chi$ , Nirmalakhandan & Speece 1988)
8325 (calculated-P/C, Yaws et al. 1991)

Octanol/Water Partition Coefficient, log $K_{OW}$:
3.48 (calculated-$\pi$ substituent constants, Hansch et al. 1968)
3.48, 3.37 (quoted, calculated- $\chi$ , Murray et al. 1975)
3.48, 3.49 (quoted, calculated-$V_M$, Wang et al. 1992)
3.52 (calculated-f const., Müller & Klein 1992)

Bioconcentration Factor, log BCF:

Sorption Partition Coefficient, log $K_{OC}$:

Half-Lives in the Environment:
Air:
Surface water:
Ground water:
Sediment:
Soil:
Biota:

Environmental Fate Rate Constant and Half-Lives:
Volatilization:
Photolysis:
Oxidation:
Hydrolysis:
Biodegradation:
Biotransformation:
Bioconcentration, Uptake ($k_1$) and Elimination ($k_2$) Rate Constants or Half-Lives:

Common Name: 1-Nonyne
Synonym:
Chemical Name: 1-nonyne
CAS Registry No: 3452-09-3
Molecular Formula: $C_9H_{16}$
Molecular Weight: 124.23
Melting Point (°C): -50
Boiling Point (°C): 150-151
Density ($g/cm^3$ at 20°C):
Molar Volume ($cm^3/mol$):
0.757 (Weast 1984)
0.7568 (20°C, Dreisbach 1959)
0.7527 (25°C, Dreisbach 1959)
Molecular Volume ($A^3$):
192.4 (calculated-LeBas method)
164.1 (20°C, calculated-density, McAuliffe 1966)
164.0 (calculated-density, Wang et al. 1992)
Total Surface Area, TSA ($A^2$):
Heat of Fusion, $\Delta H_{fus}$, (kcal/mol):
Entropy of Fusion, $\Delta S_{fus}$ (cal/mol K or e.u.):
Fugacity Ratio at 25°C, F: 1.0

Water Solubility ($g/m^3$ or mg/L at 25°C):
7.20 (shake flask-GC, McAuliffe 1966; quoted, IUPAC 1989)
7.20, 13.0 (quoted, calculated-$K_{OW}$, Hansch et al. 1968)
7.20 (quoted, Hine & Mookerjee 1975)
7.20, 5.95 (quoted, calculated-fragment solubility constants, Wakita et al. 1986)
7.20, 12.5 (quoted, calculated-$V_M$, Wang et al. 1992)
7.16 (quoted, Müller & Klein 1992)

Vapor Pressure (Pa at 25°C):
835 (Antoine eqn., Dreisbach 1959)
835 (quoted, Hine & Mookerjee 1975)

Henry's Law Constant (Pa $m^3$/mol):
14596 (calculated-$1/K_{AW}$, $C_W/C_A$, reported as exptl., Hine & Mookerjee 1975; quoted, Nirmalakhandan & Speece 1988)
11594 (calculated-group contribution, Hine & Mookerjee 1975)
13009 (calculated-bond contribution, Hine & Mookerjee 1975)
14596, 8695 (quoted, calculated- $\chi$ , Nirmalakhandan & Speece 1988)

14396 (calculated-P/C, Yaws et al. 1991)

Octanol/Water Partition Coefficient, log $K_{OW}$:

3.98 (calculated-$\pi$ substituent constants, Hansch et al. 1968)
3.98, 3.81 (quoted, calculated- $\chi$ , Murray et al. 1975)
3.98, 3.98 (quoted, calculated-$V_M$, Wang et al. 1992)
4.05 (calculated-f const., Müller & Klein 1992)

Bioconcentration Factor, log BCF:

Sorption Partition Coefficient, log $K_{OC}$:

Half-Lives in the Environment:

Air:
Surface water:
Ground water:
Sediment:
Soil:
Biota:

Environmental Fate Rate Constant and Half-Lives:

Volatilization:
Photolysis:
Oxidation:
Hydrolysis:
Biodegradation:
Biotransformation:
Bioconcentration, Uptake ($k_1$) and Elimination ($k_2$) Rate Constants or Half-Lives:

Common Name: Cyclopentene
Synonym:
Chemical Name: cyclopentene
CAS Registry No: 142-29-0
Molecular Formula: $C_5H_8$
Molecular Weight: 68.12
Melting Point (°C):
-135.08 (Dreisbach 1959; Stephenson & Malanowski 1987)
Boiling Point (°C):
44.24 (Dreisbach 1959; Stephenson & Malanowski 1987)
Density (g/cm$^3$ at 20°C):
0.774 (Weast 1984)
0.77199 (20°C, Dreisbach 1959)
0.76653 (25°C, Dreisbach 1959)
Molar Volume (cm$^3$/mol):
88.2 (20°C, calculated-density, McAuliffe 1966)
88.0 (calculated-density, Lande & Banerjee 1981; Wang et al. 1992)
88.2 (20°C, calculated-density, Stephenson & Malanowski 1987)
103.6 (calculated-LeBas method)
Molecular Volume (A$^3$):
Total Surface Area, TSA (A$^2$):
Heat of Fusion, $\Delta H_{fus}$, (kcal/mol):
Entropy of Fusion, $\Delta S_{fus}$ (cal/mol K or e.u.):
Fugacity Ratio at 25°C, F: 1.0

Water Solubility (g/m$^3$ or mg/L at 25°C):
535 (shake flask-GC, McAuliffe 1966)
541, 3625 (quoted, calculated-$K_{OW}$, Hansch et al. 1968)
611 (shake flask-titration with bromine, Natarajan & Venkatachalam 1972)
1645 (shake flask-GC, Pierotti & Liabastre 1972)
541 (quoted, Hine & Mookerjee 1975)
529 (quoted, Lande & Banerjee 1981)
540 (suggested "tentative" value, IUPAC 1989)
535, 424 (quoted, calculated-UNIFAC, Al-Sahhaf 1989)
529, 713 (quoted, calculated-$V_M$, Wang et al. 1992)
541 (quoted, Müller & Klein 1992)

Vapor Pressure (Pa at 25°C):
50690 (Antoine eqn., Dreisbach 1959)
50690 (quoted, Hine & Mookerjee 1975)
50706 (interpolated, Antoine eqn., Stephenson & Malanowski 1987)

Henry's Law Constant (Pa $m^3$/mol):

6372 (calculated-1/$K_{AW}$, $C_W/C_A$, reported as exptl., Hine & Mookerjee 1975)
3583 (calculated-group contribution, Hine & Mookerjee 1975)
9645 (calculated-bond contribution, Hine & Mookerjee 1975)
6455 (calculated-P/C, Yaws et al. 1991)

Octanol/Water Partition Coefficient, log $K_{OW}$:

1.75 (calculated-$\pi$ substituent const., Hansch et al. 1968)
1.75, 1.76 (quoted, calculated-$V_M$, Wang et al. 1992)
2.25 (calculated-f const., Müller & Klein 1992)

Bioconcentration Factor, log BCF:

Sorption Partition Coefficient, log $K_{OC}$:

Half-Lives in the Environment:

Air: photooxidation rate constant of 4.97 x$10^{-16}$ $cm^3$ $molecule^{-1}$ $sec^{-1}$ for the reaction with $O_3$ in synthetic air was determined at atmospheric pressure at 291.5 K (Bennett et al. 1987); rate constant of 4.99x$10^{-11}$ $cm^3$ $molecule^{-1}$ $sec^{-1}$ for the reaction with OH radicals in air at 298 K (Rogers 1989).

Surface water: half-lives for oxidation by OH and $RO_2$ radicals were estimated to be 320 hours and 9x$10^4$ days in aquatic system, and of 40 days, based on rate constant of 2x$10^5$ $M^{-1}$ $sec^{-1}$ for the oxidation of cyclic olefins with singlet oxygen in aquatic system (Mill & Mabey 1985).

Ground water:

Sediment:

Soil:

Biota:

Environmental Fate Rate Constant and Half-Lives:

Volatilization:

Photolysis:

Oxidation: photooxidation rate constants of 8.13x$10^{-16}$ $cm^3$ $molecule^{-1}$ $sec^{-1}$ for the gas phase reaction with $O_3$ at 298 K (Japar et al. 1974; quoted, Adeniji et al. 1981; Atkinson et al. 1983) and 9.69x$10^{-16}$ $cm^3$ $molecule^{-1}$ $sec^{-1}$ at (294 $\pm$ 2) K (Adeniji et al. 1981; Atkinson et al. 1983) and (2.75 $\pm$ 0.33)x$10^{-16}$ $cm^3$ $molecule^{-1}$ $sec^{-1}$ at (297 $\pm$ 1) K (Atkinson et al. 1983); photooxidation reaction rate constant of 6.39x$10^{-11}$ $cm^3$ $molecule^{-1}$ $sec^{-1}$ with hydroxyl radicals in air

(Atkinson et al. 1983); second-order rate constant of 4.97 $x10^{-16}$ $cm^3$ $molecule^{-1}$ $sec^{-1}$ at 291.5 K for reaction with $O_3$ in synthetic air at atmospheric pressure (Bennett et al. 1987); rate constant of $4.99x10^{-11}$ $cm^3$ $molecule^{-1}$ $sec^{-1}$ for the reaction with OH radicals in air at 298 K (Rogers 1989); photooxidation reaction rate constants of $4.0x10^{-10}$ $cm^3$ $molecule^{-1}$ $sec^{-1}$ with hydroxyl radicals and $4.6x10^{-13}$ $cm^3$ $molecule^{-1}$ $sec^{-1}$ with $NO_3$ radicals in air at 298 K (Atkinson 1990); rate constant of $6.70x10^{-11}$ $cm^3$ $molecule^{-1}$ $sec^{-1}$ for the reaction with OH radicals and $5.81x10^{-13}$ $cm^3$ $molecule^{-1}$ $sec^{-1}$ for the reaction with $NO_3$ radicals in gas phase at 298 K (Sabljic & Güsten 1990).

Hydrolysis:

Biodegradation:

Biotransformation:

Bioconcentration, Uptake ($k_1$) and Elimination ($k_2$) Rate Constants or Half-Lives:

Common Name: Cyclohexene
Synonym: 1,2,3,4-tetrahydrobenzene, tetrahydrobenzene
Chemical Name: cyclohexene
CAS Registry No: 110-83-8
Molecular Formula: $C_6H_{10}$
Molecular Weight: 82.145
Melting Point (°C): -103.512
Boiling Point (°C): 82.979
Density ($g/cm^3$ at 25°C):
0.81096 (20°C, Dreisbach 1959)
0.80609 (25°C, Dreisbach 1959)
Molar Volume ($cm^3/mol$):
101.3 (20°C, calculated-density)
101.9 (25°C, calculated-density)
101.3 (20°C, calculated-density, McAuliffe 1966)
101.0 (calculated-density, Lande & Banerjee 1981; Wang et al. 1992)
101.4 (calculated-density, Stephenson & Malanowski 1987)
125.8 (calculated-LeBas method)
Molecular Volume,TMV ($Å^3$):
Total Surface Area, TSA ($Å^2$):
217.2 (Iwase et al. 1985)
Heat of Fusion, $\Delta H_{fus}$, (kcal/mol):
0.786 (Riddick et al. 1986)
Entropy of Fusion, $\Delta S_{fus}$ (cal/mol K or e.u.):
Fugacity Ratio at 25°C, F: 1.0

Water Solubility ($g/m^3$ or mg/L at 25°C):
130 (cloud point, McBain & Lissant 1951)
160 (Farkas 1964)
213 (shake flask-GC, McAuliffe 1966; quoted, Riddick et al. 1986)
216, 1389 (quoted, calculated-$K_{OW}$, Hansch et al. 1968)
211 (quoted, Hine & Mookerjee 1975)
281, 286 (23.5°C, elution chromatography, Schwarz 1980)
206, 202 (quoted, calculated-$K_{OW}$, Yalkowsky & Morozowich 1980)
213 (quoted, Lande & Banerjee 1981)
202, 335 (quoted average, calculated-$K_{OW}$, Valvani et al. 1981)
216, 168 (quoted, calculated-fragment solubility constants, Wakita et al. 1986)
160 (recommended best value, IUPAC 1989)
213, 238 (quoted, calculated-UNIFAC, Al-Sahhaf 1989)
211, 291 (quoted, calculated-$V_M$, Wang et al. 1992)
216 (quoted, Müller & Klein 1992)

Vapor Pressure (Pa at 25°C):

11840 (Antoine eqn., Dreisbach 1959)
11840 (quoted, Hine & Mookerjee 1975)
11800 (quoted, Riddick et al. 1986)
11850 (extrapolated, Antoine eqn., Stephenson & Malanowski 1987)

Henry's Law Constant (Pa $m^3$/mol):

4020 (calculated-1/$K_{AW}$, $C_W/C_A$, reported as exptl., Hine & Mookerjee 1975)
4946 (calculated-group contribution, Hine & Mookerjee 1975)
13312 (calculated-bond contribution, Hine & Mookerjee 1975)
4568 (calculated-P/C, Yaws et al. 1991)

Octanol/Water Partition Coefficient, log $K_{OW}$:

2.16 (calculated-$\pi$ substituent constants, Hansch et al. 1968)
2.86 (shake flask-GC, Leo et al. 1975)
2.16, 2.31 (quoted, calculated- $\chi$ , Murray et al. 1975)
2.86 (calculated-f const., Yalkowsky & Morozowich 1980)
2.86 (quoted, Valvani et al. 1981)
1.90 (Canton & Wegman 1983)
2.86, 2.57 (quoted, calculated-hydrophobicity const., Iwase et al. 1985)
2.86, 2.51 (quoted, calculated-MO, Bodor et al. 1989)
2.86, 2.81 (quoted, calculated- $\pi$, Bodor et al. 1989)
2.86 (recommended, Sangster 1989)
2.16, 2.14 (quoted, calculated-$V_M$, Wang et al. 1992)
2.81 (calculated-f const., Müller & Klein 1992)

Bioconcentration Factor, log BCF:

Sorption Partition Coefficient, log $K_{OC}$:

Half-Lives in the Environment:

Air:
Surface water: half-lives for oxidation by OH and $RO_2$ radicals were estimated to be 320 hours and $9x10^4$ days in aquatic system, and 40 days, based on rate constant of $2x10^5$ $M^{-1}$ $sec^{-1}$ for oxidation of cyclic olefins by singlet oxygen in aquatic system (Mill & Mabey 1985).
Ground water:
Sediment:

Soil:
Biota:

Environmental Fate Rate Constant and Half-Lives:
Volatilization:
Photolysis:
Oxidation: room temperature rate constants of $2.20 \times 10^{-11}$ $cm^3$ $molecule^{-1}$ $sec^{-1}$ for the reaction with $O(^3P)$ (Herron & Huie 1973; quoted, Gaffney & Levine 1979) and $6.80 \times 10^{-11}$ $cm^3$ $molecule^{-1}$ $sec^{-1}$ for the reaction with OH radicals (Atkinson et al. 1979; quoted, Gaffney & Levine 1979); photooxidation reaction rate constants of $1.69 \times 10^{-16}$ $cm^3$ $molecule^{-1}$ $sec^{-1}$ with ozone at 298 K (Japar et al. 1974; quoted, Adeniji et al. 1981; Atkinson et al. 1983) and $2.04 \times 10^{-16}$ $cm^3$ $molecule^{-1}$ $sec^{-1}$ at $(294 \pm 2)$ K (Adeniji et al. 1981; quoted, Atkinson et al. 1983) and $(1.04 \pm 0.14) \times 10^{-16}$ $cm^3$ $molecule^{-1}$ $sec^{-1}$ at $(297 \pm 1)$ K (Atkinson et al. 1983); rate constant of $1.04 \times 10^{-16}$ $cm^3$ $molecule^{-1}$ $sec^{-1}$ for the reaction with ozone in air at 297 K (Atkinson & Carter 1984); second-order rate constant of $1.51 \times 10^{-16}$ $cm^3$ $molecule^{-1}$ $sec^{-1}$ for the reaction with $O_3$ in synthetic air at atmospheric pressure and 295 K (Bennett et al. 1987); rate constant of $5.40 \times 10^{-11}$ $cm^3$ $molecule^{-1}$ $sec^{-1}$ for the reaction with OH radicals at 298 K (Rogers 1989); rate constants of $1.0 \times 10^{-10}$ $cm^3$ $molecule^{-1}$ $sec^{-1}$ for the gas phase reaction with OH radicals and $5.3 \times 10^{-13}$ $cm^3$ $molecule^{-1}$ $sec^{-1}$ with $NO_3$ radicals in air at 298 K (Atkinson 1990); rate constants of $6.75 \times 10^{-11}$ $cm^3$ $molecule^{-1}$ $sec^{-1}$ for the reaction with OH radicals and of $5.26 \times 10^{-13}$ $cm^3$ $molecule^{-1}$ $sec^{-1}$ for the reaction with $NO_3$ radicals in the gas phase at 298 K (Sabljic & Güsten 1990); rate constant of $5.28 \times 10^{-13}$ $cm^3$ $molecule^{-1}$ $sec^{-1}$ for reaction with $NO_3$ in air at 295 K (Atkinson 1991).
Hydrolysis:
Biodegradation:
Biotransformation:
Bioconcentration, Uptake ($k_1$) and Elimination ($k_2$) Rate Constants or Half-Lives:

Common Name: 1-Methylcyclohexene
Synonym:
Chemical Name: 1-methylcyclohexene
CAS Registry No: 591-49-1
Molecular Formula: $C_7H_{12}$
Molecular Weight: 96.17
Melting Point (°C): -120.4
Boiling Point (°C): 110.3
Density ($g/cm^3$ at 20°C):
    0.8102 (20°C, Dreisbach 1959)
    0.8058 (25°C, Dreisbach 1959)
Molar Volume ($cm^3/mol$):
    118.7 (20°C, calculated-density, McAuliffe 1966; Stephenson & Malanowksi 1987)
    148.0 (calculated-LeBas method)
Molecular Volume ($A^3$):
Total Surface Area, TSA ($A^2$):
Heat of Fusion, $\Delta H_{fus}$, (kcal/mol):
Entropy of Fusion, $\Delta S_{fus}$ (cal/mol K or e.u.):
Fugacity Ratio at 25°C, F: 1.0

Water Solubility ($g/m^3$ or mg/L at 25°C):
    52.0 (shake flask-GC, McAuliffe 1966; quoted, IUPAC 1989)
    52.0 (quoted, Hine & Mookerjee 1975)
    52, 61.0 (quoted, calculated-fragment solubility constants, Wakita et al. 1986)
    52, 297 (quoted, calculated-$V_M$, Wang et al. 1992)

Vapor Pressure (Pa at 25°C):
    4080 (Antoine eqn., Dreisbach 1959)
    4080 (quoted, Hine & Mookerjee 1975)
    3933 (Antoine eqn., extrapolated, Stephenson & Malanowski 1987)

Henry's Law Constant (Pa $m^3/mol$):
    7660 (calculated-$1/K_{AW}$, $C_W/C_A$, reported as exptl., Hine & Mookerjee 1975; quoted, Nirmalakhandan & Speece 1988)
    7485 (calculated-group contribution, Hine & Mookerjee 1975)
    14264 (calculated-bond contribution, Hine & Mookerjee 1975)
    7660, 11072 (quoted, calculated- $\chi$ , Nirmalakhandan & Speece 1988)

Octanol/Water Partition Coefficient, log $K_{OW}$:

1.05 (calculated-regression of Lyman et al. 1982, Wang et al. 1992)

2.20 (calculated-$V_M$, Wang et al. 1992)

Bioconcentration Factor, log BCF:

Sorption Partition Coefficient, log $K_{OC}$:

Half-Lives in the Environment:

Air:

Surface water: half-lives for oxidation by OH and $RO_2$ radicals were estimated to be 320 hours and $9\times10^4$ days in aquatic system, and 40 days, based on rate constant $2\times10^5$ $M^{-1}$ $sec^{-1}$ for the oxidation of cyclic olefins by singlet oxygen in aquatic system (Mill & Mabey 1985).

Ground water:

Sediment:

Soil:

Biota:

Environmental Fate Rate Constant and Half-Lives:

Volatilization:

Photolysis:

Oxidation: photooxidation reaction rate constant of $9.45\times10^{-11}$ $cm^3$ $molecule^{-1}$ $sec^{-1}$ for the gas-phase reaction with hydroxyl radicals at 294 K (Atkinson 1985).

Hydrolysis:

Biodegradation:

Biotransformation:

Bioconcentration, Uptake ($k_1$) and Elimination ($k_2$) Rate Constants or Half-Lives:

Common Name: Cycloheptene
Synonym: suberene
Chemical Name: cycloheptene
CAS Registry No: 628-92-2
Molecular Formula: $C_7H_{12}$
Molecular Weight: 96.17
Melting Point (°C): -56.0
Boiling Point (°C): 116
Density ($g/cm^3$ at 20°C):
    0.824 (Weast 1984)
Molar Volume ($cm^3/mol$):
    116.0 (20°C, calculated-density, McAuliffe 1966)
    116.0 (calculated-density, Lande & Banerjee 1981; Wang et al. 1992)
    116.9 (20°C, calculated-density, Stephenson & Malanowski 1987)
    148.0 (calculated-LeBas method)
Molecular Volume, TMV ($Å^3$):
Total Surface Area, TSA ($Å^2$):
Heat of Fusion, $\Delta H_{fus}$, (kcal/mol):
Entropy of Fusion, $\Delta S_{fus}$ (cal/mol K or e.u.):
Fugacity Ratio at 25°C, F: 1.0

Water Solubility ($g/m^3$ or mg/L at 25°C):
    66.0 (shake flask-GC, McAuliffe 1966; quoted, IUPAC 1989)
    66.5, 516 (quoted, calculated-$K_{OW}$, Hansch et al. 1968)
    66.0 (quoted, Lande & Banerjee 1981)
    66.5, 58.2 (quoted, calculated-fragment solubility constants, Wakita et al. 1986)
    66, 59 (quoted, calculated-UNIFAC, Al-Sahhaf 1989)
    66.5, 98.4 (quoted, calculated-$V_M$, Wang et al. 1992)
    66.5 (quoted, Müller & Klein 1992)

Vapor Pressure (Pa at 25°C):
    2669 (interpolated, Antoine eqn., Stephenson & Malanowski 1987)

Henry's Law Constant (Pa $m^3$/mol):

Octanol/Water Partition Coefficient, log $K_{OW}$:
    2.57 (calculated-$\pi$ substituent constants, Hansch et al. 1968)
    2.57, 2.75 (quoted, calculated- $\chi$ , Murray et al. 1975)
    2.57, 2.58 (quoted, calculated-$V_M$, Wang et al. 1992)

3.37 (calculated-f const., Müller & Klein 1992)

Bioconcentration Factor, log BCF:

Sorption Partition Coefficient, log $K_{OC}$:

Half-Lives in the Environment:

Air:

Surface water: half-lives for oxidation by OH and $RO_2$ radicals were estimated to be 320 hours and $9x10^4$ days in aquatic system, and 40 days, based on rate constant of $2x10^5$ $M^{-1}$ $sec^{-1}$ for oxidation of cyclic olefins by singlet oxygen in aquatic system (Mill & Mabey 1985).

Ground water:

Sediment:

Soil:

Biota:

Environmental Fate Rate Constant and Half-Lives:

Volatilization:

Photolysis:

Oxidation: photooxidation reaction rate constant of $(3.19 \pm 0.36)x10^{-16}$ $cm^3$ $molecule^{-1}$ $sec^{-1}$ for the reaction with ozone in gas phase at $(297 \pm 1)$ K (Atkinson et al. 1983; quoted, Atkinson & Carter 1984); rate constants of $2.80x10^{-13}$ $cm^3$ $molecule^{-1}$ $sec^{-1}$ for the reaction with $NO_3$ radicals and of $7.13x10^{-11}$ $cm^3$ $molecule^{-1}$ $sec^{-1}$ for the reaction with OH radicals in air at 298 K (Sabljic & Güsten 1990); rate constant of $4.84x10^{-13}$ $cm^3$ $molecule^{-1}$ $sec^{-1}$ for the gas phase reaction with $NO_3$ radicals at 298 K (Atkinson 1991).

Hydrolysis:

Biodegradation:

Biotransformation:

Bioconcentration, Uptake ($k_1$) and Elimination ($k_2$) Rate Constants or Half-Lives:

Common Name: 1,4-Cyclohexadiene
Synonym:
Chemical Name: 1,4-cyclohexadiene
CAS Registry No: 628-41-1
Molecular Formula: $C_6H_8$
Molecular Weight: 80.13
Melting Point (°C): -49.2
Boiling Point (°C): 81-82
Density (g/cm³ at 20°C):

0.847 (Weast 1984)

Molar Volume (cm³/mol):

93.6 (20°C, calculated-density, McAuliffe 1966; Stephenson & Malanowski 1987)
94.0 (calculated-density, Lande & Banerjee 1981; Wang et al. 1992)
118.4 (calculated-LeBas method)

Molecular Volume, TMV (Å³):
Total Surface Area, TSA (Å²):
Heat of Fusion, $\Delta H_{fus}$, (kcal/mol):
Entropy of Fusion, $\Delta S_{fus}$ (cal/mol K or e.u.):
Fugacity Ratio at 25°C, F: 1.0

Water Solubility (g/m³ or mg/L at 25°C):

700 (shake flask-GC, McAuliffe 1966)
930 (GC, Pierotti & Liabstre 1972)
900, 715 (quoted, calculated-$K_{OW}$, Yalkowsky & Morozowich 1980)
860, 1214 (quoted, calculated-$K_{OW}$, Valvani et al. 1981)
700 (quoted, Lande & Banerjee 1981)
700, 609 (quoted, calculated-fragment solubility constants, Wakita et al. 1986)
323 (calculated-UNIFAC, Banerjee & Howard 1988)
700, 930; 800 (quoted values; recommended, IUPAC 1989)
700, 824 (quoted, calculated-UNIFAC, Al-Sahhaf 1989)
700, 518 (quoted, calculated-$V_M$, Wang et al. 1992)

Vapor Pressure (Pa at 25°C):

9009 (Antoine eqn., extrapolated, Stephenson & Malanowski 1987)

Henry's Law Constant (Pa m³/mol):

Octanol/Water Partition Coefficient, log $K_{OW}$

2.30 (calculated-f const., Yalkowsky & Morozowich 1980)
2.30, 2.48 (quoted, calculated-UNIFAC, Banerjee & Howard 1988)
2.30 (recommended, Sangster 1989)
1.49 (calculated-regression eqn. from Lyman et al. 1982, Wang et al. 1992)
1.94 (calculated-$V_M$, Wang et al. 1992)

Bioconcentration Factor, log BCF:

Sorption Partition Coefficient, log $K_{OC}$:

Half-Lives in the Environment:

Air:
Surface water: half-lives for oxidation by OH and $RO_2$ radicals were estimated to be 320 hours and $9x10^4$ days for olefins in aquatic system, and 19 hours, based on rate constant of $1.0x10^7$ $M^{-1}$ $sec^{-1}$ for oxidation of dienes by singlet oxygen in aquatic system (Mill & Mabey 1985).
Ground water:
Sediment:
Soil:
Biota:

Environmental Fate Rate Constant and Half-Lives:

Volatilization:
Photolysis:
Oxidation: rate constant of $(0.639 \pm 0.074)x10^{16}$ $cm^3$ $molecule^{-1}$ $sec^{-1}$ for reaction with $O_3$ at $(297 \pm 1)$ K (Atkinson et al. 1983); gas-phase photooxidation rate constant of $2.89x10^{-13}$ $cm^3$ $molecule^{-1}$ $sec^{-1}$ for reaction with $NO_3$ radicals at 295 K (Atkinson et al. 1984); photooxidation rate constant of $9.90x10^{-11}$ $cm^3$ $molecule^{-1}$ $sec^{-1}$ (experimental) and $1.03x10^{-10}$ $cm^3$ $molecule^{-1}$ $sec^{-1}$ (calculated value) for the reaction with OH radicals (Atkinson 1985); rate constant of $7.8x10^{-13}$ $cm^3$ $molecule^{-1}$ $sec^{-1}$ for the reaction with $NO_3$ radicals in air (Benter & Schindler 1988); rate constants of $9.91x10^{11}$ $cm^3$ $molecule^{-1}$ $sec^{-1}$ for the reaction with OH radicals and of $5.30x10^{-13}$ $cm^3$ $molecule^{-1}$ $sec^{-1}$ for the reaction with $NO_3$ radicals at nights (Sabljic & Güsten 1990); photooxidation rate constant of $5.32x10^{-13}$ $cm^3$ $molecule^{-1}$ $sec^{-1}$ and $7.80x10^{-13}$ $cm^3$ $molcule^{-1}$ $sec^{-1}$ for gas phase reaction with $NO_3$ radicals at 295 and 298 K, respectively (Atkinson 1991).

Hydrolysis:
Biodegradation:
Biotransformation:
Bioconcentration, Uptake ($k_1$) and Elimination ($k_2$) Rate Constants or Half-Lives:

Common Name: Cycloheptatriene
Synonym: tropilidene
Chemical Name: 1,3,5-cycloheptatriene
CAS Registry No: 544-25-2
Molecular Formula: $C_7H_8$
Molecular Weight: 92.14
Melting Point (°C): -73.5
Boiling Point (°C): 116-117
Density (g/cm$^3$ at 20°C):
    0.888 (Weast 1984)
Molar Volume (cm$^3$/mol):
    103.0 (20°C, calculated-density, McAuliffe 1966)
    103.0 (calculated-density, Lande & Banerjee 1981; Wang et al. 1992)
    103.8 (19°C, calculated-density, Stephenson & Malanowski 1987)
    133.2 (calculated-LeBas method)
Molecular Volume (A$^3$):
Total Surface Area, TSA (A$^2$):
Heat of Fusion, $\Delta H_{fus}$, (kcal/mol):
Entropy of Fusion, $\Delta S_{fus}$ (cal/mol K or e.u.):
Fugacity Ratio at 25°C, F: 1.0

Water Solubility (g/m$^3$ or mg/L at 25°C):
    620 (shake flask-GC, McAuliffe 1966)
    669 (shake flask-GC, Pierotti & Liabastre 1972)
    623 (quoted, Lande & Banerjee 1981)
    623, 609 (quoted, calculated-fragment solubility constants, Wakita et al. 1986)
    620, 665 (misquoted from cycloheptadiene, calculated-UNIFAC, Al-Sahhaf 1989)
    640 (recommended best value, IUPAC 1989)
    623, 285 (quoted, calculated-$V_M$, Wang et al. 1992)

Vapor Pressure (Pa at 25°C):
    2825, 3138 (Antoine equations, interpolated, Stephenson & Malanowski 1987)

Henry's Law Constant (Pa m$^3$/mol):
    432 (calculated-P/C from selected data)
    466 (calculated-P/C, Yaws et al. 1991)

Octanol/Water Partition Coefficient, log $K_{OW}$:

- 2.63 (shake flask, Eadsforth & Moser 1983)
- 3.03 (HPLC, Eadsforth & Moser 1983)
- 2.63 (recommended, Sangster 1989)
- 1.05 (calculated-regression eqn. of Lyman et al. 1982, Wang et al. 1992)
- 2.20 (calculated-$V_M$, Wang et al. 1992)

Bioconcentration Factor, log BCF:

Sorption Partition Coefficient, log $K_{OC}$:

Half-Lives in the Environment:

Air:
Surface water:
Ground water:
Sediment:
Soil:
Biota:

Environmental Fate Rate Constant and Half-Lives:

Volatilization:
Photolysis:
Oxidation: photooxidation reaction rate constant of $5.39 \times 10^{-17}$ cm$^3$ molecule$^{-1}$ sec$^{-1}$ for the gas phase reaction with ozone at 294 K (Atkinson & Carter 1984); rate constant of $9.74 \times 10^{-11}$ cm$^3$ molecule$^{-1}$ sec$^{-1}$ for the gas phase reaction with hydroxyl radicals at 294 K (Atkinson 1985); photooxidation reaction rate constants of $1.18 \times 10^{-12}$ cm$^3$ molecule$^{-1}$ sec$^{-1}$ for the reaction with $NO_3$ radicals and $9.44 \times 10^{-11}$ cm$^3$ molecule$^{-1}$ sec$^{-1}$ for the reaction with OH radicals in air at 298 K (Sabljic & Güsten 1990); rate constant of $1.19 \times 10^{-12}$ cm$^3$ molecule$^{-1}$ sec$^{-1}$ for the gas phase reaction with $NO_3$ radicals at 298 K (Atkinson 1991).
Hydrolysis:
Biodegradation:
Biotransformation:
Bioconcentration, Uptake ($k_1$) and Elimination ($k_2$) Rate Constants or Half-Lives:

Common Name: *d*-Limonene

Synonym: *d*-*p*-mentha-1,8,-diene, (R)-(+)-*p*--mentha-1,8-diene, (+)-1-methyl-4-(1-methylethenyl)cyclohexene, *p*-mentha-1,8-diene, carvene, cinene, citrene, cajeputene, kautschin

Chemical Name: *dextro*-limonene, (R)-(+)-limonene

CAS Registry No: 5989-27-5

Molecular Formula: $C_{10}H_{16}$

$$CH_2=\overset{CH_3}{\overset{|}{C}}CHCH_2CH=\overset{CH_3}{\overset{|}{C}}CH_2CH_2 \text{ (ring)}$$

Molecular Weight: 136.236

Melting Point (°C):
- -96.9 (Stull 1947)
- -74.30 (Stephenson & Malanowski 1987)

Boiling Point (°C):
- 175.0 (Riddick et al. 1986)
- 175.75 (Stephenson & Malanowski 1987)
- 175.5 (Ma et al. 1990)

Density (g/cm$^3$ at 25°C):
- 0.8392 (25°C, Massaldi & King 1973)
- 0.8403 (20°C, Riddick et al. 1986)
- 0.8383 (25°C, Riddick et al. 1986)

Molar Volume (cm$^3$/mol):
- 162.51 (calculated-density)
- 192.2 (calculated-LeBas method)

Molecular Volume (A$^3$):

Total Surface Area, TSA (A$^2$):

Heat of Fusion, $\Delta H_{fus}$, (kcal/mol):

Entropy of Fusion, $\Delta S_{fus}$ (cal/mol K or e.u.):

Fugacity Ratio at 25°C, F: 1.0

Water Solubility (g/m$^3$ or mg/L at 25°C):
- 13.49 (shake flask-GC, Massaldi & King 1973; quoted, Ma et al. 1990)
- 13.80 (quoted, Riddick et al. 1986)

Vapor Pressure (Pa at 25°C):
- 275.64 (calculated-Antoine eqn. regression, Stull 1947)
- 267.0 (quoted, Riddick et al. 1986)
- 277.0 (calculated-Antoine eqn., Stephenson & Malanowski 1987; quoted, Ma et al. 1990)

Henry's Law Constant (Pa $m^3$/mol):

2725 (calculated-P/C from selected data)

Octanol/Water Partition Coefficient, log $K_{OW}$:

5.50 (estimated, Ma et al. 1990)

Bioconcentration Factor, log BCF:

Sorption Partition Coefficient, log $K_{OC}$:

Half-Lives in the Environment:

Air:
Surface water:
Ground water:
Sediment:
Soil:
Biota:

Environmental Fate Rate Constant and Half-Lives:

Volatilization:
Photolysis:
Oxidation:
Hydrolysis:
Biodegradation:
Biotransformation:
Bioconcentration, Uptake ($k_1$) and Elimination ($k_2$) Rate Constants or Half-Lives:

Common Name: Styrene
Synonym: phenylethene, styrol, styrolene cinnamene, cinnamol, phenylethylene, vinylbenzene
Chemical Name: styrene
CAS Registry No: 100-42-5
Molecular Formula: $C_6H_5CH=CH_2$
Molecular Weight: 104.51
Melting Point (°C):
-30.60 (Stull 1947; Stephenson & Malanowski 1987; Budavari 1989)
-30.628 (Dreisbach 1955; Riddick et al. 1986)
Boiling Point (°C):
145.14 (Dreisbach 1955; Riddick et al. 1986)
145.18 (Boublik et al. 1973,84)
145.0 (Stephenson & Malanowski 1987)
145-146 (Budavari 1989)
Density (g/cm$^3$ at 20°C):
0.90600 (20°C, Dreisbach 1955; Riddick et al. 1986)
0.90122 (25°C, Dreisbach 1955; Riddick et al. 1986)
Molar Volume (cm$^3$/mol):
115.97 (20°C, calculated-density)
115.57 (25°C, Windholz 1983)
115.0 (20°C, calculated-density, Stephenson & Malanowski 1987)
133.0 (calculated-LeBas method)
Molecular Volume, TMV (Å$^3$):
Total Surface Area, TSA (Å$^2$):
Heat of Fusion, $\Delta H_{fus}$, (kcal/mol):
2.617 (Riddick et al. 1986)
Entropy of Fusion, $\Delta S_{fus}$ (cal/mol K or e.u.):
Fugacity Ratio at 25°C, F: 1.0

Water Solubility (g/m$^3$ or mg/L at 25°C):
330 (24°C, shake flask-Karl Fischer titration, Lane 1946)
310 (cloud point method, Lane 1946)
220 (shake flask-method not specified, Frilette & Hohenstein 1948)
300 (shake flask-UV, Andrews & Keefer 1950; quoted, Mackay & Shiu 1990)
160 (shake flask-HPLC/UV, Banerjee et al. 1980)
302, 262 (quoted, calculated-$K_{OW}$, Yalkowsky & Morozowich 1980)
301, 346 (quoted, calculated-$K_{OW}$, Valvani et al. 1981)
281 (quoted, Banerjee 1985)
310 (quoted, Riddick et al. 1986)
98.6 (calculated-UNIFAC, Banerjee & Howard 1988)
330, 220, 300 160; 250 (quoted; recommended best value, IUPAC 1989)
300 (quoted, Isnard & Lambert 1989)

311    (quoted, Müller & Klein 1992)

Vapor Pressure (Pa at 25°C):

969.4 (calculated-Antoine eqn. regression, Stull 1947)
807    (Antoine eqn., Dreisbach 1955)
810, 878   (Antoine eqn., Boublik et al. 1984)
1093, 1333 (quoted exptl. values, Boublik et al. 1984)
879    (extrapolated, Antoine eqn., Dean 1985)
841    (quoted, Riddick et al. 1986)
880    (interpolated, Antoine eqn., Stephenson & Malanowski 1987)
880    (quoted, Howard 1989)
853, 335 (quoted, calculated- $\chi$ , Banerjee et al. 1990)
670    (quoted, Mackay & Shiu 1990)

Henry's Law Constant (Pa $m^3$/mol):

233    (calculated-P/C, Mackay & Shiu 1990)
285, 527 (quoted, Howard et al. 1989)
267    (calculated-P/C, Yaws et al. 1991)

Octanol/Water Partition Coefficient, log $K_{OW}$:

2.95, 3.14 (quoted, calculated-f const., Rekker 1977)
2.95    (Hansch & Leo 1979, quoted, Yalkowsky & Morozowich 1980)
3.16    (shake flask-HPLC, Banerjee et al. 1980)
2.76    (HPLC Fujisawa & Masuhara 1981)
2.95    (quoted, Valvani et al. 1981)
3.39    (calculated-$\gamma$ , Arbuckle 1983)
2.90    (HPLC-RT, Wang et al. 1986)
2.95, 2.87 (quoted, calculated-UNIFAC & $\chi$ , Banerjee & Howard 1988)
3.16    (quoted, Isnard & Lambert 1989)
3.05    (recommended, Sangster 1989)
2.95    (quoted, Howard 1989)
2.87    (quoted, Fu & Alexander 1992)
2.87    (calculated-f const., Müller & Klein 1992)

Bioconcentration Factor, log BCF:

1.13    (goldfish, Ogata et al. 1984; quoted, Sabljic 1987; Howard 1989)
0.83    (goldfish, calculated- $\chi$ , Sabljic 1987)

Sorption Partition Coefficient, log $K_{OC}$:

2.43    (estimated-S, Howard 1989)
2.74    (estimated-$K_{OW}$, Howard 1989)
3.42-2.74 (quoted, Swann et al. 1983)
2.96, 2.71 (quoted exptl., calculated- $\chi$ , Meylan et al. 1992)

Half-Lives in the Environment:

Air: atmospheric half-life was estimated to be 2.4-24 hours, based on the EPA Reactivity Classification of Organics (Darnall et al. 1976); will react rapidly with both hydroxyl radicals and ozone in air with a combined, calculated half-life of 2.5 hours, the reaction half-lives of 3.5 hours with OH radicals and 9 hours with ozone (Howard 1989); atmospheric half-lives of styrene: 0.9 to 7.3 hours, based on photooxidation half-life in air (Howard et al. 1991).

Surface water: 336-672 hours, based on estimated unacclimated aqueous aerobic biodegradation half-life (Howard et al. 1991); half-life for styrene loss from surface waters was calculated to be 0.75 to 51 days (Fu & Alexander 1992).

Ground water: 672-5040 hours, based on estimated unacclimated aqueous aerobic biodegradation half-life and acclimated aqueous screening test data (Howard et al. 1991).

Sediment:

Soil: 336-672 hours, based on unacclimated grab samples of aerobic soil and acclimated aqueous screening test data (Howard et al. 1991).

Biota:

Environmental Fate Rate Constant and Half-Lives:

Volatilization: Volatilization and biodegradation may be dominant transport and transformation processes for styrene in water; calculated volatilization half-life of 3 hours from a river 1 m deep with a current speed of 1.0 m/s and wind velocity of 3 m/s (Howard 1989); volatilized rapidly from shallow layers of lake water with half-life of 1 to 3 hours, but much slower from soil (Fu & Alexander 1992).

Photolysis:

Oxidation: photooxidation reaction rate constant of $1.8 \times 10^4$ L $mole^{-1}$ $sec^{-1}$ for the reaction with ozone at 30°C in vapor phase (Bufalini & Altshuller 1965); rate constant of $5.25 \times 10^{-11}$ $cm^3$ $molecule^{-1}$ $sec^{-1}$ for the reaction with OH radicals in the gas phase (Atkinson 1985); photooxidation half-life of 0.9 to 7.3 hours, based on measured rate data for the reaction with OH radicals and $O_3$ in air (Howard et al. 1991).

Hydrolysis: no hydrolyzable groups (Howard et al. 1991).

Biodegradation: aerobic biodegradation half-life of 336-672 hours, based on unacclimated grab samples of aerobic soil and a subsurface sample (Sielicki et al. 1978; Wilson et al. 1983; quoted, Howard et al. 1991); anaerobic half-life of 1344-2688 hours, based on estimated unacclimated aqueous aerobic biodegradation half-life (Howard et al. 1991); styrene will be rapidly destroyed by biodegradation in most aerobic environments, and the rate may be slow at low concentrations in aquifers and lake waters and in environments at low pH (Fu & Alexander 1992).

Biotransformation:

Bioconcentration, Uptake ($k_1$) and Elimination ($k_2$) Rate Constants or Half-Lives:

Common Name: Methylstyrene (*m*- and *p*-methylstyrene)
Synonym: methylvinylbenzene
Chemical Name: 3- and 4-methylstyrene
CAS Registry No: 100-42-1 (*m*-methylstyrene); 622-97-9 (*p*-methylstyrene)
Molecular Formula: $CH_3C_6H_4CH=CH_2$
Molecular Weight: 118.17
Melting Point (°C): -86.34 (*m*-methylstyrene), -34.15 (*p*-methylstyrene)
Boiling Point (°C): 167.7 (*m*- and *p*-mixture, Dreisbach 1955)
Density (g/cm³ at 20°C):

0.89768, 0.89353 (20°C, 25°C, Dreisbach 1955)

Molar Volume (cm³/mol):

131.6 (calculated-density)
155.2 (calculated-LeBas method)
130.2 (calculated-density, *m*-methylstyrene, Stephenson & Malanowski 1987)
128.8 (calculated-density, *p*-methylstyrene, Stephenson & Malanowski 1987)

Molecular Volume, TMV (Å³):
Total Surface Area, TSA (Å²):
Heat of Fusion, $\Delta H_{fus}$, (kcal/mol):
Entropy of Fusion, $\Delta S_{fus}$ (cal/mol K or e.u.):
Fugacity Ratio at 25°C, F: 1.0

Water Solubility (g/m³ or mg/L at 25°C):

89.0 (quoted, *m*- and *p*-methylstyrenes, Dreisbach 1955)
89, 100 (quoted, *m*- and *p*-methylstyrenes,calculated-group contribution method, Irmann 1965)

Vapor Pressure (Pa at 25°C):

218.6 (*m*- and *p*-methylstyrenes, Dreisbach 1955)
257, 242 (*m*-methylstyrene, *p*-methylstyrene, Dreisbach 1955)
332, 216 (Antoine eqn., *m*-methylstyrene, *p*-methylstyrene, Dean 1985)
245, 242 (Antoine eqn., *m*-methylstyrene, *p*-methylstyrene, Stephenson & Malanowski 1987)

Henry's Law Constant (Pa m³/mol):

386.8 (calculated-P/C for *m*-methylstyrene, Yaws et al. 1991)
286.9 (calculated-P/C for *p*-methylstyrene, Yaws et al. 1991)

Octanol/Water Partition Coefficient, log $K_{OW}$:

3.35 (Leo et al. 1971; quoted, Ogata et al. 1984)

Bioconcentration Factor, log BCF:

1.50 (gold fish, flow-through method, *m*-methylstyrene, Ogata et al. 1984)
1.55 (gold fish, flow-through method, *p*-methylstyrene, Ogata et al. 1984)
1.50, 1.63 (gold fish, quoted, calculated-$\chi$ , *m*-methylstyrene, Sabljic 1987)
1.55, 1.62 (gold fish, quoted, calculated-$\chi$ , *p*-methylstyrene, Sabljic 1987)

Sorption Partition Coefficient, log $K_{OC}$:

Half-Lives in the Environment:

Air:
Surface water:
Ground water:
Sediment:
Soil:
Biota:

Environmental Fate Rate Constant and Half-Lives:

Volatilization:
Photolysis:
Oxidation:
Hydrolysis:
Biodegradation:
Biotransformation:
Bioconcentration, Uptake ($k_1$) and Elimination ($k_2$) Rate Constants or Half-Lives:

Common Name: Tetralin
Synonym: tetralin, naphthalene-1,2,3,4-tetrahydride
Chemical Name: 1,2,3,4-tetrahydronaphthalene
CAS Registry No: 119-64-2
Molecular Formula: $C_6H_4(CH_2)_4$, $C_{10}H_{12}$
Molecular Weight: 132.205
Melting Point (°C): -35.75
Boiling Point (°C): 207.65
Density ($g/cm^3$ at 20°C):
0.9702 (20°C, Dean 1985)
0.9695, 0.9660 (20°, 25°C, Riddick et al. 1986)
Molar Volume ($cm^3/mol$):
136.4, 136.9 (20°, 25°C, calculated-density)
177.4 (calculated-LeBas method)
Molecular Volume, TMV ($Å^3$):
Total Surface Area, TSA ($Å^2$):
Heat of Fusion, $\Delta H_{fus}$, (kcal/mol): 2.975 (Riddick et al. 1986)
Entropy of Fusion, $\Delta S_{fus}$ (cal/mol K or e.u.):
Fugacity Ratio at 25°C, F: 1.0

Water Solubility ($g/m^3$ or mg/L at 25°C):
14.94 (calculated-QSAR data base, Passino & Smith 1987)

Vapor Pressure (Pa at 25°C):
40.0 (quoted, 20°C, Verschueren 1983)
53.75 (extrapolated, Antoine eqn., Dean 1985)
53.0 (quoted, Riddick et al. 1986)
56.7 (extrapolated, Antoine eqn., Stephenson & Malanowski 1987)

Henry's Law Constant (Pa $m^3$/mol):
484 (calculated-P/C from selected data)

Octanol/Water Partition Coefficient, log $K_{OW}$:
3.83 (calculated-f const., Rekker 1977)

Bioconcentration Factor, log BCF:

Sorption Partition Coefficient, log $K_{OC}$:

Half-Lives in the Environment:

Air:
Surface water:
Ground water:
Sediment:
Soil:
Biota:

Environmental Fate Rate Constant and Half-Lives:

Volatilization:
Photolysis:
Oxidation:
Hydrolysis:
Biodegradation:
Biotransformation:
Bioconcentration, Uptake ($k_1$) and Elimination ($k_2$) Rate Constants or Half-Lives:

## 2.2 Summary Tables and QSPR Plots

Table 2.1a Summary of physical properties of hydrocarbons at 25 °C

| Compounds | CAS no. | formula | MW g/mol | M.P. °C | B.P. °C | fugacity ratio, F at 25 °C | density g/cm³ 20 °C |
|---|---|---|---|---|---|---|---|
| *Alkanes*: | | | | | | | |
| 2-Methylpropane | 78-78-4 | $C_4H_{10}$ | 58.13 | -159.4 | -11.7 | 1 | 0.55711 |
| 2,2-Dimethylpropane | 75-83-2 | $C_5H_{12}$ | 72.15 | -16.55 | 9.503 | 1 | 0.591 |
| *n*-Butane | 106-97-8 | $C_4H_{10}$ | 58.13 | -138.35 | -0.5 | 1 | 0.57861 |
| 2-Methylbutane | 78-78-4 | $C_5H_{12}$ | 72.15 | -159.9 | 27.8 | 1 | 0.6193 |
| 2,2-Dimethylbutane | 75-83-2 | $C_6H_{14}$ | 86.17 | -99.87 | 49.74 | 1 | 0.64916 |
| 2,3-Dimethylbutane | 79-29-8 | $C_6H_{14}$ | 86.17 | -128.53 | 58 | 1 | 0.66164 |
| 2,2,3-Trimethylbutane | 464-06-2 | $C_7H_{16}$ | 100.21 | -24.2 | 80.9 | 1 | 0.6901 |
| 2,2,3,3-Tetramethylbutane | 594-82-1 | $C_8H_{18}$ | 114.23 | 100.7 | 106.5 | 0.156 | 0.8242 |
| *n*-Pentane | 109-66-0 | $C_5H_{12}$ | 72.15 | -129.7 | 36.07 | 1 | 0.6262 |
| 2-Methylpentane | 107-83-5 | $C_6H_{14}$ | 86.17 | -153.67 | 60.27 | 1 | 0.6532 |
| 3-Methylpentane | 96-14-0 | $C_6H_{14}$ | 86.17 | | 83.28 | 1 | 0.66431 |
| 2,2-Dimethylpentane | 590-35-2 | $C_7H_{16}$ | 100.21 | -123.82 | 79.2 | 1 | 0.6739 |
| 2,3-Dimethylpentane | 565-59-3 | $C_7H_{16}$ | 100.21 | | 89.9 | 1 | 0.6951 |
| 2,4-Dimethylpentane | 108-08-7 | $C_7H_{16}$ | 100.21 | -119.24 | 80.5 | 1 | 0.6727 |
| 3,3-Dimethylpentane | 562-49-2 | $C_7H_{16}$ | 100.21 | -134.46 | 86.06 | 1 | 0.6936 |
| 2,2,4-Trimethylpentane | 540-84-1 | $C_8H_{18}$ | 114.23 | -107.4 | 99.2 | 1 | 0.69193 |
| 2,3,4-Trimethylpentane | 565-75-3 | $C_8H_{18}$ | 114.23 | -109.2 | 113.4 | 1 | 0.7191 |
| 2,2,3,3-Tetramethylpentane | 7154-79-2 | $C_9H_{20}$ | 128.26 | | 140.27 | 1 | |

| Compounds | CAS no. | formula | MW g/mol | mp, °C | bp, °C | fugacity ratio, F at 25 °C | density g/cm³ 20 °C |
|---|---|---|---|---|---|---|---|
| *n*-Hexane | 110-54-3 | $C_6H_{14}$ | 86.17 | -95 | 68.95 | 1 | 0.65933 |
| 2-Methylhexane | 591-76-4 | $C_7H_{16}$ | 100.21 | -118.27 | 90 | 1 | 0.67856 |
| 3-Methylhexane | 589-34-4 | $C_7H_{16}$ | 100.21 | -119 | 92 | 1 | 0.68713 |
| 2,2,5-Trimethylhexane | 3522-94-9 | $C_9H_{20}$ | 128.26 | -105.8 | 124 | 1 | 0.7072 |
| *n*-Heptane | 142-82-5 | $C_7H_{16}$ | 100.21 | -90.61 | 98.42 | 1 | 0.6837 |
| 2-Methylheptane | 562-27-6 | $C_8H_{18}$ | 114.23 | -109 | 117.6 | 1 | 0.698 |
| *n*-Octane | 111-65-9 | $C_8H_{18}$ | 114.23 | -56.8 | 125.7 | 1 | 0.70267 |
| 3-Methyloctane | 2216-33-3 | $C_9H_{20}$ | 128.26 | | 143 | 1 | 0.714 |
| *n*-Nonane | 111-84-2 | $C_9H_{20}$ | 128.26 | -51 | 150.8 | 1 | 0.71772 |
| *n*-Decane | 124-18-5 | $C_{10}H_{22}$ | 142.29 | -29.7 | 174.1 | 1 | 0.73012 |
| *n*-Undecane | 1120-21-4 | $C_{11}H_{24}$ | 156.32 | -25.59 | 195.9 | 1 | 0.74017 |
| *n*-Dodecane | 112-40-3 | $C_{12}H_{26}$ | 170.33 | -9.6 | 216.3 | 1 | 0.74869 |
| *Alkenes*: | | | | | | | |
| 2-Methylpropene | 115-117 | $C_4H_8$ | 56.11 | -140.35 | -6.9 | 1 | 0.5942 |
| 1-Butene | 106-98-9 | $C_4H_8$ | 56.11 | -185.35 | -6.26 | 1 | 0.5951 |
| 3-Methyl-1-butene | 563-45-1 | $C_5H_{10}$ | 70.14 | -168.5 | 20.1 | 1 | 0.6272 |
| 1-Pentene | 109-67-1 | $C_5H_{10}$ | 70.14 | -138 | 30.0 | 1 | 0.6405 |
| 2-Methyl-1-pentene | 76-20-3 | $C_6H_{12}$ | 84.16 | -135.7 | 60.7 | 1 | 0.6799 |
| 4-Methyl-1-pentene | 691-37-2 | $C_6H_{12}$ | 84.16 | -153.6 | 53.9 | 1 | 0.6642 |
| 1-Hexene | 592-41-6 | $C_6H_{12}$ | 84.16 | -139.8 | 63.4 | 1 | 0.6732 |
| 1-Heptene | 592-76-7 | $C_7H_{14}$ | 98.19 | -119.03 | 93.64 | 1 | 0.6970 |
| 2-Heptene | 14686-13-6 | $C_7H_{14}$ | 98.19 | -136.6 | 95.7 | 1 | 0.7012 |

| Compounds | CAS no. | formula | MW g/mol | M.P. °C | B.P. °C | fugacity ratio, F at 25 °C | density g/cm$^3$ 20 °C |
|---|---|---|---|---|---|---|---|
| 1-Octene | 111-66-0 | $C_8H_{16}$ | 112.1 | -101.7 | 121.3 | 1 | 0.7149 |
| 1-Nonene | 124-11-8 | $C_9H_{18}$ | 126.241 | -81.34 | 146.88 | 1 | 0.7292 |
| 1-Decene | 872-05-9 | $C_{10}H_{20}$ | 140.27 | -66.3 | 170.5 | 1 | 0.7408 |
| *Diolefins*: | | | | | | | |
| 1,3-Butadiene | 106-99-0 | $C_4H_6$ | 54.09 | -108.9 | -4.4 | 1 | 0.6211 |
| 2-Methyl-1,3-butadiene | 78-79-5 | $C_5H_8$ | 68.13 | -146 | 34 | 1 | 0.6810 |
| 2,3-Dimethyl-1,3-butadiene | 513-81-5 | $C_6H_{10}$ | 82.15 | -76 | 69 | 1 | 0.7267 |
| 1,4-Pentadiene | 591-93-5 | $C_5H_8$ | 68.13 | -148.3 | 26 | 1 | 0.6608 |
| 1,5-Hexadiene | 592-47-2 | $C_6H_{10}$ | 82.15 | -141 | 60 | 1 | 0.6923 |
| 1,6-Heptadiene | 3070-53-9 | $C_7H_{12}$ | 96.17 | -129 | 89-90 | 1 | |
| 1,3,5-Hexatriene | 2612-46-6 | $C_6H_8$ | 80.13 | -12 | 78.5 | 1 | 0.737 |
| *Alkynes*: | | | | | | | |
| 1-Butyne | 107-00-6 | $C_4H_6$ | 54.09 | -125.7 | 8.1 | 1 | |
| 1-Pentyne | 627-19-0 | $C_5H_8$ | 68.13 | -90 | 40.18 | 1 | 0.6901 |
| 1-Hexyne | 693-02-7 | $C_6H_{10}$ | 82.15 | -132 | 71-72 | 1 | 0.7155 |
| 1-Heptyne | 628-71-7 | $C_7H_{12}$ | 96.17 | -81 | 99.7 | 1 | 0.7328 |
| 1-Octyne | 629-05-0 | $C_8H_{14}$ | 110.2 | -79.3 | 125.2 | 1 | 0.7461 |
| 1-Nonyne | 3452-09-3 | $C_9H_{16}$ | 124.23 | -50 | 150.8 | 1 | 0.7568 |
| *Cycloalkanes*: | | | | | | | |
| Cyclopentane | 287-92-3 | $C_5H_{10}$ | 70.14 | -93.9 | 49.3 | 1 | 0.74538 |
| Methylcyclopentane | 96-37-7 | $C_6H_{12}$ | 84.16 | -142.14 | 71.8 | 1 | 0.74864 |
| Ethylcyclopentane | 1640-89-7 | $C_7H_{14}$ | 98.19 | -138 | 103.5 | 1 | 0.7665 |

| Compounds | CAS no. | formula | MW g/mol | M.P. °C | B.P. °C | fugacity ratio, F at 25 °C | density g/cm$^3$ 20 °C |
|---|---|---|---|---|---|---|---|
| Propylcyclopentane | 2040-96-2 | $C_8H_{16}$ | 112.21 | -117.3 | 101 | 1 | 0.7763 |
| Pentylcyclopentane | 3741-00-2 | $C_{10}H_{20}$ | 140.26 | -83 | | 1 | 0.7912 |
| 1,1,3-Trimethylcyclopentane | 4516-69-2 | $C_8H_{16}$ | 112.21 | -14.2 | 104.9 | 1 | 0.7703 |
| Cyclohexane | 110-83-8 | $C_6H_{12}$ | 84.16 | 6.55 | 80.7 | 1 | 0.77855 |
| Methylcyclohexane | 108-87-2 | $C_7H_{14}$ | 98.19 | -126.6 | 100.9 | 1 | 0.7694 |
| 1,2-Dimethylcyclohexane | 583-57-3 | $C_8H_{16}$ | 112.11 | | 127 | | 0.7893 |
| 1-*cis*-2-Dimethylcyclohexane | 2207-01-4 | $C_8H_{16}$ | 112.21 | -50.1 | 129.7 | 1 | 0.79627 |
| 1-*trans*-2-Dimethylcyclohexane | 6876-23-9 | $C_8H_{16}$ | 112.21 | -90 | 124 | 1 | 0.77601 |
| 1,4-*trans*-Dimethylcyclohexane | 2207-04-7 | $C_8H_{16}$ | 112.21 | -37 | 119.4 | | |
| 1,1,3-Trimethylcyclohexane | 3073-66-3 | $C_9H_{18}$ | 126.24 | -65.7 | 138.94 | 1 | 0.7664 |
| Ethylcyclohexane | 1678-91-7 | $C_8H_{16}$ | 112.2 | -111.3 | 131.8 | 1 | 0.78792 |
| Cycloheptane | 291-64-5 | $C_7H_{14}$ | 98.19 | -12 | 118.5 | 1 | 0.8001 |
| Cyclooctane | 292-64-8 | $C_8H_{18}$ | 112.22 | 14.3 | 149 | 1 | 0.8349 |
| Decahydronaphthalene, *cis*- | 493-01-6 | $C_{10}H_{18}$ | 138.23 | -42.98 | 195.82 | 1 | 0.89671 |
| Decahydronaphthalene, *trans*- | 493-02-7 | $C_{10}H_{18}$ | 138.23 | -30.38 | 187.31 | 1 | 0.86971 |
| *Cycloalkenes*: | | | | | | | |
| Cyclopentene | 142-29-0 | $C_5H_8$ | 68.12 | -135 | 44 | 1 | 0.772 |
| Cyclohexene | 110-83-8 | $C_6H_{10}$ | 82.15 | -103.5 | 82.98 | 1 | 0.8102 |
| Methylcyclohexene | 591-69-1 | $C_7H_{12}$ | 96.17 | -121 | 110.24 | 1 | 0.8102 |
| Cycloheptene | 628-82-2 | $C_7H_{12}$ | 96.17 | -56 | 112-114 | 1 | 0.8228 |

| Compounds | CAS no. | formula | MW g/mol | M.P. °C | B.P. °C | fugacity ratio, F at 25 °C | density g/cm³ 20 °C |
|---|---|---|---|---|---|---|---|
| 1,4-Cyclohexadiene | 628-41-1 | $C_6H_8$ | 80.14 | -49.2 | 85.6 | 1 | 0.8471 |
| 1,3,5-Cycloheptatriene | 544-25-2 | $C_7H_8$ | 92.14 | -79.5 | 116-117 | 1 | 0.8875 |
| *d*-Limonene | 5989-27-5 | $C_{10}H_{16}$ | 136.24 | -96.9 | 175 | 1 | 0.8403 |
| *Aromatics*: | | | | | | | |
| Styrene | 100-42-5 | $C_8H_8$ | 104.14 | -30.6 | 145.2 | 1 | 0.906 |
| *o*-Methylstyrene | 611-15-4 | $C_9H_{10}$ | 118.18 | -68.5 | 169 | 1 | 0.9036 |
| *m*-Methylstyrene | 100-80-1 | $C_9H_{10}$ | 118.18 | -86.3 | 171.6 | 1 | 0.9113 |
| *p*-Methylstyrene | 622-97-9 | $C_9H_{10}$ | 118.18 | -34.1 | 172.78 | 1 | 0.9106 |
| α-Methylstyrene | 98-83-9 | $C_9H_{10}$ | 118.18 | -23.1 | 165.38 | 1 | 0.9106 |
| *β*-Methylstyrene | 766-90-5 | $C_9H_{10}$ | 118.18 | -52.25 | 170 | 1 | 0.911 |
| Tetralin | 119-64-2 | $C_{10}H_{12}$ | 132.21 | -35.75 | 207.65 | 1 | 0.9695 |
| Diphenylene | 259-79-0 | $C_{12}H_8$ | 152.2 | 113-114 | | 0.135 | |
| Diphenylmethane | 101-81-5 | $C_{13}H_{12}$ | 168.24 | 25.3 | 264 | 0.984 | 1.006 |
| Bibenzyl | 103-29-3 | $C_{14}H_{14}$ | 182.27 | 52.2 | 285 | 0.538 | |

Table 2.1b Summary of physical properties of hydrocarbons at 25 °C

| Compounds | Molar volume | | | | | | |
|---|---|---|---|---|---|---|---|
| | $V_M$ MW/$\rho$ 20 °C | $V_M$ LeBas $cm^3/mol$ | $V_I/100$ intrinsic (a) | TSA $Å^2$ (b) | TSA $Å^2$ (c) | TMV $Å^3$ (c) | TSA $Å^2$ (d) |
| *Alkanes*: | | | | | | | |
| 2-Methylpropane | 104.34 | 96.2 | 0.458 | 249.1 | 115.64 | 75.82 | 111.1 |
| 2,2-Dimethylpropane | 122.08 | 118.4 | | 270.14 | 137.45 | 91.99 | 134.6 |
| *n*-Butane | 100.46 | 96.2 | 0.458 | 255.15 | 116.1 | 75.87 | |
| 2-Methylbutane | 116.50 | 118.4 | 0.553 | 274.58 | 130.5 | 91.12 | 132 |
| 2,2-Dimethylbutane | 132.74 | 140.6 | | 290.76 | 152.29 | 107.29 | 155.5 |
| 2,3-Dimethylbutane | 130.24 | 140.6 | | | 152.76 | 107.29 | 155.4 |
| 2,2,3-Trimethylbutane | 145.21 | 162.8 | | | 166.7 | 122.54 | |
| 2,2,3,3-Tetramethylbutane | 138.60 | 185 | | | | | |
| *n*-Pentane | 115.3 | 118.4 | 0.553 | 286.97 | 138.8 | 92.15 | 129.5 |
| 2-Methylpentane | 131.92 | 140.6 | 0.648 | 306.41 | 153.2 | 107.4 | 152.9 |
| 3-Methylpentane | 129.71 | 140.6 | | 300.08 | 153.2 | 107.4 | 152.9 |
| 2,2-Dimethylpentane | 148.70 | 162.8 | | | 174.99 | 123.57 | 176.4 |
| 2,3-Dimethylpentane | 144.17 | 162.8 | | | 171.53 | 123.14 | 176.3 |
| 2,4-Dimethylpentane | 148.97 | 162.8 | | 324.71 | 174.45 | 123.63 | 176.3 |
| 3,3-Dimethylpentane | 144.48 | 162.8 | | | 167.15 | 122.6 | |
| 2,2,4-Trimethylpentane | 165.09 | 185 | | | | | 199.8 |
| 2,3,4-Trimethylpentane | 158.85 | 185 | | | | | 199.7 |
| 2,2,3,3-Tetramethylpentane | | 207.2 | | | | | |

| Compounds | $V_M$ MW/$\rho$ 20 °C | $V_M$ LeBas $cm^3$/mol | $V_I$/100 intrinsic (a) | TSA $Å^2$ (b) | TSA $Å^2$ (c) | TMV $Å^3$ (c) | TSA $Å^2$ (d) |
|---|---|---|---|---|---|---|---|
| *n*-Hexane | 130.69 | 140.6 | 0.644 | | 161.51 | 108.44 | 150.4 |
| 2-Methylhexane | 147.68 | 162.8 | 0.745 | | 175.9 | 123.68 | 173.8 |
| 3-Methylhexane | 145.84 | 162.8 | | | 175.9 | 123.68 | 173.8 |
| 2,2,5-Trimethylhexane | 181.36 | 207.2 | | | | | 220.7 |
| *n*-Heptane | 146.57 | 162.80 | 0.745 | | 184.21 | 124.72 | 171.3 |
| 2-Methylheptane | 163.65 | 185 | | | | | |
| *n*-Octane | 162.57 | 185 | 0.842 | | 206.91 | 141 | 192.2 |
| 3-Methyloctane | 179.64 | 207.2 | | | | | |
| *n*-Nonane | 178.70 | 207.2 | | | 229.61 | 157.28 | 213.1 |
| *n*-Decane | 194.89 | 229.4 | | | | | 234 |
| *n*-Undecane | | 251.6 | | | | | |
| *n*-Dodecane | 227.5 | 273.8 | | | | | |
| *Alkenes*: | | | | | | | |
| 2-Methylpropene | 94.43 | 88.8 | | | | | |
| 1-Butene | 94.3 | 88.8 | 0.427 | | | | 102.7 |
| 3-Methyl-1-butene | 111.8 | 111 | | | | | 126.1 |
| 1-Pentene | 109.5 | 111 | | | | | |
| 2-Methyl-1-pentene | 123.8 | 133.2 | | | | | 157 |
| 4-Methyl-1-pentene | 126.7 | 133.2 | | | | | 147 |
| 1-Hexene | 125 | 133.2 | | | | | 144.5 |
| 1-Heptene | 140.9 | 155.4 | | | | | |
| 2-Heptene | 139 | 155.4 | | | | | 165.1 |

| Compounds | $V_M$ MW/$\rho$ 20 °C | $V_M$ LeBas $cm^3/mol$ | $V_I/100$ intrinsic (a) | TSA $Å^2$ (b) | TSA $Å^2$ (c) | TMV $Å^3$ (c) | TSA $Å^2$ (d) |
|---|---|---|---|---|---|---|---|
| 1-Octene | 157.8 | 177.6 | | | | | 186.3 |
| 1-Nonene | 172.93 | 199.8 | | | | | |
| 1-Decene | 189.35 | 222 | | | | | |
| *Diolefins*: | | | | | | | |
| 1,3-Butadiene | 87 | 81.4 | | | | | 96.8 |
| 2-Methyl-1,3-butadiene | 100 | 103.6 | | | | | 118.3 |
| 2,3-Dimethyl-1,3-butadiene | 113.04 | 125.8 | | | | | |
| 1,4-Pentadiene | 103 | 125.8 | | | | | |
| 1,5-Hexadiene | 119 | 148 | | | | | 138.6 |
| 1,6-Heptadiene | 134 | 170.2 | | | | | |
| 1,3,5-Hexatriene | 108.7 | 118.4 | | | | | |
| *Alkynes*: | | | | | | | |
| 1-Butyne | 83 | 81.4 | 0.415 | | | | 96.5 |
| 1-Pentyne | 99 | 103.6 | | | | | |
| 1-Hexyne | 114.81 | 125.8 | | | | | |
| 1-Heptyne | 131.24 | 148 | | | | | |
| 1-Octyne | 147.70 | 178.2 | | | | | |
| 1-Nonyne | 169.44 | 200.4 | | | | | |
| *Cycloalkanes*: | | | | | | | |
| Cyclopentane | 94.10 | 99.5 | | 255.4 | | | 104.5 |
| Methylcyclopentane | 112.42 | 121.7 | | 282.71 | | | 127.9 |
| Ethylcyclopentane | 128.10 | 142.9 | | | | | |

| Compounds | $V_M$ MW/$\rho$ 20 °C | $V_M$ LeBas $cm^3/mol$ | $V_I/100$ intrinsic (a) | TSA $Å^2$ (b) | TSA $Å^2$ (c) | TMV $Å^3$ (c) | TSA $Å^2$ (d) |
|---|---|---|---|---|---|---|---|
| Propylcyclopentane | 144.54 | 166.1 | | | | | 169.7 |
| Pentylcyclopentane | 177.28 | 207 | | | | | 211.5 |
| 1,1,3-Trimethylcyclopentane | 145.67 | 166.1 | | | | | 174.8 |
| Cyclohexane | 108.10 | 118.2 | 0.598 | 279.1 | 128.82 | 93.45 | 125.4 |
| Methylcyclohexane | 127.62 | 140.4 | | 304.85 | | | 148.8 |
| 1,2-Dimethylcyclohexane | 142.04 | 162.6 | | | | | |
| 1-*cis*-2-Dimethylcyclohexane | 140.92 | 162.6 | | 315.5 | | | 172.2 |
| 1-*trans*-2-Dimethylcyclohexane | 144.60 | 162.6 | | | | | |
| 1,4-*trans*-dimethylcyclohexane | 147.1 | 162.6 | | | | | 172.2 |
| 1,1,3-Trimethylcyclohexane | 164.72 | 184.8 | | | | | |
| Ethylcyclohexane | 142.40 | 162.6 | | | | | |
| Cycloheptane | 121 | 140.4 | | 301.94 | | | |
| Cyclooctane | 134.41 | 169.8 | | 322.28 | | | |
| Decahydronaphthalene, *cis-* | 154.15 | 184.6 | | | | | |
| Decahydronaphthalene, *trans-* | 158.94 | 184.6 | | | | | |
| *Cycloalkenes*: | | | | | | | |
| Cyclopentene | 88.24 | 77.3 | | | | | |
| Cyclohexene | 101.39 | 110.8 | | | | | |
| Methylcyclohexene | 118.70 | 133 | | | | | |
| Cycloheptene | 116.88 | 133 | | | | | |
| 1,3-Cyclohexadiene | 95.35 | 103.4 | | | | | |

| Compounds | $V_M$ MW/$\rho$ 20 °C | $V_M$ LeBas cm$^3$/mol | $V_I$/100 intrinsic (a) | TSA Å$^2$ (b) | TSA Å$^2$ (c) | TMV Å$^3$ (c) | TSA Å$^2$ (d) |
|---|---|---|---|---|---|---|---|
| 1,4-Cyclohexadiene | 94.61 | 103.4 | | | | | |
| 1,3,5-Cycloheptatriene | 103.82 | 118.2 | | | | | |
| *d*-Limonene | 162.13 | 192.2 | | | | | |
| *Aromatics*: | | | | | | | |
| Styrene (phenylethene) | 114.94 | 133 | | | | | |
| *o*-Methylstyrene | 130.8 | 155.2 | | | | | |
| *m*-Methylstyrene | 129.7 | 155.2 | | | | | |
| *p*-Methylstyrene | 129.8 | 155.2 | | | | | |
| α-Methylstyrene | 129.8 | 155.2 | | | | | |
| β-Methylstyrene | 129.7 | 155.2 | | | | | |
| Tetralin | 136.4 | 162.4 | | | | | |
| Diphenylene | | 192.0 | | | | | |
| Diphenylmethane | | 206.8 | | | 214.25 | 171.66 | |
| Bibenzyl | | 229.0 | | | 236.96 | 187.97 | |

(a) Kamlet et al. 1987, 1988
(b) Hermann 1972
(c) Pearlman 1986, Pearlman et al. 1984
(d) Okouchi et al. 1992

Table 2.2 Summary of selected physical-chemical properties of hydrocarbons at 25 °C

| Compounds | Selected properties: | | | | | | |
|---|---|---|---|---|---|---|---|
| | Vapor pressure | | Solubility | | | | Henry's law const. |
| | $P^S$, Pa | $P_L$, Pa | S, $g/m^3$ | $C^S$, $mol/m^3$ | $C_L$, $mol/m^3$ | log $K_{OW}$ | H, calc, P/C $Pa{\cdot}m^3/mol$ |
| *Alkanes*: | | | | | | | |
| 2-methylpropane (isobutane) | 357000 | 357000 | 48.9 | 0.8412 | 0.8412 | 2.8 | 120450 |
| 2,2-Dimethylpropane | 172000 | 172000 | 33.2 | 0.4602 | 0.4602 | 3.11 | 220199 |
| *n*-Butane | 243000 | 243000 | 61.4 | 1.0563 | 1.0563 | | 95929 |
| 2-Methylbutane | 91640 | 91640 | 13.8 | 0.1913 | 0.1913 | | 479118 |
| 2,2-Dimethylbutane | 42600 | 42600 | 18.4 | 0.2135 | 0.2135 | 3.82 | 199502 |
| 2,3-Dimethylbutane | 32010 | 32010 | 19.1 | 0.2217 | 0.2217 | 3.85 | 144414 |
| 2,2,3-trimethylbutane | 13652 | 13652 | 4.38 | 0.0437 | 0.0437 | | 312344 |
| 2,2,3,3-tetramethylbutane | 2776 | 2776 | | | | | |
| *n*-Pentane | 68400 | 68400 | 38.5 | 0.5336 | 0.5336 | 3.45 | 128183 |
| 2-Methylpentane | 28200 | 28200 | 13.8 | 0.1601 | 0.1601 | | 176087 |
| 3-Methylpentane | 25300 | 25300 | 12.8 | 0.1485 | 0.1485 | 3.60 | 170320 |
| 2,2-Dimethylpentane | 14000 | 14000 | 4.4 | 0.0439 | 0.0439 | | 318850 |
| 2,3-Dimethylpentane | 9180 | 9180 | 5.25 | 0.0524 | 0.0524 | | 175224 |
| 2,4-Dimethylpentane | 13100 | 13100 | 4.06 | 0.0405 | 0.0405 | | 323338 |
| 3,3-Dimethylpentane | 10940 | 10940 | 5.94 | 0.0593 | 0.0593 | | 184562 |
| 2,2,4-Trimethylpentane | 6560 | 6560 | 2.44 | 0.0214 | 0.0214 | | 307110 |
| 2,3,4-Trimethylpentane | 3600 | 3600 | 2.0 | 0.0119 | 0.0119 | | 205614 |
| 2,2,3,3-Tetramethylpentane | | | | | | | |

| Compounds | Vapor pressure | | Solubility | | | log $K_{OW}$ | Henry's law const. |
|---|---|---|---|---|---|---|---|
| | $P^S$, Pa | $P_L$, Pa | S, g/m$^3$ | $C^S$, mol/m$^3$ | $C_L$, mol/m$^3$ | | H, calc, P/C Pa·m$^3$/mol |
| *n*-Hexane | 20200 | 20200 | 9.5 | 0.1102 | 0.1102 | 4.11 | 183225 |
| 2-Methylhexane | 8780 | 8780 | 2.54 | 0.0253 | 0.0253 | | 346395 |
| 3-Methylhexane | 8210 | 8210 | 3.3 | 0.0263 | 0.0263 | | 249310 |
| 2,2,5-Trimethylhexane | 2210 | 2210 | 1.15 | 0.0090 | 0.0090 | | 246482 |
| *n*-Heptane | 6110 | 6110 | 2.93 | 0.0292 | 0.0292 | 5.0 | 208970 |
| 2-Methylheptane | 2600 | 2600 | 0.85 | 0.00744 | 0.0074 | | 349409 |
| *n*-Octane | 1800 | 1800 | 0.66 | 0.005778 | 0.0058 | 5.15 | 311536 |
| 3-methyloctane | 823 | 823 | 1.42 | 0.011071 | 0.0111 | | |
| *n*-Nonane | 571 | 571 | 0.22 | 0.001715 | 0.0017 | 5.65 | 332893 |
| *n*-Decane | 175 | 175 | 0.052 | 0.000365 | 0.00037 | 6.25 | 478861 |
| *n*-Undecane | 52.2 | 52.2 | 0.04 | 0.000256 | 0.000256 | | |
| *n*-Dodecane | 15.7 | 15.7 | 0.0037 | 0.000022 | 0.000022 | | |
| *Alkenes*: | | | | | | | |
| 2-Methylpropene | 558000 | 558000 | 263 | 4.6872 | 4.6872 | 2.35 | 21617 |
| 1-Butene | 297000 | 297000 | 222 | 3.9565 | 3.9565 | | 25610 |
| 3-Methyl-1-butene | 120000 | 120000 | 130 | 1.8534 | 1.8534 | | 54669 |
| 1-Pentene | 85000 | 85000 | 148 | 2.1101 | 2.1101 | 2.2 | 40283 |
| 2-Methyl-1-pentene | 26000 | 26000 | 78 | 0.9268 | 0.9268 | | 28053 |
| 4-Methyl-1-pentene | 36100 | 36100 | 48 | 0.5703 | 0.5703 | 2.5 | 63295 |
| 1-Hexene | 24800 | 24800 | 50 | 0.5941 | 0.5941 | 3.39 | 41743 |
| 1-Heptene | 7510 | 7510 | 18.3 | 0.1864 | 0.1864 | 3.99 | 40295 |
| 2-Heptene (*trans*-) | 6450 | 6450 | 15 | 0.1528 | 0.1528 | | 42222 |

| Compounds | Vapor pressure | | Solubility | | | log $K_{OW}$ | Henry's law const. |
|---|---|---|---|---|---|---|---|
| | $P^S$, Pa | $P_L$, Pa | S, g/m$^3$ | $C^S$, mol/m$^3$ | $C_L$, mol/m$^3$ | | H, calc, P/C Pa·m$^3$/mol |
| 1-Octene | 2320 | 2320 | 2.7 | 0.0241 | 0.0241 | 4.57 | 96323 |
| 1-Nonene | 712 | 712 | 1.12 | 0.0089 | 0.0089 | 5.15 | 80253 |
| 1-Decene | 216 | 216 | 0.1 | 0.0007 | 0.0007 | 5.31 | 302983 |
| *Diolefins*: | | | | | | | |
| 1,3-butadiene | 281000 | 281000 | 735 | 13.5885 | 13.5885 | 1.99 | 20679 |
| 2-Methyl-1,3-butadiene | 73300 | 73300 | 642 | 9.4232 | 9.4232 | | 7779 |
| 2,3-dimethyl-1,3-butadiene | 20160 | 20160 | 327 | 3.9805 | 3.9805 | | 5065 |
| 1,4-Pentadiene | 98000 | 98000 | 558 | 8.1902 | 8.1902 | 2.48 | 11965 |
| 1,5-hexadiene | | | 169 | 2.0572 | 2.0572 | 2.8 | |
| 1,6-heptadiene | | | 44 | 0.4575 | 0.4575 | | |
| 1,3,5-hexatriene | | | 620 | 7.737 | 7.737 | | |
| *Alkynes*: | | | | | | | |
| 1-Butyne | 188000 | 188000 | 2870 | 57.2969 | 57.297 | | |
| 1-Pentyne | 57600 | 57600 | 1570 | 23.0442 | 23.044 | 2.12 | 2500 |
| 1-Hexyne | 18140 | 18140 | 360 | 4.3822 | 4.382 | 2.73 | 4139 |
| 1-Heptyne | 4372 | 4372 | 94 | 0.9774 | 0.977 | | 4473 |
| 1-Octyne | 1715 | 1715 | 24 | 0.2178 | 0.218 | | 7875 |
| 1-Nonyne | | | | 0.0561 | 0.056 | | |
| *Cycloalkanes*: | | | | | | | |
| Cyclopentane | 42400 | 42400 | 156 | 2.2241 | 2.224 | 3.00 | 19064 |
| Methylcyclopentane | 18300 | 18300 | 42 | 0.4990 | 0.4990 | 3.37 | 36670 |
| Ethylcyclopentane | 5324 | 5324 | | | | | |

| Compounds | Vapor pressure | | Solubility | | | log $K_{OW}$ | Henry's law const. |
|---|---|---|---|---|---|---|---|
| | $P^S$, Pa | $P_L$, Pa | S, g/m$^3$ | $C^S$, mol/m$^3$ | $C_L$, mol/m$^3$ | | H, calc, P/C Pa·m$^3$/mol |
| Propylcyclopentane | 1640 | 1640 | 2.04 | 0.0182 | 0.0182 | | 90208 |
| Pentylcyclopentane | 152 | 152 | 0.115 | 0.0008 | 0.0008 | | 185387 |
| 1,1,3-Trimethylcyclopentane | 5300 | 5300 | 3.73 | 0.0332 | 0.0332 | | 159440 |
| Cyclohexane | 12700 | 12700 | 55 | 0.6535 | 0.6535 | 3.44 | 19433 |
| Methylcyclohexane | 6180 | 6180 | 14 | 0.1426 | 0.1426 | 3.88 | 43344 |
| 1,2-Dimethylcyclohexane | 2584 | 2584 | 6.02 | 0.0537 | 0.0537 | | 48122 |
| 1-*cis*-2-Dimethylcyclohexane | 1930 | 1930 | 6 | 0.0535 | 0.0535 | | |
| 1-*trans*-2-Dimethylcyclohexane | 2584 | 2584 | 3.73 | 0.0332 | 0.0332 | | 77735 |
| 1,4-*trans*-Dimethylcyclohexane | 3020 | 3020 | 3.84 | 0.0342 | 0.0342 | | 88248 |
| 1,1,3-Trimethylcyclohexane | 1480 | 1480 | 1.77 | 0.0140 | 0.0140 | | 105557 |
| Ethylcyclohexane | | | | | | | |
| Cycloheptane | 2930 | 2930 | 30 | 0.3055 | 0.3055 | | 9590 |
| Cyclooctane | 753 | 753 | 7.9 | 0.0704 | 0.0704 | 4.45 | 10696 |
| Decahydronaphthalene, *cis-* | 104 | 104 | 6.21 | 0.045 | 0.045 | | 2311 |
| Decahydronaphthalene, *trans-* | 168 | 168 | 6.21 | 0.045 | 0.045 | | 3644 |
| *Cycloalkenes*: | | | | | | | |
| Cyclopentene | 50706 | 50706 | 535 | 7.8538 | 7.8538 | | 6456 |
| Cyclohexene | 11850 | 11850 | 213 | 2.5928 | 2.5928 | 2.86 | 4570 |
| Methylcyclohexene | 3393 | 3393 | | | | | |
| Cycloheptene | 3363 | 3363 | 66 | 0.6863 | 0.6863 | | 4900 |
| 1,3-Cyclohexadiene | | | | | | 2.47 | |

| Compounds | Vapor pressure | | Solubility | | | log $K_{OW}$ | Henry's law const. H, calc, P/C |
|---|---|---|---|---|---|---|---|
| | $P^S$, Pa | $P_L$, Pa | S, g/m$^3$ | $C^S$, mol/m$^3$ | $C_L$, mol/m$^3$ | | Pa·m$^3$/mol |
| 1,4-Cyclohexadiene | 9010 | 9010 | 700 | 8.7347 | 8.7347 | 2.3 | 1032 |
| 1,3-Cycloheptadiene | | | 620 | 6.5845 | 6.5845 | | |
| 1,3,5-Cycloheptatriene | 3140 | 3140 | 620 | 6.7289 | 6.7289 | 2.63 | 467 |
| *d*-Limonene | 202 | 202 | 13.8 | 0.1013 | 0.1013 | | |
| *Aromatics*: | | | | | | | |
| Styrene | 880 | 800 | 300 | 2.8807 | 2.8807 | 3.05 | 305.48 |
| *o*-Methylstyrene | 240.8 | 240.8 | | | | 3.35 | |
| *m*-Methylstyrene | 257 | 257 | | | | 3.35 | |
| *p*-Methylstyrene | 241.5 | 241.5 | | | | 3.35 | |
| α-Methylstyrene | 333 | 333 | | | | 3.35 | |
| β-Methylstyrene | 267 | 267 | | | | 3.35 | |
| Tetralin | 53.0 | 53.0 | 15.0 | 0.113 | 0.113 | 3.83 | 469 |
| Diphenylene | 0.0053 | 0.0393 | | | | | |
| Diphenylmethane | 0.0885 | 0.0885 | 16.0 | 0.095 | 0.095 | 4.14 | 0.93 |
| Bibenzyl | 0.406 | 0.754 | 4.37 | 0.024 | 0.0445 | 4.70 | 16.9 |

Table 2.3 Suggested half-life classes of hydrocarbons in various environmental compartments

| Compounds | Air class | Water class | Soil class | Sediment class |
|---|---|---|---|---|
| n-Pentane | 2 | 5 | 6 | 7 |
| n-Hexane | 2 | 5 | 6 | 7 |
| n-Octane | 2 | 5 | 6 | 7 |
| n-Decane | 2 | 5 | 6 | 7 |
| n-Dodecane | 2 | 5 | 6 | 7 |
| 1-Pentene | 1 | 4 | 5 | 6 |
| 1-Octene | 1 | 4 | 5 | 6 |
| Styrene | 1 | 4 | 5 | 6 |
| 1,3-Butadiene | 1 | 4 | 5 | 6 |
| 1,4-Pentadiene | 1 | 4 | 5 | 6 |
| 1,4-Cyclohexadiene | 1 | 4 | 5 | 6 |
| 1-Hexyne | 1 | 4 | 5 | 6 |
| Cyclopentane | 2 | 5 | 5 | 6 |
| Methylcyclopentane | 2 | 5 | 5 | 6 |
| Cyclohexane | 2 | 5 | 5 | 6 |
| Methylcyclohexane | 2 | 5 | 5 | 6 |
| Cyclooctane | 2 | 5 | 5 | 6 |
| Cyclopentene | 1 | 4 | 5 | 6 |
| Cyclohexene | 1 | 4 | 5 | 6 |

where,

| Class | Mean half-life (hours) | Range (hours) |
|---|---|---|
| 1 | 5 | < 10 |
| 2 | 17 (~ 1 day) | 10-30 |
| 3 | 55 (~ 2 days) | 30-100 |
| 4 | 170 (~ 1 week) | 100-300 |
| 5 | 550 (~ 3 weeks) | 300-1,000 |
| 6 | 1700 (~ 2 months) | 1,000-3,000 |
| 7 | 5500 (~ 8 months) | 3,000-10,000 |
| 8 | 17000 (~ 2 years) | 10,000-30,000 |
| 9 | 55000 (~ 6 years) | > 30,000 |

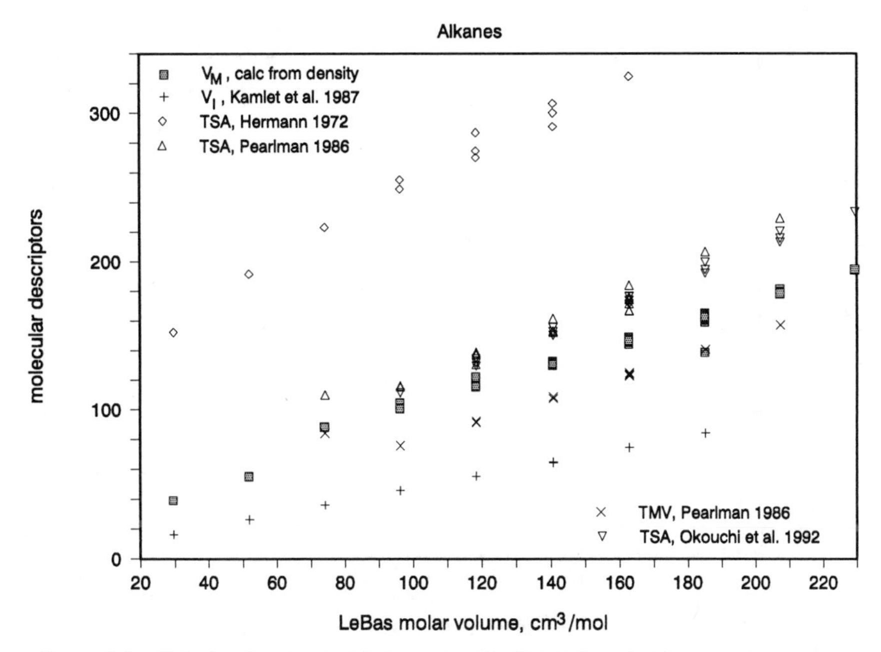


Figure 2.1a Plot of molecular descriptors versus LeBas molar volume.

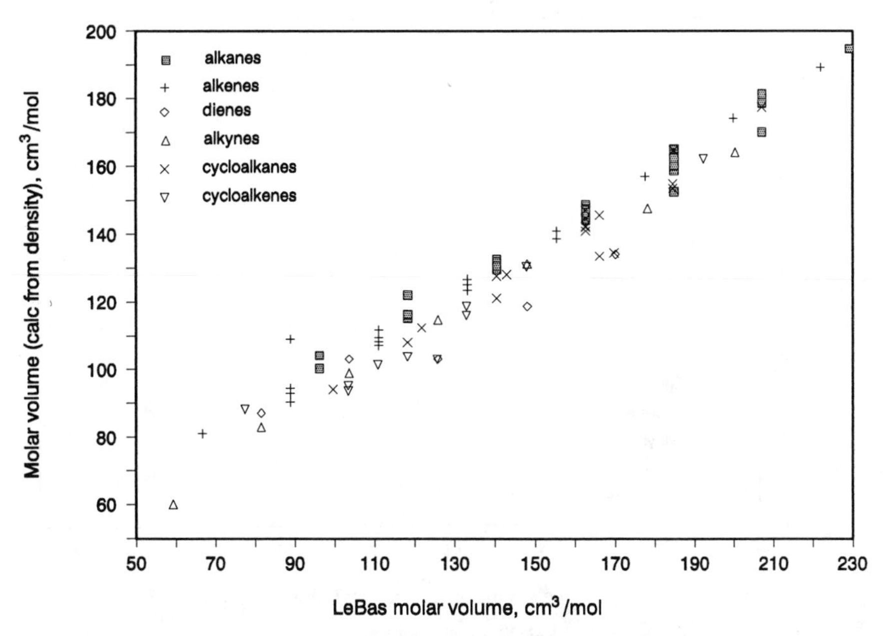


Figure 2.1b Plot of liquid molar volume (calc from density) vs LeBas molar volume.

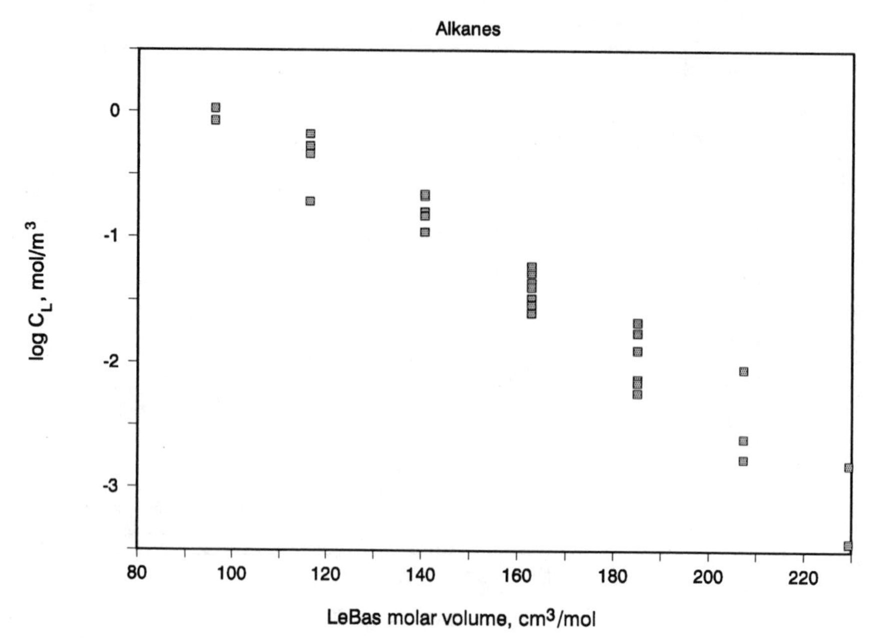


Figure 2.2a Plot of log $C_L$ versus LeBas molar volume.

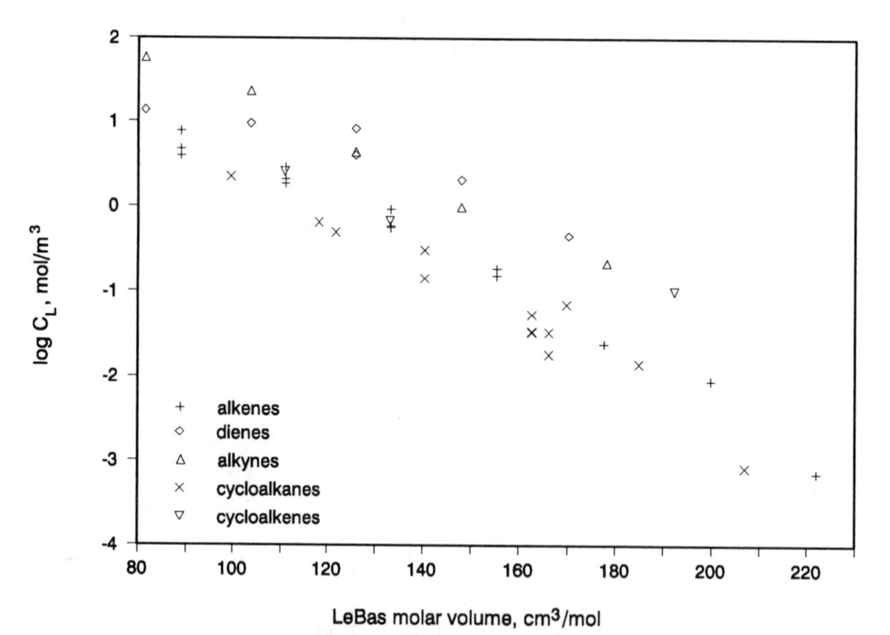


Figure 2.2b Plot of log $C_L$ versus LeBas molar volume.

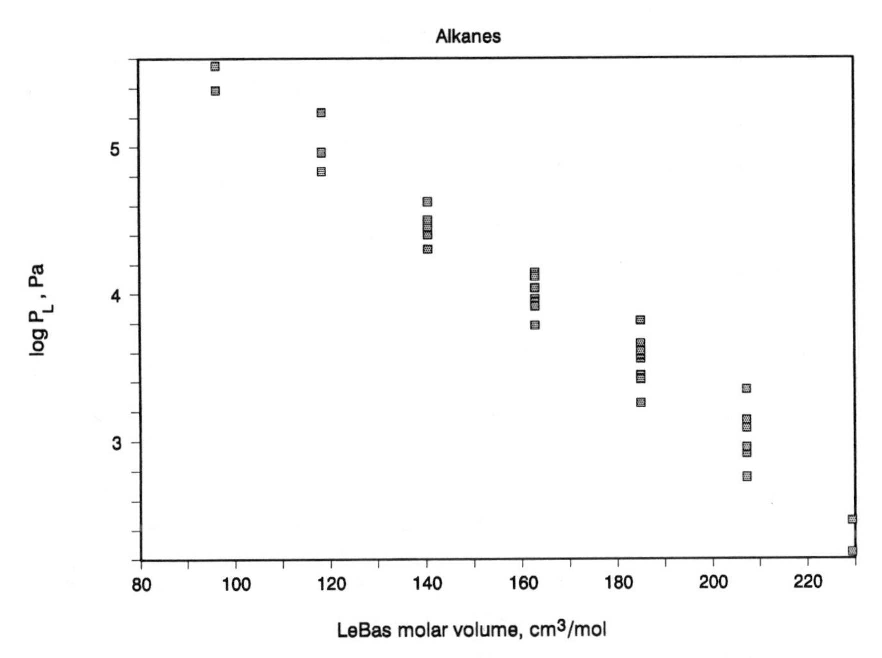

Figure 2.3a Plot of log vapor pressure versus LeBas molar volume.

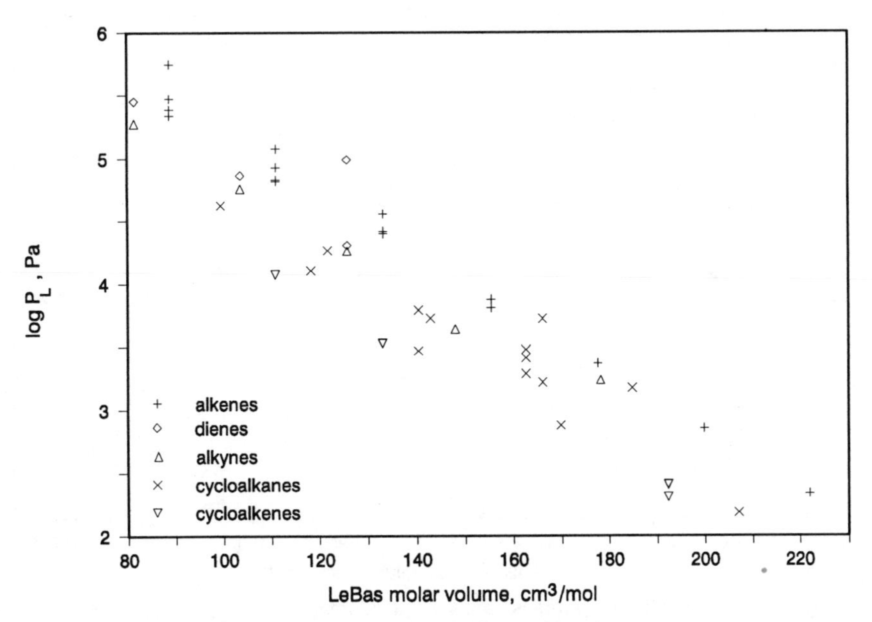

Figure 2.3b Plot of log vapor pressure versus LeBas molar volume.

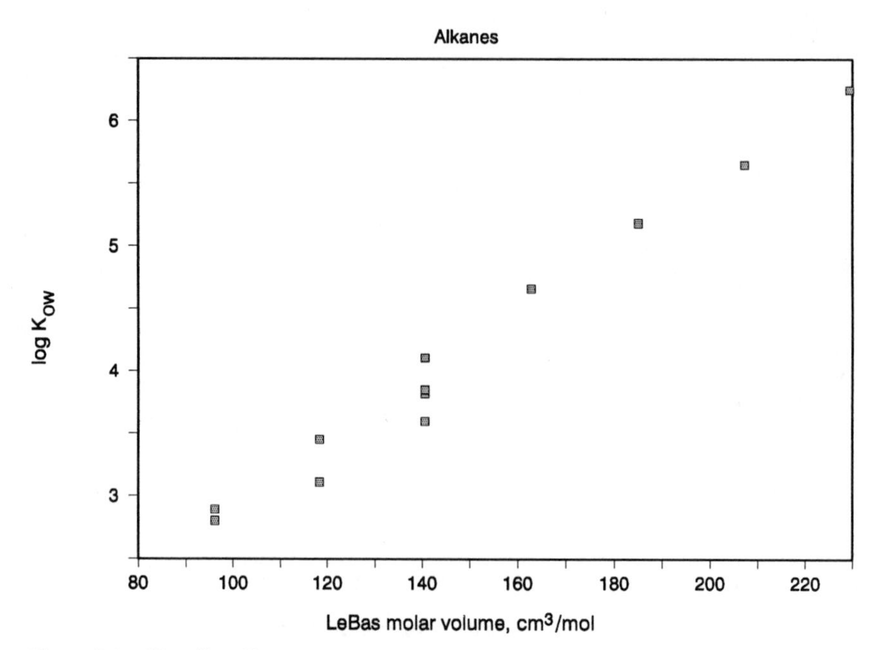


Figure 2.4a Plot of log $K_{OW}$ versus LeBas molar volume.

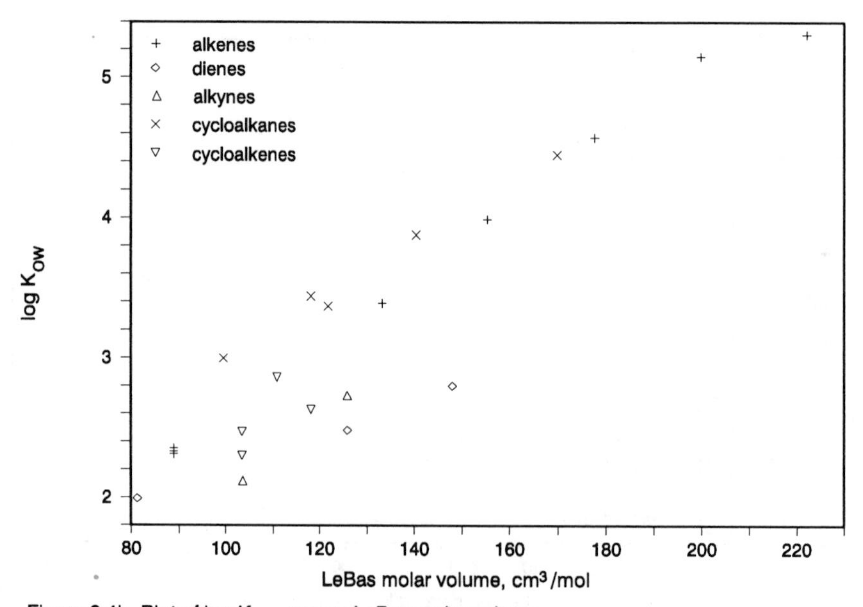


Figure 2.4b Plot of log $K_{OW}$ versus LeBas molar volume.

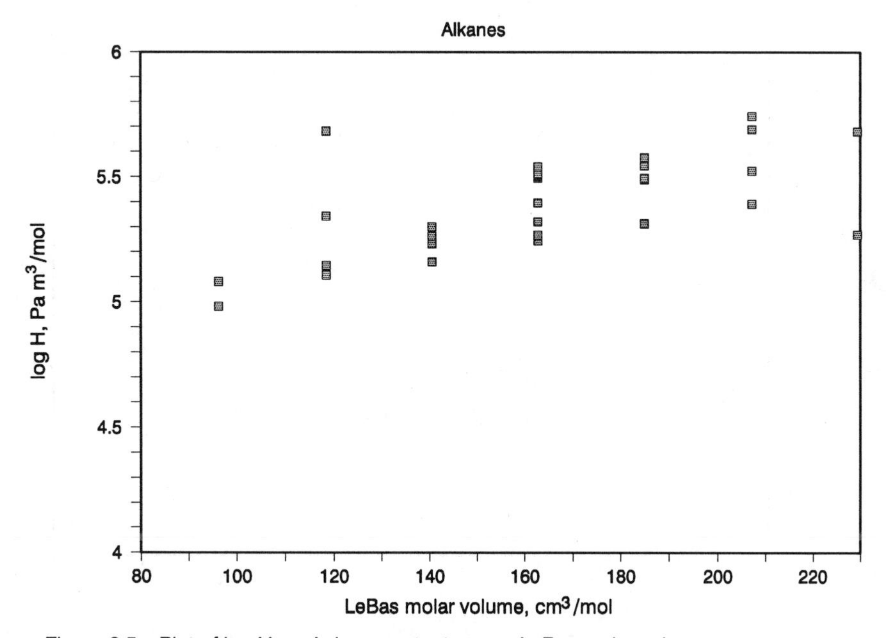


Figure 2.5a Plot of log Henry's law constant versus LeBas molar volume.

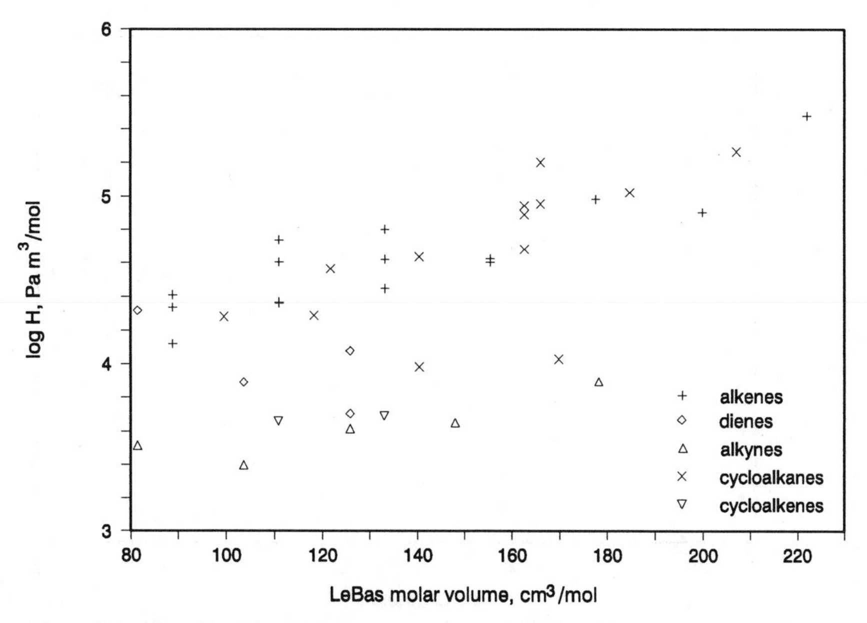


Figure 2.5a Plot of log Henry's law constant versus LeBas molar volume.

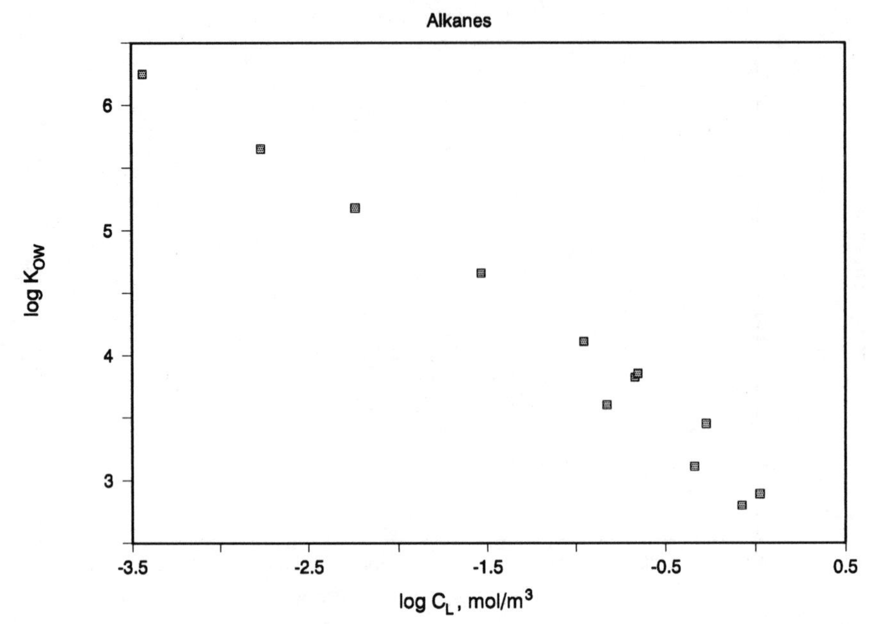

Figure 2.6a Plot of log $K_{OW}$ versus log $C_L$.

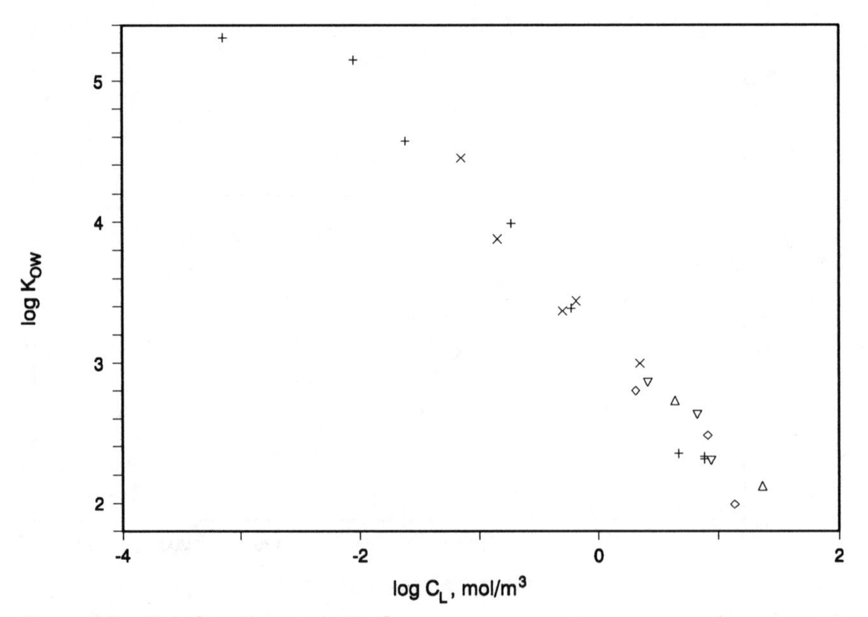

Figure 2.6b Plot of log $K_{OW}$ versus log $C_L$.

## 2.3 Illustrative Fugacity Calculations: Levels I, II and III

Chemical name: n-Pentane

Level I calculation: (six-compartment model)

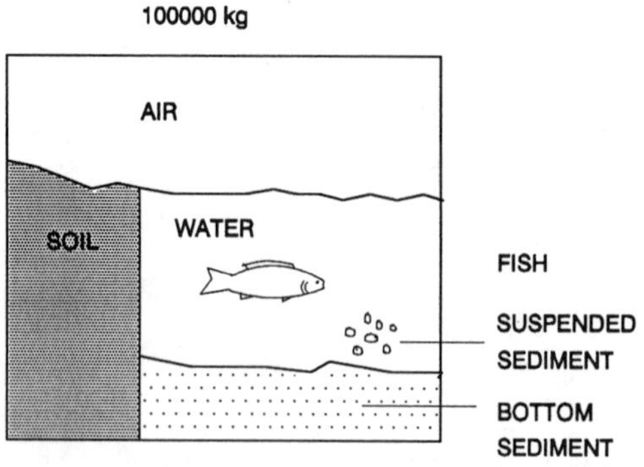

100000 kg

AIR

SOIL

WATER

FISH

SUSPENDED SEDIMENT

BOTTOM SEDIMENT

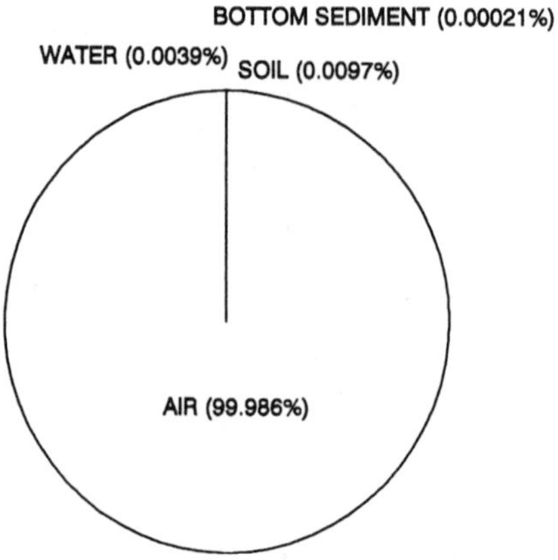

BOTTOM SEDIMENT (0.00021%)

WATER (0.0039%) SOIL (0.0097%)

AIR (99.986%)

Distribution of mass

physical-chemical properties:

MW: 72.15 g/mol

M.P.: - 129.7 °C

Fugacity ratio: 1.0

vapor pressure: 68400 Pa

solubility: 38.5 g/m $^3$

log $K_{OW}$: 3.45

| Compartment | Z | Concentration | | | Amount | Amount |
|---|---|---|---|---|---|---|
| | mol/m3 Pa | mol/m3 | mg/L (or g/m3) | ug/g | kg | % |
| Air | 4.034E-04 | 1.386E-08 | 9.999E-07 | 8.435E-04 | 99986 | 99.986 |
| Water | 7.801E-06 | 2.680E-10 | 1.934E-08 | 1.934E-08 | 3.867 | 3.87E-03 |
| Soil | 4.327E-04 | 1.486E-08 | 1.072E-06 | 4.469E-07 | 9.652 | 9.65E-03 |
| Biota (fish) | 1.099E-03 | 3.776E-08 | 2.725E-06 | 2.725E-06 | 5.45E-04 | 5.45E-07 |
| Suspended sediment | 2.704E-03 | 9.290E-08 | 6.703E-06 | 4.469E-06 | 6.70E-03 | 6.70E-06 |
| Bottom sediment | 8.654E-04 | 2.973E-08 | 2.145E-06 | 8.937E-07 | 0.2145 | 2.14E-04 |
| | Total | | | | 100000 | 100 |

f = 3.435E-05 Pa

Chemical name: n-Pentane

Level II calculation: (six-compartment model)

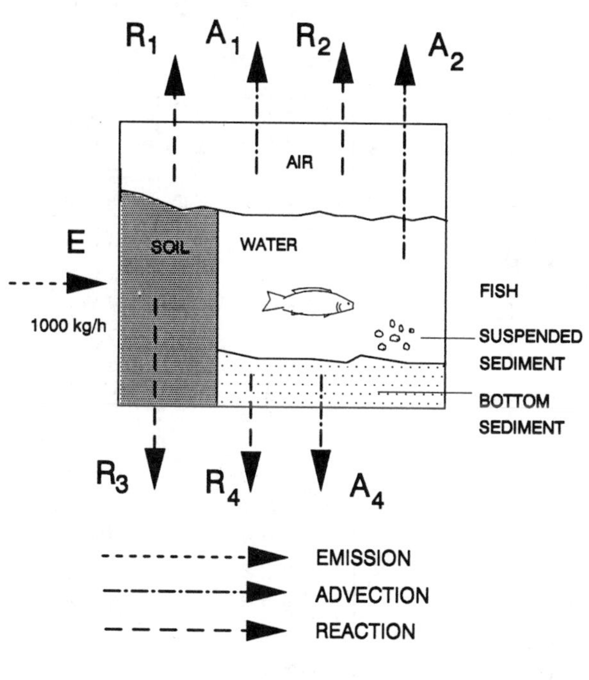


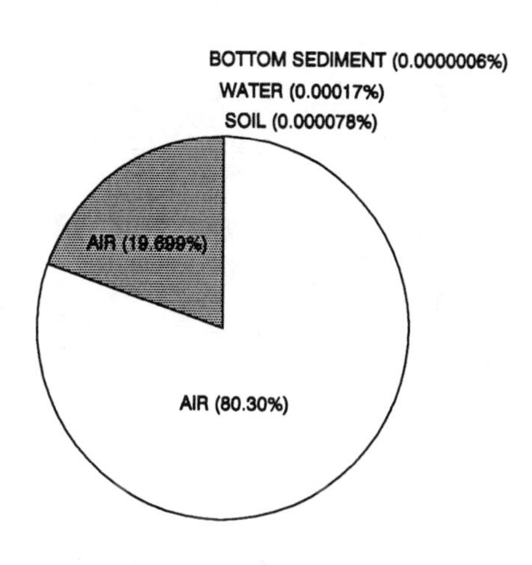


Distribution of removal rates

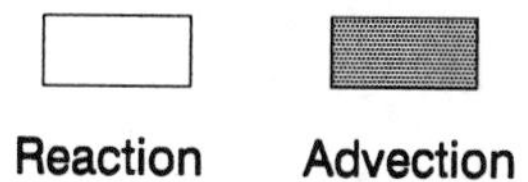


| Compartment | Half-life h | D Values Reaction mol/Pa h | D Values Advection mol/Pa h | Conc'n mol/m3 | Loss Reaction kg/h | Loss Advection kg/h | Removal % |
|---|---|---|---|---|---|---|---|
| Air | 17 | 1.64E+09 | 4.03E+08 | 2.73E-09 | 803.011 | 196.987 | 99.9997 |
| Water | 550 | 1.97E+03 | 1.56E+03 | 5.28E-11 | 9.60E-04 | 7.62E-04 | 1.72E-04 |
| Soil | 1700 | 1.59E+03 | | 2.93E-09 | 7.75E-04 | | 7.75E-05 |
| Biota (fish) | | | | 7.44E-09 | | | |
| Suspended sediment | | | | 1.83E-08 | | | |
| Bottom sediment | 5500 | 1.09E+01 | 1.73E+00 | 5.86E-09 | 5.32E-06 | 8.45E-07 | 6.17E-07 |
| | Total | 1.64E+09 | 4.03E+08 | | 803.01 | 196.99 | 100 |
| | R + A | | 2.05E+09 | | | 1000 | |

f = 6.768E-06 Pa

Total Amt= 19701 kg

Overall residence time = 19.70 h

Reaction time = 24.53 h

Advection time = 100.01 h

## Fugacity Level III calculations: (four-compartment model)

Chemical name: n-Pentane

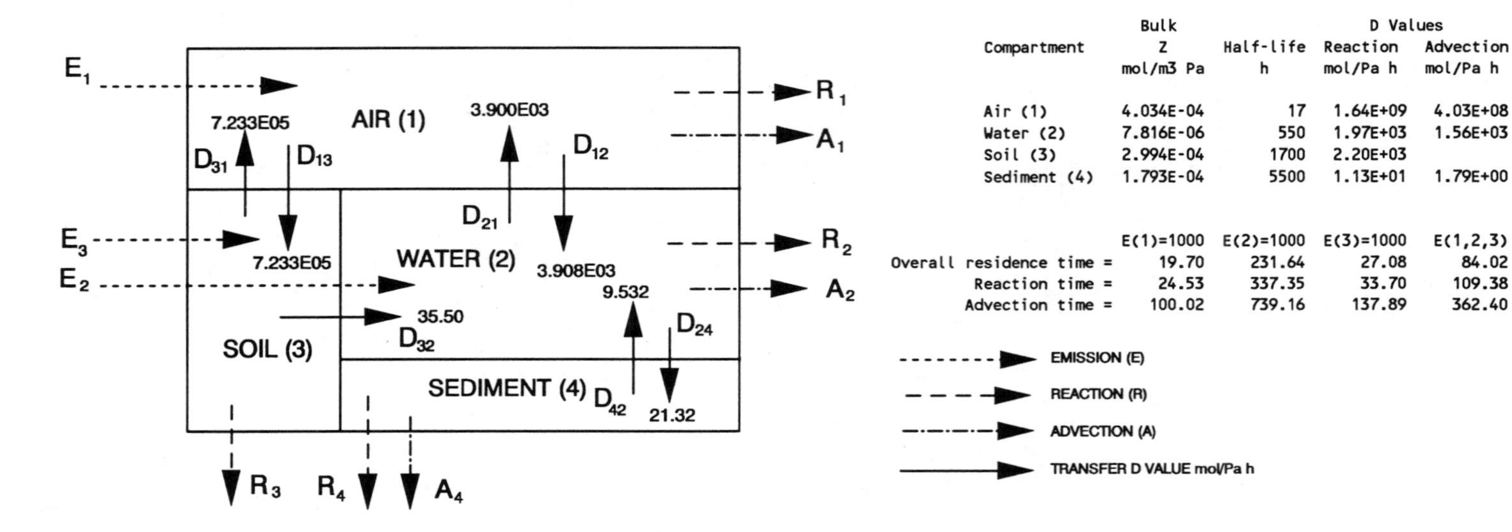


Phase Properties and Rates:

| Compartment | Bulk Z mol/m3 Pa | Half-life h | D Values Reaction mol/Pa h | D Values Advection mol/Pa h |
|---|---|---|---|---|
| Air (1) | 4.034E-04 | 17 | 1.64E+09 | 4.03E+08 |
| Water (2) | 7.816E-06 | 550 | 1.97E+03 | 1.56E+03 |
| Soil (3) | 2.994E-04 | 1700 | 2.20E+03 | |
| Sediment (4) | 1.793E-04 | 5500 | 1.13E+01 | 1.79E+00 |

| | E(1)=1000 | E(2)=1000 | E(3)=1000 | E(1,2,3) |
|---|---|---|---|---|
| Overall residence time = | 19.70 | 231.64 | 27.08 | 84.02 h |
| Reaction time = | 24.53 | 337.35 | 33.70 | 109.38 h |
| Advection time = | 100.02 | 739.16 | 137.89 | 362.40 h |



Phase Properties, Compositions, Transport and Transformation Rates:

| Emission, kg/h E(1) | E(2) | E(3) | Fugacity, Pa f(1) | f(2) | f(3) | f(4) | Concentration, g/m3 C(1) | C(2) | C(3) | C(4) | Amounts, kg m(1) | m(2) | m(3) | m(4) | Total Amount, kg |
|---|---|---|---|---|---|---|---|---|---|---|---|---|---|---|---|
| 1000 | 0 | 0 | 6.768E-06 | 3.585E-06 | 6.748E-06 | 3.379E-06 | 1.970E-07 | 2.021E-09 | 1.457E-07 | 4.371E-08 | 1.970E+04 | 4.043E-01 | 2.623E+00 | 2.186E-02 | 1.970E+04 |
| 0 | 1000 | 0 | 3.545E-06 | 1.862E+00 | 3.535E-06 | 1.755E+00 | 1.032E-07 | 1.050E-03 | 7.635E-08 | 2.270E-02 | 1.032E+04 | 2.100E+05 | 1.374E+00 | 1.135E+04 | 2.316E+05 |
| 0 | 0 | 1000 | 6.747E-06 | 9.466E-05 | 1.911E-02 | 8.922E-05 | 1.964E-07 | 5.338E-08 | 4.128E-04 | 1.154E-06 | 1.964E+04 | 1.068E+01 | 7.430E+03 | 5.772E-01 | 2.708E+04 |
| 600 | 300 | 100 | 5.799E-06 | 5.585E-01 | 1.916E-03 | 5.264E-01 | 1.688E-07 | 3.150E-04 | 4.139E-05 | 6.811E-03 | 1.688E+04 | 6.299E+04 | 7.450E+02 | 3.406E+03 | 8.402E+04 |

| Emission, kg/h E(1) | E(2) | E(3) | Loss, Reaction, kg/h R(1) | R(2) | R(3) | R(4) | Loss, Advection, kg/h A(1) | A(2) | A(4) | Intermedia Rate of Transport, kg/h T12 air-water | T21 water-air | T13 air-soil | T31 soil-air | T32 soil-water | T24 water-sed | T42 sed-water |
|---|---|---|---|---|---|---|---|---|---|---|---|---|---|---|---|---|
| 1000 | 0 | 0 | 8.030E+02 | 5.094E-04 | 1.07E-03 | 2.754E-06 | 1.970E+02 | 4.043E-04 | 4.371E-07 | 1.908E-03 | 1.009E-03 | 3.532E-01 | 3.521E-01 | 1.728E-05 | 5.515E-06 | 2.324E-06 |
| 0 | 1000 | 0 | 4.206E+02 | 2.646E+02 | 5.60E-04 | 1.430E+00 | 1.032E+02 | 2.100E+02 | 2.270E-01 | 9.996E-04 | 5.238E+02 | 1.850E-01 | 1.844E-01 | 9.052E-06 | 2.864E+00 | 1.207E+00 |
| 0 | 0 | 1000 | 8.006E+02 | 1.345E-02 | 3.03E+00 | 7.272E-05 | 1.964E+02 | 1.068E-02 | 1.154E-05 | 1.902E-03 | 2.663E-02 | 3.521E-01 | 9.973E+02 | 4.894E-02 | 1.456E-04 | 6.136E-05 |
| 600 | 300 | 100 | 6.881E+02 | 7.937E+01 | 3.04E-01 | 4.291E-01 | 1.688E+02 | 6.299E+01 | 6.811E-02 | 1.635E-03 | 1.572E+02 | 3.026E-01 | 9.999E+01 | 4.907E-03 | 8.592E-01 | 3.620E-01 |

# Level III Distribution

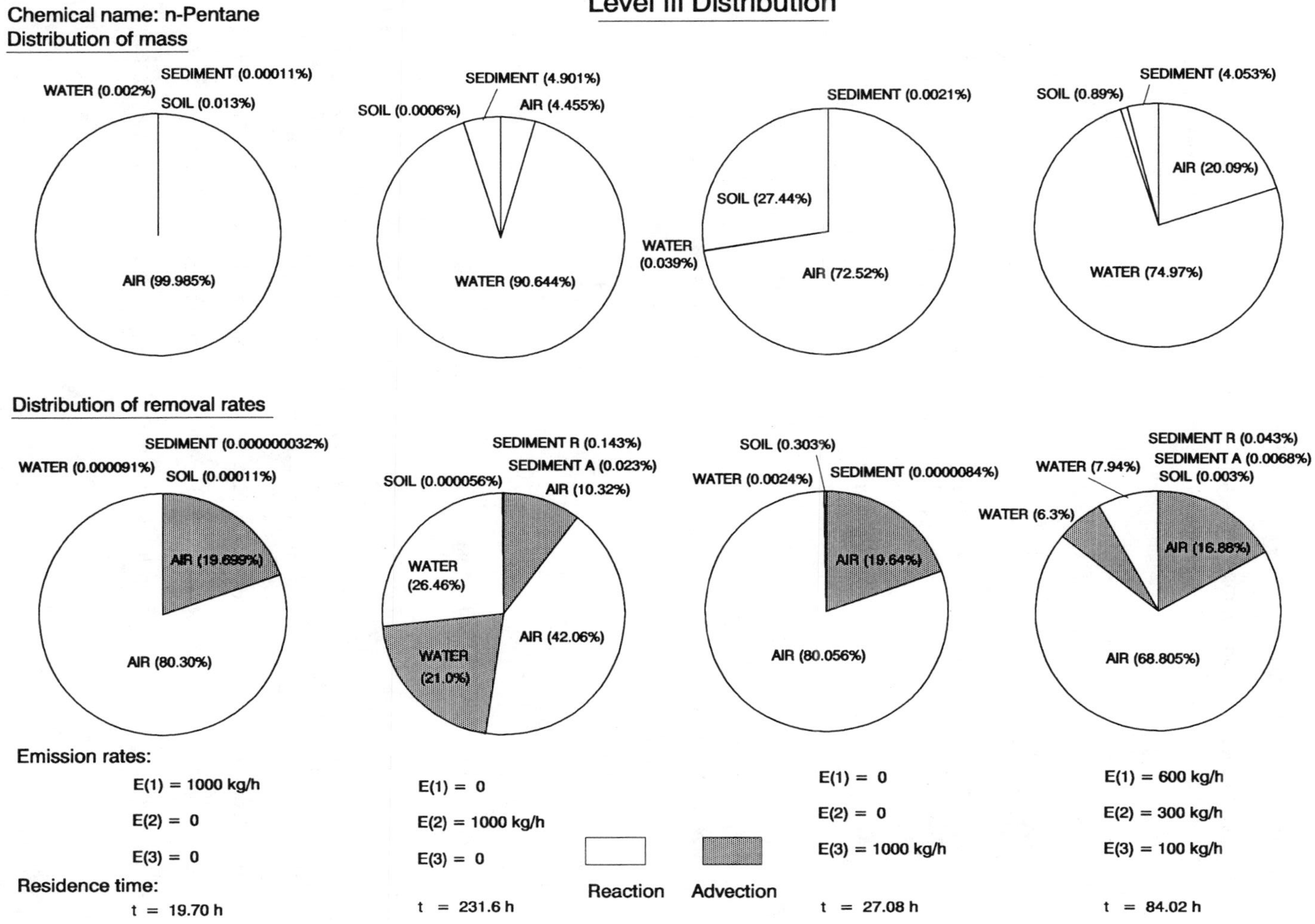

Chemical name: n-Pentane

Distribution of mass

WATER (0.002%)
SEDIMENT (0.00011%)
SOIL (0.013%)
AIR (99.985%)

SEDIMENT (4.901%)
SOIL (0.0006%)
AIR (4.455%)
WATER (90.644%)

SEDIMENT (0.0021%)
SOIL (27.44%)
WATER (0.039%)
AIR (72.52%)

SEDIMENT (4.053%)
SOIL (0.89%)
AIR (20.09%)
WATER (74.97%)

Distribution of removal rates

SEDIMENT (0.000000032%)
WATER (0.000091%)
SOIL (0.00011%)
AIR (19.699%)
AIR (80.30%)

SEDIMENT R (0.143%)
SEDIMENT A (0.023%)
SOIL (0.000056%)
AIR (10.32%)
WATER (26.46%)
AIR (42.06%)
WATER (21.0%)

SOIL (0.303%)
WATER (0.0024%)
SEDIMENT (0.0000084%)
AIR (19.64%)
AIR (80.056%)

SEDIMENT R (0.043%)
SEDIMENT A (0.0068%)
WATER (7.94%)
SOIL (0.003%)
WATER (6.3%)
AIR (16.88%)
AIR (68.805%)

Reaction | Advection

Emission rates:

| | | | |
|---|---|---|---|
| E(1) = 1000 kg/h | E(1) = 0 | E(1) = 0 | E(1) = 600 kg/h |
| E(2) = 0 | E(2) = 1000 kg/h | E(2) = 0 | E(2) = 300 kg/h |
| E(3) = 0 | E(3) = 0 | E(3) = 1000 kg/h | E(3) = 100 kg/h |

Residence time:

| | | | |
|---|---|---|---|
| t = 19.70 h | t = 231.6 h | t = 27.08 h | t = 84.02 h |

Chemical name: n-Hexane

Level I calculation: (six-compartment model)

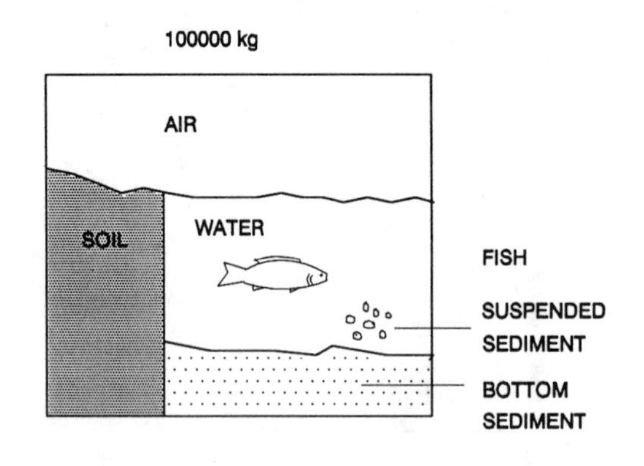


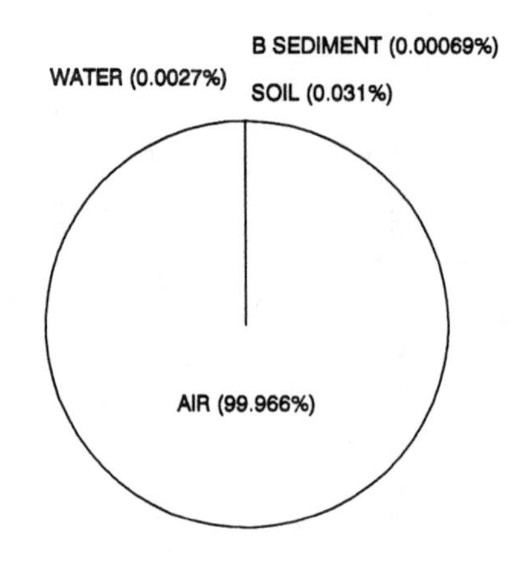


Distribution of mass

physical-chemical properties:

MW: 86.17 g/mol

M.P.: - 95.0 °C

Fugacity ratio: 1.0

vapor pressure: 20200 Pa

solubility: 9.5 g/m $^3$

log $K_{OW}$: 4.11

| Compartment | Z | Concentration | | | Amount | Amount |
|---|---|---|---|---|---|---|
| | mol/m3 Pa | mol/m3 | mg/L (or g/m3) | ug/g | kg | % |
| Air | 4.034E-04 | 1.160E-08 | 9.997E-07 | 8.433E-04 | 99966 | 99.966 |
| Water | 5.458E-06 | 1.569E-10 | 1.352E-08 | 1.352E-08 | 2.705 | 2.70E-03 |
| Soil | 1.384E-03 | 3.979E-08 | 3.429E-06 | 1.429E-06 | 30.859 | 3.09E-02 |
| Biota (fish) | 3.515E-03 | 1.011E-07 | 8.711E-06 | 8.711E-06 | 1.74E-03 | 1.74E-06 |
| Suspended sediment | 8.648E-03 | 2.487E-07 | 2.143E-05 | 1.429E-05 | 2.14E-02 | 2.14E-05 |
| Bottom sediment | 2.767E-03 | 7.958E-08 | 6.858E-06 | 2.857E-06 | 0.6858 | 6.86E-04 |
| | Total | | | | 100000 | 100 |

f = 2.876E-05 Pa

Chemical name: n-Hexane

Level II calculation: (six-compartment model)

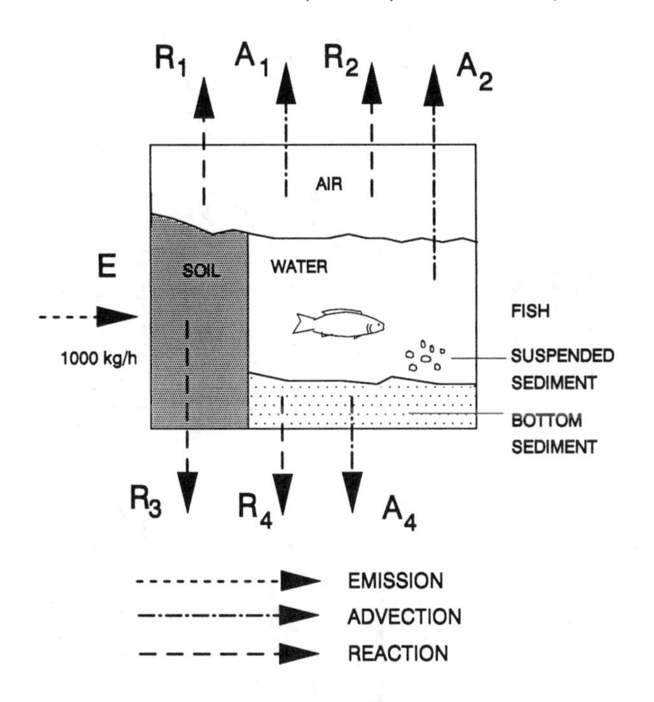


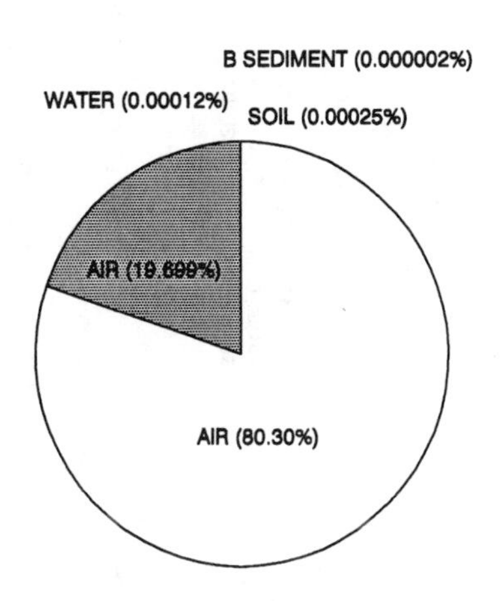


Distribution of removal rates

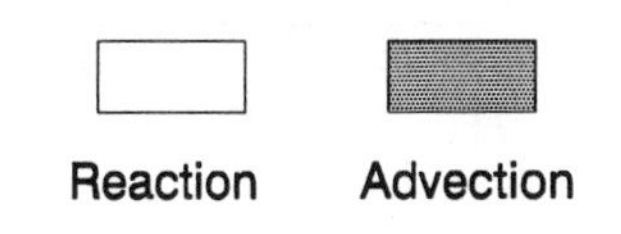


| Compartment | Half-life h | D Values Reaction mol/Pa h | D Values Advection mol/Pa h | Conc'n mol/m3 | Loss Reaction kg/h | Loss Advection kg/h | Removal % |
|---|---|---|---|---|---|---|---|
| Air | 17 | 1.64E+09 | 4.03E+08 | 2.29E-09 | 803.010 | 196.987 | 99.9996 |
| Water | 550 | 1.38E+03 | 1.09E+03 | 3.09E-11 | 6.72E-04 | 5.33E-04 | 1.20E-04 |
| Soil | 1700 | 5.08E+03 | | 7.84E-09 | 2.48E-03 | | 2.48E-04 |
| Biota (fish) | | | | 1.99E-08 | | | |
| Suspended sediment | | | | 4.90E-08 | | | |
| Bottom sediment | 5500 | 3.49E+01 | 5.53E+00 | 1.57E-08 | 1.70E-05 | 2.70E-06 | 1.97E-06 |
| | Total | 1.64E+09 | 4.03E+08 | | 803.01 | 196.99 | 100 |
| | R + A | | 2.05E+09 | | | 1000 | |

f = 5.667E-06 Pa
Total Amt= 19705 kg

Overall residence time = 19.71 h
Reaction time = 24.54 h
Advection time = 100.03 h

Fugacity Level III calculations: (four-compartment model)

Chemical name: n-Hexane

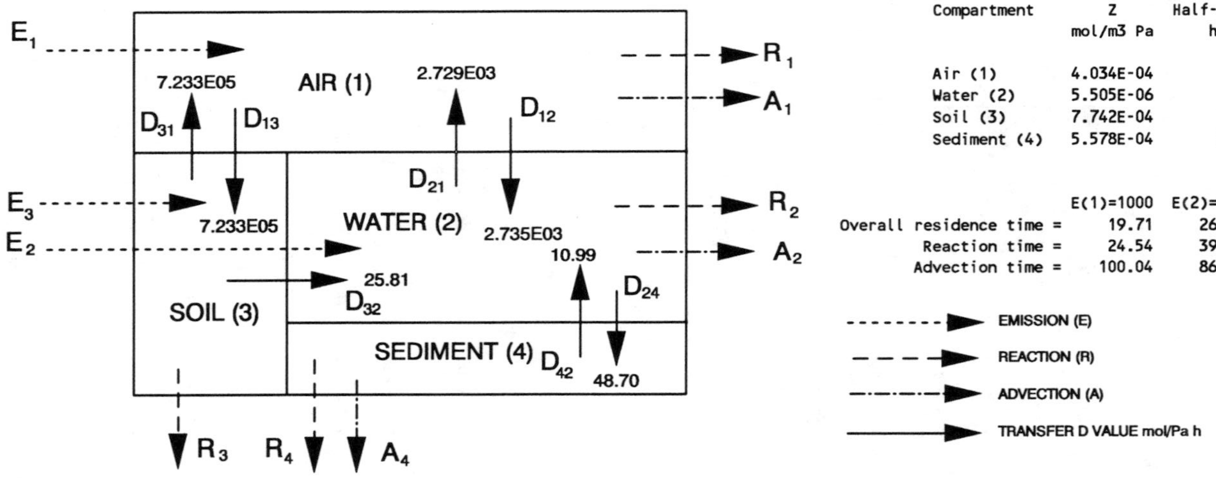

Phase Properties and Rates:

| Compartment | Bulk Z mol/m3 Pa | Half-life h | D Values Reaction mol/Pa h | Advection mol/Pa h |
|---|---|---|---|---|
| Air (1) | 4.034E-04 | 17 | 1.64E+09 | 4.03E+08 |
| Water (2) | 5.505E-06 | 550 | 1.39E+03 | 1.10E+03 |
| Soil (3) | 7.742E-04 | 1700 | 5.68E+03 | |
| Sediment (4) | 5.578E-04 | 5500 | 3.51E+01 | 5.58E+00 |

| | E(1)=1000 | E(2)=1000 | E(3)=1000 | E(1,2,3) |
|---|---|---|---|---|
| Overall residence time = | 19.71 | 269.71 | 38.68 | 96.61 h |
| Reaction time = | 24.54 | 392.47 | 48.07 | 125.72 h |
| Advection time = | 100.04 | 862.31 | 197.88 | 417.17 h |

Phase Properties, Compositions, Transport and Transformation Rates:

| Emission, kg/h | | | Fugacity, Pa | | | | Concentration, g/m3 | | | | Amounts, kg | | | | Total Amount, kg |
|---|---|---|---|---|---|---|---|---|---|---|---|---|---|---|---|
| E(1) | E(2) | E(3) | f(1) | f(2) | f(3) | f(4) | C(1) | C(2) | C(3) | C(4) | m(1) | m(2) | m(3) | m(4) | |
| 1000 | 0 | 0 | 5.667E-06 | 2.977E-06 | 5.623E-06 | 2.803E-06 | 1.970E-07 | 1.412E-09 | 3.751E-07 | 1.347E-07 | 1.970E+04 | 2.824E-01 | 6.752E+00 | 6.737E-02 | 1.971E+04 |
| 0 | 1000 | 0 | 2.942E-06 | 2.208E+00 | 2.919E-06 | 2.080E+00 | 1.023E-07 | 1.048E-03 | 1.948E-07 | 9.996E-02 | 1.023E+04 | 2.095E+05 | 3.506E+00 | 4.998E+04 | 2.697E+05 |
| 0 | 0 | 1000 | 5.622E-06 | 8.113E-05 | 1.593E-02 | 7.640E-05 | 1.954E-07 | 3.848E-08 | 1.062E-03 | 3.672E-06 | 1.954E+04 | 7.696E+00 | 1.912E+04 | 1.836E+00 | 3.868E+04 |
| 600 | 300 | 100 | 4.845E-06 | 6.625E-01 | 1.597E-03 | 6.239E-01 | 1.684E-07 | 3.143E-04 | 1.065E-04 | 2.999E-02 | 1.684E+04 | 6.285E+04 | 1.917E+03 | 1.499E+04 | 9.661E+04 |

| Emission, kg/h | | | Loss, Reaction, kg/h | | | | Loss, Advection, kg/h | | | Intermedia Rate of Transport, kg/h T12 | T21 | T13 | T31 | T32 | T24 | T42 |
|---|---|---|---|---|---|---|---|---|---|---|---|---|---|---|---|---|
| E(1) | E(2) | E(3) | R(1) | R(2) | R(3) | R(4) | A(1) | A(2) | A(4) | air-water | water-air | air-soil | soil-air | soil-water | water-sed | sed-water |
| 1000 | 0 | 0 | 8.030E+02 | 3.558E-04 | 2.75E-03 | 8.488E-06 | 1.970E+02 | 2.824E-04 | 1.347E-06 | 1.335E-03 | 6.998E-04 | 3.532E-01 | 3.504E-01 | 1.250E-05 | 1.249E-05 | 2.655E-06 |
| 0 | 1000 | 0 | 4.169E+02 | 2.640E+02 | 1.43E-03 | 6.298E+00 | 1.023E+02 | 2.095E+02 | 9.996E-01 | 6.933E-04 | 5.192E+02 | 1.834E-01 | 1.820E-01 | 6.492E-06 | 9.267E+00 | 1.970E+00 |
| 0 | 0 | 1000 | 7.967E+02 | 9.697E-03 | 7.80E+00 | 2.314E-04 | 1.954E+02 | 7.696E-03 | 3.672E-05 | 1.325E-03 | 1.908E-02 | 3.504E-01 | 9.925E+02 | 3.541E-02 | 3.404E-04 | 7.237E-05 |
| 600 | 300 | 100 | 6.866E+02 | 7.919E+01 | 7.82E-01 | 1.889E+00 | 1.684E+02 | 6.285E+01 | 2.999E-01 | 1.142E-03 | 1.558E+02 | 3.020E-01 | 9.952E+01 | 3.551E-03 | 2.780E+00 | 5.910E-01 |

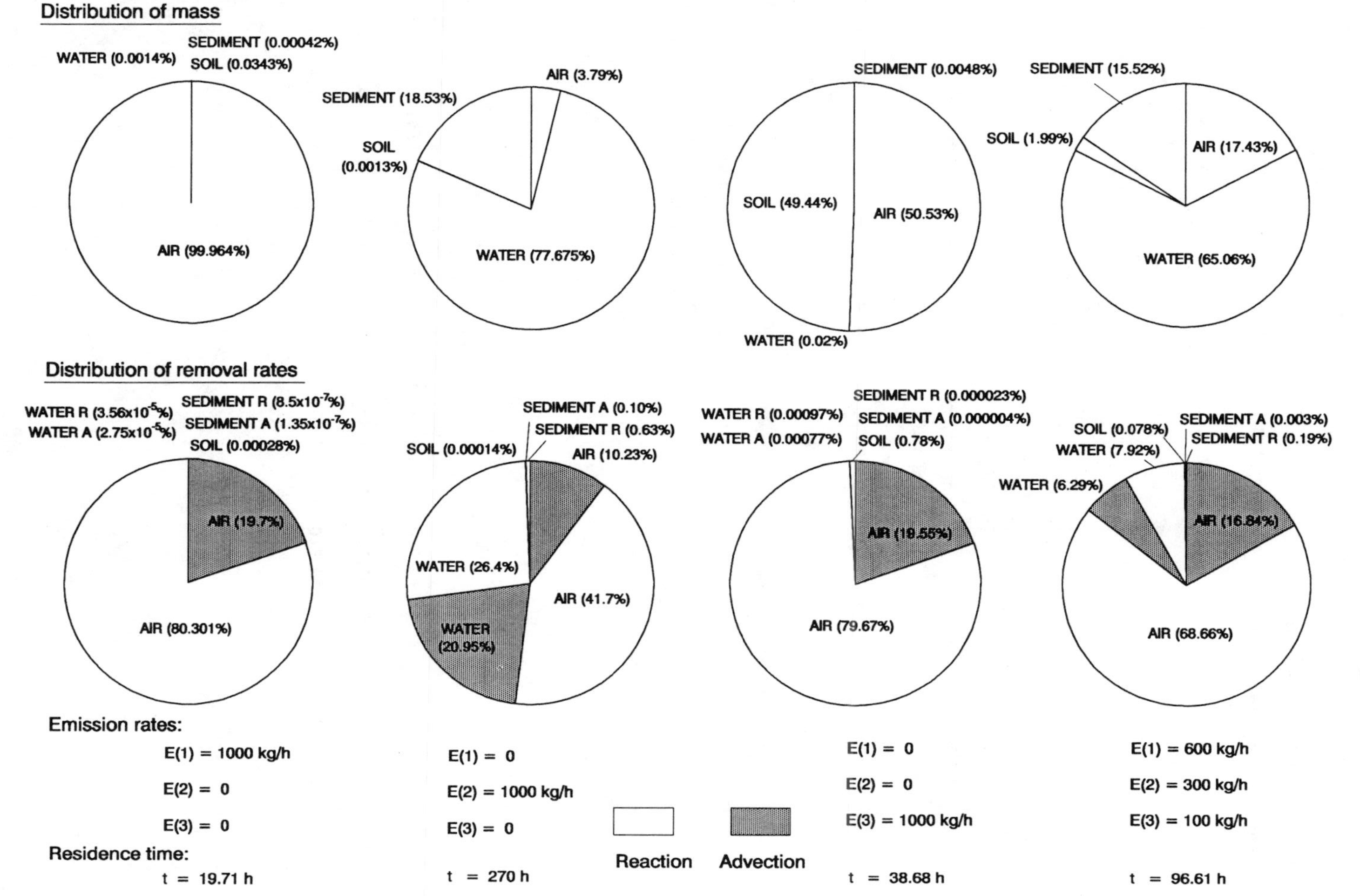

Level III Distribution
Chemical name: n-Hexane
Distribution of mass
WATER (0.0014%)
SEDIMENT (0.00042%)
SOIL (0.0343%)
AIR (99.964%)
SEDIMENT (18.53%)
AIR (3.79%)
SOIL (0.0013%)
WATER (77.675%)
SEDIMENT (0.0048%)
SOIL (49.44%)
AIR (50.53%)
WATER (0.02%)
SEDIMENT (15.52%)
SOIL (1.99%)
AIR (17.43%)
WATER (65.06%)
Distribution of removal rates
WATER R (3.56x10^-5%)
WATER A (2.75x10^-5%)
SEDIMENT R (8.5x10^-7%)
SEDIMENT A (1.35x10^-7%)
SOIL (0.00028%)
AIR (19.7%)
AIR (80.301%)
SEDIMENT A (0.10%)
SEDIMENT R (0.63%)
SOIL (0.00014%)
AIR (10.23%)
WATER (26.4%)
AIR (41.7%)
WATER (20.95%)
SEDIMENT R (0.000023%)
WATER R (0.00097%)
SEDIMENT A (0.000004%)
WATER A (0.00077%)
SOIL (0.78%)
AIR (19.55%)
AIR (79.67%)
SOIL (0.078%)
SEDIMENT A (0.003%)
SEDIMENT R (0.19%)
WATER (7.92%)
WATER (6.29%)
AIR (16.84%)
AIR (68.66%)
Emission rates:
E(1) = 1000 kg/h
E(2) = 0
E(3) = 0
Residence time:
t = 19.71 h
E(1) = 0
E(2) = 1000 kg/h
E(3) = 0
t = 270 h
Reaction
Advection
E(1) = 0
E(2) = 0
E(3) = 1000 kg/h
t = 38.68 h
E(1) = 600 kg/h
E(2) = 300 kg/h
E(3) = 100 kg/h
t = 96.61 h

Chemical name: n-Octane

Level I calculation: (six-compartment model)

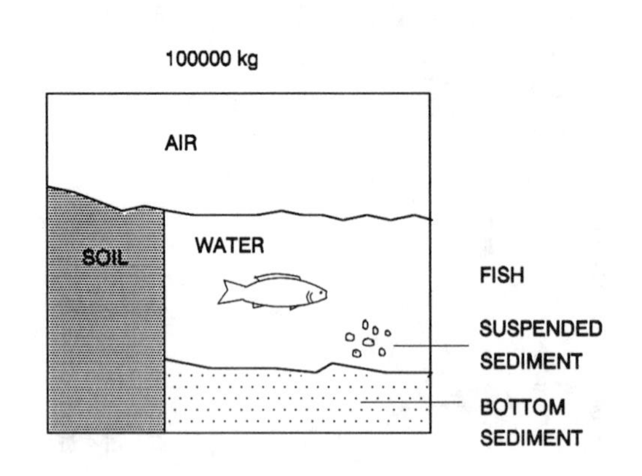


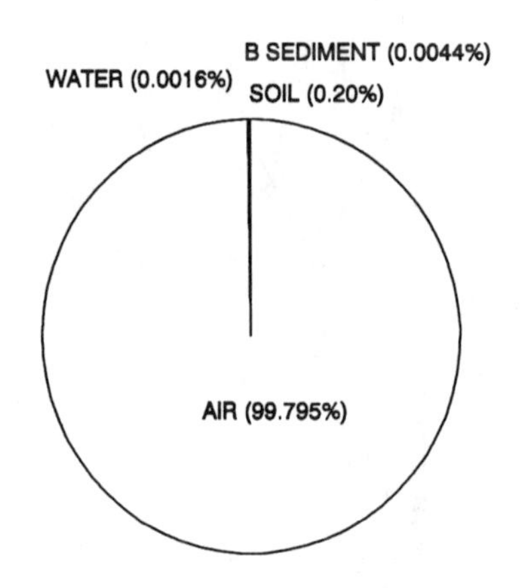


Distribution of mass

physical-chemical properties:

MW: 114.23 g/mol

M.P.: - 56.8 °C

Fugacity ratio: 1.0

vapor pressure: 1800 Pa

solubility: 0.66 g/m$^3$

log $K_{OW}$: 5.15

| Compartment | Z | Concentration | | | Amount | Amount |
|---|---|---|---|---|---|---|
| | mol/m3 Pa | mol/m3 | mg/L (or g/m3) | ug/g | kg | % |
| Air | 4.034E-04 | 8.736E-09 | 9.980E-07 | 8.419E-04 | 99795 | 99.795 |
| Water | 3.210E-06 | 6.951E-11 | 7.940E-09 | 7.940E-09 | 1.588 | 1.59E-03 |
| Soil | 8.923E-03 | 1.932E-07 | 2.207E-05 | 9.197E-06 | 198.661 | 0.1987 |
| Biota (fish) | 2.267E-02 | 4.909E-07 | 5.608E-05 | 5.608E-05 | 1.12E-02 | 1.12E-05 |
| Suspended sediment | 5.577E-02 | 1.208E-06 | 1.380E-04 | 9.197E-05 | 0.1380 | 1.38E-04 |
| Bottom sediment | 1.785E-02 | 3.865E-07 | 4.415E-05 | 1.839E-05 | 4.4147 | 4.41E-03 |
| | Total | | | | 100000 | 100 |

f = 2.166E-05 Pa

Chemical name: n-Octane

Level II calculation: (six-compartment model)

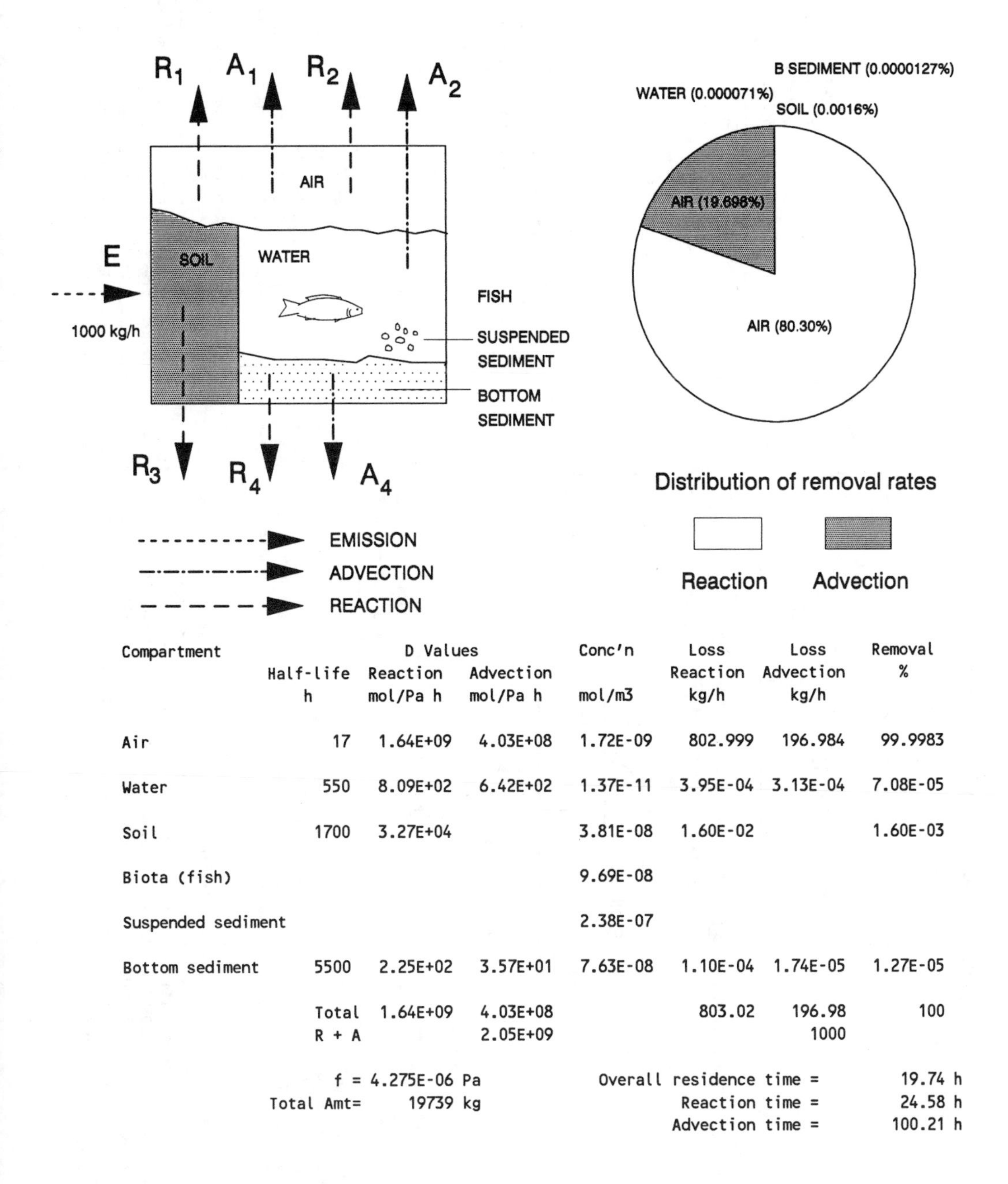


| Compartment | Half-life h | D Values Reaction mol/Pa h | D Values Advection mol/Pa h | Conc'n mol/m3 | Loss Reaction kg/h | Loss Advection kg/h | Removal % |
|---|---|---|---|---|---|---|---|
| Air | 17 | 1.64E+09 | 4.03E+08 | 1.72E-09 | 802.999 | 196.984 | 99.9983 |
| Water | 550 | 8.09E+02 | 6.42E+02 | 1.37E-11 | 3.95E-04 | 3.13E-04 | 7.08E-05 |
| Soil | 1700 | 3.27E+04 | | 3.81E-08 | 1.60E-02 | | 1.60E-03 |
| Biota (fish) | | | | 9.69E-08 | | | |
| Suspended sediment | | | | 2.38E-07 | | | |
| Bottom sediment | 5500 | 2.25E+02 | 3.57E+01 | 7.63E-08 | 1.10E-04 | 1.74E-05 | 1.27E-05 |
| | Total | 1.64E+09 | 4.03E+08 | | 803.02 | 196.98 | 100 |
| | R + A | | 2.05E+09 | | | 1000 | |

f = 4.275E-06 Pa
Total Amt= 19739 kg

Overall residence time = 19.74 h
Reaction time = 24.58 h
Advection time = 100.21 h

Fugacity Level III calculations: (four-compartment model)

Chemical name: n-Octane

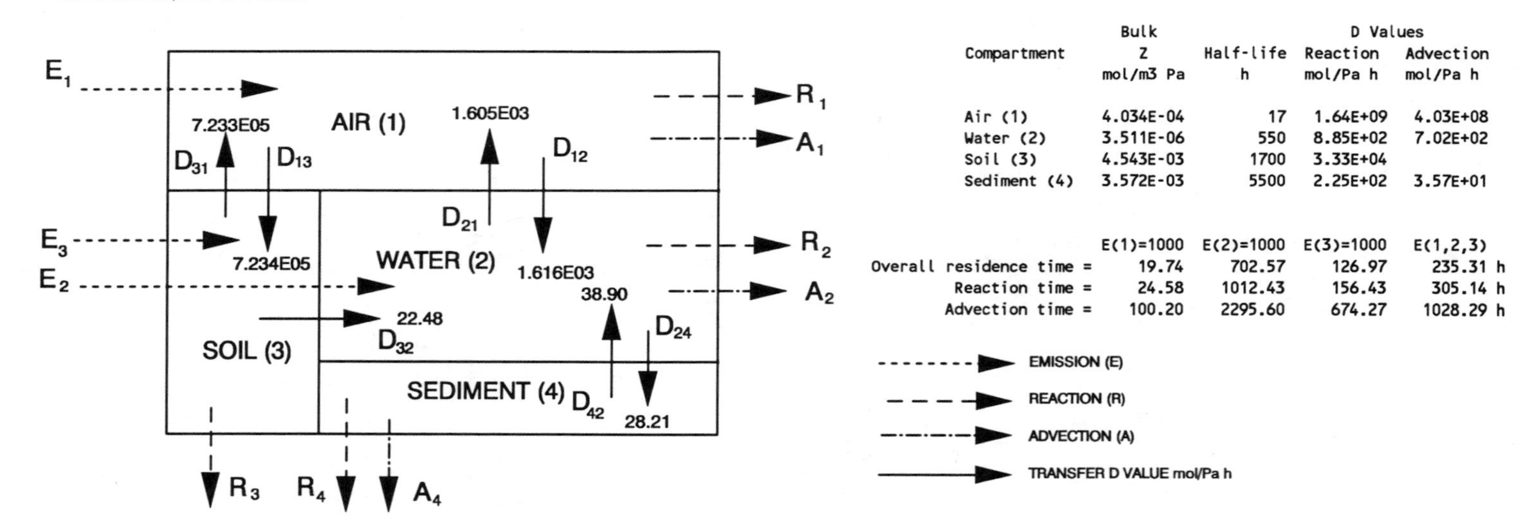


Phase Properties and Rates:

| Compartment | Bulk Z mol/m3 Pa | Half-life h | D Values Reaction mol/Pa h | D Values Advection mol/Pa h |
|---|---|---|---|---|
| Air (1) | 4.034E-04 | 17 | 1.64E+09 | 4.03E+08 |
| Water (2) | 3.511E-06 | 550 | 8.85E+02 | 7.02E+02 |
| Soil (3) | 4.543E-03 | 1700 | 3.33E+04 | |
| Sediment (4) | 3.572E-03 | 5500 | 2.25E+02 | 3.57E+01 |

| | E(1)=1000 | E(2)=1000 | E(3)=1000 | E(1,2,3) |
|---|---|---|---|---|
| Overall residence time = | 19.74 | 702.57 | 126.97 | 235.31 h |
| Reaction time = | 24.58 | 1012.43 | 156.43 | 305.14 h |
| Advection time = | 100.20 | 2295.60 | 674.27 | 1028.29 h |

Phase Properties, Compositions, Transport and Transformation Rates:

| Emission, kg/h | | | Fugacity, Pa | | | | Concentration, g/m3 | | | | Amounts, kg | | | | Total |
|---|---|---|---|---|---|---|---|---|---|---|---|---|---|---|---|
| E(1) | E(2) | E(3) | f(1) | f(2) | f(3) | f(4) | C(1) | C(2) | C(3) | C(4) | m(1) | m(2) | m(3) | m(4) | Amount, kg |
| 1000 | 0 | 0 | 4.275E-06 | 2.036E-06 | 4.087E-06 | 1.917E-06 | 1.970E-07 | 8.168E-10 | 2.121E-06 | 7.821E-07 | 1.970E+04 | 1.634E-01 | 3.818E+01 | 3.911E-01 | 1.974E+04 |
| 0 | 1000 | 0 | 1.996E-06 | 2.547E+00 | 1.908E-06 | 2.397E+00 | 9.197E-08 | 1.022E-03 | 9.902E-07 | 9.781E-01 | 9.197E+03 | 2.043E+05 | 1.782E+01 | 4.891E+05 | 7.026E+05 |
| 0 | 0 | 1000 | 4.086E-06 | 7.760E-05 | 1.157E-02 | 7.304E-05 | 1.883E-07 | 3.113E-08 | 6.007E-03 | 2.980E-05 | 1.883E+04 | 6.225E+00 | 1.081E+05 | 1.490E+01 | 1.270E+05 |
| 600 | 300 | 100 | 3.572E-06 | 7.640E-01 | 1.160E-03 | 7.192E-01 | 1.646E-07 | 3.065E-04 | 6.022E-04 | 2.934E-01 | 1.646E+04 | 6.129E+04 | 1.084E+04 | 1.467E+05 | 2.353E+05 |

| Emission, kg/h | | | Loss, Reaction, kg/h | | | | Loss, Advection, kg/h | | | Intermedia Rate of Transport, kg/h T12 | T21 | T13 | T31 | T32 | T24 | T42 |
|---|---|---|---|---|---|---|---|---|---|---|---|---|---|---|---|---|
| E(1) | E(2) | E(3) | R(1) | R(2) | R(3) | R(4) | A(1) | A(2) | A(4) | air-water | water-air | air-soil | soil-air | soil-water | water-sed | sed-water |
| 1000 | 0 | 0 | 8.030E+02 | 2.058E-04 | 1.56E-02 | 4.927E-05 | 1.970E+02 | 1.634E-04 | 7.821E-06 | 7.891E-04 | 3.733E-04 | 3.532E-01 | 3.376E-01 | 1.049E-05 | 6.561E-05 | 8.518E-06 |
| 0 | 1000 | 0 | 3.749E+02 | 2.574E+02 | 7.27E-03 | 6.162E+01 | 9.197E+01 | 2.043E+02 | 9.781E+00 | 3.684E-04 | 4.669E+02 | 1.649E-01 | 1.576E-01 | 4.898E-06 | 8.205E+01 | 1.065E+01 |
| 0 | 0 | 1000 | 7.676E+02 | 7.844E-03 | 4.41E+01 | 1.878E-03 | 1.883E+02 | 6.225E-03 | 2.980E-04 | 7.543E-04 | 1.423E-02 | 3.376E-01 | 9.562E+02 | 2.971E-02 | 2.500E-03 | 3.246E-04 |
| 600 | 300 | 100 | 6.710E+02 | 7.723E+01 | 4.42E+00 | 1.849E+01 | 1.646E+02 | 6.129E+01 | 2.934E+00 | 6.594E-04 | 1.401E+02 | 2.952E-01 | 9.587E+01 | 2.979E-03 | 2.462E+01 | 3.196E+00 |

# Level III Distribution

Chemical name: n-Octane

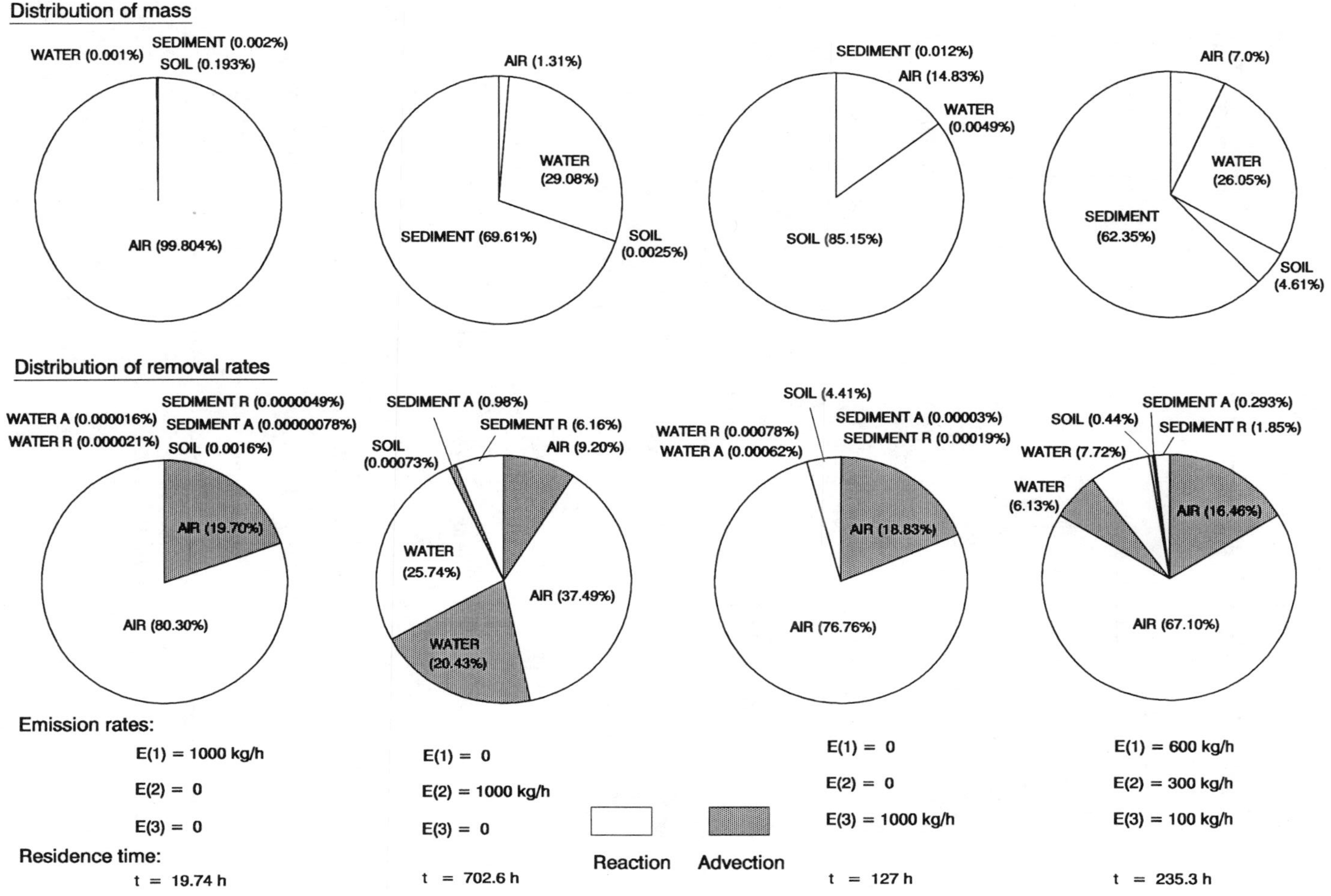

## Distribution of mass



## Distribution of removal rates



| Emission rates: | | | | |
|---|---|---|---|---|
| | E(1) = 1000 kg/h | E(1) = 0 | E(1) = 0 | E(1) = 600 kg/h |
| | E(2) = 0 | E(2) = 1000 kg/h | E(2) = 0 | E(2) = 300 kg/h |
| | E(3) = 0 | E(3) = 0 | E(3) = 1000 kg/h | E(3) = 100 kg/h |
| Residence time: | t = 19.74 h | t = 702.6 h | t = 127 h | t = 235.3 h |

Chemical name: n-Decane

Level I calculation: (six-compartment model)

100000 kg

AIR

SOIL

WATER

FISH

SUSPENDED SEDIMENT

BOTTOM SEDIMENT

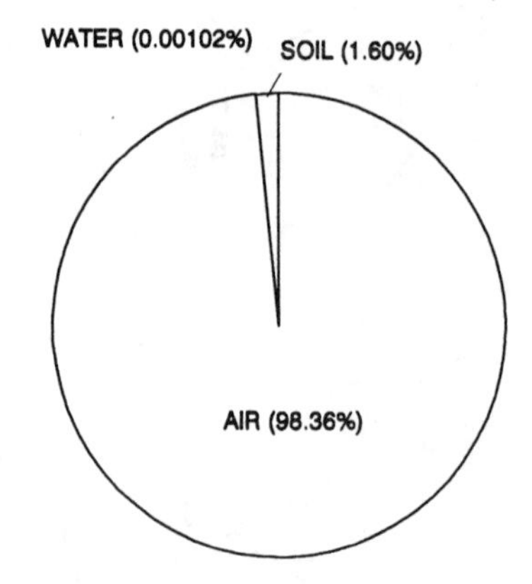


Distribution of mass

physical-chemical properties:

MW: 142.29 g/mol

M.P.: -29.7 °C

Fugacity ratio: 1.0

vapor pressure: 175 Pa

solubility: 0.052 g/m $^3$

log $K_{OW}$: 6.25

| Compartment | Z | Concentration | | | Amount | Amount |
|---|---|---|---|---|---|---|
| | mol/m3 Pa | mol/m3 | mg/L (or g/m3) | ug/g | kg | % |
| Air | 4.034E-04 | 6.913E-09 | 9.836E-07 | 8.297E-04 | 98358 | 98.358 |
| Water | 2.088E-06 | 3.578E-11 | 5.092E-09 | 5.092E-09 | 1.018 | 1.02E-03 |
| Soil | 7.308E-02 | 1.252E-06 | 1.782E-04 | 7.424E-05 | 1603.670 | 1.6037 |
| Biota (fish) | 1.857E-01 | 3.182E-06 | 4.527E-04 | 4.527E-04 | 9.05E-02 | 9.05E-05 |
| Suspended sediment | 4.568E-01 | 7.827E-06 | 1.114E-03 | 7.424E-04 | 1.1137 | 1.11E-03 |
| Bottom sediment | 1.462E-01 | 2.505E-06 | 3.564E-04 | 1.485E-04 | 35.6371 | 3.56E-02 |
| | Total | | | | 100000 | 100 |

f = 1.713E-05 Pa

Chemical name: n-Decane

Level II calculation: (six-compartment model)

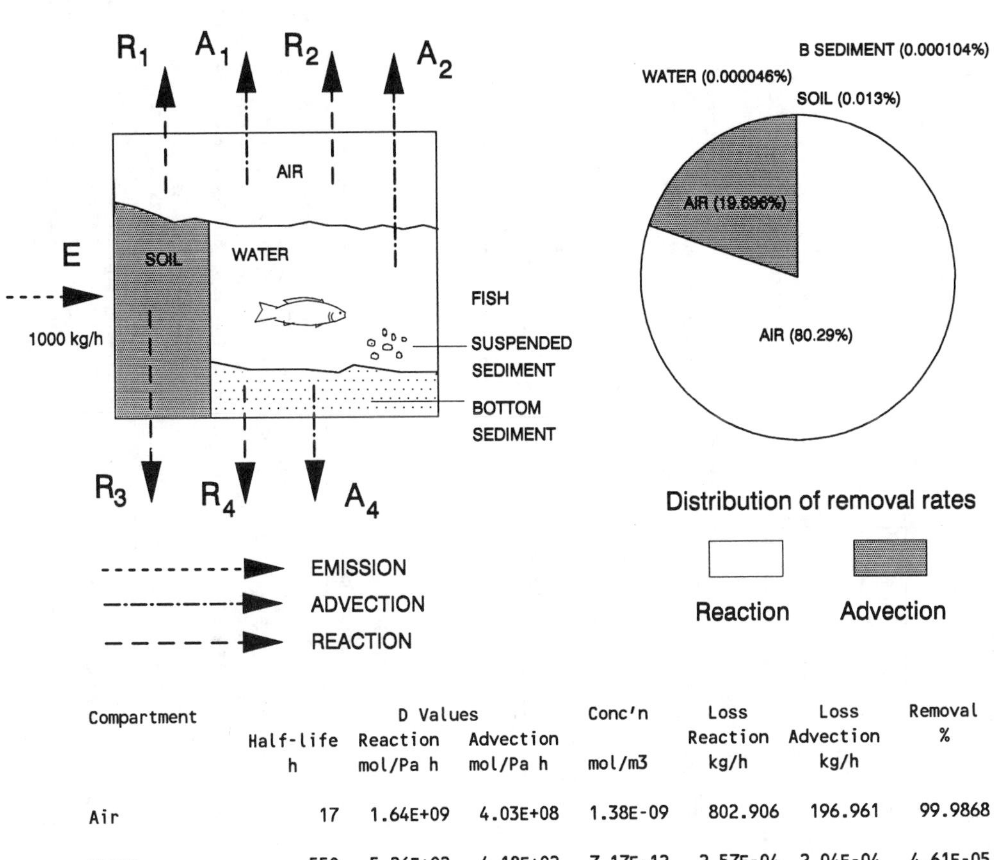


| Compartment | Half-life h | D Values Reaction mol/Pa h | D Values Advection mol/Pa h | Conc'n mol/m3 | Loss Reaction kg/h | Loss Advection kg/h | Removal % |
|---|---|---|---|---|---|---|---|
| Air | 17 | 1.64E+09 | 4.03E+08 | 1.38E-09 | 802.906 | 196.961 | 99.9868 |
| Water | 550 | 5.26E+02 | 4.18E+02 | 7.17E-12 | 2.57E-04 | 2.04E-04 | 4.61E-05 |
| Soil | 1700 | 2.68E+05 | | 2.51E-07 | 0.1309 | | 1.31E-02 |
| Biota (fish) | | | | 6.37E-07 | | | |
| Suspended sediment | | | | 1.57E-06 | | | |
| Bottom sediment | 5500 | 1.84E+03 | 2.92E+02 | 5.02E-07 | 8.99E-04 | 1.43E-04 | 1.04E-04 |
| | Total | 1.64E+09 | 4.03E+08 | | 803.04 | 196.96 | 100 |
| | R + A | | 2.05E+09 | | | 1000 | |

f = 3.431E-06 Pa
Total Amt= 20025 kg

Overall residence time = 20.02 h
Reaction time = 24.94 h
Advection time = 101.67 h

## Fugacity Level III calculations: (four- compartment model)

Chemical name: n-Decane

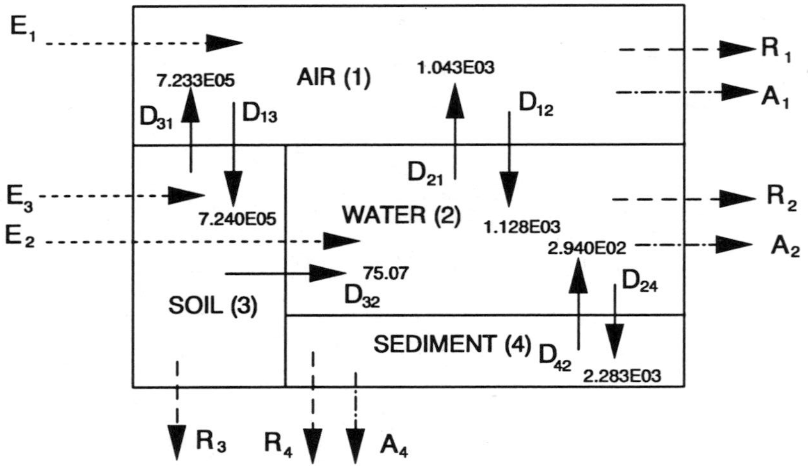

E1, E2, E3, AIR (1), WATER (2), SOIL (3), SEDIMENT (4), R1, A1, R2, A2, R3, R4, A4
D31 7.233E05, D13 7.240E05, D21 1.043E03, D12 1.128E03, D32 75.07, D42 2.940E02, D24 2.283E03

Phase Properties and Rates:

| Compartment | Bulk Z mol/m3 Pa | Half-life h | D Values Reaction mol/Pa h | D Values Advection mol/Pa h |
|---|---|---|---|---|
| Air (1) | 4.034E-04 | 17 | 1.64E+09 | 4.03E+08 |
| Water (2) | 4.558E-06 | 550 | 1.15E+03 | 9.12E+02 |
| Soil (3) | 3.662E-02 | 1700 | 2.69E+05 | |
| Sediment (4) | 2.923E-02 | 5500 | 1.84E+03 | 2.92E+02 |

| | E(1)=1000 | E(2)=1000 | E(3)=1000 | E(1,2,3) |
|---|---|---|---|---|
| Overall residence time = | 19.93 | 2873.31 | 679.24 | 941.88 h |
| Reaction time = | 24.82 | 3948.60 | 793.15 | 1198.68 h |
| Advection time = | 101.20 | 10551.11 | 4729.38 | 4396.36 h |

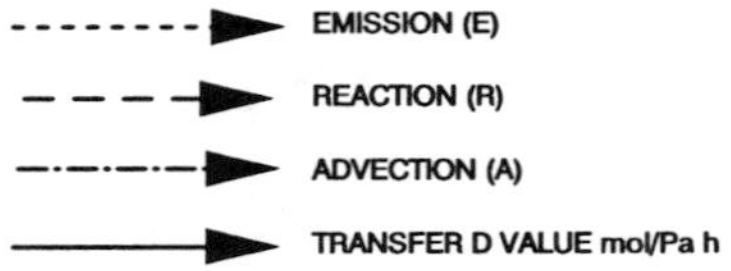

EMISSION (E)
REACTION (R)
ADVECTION (A)
TRANSFER D VALUE mol/Pa h

Phase Properties, Compositions, Transport and Transformation Rates:

| Emission, kg/h | | | Fugacity, Pa | | | | Concentration, g/m3 | | | | Amounts, kg | | | | Total Amount, kg |
|---|---|---|---|---|---|---|---|---|---|---|---|---|---|---|---|
| E(1) | E(2) | E(3) | f(1) | f(2) | f(3) | f(4) | C(1) | C(2) | C(3) | C(4) | m(1) | m(2) | m(3) | m(4) | |
| 1000 | 0 | 0 | 3.431E-06 | 7.946E-07 | 2.504E-06 | 7.479E-07 | 1.970E-07 | 5.153E-10 | 1.305E-05 | 3.111E-06 | 1.970E+04 | 1.031E-01 | 2.349E+02 | 1.556E+00 | 1.993E+04 |
| 0 | 1000 | 0 | 7.007E-07 | 1.375E+00 | 5.114E-07 | 1.294E+00 | 4.022E-08 | 8.914E-04 | 2.665E-06 | 5.382E+00 | 4.022E+03 | 1.783E+05 | 4.797E+01 | 2.691E+06 | 2.873E+06 |
| 0 | 0 | 1000 | 2.502E-06 | 1.047E-04 | 7.086E-03 | 9.858E-05 | 1.436E-07 | 6.792E-08 | 3.693E-02 | 4.101E-04 | 1.436E+04 | 1.358E+01 | 6.647E+05 | 2.050E+02 | 6.792E+05 |
| 600 | 300 | 100 | 2.519E-06 | 4.124E-01 | 7.103E-04 | 3.881E-01 | 1.446E-07 | 2.674E-04 | 3.701E-03 | 1.615E+00 | 1.446E+04 | 5.349E+04 | 6.662E+04 | 8.073E+05 | 9.419E+05 |

| Emission, kg/h | | | Loss, Reaction, kg/h | | | | Loss, Advection, kg/h | | | Intermedia Rate of Transport, kg/h T12 | T21 | T13 | T31 | T32 | T24 | T42 |
|---|---|---|---|---|---|---|---|---|---|---|---|---|---|---|---|---|
| E(1) | E(2) | E(3) | R(1) | R(2) | R(3) | R(4) | A(1) | A(2) | A(4) | air-water | water-air | air-soil | soil-air | soil-water | water-sed | sed-water |
| 1000 | 0 | 0 | 8.029E+02 | 1.299E-04 | 9.58E-02 | 1.960E-04 | 1.970E+02 | 1.031E-04 | 3.111E-05 | 5.513E-04 | 1.180E-04 | 3.535E-01 | 2.577E-01 | 2.679E-05 | 2.585E-04 | 3.133E-05 |
| 0 | 1000 | 0 | 1.640E+02 | 2.246E+02 | 1.96E-02 | 3.391E+02 | 4.022E+01 | 1.783E+02 | 5.382E+01 | 1.126E-04 | 2.042E+02 | 7.219E-02 | 5.263E-02 | 5.470E-06 | 4.471E+02 | 5.420E+01 |
| 0 | 0 | 1000 | 5.854E+02 | 1.712E-02 | 2.71E+02 | 2.583E-02 | 1.436E+02 | 1.358E-02 | 4.101E-03 | 4.019E-04 | 1.556E-02 | 2.577E-01 | 7.292E+02 | 7.579E-02 | 3.407E-02 | 4.130E-03 |
| 600 | 300 | 100 | 5.895E+02 | 6.739E+01 | 2.72E+01 | 1.017E+02 | 1.446E+02 | 5.349E+01 | 1.615E+01 | 4.048E-04 | 6.126E+01 | 2.595E-01 | 7.309E+01 | 7.597E-03 | 1.341E+02 | 1.626E+01 |

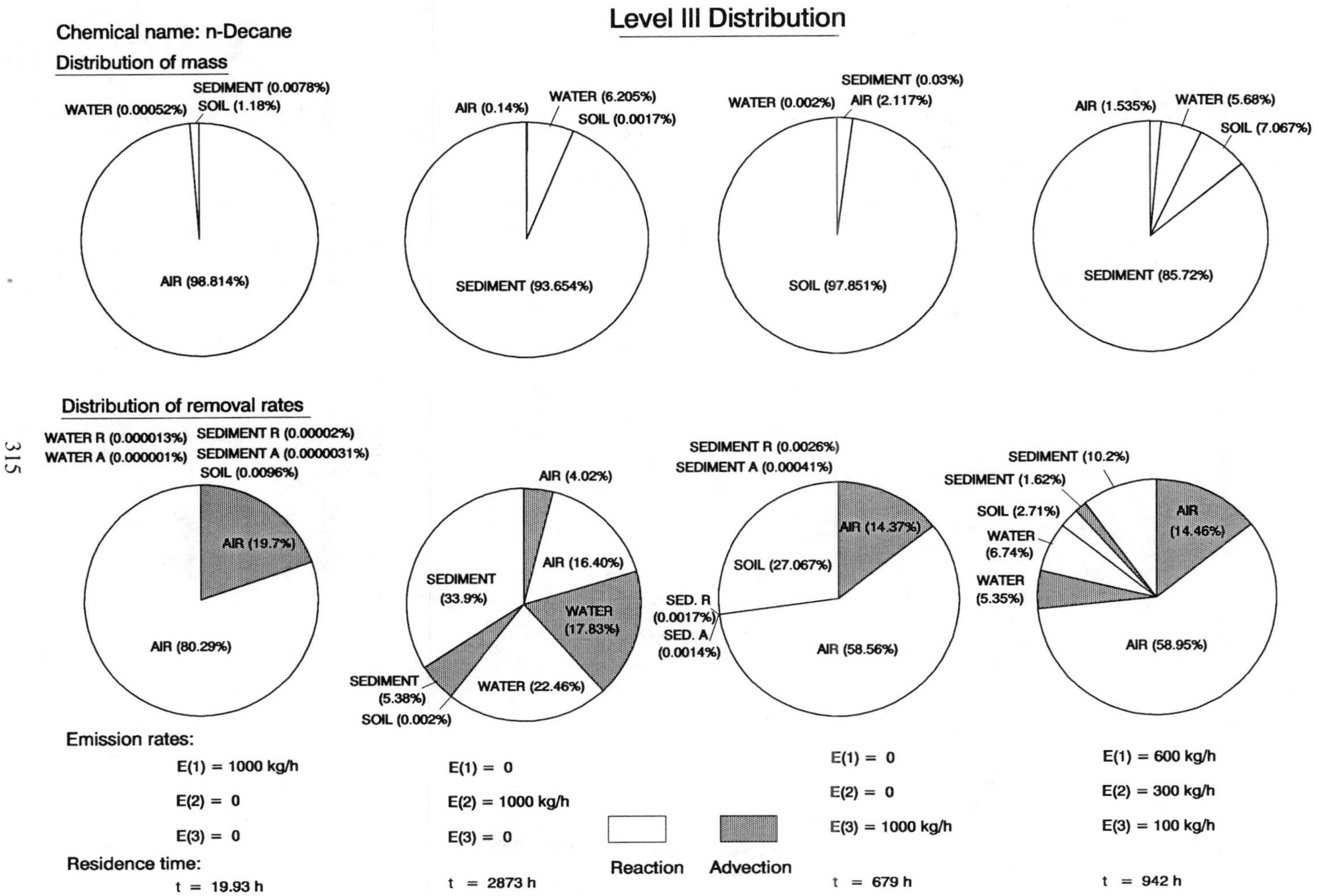

Level III Distribution
Chemical name: n-Decane
Distribution of mass
WATER (0.00052%)
SEDIMENT (0.0078%)
SOIL (1.18%)
AIR (98.814%)
AIR (0.14%)
WATER (6.205%)
SOIL (0.0017%)
SEDIMENT (93.654%)
WATER (0.002%)
SEDIMENT (0.03%)
AIR (2.117%)
SOIL (97.851%)
AIR (1.535%)
WATER (5.68%)
SOIL (7.067%)
SEDIMENT (85.72%)
Distribution of removal rates
WATER R (0.000013%)
WATER A (0.000001%)
SEDIMENT R (0.00002%)
SEDIMENT A (0.0000031%)
SOIL (0.0096%)
AIR (19.7%)
AIR (80.29%)
AIR (4.02%)
AIR (16.40%)
SEDIMENT (33.9%)
WATER (17.83%)
WATER (22.46%)
SEDIMENT (5.38%)
SOIL (0.002%)
SEDIMENT R (0.0026%)
SEDIMENT A (0.00041%)
AIR (14.37%)
SOIL (27.067%)
SED. R (0.0017%)
SED. A (0.0014%)
AIR (58.56%)
SEDIMENT (10.2%)
SEDIMENT (1.62%)
SOIL (2.71%)
WATER (6.74%)
WATER (5.35%)
AIR (14.46%)
AIR (58.95%)
Emission rates:
E(1) = 1000 kg/h
E(2) = 0
E(3) = 0
Residence time:
t = 19.93 h
E(1) = 0
E(2) = 1000 kg/h
E(3) = 0
t = 2873 h
Reaction
Advection
E(1) = 0
E(2) = 0
E(3) = 1000 kg/h
t = 679 h
E(1) = 600 kg/h
E(2) = 300 kg/h
E(3) = 100 kg/h
t = 942 h

Chemical name: n-Dodecane

Level I calculation: (six-compartment model)

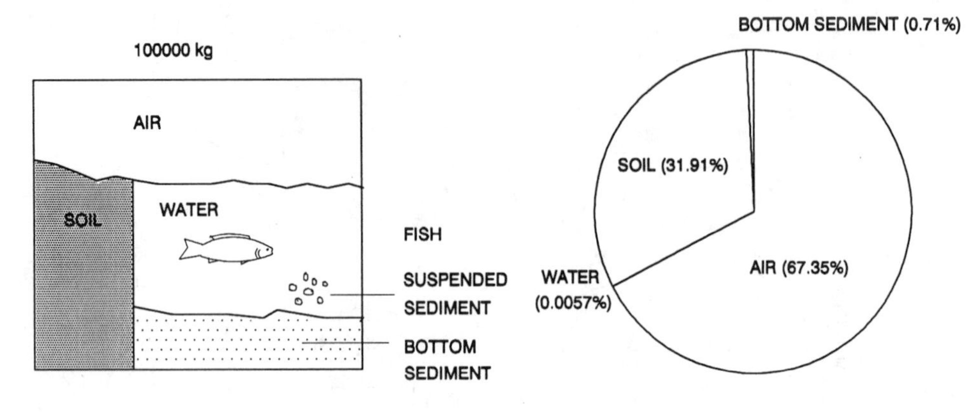


Distribution of mass

physical-chemical properties:

MW: 170.33 g/mol

M.P.: -9.6 ℃

Fugacity ratio: 1.0

vapor pressure: 1.27 Pa

solubility: 0.0037 g/m$^3$

log $K_{OW}$: 6.80

| Compartment | Z | Concentration | | | Amount | Amount |
|---|---|---|---|---|---|---|
| | mol/m3 Pa | mol/m3 | mg/L (or g/m3) | ug/g | kg | % |
| Air | 4.034E-04 | 3.954E-09 | 6.735E-07 | 5.682E-04 | 67349 | 67.349 |
| Water | 1.710E-05 | 1.676E-10 | 2.856E-08 | 2.856E-08 | 5.711 | 5.71E-03 |
| Soil | 2.124E+00 | 2.082E-05 | 3.546E-03 | 1.477E-03 | 31912 | 31.912 |
| Biota (fish) | 5.396E+00 | 5.289E-05 | 9.009E-03 | 9.009E-03 | 1.8017 | 1.80E-03 |
| Suspended sediment | 1.327E+01 | 1.301E-04 | 2.216E-02 | 1.477E-02 | 22.161 | 2.22E-02 |
| Bottom sediment | 4.248E+00 | 4.163E-05 | 7.092E-03 | 2.955E-03 | 709.15 | 0.7092 |
| | Total | | | | 100000 | 100 |

f = 9.801E-06 Pa

Chemical name: n-Dodecane

Level II calculation: (six-compartment model)

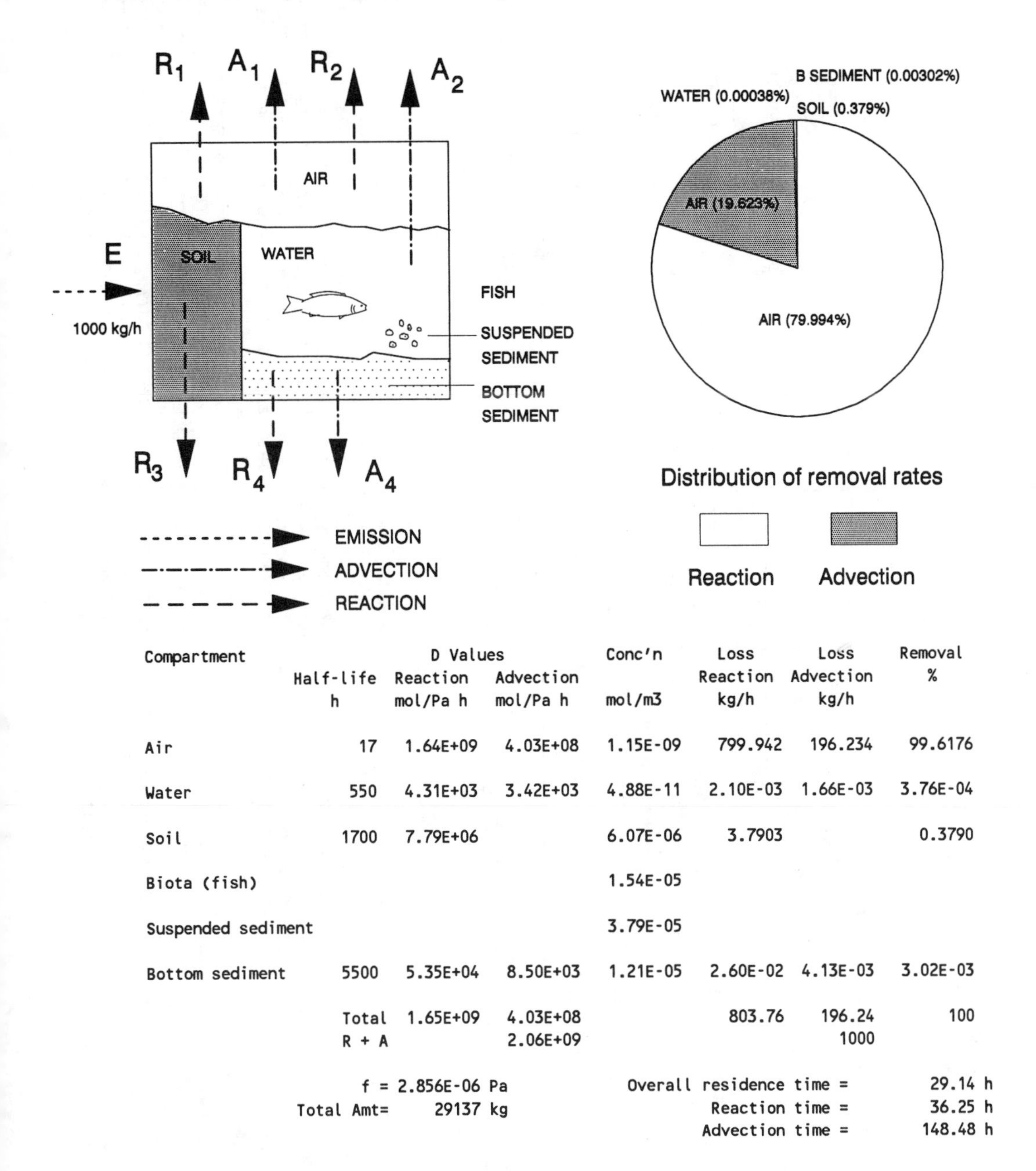


| Compartment | Half-life h | D Values Reaction mol/Pa h | D Values Advection mol/Pa h | Conc'n mol/m3 | Loss Reaction kg/h | Loss Advection kg/h | Removal % |
|---|---|---|---|---|---|---|---|
| Air | 17 | 1.64E+09 | 4.03E+08 | 1.15E-09 | 799.942 | 196.234 | 99.6176 |
| Water | 550 | 4.31E+03 | 3.42E+03 | 4.88E-11 | 2.10E-03 | 1.66E-03 | 3.76E-04 |
| Soil | 1700 | 7.79E+06 | | 6.07E-06 | 3.7903 | | 0.3790 |
| Biota (fish) | | | | 1.54E-05 | | | |
| Suspended sediment | | | | 3.79E-05 | | | |
| Bottom sediment | 5500 | 5.35E+04 | 8.50E+03 | 1.21E-05 | 2.60E-02 | 4.13E-03 | 3.02E-03 |
| | Total | 1.65E+09 | 4.03E+08 | | 803.76 | 196.24 | 100 |
| | R + A | | 2.06E+09 | | | 1000 | |

f = 2.856E-06 Pa
Total Amt= 29137 kg

Overall residence time = 29.14 h
Reaction time = 36.25 h
Advection time = 148.48 h

## Fugacity Level III calculations: (four compartment model)

Chemical name: n-Dodecane

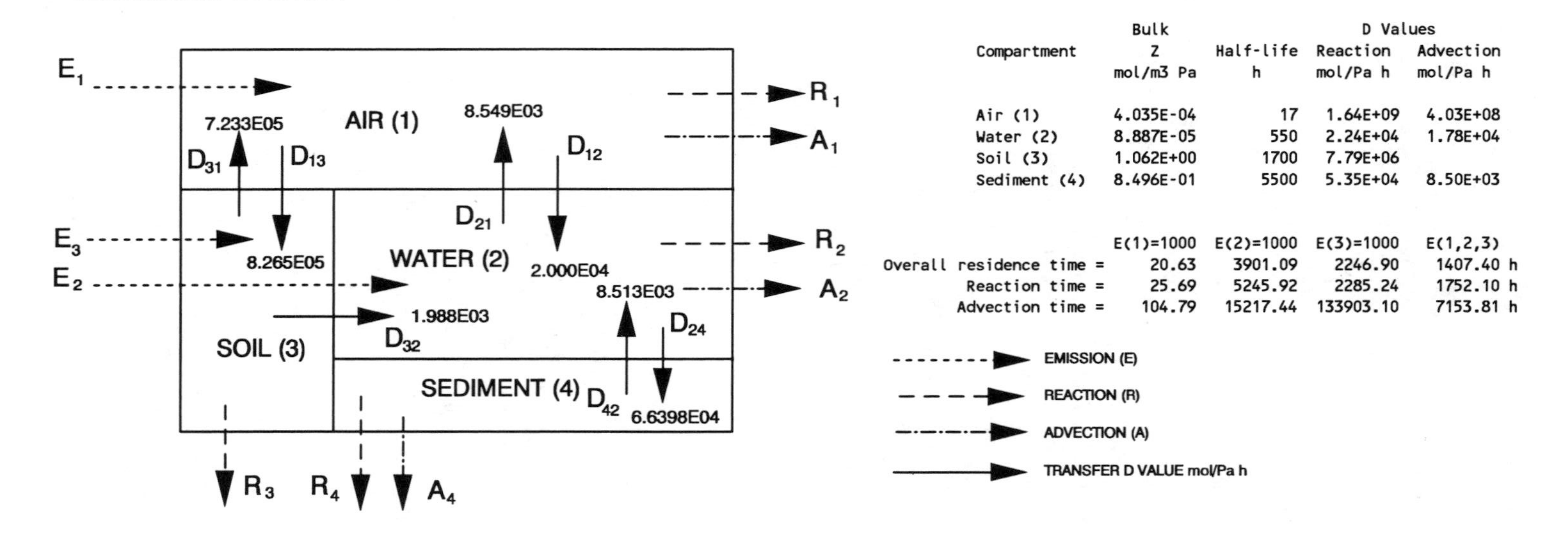


Phase Properties and Rates:

| Compartment | Bulk Z mol/m3 Pa | Half-life h | D Values Reaction mol/Pa h | D Values Advection mol/Pa h |
|---|---|---|---|---|
| Air (1) | 4.035E-04 | 17 | 1.64E+09 | 4.03E+08 |
| Water (2) | 8.887E-05 | 550 | 2.24E+04 | 1.78E+04 |
| Soil (3) | 1.062E+00 | 1700 | 7.79E+06 | |
| Sediment (4) | 8.496E-01 | 5500 | 5.35E+04 | 8.50E+03 |

| | E(1)=1000 | E(2)=1000 | E(3)=1000 | E(1,2,3) | |
|---|---|---|---|---|---|
| Overall residence time = | 20.63 | 3901.09 | 2246.90 | 1407.40 | h |
| Reaction time = | 25.69 | 5245.92 | 2285.24 | 1752.10 | h |
| Advection time = | 104.79 | 15217.44 | 133903.10 | 7153.81 | h |

Phase Properties, Compositions, Transport and Transformation Rates:

| Emission, kg/h | | | Fugacity, Pa | | | | Concentration, g/m3 | | | | Amounts, kg | | | | Total Amount, kg |
|---|---|---|---|---|---|---|---|---|---|---|---|---|---|---|---|
| E(1) | E(2) | E(3) | f(1) | f(2) | f(3) | f(4) | C(1) | C(2) | C(3) | C(4) | m(1) | m(2) | m(3) | m(4) | |
| 1000 | 0 | 0 | 2.865E-06 | 5.403E-07 | 2.780E-07 | 5.086E-07 | 1.969E-07 | 8.179E-09 | 5.028E-05 | 7.359E-05 | 1.969E+04 | 1.636E+00 | 9.051E+02 | 3.680E+01 | 2.063E+04 |
| 0 | 1000 | 0 | 2.287E-07 | 5.482E-02 | 2.219E-08 | 5.160E-02 | 1.572E-08 | 8.298E-04 | 4.014E-06 | 7.467E+00 | 1.572E+03 | 1.660E+05 | 7.225E+01 | 3.733E+06 | 3.901E+06 |
| 0 | 0 | 1000 | 2.434E-07 | 1.284E-05 | 6.893E-04 | 1.209E-05 | 1.672E-08 | 1.944E-07 | 1.247E-01 | 1.749E-03 | 1.672E+03 | 3.888E+01 | 2.244E+06 | 8.747E+02 | 2.247E+06 |
| 600 | 300 | 100 | 1.812E-06 | 1.645E-02 | 6.910E-05 | 1.548E-02 | 1.245E-07 | 2.490E-04 | 1.250E-02 | 2.240E+00 | 1.245E+04 | 4.980E+04 | 2.250E+05 | 1.120E+06 | 1.407E+06 |

| Emission, kg/h | | | Loss, Reaction, kg/h | | | | Loss, Advection, kg/h | | | Intermedia Rate of Transport, kg/h T12 | T21 | T13 | T31 | T32 | T24 | T42 |
|---|---|---|---|---|---|---|---|---|---|---|---|---|---|---|---|---|
| E(1) | E(2) | E(3) | R(1) | R(2) | R(3) | R(4) | A(1) | A(2) | A(4) | air-water | water-air | air-soil | soil-air | soil-water | water-sed | sed-water |
| 1000 | 0 | 0 | 8.027E+02 | 2.061E-03 | 3.69E-01 | 4.636E-03 | 1.969E+02 | 1.636E-03 | 7.359E-04 | 9.762E-03 | 7.867E-04 | 4.033E-01 | 3.425E-02 | 9.415E-05 | 6.110E-03 | 7.374E-04 |
| 0 | 1000 | 0 | 6.407E+01 | 2.091E+02 | 2.95E-02 | 4.704E+02 | 1.572E+01 | 1.660E+02 | 7.467E+01 | 7.792E-04 | 7.982E+01 | 3.219E-02 | 2.734E-03 | 7.515E-06 | 6.199E+02 | 7.482E+01 |
| 0 | 0 | 1000 | 6.817E+01 | 4.899E-02 | 9.15E+02 | 1.102E-01 | 1.672E+01 | 3.888E-02 | 1.749E-02 | 8.291E-04 | 1.870E-02 | 3.425E-02 | 8.491E+01 | 2.334E-01 | 1.452E-01 | 1.753E-02 |
| 600 | 300 | 100 | 5.077E+02 | 6.274E+01 | 9.17E+01 | 1.411E+02 | 1.245E+02 | 4.980E+01 | 2.240E+01 | 6.174E-03 | 2.395E+01 | 2.551E-01 | 8.513E+00 | 2.340E-02 | 1.860E+02 | 2.245E+01 |

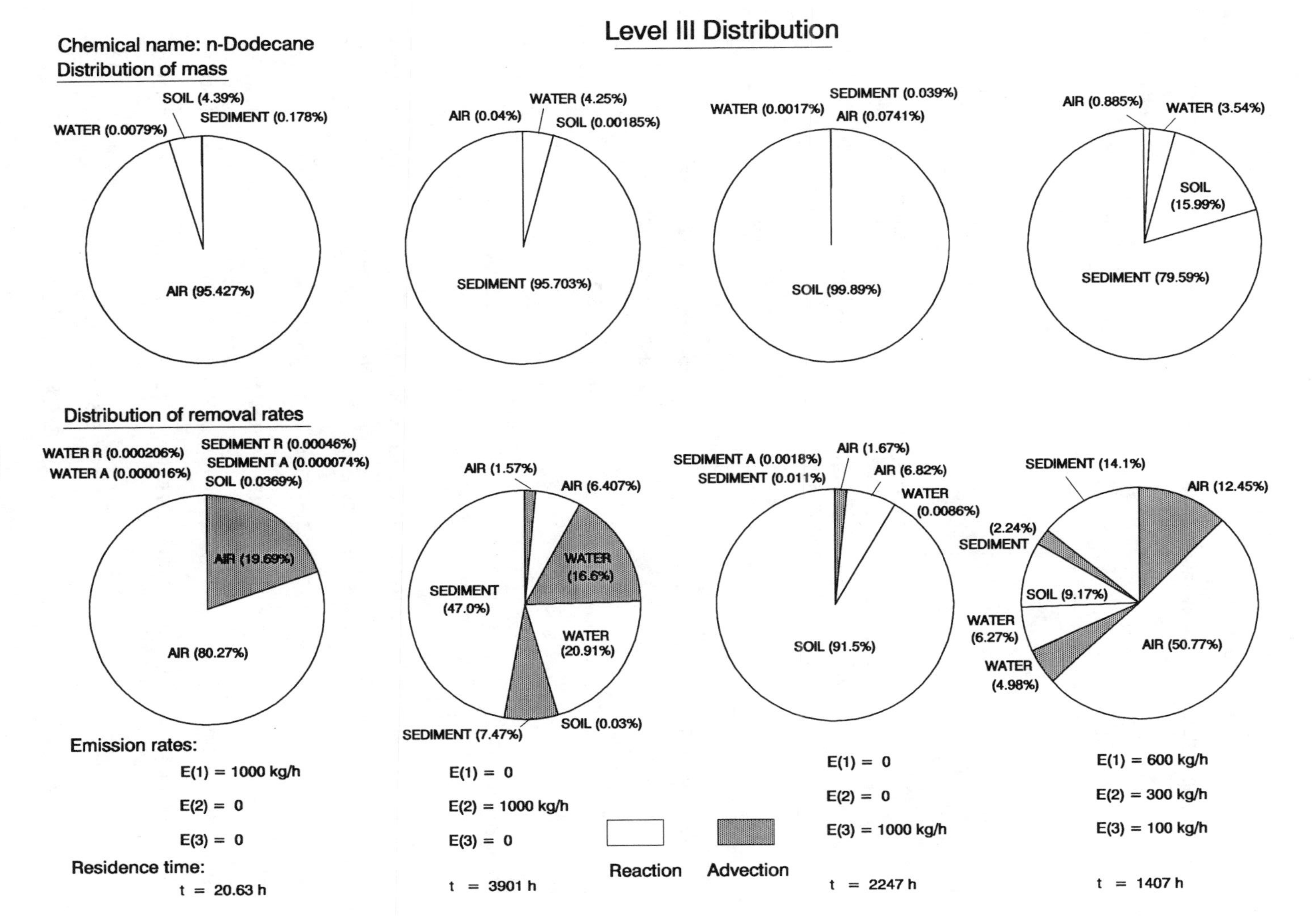
Level III Distribution
Chemical name: n-Dodecane
Distribution of mass
WATER (0.0079%)
SOIL (4.39%)
SEDIMENT (0.178%)
AIR (95.427%)
AIR (0.04%)
WATER (4.25%)
SOIL (0.00185%)
SEDIMENT (95.703%)
WATER (0.0017%)
SEDIMENT (0.039%)
AIR (0.0741%)
SOIL (99.89%)
AIR (0.885%)
WATER (3.54%)
SOIL (15.99%)
SEDIMENT (79.59%)
Distribution of removal rates
WATER R (0.000206%)
WATER A (0.000016%)
SEDIMENT R (0.00046%)
SEDIMENT A (0.000074%)
SOIL (0.0369%)
AIR (19.69%)
AIR (80.27%)
AIR (1.57%)
AIR (6.407%)
WATER (16.6%)
SEDIMENT (47.0%)
WATER (20.91%)
SEDIMENT (7.47%)
SOIL (0.03%)
SEDIMENT A (0.0018%)
SEDIMENT (0.011%)
AIR (1.67%)
AIR (6.82%)
WATER (0.0086%)
SOIL (91.5%)
SEDIMENT (14.1%)
AIR (12.45%)
(2.24%) SEDIMENT
SOIL (9.17%)
WATER (6.27%)
WATER (4.98%)
AIR (50.77%)
Emission rates:
E(1) = 1000 kg/h
E(2) = 0
E(3) = 0
Residence time:
t = 20.63 h
E(1) = 0
E(2) = 1000 kg/h
E(3) = 0
t = 3901 h
Reaction
Advection
E(1) = 0
E(2) = 0
E(3) = 1000 kg/h
t = 2247 h
E(1) = 600 kg/h
E(2) = 300 kg/h
E(3) = 100 kg/h
t = 1407 h

Chemical name: Cyclopentane

Level I calculation: (six-compartment model)

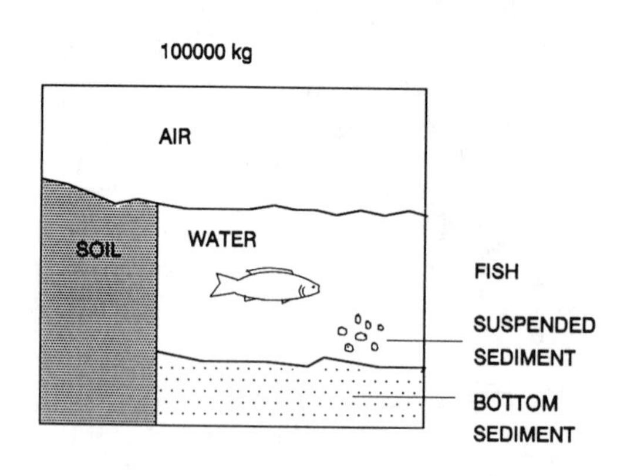


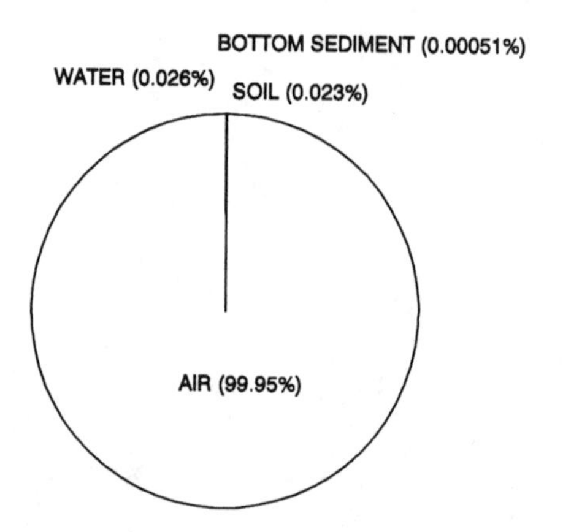


Distribution of mass

physical-chemical properties:

MW: 70.14 g/mol

M.P.: - 93.9 °C

Fugacity ratio: 1.0

vapor pressure: 42400 Pa

solubility: 156 g/m$^3$

log $K_{OW}$: 3.0

| Compartment | Z | Concentration | | | Amount | Amount |
|---|---|---|---|---|---|---|
| | mol/m3 Pa | mol/m3 | mg/L (or g/m3) | ug/g | kg | % |
| Air | 4.034E-04 | 1.425E-08 | 9.995E-07 | 8.432E-04 | 99950 | 99.950 |
| Water | 5.246E-05 | 1.853E-09 | 1.300E-07 | 1.300E-07 | 25.993 | 2.60E-02 |
| Soil | 1.032E-03 | 3.647E-08 | 2.558E-06 | 1.066E-06 | 23.019 | 0.023 |
| Biota (fish) | 2.623E-03 | 9.265E-08 | 6.498E-06 | 6.498E-06 | 1.30E-03 | 1.30E-06 |
| Suspended sediment | 6.452E-03 | 2.279E-07 | 1.599E-05 | 1.066E-05 | 0.0160 | 1.60E-05 |
| Bottom sediment | 2.065E-03 | 7.293E-08 | 5.115E-06 | 2.131E-06 | 0.5115 | 5.12E-04 |
| | Total | | | | 100000 | 100 |

f = 3.532E-05 Pa

Chemical name: Cyclopentane

Level II calculation: (six-compartment model)

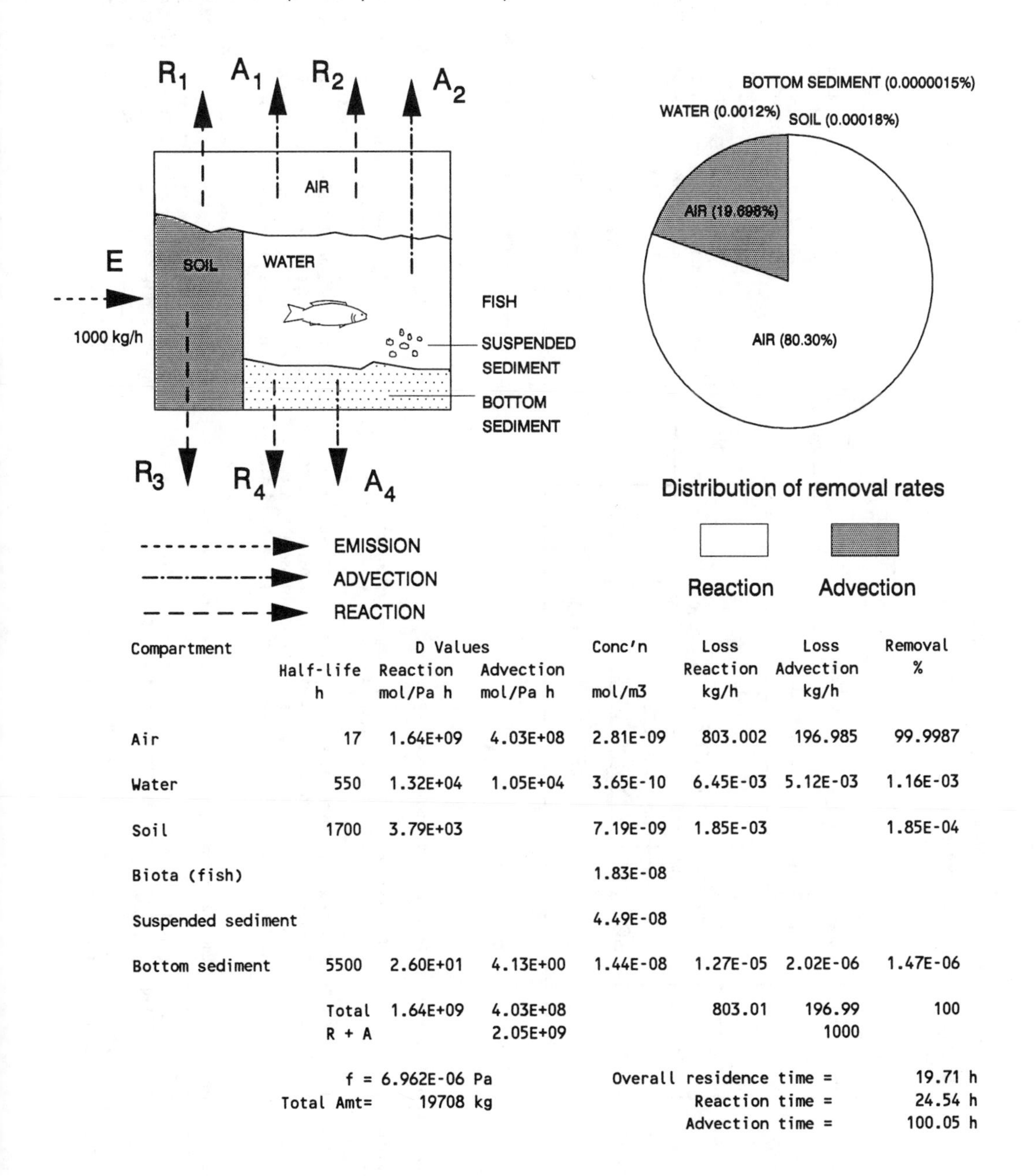


| Compartment | Half-life h | D Values Reaction mol/Pa h | D Values Advection mol/Pa h | Conc'n mol/m3 | Loss Reaction kg/h | Loss Advection kg/h | Removal % |
|---|---|---|---|---|---|---|---|
| Air | 17 | 1.64E+09 | 4.03E+08 | 2.81E-09 | 803.002 | 196.985 | 99.9987 |
| Water | 550 | 1.32E+04 | 1.05E+04 | 3.65E-10 | 6.45E-03 | 5.12E-03 | 1.16E-03 |
| Soil | 1700 | 3.79E+03 | | 7.19E-09 | 1.85E-03 | | 1.85E-04 |
| Biota (fish) | | | | 1.83E-08 | | | |
| Suspended sediment | | | | 4.49E-08 | | | |
| Bottom sediment | 5500 | 2.60E+01 | 4.13E+00 | 1.44E-08 | 1.27E-05 | 2.02E-06 | 1.47E-06 |
| | Total | 1.64E+09 | 4.03E+08 | | 803.01 | 196.99 | 100 |
| | R + A | | 2.05E+09 | | | 1000 | |

f = 6.962E-06 Pa

Total Amt= 19708 kg

Overall residence time = 19.71 h

Reaction time = 24.54 h

Advection time = 100.05 h

Fugacity Level III calculations: (four-compartment model)

Chemical name: Cyclopentane

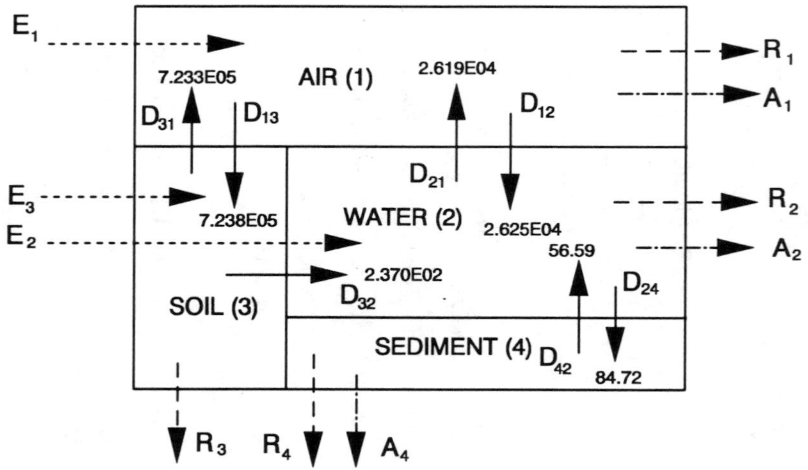

Phase Properties and Rates:

| Compartment | Bulk Z mol/m3 Pa | Half-life h | D Values Reaction mol/Pa h | D Values Advection mol/Pa h |
|---|---|---|---|---|
| Air (1) | 4.034E-04 | 17 | 1.64E+09 | 4.03E+08 |
| Water (2) | 5.249E-05 | 550 | 1.32E+04 | 1.05E+04 |
| Soil (3) | 6.126E-04 | 1700 | 4.49E+03 | |
| Sediment (4) | 4.549E-04 | 5500 | 2.87E+01 | 4.55E+00 |

| | E(1)=1000 | E(2)=1000 | E(3)=1000 | E(1,2,3) |
|---|---|---|---|---|
| Overall residence time = | 19.71 | 224.80 | 34.80 | 82.74 h |
| Reaction time = | 24.54 | 327.48 | 43.27 | 107.72 h |
| Advection time = | 100.04 | 716.94 | 177.71 | 356.90 h |

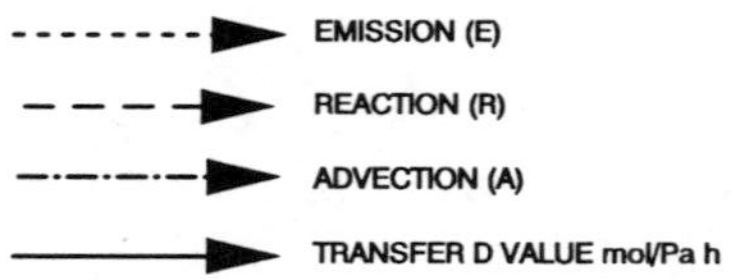

Phase Properties, Compositions, Transport and Transformation Rates:

| Emission, kg/h | | | Fugacity, Pa | | | | Concentration, g/m3 | | | | Amounts, kg | | | | Total |
|---|---|---|---|---|---|---|---|---|---|---|---|---|---|---|---|
| E(1) | E(2) | E(3) | f(1) | f(2) | f(3) | f(4) | C(1) | C(2) | C(3) | C(4) | m(1) | m(2) | m(3) | m(4) | Amount, kg |
| 1000 | 0 | 0 | 6.962E-06 | 3.691E-06 | 6.921E-06 | 3.482E-06 | 1.970E-07 | 1.359E-08 | 2.974E-07 | 1.111E-07 | 1.970E+04 | 2.718E+00 | 5.353E+00 | 5.555E-02 | 1.971E+04 |
| 0 | 1000 | 0 | 3.651E-06 | 2.854E-01 | 3.629E-06 | 2.693E-01 | 1.033E-07 | 1.051E-03 | 1.559E-07 | 8.592E-03 | 1.033E+04 | 2.102E+05 | 2.807E+00 | 4.296E+03 | 2.248E+05 |
| 0 | 0 | 1000 | 6.918E-06 | 9.657E-05 | 1.959E-02 | 9.111E-05 | 1.957E-07 | 3.556E-07 | 8.417E-04 | 2.907E-06 | 1.957E+04 | 7.111E+01 | 1.515E+04 | 1.454E+00 | 3.480E+04 |
| 600 | 300 | 100 | 5.964E-06 | 8.564E-02 | 1.964E-03 | 8.080E-02 | 1.688E-07 | 3.153E-04 | 8.440E-05 | 2.578E-03 | 1.688E+04 | 6.306E+04 | 1.519E+03 | 1.289E+03 | 8.274E+04 |

| Emission, kg/h | | | Loss, Reaction, kg/h | | | | Loss, Advection, kg/h | | | Intermedia Rate of Transport, kg/h | | | | | | |
|---|---|---|---|---|---|---|---|---|---|---|---|---|---|---|---|---|
| | | | | | | | | | | T12 | T21 | T13 | T31 | T32 | T24 | T42 |
| E(1) | E(2) | E(3) | R(1) | R(2) | R(3) | R(4) | A(1) | A(2) | A(4) | air-water | water-air | air-soil | soil-air | soil-water | water-sed | sed-water |
| 1000 | 0 | 0 | 8.030E+02 | 3.424E-03 | 2.18E-03 | 7.000E-06 | 1.970E+02 | 2.718E-03 | 1.111E-06 | 1.282E-02 | 6.781E-03 | 3.534E-01 | 3.511E-01 | 1.150E-04 | 2.193E-05 | 1.382E-05 |
| 0 | 1000 | 0 | 4.211E+02 | 2.648E+02 | 1.14E-03 | 5.413E-01 | 1.033E+02 | 2.102E+02 | 8.592E-02 | 6.721E-03 | 5.244E+02 | 1.853E-01 | 1.841E-01 | 6.033E-05 | 1.696E+00 | 1.069E+00 |
| 0 | 0 | 1000 | 7.979E+02 | 8.960E-02 | 6.18E+00 | 1.831E-04 | 1.957E+02 | 7.111E-02 | 2.907E-05 | 1.273E-02 | 1.774E-01 | 3.512E-01 | 9.938E+02 | 3.256E-01 | 5.738E-04 | 3.616E-04 |
| 600 | 300 | 100 | 6.879E+02 | 7.946E+01 | 6.19E-01 | 1.624E-01 | 1.688E+02 | 6.306E+01 | 2.578E-02 | 1.098E-02 | 1.573E+02 | 3.028E-01 | 9.965E+01 | 3.265E-02 | 5.089E-01 | 3.207E-01 |

Chemical name: Cyclopentane

# Level III Distribution

## Distribution of mass

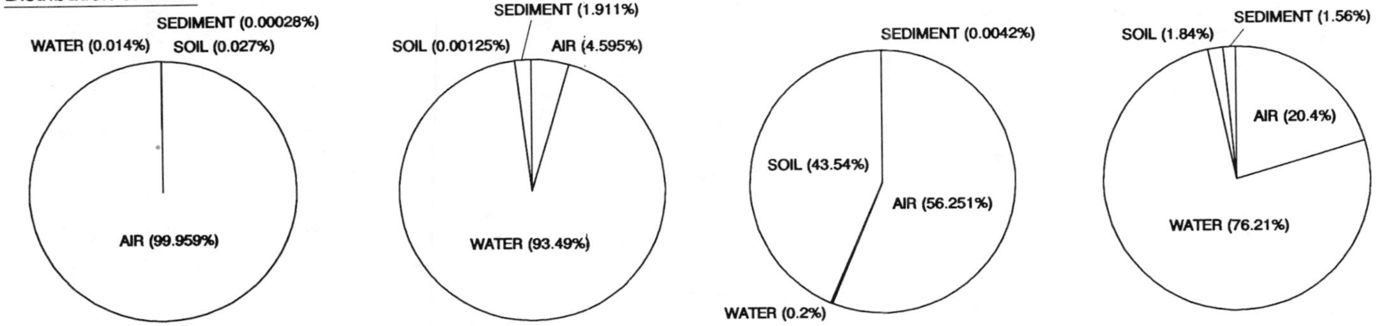


## Distribution of removal rates

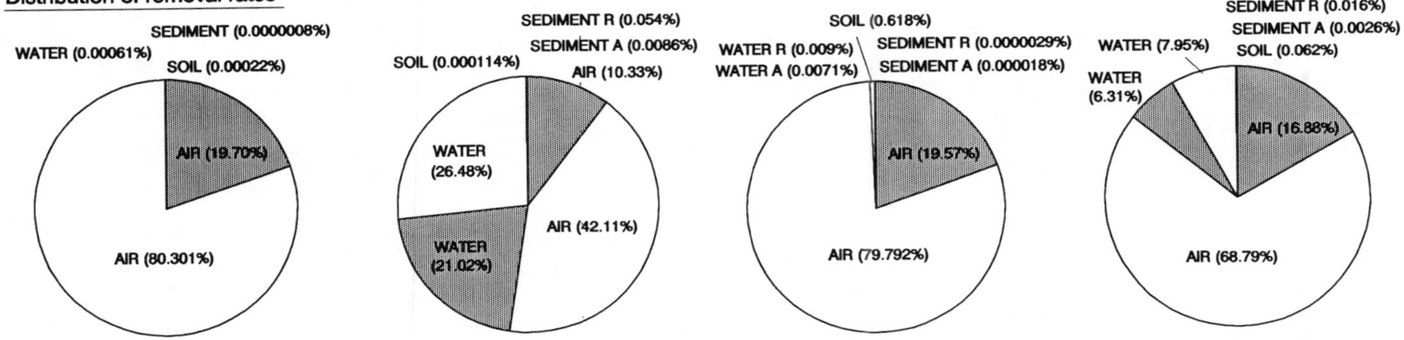


Reaction Advection

Emission rates:

| | | | |
|---|---|---|---|
| E(1) = 1000 kg/h | E(1) = 0 | E(1) = 0 | E(1) = 600 kg/h |
| E(2) = 0 | E(2) = 1000 kg/h | E(2) = 0 | E(2) = 300 kg/h |
| E(3) = 0 | E(3) = 0 | E(3) = 1000 kg/h | E(3) = 100 kg/h |

Residence time:

| | | | |
|---|---|---|---|
| t = 19.71 h | t = 224.8 h | t = 34.8 h | t = 82.74 h |

Chemical name: Cyclohexane

Level I calculation: (six-compartment model)

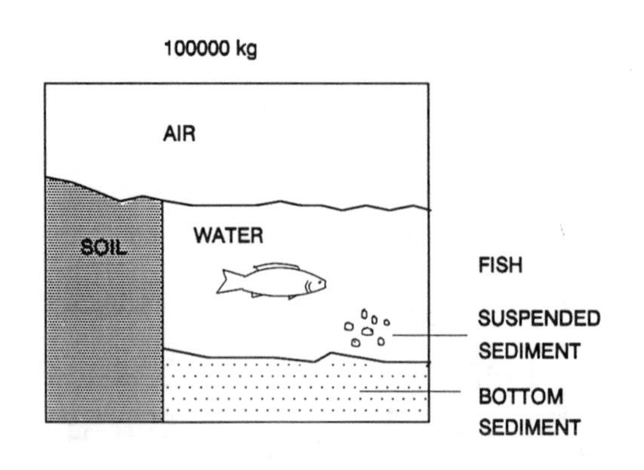


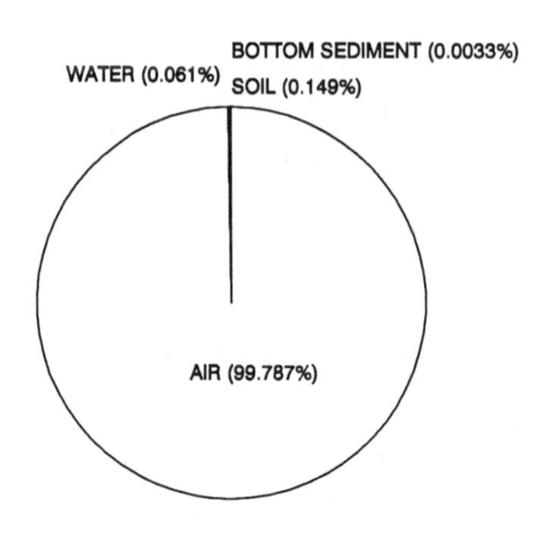


Distribution of mass

physical-chemical properties:

MW: 84.16 g/mol

M.P.: 6.55°C

Fugacity ratio: 1.0

vapor pressure: 5300 Pa

solubility: 55 $g/m^3$

log $K_{OW}$: 3.44

| Compartment | Z | Concentration | | | Amount | Amount |
|---|---|---|---|---|---|---|
| | mol/m3 Pa | mol/m3 | mg/L (or g/m3) | ug/g | kg | % |
| Air | 4.034E-04 | 1.186E-08 | 9.979E-07 | 8.418E-04 | 99787 | 99.787 |
| Water | 1.233E-04 | 3.624E-09 | 3.050E-07 | 3.050E-07 | 61.000 | 6.10E-02 |
| Soil | 6.684E-03 | 1.964E-07 | 1.653E-05 | 6.888E-06 | 148.788 | 0.1488 |
| Biota (fish) | 1.698E-02 | 4.991E-07 | 4.200E-05 | 4.200E-05 | 8.40E-03 | 8.40E-06 |
| Suspended sediment | 4.177E-02 | 1.228E-06 | 1.033E-04 | 6.888E-05 | 0.1033 | 1.03E-04 |
| Bottom sediment | 1.337E-02 | 3.929E-07 | 3.306E-05 | 1.378E-05 | 3.3064 | 3.31E-03 |
| | Total | | | | 100000 | 100 |

f = 2.939E-05 Pa

Chemical name: Cyclohexane

Level II calculation: (six-compartment model)

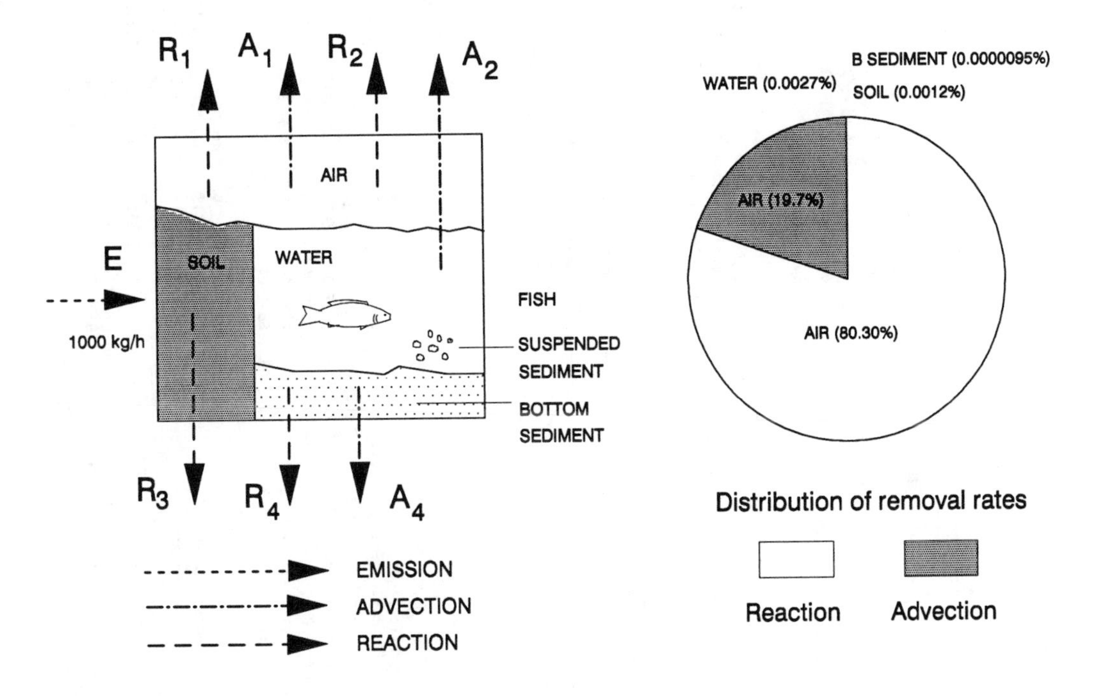


| Compartment | Half-life h | D Values Reaction mol/Pa h | D Values Advection mol/Pa h | Conc'n mol/m3 | Loss Reaction kg/h | Loss Advection kg/h | Removal % |
|---|---|---|---|---|---|---|---|
| Air | 17 | 1.64E+09 | 4.03E+08 | 2.34E-09 | 802.981 | 196.980 | 99.9961 |
| Water | 550 | 3.11E+04 | 2.47E+04 | 7.15E-10 | 0.0152 | 0.0120 | 2.72E-03 |
| Soil | 1700 | 2.45E+04 | | 3.88E-08 | 0.0120 | | 1.20E-03 |
| Biota (fish) | | | | 9.85E-08 | | | |
| Suspended sediment | | | | 2.42E-07 | | | |
| Bottom sediment | 5500 | 1.68E+02 | 2.67E+01 | 7.76E-08 | 8.22E-05 | 1.31E-05 | 9.53E-06 |
| | Total | 1.64E+09 | 4.03E+08 | | 803.01 | 196.99 | 100 |
| | R + A | | 2.05E+09 | | | 1000 | |

f = 5.802E-06 Pa

Total Amt= 19740 kg

Overall residence time = 19.74 h

Reaction time = 24.58 h

Advection time = 100.21 h

## Fugacity Level III calculations: (four-compartment model)

Chemical name: Cyclohexane

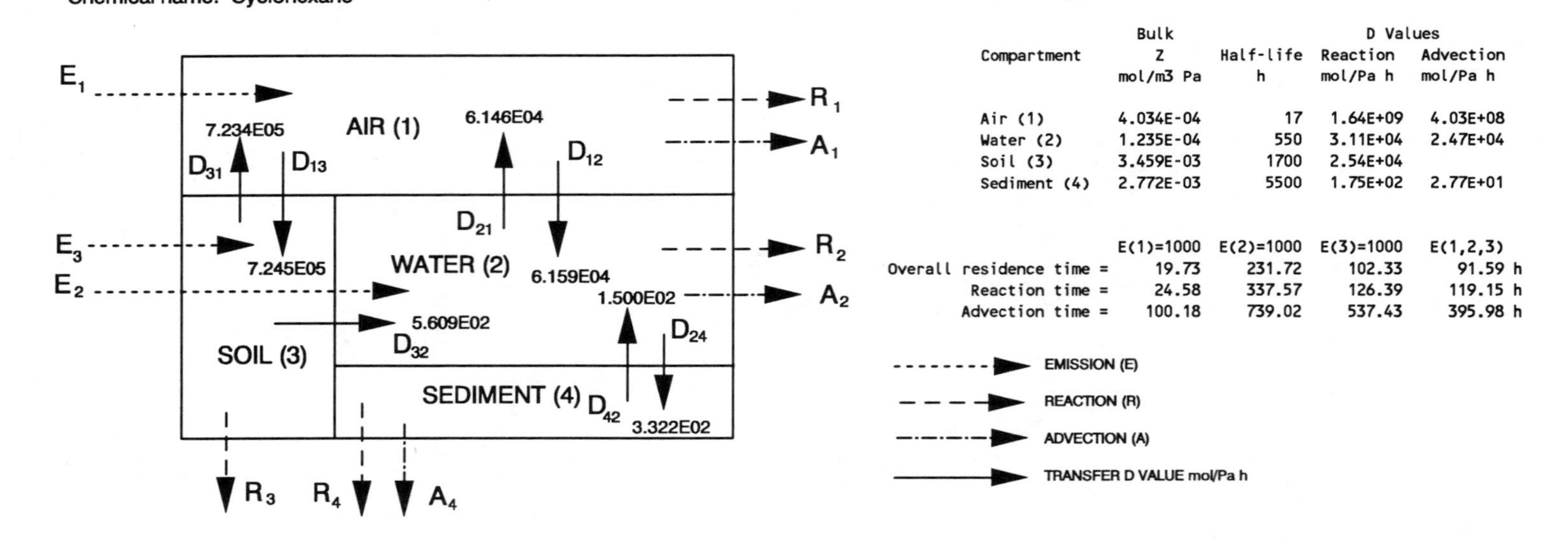


Phase Properties and Rates:

| Compartment | Bulk Z mol/m3 Pa | Half-life h | D Values Reaction mol/Pa h | D Values Advection mol/Pa h |
|---|---|---|---|---|
| Air (1) | 4.034E-04 | 17 | 1.64E+09 | 4.03E+08 |
| Water (2) | 1.235E-04 | 550 | 3.11E+04 | 2.47E+04 |
| Soil (3) | 3.459E-03 | 1700 | 2.54E+04 | |
| Sediment (4) | 2.772E-03 | 5500 | 1.75E+02 | 2.77E+01 |

| | E(1)=1000 | E(2)=1000 | E(3)=1000 | E(1,2,3) |
|---|---|---|---|---|
| Overall residence time = | 19.73 | 231.72 | 102.33 | 91.59 h |
| Reaction time = | 24.58 | 337.57 | 126.39 | 119.15 h |
| Advection time = | 100.18 | 739.02 | 537.43 | 395.98 h |

Phase Properties, Compositions, Transport and Transformation Rates:

| Emission, kg/h | | | Fugacity, Pa | | | | Concentration, g/m3 | | | | Amounts, kg | | | | Total Amount, kg |
|---|---|---|---|---|---|---|---|---|---|---|---|---|---|---|---|
| E(1) | E(2) | E(3) | f(1) | f(2) | f(3) | f(4) | C(1) | C(2) | C(3) | C(4) | m(1) | m(2) | m(3) | m(4) | |
| 1000 | 0 | 0 | 5.802E-06 | 3.068E-06 | 5.610E-06 | 2.892E-06 | 1.970E-07 | 3.190E-08 | 1.633E-06 | 6.747E-07 | 1.970E+04 | 6.380E+00 | 2.940E+01 | 3.373E-01 | 1.973E+04 |
| 0 | 1000 | 0 | 3.035E-06 | 1.011E-01 | 2.935E-06 | 9.533E-02 | 1.030E-07 | 1.051E-03 | 8.544E-07 | 2.224E-02 | 1.030E+04 | 2.103E+05 | 1.538E+01 | 1.112E+04 | 2.317E+05 |
| 0 | 0 | 1000 | 5.603E-06 | 7.866E-05 | 1.586E-02 | 7.415E-05 | 1.902E-07 | 8.178E-07 | 4.618E-03 | 1.730E-05 | 1.902E+04 | 1.636E+02 | 8.313E+04 | 8.649E+00 | 1.023E+05 |
| 600 | 300 | 100 | 4.952E-06 | 3.035E-02 | 1.591E-03 | 2.861E-02 | 1.681E-07 | 3.155E-04 | 4.631E-04 | 6.674E-03 | 1.681E+04 | 6.311E+04 | 8.335E+03 | 3.337E+03 | 9.159E+04 |

| Emission, kg/h | | | Loss, Reaction, kg/h | | | | Loss, Advection, kg/h | | | Intermedia Rate of Transport, kg/h T12 | T21 | T13 | T31 | T32 | T24 | T42 |
|---|---|---|---|---|---|---|---|---|---|---|---|---|---|---|---|---|
| E(1) | E(2) | E(3) | R(1) | R(2) | R(3) | R(4) | A(1) | A(2) | A(4) | air-water | water-air | air-soil | soil-air | soil-water | water-sed | sed-water |
| 1000 | 0 | 0 | 8.030E+02 | 8.038E-03 | 1.20E-02 | 4.251E-05 | 1.970E+02 | 6.380E-03 | 6.747E-06 | 3.007E-02 | 1.587E-02 | 3.538E-01 | 3.415E-01 | 2.648E-04 | 8.577E-05 | 3.652E-05 |
| 0 | 1000 | 0 | 4.201E+02 | 2.650E+02 | 6.27E-03 | 1.401E+00 | 1.030E+02 | 2.103E+02 | 2.224E-01 | 1.573E-02 | 5.232E+02 | 1.851E-01 | 1.787E-01 | 1.385E-04 | 2.827E+00 | 1.204E+00 |
| 0 | 0 | 1000 | 7.755E+02 | 2.061E-01 | 3.39E+01 | 1.090E-03 | 1.902E+02 | 1.636E-01 | 1.730E-04 | 2.904E-02 | 4.069E-01 | 3.417E-01 | 9.657E+02 | 7.488E-01 | 2.199E-03 | 9.363E-04 |
| 600 | 300 | 100 | 6.854E+02 | 7.951E+01 | 3.40E+00 | 4.205E-01 | 1.681E+02 | 6.311E+01 | 6.674E-02 | 2.567E-02 | 1.570E+02 | 3.019E-01 | 9.683E+01 | 7.508E-02 | 8.484E-01 | 3.612E-01 |

# Level III Distribution

Chemical name: Cyclohexane

## Distribution of mass

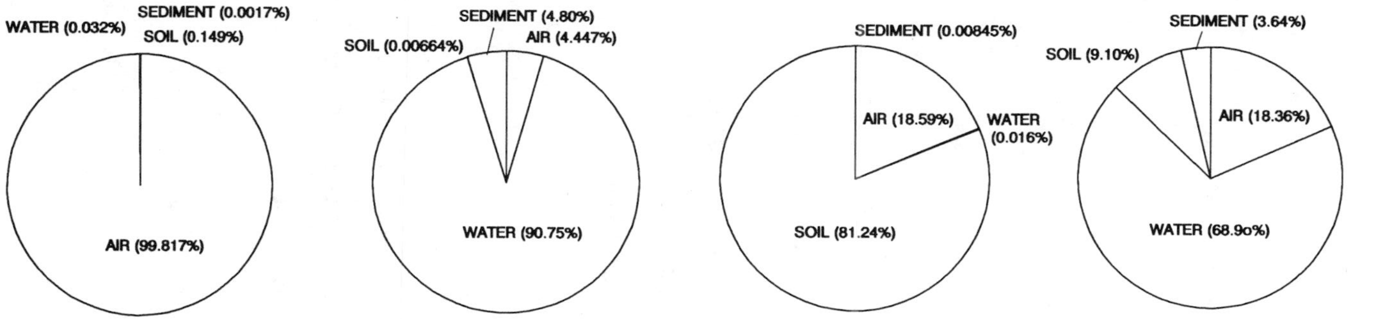


## Distribution of removal rates

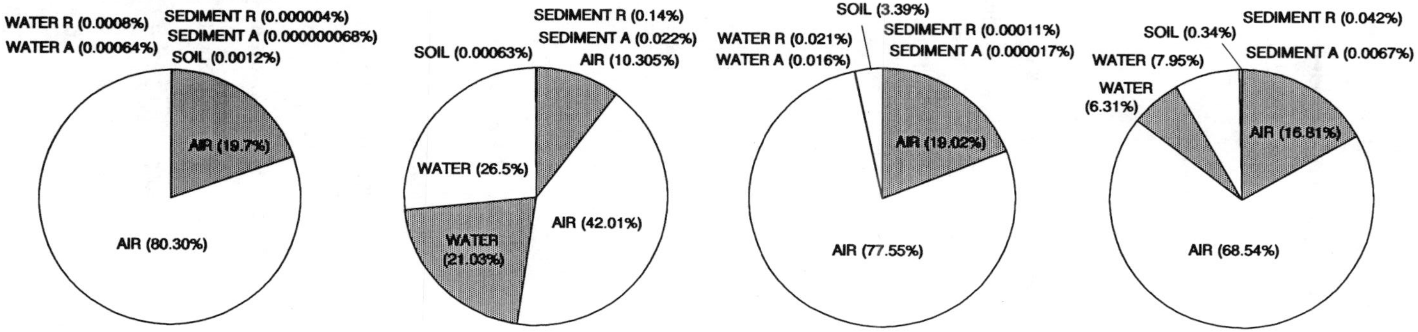


Reaction  Advection

| | | | | |
|---|---|---|---|---|
| Emission rates: | E(1) = 1000 kg/h | E(1) = 0 | E(1) = 0 | E(1) = 600 kg/h |
| | E(2) = 0 | E(2) = 1000 kg/h | E(2) = 0 | E(2) = 300 kg/h |
| | E(3) = 0 | E(3) = 0 | E(3) = 1000 kg/h | E(3) = 100 kg/h |
| Residence time: | t = 19.73 h | t = 231.7 h | t = 102.3 h | t = 91.6 h |

Chemical name: Methylcyclohexane

Level I calculation: (six-compartment model)

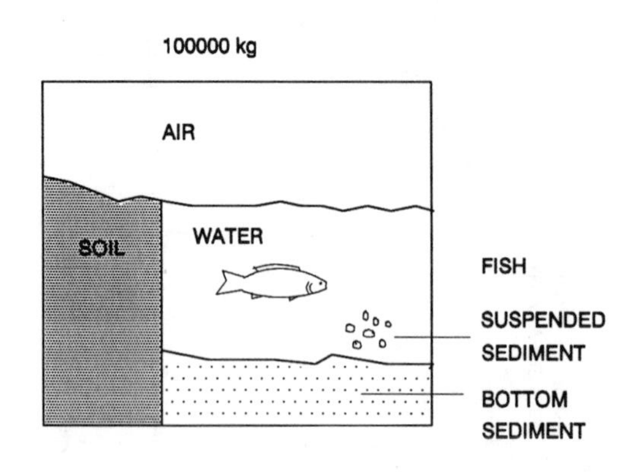


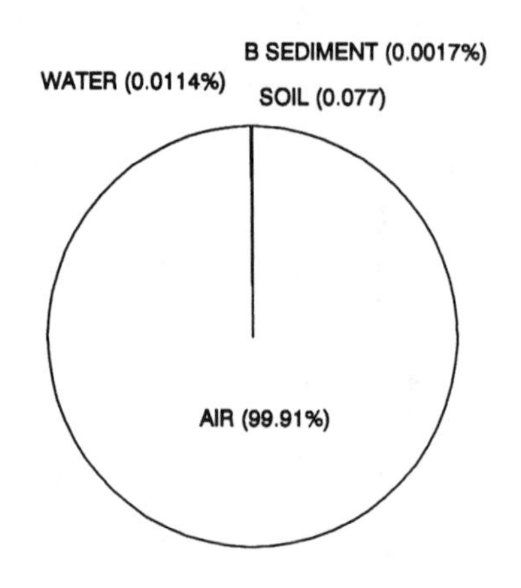


Distribution of mass

physical-chemical properties:

MW: 98.19 g/mol

M.P.: -126.6°C

Fugacity ratio: 1.0

vapor pressure: 6180 Pa

solubility: 14.0 g/m$^3$

log $K_{OW}$: 3.88

| Compartment | Z | Concentration | | | Amount | Amount |
|---|---|---|---|---|---|---|
| | mol/m3 Pa | mol/m3 | mg/L (or g/m3) | ug/g | kg | % |
| Air | 4.034E-04 | 1.018E-08 | 9.991E-07 | 8.428E-04 | 99910 | 99.910 |
| Water | 2.307E-05 | 5.819E-10 | 5.714E-08 | 5.714E-08 | 11.428 | 1.14E-02 |
| Soil | 3.444E-03 | 8.687E-08 | 8.530E-06 | 3.554E-06 | 76.770 | 0.0768 |
| Biota (fish) | 8.751E-03 | 2.207E-07 | 2.167E-05 | 2.167E-05 | 4.33E-03 | 4.33E-06 |
| Suspended sediment | 2.153E-02 | 5.430E-07 | 5.331E-05 | 3.554E-05 | 0.0533 | 5.33E-05 |
| Bottom sediment | 6.889E-03 | 1.737E-07 | 1.706E-05 | 7.108E-06 | 1.7060 | 1.71E-03 |
| | Total | | | | 100000 | 100 |

f = 2.522E-05 Pa

Chemical name: Methylcyclohexane

Level II calculation: (six-compartment model)

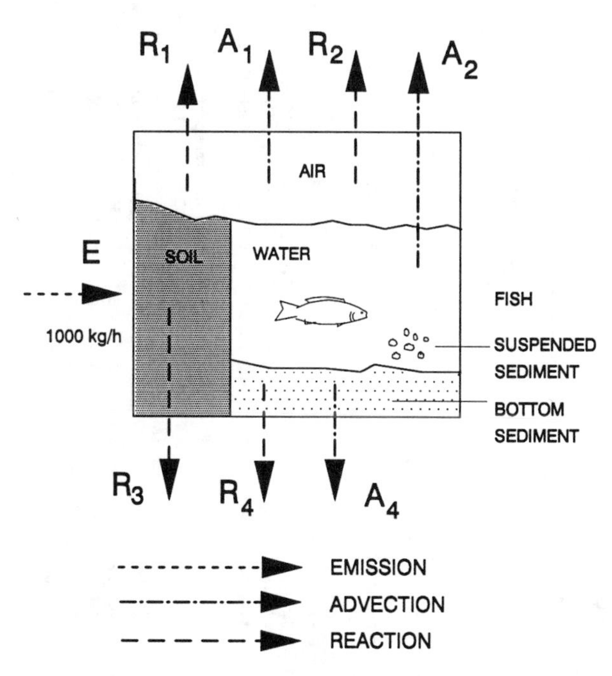


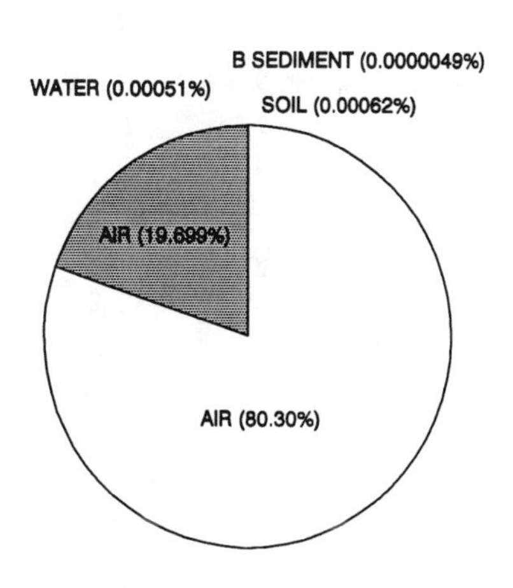


Distribution of removal rates

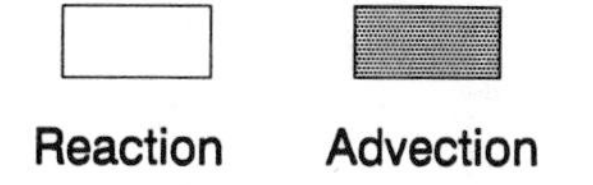


| Compartment | Half-life h | D Values Reaction mol/Pa h | D Values Advection mol/Pa h | Conc'n mol/m3 | Loss Reaction kg/h | Loss Advection kg/h | Removal % |
|---|---|---|---|---|---|---|---|
| Air | 17 | 1.64E+09 | 4.03E+08 | 2.01E-09 | 803.004 | 196.985 | 99.9989 |
| Water | 550 | 5.81E+03 | 4.61E+03 | 1.15E-10 | 2.84E-03 | 2.25E-03 | 5.09E-04 |
| Soil | 1700 | 1.26E+04 | | 1.71E-08 | 6.17E-03 | | 6.17E-04 |
| Biota (fish) | | | | 4.35E-08 | | | |
| Suspended sediment | | | | 1.07E-07 | | | |
| Bottom sediment | 5500 | 8.68E+01 | 1.38E+01 | 3.43E-08 | 4.24E-05 | 6.73E-06 | 4.91E-06 |
| | Total | 1.64E+09 | 4.03E+08 | | 803.01 | 196.99 | 100 |
| | R + A | | 2.05E+09 | | | 1000 | |

f = 4.973E-06 Pa
Total Amt= 19716 kg

Overall residence time = 19.72 h
Reaction time = 24.55 h
Advection time = 100.09 h

## Fugacity Level III calculations: (four-compartment model)

Chemical name: Methylcyclohexane

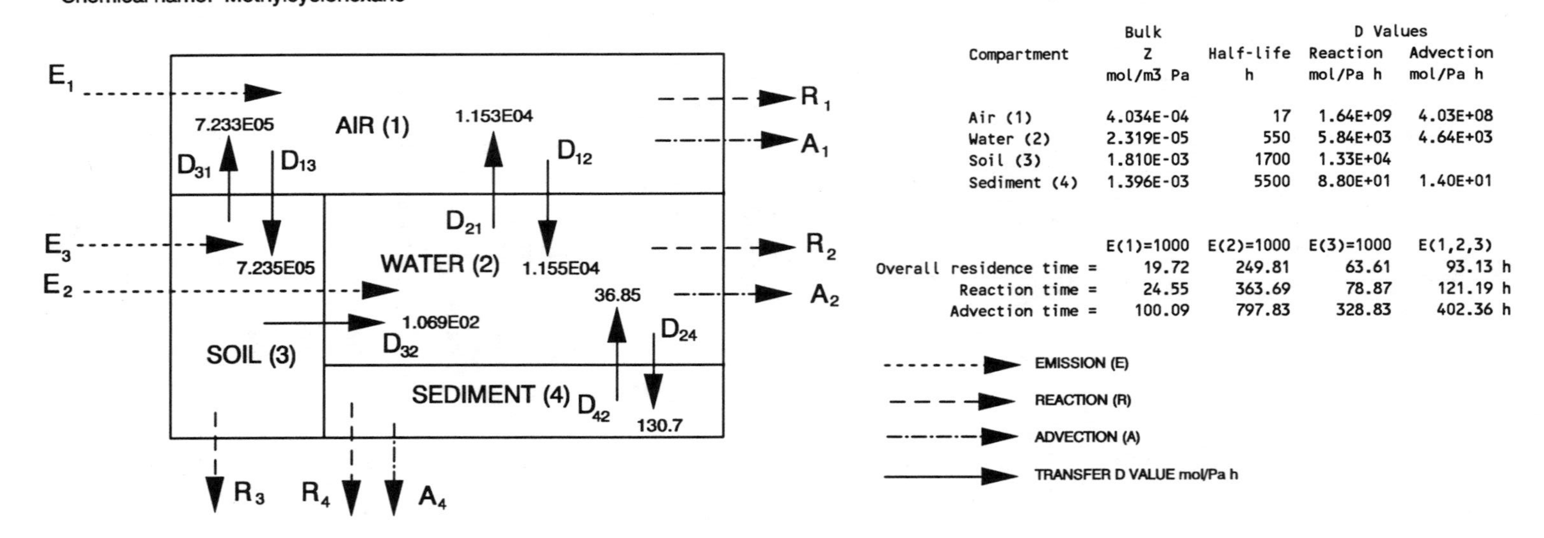


Phase Properties and Rates:

| Compartment | Bulk Z mol/m3 Pa | Half-life h | D Values Reaction mol/Pa h | D Values Advection mol/Pa h |
|---|---|---|---|---|
| Air (1) | 4.034E-04 | 17 | 1.64E+09 | 4.03E+08 |
| Water (2) | 2.319E-05 | 550 | 5.84E+03 | 4.64E+03 |
| Soil (3) | 1.810E-03 | 1700 | 1.33E+04 | |
| Sediment (4) | 1.396E-03 | 5500 | 8.80E+01 | 1.40E+01 |

| | E(1)=1000 | E(2)=1000 | E(3)=1000 | E(1,2,3) |
|---|---|---|---|---|
| Overall residence time = | 19.72 | 249.81 | 63.61 | 93.13 h |
| Reaction time = | 24.55 | 363.69 | 78.87 | 121.19 h |
| Advection time = | 100.09 | 797.83 | 328.83 | 402.36 h |

Phase Properties, Compositions, Transport and Transformation Rates:

| Emission, kg/h | | | Fugacity, Pa | | | | Concentration, g/m3 | | | | Amounts, kg | | | | Total Amount, kg |
|---|---|---|---|---|---|---|---|---|---|---|---|---|---|---|---|
| E(1) | E(2) | E(3) | f(1) | f(2) | f(3) | f(4) | C(1) | C(2) | C(3) | C(4) | m(1) | m(2) | m(3) | m(4) | |
| 1000 | 0 | 0 | 4.973E-06 | 2.623E-06 | 4.884E-06 | 2.470E-06 | 1.970E-07 | 5.972E-09 | 8.679E-07 | 3.387E-07 | 1.970E+04 | 1.194E+00 | 1.562E+01 | 1.693E-01 | 1.972E+04 |
| 0 | 1000 | 0 | 2.594E-06 | 4.607E-01 | 2.547E-06 | 4.339E-01 | 1.027E-07 | 1.049E-03 | 4.526E-07 | 5.949E-02 | 1.027E+04 | 2.098E+05 | 8.148E+00 | 2.974E+04 | 2.498E+05 |
| 0 | 0 | 1000 | 4.883E-06 | 6.944E-05 | 1.383E-02 | 6.541E-05 | 1.934E-07 | 1.581E-07 | 2.458E-03 | 8.967E-06 | 1.934E+04 | 3.162E+01 | 4.424E+04 | 4.483E+00 | 6.361E+04 |
| 600 | 300 | 100 | 4.250E-06 | 1.382E-01 | 1.387E-03 | 1.302E-01 | 1.684E-07 | 3.147E-04 | 2.464E-04 | 1.785E-02 | 1.684E+04 | 6.294E+04 | 4.435E+03 | 8.924E+03 | 9.313E+04 |

| Emission, kg/h | | | Loss, Reaction, kg/h | | | | Loss, Advection, kg/h | | | Intermedia Rate of Transport, kg/h T12 | T21 | T13 | T31 | T32 | T24 | T42 |
|---|---|---|---|---|---|---|---|---|---|---|---|---|---|---|---|---|
| E(1) | E(2) | E(3) | R(1) | R(2) | R(3) | R(4) | A(1) | A(2) | A(4) | air-water | water-air | air-soil | soil-air | soil-water | water-sed | sed-water |
| 1000 | 0 | 0 | 8.030E+02 | 1.505E-03 | 6.37E-03 | 2.134E-05 | 1.970E+02 | 1.194E-03 | 3.387E-06 | 5.642E-03 | 2.969E-03 | 3.533E-01 | 3.469E-01 | 5.128E-05 | 3.366E-05 | 8.939E-06 |
| 0 | 1000 | 0 | 4.188E+02 | 2.643E+02 | 3.32E-03 | 3.748E+00 | 1.027E+02 | 2.098E+02 | 5.949E-01 | 2.942E-03 | 5.215E+02 | 1.843E-01 | 1.809E-01 | 2.674E-05 | 5.913E+00 | 1.570E+00 |
| 0 | 0 | 1000 | 7.885E+02 | 3.984E-02 | 1.80E+01 | 5.649E-04 | 1.934E+02 | 3.162E-02 | 8.967E-05 | 5.540E-03 | 7.861E-02 | 3.469E-01 | 9.822E+02 | 1.452E-01 | 8.912E-04 | 2.367E-04 |
| 600 | 300 | 100 | 6.863E+02 | 7.930E+01 | 1.81E+00 | 1.124E+00 | 1.684E+02 | 6.294E+01 | 1.785E-01 | 4.822E-03 | 1.565E+02 | 3.019E-01 | 9.848E+01 | 1.456E-02 | 1.774E+00 | 4.710E-01 |

# Level III Distribution

Chemical name: Cyclohexane

## Distribution of mass

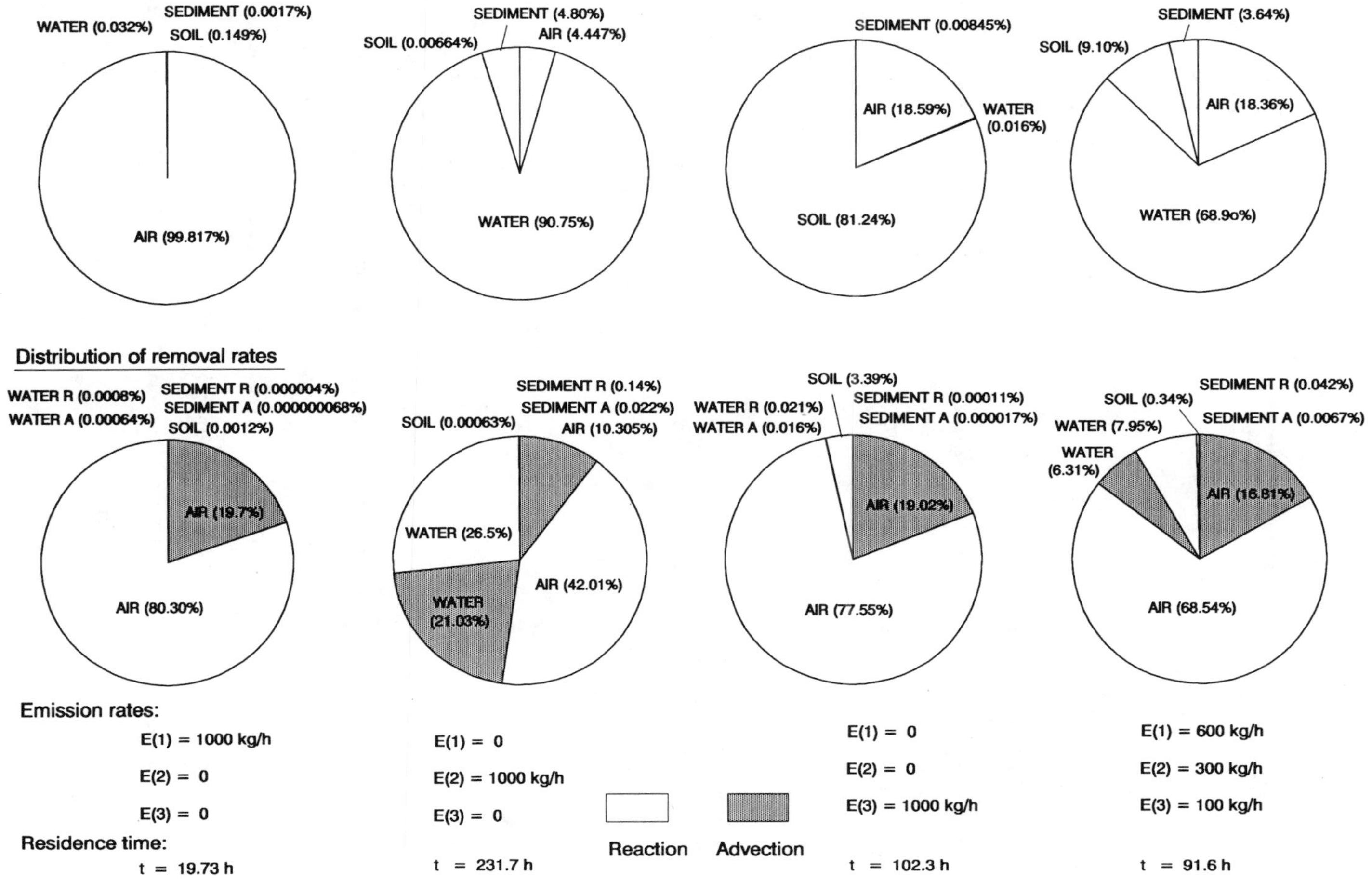

WATER (0.032%)
SEDIMENT (0.0017%)
SOIL (0.149%)
AIR (99.817%)

SOIL (0.00664%)
SEDIMENT (4.80%)
AIR (4.447%)
WATER (90.75%)

SEDIMENT (0.00845%)
AIR (18.59%)
WATER (0.016%)
SOIL (81.24%)

SOIL (9.10%)
SEDIMENT (3.64%)
AIR (18.36%)
WATER (68.9o%)

## Distribution of removal rates

WATER R (0.0008%)
WATER A (0.00064%)
SEDIMENT R (0.000004%)
SEDIMENT A (0.000000068%)
SOIL (0.0012%)
AIR (19.7%)
AIR (80.30%)

SOIL (0.00063%)
SEDIMENT R (0.14%)
SEDIMENT A (0.022%)
AIR (10.305%)
WATER (26.5%)
WATER (21.03%)
AIR (42.01%)

WATER R (0.021%)
WATER A (0.016%)
SOIL (3.39%)
SEDIMENT R (0.00011%)
SEDIMENT A (0.000017%)
AIR (19.02%)
AIR (77.55%)

SOIL (0.34%)
WATER (7.95%)
WATER (6.31%)
SEDIMENT R (0.042%)
SEDIMENT A (0.0067%)
AIR (16.81%)
AIR (68.54%)

Reaction
Advection

| | | | | |
|---|---|---|---|---|
| Emission rates: | E(1) = 1000 kg/h | E(1) = 0 | E(1) = 0 | E(1) = 600 kg/h |
| | E(2) = 0 | E(2) = 1000 kg/h | E(2) = 0 | E(2) = 300 kg/h |
| | E(3) = 0 | E(3) = 0 | E(3) = 1000 kg/h | E(3) = 100 kg/h |
| Residence time: | t = 19.73 h | t = 231.7 h | t = 102.3 h | t = 91.6 h |

Chemical name: Cyclooctane

Level I calculation: (six-compartment model)

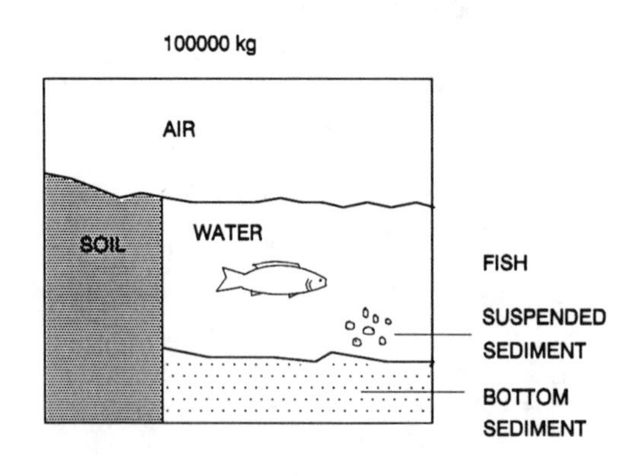


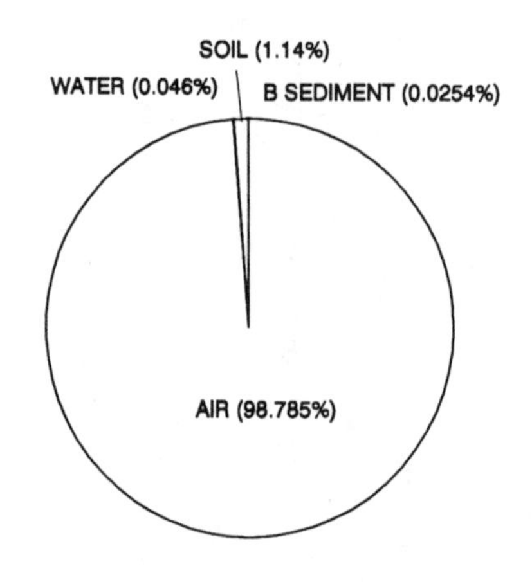


Distribution of mass

physical-chemical properties:

soMW: 112.2 g/mol

M.P.: 14.3°C

Fugacity ratio: 1.0

vapor pressure: 753 Pa

solubility: 7.90 g/m$^3$

log $K_{OW}$: 4.45

| Compartment | Z | Concentration | | | Amount | Amount |
|---|---|---|---|---|---|---|
| | mol/m3 Pa | mol/m3 | mg/L (or g/m3) | ug/g | kg | % |
| Air | 4.034E-04 | 8.803E-09 | 9.879E-07 | 8.333E-04 | 98785 | 98.785 |
| Water | 9.349E-05 | 2.040E-09 | 2.289E-07 | 2.289E-07 | 45.786 | 4.58E-02 |
| Soil | 5.185E-02 | 1.131E-06 | 1.270E-04 | 5.291E-05 | 1142.79 | 1.1428 |
| Biota (fish) | 1.317E-01 | 2.875E-06 | 3.226E-04 | 3.226E-04 | 6.45E-02 | 6.45E-05 |
| Suspended sediment | 3.241E-01 | 7.072E-06 | 7.936E-04 | 5.291E-04 | 0.7936 | 7.94E-04 |
| Bottom sediment | 1.037E-01 | 2.263E-06 | 2.540E-04 | 1.058E-04 | 25.3953 | 2.54E-02 |
| | Total | | | | 100000 | 100 |

f = 2.182E-05 Pa

Chemical name: Cyclooctane

Level II calculation: (six-compartment model)

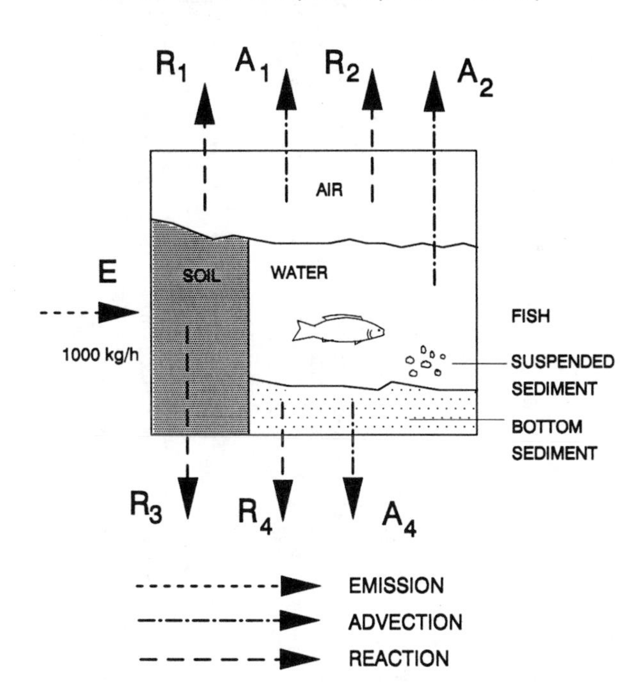


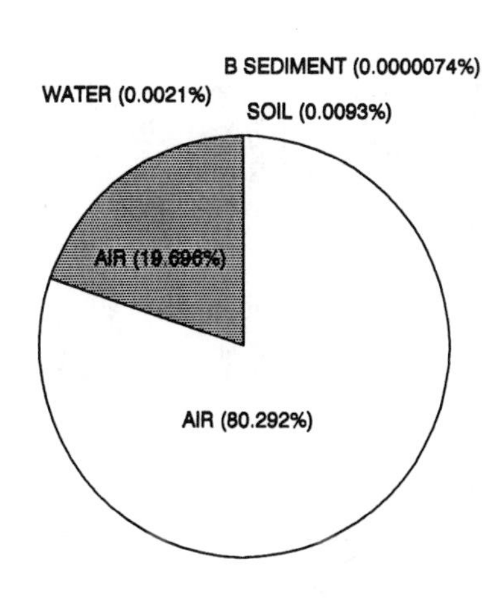


Distribution of removal rates

| Compartment | Half-life h | D Values Reaction mol/Pa h | D Values Advection mol/Pa h | Conc'n mol/m3 | Loss Reaction kg/h | Loss Advection kg/h | Removal % |
|---|---|---|---|---|---|---|---|
| Air | 17 | 1.64E+09 | 4.03E+08 | 1.76E-09 | 802.921 | 196.965 | 99.9886 |
| Water | 550 | 2.36E+04 | 1.87E+04 | 4.07E-10 | 1.15E-02 | 9.13E-03 | 2.06E-03 |
| Soil | 1700 | 1.90E+05 | | 2.26E-07 | 9.29E-02 | | 9.29E-03 |
| Biota (fish) | | | | 5.73E-07 | | | |
| Suspended sediment | | | | 1.41E-06 | | | |
| Bottom sediment | 5500 | 1.31E+03 | 2.07E+02 | 4.51E-07 | 6.38E-04 | 1.01E-04 | 7.39E-05 |
| | Total | 1.64E+09 | 4.03E+08 | | 803.03 | 196.97 | 100 |
| | R + A | | 2.05E+09 | | | 1000 | |

f = 4.351E-06 Pa

Total Amt= 19939 kg

Overall residence time = 19.94 h

Reaction time = 24.83 h

Advection time = 101.23 h

## Fugacity Level III calculations: (four-compartment model)

Chemical name: Cyclooctane

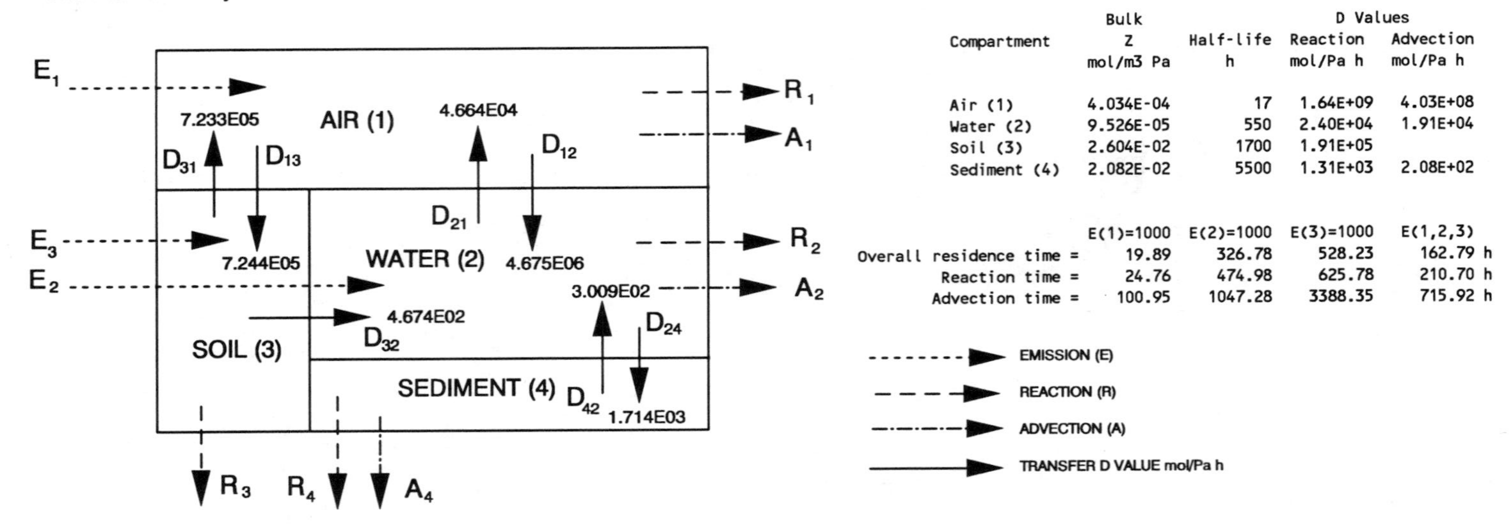


Phase Properties and Rates:

| Compartment | Bulk Z mol/m3 Pa | Half-life h | D Values Reaction mol/Pa h | D Values Advection mol/Pa h |
|---|---|---|---|---|
| Air (1) | 4.034E-04 | 17 | 1.64E+09 | 4.03E+08 |
| Water (2) | 9.526E-05 | 550 | 2.40E+04 | 1.91E+04 |
| Soil (3) | 2.604E-02 | 1700 | 1.91E+05 | |
| Sediment (4) | 2.082E-02 | 5500 | 1.31E+03 | 2.08E+02 |

| | E(1)=1000 | E(2)=1000 | E(3)=1000 | E(1,2,3) |
|---|---|---|---|---|
| Overall residence time = | 19.89 | 326.78 | 528.23 | 162.79 h |
| Reaction time = | 24.76 | 474.98 | 625.78 | 210.70 h |
| Advection time = | 100.95 | 1047.28 | 3388.35 | 715.92 h |

EMISSION (E)

REACTION (R)

ADVECTION (A)

TRANSFER D VALUE mol/Pa h

Phase Properties, Compositions, Transport and Transformation Rates:

| Emission, kg/h E(1) | E(2) | E(3) | Fugacity, Pa f(1) | f(2) | f(3) | f(4) | Concentration, g/m3 C(1) | C(2) | C(3) | C(4) | Amounts, kg m(1) | m(2) | m(3) | m(4) | Total Amount, kg |
|---|---|---|---|---|---|---|---|---|---|---|---|---|---|---|---|
| 1000 | 0 | 0 | 4.352E-06 | 2.250E-06 | 3.445E-06 | 2.119E-06 | 1.970E-07 | 2.405E-08 | 1.007E-05 | 4.949E-06 | 1.970E+04 | 4.810E+00 | 1.812E+02 | 2.475E+00 | 1.989E+04 |
| 0 | 1000 | 0 | 2.227E-06 | 9.780E-02 | 1.763E-06 | 9.207E-02 | 1.008E-07 | 1.045E-03 | 5.152E-06 | 2.151E-01 | 1.008E+04 | 2.091E+05 | 9.274E+01 | 1.075E+05 | 3.268E+05 |
| 0 | 0 | 1000 | 3.442E-06 | 5.175E-05 | 9.745E-03 | 4.872E-05 | 1.558E-07 | 5.531E-07 | 2.847E-02 | 1.138E-04 | 1.558E+04 | 1.106E+02 | 5.125E+05 | 5.691E+01 | 5.282E+05 |
| 600 | 300 | 100 | 3.623E-06 | 2.935E-02 | 9.770E-04 | 2.763E-02 | 1.640E-07 | 3.137E-04 | 2.855E-03 | 6.454E-02 | 1.640E+04 | 6.273E+04 | 5.138E+04 | 3.227E+04 | 1.628E+05 |

| Emission, kg/h E(1) | E(2) | E(3) | Loss, Reaction, kg/h R(1) | R(2) | R(3) | R(4) | Loss, Advection, kg/h A(1) | A(2) | A(4) | Intermedia Rate of Transport, kg/h T12 air-water | T21 water-air | T13 air-soil | T31 soil-air | T32 soil-water | T24 water-sed | T42 sed-water |
|---|---|---|---|---|---|---|---|---|---|---|---|---|---|---|---|---|
| 1000 | 0 | 0 | 8.029E+02 | 6.061E-03 | 7.39E-02 | 3.118E-04 | 1.970E+02 | 4.810E-03 | 4.949E-05 | 2.283E-02 | 1.178E-02 | 3.537E-01 | 2.796E-01 | 1.807E-04 | 4.328E-04 | 7.154E-05 |
| 0 | 1000 | 0 | 4.110E+02 | 2.634E+02 | 3.78E-02 | 1.355E+01 | 1.008E+02 | 2.091E+02 | 2.151E+00 | 1.169E-02 | 5.118E+02 | 1.810E-01 | 1.431E-01 | 9.249E-05 | 1.881E+01 | 3.109E+00 |
| 0 | 0 | 1000 | 6.350E+02 | 1.394E-01 | 2.09E+02 | 7.170E-03 | 1.558E+02 | 1.106E-01 | 1.138E-03 | 1.806E-02 | 2.708E-01 | 2.797E-01 | 7.909E+02 | 5.111E-01 | 9.953E-03 | 1.645E-03 |
| 600 | 300 | 100 | 6.686E+02 | 7.904E+01 | 2.09E+01 | 4.066E+00 | 1.640E+02 | 6.273E+01 | 6.454E-01 | 1.901E-02 | 1.536E+02 | 2.945E-01 | 7.930E+01 | 5.124E-02 | 5.644E+00 | 9.330E-01 |

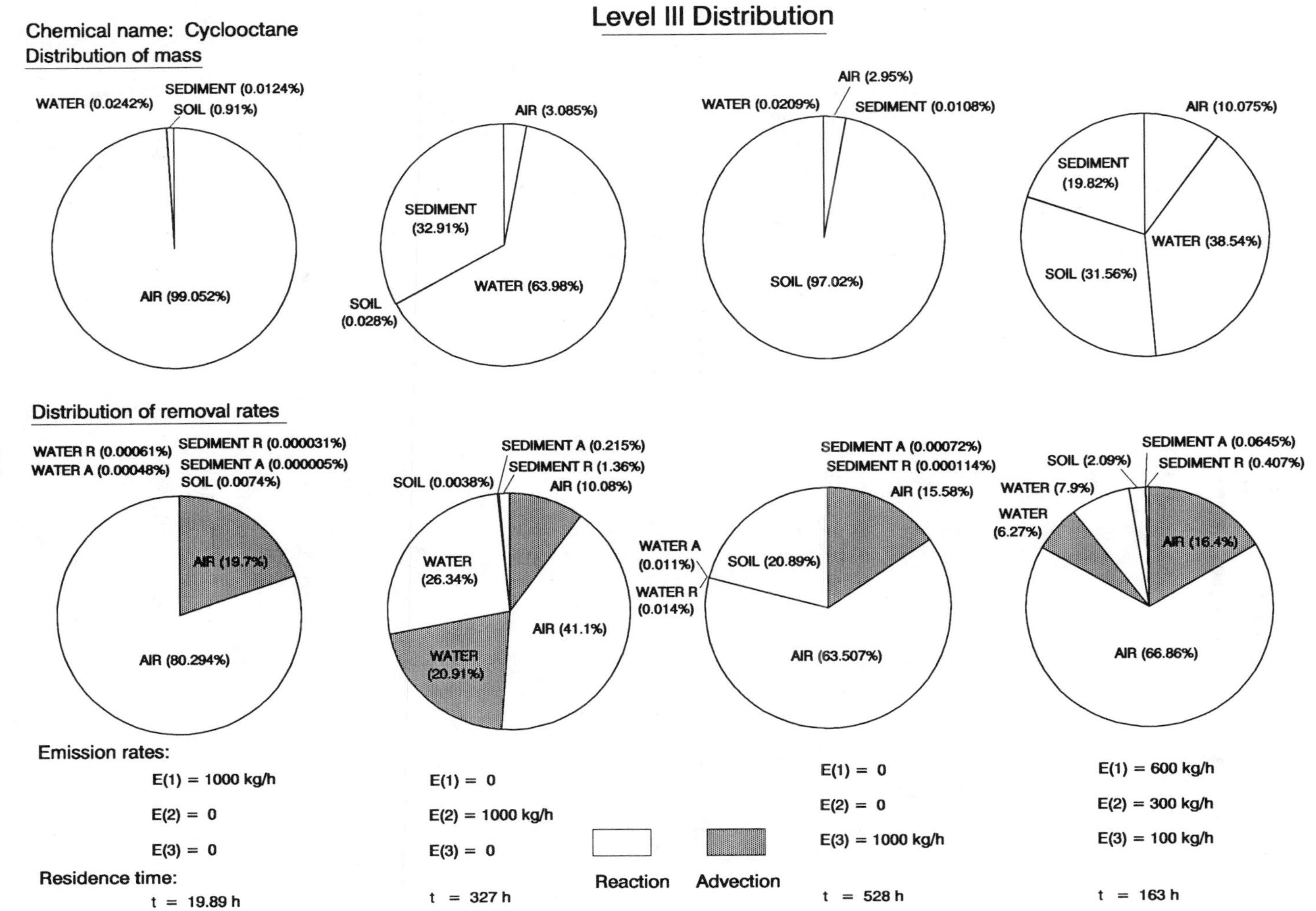
Level III Distribution
Chemical name: Cyclooctane
Distribution of mass
WATER (0.0242%)
SEDIMENT (0.0124%)
SOIL (0.91%)
AIR (99.052%)
AIR (3.085%)
SEDIMENT (32.91%)
WATER (63.98%)
SOIL (0.028%)
AIR (2.95%)
WATER (0.0209%)
SEDIMENT (0.0108%)
SOIL (97.02%)
AIR (10.075%)
SEDIMENT (19.82%)
WATER (38.54%)
SOIL (31.56%)
Distribution of removal rates
WATER R (0.00061%)
WATER A (0.00048%)
SEDIMENT R (0.000031%)
SEDIMENT A (0.000005%)
SOIL (0.0074%)
AIR (19.7%)
AIR (80.294%)
SEDIMENT A (0.215%)
SEDIMENT R (1.36%)
SOIL (0.0038%)
AIR (10.08%)
WATER (26.34%)
WATER (20.91%)
AIR (41.1%)
SEDIMENT A (0.00072%)
SEDIMENT R (0.000114%)
AIR (15.58%)
WATER A (0.011%)
WATER R (0.014%)
SOIL (20.89%)
AIR (63.507%)
SEDIMENT A (0.0645%)
SEDIMENT R (0.407%)
SOIL (2.09%)
WATER (7.9%)
WATER (6.27%)
AIR (16.4%)
AIR (66.86%)
Emission rates:
E(1) = 1000 kg/h
E(2) = 0
E(3) = 0
Residence time:
t = 19.89 h
E(1) = 0
E(2) = 1000 kg/h
E(3) = 0
t = 327 h
Reaction
Advection
E(1) = 0
E(2) = 0
E(3) = 1000 kg/h
t = 528 h
E(1) = 600 kg/h
E(2) = 300 kg/h
E(3) = 100 kg/h
t = 163 h

Chemical name: Cyclohexene

Level I calculation: (six-compartment model)

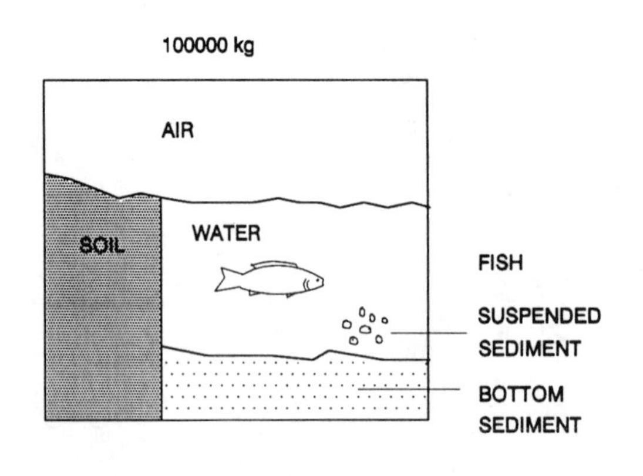


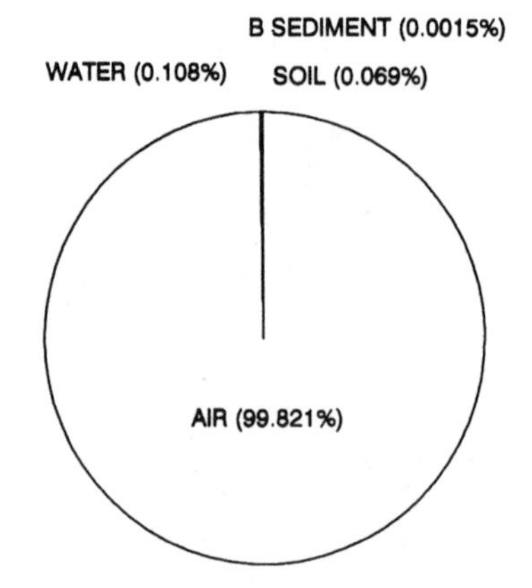


Distribution of mass

physical-chemical properties:

MW: 82.15 g/mol

M.P.: - 103.5°C

Fugacity ratio: 1.0

vapor pressure: 11850 Pa

solubility: 213 g/m$^3$

log $K_{OW}$: 2.86

| Compartment | Z | Concentration | | | Amount | Amount |
|---|---|---|---|---|---|---|
| | mol/m3 Pa | mol/m3 | mg/L (or g/m3) | ug/g | kg | % |
| Air | 4.034E-04 | 1.215E-08 | 9.982E-07 | 8.421E-04 | 99821 | 99.821 |
| Water | 2.188E-04 | 6.590E-09 | 5.414E-07 | 5.414E-07 | 108.280 | 1.08E-01 |
| Soil | 3.119E-03 | 9.396E-08 | 7.719E-06 | 3.216E-06 | 69.468 | 0.0695 |
| Biota (fish) | 7.925E-03 | 2.387E-07 | 1.961E-05 | 1.961E-05 | 3.92E-03 | 3.92E-06 |
| Suspended sediment | 1.950E-02 | 5.872E-07 | 4.824E-05 | 3.216E-05 | 0.0482 | 4.82E-05 |
| Bottom sediment | 6.239E-03 | 1.879E-07 | 1.544E-05 | 6.432E-06 | 1.5437 | 1.54E-03 |
| | Total | | | | 100000 | 100 |

f = 3.012E-05 Pa

Chemical name: Cyclohexene

Level II calculation: (six-compartment model)

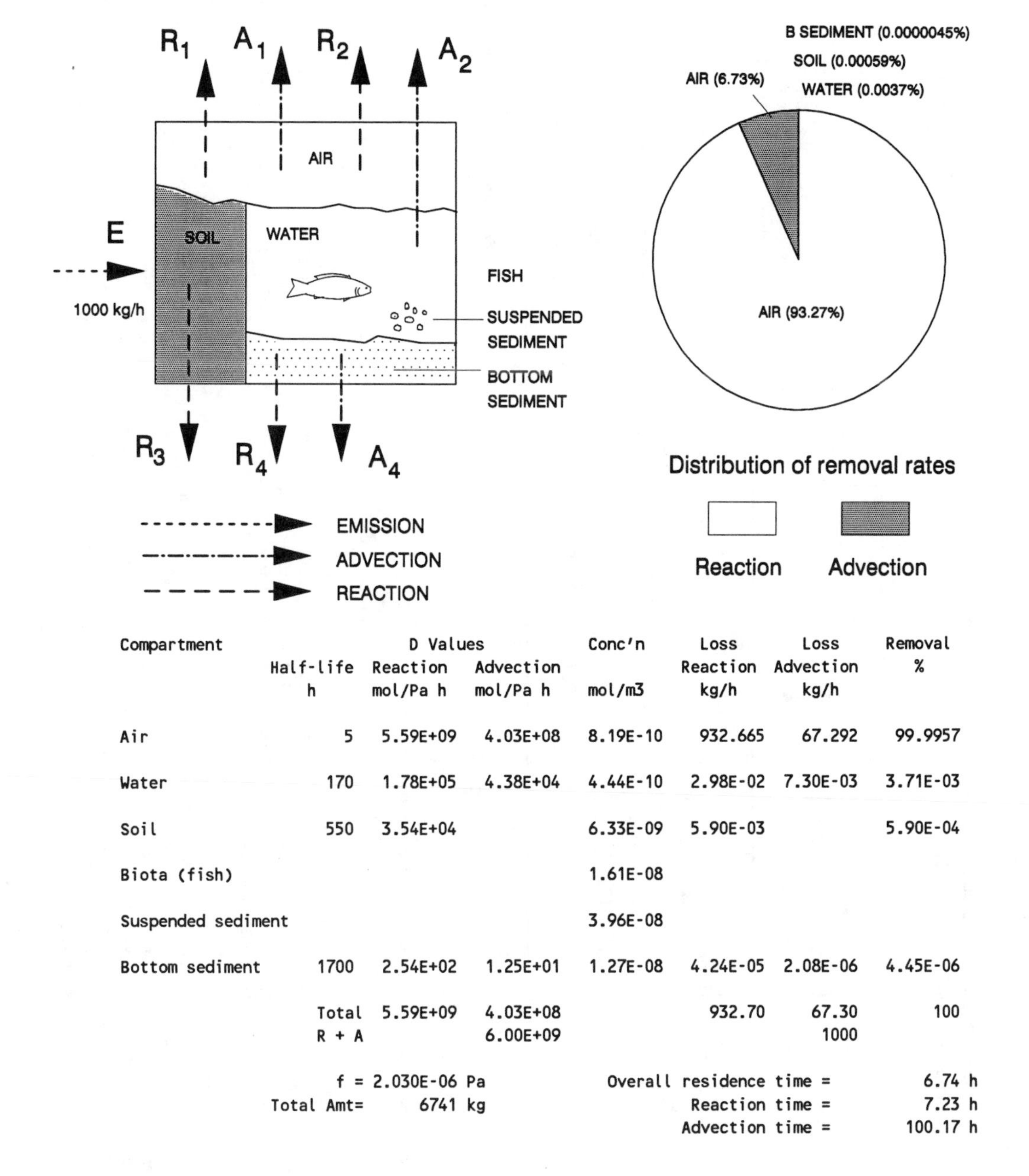


| Compartment | Half-life h | D Values Reaction mol/Pa h | D Values Advection mol/Pa h | Conc'n mol/m3 | Loss Reaction kg/h | Loss Advection kg/h | Removal % |
|---|---|---|---|---|---|---|---|
| Air | 5 | 5.59E+09 | 4.03E+08 | 8.19E-10 | 932.665 | 67.292 | 99.9957 |
| Water | 170 | 1.78E+05 | 4.38E+04 | 4.44E-10 | 2.98E-02 | 7.30E-03 | 3.71E-03 |
| Soil | 550 | 3.54E+04 | | 6.33E-09 | 5.90E-03 | | 5.90E-04 |
| Biota (fish) | | | | 1.61E-08 | | | |
| Suspended sediment | | | | 3.96E-08 | | | |
| Bottom sediment | 1700 | 2.54E+02 | 1.25E+01 | 1.27E-08 | 4.24E-05 | 2.08E-06 | 4.45E-06 |
| | Total | 5.59E+09 | 4.03E+08 | | 932.70 | 67.30 | 100 |
| | R + A | | 6.00E+09 | | | 1000 | |

f = 2.030E-06 Pa
Total Amt= 6741 kg

Overall residence time = 6.74 h
Reaction time = 7.23 h
Advection time = 100.17 h

## Fugacity Level III calculations: (four-compartment model)

Chemical name: Cyclohexene

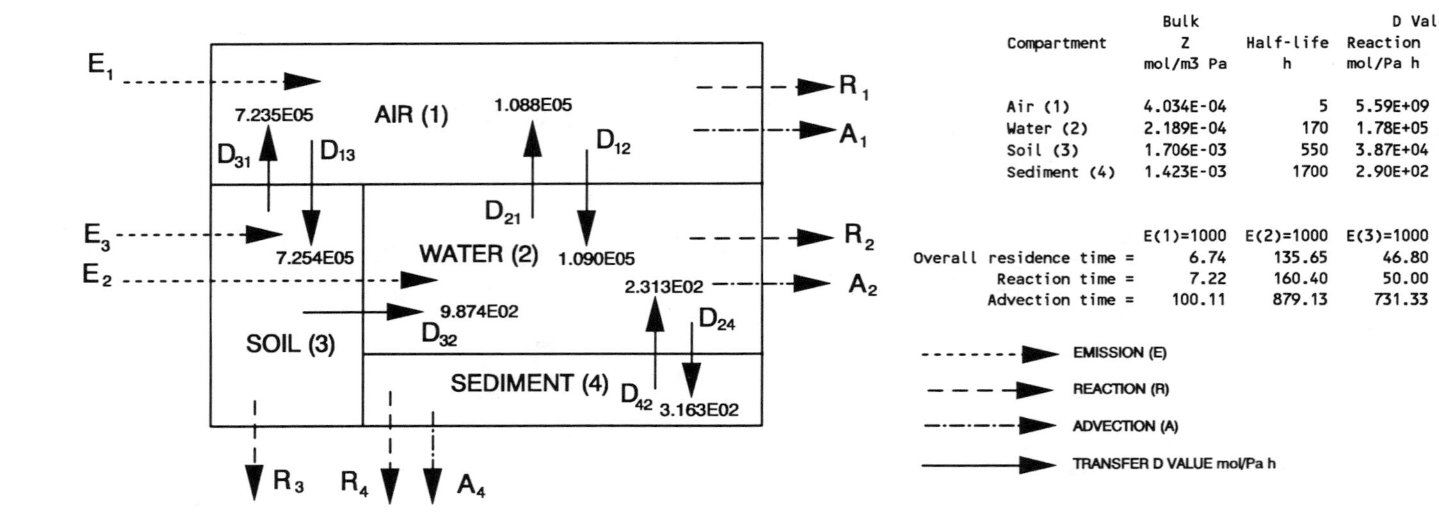


Phase Properties and Rates:

| Compartment | Bulk Z mol/m3 Pa | Half-life h | D Values Reaction mol/Pa h | D Values Advection mol/Pa h |
|---|---|---|---|---|
| Air (1) | 4.034E-04 | 5 | 5.59E+09 | 4.03E+08 |
| Water (2) | 2.189E-04 | 170 | 1.78E+05 | 4.38E+04 |
| Soil (3) | 1.706E-03 | 550 | 3.87E+04 | |
| Sediment (4) | 1.423E-03 | 1700 | 2.90E+02 | 1.42E+01 |

| | E(1)=1000 | E(2)=1000 | E(3)=1000 | E(1,2,3) |
|---|---|---|---|---|
| Overall residence time = | 6.74 | 135.65 | 46.80 | 49.42 h |
| Reaction time = | 7.22 | 160.40 | 50.00 | 54.49 h |
| Advection time = | 100.11 | 879.13 | 731.33 | 530.99 h |

Phase Properties, Compositions, Transport and Transformation Rates:

| Emission, kg/h | | | Fugacity, Pa | | | | Concentration, g/m3 | | | | Amounts, kg | | | | Total |
|---|---|---|---|---|---|---|---|---|---|---|---|---|---|---|---|
| E(1) | E(2) | E(3) | f(1) | f(2) | f(3) | f(4) | C(1) | C(2) | C(3) | C(4) | m(1) | m(2) | m(3) | m(4) | Amount, kg |
| 1000 | 0 | 0 | 2.031E-06 | 6.741E-07 | 1.930E-06 | 3.981E-07 | 6.729E-08 | 1.212E-08 | 2.705E-07 | 4.654E-08 | 6.729E+03 | 2.425E+00 | 4.869E+00 | 2.327E-02 | 6.737E+03 |
| 0 | 1000 | 0 | 6.670E-07 | 3.675E-02 | 6.341E-07 | 2.170E-02 | 2.211E-08 | 6.609E-04 | 8.886E-08 | 2.537E-03 | 2.211E+03 | 1.322E+05 | 1.600E+00 | 1.268E+03 | 1.357E+05 |
| 0 | 0 | 1000 | 1.926E-06 | 4.819E-05 | 1.595E-02 | 2.846E-05 | 6.382E-08 | 8.666E-07 | 2.236E-03 | 3.327E-06 | 6.382E+03 | 1.733E+02 | 4.025E+04 | 1.663E+00 | 4.680E+04 |
| 600 | 300 | 100 | 1.611E-06 | 1.103E-02 | 1.597E-03 | 6.514E-03 | 5.339E-08 | 1.984E-04 | 2.238E-04 | 7.614E-04 | 5.339E+03 | 3.967E+04 | 4.028E+03 | 3.807E+02 | 4.942E+04 |

| Emission, kg/h | | | Loss, Reaction, kg/h | | | | Loss, Advection, kg/h | | | Intermedia Rate of Transport, kg/h T12 | T21 | T13 | T31 | T32 | T24 | T42 |
|---|---|---|---|---|---|---|---|---|---|---|---|---|---|---|---|---|
| E(1) | E(2) | E(3) | R(1) | R(2) | R(3) | R(4) | A(1) | A(2) | A(4) | air-water | water-air | air-soil | soil-air | soil-water | water-sed | sed-water |
| 1000 | 0 | 0 | 9.327E+02 | 9.884E-03 | 6.14E-03 | 9.485E-06 | 6.729E+01 | 2.425E-03 | 4.654E-07 | 1.819E-02 | 6.026E-03 | 1.210E-01 | 1.147E-01 | 1.566E-04 | 1.752E-05 | 7.565E-06 |
| 0 | 1000 | 0 | 3.064E+02 | 5.388E+02 | 2.02E-03 | 5.171E-01 | 2.211E+01 | 1.322E+02 | 2.537E-02 | 5.974E-03 | 3.285E+02 | 3.975E-02 | 3.768E-02 | 5.143E-05 | 9.548E-01 | 4.124E-01 |
| 0 | 0 | 1000 | 8.846E+02 | 7.065E-01 | 5.07E+01 | 6.781E-04 | 6.382E+01 | 1.733E-01 | 3.327E-05 | 1.725E-02 | 4.307E-01 | 1.148E-01 | 9.481E+02 | 1.294E+00 | 1.252E-03 | 5.408E-04 |
| 600 | 300 | 100 | 7.400E+02 | 1.617E+02 | 5.08E+00 | 1.552E-01 | 5.339E+01 | 3.967E+01 | 7.614E-03 | 1.443E-02 | 9.859E+01 | 9.601E-02 | 9.489E+01 | 1.295E-01 | 2.866E-01 | 1.238E-01 |

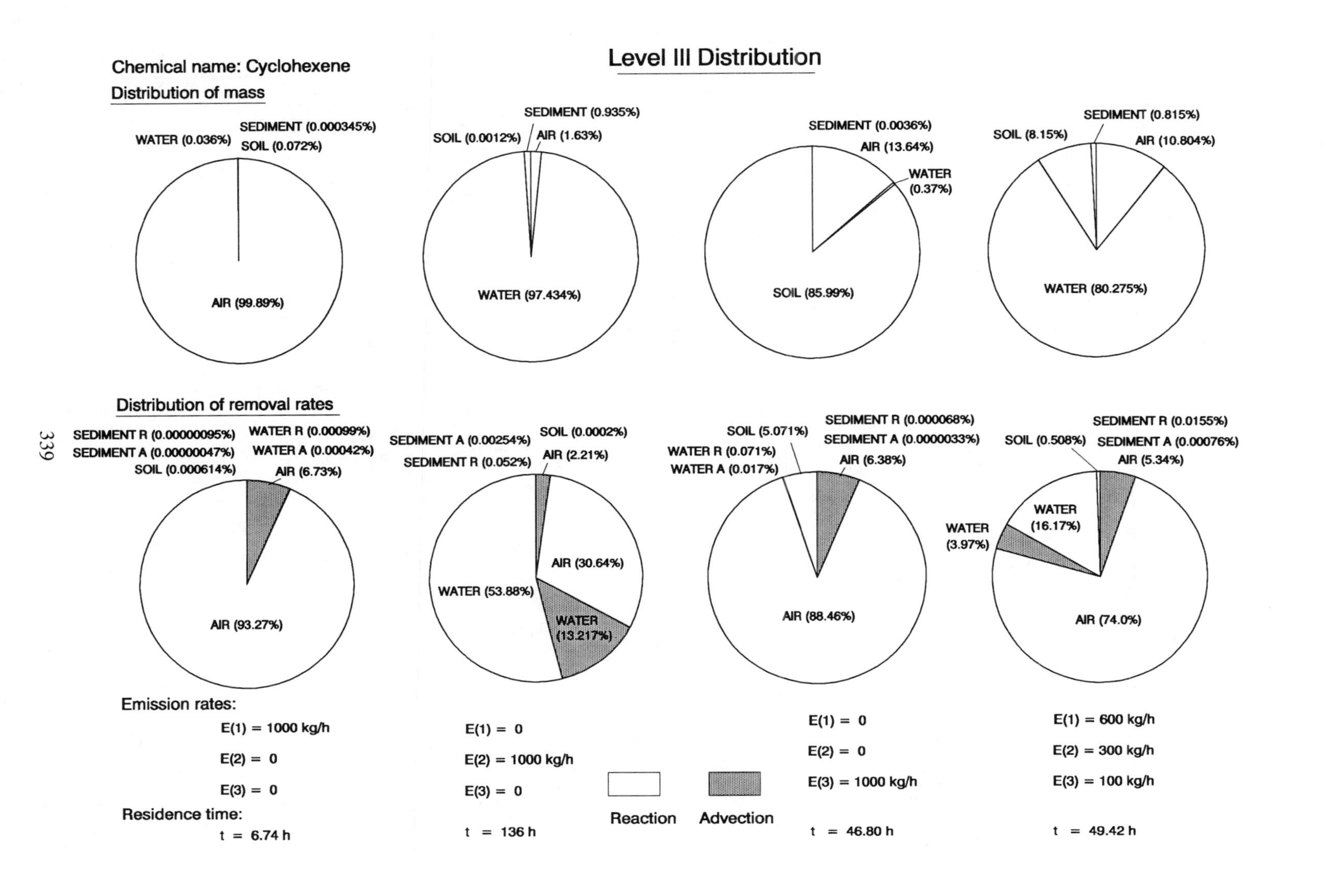

Level III Distribution
Chemical name: Cyclohexene
Distribution of mass
WATER (0.036%)
SEDIMENT (0.000345%)
SOIL (0.072%)
AIR (99.89%)
SEDIMENT (0.935%)
SOIL (0.0012%)
AIR (1.63%)
WATER (97.434%)
SEDIMENT (0.0036%)
AIR (13.64%)
WATER (0.37%)
SOIL (85.99%)
SEDIMENT (0.815%)
SOIL (8.15%)
AIR (10.804%)
WATER (80.275%)
Distribution of removal rates
SEDIMENT R (0.00000095%)
SEDIMENT A (0.00000047%)
SOIL (0.000614%)
WATER R (0.00099%)
WATER A (0.00042%)
AIR (6.73%)
AIR (93.27%)
SEDIMENT A (0.00254%)
SEDIMENT R (0.052%)
SOIL (0.0002%)
AIR (2.21%)
AIR (30.64%)
WATER (53.88%)
WATER (13.217%)
SOIL (5.071%)
WATER R (0.071%)
WATER A (0.017%)
SEDIMENT R (0.000068%)
SEDIMENT A (0.0000033%)
AIR (6.38%)
AIR (88.46%)
SEDIMENT R (0.0155%)
SOIL (0.508%)
SEDIMENT A (0.00076%)
AIR (5.34%)
WATER (16.17%)
WATER (3.97%)
AIR (74.0%)
Emission rates:
E(1) = 1000 kg/h
E(2) = 0
E(3) = 0
Residence time:
t = 6.74 h
E(1) = 0
E(2) = 1000 kg/h
E(3) = 0
t = 136 h
Reaction
Advection
E(1) = 0
E(2) = 0
E(3) = 1000 kg/h
t = 46.80 h
E(1) = 600 kg/h
E(2) = 300 kg/h
E(3) = 100 kg/h
t = 49.42 h

Chemical name: 1-Pentene

Level I calculation: (six-compartment model)

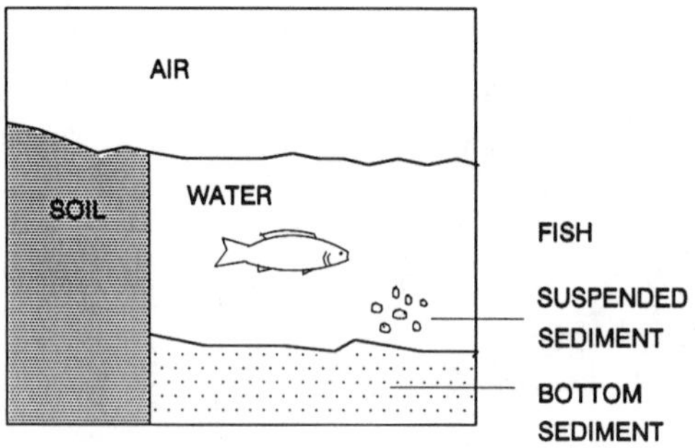

100000 kg

AIR

SOIL

WATER

FISH

SUSPENDED SEDIMENT

BOTTOM SEDIMENT

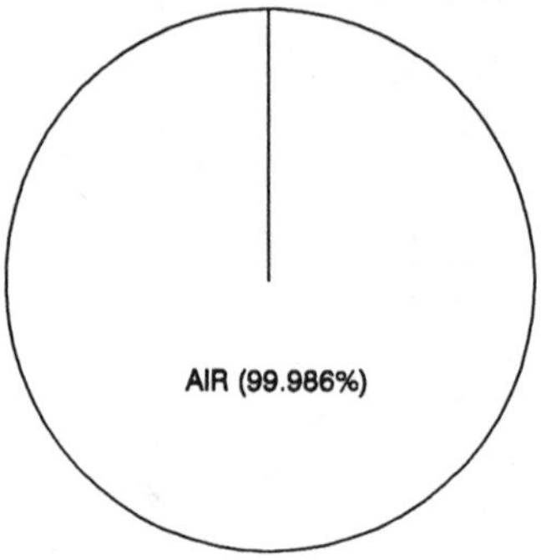

BOTTOM SEDIMENT (0.000038%)

SOIL (0.0017%)

WATER (0.0123%)

AIR (99.986%)

Distribution of mass

physical-chemical properties:

MW: 70.14 g/mol

M.P.: - 138°C

Fugacity ratio: 1.0

vapor pressure: 85000 Pa

solubility: 148 g/m$^3$

log $K_{OW}$: 2.20

| Compartment | Z | Concentration | | | Amount | Amount |
|---|---|---|---|---|---|---|
| | mol/m3 Pa | mol/m3 | mg/L (or g/m3) | ug/g | kg | % |
| Air | 4.034E-04 | 1.426E-08 | 9.999E-07 | 8.435E-04 | 99986 | 99.986 |
| Water | 2.482E-05 | 8.772E-10 | 6.153E-08 | 6.153E-08 | 12.305 | 0.0123 |
| Soil | 7.743E-05 | 2.736E-09 | 1.919E-07 | 7.996E-08 | 1.7271 | 1.73E-03 |
| Biota (fish) | 1.967E-04 | 6.951E-09 | 4.876E-07 | 4.876E-07 | 9.75E-05 | 9.75E-08 |
| Suspended sediment | 4.839E-04 | 1.710E-08 | 1.199E-06 | 7.996E-07 | 1.20E-03 | 1.20E-06 |
| Bottom sediment | 1.549E-04 | 5.472E-09 | 3.838E-07 | 1.599E-07 | 0.0384 | 3.84E-05 |
| | Total | | | | 100000 | 100 |

f = 3.534E-05 Pa

Chemical name: 1-Pentene

Level II calculation: (six-compartment model)

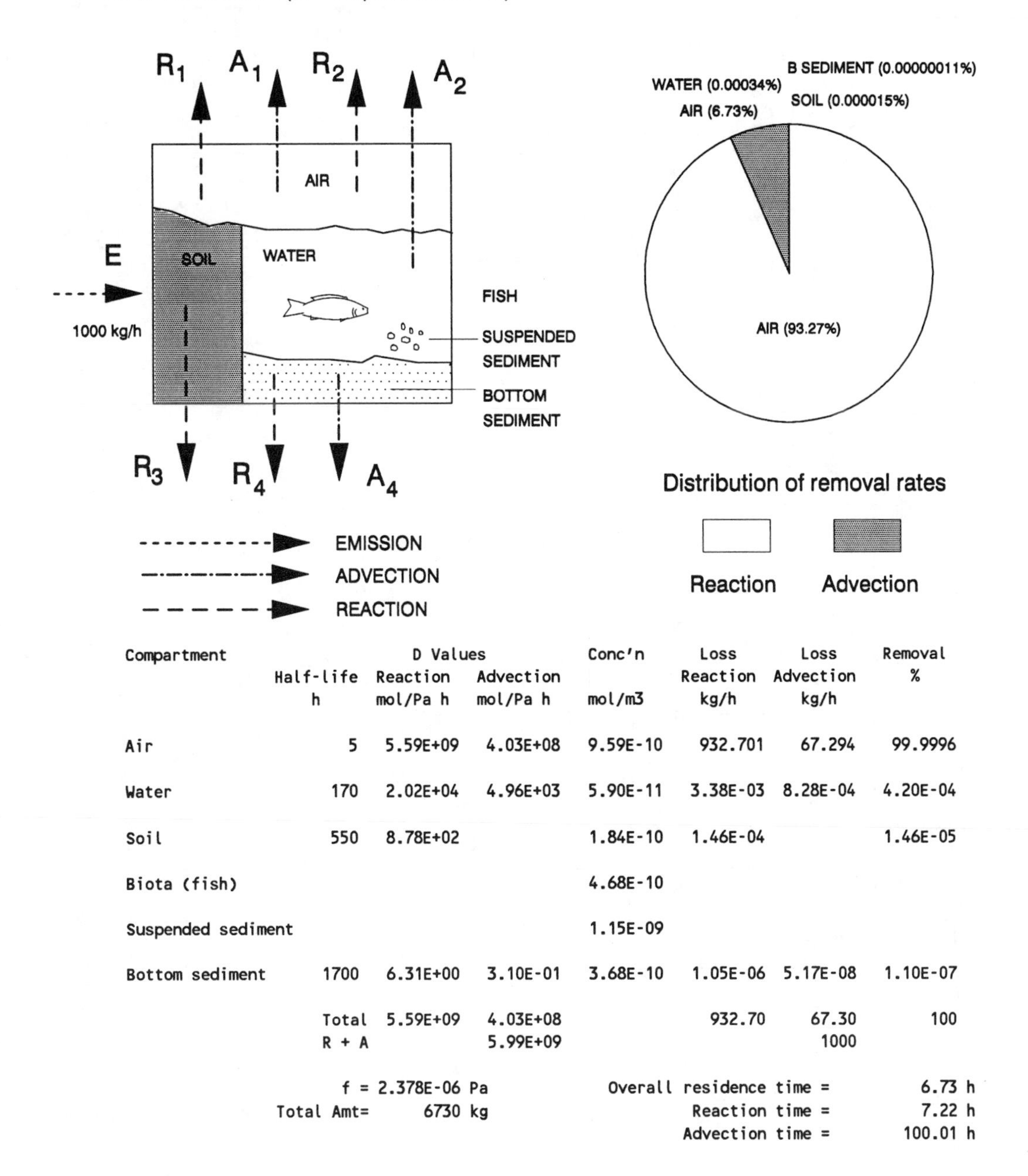


| Compartment | Half-life h | D Values Reaction mol/Pa h | D Values Advection mol/Pa h | Conc'n mol/m3 | Loss Reaction kg/h | Loss Advection kg/h | Removal % |
|---|---|---|---|---|---|---|---|
| Air | 5 | 5.59E+09 | 4.03E+08 | 9.59E-10 | 932.701 | 67.294 | 99.9996 |
| Water | 170 | 2.02E+04 | 4.96E+03 | 5.90E-11 | 3.38E-03 | 8.28E-04 | 4.20E-04 |
| Soil | 550 | 8.78E+02 | | 1.84E-10 | 1.46E-04 | | 1.46E-05 |
| Biota (fish) | | | | 4.68E-10 | | | |
| Suspended sediment | | | | 1.15E-09 | | | |
| Bottom sediment | 1700 | 6.31E+00 | 3.10E-01 | 3.68E-10 | 1.05E-06 | 5.17E-08 | 1.10E-07 |
| | Total | 5.59E+09 | 4.03E+08 | | 932.70 | 67.30 | 100 |
| | R + A | | 5.99E+09 | | | 1000 | |

f = 2.378E-06 Pa

Total Amt= 6730 kg

Overall residence time = 6.73 h

Reaction time = 7.22 h

Advection time = 100.01 h

## Fugacity Level III calculations: (four-compartment model)

Chemical name: 1-Pentene

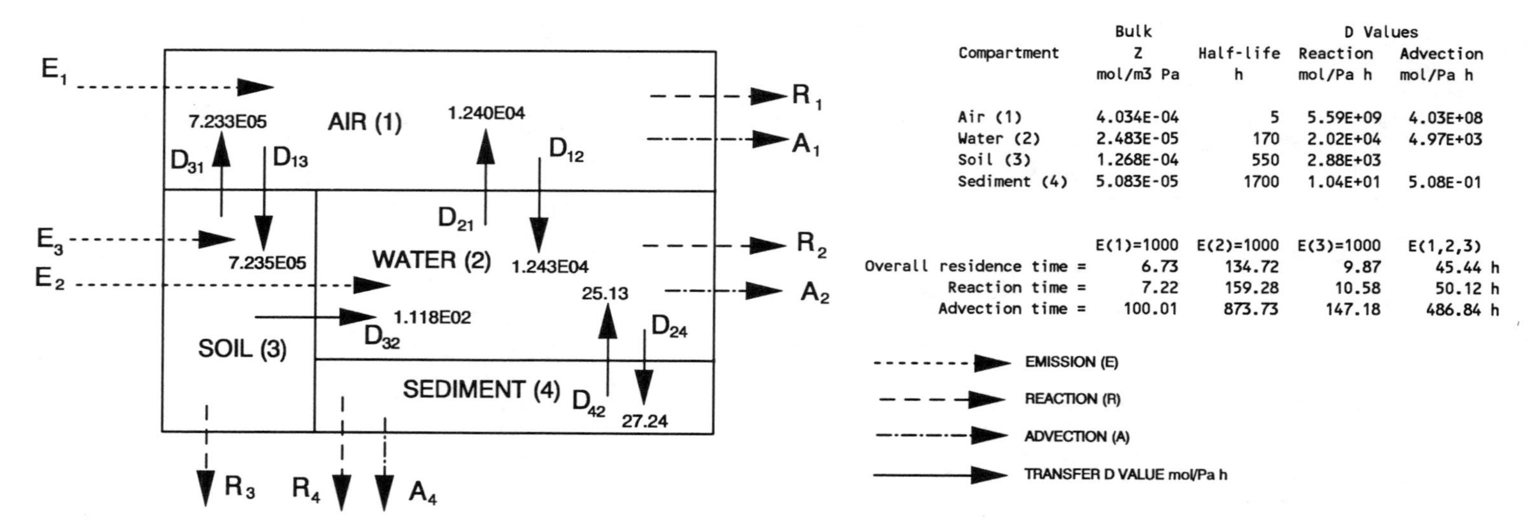


Phase Properties and Rates:

| Compartment | Bulk Z mol/m3 Pa | Half-life h | D Values Reaction mol/Pa h | D Values Advection mol/Pa h |
|---|---|---|---|---|
| Air (1) | 4.034E-04 | 5 | 5.59E+09 | 4.03E+08 |
| Water (2) | 2.483E-05 | 170 | 2.02E+04 | 4.97E+03 |
| Soil (3) | 1.268E-04 | 550 | 2.88E+03 | |
| Sediment (4) | 5.083E-05 | 1700 | 1.04E+01 | 5.08E-01 |

| | E(1)=1000 | E(2)=1000 | E(3)=1000 | E(1,2,3) |
|---|---|---|---|---|
| Overall residence time = | 6.73 | 134.72 | 9.87 | 45.44 h |
| Reaction time = | 7.22 | 159.28 | 10.58 | 50.12 h |
| Advection time = | 100.01 | 873.73 | 147.18 | 486.84 h |

EMISSION (E)

REACTION (R)

ADVECTION (A)

TRANSFER D VALUE mol/Pa h

Phase Properties, Compositions, Transport and Transformation Rates:

| Emission, kg/h | | | Fugacity, Pa | | | | Concentration, g/m3 | | | | Amounts, kg | | | | Total |
|---|---|---|---|---|---|---|---|---|---|---|---|---|---|---|---|
| E(1) | E(2) | E(3) | f(1) | f(2) | f(3) | f(4) | C(1) | C(2) | C(3) | C(4) | m(1) | m(2) | m(3) | m(4) | Amount, kg |
| 1000 | 0 | 0 | 2.378E-06 | 7.928E-07 | 2.369E-06 | 5.999E-07 | 6.729E-08 | 1.381E-09 | 2.108E-08 | 2.139E-09 | 6.729E+03 | 2.761E-01 | 3.794E-01 | 1.069E-03 | 6.730E+03 |
| 0 | 1000 | 0 | 7.842E-07 | 3.790E-01 | 7.812E-07 | 2.868E-01 | 2.219E-08 | 6.600E-04 | 6.950E-09 | 1.022E-03 | 2.219E+03 | 1.320E+05 | 1.251E-01 | 5.112E+02 | 1.347E+05 |
| 0 | 0 | 1000 | 2.369E-06 | 5.912E-05 | 1.963E-02 | 4.474E-05 | 6.702E-08 | 1.029E-07 | 1.747E-04 | 1.595E-07 | 6.702E+03 | 2.059E+01 | 3.144E+03 | 7.975E-02 | 9.867E+03 |
| 600 | 300 | 100 | 1.899E-06 | 1.137E-01 | 1.965E-03 | 8.604E-02 | 5.374E-08 | 1.980E-04 | 1.748E-05 | 3.068E-04 | 5.374E+03 | 3.960E+04 | 3.147E+02 | 1.534E+02 | 4.544E+04 |

| Emission, kg/h | | | Loss, Reaction, kg/h | | | | Loss, Advection, kg/h | | | Intermedia Rate of Transport, kg/h | | | | | | |
|---|---|---|---|---|---|---|---|---|---|---|---|---|---|---|---|---|
| | | | | | | | | | | T12 | T21 | T13 | T31 | T32 | T24 | T42 |
| E(1) | E(2) | E(3) | R(1) | R(2) | R(3) | R(4) | A(1) | A(2) | A(4) | air-water | water-air | air-soil | soil-air | soil-water | water-sed | sed-water |
| 1000 | 0 | 0 | 9.327E+02 | 1.126E-03 | 4.78E-04 | 4.360E-07 | 6.729E+01 | 2.761E-04 | 2.139E-08 | 2.073E-03 | 6.898E-04 | 1.207E-01 | 1.202E-01 | 1.858E-05 | 1.515E-06 | 1.058E-06 |
| 0 | 1000 | 0 | 3.075E+02 | 5.381E+02 | 1.58E-04 | 2.084E-01 | 2.219E+01 | 1.320E+02 | 1.022E-02 | 6.837E-04 | 3.297E+02 | 3.980E-02 | 3.963E-02 | 6.125E-06 | 7.242E-01 | 5.056E-01 |
| 0 | 0 | 1000 | 9.289E+02 | 8.393E-02 | 3.96E+00 | 3.251E-05 | 6.702E+01 | 2.059E-02 | 1.595E-06 | 2.065E-03 | 5.144E-02 | 1.202E-01 | 9.960E+02 | 1.539E-01 | 1.130E-04 | 7.887E-05 |
| 600 | 300 | 100 | 7.448E+02 | 1.614E+02 | 3.96E-01 | 6.252E-02 | 5.374E+01 | 3.960E+01 | 3.068E-03 | 1.656E-03 | 9.893E+01 | 9.637E-02 | 9.968E+01 | 1.541E-02 | 2.173E-01 | 1.517E-01 |

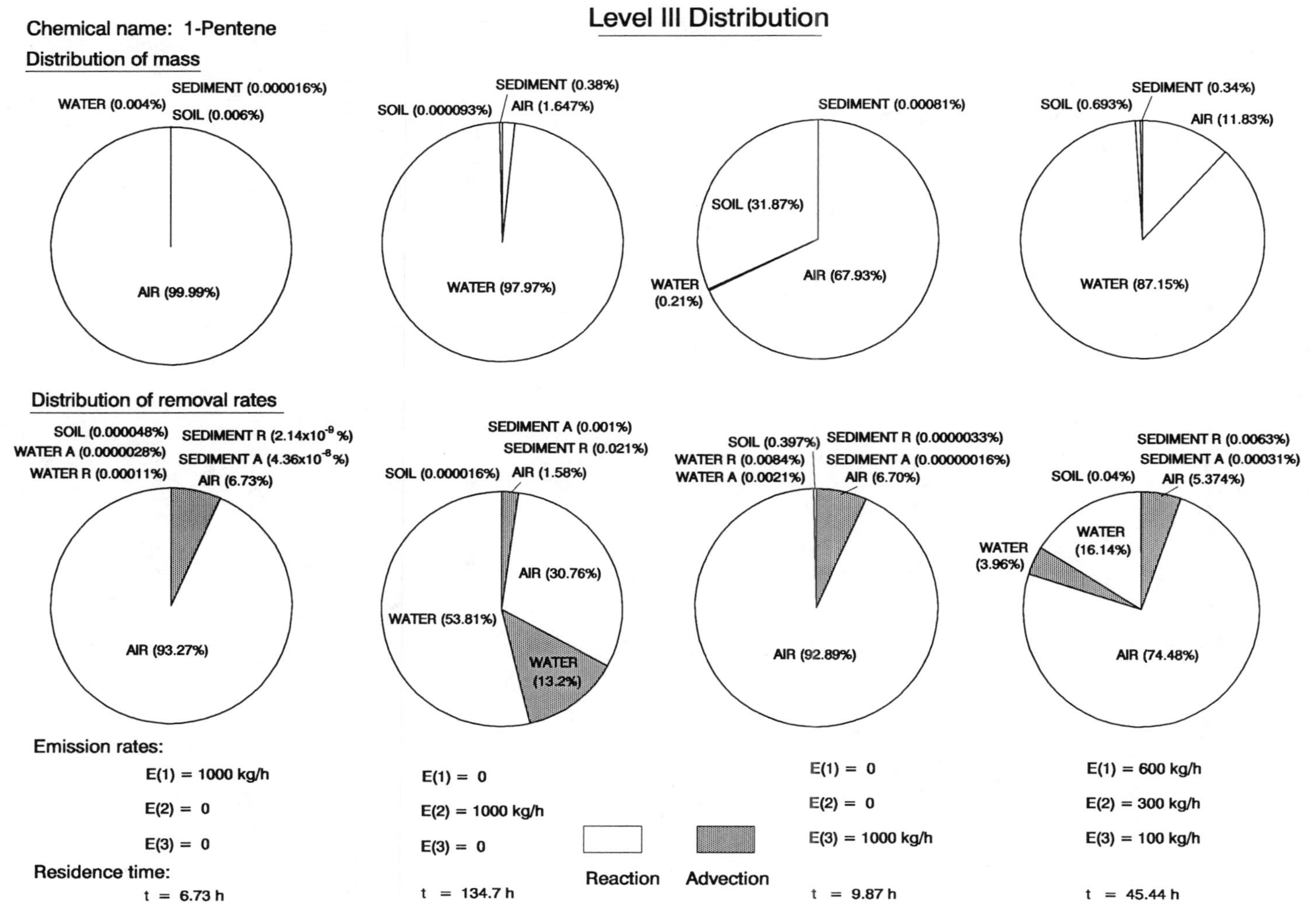

Level III Distribution
Chemical name: 1-Pentene
Distribution of mass
WATER (0.004%)
SEDIMENT (0.000016%)
SOIL (0.006%)
AIR (99.99%)
SOIL (0.000093%)
SEDIMENT (0.38%)
AIR (1.647%)
WATER (97.97%)
SEDIMENT (0.00081%)
SOIL (31.87%)
WATER (0.21%)
AIR (67.93%)
SOIL (0.693%)
SEDIMENT (0.34%)
AIR (11.83%)
WATER (87.15%)
Distribution of removal rates
SOIL (0.000048%)
WATER A (0.0000028%)
WATER R (0.00011%)
SEDIMENT R (2.14x10$^{-9}$ %)
SEDIMENT A (4.36x10$^{-8}$ %)
AIR (6.73%)
AIR (93.27%)
SEDIMENT A (0.001%)
SEDIMENT R (0.021%)
SOIL (0.000016%)
AIR (1.58%)
AIR (30.76%)
WATER (53.81%)
WATER (13.2%)
SOIL (0.397%)
WATER R (0.0084%)
WATER A (0.0021%)
SEDIMENT R (0.0000033%)
SEDIMENT A (0.00000016%)
AIR (6.70%)
AIR (92.89%)
SEDIMENT R (0.0063%)
SEDIMENT A (0.00031%)
SOIL (0.04%)
AIR (5.374%)
WATER (16.14%)
WATER (3.96%)
AIR (74.48%)
Emission rates:
E(1) = 1000 kg/h
E(2) = 0
E(3) = 0
Residence time:
t = 6.73 h
E(1) = 0
E(2) = 1000 kg/h
E(3) = 0
t = 134.7 h
Reaction
Advection
E(1) = 0
E(2) = 0
E(3) = 1000 kg/h
t = 9.87 h
E(1) = 600 kg/h
E(2) = 300 kg/h
E(3) = 100 kg/h
t = 45.44 h

Chemical name: 1-Octene

Level I calculation: (six-compartment model)

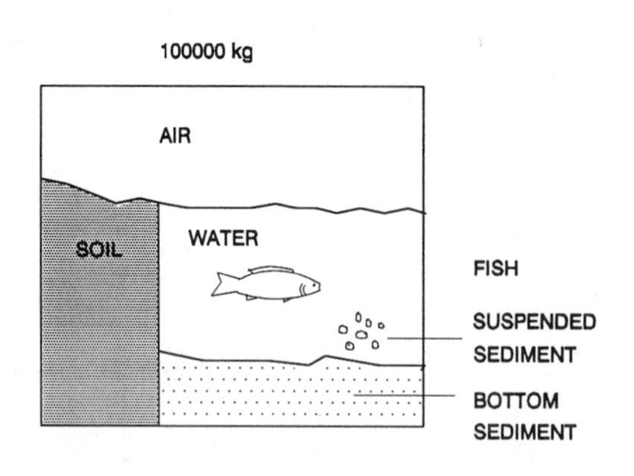


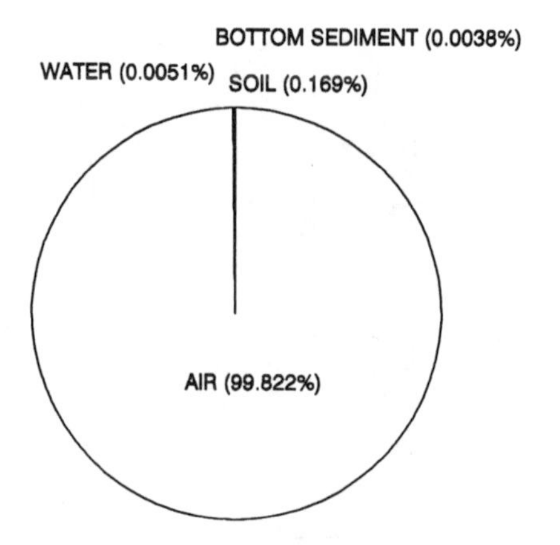


Distribution of mass

physical-chemical properties:

MW: 112.1 g/mol

M.P.: - 101.7°C

Fugacity ratio: 1.0

vapor pressure: 2320 Pa

solubility: 2.70 g/m$^3$

log $K_{OW}$: 4.57

| Compartment | Z | Concentration | | | Amount | Amount |
|---|---|---|---|---|---|---|
| | mol/m3 Pa | mol/m3 | mg/L (or g/m3) | ug/g | kg | % |
| Air | 4.034E-04 | 8.905E-09 | 9.982E-07 | 8.421E-04 | 99822 | 99.822 |
| Water | 1.038E-05 | 2.292E-10 | 2.569E-08 | 2.569E-08 | 5.138 | 5.14E-03 |
| Soil | 7.591E-03 | 1.676E-07 | 1.878E-05 | 7.826E-06 | 169.05 | 0.1690 |
| Biota (fish) | 1.929E-02 | 4.257E-07 | 4.772E-05 | 4.772E-05 | 9.54E-03 | 9.54E-06 |
| Suspended sediment | 4.744E-02 | 1.047E-06 | 1.174E-04 | 7.826E-05 | 0.1174 | 1.17E-04 |
| Bottom sediment | 1.518E-02 | 3.351E-07 | 3.757E-05 | 1.565E-05 | 3.7566 | 3.76E-03 |
| | Total | | | | 100000 | 100 |

f = 2.207E-05 Pa

Chemical name: 1-Octene

Level II calculation: (six-compartment model)

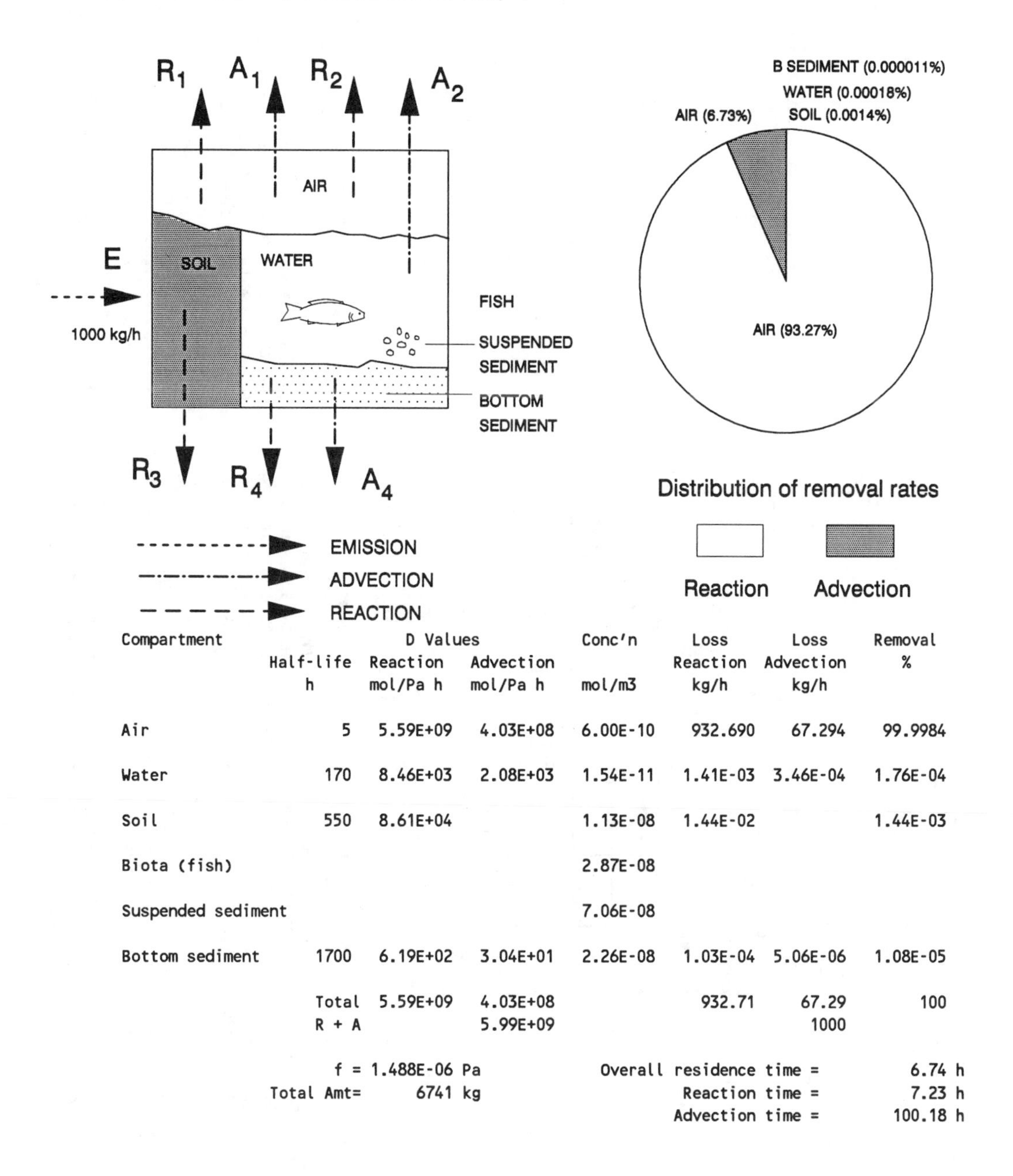


| Compartment | Half-life h | D Values Reaction mol/Pa h | D Values Advection mol/Pa h | Conc'n mol/m3 | Loss Reaction kg/h | Loss Advection kg/h | Removal % |
|---|---|---|---|---|---|---|---|
| Air | 5 | 5.59E+09 | 4.03E+08 | 6.00E-10 | 932.690 | 67.294 | 99.9984 |
| Water | 170 | 8.46E+03 | 2.08E+03 | 1.54E-11 | 1.41E-03 | 3.46E-04 | 1.76E-04 |
| Soil | 550 | 8.61E+04 | | 1.13E-08 | 1.44E-02 | | 1.44E-03 |
| Biota (fish) | | | | 2.87E-08 | | | |
| Suspended sediment | | | | 7.06E-08 | | | |
| Bottom sediment | 1700 | 6.19E+02 | 3.04E+01 | 2.26E-08 | 1.03E-04 | 5.06E-06 | 1.08E-05 |
| | Total | 5.59E+09 | 4.03E+08 | | 932.71 | 67.29 | 100 |
| | R + A | | 5.99E+09 | | | 1000 | |

f = 1.488E-06 Pa
Total Amt= 6741 kg

Overall residence time = 6.74 h
Reaction time = 7.23 h
Advection time = 100.18 h

## Fugacity Level III calculations: (four-compartment model)

Chemical name: 1-Octene

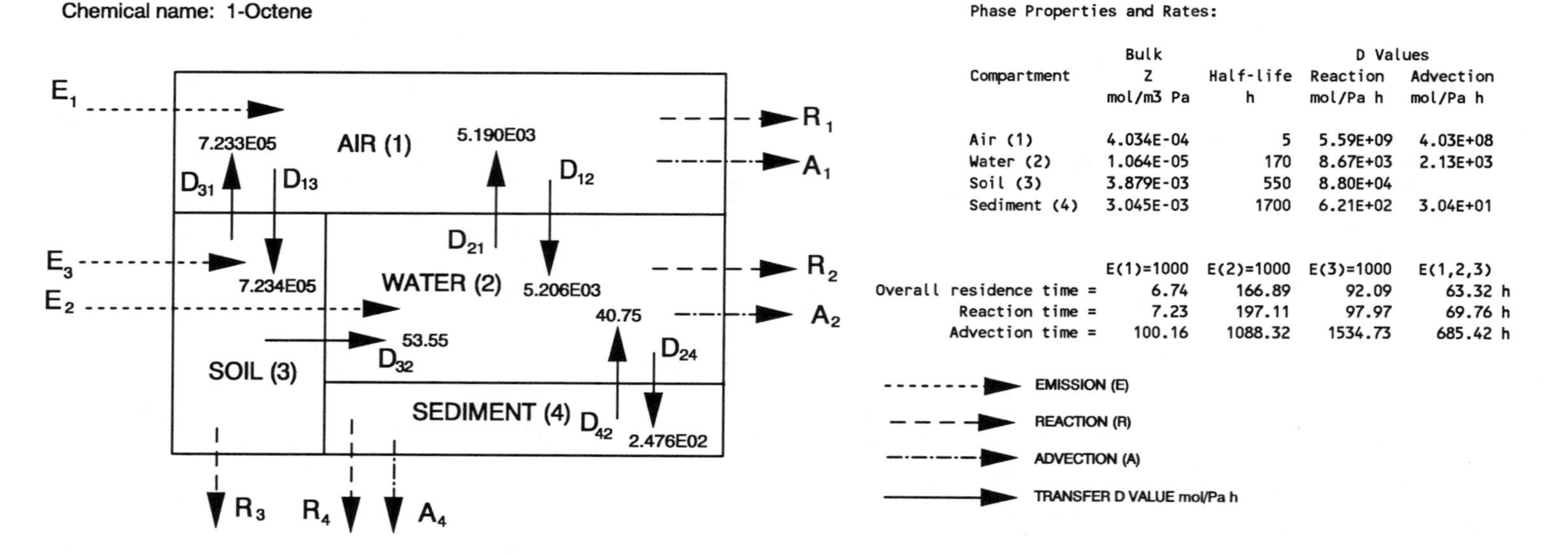


Phase Properties and Rates:

| Compartment | Bulk Z mol/m3 Pa | Half-life h | D Values Reaction mol/Pa h | D Values Advection mol/Pa h |
|---|---|---|---|---|
| Air (1) | 4.034E-04 | 5 | 5.59E+09 | 4.03E+08 |
| Water (2) | 1.064E-05 | 170 | 8.67E+03 | 2.13E+03 |
| Soil (3) | 3.879E-03 | 550 | 8.80E+04 | |
| Sediment (4) | 3.045E-03 | 1700 | 6.21E+02 | 3.04E+01 |

| | E(1)=1000 | E(2)=1000 | E(3)=1000 | E(1,2,3) |
|---|---|---|---|---|
| Overall residence time = | 6.74 | 166.89 | 92.09 | 63.32 h |
| Reaction time = | 7.23 | 197.11 | 97.97 | 69.76 h |
| Advection time = | 100.16 | 1088.32 | 1534.73 | 685.42 h |

Phase Properties, Compositions, Transport and Transformation Rates:

| Emission, kg/h | | | Fugacity, Pa | | | | Concentration, g/m3 | | | | Amounts, kg | | | | Total Amount, kg |
|---|---|---|---|---|---|---|---|---|---|---|---|---|---|---|---|
| E(1) | E(2) | E(3) | f(1) | f(2) | f(3) | f(4) | C(1) | C(2) | C(3) | C(4) | m(1) | m(2) | m(3) | m(4) | |
| 1000 | 0 | 0 | 1.488E-06 | 4.819E-07 | 1.327E-06 | 1.725E-07 | 6.729E-08 | 5.747E-10 | 5.770E-07 | 5.887E-08 | 6.729E+03 | 1.149E-01 | 1.039E+01 | 2.943E-02 | 6.740E+03 |
| 0 | 1000 | 0 | 4.760E-07 | 5.499E-01 | 4.244E-07 | 1.968E-01 | 2.153E-08 | 6.557E-04 | 1.846E-07 | 6.717E-02 | 2.153E+03 | 1.311E+05 | 3.322E+00 | 3.359E+04 | 1.669E+05 |
| 0 | 0 | 1000 | 1.327E-06 | 3.672E-05 | 1.100E-02 | 1.314E-05 | 5.999E-08 | 4.379E-08 | 4.782E-03 | 4.486E-06 | 5.999E+03 | 8.759E+00 | 8.608E+04 | 2.243E+00 | 9.209E+04 |
| 600 | 300 | 100 | 1.168E-06 | 1.650E-01 | 1.101E-03 | 5.904E-02 | 5.283E-08 | 1.967E-04 | 4.786E-04 | 2.015E-02 | 5.283E+03 | 3.934E+04 | 8.615E+03 | 1.008E+04 | 6.332E+04 |

| Emission, kg/h | | | Loss, Reaction, kg/h | | | | Loss, Advection, kg/h | | | Intermedia Rate of Transport, kg/h T12 | T21 | T13 | T31 | T32 | T24 | T42 |
|---|---|---|---|---|---|---|---|---|---|---|---|---|---|---|---|---|
| E(1) | E(2) | E(3) | R(1) | R(2) | R(3) | R(4) | A(1) | A(2) | A(4) | air-water | water-air | air-soil | soil-air | soil-water | water-sed | sed-water |
| 1000 | 0 | 0 | 9.327E+02 | 4.685E-04 | 1.31E-02 | 1.200E-05 | 6.729E+01 | 1.149E-04 | 5.887E-07 | 8.684E-04 | 2.803E-04 | 1.207E-01 | 1.076E-01 | 7.965E-06 | 1.338E-05 | 7.878E-07 |
| 0 | 1000 | 0 | 2.983E+02 | 5.346E+02 | 4.19E-03 | 1.369E+01 | 2.153E+01 | 1.311E+02 | 6.717E-01 | 2.778E-04 | 3.199E+02 | 3.860E-02 | 3.441E-02 | 2.548E-06 | 1.526E+01 | 8.989E-01 |
| 0 | 0 | 1000 | 8.315E+02 | 3.570E-02 | 1.08E+02 | 9.144E-04 | 5.999E+01 | 8.759E-03 | 4.486E-05 | 7.742E-04 | 2.136E-02 | 1.076E-01 | 8.916E+02 | 6.601E-02 | 1.019E-03 | 6.004E-05 |
| 600 | 300 | 100 | 7.323E+02 | 1.604E+02 | 1.09E+01 | 4.107E+00 | 5.283E+01 | 3.934E+01 | 2.015E-01 | 6.818E-04 | 9.597E+01 | 9.474E-02 | 8.923E+01 | 6.607E-03 | 4.579E+00 | 2.697E-01 |

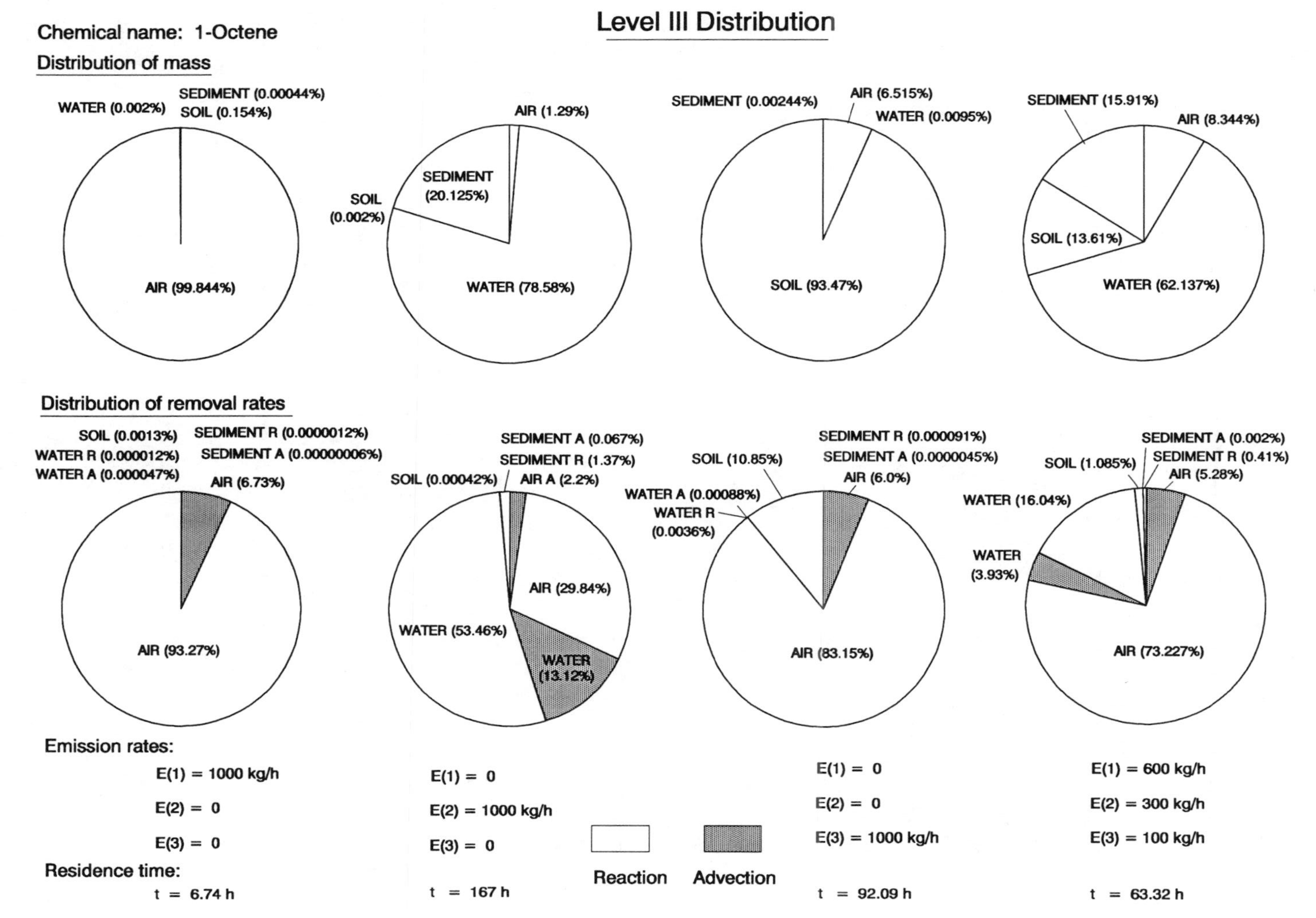

Level III Distribution
Chemical name: 1-Octene
Distribution of mass
WATER (0.002%)
SEDIMENT (0.00044%)
SOIL (0.154%)
AIR (99.844%)
AIR (1.29%)
SOIL (0.002%)
SEDIMENT (20.125%)
WATER (78.58%)
SEDIMENT (0.00244%)
AIR (6.515%)
WATER (0.0095%)
SOIL (93.47%)
SEDIMENT (15.91%)
AIR (8.344%)
SOIL (13.61%)
WATER (62.137%)
Distribution of removal rates
SOIL (0.0013%)
WATER R (0.000012%)
WATER A (0.000047%)
SEDIMENT R (0.0000012%)
SEDIMENT A (0.00000006%)
AIR (6.73%)
AIR (93.27%)
SEDIMENT A (0.067%)
SEDIMENT R (1.37%)
SOIL (0.00042%)
AIR A (2.2%)
AIR (29.84%)
WATER (53.46%)
WATER (13.12%)
SOIL (10.85%)
SEDIMENT R (0.000091%)
SEDIMENT A (0.0000045%)
AIR (6.0%)
WATER A (0.00088%)
WATER R (0.0036%)
AIR (83.15%)
SEDIMENT A (0.002%)
SEDIMENT R (0.41%)
SOIL (1.085%)
AIR (5.28%)
WATER (16.04%)
WATER (3.93%)
AIR (73.227%)
Emission rates:
E(1) = 1000 kg/h
E(2) = 0
E(3) = 0
Residence time:
t = 6.74 h
E(1) = 0
E(2) = 1000 kg/h
E(3) = 0
t = 167 h
Reaction
Advection
E(1) = 0
E(2) = 0
E(3) = 1000 kg/h
t = 92.09 h
E(1) = 600 kg/h
E(2) = 300 kg/h
E(3) = 100 kg/h
t = 63.32 h

Chemical name: 1-Hexyne

Level I calculation: (six-compartment model)

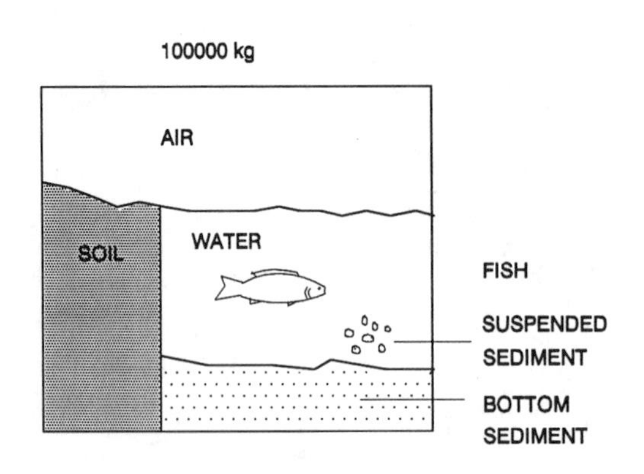


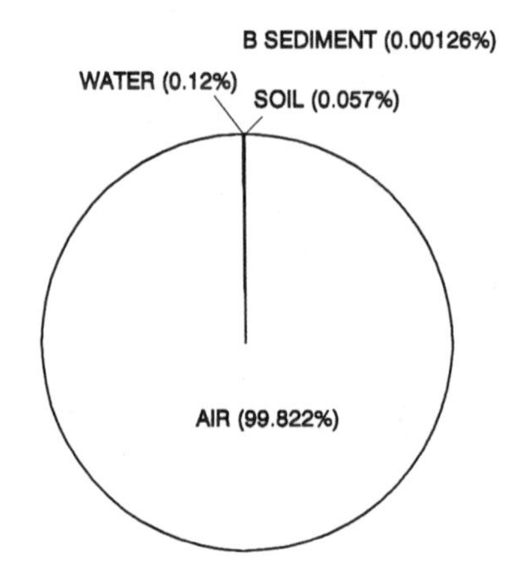


Distribution of mass

physical-chemical properties:

MW: 82.15 g/mol

M.P.: -132 °C

Fugacity ratio: 1.0

vapor pressure: 18140 Pa

solubility: 360 g/m$^3$

log $K_{OW}$: 2.73

| Compartment | Z | Concentration | | | Amount | Amount |
|---|---|---|---|---|---|---|
| | mol/m3 Pa | mol/m3 | mg/L (or g/m3) | ug/g | kg | % |
| Air | 4.034E-04 | 1.215E-08 | 9.982E-07 | 8.421E-04 | 99822 | 99.822 |
| Water | 2.416E-04 | 7.276E-09 | 5.978E-07 | 5.978E-07 | 119.553 | 0.1196 |
| Soil | 2.553E-03 | 7.690E-08 | 6.318E-06 | 2.632E-06 | 56.859 | 0.0569 |
| Biota (fish) | 6.487E-03 | 1.954E-07 | 1.605E-05 | 1.605E-05 | 3.21E-03 | 3.21E-06 |
| Suspended sediment | 1.596E-02 | 4.806E-07 | 3.949E-05 | 2.632E-05 | 0.0395 | 3.95E-05 |
| Bottom sediment | 5.106E-03 | 1.538E-07 | 1.264E-05 | 5.265E-06 | 1.2635 | 1.26E-03 |
| | Total | | | | 100000 | 100 |

f = 3.012E-05 Pa

Chemical name: 1-Hexyne

Level II calculation: (six-compartment model)

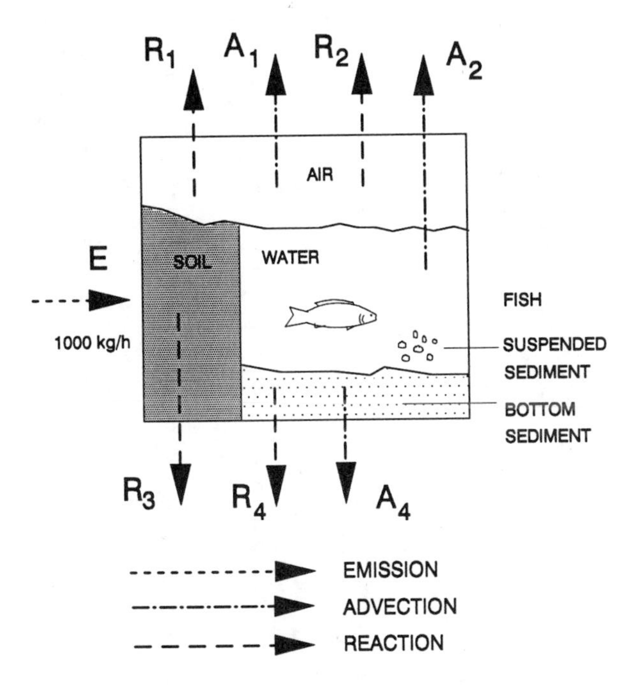


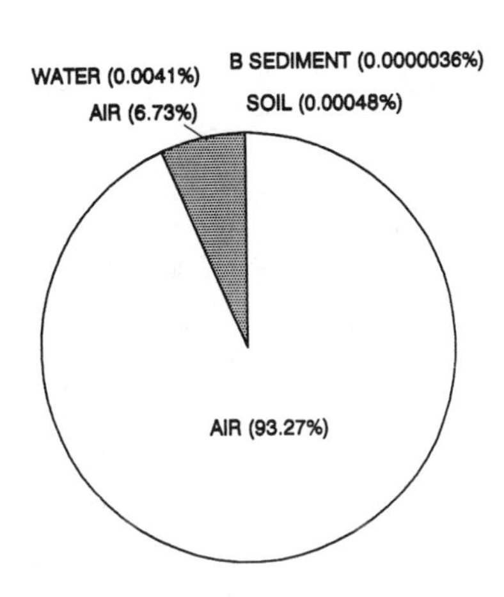


Distribution of removal rates

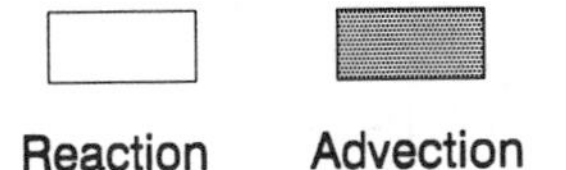


| Compartment | Half-life h | D Values Reaction mol/Pa h | D Values Advection mol/Pa h | Conc'n mol/m3 | Loss Reaction kg/h | Loss Advection kg/h | Removal % |
|---|---|---|---|---|---|---|---|
| Air | 5 | 5.59E+09 | 4.03E+08 | 8.19E-10 | 932.663 | 67.292 | 99.9954 |
| Water | 170 | 1.97E+05 | 4.83E+04 | 4.91E-10 | 3.29E-02 | 8.06E-03 | 4.09E-03 |
| Soil | 550 | 2.90E+04 | | 5.18E-09 | 4.83E-03 | | 4.83E-04 |
| Biota (fish) | | | | 1.32E-08 | | | |
| Suspended sediment | | | | 3.24E-08 | | | |
| Bottom sediment | 1700 | 2.08E+02 | 1.02E+01 | 1.04E-08 | 3.47E-05 | 1.70E-06 | 3.64E-06 |
| | Total | 5.59E+09 | 4.03E+08 | | 932.70 | 67.30 | 100 |
| | R + A | | 6.00E+09 | | | 1000 | |

f = 2.030E-06 Pa

Total Amt= 6741 kg

Overall residence time = 6.74 h

Reaction time = 7.23 h

Advection time = 100.17 h

## Fugacity Level III calculations: (four-compartment model)

Chemical name: 1-Hexyne

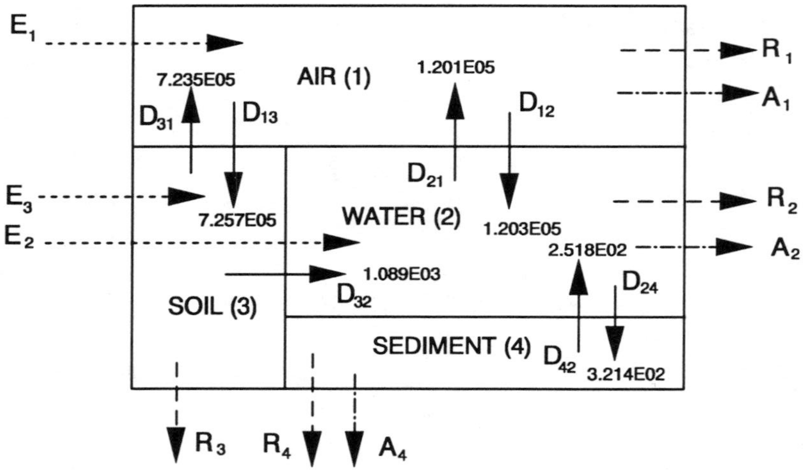

E1 AIR (1) R1 A1
7.235E05 D31 D13 1.201E05 D21 D12
E3 E2 WATER (2) 7.257E05 1.203E05 R2 A2
SOIL (3) 1.089E03 D32 2.518E02 D24
SEDIMENT (4) D42 3.214E02
R3 R4 A4

EMISSION (E)
REACTION (R)
ADVECTION (A)
TRANSFER D VALUE mol/Pa h

Phase Properties and Rates:

| Compartment | Bulk Z mol/m3 Pa | Half-life h | D Values Reaction mol/Pa h | D Values Advection mol/Pa h |
|---|---|---|---|---|
| Air (1) | 4.034E-04 | 5 | 5.59E+09 | 4.03E+08 |
| Water (2) | 2.417E-04 | 170 | 1.97E+05 | 4.83E+04 |
| Soil (3) | 1.430E-03 | 550 | 3.24E+04 | |
| Sediment (4) | 1.215E-03 | 1700 | 2.48E+02 | 1.21E+01 |

| | E(1)=1000 | E(2)=1000 | E(3)=1000 | E(1,2,3) |
|---|---|---|---|---|
| Overall residence time = | 6.74 | 135.46 | 40.63 | 48.74 h |
| Reaction time = | 7.22 | 160.18 | 43.43 | 53.75 h |
| Advection time = | 100.10 | 877.75 | 629.54 | 523.39 h |

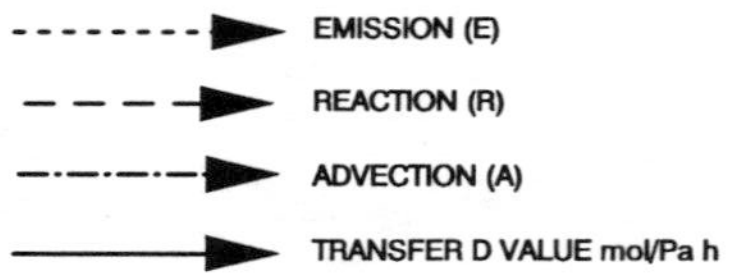

Phase Properties, Compositions, Transport and Transformation Rates:

| Emission, kg/h | | | Fugacity, Pa | | | | Concentration, g/m3 | | | | Amounts, kg | | | | Total |
|---|---|---|---|---|---|---|---|---|---|---|---|---|---|---|---|
| E(1) | E(2) | E(3) | f(1) | f(2) | f(3) | f(4) | C(1) | C(2) | C(3) | C(4) | m(1) | m(2) | m(3) | m(4) | Amount, kg |
| 1000 | 0 | 0 | 2.031E-06 | 6.740E-07 | 1.946E-06 | 4.235E-07 | 6.729E-08 | 1.338E-08 | 2.286E-07 | 4.225E-08 | 6.729E+03 | 2.676E+00 | 4.115E+00 | 2.113E-02 | 6.736E+03 |
| 0 | 1000 | 0 | 6.669E-07 | 3.330E-02 | 6.393E-07 | 2.092E-02 | 2.210E-08 | 6.610E-04 | 7.509E-08 | 2.087E-03 | 2.210E+03 | 1.322E+05 | 1.352E+00 | 1.044E+03 | 1.355E+05 |
| 0 | 0 | 1000 | 1.942E-06 | 4.856E-05 | 1.608E-02 | 3.051E-05 | 6.435E-08 | 9.641E-07 | 1.889E-03 | 3.044E-06 | 6.435E+03 | 1.928E+02 | 3.400E+04 | 1.522E+00 | 4.063E+04 |
| 600 | 300 | 100 | 1.613E-06 | 9.994E-03 | 1.610E-03 | 6.279E-03 | 5.344E-08 | 1.984E-04 | 1.891E-04 | 6.265E-04 | 5.344E+03 | 3.968E+04 | 3.403E+03 | 3.133E+02 | 4.874E+04 |

| Emission, kg/h | | | Loss, Reaction, kg/h | | | | Loss, Advection, kg/h | | | Intermedia Rate of Transport, kg/h T12 | T21 | T13 | T31 | T32 | T24 | T42 |
|---|---|---|---|---|---|---|---|---|---|---|---|---|---|---|---|---|
| E(1) | E(2) | E(3) | R(1) | R(2) | R(3) | R(4) | A(1) | A(2) | A(4) | air-water | water-air | air-soil | soil-air | soil-water | water-sed | sed-water |
| 1000 | 0 | 0 | 9.327E+02 | 1.091E-02 | 5.19E-03 | 8.612E-06 | 6.729E+01 | 2.676E-03 | 4.225E-07 | 2.007E-02 | 6.648E-03 | 1.210E-01 | 1.157E-01 | 1.742E-04 | 1.779E-05 | 8.760E-06 |
| 0 | 1000 | 0 | 3.063E+02 | 5.389E+02 | 1.70E-03 | 4.254E-01 | 2.210E+01 | 1.322E+02 | 2.087E-02 | 6.591E-03 | 3.284E+02 | 3.975E-02 | 3.799E-02 | 5.721E-05 | 8.790E-01 | 4.327E-01 |
| 0 | 0 | 1000 | 8.918E+02 | 7.860E-01 | 4.28E+01 | 6.205E-04 | 6.435E+01 | 1.928E-01 | 3.044E-05 | 1.919E-02 | 4.790E-01 | 1.157E-01 | 9.558E+02 | 1.439E+00 | 1.282E-03 | 6.311E-04 |
| 600 | 300 | 100 | 7.407E+02 | 1.618E+02 | 4.29E+00 | 1.277E-01 | 5.344E+01 | 3.968E+01 | 6.265E-03 | 1.594E-02 | 9.858E+01 | 9.613E-02 | 9.566E+01 | 1.441E-01 | 2.638E-01 | 1.299E-01 |

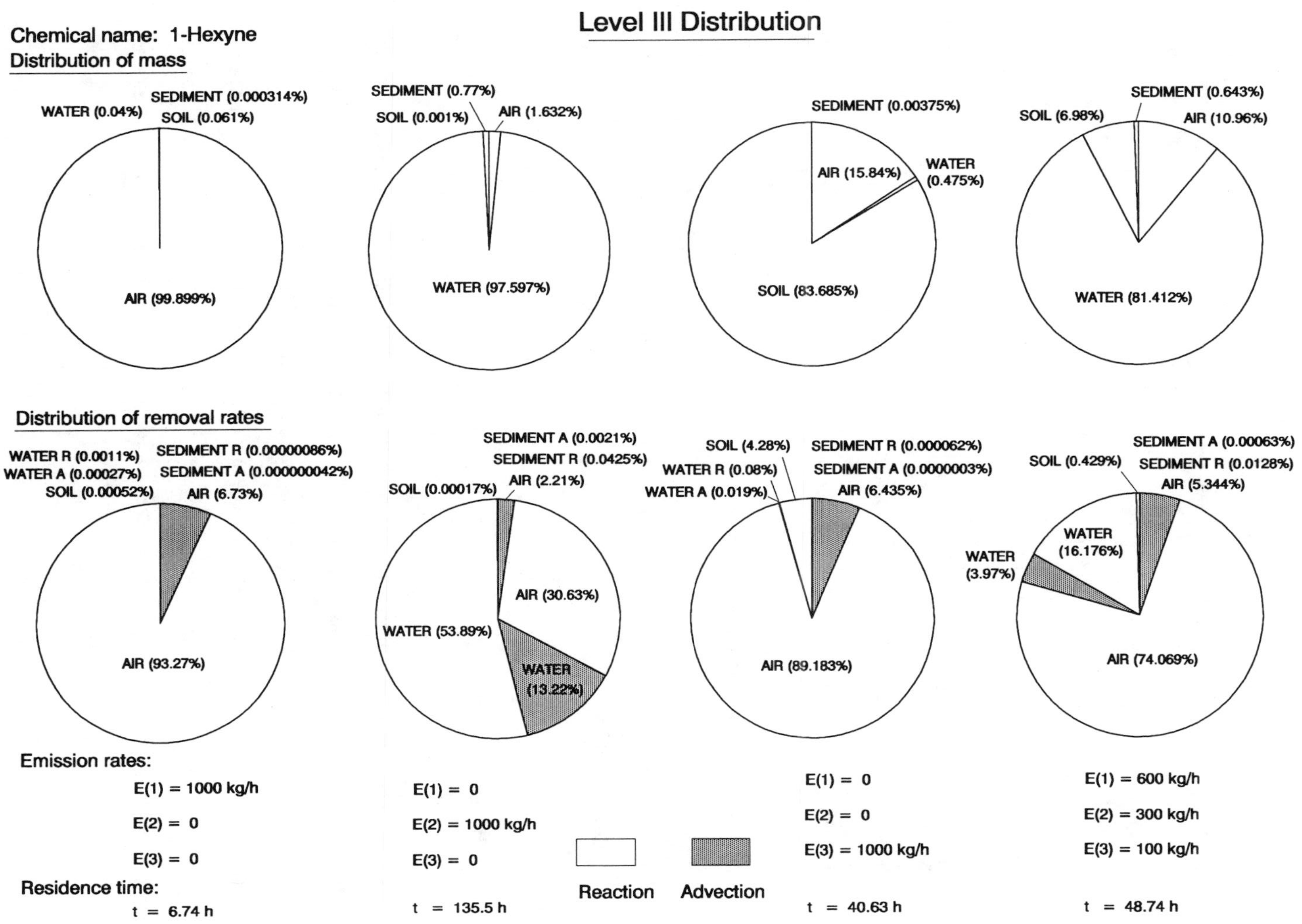

# Level III Distribution

Chemical name: 1-Hexyne

Distribution of mass

WATER (0.04%)
SEDIMENT (0.000314%)
SOIL (0.061%)
AIR (99.899%)

SEDIMENT (0.77%)
SOIL (0.001%)
AIR (1.632%)
WATER (97.597%)

SEDIMENT (0.00375%)
AIR (15.84%)
WATER (0.475%)
SOIL (83.685%)

SEDIMENT (0.643%)
SOIL (6.98%)
AIR (10.96%)
WATER (81.412%)

Distribution of removal rates

WATER R (0.0011%)
WATER A (0.00027%)
SOIL (0.00052%)
SEDIMENT R (0.00000086%)
SEDIMENT A (0.000000042%)
AIR (6.73%)
AIR (93.27%)

SEDIMENT A (0.0021%)
SEDIMENT R (0.0425%)
AIR (2.21%)
SOIL (0.00017%)
AIR (30.63%)
WATER (53.89%)
WATER (13.22%)

SOIL (4.28%)
WATER R (0.08%)
WATER A (0.019%)
SEDIMENT R (0.000062%)
SEDIMENT A (0.0000003%)
AIR (6.435%)
AIR (89.183%)

SEDIMENT A (0.00063%)
SOIL (0.429%)
SEDIMENT R (0.0128%)
AIR (5.344%)
WATER (16.176%)
WATER (3.97%)
AIR (74.069%)

Reaction
Advection

| | | | | |
|---|---|---|---|---|
| Emission rates: | E(1) = 1000 kg/h | E(1) = 0 | E(1) = 0 | E(1) = 600 kg/h |
| | E(2) = 0 | E(2) = 1000 kg/h | E(2) = 0 | E(2) = 300 kg/h |
| | E(3) = 0 | E(3) = 0 | E(3) = 1000 kg/h | E(3) = 100 kg/h |
| Residence time: | t = 6.74 h | t = 135.5 h | t = 40.63 h | t = 48.74 h |

Chemical name: 1,3-Butadiene

Level I calculation: (six-compartment model)

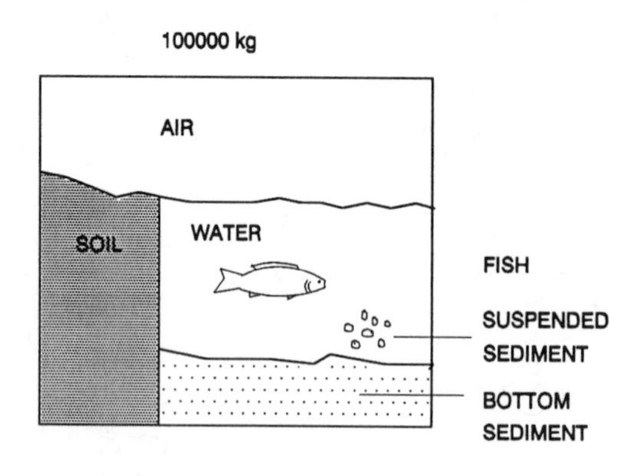


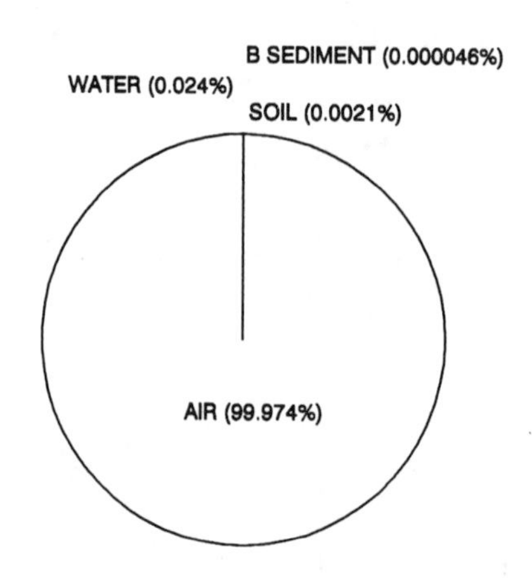


Distribution of mass

physical-chemical properties:

MW: 54.09 g/mol

M.P.: - 108.9°C

Fugacity ratio: 1.0

vapor pressure: 281000 Pa

solubility: 735 g/m³

log $K_{OW}$: 1.99

| Compartment | Z | Concentration | | | Amount | Amount |
|---|---|---|---|---|---|---|
| | mol/m3 Pa | mol/m3 | mg/L (or g/m3) | ug/g | kg | % |
| Air | 4.034E-04 | 1.848E-08 | 9.997E-07 | 8.434E-04 | 99974 | 99.974 |
| Water | 4.836E-05 | 2.216E-09 | 1.198E-07 | 1.198E-07 | 23.968 | 0.0240 |
| Soil | 9.300E-05 | 4.261E-09 | 2.305E-07 | 9.603E-08 | 2.074 | 0.0021 |
| Biota (fish) | 2.363E-04 | 1.083E-08 | 5.856E-07 | 5.856E-07 | 1.17E-04 | 1.17E-07 |
| Suspended sediment | 5.813E-04 | 2.663E-08 | 1.440E-06 | 9.603E-07 | 1.44E-03 | 1.44E-06 |
| Bottom sediment | 1.860E-04 | 8.522E-09 | 4.609E-07 | 1.921E-07 | 0.0461 | 4.61E-05 |
| | Total | | | | 100000 | 100 |

f = 4.582E-05 Pa

Chemical name: 1,3-Butadiene

Level II calculation: (six-compartment model)

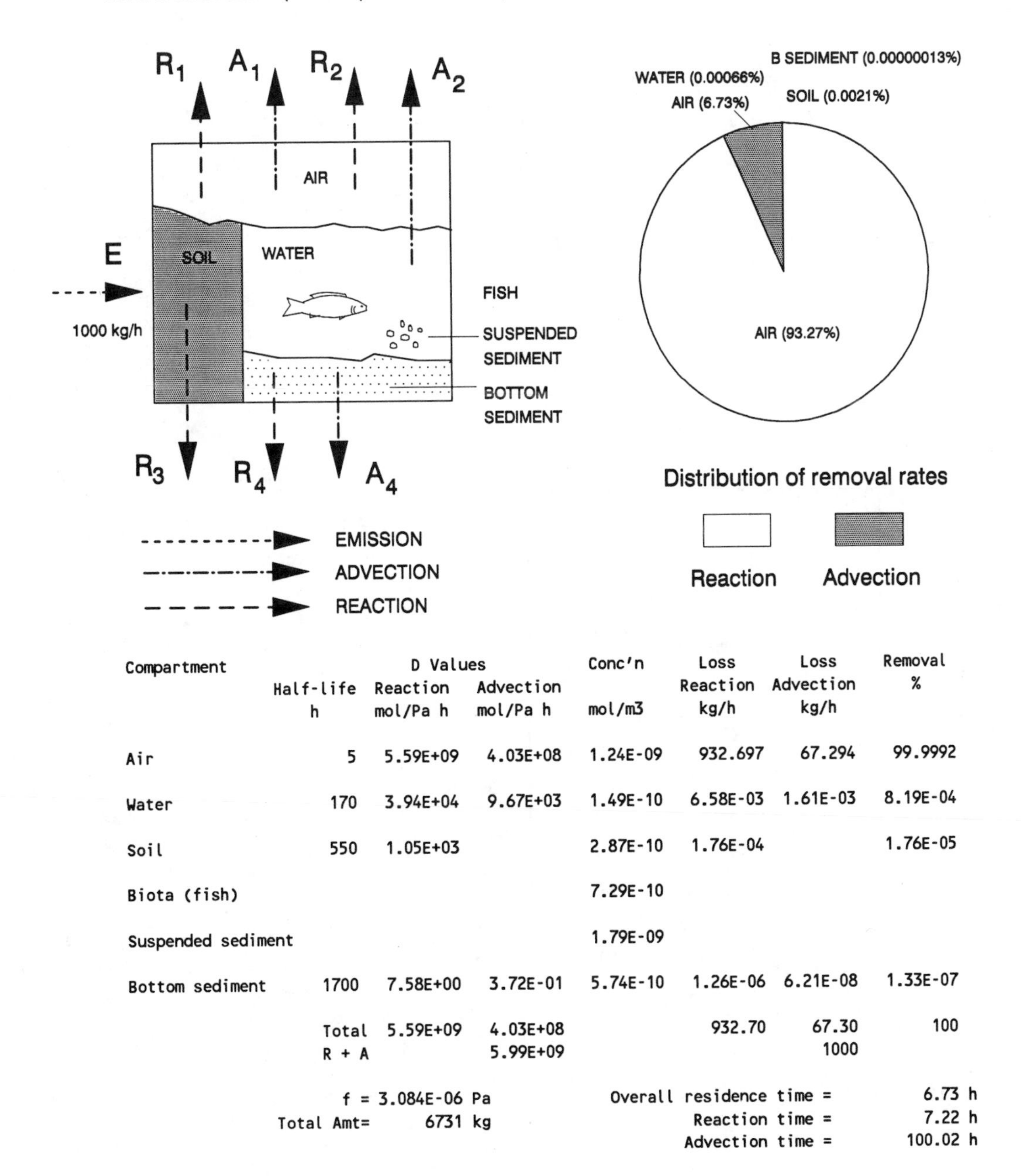


| Compartment | Half-life h | D Values Reaction mol/Pa h | D Values Advection mol/Pa h | Conc'n mol/m3 | Loss Reaction kg/h | Loss Advection kg/h | Removal % |
|---|---|---|---|---|---|---|---|
| Air | 5 | 5.59E+09 | 4.03E+08 | 1.24E-09 | 932.697 | 67.294 | 99.9992 |
| Water | 170 | 3.94E+04 | 9.67E+03 | 1.49E-10 | 6.58E-03 | 1.61E-03 | 8.19E-04 |
| Soil | 550 | 1.05E+03 | | 2.87E-10 | 1.76E-04 | | 1.76E-05 |
| Biota (fish) | | | | 7.29E-10 | | | |
| Suspended sediment | | | | 1.79E-09 | | | |
| Bottom sediment | 1700 | 7.58E+00 | 3.72E-01 | 5.74E-10 | 1.26E-06 | 6.21E-08 | 1.33E-07 |
| | Total | 5.59E+09 | 4.03E+08 | | 932.70 | 67.30 | 100 |
| | R + A | | 5.99E+09 | | | 1000 | |

f = 3.084E-06 Pa
Total Amt= 6731 kg

Overall residence time = 6.73 h
Reaction time = 7.22 h
Advection time = 100.02 h

## Fugacity Level III calculations: (four-compartment model)

Chemical name: 1,3-Butadiene

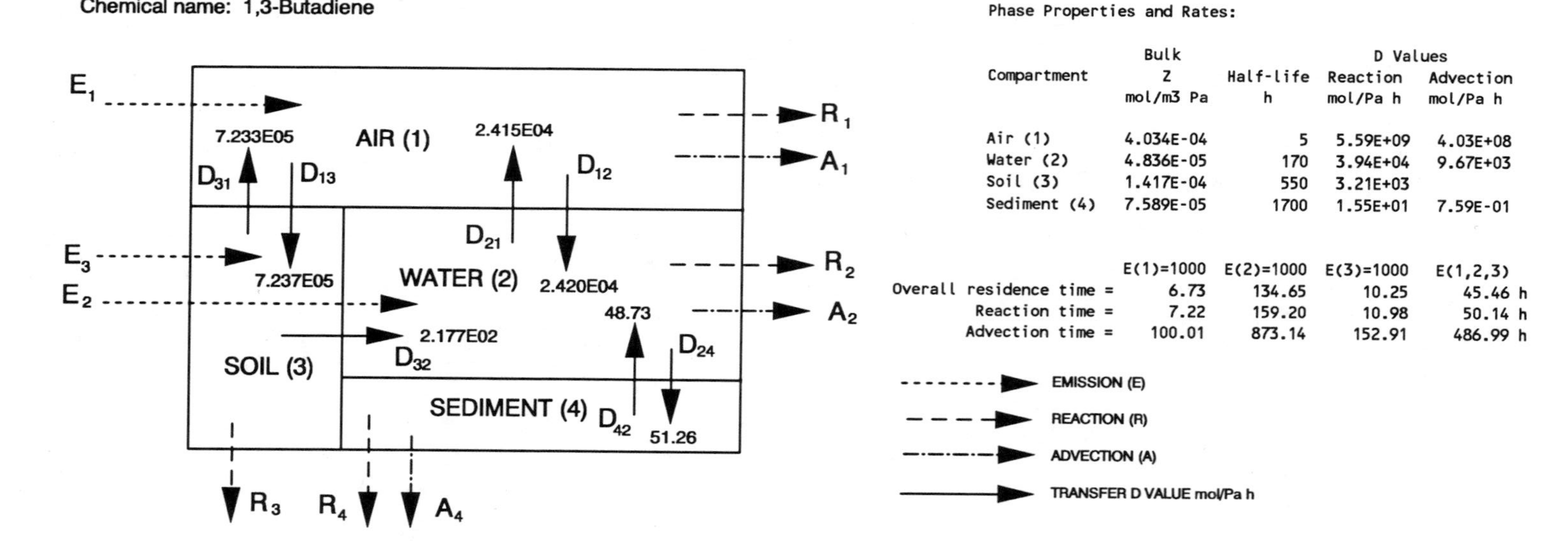


Phase Properties and Rates:

| Compartment | Bulk Z mol/m3 Pa | Half-life h | D Values Reaction mol/Pa h | D Values Advection mol/Pa h |
|---|---|---|---|---|
| Air (1) | 4.034E-04 | 5 | 5.59E+09 | 4.03E+08 |
| Water (2) | 4.836E-05 | 170 | 3.94E+04 | 9.67E+03 |
| Soil (3) | 1.417E-04 | 550 | 3.21E+03 | |
| Sediment (4) | 7.589E-05 | 1700 | 1.55E+01 | 7.59E-01 |

| | E(1)=1000 | E(2)=1000 | E(3)=1000 | E(1,2,3) |
|---|---|---|---|---|
| Overall residence time = | 6.73 | 134.65 | 10.25 | 45.46 h |
| Reaction time = | 7.22 | 159.20 | 10.98 | 50.14 h |
| Advection time = | 100.01 | 873.14 | 152.91 | 486.99 h |

Phase Properties, Compositions, Transport and Transformation Rates:

| Emission, kg/h | | | Fugacity, Pa | | | | Concentration, g/m3 | | | | Amounts, kg | | | | Total Amount, kg |
|---|---|---|---|---|---|---|---|---|---|---|---|---|---|---|---|
| E(1) | E(2) | E(3) | f(1) | f(2) | f(3) | f(4) | C(1) | C(2) | C(3) | C(4) | m(1) | m(2) | m(3) | m(4) | |
| 1000 | 0 | 0 | 3.084E-06 | 1.028E-06 | 3.071E-06 | 8.111E-07 | 6.729E-08 | 2.688E-09 | 2.354E-08 | 3.329E-09 | 6.729E+03 | 5.377E-01 | 4.237E-01 | 1.665E-03 | 6.730E+03 |
| 0 | 1000 | 0 | 1.017E-06 | 2.523E-01 | 1.012E-06 | 1.992E-01 | 2.218E-08 | 6.601E-04 | 7.759E-09 | 8.175E-04 | 2.218E+03 | 1.320E+05 | 1.397E-01 | 4.087E+02 | 1.346E+05 |
| 0 | 0 | 1000 | 3.070E-06 | 7.661E-05 | 2.544E-02 | 6.046E-05 | 6.698E-08 | 2.004E-07 | 1.950E-04 | 2.482E-07 | 6.698E+03 | 4.008E+01 | 3.510E+03 | 1.241E-01 | 1.025E+04 |
| 600 | 300 | 100 | 2.462E-06 | 7.571E-02 | 2.546E-03 | 5.975E-02 | 5.373E-08 | 1.981E-04 | 1.952E-05 | 2.453E-04 | 5.373E+03 | 3.961E+04 | 3.513E+02 | 1.226E+02 | 4.546E+04 |

| Emission, kg/h | | | Loss, Reaction, kg/h | | | | Loss, Advection, kg/h | | | Intermedia Rate of Transport, kg/h T12 | T21 | T13 | T31 | T32 | T24 | T42 |
|---|---|---|---|---|---|---|---|---|---|---|---|---|---|---|---|---|
| E(1) | E(2) | E(3) | R(1) | R(2) | R(3) | R(4) | A(1) | A(2) | A(4) | air-water | water-air | air-soil | soil-air | soil-water | water-sed | sed-water |
| 1000 | 0 | 0 | 9.327E+02 | 2.192E-03 | 5.34E-04 | 6.786E-07 | 6.729E+01 | 5.377E-04 | 3.329E-08 | 4.037E-03 | 1.342E-03 | 1.207E-01 | 1.202E-01 | 3.616E-05 | 2.850E-06 | 2.138E-06 |
| 0 | 1000 | 0 | 3.074E+02 | 5.382E+02 | 1.76E-04 | 1.666E-01 | 2.218E+01 | 1.320E+02 | 8.175E-03 | 1.331E-03 | 3.296E+02 | 3.980E-02 | 3.961E-02 | 1.192E-05 | 6.997E-01 | 5.249E-01 |
| 0 | 0 | 1000 | 9.284E+02 | 1.634E-01 | 4.42E+00 | 5.059E-05 | 6.698E+01 | 4.008E-02 | 2.482E-06 | 4.018E-03 | 1.001E-01 | 1.202E-01 | 9.954E+02 | 2.996E-01 | 2.124E-04 | 1.594E-04 |
| 600 | 300 | 100 | 7.447E+02 | 1.615E+02 | 4.43E-01 | 4.999E-02 | 5.373E+01 | 3.961E+01 | 2.453E-03 | 3.223E-03 | 9.890E+01 | 9.639E-02 | 9.962E+01 | 2.998E-02 | 2.099E-01 | 1.575E-01 |

# Level III Distribution

Chemical name: 1,3-Butadiene

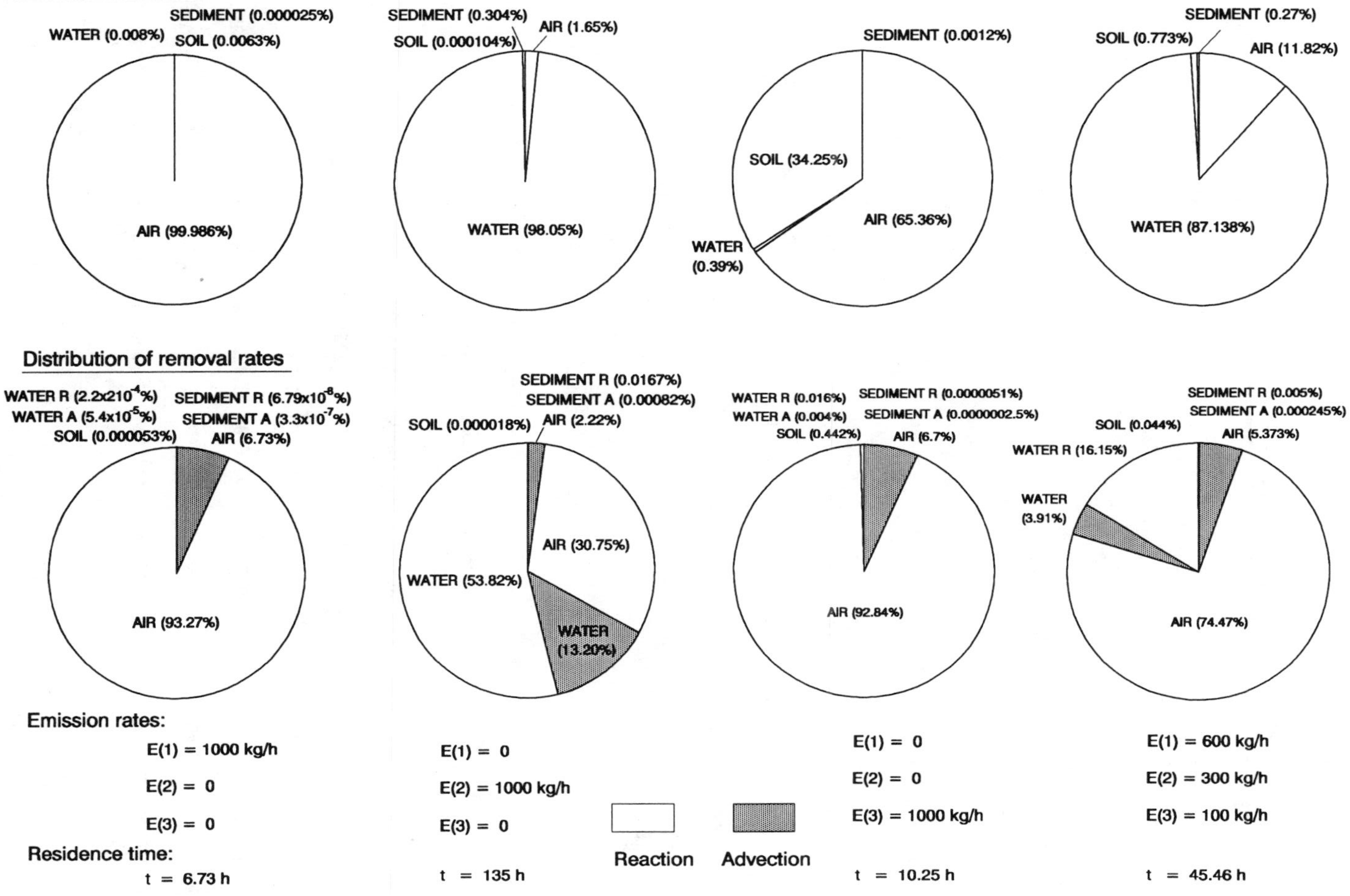

Distribution of mass



Distribution of removal rates



Emission rates:

| | | | |
|---|---|---|---|
| E(1) = 1000 kg/h | E(1) = 0 | E(1) = 0 | E(1) = 600 kg/h |
| E(2) = 0 | E(2) = 1000 kg/h | E(2) = 0 | E(2) = 300 kg/h |
| E(3) = 0 | E(3) = 0 | E(3) = 1000 kg/h | E(3) = 100 kg/h |

Residence time:

| | | | |
|---|---|---|---|
| t = 6.73 h | t = 135 h | t = 10.25 h | t = 45.46 h |

Chemical name: 1,4-Pentadiene

Level I calculation: (six-compartment model)

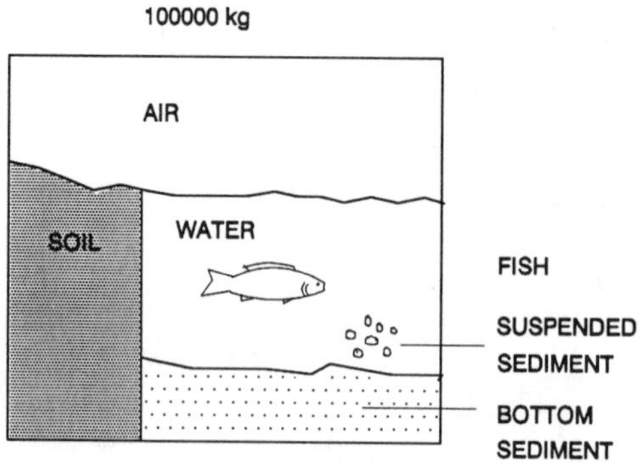

100000 kg

AIR

SOIL

WATER

FISH

SUSPENDED SEDIMENT

BOTTOM SEDIMENT

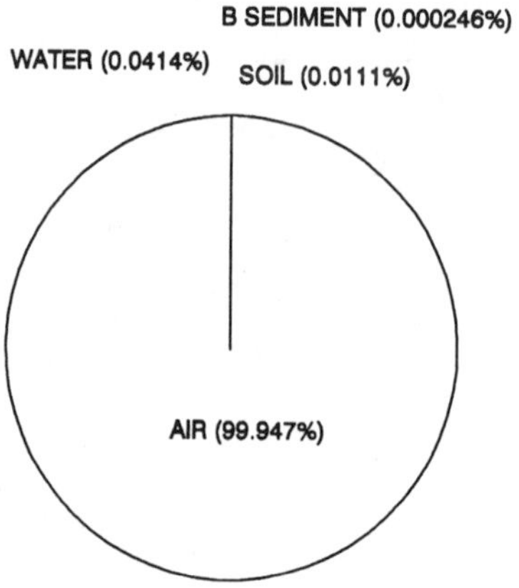

B SEDIMENT (0.000246%)

WATER (0.0414%)

SOIL (0.0111%)

AIR (99.947%)

Distribution of mass

physical-chemical properties:

MW: 68.13 g/mol

M.P.: - 148.3 °C

Fugacity ratio: 1.0

vapor pressure: 98000 Pa

solubility: 558 g/m$^3$

log $K_{OW}$: 2.48

| Compartment | Z | Concentration | | | Amount | Amount |
|---|---|---|---|---|---|---|
| | mol/m3 Pa | mol/m3 | mg/L (or g/m3) | ug/g | kg | % |
| Air | 4.034E-04 | 1.467E-08 | 9.995E-07 | 8.431E-04 | 99947 | 99.947 |
| Water | 8.357E-05 | 3.039E-09 | 2.071E-07 | 2.071E-07 | 41.411 | 0.0414 |
| Soil | 4.967E-04 | 1.806E-08 | 1.231E-06 | 5.127E-07 | 11.075 | 1.11E-02 |
| Biota (fish) | 1.262E-03 | 4.589E-08 | 3.126E-06 | 3.126E-06 | 6.25E-04 | 6.25E-07 |
| Suspended sediment | 3.104E-03 | 1.129E-07 | 7.691E-06 | 5.127E-06 | 7.69E-03 | 7.69E-06 |
| Bottom sediment | 9.934E-04 | 3.612E-08 | 2.461E-06 | 1.025E-06 | 0.2461 | 2.46E-04 |
| | Total | | | | 100000 | 100 |

f = 3.636E-05 Pa

Chemical name: 1,4-Pentadiene

Level II calculation: (six-compartment model)

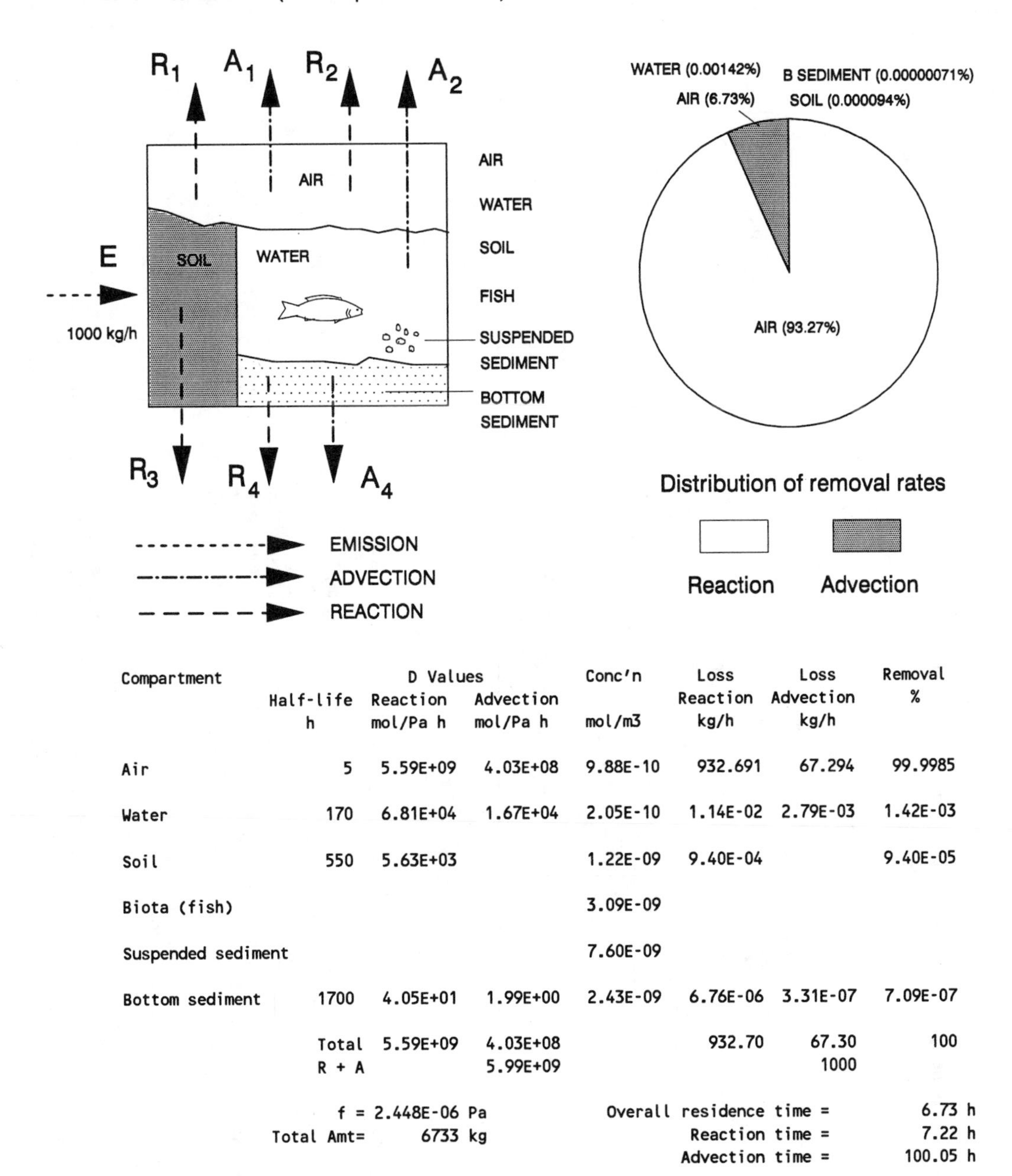


| Compartment | Half-life h | D Values Reaction mol/Pa h | D Values Advection mol/Pa h | Conc'n mol/m3 | Loss Reaction kg/h | Loss Advection kg/h | Removal % |
|---|---|---|---|---|---|---|---|
| Air | 5 | 5.59E+09 | 4.03E+08 | 9.88E-10 | 932.691 | 67.294 | 99.9985 |
| Water | 170 | 6.81E+04 | 1.67E+04 | 2.05E-10 | 1.14E-02 | 2.79E-03 | 1.42E-03 |
| Soil | 550 | 5.63E+03 | | 1.22E-09 | 9.40E-04 | | 9.40E-05 |
| Biota (fish) | | | | 3.09E-09 | | | |
| Suspended sediment | | | | 7.60E-09 | | | |
| Bottom sediment | 1700 | 4.05E+01 | 1.99E+00 | 2.43E-09 | 6.76E-06 | 3.31E-07 | 7.09E-07 |
| | Total | 5.59E+09 | 4.03E+08 | | 932.70 | 67.30 | 100 |
| | R + A | | 5.99E+09 | | | 1000 | |

f = 2.448E-06 Pa
Total Amt= 6733 kg

Overall residence time = 6.73 h
Reaction time = 7.22 h
Advection time = 100.05 h

Fugacity Level III calculations: (four-compartment model)

Chemical name: 1,4-Pentadiene

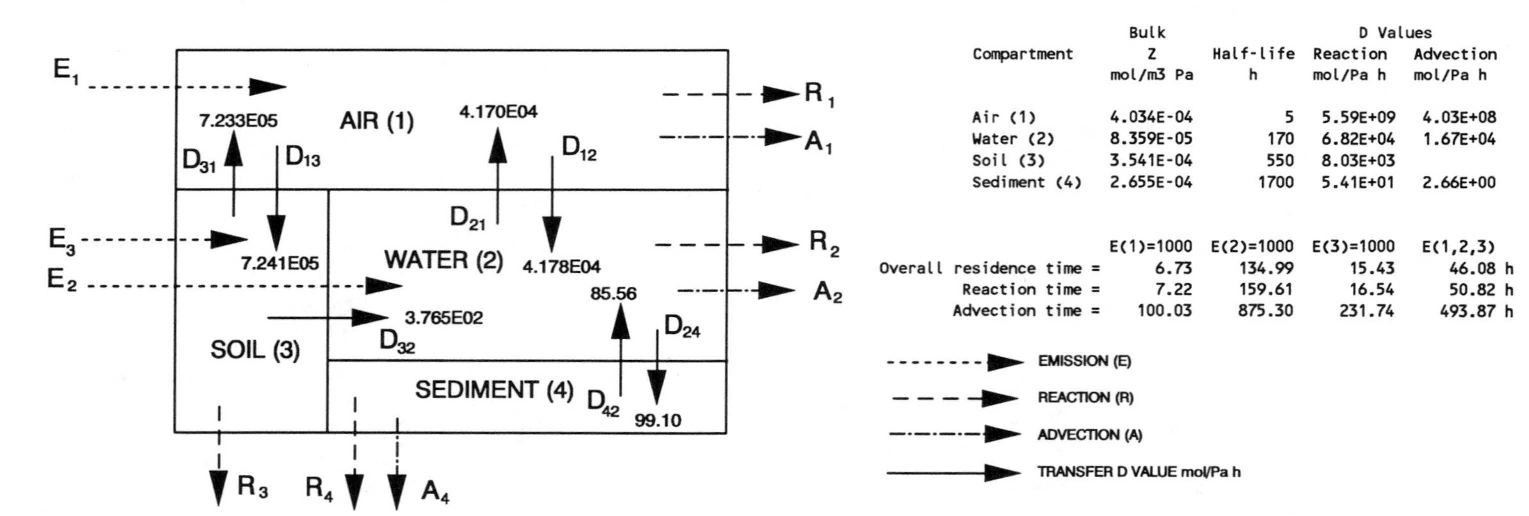


Phase Properties and Rates:

| Compartment | Bulk Z mol/m3 Pa | Half-life h | D Values Reaction mol/Pa h | D Values Advection mol/Pa h |
|---|---|---|---|---|
| Air (1) | 4.034E-04 | 5 | 5.59E+09 | 4.03E+08 |
| Water (2) | 8.359E-05 | 170 | 6.82E+04 | 1.67E+04 |
| Soil (3) | 3.541E-04 | 550 | 8.03E+03 | |
| Sediment (4) | 2.655E-04 | 1700 | 5.41E+01 | 2.66E+00 |

| | E(1)=1000 | E(2)=1000 | E(3)=1000 | E(1,2,3) | |
|---|---|---|---|---|---|
| Overall residence time = | 6.73 | 134.99 | 15.43 | 46.08 | h |
| Reaction time = | 7.22 | 159.61 | 16.54 | 50.82 | h |
| Advection time = | 100.03 | 875.30 | 231.74 | 493.87 | h |

Phase Properties, Compositions, Transport and Transformation Rates:

| Emission, kg/h | | | Fugacity, Pa | | | | Concentration, g/m3 | | | | Amounts, kg | | | | Total |
|---|---|---|---|---|---|---|---|---|---|---|---|---|---|---|---|
| E(1) | E(2) | E(3) | f(1) | f(2) | f(3) | f(4) | C(1) | C(2) | C(3) | C(4) | m(1) | m(2) | m(3) | m(4) | Amount, kg |
| 1000 | 0 | 0 | 2.448E-06 | 8.152E-07 | 2.423E-06 | 5.676E-07 | 6.729E-08 | 4.643E-09 | 5.845E-08 | 1.027E-08 | 6.729E+03 | 9.286E-01 | 1.052E+00 | 5.134E-03 | 6.731E+03 |
| 0 | 1000 | 0 | 8.064E-07 | 1.159E-01 | 7.980E-07 | 8.071E-02 | 2.216E-08 | 6.602E-04 | 1.925E-08 | 1.460E-03 | 2.216E+03 | 1.320E+05 | 3.465E-01 | 7.301E+02 | 1.350E+05 |
| 0 | 0 | 1000 | 2.421E-06 | 6.046E-05 | 2.006E-02 | 4.209E-05 | 6.653E-08 | 3.443E-07 | 4.840E-04 | 7.615E-07 | 6.653E+03 | 6.886E+01 | 8.712E+03 | 3.807E-01 | 1.543E+04 |
| 600 | 300 | 100 | 1.953E-06 | 3.479E-02 | 2.008E-03 | 2.422E-02 | 5.368E-08 | 1.981E-04 | 4.844E-05 | 4.381E-04 | 5.368E+03 | 3.962E+04 | 8.719E+02 | 2.191E+02 | 4.608E+04 |

| Emission, kg/h | | | Loss, Reaction, kg/h | | | | Loss, Advection, kg/h | | | Intermedia Rate of Transport, kg/h T12 | T21 | T13 | T31 | T32 | T24 | T42 |
|---|---|---|---|---|---|---|---|---|---|---|---|---|---|---|---|---|
| E(1) | E(2) | E(3) | R(1) | R(2) | R(3) | R(4) | A(1) | A(2) | A(4) | air-water | water-air | air-soil | soil-air | soil-water | water-sed | sed-water |
| 1000 | 0 | 0 | 9.327E+02 | 3.785E-03 | 1.33E-03 | 2.093E-06 | 6.729E+01 | 9.286E-04 | 1.027E-07 | 6.970E-03 | 2.316E-03 | 1.208E-01 | 1.194E-01 | 6.215E-05 | 5.504E-06 | 3.308E-06 |
| 0 | 1000 | 0 | 3.072E+02 | 5.383E+02 | 4.37E-04 | 2.976E-01 | 2.216E+01 | 1.320E+02 | 1.460E-02 | 2.296E-03 | 3.294E+02 | 3.978E-02 | 3.933E-02 | 2.047E-05 | 7.827E-01 | 4.705E-01 |
| 0 | 0 | 1000 | 9.221E+02 | 2.807E-01 | 1.10E+01 | 1.552E-04 | 6.653E+01 | 6.886E-02 | 7.615E-06 | 6.891E-03 | 1.718E-01 | 1.194E-01 | 9.886E+02 | 5.146E-01 | 4.082E-04 | 2.454E-04 |
| 600 | 300 | 100 | 7.440E+02 | 1.615E+02 | 1.10E+00 | 8.930E-02 | 5.368E+01 | 3.962E+01 | 4.381E-03 | 5.560E-03 | 9.883E+01 | 9.635E-02 | 9.895E+01 | 5.151E-02 | 2.349E-01 | 1.412E-01 |

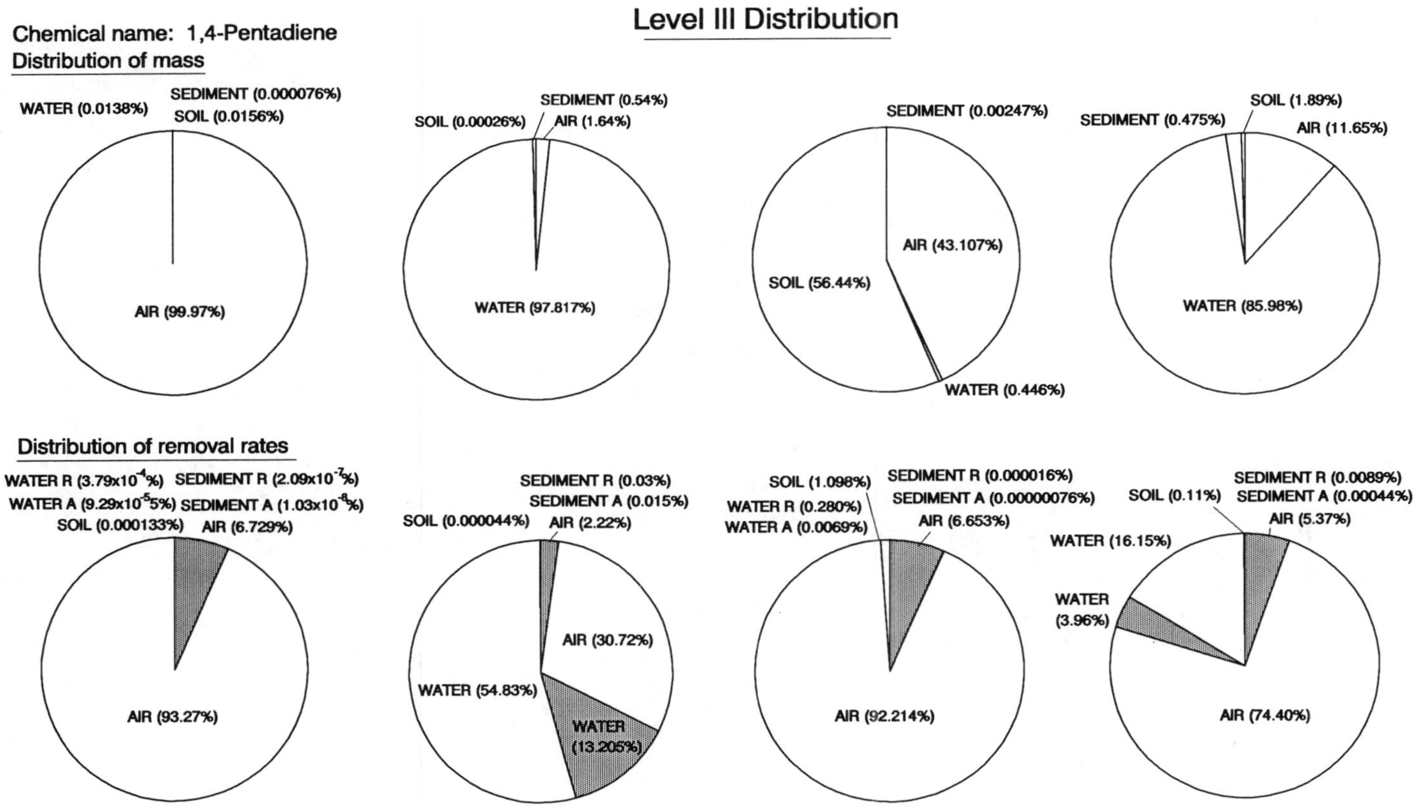

## Level III Distribution

Chemical name: 1,4-Pentadiene

Distribution of mass



Distribution of removal rates



Emission rates:

| | | | |
|---|---|---|---|
| E(1) = 1000 kg/h | E(1) = 0 | E(1) = 0 | E(1) = 600 kg/h |
| E(2) = 0 | E(2) = 1000 kg/h | E(2) = 0 | E(2) = 300 kg/h |
| E(3) = 0 | E(3) = 0 | E(3) = 1000 kg/h | E(3) = 100 kg/h |

Residence time:

| | | | |
|---|---|---|---|
| t = 6.73 h | t = 135 h | t = 15.43 h | t = 46.08 h |

Reaction Advection

Chemical name: 1,4-Cyclohexadiene

Level I calculation: (six-compartment model)

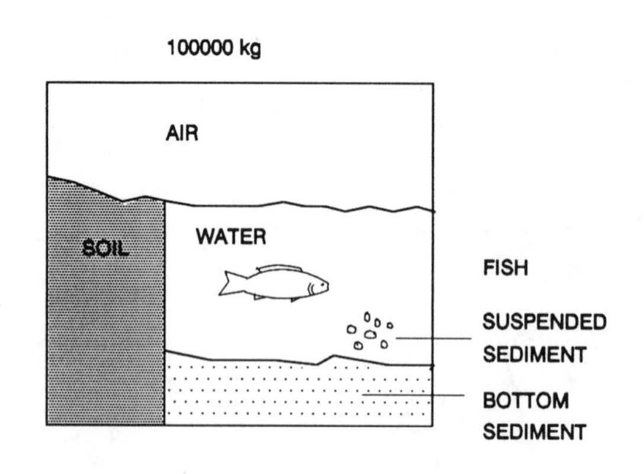


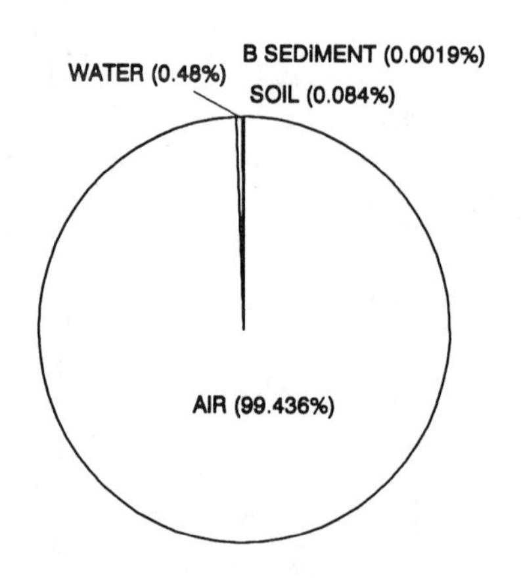


Distribution of mass

physical-chemical properties:

MW: 80.14 g/mol

M.P.: - 49.3 °C

Fugacity ratio: 1.0

vapor pressure: 9010 Pa

solubility: 700 g/m$^3$

log $K_{OW}$: 2.30

| Compartment | Z | Concentration | | | Amount | Amount |
|---|---|---|---|---|---|---|
| | mol/m3 Pa | mol/m3 | mg/L (or g/m3) | ug/g | kg | % |
| Air | 4.034E-04 | 1.241E-08 | 9.944E-07 | 8.388E-04 | 99436 | 99.436 |
| Water | 9.694E-04 | 2.982E-08 | 2.390E-06 | 2.390E-06 | 477.904 | 0.4779 |
| Soil | 3.807E-03 | 1.171E-07 | 9.383E-06 | 3.910E-06 | 84.446 | 8.44E-02 |
| Biota (fish) | 9.672E-03 | 2.975E-07 | 2.384E-05 | 2.384E-05 | 4.77E-03 | 4.77E-06 |
| Suspended sediment | 2.379E-02 | 7.318E-07 | 5.864E-05 | 3.910E-05 | 5.86E-02 | 5.86E-05 |
| Bottom sediment | 7.613E-03 | 2.342E-07 | 1.877E-05 | 7.819E-06 | 1.8766 | 1.88E-03 |
| | Total | | | | 100000 | 100 |

f = 3.076E-05 Pa

Chemical name: 1,4-Cyclohexadiene

Level II calculation: (six-compartment model)

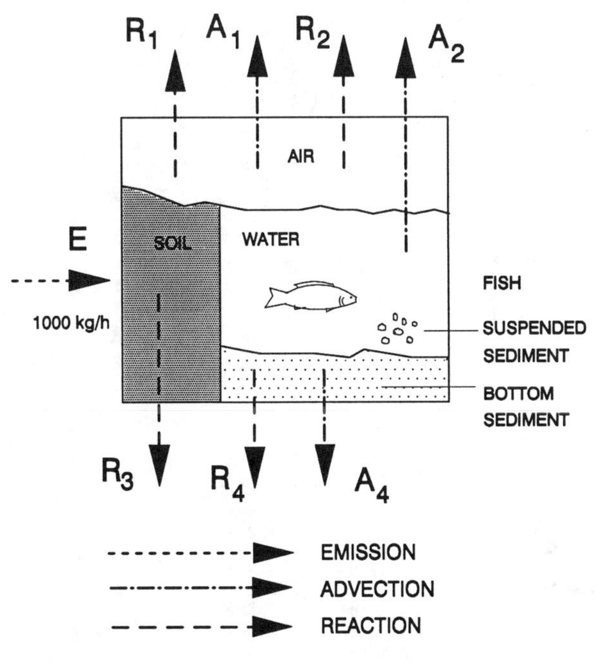


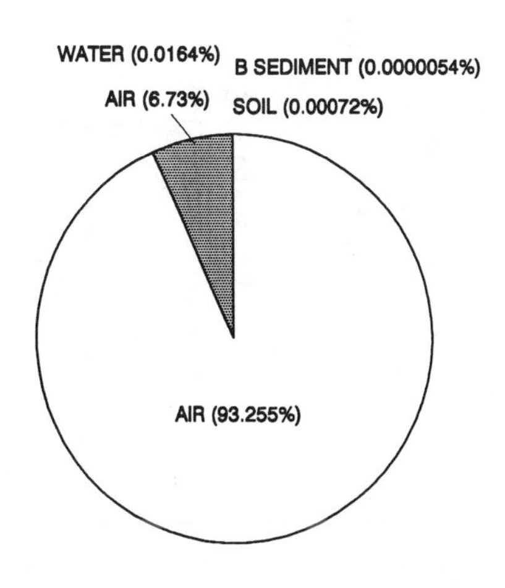


Distribution of removal rates

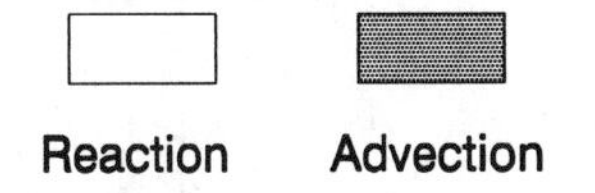


| Compartment | Half-life h | D Values Reaction mol/Pa h | D Values Advection mol/Pa h | Conc'n mol/m3 | Loss Reaction kg/h | Loss Advection kg/h | Removal % |
|---|---|---|---|---|---|---|---|
| Air | 5 | 5.59E+09 | 4.03E+08 | 8.40E-10 | 932.545 | 67.283 | 99.9829 |
| Water | 170 | 7.90E+05 | 1.94E+05 | 2.02E-09 | 0.1318 | 0.0323 | 0.0164 |
| Soil | 550 | 4.32E+04 | | 7.92E-09 | 7.20E-03 | | 7.20E-04 |
| Biota (fish) | | | | 2.01E-08 | | | |
| Suspended sediment | | | | 4.95E-08 | | | |
| Bottom sediment | 1700 | 3.10E+02 | 1.52E+01 | 1.58E-08 | 5.18E-05 | 2.54E-06 | 5.43E-06 |
| | Total | 5.59E+09 | 4.04E+08 | | 932.68 | 67.32 | 100 |
| | R + A | | 6.00E+09 | | | 1000 | |

f = 2.081E-06 Pa

Total Amt= 6767 kg

Overall residence time = 6.77 h

Reaction time = 7.25 h

Advection time = 100.52 h

## Fugacity Level III calculations: (four-compartment model)

Chemical name: 1,4-Cyclohexadiene

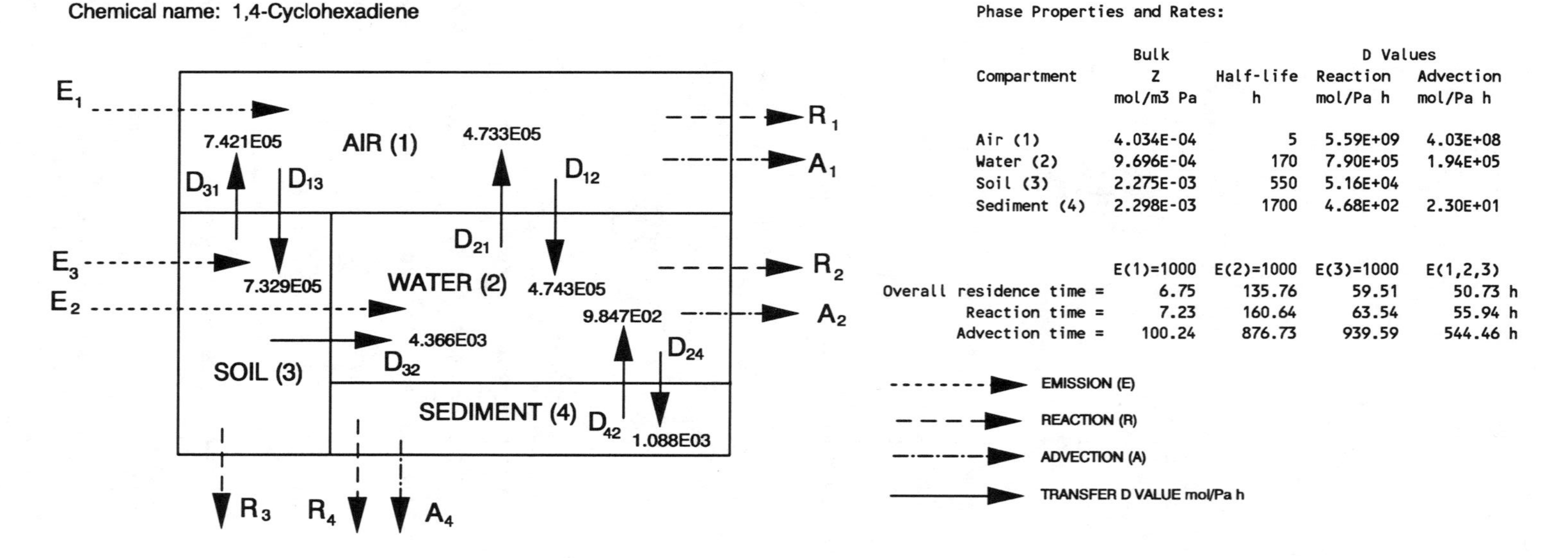


Phase Properties and Rates:

| Compartment | Bulk Z mol/m3 Pa | Half-life h | D Values Reaction mol/Pa h | D Values Advection mol/Pa h |
|---|---|---|---|---|
| Air (1) | 4.034E-04 | 5 | 5.59E+09 | 4.03E+08 |
| Water (2) | 9.696E-04 | 170 | 7.90E+05 | 1.94E+05 |
| Soil (3) | 2.275E-03 | 550 | 5.16E+04 | |
| Sediment (4) | 2.298E-03 | 1700 | 4.68E+02 | 2.30E+01 |

| | E(1)=1000 | E(2)=1000 | E(3)=1000 | E(1,2,3) |
|---|---|---|---|---|
| Overall residence time = | 6.75 | 135.76 | 59.51 | 50.73 h |
| Reaction time = | 7.23 | 160.64 | 63.54 | 55.94 h |
| Advection time = | 100.24 | 876.73 | 939.59 | 544.46 h |

Phase Properties, Compositions, Transport and Transformation Rates:

| Emission, kg/h | | | Fugacity, Pa | | | | Concentration, g/m3 | | | | Amounts, kg | | | | Total |
|---|---|---|---|---|---|---|---|---|---|---|---|---|---|---|---|
| E(1) | E(2) | E(3) | f(1) | f(2) | f(3) | f(4) | C(1) | C(2) | C(3) | C(4) | m(1) | m(2) | m(3) | m(4) | Amount, kg |
| 1000 | 0 | 0 | 2.081E-06 | 6.829E-07 | 1.955E-06 | 5.036E-07 | 6.729E-08 | 5.306E-08 | 3.565E-07 | 9.275E-08 | 6.729E+03 | 1.061E+01 | 6.417E+00 | 4.637E-02 | 6.746E+03 |
| 0 | 1000 | 0 | 6.757E-07 | 8.558E-03 | 6.348E-07 | 6.310E-03 | 2.184E-08 | 6.650E-04 | 1.157E-07 | 1.162E-03 | 2.184E+03 | 1.330E+05 | 2.083E+00 | 5.811E+02 | 1.358E+05 |
| 0 | 0 | 1000 | 1.936E-06 | 4.853E-05 | 1.600E-02 | 3.578E-05 | 6.259E-08 | 3.771E-06 | 2.917E-03 | 6.591E-06 | 6.259E+03 | 7.542E+02 | 5.250E+04 | 3.295E+00 | 5.951E+04 |
| 600 | 300 | 100 | 1.645E-06 | 2.573E-03 | 1.601E-03 | 1.897E-03 | 5.319E-08 | 1.999E-04 | 2.919E-04 | 3.494E-04 | 5.319E+03 | 3.998E+04 | 5.254E+03 | 1.747E+02 | 5.073E+04 |

| Emission, kg/h | | | Loss, Reaction, kg/h | | | | Loss, Advection, kg/h | | | Intermedia Rate of Transport, kg/h T12 | T21 | T13 | T31 | T32 | T24 | T42 |
|---|---|---|---|---|---|---|---|---|---|---|---|---|---|---|---|---|
| E(1) | E(2) | E(3) | R(1) | R(2) | R(3) | R(4) | A(1) | A(2) | A(4) | air-water | water-air | air-soil | soil-air | soil-water | water-sed | sed-water |
| 1000 | 0 | 0 | 9.326E+02 | 4.326E-02 | 8.09E-03 | 1.890E-05 | 6.729E+01 | 1.061E-02 | 9.275E-07 | 7.912E-02 | 2.591E-02 | 1.222E-01 | 1.135E-01 | 6.842E-04 | 5.957E-05 | 3.974E-05 |
| 0 | 1000 | 0 | 3.028E+02 | 5.421E+02 | 2.62E-03 | 2.369E-01 | 2.184E+01 | 1.330E+02 | 1.162E-02 | 2.568E-02 | 3.246E+02 | 3.968E-02 | 3.684E-02 | 2.221E-04 | 7.465E-01 | 4.980E-01 |
| 0 | 0 | 1000 | 8.674E+02 | 3.074E+00 | 6.61E+01 | 1.343E-03 | 6.259E+01 | 7.542E-01 | 6.591E-05 | 7.359E-02 | 1.841E+00 | 1.137E-01 | 9.284E+02 | 5.597E+00 | 4.233E-03 | 2.824E-03 |
| 600 | 300 | 100 | 7.372E+02 | 1.630E+02 | 6.62E+00 | 7.121E-02 | 5.319E+01 | 3.998E+01 | 3.494E-03 | 6.253E-02 | 9.759E+01 | 9.662E-02 | 9.292E+01 | 5.602E-01 | 2.244E-01 | 1.497E-01 |

## Level III Distribution

Chemical name: 1,4-Cyclohexadiene

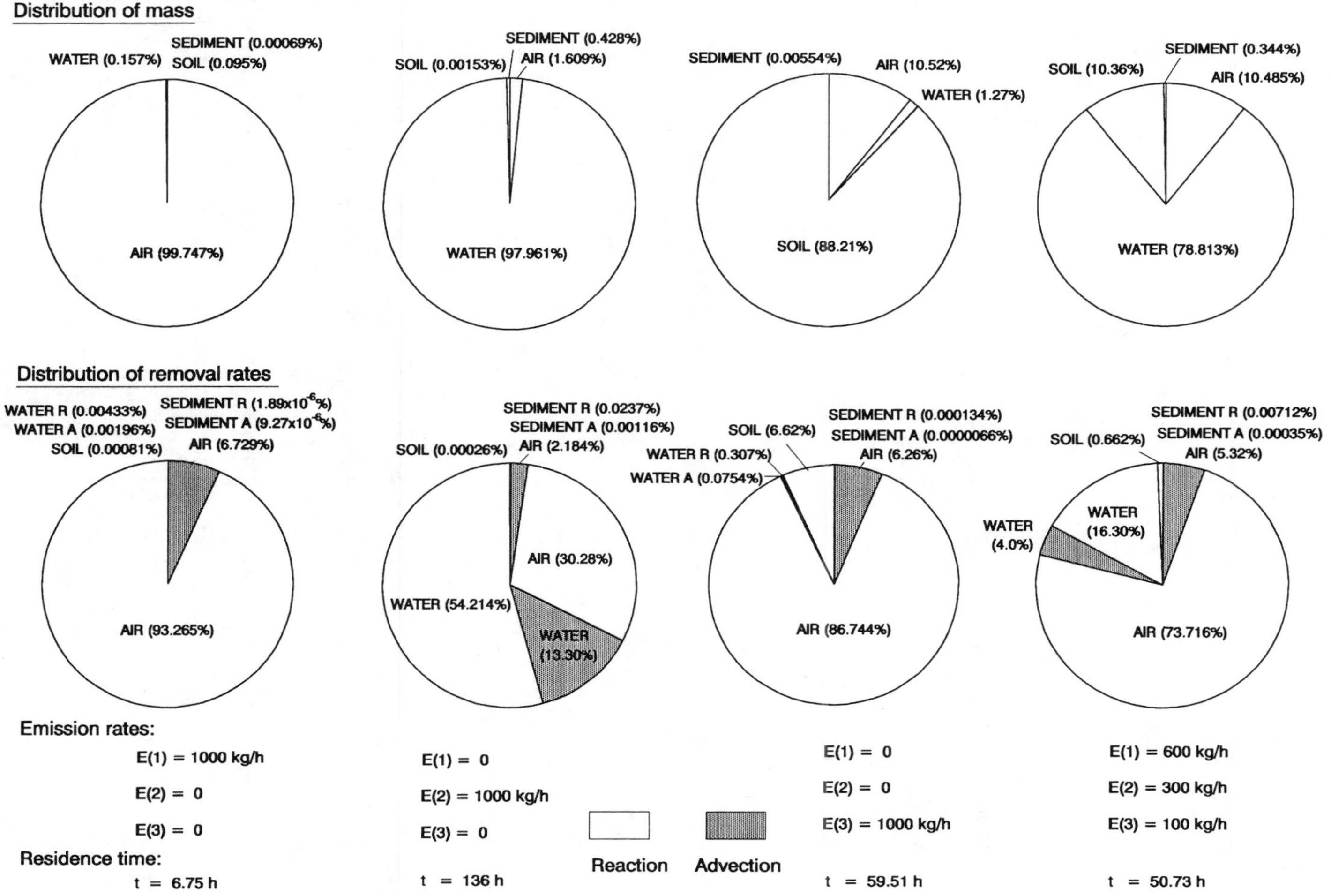

Distribution of mass

WATER (0.157%) SEDIMENT (0.00069%) SOIL (0.095%) AIR (99.747%)

SOIL (0.00153%) SEDIMENT (0.428%) AIR (1.609%) WATER (97.961%)

SEDIMENT (0.00554%) AIR (10.52%) WATER (1.27%) SOIL (88.21%)

SOIL (10.36%) SEDIMENT (0.344%) AIR (10.485%) WATER (78.813%)

Distribution of removal rates

WATER R (0.00433%) WATER A (0.00196%) SOIL (0.00081%) SEDIMENT R ($1.89 \times 10^{-6}$%) SEDIMENT A ($9.27 \times 10^{-6}$%) AIR (6.729%) AIR (93.265%)

SOIL (0.00026%) SEDIMENT R (0.0237%) SEDIMENT A (0.00116%) AIR (2.184%) AIR (30.28%) WATER (54.214%) WATER (13.30%)

SOIL (6.62%) WATER R (0.307%) WATER A (0.0754%) SEDIMENT R (0.000134%) SEDIMENT A (0.0000066%) AIR (6.26%) AIR (86.744%)

SOIL (0.662%) SEDIMENT R (0.00712%) SEDIMENT A (0.00035%) AIR (5.32%) WATER (16.30%) WATER (4.0%) AIR (73.716%)

Reaction Advection

| | | | | |
|---|---|---|---|---|
| Emission rates: | E(1) = 1000 kg/h | E(1) = 0 | E(1) = 0 | E(1) = 600 kg/h |
| | E(2) = 0 | E(2) = 1000 kg/h | E(2) = 0 | E(2) = 300 kg/h |
| | E(3) = 0 | E(3) = 0 | E(3) = 1000 kg/h | E(3) = 100 kg/h |
| Residence time: | t = 6.75 h | t = 136 h | t = 59.51 h | t = 50.73 h |

Chemical name: Styrene

Level I calculation: (six-compartment model)

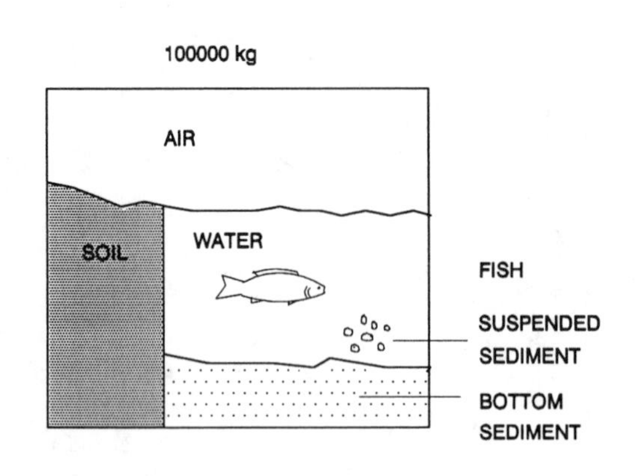


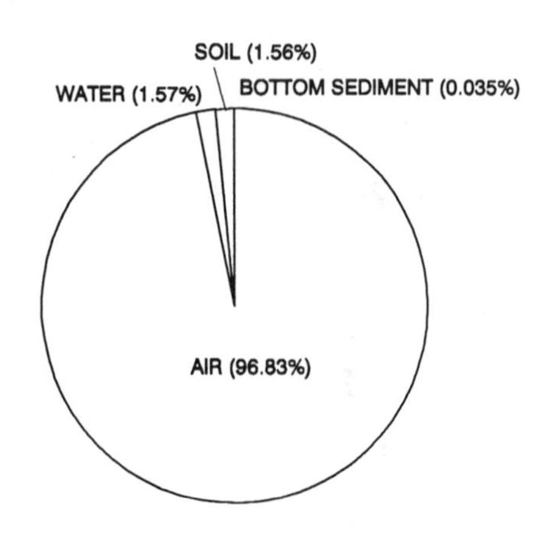


Distribution of mass

physical-chemical properties:

MW: 104.16 g/mol

M.P.: - 96.0 °C

Fugacity ratio: 1.0

vapor pressure: 880 Pa

solubility: 300 g/m$^3$

log $K_{OW}$: 3.05

| Compartment | Z | Concentration | | | Amount | Amount |
|---|---|---|---|---|---|---|
| | mol/m3 Pa | mol/m3 | mg/L (or g/m3) | ug/g | kg | % |
| Air | 4.034E-04 | 9.296E-09 | 9.683E-07 | 8.169E-04 | 96832 | 96.832 |
| Water | 3.273E-03 | 7.542E-08 | 7.856E-06 | 7.856E-06 | 1571.19 | 1.5712 |
| Soil | 7.227E-02 | 1.665E-06 | 1.735E-04 | 7.228E-05 | 1561.23 | 1.5612 |
| Biota (fish) | 1.836E-01 | 4.231E-06 | 4.407E-04 | 4.407E-04 | 0.0881 | 8.81E-05 |
| Suspended sediment | 4.517E-01 | 1.041E-05 | 1.084E-03 | 7.228E-04 | 1.0842 | 1.08E-03 |
| Bottom sediment | 1.445E-01 | 3.331E-06 | 3.469E-04 | 1.446E-04 | 34.694 | 3.47E-02 |
| | Total | | | | 100000 | 100 |

f = 2.304E-05 Pa

Chemical name: Styrene

Level II calculation: (six-compartment model)

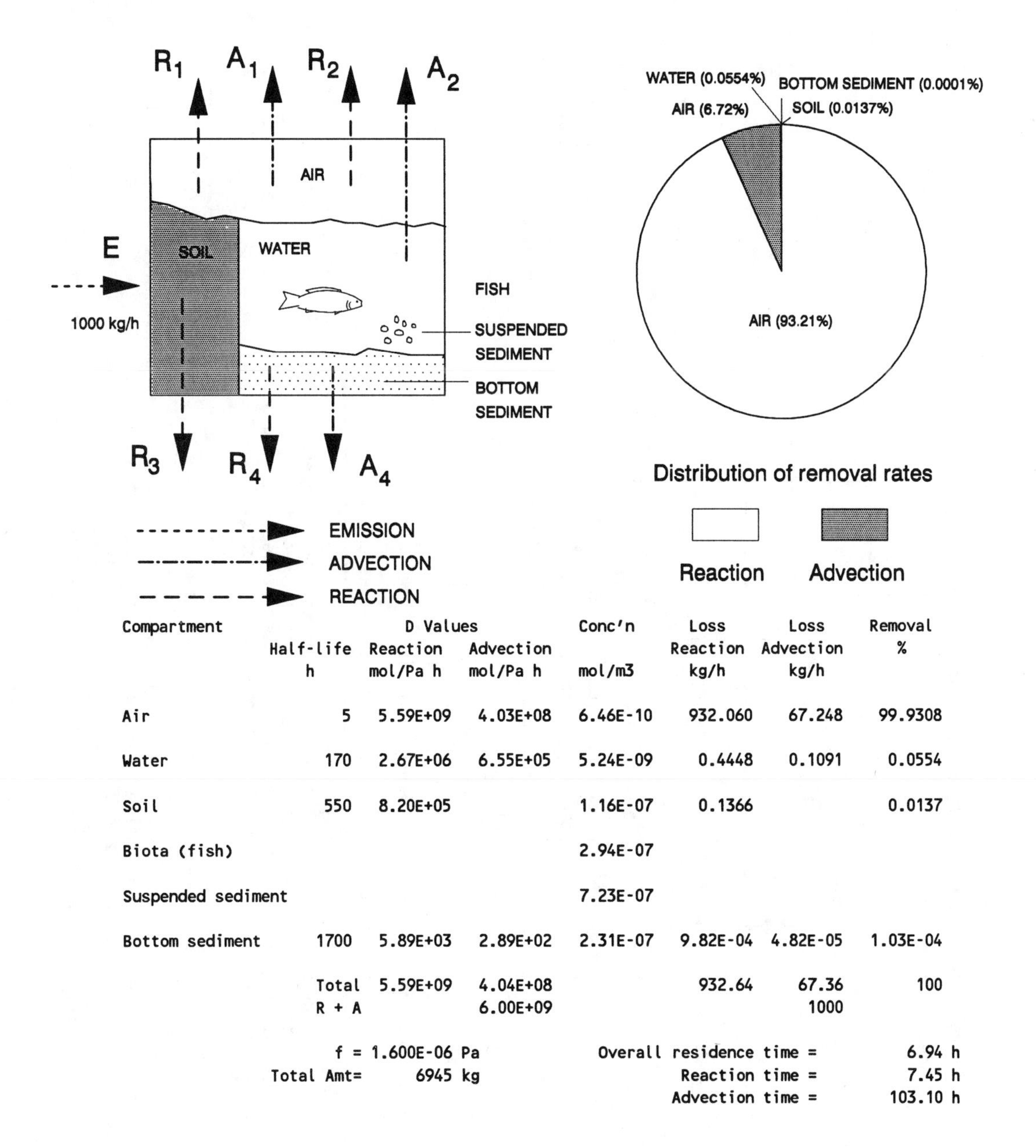


| Compartment | Half-life h | D Values Reaction mol/Pa h | D Values Advection mol/Pa h | Conc'n mol/m3 | Loss Reaction kg/h | Loss Advection kg/h | Removal % |
|---|---|---|---|---|---|---|---|
| Air | 5 | 5.59E+09 | 4.03E+08 | 6.46E-10 | 932.060 | 67.248 | 99.9308 |
| Water | 170 | 2.67E+06 | 6.55E+05 | 5.24E-09 | 0.4448 | 0.1091 | 0.0554 |
| Soil | 550 | 8.20E+05 | | 1.16E-07 | 0.1366 | | 0.0137 |
| Biota (fish) | | | | 2.94E-07 | | | |
| Suspended sediment | | | | 7.23E-07 | | | |
| Bottom sediment | 1700 | 5.89E+03 | 2.89E+02 | 2.31E-07 | 9.82E-04 | 4.82E-05 | 1.03E-04 |
| | Total | 5.59E+09 | 4.04E+08 | | 932.64 | 67.36 | 100 |
| | R + A | | 6.00E+09 | | | 1000 | |

f = 1.600E-06 Pa
Total Amt= 6945 kg

Overall residence time = 6.94 h
Reaction time = 7.45 h
Advection time = 103.10 h

**Fugacity Level III calculations:** (four-compartment model)

Chemical name: Styrene

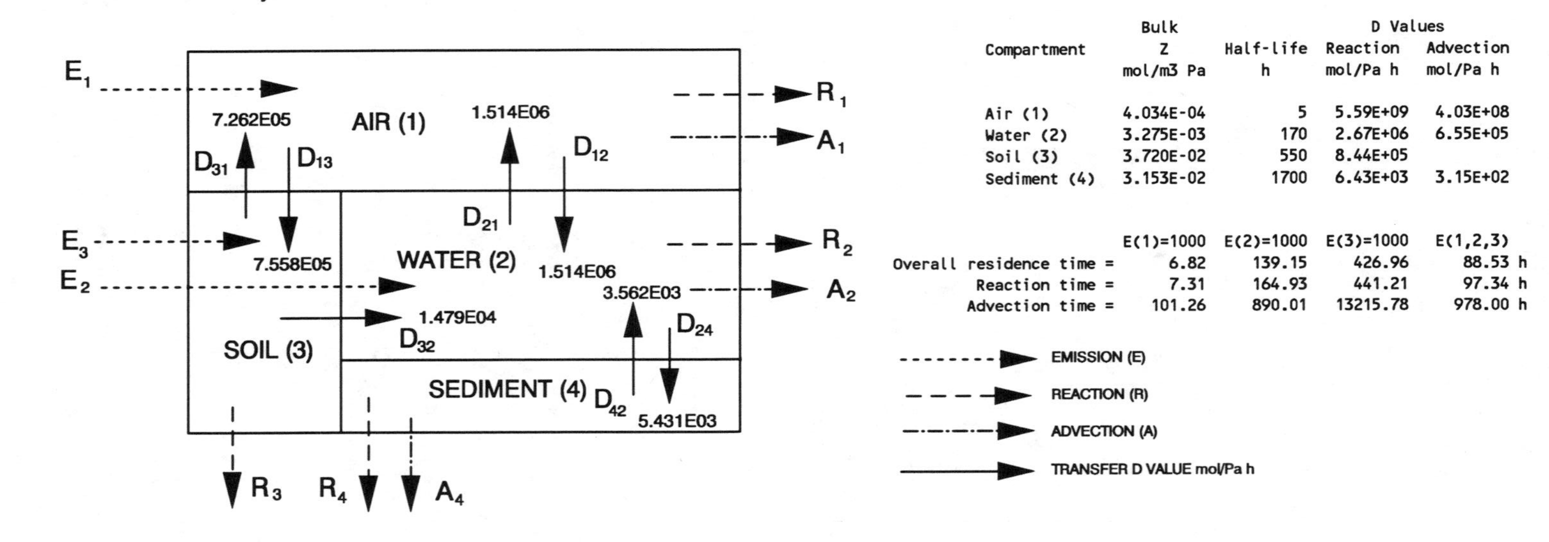


Phase Properties and Rates:

| Compartment | Bulk Z mol/m3 Pa | Half-life h | D Values Reaction mol/Pa h | D Values Advection mol/Pa h |
|---|---|---|---|---|
| Air (1) | 4.034E-04 | 5 | 5.59E+09 | 4.03E+08 |
| Water (2) | 3.275E-03 | 170 | 2.67E+06 | 6.55E+05 |
| Soil (3) | 3.720E-02 | 550 | 8.44E+05 | |
| Sediment (4) | 3.153E-02 | 1700 | 6.43E+03 | 3.15E+02 |

| | E(1)=1000 | E(2)=1000 | E(3)=1000 | E(1,2,3) |
|---|---|---|---|---|
| Overall residence time = | 6.82 | 139.15 | 426.96 | 88.53 h |
| Reaction time = | 7.31 | 164.93 | 441.21 | 97.34 h |
| Advection time = | 101.26 | 890.01 | 13215.78 | 978.00 h |

Phase Properties, Compositions, Transport and Transformation Rates:

| Emission, kg/h | | | Fugacity, Pa | | | | Concentration, g/m3 | | | | Amounts, kg | | | | Total Amount, kg |
|---|---|---|---|---|---|---|---|---|---|---|---|---|---|---|---|
| E(1) | E(2) | E(3) | f(1) | f(2) | f(3) | f(4) | C(1) | C(2) | C(3) | C(4) | m(1) | m(2) | m(3) | m(4) | |
| 1000 | 0 | 0 | 1.601E-06 | 5.039E-07 | 7.636E-07 | 2.705E-07 | 6.728E-08 | 1.719E-07 | 2.959E-06 | 8.883E-07 | 6.728E+03 | 3.438E+01 | 5.326E+01 | 4.441E-01 | 6.816E+03 |
| 0 | 1000 | 0 | 5.004E-07 | 1.983E-03 | 2.387E-07 | 1.064E-03 | 2.103E-08 | 6.764E-04 | 9.248E-07 | 3.495E-03 | 2.103E+03 | 1.353E+05 | 1.665E+01 | 1.748E+03 | 1.391E+05 |
| 0 | 0 | 1000 | 7.384E-07 | 1.874E-05 | 6.059E-03 | 1.006E-05 | 3.103E-08 | 6.393E-06 | 2.348E-02 | 3.304E-05 | 3.103E+03 | 1.279E+03 | 4.226E+05 | 1.652E+01 | 4.270E+05 |
| 600 | 300 | 100 | 1.185E-06 | 5.970E-04 | 6.064E-04 | 3.205E-04 | 4.978E-08 | 2.037E-04 | 2.350E-03 | 1.052E-03 | 4.978E+03 | 4.073E+04 | 4.229E+04 | 5.262E+02 | 8.853E+04 |

| Emission, kg/h | | | Loss, Reaction, kg/h | | | | Loss, Advection, kg/h | | | Intermedia Rate of Transport, kg/h T12 | T21 | T13 | T31 | T32 | T24 | T42 |
|---|---|---|---|---|---|---|---|---|---|---|---|---|---|---|---|---|
| E(1) | E(2) | E(3) | R(1) | R(2) | R(3) | R(4) | A(1) | A(2) | A(4) | air-water | water-air | air-soil | soil-air | soil-water | water-sed | sed-water |
| 1000 | 0 | 0 | 9.325E+02 | 1.401E-01 | 6.71E-02 | 1.811E-04 | 6.728E+01 | 3.438E-02 | 8.883E-06 | 2.530E-01 | 7.944E-02 | 1.260E-01 | 5.776E-02 | 1.177E-03 | 2.903E-04 | 1.004E-04 |
| 0 | 1000 | 0 | 2.915E+02 | 5.515E+02 | 2.10E-02 | 7.124E-01 | 2.103E+01 | 1.353E+02 | 3.495E-02 | 7.907E-02 | 3.126E+02 | 3.940E-02 | 1.805E-02 | 3.678E-04 | 1.142E+00 | 3.949E-01 |
| 0 | 0 | 1000 | 4.300E+02 | 5.212E+00 | 5.32E+02 | 6.734E-03 | 3.103E+01 | 1.279E+00 | 3.304E-04 | 1.167E-01 | 2.955E+00 | 5.813E-02 | 4.583E+02 | 9.336E+00 | 1.080E-02 | 3.733E-03 |
| 600 | 300 | 100 | 6.899E+02 | 1.660E+02 | 5.33E+01 | 2.145E-01 | 4.978E+01 | 4.073E+01 | 1.052E-02 | 1.872E-01 | 9.412E+01 | 9.326E-02 | 4.587E+01 | 9.344E-01 | 3.439E-01 | 1.189E-01 |

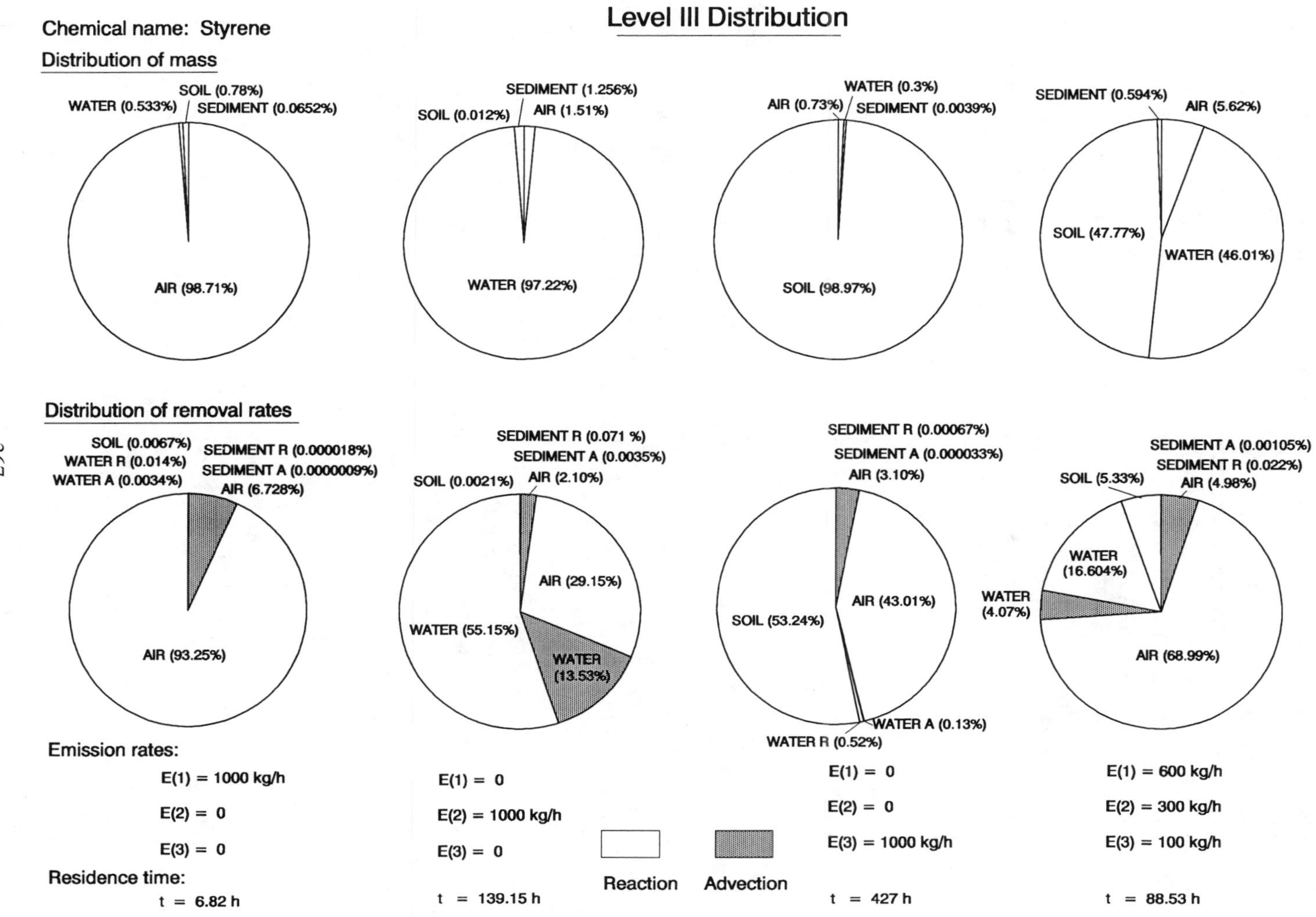
Level III Distribution
Chemical name: Styrene
Distribution of mass
WATER (0.533%)
SOIL (0.78%)
SEDIMENT (0.0652%)
AIR (98.71%)
SOIL (0.012%)
SEDIMENT (1.256%)
AIR (1.51%)
WATER (97.22%)
AIR (0.73%)
WATER (0.3%)
SEDIMENT (0.0039%)
SOIL (98.97%)
SEDIMENT (0.594%)
AIR (5.62%)
SOIL (47.77%)
WATER (46.01%)
Distribution of removal rates
SOIL (0.0067%)
WATER R (0.014%)
WATER A (0.0034%)
SEDIMENT R (0.000018%)
SEDIMENT A (0.0000009%)
AIR (6.728%)
AIR (93.25%)
SEDIMENT R (0.071 %)
SEDIMENT A (0.0035%)
SOIL (0.0021%)
AIR (2.10%)
AIR (29.15%)
WATER (55.15%)
WATER (13.53%)
SEDIMENT R (0.00067%)
SEDIMENT A (0.000033%)
AIR (3.10%)
SOIL (53.24%)
AIR (43.01%)
WATER R (0.52%)
WATER A (0.13%)
SEDIMENT A (0.00105%)
SEDIMENT R (0.022%)
SOIL (5.33%)
AIR (4.98%)
WATER (16.604%)
WATER (4.07%)
AIR (68.99%)
Emission rates:
E(1) = 1000 kg/h
E(2) = 0
E(3) = 0
Residence time:
t = 6.82 h
E(1) = 0
E(2) = 1000 kg/h
E(3) = 0
t = 139.15 h
Reaction
Advection
E(1) = 0
E(2) = 0
E(3) = 1000 kg/h
t = 427 h
E(1) = 600 kg/h
E(2) = 300 kg/h
E(3) = 100 kg/h
t = 88.53 h

## 2.4 COMMENTARY ON THE PHYSICAL-CHEMICAL PROPERTIES AND ENVIRONMENTAL FATE

**Reactivity**

The half-life classes in Table 2.3 reflect the relatively rapid reaction of hydrocarbons in the atmosphere, mainly with hydroxyl (OH) radicals. Although there are extensive compilations of rate constants for individual species [notably by Atkinson and coworkers (1984, 1985, 1987, 1990, 1991) as discussed in Chatper 1], we have assigned class 2 (half-life 17 hours) to all saturated compounds. The unsaturated compounds are more reactive by one class (half-life 5 hours), thus in all cases class 1 or 2 applies. These reaction rates are so rapid that they entirely dominate the chemicals' fate when viewed in a multimedia context. When discharged to soil or water it is often the rate of evaporation which controls the overall residence time, because once evaporated, these substances rapidly decay. The water class was selected to be 3 more than that of air (factor of 30 slower), with soil a further one class slower, and with bottom sediment 2 classes slower than water. The net result is that the half-life of saturated hydrocarbons in sediments can approach one year; under anaerobic conditions these substances probably persist indefinitely.

**QSPR Plots**

The QSPR plots show the generally "well-behaved" character of this group of chemicals, but there are significant differences between alkanes, alkenes, dienes, alkynes and the cyclic and acyclic species.

Figures 2.1a and 2.1b show that the LeBas molar volume is an adequate descriptor of molecular size, being well correlated with other molecular volumes and areas.

The solubility data for alkanes (Figure 2.2a) show the steady drop in solubility with molecular size which corresponds to a factor of 3.4 (0.53 log units) per carbon (methylene group) added. The other hydrocarbons show a very similar trend, but there are obvious differences in the absolute solubilities. The alkenes and alkynes are less hydrophobic than the alkanes since there is apparently enhanced interaction between water and the unsaturated groups. This is partly, but not totally, taken into account by the reduced molar volume; thus when viewed in terms of LeBas molar volume, the alkenes are generally a factor of about 2 more soluble than the corresponding alkanes, the alkynes a factor of about 10 more soluble, and the dienes a factor of about 5 more soluble. There is an obvious need to correlate solubility separately for each class since molar volume is not successful as a single descriptor. This type of relationship was originally by McAuliffe (1966) who plotted log solubility versus liquid molar volume for various homologous groups.

The vapor pressure data in Figures 2.3a and 2.3b show a similar trend with each methylene addition causing a drop in vapor pressure by a factor of 3.1 or 0.50 log units. Again, the unsaturation and cyclic character of the molecule influence vapor pressure, there being about a factor of 8 spread at any given molar volume.

The log $K_{OW}$ data in Figures 2.4a and 2.4b show the expected increase in hydrophobicity which is largely attributable to the drop in solubility in water, or equivalently the increase in activity coefficient in the water phase. The slope corresponds to a factor of about 4.5 or 0.65 log units per methylene added. The similarity in slope to that of water solubility suggests that these hydrocarbons have approximately equal solubilities in, or affinities for, the octanol phase and the variation in $K_{OW}$ is almost entirely due to changes in water solubility.

The Henry's law constants (H) in Figures 2.5a and 2.5b show that the alkanes have high values averaging 5.3 in log H, or 200,000 Pa·m$^3$/mol in H. This corresponds to an air-water partition coefficient of about 80, thus the alkanes have a very strong partitioning tendency into air. The unsaturated and cyclic hydrocarbons have lower values of H, mainly because of their greater solubility, but the range from 2500 to over 100,000 represents partition coefficients of 1 to 40. The lowest coefficients occur for the most unsaturated compounds. There is a tendency for H to increase with molar volume because the solubility drops more than the vapor pressure for each carbon added.

The log $K_{OW}$ - solubility plots in Figures 2.6a and 2.6b show slopes of about 0.95 confirming the almost reciprocal linear relationship between $K_{OW}$ and solubility.

No attempt has been made to examine the properties of the alkanes above *n*-dodecane, but it is expected that they will maintain the trends exhibited by the lower, more volatile alkanes. The solubility of *n*-decane is 0.052 g/m$^3$ or mg/L. If the slope of the solubility - molar volume line is maintained, a solubility of 0.004 g/m$^3$ is expected for *n*-tetradecane [with experimental measured values of 0.00694 g/m$^3$ by Franks (1966), 0.0022 g/m$^3$ by Sutton and Calder (1974), 0.0259 g/m$^3$ by Mackay et al. (1975), all considered to be uncertain by IUPAC Solubility Data Series (1989)]; and a value of 0.00003 g/m$^3$ for *n*-hexadecane [experimental reported values, 0.0063 g/m$^3$ by Franks (1966), and 0.00009 g/m$^3$ by Sutton and Calder (1974)]. These solubilities rapidly fall below levels at which experimental determination is reliable. The experimentally measured values are all greater than those predicted by correlation as demonstrated by McAuliffe (1969) and Peake and Hodgson (1966, 1967); the discrepancy has been largely attributed to emulsion formation for higher alkanes during preparation of saturated aqueous solutions. The few reported measured solubility values of higher alkanes in fresh water and seawater (Franks 1966, Peake and Hodgson 1967, McAuliffe 1969 and Sutton and Calder 1974) were too scattered to permit a reliable value to be selected. The $K_{OW}$ values also become suspect beyond *n*-decane (Coates et al. 1985, Sangster 1989).

These plots can be used, with appropriate caution, to deduce the properties of other volatile hydrocarbons. The preferred procedure is to plot the property as a function of molar volume for a group of structurally similar compounds and to interpolate. Extrapolation brings a high degree of uncertainty.

**Evaluative Calculations**

Rather than to give diagrams for all hydrocarbons, only a selection is presented. The fate of the others can be inferred from those of the closest homolog for which a diagram is given.

The Level I calculations show almost total partitioning into the atmosphere with only minor amounts in the other compartments. The chemicals with higher $K_{OW}$ partition slightly into soil.

The Level II diagrams show the dominance of atmospheric reaction and advection. Even if soil, water, or sediment rate constants were increased by a factor of 1000 (which is inconceivable), the rates of reaction in these media would remain negligible.

The Level III diagrams show that the chemical discharged into air experiences very little tendency to enter soil or water because the rates of loss by reaction and advection in air greatly exceed the rates of absorption or deposition. The overall residence times are generally about 1 day.

When discharged to water there is usually competition between evaporation, advection from water and reaction in water. The more hydrophobic chemicals also tend to partition into sediments. The overall residence times are much longer (e.g., 1 month for *n*-octane) as a result of slower reactions and advection in water and the constraint imposed by evaporation. Essentially, it is diffusion through the boundary layer in the water at the air-water interface which limits the evaporation rate.

When discharged to soil, most hydrocarbon is destined to evaporate. and again the rate of loss is limited by diffusion processes, but in this case, at the soil-air interface.

The Level III pie charts show these trends pictorially.

In general, the unsaturated hydrocarbons are more susceptible to loss by reaction. Styrene is somewhat anomalous because its aromaticity imparts a high water solubility. It is probably more appropriately assessed along with the aromatic hydrocarbons which were reviewed in Volume I of this series.

From the perspective of the environmental toxicology of this series, it is apparent that regardless of how the hydrocarbon is discharged, much of it will eventually enter the atmosphere and will participate in the complex series of chemical reactions which ultimately result in conversion to carbon dioxide and water. It is often the intermediates, such as aldehydes, and by-products, such as ozone, which are of primary concern. The aquatic and soil environments are susceptible only when the hydrocarbon is discharged directly to these media. Of particular concern are situations in which evaporation is impeded, as occurs to hydrocarbons deep in soils, or under ice in lakes. The overwhelming importance of the medium of discharge can be exemplified by *n*-octane in which the prevailing concentration in air, water and soil can be compared for the four discharge scenarios. In all cases the concentration in air is about $10^{-7}$ g/m$^3$ or 0.1 $\mu$g/m$^3$. If the discharge of the 1000 kg/h is to water, the concentration in water is $10^{-3}$ g/m$^3$, whereas it becomes $10^{-9}$ g/m$^3$, a factor of a million lower, when the discharge is to air. Similarly, the soil concentration is about $10^{-6}$ g/m$^3$ when discharge is to air or water, but it becomes $6\times10^{-3}$ g/m$^3$, a factor of 6000 higher when discharge is to soil directly. It can thus be argued that any direct ecotoxicological effects of these compounds in the soil or water ecosystems will usually be the result of direct discharges to these ecosystems, and not the result of air emissions. This does not, of course, necessarily apply to the by-products of transformation in the atmosphere - it applies only to the parent compound.

## 2.5 REFERENCES


Abernethy, S., Bobra, A.M., Shiu, W.Y., Wells. P.G., Mackay, D. (1986) Acute lethal toxicity of hydrocarbon to two planktonic crustaceans: the key role of organism-water partitioning. *Aquat. Toxicol.* 8, 163-174.

Abernethy, S., Mackay, D. (1987) A discussion of correlations for narcosis in aquatic species. In: *QSAR in Environmental Toxicology - II*, Kaiser, K.L.E., Ed., pp.1-16, D. Reidel Publ. Co., Dordrecht, Holland.

Abernethy, S., Mackay, D., McCarty, L.S. (1988) "Volume fraction" correlation for narcosis in aquatic organisms: the key role of partitioning. *Environ. Toxicol. Chem.* 7, 469-481.

Adeniji, S.A., Kerr, J.A., Williams, M.R. (1981) Rate constants for ozone-alkene reactions under atmospheric conditions. *Int'l. J. Chem. Kinet.* 13, 209-217.

Altshuller, A.P. (1991) Chemical reactions and transport of alkanes and their products in the troposphere. *J. Atmos. Chem.* 12, 19-61.

Altshuller, A.P., Bufalini, J.J. (1971) Photochemical aspects of air pollution: A review. *Environ. Sci. Technol.* 5(1), 39-64.

Al-Sahhaf, T.A. (1989) Prediction of the solubility of hydrocarbons in water using UNIFAC. *J. Environ. Sci. Health* A24, 49-56.

Amidon, G.L., Anik, S.T. (1976) Comparison of several molecular topological indices with molecular surface area in aqueous solubility estimation. *J. Pharm. Sci.* 65, 801-805.

Amidon, G.L., Anik, S.T. (1980) Hydrophobicity of polycyclic aromatic compound, thermodynamic partitioning analysis. *J. Phys. Chem.* 84, 970-974.

Amidon, G.L., Anik, S.T. (1981) Application of the surface area approach to the correlation and estimation of aqueous solubility and vapor pressure. Alkyl aromatic hydrocarbons. *J. Chem. Eng. Data* 26(1), 28-33.

Amidon, G.L., Yalkowsky, S.H., Leung, S. (1974) Solubility of nonelectrolytes in polar solvents II: Solubility of aliphatic alcohols in water. *J. Pharm. Sci.* 63, 1858-1866.

Amidon, G.L., Yalkowsky, S.H., Anik, S.T., Leung, S. (1975) Solubility of nonelectrolytes in polar solvents. V. Estimation of the solubility of aliphatic monofunctional compounds in water using a molecular surface area approach. *J. Phys. Chem.* 9, 2239-2245.

Amidon, G.L., Pearlman, R.S., Anik, S.T. (1979) The solvent contribution to the free energy of protein-ligand interactions. *J. Theor. Biol.* 77, 161-170.

Amidon, G.L., Williams, N.A. (1982) A solubility equation for non-electrolytes in water. *Intl. J. Pharm.* 11, 249-256.

Andrews L.J., Keefer, R.M. (1949) Cation complexed of compounds containing carbon-carbon double bonds. IV. The argentation of aromatic hydrocarbons. *J. Am. Chem. Soc.* 71, 3644-3647.

Andrews, L.J., Keefer, R.M. (1950) Cation complexes of compounds containing carbon-carbon double bonds. VII. Further studies on the argentation of substituted benzenes. *J. Am. Chem. Soc.* 72, 5034-5037.

API (1985) Literature survey: hydrocarbon solubilities and attenuation mechanisms. *Health and Environmental Sci. Dept. API Publication* No. 4414, Am. Petroleum Institute, Washington, D.C.

Aquan-Yuen, M., Mackay, D., Shiu, W.Y. (1979) Solubility of hexane, phenanthrene, chlorobenzene, and p-dichlorobenzene in aqueous electrolyte solutions. *J. Chem. Eng. Data* 24, 30-34.

Arbuckle, W.B. (1983) Estimating acitivity coefficients for use in calculating environmental parameters. *Environ. Sci. Technol.* 17, 537-542.

Ashworth, R.A., Howe, G.B., Mullins, M.E., Rogers, T.N. (1988) Air-water partitioning coefficients of organics in dilute aqueous solutions. *J. Hazard. Materials* 18, 25-36.

Atkinson, R. (1985) Kinetics and mechanisms of the gas phase reaction of hydroxy radicals with organic compounds under atmospheric conditions. *Chem. Rev.* 85, 69-201.

Atkinson, R. (1987) Structure-activity relationship for the estimation of the rate constants for the gas phase reactions of OH radicals with organic compounds. *Int. J. Chem. Kinetics* 19, 799-828.

Atkinson, R. (1990) Gas-phase tropospheric chemistry of organic compounds, a review. *Atmos. Environ.* 24A, 1-41.

Atkinson, R. (1991) Kinetics and mechanisms of the gas-phase reactions of the $NO_3$ radical with organic compounds. *J. Phys. Chem. Data* 20, 450-507.

Atkinson, R., Aschmann, S.M. (1984) Rate constants for the reactions of $O_3$ and OH radicals with a series of alkynes. *Int'l. J. Chem. Kinet.* 16, 259-268.

Atkinson, R., Carter, W.L. (1984) Kinetics and mechanisms of the gas-phase reactions of ozone with organic compounds under atmospheric conditions. *Chem. Rev.* 84, 437-470.

Atkinson, R., Aschmann, S.M., Carter, W.L., Pitts, Jr., J.N. (1983) Effects of ring strain on gas-phase rate constants. 1. Ozone reactions with cycloalkenes. *Int'l. J. Chem. Kinet.* 15, 721-731.

Atkinson, R., Aschmann, S.M., Winer, A.M., Pitts, Jr., J.N. (1984) Kinetics of the gas-phase reactions of $NO_3$ radicals with a series of dialkenes, cycloalkenes, and monoterpenes at 295 $\pm$ 1 K. *Environ. Sci. Technol.* 18, 370-375.

Atkinson, R., Darnall, K.R., Lloyd, A.C., Winer, A.M., Pitts, Jr., J.N. (1979) Kinetics and mechanisms of reaction of hydroxyl radicals with organic compounds in the gas phase. *Adv. Photochem.* 11, 375-488.

Atkinson, R., Pitts, J.N., Jr. (1975) Rate constants for the reaction of OH radicals with propylene and the butenes over the temperature range 297 - 425 K. *J. Chem. Phys.* 63, 3591-3595.

Atkinson, R., Pitts, J.N., Jr. (1977) Absolute rate constants for the reaction of oxygen ($^3$P) atoms with a series of olefins over temperature range 293-439 K. *J. Phys. Chem.* 67, 38-43.

Atkinson, R., Pitts, J.N., Jr. (1977) Absolute rate constants for the reaction of oxygen ($^3$P) atoms with allene, 1,3-butadiene and vinyl methyl ether over temperature range 297-439 K. *J. Phys. Chem.* 67, 2492-2495.

Atkinson, R., Pitts, J.N., Jr. (1978) Kinetics for the reaction of oxygen ($^3$P) atoms and hydroxyl radicals with 2-methyl-2-butene. *J. Phys. Chem.* 68, 2992.

Baker, E.G. (1958) American Chemical Society, Division of Petroleum Chemistry, Preprints 3, No.4, C61-68.

Baker, E.G. (1959) Origin and migration of oil. *Science* 129, 871-874.

Baker, E.G. (1960) A hypothesis concerning the accumlation of sediment hydrocarbons to form crude oil. *Geochim. Cosmochim. Acta* 19, 309-317.

Baker, E.G. (1967) A geochemical evaluation of petroleum migration and accumulation. In: *Fundamental Aspects of Petroleum Geochemistry.* Nagy, B., Colombo, V. Eds., pp. 299-330, Elsevier, New York, New York.

Banerjee, S. (1984) Solubility of organic mixture in water. *Environ. Sci. Technol.* 18, 587-591.

Banerjee S. (1985) Calculation of water solubility of organic compounds with UNIFAC-derived parameters. *Environ. Sci. Technol.* 19, 369-370.

Banerjee, S., Howard, P.H. (1988) Improved estimation of solubility and partitioning through correction of UNIFAC-derived activity coefficients. *Environ. Sci. Technol.* 22, 839-841.

Banerjee, S., Howard, P.H., Lande, S.S. (1990) General structure vapor pressure relationship for organics. *Chemosphere* 21, 1173-1180.

Banerjee, S., Yalkowsky, S.H., Valvani, S.C. (1980) Water solubility and octanol/water partition coefficient of organics. Limitations of solubility-partition coefficient correlation. *Environ. Sci. Technol.* 14, 1227-1229.

Barone, G., Crescenzi, V., Pispisa, B., Quadrifoglio, B. (1966) Hydrophobic interactions in polyelectrolytes solutions. II. Solubility of some $C_3$-$C_6$ alkanes in poly(methacrylic acid) aqueous solutions. *J. Macromol. Chem.* 1, 761-771.

Becke, A., Quitzsch, G. (1977) Das phasengleichgewichtsverhalten ternärer systeme der art $C_4$-alkohol-wasser-kohlenwasserstoff. *Chem. Techn.* 29, 49-51.

Bennet, P.J., Harris, S.J., Kerr, J. A. (1987) A reinvestigation of the rate constants for the reactions of ozone with cyclopentene and cyclohexene under atmospheric conditions. *Int. J. Chem. Kinet.* 19, 609-614.

Benter, Th., Schindler, R.N. (1988) Absolute rate coefficients for the reaction of $NO_3$ radicals with simple dienes, *Chem. Phys. Lett.* 145, 67-70.

Berti, P., Cabani, S. Conti, G., Mollica, V. (1986) The thermodynamic study of organic compounds in octan-1-ol. *J. Chem. Soc., Faraday, Trans. 1,* 82, 2547.

Bissell, T.G., Williamson, A.G. (1975) Vapour pressures and excess Gibbs energies of n-hexane and of n-heptane + carbon tetrachloride and + chloroform at 298.15 K. *J. Chem. Thermodyn.* 7, 131-136.

Bittrich, H.J., Gedan, H., Feix, G. (1979) Zur löslichkeitsbeeinflussung von kohlenwasserstoffen in wasser. *Z. Phys. Chem.* (Leibzig) 260, 1009-1013.

Black, C., Joris, G.G., Taylor, H.S. (1948) The solubility of water in hydrocarbons. *J. Chem. Phys,* 16, 537-548.

Bobra, A.M., Shiu, W.Y., Mackay, D. (1979) Distribution of hydrocarbons among oil, water and vapor phases during oil dispersant toxicity tests. *Bull. Environ. Contam. Toxicol.* 4, 297-305.

Bobra, A.M., Shiu, W.Y., Mackay, D. (1984) Structure-activity relationships for toxicity of hydrocarbons, chlorinated hydrocarbons and oils to *daphnia magna.* In: *QSAR in Environmental Toxicology.* Kaiser, K.L.E., Editor, D. Reidel Publishing Company, Dordrecht, Holland. pp.3-16.

Bodor, N., Gabanyi, Z., Wong, C.-K. (1989) A new method for the estimation of partition coefficient. *J. Am. Chem. Soc.* 111, 3783-3786.

Booth, H.S., Everson, H.E. (1948) Hydrotropic solubilities: solubilities in 40 per cent sodium xylenesulfonate. *Ind. Eng. Chem.* 40(8), 1491-1493.

Boublik, T., Fried, V., Hala, E. (1973) *The Vapour Pressure of Pure Substances.* Elsevier, Amsterdam.

Boublik, T., Fried, V., Hala, E. (1984) *The Vapour Pressures of Pure Substances.* (second revised edition), Elsevier, Amsterdam.

Brady, A.P., Huff, H. (1958) Vapor benzene over aqueous detergent solutions. *J. Phys. Chem.* 62, 644-649.

Brice, K.A., Derwent, R.G. (1978) Emissions inventory for hydrocarbons in United Kingdom. *Atom. Environ.* 12, 2045-2054.

Bridie, A.L., Wolff, C.J.M., Winter, M. (1979) BOD and COD of some petrochemicals. *Water Res.* 13, 627-630.

Bright, N.F.H. (1951) The vapor pressure of diphenyl, dibenzyl and diphenylmethane. *J. Chem. Soc.* Part I, 624-625.

Brookman, G.T., Flanagan, M., Kebe, J.O. (1985) Literature Survey: Hydrocarbon solubilities and attenuation mechanisms, prepared for Environmental Affairs Dept. of American Petroleum Institute. *API Publication* No. 4414, August, 1985, Washington D.C.

Brown, R.L., Wasik, S.P. (1974) A method of measuring the solubilities of hydrocarbons in aqueous solutions. *J. Res. Nat. Bur. Std.* 78A, 453-460.

Budavari, S., Ed. (1989) The Merck Index. *An Encyclopedia of Chemicals, Drugs and Biologicals.* 11th edition, Merck & Co. Inc., Rahway, New Jersey.

Buffalini, J.J., Altshuller, A.P. (1965) Kinetics of vapor-phase hydrocarbon-ozone reactions. *Can. J. Chem.* 43, 2243-2250.

Burkhard, L.P., Kuehl, D.W., Veith, G.D. (1985) Evaluation of reverse phase liquid chromatograph/mass spectrometry for estimation of *n*-octanol'water partition coefficients. *Chemosphere* 14(10), 1551-1560.

Burris, D.R., MacIntyre, W.G. (1986) A thermodynamic study of solutions of liquid hydrocarbon mixtures in water. *Geochim. Cosmochim. Acta* 50, 1545-1549.

Cadle, R.D., Schadt, C. (1952) Kinetics of the gas-phase reaction of olefins with ozone. *J. Am. Chem. Soc.* 74, 6002-6004.

Campbell, A.N., Kartzmark, E.M., Anand. S.C., Cheng, Y., Dzikowski, H.P., Skrynyk, S.M. (1968) Partially miscible liquid systems:the density, change of volume on mixing, vapor pressure, surface tension, and viscosity in the system: aniline-hexane. *Can. J. Chem.* 46, 2399.

Canton, J.H., Wegman, R.C.C. (1983) Studies on the toxicity of tribromomethane, cyclohexene, and bromocyclohexane to different fresh water organisms. *Water Res.* 17, 743-747.

Carruth, G.F., Kobayashi, R. (1973) Vapor pressure of normal paraffins ethane through *n*-decane from their triple points to about 10 mm mercury. *J. Chem. Eng. Data* 18(2), 115-126.
Castello, G., D'Amato, G. (1976) Influence of vapour pressure, activity coefficient and structure on the retention volumes of branched-chain nonanes. *J. Chromatogr.* 116, 249-255.
Chou, J.T., Jurs, P.C. (1979) Computation of partition coefficients from molecular structures by a fragment addition method. In: *Physical Chemical Properties of Drugs. Medical Research Series, Volume 10.* Yalkowsky, S.H., Sinkula, A.A., Valvani, S.C., Editors, Marcel Dekker, Inc., New York. pp.163-199.
Claussen, W.F., Polglase, M.F. (1952) Solubilities and structure in aqueous aliphatic hydrocarbons solution. *J. Am. Chem. Soc.* 74, 4817-4819.
Coates, M., Connell, D.W., Barron, D.M. (1985) Aqueous solubility and octan-1-ol to water partition coefficients of aliphatic hydrocarbons. *Environ. Sci. Technol.* 19, 628-632.
Connolly, J.F. (1966) Solubility of hydrocarbons in water near the critical solution temperatures. *J. Chem. Eng. Data* 11, 13-16.
Coutant, R.W., Keigley, G.W. (1988) An alternative method for gas chromatographic determination of volatile organic compounds in water. *Anal. Chem.* 60, 2436-2537.
Cox, R.A., Penkett, S.A. (1972) Aerosol formation from sulfur dioxide in the presence of ozone and olefinic hydrocarbons. *J. Chem. Soc. Faraday Trans.* 1, 68, 1735-1753.
Cramer III, R.D. (1977) "Hydrophobic interaction" and solvation energies: Discrepancies between theory and experimental data. *J. Am. Chem. Soc.* 99, 5408-5412.
Cruickshank, A.J.B., Cutler, A.J.B. (1966) Vapor pressure of cyclohexane, 25-75°C. *J. Chem. Eng. Data* 12, 326-329.
D'Amboise, M., Hanai, T. (1982) Hydrophobicity and retention in reverse phase liquid chromatography. *J. Liq. Chromatogr.* 5, 229-244.
Darnall, K.R., Atkinson, R., Pitts, J.N., Jr. (1978) Rate constants for the reaction of the OH radical with selected alkanes at 300 K. *J. Phys. Chem.* 82, 1581-1584.
Darnall, K.R., Carter, W.P.L., Winer, A.M., Lloyd, A.C., Pitts, J.N., Jr. (1976) Importance of $RO_2$ + NO in allyl nitrate formation from $C_4$-$C_6$ alkane photooxidations under simulated atmospheric conditions. *J. Phys. Chem.* 80, 1948-1950.
Darnall, K.R., Lloyd, A.C., Winer, A.M., Pitts, J.N. (1976) Reactivity scale for atmospheric hydrocarbons based on reaction with hydroxyl radicals. *Environ. Sci. Technol.* 10, 692-696.
Daubert, T.E., Danner, R.P. (1985) *Data Compilation Tables of Properties of Pure Compounds.* Am. Institute of Chem. Engineers. pp 450.
Dean, J.D., Ed. (1979) *Lange's Handbook of Chemistry.* 12th ed. McGraw-Hill, Inc., New York.
Dean, J.D., Ed. (1985) *Lange's Handbook of Chemistry.* 13th ed. McGraw-Hill, Inc., New York.
DeMore, W.B. (1971) Rates and mechanism of alkyne ozonation. *Int'l. J. Chem. Kinet.* 3, 161-173.
Deno, N.C., Berkheimer, H.E. (1960) Phase equilibria molecular transport thermodynamics: activity coefficients as a function of structure and media. *J. Chem. Eng. Data* 5, 1-5.

DeVoe, H., Miller, M.M., Wasik, S.P. (1981) Generator columns and high pressure liquid chromatography for determining aqueous solubilities and octanol-water partition coefficients of hydrophobic substances. *J. Res. Natl. Bur. Std.* 86, 361-.

Dillemuth, F.J., Schubert, C.C., Skidmore, D.R. (1963) The reaction of $O_3$ with acetylenic hydrocarbons. *Combust. Flame* 6(3), 211-212.

Dilling, W.L., Gonsior, S.J., Boggs, G.U., Mendoza, C.G. (1988) Organic photochemistry. 20. A method for estimating gas-phase rate for reactions of hydroxyl radicals with organic compounds from their relative rates of reaction with hydrogen peroxide under photolysis in 1,1,2-trichlorotrifluoroethane solution. *Environ. Sci. Technol.* 22, 1447-1453.

Dorfman, L.M., Adams, G.E. (1973) Reactivity of the hydroxy radical in aqueous solution. NSRD-NDB-46. NTIS COM-73-50623. National Bureau Standards, Washington D.C. pp. 51.

Doyle, G.J., Lloyd, A.C., Darnall, K.R., Winer, A.M., Pitts, J.N. Jr. (1975) Gas phase kinetic study of relative rates of reaction of selected aromatic compounds with hydroxy radicals in environmental chamber. *Environ. Sci. Technol.* 9(3), 237-241.

Dreisbach, R.R. (1955) *Physical Properties of Chemical Compounds. Adv. Chem. Ser.* 15, American Chemical Society, Washington D.C.

Dreisbach, R.R. (1959) *Physical Properties of Chemical Compounds II. Adv. Chem. Ser.* 22, American Chemical Society, Washington D.C.

Durand, R. (1948) Investigations on hydrotropy. The solubility of benzene, hexane and cyclohexane in aqueous solutions of fatty acid salts. *Compt. Rend.* 226, 409-410.

Eadsforth, C.V. (1986) Application of reverse phase HPLC for the determination of partition coefficients. *Pest. Sci.* 17, 311-325.

Eadsforth, C.V., Moser, P. (1983) Assessments of reversed phase chromatographic methods for determining partition coefficients. *Chemosphere* 12, 1459-1475.

Eastcott, L., Shiu, W.Y., Mackay, D. (1988) Environmentally relevant physical-chemical properties of hydrocarbons: A review of data and development of simple correlations. *Oil & Chem. Pollut.* 4, 191-216.

Edney, E.O., Kleindienst, T.E., Corse, E.W. (1986) Room temperature rate constants for the reaction of OH with selected chlorinated and oxygenated hydrocarbons. *Int'l J. Chem. Kinet.* 18, 1355-1371.

Farkas, E.J. (1964) *Anal. Chem.* 37, 1173-1175.

Forziati, A.F., Norris, W.R., Rossini, F.D. (1949) Vapor pressures and boiling points of sixty API-NBS hydrocarbons. *J. Res. Natl. Bur. Std.* 43, 555-563.

Forziati, A.F., Camin, D.L., Rossini, F.D. (1950) Density, refractive index, boiling point, and vapor pressure of eight monoolefin (1-alkene), six pentadiene, and two cyclomonoolefin hydrocarbons. *J. Res. Natl. Bur. Std.* 45, 406-410.

Franks, F. (1966) Solute-water interactions and the solubility behaviour of long-chain paraffin hydrocarbons. *Nature* 210, 87-88.

Frilette, V.J., Hohenstein, W.P. (1948) Polymerization of styrene in soap solutions. *J. Polymer Sci.* 3, 22-31.

Fu, M.H., Alexander, M. (1992) Biodegradation of styrene in samples of natural environments. *Environ. Sci. Technol.* 26, 1540-1546.

Führer, H. (1924) Die wasserlöslichkeit in homologen reihen. *Chem. Ber.* 57, 510-515.

Fujita, T., Iwasa, Hansch, C. (1964). A new substituent constant, "pi" derived from partition coefficients. *J. Am. Chem. Soc.* 86, 5175-5180.

Fujisawa, S., Masuhara, E. (1981) Determination of partition coefficients of acrylates, methacrylates and vinyl monomers using high performance liquid chromatograph (HPLC). *J. Biomed. Mater. Res.* 15, 787-793.

Furuyama, S., Atkinson, R., Colussi, A.J., Cvetanovic, R.J. (1974) Determination by the phase shift method of the absolute rate constants of reactions of oxygen ($^3P$) atoms with olefins at 25°C. *Int'l. J. Chem. Kinet.* 6, 741.

Gaffney, J.S., Levine, S.Z. (1979) Predicting gas phase organic molecule reaction rates using linear free-energy correlations. I. O($^3P$) and OH addition and abstraction reactions. *Int'l J. Chem. Kinet.* 11, 1197-1209.

Gavezzotti, A. (1983) The calculation of molecular volumes and the use volume analysis in the investigation of structured media and solid-state organic reactivity. *J. Am. Chem. Soc.* 105, 5220-5225.

Glew, D.N., Roberson, R.E. (1956) The spectrophotometric determination of the solubility of cumene in water by a kinetic method. *J. Phys. Chem.* 60, 332-337.

Greiner, N.R. (1967) Hydroxyl-radical kinetics by kinetic spectroscopy. II. Reactions with $C_2H_6$, $C_3H_8$, and *iso*-$C_4H_{10}$ at 300 K. *J. Chem. Phys.* 46, 3389-3392.

Greiner, (1970) Hydroxyl radical kinetics by kinetic spectroscopy. VI. Reactions with alkanes in the range 300-500 K. *J. Chem. Phys.* 53, 1070-1076.

Gross, P.M., Saylor, J.H. (1931) The solubilities of certain slightly soluble organic compounds in water. *J. Am. Chem. Soc.* 53, 1744-1751.

Groves, Jr, F.R. (1988) Solubility of cycloparaffins in distilled water and salt water. *J. Chem. Eng. Data* 33, 136-138.

Güesten, H., Filby, W.G., Schoop, S. (1981) Prediction of hydroxy radical reaction rates with organic compounds in the gas phase. *Atmos. Environ.* 15, 1763-1765.

Haag, W.R., Yao, C.C.D. (1992) Rate constants for reaction of hydroxyl radicals with several drinking water contaminants. *Environ. Sci. Technol.* 26, 1005-1013.

Hafkenscheid, T.L., Tomlinson, E. (1983) Correlations between alkane/water and octan-1-ol/water distribution coefficients and isocratic reversed-phase liquid chromatographic capacity factors of acids, bases and neutrals. *Int'l. J. Pharmaceu.* 16, 225-239.

Hall, L.H., Kier, L.B., Murray, W.J. (1975) Molecular connectivity II: Relationship to water solubility and boiling point. *J. Pharm. Sci.* 64, 1974-1977.

Hammers, W.E., Meurs, G.J., De Ligny, C.L. (1982) Correlations between liquid chromatographic capacity ratio data on Lichrosorb RP-18 and partition coefficients in the octanol-water system. *J. Chromatogr.* 247, 1-13.

Hansch C., Anderson, S. (1967) The effect of intramolecular hydrophobic bonding on partition coefficients. *J. Org. Chem.* 32, 2583.

Hansch, C., Leo, A. (1979) *Substituent Constants for Correlation Analysis in Chemistry and Biology*. Wiley, New York.

Hansch, C. & Leo, A. (1985) Medichem Project. Pomona College, Claremont, California.

Hansch, C., Leo, A., Nickaitani, D. (1972) On the additive - constitutive character of partition coefficients. *J. Org. Chem.* 37, 3090-3092.

Hansch, C., Quinlan, J.E., Lawrance, G.L. (1968) The linear free-energy relationship between partition coefficients and the aqueous solubility of organic liquids. *J. Am. Chem. Soc.* 33, 345-350.

Hansen, D.A., Atkinson, R., Pitts, J.N. Jr. (1975) Rate constants for the reaction of OH radicals with a series of aromatic hydrocarbons. *J. Phys. Chem.* 79(17), 1763-1766.

Hanst, P.L., Stephens, E.R., Scott, W.E., Doerr, R.C. (1958) *Atmospheric Ozone-Olefin Reactions*. Franklin Institute, Philadelphia, Pa.

Harris, K.R., Dunlop, P.J. (1970) Vapor pressures and excess Gibbs energies of mixtures of benzene with chlorobenzene, *n*-hexane and *n*-heptane at 25°C. *J. Chem. Thermodyn.* 2, 805.

Hayashi, M., Sasaki, T. (1956) Measurements of solubilities of sparingly soluble liquids in water and aqueous detergent solutions using nonionic surfactant. *Bull. Chem. Soc, Jpn.* 29, 857.

Hendry, D.G., Mill, T., Piszkiewicz, L., Howard, J.A., Eigenman, H.K. (1974) A critical review of H-atom transfer in the liquid phase: chlorine atom, alkyl, trichloromethyl, alkoxy and alkylperoxy radicals. *J. Phys. Chem. Ref. Data* 3, 944-978.

Hermann, R.B. (1972) Theory of hydrophobic bonding. II. The correlation of hydrocarbon solubility in water with solvent cavity surface area. *J. Phys. Chem.* 76, 2754-2758.

Herron, J.T., Huie, R.E. (1973) Rate constants for the reactions of atomic oxygen ($^3P$) with organic compounds in the gas phase. *J. Phys. Chem. Ref. Data* 2, 467.

Herron, J.T., Huie, R.E. (1974) Rate constants for the reactions of ozone with ethene and propene from 235-362 K. *J. Phys. Chem.*78, 2085.

Hine, J., Mookerjee, P.K. (1975) The intrinsic hydrophilic character of organic compounds. Correlations in terms of structural contributions. *J. Org. Chem.* 40(3), 292-298.

Hoigné, J., Bader, H. (1983) Rate constants of reactions of ozone with organic and inorganic compounds in water-1. *Water Res. 17, 173-183.*

Hollifield, H.C. (1979) Rapid nephelometric estimate of water solubility of highly insoluble organic chemicals of environmental interest. *Bull. Environ. Contam. Toxicol.* 23, 579-586.

Horvath, A.L. (1982) *Halogenated Hydrocarbons, Solubility-Miscibility with Water.* Marcel Dekker, Inc., New York, N.Y.

Howard, P.H., Ed. (1989) *Handbook of Fate and Exposure Data for Organic Chemicals.* Vol.I - Large Production and Priority Pollutants. Lewis Publishers, Chelsea, Michigan.

Howard, P.H., Ed., (1990) *Handbook of Fate and Exposure Data for Organic Chemicals.* Vol. II - Solvents. Lewis Publ., Inc., Chelsea, Michigan.

Howard, P.H., Boethling, R.S., Jarvis, W.F., Meylan, W.M., Michalenko, E.M. (1991) *Handbook of Environmental Degradation Rates*. Lewis Publishers, Chelsea, MI.

Hoy, K.L. (1970) New values of the solubility parameters from vapor pressure data. *J. Paint Technol.* 42, 76-118.

Huie, R.E., Herron, J.T. (1975) Temperature dependence of the rate constants for reaction of ozone with some olefins. *Int'l. J. Chem. Kinet.* S1, 165.

Hussam, A., Carr, P.W. (1985) Rapid and precise method for the measurement of vapor/liquid equilibria by headspace gas chromatography. *Anal. Chem.* 57, 793-801.

Hustert, K., Mansour, M., Korte, F. (1981) The EPA Test-a method to determine the photochemical degradation of organic compounds in aqueous systems. *Chemosphere* 10, 995-998.

Hutchinson, T.C., Hellebust, J.A., Tam, D., Mackay, D., Mascarenhas, R.A., Shiu, W.Y. (1980) The correlation of the toxicity to algae of hydrocarbons and halogenated hydrocarbons with their physical-chemical properties. In: *Hydrocarbons and Halogenated Hydrocarbons in the Aquatic Envrionment*. Afghan, B.K., Mackay, D., Eds., pp. 577-586. Plenum Press, New York.

Irmann, F. (1965) Eine einfache korrelation zwischen wasserlöslichkeit und struktur von kohlenwasserstoffen un halogenkohlenwasserstoffen. *Chem. Ing.-Techn.* 37, 789-798.

Isnard, P., Lambert, S. (1989) Aqueous solubility/n-octanol water partition coefficient correlations. *Chemosphere* 18, 1837-1853.

IUPAC Solubility Data Series (1989a) Vol. 37: *Hydrocarbons ($C_5$-$C_7$) with Water and Seawater.* Shaw, D.G., Ed., Pergamon Press, Oxford, England.

IUPAC Solubility Data Series (1989b) Vol. 38: *Hydrocarbons ($C_8$-$C_{36}$) with Water and Seawater.* Shaw, D.G., Ed., Pergamon Press, Oxford, England.

Iwasa, J., Fujita, T., Hansch, C. (1965) Substituent constants for aliphatic functions obtained from partition coefficients. *J. Med. Chem.* 8, 150-153.

Iwase, K., Komatsu, K., Hirono, S., Nakagawa, S., Moriguchi, I. (1985) Estimation of hydrophobicity based on the solvent-accessible surface area of molecules. *Chem. Pharm. Bull.* 33, 2114-2121.

Japar, S.M., Wu,. C.H., Niki, H. (1974) Rate constants for the reaction of ozone with olefins in the gas phase. *J. Phys. Chem.* 23, 2318-2320.

Jönsson, J.A., Vejrosta, J., Novak, J. (1982) Air/water partition coefficients for normal alkanes (n-pentane to n-nonane) *Fluid Phase Equil.* 9, 279-286.

Kabadi, V.N., Danner, R.P. (1979) Nomograph solves for solubilities of hydrocarbons in water. *Hydrocarbon Processing* 58, 245-246.

Kamlet, M.J., Abraham, D.J., Doherty, R.M., Taft, R.W., Abraham, M.H. (1987) Solubility properties in polymers and biological media: 6. An equation for correlation and prediction of solubilities of liquid organic nonelectrolytes in blood. *J. Pharm. Sci.* 75, 350-355.

Kamlet, M.J., Doherty, R.M., Abboud, J-L M., Abrahm, M.H. (1986) Linear solvation energy relationships: 36. Molecular properties governing solubilities of organic nonelectrolytes in water. *J. Pharm. Sci.*75(4), 338-349.

Kamlet, M.J., Doherty, R.M., Abrahm, M.H., Carr, P.W., Doherty, R.F., Taft, R.W. (1987) Linear solvation energy relationships. 41. Important differences between aqueous solubility relationships for aliphatic and aromatic solutes. *J. Phys. Chem.* 91, 1996-2004.

Kamlet, M.J., Doherty, R.M., Abraham, M.H., Marcus, Y., Tagft, R.W. (1988) Linear solvation energy relationships. 46. An improved equation for correlation and prediction of octanol/water partition coefficients of organic nonelectrolytes (including strong hydrogen bond donor solutes). *J. Phys. Chem.* 92, 5244-5255.

Kier, L.B., Hall, L.H. (1976) Molar properties and molecular connectivity. In: *Molecular Connectivity in Chemistry and Drug Design.* Medicinal Chem. Vol.14, pp. 123-167. Academic Press, New York.

Kier, L.B., Hall, L.H. (1986) *Molecular Connectivity in Structure-Activity Analysis.* Wiley, New York.

Kier, L.B., Hall, L.H., Murray, W.J., Randic, M. (1975) Molecular connectivity I: Relationship to nonspecific local anesthesia. *J. Pharm. Sci.* 64, 1971-1974.

Könemann, W.H. (1979) Quantitative Structure Activity Relationship for Kinetics and Toxicity of Aquatic Pollutants and their Mixtures in Fish. University Ultrecht, Netherlands.

Könemann, H. (1981) Quantitative structure-activity relationships in fish toxicity studies. Part 1: Relationship for 50 industrial pollutants. *Toxicology* 19, 209-221.

Kopcynski, S.L., Lonneman. W.A., Sutterfield, F.D., Darley, P.E. (1972) Photochemistry of atmospheric samples in Los Angeles. *Environ. Sci. Technol.* 6, 342.

Korn, S. et al. (1977) The uptake, distribution and depuration of carbon-14 labelled benzene and carbon-14 labelled toluene in pacific herring. *Fish Bull. Natl. Marine Fish Ser.* 75, 633-636.

Krasnoshchekova, R.Ya., Gubertritis, M.Ya. (1973) Solubility of paraffin hydrocarbons in fresh and saltwater. *Neftekhimiya* 13, 885-887.

Krasnoshchekova, R.Ya., Gubertritis, M.Ya. (1975) Solubility of alkylbenzenes in fresh and salt waters. *Vodnye. Resursy.* 2, 170-173.

Krzyzanowska, T., Szeliga, J. (1978) A method for determining the solubility of individual hydrocarbons. *Nafta (Katowice)* 28, 414-417.

Lande, S.S., Banerjee, S. (1981) Predicting aqueous solubility of organic nonelectrolytes from molar volume. *Chemosphere* 10, 751-759.

Lande, S.S., Hagen, D.F., Seaver, A.E. (1985) Computation of total molecular surface area from gas phase ion mobility data and its correlation with aqueous solubilities of hydrocarbons. *Environ. Toxicol. Chem.* 4, 325-334.

Lane, W.H. (1946) Determination of the solubility of styrene in water and of water in styrene. *Ind. Eng. Chem. Anal. Ed.* 18, 295-296.

Leahy, D.E. (1986) Intrinsic molecular volume as a measure of the cavity term in linear solvation energy relationships: octanol-water partition coefficients and aqueous solubilities. *J. Pharm. Sci.* 75, 629-636.

Leighton, D.T., Calo, J.M. (1981) Distribution coefficients of chlorinated hydrocarbons in dilute air-water systems for groundwater contamination applications. *J. Chem. Eng. Data* 26, 381-385.

Leinonen, P.J., Mackay, D. (1973) The multicomponent solubility of hydrocarbons in water. *Can. J. Chem. Eng.* 51, 230-233.

Leo, A. (1985) Medchem. Proj. Issue No.26, Pomona College, Claremont, CA.

Leo, A., Hansch, C., Elkins, D. (1971) Partition coefficients and their uses. *Chemical Reviews* 71, 525-616.

Leo, A., Jow, P.Y.C., Silipo, C., Hansch, C. (1975) Calculation of hydrophobic constant (Log *P*) from $\pi$ and *f* constants. *J. Med. Chem.* 18(9), 865-868.

Lloyd, A.C., Darnall, K.R., Winer, A.M., Pitts, Jr., J.N. (1976) Relative rate constants for reaction of the hydroxyl radical with a series of alkanes, alkenes, and aromatic hydrocarbons. *J. Phys. Chem.* 80, 189-794.

LOGP and Related Computerized Data Base, Pomona College Med-Chem. Project, Pomona College, Claremont, CA. Technical Data Base (TDS) Inc.

Lu, P.Y., Metcalf, R.L., Calson, E.M. (1978) Environmental fate of five radiolabelled coal conversion by-products evaluated in a laboratory model ecosystems. *Environ. Health Pespectives* 24, 201.

Lyman, W.J. (1982) Adsorption coefficients for soil and sediments. Chapter 4, In: *Handbook of Chemical Property Estimation Methods*, W.J. Lyman, W.F. Reehl, D.H. Rosenblatt, Eds., McGraw-Hill, New York.

Lyman, W.J., Reehl, W.F., Rosenblatt, D.H. (1982) *Handbook of Chemical Property Estimation Methods*, McGraw-Hill, New York.

Ma, K.C., Shiu, W.Y., Mackay, D. (1990) *A Critically Reviewed Compilation of Physical and Chemical and Persistence Data for 110 Selected EMPPL Substances.* A report prepared for the Ontario Ministry of Environment, Water Resources Branch, Toronto, Canada.

Mabey, W., Smith, , J.H., Podoll, R.T., Johnson, H.L., Mill, T., Chou, T.W., Gates, J., Waight-Partridge, I., Vanderberg, D. (1982) Aquatic Fate Process for Organic Priority Pollutants. EPA Report No. 440/4-81-014.

Mackay, D., Shiu, W.Y., Wolkoff, A.W. (1975) Gas chromatograph determination of low concentration hydrocarbons in water by vapor phase extraction. *ASTM Special Tech. Publication* No. 573, pp. 251-258, Philadelphia.

Mackay, D. (1981) Environmental and laboratory rates of volatilization of toxic chemicals from water. In: *Hazardous Assessment of Chemicals, Current Development*. Volume 1, Academic Press.

Mackay, D. (1982) Correlation of bioconcentration factors. *Environ. Sci. Technol.* 16, 274-278.

Mackay, D. (1988) The chemistry and modelling of soil contamination with petroleum. In: *Soils Contaminated by Petroleum: Environmental and Public Health Effects*. Calabrese, E.J., Kostecki, P.T., Editors, John Wiley & Sons, New York.

Mackay, D., Bobra, A.M., Chan, D.W., Shiu, W.Y. (1982) Vapor pressure correlation for low-volatility environmental chemicals. *Environ. Sci. Technol.* 16, 645-649.

Mackay, D., Bobra, A., Shiu, W.Y., Yalkowsky, S.H. (1980) Relationships between aqueous solubility and octanol-water partition coefficient. *Chemosphere* 9, 701-711.

Mackay, D., Leinonen, P.J. (1975) Rate of evaporation of low-solubility contaminants from water bodies to atmosphere. *Environ. Sci. Technol.* 7, 1178-1180.

Mackay, D. Paterson, S., Chung, B., Neely W.B. (1985) Evaluation of the environmental behavior of chemicals with a level III fugacity model. *Chemosphere* 14(3/4), 335-374.

Mackay, D., Shiu, W.Y. (1975) The aqueous solubility and air-water exchange characteristics of hydrocarbons under environmental conditions. In: *Chemistry and Physics of Aqueous Gas Solutions*. Adams, W.A., Greer, G., Desnoyers, J.E., Atkinson, G., Kell, K.B., Oldham, K.B., Walkey, J., Eds., pp. 93-110, Electrochem. Soc. , Inc., Princeton, N.J.

Mackay, D., Shiu, W.Y. (1981) A critical review of Henry's law constants for chemicals of environmental interest. *J. Phys. Chem. Ref. Data* 10, 1175-1199.

Mackay, D., Shiu, W.Y. (1990) Physical-chemical properties and fate of volatile organic compounds: an application of the fugacity approach. In: *Significance and Treatment of Volatile Organic Compounds in Water Supplies*. Ram, N.M., Christman, R.F., Cantor, K.P., Eds., pp.183-203, Lewis Publishers, Inc., Chelsea, Michigan.

Mackay, D., Shiu, W.Y. (1991) Chapter 9, Estimating the multimedia partitioning of hydrocarbons: The effective solubility approach. In: *Hydrocarbon Contaminated Soils and Groundwater*. Volume 2, Calabrese, E.J., Kostecki, P.T., Editors, Lewis Publishers, Inc., Chelsea, Michigan.

Mackay, D., Shiu, W.Y., Sutherland, R.P. (1979) Determination of air-water Henry's law constants for hydrophobic pollutants. *Environ. Sci. Technol.* 13, 333-337.

Mackay, D., Shiu, W.Y., Wolkoff, A.W. (1975) Gas chromatographic determination of low concentrations of hydrocarbons in water by vapor phase extraction. *ASTM STP 573*, pp. 251-258, Am. Soc. Testing and Materials, Philadelphia.

Mackay, D., Wolkoff, A.W. (1973) Rate of evaporation of low-solubility contaminants from water bodies to atmosphere. *Environ. Sci. Technol.* 7, 611-614.

Mailhot, H., Peters, R.H. (1988) Empirical relationships between the 1-octanol/water partition coefficient and nine physicochemical properties. *Environ. Sci. Technol.* 22, 1479-1488.

Marion, C.V., Malaney, G.W. (1964) Ability of activated sludge microorganisms to oxidize aromatic organic compounds. In: *Proc. Ind. Waste Conf.*, Eng. Bull., Purdue Univ., Eng. Ext. Ser., pp. 297-308.

Massaldi, H.A., King, C.J. (1973) Simple technique to determine solubilities of sparingly soluble organics: solubility and activity coefficients of d-limonene, butylbenzene, and n-hexyl acetate in water and sucrose solutions. *J. Chem. Eng. Data* 18, 393-397.

Matter-Müller, C., Gujer, W., Giger, W. (1981) Transfer of volatile substances from water to the atmosphere. *Water Res.* 15, 1271-1279.

McAuliffe, C. (1963) Solubility in water of $C_1$ - $C_9$ hydrocarbons. *Nature* (London) 200, 1092-1093.

McAuliffe, C. (1966) Solubility in water of paraffin, cycloparaffin, olefin, acetylene, cycloolefin and aromatic hydrocarbons. *J. Phys. Chem.* 76, 1267-1275.

McAuliffe, C. (1969) Solubility in water of normal $C_9$ and $C_{10}$ alkane hydrocarbons. *Science*, 163, 478-479.

McBain, J.W., Lissant, K.J. (1951) The solubilization of four typical hydrocarbons in aqueous solution by three typical detergents. *J. Phys. Colloid Chem.* 55, 655-662.

The Merck Index. *An Encyclopedia of Chemicals, Drugs and Biologicals* (1983). Windholz, M. (Ed.), Merck and Co., Inc., Rahway, N.J., U.S.A., 10th ed.

The Merck Index. *An Encyclopedia of Chemicals, Drugs and Biologicals.* (1989). Budavari, S., Editor, Merck & Co., Inc., Rahway, N.J., 11th ed.

Meylan, W.M., Howard, P.H., Boethling, R.S. (1992) Molecular topograph/fragment contribution method for predicting soil sorption coefficients. *Environ. Sci. Technol.* 26, 1560-1567.

Mill, T. (1982) Hydrolysis and oxidation processes in the environment. *Environ. Toxicol. Chem.* 1, 135-141.

Mill, T., Hendry, D.G., Richardson, H. (1980) Free radical oxidants in natural waters. *Science* 207, 886-887.

Mill, T., Mabey, W. (1985) Photochemical transformations. In: *Environmental Exposure from Chemicals.* Vol. I, Neely, W.B., Blau, G.E., Eds., Chap. 8, pp. 175-216. CRC Press, Boca Raton, Florida.

Miller, M.M., Ghodbane, S., Wasik, S.P., Tewari, Y.B., Martire, D.E. (1984) Aqueous solubilities, octanol/water partition coefficients and entropies of melting of chlorinated benzenes and biphenyls. *J. Chem. Eng. Data* 29, 184-190.

Miller, M.M., Wasik, S.P., Huang, G.L., Shiu, W.Y., Mackay, D. (1985) Relationships between octanol-water partition coefficient and aqueous solubility. *Environ. Sci. Technol.* 19, 522-529.

Mills, W.B., Dean, J.D., Porcella, D.B., Gherini, S.A., Hudson, R.J.M., Frick, W.E., Rupp, G.L., Bowie, G.L. (1982) *Water Quality Assessment: A Screening Procedure for Toxic and Conventional Pollutants.* Part 1, U.S. EPA, EPA-600/6-82-004a.

Miyake, K., Terada, H. (1982) Determination of partition coefficients of very hydrophobic compounds by high performance liquid chromatography on glyceryl-coated controlled-pore glass. *J. Chromatogr.* 240, 9-20.

Moriguchi, I., Kanada, Y., Komatsu, K. (1976) Van der Waals volume and the related parameters for hydrophobicity in structure-activity studies. *Chem. Pharm. Bull.* 24(8), 1799-1806.

Morrison, T.J., Bilett, F. (1952) The salting out of non-electrolytes. Part II. The effect of variation in non-electrolyte. *J. Chem. Soc.* 3819-3822.

Müller, M., Klein, W. (1992) Comparative evaluation of methods predicting water solubility for organic compounds. *Chemosphere* 25(6), 769-782.

Munz, C., Robert, P.V. (1989) Gas- and liquid-phase mass transfer resistances of organic compounds during mechanical surface aeration. *Water Res.* 23, 589-601.

Murray, W.J., Hall, L.H., Kier, LB. (1975) Molecular connectivity III: Relationship to partition coefficients. *J. Pharm. Sci.* 64, 1978-1981.

Myrdal, P., Ward, G.H., Dannenfelser, R.-M., Mishra, D., Yalkowsky, S.H. (1992) AQUAFAC 1: Aqueous functional group activity coefficients; application to hydrocarbons. *Chemosphere* 24, 1047-1061.

Nahum, A., Horvath, C. (1980) Evaluation of octanol-water partition coefficients by using high-performance liquid chromatography. *J. Chromatogr.* 192, 315-322.

Natarajan, G.S., Venkatachalam, K.A. (9172) Solubilities of some olefins in aqueous solutions. *J. Chem. Eng. Data* 17, 328-329.

Nelson, H.D., De Ligny, C.L. (1968) The determination of the solubilities of some *n*-alkanes in water at different temperatures, by means of gas chromatography. *Rec. Trav. Chim. Payus-Bae (Recueil)* 87, 528-544.

Nicolini, E., Laffitte, P. (1949) Vapor pressure of some pure organic liquids. *Comept. Rend.* 229, 757-759.

Nirmalakhandan, N.N., Speece, R.E. (1988a) Prediction of aqueous solubility of organic chemicals based on molecular structure. *Environ. Sci. Technol.* 22, 328-338.

Nirmalakhandan, N.N., Speece, R.E. (1988b) QSAR model for predicting Henry's law constant. *Environ. Sci. Technol.* 22, 1349-1357.

Nunes, P., Benville, P.E., Jr. (1979) Uptake and depuration of petroleum hydrocarbons in the Manila clams, Tapes semidecussata Reeve. *Bull Environ. Contam. Toxicol.* 21, 719-724.

Ogata, M., Miyake, Y. (1978) Disappearance of aromatic hydrocarbons and organic sulfur compounds from fish flesh reared in crude oil suspension. *Water Res.* 12, 1041-1044.

Ogata, M., Fujisawa, K., Ogino, Y., Mano, E. (1984) Partition coefficients as a measure of bioconcentration potential of crude oil compounds in fish and shellfish. *Bull. Environ. Contam. Toxicol.* 33, 561-567.

Ohta, T., Ohyama, T. (1985) *Bull. Chem. Soc. Jpn.* 89, 3556.

Okouchi, S., Saegusa, H., Nojima, O. (1992) Prediction of environmental parameters by adsorbability index: Water solubilities of hydrophobic organic pollutants. *Environ. Intl.* 18, 249-261.

Osborn, A.G., Douslin, D.R. (1974) Vapor pressure relations for 15 hydrocarbons. *J. Chem. Eng. Data* 19, 114-117.

Pankow, J.F. (1986) Magnitude of artifacts caused by bubbles and head-space in the determination of volatile compounds in water. *Anal. Chem.* 58, 1822-1826.

Pankow, J.F. (1990) Minimization of volatilization losses during sampling and analysis of volatile organic compounds in water. In: *Significance, Treatment of Volatile Organic Compounds in Water Supplies*. Ram, N.M., Christman, R.F., Cantor, K.P., Eds., pp. 73-86, Lewis Publishers, Inc., Chelsea, Michigan.

Pankow, J.F., Rosen, M.E. (1988) Determination of volatile compounds in water by purging directly to a capillary column with whole column cryotrapping. *Environ. Sci. Technol.* 22, 398-405.

Pankow, J.F., Isabelle, L.M., Asher, W.E. (1984) Trace organic compounds in rain. 1. Sample design and analysis by adsorption/thermal desorption (ATD). *Environ. Sci. Technol.* 18, 310-318.

Parks, G.S., Huffman, H.M. (1931) Some fusion and transition data for hydrocarbons. *Ind. Eng. Chem.* 23, 1138-1139.

Passino, D.R.M., Smith, S.B. (1987) Quantitative structure-activity relationships (QSAR) and toxicity data in hazard assessment. In: *QSAR in Environmental Toxicology-II.* Kaiser, K.L.E., Editor, D. Reidel Publishing Co., Dordrecht, Holland. pp.261-270.

Paulson, S.E., Seinfeld, J.H. (1992) Atmospheric photochemical oxidation of 1-octene: OH, $O_3$, $O(^3P)$ reactions. *Environ. Sci. Technol.* 26, 1165-1173.

Pavlova, S.P., Pavlov, S.Yu., Serafimov, L.A., Kofman, L.S. (1966) *Promyshlennost. Sinteticheskogo Kouchuka* 3, 18-20.

Pearlmam, R.S. (1986) Molecular surface area and volume: their calculation and use in predicting solubilities and free energies of dissolution. In: *Partition Coefficient, Determination and Estimation.* Dunn, III, W.J., Block, J.H., Pearlman, R.S., Eds., pp. 3-20. Pergamon Press, New York.

Pearlman, R.S., Yalkowsky, S.H., Banerjee, S. (1984) Water solubilities of polynuclear aromatic and heteroaromatic compounds. *J. Phys. Chem. Ref. Data* 13, 555-562.

Pierotti, R.A., Liabastre, A.A. (1972) *Structure and Properties of Water Solutions.* U.S. Natl. Tech. Inform. Ser., PB rep. no. 21163, 113 pp.

Platford, R.F. (1979) Glyceryl trioleate-water partition coefficients for three simple organic compounds. *Bull. Environ. Contam. Toxicol.* 21, 68.

Platford, R.F. (1983) The octanol-water partitioning of some hydrophobic and hydrophilic compounds. *Chemosphere* 12(7/8), 1107-1111.

Polak, J., Lu, B.C.Y. (1973) Mutural solubilities of hydrocarbons and water at 0° and 25°C. *Can. J. Chem.* 51, 4018-4023.

Prausnitz, J.M. (1969) *Molecular Thermodynamic of Fluid-Phase Equilibria.* Prentice Hall, Englewood Cliffs, N.J.

Price, L.C. (1976) Aqueous solubility of petroleum as applied to its origin and primary migration. *Am. Assoc. Petrol. Geol. Bull.* 60, 213-244.

Radding, S.B., Liu, D.H., Johnson, H.L., Mill, T. (1977) Review of the Environmental Fate of Selected Chemicals. U.S. Environmental Protection Agency Report No. EPA-560/5-77-003.

Reid, R.C., Prausnitz, J.M., Sherwood, T.K. (1977) *The Properties of Gases and Liquids*, 3rd ed. McGraw Hill, New York.

Reid, R.C., Prausnitz, J.M., Poling, B.E. (1987) *The Properties of Gases and Liquids.* 3rd ed., McGraw-Hill, New York.

Rekker, R.F. (1977) *The Hydrophobic Fragmental Constants. Its Derivation and Application, A Means of Characterizing Membrane Systems.* Elsevier Sci. Publ. Co., Oxford, England.

Riddick, J.A., Bunger, W.B., Sakano, T.K. (1986) *Organic Solvents.* Wiley Interscience, New York.

Rogers, J.D. (1989) Rate constant measurements for the reaction of the hydroxyl radical with cyclohexene, cyclopentene, and glutaraldehyde. *Environ. Sci. Technol.* 23, 177-181.

Rudakov, E.S., Lutsyk, A.I. (1979) *Zh. Fiz. Khim.* 53, 1298-1300.

Sabljic, A. (1984) Predictions of the nature and strength of soil sorption of organic pollutants by molecular topology. *J. Agric. Food Chem.* 32, 243-246.

Sabljic, A. (1987a) On the prediction of soil sorption coefficients of organic pollutants from molecular structure: Application of molecular topology model. *Environ. Sci. Technol.* 27, 358-366.

Sabljic, A. (1987b) Nonempirical modeling of environmental distribution and toxicity of major organic pollutants. In: *QSAR in Environmental Toxicology - II.* Kaiser, K.L.E., Ed., pp. 309-332, D. Reidel Publ. Co., Dordrecht, Netherlands.

Sabljic, A., Güsten, H. (1990) Predicting the night-time $NO_3$ radical reactivity in the troposphere. *Atmos. Environ.* 24A, 73-78.

Saltzmann, B.E., Gilbert, N. (1959) Ozone reaction with 1-hexene: Clue to smog formation. *Ind. Eng. Chem.* 51, 1415.

Sangster, J. (1989) Octanol-water partition coefficients of simple organic compounds. *J. Phys. Chem. Ref. Data* 18, 1111-1230.

Schantz, M.M., Martire, D.E. (1987) Determination of hydrocarbon-water partition coefficients from chromatographic data and based on solution thermodynamics and theory. *J. Chromatogr.* 391, 35-51.

Schuck, E.A., Doyle, G.J., Endow, E. (1960) Air Pollution Foundation, San Marino, California. Sept. 31, 1960.

Schumann, S.C., Aston, J.S., Sagenkahn, M. (1942) The heat capacity and entropy, heats of fusion and vaporization and the vapor pressures of isopentane. *J. Am. Chem. Soc.* 64, 1039-1043.

Schwarz, F.P. (1977) Determination of temperature dependence of solubilities of polycyclic aromatic hydrocarbons in aqueous solutions by a fluoresence method. *J. Chem. Eng. Data* 22, 273-277.

Schwarz, F.P. (1980) Measurement of the solubilities of slightly soluble organic liquids in water by elution chromatography. *Anal. Chem.* 52, 10-15.

Schwarz, F.P., Miller, J. (1980) Measurement of the solubilities of slightly soluble organic liquids in water by elution chromatography. *Anal. Chem.* 52, 2161-2164.

Sielicki,M., Focht, D.D., Martin, J.P. (1978) Microbial transformations of styrene and $^{14}C$-styrene in soil and enrichment cultures. *Appl. Environ. Microbiol.* 35, 124-128.

Singleton, D.L., Cvetanovic, R.J. (1976) Temperature dependence of oxygen atoms with olefins. *J. Am. Chem. Soc.* 98, 6812.

Smith, J.H., Mabey, W.R., Bahonos, N., Holt, B.R., Lee, S.S., Chou, T.W., Bomberger, D.C., Mill, T. (1978) *Environmental Pathways of Selected Chemicals in Freshwater Systems: Part II. Laboratory Studies.* Interagency Energy-environment Research and Development Program Report. EPA-600/7-78-074. Environmental Research Laboratory Office of Research and Development. U.S. Environmental Protection Agency, Athens, Georgia 30605, p. 304.

Stearns, R.S., Oppenheimer, H., Simon, E., Harkins, W.D. (1947) Solubilization by solutions of long chain colloidal electrolytes. *J. Chem. Phys.* 15, 496-507.

Stedman, D.H., Wu, C.H., Niki, H. (1973) Kinetics of gas-phase reactions of ozone with some olefins. *J. Phys. Chem.* 77, 2511.

Stephen, H., Stephen, Y. (1963) Solubilities of Inorganic and Organic Compounds. Vol. 1 & 2, Pergamon Press, Oxford.

Stephenson, R.M., Malanowski, S. (1987) *Handbook of the Thermodynamic of Organic Compounds.* Elsevier Science publishing Co. Inc., New York, N.Y.

Stull, D.R. (1947) Vapor pressure of pure substances organic compounds. *Ind. Eng. & Chem.* 39(4), 517-560.

Sutton, C., Calder, J.A. (1974) Solubility of higher-molecular-weight-paraffins in distilled water and seawater. *Environ. Sci. Technol.* 8, 654.

Swann, R.L., Laskowski, D.A., McCall, P.J., Vender Kuy, K., Dishburger, J.J. (1983) A rapid method for the estimation of the environmental parameters octanol/water partition coefficient, soil sorption constant, water to air ratio, and water solubility. *Res. Rev.* 85, 17-28.

Tabak, H.H., Quave, S.A., Moshni, C.I., Barth, E.F. (1981) Biodegradability studies with organic priority pollutant compounds. *J. Water Pollut. Control Fed.* 53, 1503-1518.

Taft, R.W., Abrahm, M.H., Famini, G.R., Doherty, R.M., Abboud, J.L., Kamlet, M.J. (1985a) Solubility properties in polymers and biological media 5: An analysis of physicochemical properties which influence octanol-water partition coefficients of aliphatic and aromatic solutes. *J. Pharm. Sci.* 74(8), 807-814.

Taft, R.W., Abraham, M.H., Doherty, R.M., Kamlet, M.J. (1985b) The molecular properties governing solubilities of ofganic nonelectrolytes in water. *Nature* 313, 384-386.

Taha, A.A., Grisby, R.D., Johnson, J.R., Christian, S.D., Affsprung, H.E. (1966) Manometric for vapor and solution studies. *J. Chem. Education* 43, 432.

Tewari, Y.B., Martire, D.E., Wasik, S.P., Miller, M.M. (1982a) Aqueous solubilities and octanol-water partition coefficients of binary liquid mixtures of organic compounds at 25°C. *J. Solution Chem.* 11, 435-445.

Tewari, Y.B., Miller, M.M., Wasik, S.P. (1982b) Calculation of aqueous solubilities of organic compounds. *NBS J. Res.* 87, 155-158.

Tewari, Y.B., Miller, M.M., Wasik, S.P., Martire, D.E. (1982c) Aqueous solubility and octanol/water partition coefficient of organic compounds at 25.0 °C. *J. Chem. Eng. Data* 27, 451-454.

Thoms, S.R., Lion, L.W. (1992) Vapor-phase partitioning of volatile organic compounds: a regression approach. *Environ. Toxicol. Chem.* 11, 1377-1388.

Tsonopoulos, C., Prausnitz, J.M. (1971) Activity coefficients of aromatic solutes in dilute aqueous solutions. *I & EC Fundum.* 593-600.

Trzilova, B., Horska, E. (1988) Biodegradation of amines and alkanes in aquatic environment. *Biologia (Bratislava)* 43, 209-218.

Vaishnav, D.D., Babeu, L. (1987) Comparison of occurrence and rates of chemical biodegradation in natural waters. *Bull. Environ. Contam. Toxicol.* 39, 237-244.

Valsaraj, K.T. (1988) On the physico-chemical aspects of partitioning of non-polar hydrophobic organics at the air-water interface. *Chemosphere* 17, 875-887.

Valvani, S.C., Yalkowsky, S.H. (1980) Solubility and partitioning in drug design. In: *Physical Chemical Properties of Drugs.* Med. Res. Ser. vol 10, Yalkowsky, S.H., Sinkula, A.A., Valvani, S.C., Eds., pp. 201-229, Marcel Dekker Inc., New York.

Valvani, S.C., Yalkowsky, S.H., Amidon, G.L. (1976) Solubility of nonelectrolytes in polar solvents. VI. Refinements in molecular surface area computations. *J. Phys. Chem.* 80, 829-835.

Valvani, S.C., Yalkowsky, S.H., Roseman, T.J. (1981) Solubility and partitioning IV. Aqueous solubility and octanol-water partition coefficient of liquid electrolytes. *J. Pharm. Sci.* 70, 502-507.

Van der Linden, A.C. (1978) Degradation of oil in the marine environment. *Dev. Biodegrad. Hydrocarbons* 1, 165-200.

Verschueren, K. (1977) *Handbook of Environmental Data on Organic Chemicals.* Van Nostrand Reinhold, New York.

Verschueren, K. (1983) *Handbook of Environmental Data on Organic Chemicals*, 2nd edition, Van Nostrand Reinhold, New York.

Vowles, P.D., Mantoura, R.F.C. (1987) Sediment-water partition coerfficients and HPLC retention factors of aromatic hydrocarbons. *Chemosphere* 16, 109-116.

Wakeham, S.G., Davis, A.C., Karas, J.L. (1983) Microcosm experiments to determine the fate and persistence of volatile organic compounds in coastal seawater. *Environ. Sci. Technol.* 17, 611-617.

Wakita, K., Yoshimoto, M., Miyamoto, S., Watanabe, H. (1986) A method for calculation of the aqueous solubility of organic compounds by using new fragment solubility constants. *Chem. Pharm. Bull.* 34, 4663-4681.

Wang, L., Wang, X., Xu, O., Tian, L. (1986) *Huanjing Kexue Xuebao* 6, 491.

Wang, L., Zhao, Y., Hong, G. (1992) Predicting aqueous solubility and octanol/water partition coefficients of organic chemicals from molar volume. *Environ. Chem.* 11, 55-70.

Warne, M.St.J., Connell, D.W., Hawker, D.W., Schuurmann, G. (1990) Prediction of aqueous solubility and the octanol-water partition coefficient for lipophilic organic compounds using molecular descriptors and physicochemical properties. *Chemosphere* 21, 877-888.

Warne, M.St.J., Connell, D.W., Hawker, D.W. (1991) Comparison of the critical concentration and critical volume hypotheses to model non-specific toxicity of individual compounds. *Toxicology* 66, 187-195.

Warner, M.P., Cohen, J.M., Ireland, J.C. (1987) Determination of Henry's law constants of selected priority pollutants. EPA/600/D-87/229, U.S.EPA, Cincinnati, OH 45268.

Wasik, S.P., Tewari, Y.B., Miller, M.M., Martire, D.E. (1981) Octanol/water partition coefficients and aqueous solubilities of organic compounds. *NBSIR* 81-2406, report prepared for Office of Toxic Substances, Environmental Protection Agency, Washington, DC.

Wasik, S.P., Miller, M.M., Tewari, Y.B., May, W.E., Sonnefeld, W.J., DeVoe, H., Zoller, W.H. (1983) Determination of the vapor pressure, aqueous solubility, and octanol/water partition coefficient of hydrophobic substances by coupled generator column/liquid chromatographic methods. *Residue Rev.* 85, 29-42.

Wasik, S.P., Schwarz, F.P., Tewari, Y.B., Miller, M.M., Purnell, J.H. (1984) A head-space method for measuring activity coefficients, partition coefficients and solubilities of hydrocarbons in saline solutions. *J. Res. Natl. Bur. Std.* 89, 273-277.

Wasik, S.P., Tewari, Y.B., Miller, M.M. (1982) Measurements of octanol/water partition coefficient by chromatographic method. *J. Res. Natl. Bur. Std.* 87, 311-315.

Weast, R.C., Ed. (1972-73). *Handbook of Chemistry and Physics*, 53th ed. CRC Press, Cleveland.

Weast, R.C. (1983-84). *Handbook of Chemistry and Physics*, 64th ed., CRC Press, Florida.

Webster, G.R.B., Sarna, L.P., Muir, D.C.G. (1985) Octanol-water partition coefficient of 1,3,6,8-tetrachlorodibenzo-*p*-dioxin and octachlorodibenzo-*p*-dioxin. In: *Chlorinated Dioxins and Dibenzofurans in the Total Environment. Vol. 2*, Keith, L.H., Rappe, C., Choudhary, G., Editors, Butterworths, London.

Whitehouse, B.G., Cooke, R.C. (1982) Estimating the aqueous solubility of aromatic hydrocarbons by high performance liquid chromatography. *Chemosphere* 11, 689-699.

Willingham, C.B., Taylor, W.J., Pignocco, J.M., Rossini, F.D. (1945) Vapor pressure and boiling points of some paraffin, alkylcyclopentane, alkylcyclohexane, and alkylbenzene hydrocarbons. *J. Res. Natl. Bur. Std.* 34, 219.

Wilson, B.H., Smith, G.B., Rees, J.F. (1986) Biotransformations of selected alkylbenzenes and halogenated aliphatic hydrocarbons in methanogenic aquifer material: A Microcosm Study. *Environ. Sci. Technol.* 20, 997-1002.

Wilson, J.T., Enfield, C.G., Dunlap, W.J., Crosby, R.L., Foster, D.A., Baskin, L.B. (1981) Transport and fate of selected organic pollutants in a sandy soil. *J. Environ. Qual.* 10, 501-506.

Wilson, J.T., McNabb, J.F., Balkwill, D.L., Ghiorse, W.C. (1983) Enumeration and characterization of bacteria indigenous to a shallow water-table aquifer. *Ground Water* 21, 134-142.

Wilson, J.T., McNabb, J.F., Wilson, R.H., Noonan, M.J. (1983) Biotransformation of selected organic pollutants in ground water. *Dev. Ind. Microbiol.* 24, 225-233.

Winer, A.M., Darnall, K.R., Atkinson, R. Pitts, Jr., J.N. (1979) Smog chamber study of the correlation of hydroxyl radical rate constants with ozone formation. *Environ. Sci. Technol.* 7, 622-626.

Winer, A.M., Atkinson, R., Pitts, J.N., Jr. (1984) Gasous nitrate radical: possible nightime atmospheric sink for biogenic organic compounds. *Science* 224, 156-159.

Windholz, M., Ed. (1983) *The Merck Index, An Encyclopedia of Chemicals, Drugs and Biologicals.* 10th edition, Merck & Co., Inc., Rahway, New Jersey.

Wolfe, N.L., Zepp, R.G., Schlotzhauer, P., Sink, M. (1982) Transformation pathways of hexachlorocyclopentadiene in the equatic environment. *Chemosphere* 11, 91-101.

Wong, P.T.S., Chau, Y.K., Rhamey, J.S., Docker, M. (1984) Relationship between water solubility of chlorobenzenes and their effects on a fresh water green algae. *Chemosphere* 13(9), 991-996.

Yalkowsky, S.H., Valvani, S.C. (1976) Partition coefficients and surface areas of some alkylbenzenes. *J. Med. Chem.* 19, 727-728.

Yalkowsky, S.H. (1979) Estimation of entropies of fusion of organic compounds. *Ind. Eng. Chem. Fundam.* 18, 108-111.

Yalkowsky, S.H., Morozowich, W. (1980) A physical chemical basis for the design of orally active prodrugs. In: *Drug Design*, Vol IX, pp. 121-185, Academic Press Inc., New York.

Yalkowsky, S.H., Valvani, S.C., Mackay, D. (1983) Estimation of the aqueous solubility os some aromatic compounds. *Res. Rev.* 85, 43-55.

Yalkowsky, S.H., Valvani, S.C., Roseman, T.J. (1983) Solubility and partitioning VI: Octanol solubility and octanol-water partition coefficients. *J. Pharmaceut. Sci.* 72, 866-870.

Yalkowsky, S.H., Valvani, S.C., Kuu, W.-Y., Dannenfelser, R.M., Eds. (1987) Arizona Database of Aqueous for Organic Compounds. College of Pharmacy, University of Arizona, Tucson, Arizona.

Yoshida, K., Shigeoka, T., Yamauchi, F. (1983) Relationship between molar refraction and n-octanol/water partition coefficient. *Ecotox. Environ. Saf.* 7, 558-565.

Yurteri, C. Ryan, D.F., Callow, J.J., Gurol, J.J. (1987) The effect of chemical composition of water on Henry's law constant. *J. WPCF* 59, 950-956.

Zafonte, L., Bonamassa, F. (1977) Relative photochemical reactivity of propane and *n*-butane. *Environ. Sci. Technol.* 11, 1015-1017.

Zoeteman, B.C.J., Harmsen, K., Linders, J.B.H. (1980) Persistant organic pollutants in river water and ground water of the Netherlands. *Chemosphere* 9, 231-249.

Zwolinski, B.J., Wilhoit, R.C. (1971) *Handbook of Vapor Pressures and Heats of Vaporization of Hydrocarbons and Related Compounds*. API-44 TRC Publication No.101, Texas A. & M. University, Evans Press, Fort Worth, Texas.

# 3. Halogenated Hydrocarbons

3.1 List of Chemicals and Data Compilations:


Methyl chloride . . . . 395
Dichloromethane . . . . 400
Chloroform . . . . 407
Carbon tetrachloride . . . . 416
Chloroethane (Ethyl chloride) . . . . 426
1,1-Dichloroethane . . . . 431
1,2-Dichloroethane . . . . 436
1,1,1-Trichloroethane . . . . 443
1,1,2-Trichloroethane . . . . 451
1,1,1,2-Tetrachloroethane . . . . 456
1,1,2,2-Tetrachloroethane . . . . 459
Pentachloroethane . . . . 465
Hexachloroethane . . . . 469
1-Chloropropane (*n*-Propyl chloride) . . . . 473
2-Chloropropane . . . . 476
1,2-Dichloropropane . . . . 479
1,2,3-Trichloropropane . . . . 483
1-Chlorobutane (*n*-Butyl chloride) . . . . 486
2-Chlorobutane . . . . 489
1-Chloropentane . . . . 491
Vinyl chloride . . . . 493
1,1-Dichloroethene . . . . 498
*cis*-1,2-Dichloroethene . . . . 503
*trans*-1,2-Dichloroethene . . . . 507
Trichloroethylene . . . . 512
Tetrachloroethylene . . . . 522
1,3-Dichloropropene . . . . 531
Chloroprene . . . . 534
Hexachlorobutadiene . . . . 536
Hexachlorocyclopentadiene . . . . 538
Bromomethane . . . . 541
Dibromomethane . . . . 545
Tribromomethane . . . . 548
Bromoethane (Ethyl bromide) . . . . 552
1,2-Dibromoethane . . . . 555
1-Bromopropane (*n*-Propyl bromide) . . . . 558
2-Bromopropane . . . . 561
1,2-Dibromopropane . . . . 564
1-Bromobutane . . . . 566
1-Bromopentane . . . . 569
Vinyl bromide . . . . 571

Methyl iodide . . . . . 573
Iodoethane (Ethyl iodide) . . . . . 576
1-Iodopropane (*n*-Propyl iodide) . . . . . 579
2-Iodopropane . . . . . 581
1-Iodobutane . . . . . 583
Bromochloromethane . . . . . 585
Bromodichloromethane . . . . . 588
Dibromochloromethane . . . . . 591
Chlorodifluoromethane . . . . . 594
Dichlorodifluoromethane . . . . . 597
Trichlorofluoromethane . . . . . 600
1,1,2-Trichloro-1,2,2-trifluoroethane . . . . . 604
1,1,2,2-Tetrachloro-1,2-difluoroethane . . . . . 607
Fluorobenzene . . . . . 609
Bromobenzene . . . . . 611
Iodobenzene . . . . . 614
Chlorobenzenes (please see Volume I)
3.2 Summary Tables and QSPR Plots . . . . . 616
3.3 Illustrative Fugacity Calculations: Levels I, II and III . . . . . 633
Dichloromethane . . . . . 634
Chloroform . . . . . 638
Carbon tetrachloride . . . . . 642
1,2-Dichloroethane . . . . . 646
1,1,2,2-Tetrachloroethane . . . . . 650
Pentachloroethane . . . . . 654
Hexachloroethane . . . . . 658
Vinyl chloride . . . . . 662
Trichloroethylene . . . . . 666
Tetrachloroethylene . . . . . 670
1,2-Dichloropropane . . . . . 674
1,2,3-Trichloropropane . . . . . 678
Chloroprene . . . . . 682
Tribromomethane . . . . . 686
Bromochloromethane . . . . . 690
Bromodichloromethane . . . . . 694
Trichlorofluoromethane . . . . . 698
Fluorobenzene . . . . . 702
Bromobenzene . . . . . 706
Iodobenzene . . . . . 710
3.4 Commentary on the Physical-Chemical Properties and Environmental Fate . . . . . 714
3.5 References . . . . . 717

## 3.1 List of Chemicals and Data Compilations

Common Name: Methyl chloride
Synonym: chloromethane, monochloromethane
Chemical Name: chloromethane
CAS Registry No: 74-87-3
Molecular Formula: $CH_3Cl$
Molecular Weight: 50.49
Melting Point (°C):
-97.7 (Stull 1947; Mackay & Shiu 1981,1990; Verschueren 1983; Dean 1985; Riddick et al. 1986; Stephenson & Malanowski 1987)
-97.72 (Dreisbach 1959)
-97.73 (Weast 1977; quoted, Callahan et al. 1979; Horvath 1982; Mabey et al. 1982)
-97.1 (Weast 1982-83)
Boiling Point (°C):
-23.76 (McGovern 1943)
-24.22 (Dreisbach 1959; Dean 1985)
-23.84, -24.16 (Boublik et al. 1973)
-24.2 (McConnell et al. 1975; Pearson & McConnell 1975; Weast 1977; Weast 1982-83; quoted, Callahan et al. 1979; Mackay & Shiu 1981,1990; Horvath 1982; Mabey et al. 1982; Riddick et al. 1986)
-24.0 (Verschueren 1983)
-23.73 (Stephenson & Malanowski 1987)
Density ($g/cm^3$ at 20°C):
0.920 (McGovern 1943)
0.9159 (Horvath 1982; Weast 1982-83)
0.920 (Dean 1985)
0.9214 (Riddick et al. 1986)
Molar Volume ($cm^3/mol$):
50.6, 48.8, 54.3, 50.4 (exptl., Tyn and Calus method, Schroeder method, LeBas method, Reid et al. 1987)
50.5 (LeBas method, Mackay & Shiu 1990)
55.0 (calculated from density, Wang et al. 1992)
Molecular Volume ($A^3$):
Total Surface Area, TSA ($A^2$):
157.8 (Iwase et al. 1985)
62.50 (Okouchi et al. 1992)
Heat of Fusion, $\Delta H_{fus}$, (kcal/mol):
1.537 (Riddick et al. 1986)
Entropy of Fusion, $\Delta S_{fus}$ (cal/mol K or e.u.):
Fugacity Ratio at 25 °C, F: 1.0

Water Solubility (g/m³ or mg/L at 25°C):

7400 (McGovern 1943; quoted, Horvath 1982)
5350 (gravitational method, Glew & Moelwyn-Hughes 1953; quoted, Mackay & Shiu 1981)
7400 (30°C, quoted, Gladis 1960)
5100 (quoted, Irmann 1965)
6450 (20°C, Dean 1973; quoted, Callahan et al. 1979; Mabey et al. 1982; Mills et al. 1982)
5049 (quoted, Hine & Mookerjee 1975)
7250 (20°C, McConnell et al. 1975; Pearson & McConnell 1975; quoted, Callahan et al. 1979; Mackay & Shiu 1981,1990; Mills et al. 1982)
7400 (30°C, Neely 1976; quoted, Mackay & Shiu 1981)
6270 (20°C, Dilling 1977; quoted, Mackay & Shiu 1981)
5380 (quoted average, Dilling 1977; quoted, Horvath 1982)
4600 (20°C, selected, Nathan 1978)
5325 (recommended, Horvath 1982)
7400 (quoted, Thomas 1982)
4800 (Dean 1985)
6480 (30°C, quoted, Riddick et al. 1986)
6226 (calculated-AI, Okouchi et al. 1992)
6504, 19196 (quoted exptl., calculated-molar volume, Wang et al. 1992)

Vapor Pressure (Pa at 25°C):

559860 (McGovern 1943)
100788 (Glew & Moelwyn-Hughes 1953)
574483 (calculated-Antoine eqn., Dreisbach 1959)
539270 (20-25°C, calculated-Antoine eqn., Weast 1972-73; quoted, Mackay & Shiu 1990)
539272 (calculated-Antoine eqn., Weast 1972-73)
570000 (Boublik et al. 1973; quoted, Mackay & Shiu 1981)
567934, 576309 (calculated values, Boublik et al. 1973)
500675 (20°C, McConnell et al. 1975; quoted, Callahan et al. 1979; Mabey et al. 1982)
480000 (20°C, Neely 1976; quoted, Mackay & Shiu 1981)
499000 (20°C, Pearson & McConnell 1975; quoted, Mackay & Shiu 1981)
536932 (selected, Nathan 1978)
493210 (20°C, selected, Mills et al. 1982)
572825 (calculated-Antoine eqn., Stephenson & Malanowski 1987)

Henry's Law Constant (Pa $m^3$/mol):

953, 957 (exptl., calculated-P/C, Glew & Moelwyn-Hughes 1953)

1010 (calculated as $1/K_{AW}$, $C_W/C_A$, reported as exptl., Hine & Mookerjee 1975; quoted, Nirmaklahanden & Speece 1988)

744, 892 (exptl. as per McConnell et al. 1975, calculated-P/C, Neely 1976)

942 (calculated-P/C, Dilling 1977)

951 (calculated-P/C, Mackay & Shiu 1981)

950 (recommended, Mackay & Shiu 1981)

4052 (20°C, calculated-P/C, Mabey et al. 1982)

2431 (calculated-P/C, Thomas 1982)

894 (EPICS, Gossett 1987; quoted, Mackay & Shiu 1990)

2367 (calculated-QSAR, Nirmalakhandan & Speece 1988)

669 (20-25°C and low ionic strength, quoted, Pankow & Rosen 1988; Pankow 1990)

39661 (WERL Treatability Database, quoted, Ryan et al. 1988)

4015 (selected, Jury et al. 1990)

706 (calculated-P/C, Mackay & Shiu 1990)

Octanol/Water Partition Coefficient, log $K_{OW}$:

0.91 (shake flask-GC, Hansch et al. 1975; quoted, Callahan et al. 1979)

0.91 (Hansch & Leo 1979; quoted, Callahan et al. 1979; Ryan et al. 1988; Bodor et al. 1989; Sangster 1989; Mackay & Shiu 1990)

0.95 (calculated-f. const., Mabey et al. 1982)

0.903 (selected, Mills et al. 1982)

0.91, 1.03 (quoted, calculated-molar refraction, Yoshida et al. 1983)

0.91, 0.81 (quoted, calculated-hydrophobicity const., Iwase et al. 1985)

0.91, 0.89 (quoted, calculated-MO, Bodor et al. 1989)

0.91, 0.936 (quoted, calculated-$\pi$ substituent const., Bodor et al. 1989)

0.91 (recommended, Sangster 1989)

0.94, 1.06 (quoted, calculated-molar volume, Wang et al. 1992)

Bioconcentration Factor, log BCF:

0.505 (microorganism-water, calculated from $K_{OW}$, Mabey et al. 1982)

Sorption Partition Coefficient, log $K_{OC}$:

0.633 (calculated-$K_{OW}$, Mabey et al. 1982)

0.778 (selected, Jury et al. 1990)

Half-Lives in the Environment:

Air: disappearance half-life of 2.4-24 hours from air for the reaction with OH radicals (EPA 1974; quoted, Darnall et al. 1976); estimated residence time to be about 2 years for the reaction with OH radicals in troposphere (Singh et al. 1979); estimated residence time in troposphere to be one year (Lyman 1982); 231 days, estimated as toxic chemical residence time with rate constant of $5 \times 10^{-14}$ $cm^3$ $molecule^{-1}$ $sec^{-1}$ at 300 K for the reaction with OH radicals (Singh et al. 1981); 1472-14717 hours, based on photooxidation half-life in air from measured rate constants for reaction with hydroxyl radicals in air (Atkinson 1985; quoted, Howard et al. 1991); estimated tropospheric lifetime of 1.3 year and 1.54 year by rigorous calculation (Nimitz & Skaggs 1992).

Surface water: 168-672 hours, based on estimated aerobic biodegradation half-life (Howard et al. 1991).

Ground water: 336-1344 hours, based on estimated aerobic biodegradation half-life (Howard et al. 1991).

Sediment:

Soil: >50 days (Ryan et al. 1988); estimated half-life of 120 days for volatilization from soil (Jury et al. 1990); 168-672 hours, based on estimated aerobic biodegradation half-life (Howard et al. 1991).

Biota: >50 days, subject to plant uptake from soil via volatilization (Ryan et al. 1988).

Environmental Fate Rate Constant and Half-Lives:

Volatilization: half-life of 27-28 minutes for initial concentration of 1 mg/liter in an open container stirred at 200 rpm (Dilling et al. 1975; Dilling 1977; quoted, Callahan et al. 1979; Mills et al. 1982); calculated half-lives: 0.599 minute, 14.9 minutes (Dilling 1977); estimated half-life from water to be 2.4 hours (Thomas 1982); estimated half-life from soil to be 120 days (Jury et al. 1990).

Photolysis:

Oxidation: bimolecular rate constant for reaction with hydroxyl radicals was reported to be $4.7 \times 10^{-14}$ $cm^3$ $sec^{-1}$ (Yung et al. 1975; quoted, Callahan et al. 1979) and $8.5 \times 10^{-14}$ $cm^3$ $sec^{-1}$ with a lifetime of 0.37 years in the troposphere (Cox et al. 1976; quoted, Callahan et al. 1979); calculated rate constant of $4.6 \times 10^{-14}$ $cm^3$ $molecule^{-1}$ $sec^{-1}$ for the reaction with OH radicals at 298 K with lifetime of 3.6 years in troposphere (Davis et al. 1976; quoted, Altshuller 1980); calculated rate constant of $4.3 \times 10^{-14}$ $cm^3$ $molecule^{-1}$ $sec^{-1}$ for the reaction with OH radicals at 298 K with lifetime of 4.4 years in troposphere (Perry et al. 1976; quoted, Altshuller 1980); $2.4 \times 10^{10}$ $cm^3$ $mol^{-1}$ $sec^{-1}$, estimated for the reaction with OH radicals at 300 K (Lyman 1982); calculated rate constant of $<<360$ $M^{-1}$ $hour^{-1}$ for singlet oxygen and 0.05 $M^{-1}$ $hour^{-1}$ for peroxy radical (Mabey et al. 1982).

Hydrolysis: $6.8 \times 10^{-5}$ $hour^{-1}$ with hydrolytic half-life of 417 days at pH 7 and 25°C (Radding et al. 1977; Mabey & Mill 1978; quoted, Callahan et al. 1979; Mabey et al. 1982); half-life of 7000 hours, based on neutral and base catalyzed hydrolysis rate constants at 25°C extrapolated from data obtained at higher temperatures (Mabey & Mill 1978; quoted, Howard et al. 1991).

Biodegradation: aqueous aerobic half-life: 168-672 hours, based on unacclimated aerobic aqueous screening test data for dichloromethane from experiments utilizing selected domestic waste water inoculum (Tabak et al. 1981; quoted, Howard et al. 1991) and activated sludge inoculum (Klecka 1982; quoted, Howard et al. 1991); aqueous anaerobic half-life: 672-2688 hours, based on estimated aerobic biodegradation half-life (Howard et al. 1991).

Biotransformation:

Bioconcentration and Uptake and Elimination Rate Constants ($k_1$ and $k_2$):

Common Name: Dichloromethane
Synonym: methylene chloride, methylene dichloride, methane dichloride, methylene bichloride
Chemical Name: dichloromethane
CAS Registry No: 75-09-2
Molecular Formula: $CH_2Cl_2$
Molecular Weight: 84.94
Melting Point (°C):

- -96.7 (Stull 1947; Dean 1985; Stephenson & Malanowski 1987)
- -95.14 (Dreisbach 1959)
- -95.0 (Weast 1977; Callahan et al. 1979; Mabey et al. 1982)
- -95.1 (Mackay & Shiu 1981; Horvath 1982; Weast 1982-83; Suntio et al. 1988; Howard 1990; Mackay & Shiu 1990)
- -97.0 (Verschueren 1983)
- -94.92 (Riddick et al. 1986)

Boiling Point (°C):

- 41.0 (Rex 1906)
- 39.80 (McGovern 1943; Stephenson & Malanowski 1987)
- 39.75 (Dreisbach 1959; Weast 1977; Callahan et al. 1979; Mabey et al. 1982; Howard 1990)
- 39.76, 40.19 (Boublik et al. 1973)
- 40.1 (McConnell et al. 1975; Pearson & McConnell 1975)
- 39.70 (Mackay & Shiu 1981,1990)
- 40.0 (Horvath 1982; Weast 1982-83)
- 42.0 (Verschueren 1983)
- 40.5 (Dean 1985)
- 39.64 (Riddick et al. 1986)

Density ($g/cm^3$ at 20°C):

- 1.3260 (McGovern 1943)
- 1.32554 (Dreisbach 1959)
- 1.3163 (25°C, Dreisbach 1959)
- 1.3266 (Horvath 1982; Weast 1982-83)
- 1.3255 (Dean 1985)
- 1.3256 (Riddick et al. 1986)
- 1.330 (Gillham & Rao 1990)

Molar Volume ($cm^3/mol$):

- 64.51 (Hoy 1970; Amidon & Williams 1982)
- 0.624 ($V_M/100$, Taft et al. 1985; Kamlet et al. 1986,1987; Leahy 1986)
- 0.336 (calculated intrinsic volume: $V_I/100$, Leahy 1986; Kamlet et al. 1988; Thoms & Lion 1992)
- 71.0 (LeBas method, Abernethy & Mackay 1987; Abernethy et al. 1988; Mackay & Shiu 1990)
- 34.0 (intrinsic volume-van der Waals method, Abernethy et al. 1988)
- 62.0 (calculated from density, Abernethy et al. 1988)

64.0 (calculated from density, Wang et al. 1992)

Molecular Volume ($A^3$):

Total Surface Area, TSA ($A^2$):

213.49 (Amidon 1976; Horvath 1982)

82.0 (Mackay et al. 1982; Valsaraj 1988; Valsaraj & Thibodeaux 1989)

191.4 (Iwase et al. 1985)

78.90 (Okouchi et al. 1992)

Heat of Fusion, $\Delta H_{fus}$, (kcal/mol):

1.435 (calculated, Dreisbach 1959)

1.472 (quoted, Riddick et al. 1986)

Entropy of Fusion, $\Delta S_{fus}$ (cal/mol K or e.u.):

Fugacity Ratio at 25 °C, F:

1.0 (Suntio et al. 1988)

Water Solubility (g/$m^3$ or mg/L at 25°C):

20000 (20°C, volumetric, Rex 1906)

19690 (30°C, volumetric, Rex 1906)

19912 (Seidell 1940; quoted, Deno & Berkheimer 1960)

13200 (McGovern 1943; quoted, Horvath 1982)

34476 (Booth & Everson 1948; quoted, Horvath 1982)

13200 (quoted, Gladis 1960; Irmann 1965)

16200 (quoted, Irmann 1965)

20000 (Pechiney-Saint-Gobain 1971; quoted, Horvath 1982)

20000 (Dean 1973; quoted, Callahan et al. 1979; Mabey et al. 1982; Mills et al. 1982)

19800 (quoted, Dilling et al. 1975)

19016 (quoted, Hine & Mookerjee 1975)

13200 (20°C, McConnell et al. 1975; Pearson & McConnell 1975; quoted, Callahan et al. 1979; Mackay & Shiu 1981,1990; Mills et al. 1982; Neely 1984; Dean 1985; Suntio et al. 1988; Thoms & Lion 1992)

13200 (20°C, Neely 1976; quoted, Mackay & Shiu 1981; Suntio et al. 1988)

20000 (Archer & Sterns 1977; quoted, Horvath 1982)

19400 (Dilling 1977; quoted, Afghan & Mackay 1980; Hutchinson et al. 1980; Mackay & Shiu 1981; Bobra et al. 1984; Suntio et al. 1988; Thoms & Lion 1992)

20000 (20°C, Verschueren 1977,1983; quoted, Suntio et al. 1988)

20000 (Andelman 1978; quoted, Horvath 1982)

19600 (20°C, selected, Nathan 1978)

13200 (Selenka & Bauer 1978; quoted, Horvath 1982)

19680 (20°C, recommended, Sørenson & Arit 1979; quoted, Wright et al. 1992)

16700 (quoted, Cowen & Baynes 1980; Warner et al. 1987)

19016, 15105 (quoted average, calculated-$K_{OW}$, Valvani et al. 1981)

18580, 29450 (quoted, calculated-$K_{OW}$, molar volume & M.P., Amidon & Williams 1982)

16200, 14100 (quoted exptl., calculated-group contribution as per Irmann 1965, Horvath 1982)
13030 (recommended, Horvath 1982)
13000-20000 (selected, Mills et al. 1982)
13000 (quoted, Thomas 1982)
16700 (quoted, Verschueren 1983)
19016 (calculated-UNIFAC, Banerjee 1985)
16000 (selected, Daniels et al. 1985)
19016, 20850 (quoted, calculated-molar volume & solvatochromic p., Leahy 1986)
13000 (Riddick et al. 1986; quoted, Howard 1990)
19912, 7747 (quoted, calculated-fragment consts., Wakita et al. 1986)
19016, 26860 (quoted, calculated-molar volume & solvatochromic p., Kamlet et al. 1987)
13031; 19952, 13305 (quoted; predicted- $\chi$ & polarizability, Nirmalakhandan & Speece 1988)
18000 (selected, Suntio et al. 1988)
16700 (selected, Valsaraj 1988)
10162 (quoted from Mackay et al. 1982, Valsaraj & Thibodeaux 1989)
20000 (selected, Gillham & Rao 1990)
13035 (calculated-AI, Okouchi et al. 1992)
18000 (17.5°C, shake flask-GC/TC, Stephenson 1992)
17200 (26.8°C, shake flask-GC/TC, Stephenson 1992)
19912, 15457 (quoted exptl., calculated-molar volume, Wang et al. 1992)
18780, 18680 (20°C, calculated-activity coefficients, Wright et al. 1992)

Vapor Pressure (Pa at 25°C):

46508 (20°C, Rex 1906)
68170 (30°C, Rex 1906)
57120 (McGovern 1943)
57476 (Antoine eqn. regression, Stull 1947)
58100 (calculated-Antoine eqn., Dreisbach 1959)
57388 (calculated-Antoine eqn., Weast 1972-73)
58275, 57267 (calculated-Antoine eqn., Boublik et al. 1973)
56786 (quoted, Dilling et al. 1975)
54813 (quoted, Hine & Mookerjee 1975)
48200 (20°C, McConnell et al. 1975; Pearson & McConnell 1975; quoted, Callahan et al. 1979; Mackay & Shiu 1981; Mabey et al. 1982; Suntio et al. 1988)
46530 (20°C, Neely 1976; quoted, Mackay & Shiu 1981; Thomas 1982; Suntio et al. 1988)
58400 (Dilling 1977; quoted, Mackay & Shiu 1981; Suntio et al. 1988)
55986 (selected, Nathan 1978)
55314 (quoted, Cowen & Baynes 1980)

48255 (20°C, selected, Mills et al. 1982)
57972 (calculated-Antoine eqn., Boublik et al. 1984; quoted, Howard 1990)
57953, 57983 (calculated-Antoine eqn., Boublik et al. 1984)
47988 (selected, Daniels et al. 1985)
57365 (quoted, Kamlet et al. 1986)
57989 (calculated-Antoine eqn., Stephenson & Malanowski 1987)
60694 (interpolated between two data points, Warner et al. 1987)
52000 (selected, Suntio et al. 1988)
60683 (selected, Valsaraj 1988)
46522 (20°C, selected, Gillham & Rao 1990)
58400 (quoted from Weast 1972-73, Mackay & Shiu 1990)
47656 (20°C, quoted from DIPPR, Tse et al. 1992)
71003 (30°C, quoted from DIPPR, Tse et al. 1992)

Henry's Law Constant (Pa $m^3$/mol):
231.3 (calculated as $1/K_{AW}$, $C_W/C_A$, reported as exptl., Hine & Mookerjee 1975; quoted, Nirmalakhandan & Speece 1988)
301.0 (McConnell et al. 1975; quoted, Mackay & Shiu 1981)
297.5, 322.3 (quoted, calculated, Neely 1976)
271.5 (Dilling 1977; quoted, Mackay & Shiu 1981; Suntio et al. 1988; Howard 1990)
247.9 (calculated-P/C, Dilling 1977)
327.0 (concentration ratio-GC, Leighton & Calo 1981)
205.7 (calculated-P/C, Mabey et al. 1982)
303.9 (calculated-P/C, Thomas 1982)
229.1 (20°C, EPICS, Lincoff & Gossett 1983; quoted, Yurteri et al. 1987)
227.9 (20°C, EPICS, Lincoff & Gossett 1984)
199.6 (20°C, batch-stripping, Lincoff & Gossett 1984)
247.9 (selected, Daniels et al. 1985)
222.0 (EPICS, Gossett 1987; quoted, Suntio et al. 1988; Mackay & Shiu 1990; Thoms & Lion 1992)
173.0 (20°C, EPICS, Gossett 1987; quoted, Yurteri et al. 1987; Tse et al. 1987)
187.7 (20°C, EPICS, Yurteri et al. 1987)
323, 308 (quoted, calculated-P/C, Warner et al. 1987)
451.1 (calculated-QSAR, Nirmalakhandan & Speece 1988)
202.6 (20-25°C and low ionic strength, quoted, Pankow & Rosen 1988; Pankow 1990)
210.7 (quoted, Ryan et al. 1988)
254.4 (calculated-P/C, Suntio et al. 1988)
323.2 (selected, Valsaraj 1988)
317.3 (selected, Jury et al. 1990)
376.0 (calculated-P/C, Mackay & Shiu 1990)
247.9 (quoted, Merlin et al. 1992)
212.7 (20°C, Tse et al. 1992)

314.1 (30°C, Tse et al. 1992)

Octanol/Water Partition Coefficient, log $K_{OW}$:

1.25 (shake flask-GC, Hansch et al. 1975; quoted, Callahan et al. 1979; Suntio et al. 1988)
1.25 (Hansch & Leo 1979; quoted, Iwase et al. 1985; Ryan et al. 1988; Bodor et al. 1989; Sangster 1989; Mackay & Shiu 1990)
1.51 (Hansch & Leo 1979; quoted, Suntio et al. 1988)
1.25 (quoted, Valvani et al. 1981)
1.26 (calculated-f const., Mabey et al. 1982; Dobbs et al. 1989)
1.301 (selected, Mills et al. 1982)
1.25, 1.22 (quoted, calculated-HPLC-k', Hafkenscheid & Tomlinson 1983)
1.25, 1.63 (quoted, calculated-molar refraction, Yoshida et al. 1983)
1.30 (selected, Daniels et al. 1985)
1.25 (calculated, Hansch & Leo 1985; quoted, Howard 1990; Thoms & Lion 1992)
1.25, 1.45 (quoted, calculated-hydrophobicity const., Iwase et al. 1985)
1.15, 1.01 (quoted, calculated-molar volume & solvatochromic p., Taft et al. 1985)
1.51 (Afghan & Mackay 1980; Abernethy et al. 1988)
1.25 (Valvani et al. 1981; quoted, Amidon & Williams 1982)
1.15, 1.02 (quoted, calculated-molar volume & solvatochromic p., Leahy 1986)
1.25 (estimated-HPLC-k', Tomlinson & Hafkenscheid 1986)
1.15 (Abernethy et al. 1988)
1.15, 1.07 (quoted, calculated-molar volume & solvatochromic p., Kamlet et al. 1988)
1.51 (selected, Suntio et al. 1988)
1.25 (selected, Valsaraj 1988)
1.25, 1.50 (quoted, calculated-MO, Bodor et al. 1989)
1.25, 1.249 (quoted, calculated-$\pi$ substituent const., Bodor et al. 1989)
1.25 (recommended, Sangster 1989)
1.25 (quoted from Chiou 1981, Valsaraj & Thibodeaux 1989)
1.25 (quoted, Howard 1990; Thoms & Lion 1992; Van Leeuwen et al. 1992; Verhaar et al. 1992)
1.26, 1.32 (quoted, calculated-molar volume, Wang et al. 1992)

Bioconcentration Factor, log BCF:

0.699 (calculated as per Lyman et al. 1982, Howard 1990)
0.778 (microorganism-water, calculated from $K_{OW}$, Mabey et al. 1982)
0.60 (calculated- $\chi$ , Koch 1983)
0.362 (selected, Daniels et al. 1985)

Sorption Partition Coefficient, log $K_{OC}$:

0.944 (calculated-$K_{OW}$, Mabey et al. 1982)

1.44 (calculated- $\chi$ , Koch 1983; quoted, Bahnick & Doucette 1988)

1.0 (selected, Daniels et al. 1985)

1.39 (calculated- $\chi$ , Bahnick & Doucette 1988)

Sorption Partition Coefficient, log $K_{OM}$:

1.44, 1.23 (quoted, calculated- $\chi$ , Sabljic 1984)

Half-Lives in the Environment:

Air: disappearance half-life of 2.4-24 hours from air for the reaction with OH radicals (EPA 1974; quoted, Darnall et al. 1976); estimated disappearance time to be 12 hours to 1.0 year in simulated troposphere chamber (Dilling & Goersch 1979; quoted, Lyman 1982); 77 days, estimated as toxic chemical residence time with rate constant of $1.5x10^{-13}$ $cm^3$ $molecule^{-1}$ $sec^{-1}$ at 300 K for the reaction with OH radicals (Singh et al. 1981); estimated residence time in troposphere to be 160-250 days (Lyman 1982); 458-4584 hours, based on photooxidation half-life in air (Atkinson 1985; quoted, Howard et al. 1991); estimated tropospheric lifetime of 0.47 year (Nimitz & Skaggs 1992).

Surface water: 168-672 hours, based on estimated unacclimated aqueous aerobic biodegradation half-life (Howard et al. 1991); half-life of disappearance from water calculated from a pseudo first order equation of mesocosms experiment: 1-1.2 day for the first days up to 4 days and 2.71-2.80 days for a period of up to 14 days (Merlin et al. 1992).

Ground water: estimated half-life of 10 years in the ground water of the Netherlands (Zoeteman et al. 1981); 336-1344 hours, based on estimated unacclimated aqueous aerobic biodegradation half-life (Howard et al. 1991).

Sediment:

Soil: 10-50 days (Ryan et al. 1988); estimated volatilization half-life of 100 days from soil (Jury et al. 1990); aerobic half-lives : 1.3 day (0.16 ppm), 9.4 days (0.5 ppm), 191.4 days (5 ppm) all in sandy loam soil; 54.8 days in sand (0.5 ppm); 12.7 days (0.5 ppm) in sandy clay loam soil; 7.2 days (0.5 ppm) in clay with 50 days lag and anaerobic half-life of 21.5 days (5 ppm) with 70 days lag (Davis & Madsen 1991); 168-672 hours, based on estimated unacclimated aqueous aerobic biodegradation half-life (Howard et al. 1991).

Biota: 10-50 days, subject to plant uptake in soil via volatilization (Ryan et al. 1988); half-life of 0.4-0.5 day to eliminate from small fish (McCarty et al. 1992).

Environmental Fate Rate Constant and Half-Lives:

Volatilization: calculated half-life, 2.23 minutes (Mackay & Wolkoff 1973; quoted, Dilling et al. 1975; Dilling 1977; Callahan et al. 1979); calculated half-life, 20.7 minutes (Mackay & Leinonen 1975; quoted, Dilling 1977; Callahan et al. 1979);

estimated experimental half-life for 1 ppm in water at 25°C to be 21 ± 3 minutes when stirred at 200 rpm in water (Dilling et al. 1975; Dilling 1977; quoted, Callahan et al. 1979; Mills et al. 1982); half-life from water estimated to be 3.0 hours (Thomas 1982); half-life from soil estimated to be 100 days (Jury et al. 1990).

Photolysis: estimated photodecomposition half-life greater than 250 hours from a simulated environmental sunlight exposure ( >290 nm at 27 ± 1°C) study (Dilling et al. 1976; quoted, Callahan et al. 1979); photodegradation half-life of 30-120 days (Darnall et al. 1976; quoted, Daniels et al. 1985); photocatalyzed mineralization by the presence of $TiO_2$ with a rate of 1.6 ppm/min per gram of catalyst (Ollis 1985).

Oxidation: the reaction rate with hydroxy radicals is reported to be $1.04 \times 10^{-13}$ $cm^3$ $molecule^{-1}$ $sec^{-1}$ corresponding to a lifetime of 0.30 year (Cox et al. 1976; quoted, Callahan et al. 1979); calculated rate constants of $1.1 \times 10^{-13}$ $cm^3$ $molecule^{-1}$ $sec^{-1}$ for the reaction with OH radicals at 298 K with lifetime of 1.4 year in the troposphere (Davis et al. 1976; quoted, Altshuller 1980); calculated rate constants of $<<360$ $M^{-1}$ $hour^{-1}$ for singlet oxygen and 0.2 $M^{-1}$ $hour^{-1}$ for peroxy radical at 25°C (Mabey et al. 1982); photooxidation half-life of 458-4584 hours based on measured rate data for the vapor phase reaction with hydroxy radicals in air (Atkinson 1985; quoted, Howard et al. 1991); rate constant of $(9 \pm 6) \times 10^7$ $M^{-1}$ $sec^{-1}$ for the reaction with OH radicals in aqueous solution at pH 8.5 (Haag & Yao 1992) with reference to $CH_3CCl_3$ having rate constant of $4.0 \times 10^7$ $M^{-1}$ $sec^{-1}$ (Buxton et al. 1988; quoted, Haag & Yao 1992).

Hydrolysis: a minimum hydrolysis half-life of 18 months was estimated from aqueous reactivity experiments (Dilling et al. 1975; quoted, Callahan et al. 1979; Howard 1990); rate constant of $3.2 \times 10^{-11}$ $sec^{-1}$ with a maximum half-life of 704 years for hydrolysis at pH 7 and 25°C was reported from the extrapolated experimental data obtained at 100-150°C (Radding et al. 1977; quoted, Callahan et al. 1979; Mabey et al. 1982).

Biodegradation: completely biodegradable under aerobic conditions with sewage seed or activated sludge between 6 hours to 7 days (Rittman & McCarty 1980; Davis et al. 1981; Tabak et al. 1981; Klecka 1982; Stover & Kincannon 1983; quoted, Howard 1990); aqueous aerobic biodegradation half-life of 168-672 hours based on unacclimated aerobic screening test data (Kawasaki 1980; Tabak et al. 1981; quoted, Howard et al. 1991); aqueous anaerobic biodegradation half-life of 672-2688 hours based on unacclimated aerobic biodegradation half-life (Howard et al. 1991).

Biotransformation:

Bioconcentration and Uptake and Elimination Rate Constants ($k_1$ and $k_2$):

Common Name: Chloroform
Synonym: trichloromethane
Chemical Name: trichloromethane, chloroform
CAS Registry No: 67-66-3
Molecular Formula: $CHCl_3$
Molecular Weight: 119.38
Melting Point (°C):
-63.5 (Stull 1947; Weast 1982-83; Callahan et al. 1979; Horvath 1982; Mabey et al. 1982; Suntio et al. 1988; Mackay & Shiu 1990)
-63.59 (Dreisbach 1959; Dean 1985)
-64.0 (Verschueren 1983; Mailhot 1987)
-63.52 (Riddick et al. 1986)
-63.2 (Stephenson & Malanowski 1987)
Boiling Point (°C):
61.7 (Rex 1906; Weast 1982-83; Callahan et al. 1979; Horvath 1982; Mabey et al. 1982; McNally & Grob 1984; Dean 1985)
61.73 (Dreisbach 1959)
61.20, 61.25 (Boublik et al. 1973)
61.3 (Pearson & McConnell 1975)
61.0 (Jones et al. 1977/1978; Banerjee et al. 1990)
62.0 (Verschueren 1983; Mailhot 1987)
61.18 (Riddick et al. 1986)
71.3 (Stephenson & Malanowski 1987)
Density ($g/cm^3$ at 20°C):
1.4832 (Dreisbach 1959; Horvath 1982; Weast 1982-83; McNally & Grob 1984)
1.4799 (25°C, Dreisbach 1959)
1.489 (Verschueren 1983; quoted, Grathwohl 1990)
1.4985 (15°C, Dean 1985)
1.4891 (Riddick et al. 1986)
1.490 (Mailhot 1987; Gillham & Rao 1990)
Molar Volume ($cm^3/mol$):
80.64 (Hoy 1970; Amidon & Williams 1982)
0.805 ($V_M/100$, Taft et al. 1985; Kamlet et al. 1986,1987; Leahy 1986)
92.0 (LeBas method, Abernethy & Mackay 1987; Abernethy et al. 1988; Mackay & Shiu 1990)
0.427 (calculated intrinsic volume: $V_I/100$, Kamlet et al. 1987,1988; Thoms & Lion 1992)
80 (calculated from density, Mailhot 1987; Wang et al. 1992)
43.0 (intrinsic volume-van der Waals method, Abernethy et al. 1988)
81.0 (calculated from density, Abernethy et al. 1988)
Molecular Volume, ($Å^3$):
66.3 (Moriguchi et al. 1976)
61.3 (corrected for branching, Moriguchi et al. 1976)

Total Surface Area, TSA ($Å^2$):

238.68 (Amidon 1976; Horvath 1982)
95.9 (Moriguchi et al. 1976)
99.0 (Mackay et al. 1982; Valsaraj 1988; Valsaraj & Thibodeaux 1989)
224.0 (Iwase et al. 1985)
97.9 (Okouchi et al. 1992)

Heat of Fusion, $\Delta H_{fus}$, (kcal/mol):

2.104 (calculated, Dreisbach 1959)
2.280 (quoted, Riddick et al. 1986)

Fugacity Ratio, F:

1.0 (Suntio et al. 1988)

Water Solubility ($g/m^3$ or mg/L at 25 °C):

8220 (20°C, volumetric, Rex 1906)
7760 (30°C, volumetric, Rex 1906)
7710 (30°C, shake flask-interferometer, Gross & Saylor 1931; quoted, Shiu et al. 1990)
8520 (15°C, shake flask-interferometer, Gross & Saylor 1931; quoted, Shiu et al. 1990)
8000 (Wright & Schaffer 1932; quoted, Horvath 1982)
7361 (Seidell 1940; quoted, Deno & Berkheimer 1960; Hine & Mookerjee 1975)
7700 (Seidell 1941; quoted, Irmann 1965)
7900 (McGovern 1943; quoted, Shiu et al. 1990)
13320 (Booth & Everson 1948; quoted, Horvath 1982)
8000 (quoted, Gladis 1960)
7950 (Marsden & Mann 1962; quoted, Shiu et al. 1990)
8000 (20°C, Stephen & Stephen 1963)
14352, 3435 (quoted, calculated-$K_{OW}$, Hansch et al. 1968)
8150 (20°C, Riddick & Bunger 1970; quoted, Lo et al. 1986)
7950 (quoted, Dilling et al. 1975; Hutchinson et al. 1980; Mackay et al. 1980; Bobra et al. 1984; Howard 1990)
7361 (quoted, Hine & Mookerjee 1975)
7950 (Kenaga 1975,1980; quoted, Freed et al. 1977; Afghan & Mackay 1980; Mackay et al. 1980; Horvath 1982; Suntio et al. 1988)
8200 (20°C, shake flask-GC, Pearson & McConnell 1975; quoted, Jones et al. 1977/1978; Callahan et al. 1979; Mackay & Shiu 1981; Mabey et al. 1982; Mills et al. 1982; Mackay & Shiu 1990; Shiu et al. 1990; Thoms & Lion 1992)
8000 (20°C, Neely 1976; quoted, Mackay & Shiu 1981; Thoms & Lion 1992)
7950 (quoted, Chiou et al. 1977; Freed et al. 1977, Kenaga 1980; Horvath 1982)
8100 (20°C, quoted, Chiou & Freed 1977)
7840 (quoted, Dilling 1977; quoted, Smith et al. 1980; Horvath 1982)
8000 (20°C, Verschueren 1977; quoted, Wilson et al. 1981; Suntio et al. 1988)

7800 (quoted, Andelman 1978; Horvath 1982)
8000 (20°C, selected, Nathan 1978)
8160 (20°C, recommended, Sørensen & Arit 1979; quoted, Wright et al. 1992)
7230 (shake flask-LSC, Banerjee et al. 1980; quoted, Banerjee 1985; Shiu et al. 1990)
9300 (quoted, Cowen & Baynes 1980)
7361, 3863 (shake flask-LSC, calculated-f const., Veith et al. 1980)
9056, 3953 (quoted, calculated-$K_{OW}$, Valvani et al. 1981)
7360, 6869 (quoted, calculated-$K_{OW}$, molar volume, & M.P., , Amidon & Williams 1982)
7700, 7940 (quoted exptl., calculated-group contribution as per Irmann 1965, Horvath 1982)
7920 (recommended, Horvath 1982; quoted, Broholm et al. 1992)
8200 (recommended, Horvath 1982; quoted, Munz & Roberts 1986)
8000 (quoted from Mackay & Yuen 1979, Thomas 1982; quoted, Jury et al. 1984)
7223 (quoted actual exptl. value from Banerjee et al. 1980, Arbuckle 1983)
7802 (calculated-UNIFAC, Arbuckle 1983)
9300 (quoted, Verschueren, 1983; Shiu et al. 1990)
2525 (30°C, headspace-GC, McNally & Grob 1984)
7193 (calculated-UNIFAC, Banerjee 1985)
8200 (Dean 1985)
9056, 5211 (quoted from Hine & Mookerjee 1975, calculated-vapor pressure, molar volume & $\pi$, Kamlet et al. 1986)
9056, 4977 (quoted, calculated-molar volume & solvatochromic p., Leahy 1986)
8200 (radiometric method, Lo et al. 1986)
8150 (20°C, quoted, Riddick et al. 1986)
14353, 2552 (quoted, calculated-fragment consts., Wakita et al. 1986)
9056, 6123 (quoted, calculated-molar volume & solvatochromic p., Kamlet et al. 1987)
8451 (quoted, Mailhot 1987)
9600 (quoted, Warner et al. 1987)
8128 (quoted, Isnard & Lambert 1988,1989)
7925; 6266, 4436 (quoted; predicted- $\chi$ & polarizability, Nirmalakhandan & Speece 1988)
8000 (selected, Suntio et al. 1988)
9600 (selected, Valsaraj 1988)
9560 (quoted from Mackay et al. 1982, Valsaraj & Thibodeaux 1989)
7925 (20°C, quoted from Horvath 1982, Grathwohl 1990)
8200 (20°C, selected, Gillham & Rao 1990)
8668 (23-24°C, shake flask-GC, Broholm et al. 1992)
7924 (calculated-AI, Okouchi et al. 1992)
8200 (19.6°C, shake flask-GC/TC, Stephenson 1992)
7900 (29.5°C, shake flask-GC/TC, Stephenson 1992)
8259, 5847 (quoted exptl., calculated-molar volume, Wang et al. 1992)
8090, 8220 (20°C, calculated-activity coefficients, Wright et al. 1992)

Vapor Pressure (Pa at 25 °C):

21115 (20°C, Rex 1906)
31992 (30°C, Rex 1906)
25700 (Antoine eqn. regression, Stull 1947)
27033 (Stull 1947; Jorden 1954; quoted, Hine & Mookerjee 1975)
26241 (McGlashan et al. 1954; quoted, Bissell & Williamson 1975)
26126 (Moelwyn-Hughes & Missen 1957; quoted, Bissell & Williamson 1975)
26271 (Mueller & Kearns 1958; quoted, Bissell & Williamson 1975)
26313 (calculated-Antoine eqn., Dreisbach 1959)
32792 (Gallant, 1966; Howard 1990)
23080 (interpolated from Antoine eqn., Weast 1972-73)
26244, 18946 (calculated-Antoine eqn., Boublik et al. 1973)
26116 (static method, Bissell & Williamson 1975)
26660 (quoted, Dilling et al. 1975)
27033 (quoted, Hine & Mookerjee 1975)
20000 (20°C, Pearson & McConnell 1975; quoted, Callahan et al. 1979; Mabey et al. 1982)
32792 (20°C, Neely 1976; quoted, Thomas 1982)
25594 (quoted average, Dilling 1977)
25954 (selected, Nathan 1978)
24661 (23.4°C, quoted, Chiou et al. 1980)
25631 (quoted, Cowen & Baynes 1980)
11997 (20°C, selected, Mills et al. 1982)
2603.62 (estimated as per Perry & Chilton 1973, Arbuckle 1983)
21328 (20°C, Verschueren 1983; quoted, Suntio et al. 1988; Mackay & Shiu 1990)
25780, 25774 (quoted exptl., calculated-Antoine eqn., Boublik et al. 1984)
26222, 32084 (calculated-Antoine eqn., Boublik et al. 1984)
21328 (20°C, quoted, McNally & Grob 1984)
26221 (quoted, Kamlet et al. 1986)
25970 (quoted, Riddick et al. 1986)
26221.5 (calculated-Antoine eqn., Stephenson & Malanowski 1987)
26344 (quoted, Warner et al. 1987)
26340 (selected, Valsaraj 1988)
26260, 18262 (quoted, calculated-solvatochromic p. & UNIFAC, Banerjee et al. 1990)
20128 (selected, Gillham & Rao 1990)

Henry's Law Constant (Pa $m^3$/mol):

314.1 (20°C, Dilling et al. 1975; quoted, Nicholson et al. 1984)
440.8 (calculated as $1/K_{AW}$, $C_W/C_A$, Hine & Mookerjee 1975; Nirmalakhandan & Speece 1988; Howard 1990)
283.4 (20°C, McConnell et al. 1975; quoted, Jones et al. 1977/1978)
297.5, 297.5 (quoted as exptl., calculated-P/C, Neely 1976; quoted, Ryan et al. 1988)

322, 396.6 (quoted, calculated, Dilling 1977; quoted, Suntio et al. 1988)
486.3 (20°C, ESE 1980; quoted, Nicholson et al. 1984)
364.7 (20°C, Symons et al. 1981; quoted, Nicholson et al. 1984)
401 (concentration ratio-GC, Leighton & Calo 1981)
292 (20°C, calculated-P/C, Mabey et al. 1982; Mills et al. 1982)
486.3 (calculated-P/C, Thomas 1982)
536.9 (batch stripping, Munz & Roberts 1982; quoted, Roberts & Dändliker 1983; Roberts et al. 1985)
313.8 (quoted actual value from Kavanaugh & Trussell 1980, Arbuckle 1983)
432.2 (calculated-UNIFAC, Arbuckle 1983)
297.5 (calculated, Jury et al. 1984)
237.4 (20°C, EPICS, Lincoff & Gossett 1984)
308.0 (20°C, batch stripping, Lincoff & Gossett 1984)
303.9 (20°C, batch stripping, Nicholson et al. 1984)
314.1 (20°C, calculated-P/C, Nicholson et al. 1984)
343.4 (adsorption isotherm, Urano & Murata 1985)
372 (EPICS, Grossett 1987; quoted, Suntio et al. 1988; Mackay & Shiu 1990; Thoms & Lion 1992)
319.9 (20°C, calculated-P/C, McKone 1987)
347 (Munz & Roberts 1987; quoted, Tancréde et al. 1992)
343, 327 (quoted, calculated-P/C, Warner et al. 1987)
427 (EPICS, Ashworth et al. 1988)
298 (20°C, interpolated of Gossett 1987 data, Grathwohl 1990)
259 (calculated-QSAR, Nirmalakhandan & Speece 1988)
293.8 (20-25°C and low ionic strength, quoted, Pankow & Rosen 1988; Pankow 1990)
318 (calculated-P/C, Suntio et al. 1988; Mackay & Shiu 1990)
327.2 (selected, Valsaraj 1988)
297.5 (selected, Jury et al. 1990)
198 (quoted from Tancréde & Yanagisawa 1990, Tancréde et al. 1992)
283.7 (20°C, quoted from Gossett 1987, Tse et al. 1992)

Octanol/Water Partition Coefficient, log $K_{OW}$:

1.97 (Hansch & Anderson 1967; quoted, Freed et al. 1977; Mackay et al. 1980; Sangster 1989)
1.97 (shake flask-AS, Hansch et al. 1968)
1.97 (Leo et al. 1971; quoted, Moriguchi et al. 1976; Hansch & Leo 1979; Chiou et al. 1977; Callahan et al. 1979; McDuffie 1981; Schwarzenbach et al. 1983; Iwase et al. 1985; Munz & Roberts 1986; Hodson et al. 1988; Suntio et al. 1988; Bodor et al. 1989; Mackay & Shiu 1990)
1.97 (shake flask-GC, Hansch et al. 1975)
1.97, 1.87 (quoted, calculated-hydrophobicity const., Moriguchi et al. 1976)
2.00 (Chiou & Freed 1977)

1.97, 2.11 (quoted observed & calculated values, Chou & Jurs 1979)
1.67, 1.96 (calculated-$\pi$, f consts., Chou & Jurs 1979)
1.94 (Hansch & Leo 1979; quoted, Suntio et al. 1988; Sangster 1989)
1.95, 1.43 (quoted, calculated-S, Mackay et al. 1980; quoted, Sangster 1989)
1.90 (shake flask-LSC, Banerjee et al. 1980; quoted, Davies & Dobbs 1984; Banerjee & Howard 1988; Suntio et al. 1988; Sangster 1989)
1.96, 1.90 (quoted, shake flask-LSC, Veith et al. 1980)
1.97, 1.81 (quoted, calculated-HPLC-k', McDuffie 1981)
1.96 (Valvani et al. 1981; quoted, Amidon & Williams 1982; Valsaraj 1988)
2.15 (HPLC-k', Wells et al. 1981)
1.96 (calculated-f constant, Mabey et al. 1982)
1.95 (Mackay 1982; quoted, Schüürmann & Klein 1988)
1.97 (selected, Mills et al. 1982)
1.90 (Veith & Kosian 1982; quoted, Saito et al. 1992)
1.90, 1.91 (quoted, calculated-UNIFAC, Arbuckle 1983; quoted, Sangster 1989)
1.95, 1.90 (quoted, calculated-HPLC-k', Hafkenscheid & Tomlinson 1983)
1.97 (quoted, Verschueren 1983; Yoshida et al. 1983)
2.22 (calculated-molar fraction, Yoshida et al. 1983)
1.97, 2.07 (quoted, calculated-hydrophobicity const., Iwase et al. 1985)
1.94, 1.74 (quoted, calculated-molar volume & solvatochromic p., Taft et al. 1985)
2.02 (Abernethy & Mackay 1987; Abernethy et al. 1988)
1.94, 1.80 (selected, calculated-molar volume & solvatochromic p., Leahy 1986)
2.14, 2.13, 2.03 (HPLC-k', Tomlinson & Hafkenscheid 1986)
1.94 (estimated-HPLC-k', Tomlinson & Hafkenscheid 1986)
1.94 (quoted, Mailhot 1987)
1.97, 1.66 (quoted, calculated-UNIFAC, Banerjee & Howard 1988)
1.90 (quoted, Isnard & Lambert 1988,1989; Ryan et al. 1988)
1.94, 1.80 (quoted, calculated-molar volume & solvatochromic p., Kamlet et al. 1988)
1.97, 1.95 (quoted of THOR 1986, calculated, Schüürmann & Klein 1988)
1.95 (selected, Suntio et al. 1988)
1.96, 2.20 (quoted, calculated-MO, Bodor et al. 1989)
1.97, 1.952 (quoted, calculated-$\pi$ subtituent const., Bodor et al. 1989)
1.97 (recommended, Sangster 1989)
1.96 (quoted from Chiou 1981, Valsaraj & Thibodeaux 1989)
1.94 (quoted from Hansch & Leo 1979, Grathwohl 1990)
1.97 (quoted from Hansch et al. 1975, Howard 1990; quoted, Thoms & Lion 1992)
1.97 (quoted, Van Leeuwen et al. 1992; Verhaar et al. 1992)
1.97, 1.78 (quoted, calculated-molar volume, Wang et al. 1992)

Bioconcentration Factor, log BCF:

0.92 (calculate-$K_{OW}$, Veith et al. 1979; quoted, Veith et al. 1980)
0.78 (bluegill sunfish, Barrows et al. 1980)
0.78 (bluegill sunfish, Veith et al. 1980; Veith & Kosian 1982; quoted, Davies & Dobbs 1984; Suntio et al. 1988; Saito et al. 1992)
0.778 (quoted, bluegill sunfish, Bysshe 1982)
0.63 (Mackay 1982)
1.41 (microorganisms-water, calculated from $K_{OW}$, Mabey et al. 1982)
0.97 (calculated- $\chi$ , Koch 1983)
0.78 (bluegill sunfish, LSC, Davies & Dobbs 1984)
2.84 (green alga, Mailhot 1987)
0.52-1.01 (rainbow trout, Howard 1990)
0.204-0.4 (bluegill sunfish, Howard 1990)
0.46-0.49 (large mouth bass, Howard 1990)
0.52-0.57 (catfish, Howard 1990)
0.78 (bluegill sunfish, Davies & Dobbs 1984)
2.84 (*selenastrum capricornutum*, Mailhot 1987)
0.78 (quoted, Isnard & Lambert 1988)

Sorption Partition Coefficient, log $K_{OC}$:

1.64 (calculated-$K_{OW}$, Mabey et al. 1982)
1.65 (calculated- $\chi$ , Koch 1983)
2.15 (wastewater solids with correlation to $K_{OW}$ Dobbs et al. 1989)
1.44 (20°C, soil, sand & loess, Grathwohl 1990)
1.98 (20°C, weathered shale, mudrock, Grathwohl 1990)
2.79 (20°C, unweathered shale & mudrock, Grathwohl 1990)
1.85, 1.92 (20°C, calculated, Grathwohl 1990)
1.53 (soil, Howard 1990)
1.462 (selected, Jury et al. 1990)

Sorption Partition Coefficient, log $K_{OM}$:

1.65, 1.40 (quoted, calculated- $\chi$ , Sabljic 1984)

Half-Lives in the Environment:

Air: disappearance half-life of 2.4-24 hours from air for the reaction with OH radicals (EPA 1974; quoted, Darnall et al. 1976); residence time of 1.7 year in troposphere, based on one compartment nonsteady state model (Singh et al. 1978; quoted, Lyman 1982); 116 days, estimated as toxic chemical residence time with rate constant of $1.0 \times 10^{-13}$ $cm^3$ $molecule^{-1}$ $sec^{-1}$ at 300 K for the reaction with OH radicals (Singh et al. 1981); estimated residence time in N. troposphere to be 100 days (Lyman 1982); 623-6231 hours, based on photooxidation half-life in air from measured reaction data for the vapor phase reaction with hydroxyl radicals in air (Atkinson 1985; quoted, Howard et al. 1991); estimated tropospheric lifetime of 0.17 year (Nimitz & Skaggs 1992).

Surface water: 672-4320 hours, based on estimated aqueous aerobic biodegradation half-life (Howard et al. 1991).

Ground water: 1344-43200 hours, based on unacclimated aqueous aerobic biodegradation and grab sample data of aerobic soil from a ground water aquifer (Wilson et al. 1983; quoted, Howard et al. 1991).

Soil: 10-50 days (Ryan et al. 1988); 100 days, estimated volatilization loss from soil (Jury et al. 1990); disappearance half-life of 4.1 days was calculated from 1st-order kinetic for volatilization loss from soil mixtures (Anderson et al. 1991); 672-4320 hours, based on estimated aqueous aerobic biodegradation (Howard et al. 1991).

Biota: $<1$ day in tissues of bluegill sunfish (Barrows et al. 1980); 10-50 days, subject plant uptake via volatilization (Ryan et al. 1988).

Environmental Fate Rate Constants or Half-Lives:

Volatilization: calculated half-life of 1.4 minute (Mackay & Wolkoff 1973; quoted, Dilling et al. 1975; Dilling 1977); experimental evaporation rate of $13.2 \times 10^{-5}$ g/cm$^2$ sec$^{-1}$ to still air (Chiou & Freed 1977; Chiou et al. 1980); half-life of 29 miunutes to 11.3 days from rivers & streams (calculated using Langbein & Durum 1967 published $O_2$ reaeration values, Kaczmar et al. 1984); calculated half-life of 31.2 hours from rivers & streams (Kaczmar et al. 1984); the primary transport process from the aquatic environment with half-life of 21-26 minutes at pH 7 and 25°C stirred at 200 rpm in an open container (Dilling et al. 1975; Dilling 1977; quoted, Callahan et al. 1979; Mills et al. 1982); the ratio of evaporation rate constant to the oxygen reaeration rate constant: measured to be 0.52 as compared to an estimated ratio of 0.47 (Smith et al. 1980); different laboratory studies of evaporation from water given half-lives of 3-5.6 hours with moderate mixing conditions (Smith et al. 1980; Rathbun & Tai 1981; Lyman et al. 1982; quoted, Howard 1990); estimated half-life from water to be 3.7 hours (Thomas 1982); 29 minutes-11.3 days, calculated using published $O_2$ reaeration values; and 31.2 hours, calculated using estimated $O_2$ reaeration rate constant (Kaczmar et al. 1984); a modeling predicted half-lives of 36 hours in a river, 40 hours in a pond and 9-10 days in a lake (USEPA 1984; quoted, Howard 1990); half-life of 4 hours at 20°C was predicted from a model river of 1 m deep at flowing speed of 1 m/sec with a wind velocity of 3 m/sec and its Henry's law constant (Smith et al. 1980; quoted, Howard 1990); half-life of 1.2 day in Rhine River and 31 days in a lake in Rhine Basin. (Zoeteman et al. 1980; quoted, Howard 1990).

Photolysis: not important only by UV in the stratosphere (Robbins 1976); probably not significant in aquatic systems (Callahan et al. 1979); photocatalyzed mineralization by the presence of $TiO_2$ with a rate of 4.4 ppm/min per gram of catalyst (Ollis 1985).

Oxidation: not important for aqueous phase (Dilling et al. 1975); calculated rate constant of $1.0 \times 10^{-13}$ $cm^3$ $molecule^{-1}$ $sec^{-1}$ for the reaction with OH radicals at 298 K with lifetime of 1.5 year in the troposphere (Davis et al. 1976; quoted, Altshuller 1980); the primary fate this compound is attacked by hydroxyl radicals in the troposphere with half-life of 0.19-0.32 year (Callahan et al. 1979); approximate $6.6 \times 10^{10}$ $cm^3$ $mol^{-1}$ $sec^{-1}$ of estimated rate constant for the reaction with OH radicals at 300 K (Lyman 1982); estimated half-life of 78.5-3140 years from the rate constant for the reaction with OH radicals (Dorfman & Adams 1973); rate constant of $<<$ 360 $M^{-1}$ $hour^{-1}$ for singlet oxygen and 0.7 $M^{-1}$ $hour^{-1}$ for peroxy radical both at 25°C (Mabey et al. 1982); photooxidation half-life of 26-260 days, based on measured data for the vapor phase reaction with hydroxyl radicals in air (Atkinson 1985; quoted, Howard et al. 1991); rate constant of $(5.4 \pm 3.0) \times 10^7$ $M^{-1}$ $sec^{-1}$ for the reaction with OH radicals in aqueous solution at pH 8.5 (Haag & Yao 1992) with reference to $CH_3CCl_3$ having rate constant of $4.0 \times 10^7$ $M^{-1}$ $sec^{-1}$ (Buxton et al. 1988; quoted, Haag & Yao 1992).

Hydrolysis: first-order rate constant of 0.045 $month^{-1}$ with a half-life about 15 months (Dilling et al. 1975; quoted, Callahan et al. 1979); $6.9 \times 10^{-12}$ $sec^{-1}$ (Radding 1976); no hydrolysis in acidic aqueous solutions, and rate constant in alkaline aqueous solution, 0.23 $M^{-1}$ $hour^{-1}$ at 25°C, and $2.5 \times 10^{-9}$ $M^{-1}$ $hour^{-1}$ at 25°C in neutral aqueous solutions (Mabey & Mill 1978; Mabey et al. 1982; Mills et al. 1982); probably not a significant fate process with half-life of 3500 years, based on reported rate constant of $6.9 \times 10^{-12}$ $sec^{-1}$ at pH 7 and 25°C (Mabey & Mill 1978; quoted, Callahan et al. 1979; Haque et al. 1980).

Biodegradation: very slow by BOD bottle experiments (Pearson & McConnell 1975);
Aerobic half-life: 4 weeks to 6 months by unacclimated screening tests (Kawasaki 1980; Flathman & Dahlgran 1982; quoted, Howard et al. 1991);
Anaerobic half-life: 1 to 4 weeks by unacclimated unaerobic screening tests (Bouwer et al. 1981; Bouwer & McCarty 1983; quoted, Howard et al. 1991);
0.5 $day^{-1}$ (Tabak et al. 1981; Mills et al. 1982).

Bioaccumulation: weak to moderate bioaccumulation; no evidence of biomagnification of trichloromethane in marine food chain (Callahan et al. 1979).

Bioconcentration, Uptake($k_1$) and Elimination($k_2$) Rate Constants or Half-Lives:

Common Name: Carbon tetrachloride
Synonym: tetrachloromethane, methane tetrachloride, perchloromethane, benzinoform
Chemical Name: carbon tetrachloride
CAS Registry No: 56-23-5
Molecular Formula: $CCl_4$
Molecular Weight: 153.82
Melting Point (°C):
-22.6 (Stull 1947)
-22.99 (Dreisbach 1959; Horvath 1982)
-22.9 (Weast 1977; quoted, Mackay & Shiu 1981; Callahan et al. 1979; Ma et al. 1990)
-23.0 (Weast 1982-83; Verschueren 1983; Suntio et al. 1988; Howard 1990; Mackay & Shiu 1990)
-22.82 (Riddick et al. 1986)
-21.2 (Stephenson & Malanowski 1987)
Boiling Point (°C):
76.7 (Kahlbaum & Arndt 1898; Rex 1906; Boublik et al. 1973; Verschueren 1983; Stephenson & Malanowski 1987)
76.54 (Dreisbach 1959; Weast 1977; quoted, Callahan et al. 1979; Horvath 1982; Mabey et al. 1982; Howard 1990)
76.8 (McConnell et al. 1975; Pearson & McConnell 1975)
76.50 (Weast 1982-83; Mackay & Shiu 1981,1990)
76.64 (Riddick et al. 1986)
77.0 (Jones et al. 1977/1978; Banerjee et al. 1990)
Density ($g/cm^3$ at 20°C):
1.591 (Kahlbaum & Arndt 1898)
1.594 (McGovern 1943; Horvath 1982)
1.5940 (Dreisbach 1959; Weast 1982-83; Riddick et al. 1986)
1.5843 (25°C, Dreisbach 1959)
1.590 (Verschueren 1983; Gillham & Rao 1990)
Molar Volume ($cm^3/mol$):
97.10 (Hoy 1970; Pierotti 1976; Amidon & Williams 1982)
0.968 ($V_M/100$, Taft et al. 1985; Kamlet et al. 1986,1987; Leahy 1986)
0.514 (calculated intrinsic volume: $V_I/100$, Leahy 1986; Kamlet et al. 1987; Hawker 1990)
113.0 (LeBas method, Abernethy & Mackay 1987; Abernethy et al. 1988; Mackay & Shiu 1990)
102.0, 101.0, 104.8, 113.0 (exptl., Tyn & Calus method, Schroeder method, LeBas method, Reid et al. 1987)
51.0 (intrinsic volume-van der Waals method, Abernethy et al. 1988)
97.0 (calculated from density, Abernethy et al. 1988)
96.0 (calculated from density, Wang et al. 1992)
Molecular Volume ($A^3$):

Total Surface Area, TSA ($A^2$):

260.69 (Amidon 1976; quoted, Horvath 1982)

119.0 (Mackay et al. 1982; Valsaraj 1988; Valsaraj & Thibodeaux 1989)

248.2 (Iwase et al. 1985)

117.0 (Okouchi et al. 1992)

Heat of Fusion, $\Delta H_{fus}$, (kcal/mol):

0.775 (calculated, Dreisbach 1959)

0.581 (quoted, Riddick et al. 1986)

Entropy of Fusion, $\Delta S_{fus}$ (cal/mol K, or e.u.):

Fugacity Ratio at 25 °C, F:

1.0 (Suntio et al. 1988)

Water Solubility (g/m³ or mg/L at 25°C):

800 (20°C, volumetric, Rex 1906)

850 (30°C, volumetric, Rex 1906)

770 (Gross 1929a,b; quoted, Horvath 1982)

770 (15°C, shake flask-interferometer, Gross & Saylor 1931; quoted, Horvath 1982; Shiu et al. 1990)

810 (30°C, shake flask-interferometer, Gross & Saylor 1931)

800 (20°C, Smith 1932; quoted, Horvath 1982)

771 (Seidell 1940; quoted, Deno & Berkheimer 1960; Hine & Mookerjee 1975; Chiou 1981)

780 (Seidell 1941; quoted, Irmann 1965)

800 (McGovern 1943; quoted, Dilling 1977; Mackay & Shiu 1981; Horvath 1982; Love Jr. & Eilers 1982; Chiou et al. 1988; Suntio et al. 1988; Shiu et al. 1990; Rutherford & Chiou 1992)

800 (Hodgman 1952; quoted, Kenaga 1975; Freed et al. 1977)

770 (15°C, Jones et al. 1957; quoted, Horvath 1982)

800 (quoted, Gladis 1960)

762 (Liu & Huang 1961; quoted, Horvath 1982)

800 (Marsden & Mann 1962; quoted, Ma et al. 1990)

800 (20°C, Metcalf 1962; quoted, Horvath 1982)

800 (Günther et al. 1968; quoted, Kenaga 1980; Kenaga & Goring 1980; Horvath 1982; Shiu et al. 1990)

770 (Riddick & Bunger 1970)

810 (15°C, Svetlanov et al. 1971; quoted, Horvath 1982)

800 (20°C, Hancock 1973; quoted, Horvath 1982)

762 (Gmelins 1974; quoted, Horvath 1982)

800 (20°C, Neely et al. 1974; quoted, Suntio et al. 1988)

771 (quoted, Hine & Mookerjee 1975)

785 (20°C, shake flask-GC, McConnell et al. 1975; Pearson & McConnell 1975; quoted, Jones et al. 1977/1978; Callahan et al. 1979; Mackay & Shiu 1981;

Mabey et al. 1982; Mills et al. 1982; Neely 1984; Ma et al. 1990; Shiu et al. 1990)

769, 1009 (exptl., calculated-TSA, Amidon 1976; quoted, Horvath 1982)

800 (20°C, Neely 1976; quoted, Mackay & Shiu 1981,1990; Shiu et al. 1990; Rutherford & Chiou 1992)

800 (20°C, Chiou et al. 1977; quoted, Horvath 1982)

800 (GC/ECD, Dilling 1977; quoted, Smith et al. 1980; Horvath 1982; Warner et al. 1987)

1160 (Verschueren 1977,1983; quoted, Mackay & Shiu 1981; Suntio et al. 1988; Shiu et al. 1990)

1600 (selected, Nathan 1978)

788 (20°C, Selenka & Bauer 1978; quoted, Horvath 1982)

870 (Aref'eva et al. 1979; quoted, Horvath 1982)

778 (20°C, recommended, Sørensen & Arit 1979; quoted, Wright et al. 1992)

757 (shake flask-LSC, Banerjee et al. 1980; quoted, Davies & Dobbs 1984; Ma et al. 1990; Shiu et al. 1990)

1160 (quoted, Cowen & Baynes 1980)

800 (Kenaga 1980; quoted, Ma et al. 1990)

771, 486 (quoted exptl., calculated, Veith et al. 1980)

927, 671 (quoted exptl., calculated-$K_{OW}$, Valvani et al. 1981)

927, 865 (quoted exptl., calculated-molar volume, Amidon & Williams 1982)

780, 1000 (quoted exptl., calculated-group contribution as per Irmann 1965, Horvath 1982)

793.4 (recommended, Horvath 1982)

800 (quoted, Thomas 1982)

800 (quoted from Mackay & Yuen 1979, Thomas 1982; quoted, Jury et al. 1984)

757 (quoted actual exptl. value from Banerjee et al. 1980, Arbuckle 1983)

831 (calculated-UNIFAC, Arbuckle 1983)

753.4 (calculated-UNIFAC, Banerjee 1985; quoted, Farrington 1991)

800 (quoted, Urano & Murata 1985)

927, 1436 (quoted from Hine & Mookerjee 1975, calculated-vapor pressure, molar volume & solvatochromic p., Kamlet et al. 1986)

927, 1016 (quoted, calculated- molar volume & solvatochromic p., Leahy 1986)

800 (radiometric method, Lo et al. 1986)

803 (recommended value quoted from Horvath 1982, Munz & Roberts 1986)

770 (quoted, Riddick et al. 1986)

807, 789 (quoted, calculated-fragment const., Wakita et al. 1986)

927, 1140 (quoted, calculated- molar volume & solvatochromic p., Kamlet et al. 1987)

794 (quoted, Isnard & Lambert 1988,1989)

791; 895,1552 (quoted exptl.; predicted- $\chi$ & polarizability, Nirmalakhandan & Speece 1988; quoted, Ma et al. 1990)

900 (selected, Suntio et al. 1988)

800 (selected, Valsaraj 1988)

800 (quoted from Mackay et al. 1982, Valsaraj & Thibodeaux 1989)
785 (20°C, selected, Gillham & Rao 1990)
805 (20°C, quoted from Horvath 1982, Howard 1990; Ma et al. 1990)
800 (selected, Ma et al. 1990)
793 (quoted, Broholm et al. 1992)
780 (23-24°C, shake flask-GC, Broholm et al. 1992)
791 (calculated-AI, Okouchi et al. 1992)
600 (20.5°C, shake flask-GC/TC, Stephenson 1992)
720 (31.0°C, shake flask-GC/TC, Stephenson 1992)
789, 2075 (quoted exptl., calculated-molar volume, Wang et al. 1992)
701, 702 (20°C, calculated-activity coefficients, Wright et al. 1992)

Vapor Pressure (Pa at 25°C):
17170 (20°C, Rex 1906)
18809 (30°C, Rex 1906)
15184 (Scatchard et al. 1939; quoted, Bissell & Williamson 1975)
14530 (McGovern 1943)
14339 (Antoine eqn. regression, Stull 1947)
15193 (McGlashan et al. 1954; quoted, Bissell & Williamson 1975)
15096 (Moelwyn-Hughes & Missen 1957; quoted, Bissell & Williamson 1975)
15356 (calculated-Antoine eqn., Dreisbach 1959)
15220 (Hildenbrand & McDonald 1959; quoted, Bissell & Williamson 1975)
15227 (Marsh 1968; quoted, Bissell & Williamson 1975)
13188 (calculated-Antoine eqn., Weast 1972-73)
15249, 15240 (calculated-Antoine eqn., Boublik et al. 1973)
13200 (Weast 1972-73; quoted, Chiou et al. 1988; Ma et al. 1990)
15190 (static method, Bissell & Williamson 1975)
14610 (quoted, Hine & Mookerjee 1975; Kamlet et al. 1986)
12000 (20°C, Pearson & McConnell 1975; quoted, Callahan et al. 1979; Mackay & Shiu 1981; Mabey et al. 1982; Mills et al. 1982; Suntio et al. 1988; Ma et al. 1990)
12130 (20°C, Neely 1976; quoted, Mackay & Shiu 1981; Thomas 1982; Suntio et al. 1988)
15060 (quoted average, Dilling 1977; quoted, Mackay & Shiu 1981; Love Jr. & Eilers 1982; Suntio et al. 1988; Ma et al. 1990)
14000 (Verschueren 1977; quoted, Ma et al. 1990)
15363 (selected, Nathan 1978)
14690 (quoted, Cowen & Baynes 1980)
11853 (estimated as per Perry & Chilton 1973, Arbuckle 1983)
15063 (quoted, Verschueren 1983)
15226 (calculated-Antoine eqn., Boublik et al. 1984)
15372, 15304 (25.2°C, quoted exptl., calculated-Antoine eqn., Boublik et al. 1984)

15170 (Daubert & Danner 1985; quoted, Howard 1990; Ma et al. 1990)
14610 (quoted, Kamlet et al. 1986)
15360 (quoted, Riddick et al. 1986)
15060 (EPICS, Gossett 1987; quoted, Mackay & Shiu 1990)
15214 (calculated-Antoine eqn., Stephenson & Malanowski 1987)
14000 (selected, Suntio et al. 1988)
15097 (quoted, Warner et al. 1987)
15095 (selected, Valsaraj 1988)
15330, 9158 (quoted, calculated-solvatochromic p. & UNIFAC, Banerjee et al. 1990)
11997 (20°C, selected, Gillham & Rao 1990)
14000 (selected, Ma et al. 1990)
12138 (20°C, quoted from DIPPR, Tse et al. 1992)
18898 (30°C, quoted from DIPPR, Tse et al. 1992)

Henry's Law Constant (Pa $m^3$/mol):

2912 (calculated as $1/K_{AW}$, $C_W/C_A$, reported as exptl., Hine & Mookerjee 1975; quoted, Nirmalakhandan & Speece 1988)
152.84 (calculated-bond method, Hine & Mookerjee 1975)
2216 (20°C, McConnell et al. 1975; quoted, Jones et al. 1977/1978)
2256, 2677; 2454 (quoted exptl.; calculated-P/C, Neely 1976)
2157, 2975 (quoted, calculated-P/C, Dilling 1977; quoted, Love Jr. & Eilers 1982)
2000 (Dilling 1977; quoted, Mackay & Shiu 1981)
2776 (batch stripping, Mackay et al. 1979; quoted, Yurteri et al. 1987)
2774 (concentration ratio, Leighton & Calo 1981)
2160 (calculated-P/C, Mackay & Shiu 1981; quoted, Suntio et al. 1988; Ma et al. 1990)
2330 (20°C, calculated-P/C, Mabey et al. 1982)
2454 (20°C, batch stripping, Munz & Roberts 1982; quoted, Roberts & Dändliker 1983; Roberts et al. 1985; Yurteri et al. 1987)
2348 (quoted actual value from Kavanaugh & Trussell 1980, Arbuckle 1983)
2418 (20°C, predicted-UNIFAC, Arbuckle 1983; quoted, Yurteri et al. 1987)
3081 (20°C, EPICS, Lincoff & Gossett 1983; Gossett 1985; quoted, Yurteri et al. 1987)
2330 (calculated-P/C, Thomas 1982; quoted, Jury et al. 1984)
3060 (adsorption isotherm, Urano & Murata 1985)
2930 (20°C, EPICS, Ashworth et al. 1986; quoted, Yurteri et al. 1987)
2369 (20°C, multiple equilibration, Roberts 1986; quoted, Yurteri et al. 1987)
3080 (EPICS, Gossett 1987; quoted, Suntio et al. 1988; Howard 1990; Ma et al. 1990; Mackay & Shiu 1990; Tancréde & Yanagisawa 1990)
2367 (20°C, EPICS, Gossett 1987; quoted, Yurteri et al. 1987; Tse et al. 1992)
2266 (20°C, calculated-P/C, McKone 1987)
3027 (Munz & Roberts 1987; quoted, Tancréde & Yanagisawa 1990)

3060, 2898 (quoted, calculated-P/C, Warner et al. 1987)
2281 (20°C, EPICS, Yurteri et al. 1987)
2989 (EPICS, Ashworth et al. 1988; quoted, Ma et al. 1990)
2330 (20-25°C and low ionic strength, quoted, Pankow & Rosen 1988; Ma et al. 1990; Pankow 1990)
2912, 2846 (quoted, calculated-QSAR, Nirmalakhandan & Speece 1988)
2379.7 (quoted, Ryan et al. 1988)
293.3 (calculated-P/C, Suntio et al. 1988)
3060 (selected, Valsaraj 1988)
2895 (calculated-P/C, Mackay & Shiu 1990)
2875 (purge & trap-GC/ECD, Tancréde & Yanagisawa 1990)
2900 (quoted from Tancréde & Yanagisawa 1990, Tancréde et al. 1992)
2067 (20°C, Tse et al. 1992)
3414 (30°C, Tse et al. 1992)

Octanol/Water Partition Coefficient, log $K_{OW}$:
2.64 (Macy 1948; quoted, Kenaga 1975; Chiou et al. 1977; Freed et al. 1977; Mackay et al. 1980)
2.64 (Hansch & Elkins 1971; Leo et al. 1971; Neely et al. 1974; Chiou et al. 1977; quoted, Callahan et al. 1979; Neely 1979; Green et al. 1983; Ryan et al. 1988; Dobbs et al. 1989; Thomann 1989; Ma et al. 1990)
2.62 (shake flask-GC, Chiou et al. 1977; quoted, Suntio et al. 1988; Sangster 1989)
2.70 (Verschueren 1977; quoted, Ma et al. 1990)
2.83 (Hansch & Leo 1979; quoted, Valvani et al. 1981; Sangster 1989; Ma et al. 1990; Mackay & Shiu 1990; Rutherford & Chiou 1992)
2.83, 2.96 (quoted, calculated-f const., Chou & Jurs 1979)
2.83, 2.99 (quoted, calculated-f const., Rekker & Kort 1979)
2.64 (Veith et al. 1979; Veith & Kosian 1982; quoted, Ma et al. 1990; Saito et al. 1992)
2.64 (quoted, Kenaga & Goring 1980)
2.54 (calculated-S, Mackay et al. 1980; quoted, Sangster 1989)
2.73 (shake flask-LSC, Banerjee et al. 1980; quoted, McDuffie 1981; Arbuckle 1983; Davies & Dobbs 1984; Sangster 1989; Ma et al. 1990)
2.73 (shake flask-LSC, Veith et al. 1980; Veith & Kosian 1982; quoted, Ma et al. 1990; Saito et al. 1992)
2.96 (calculated-f const., Veith et al. 1980)
2.94 (calculated-HPLC-k', McDuffie 1981)
2.96 (calculated-f const., Mabey et al. 1982)
2.64 (Mackay 1982; quoted, Schüürmann & Klein 1988)
2.60 (selected, Mills et al. 1982)
2.81 (calculated-UNIFAC with octanol & water mutual solubility considered, Arbuckle 1983; quoted, Sangster 1989)

2.90 (calculated-UNIFAC with octanol & water mutual solubility not considered, Arbuckle 1983)
2.73, 2.83 (quoted, calculated-HPLC-k', Hafkenscheid & Tomlinson 1983)
2.64 (quoted of calculated value, Verschueren 1983)
2.64, 2.83 (quoted, calculated-molar refraction, Yoshida et al. 1983; quoted, Ma et al. 1990)
2.73, 2.83 (Geyer et al. 1984)
2.64 (quoted, Hawker & Connell 1985)

2.62, 2.53 (quoted, calculated-hydrophobicity const., Iwase et al. 1985)
2.83, 2.46 (quoted, calculated-molar volume & solvatochromic p., Taft et al. 1985)
2.03 (estimated-HPLC, Eadsforth 1986; quoted, Sangster 1989)
2.83, 2.63 (quoted, calculated-molar volume & solvatochromic p., Leahy 1986)
2.88 (calculated, Leo 1986; quoted, Schüürmann & Klein 1988)
2.83 (THOR 1986; quoted, Schüürmann & Klein 1988)
2.73 (estimated-HPLC-k', Tomlinson & Hafkenscheid 1986)
2.79 (Abernethy & Mackay 1987; Abernethy et al. 1988)
2.83, 2.38 (quoted, calculated-UNIFAC, Banerjee & Howard 1988)
2.64 (Connell & Hawker 1988; Hawker & Connell 1988; Hawker 1990; quoted, Ma et al. 1990)
2.64 (quoted, Hodson et al. 1988; Isnard & Lambert 1988,1989; Banerjee & Baughman 1991)
2.63, 2.55 (quoted, calculated-molar volume & solvatochromic p., Kamlet et al. 1988)
2.70 (selected, Suntio et al. 1988)
2.83 (selected, Valsaraj 1988)
2.73, 3.00 (quoted, calculated-MO, Bodor et al. 1989)
2.83, 2.88 (quoted, calculated-$\pi$ substituent const., Bodor et al. 1989)
2.83 (quoted from Chiou 1981, Valsaraj & Thibodeaux 1989)
2.83 (recommended, Sangster 1989; quoted, Ma et al. 1990)
2.80 (selected, Ma et al. 1990)
2.83 (quoted, Van Leeuwen et al. 1992; Verhaar et al. 1992)
2.64, 2.24 (quoted, calculated-molar volume, Wang et al. 1992)

Bioconcentration Factor, log BCF:
1.24 (rainbow trout, Neely et al. 1974; Veith et al. 1979; quoted, Verschueren 1983; Suntio et al. 1988; Howard 1990; Ma et al. 1990)
1.25 (trout muscle, calculated-$k_1/k_2$, Neely et al. 1974; quoted, Kenaga & Goring 1980; Kenaga 1980; Chiou 1981)
1.62 (calculated-$K_{OW}$, Veith et al. 1979,1980)
1.48 (bluegill sunfish, Barrows et al. 1980; Veith et al. 1980; Veith & Kosian 1982; quoted, Davies & Dobbs 1984; Suntio et al. 1988; Howard 1990; Ma et al. 1990; Saito et al. 1992)

1.477 (quoted, bluegill sunfish, Bysshe 1982)
1.241 (quoted, rainbow trout, Bysshe 1982)
2.32 (microorganisms-water, calculated-$K_{OW}$, Mabey et al. 1982; quoted, Ma et al. 1990)
1.24 (Mackay 1982; quoted, Schüürmann & Klein 1988)
1.72 (rainbow trout, Veith & Kosian 1982; quoted, Saito et al. 1992)
1.30 (calculated- $\chi$ , Koch 1983)
2.68 (activated sludge, Freitag et al. 1984)
2.477 (*alga chlorella fusca*-LSC, Geyer et al. 1984)
2.057 (calculated-$K_{OW}$, Geyer et al. 1984)
2.68, 2.48, <1.0 (activated sludge, *chlorella fusca*, golden ide, Freitag et al. 1985; quoted, Ma et al. 1990)
1.24 (fish, Connell & Hawker 1988; Hawker 1990; quoted, Ma et al. 1990)
1.24 (Isnard & Lambert 1988; quoted, Ma et al. 1990; Banerjee & Baughman 1991)
2.06 (calculated-$K_{OW}$, $S_{OCTANOL}$, & M.P., Banerjee & Baughman 1991)

Sorption Partition Coefficient, log $K_{OC}$:
2.041 (calculated-S as per Kenaga & Goring 1978, Kenaga 1980; quoted, Howard 1990; Ma et al. 1990)
2.642 (calculated-$K_{OW}$, Mabey et al. 1982; quoted, Ma et al. 1990)
1.85 (calculated- $\chi$ , Koch 1983; quoted, Bahnick & Doucette 1988)
1.70 (calculated- $\chi$ , Bahnick & Doucette 1988)
1.26 (DTMA-clay, Smith et al. 1990)
1.34 (TTMA-clay, Smith et al. 1990)
1.70 (HTMA-clay, Smith et al. 1990)
1.96 (BDHA-clay, Smith et al. 1990)
2.07 (DDPA-clay, Smith et al. 1990)
1.69 (80%DTMA-clay at 20°C, Smith & Jaffé 1991)

Sorption Partition Coefficient, log $K_{OM}$:
1.85, 1.55 (quoted, calculated- $\chi$ , Sabljic 1984)
1.65, 1.44 ( peat, muck, Rutherford & Chiou 1992; Rutherford et al. 1992)
0.243, 1.866 ( cellulose, extracted peat, Rutherford et al. 1992)

Half-Lives in the Environment:
Air: disappearance half-life larger than 10 days for the reaction with OH radicals in air (EPA 1974; quoted, Darnall et al. 1976); 30-50 years residence time in the troposphere (Molina & Rowland 1974; quoted, Howard 1990); estimated

residence time in troposphere, >330 years (CEQ 1975); estimated residence time of 100 years in troposphere (Singh et al. 1979); >11600 days, estimated as toxic chemical residence time with rate constant of $<1x10^{-15}$ $cm^3$ $molecule^{-1}$ $sec^{-1}$ at 300 K for the reaction with OH radicals (Singh et al. 1981); estimated residence time of 4-20 years in troposphere (Lyman 1982); $1.6x10^4$-$1.6x10^5$ hours, based on photooxidation half-life in air from measured rate data for the vapor phase reaction with hydroxyl radicals in air (Atkinson 1985; quoted, Howard et al. 1991).

Surface water: 3-30 days, estimated half-life in rivers and 3-300 days in lakes (Zoeteman et al. 1980; quoted, Howard 1990); 4320-8640 hours, based on estimated aqueous aerobic biodegradation half-life (Howard et al. 1991).

Ground water: estimated half-life to be 3-300 days (Zoeteman et al. 1980; quoted, Howard 1990); 168-8640 hours, based on estimated aqueous aerobic biodegradation half-life and acclimated anaerobic sediment/aquifer grab sample data (Parsons et al. 1985; quoted, Howard et al. 1991).

Sediment:

Soil: 4320-8640 hours, based on estimated aqueous aerobic biodegradation half-life (Howard et al. 1991); calculated half-life of 50 days in soil mixtures from first-order kinetic (Anderson et al. 1991).

Biota: <1 day in tissues of bluegill sunfish (Barrows et al. 1980).

Environmental Fate Rate Constant and Half-Lives:

Volatilization: half-life of 29 minutes was determined in an open container with initial concentration of 1 mg/liter when stirred at 100 rpm (Dilling et al. 1975 & Dilling 1977; quoted, Callahan et al. 1979; Mills et al. 1982; Verschueren 1983); calculated half-lives: 0.2 minute, 25.5 minutes (Dilling 1977); the ratio of evaporation rate constant to that of oxygen reaeration: 0.47 as measured value compared to 0.43 as predicted (Smith et al. 1980); $< 2.4x10^8$ $cm^3$ $mol^{-1}$ $sec^{-1}$, estimated rate constant for the reaction with OH radicals at 300 K (Lyman 1982); half-life from a model river of 1 m deep for water flowing at 1 m/sec with wind speed of 3 m/sec and Henry's law constant was estimated to be 3.7 hours (Thomas 1982; Lyman et al. 1982; quoted, Howard 1990).

Photolysis: photocatalyzed mineralization by the presence of $TiO_2$ with the rate of 0.18 ppm/min per gram of catalyst (Ollis 1985).

Oxidation: bimolecular reaction rate with hydroxyl radicals was reported to be less than $1x10^{-16}$ $cm^3$ $sec^{-1}$ with a reported tropospheric lifetime of greater than 330 years (Cox et al. 1976; quoted, Callahan et al. 1979); calculated rate constants at 25°C: $<<360$ $M^{-1}$ $hour^{-1}$ for singlet oxygen and $<<1$ $M^{-1}$ $hour^{-1}$ for peroxy radical (Mabey et al. 1982); photooxidation half-life in air: $1.6x10^4$-$1.6x10^5$ hours, based on measured rate data for the vapor phase reaction with hydroxyl radicals in air (Atkinson 1985; quoted, Howard et al. 1991); calculated rate constant of $<2 \times 10^6$ $M^{-1}$ $sec^{-1}$ for the reaction with OH radicals in aqueous solution (Haag & Yao 1992).

Hydrolysis: calculated first-order half-life of 7000 years, based on reported rate constant of $4.8 \times 10^{-7}$ $mol^{-1}$ $sec^{-1}$ for 1 mg/liter concentration at pH 7 and 25°C (Mabey & Mill 1978; quoted, Callahan et al. 1979; Howard 1990; Howard et al. 1991).

Biodegradation: aerobic half-life: 4320-8640 hours, based on acclimated aerobic screening test data (Tabak et al. 1981; quoted, Howard et al. 1991); anaerobic half-life: 168-672 hours, based on unacclimated anaerobic screening test data (Bouwer & McCarty 1983; quoted, Howard et al. 1991) and acclimated anaerobic sediment/aquifer grab sample data (Parsons et al. 1985; quoted, Howard et al. 1991).

Biotransformation: estimated rate constant of $1 \times 10^{-10}$ ml $cell^{-1}$ $hour^{-1}$ for bacteria in water (Mabey et al. 1982).

Bioconcentration and Uptake and Elimination Rate Constants ($k_1$ and $k_2$):

$k_1$: 4.05 $hour^{-1}$ (trout muscle, Neely et al. 1974; quoted, Thomann 1989)

$k_2$: 0.229 $hour^{-1}$ (trout muscle, Neely et al. 1974; quoted, Hawker & Connell 1988; Thomann 1989)

$k_1$: 4.05 mL $gram^{-1}$ $hour^{-1}$ (10°C, trout gill, Neely 1979)

$k_1$: 4.10 $hour^{-1}$ (trout, Hawker & Connell 1985)

$k_2$: 0.25 $hour^{-1}$ (trout, Hawker & Connell 1985)

Common Name: Ethyl chloride
Synonym: chloroethane, monochloroethane
Chemical Name: chloroethane, ethyl chloride, monochloroethane, hydrochloric ether, muriatic ether
CAS Registry No: 75-00-3
Molecular Formula: $C_2H_5CL$
Molecular Weight: 64.52
Melting Point (°C):
- -139.0 (Stull 1947)
- -136.4 (Callahan et al. 1979; Mackay & Shiu 1981,1990; Mabey et al. 1982; Weast 1982-83; Riddick et al. 1986)
- -138.3 (Verschueren 1983; Stephenson & Malanowski 1987)
- -136 to -138 (Dean 1985)
- -138.7 (Howard 1990)

Boiling Point (°C):
- 12.3 (McGovern 1943; Mackay & Shiu 1981,1990; Weast 1982-83; Dean 1985; Howard 1990)
- 12.27 (Dreisbach 1959; Callahan et al. 1979; Mabey et al. 1982; Riddick et al. 1986)
- 12.26 (Boublik et al. 1973)
- 12.40 (Verschueren 1983; Stephenson & Malanowski 1987)

Density ($g/cm^3$ at 20°C):
- 0.8970 (McGovern 1943)
- 0.8706 (25°C, Dreisbach 1959)
- 0.8978 (Weast 1982-83)
- 0.9214 (0°C, Dean 1985)
- 0.8960 (Riddick et al. 1986)

Molar Volume ($cm^3/mol$):
- 0.719 ($V_M/100$, Kamlet et al. 1987)
- 0.352 (calculated intrinsic volume: $V_I/100$, Kamlet et al. 1987)
- 72.7 (LeBas method, Mackay & Shiu 1990)
- 72.0 (calculated from density, Wang et al. 1992)

Molecular Volume ($A^3$):
Total Surface Area, TSA ($A^2$):
- 185.0 (Iwase et al. 1985)
- 83.3 (Okouchi et al. 1992)

Heat of Fusion, $\Delta H_{fus}$, (kcal/mol):
- 1.064 (Riddick et al. 1986)

Entropy of Fusion, $\Delta S_{fus}$ (cal/mol K or e.u.):
Fugacity Ratio at 25 °C, F:

Water Solubility ($g/m^3$ or mg/L at 25°C):

5735 (Führer 1924; quoted, Hansch et al. 1968)
7580 (Seidell 1940; quoted, Deno & Berkheimer 1960)
5100 (18.9°C, quoted, Gladis 1960)
4700 (20°C, Marsden & Mann 1962; quoted, Mackay & Shiu 1981)
4100 (quoted, Irmann 1965)
9390 (calculated-$K_{OW}$, Hansch et al. 1968)
7239 (quoted, Hine & Mookerjee 1975)
5710 (20°C, Neely 1976; quoted, Mackay & Shiu 1981,1990; Howard 1990)
5700 (20°C, quoted, Dilling 1977)
5740 (20°C, Verschueren 1977,1983; quoted, Callahan et al. 1979; Mabey et al. 1982; Mills et al. 1982)
5740 (20°C, selected, Nathan 1978; Haque et al. 1980)
6162, 7408 (quoted average, calculated-$K_{OW}$, Valvani et al. 1981)
4500 (0°C, Dean 1985)
5750, 6913 (quoted, calculated-molar volume & solvatochromic p., Kamlet et al. 1987)
5675; 11748, 14962 (quoted; calculated- $\chi$ & polarizability, Nirmalakhandan & Speece 1988)
5685 (calculated-AI, Okouchi et al. 1992)
5750, 6162 (quoted exptl., calculated-molar volume, Wang et al. 1992)

Vapor Pressure (Pa at 25°C):

136633 (McGovern 1943)
186787 (Antoine eqn. regression, Stull 1947)
159827 (calculated-Antoine eqn., Dreisbach 1959)
159989 (calculated-Antoine eqn., Boublik et al. 1973)
101325 (quoted, Hine & Mookerjee 1975)
100700 (20°C, Neely 1976; quoted, Mackay & Shiu 1981,1990)
133300 (20°C, Verschueren 1977,1983; quoted, Callahan et al. 1979; Mabey et al. 1982; Mills et al. 1982)
159827 (selected, Nathan 1978)
102108 (12.5°C, Boublik et al. 1984; quoted, Howard 1990)
159880 (quoted, Riddick et al. 1986)
159818 (calculated-Antoine eqn., Stephenson & Malanowski 1987)

Henry's Law Constant (Pa $m^3$/mol):

859.5 (calculated as $1/K_{AW}$, $C_W/C_A$, reported as exptl., Hine & Mookerjee 1975; quoted, Nirmalakhandan & Speece 1988)
1140.3 (calculated, Dilling 1977)
1145 (calculated-P/C, Mackay & Shiu 1981,1990)
14994 (20°C, calculated-P/C, Mabey et al. 1982)

1125 (EPICS, Gossett 1987; quoted, Mackay & Shiu 1990)
1271.3 (calculated-QSAR, Nirmalakhandan & Speece 1988)
942.2 (20-25°C and low ionic strength, quoted, Pankow & Rosen 1988; Pankow 1990)
1512 (quoted, Ryan et al. 1988)
15245 (selected, Jury et al. 1990)

Octanol/Water Partition Coefficient, log $K_{OW}$:
1.54 (Leo et al. 1971; quoted, Callahan et al. 1979; Mills et al. 1982; Ryan et al. 1988)
1.39 (calculated-$\pi$, Hansch et al. 1968)
1.43 (shake flask-GC, Hansch et al. 1975)
1.39 (Hansch & Leo 1979; quoted, Mackay & Shiu 1990)
1.43 (quoted, Valvani et al. 1981)
1.49 (calculated-f const., Valvani et al. 1981)
1.49 (calculated-f const., Mabey et al. 1982)
1.54 (calculated, Verschueren 1983)
1.43 (Hansch & Leo 1985; quoted, Howard 1985; Bodor et al. 1989; Sangster 1989)
1.43, 1.33 (quoted, calculated-hydrophobicity const., Iwase et al. 1985)
1.39 (selected, Mailhot & Peters 1988)
1.43, 1.20 (quoted, calculated-MO, Bodor et al. 1989)
1.43, 1.465 (quoted, calculated-$\pi$ substituent const., Bodor et al. 1989)
1.43 (recommended, Sangster 1989)
1.39, 1.55 (quoted, calculated-molar volume, Wang et al. 1992)

Bioconcentration Factor, log BCF:
0.86, 0.67 (estimated-$K_{OW}$, S, Lyman et al. 1982; quoted, Howard et al. 1990)
0.991 (microorganisms-water, calculated-$K_{OW}$, Mabey et al. 1982)

Sorption Partition Coefficient, log $K_{O}C$:
2.16, 1.52 (estimated-$K_{OW}$, S, Lyman et al. 1982; quoted, Howard 1990)
1.173 (sediment-water, calculated-$K_{OW}$, Mabey et al. 1982)
1.398 (soil, selected, Jury et al. 1990)

Half-Lives in the Environment:
Air: disappearance half-life of 2.4-24 hours from air for the reaction with OH radicals (EPA 1974; quoted, Darnall et al. 1976); 160-1604 hours, based on photooxidation half-life in air from measured rate constants for the gas phase reaction with hydroxyl radicals in air (Atkinson 1985; quoted, Howard et al. 1991); estimated tropospheric lifetime of 0.04 year (Nimitz & Skaggs 1992).

Surface water: 168-672 hours, based on estimated unacclimated aerobic aqueous biodegradation half-life (Howard et al. 1991).

Ground water: 336-1344 hours, based on estimated unacclimated aerobic aqueous biodegradation half-life (Howard et al. 1991).

Sediment:

Soil: 10-50 days (Ryan et al. 1988); 30 days, volatilization loss from soil (Jury et al. 1990); 168-672 hours, based on estimated unacclimated aerobic aqueous biodegradation half-life (Howard et al. 1991).

Biota: 10-50 days, subject to plant uptake by soil through volatilization (Ryan et al. 1988).

Environmental Fate Rate Constant and Half-Lives:

Volatilization: estimated experimental half-life of 21 minutes for 1 mg/liter to evaporate from aqueous solution stirred at 200 rpm in an open container of depth 65 mm at 25°C (Dilling et al. 1975; Dilling 1977; quoted, Callahan et al. 1979; Mills et al. 1982); calculated half-lives: 0.5 minute, 16.7 minutes (Dilling 1977); based on Henry's law constant, half-life of volatilization from a model river of 1 m deep can be estimated to be about 2.5 hours (Lyman et al. 1982; quoted, Howard 1990); based on oxygen reaeration ratio of 0.645, half-lives of 5.6, 1.1, and 4.5 days from representative pond, river and lake were estimated (Mabey et al. 1982; quoted, Howard 1990); half-life of 30 days for volatilization loss from soil (Jury et al. 1990).

Photolysis:

Oxidation: $3.9 \times 10^{-13}$ $cm^3$ $molecule^{-1}$ $sec^{-1}$ for reaction with hydroxyl radicals of concentration $10^6$ $cm^{-3}$ at 23°C having corresponding lifetime of one month (Howard & Evenson 1976; quoted, Callahan et al. 1979; Altshuller 1980); $2.3 \times 10^{11}$ $cm^3$ $mol^{-1}$ $sec^{-1}$, estimated rate constant for the reaction with OH radicals at 300 K (Lyman 1982); calculated rate constants at 25°C: $<<360$ $M^{-1}$ $hour^{-1}$ for singlet oxygen and $<<1$ $M^{-1}$ $hour^{-1}$ for peroxy radicals (Mabey et al. 1982); photooxidation half-life of 160-1604 hours, based on measured rate constants for the reaction with hydroxyl radicals in air (Atkinson 1985; quoted, Howard 1990; Howard et al. 1991).

Hydrolysis: $2 \times 10^{-7}$ $sec^{-1}$ was estimated for reaction at pH 7 and 25°C with a maximum half-life of 40 days (Radding et al. 1977; quoted, Callahan et al. 1979; Mabey et al. 1982); estimated half-life of 38 days at 25°C was based on an exptl. half-life of 1.68 hour at 100°C with ethanol and HCl being the hydrolysis product (Mabey & Mill 1978; quoted, Haque et al. 1980; Howard 1990; Howard et al. 1991); $4.5 \times 10^{-5}$ $hour^{-1}$ at pH 7 and 25°C with a calculated half-life of 1.8 year (Jeffers 1989; quoted, Ellington 1989); abiotic hydrolysis or dehydrohalogenation half-life of 1.3 month (Olsen & Davis 1990).

Biodegradation: aerobic aqueous biodegradation half-life: 168-672 hours, based on aqueous aerobic screening test data for 1-chloropropane and 1-chlorobutane (Gerhold & Malaney 1966; quoted, Howard et al. 1991); aqueous anaerobic half-life: 672-2688 hours, based on estimated unacclimated aqueous aerobic half-life (Howard et al. 1991); 0.02 $year^{-1}$ with half-life of 10 days and 0.001 $year^{-1}$ with half-life of 700 days (Olsen & Davis 1990).

Biotransformation:

Bioconcentration, Uptake ($k_1$) and Elimination ($k_2$) Rate Constants or Half-Lives:

Common Name: 1,1-Dichloroethane
Synonym: ethylidene chloride, ethylidene dichloride
Chemical Name: 1,1-dichloroethane
CAS Registry No: 75-34-3
Molecular Formula: $CH_3CHCl_2$
Molecular Weight: 98.96
Melting Point (°C):
-96.7 (Stull 1947; Stephenson & Malanowski 1987)
-96.98 (Weast 1977; Callahan et al. 1979; Mackay & Shiu 1981; Horvath 1982; Mabey et al. 1982; Howard 1990; Ma et al. 1990; Mackay & Shiu 1990)
-97.0 (Weast 1982-83; Dean 1985)
-97.40 (Verschueren 1983)
Boiling Point (°C):
57.0 (Rex 1906)
57.28 (Weast 1977; Horvath 1982; Mabey et al. 1982; Ma et al. 1990)
57.30 (Weast 1982-83; Verschueren 1983; McNally & Grob 1984; Dean 1985; Riddick et al. 1986; Stephenson & Malanowski 1987; Howard 1990)
57.50 (Mackay & Shiu 1981,1990)
Density ($g/cm^3$ at 20°C):
1.175 (McNally & Grob 1984)
1.1757 (Horvath 1982; Weast 1982-83; Dean 1985)
1.174 (Verschueren 1983)
1.1755 (Riddick et al. 1986)
Molar Volume ($cm^3/mol$):
94.0 (LeBas method, Abernethy et al. 1988)
44.0 (intrinsic volume-van der Waals method, Abernethy et al. 1988)
84.0 (calculated from density, Abernethy et al. 1988; Wang et al. 1992)
Molecular Volume ($A^3$):
Total Surface Area, TSA ($A^2$):
242.97 (Amidon 1976; Horvath 1982)
214.50 (Iwase et al. 1985)
102.30 (Okouchi et al. 1992)
Heat of Fusion, $\Delta H_{fus}$, (kcal/mol):
1.881 (Riddick et al. 1986)
Entropy of Fusion, $\Delta S_{fus}$ (cal/mol K or e.u.):
Fugacity Ratio at 25°C, F:

Water Solubility ($g/m^3$ or mg/L at 25°C):
5500 (20°C, volumetric, Rex 1906)
5400 (30°C, volumetric, Rex 1906)
5060 (Gross 1929; quoted, Horvath 1982)
5555 (Wright & Schaffer 1932; quoted, Horvath 1982)

5075 (Seidell 1940; quoted, Deno & Berkheimer 1960)
5060 (Seidell 1941; quoted, Howard 1990; Ma et al. 1990)
5075 (quoted, Hine & Mookerjee 1975)
5500 (20°C, Verschueren 1977,1983; Dilling 1977; quoted, Callahan et al. 1979; Mackay & Shiu 1981; Mabey et al. 1982; Mills et al. 1982; Warner et al. 1987)
5100 (Neely 1976; quoted, Mackay & Shiu 1981; Dean 1985; Mackay & Shiu 1990)
5100 (quoted average, Dilling 1977; quoted, Smith et al. 1980; Horvath 1982)
5370 (30°C, selected, Nathan 1978)
5190 (20°C, recommended, Sørensen & Arit 1979; quoted, Wright et al. 1992)
5500 (quoted, Cowen & Baynes 1980; McNally & Grob 1984)
5315, 4960 (quoted average, calculated-$K_{OW}$, Valvani et al. 1981)
4767 (recommended, Horvath 1982)
5000, 4470 (quoted exptl., calculated-group contribution as per Irmann 1965, Horvath 1982)
5000, 4834.4 (quoted, headspace-GC analysis. McNally & Grob 1983)
5000; 4588.5 (quoted; 30°C, headspace-GC analysis, McNally & Grob 1984)
5500 (20°C, quoted, Verschueren 1983)
5030 (20°C, quoted, Riddick et al. 1986)
5075, 2789 (quoted, calculated-fragment consts., Wakita et al. 1986)
4775; 4842, 4984 (quoted; predicted- $\chi$ & polarizability, Nirmaklahanden & Speece 1988)
4842 (Nirmalakhanden & Speece 1988; quoted, Ma et al. 1990)
5495 (Isnard & Lambert 1989; quoted, Ma et al. 1990)
5500 (selected, Ma et al. 1990)
5056 (calculated-$K_{OW}$ with regression, Müller & Klein 1992)
4780 (calculated-AI, Okouchi et al. 1992)
5438, 3511 (quoted exptl., calculated-molar volume, Wang et al. 1992)
4997 (20°C, calculated-activity coefficients, Wright et al. 1992)

Vapor Pressure (Pa at 25°C):
24274 (20°C, Rex 1906)
36951 (30°C, Rex 1906)
29811 (Antoine eqn. regression, Stull 1947)
25931 (calculated-Antoine eqn., Weast 1972-73)
30261 (calculated-Antoine eqn., Boublik et al. 1973)
30632 (quoted, Hine & Mookerjee 1975)
30100 (Neely 1976; quoted, Mackay & Shiu 1981)
23994 (20°C, Verschueren 1977,1983; quoted, Callahan et al. 1979; Mabey et al. 1982; Mills et al. 1982)
30126 (quoted average, Dilling 1977)
30100 (Dilling 1977; quoted, Mackay & Shiu 1990)
29993 (selected, Nathan 1978)

29683 (quoted, Cowen & Baynes 1980)
31192 (quoted, Verschueren 1983)
30259 (Boublik et al. 1984; quoted, Howard 1990; Ma et al. 1990)
30659 (quoted, McNally & Grob 1984)
30360 (quoted, Riddick et al. 1986)
30358 (calculated-Antoine eqn., Stephenson & Malanowski 1987)
30700 (quoted, Warner et al. 1987)
30000 (selected, Ma et al. 1990)
24440 (20°C, quoted from DIPPR, Tse et al. 1992)
37242 (30°C, quoted from DIPPR, Tse et al. 1992)

Henry's Law Constant (Pa $m^3$/mol):
594.6 (calculated-1/$K_{AW}$, $C_W/C_A$, reported as exptl., Hine & Mookerjee 1975; quoted, Nirmalakhandan & Speece 1988; Howard 1990; Ma et al. 1990)
594.9 (calculated, Dilling 1977)
439 (calculated-P/C, Mackay & Shiu 1981)
432 (20°C, calculated-P/C, Mabey et al. 1982; quoted, Ma et al. 1982)
569 (EPICS, Gossett 1987; quoted, Ma et al. 1990; Mackay & Shiu 1990)
552.0, 561.0 (quoted, calculated-P/C, Warner et al. 1987)
594.6, 714.9 (quoted, calculated-QSAR, Nirmalakhandan & Speece 1988)
436 (20-25°C and low ionic strength, quoted, Pankow & Rosen 1988; Ma et al. 1990; Pankow 1990)
421.4 (quoted, Ryan et al. 1988)
438.8 (selected, Jury et al. 1990)
584 (calculated-P/C, Mackay & Shiu 1990)
435.6 (20°C, quoted from Gossett 1987, Tse et al. 1992)
466.0 (20°C, Tse et al. 1992)
709.2 (30°C, Tse et al. 1992)

Octanol/Water Partition Coefficient, log $K_{OW}$:
1.79 (shake flask-GC, Hansch et al. 1975; quoted, Callahan et al. 1979; Mabey et al. 1982; Hansch & Leo 1985; Ryan et al. 1988; Howard 1990; Ma et al. 1990)
1.92 (Hansch & Leo 1979; quoted, Abernethy & Mackay 1987; Abernethy et al. 1988; Howard 1990; Ma et al. 1990; Mackay & Shiu 1990)
1.79 (quoted, Valvani et al. 1981)
1.80 (calculated-f const., Mabey et al. 1982)
1.78 (selected, Mills et al. 1982)
1.79, 1.89 (quoted, calculated-hydrophobicity const., Iwase et al. 1985)
1.79 (selected, Ma et al. 1990)
1.79 (quoted, Olsen & Davis 1990)
1.78 (calculated, Müller & Klein 1992)

1.79 (quoted, Van Leeuwen et al. 1992; Verhaar et al. 1992)
1.79, 1.90 (quoted, calculated-molar volume, Wang et al. 1992)

Bioconcentration Factor, log BCF:
1.2 (estimated-S, Lyman et al. 1982; quoted, Howard 1990; Ma et al. 1990)
1.28 (microorganisms-water, calculated-$K_{OW}$, Mabey et al. 1982; quoted, Ma et al. 1990)

Sorption Partition Coefficient, log $K_{OC}$:
1.60 (estimated-S, Lyman et al. 1982; quoted, Howard 1990)
1.477 (sediment-water, calculated-$K_{OW}$, Mabey et al. 1982; quoted, Ma et al. 1990)
1.663 (soil, selected, Jury et al. 1990)

Half-Lives in the Environment:

Air: disappearance half-life of 2.4-24 hours from air for the reaction with OH radicals (EPA 1974; quoted, Darnall et al. 1976); 44 days, estimated as toxic chemical residence time with rate constant of $2.6 \times 10^{-13}$ $cm^{-3}$ $molecule^{-1}$ $sec^{-1}$ at 300 K for the reaction with OH radicals (Singh et al. 1981); 247-2468 hours, based on photooxidation half-life in air (Atkinson 1985; quoted, Howard et al. 1991); estimated tropospheric lifetime of 0.02 year (Nimitz & Skaggs 1992).

Surface water: 768-3696 hours, based on estimated aqueous aerobic biodegradation half-life (Howard et al. 1991).

Groundwater: 1344-8640 hours, based on estimated aqueous aerobic biodegradation half-life and sub-soil grab sample data from a groundwater aquifer (Wilson et al. 1983; quoted, Howard et al. 1991).

Sediment:

Soil: 10-50 days (Ryan et al. 1988); 768-3696 hours, based on methane acclimated soil grab sample data (Henson et al. 1989; quoted, Howard et al. 1991) and sub-soil grab sample data from a groundwater aquifer (Wilson et al. 1983; quoted, Howard et al. 1991); 45 days, volatilization loss from soil (Jury et al. 1990).

Biota: < 2 days of elimination from whole body of bluegill sunfish (USEPA 1980; quoted, Howard 1990); 10-50 days, subject to plant uptake via volatilization(Ryan et al. 1988).

Environmental Fate Rate Constant and Half-Lives:

Volatilization: estimated experimental half-life of 22 minutes for initial concentration of 1.0 mg/liter when stirred at 200 rpm in water at appr. 25°C in an open container of depth 65 mm (Dilling et al. 1975; quoted, Callahan et al. 1979; Mills et al. 1982); an average half-life of 32.2 minutes was found by repeating Dilling et al.

1975 experiment (Dilling 1977; quoted, Howard 1990); calculated half-life of 0.98 minute (Mackay & Wolkoff 1973; quoted, Dilling 1977); calculated half-life of 21.2 minutes (Mackay & Leinonen 1975; quoted, Dilling 1977); half-lives: 6-9 days in a typical pond, 5-8 days in a typical lake, and 24-32 hours in a typical river (Smith et al. 1980; quoted, Howard 1990); the ratio of rate of evaporation to that of oxygen reaeration: measured value of 0.57 compared to that predicted value of 0.47 (Smith et al. 1980); half-life of 45 days for the volatilization loss from soil (Jury et al. 1990).

Photolysis:

Oxidation: rate constant of $(2.6 \pm 0.6) \times 10^{-13}$ $cm^3$ $molecule^{-1}$ $sec^{-1}$ was found for the photooxidation in the troposphere with $10^6$ $cm^{-3}$ hydroxyl radicals at 23°C with corresponding lifetime of 1.5 months (Howard & Evenson 1976; quoted, Callahan et al. 1979; Altshuller 1980; Howard 1990); estimated rate constant of $<1.6 \times 10^{11}$ $cm^3$ $mol^{-1}$ $sec^{-1}$ for the reaction with OH radicals at 300 K (Lyman 1982); calculated rate constants of $<<360$ $M^{-1}$ $hour^{-1}$ for singlet oxygen and 1.0 $M^{-1}$ $hour^{-1}$ for peroxy radicals both at 25°C (Mabey et al. 1982; quoted, Ma et al. 1990); photooxidation half-life in air of 247-2468 hours, based on measured rate data for the vapor phase reaction with hydroxyl radicals in air (Atkinson 1985; quoted, Howard et al. 1991).

Hydrolysis: rate constant of $1.15 \times 10^{-7}$ $M^{-1}$ $hour^{-1}$ for neutral process by analogy to dichloromethane (Mabey et al. 1982; quoted, Ma et al. 1990); $1.29 \times 10^{-6}$ $hour^{-1}$ at pH 7 and 25°C with calculated half-life of 61 years (Jeffers et al. 1989; quoted, Ellington 1989).

Biodegradation: aqueous aerobic half-life of 768-3696 hours, based on estimated methane acclimated soil grab sample data (Henson et al. 1989; quoted, Howard et al. 1991) and sub-soil grab sample data from a ground water aquifer (Wilson et al. 1983; quoted, Howard et al. 1991); aqueous anaerobic half-life of 3072-14784 hours, based on estimated aqueous aerobic biodegradation half-life (Howard et al. 1991); half-life of >60 days (Olsen & Davis 1990).

Biotransformation:

Bioconcentration, Uptake ($k_1$) and Elimination ($k_2$) Rate Constants or Half-Lives:

Common Name: 1,2-Dichloroethane
Synonym: ethylene chloride, ethylene dichloride, glycol dichloride, sym-dichloroethane, Dutch oil, EDC
Chemical Name: 1,2-dichloroethane
CAS Registry No: 107-06-2
Molecular Formula: $CH_2ClCH_2Cl$
Molecular Weight: 98.96
Melting Point (°C):
-35.5 (McGovern 1943; Coca & Diaz 1980)
-35.3 (Stull 1947; Stephenson & Malanowski 1987)
-35.66 (Dreisbach 1959; Riddick et al. 1986)
-35.36 (Callahan et al. 1979; Mackay & Shiu 1981; Horvath 1982; Mabey et al. 1982; Environment Canada 1984; Suntio et al. 1988; Howard 1990; Ma et al. 1990)
-35.40 (Verschueren 1983; Mackay & Shiu 1990)
-35.70 (Dean 1985)
Boiling Point (°C):
83.7-84.3 (Rex 1906)
83.50 (McGovern 1943; Verschueren 1983; McNally & Grob 1984; Mackay & Shiu 1990)
83.47 (Dreisbach 1959; Callahan et al. 1979; Mackay & Shiu 1981; Horvath 1982; Mabey et al. 1982; Environment Canada 1984; Ma et al. 1990)
83.81 (Boublik et al. 1973)
83.48 (Riddick et al. 1986)
83.70 (Stephenson & Malanowski 1987)
83.0 (Banerjee et al. 1990)
Density ($g/cm^3$ at 20°C):
1.253 (McGovern 1943)
1.2531 (Dreisbach 1959; Dean 1985)
1.2458 (25°C, Dreisbach 1959)
1.252 (Ullmann 1975; Environment Canada 1984)
1.2351 (Horvath 1982)
1.250 (Verschueren 1983)
1.258 (McNally & Grob 1984)
1.2521 (Riddick et al. 1986)
1.260 (Gillham & Rao 1990)
Molar Volume ($cm^3/mol$):
0.787 ($V_M/100$, Taft et al. 1985; Kamlet et al. 1986,1987; Leahy 1986)
0.442 (calculated intrinsic volume: $V_I/100$, Leahy 1986; Kamlet et al. 1987,1988)
44.0 (intrinsic volume-van der Waals method, Abernethy et al. 1988)
94.0 (LeBas method, Abernethy et al. 1988)
79.0 (calculated from density, Abernethy et al. 1988)
78.0 (calculated from density, Wang et al. 1992)
Molecular Volume ($A^3$):

Total Surface Area, TSA ($A^2$):

246.27 (Amidon 1976; Horvath 1982)

99.80 (Okouchi et al. 1992)

Heat of Fusion, $\Delta H_{fus}$, (kcal/mol):

1.982 (calculated, Dreisbach 1959)

2.112 (quoted, Riddick et al. 1986)

Entropy of Fusion, $\Delta S_{fus}$ (cal/mol K or e.u.):

Fugacity Ratio at 25°C, F:

1.0 (Suntio et al. 1988)

Water Solubility ($g/m^3$ or mg/L at 25°C):

8690 (20°C, volumetric, Rex 1906)

8940 (30°C, volumetric, Rex 1906)

8650 (Gross 1929; quoted, Horvath 1982)

9000 (30°C, shake flask-interferometer, Gross & Saylor 1931)

8720 (15°C, shake flask-interferometer, Gross & Saylor 1931)

8696 (Wright & Schaffer 1932; quoted, Horvath 1982)

8620 (Seidell 1940; quoted, Deno & Berkheimer 1960)

8400 (quoted, McGovern 1943; Horvath 1982)

8690 (CRC 1962, quoted, McNally & Grob 1984)

8600 (quoted, Irmann 1965)

8619 (quoted, Hine & Mookerjee 1975)

8800 (shake flask-GC, McConnell et al. 1975; quoted, Mackay & Shiu 1981; Suntio et al. 1988; Shiu et al. 1990)

8000 (20°C, Pearson & McConnell 1975; quoted, Mackay & Shiu 1981; Suntio et al. 1988)

8000 (20°C, Neely 1976; quoted, Suntio et al. 1988; Ma et al. 1990; Mackay & Shiu 1981,1990)

8700 (Dilling 1977; quoted, Mackay & Shiu 1981; Suntio et al. 1988; Ma et al. 1990; Mackay & Shiu 1990; Shiu et al. 1990)

8690 (20°C, Verschueren 1977; quoted, Callahan et al. 1979; Wilson et al. 1981; Mabey et al. 1982; Mills et al. 1982; Suntio et al. 1988; Ma et al. 1990; Shiu et al. 1990)

8630 (20°C, recommended, Sørensen & Arit 1979; quoted, Wright et al. 1992)

7987 (shake flask-LSC, Banerjee et al. 1980; quoted, Karickhoff 1981; Banerjee 1985; Shiu et al. 1990)

8690 (Cowen & Baynes 1980; quoted, McNally & Grob 1984)

8600, 8910 (quoted exptl., calculated-group contribution as per Irmann 1965, Horvath 1982)

8524 (Horvath 1982; quoted, Howard 1990; Ma et al. 1990)

8608 (recommended, Horvath 1982)

8000 (quoted, Thomas 1982)

8690 (20°C, quoted, Verschueren 1983)
9000; 3506 (quoted; 30°C, headspace-GC, McNally & Grob 1984)
8044 (calculated-UNIFAC, Banerjee 1985; quoted, Farrington 1991)
8100 (Dean 1985)
8820, 6389 (quoted from Hine & Mookerjee 1975, calculated-molar volume & solvatochromic p., Kamlet et al. 1986)
8820, 5695 (quoted, calculated-molar volume & solvatochromic p., Leahy 1986)
8100 (20°C, quoted, Riddick et al. 1986)
9025, 10362 (quoted, calculated-fragment consts., Wakita et al. 1986)
8820, 7336 (quoted, calculated-molar volume & solvatochromic p., Kamlet et al. 1987)
8300 (quoted, Warner et al. 1987)
8610; 6383, 4395 (quoted; predicted- $\chi$ & polarizability, Nirmalakahandan & Speece 1988)
8500 (selected, Suntio et al. 1988)
8511 (Isnard & Lambert 1988,1989; quoted, Ma et al. 1990)
8690 (20°C, selected, Gillham & Rao 1990)
8500 (selected, Ma et al. 1990)
8519 (calculated-$K_{OW}$ with regression, Müller & Klein 1992)
8619 (calculated-AI, Okouchi et al. 1992)
7200 (19.7°C, shake flask-GC/TC, Stephenson 1992)
8100 (29.7°C, shake flask-GC/TC, Stephenson 1992)
8619, 5695 (quoted exptl., calculated-molar volume, Wang et al. 1992)
9406, 9621 (20°C, calculated-activity coefficients, Wright et al. 1992)

Vapor Pressure (Pa at 25°C):
8131 (20°C, Rex 1906)
12983 (30°C, Rex 1906)
10664 (quoted, McGovern 1943)
10740 (Antoine eqn. regression, Stull 1947)
10704 (calculated-Antoine eqn., Dreisbach 1959)
10154 (calculated-Antoine eqn., Weast 1972-73)
10536 (calculated-Antoine eqn., Boublik et al. 1973)
11413 (quoted, Hine & Mookerjee 1975)
8520 (McConnell et al. 1975; quoted, Mackay & Shiu 1981; Suntio et al. 1988)
8400 (Pearson & McConnell 1975; quoted, Mackay & Shiu 1981; Suntio et al. 1988)
8500 (20°C, Ullmann 1975; quoted, Environment Canada 1984)
8930 (20°C, Neely 1976; quoted, Mackay & Shiu 1981; Thomas 1982; Suntio et al. 1988)
10930 (quoted average, Dilling 1977; quoted, Mackay & Shiu 1981,1990; Love Jr. & Eilers 1982; Suntio et al. 1988; Ma et al. 1990)
8132 (20°C, Verschueren 1977,1983; quoted, Callahan et al. 1979; Mabey et al. 1982; Mills et al. 1982; Suntio et al. 1988; Ma et al. 1990)

10931 (selected, Nathan 1978)
10490 (Boublik et al. 1984; quoted, Howard 1990; Ma et al. 1990)
11331 (quoted, McNally & Grob 1984)
10560 (quoted, Kamlet et al. 1986)
11109 (calculated-Antoine eqn., Stephenson & Malanowski 1987)
11450 (quoted, Warner et al. 1987)
9200 (selected, Suntio et al. 1988)
10931, 13863 (quoted, calculated- solvatochromic p. & UNIFAC, Banerjee et al. 1990)
8131 (20°C, selected, Gillham & Rao 1990)
9000 (selected, Ma et al. 1990)
10462 (resistance measurements-Antoine eqn., Foco et al. 1992)
8272 (20°C, quoted from DIPPR, Tse et al. 1992)
13321 (30°C, quoted from DIPPR, Tse et al. 1992)

Henry's Law Constant (Pa $m^3$/mol):
133.1 (calculated as $1/K_{AW}$, $C_W/C_A$, reported as exptl., Hine & Mookerjee 1975; quoted, Nirmalakhandan & Speece 1988; Brüggemann et al. 1991)
92.3 (calculated-$C_W/C_A$, McConnell et al. 1975)
123.9 (calculated, Dilling 1977; quoted, Love Jr. & Eilers 1982)
99.0 (Dilling 1977; quoted, Mackay & Shiu 1981; Jafvert & Wolfe 1987; Suntio et al. 1988; Howard 1990; Ma et al. 1990)
99, 110 (quoted, recommended, Mackay & Shiu 1981; quoted, Ma et al. 1990; Mackay & Shiu 1990)
124 (calculated-P/C, Mackay & Shiu 1981,1990)
92.6 (20°C, calculated-P/C, Mabey et al. 1982; quoted, Ma et al. 1990)
111.4 (calculated-P/C, Thomas 1982)
112.5 (EPICS, Gossett 1987; quoted, Suntio et al. 1988)
111.5, 136.8 (quoted, calculated-P/C, Warner et al. 1987)
143 (EPICS, Ashworth et al. 1988; quoted, Ma et al. 1990)
92 (20-25°C and low ionic strength, quoted, Pankow & Rosen 1988; Ma et al. 1990; Pankow 1990)
133.1, 567.9 (calculated-QSAR, Nirmalakhandan & Speece 1988; quoted, Brüggemann et al. 1991)
94.2 (quoted, Ryan et al. 1988)
1107.1 (calculated-P/C, Suntio et al. 1988)
94.2 (selected, Jury et al. 1990)
101.3 (20°C, Tse et al. 1992)
152.0 (30°C, Tse et al. 1992)

Octanol/Water Partition Coefficient, log $K_{OW}$:

1.48 (shake flask-GC, Leo et al. 1975)
1.48 (Radding et al. 1977; quoted, Callahan et al. 1979; Mills et al. 1982; Ryan et al. 1988)
1.48 (Hansch & Leo 1979; quoted, Suntio et al. 1988; Howard 1990; Ma et al. 1990)
1.45 (shake flask-LSC, Banerjee et al. 1980; quoted, Karickhoff 1981; Banerjee & Howard 1988; Suntio et al. 1988; Olsen & Davis 1990)
1.45 (Veith et al. 1980; Veith & Kosian 1982; quoted, Ma et al. 1990; Saito et al. 1992)
1.48 (calculated-f const., Mabey et al. 1982; quoted, Ma et al. 1990)
1.45 (actual value from Banerjee et al. 1980, Arbuckle 1983)
1.58 (octanol & water mutual solubility not considered, Arbuckle 1983)
1.54 (octanol & water mutual solubility considered, Arbuckle 1983)
1.79 (quoted measured value from Hansch & Leo 1979, Veith et al. 1983)
1.48, 1.44 (quoted, calculated-UNIFAC, Banerjee & Howard 1988)
1.48, 1.30 (quoted, calculated-molar volume & solvatochromic p., Taft et al. 1985)
1.48, 1.58 (quoted, calculated-molar volume & solvatochromic p., Leahy 1986)
1.451 (BUA 1987; quoted, Brüggemann et al. 1991)
1.76 (Abernethy et al. 1988; Mackay & Shiu 1990)
1.48, 1.63 (quoted, calculated-molar volume & solvatochromic p., Kamlet et al. 1988)
1.48 (quoted, Ryan et al. 1988)
1.47 (selected, Mailhot & Peters 1988; Ryan et al. 1988)
1.45 (quoted, Isnard & Lambert 1988,1989)
1.48 (selected, Ma et al. 1990)
1.46 (calculated, Müller & Klein 1992)
1.48 (quoted, Van Leeuwen et al. 1992; Verhaar et al. 1992)
1.48, 1.72 (quoted, calculated-molar volume, Wang et al. 1992)

Bioconcentration Factor, log BCF:

0.30 (bluegill sunfish, Barrows et al. 1980; quoted, Howard 1990; Ma et al. 1990)
0.301 (bluegill sunfish, Veith et al. 1980; Veith & Kosian 1982; quoted, Bysshe 1982; Suntio et al. 1988; Ma et al. 1990; Saito et al. 1992)
0.954 (microorganisms-water, calculated-$K_{OW}$, Mabey et al. 1982; quoted, Ma et al. 1990)
0.30 (bluegill sunfish, Davies & Dobbs; quoted, Sabljic 1987; Ma et al. 1990)
0.30, 0.10 (quoted, calculated, Sabljic 1987)
0.30 (quoted, Isnard & Lambert 1988)

Sorption Partition Coefficient, log $K_{OC}$:

1.51 (Karickhoff 1981; quoted, Ma et al. 1990)
1.06 (estimated-$K_{OW}$, Karickhoff 1981)

1.19 (estimated-S & M.P., Karickhoff 1981)
1.48 (estimated-S, Karickhoff 1981)
2.18 (estimated-$K_{OW}$, Lyman et al. 1982; quoted, Howard 1990; Ma et al. 1990)
1.146 (sediment-water, calculated-$K_{OW}$, Mabey et al. 1982; quoted, Ma et al. 1990)
1.09 (BUA 1987; quoted, Brüggemann et al. 1991)
1.342 (soil, selected, Jury et al. 1990)

Sorption Partition Coefficient, log $K_{OM}$:
1.28, 1.50 (quoted, calculated- $\chi$ , Sabljic 1984)

Half-Lives in the Environment:

Air: disappearance half-life of 2.4-24 hours from air for the reaction with OH radicals (EPA 1974; quoted, Darnall et al. 1976); 53 days, estimated as toxic chemical residence time with rate constant of $2.2 x 10^{-13}$ $cm^3$ $molecule^{-1}$ $sec^{-1}$ at 300 K for the reaction with OH radicals (Singh et al. 1981); 292-2917 hours, based on photooxidation half-life in air from measured rate data for the vapor phase reaction with hydroxyl radicals in air (Atkinson 1985; quoted, Howard et al. 1991); estimated tropospheric lifetime of 0.09 year (Nimitz & Skaggs 1992).

Surface water: 2400-4320 hours, based on unacclimated grab sample of aerobic soil from ground water aquifers and acclimated river die-away rate data (Wilson et al. 1983; Mudder 1981; quoted, Howard et al. 1991).

Ground water: 2400-8640 hours, based on unacclimated grab sample of aerobic soil from ground water aquifers and estimated aqueous aerobic biodegradation half-life (Wilson et al. 1983; Howard et al. 1991).

Sediment:

Soil: 10-50 days (Ryan et al. 1988); 90 days, estimated volatilization loss from soil (Jury et al. 1990); 2400-4320 hours, based on estimated aqueous aerobic biodegradation half-life (Howard et al. 1991).

Biota: $>1<2$ days in tissues of bluegill sunfish (Barrows et al. 1980); 10-50 days, subject to plant uptake via volatilization (Ryan et al. 1988).

Environmental Fate Rate Constant and Half-Lives:

Volatilization: estimated experimental half-life of volatilization from aqueous solution of 1 mg/liter to be 28 $\pm$ 1 minutes when stirred at 200 rpm in water at appr. 25°C in an open container (Dilling et al. 1975; Dilling 1977; quoted, Callahan et al. 1979; Mills et al. 1982); calculated half-lives: 4.5, 24.5 minutes (Dilling 1977); rate of evaporation of 2.4 gram $m^{-2}$ $sec^{-1}$ (Environment Canada 1984); half-life of 90 days, estimated volatilization loss from soil (Jury et al. 1990).

Photolysis: photooxidation half-life of 292-2917 hours, based on measured rate data for the vapor phase reaction with hydroxyl radicals in air (Atkinson 1985; quoted, Howard et al. 1991); photocatalyzed mineralization by the presence of $TiO_2$ with the rate of 1.1 ppm/min per gram of catalyst (Ollis 1985).

Oxidation: bimolecular rate reaction with OH radicals of concentration of $10^6$ $cm^{-1}$ at 23°C was $(2.2 \pm 0.5)x10^{-13}$ $cm^3$ $molecule^{-1}$ $sec^{-1}$ with a lifetime of 1.7 months (Howard & Everson 1976; quoted, Callahan et al. 1979) and same reaction with rate constant of $0.1x10^{9}$ $M^{-1}$ $sec^{-1}$ and half-life of 234 hours (Radding et al. 1977; quoted, Callahan et al. 1979); $1.3x10^{11}$ $cm^3$ $mol^{-1}$ $sec^{-1}$, estimated rate constant for the reaction with OH radicals at 300 K (Lyman 1982); calculated rate constants at 25°C: $<360$ $M^{-1}$ $hour^{-1}$ for singlet oxygen and $<1$ $M^{-1}$ $hour^{-1}$ for peroxy radicals (Mabey et al. 1982; quoted, Ma et al. 1990).

Hydrolysis: estimated first-order rate constant of $5x10^{-13}$ $sec^{-1}$ with a max. half-life of 50,000 years at pH 7 and 25°C from experimental data at 100-150°C (Radding et al. 1977; quoted, Callahan et al. 1979); $1.80x10^{-9}$ $hour^{-1}$ for neutral process (Mabey et al. 1982; quoted, Ma et al. 1990); abiotic hydrolysis or dehydrohalogenation half-life of 50 months (Mabey et al. 1983; quoted, Olsen & Davis 1990); rate constant at neutral pH of 0.63 $year^{-1}$ with first-order hydrolysis half-life of 1.1 year (Kollig et al. 1987); $1.1x10^{-6}$ $hour^{-1}$ at pH 7 and 25°C with calculated hydrolysis half-life of 72 years (Jeffers et al. 1989; quoted, Ellington 1989; Brüggemann et al. 1991).

Biodegradation: relatively undegradable (Lyman et al. 1982; quoted, Brüggmann et al. 1991); aqueous aerobic degradation half-life of 2400-4320 hours, based on unacclimated grab sample of aerobic soil from ground water aquifers and acclimated river die-away rate data (Wilson et al. 1983; Mudder 1981; quoted, Howard et al. 1991); anaerobic aqueous degradation half-life estimated to be 9600-17280 hours (Howard et al. 1991); half-life of $>60$ days (Olsen & Davis 1990).

Biotransformation: estimated to be $1x10^{-10}$ ml $cell^{-1}$ $hour^{-1}$ for bacteria (Mabey et al. 1982).

Bioconcentration, Uptake ($k_1$) and Elimination ($k_2$) Rate Constants or Half-Lives:

Common Name: 1,1,1-Trichloroethane
Synonym: methyl chloroform, chlorotene, Genklene, Baltana
Chemical Name: 1,1,1-trichloroethane
CAS Registry No: 71-55-6
Molecular Formula: $CH_3CCl_3$
Molecular Weight: 133.41
Melting Point (°C):
-30.6 (Stull 1947)
-32.5 (Windholz 1976; Neely & Blau 1985; Stephenson & Malanowski 1987)
-30.41 (Dreisbach 1959; Weast 1977; Callahan et al. 1979; Horvath 1982; Mabey et al. 1982)
-30.4 (Weast 1982-83; Dean 1985; Riddick et al. 1986)
-32.0 (Verschueren 1983; Suntio et al. 1988)
Boiling Point (°C):
74.1 (Dreisbach 1959; Pearson & McConnell 1975; Windholz 1976; Weast 1977,1982-83; Callahan et al. 1979; Horvath 1982; Mabey et al. 1982; McNally & Grob 1984; Neely & Blau 1985; Stephenson & Malanowski 1987)
74.0 (Jones et al. 1977/1978; Dean 1985)
71/81 (Verschueren 1983)
74.08 (Riddick et al. 1986)
75.0 (Banerjee et al. 1990)
Density (g/cm³ at 20°C):
1.3390 (Dreisbach 1959; Horvath 1982; Weast 1982-83; McNally & Grob 1984)
1.3306 (25°C, Dreisbach 1959)
1.3303 (Coca & Diaz 1980)
1.350 (Verschueren 1983; Gillham & Rao 1990)
1.3376 (Dean 1985)
1.3381 (Riddick et al. 1986)
Molar Volume (cm³/mol):
0.989 ($V_M$/100, Taft et al. 1985; Leahy 1986; Kamlet et al. 1987)
0.996 ($V_M$/100, Kamlet et al. 1986,1987)
0.519 (calculated intrinsic volume, Leahy 1986; Kamlet et al. 1987,1988; Thoms & Lion 1992)
115.0 (LeBas method, Abernethy et al. 1988)
52.0 (intrinsic volume-van der Waals method, Abernethy et al. 1988)
99.0 (calculated from density, Abernethy et al. 1988)
100.0 (calculated from density, Wang et al. 1992)
Molecular Volume (A³):
Total Surface Area, TSA (A²):
263.93 (Amidon 1976; Horvath 1982)
241.30 (Iwase et al. 1985)
121.40 (Okouchi et al. 1992)

Heat of Fusion, $\Delta H_{fus}$, (kcal/mol):

0.654 (calculated, Dreisbach 1959)

0.5617 (Riddick et al. 1986)

Entropy of Fusion, $\Delta S_{fus}$ (cal/mol K or e.u.):

Fugacity Ratio at 25°C, F:

1.0 (Suntio et al. 1988)

Water Solubility (g/m³ or mg/L at 25°C):

1320 (Van Arkel & Vles 1936; quoted, Schwarz & Miller 1980)

1304 (Seidell 1040; quoted, Deno & Berkheimer 1960)

1300 (O'Connell 1963; quoted, Horvath 1982)

1300 (quoted, Irmann 1965)

700 (Dow Chemical 1972; quoted, Horvath 1982)

1490 . (Walraevens et al. 1974; quoted, Horvath 1982)

1334 (Amidon et al. 1975; quoted, Kamlet et al. 1986)

1300 (quoted, Dilling et al. 1975)

1304 (quoted, Hine & Mookerjee 1975)

480 (20°C, shake flask-GC, Pearson & McConnell 1975; McConnell et al. 1975; quoted, Jones et al. 1977/1978; Callahan et al. 1979; Hutchinson et al. 1980; Suntio et al. 1988; Shiu et al. 1990)

260 (Aviado et al. 1976; quoted, Horvath 1982)

700 (Archer & Stevens 1977; quoted, Horvath 1982)

730 (Dilling 1977; quoted, Hutchinson et al. 1980; Mackay & Shiu 1981; Horvath 1982; Mabey et al. 1982; Bobra et al. 1984; Neely 1984; Chiou et al. 1988; Suntio et al. 1988; Shiu et al. 1990; Thoms & Lion 1992)

4400 (20°C, Verschueren 1977; quoted, Callahan et al. 1979; Love Jr. & Eilers 1982)

1320 (20°C, recommended, Sørensen & Arit 1979; quoted, Wright et al. 1992)

720 (selected, Afghan & Mackay 1980; Horvath 1982; Suntio et al. 1988)

1334 (shake flask-LSC, Banerjee et al. 1980; quoted, Karickhoff 1981,1985; Davies & Dobbs 1984; Banerjee 1985; Lyman 1985; Mackay 1985; Shiu et al. 1990)

100 (Coca & Diaz 1980; quoted, Horvath 1982)

4400 (quoted, Cowen & Baynes 1980)

1175 (elution chromatography, Schwarz 1980)

1320 (20°C, quoted, Schwarz 1980)

1850 (20°C, elution chromatography, Schwarz & Miller 1980; quoted, Shiu et al. 1990)

1334 (shake flask-LSC, Veith et al. 1980)

278.7 (calculated-f const., Veith et al. 1980)

1300, 1260 (quoted exptl., calculated-group contribution as per Irmann 1965, Horvath 1982)

1485 (recommended, Horvath 1982)

440-4400 (20°C, selected, Mills et al. 1982)

480 (selected, Neely 1982)
950 (quoted, Thomas 1982)
1334 (quoted actual exptl. value from Banerjee et al. 1980, Arbuckle 1983)
3200 (calculated-UNIFAC, Arbuckle 1983)
479.8 (30°C, headspace-GC, McNally & Grob 1984)
1334 (calculated-UNIFAC, Banerjee 1985)
1300 (Dean 1985)
1466 (estimated-$K_{OW}$ and M.P., Lyman 1985)
1334 (quoted extl. average, Neely & Blau 1985; quoted, Karickhoff 1985)
1334, 1245 (quoted, calculated-vapor pressure, molar volume & solvatochromic p., Kamlet et al. 1986)
1334, 1334 (quoted, calculated-molar volume & solvatochromic p., Leahy 1986)
1320 (20°C, quoted, Riddick et al. 1986)
1304, 881 (quoted, calculated-fragment const., Wakita et al. 1986)
1334, 1567 (quoted, calculated-molar volume & solvatochromic p., Kamlet et al. 1987)
5497 (interpolated between two data points, Warner et al. 1987)
1549 (quoted, Isnard & Lambert 1988,1989)
1500; 1074, 1746 (quoted; calculated- $\chi$ & polarizability, Nirmalakhanden & Speece 1988)
720 (20°C, selected, Gillham & Rao 1990)
1500 (quoted from Horvath 1982, Grathwohl 1990)
1495 (quoted, Broholm et al. 1992)
1252 (23-24°C, shake flask-GC, Broholm et al. 1992)
1494 (calculated-$K_{OW}$ with regression, Müller & Klein 1992)
1500 (calculated-AI, Okouchi et al. 1992)
700 (20.2°C, shake flask-GC/TC, Stephenson 1992)
760 (31.6°C, shake flask-GC/TC, Stephenson 1992)
1260, 1267 (20°C, calculated-activity coefficients, Wright et al. 1992)
716.5, 1304 (quoted exptl., calculated-molar volume, Wang et al. 1992)

Vapor Pressure (Pa at 25°C):
16191 (Antoine eqn. regression, Stull 1947)
16445 (calculated-Antoine eqn., Dreisbach 1959)
16169 (calculated-Antoine eqn., Weast 1972-73)
15330 (24°C, Weast 1972-73; quoted, Chiou et al. 1988)
17770 (calculated-Antoine eqn., Boublik et al. 1973,1984)
16500 (quoted, Dilling et al. 1975; Dilling 1977; quoted, Mabey et al. 1982; Mackay et al. 1982; Lyman 1985; Mackay 1985; Neely & Blau 1985)
16089 (quoted, Hine & Mookerjee 1975)
12800 (20°C, Pearson & McConnell 1975; McConnell et al. 1975; quoted, Callahan et al. 1979; Suntio et al. 1988)
13330 (20°C, Neely 1976; quoted, Thomas 1982; Suntio et al. 1988)

16529 (quoted, Dilling 1977; Suntio et al. 1988)
16663 (selected, Nathan 1978)
16615 (quoted, Cowen & Baynes 1980)
12797 (selected, Mills et al. 1982; Neely 1982)
13373 (estimated as per Perry & Chilton 1973, Arbuckle 1983)
13330 (20°C, quoted, Verschueren 1983)
12797 (20°C, quoted, McNally & Grob 1984)
17768 (quoted, Kamlet et al. 1986)
16490 (quoted, Riddick et al. 1986)
16491 (calculated-Antoine eqn., Stephenson & Malanowski 1987)
17023 (interpolated between two data points, Warner et al. 1987)
13300 (selected, Suntio et al. 1988)
16529, 960 (quoted, calculated-solvatochromic p. & UNIFAC, Banerjee et al. 1990)
13330 (20°C, selected, Gillham & Rao 1990)
13149 (20°C, quoted from DIPPR, Tse et al. 1992)
20519 (30°C, quoted from DIPPR, Tse et al. 1992)

Henry's Law Constant (Pa $m^3$/mol):
1638 (calculated as $1/K_{AW}$, $C_W/C_A$, reported as exptl., Hine & Mookerjee 1975; quoted, Nirmalakhandan & Speece 1988)
3433 (20°C, McConnell et al. 1975; quoted, Jones et al. 1977/1978)
3495 (calculated as per McConnell et al. 1975 reported as exptl., Neely 1976)
3473 (calculated-P/C, Neely 1976)
2800 (exptl., Dilling 1977; quoted, Suntio et al. 1988)
2975 (calculated, Dilling 1977; quoted, Love Jr. & Eilers 1982)
2025 (20°C, batch stripping, Mackay et al. 1979; quoted, Yurteri et al. 1987)
1996 (concentration ratio, Leighton & Calo 1981)
3039 (calculated-P/C, Mabey et al. 1982)
1520 (20°C, batch stripping, Munz & Roberts 1982; quoted, Roberts & Dändliker 1983; Roberts et al. 1985; Yurteri et al. 1987)
1824 (calculated-P/C, Thomas 1982)
728.9 (quoted actual value from Kavanaugh & Trussell 1980, Arbuckle 1983)
606.9 (20°C, predicted-UNIFAC, Arbuckle 1983; quoted, Yurteri et al. 1987)
1743 (20°C, EPICS, Lincoff & Gossett 1983; Gossett 1985; quoted, Yurteri et al. 1987)
1337 (20°C, EPICS, Lincoff & Gossett 1984)
1358 (20°C, batch stripping, Lincoff & Gossett 1984)
1652 (estimated-P/C, Lyman 1985; Mackay 1985)
1621 (estimated as per Hine & Mookerjee 1975, Lyman 1985)
1013 (estimated as per Cramer 1980, Lyman 1985)
498.5 (adsorption isotherm, Urano & Murata 1985)
1735 (20°C, EPICS, Ashworth et al. 1986; quoted, Yurteri et al. 1987)

1360 (20°C, multiple equilibration, Munz & Roberts 1986; quoted, Yurteri et al. 1987)
399.9 (20°C, calculated-P/C, McKone 1987)
498.5, 413.4 (quoted, calculated-P/C, Warner et al. 1987)
3223 (quoted, Ryan et al. 1988)
1345 (20°C, EPICS, Gossett 1987; quoted, Yurteri et al. 1987; Tse et al.1992)
1572 (20°C, EPICS, Yurteri et al. 1987)
1735 (EPICS, Ashworth et al. 1988)
1743 (EPICS, quoted from Gossett 1987, Suntio et al. 1988; Thoms & Lion 1992)
2464 (calculated-P/C, Suntio et al. 1988)
1413 (EPICS, Gossett 1987; quoted, Grathwohl 1990)
1638, 430.8 (quoted, calculated-QSAR, Nirmalakhandan & Speece 1988)
1317 (20-25°C and low ionic strength, quoted, Pankow & Rosen 1988; Pankow 1990)
3619 (selected, Jury et al. 1990)
1276 (20°C, Tse et al. 1992)
2026 (30°C, Tse et al. 1992)

Octanol/Water Partition Coefficient, log $K_{OW}$:
2.17 (Tute 1971; quoted, Callahan et al. 1979; Schwarzenbach et al. 1983)
2.49 (Hansch & Leo 1979; quoted, Iwase et al. 1985)
2.47 (shake flask-LSC, Banerjee et al. 1980; quoted, Karickhoff 1981,1985; Davies & Dobbs 1984; Neely & Blau 1985; Banerjee & Howard 1988; Suntio et al. 1988; Olsen & Davis 1990)
2.47 (shake flask-LSC, Veith et al. 1980)
2.51 (calculated-f const., Mabey et al. 1982)
2.18 (selected, Mills et al. 1982)
2.6 (selected, Neely 1982)
2.47 (Veith & Kosian 1982; quoted, Saito et al. 1992)
2.47 (quoted actual exptl. value from Banerjee et al. 1980, Arbuckle 1983)
2.35 (calculated-UNIFAC with octanol & water mutual solubility not considered, Arbuckle 1983)
2.29 (calculated-UNIFAC with octanol & water mutual solubility considered, Arbuckle 1983)
2.49 (calculated, Hansch & Leo 1985; quoted, Howard 1990; Thoms & Lion 1992)
2.49, 2.39 (quoted, calculated-hydrophobicity const., Iwase et al. 1985)
2.49, 2.39 (quoted, calculated-molar volume & solvatochromic p., Leahy 1986)
2.49 (Abernethy et al. 1988)
2.49, 1.96 (quoted, calculated-UNIFAC, Banerjee & Howard 1988)
2.47 (quoted, Isnard & Lambert 1988,1989)
2.49, 2.36 (quoted, calculated-molar volume & solvatochromic p., Kamlet et al. 1988)
2.48 (selected, Suntio et al. 1988)
2.49 (quoted from Hansch & Leo 1979, Grathwohl 1990)
2.48 (calculated, Müller & Klein 1992)

2.49 (quoted, Van Leeuwen et al. 1992; Verhaar et al. 1992)
2.47, 2.36 (quoted, calculated-molar volume, Wang et al. 1992)

Bioconcentration Factor, log BCF:
1.40 (calculated-$K_{OW}$, Veith et al. 1979; quoted, Veith et al. 1980)
0.954 (bluegill sunfish, Veith et al. 1980; quoted, Bysshe 1982; Davies & Dobbs 1984)
0.954 (bluegill sunfish, Barrows et al. 1980)
1.908 (microorganisms-water, calculated-$K_{OW}$, Mabey et al. 1982)
0.95 (bluegill sunfish, Veith & Kosian 1982; quoted, Saito et al. 1992)
1.70 (calculated- $\chi$ , Koch 1983)
0.95 (quoted, Isnard & Lambert 1988)

Sorption Partition Coefficient, log $K_{OC}$:
2.25 (quoted from Chiou et al. 1979, Karickhoff 1981,1985)
2.08 (estimated-$K_{OW}$, Karickhoff 1981)
2.04 (estimated-S & M. P., Karickhoff 1981)
2.02 (estimated-S, Karickhoff 1981)
2.182 (sediment-water, calculated-$K_{OW}$, Mabey et al. 1982)
2.27, 2.36, 2.08, 2.15 (estimated-$K_{OW}$, Karickhoff 1985)
2.04, 2.28, 2.17, 2.40 (estimated-S, Karickhoff 1985)
2.20 (best estimate, Karickhoff 1985)
2.11 (average from estimates of $K_{OW}$ and S, Karickhoff 1985; quoted, Neely & Blau 1985)
2.26 (quoted from Chiou et al. 1979, Bahnick & Doucette 1988)
1.70 (calculated- $\chi$ , Bahnick & Doucette 1988)
2.26 (20°C, soil, Chiou et al. 1988; quoted, Grathwohl 1990)
1.65 (20°C, soil, sand & loess, Grathwohl 1990)
2.22 (20°C, weathered shale & mudrock, Grathwohl 1990)
3.02 (20°C, unweathered shale & mudrock, Grathwohl 1990)
2.08, 2.29 (20°C, calculated, Grathwohl 1990)
2.053 (soil, selected, Jury et al. 1990)

Sorption Partition Coefficient, log$K_{OM}$:
2.02, 1.55 (quoted, calculated- $\chi$ , Sabljic 1984)

Half-Lives in the Environment:
Air: disappearance half-life of 2.4-24 hours from air for the reaction with OH radicals (EPA 1974; quoted, Darnall et al. 1976); estimated troposphere residence time of 1.1 year for reaction with OH radicals (CEQ 1975); estimated N. troposphere residence time of 7.2 years by one compartment nonsteady state model (Singh

et al. 1978); estimated troposphere residence time of 8-10 years by two compartment nonsteady state model (Singh 1977; Singh et al. 1979); 1160 days, estimated as toxic chemical residence time with rate constant of $1.0 \times 10^{-14}$ $cm^3$ $molecule^{-1}$ $sec^{-1}$ at 300 K for the reaction with OH radicals (Singh et at. 1981); 5393-53929 hours, based on measured rate data for the vapor phase reaction with hydroxyl radicals in air (Atkinson 1985; quoted, Howard et al. 1991); estimated tropospheric lifetime of 7.8 years and 6.1 years by rigorous calculation (Nimitz & Skaggs 1992).

Surface water: 3360-6552 hours, based on estimated unacclimated aqueous aerobic biodegradation half-life (Howard et al. 1991).

Ground water: estimated half-life of 1.0 year in the groundwater of the Netherlands (Zoeteman et al. 1981); 3360-13104 hours, based on estimated unacclimated aqueous aerobic biodegradation half-life and sub-soil grab sample data from a ground water aquifer (Howard et al. 1991; Wilson et al. 1983).

Sediment: measured half-life of 450 days at 25°C, based on neutral and base-catalyzed hydrolysis rates studied in pure water and in barely saturated subsurface sediment at 25-60°C (Haag & Mill 1988).

Soil: 365 days, estimated volatilization loss (Jury et al. 1990); 3360-6552 hours, based on estimated unacclimated aqueous aerobic biodegradation half-life (Howard et al. 1991).

Biota: $< 1$ day in tissues of bluegill sunfish (Barrows et al. 1980).

Environmental Fate Rate Constant and Half-Lives:

Volatilization: calculated half-life of 0.34 minute (Mackay & Wolkoff 1973; quoted, Dilling et al. 1975); estimated exptl. half-life of $20 \pm 3$ minutes at 25°C for an aqueous solution of 1 mg/liter when stirred at 200 rpm in an open container of depth 65 mm (Dilling et al. 1975; quoted, Callahan et al. 1979; Mills et al. 1982); half-life of 24.9 minutes was obtained under same exptl. conditions with 0.97 mg/liter in water (Dilling 1977; quoted, Callahan et al. 1979); calculated half-lives: 0.19 minute, 23.7 minutes (Dilling 1977); estimated half-life from water to be 3.7 hours (Thomas 1982); half-lives from water column of simulated marine mesocosm: 24 days at 8-16°C in the spring, 12 days at 20-22°C in the summer and 11 days at 3-7°C in the winter (Wakeham et al. 1983); estimated half-life from volatilization loss from soil to be 365 days (Jury et al. 1990).

Photolysis: not expected to be important (Howard et al. 1991).

Oxidation: bimolecular rate constant of $8.2 \times 10^{-13}$ $cm^3$ $molecule^{-1}$ $sec^{-1}$ with a max. tropospohereic life-time of 3.1 years (Yung et al. 1975; quoted, Callahan et al. 1979); based on a bimolecular rate constant of $2.8 \times 10^{-14}$ $cm^3$ $molecule^{-1}$ $sec^{-1}$ for reaction with hydroxyl radicals with a troposphere life time of 1.1 year (Cox et al. 1976; quoted, Callahan et al. 1979) and rate constant of $(1.59 \pm 0.16) \times 10^{-14}$ $cm^3$ $molecule^{-1}$ $sec^{-1}$ at 298 K with lifetime of 13 years in troposphere was reported for the same reaction (Watson et al. 1977; quoted, Callahan et al.

1979; Altshuller 1980); calculated rate constant of $2.2x10^{-14}$ $cm^3$ $molecule^{-1}$ $sec^{-1}$ at 298 K with lifetime of 8 years in troposphere (Chang & Kaufman 1977; quoted, Altshuller 1980); $9.0x10^9$ $cm^3$ $mol^{-1}$ $sec^{-1}$, estimated rate constant for the reaction with OH radicals at 300 K (Lyman 1982); calculated rate constants at 25°C: << 360 $M^{-1}$ $hour^{-1}$ for singlet oxygen and 1.0 $M^{-1}$ $hour^{-1}$ for peroxy radicals (Mabey et al. 1982); photooxidation half-life in air of 5393-53929 hours, based on measured rate data for the vapor phase reaction with OH radicals in air (Atkinson 1985; quoted, Howard et al. 1991); experimentally determined rate constant of $1.19x10^{-14}$ $cm^3$ $molecule^{-1}$ $sec^{-1}$ for the reaction with hydroxyl radicals in air at room temperature with an estimated rate constant of $1.3x10^{-14}$ $cm^3$ $molecule^{-1}$ $sec^{-1}$ under same conditions (Atkinson 1987); $1.19x10^{-14}$ $cm^3$ $molecule^{-1}$ $sec^{-1}$ for the reaction with hydroxyl radicals with a troposphere life time of 1.7 year for a global average concentration of OH radicals (Prinn et al. 1987; quoted, Bunce et al. 1991).

Hydrolysis: first-order rate constant of 0.12 $month^{-1}$ with an exptl. half-life of about six months at pH 7 and 25°C (Dilling et al. 1975; quoted, Callahan et al. 1979; Mabey et al. 1979; Neely 1985); 0.96 $year^{-1}$ at pH 7 and 25°C with a first-order hydrolysis half-life of 0.73 year (Kollig et al. 1987); measured rate constants at 25°C: $(1.93 \pm 0.40)x10^{-8}$ $sec^{-1}$ in distilled water contained 0.1% w/w $CH_2O$ as sterilant with 39% conversion, $(2.04 \pm 0.47)x10^{-8}$ $sec^{-1}$ in autoclaved distilled water with 40% conversion and $(1.80 \pm 0.90)x10^{-8}$ $sec^{-1}$ in sediment contained 0.1% w/w $CH_2O$ as sterilant with 25% conversion (Haag & Mill 1988); abiotic hydrolysis or dehydrohalogenation half-life of 6 months (Olsen & Davis 1990).

Biodegradation: 0.043 $day^{-1}$ in fresh water plus sediment incubated under anaerobic conditions (Wood et al. 1981; quoted, Klečka 1985); aqueous aerobic half-life of 3360-6552 hours, based on unacclimated aerobic seawater grab sample data and sub-soil sample data from a ground water aquifer (Pearson & McConnell 1975; Wilson et al. 1983; quoted, Howard et al. 1991); anaerobic half-life of 13440-26208 hours, based on estimated unacclimated aerobic biodegradation half-life (Howard et al. 1991).

Biotransformation:

Bioconcentration, Uptake ($k_1$) and Elimination ($k_2$) Rate Constants or Half-Lives:

Common Name: 1,1,2-Trichloroethane
Synonym: vinyl trichloride
Chemical Name: 1,1,2-trichloroethane
CAS Registry No: 79-00-5
Molecular Formula: $CH_2ClCHCl_2$
Molecular Weight: 133.41
Melting Point (°C):
- -35.5 (McGovern 1943; Stephenson & Malanowski 1987)
- -36.7 (Stull 1947; Dean 1985)
- -36.59 (Dreisbach 1959)
- -36.5 (Weast 1977; Callahan et al. 1979; Mackay & Shiu 1981,1990; Horvath 1982; Mabey et al. 1982; Weast 1982-83; Howard 1990)
- -35/-36.7 (Verschueren 1983)
- -36.53 (Riddick et al. 1986)

Boiling Point (°C):
- 133.7 (McGovern 1943; Verschueren 1983)
- 113.77 (Dreisbach 1959)
- 113.67 (Boublik et al. 1973)
- 133.77 (Weast 1977; Callahan et al. 1979; Horvath 1982; Mabey et al. 1982)
- 113.8 (Mackay & Shiu 1981,1990; McNally & Grob 1984; Howard 1990)
- 133.8 (Weast 1982-83)
- 113.5 (Dean 1985)
- 113.85 (Riddick et al. 1986)
- 113.0 (Michael et al. 1988; Stephenson & Malanowski 1987)
- 147.0 (Banerjee et al. 1990)

Density (g/cm³ at 20°C):
- 1.438 (McGovern 1943)
- 1.4397 (Dreisbach 1959; Coca & Diaz 1980; Horvath 1982; Weast 1982-83; McNally & Grob 1984)
- 1.4319 (25°C, Dreisbach 1959)
- 1.4401 (Coca & Diaz 1980)
- 1.440 (Verschueren 1983)
- 1.4416 (Dean 1985)
- 1.4393 (Riddick et al. 1986)
- 1.440 (Gillham & Rao 1990)

Molar Volume (cm³/mol):
- 52.0 (intrinsic volume-van der Waals method, Abernethy et al. 1988)
- 99.0 (calculated from density, Abernethy et al. 1988)
- 115.0 (LeBas method, Abernethy & Mackay 1987; Abernethy et al. 1988)
- 93.0 (calculated from density, Wang et al. 1992)

Molecular Volume (A³):

Total Surface Area, TSA ($A^2$):

266.48 (Amidon 1976; Horvath 1982)

118.80 (Okouchi et al. 1992)

Heat of Fusion, $\Delta H_{fus}$, (kcal/mol):

2.759 (calculated, Dreisbach 1959; quoted, Riddick et al. 1986)

Entropy of Fusion, $\Delta S_{fus}$ (cal/mol K or e.u.):

Fugacity Ratio at 25°C, F:

Water Solubility (g/m³ or mg/L at 25°C):

3704 (Wright & Schaffer 1932; quoted, Horvath 1982)
4580 (Van Arkel & Vles 1936; quoted, Horvath 1982)
4418 (Seidell 1940; quoted, Deno & Berkheimer 1960)
4400 (McGovern 1943; quoted, Horvath 1982)
4400 (Lange's 1956; quoted, McNally & Grob 1984)
4500 (18.9°C, Gladis 1960)
4600 (quoted, Irmann 1965)
4380 (Walraevens et al. 1974; quoted, Horvath 1982)
4626 (quoted, Hine & Mookerjee 1975)
4420 (quoted average, Dilling 1977; quoted, Mackay & Shiu 1981; Howard 1990)
4500 (20°C, Verschueren 1977; quoted, Callahan et al. 1979; Wilson et al. 1981; Mabey et al. 1982; Mills et al. 1982)
4370 (20°C, recommended, Sørensen & Arit 1979; quoted, Wright et al. 1992)
4500 (quoted, Cowen & Baynes 1980; McNally & Grob 1984)
4600, 5010 (quoted exptl., calculated-group contribution as per Irmann 1965, Horvath 1982)
4394 (recommended, Horvath 1982)
4500 (20°C, Verschueren 1983; quoted, Mackay & Shiu 1990; Shiu et al. 1990)
4365 (30°C, headspace-GC, McNally & Grob 1984)
4800 (Dean 1985)
4400 (20°C, quoted, Riddick et al. 1986)
4626, 6534 (quoted, calculated-fragment const., Wakita et al. 1986)
4467 (quoted, Isnard & Lambert 1988,1989)
4385; 2858, 1469 (quoted; predicted- $\chi$ & polarizability, Nirmaklahanden & Speece 1988)
4500 (20°C, selected, Gillham & Rao 1990)
4416 (calculated-$K_{OW}$ with regression, Müller & Klein 1992)
4387 (calculated-AI, Okouchi et al. 1992)
4580 (31.3°C, shake flask-GC/TC, Stephenson 1992)
4877, 4921 (20°C, calculated-activity coefficients, Wright et al. 1992)
4521, 2318 (quoted exptl., calculated-molar volume, Wang et al. 1992)

Vapor Pressure (Pa at 25°C):

3091 (Antoine eqn. regression, Stull 1947)
2998 (calculated-Antoine eqn., Dreisbach 1959)
3088 (calculated-Antoine eqn., Weast 1972-73)
2913 (calculated-Antoine eqn., Boublik et al. 1973)
3217 (quoted, Hine & Mookerjee 1975)
3066 (quoted average, Dilling 1977)
4040 (Dilling 1977; quoted, Mackay & Shiu 1981,1990; Howard 1990)
2533 (20°C, Verschueren 1977,1983; quoted, Callahan et al. 1979; Mabey et al. 1982; Mills et al. 1982; Gillham & Rao 1990)
3181 (quoted, Cowen & Baynes 1980)
4266 (30°C, quoted, Verschueren 1983)
4266 (30°C, quoted, McNally & Grob 1984)
2998 (quoted, Riddick et al. 1986)
3218 (calculated-Antoine eqn., Stephenson & Malanowski 1987)
3066, 3719 (quoted, calculated-UNIFAC, Banerjee et al. 1990)
2368.9 (20°C, quoted from DIPPR, Tse et al. 1992)
4011.8 (30°C, quoted from DIPPR, Tse et al. 1992)

Henry's Law Constant (Pa $m^3$/mol):

92.10 (calculated as $1/K_{AW}$, $C_W/C_A$, reported as exptl., Hine & Mookerjee 1975; quoted, Nirmalakhandan & Speece 1988)
94.2 (calculated-P/C, Dilling 1977)
122, 120 (calculated-P/C, recommended, Mackay & Shiu 1981,1990)
83 (concentration ratio-GC, Leighton & Calo 1981)
75.2 (20°C, calculated-P/C, Mabey et al. 1982)
81.1 (20°C, EPICS, Gossett 1987; quoted, Tse et al. 1992)
92.2 (EPICS, Ashworth et al. 1988; quoted, Mackay & Shiu 1990)
92.1, 305 (quoted, calculated-QSAR, Nirmalakhandan & Speece 1988)
768.4 (quoted, Ryan et al. 1988)
74.97 (20-25°C and low ionic strength, quoted, Pankow & Rosen 1988; Pankow 1990)
70.92 (20°C, Tse et al. 1992)
115.5 (30°C, Tse et al. 1992)

Octanol/Water Partition Coefficient, log $K_{OW}$:

2.17 (calculated as per Tute 1971, Callahan et al. 1979; Ryan et al. 1988)
2.38 (Hansch & Leo 1979; quoted, Abernethy & Mackay 1987; Abernethy et al. 1988; Mackay & Shiu 1990)
2.07 (calculated-f const., Mabey et al. 1982)
2.18 (selected, Mills et al. 1982)
2.07 (calculated-f const. as per Hansch & Leo 1979, Veith et al. 1983)

2.07 (calculated, Hansch & Leo 1985; quoted, Howard 1990)
2.42 (quoted, Isnard & Lambert 1988,1989)
2.05 (quoted calculated value, Müller & Klein 1992)
1.89 (quoted, Van Leeuwen et al. 1992; Verhaar et al. 1992)
2.07, 2.15 (quoted, calculated-molar volume, Wang et al. 1992)

Bioconcentration Factor, log BCF:
$<1.0$ (Kawasaki 1980; quoted, Howard 1990)
1.519 (microorganisms-water, calculated-$K_{OW}$, Mabey et al. 1982)
1.23 (quoted, Isnard & Lambert 1988)
1.049 (calculated-$K_{OW}$, McCarty et al. 1992)

Sorption Partition Coefficient, log $K_{OC}$:
1.845 (sandy soil column, Wilson et al. 1981; quoted, Howard 1990)
1.748 (sediment-water, calculated-$K_{OW}$, Mabey et al. 1982)

Half-Lives in the Environment:

Air: disappearance half-life of 2.4-24 hours for the reaction with OH radicals in air (EPA 1974; quoted, Darnall et al. 1976); 35 days, estimated as toxic chemical residence time with rate constant of $3.3 \times 10^{-13}$ $cm^3$ $molecule^{-1}$ $sec^{-1}$ at 300 K for the reaction with OH radicals (Singh et al. 1981); 196-1956 hours, based on measured rate constants for reaction with hydroxyl radicals in air (Atkinson 1985; quoted, Howard et al. 1991).

Surface water: 3263-8760 hours, based on estimated hydrolysis half-life at pH 9 and 25°C (Mabey et al. 1983; quoted, Howard et al. 1991) and estimated unacclimated aerobic aqueous degradation half-life (Howard et al. 1991).

Ground water: 3263-117520 hours, based on estimated hydrolysis half-life pH 9 and 25°C (Mabey et al. 1983; quoted, Howard et al. 1991) and data from estimated unacclimated aerobic aqueous biodegradation half-life as well as a ground water die-away study in which no biodegradation was observed (Wilson et al. 1984; quoted, Howard et al. 1991).

Sediment:

Soil: 3263-8760 hours, based on estimated hydrolysis half-life at pH 9 and 25°C (Mabey et al. 1983; quoted, Howard et al. 1991) and estimated unacclimated aerobic aqueous biodegradation half-life (Howard et al. 1991) and a soil column test in which no biodegradation was observed (Wilson et al 1981; quoted, Howard et al. 1991).

Biota:

Environmental Fate Rate Constant and Half-Lives:

Volatilization: half-life of 21 minutes, estimated from lab. experiment of initial 1 mg/liter in water stirred at 200 rpm at 25°C in an open container of depth 65 mm (Dilling et al. 1975; quoted, Callahan et al. 1979; Mills et al. 1982; Howard 1990) and 35.1 minutes under same lab. experimental conditions (Dilling 1977; quoted, Callahan et al. 1979); calculated half-lives 6.1, 30.1 minutes (Dilling 1977).

Photolysis:

Oxidation: calculated rate constants at 25°C: $<<360$ $M^{-1}$ $hour^{-1}$ for singlet oxygen and 3.0 $M^{-1}$ $hour^{-1}$ for peroxy radicals (Mabey et al. 1982); photooxidation half-life in air was estimated to be 196-1956 hours, based on measured rate constants for reaction with hydroxyl radicals in air (Atkinson 1985; quoted, Howard et al. 1991); rate constant of $1.1 \times 10^{8}$ $M^{-1}$ $sec^{-1}$ for the reaction with OH radicals in aqueous solution (Haag & Yao 1992).

Hydrolysis: estimated rate constant at 25°C and pH 7 of $1.2 \times 10^{-7}$ $hour^{-1}$ (Mabey et al. 1982); $5.9 \times 10^{-3}$ $M^{-1}$ $sec^{-1}$ with half-life of 3263 hours, based on alkaline catalyzed hydrolysis reaction at pH 9 and 25°C (Mabey et al. 1983; quoted, Howard et al. 1991); first-order half-life of 37 years, based on hydrolysis rate constant measured at pH 7 and 25°C (Mabey et al. 1983; quoted, Howard 1990; Howard et al. 1991); abiotic hydrolysis or dehydrohalogenation half-life of 170 months (Mabey et al. 1983; quoted, Olsen & Davis 1990).

Biodegradation: aqueous aerobic half-life of 4320-8760 hours, based on the extremely low or no biodegradation which was observed in screening tests and a river die-away test (Tabak et al. 1981; Kawasaki 1980; Mudder & Musterman 1982; quoted, Howard 1990; Howard et al. 1991); aqueous anaerobic half-life of 17280-35040 hours, based on estimated unacclimated aerobic aqueous biodegradation half-life (Howard et al. 1991); 0.04 $year^{-1}$ with half-life of 24 days (Wood et al. 1985; quoted, Olsen & Davis 1990).

Biotransformation: estimated rate constant of $3 \times 10^{-12}$ ml $cell^{-1}$ $hour^{-1}$ for bacteria (Mabey et al. 1982).

Bioconcentration, Uptake ($k_1$) and Elimination ($k_2$) Rate Constants or Half-Lives:

$k_1$: 0.763 $hour^{-1}$ (flagfish, calculated-BCFx$k_2$, McCarty et al. 1992)

$k_2$: 0.0676 $hour^{-1}$ (flagfish, estimated from one compartment first-order kinetic, McCarty et al. 1992)

Common Name: 1,1,1,2-Tetrachloroethane
Synonym:
Chemical Name: 1,1,1,2-tetrachloroethane
CAS Registry No: 630-20-6
Molecular Formula: $CH_2ClCCl_3$
Molecular Weight: 167.85
Melting Point (°C):
-68.7 (Stull 1947; Stephenson & Malanowski 1987)
-70.2 (Mackay & Shiu 1981; Horvath 1982; Weast 1982-83)
Boiling Point (°C):
130.2 (Boublik et al. 1973)
130.5 (Mackay & Shiu 1981; Horvath 1982; Weast 1982-83; Stephenson & Malanowski 1987)
138 (Dean 1985)
Density (g/cm³ at 20°C):
1.5406 (Horvath 1982; Weast 1982-83)
1.4819 (Dean 1985)
Molar Volume (cm³/mol):
135.0 (LeBas method, Mackay & Shiu 1990)
108.95 (calculated from density)
Molecular Volume (A³):
Total Surface Area, TSA (A²):
284.48 (Amidon 1976; Horvath 1982)
137.90 (Okouchi et al. 1992)
Heat of Fusion, $\Delta H_{fus}$, (kcal/mol):
Entropy of Fusion, $\Delta S_{fus}$ (cal/mol K or e.u.):
Fugacity Ratio at 25°C, F: 1.0

Water Solubility (g/m³ or mg/L at 25°C):
1110 (Seidell 1940; quoted, Deno & Berkheimer 1960)
1100 (Walraevens et al. 1974; quoted, Horvath 1982)
1100 (Dilling 1977; quoted, Mackay & Shiu 1981; Horvath 1982)
1090 (20°C, recommended, Sørensen & Arit 1979; quoted, Wright et al. 1992)
1100, 1410 (quoted exptl., calculated-group contribution as per Irmann 1965, Horvath 1982)
1110 (recommended, Horvath 1982)
200 (Dean 1985)
1099; 710, 494 (quoted; predicted- $\chi$ & polarizability, Nirmalakhandan & Speece 1988)
1099 (calculated-$K_{OW}$ with regression, Müller & Klein 1992)
1099 (calculated-AI, Okouchi et al. 1992)
1007 (20°C, calculated-activity coefficients, Wright et al. 1992)

Vapor Pressure (Pa at 25°C):

1767 (Antoine eqn. regression, Stull 1947)
1778 (calculated-Antoine eqn., Weast 1972-73)
1604 (calculated-Antoine eqn., Boublik et al. 1973)
1853 (20°C, Dilling 1977; quoted, Mackay & Shiu 1981)
1201 (20°C, quoted from DIPPR, Tse et al. 1992)
2117.3 (30°C, quoted from DIPPR, Tse et al. 1992)
1577.6 (calculated-Antoine eqn., Stephenson & Malanowski 1987)

Henry's Law Constant (Pa $m^3$/mol):

272.7 (calculated, Dilling 1977)
283, 280 (calculated-P/C, recommended, Mackay & Shiu 1981)
172.2 (20°C, Tse et al. 1992)
283.7 (30°C, Tse et al. 1992)

Octanol/Water Partition Coefficient, log $K_{OW}$:

3.03 (calculated, Müller & Klein 1992)

Bioconcentration Factor, log BCF:

Sorption Partition Coefficient, log $K_{OC}$:

Half-Lives in the Environment:

Air: disappearance half-life of 2.4-24 hours for the reaction with OH radicals in air (EPA 1974; quoted, Darnall et al. 1976); >1160 days, estimated as toxic chemical residence time with rate constant of $<1x10^{-14}$ $cm^3$ $molecule^{-1}$ $sec^{-1}$ at 300 K for the reaction with OH radicals (Singh et al. 1981); 2236-22361 hours, based on an estimated rate constant for the vapor phase reaction with hydroxyl radicals in air (Atkinson 1987; quoted, Howard et al. 1991); estimated tropospheric lifetime of 1.9 year (Nimitz & Skaggs 1992).

Surface water: 16-1604 hours, based on hydrolysis half-lives at pH 7 & 9 (Mabey et al. 1983; quoted, Howard et al. 1991).

Ground water: 16-1604 hours, based on hydrolysis half-lives at pH 7 & 9 (Mabey et al. 1983; quoted, Howard et al. 1991).

Sediment:

Soil: 16-1604, based on hydrolysis half-lives at pH 7 and 9 (Mabey et al. 1983; quoted, Howard et al. 1991).

Biota:

Environmental Fate Rate Constant and Half-Lives:

Volatilization: half-life for the evaporation from dilute aqueous solution was estimated to be 43 minutes (Dilling et al. 1975) and 42.3 minutes under same experimental conditions (Dilling 1977); calculated half-lives: 2.01, 28.8 minutes (Dilling 1977).

Photolysis:

Oxidation: photooxidation half-life of 2236-22361 hours, based on an estimated rate constant for the vapor phase reaction with hydroxyl radicals in air (Atkinson 1987; quoted, Howard et al. 1991).

Hydrolysis: 1.2 $M^{-1}$ $sec^{-1}$ for reaction at pH 7 and 25°C (Mabey et al. 1983; quoted, Howard et al. 1991); 4320 $M^{-1}$ $hour^{-1}$ for base reaction at pH 9 and 25°C (Mabey et al. 1983; quoted, Howard et al. 1991); abiotic hydrolysis or dehydrohalogenation half-life of 384 months (Mabey et al. 1983; quoted, Olsen & Davis 1990).

Biodegradation: aqueous aerobic half-life of 672-4320 hours, based on acclimated river die-away rate data for 1,1,2,2-tetrachloroethane (Mudder 1981; quoted, Howard et al. 1991), unacclimated sea water (Pearson & McConnell 1975; quoted, Howard et al. 1991) and sub-soil grab sample data for a ground water aquifer for 1,1,1-trichloroethane (Wilson et al. 1983; quoted, Howard et al. 1991); aqueous anaerobic half-life of 2688-17280 hours, based on aerobic biodegradation half-life (Howard et al. 1991).

Biotransformation:

Bioconcentration, Uptake ($k_1$) and Elimination ($k_2$) Rate Constants or Half-Lives:

Common Name: 1,1,2,2-Tetrachloroethane
Synonym: sym-tetrachloroethane, acetylene tetrachloride
Chemical Name: 1,1,2,2-tetrachloroethane
CAS Registry No: 79-34-5
Molecular Formula: $CHCl_2CHCl_2$
Molecular Weight: 167.85
Melting Point (°C):
- -42.5 (McGovern 1943; Verschueren 1983; Stephenson & Malanowski 1987)
- -36.0 (Stull 1947; Weast 1977; Callahan et al. 1979; Mackay & Shiu 1981,1990; Horvath 1982; Mabey et al. 1982; Weast 1982-83)
- -43.8 (Verschueren 1983; Riddick et al. 1986)
- -46.8 (Dean 1985)
- -44.0 (Suntio et al. 1988)

Boiling Point (°C):
- 146.2 (Weast 1977; Callahan et al. 1979; Mackay & Shiu 1981,1990; Horvath 1982; Mabey et al. 1982; Weast 1982-83)
- 146.3 (McGovern 1943; Dean 1985; Stephenson & Malanowski 1987)
- 146.4 (Verschueren 1983)
- 146.5 (McNally & Grob 1984)
- 145.1 (Riddick et al. 1986)
- 147.0 (Banerjee et al. 1990)

Density ($g/cm^3$ at 20°C):
- 1.596 (McGovern 1943)
- 1.5953 (Horvath 1982; Weast 1982-83)
- 1.60 (Verschueren 1983; McNally & Grob 1984)
- 1.5945 (Riddick et al. 1986)
- 1.5867 (25°C, Riddick et al. 1986)
- 1.60 (Gillham & Rao 1990)

Molar Volume ($cm^3/mol$):
- 1.052 (intrinsic volume: $V_M/100$, Kamlet et al. 1986,1987; Leahy 1986)
- 0.617 (calculated intrinsic volume, Leahy 1986; Kamlet et al. 1987,1988)
- 135.0 (LeBas method, Abernethy & Mackay 1987; Abernethy et al. 1988)
- 62.0 (intrinsic volume-van der Waals method, Abernethy et al. 1988)
- 105.0 (calculated from density, Abernethy et al. 1988; Wang et al. 1992)

Molecular Volume ($A^3$):
Total Surface Area, TSA ($A^2$):
- 286.73 (Amidon 1976; Horvath 1982)
- 137.80 (Okouchi et al. 1992)

Heat of Fusion, $\Delta H_{fus}$, (kcal/mol):
Entropy of Fusion, $\Delta S_{fus}$ (cal/mol K or e.u.):
Fugacity Ratio at 25°C, F:
- 1.0 (Suntio et al. 1988)

Water Solubility (g/m$^3$ or mg/L at 25°C):

2857 (Wright & Schaffer 1932; quoted, Horvath 1982)
2880 (20°C, Van Arkel & Vles 1936; quoted, Schwarz & Miller 1980; Horvath 1982)
2850 (Seidell 1940; quoted, Deno & Berkheimer 1960)
2900 (McGovern 1943; quoted, Horvath 1982; Shiu et al. 1990)
2900 (quoted, Gladis 1960)
3200 (Neely et al. 1974; Neely 1976; quoted, Mackay & Shiu 1981; Suntio et al. 1988)
2917 (quoted, Hine & Mookerjee 1975)
3000 (quoted, Dilling 1977; quoted, Mackay & Shiu 1981,1990; Horvath 1982; Suntio et al. 1988; Shiu et al. 1990)
2900 (20°C, Verschueren 1977,1983; quoted, Callahan et al. 1979; Mabey et al. 1982; Mills et al. 1982; Gillham & Rao 1990; Shiu et al. 1990)
2880 (20°C, recommended, Sørensen & Arit 1979; quoted, Wright et al. 1992)
480 (Afghan & Mackay 1980; quoted, Horvath 1982)
2970 (shake flask-LSC, Banerjee et al. 1980; quoted, Karickhoff 1981; Horvath 1982; Banerjee 1985; Shiu et al. 1990)
2900 (quoted, Cowen & Baynes 1980)
2960 (23.5°C, elution chromatography, Schwarz 1980; quoted, Shiu et al. 1990)
2870 (20°C, quoted, Schwarz 1980)
3850 (20°C, elution chromatography, Schwarz & Miller 1980; quoted, Shiu et al. 1990)
3670 (30°C, elution chromatography, Schwarz & Miller 1980)
2985 (shake flask-LSC, Veith et al. 1980)
1364 (calculated-f const., Veith et al. 1980)
2962 (calculated-$K_{OW}$, Valvani et al. 1981)
2962 (recommended, Horvath 1982; quoted, Howard 1990)
2900, 2820 (quoted exptl., calculated-group contribution as per Irmann 1965, Horvath 1982)
2971 (quoted actual exptl. value from Banerjee et al. 1980, Arbuckle 1983)
866.4 (calculated-UNIFAC, Arbuckle 1983)
2915 (30°C, headspace-GC, McNally & Grob 1984)
2985 (calculated-UNIFAC, Banerjee 1985)
2900 (Dean 1985)
2985, 1640 (quoted, calculated-molar volume & solvatochromic p., Kamlet et al. 1986)
2985, 1273 (quoted, calculated-molar volume & solvatochromic p., Leahy 1986)
2870 (20°C, quoted, Riddick et al. 1986)
2985, 3845 (quoted, calculated-fragment consts., Wakita et al. 1986)
2985, 1679 (quoted, calculated-molar volume & solvatochromic p., Kamlet et al. 1987)
2884 (quoted, Isnard & Lambert 1988,1989)
2958; 1371, 7413 (quoted; predicted- $\chi$ & polarizability, Nirmalakhandan & Speece 1988)
3000 (selected, Suntio et al. 1988)

2954 (calculated-$K_{OW}$ with regression, Müller & Klein 1992)
2958 (calculated-AI, Okouchi et al. 1992)
2910 (20°C, shake flask-GC/TC, Stephenson 1992)
2920 (29.7°C, shake flask-GC/TC, Stephenson 1992)
2917, 1109 (quoted exptl., calculated-molar volume, Wang et al. 1992)
2424, 2434 (20°C, calculated-activity coefficients, Wright et al. 1992)

Vapor Pressure (Pa at 25°C):
850 (Antoine eqn. regression, Stull 1947)
851 (calculated-Antoine eqn., Weast 1972-73)
582 (calculated-Antoine eqn., Boublik et al. 1973)
793 (quoted, Hine & Mookerjee 1975)
867 (Dilling 1977; quoted, Mackay & Shiu 1981; Suntio et al. 1988)
800 (Dilling 1977; quoted, Mackay & Shiu 1990)
667 (20°C, Verschueren 1977,1983; quoted, Callahan et al. 1979; Mabey et al. 1982; Mills et al. 1982; Gillham & Rao 1990)
853 (quoted, Cowen & Baynes 1980)
1133 (30°C, quoted, Verschueren 1983)
2133 (quoted, McNally & Grob 1984)
849 (quoted, Kamlet et al. 1986)
793 (quoted, Riddick et al. 1986)
793 (calculated-Antoine eqn., Stephenson & Malanowski 1987)
800 (selected, Suntio et al. 1988)
813.3, 1056 (quoted, calculated-UNIFAC, Banerjee et al. 1990)
440.1 (20°C, quoted from DIPPR, Tse et al. 1992)
848.2 (30°C, quoted from DIPPR, Tse et al. 1992)

Henry's Law Constant (Pa $m^3$/mol):
46.2 (calculated as $1/K_{AW}$, $C_W/C_A$, reported as exptl., Hine & Mookerjee; quoted, Nirmalakhandan & Speece 1988)
47.1 (calculated-P/C, Dilling 1977; quoted, Jafvert & Wolfe 1987)
48.5, 48.0 (calculated-P/C, recommended, Mackay & Shiu 1981)
38.5 (20°C, calculated-P/C, Mabey et al. 1982)
48.6 (quoted, Pankow et al. 1984)
46.2, 163.8 (quoted, calculated-QSAR, Nirmalakhandan & Speece 1988)
38.5 (20-25°C and low ionic strength, quoted, Pankow & Rosen 1988; Pankow 1990)
29.75 (quoted, Ryan et al. 1988)
44.76 (calculated-P/C, Suntio et al. 1988)
46.1 (calculated-P/C, Howard 1990)
45.0 (calculated-P/C, Mackay & Shiu 1990)
25.33 (Nicholson et al. 1984; quoted, Mackay & Shiu 1990)

30.39 (20°C, Tse et al. 1992)
50.65 (30°C, Tse et al. 1992)

Octanol/Water Partition Coefficient, log $K_{OW}$:

2.56 (calculated as per Tute 1971, Callahan et al. 1979; Ryan et al. 1988; Mackay & Shiu 1990)
2.39 (shake flask-LSC, Banerjee et al. 1980; quoted, Karickhoff 1981; Banerjee & Howard 1988; Suntio et al. 1988; Olsen & Davis 1990)
2.39 (shake flask-LSC, Veith et al. 1980; quoted, Veith et al. 1983)
2.66 (calculated, Veith et al. 1980)
2.389 (calculated-f const., Mabey et al. 1982)
2.56 (selected, Mills et al. 1982)
2.39 (Veith & Kosian 1982; quoted, Saito et al. 1992)
2.39 (quoted actual value from Banerjee et al. 1980, Arbuckle 1983)
2.84 (calculated-UNIFAC with octanol & water mutual solubility not considered, Arbuckle 1983)
2.79 (calculated-UNIFAC with octanol & water mutual solubility considered, Arbuckle 1983)
2.39 (quoted measured value from Veith et al. 1980, Veith et al. 1983)
2.39, 3.27 (quoted, calculated-molar refraction, Yoshida et al. 1983)
2.39, 2.30 (quoted, calculated-molar volume & solvatochromic p., Leahy 1986)
3.01 (Abernethy & Mackay 1987; Abernethy et al. 1988)
2.39, 2.39 (quoted, calculated-UNIFAC, Banerjee & Howard 1988)
2.39 (quoted, Isnard & Lambert 1988,1989)
2.39, 2.43 (quoted, calculated-molar volume & solvatochromic p., Kamlet et al. 1988)
2.39 (selected, Mailhot & Peters 1988)
2.46 (selected, Suntio et al. 1988)
2.39 (calculated, Hansch & Leo 1985; quoted, Howard 1990)
2.64 (calculated, Müller & Klein 1992)
2.39 (quoted, Van Leeuwen et al. 1992; Verhaar et al. 1992)
2.38, 2.50 (quoted, calculated-molar volume, Wang et al. 1992)

Bioconcentration Factor, log BCF:

0.9-1.0 (bluegill sunfish, Barrows et al. 1980; Kawasaki 1980; quoted, Howard 1990)
0.90 (bluegill sunfish, Veith et al. 1980; Veith & Kosian 1982; quoted, Suntio et al. 1988; Saito et al. 1992)
1.959 (microorganisms-water, calculated-$K_{OW}$, Mabey et al. 1982)
0.91 (fathead minnow, Veith & Kosian 1982; quoted, Saito et al. 1992)
0.90 (quoted, Isnard & Lambert 1988)
2.029 (calculated-$K_{OW}$, McCarty et al. 1992)

Sorption Partition Coefficient, log $K_{OC}$:

1.90 (silt loam, Chiou et al. 1979; quoted, Karickhoff 1981; Howard 1990)
2.00 (estimated-$K_{OW}$, Karickhoff 1981)
1.80 (estimated-S & M.P., Karickhoff 1981)
1.87 (estimated-S, Karickhoff 1981)
2.072 (sediment-water, calculated-$K_{OW}$, Mabey et al. 1982)

Sorption Partition Coefficient, log $K_{OM}$:

1.66, 1.90 (quoted, calculated- $\chi$ , Sabljic 1984)

Half-Lives in the Environment:

Air: disappearance half-life of 2.4-24 hours for the reaction with OH radicals in air (EPA 1974; quoted, Darnall et al. 1976); >1160 days, estimated as toxic chemical residence time with rate constant of $1.0 \times 10^{-14}$ $cm^3$ $molecule^{-1}$ $sec^{-1}$ at 300 K for the reaction with OH radicals (Singh et al. 1981); 213-2131 hours, based on an estimated rate constant for the vapor phase reaction with hydroxyl radicals in air (Atkinson 1987; quoted, Howard et al. 1991); estimated tropospheric lifetime of 0.09 year (Nimitz & Skaggs 1992).

Surface water: 10.7-1056 hours, based on hydrolysis half-lives at pH 7 and 9 (Haag & Mill 1988; quoted, Howard et al. 1991).

Ground water: 10.7-1056 hours, based on hydrolysis half-lives at pH 7 and 9 (Haag & Mill 1988; quoted, Howard et al. 1991).

Sediment: measured half-life of 29 days at 25°C, based on neutral and base-catalyzed hydrolysis rates studies in pure water and in barely saturated subsurface sediment at 25-60°C (Haag & Mill 1988).

Soil: <10 days (Ryan et al. 1988); 10.7-1056 hours, based on hydrolysis half-lives at pH 7 and 9 (Haag & Mill 1988; quoted, Howard et al. 1991).

Biota: <1 day in tissues of bluegill sunfish (Barrows et al. 1980); <10 days, subject to plant uptake via volatilization (Ryan et al. 1988).

Environmental Fate Rate Constant and Half-Lives:

Volatilization: estimated experimental half-life for initial concentration of 1 mg/liter to be 56 minutes when stirred at 200 rpm in water at approximately 25°C in an open container of depth 65 mm (Dilling et al. 1975; quoted, Callahan et al. 1979; Mills et al. 1982) and 55.2 minutes was obtained again with an initial concentration of 0.92 mg/liter under same experimental condition (Dilling 1977; quoted, Callahan et al. 1979); calculated half-lives: 12, 40.5 minutes (Dilling 1977); estimated half-life of 6.3 hours for a model river of 1 meter deep flowing at 1 m/sec with a wind speed of 3 m/sec, based on calculated Henry's law constant (Lyman et al. 1982; quoted, Howard 1990); estimated half-life of 3.5 days for a model pond was based on the effect of adsorption (USEPA 1987; quoted, Howard 1990).

Photolysis:

Oxidation: calculated rate constants at 25°: $<<360\ M^{-1}\ hour^{-1}$ for singlet oxygen and 2.0 $M^{-1}\ hour^{-1}$ for peroxy radicals (Mabey et al. 1982); photooxidation half-life in air of 213-2131 hours, based on an estimated rate constant for the vapor phase reaction with hydroxyl radicals in air (Atkinson 1987; quoted, Howard et al. 1991).

Hydrolysis: estimated rate constant at 25°C and pH 7 of $1.2x10^{-7}\ hour^{-1}$ (Mabey et al. 1982); abiotic hydrolysis or dehydrohalogenation half-life of 10 months (Mabey et al. 1983; quoted, Olsen & Davis 1990) and 3.3 months (Cooper et al. 1987; quoted, Olsen & Davis 1990); hydrolysis in alkaline soil of $2.3x10^{7}\ M^{-1}\ year^{-1}$ at pH 9 nd 25°C with half-life of 1.1 day and 111 days at pH 7 (Kollig et al. 1987; quoted, Howard 1990); 1.8 $M^{-1}\ sec^{-1}$ for reaction at 25°C and pH 7 with first-order half-life of 1056 hours (Haag & Mill 1988; quoted, Howard et al. 1991); 6480 $M^{-1}\ hour^{-1}$ for base reaction at pH 9 and 25°C (Haag & Mill 1988; quoted, Howard et al. 1991); $(27.6 \pm 4.0)x10^{-8}\ sec^{-1}$ in sediment with 68% conversion (Haag & Mill 1988).

Biodegradation: abiotic degradation half-life of >800 days for the reaction with photochemically produced hydroxyl radicals (Singh et al. 1981; quoted, Howard 1990); aqueous aerobic half-life of 672-4320 hours, based on acclimated river die-away rate data (Mudder 1981; quoted, Howard et al. 1991); aqueous anaerobic half-life of 168-672 hours, based on anaerobic sediment grab sample data (Jafvert & Wolfe 1987; quoted, Howard et al. 1991) and anaerobic screening test data (Hallen et al. 1986; quoted, Howard et al. 1991).

Biotransformation: estimated rate constant of $3x10^{-12}$ ml $cell^{-1}\ hour^{-1}$ for bacteria (Mabey et al. 1982).

Bioconcentration, Uptake ($k_1$) and Elimination ($k_2$) Rate Constants or Half-Lives:

$k_1$: 13.2 $hour^{-1}$ (flagfish, calculated-BCFx$k_2$, McCarty et al. 1992)

$k_2$: 0.123 $hour^{-1}$ (flagfish, estimated-one compartment first-order kinetic, McCarty et al. 1992)

Common Name: Pentachloroethane
Synonym: 1,1,1,2,2-pentachloroethane, pentalin
Chemical Name: pentachloroethane
CAS Registry No: 76-01-7
Molecular Formula: $CHCl_2CCl_3$
Molecular Weight: 202.30
Melting Point (°C):
-29.0 (Horvath 1982; Weast 1982-83; Verschueren 1983; Dean 1985; Riddick et al. 1986; Stephenson & Malanowski 1987; Howard 1990)
Boiling Point (°C):
159.72 (Boublik et al. 1973)
162.0 (Horvath 1982; Weast 1982-83; Verschueren 1983)
160.5 (Dean 1985)
159.88 (Riddick et al. 1986)
161.95 (Stephenson & Malanowski 1987)
161.0 (Banerjee et al. 1990)
Density (g/cm$^3$ at 20°C):
1.6796 (Horvath 1982; Weast 1982-83)
1.670 (Verschueren 1983)
1.6808 (Riddick et al. 1986)
1.6732 (25°C, Riddick et al. 1986)
Molar Volume (cm$^3$/mol):
1.204 ($V_M$/100, Kamlet et al. 1986,1987; Leahy 1986)
0.700 (calculated intrinsic volume: $V_I$/100, Leahy 1986; Kamlet et al. 1987,1988)
70.0 (intrinsic volume-van der Waals method, Abernethy et al. 1988)
156.0 (LeBas method, Abernethy et al. 1988)
120.0 (calculated from density, Abernethy et al. 1988)
Molecular Volume (A$^3$):
Total Surface Area, TSA (A$^2$):
302.47 (Amidon 1976; Horvath 1982)
156.90 (Okouchi et al. 1992)
Heat of Fusion, $\Delta H_{fus}$, (kcal/mol):
2.710 (Riddick et al. 1986)
Entropy of Fusion, $\Delta S_{fus}$ (cal/mol K or e.u.):
Fugacity Ratio at 25°C, F: 1.0

Water Solubility (g/m$^3$ or mg/L at 25°C):
345 (Wright & Schaffer 1932; quoted, Horvath 1982)
463 (Seidell 1940; quoted, Deno & Berkheimer 1960)
500 (McGovern 1943; quoted, Horvath 1982)
500 (quoted, Gladis 1960)
470 (O'Connell 1963; quoted, Horvath 1982)

500 (Walraevens et al. 1974; quoted, Horvath 1982)
480 (Dilling et al. 1975; Dilling 1977; quoted, Horvath 1982; Bobra et al. 1984; Howard 1990)
497 (quoted, Hine & Mookerjee 1975)
776 (shake flask-LSC, Veith et al. 1980; quoted, Davies & Dobbs 1984)
403.6 (calculated-f const., Veith et al. 1980)
500, 790 (quoted exptl., calculated-group contribution as per Irmann 1965, Horvath 1982)
499.5 (recommended, Horvath 1982)
500 (Dean 1985)
496.6, 442.6 (quoted, calculated-molar volume & solvatochromic p., Kamlet et al. 1986)
496.6, 243.2 (quoted, calculated-molar volume & solvatochromic p., Leahy 1986)
500 (quoted, Riddick et al. 1986)
497, 1086 (quoted, calculated-fragment const., Wakita et al. 1986)
496.6, 292.4 (quoted, calculated-molar volume & solvatochromic p., Kamlet et al. 1987)
776 (quoted, Isnard & Lambert 1988,1989)
500; 365, 165 (quoted; predicted- $\chi$ & polarizability, Nirmalakhandan & Speece 1988)
500 (calculated-AI, Okouchi et al. 1992)

Vapor Pressure (Pa at 25°C):
596.3 (Antoine eqn. regression, Stull 1947)
595.8 (calculated-Antoine eqn., Weast 1972-73)
465 (calculated-Antoine eqn., Boublik et al. 1973)
605.2 (quoted, Hine & Mookerjee 1975)
599.9 (quoted average, Dilling 1977)
453.2 (20°C, quoted, Verschueren 1983)
800 (30°C, quoted, Verschueren 1983)
467 (Boublik et al. 1984; quoted, Howard 1990)
595 (quoted, Kamlet et al. 1986)
590 (quoted, Riddick et al. 1986)
624.9 (calculated-Antoine eqn., Stephenson & Malanowski 1987)
466.6, 453.2 (quoted, calculated-UNIFAC, Banerjee et al. 1990)

Henry's Law Constant (Pa m$^3$/mol):
247.9 (calculated-1/$K_{AW}$, $C_W/C_A$, reported as exptl., Hine & Mookerjee 1975; quoted, Nirmalakhandan & Speece 1988)
247.9 (calculated-P/C, Dilling 1977)
247.9, 96.44 (quoted, calculated-QSAR, Nirmalakhandan & Speece 1988)
196.5 (calculated-P/C, Howard 1990)

Octanol/Water Partition Coefficient, log $K_{OW}$:

2.89 (shake flask-LSC, Veith et al. 1980; quoted, Davies & Dobbs 1984)
3.21 (Veith & Kosian 1982; quoted, Saito et al. 1992)
3.64 (calculated-f const. as per Hansch & Leo 1979, Veith et al. 1983)
3.05 (calculated, Hansch & Leo 1985; quoted, Howard 1990)
3.05, 3.15 (quoted, calculated-molar volume & solvatochromic p., Leahy 1986)
3.58 (Abernethy et al. 1988)
2.89 (selected, Mailhot & Peters 1988)
2.89 (quoted, Isnard & Lambert 1988,1989)
3.05, 3.20 (quoted, calculated-molar volume & solvatochromic p., Kamlet et al. 1988)
3.63 (quoted, Van Leeuwen et al. 1992; Verhaar et al. 1992)

Bioconcentration Factor, log BCF:

1.75 (calculated-$K_{OW}$, Veith et al. 1979,1980)
1.826 (bluegill sunfish, Barrows et al. 1980)
1.83 (bluegill sunfish, Veith et al. 1980; Veith & Kosian 1982; quoted, Davies & Dobbs 1984; Saito et al. 1992)
1.826 (bluegill sunfish, quoted, Bysshe 1982)
1.78 (fathead minnow, Veith & Kosian 1982; quoted, Saito et al. 1992)
1.83 (quoted, Isnard & Lambert 1988)

Sorption Partition Coefficient, log $K_{OC}$:

Half-Lives in the Environment:

Air: disappearance half-life of 2.4-24 hours from air for the reaction with OH radicals (EPA 1974; quoted, Darnall et al. 1976); estimated tropospheric lifetime of 0.58 hour (Nimitz & Skaggs 1992).
Surface water:
Ground water:
Sediment:
Soil:
Biota: <1 day in tissues of bluegill sunfish (Barrows et al. 1980).

Environmental Fate Rate Constant and Half-Lives:

Volatilization: estimated experimental half-life for 1 mg/liter aqueous solution to be 48 minutes when stirred at 200 rpm at appr. 25°C in an open container of depth 65 mm (Dilling et al. 1975) and 46.5 minutes for 0.9 mg/liter aqueous solution under same experimental conditions (Dilling 1977); calculated half-lives: 2.3, 32 minutes (Dilling 1977).

Photolysis:
Oxidation:
Hydrolysis:
Biodegradation:
Biotransformation:
Bioconcentration, Uptake ($k_1$) and Elimination ($k_2$) Rate Constants or Half-Lives:

Common Name: Hexachloroethane
Synonym: carbon hexachloride, perchloroethane, phenohep, HCE
Chemical Name: hexachloroethane
CAS Registry No: 67-72-1
Molecular Formula: $CCl_3CCl_3$
Molecular Weight: 236.74
Melting Point (°C):
    186.8 (Horvath 1982)
    187.0 (Verschueren 1983)
    187.5 (Dean 1985)
    185 (Stephenson & Malanowski 1987)
    192 (Banerjee & Baughman 1991)
Boiling Point (°C):
    186.8 (McGovern 1943)
    186.0 (Callahan et al. 1979; Mackay & Shiu 1981; Horvath 1982; Weast 1982-83; Stephenson & Malanowski 1987)
Density (g/cm$^3$ at 20°C):
    2.091 (McGovern 1943; Horvath 1982; Weast 1982-83; Dean 1985)
    2.090 (Verschueren 1983)
Molar Volume (cm$^3$/mol):
    79.0 (intrinsic volume-van der Waals method, Abernethy et al. 1988)
    177.0 (LeBas method, Abernethy et al. 1988)
    113.0 (calculated from density, Abernethy et al. 1988; Wang et al. 1992)
Molecular Volume (A$^3$):
Total Surface Area, TSA (A$^2$):
    318.07 (Amidon 1976; Horvath 1982)
Heat of Fusion, $\Delta H_{fus}$, (kcal/mol):
Entropy of Fusion, $\Delta S_{fus}$ (cal/mol K or e.u.):
Fugacity Ratio at 25°C, F: 1.0

Water Solubility (g/m$^3$ or mg/L at 25°C):
    50 (22.3°C, Van Arkel & Vles 1936; quoted, Horvath 1982)
    50 (22.3°C, McGovern 1943; quoted, Horvath 1982)
    50.6 (quoted, Hine & Mookerjee 1975)
    50 (22°C, Neely 1976; quoted; Mackay & Shiu 1981; Mills et al. 1982)
    8.0 (Dilling 1977; quoted, Mackay & Shiu 1981; Horvath 1982)
    50 (20°C, Verschueren 1977; quoted, Callahan et al. 1979; Mabey et al. 1982; Neuhauser et al. 1985; Warner et al. 1987)
    50 (quoted, Cowen & Baynes 1980)
    27.2 (shake flask-LSC, Veith et al. 1980; quoted,Davies & Dobbs 1984)
    3.67 (calculated-f const., Veith et al. 1980; quoted, Davies & Dobbs 1984)
    50 (22.3°C, recommended, Horvath 1982)

50 (22°C, Verschueren 1983)
50.12 (quoted, Isnard & Lambert 1988,1989)
50; 106.9, 57.7 (quoted; predicted- $\chi$ & polarizability, Nirmalakhanden & Speece 1988)
802 (calculated-molar volume, Wang et al. 1992)

Vapor Pressure (Pa at 25°C):
75.4 (quoted, Hine & Mookerjee 1975)
28 (20°C, Neely 1976; quoted, Mackay & Shiu 1981)
44 (Dilling 1977; quoted, Mackay & Shiu 1981)
53 (20°C, Verschueren 1977; quoted, Callahan et al. 1979; Mabey et al. 1982; Mills et al. 1982; Neuhauser et al. 1985)
61 (quoted, Cowen & Baynes 1980)
53.3 (22°C, Verschueren 1983)
49.5 (calculated-Antoine eqn., Stephenson & Malanowski 1987)

Henry's Law Constant (Pa m$^3$/mol):
231.3 (calculated as $1/K_{AW}$, $C_W/C_A$, reported as exptl., Hine & Mookerjee 1975; quoted, Nirmalakhandan & Speece 1988)
123.9 (calculated-P/C, Dilling 1977; quoted, Jafvert & Wolfe 1987)
1302 (calculated-P/C, Mackay & Shiu 1981)
252.3 (22°C, calculated-P/C, Mabey et al. 1982)
253.3 (21°C, quoted, Pankow et al. 1984)
998.1 (quoted, Warner et al. 1987)
231.3, 986.8 (quoted, calculated-QSAR, Nirmalakhandan & Speece 1988)
253.3 (20-25°C and low ionic strength, quoted, Pankow & Rosen 1988; Pankow 1990)
1016.3 (quoted, Ryan et al. 1988)

Octanol/Water Partition Coefficient, log $K_{OW}$:
3.34 (calculated as per Tute 1971; quoted, Callahan et al. 1979; Neuhauser et al. 1985)
3.82 (calculated-HPLC-k', Könemann et al. 1979)
3.58 (calculated-f const., Könemann et al. 1979)
3.93 (shake flask-LSC, Veith et al. 1980; quoted, Davies & Dobbs 1984)
4.05 (calculated-RPHPLC, Veith et al. 1980; quoted, Davies & Dobbs 1984)
4.62 (calculated-f. const. as per Hansch & Leo 1979, Veith et al. 1980,1983; Mabey et al. 1982; quoted, Abernethy et al. 1988; Ryan et al. 1988)
3.34 (selected, Mills et al. 1982)
3.93 (Veith & Kosian 1982; quoted, Saito et al. 1992)
3.93 (quoted, Isnard & Lambert 1988,1989; quoted, Banerjee & Baughman 1991)
3.58 (selected, Thomann 1989)

4.14 (quoted, Van Leeuwen et al. 1992)
2.73 (calculated-molar volume, Wang et al. 1992)

Bioconcentration Factor, log BCF:
0.92-3.23 (calculated-$K_{OW}$, Veith et al. 1979,1980)
2.143 (bluegill sunfish, Barrows et al. 1980)
2.14 (bluegill sunfish, Veith et al. 1980; quoted, Davies & Dobbs 1984)
2.143 (bluegill sunfish, quoted, Bysshe 1982)
3.83 (microorganisms-water, calculated-$K_{OW}$, Mabey et al. 1982)
2.14 (bluegill sunfish, Veith & Kosian 1982; quoted, Saito et al. 1992)
2.85 (fathead minnow, Veith & Kosian 1982; quoted, Saito et al. 1992)
2.14 (quoted, Isnard & Lambert 1988; quoted, Banerjee & Baughman 1991)
4.57 (calculated-$K_{OW}$, Thomann 1989)

Sorption Partition Coefficient, log $K_{OC}$:
4.3 (sediment-water, calculated-$K_{OW}$, Mabey et al. 1982)

Half-Lives in the Environment:
Air: disappearance half-life of larger than 10 days from air for the reaction with OH radicals (EPA 1974; quoted, Darnall et al. 1976); >7.3-73 years, based on estimated maximum rate constant for reaction with hydroxyl radicals (Singh et al. 1980; quoted, Howard et al. 1991); >11600 days, estimated as toxic chemical residence time with rate constant of $<1.0\times10^{-15}$ $cm^3$ $molecule^{-1}$ $sec^{-1}$ at 300 K for the reaction with OH radicals (Singh et al. 1981).
Surface water: 672-4320 hours, based on estimated aqueous aerobic biodegradation half-life (Howard et al. 1991).
Ground water: 1344-8640 hours, based on estimated aqueous aerobic biodegradation half-life (Howard et al. 1991).
Sediment:
Soil: 672-4320 hours, based on estimated aqueous aerobic biodegradation half-life (Howard et al. 1991).
Biota: <1 day in tissues of bluegill sunfish (Barrows et al. 1980).

Environmental Fate Rate Constant and Half-Lives:
Volatilization: estimated experimental half-life of 40-45 minutes from dilute aqueous solution in open container of depth 65mm and stirring at 200 rpm (Dilling et al. 1975; Dilling 1977; quoted, Callahan et al. 1979; Mills et al. 1982); calculated half-lives: 4, 38 minutes (Dilling 1977).
Photolysis:

Oxidation: >7.3-73 years, based on estimated maximum rate constant for the reaction with hydroxyl radicals in air (Singh et al. 1980; quoted, Howard et al. 1991).
Hydrolysis: not significant, based on hydrolysis studies after 13 days at 85°C and pH 3, 7, and 11 (Ellington et al. 1987; quoted, Howard et al. 1991).
Biodegradation:
Biotransformation: estimated rate constant of $1x10^{-10}$ ml $cell^{-1}$ $hour^{-1}$ for bacteria (Mabey et al. 1982).
Bioconcentration, Uptake ($k_1$) and Elimination ($k_2$) Rate Constants or Half-Lives:

Common Name: 1-Chloropropane
Synonym: *n*-Propyl chloride, propyl chloride
Chemical Name: 1-chloropropane, n-Propyl chloride, propyl chloride
CAS Registry No: 540-54-5
Molecular Formula: $CH_3CH_2CH_2Cl$
Molecular Weight: 78.54
Melting Point (°C):
-122.8 (Horvath 1982; Verschueren 1983; Dean 1985; Riddick et al. 1986)
Boiling Point (°C):
46.20 (Rex 1906)
46.60 (Horvath 1982; Dean 1985)
47.20 (Verschueren 1983)
46.52 (Riddick et al. 1986)
Density ($g/cm^3$ at 20°C):
0.8909 (Horvath 1982)
0.8899 (Verschueren 1983; Riddick et al. 1986)
0.8830 (25°C, Riddick et al. 1986)
Molar Volume ($cm^3/mol$):
94.9 (calculated-LeBas method)
88.25 (calculated from density)
Molecular Volume ($A^3$):
Total Surface Area, TSA ($A^2$):
104.2 (Okouchi et al. 1992)
Heat of Fusion, $\Delta H_{fus}$, (kcal/mol):
1.325 (Riddick et al. 1986)
Entropy of Fusion, $\Delta S_{fus}$ (cal/mol K or e.u.):
Fugacity Ratio at 25°C, F:

Water Solubility ($g/m^3$ or mg/L at 25°C):
2720 (20°C, volumetric, Rex 1906; quoted, Horvath 1982)
2770 (30°C, volumetric, Rex 1906; quoted, Horvath 1982)
2500 (Wright & Schaffer 1932; quoted, Irmann 1965; Horvath 1982)
2772 (30°C, Van Arkel & Vles 1936; quoted, Horvath 1982)
2723 (Seidell 1940; quoted, Deno & Berkheimer 1960)
2334 (Seidell 1941; quoted, Hansch et al. 1968)
2333 (20°C, Saracco & Spaccamela Marchetti 1958; quoted, Horvath 1982)
2700 (20°C, Patty 1962; Lange 1967; quoted, Horvath 1982)
2510 (calculated-group contribution, Irmann 1965; quoted, Horvath 1982)
2334, 2825 (quoted, calculated-$K_{OW}$, Hansch et al. 1968)
2850, 2850 (exptl., calculated-group contribution, Korenman et al. 1971; quoted, Horvath 1982)

2723 (quoted, Hine & Mookerjee 1975)
2950 (selected exptl., Horvath 1982)
2700 (20°C, quoted, Verschueren 1983)
2710 (20°C, quoted, Riddick et al. 1986)
2318, 1162 (quoted, calculated-fragment const., Wakita et al. 1986)
2649 (calculated-AI, Okouchi et al. 1992)

Vapor Pressure (Pa at 25°C):
37364 (20°C, Rex 1906)
55240 (30°C, Rex 1906)
46495 (quoted, Hine & Mookerjee 1975)
46655 (quoted, Verschueren 1983)
45920 (quoted, Riddick et al. 1986)

Henry's Law Constant (Pa $m^3$/mol):
1341 (calculated-P/C, using Hine & Mookerjee 1975 data)
1362.2, 1600.5 (quoted, calculated-QSAR, Nirmalakhandan & Speece 1988)

Octanol/Water Partition Coefficient, log $K_{OW}$:
2.04 (quoted exptl., Sangster 1989)
2.04 (recommended, Sangster 1989)

Bioconcentration Factor, log BCF:

Sorption Partition Coefficient, log $K_{OC}$:

Half-Lives in the Environment:
Air: disappearance half-life of 2.4-24 hours from air for the reaction with OH radicals (EPA 1974; quoted, Darnall et al. 1976).
Surface water:
Ground water:
Sediment:
Soil:
Biota:

Environmental Fate Rate Constant and Half-Lives:

Volatilization:

Photolysis:

Oxidation:

Hydrolysis:

Biodegradation:

Biotransformation:

Bioconcentration, Uptake ($k_1$) and Elimination ($k_2$) Rate Constants or Half-Lives:

Common Name: 2-Chloropropane
Synonym: isopropyl chloride
Chemical Name: 2-chloropropane, isopropyl chloride
CAS Registry No: 75-29-6
Molecular Formula: $CH_3CHClCH_3$
Molecular Weight: 78.54
Melting Point (°C):
-117.18 (Dreisbach 1959; Horvath 1982; Riddick et al. 1986)
-117.0 (Verschueren 1983)
-117.20 (Dean 1985)
Boiling Point (°C):
34.80 (Rex 1906; Dreisbach 1959)
35.74 (Horvath 1982; Riddick et al. 1986)
36.5 (Verschueren 1983)
35.0 (Dean 1985)
Density ($g/cm^3$ at 20°C):
0.8626 (Dreisbach 1959)
0.8560 (25°C, Dreisbach 1959)
0.8617 (Horvath 1982; Riddick et al. 1986)
0.859 (Verschueren 1983)
0.8563 (25°C, Riddick et al. 1986)
Molar Volume ($cm^3/mol$):
94.90 (calculated-LeBas method)
91.33 (calculated from density)
Molecular Volume ($A^3$):
Total Surface Area, TSA ($A^2$):
246.38 (Amidon 1976; quoted, Horvath 1982)
106.70 (Okouchi et al. 1992)
Heat of Fusion, $\Delta H_{fus}$, (kcal/mol):
1.766 (calculated, Dreisbach 1959)
Entropy of Fusion, $\Delta S_{fus}$ (cal/mol K or e.u.):
Fugacity Ratio at 25°C, F: 1.0

Water Solubility ($g/m^3$ or mg/L at 25°C):
3050 (20°C, volumetric, Rex 1906; quoted, Horvath 1982)
3040 (30°C, volumetric, Rex 1906; quoted, Horvath 1982)
3444 (Fühner 1924; quoted, Hansch et al. 1968)
3056 (Seidell 1940; quoted, Deno & Berkheimer 1960)
3000 (Seidell 1941; quoted, Irmann 1965)
3100 (20°C, Patty 1962; Lange 1967; quoted, Horvath 1982)
3000 (quoted exptl., Irmann 1965)
3160 (calculated-group contribution, Irmann 1965; quoted, Horvath 1982)

3444, 4944 (quoted, calculated-$K_{OW}$, Hansch et al. 1968)
3130 (calculated-group contribution, Korenman et al. 1971; quoted, Horvath 1982)
3055 (quoted, Hine & Mookerjee 1975)
3000, 3713 (quoted, calculated-TSA, Amidon 1976; quoted, Horvath 1982)
2950 (selected exptl., Horvath 1982)
3100 (20°C, quoted, Verschueren 1983)
3400 (Dean 1985)
3428, 1462 (quoted, calculated-fragment const., Wakita et al. 1986)
2945 (calculated-AI, Okouchi et al. 1992)

Vapor Pressure (Pa at 25°C):
58052 (20°C, Rex 1906)
83939 (30°C, Rex 1906)
70582 (calculated-Antoine eqn., Dreisbach 1959)
64264 (quoted, Hine & Mookerjee 1975)
69716 (quoted, Verschueren 1983)
68700 (quoted, Riddick et al. 1986)

Henry's Law Constant (Pa $m^3$/mol):
1652 (calculated-P/C, using Hine & Mookerjee 1975 data)
1637.7, 247.9 (quoted, calculated-QSAR, Nirmalakhandan & Speece 1988)

Octanol/Water Partition Coefficient, log $K_{OW}$:
1.69 (calculated-$\pi$, Hansch et al. 1968)
1.90 (quoted exptl., Sangster 1989)
1.90 (recommended, Sangster 1989)

Bioconcentration Factor, log BCF:

Sorption Partition Coefficient, log $K_{OC}$:

Half-Lives in the Environment:
Air: disappearance half-life of 2.4-24 hours from air for the reaction with OH radicals (USEPA 1974; quoted, Darnall et al. 1976).
Surface water:
Ground water:
Sediment:

Soil:
Biota:

Environmental Fate Rate Constant and Half-Lives:
Volatilization:
Photolysis:
Oxidation:
Hydrolysis:
Biodegradation:
Biotransformation:
Bioconcentration, Uptake ($k_1$) and Elimination ($k_2$) Rate Constants or Half-Lives:

Common Name: 1,2-Dichloropropane
Synonym: propylene chloride, propylene dichloride
Chemical Name: 1,2-dichloropropane
CAS Registry No: 78-87-5
Molecular Formula: $CH_3CHClCH_2Cl$
Molecular Weight: 112.99
Melting Point (°C):
-100.44 (Dreisbach 1959; Horvath 1982; Riddick et al. 1986)
-100.4 (Weast 1982-83; Dean 1985)
-100/-80 (Verschueren 1983)
Boiling Point (°C):
96.40 (McGovern 1943; Timmermans 1950; McNally & Grob 1984)
96.37 (Dreisbach 1959; Horvath 1982; Riddick et al. 1986)
96.22 (Boublik et al. 1973)
94.0 (Weast 1973-74; Jones et al. 1977/1978)
96.4 (Weast 1982-83; Dean 1985)
96.8 (Verschueren 1983)
96.0 (Banerjee et al. 1990)
Density ($g/cm^3$ at 20°C):
1.155 (McGovern 1943)
1.1560 (Dreisbach 1959; Horvath 1982; Weast 1982-83; McNally & Grob 1984)
1.1494 (25°C, Dreisbach 1959)
1.160 (Verschueren 1983)
1.1558 (Dean 1985)
1.15597 (Riddick et al. 1986)
Molar Volume ($cm^3/mol$):
96.90 (14°C, calculated from density, Stephenson & Malanowski 1987)
116 (LeBas method, Abernethy et al. 1988)
54.0 (intrinsic volume-van der Waals method, Abernethy et al. 1988)
98.0 (calculated from density, Abernethy et al. 1988; Wang et al. 1992)
Molecular Volume ($A^3$):
Total Surface Area, TSA ($A^2$):
268.64 (Amidon 1976; quoted, Horvath 1982)
123.2 (Okouchi et al. 1992)
Heat of Fusion, $\Delta H_{fus}$, (kcal/mol):
1.529 (calculated, Dreisbach; quoted, Riddick et al. 1986)
Entropy of Fusion, $\Delta S_{fus}$ (cal/mol K or e.u.):
Fugacity Ratio at 25°C, F: 1.0

Water Solubility ($g/m^3$ or mg/L at 25°C):
2800 (Gross 1929; Stephen & Stephen 1963; quoted, Horvath 1982)
2773 (Seidell 1940; quoted, Deno & Berkheimer 1960)

2700 (20°C, McGovern 1943)
2750 (measured by Dow Chemical, Dreisbach 1955-1961; quoted, Horvath 1982)
2800 (quoted, Irmann 1965)
2774 (quoted, Hine & Mookerjee 1975)
2802, 1409 (quoted, calculated-TSA, Amidon 1976; quoted, Horvath 1982)
2096 (shake flask-GC, Jones et al. 1977/1978)
2700 (20°C, quoted, Nathan 1978)
2700 (quoted, Cowen & Baynes 1980)
2800 (recommended, Horvath 1982)
2800, 2820 (quoted exptl., calculated-group contribution as per Irmann 1965, Horvath 1982)
2700 (selected, Mills et al. 1982)
2700 (20°C, quoted, Verschueren 1983; Warner et al. 1987)
2700, 2420.4 (quoted, headspace/GC, McNally & Grob 1983)
2700; 2069.5 (quoted; 30°C, headspace/GC, McNally & Grob 1984)
2600 (Dean 1985)
2740 (quoted, Riddick et al. 1986; quoted, Howard 1990)
2799, 1648 (quoted, predicted- $\chi$ & polarizability, Nirmalakhandan & Speece 1988)
2746 (calculated-$K_{OW}$ with regression, Müller & Klein 1992)
2799 (calculated-AI, Okouchi et al. 1992)
3000 (20°C, shake flask-GC/TC, Stephenson 1992)
2900 (29.7°C, shake flask-GC/TC, Stephenson 1992)
2680, 2699 (20°C, calculated-activity coefficients, Wright et al. 1992)
2649, 1297 (quoted experimental, calculated-molar volume, Wang et al. 1992)

Vapor Pressure (Pa at 25°C):
7198 (Nelson & Young 1933; quoted, Timmermans 1950)
6932 (McGovern 1943)
6777 (Antoine eqn. regression, Stull 1947)
6621 (calculated-Antoine eqn., Dreisbach 1959)
6724 (calculated-Antoine eqn., Boublik et al. 1973)
7146 (quoted, Hine & Mookerjee 1975)
6665 (quoted, Nathan 1978; Verschueren 1983)
6930 (quoted, Cowen & Baynes 1980)
5600 (20°C, selected, Mills et al. 1982)
5332 (19.4°C, quoted, McNally & Grob 1984)
6622 (quoted, Riddick et al. 1986; quoted, Howard 1990)
6617 (calculated-Antoine eqn., Stephenson & Malanowski 1987)
6667 (quoted, Warner et al. 1987)
7105, 4466 (quoted, calculated-UNIFAC, Banerjee et al. 1990)
5410 (20°C, quoted from DIPPR, Tse et al. 1992)
8791 (30°C, quoted from DIPPR, Tse et al. 1992)

Henry's Law Constant (Pa $m^3$/mol):

298 (calculated-1/$K_{AW}$, $C_W/C_A$, reported as exptl., Hine & Mookerjee; quoted, Nirmalakhandan & Speece 1988)

31.21 (calculated-group contribution method, Hine & Mookerjee 1975)

2261 (calculated-bond contribution method, Hine & Mookerjee 1975)

287 (concentration ratio-GC, Leighton & Calo 1981)

209.7 (calculated-P/C, Mackay & Yuen 1983; quoted, Howard 1990)

285.7, 278.6 (quoted, calculated-P/C, Warner et al. 1987)

298.0, 839.9 (quoted, calculated-QSAR, Nirmalakhandan & Speece 1988)

233.0 (20-25°C and low ionic strength, quoted, Pankow & Rosen 1988; Pankow 1990)

Octanol/Water Partition Coefficient, log $K_{OW}$:

2.00 (Hansch & Leo 1979)

2.28 (selected, Mills et al. 1982)

1.99 (USEPA 1987; quoted, Howard 1990)

2.16 (selected, Abernethy et al. 1988)

1.99 (quoted, Van Leeuwen et al. 1992; Verhaar et al. 1992)

2.02, 2.30 (quoted, calculated-molar volume, Wang et al. 1992)

1.99 (calculated, Müller & Klein 1992)

Bioconcentration Factor, log BCF:

$<1.0$ (fish, Kawasaki 1980; quoted, Howard 1990)

1.29 (calculated-$K_{OW}$, USEPA 1986; quoted, Howard 1990)

Sorption Partition Coefficient, log $K_{OC}$:

1.672 (silt loam soil, Chiou et al. 1979; quoted, Howard 1990)

Sorption Partition Coefficient, log $K_{OM}$:

1.43, 1.70 (quoted, calculated- $\chi$ , Sabljic 1984)

Half-Lives in the Environment:

Air: disappearance half-life of 2.4-24 hours from air for the reaction with OH radicals (EPA 1974; quoted, Darnall et al. 1976); 65-646 hours, based on an estimated rate constant for the vapor phase reaction with hydroxyl radicals in air (Atkinson 1987; quoted, Howard et al. 1991); $>23$ days, reaction with photochemically produced OH radicals (Howard 1990) .

Surface water: 4008-30936 hours, based on estimated aqueous aerobic biodegradation half-life (Howard et al. 1991).

Ground water: 8016-61872 hours, based on estimated aqueous aerobic biodegradation half-life (Howard et al. 1991).

Sediment:

Soil: 4008-30936 hours, based on estimated aqueous aerobic biodegradation half-life (Howard et al. 1991).

Biota:

Environmental Fate Rate Constant and Half-Lives:

Volatilization: half-life of less than 50 minutes in water from stirring in an open container of depth 65 mm at 200 rpm (Dilling et al. 1975; quoted, Mills et al. 1982); half-life of 8 minutes from a stirred solution 1.6 cm deep (Chiou et al. 1980; quoted, Howard 1990); half-lives of 1.7-10 days, estimated from an EXAMS model of the fate of 1,2-dichloropropane in a pond, river, and two lakes (Burns 1981; quoted, Howard 1990); half-life of 8.3 hours at 1 m depth of stirred aqueous solution (Lyman et al. 1982; quoted, Howard 1990); estimated half-life of 5.5 hours for the removal from a stream of 1 m depth with 1 m/sec current, based on laboratory-determined relative transfer coefficients (Cadena et al. 1984; quoted, Howard 1990).

Photolysis:

Oxidation: photooxidation half-life in air of 65-646 hours, based on an estimated rate constant for the vapor phase reaction with hydroxyl radicals in air (Atkinson 1987; quoted, Howard et al. 1991).

Hydrolysis: $5.0 \times 10^{-6}$ $hour^{-1}$ at pH 7 to 9 and 25°C with a calculated half-life of 15.8 years (Ellington et al. 1987; quoted, Ellington 1989; Howard et al. 1991).

Biodegradation: aqueous aerobic half-life of 4008-30936 hours, based on acclimated aerobic soil grab sample data (Roberts & Stoydin 1976; quoted, Howard et al. 1991); aqueous anaerobic half-life of 16032-123744 hours, based on estimated aqueous aerobic biodegradation half-life (Howard et al. 1991).

Biotransformation:

Bioconcentration, Uptake ($k_1$) and Elimination ($k_2$) Rate Constants or Half-Lives:

Common Name: 1,2,3-Trichloropropane
Synonym:
Chemical Name: 1,2,3-trichloropropane
CAS Registry No: 96-18-4
Molecular Formula: $CH_2ClCHClCH_2Cl$
Molecular Weight: 147.43
Melting Point (°C):
-14.7 (Stull 1947; Horvath 1982; Weast 1982-83; Dean 1985; Stephenson & Malanowski 1987)
-14.0 (Verschueren 19983)
Boiling Point (°C):
156.85 (Horvath 1982)
156.8 (Weast 1982-83)
156.0 (Verschueren 19983)
156.9 (Dean 1985)
158.0 (Stephenson & Malanowski 1987)
Density (g/cm$^3$ at 20°C):
1.3889 (Horvath 1982; Weast 1982-83)
1.3880 (Dean 1985)
1.3832 (25°C, Riddick et al. 1986)
Molar Volume (cm$^3$/mol):
103.8 (15°C, calculated from density, Stephenson & Malanowski 1987)
63.0 (intrinsic volume, Abernethy et al. 1988)
137 (LeBas method, Abernethy et al. 1988)
106 (calculated from density, Wang et al. 1992)
Molecular Volume (A$^3$):
Total Surface Area, TSA (A$^2$):
287.51 (Amidon 1976; Horvath 1982)
139.7 (Okouchi et al. 1992)
Heat of Fusion, $\Delta H_{fus}$, (kcal/mol):
Entropy of Fusion, $\Delta S_{fus}$ (cal/mol K or e.u.):
Fugacity Ratio at 25°C, F: 1.0

Water Solubility (g/m$^3$ or mg/L at 25°C):
1900, 1016 (quoted, calculated-TSA, Amidon 1976; quoted, Horvath 1982)
1900 (Dilling 1977; quoted, Horvath 1982)
1900 (Afghan & Mackay 1980; quoted, Horvath 1982)
1896 (recommended, Horvath 1982)
1900 (quoted, Riddick et al. 1986)
1900, 485.3 (quoted, predicted- $\chi$ & polarizability, Nirmalakhandan & Speece 1988)
1899 (calculated-AI, Okouchi et al. 1992)
1772, 868.1 (quoted, calculated-molar volume, Wang et al. 1992)

Vapor Pressure (Pa at 25°C):

266.6 (20°C, quoted, Verschueren 1983)

533.2 (30°C, quoted, Verschueren 1983)
19.77 (calculated-Antoine eqn., Boublik et al. 1984)
492 (quoted, Riddick et al. 1986)
492 (calculated-Antoine eqn., Stephenson & Malanowski 1987)
413, 960 (quoted, calculated-solvatochromic p. & UNIFAC, Banerjee et al. 1990)

Henry's Law Constant (Pa $m^3$/mol):

32.22 (selected from Dilling 1977 & Mackay & Shiu 1981, Tancréde & Yanagisawa 1990)
36.0 (concn. ratio-GC, Leighton & Calo 1981)
9.92 (purge & trap/GC/ECD, Tancréde & Yanagisawa 1990)
22.31 (quoted from Tancréde & Yanagisawa 1990, Tancréde et al. 1992)

Octanol/Water Partition Coefficient, log $K_{OW}$:

2.63 (selected, Abernethy et al. 1988)
1.98 (quoted, Van Leeuwen et al. 1992; Verhaar et al. 1992)
2.25, 2.36 (quoted, calculated-molar volume, Wang et al. 1992)

Bioconcentration Factor, log BCF:

Sorption Partition Coefficient, log $K_{OC}$:

Half-Lives in the Environment:

Air: disappearance half-life of 2.4-24 hours from air for the reaction with OH radicals (EPA 1974; quoted, Darnall et al. 1976); 61-613 hours, based on an estimated rate constant for the vapor phase reaction with hydroxyl radicals in air (Atkinson 1987; quoted, Howard et al. 1991).

Surface water: 4320-8640 hours, based on estimated aqueous aerobic biodegradation half-life (Howard et al. 1991).

Ground water: 8640-17280 hours, based on estimated aqueous aerobic biodegradation half-life (Howard et al. 1991).

Sediment:

Soil: disappearance half-life of 2.7 days was calculated from first-order kinetic for volatilization loss of mixtures in soils (Anderson et al. 1991); 4320-8640 hours, based on estimated aqueous aerobic biodegradation half-life (Howard et al. 1991).

Biota:

Environmental Fate Rate Constant and Half-Lives:

Volatilization:

Photolysis:

Oxidation: photooxidation half-life of 61-613 hours, based on an estimated rate constant for the vapor phase reaction with hydroxyl radicals in air (Atkinson 1987; quoted, Howard et al. 1991).

Hydrolysis: $1.8 \times 10^{-6}$ $hour^{-1}$ at pH 7 to 9 at 25°C with a calculated half-life of 44 years (Ellington et al. 1987; quoted, Ellington 1989; Howard et al. 1991).

Biodegradation: aqueous aerobic half-life of 4320-8640 hours, based on acclimated aerobic grab sample test data for 1,3-dichloropropane (Roberts & Stoydin 1976; quoted, Howard et al. 1991); aqueous anaerobic half-life of 17280-34560 hours, based on aqueous aerobic biodegradation half-life (Howard et al. 1991).

Biotransformation:

Bioconcentration, Uptake ($k_1$) and Elimination ($k_2$) Rate Constants or Half-Lives:

Common Name: 1-Chlorobutane
Synonym: *n*-butyl chloride, butyl chloride
Chemical Name: 1-chlorobutane, n-butyl chloride
CAS Registry No: 109-69-3
Molecular Formula: $CH_3CH_2CH_2CH_2Cl$
Molecular Weight: 92.57
Melting Point (°C):
-123.2 (Dreisbach 1959)
-123.1 (Horvath 1982; Dean 1985; Riddick et al. 1986)
-123.0 (Verschueren 1983; Miller et al. 1985)
Boiling Point (°C):
78.44 (Dreisbach 1959; Horvath 1982; Dean 1985)
78.0 (Verschueren 1983)
78.43 (Riddick et al. 1986)
Density ($g/cm^3$ at 20°C):
0.8862 (Dreisbach 1959; Horvath 1982)
0.8808 (25°C, Dreisbach 1959)
0.8840 (Verschueren 1983)
0.8864 (Dean 1985)
0.8857 (Riddick et al. 1986)
0.8810 (25°C, Riddick et al. 1986)
Molar Volume ($cm^3$/mol):
66.04 (Miller et al. 1985)
55.0 (intrinsic volume, Abernethy et al. 1988)
117.0 (LeBas method, Abernethy et al. 1988)
104.47 (calculated from density)
Molecular Volume ($A^3$):
85.5 (Moriguchi et al. 1976)
Total Surface Area, TSA ($A^2$):
282.62 (Amidon 1976; quoted, Horvath 1982)
121.5 (Moriguchi et al. 1976)
125.1 (Okouchi et al. 1992)
Heat of Fusion, $\Delta H_{fus}$, (kcal/mol):
Entropy of Fusion, $\Delta S_{fus}$ (cal/mol K or e.u.):
Fugacity Ratio at 25°C, F: 1.0

Water Solubility ($g/m^3$ or mg/L at 25°C):
666 (Fühner 1924; quoted, Hansch et al. 1968)
640 (Seidell 1940; quoted, Deno & Berkheimer 1960)
741 (Kakovsky 1957; quoted, Horvath 1982)
667 (17.5°C, Saracco & Spaccamela Marchetti 1958; quoted, Horvath 1982)

660 (quoted, Irmann 1965)
794 (calculated-group contribution, Irmann 1965; quoted, Horvath 1982)
666, 823 (quoted, calculated-$K_{OW}$, Hansch et al. 1968; quoted, Tewari et al. 1982)
671, 930 (20°C, exptl., calculated-group contribution, Korenman et al. 1971; Horvath 1982)
671 (quoted, Hine & Mookerjee 1975)
659, 686 (quoted, calculated-TSA, Amidon 1976; quoted, Horvath 1982)
700 (quoted, Nathan 1978)
702, 640 (quoted, calculated-$K_{OW}$, Valvani et al. 1981)
615 (quoted exptl., Horvath 1982)
872 (gen. col.-HPLC, Tewari et al. 1982; quoted, Miller et al. 1985)
1100 (Dean 1985)
1100 (20°C, quoted, Riddick et al. 1986)
671, 336 (quoted, calculated-fragment const., Wakita et al. 1986)
702 (quoted, Isnard & Lambert 1989)
614 (calculated-AI, Okouchi et al. 1992)

Vapor Pressure (Pa at 25°C):
13654 (calculated-Antoine eqn., Dreisbach 1959)
14170 (quoted, Hine & Mookerjee 1975)
13997 (quoted, Nathan 1978)
10677 (20°C, quoted, Verschueren 1983)
13300 (24.47°C, quoted, Riddick et al. 1986)

Henry's Law Constant (Pa $m^3$/mol):
1969 (calculated-$1/K_{AW}$, $C_W/C_A$, reported as exptl., Hine & Mookerjee 1975; quoted, Nirmalakhandan & Speece 1988)
1796 (calculated-group contribution method, Hine & Mookerjee 1975)
8595 (calculated-bond contribution method, Hine & Mookerjee 1975)
1969, 2015 (quoted, calculated-QSAR, Nirmalakhandan & Speece 1988)

Octanol/Water Partition Coefficient, log $K_{OW}$:
2.39 (shake flask-GC, Fujita et al. 1964; Hansch et al. 1968; quoted, Tewari et al. 1982; Sangster 1989)
2.39 (Leo et al. 1971; quoted, Moriguchi et al. 1976)
2.39, 2.44 (quoted, calculated-hydrophobicity const., Moriguchi et al. 1976)
2.55 (gen. col.-GC, DeVoe et al. 1981; quoted, Sangster 1989)
2.64 (quoted, Valvani et al. 1981)
2.55 (gen. col.-HPLC, Wasik et al. 1981; quoted, Sangster 1989)
2.55 (gen. col.-HPLC, Tewari et al. 1982; quoted, Miller et al. 1985; Sangster 1989)

2.39 (quoted, Verschueren 1983)
2.35 (selected, Abernethy et al. 1988)
2.55 (quoted, Isnard & Lambert 1989)
2.64 (quoted exptl., Sangster 1989)
2.64 (recommended, Sangster 1989)
2.64 (quoted, Van Leeuwen et al. 1992)

Bioconcentration Factor, log BCF:

Sorption Partition Coefficient, log $K_{OC}$:

Half-Lives in the Environment:

Air: disappearance half-life of 2.4-24 hours from air for the reaction with OH radicals (USEPA 1974; quoted, Darnall et al. 1976).

Surface water:

Ground water:

Sediment:

Soil:

Biota:

Environmental Fate Rate Constant and Half-Lives:

Volatilization:

Photolysis:

Oxidation:

Hydrolysis:

Biodegradation:

Biotransformation:

Bioconcentration, Uptake ($k_1$) and Elimination ($k_2$) Rate Constants or Half-Lives:

Common Name: 2-Chlorobutane
Synonym: methylethylchloromethane, *sec*-butyl chloride
Chemical Name: 2-chlorobutane, methylethylchloromethane, *sec*-butyl chloride
CAS Registry No: 78-86-4
Molecular Formula: $CH_3CH_2CHClCH_3$
Molecular Weight: 92.57
Melting Point (°C):
-131.3 (Horvath 1982)
-113.3 (racemic, Riddick et al. 1986)
-140.5 (active, Riddick et al. 1986)
Boiling Point (°C):
68.25 (Horvath 1982; Riddick et al. 1986)
Density ($g/cm^3$ at 20°C):
0.8732 (Horvath 1982; Dean 1985)
0.87323 (racemic, Riddick et al. 1986)
Molar Volume ($cm^3/mol$):
Molecular Volume ($A^3$):
Total Surface Area, TSA ($A^2$):
127.6 (Okouchi et al. 1992)
Heat of Fusion, $\Delta H_{fus}$, (kcal/mol):
Entropy of Fusion, $\Delta S_{fus}$ (cal/mol K or e.u.):
Fugacity Ratio at 25°C, F: 1.0

Water Solubility ($g/m^3$ or mg/L at 25°C):
1000 (Wright & Schaffer 1932; quoted, Irmann 1965; Horvath 1982)
1000 (quoted, Nathan 1978)
1000 (Dean 1985)
1000 (quoted, Riddick et al. 1986)
1001 (calculated-AI, Okouchi et al. 1992)

Vapor Pressure (Pa at 25°C):
20662 (quoted, Nathan 1978)
20210 (quoted, Riddick et al. 1986)

Henry's Law Constant (Pa $m^3$/mol):
1871 (calculated-P/C, using Riddick et al. 1986 data)

Octanol/Water Partition Coefficient, log $K_{OW}$:

Bioconcentration Factor, log BCF:

Sorption Partition Coefficient, log $K_{OC}$:

Half-Lives in the Environment:

Air: disappearance half-life of 2.4-24 hours from air for the reaction with OH radicals (EPA 1974; quoted, Darnall et al. 1976).

Surface water:

Ground water:

Sediment:

Soil:

Biota:

Environmental Fate Rate Constant and Half-Lives:

Volatilization:

Photolysis:

Oxidation:

Hydrolysis:

Biodegradation:

Biotransformation:

Bioconcentration, Uptake ($k_1$) and Elimination ($k_2$) Rate Constants or Half-Lives:

Common Name: 1-Chloropentane
Synonym: *n*-amyl chloride, monochloropentane, pentyl chloride
Chemical Name: n-amyl chloride, 1-chloropentane, pentyl chloride
CAS Registry No: 543-59-9
Molecular Formula: $CH_3CH_2CH_2CH_2CH_2Cl$
Molecular Weight: 106.60
Melting Point (°C):
-99.0 (Horvath 1982; Riddick et al. 1982; Dean 1985)
Boiling Point (°C):
107.80 (Horvath 1982)
98.30 (Dean 1985)
108.39 (Riddick et al. 1986)
Density (g/cm$^3$ at 20°C):
0.8818 (Horvath 1982)
0.8824 (Dean 1985)
0.8820 (Riddick et al. 1986)
0.8770 (25°C, Riddick et al. 1986)
Molar Volume (cm$^3$/mol):
139.3 (calculated-LeBas method)
120.86 (calculated from density)
Molecular Volume (A$^3$):
Total Surface Area, TSA (A$^2$):
146.0 (Okouchi et al. 1992)
Heat of Fusion, $\Delta H_{fus}$, (kcal/mol):
Entropy of Fusion, $\Delta S_{fus}$ (cal/mol K or e.u.):
Fugacity Ratio at 25°C, F: 1.0

Water Solubility (g/m$^3$ or mg/L at 25°C):
200 (Wright & Schaffer 1932; quoted, Irmann 1965; Horvath 1982)
789 (Booth & Everson 1948; quoted, Horvath 1982)
199 (quoted, Hine & Mookerjee 1975)
200 (Dean 1985)
200 (quoted, Riddick et al. 1986)
199, 95 (quoted, calculated-fragment const., Wakita et al. 1986)
199 (calculated-AI, Okouchi et al. 1992)

Vapor Pressure (Pa at 25°C):
4142 (quoted, Hine & Mookerjee 1975)
4142 (quoted, Riddick et al. 1986)

Henry's Law Constant (Pa $m^3$/mol):

2209 (calculated as $1/K_{AW}$, $C_W/C_A$, reported as exptl., Hine & Mookerjee 1975)
2479 (calculated-group contribution method, Hine & Mookerjee 1975)
13009 (calculated-bond contribution method, Hine & Mookerjee 1975)

Octanol/Water Partition Coefficient, log $K_{OW}$:

Bioconcentration Factor, log BCF:

Sorption Partition Coefficient, log $K_{OC}$:

Half-Lives in the Environment:

Air: disappearance half-life of 2.4-24 hours from air for the reaction with OH radicals (EPA 1974; quoted, Darnall et al. 1976).
Surface water:
Ground water:
Sediment:
Soil:
Biota:

Environmental Fate Rate Constant and Half-Lives:

Volatilization:
Photolysis:
Oxidation:
Hydrolysis:
Biodegradation:
Biotransformation:
Bioconcentration, Uptake ($k_1$) and Elimination ($k_2$) Rate Constants or Half-Lives:

Common Name: Vinyl Chloride
Synonym: chloroethene, chloroethylene, monochloroethylene, monovinylchloride, MVC
Chemical Name: chloroethylene, vinyl chloride, chloroethene
CAS Registry No: 75-01-4
Molecular Formula: $H_2C=CHCl$
Molecular Weight: 62.5
Melting Point (°C):

- -153.7 (Stull 1947)
- -153.79 (Dreisbach 1959; Riddick et al. 1986)
- -153.8 (Weast 1977; Callahan et al. 1979; Mackay & Shiu 1981; Horvath 1982; Mabey et al. 1982; Stephenson & Malanowski 1987; Howard 1989; Ma et al. 1990; Mackay & Shiu 1990)
- -153/-160 (Verschueren 1983)
- -159.7 (Dean 1985)

Boiling Point (°C):

- -13.37 (Dreisbach 1959; Weast 1977; Callahan et al. 1979; Horvath 1982; Mabey et al. 1982; Howard 1989)
- -13.81 (Boublik et al. 1973)
- -13.9 (Pearson & McConnell 1975; Verschueren 1983)
- -13.40 (Mackay & Shiu 1981,1990; Stephenson & Malanowski 1987)
- -13.80 (Riddick et al. 1986)

Density ($g/cm^3$ at 20°C):

- 0.9106 (Riddick et al. 1986)

Molar Volume ($cm^3/mol$):

- 65.3 (calculated-LeBas method)
- 68.64 (calculated from density)

Molecular Volume ($A^3$):
Total Surface Area, TSA ($A^2$):

- 77.40 (Okouchi et al. 1992)

Heat of Fusion, $\Delta H_{fus}$, (kcal/mol):

- 1.134 (calculated, Dreisbach; quoted, Riddick et al. 1986)

Entropy of Fusion, $\Delta S_{fus}$ (cal/mol K or e.u.):
Fugacity Ratio at 25°C, F: 1.0

Water Solubility ($g/m^3$ or mg/L at 25°C):

- 1100 (quoted, Irmann 1965)
- 2700 (Hayduk & Laudie 1974; quoted, Mackay & Shiu 1981; Riddick et al 1986; Ma et al. 1990)
- 1111 (quoted, Hine & Mookerjee 1975)
- 2700 (Dilling 1977; quoted, Horvath 1982; Mabey et al. 1982)
- 90 (20°C, Neely 1976; quoted, Mackay & Shiu 1981; Thomas 1982; Jury et al. 1984)
- 1100 (Verschueren 1977,1983; quoted, Ma et al. 1990; Mackay & Shiu 1990)

6800 (20°C, selected, Nathan 1978)
8700 (restatement of Hayduk & Laudie 1974, DeLassus & Schmidt 1981)
8800 (solubility bomb-GC head space, DeLassus & Schmidt 1981)
2763 (Horvath 1982; quoted, Howard 1989; Ma et al. 1990)
1100 (Horvath 1982; quoted, Ma et al. 1990)
2763 (recommended, Horvath 1982)
60 (10°C, selected, Mills et al. 1982)
2700 (quoted, Riddick et al. 1986)
2761, 3467 (quoted, predicted- $\chi$ & polarizability, Nirmalakhandan & Speece 1988)
2000 (selected, Ma et al. 1990)
2760 (calculated-AI, Okouchi et al. 1992)

Vapor Pressure (Pa at 25°C):

538000 (Antoine eqn. regression, Stull 1947)
354578 (calculated-Antoine eqn., Dreisbach 1959; quoted, Riddick et al. 1986)
546801 (calculated-Antoine eqn., Weast 1972-73)
392798 (calculated-Antoine eqn., Boublik et al. 1973)
309300 (Pearson & McConnell 1975; quoted, Ma et al. 1990)
308000 (20°C, Pearson & McConnell 1975; quoted, Mackay & Shiu 1981)
344000 (20°C, Neely 1976; quoted, Mackay & Shiu 1981; Thomas 1982; Ma et al. 1990; Mackay & Shiu 1990)
354578 (Verschueren 1977,1983; quoted, Callahan et al. 1979; Mabey et al. 1982; Mills et al. 1982; Ma et al. 1990)
354578 (selected, Nathan 1978)
354578 (20°C, selected, Mills et al. 1982)
354600 (Riddick et al. 1986; quoted, Howard 1989; Ma et al. 1990)
350000 (selected, Ma et al. 1990)

Henry's Law Constant (Pa $m^3$/mol):

5679 (calculated as $1/K_{AW}$, $C_W/C_A$, reported as exptl., Hine & Mookerjee 1975; quoted, Nirmalakhandan & Speece 1988)
4723 (calculated-bond contribution method, Hine & Mookerjee 1975)
123941, 39661 (quoted exptl., calculated-P/C, Neely 1976)
123941, 106589 (quoted, calculated-P/C, Dilling 1977)
2350 (calculated-P/C, Mackay & Shiu 1981)
8246 (calculated-P/C, Mabey et al. 1982; quoted, Ma et al. 1990)
243139 (calculated-P/C, Thomas 1982)
2817 (EPICS, Gossett 1987; quoted, Ma et al. 1990; Mackay & Shiu 1990)
2685 (EPICS, Ashworth et al. 1988; quoted, Ma et al. 1990)
5679, 494.6 (quoted, calculated-QSAR, Nirmalakhandan & Speece 1988)
2230 (Pankow & Rosen 1988; quoted, Ma et al. 1990; Pankow 1990)

1537 (quoted, Ryan et al. 1988)
377000 (selected, Jury et al. 1990)
5757 (calculated-P/C, Mackay & Shiu 1990)

Octanol/Water Partition Coefficient, log $K_{OW}$:
1.39 (calculated-$\pi$, Hansch et al. 1968)
0.6 (Callahan et al. 1979; quoted, Ryan et al. 1988; Mackay & Shiu 1990)
1.23 (calculated-f const., Mabey et al. 1982)
0.602 (selected, Mills et al. 1982)
1.38 (Hansch & Leo 1985; quoted, Howard 1989; Ma et al. 1990)
1.00 (selected, Ma et al. 1990)

Bioconcentration Factor, log BCF:
0.845 (estimated-S, Lyman et al. 1982; quoted, Howard 1989)
0.756 (microorganisms-water, calculated-$K_{OW}$, Mabey et al. 1982; quoted, Ma et al. 1990)
3.04 (activated sludge, Freitag et al. 1984)
1.6 (*chlorella fusca*, Freitag et al. 1985; quoted, Ma et al. 1990)
3.04 (activated sludge, Freitag et al. 1985; quoted, Ma et al. 1990)
<1.0 (golden ide, Freitag et al. 1985; quoted, Ma et al. 1990)
0.068 (quoted from USEPA 86, Yeh & Kastenberg 1991)

Sorption Partition Coefficient, log $K_{OC}$:
0.9138 (sediment-water, calculated-$K_{OW}$, Mabey et al. 1982; quoted, Ma et al. 1990)
1.748 (estimated-S, Lyman et al. 1982; quoted, Howard 1989; Jury et al. 1992)
0.477 (soil, selected, Jury et al. 1990)
1.756 (quoted from USEPA 86, Yeh & Kastenberg 1991)

Half-Lives in the Environment:
Air: disappearance half-life of 0.24-2.4 hours from air for the reaction with OH radicals (EPA 1974; quoted, Darnall et al. 1976); estimated half-life of 1.5 days, based on its photochemical reaction with hydroxyl radicals in air (Perry et al. 1977; quoted, Howard 1989); 1.8 day, estimated as toxic chemical residence time with rate constant of $6.6 \times 10^{-12}$ $cm^3$ $molecule^{-1}$ $sec^{-1}$ at 300 K for the reaction with OH radicals (Singh et al. 1981); 9.7-97 hours, based on measured rate for the reaction with hydroxyl radicals in air (Atkinson 1985; quoted, Howard et al. 1991); atmospheric lifetimes: 42 days for the reaction with $NO_3$ for a 12-hour nighttime average concn. of $2.4 \times 10^9$ molecule $cm^{-3}$, 3.5 days for the reaction

with OH for a 12-hour average concn. of $1.0 \times 10^6$ molecule $cm^{-3}$, and 66 days for the reaction with $O_3$ for a 24-hour average concn. of $7 \times 10^{11}$ molecule $cm^{-3}$ (quoted, Atkinson et al. 1987).

Surface water: estimated half-life of 0.805 hour for volatilization from a river of 1 meter deep with a current of 3 m/sec and a wind velocity of 3 m/sec (Lyman et al. 1982; quoted, Howard 1989); 672-4320 hours, based on aqueous screening test data (Freitag et al. 1984; Helfgott et al. 1977; quoted, Howard et al. 1991).

Ground water: 1344-69000 hours, based on estimated unacclimated aqueous aerobic biodegradation half-life and an estimated half-life for anaerobic biodegradation from a ground water field study of chlorinated ethenes (Silka & Wallen 1988; quoted, Howard et al. 1991).

Sediment:

Soil: <10 days (Ryan et al. 1988); 0.2 to 0.5 day for volatilization from soil at 1 and 10 cm incorporation (Jury et al. 1984; quoted, Howard 1989); 30 days, estimated volatilization loss from soil (Jury et al. 1990); 30-180 days (Howard et al. 1991; quoted, Jury et al. 1992); 672-4320 hours, based on estimated unacclimated aqueous aerobic biodegradation half-life (Howard et al. 1991).

Biota: <10 days, subject to plant uptake via volatilization (Ryan et al. 1988; quoted, Jury et al. 1992).

Environmental Fate Rate Constant and Half-Lives:

Volatilization: half-life from water of 26 minutes, obtained by rapidly stirring aqueous solutions in an open container of depth 65 mm at 200 rpm (Dilling et al. 1975; quoted, Callahan et al. 1979; Mills et al 1982) and 27.6 minutes under same experimental conditions (Dilling 1977); calculated half-lives: 0.0054 minute, 16.1 minutes (Dilling 1977); estimated half-life of 0.805 hour for volatilization from a river of 1 meter deep with a current of 3 m/sec and wind velocity of 3 m/sec (Lyman et al. 1982; quoted, Howard 1989); estimated half-life from water to be 2.5 hours (Thomas 1982); volatilization half-life of 0.2 to 0.5 day from soil at 1 and 10 cm incorporation (Jury et al. 1984; quoted, Howard 1989); half-life of 30 days, estimated volatilization from soil (Jury et al. 1990).

Photolysis: degrade rapidly in air by reaction with photochemically produced hydroxyl radicals with an estimated half-life of 1.5 day (Perry et al. 1977; quoted, Howard 1989).

Oxidation: $6.6 \times 10^{-12}$ $cm^3$ $molecule^{-1}$ $sec^{-1}$ at 26°C for the reaction with OH radicals in air (Perry et al. 1977); half-life of photooxidation in the troposphere was reported to be a few hours (Callahan et al. 1979); $6.6 \times 10^{-12}$ $cm^3$ $molecule^{-1}$ $sec^{-1}$ for the reaction with hydroxyl radicals at room temperature (Atkinson et al. 1979,1982; quoted, Atkinson & Carter 1984; Tuazon et al. 1984); rate constants for the gas-phase reaction with ozone: $6.5 \times 10^{-21}$ $cm^3$ $molecule^{-1}$ $sec^{-1}$ at 295 K (Sanhueza et al. 1976; quoted, Atkinson & Carter 1984), $2.3 \times 10^{-19}$ $cm^3$ $molecule^{-1}$ $sec^{-1}$ at room temperature (Gay et al. 1976; quoted, Atkinson & Carter 1984), $2.5 \times 10^{-19}$

$cm^3$ $molecule^{-1}$ $sec^{-1}$ at room temperature (Atkinson et al. 1979,1982; quoted, Atkinson & Carter 1984; Tuazon et al. 1984; Atkinson et al. 1987) and (2.45 ± 0.45)x$10^{-19}$ $cm^3$ $molecule^{-1}$ $sec^{-1}$ at 298 K (Zhang et al. 1983; quoted, Atkinson & Carter 1984); 3.9x$10^{12}$ $cm^3$ $mol^{-1}$ $sec^{-1}$ estimated rate constant for the reaction with OH radicals and 1.2x$10^6$ $cm^3$ $mol^{-1}$ $sec^{-1}$ for ozone both at 300 K (Lyman 1982); calculated rate constants at 25°C: < $10^8$ $M^{-1}$ $hour^{-1}$ for singlet oxygen and 3.0 $M^{-1}$ $hour^{-1}$ for peroxy radicals (Mabey et al. 1982); photooxidation half-life of 9.7-97 hours, based on measured rate constant for the reaction with hydroxyl radicals in air (Atkinson 1985; quoted, Howard et al. 1991); rate constant of 6.61x$10^{-12}$ $cm^3$ $molecule^{-1}$ $sec^{-1}$ for the gas-phase reaction with OH radicals at 298 K (Atkinson 1986; quoted, Atkinson et al. 1987; Sabljic & Güsten 1990); rate constant of (2.30 ± 1.10)x$10^{-16}$ $cm^3$ $molecule^{-1}$ $sec^{-1}$ for the gas-phase reaction with $NO_3$ radicals at 298 ± 2 K (Atkinson et al. 1987) and (4.26 ± 0.19)x$10^{-16}$ $cm^3$ $molecule^{-1}$ $sec^{-1}$ relative to ethene for the gas-phase reaction with $NO_3$ radicals at 298 ± 2 K (Atkinson et al. 1987; quoted, Atkinson 1991) while (3.30 ± 1.66)x$10^{-16}$ $cm^3$ $molecule^{-1}$ $sec^{-1}$ for the same reaction at 296 ± 1 K (Anderson & Ljungström 1989; quoted, Atkinson 1991); rate constant of 4.45x$10^{-16}$ $cm^3$ $molecule^{-1}$ $sec^{-1}$ for the gas-phase reaction with $NO_3$ radicals at 298 K (Atkinson et al. 1988; quoted, Sabljic & Güsten 1990); estimated rate constant of 5.25x$10^{-12}$ $cm^3$ $molecule^{-1}$ $sec^{-1}$ for the reaction with OH radicals (as per Atkinson 1987 and 1988, Müller & Klein 1991).

Hydrolysis: half-life < 10 years (Callahan et al. 1979); estimated acid-catalyzed rate constant of 3.30x$10^{-12}$ $mol^{-1}$ $sec^{-1}$ at pH 5 with calculated half-life of 2x$10^{11}$ days (Wolfe 1980; quoted, Ma et al. 1990); abiotic hydrolysis or dehydrohalogenation half-life of <120 months (Olsen & Davis 1990)..

Biodegradation: abiotic degradation rate constant of 6.6x$10^{-12}$ $cm^3$ $molecule^{-1}$ $sec^{-1}$ at 26°C determined in laboratory experiments for the vapor phase reaction with hydroxyl radicals (Perry et al. 1977; quoted, Howard 1989); aqueous aerobic half-life of 672-4320 hours, based on aqueous screening test data (Freitag et al. 1984; Helfgott et al. 1977; quoted, Howard et al. 1991); under aerobic conditions $^{14}C$-labelled vinyl chloride in samples taken from a shallow aquifer was readily degraded with greater than 99% after 108 days and approximately 65% being mineralized to $^{14}CO_2$ (Davis & Carpenter 1990); aqueous anaerobic half-life of 2688-17280 hours, based on estimated unacclimated aqueous aerobic biodegradation half-life (Howard et al. 1991); half-life of >60 days (Olsen & Davis 1990).

Biotransformation:

Bioconcentration, Uptake ($k_1$) and Elimination ($k_2$) Rate Constants or Half-Lives:

Common Name: 1,1-Dichloroethene
Synonym: 1,1-dichloroethylene, vinylidene chloride, vinylidine chloride, 1,1-DCE
Chemical Name: 1,1-dichloroethene, 1,1-dichloroethylene
CAS Registry No: 75-35-4
Molecular Formula: $CH_2=CCl_2$
Molecular Weight: 96.94
Melting Point (°C):

-122.5 (Stull 1947; Verschueren 1983; Howard 1989)

-122.1 (Weast 1977,1982-83; Callahan et al. 1979; Mackay & Shiu 1981,1990; Horvath 1982; Mabey et al. 1982)

-122.6 (Dean 1985)

-122.56 (Riddick et al. 1986; Stephenson & Malanowski 1987)

Boiling Point (°C):

31.56 (Boublik et al. 1973; Riddick et al. 1986; Stephenson & Malanowski 1987)

31.9 (McConnell et al. 1975; Pearson & McConnell 1975; Verschueren 1983)

37.0 (Weast 1977,1982-83; Callahan et al. 1979; Mackay & Shiu 1981,1990; Horvath 1982; Mabey et al. 1982)

31.6 (Dean 1985)

31.7 (Howard 1989)

Density (g/cm³ at 20°C):

1.218 (Horvath 1982; Weast 1982-83; Verschueren 1983)

1.2129 (Dean 1985)

1.2132 (Riddick et al. 1986)

1.220 (Gillham & Rao 1990)

Molar Volume (cm³/mol):

0.402 (calculated intrinsic volume, Kamlet et al. 1988; Thoms & Lion 1992)

86.20 (calculated-LeBas method)

79.46 (calculated from density)

Molecular Volume (A³):
Total Surface Area, TSA (A²):

94.5 (Okouchi et al. 1992)

Heat of Fusion, $\Delta H_{fus}$, (kcal/mol):

1.557 (Riddick et al. 1986)

Entropy of Fusion, $\Delta S_{fus}$ (cal/mol K or e.u.):
Fugacity Ratio at 25°C, F: 1.0

Water Solubility (g/m³ or mg/L at 25°C):

400 (20°C, McConnell et al. 1975; Pearson & McConnell 1975; quoted, Callahan et al. 1979; Mackay & Shiu 1981,1990; Mabey et al. 1982; Mills et al. 1982; Gillham & Rao 1990; Thoms & Lion 1992)

400 (20°C, Dilling 1977; quoted, Mackay & Shiu 1981; Horvath 1982; Thoms & Lion 1992)

400 (quoted, Cowen & Baynes 1980)
2250 (solubility bomb-GC head space, DeLassus & Schmidt 1981)
210 (at satn. pressure, Horvath 1982)
3344 (at atm. pressure, Horvath 1982)
2640 (20°C, saturation concn., quoted, Verschueren 1983)
3675 (30°C, saturation concn., quoted, Verschueren 1983)
2500; 2232 (quoted literature best; 30°C, headspace-GC, McNally & Grob 1984)
2600 (quoted, Neely 1984)
210 (Dean 1985)
210 (quoted, Riddick et al. 1986)
5000 (quoted, Warner et al. 1987)
3342; 1592, 3365 (quoted; predicted- $\chi$ & polarizability, Nirmalakhandan & Speece 1988)
2802 (calculated-$K_{OW}$ with regression, Müller & Klein 1992)
1594 (calculated-AI, Okouchi et al. 1992)

Vapor Pressure (Pa at 25°C):
86433 (Antoine eqn. regression, Stull 1947)
84500 (calculated-Antoine eqn., Weast 1972-73)
80042 (calculated-Antoine eqn., Boublik et al. 1973)
66195 (20°C, McConnell et al. 1975)
65900 (20°C, Pearson & McConnell 1975; quoted, Neely 1976; Mackay & Shiu 1981)
79713 (Dilling 1977; quoted, Mackay & Shiu 1981,1990)
78780 (Verschueren 1977,1983; quoted, Callahan et al. 1979; Mabey et al. 1982; Mills et al. 1982)
79122 (quoted, Cowen & Baynes 1980)
73400 (23°C, Schmidt-Bleek et al. 1982; quoted, Politzki et al. 1982)
78780 (IARC 1986; quoted, Howard 1989)
79860 (quoted, Riddick et al. 1986)
80063 (calculated-Antoine eqn., Stephenson & Malanowski 1987)
78831 (quoted, Warner et al. 1987)
78647 (20°C, selected, Gillham & Rao 1990)
66188 (20°C, quoted from DIPPR, Tse et al. 1992)
95743 (30°C, quoted from DIPPR, Tse et al. 1992)

Henry's Law Constant (Pa $m^3$/mol):
19335 (calculated, Dilling 1977)
3729 (20°C, batch stripping, Mackay et al. 1979; quoted, Yurteri et al. 1987)
19380 (calculated-P/C, Mackay & Shiu 1981)
19249 (20°C, calculated-P/C, Mabey et al. 1982)

2649 (20°C, EPICS, Linoff & Gossett 1983; Gossett 1985; quoted, Yurteri et al. 1987)
2569 (20°C, EPICS, Ashworth et al. 1986; quoted, Yurteri et al. 1987)
2645 (EPICS, Gossett 1987; quoted, Mackay & Shiu 1990; Thoms & Lion 1992)
2101 (20°C, EPICS, Gossett 1987; quoted, Yurteri et al. 1987; Tse et al. 1992)
1520, 1530 (quoted, calculated-P/C, Warner et al. 1987)
7529 (20°C, EPICS, Yurteri et al. 1987)
19249 (20-25°C and low ionic strength, quoted, Pankow & Rosen 1988; Pankow 1990)
446.2 (quoted, Ryan et al. 1988)
3049 (calculated-P/C, Howard 1989)
2320 (20°C, Tse et al. 1992)
3414 (30°C, Tse et al. 1992)

Octanol/Water Partition Coefficient, log $K_{OW}$:
1.48 (calculated as per Tute 1971, Callahan et al. 1979; quoted, Ryan et al. 1988; Mackay & Shiu 1990)
2.13 (estimated-f const., Lyman 1982; quoted, Thoms & Lion 1992)
2.13 (calculated-f const., Mabey et al. 1982; quoted, Dobbs et al. 1989)
1.48 (selected, Mills et al. 1982)
2.13 (Hansch & Leo 1985; quoted, Howard 1989)
1.48 (quoted, Olsen & Davis 1990)
2.11 (calculated, Müller & Klein 1992)

Bioconcentration Factor, log BCF:
1.724 (microorganisms-water, calculated-$K_{OW}$, Mabey et al. 1982)

Sorption Partition Coefficient, log $K_{OC}$:
2.176 (calculated-$K_{OW}$, Kenaga & Goring 1980; quoted, Howard 1989)
1.813 (sediment-water, calculated-$K_{OW}$, Mabey et al. 1982)

Half-Lives in the Environment:
Air: disappearance half-life of 0.24-2.4 hours from air for the reaction with OH radicals (EPA 1974; quoted, Darnall et al. 1976); 2.9 days, estimated as toxic chemical residence time with rate constant of $4x10^{-12}$ $cm^3$ $molecule^{-11}$ $sec^{-1}$ at 300 K for the reaction with OH radicals (Singh et al. 1981); photooxidation half-life of 11 hours in relatively clean air (Edney et al. 1983; quoted, Howard 1989) or under 2 hours in polluted air (Gay et al. 1976; quoted, Howard 1989); lifetime of 0.75 day, based on measured rate data for the vapor phase reaction with hydroxyl radicals in air at (22.3 ± 1.2)°C (Edney et al. 1986); 9.9-98.7 hours, based on

measured rate data for the vapor phase reaction with hydroxyl radicals in air (Goodman et al. 1986; quoted, Howard et al. 1991); atmospheric lifetimes: 15 days for the reaction with $NO_3$ for a 12-hour nighttime average concn. of $2.4 \times 10^9$ molecule $cm^{-3}$, 3.4 days for the reaction with OH for a 12-hour average concn. of $1.0 \times 10^6$ molecule $cm^{-3}$, and 12 years for the reaction with $O_3$ for a 24-hour average concn. of $7 \times 10^{11}$ molecule $cm^{-3}$ (quoted, Atkinson et al. 1987).

Surface water: 672-4320 hours, based on acclimated aerobic soil screening test data (Tabak et al. 1981; quoted, Howard et al. 1991).

Ground water: 1344-3168 hours, based on estimated aqueous aerobic biodegradation half-life and anaerobic grab sample data for soil from ground water aquifer receiving landfill leachate (Wilson et al. 1986; quoted, Howard et al. 1991).

Sediment:

Soil: 672-4320 hours, based on acclimated aerobic soil screening test data (Tabak et al. 1981; quoted, Howard et al. 1991); < 10 days (Ryan et al. 1988).

Biota: < 10 days, subject to plant uptake via volatilization (Ryan et al. 1988).

Environmental Fate Rate Constant and Half-Lives:

Volatilization: estimated experimental half-life of 22 minutes at 25°C for volatilization of 1 mg/liter in water in an open container of 65 mm in depth stirring at 200 rpm (Dilling et al. 1975; quoted, Callahan et al. 1979; Mills et al. 1982) and 27.2 minutes under same experimental conditions (Dilling 1977); calculated half-lives: 0.069 minute, 20.1 minutes (Dilling 1977); using data for the oxygen reaeration rate of typical bodies of water, one can calculate the half-life for evaporation to be 5.9, 1.2, and 4.7 days from a pond, river, and lake, respectively (Mill et al. 1982; quoted, Howard 1989).

Photolysis:

Oxidation: rate constant of $(3.7 \pm 1.0) \times 10^{-21}$ $cm^3$ $molecule^{-1}$ $sec^{-1}$ for the gas-phase reaction with ozone at 298 K (Hull et al. 1973; quoted, Atkinson & Carter 1984; Tuazon et al. 1984; Atkinson et al. 1987); calculated rate constants at 25°C: $< 10^8$ $M^{-1}$ $hour^{-1}$ for singlet oxygen and 3.0 $M^{-1}$ $hour^{-1}$ for peroxy radicals (Mabey et al. 1982); photooxidation half-life of 11 hours in relatively clean air (Edney et al. 1983; quoted, Howard 1989) or under 2 hours in polluted air (Gay et al. 1976; quoted, Howard 1989); rate constant of $(14.9 \pm 2.1) \times 10^{12}$ $cm^3$ $molecule^{-1}$ $sec^{-1}$ with reference to *n*-butane for the gas-phase reaction with OH radicals at $(22.3 \pm 1.2)$°C with a lifetime of 0.75 day in the atmosphere and $15.0 \times 10^{-12}$ $cm^3$ $molecule^{-1}$ $sec^{-1}$ with reference to *n*-pentane for the gas-phase reaction with OH radicals at $(24.4 \pm 0.4)$°C (Edney et al. 1986); photooxidation half-life of 9.9-98.7 hours, based on measured rate data for the vapor phase reaction with hydroxyl radicals in air (Goodman et al. 1986; quoted, Howard et al. 1991); rate constant of $8.09 \times 10^{12}$ $cm^3$ $molecule^{-1}$ $sec^{-1}$ for the gas-phase reaction with OH radicals at 298 K (Atkinson 1987; Atkinson et al. 1987;

quoted, Sabljic & Güsten 1990); rate constant of $(6.60 \pm 3.10) \times 10^{-16}$ (Atkinson et al. 1987) and $(1.23 \pm 0.15) \times 10^{-15}$ $cm^3$ $molecule^{-1}$ $sec^{-1}$ relative to ethene for the gas-phase reaction with $NO_3$ radicals at $298 \pm 2$ K (Atkinson et al. 1987; quoted, Atkinson 1991) while rate constant of $1.28 \times 10^{-15}$ $cm^3$ $molecule^{-1}$ $sec^{-1}$ for the gas-phase reaction with $NO_3$ radicals at 298 K (Atkinson et al. 1988; quoted, Sabljic & Güsten 1990); estimated rate constant of $2.04 \times 10^{-12}$ $cm^3$ $molecule^{-1}$ $sec^{-1}$ for the reaction with OH radicals (as per Atkinson 1987 and 1988, Müller & Klein 1991).

Hydrolysis: estimated acid-catalyzed rate constant of $1.4 \times 10^{-13}$ $mol^{-1}$ $sec^{-1}$ at pH 5 with calculated half-life of $6 \times 10^{12}$ days at pH 5 (Wolfe 1980); estimated half-life of 2.0 years at pH 7.0 (Schmidt-Bleek et al. 1982; quoted, Howard 1989); half-life of 6-9 months has been observed with no significant difference in hydrolysis rate between pH 4.5 and 8.5 (Cline & Delfino 1987; quoted, Howard 1989); abiotic hydrolysis or dehydrohalogenation half-life of 12 months (Olsen & Davis 1990).

Biodegradation: aqueous aerobic half-life of 672-4320 hours, based on acclimated aerobic soil screening test data (Tabak et al. 1981; quoted, Howard et al. 1991); aqueous anaerobic half-life of 1944-4152 hours, based on anaerobic sediment grab sample data (Barrio-Lage et al. 1986; quoted, Howard 1989; Olsen & Davis 1990; Howard et al. 1991).

Biotransformation:

Bioconcentration, Uptake ($k_1$) and Elimination ($k_2$) Rate Constants or Half-Lives:

Common Name: *cis*-1,2-Dichloroethylene
Synonym: *cis*-acetylene dichloride, *cis*-1,2-dichloroethene, *cis*-1,2-dichloroethylene, (Z)-1,2-dichloroethene
Chemical Name: *cis*-1,2-dichloroethene, *cis*-1,2-dichloroethylene
CAS Registry No: 156-59-2
Molecular Formula: CHCl=CHCl
Molecular Weight: 96.94
Melting Point (°C):
-80.5 (Stull 1947; Horvath 1982; Weast 1982-83; Howard 1990)
-81.0 (Verschueren 1983)
-80.1 (Dean 1985)
-80.0 (Riddick et al. 1986)
-81.47 (Stephenson & Malanowski 1987)
Boiling Point (°C):
60.44 (Boublik et al. 1973)
60.3 (Horvath 1982; Weast 1982-83; Howard 1990)
60.0 (Verschueren 1983; Banerjee et al. 1990)
60.7 (Dean 1985)
60.63 (Riddick et al. 1986)
60.2 (Stephenson & Malanowski 1987)
Density ($g/cm^3$ at 20°C):
1.2837 (Horvath 1982; Weast 1982-83; Riddick et al. 1986)
1.280 (Verschueren 1983)
1.2818 (Dean 1985)
Molar Volume ($cm^3/mol$):
86.20 (calculated-LeBas method)
75.52 (calculated from density)
Molecular Volume ($A^3$):
Total Surface Area, TSA ($A^2$):
93.6 (Okouchi et al. 1992)
Heat of Fusion, $\Delta H_{fus}$, (kcal/mol):
1.722 (Riddick et al. 1986)
Entropy of Fusion, $\Delta S_{fus}$ (cal/mol K or e.u.):
Fugacity Ratio at 25°C, F: 1.0

Water Solubility ($g/m^3$ or mg/L at 25°C):
3520 (Seidell 1940; quoted, Deno & Berkheimer 1960)
3500 (McGovern 1943; quoted, Dilling 1977; Horvath 1982; Love Jr. & Eilers 1982)
7700, 5620 (quoted exptl., calculated-group contribution as per Irmann 1965, Horvath 1982)
7700 (quoted, Hine & Mookerjee 1975)

3500 (recommended, Horvath 1982)
800 (20°C, quoted, Verschueren 1983)
7700 (Dean 1985)
3500 (quoted, Riddick et al. 1986; quoted, Howard 1990)
3499; 1884, 3236 (quoted; predicted- $\chi$ & polarizability, Nirmalakhandan & Speece 1988)
3500 (calculated-$K_{OW}$ with regression, Müller & Klein 1992)
3495 (calculated-AI, Okouchi et al. 1992)
6301, 6409 (20°C, calculated-activity coefficients, Wright et al. 1992)

Vapor Pressure (Pa at 25°C):
27261 (Antoine eqn. regression, Stull 1947)
23542 (calculated-Antoine eqn., Weast 1972-73)
27007 (calculated-Antoine eqn., Boublik et al. 1973)
27020 (quoted, Hine & Mookerjee 1975)
27460 (quoted average, Dilling 1977; quoted, Love Jr. & Eilers 1982)
26660 (quoted, Verschueren 1983)
26700 (quoted, Riddick et al. 1986)
26978 (calculated-Antoine eqn., Stephenson & Malanowski 1987)
26660 (35°C, Riddick et al. 1986; quoted, Howard 1990)
26793, 33592 (quoted, calculated-solvatochromic p. & UNIFAC, Banerjee et al. 1990)
21767 (20°C, estimated, Tse et al. 1992)
33305 (30°C, estimated, Tse et al. 1992)

Henry's Law Constant (Pa $m^3$/mol):
342.2 (calculated $1/K_{AW}$, $C_W/C_A$, reported as exptl., Hine & Mookerjee 1975; quoted, Howard 1990)
1969 (calculated-group contribution method, Hine & Mookerjee 1975)
2367 (calculated-bond contribution method, Hine & Mookerjee 1975)
443.6 (20°C, EPICS, Lincoff & Gossett 1983; Gossett 1985; quoted, Yurteri et al. 1987)
453.3 (20°C, EPICS, Ashworth et al. 1986; quoted, Yurteri et al. 1987)
299.8 (20°C, EPICS, Gossett 1987; quoted, Yurteri et al. 1987; Tse et al. 1992)
413 (EPICS, Gossett 1987)
441.1 (20°C, EPICS, Yurteri et al. 1987)
324.2 (20°C, Tse et al. 1992)
496.4 (30°C, Tse et al. 1992)

Octanol/Water Partition Coefficient, log $K_{OW}$:
1.86 (Hansch & Leo 1985; quoted, Howard 1990)
1.51 (calculated, Müller & Klein 1992)

Bioconcentration Factor, log BCF:
1.176 (calculated-$K_{OW}$, Lyman et al. 1982; quoted, Howard 1990)

Sorption Partition Coefficient, log $K_{OC}$:
1.69 (calculated-$K_{OW}$, Lyman et al. 1982; quoted, Howard 1990)

Half-Lives in the Environment:

Air: disappearance half-life of 0.24-2.4 hours from air for the reaction with OH radicals (EPA 1974; quoted, Darnall et al. 1976); 2.9 days, estimated as toxic chemical residence time with rate constant of $4 \times 10^{-12}$ $cm^3$ $molecule^{-1}$ $sec^{-1}$ for the reaction with OH radicals (Singh et al. 1981); half-life of 129 days resulting from the ozone attacking of the double bond (Tuazon et al. 1984; quoted, Howard 1990); 8.0 days in the atmosphere for the reaction with photochemically produced hydroxyl radicals (Goodman et al. 1986; quoted, Howard 1990); 286 hours, based on estimated rate constant for the reaction with hydroxyl radicals in air (Atkinson 1987; quoted, Howard et al. 1991); atmospheric lifetimes: 130 days for the reaction with $NO_3$ for a 12-hour nighttime average concn. of $2.4 \times 10^9$ molecule $cm^{-3}$, 12 days for the reaction with OH for a 12-hour average concn. of $1.0 \times 10^6$ molecule $cm^{-3}$, and >9.0 years for the reaction with $O_3$ for a 24-hour average concn. of $7 \times 10^{11}$ molecule $cm^{-3}$ (quoted, Atkinson et al. 1987).

Surface water: 672-4320 hours, based on estimated unacclimated aqueous aerobic biodegradation half-life (Howard et al. 1991).

Ground water: 1344-69000 hours, based on estimated aqueous aerobic biodegradation half-life (Howard et al. 1991) and an estimated half-life for anaerobic biodegradation from a ground water field studies of chlorinated ethylenes (Silka & Wallen 1988; quoted, Howard et al. 1991).

Sediment:

Soil: 672-4320 hours, based on estimated unacclimated aqueous aerobic biodegradation half-life (Howard et al. 1991).

Biota:

Environmental Fate Rate Constant and Half-Lives:

Volatilization: half-life of 19.4 minutes from a slowly stirred beaker 6.5 cm deep equivalent to half-life of 5.0 hours in a body of water 1 meter deep (Dilling 1977; quoted, Verschueren 1983; Howard 1990); calculated half-lives: 0.75

minute, 20.7 minutes (Dilling 1977); half-life of 3.1 hours was estimated from Henry's law constant for a model river 1 meter deep with 1 m/sec current and 3 m/sec wind (Lyman et al. 1982; quoted, Howard 1990).

Photolysis:

Oxidation: rate constants for the gas-phase reaction with ozone: $6.2 \times 10^{-20}$ $cm^3$ $molecule^{-1}$ $sec^{-1}$ at 296 K (Blume et al. 1976; quoted, Atkinson & Carter 1984; Tuazon et al. 1984) and $< 5.0 \times 10^{-21}$ $cm^3$ $molecule^{-1}$ $sec^{-1}$ at room temperature (Niki et al. 1983; quoted, Atkinson & Carter 1984; Tuazon et al. 1984; Atkinson et al. 1987), while $3.7 \times 10^4$ $cm^3$ $mol^{-1}$ $sec^{-1}$, was estimated rate constant for the reaction with ozone at 300 K (Lyman 1982); photoxidation half-life of 286 hours, based on estimated rate constant for the reaction with hydroxyl radicals in air (Atkinson 1987; quoted, Howard et al. 1991); rate constant of $2.38 \times 10^{12}$ $cm^3$ $molecule^{-1}$ $sec^{-1}$ for the gas-phase reaction with OH radicals at 298 K (Atkinson 1986; quoted, Atkinson et al. 1987; Sabljic & Güsten 1990); rate constant of $(0.75 \pm 0.35) \times 10^{-16}$ $cm^3$ $molecule^{-1}$ $sec^{-1}$ for the gas-phase reaction with $NO_3$ radicals at $298 \pm 2$ K (Atkinson et al. 1987); rate constant of $(1.39 \pm 0.13) \times 10^{-16}$ $cm^3$ $molecule^{-1}$ $sec^{-1}$ relative to ethene for the gas-phase reaction with $NO_3$ radicals at $298 \pm 2$ K (Atkinson et al. 1987; quoted, Atkinson 1991) while rate constant of $1.46 \times 10^{-16}$ $cm^3$ $molecule^{-1}$ $sec^{-1}$ for the gas-phase reaction with $NO_3$ radicals at 298 K (Atkinson et al. 1988; quoted, Sabljic & Güsten 1990); estimated rate constant of $2.24 \times 10^{-12}$ $cm^3$ $molecule^{-1}$ $sec^{-1}$ for the reaction with OH radicals (as per Atkinson 1987 and 1988, Müller & Klein 1991).

Hydrolysis:

Biodegradation: aqueous aerobic half-life of 672-4320 hours, based on unacclimated aerobic aqueous screening test data (Tabak et al. 1981; quoted, Howard et al. 1991); anaerobic half-life of 2688-17280 hours, based on estimated unacclimated aqueous aerobic biodegradation half-life (Howard et al. 1991); half-life of >60 days (Wood et al. 1985; quoted, Olsen & Davis 1990); 0.74 $year^{-1}$ with half-life of 88-339 days (Barrio-Lage et al. 1986; quoted, Olsen & Davis 1990).

Biotransformation:

Bioconcentration, Uptake ($k_1$) and Elimination ($k_2$) Rate Constants or Half-Lives:

Common Name: *trans*-1,2-Dichloroethylene
Synonym: *trans*-1,2-dichloroethene, *trans*-1,2-dichloroethylene, *trans*-acetylene dichloride, Dioform, (E)-1,2-dichloroethene
Chemical Name: *trans*-1,2-dichloroethylene, *trans*-1,2-dichloroethene
CAS Registry No: 156-60-5
Molecular Formula: ClCH=CHCl
Molecular Weight: 96.94
Melting Point (°C):
- -50 (Stull 1947; Weast 1977,1982-83; Mackay & Shiu 1981; Horvath 1982; Verschueren 1983; Howard 1990)
- -49.8 (Dean 1985; Riddick et al. 1986)
- -49.44 (Stephenson & Malanowski 1987)

Boiling Point (°C):
- 47.5 (Weast 1977,1982-83; Mackay & Shiu 1981; Horvath 1982)
- 47.7 (Boublik et al. 1973; Dean 1985; Stephenson & Malanowski 1987)
- 48.0 (Verschueren 1983; Hewitt et al. 1992)
- 47.67 (Riddick et al. 1986)

Density ($g/cm^3$ at 20°C):
- 1.2565 (Horvath 1982; Weast 1982-83)
- 1.260 (Verschueren 1983)
- 1.2546 (Dean 1985)

Molar Volume ($cm^3/mol$):
- 86.20 (calculated-LeBas method)
- 77.20 (calculated from density)
- 0.772 ($V_M/100$, Kamlet et al. 1986,1987; Leahy 1986)
- 0.406 (calculated intrinsic volume $V_I/100$, Leahy 1986; Kamlet et al. 1987,1988)

Molecular Volume ($A^3$):
Total Surface Area, TSA ($A^2$):
Heat of Fusion, $\Delta H_{fus}$, (kcal/mol):
- 2.864 (Riddick et al. 1986)

Entropy of Fusion, $\Delta S_{fus}$ (cal/mol K or e.u.):
Fugacity Ratio at 25°C, F: 1.0

Water Solubility ($g/m^3$ or mg/L at 25°C):
- 6259 (Seidell 1940; quoted, Deno & Berkheimer 1960)
- 6300 (McGovern 1943; quoted, Dilling 1977; Mackay & Shiu 1981; Horvath 1982; Warner et al. 1987)
- 600 (20°C, Verschueren 1977,1983; quoted, Callahan et al. 1979; Mabey et al. 1982; Mills et al. 1982; Gillham & Rao 1990)
- 600 (quoted, Cowen & Baynes 1980)

6300, 5620 (quoted exptl., calculated-group contribution as per Irmann 1965, Horvath 1982)
6259 (quoted, Hine & Mookerjee 1975)
6260 (recommended, Horvath 1982)
6300 (Dean 1985)
6300 (quoted, Riddick et al. 1986; quoted, Howard 1990)
6259, 4859 (quoted, calculated-molar volume & solvatochromic p., Kamlet et al. 1986)
6259, 3859 (quoted, calculated-molar volume & solvatochromic p., Leahy 1986)
6259, 4640 (quoted, calculated-molar volume & solvatochromic p., Kamlet et al. 1987)
6000 (20°C, quoted from McGovern 1943, Hewitt et al. 1992)
6300 (calculated-$K_{OW}$ with regression, Müller & Klein 1992)
4486, 4540 (20°C, calculated-activity coefficients, Wright et al. 1992)

Vapor Pressure (Pa at 25°C):
43177 (Antoine eqn. regression, Stull 1947)
43470 (Hardie 1964; quoted, Mackay & Shiu 1981)
36743 (calculated-Antoine eqn., Weast 1972-73)
44188 (calculated-Antoine eqn., Boublik et al. 1973)
43989 (quoted, Hine & Mookerjee 1975)
34650 (20°C, Neely 1976; quoted, Mackay & Shiu 1981)
43456 (Dilling 1977; quoted, Mabey et al. 1982)
26660 (20°C, Verschueren 1977; quoted, Callahan et al. 1979; Mills et al. 1982)
42144 (quoted, Cowen & Baynes 1980)
26660 (14°C, quoted, Verschueren 1983)
45300 (quoted, Riddick et al. 1986)
36698 (quoted, Kamlet et al. 1986)
44403 (calculated-Antoine eqn., Stephenson & Malanowski 1987)
26698 (14°C, quoted, Warner et al. 1987)
43456 (20°C, selected, Gillham & Rao 1990)
36247 (20°C, quoted from DIPPR, Tse et al. 1992)
53974 (30°C, quoted from DIPPR, Tse et al. 1992)

Henry's Law Constant (Pa $m^3$/mol):
682.7 (calculated as $1/K_{AW}$, $C_W/C_A$, reported as exptl., Hine & Mookerjee 1975; quoted, Howard 1990; Hewitt et al. 1992)
1969 (calculated-group contribution method, Hine & Mookerjee 1975)
2367 (calculated-bond contribution method, Hine & Mookerjee 1975)
669.3 (calculated-P/C, Dilling 1977; quoted, Ryan et al. 1988)
669 (calculated-P/C, Mackay & Shiu 1981)
6788 (20°C, calculated-P/C, Mabey et al. 1982)

950.5 (20°C, EPICS, Lincoff & Gossett 1983; Gossett 1985; quoted, Yurteri et al. 1987)
940.8 (20°C, EPICS, Ashworth et al. 1986; quoted, Yurteri et al. 1987)
950 (EPICS, Gossett 1987)
914.0 (20°C, EPICS, Yurteri et al. 1987)
739.5 (20°C, quoted from Gossett 1987, Tse et al. 1992)
539.0, 410.4 (quoted, calculated-P/C, Warner et al. 1987)
729.4 (20-25°C and low ionic strength, quoted, Pankow & Rosen 1988; Pankow 1990)
800.3 (20°C, Tse et al. 1992)
1195 (30°C, Tse et al. 1992)

Octanol/Water Partition Coefficient, log $K_{OW}$:
1.48 (calculated as per Tute 1971, Callahan et al. 1979; quoted, Ryan et al. 1988)
2.09 (calculated-f const., Mabey 1982)
1.48 (selected, Mills et al. 1982)
2.06 (Hansch & Leo 1985; quoted, Howard 1990)
2.09, 1.87 (quoted, calculated-molar volume & solvatochromic p., Leahy 1986)
2.09, 2.00 (quoted, calculated-molar volume & solvatochromic p., Kamlet et al. 1988)
1.48 (quoted, Olsen & Davis 1990)
1.93 (quoted from Hansch & Leo 1979, Hewitt et al. 1992)
1.51 (calculated, Müller & Klein 1992)

Bioconcentration Factor, log BCF:
1.34 (calculated-$K_{OW}$, Lyman et al. 1982; quoted, Howard 1990)
1.68 (microorganisms-water, calculated-$K_{OW}$, Mabey et al. 1982)

Sorption Partition Coefficient, log $K_{OC}$:
1.56 (calculated-$_{OW}$, Lyman et al. 1982; quoted, Howard 1990)
1.77 (sediment-water, calculated-$K_{OW}$, Mabey et al. 1982)

Half-Lives in the Environment:
Air: disappearance half-life of 0.24-2.4 hours from air for the reaction with OH radicals (EPA 1974; quoted, Darnall et al. 1976); 2.9 days, estimated as toxic chemical residence time with rate constant of $4x10^{-12}$ $cm^3$ $molecule^{-1}$ $sec^{-1}$ at 300 K for the reaction with hydroxyl radicals (Singh et al. 1981); 44 days resulting from ozone attacking of the double bond (Tuazon et al. 1984; quoted, Howard 1990); 3.6 days in the atmosphere for the reaction with photochemically produced hydroxyl radicals (Goodman et al. 1986; quoted, Howard 1990); 25.2 hours, based on estimated rate constant for reaction with hydroxyl radicals in air (Atkinson 1987;

quoted, Howard et al. 1991); atmospheric lifetimes: 170 days for the reaction with $NO_3$ for a 12-hour nighttime average concn. of $2.4x10^9$ molecule $cm^{-3}$, 5.1 days for the reaction with OH for a 12-hour average concn. of $1.0x10^6$ molecule $cm^{-3}$, and 110 days for the reaction with $O_3$ for a 24-hour average concn. of $7x10^{11}$ molecule $cm^{-3}$ (quoted, Atkinson et al. 1987).

Surface water: 672-4320 hours, based on unacclimated aerobic aqueous screening test data (Tabak et al. 1981; quoted, Howard et al. 1991).

Ground water: 1344-69000 hours, based on estimated unacclimated aqueous aerobic biodegradation half-life and estimated half-life for anaerobic biodegradation of chlorinated ethylenes from a ground water field study (Silka & Wallen 1988; quoted, Howard et al. 1991).

Sediment:

Soil: 672-4320 hours, based on unacclimated aerobic aqueous screening test data (Tabak et al. 1981; quoted, Howard et al. 1991); $<10$ days (Ryan et al. 1988).

Biota: $<10$ days, subject to plant uptake via volatilization (Ryan et al. 1988).

Environmental Fate Rate Constant and Half-Lives:

Volatilization: experimental half-life for 1 mg/liter from water to be 24 minutes when stirred at 200 rpm at approximately 25°C in an open container of depth 65 mm (Dilling et al. 1975; Dilling 1977; quoted, Callahan et al. 1979; Mills et al. 1982); calculated half-lives: 0.85 minute, 25.8 minutes (Dilling 1977); half-life of volatilization from a model river 1 m deep with 1 m/sec current and a 3 m/sec wind is 3.0 hours (Lyman et al. 1982; quoted, Howard 1990).

Photolysis:

Oxidation: rate constants for the gas-phase reaction with ozone: $3.8x10^{-19}$ $cm^3$ $molecule^{-1}$ $sec^{-1}$ at 296 K (Blume et al. 1976; quoted, Atkinson & Carter 1984; Tuazon et al. 1984), $(1.8 \pm 0.29)x10^{-19}$ $cm^3$ $molecule^{-1}$ $sec^{-1}$ at 298 K (Zhang et al. 1983; quoted, Atkinson & Carter 1984; Tuazon et al. 1984; Atkinson et al. 1987), $1.2x10^{-19}$ $cm^3$ $molecule^{-1}$ $sec^{-1}$ at room temp. (Niki et al. 1983; quoted, Atkinson & Carter 1984; Tuazon et al. 1984) and $(1.2 \pm 0.24)x10^{-19}$ $cm^3$ $molecule^{-1}$ $sec^{-1}$ at 300 K (Niki et al. 1984; quoted, Atkinson & Carter 1984; Tuazon et al. 1984), while $2.3x10^5$ $cm^3$ $mol^{-1}$ $sec^{-1}$, was estimated rate constant for the reaction with ozone at 300 K (Lyman 1982); calculated rate constants at 25°C: $<10^5$ $M^{-1}$ $hour^{-1}$ for singlet oxygen and 6.0 $M^{-1}$ $hour^{-1}$ for peroxy radicals (Mabey et al. 1982); photooxidation half-life of 25.2 hours, based on estimated rate constant for the reaction with hydroxyl radicals in air (Atkinson 1987; quoted, Howard et al. 1991); rate constant of $1.79x10^{-12}$ $cm^3$ $molecule^{-1}$ $sec^{-1}$ for the gas-phase reaction with OH radicals at 298 K (Atkinson 1986; quoted, Atkinson et al. 1987; Sabljic & Güsten 1990); rate constant of $(0.57 \pm 0.27)x10^{-16}$ $cm^3$ $molecule^{-1}$ $sec^{-1}$ for the gas-phase reaction with $NO_3$ radicals at $298 \pm 2$ K (Atkinson et al. 1987); rate constant of $(1.07 \pm 0.11)x10^{16}$ $cm^3$ $molecule^{-1}$ $sec^{-1}$ relative to ethene for the gas-phase reaction with $NO_3$ radicals at $298 \pm 2$ K

(Atkinson et al. 1987; quoted, Atkinson 1991) while rate constant of $1.11 \times 10^{-16}$ $cm^3$ $molecule^{-1}$ $sec^{-1}$ for the gas-phase reaction with $NO_3$ radicals at 298 K (Atkinson et al. 1988; quoted, Sabljic & Güsten 1990); estimated rate constant of $2.57 \times 10^{-12}$ $cm^3$ $molecule^{-1}$ $sec^{-1}$ for the reaction with OH radicals (as per Atkinson 1987 and 1988, Müller & Klein 1991).

Hydrolysis:

Biodegradation: aqueous aerobic half-life of 672-4320 hours, based on unacclimated aerobic aqueous screening test data (Tabak et al. 1981; quoted, Howard et al. 1991); aqueous anaerobic half-life of 2688-17280 hours, based on estimated unacclimated aqueous aerobic degradation half-life (Howard et al. 1991); 0.1 $year^{-1}$ with half-life of 53 days (Wood et al. 1985; quoted, Olsen & Davis 1990); 1.8 $year^{-1}$ with half-life of 132-147 days (Barrio-Lage et al. 1986; quoted, Olsen & Davis 1990).

Biotransformation:

Bioconcentration, Uptake ($k_1$) and Elimination ($k_2$) Rate Constants or Half-Lives:

Common Name: Trichloroethylene

Synonym: acetylene trichloride, Algylen, Blacosolv, Chlorylen, Circosolv, Dow-Tri, ethinyl trichloride, ethylene trichloride, Fleck-Flip, Gemalgene, Lanadin, Lethurin, Nialk, Perm-A-Clor, Petzinol, Philex, TCE, Threthylen, Trethylene, Trial, TRI, Triad, Triasol, Trichloran, Trichloren, trichloroethene, Triclene, Tri-Clene, Trielene, Trielin, Triklone, Trilene, Triline, Trimar, Vestrol, Vitran, Westrosol

Chemical Name: 1,1,2-trichloroethylene

CAS Registry No: 79-01-6

Molecular Formula: $CHCl=CCl_2$

Molecular Weight: 131.39

Melting Point (°C):

- -86.4 (McGovern 1943; Riddick et al. 1986)
- -73.0 (Stull 1947; Weast 1977; Callahan et al. 1979; Mackay & Shiu 1981,1990; Horvath 1982; Mabey et al. 1982; Weast 1982-83; Schmidt-Bleek et al. 1982; Suntio et al. 1988)
- -87.0 (Verschueren 1983; Olsen & Davis 1990)
- -84.8 (Dean 1985)
- -87.15 (Miller et al. 1985)
- -87.10 (Stephenson & Malanowski 1987)

Boiling Point (°C):

- 86.9 (McGovern 1943; Verschueren 1983; Pavlostathis & Mathavan 1992)
- 87.08 (Dreisbach 1959)
- 87.30 (Boublik et al. 1973)
- 87.0 (Pearson & McConnell 1975; Jones et al. 1977/1978; Weast 1977; Callahan et al. 1979; Mackay & Shiu 1981,1990; Horvath 1982; Mabey et al. 1982; Weast 1982-83; McNally & Grob 1984)
- 86.0 (Schmidt-Bleek et al. 1982)
- 86.7 (Verschueren 1983; Dean 1985; Stephenson & Malanowski 1987; Hewitt et al. 1992)
- 87.19 (Riddick et al. 1986)

Density ($gm/cm^3$ at 20°C):

- 1.464 (McGovern 1943; Schmidt-Bleek et al. 1982)
- 1.4642 (Dreisbach 1959; Coca & Diaz 1980; Horvath 1982; Weast 1982-83; McNally & Grob 1984)
- 1.4554 (25°C, Dreisbach 1959)
- 1.460 (Verschueren 1983; Windholz et al. 1983; Gillham & Rao 1990; Pinal et al. 1990)
- 1.4649 (Dean 1985)

Molar Volume ($cm^3/mol$):

- 89.02 (Miller et al. 1985)
- 0.897 ($V_M/100$, Taft et al. 1985; Kamlet et al. 1986,1987)
- 0.492 (calculated intrinsic volume: $V_I/100$, Leahy 1986; Kamlet et al. 1988; Thoms & Lion 1992)

107.0 (LeBas method, Abernethy & Mackay 1987; Abernethy et al. 1988)
49.0 (intrinsic volume-van der Waals method, Abernethy et al. 1988)
90.0 (calculated from density, Abernethy et al. 1988)

Total Surface Area, TSA ($Å^2$):
110.70 (Okouchi et al. 1992)

Fugacity Ratio at 25°C, F:
1.0 (Suntio et al. 1988)

Water Solubility ($g/m^3$ or mg/L at 25°C):
1818 (Wright & Schaffer 1932; quoted, Horvath 1982)
997 (Seidell 1940; quoted, Deno & Berkheimer 1960)
1100 (McGovern 1943)
1000 (Lange 1956; quoted, Horvath 1982; McNally & Grob 1984)
1288 (Vallaud et al. 1957; quoted, Horvath 1982)
1100 (quoted, Gladis 1960)
1100 (Sconce 1962; quoted, Horvath 1982)
1100 (O'Connell 1963; quoted, Horvath 1982)
1100 (quoted, Irmann 1965)
1051 (Hansch et al. 1968; quoted, Tewari et al. 1982)
1000 (Kay 1973; quoted, Horvath 1982)
1000 (Neely et al. 1974; Neely 1976; quoted, Suntio et al. 1988)
1100 (quoted, Dilling et al. 1975)
1093 (quoted, Hine & Mookerjee 1975)
1100 (20°C, McConnell et al. 1975; Pearson & McConnell 1975; quoted, Jones et al. 1977/1978; Callahan et al. 1979; Mabey et al. 1982; Mills et al. 1982; Gillham & Rao 1990)
1000 (Aviado et al. 1976; quoted, Horvath 1982)
1500 (20°C, shake flask-GC/ECD, Chiou & Freed 1977)
1100 (20°C, GC/ECD, Dilling 1977; quoted, Smith et al. 1980; Mackay & Shiu 1981; Horvath 1982; Love Jr. & Eilers 1982; Bobra et al. 1984; Neely 1984; Cohen & Ryan 1985; Chiou et al. 1988; Suntio et al. 1988; Rutherford & Chiou 1992; Thoms & Lion 1992)
1100 (quoted, Verschueren 1977,1983; Smith et al. 1980; Mackay & Shiu 1981; Wilson et al. 1981; Klöpffer et al. 1982; Mackay et al. 1985; Warner et al. 1987; Pinal et al. 1990; Mackay & Paterson 1991; Hewitt et al. 1992; Pavlostathis & Mathavan 1992)
1100 (selected, Nathan 1978)
1472 (shake flask-LSC, Banerjee et al. 1980; quoted, Horvath 1982; Davies & Dobbs 1984)
1100 (quoted, Cowen & Baynes 1980; McNally & Grob 1984)
1474 (shake flask-LSC, Veith et al. 1980)
997 (calculated-f const., Veith et al. 1980)

829, 1987 (quoted average, calculated-$K_{OW}$, Valvani et al. 1981)
1100, 1590 (quoted exptl., calculated-group contribution as per Irmann 1965, Horvath 1982)
1100 (recommended, Horvath 1982; quoted, Grathwohl 1990; Broholm et al. 1992)
1100 (quoted, Klöpffer et al. 1982)
1100 (20°C, selected, Mills et al. 1982)
1016, 2850 (20°C, quoted, Schmidt-Bleek et al. 1982)
1366 (gen. col.-HPLC, Tewari et al. 1982; quoted, Miller et al. 1985)
1000 (quoted, Thomas 1982)
743.1 (30°C, headspace-GC, McNally & Grob 1984)
1000 (quoted, McNally & Grob 1984; Dean 1985)
1367 (quoted, Miller et al. 1985)
1000-1100 (quoted, Urano & Murata 1985)
1474, 2745 (quoted from Hine & Mookerjee 1975, calculated-molar volume & solvatochromic p., Kamlet et al. 1986)
1474, 2035 (quoted, calculated-molar volume & solvatochromic p., Leahy 1986)
1370 (quoted, Riddick et al. 1986)
1474, 659 (quoted, calculated-fragment const., Wakita et al. 1986)
1474, 2447 (quoted, calculated-molar volume & solvatochromic p., Kamlet et al. 1987)
1380 (quoted, Isnard & Lambert 1988,1989)
1099, 1122 (quoted, predicted- $\chi$ & polarizability, Nirmalakhandan & Speece 1988)
1050 (selected, Suntio et al. 1988)
7731 (20°C, selected, Gillham & Rao 1990)
1100 (quoted, Olsen & Davis 1990; Broholm et al. 1992)
1421 (23-24°C, shake flask-GC, Broholm et al. 1992)
1098 (calculated-AI, Okouchi et al. 1992)
1350 (20°C, calculated-activity coefficients, Wright et al. 1992)

Vapor Pressure (Pa at 25°C):
9331 (McGovern 1943)
9735 (Antoine eqn. regression, Stull 1947)
9906 (calculated-Antoine eqn., Dreisbach 1959)
9723 (calculated-Antoine eqn., Weast 1972-73)
9464 (Weast 1972-73; quoted, Chiou et al. 1988)
9224 (calculated-Antoine eqn., Boublik et al. 1973)
7998 (20°C, Perry & Chilton 1973; quoted, Cohen & Ryan 1985)
9680 (Weast 1973-74; quoted, Klöpffer et al. 1982)
9906 (quoted, Hine & Mookerjee 1975)
9864 (quoted, Dilling et al. 1975)
7700 (20°C, McConnell et al. 1975; Pearson & McConnell 1975; quoted, Callahan et al. 1979; Mabey et al. 1982; Mills et al. 1982)
8000 (20°C, Neely 1976; quoted, Mackay & Shiu 1981; Thomas 1982)

9870 (Dilling 1977; quoted, Mackay & Shiu 1981,1990; Mackay et al. 1985; Suntio et al. 1988; Mackay & Paterson 1991)
9904 (selected, Nathan 1978)
10029 (quoted, Cowen & Baynes 1980)
9680 (quoted, Klöpffer et al. 1982)
8450 (22°C, Schmidt-Bleek et al. 1982; quoted, Politzki et al. 1982)
8580 (20°C, quoted, Schmidt-Bleek et al. 1982)
7998 (20°C, quoted, Verschueren 1983)
12664 (30°C, quoted, Verschueren 1983)
9198 (interpolated-Antoine eqn., Boublik et al. 1984)
10264 (quoted, McNally & Grob 1984)
9218 (quoted, Kamlet et al. 1986)
6307 (quoted, Riddick et al. 1986)
9911 (calculated-Antoine eqn., Stephenson & Malanowski 1987)
9910 (quoted, Warner et al. 1987)
8000 (selected, Suntio et al. 1988)
9691 (resistance measurement-Antoine eqn., Foco et al. 1992)
7998 (20°C, quoted from Verschueren 1983, Hewitt et al. 1992)
7753 (20°C, quoted from DIPPR, Tse et al. 1992)
12347 (30°C, quoted from DIPPR, Tse et al. 1992)

Henry's Law Constant (Pa $m^3$/mol):
1186 (calculated as $1/K_{AW}$, $C_W/C_A$, reported as exptl., Hine & Mookerjee 1975; quoted, Nirmalakhandan & Speece 1988)
247.9 (calculated-group contribution method, Hine & Mookerjee 1975)
1186 (calculated-bond contribution method, Hine & Mookerjee 1975)
892.4 (calculated as per McConnell et al. 1975 & reported as exptl., Neely 1976; quoted, Jones et al. 1977/1978)
966.7 (calculated-P/C, Neely 1976)
1215 (calculated-P/C, Dilling 1977; quoted, Love Jr. & Eilers 1982)
904 (20°C, Dilling 1977; quoted, Mackay & Shiu 1981)
984.7 (20°C, batch stripping, Mackay et al. 1979; quoted, Yurteri et al. 1987)
989 (concentration ratio-GC, Leighton & Calo 1981)
939 (20°C, calculated-P/C, Mackay & Shiu 1981)
1179 (calculated-P/C, Mackay & Shiu 1981,1990)
922 (20°C, calculated-P/C, Mabey et al. 1982)
999.3 (20°C, batch stripping, Munz & Roberts 1982; quoted, Roberts & Dändliker 1983; Roberts et al. 1985; Yurteri et al. 1987)
1013 (calculated-P/C, Thomas 1982)
970.0 (20°C, EPICS, Lincoff & Gossett 1983; Gossett 1985; quoted, Yurteri et al. 1987)
774.0 (20°C, EPICS, Lincoff & Gossett 1984)

682.8 (20°C, batch stripping, Lincoff & Gossett 1984)
762.9 (20°C, calculated-P/C, Cohen & Ryan 1985)
1185 (adsorption isotherm, Urano & Murata 1985)
1016 (20°C, EPICS, Ashworth et al. 1986; quoted, Yurteri et al. 1987)
797.0 (20°C, multiple equilibration, Munz & Roberts 1986; quoted, Yurteri et al. 1987)
1178 (quoted from USEPA 1986, Yeh & Kastenberg 1991)
728.7 (20°C, EPICS, Gossett 1987; quoted, Yurteri et al. 1987; Grathwohl 1990; Tse et al. 1992)
1034 (EPICS, Gossett 1987; quoted, Tancrède & Yanagisawa 1990; Thoms & Lion 1992)
1066 (20°C, calculated-P/C, McKone 1987)
949.4 (Munz & Roberts 1987; quoted, Tancrède & Yanagisawa 1990)
1185, 1185 (quoted, calculated-P/C, Warner et al. 1987)
1048 (20°C, EPICS, Yurteri et al. 1987)
119 (calculated-QSAR, Nirmaklahanden & Speece 1988)
921.9 (20-25°C and low ionic strength, quoted, Pankow & Rosen 1988; Pankow 1990)
942 (quoted, Ryan et al. 1988)
971 (EPICS, quoted from Gossett 1987, Mackay & Shiu 1990)
768.4 (purge & trap method, Tancrède & Yanagisawa 1990)
809.9 (22°C, shower spray data, Giardino et al. 1992)
678.8 (quoted from Roberts & Dändliker 1983, Hewitt et al. 1992)
925 (quoted from Lyman et al. 1982 & EPA 1986, Pavlostathis & Mathavan 1992)
744 (quoted from Tancréde & Yanagisawa 1990, Tancréde et al. 1992)
709.2 (20°C, Tse et al. 1992)
1155 (30°C, Tse et al. 1992)
1246 (30°C, quoted from Gossett 1987, Tse et al. 1992)

Octanol/Water Partition Coefficient, log $K_{OW}$:
2.37 (Hansch & Elkins 1971; quoted, Green et al. 1983)
2.29 (Leo et al. 1971; Hansch & Leo 1979; quoted, Callahan et al. 1979; Schwarzenbach et al. 1983; Mackay et al. 1985; Ryan et al. 1988; Grathwohl 1990; Mackay & Paterson 1991; Hewitt et al. 1992)
2.61 (shake flask-GC/ECD, Chiou & Freed 1977; quoted, Rutherford & Chiou 1992)
2.42 (shake flask-LSC, Banerjee et al. 1980; quoted, McDuffie 1981; Arbuckle 1983; Suntio et al. 1988; Olsen & Davis 1990)
2.28, 2.42 (quoted, shake flask-LSC, Veith et al. 1980)
2.42, 2.86 (quoted, calculated-HPLC-k', McDuffie 1981)
2.29 (quoted, Valvani et al. 1981)
2.42 (calculated, Mabey et al. 1982)
2.30 (selected, Mills et al. 1982)
2.31, 2.88 (quoted, Schmidt-Bleek et al. 1982)
2.53 (gen. col.-HPLC, Tewari et al. 1982)

2.42 (Veith & Kosian 1982; quoted, Saito et al. 1992)
2.42 (quoted actual value from Banerjee et al. 1980; Arbuckle 1983)
2.28 (calculated as per Rekker 1977, Harnisch et al. 1983)
2.71, 2.79, 3.49, 3.57 (quoted values used for OECD Lab. comparison tests, Harnisch et al. 1983)
2.42 (quoted measured value from Veith et al. 1980, Veith et al. 1983)
2.29, 2.32 (quoted, calculated-molar refraction, Yoshida et al. 1983)
2.42 (Pavlou & Weston, 1983,1984)
3.30 (quoted, Davies & Dobbs 1984)
3.24, 3.30 (quoted, Geyer et al. 1984)
2.332 (quoted from Mackay et al. 1980, Cohen & Ryan 1985)
2.42 (Hansch & Leo 1985; quoted, Howard 1990; Thoms & Lion 1992)
2.53 (quoted, Miller et al. 1985)
2.29, 2.04 (quoted, calculated-molar volume & solvatochromic p., Taft et al. 1985)
2.29, 2.20 (quoted, calculated-molar volume & solvatochromic p., Leahy 1986)
2.42 (Abernethy & Mackay 1987,1988; Abernethy et al. 1988; quoted, Pinal et al. 1990)
2.29 (quoted, Hodson & Williams 1988)
2.36 (selected, Mailhot & Peters 1988)
2.53 (quoted, Isnard & Lambert 1988,1989)
2.35, 2.27 (quoted, calculated-molar volume & solvatochromic p., Kamlet et al. 1988)
2.61 (quoted, Lee et al. 1989)
2.42 (quoted from Hansch & Leo 1979, Mackay & Shiu 1990)
2.38 (quoted from Lyman et al. 1982 & USEPA 1986, Pavlostathis & Mathavan 1992)
2.42 (quoted, Van Leeuwen et al. 1992; Verhaar et al. 1992)

Bioconcentration Factor, log BCF:
1.59 (rainbow trout, Neely et al. 1974; quoted, Suntio et al. 1988)
1.23-1.36 (calculated-$K_{OW}$, Veith et al. 1979; quoted, Veith et al. 1980)
1.23 (bluegill sunfish, Veith et al. 1980; Veith & Kosian 1982; quoted, Bysshe 1982; Dobbs & Williams 1983; Davies & Dobbs 1984; Sabljic 1987; Saito et al. 1992)
1.23-1.59 (bluegill sunfish & rainbow trout, Barrows et al. 1980; Lyman 1981)
1.591 (rainbow trout, quoted, Bysshe 1982)
1.987 (microorganisms-water, calculated-$K_{OW}$, Mabey et al. 1982)
1.20 (calculated- $\chi$ , Koch 1983; quoted, Sabljic 1987)
2.996 (activated sludge, Freitag et al. 1984)
3.06 (*chlorella fusca*, Geyer et al. 1984)
2.391 (calculated-$K_{OW}$, Geyer et al. 1984)
1.20, 1.05 (quoted, calculated, Sabljic 1987)
1.23 (quoted, Isnard & Lambert 1988)
1.025 (quoted from USEPA 1986, Yeh & Kastenberg 1991)

1.420 (calculated-$K_{OW}$, McCarty et al. 1992)

Sorption Partition Coefficient, log $K_{OC}$:

2.1 (sediment-water, calculated-$K_{OW}$, Mabey et al. 1982; USEPA 1986; Yeh & Kastenberg 1991; Pavlostathis & Mathavan 1992)
2.20 (Pavlou & Weston 1983,1984)
1.76 (ICN humic acid-coated $Al_2O_3$, Garbarini & Lion 1985)
2.20 (ICN humic acid, Garbarini & Lion 1985)
1.66 (predicted-S, Garbarini & Lion 1985)
2.14, 2.40 (predicted-$K_{OW}$, Garbarini & Lion 1985)
1.76 (Sapsucker Woods humic acid, Garbarini & Lion 1986; quoted, Grathwohl 1990)
0.616 (Sapsucker Woods fulvic acid, Garbarini & Lion 1986)
1.238, 2.079, 2.045, 0.30, 1.827 (tannic acid, lignin, zein, cellulose, Aldrich humic acid, Garbarini & Lion 1986)
2.025 (Sapsucker Woods soil, Garbarini & Lion 1986)
2.086 (Sapsucker Woods ether-extracted soil, Gabarini & Lion 1986)
2.161 (humin, Garbarini & Lion 1986)
2.458 (oxidized humin, Garbarini & Lion 1986)
2.663 (fats, waxes, resins, Garbarini & Lion 1986)
2.00 (quoted from Seip et al. 1986, Bahnick & Doucette 1988)
1.84 (calculated- $\chi$, Bahnick & Doucette 1988)
2.03 (soil, Chiou et al. 1988; quoted, Grathwohl 1990)
2.02, 2.11 (soil: quoted, HPLC, Hodson & Williams 1988)
1.79 (20°C, humic acid, Peterson et al. 1988; quoted, Grathwohl 1990)
2.09, 2.56, 3.43 (soil, weathered shale, unweathered shale at 20°C, Grathwohl 1990)
1.97, 2.15 (20°C, calculated, Grathwohl 1990)
2.01 (calculated average, Olsen & Davis 1990)
2.20, 1.78 (humic acid, humic acid-coated $Al_2O_3$, Pavlostathis & Jaglal 1991)
2.03 (surface soil, Pavlostathis & Jaglal 1991)
3.39, 2.00 (organic carbon soil, Doust & Huang 1992)
1.60, 1.60, 2.15 (bentonite, green & tan clay, Doust & Huang 1992)
0.35, 1.40, 1.90, 2.20 (Barnwell, Congaree, McBean I & II sands, Doust & Huang 1992)
1.66, 2.64, 2.83 (calculated-equilibrium desorption data, Pavlostathis & Mathavan 1992)
1.52 (reported as log $K_{OM}$ of peat from water normalized for organic matter content, Rutherford & Chiou 1992)

Sorption Partition Coefficient, log $K_{OM}$:

1.23 (A horizon untreated soil, Lee et al. 1989)
2.42, 1.89 (A horizon soil treated with HDTMA, DDTMA, Lee et al. 1989)
2.78, 2.36 ($B_t$ horizon soil treated with HDTMA, DDTMA, Lee et al. 1989)

Environmental Half-Lives:

Air: disappearance half-life of 0.24-2.4 hours from air for the reaction with OH radicals (EPA 1974; quoted, Darnall et al. 1976); 120 minutes for irradiation with UV light (Gay et al. 1976); 5.3 days, estimated as toxic chemical residence time with rate constant of $2.2 \times 10^{-12}$ $cm^3$ $molecule^{-1}$ $sec^{-1}$ at 300 K for the reaction with OH radicals (Singh et al. 1981); 27-272 hours, based on photooxidation half-life in air (Atkinson 1985; quoted, Howard et al. 1991); atmospheric lifetime of 4.0 days, based on the photooxidation rate constant in the gas phase at (23.2 ± 1.1)°C (Edney et al. 1986); atmospheric lifetimes: 64 days for the reaction with $NO_3$ for a 12-hour nighttime average concn. of $2.4 \times 10^9$ molecule $cm^{-3}$, 9.6 days for the reaction with OH radicals for a 12-hour average concn. of $1.0 \times 10^6$ molecule $cm^{-3}$, and >1.5 years for the reaction with $O_3$ for a 24-hour average concn. of $7 \times 10^{11}$ molecule $cm^{-3}$ (quoted, Atkinson et al. 1987); estimated to be 3.7 days (Yeh & Kastenberg 1991).

Surface water: 4320-8640 hours, based on estimated aqueous aerobic biodegradation half-life (Howard et al. 1991); estimated to be 90 days (Yeh & Kastenberg 1991).

Ground water: 7704-39672 hours, based on hydrolysis half-life (Dilling et al. 1975; quoted, Howard et al. 1991) and anaerobic sediment grab sample data (Barrio-Lage et al. 1986; quoted, Howard et al. 1991); estimated half-life to be 2.0 years in the ground water in the Netherlands (Zoeteman et al. 1981); estimated to be 86 days (Yeh & Kastenberg 1991).

Soil: <10 days (Ryan et al. 1988); 4320-8640 hours, based on estimated aqueous aerobic biodegradation half-life (Howard et al. 1991); estimated to be 43 days in upper soil and 86 days in lower soil (Yeh & Kastenberg 1991).

Sediment: estimated to be 43 days (Yeh & Kastenberg 1991).

Biota: <1 day in tissues of bluegill sunfish (Barrows et al. 1980); <10 days, subject to plant uptake via volatilization (Ryan et al. 1988).

Environment Fate Rate Constants or Half-Lives:

Volatilization: calculated half-life of 0.48 minute (Mackay & Wolkoff 1973; quoted, Dilling et al. 1975); half-life of evaporation from dilute aqueous solution of 1 ppm initial concn. in a 250 ml beaker with constant stirring at room temperature was found to be 21 ± 3 minutes compared with a calculated half-life of 0.48 minute (Dilling et al. 1975; quoted, Mills et al. 1982); volatilization into the atmosphere with rate of 2.79, 5.07 g/($cm^2$ sec) at 23.7 ± 0.5°C (Chiou & Freed 1977); calculated half-lives: 0.47 minute, 23.8 minutes and experimental half-lives: 17.7, 18.5, 23.5 minutes (Dilling 1977); the ratio of evaporation rate constant to that of oxygen reaeration rate constant: measured as 0.49 as compared to the predicted 0.44 (Smith et al. 1980); half-lives of $1.42 \times 10^4$ seconds (exptl.) and $5.1 \times 10^3$ seconds (calculated) for water body of depth of 22.5 m (Klöpffer et al. 1982); estimated half-life from water to be 3.4 hours (Thomas 1982); half-lives in a simulated marine mesocosm: 28 days in the spring, 13 days in the summer and 15 days in the winter (Wakeham et al. 1983); $7.22x10^{-3}$ hour$^{-1}$ (Mackay et al. 1985).

Photolysis: half-life of 3.5 hours estimated from lab. simulated UV photolysis (light intensity about 6 times of natural sunlight at noon on a summer day in Freeport) for 10 ppm to react with 5 ppm NO at 27 ± 1 °C (Dilling et al. 1976); probably would not occur (Callahan et al., 1979); photocatalyzed mineralization by the presence of $TiO_2$ with the rate of 830.0 ppm/min per gram of catalyst (Ollis 1985).

Oxidation: $2x10^{-12}$ cm$^3$ molecule$^{-1}$ sec$^{-1}$ at 23°C for the reaction with OH radicals in air (Howard 1976); $2.4x10^{-12}$ cm$^3$ molecule$^{-1}$ sec$^{-1}$ for room temperature reaction with OH radicals (Atkinson et al. 1979,1982; quoted, Tuazon et al. 1984; Atkinson et al. 1987); $1.2x10^{12}$ cm$^3$ mol$^{-1}$ sec$^{-1}$ estimated for the reaction with hydroxyl radicals and $3.6x10^3$ cm$^3$ mol$^{-1}$ sec$^{-1}$ for ozone both at 300 K (Lyman 1982); < $3x10^{-20}$ cm$^3$ molecule$^{-1}$ sec$^{-1}$ for the gas-phase reaction with ozone at 296 ± 2 K (Atkinson et al. 1982; quoted, Atkinson & Carter 1984; Tuazon et al. 1984; Atkinson et al. 1987); for singlet oxygen, < 1000 M$^{-1}$ hour$^{-1}$, and $RO_2$ radical, 6 M$^{-1}$ hour$^{-1}$ (Mabey et al. 1982); 66 % reacted after 140 minutes in long-path infrared chamber in the presence of $NO_2$ (Gay et al. 1976); rate constant of $2.36x10^{-12}$ cm$^3$ molecule$^{-1}$ sec$^{-1}$ for the gas-phase reaction with OH radicals in air at 298 K (Atkinson 1985; quoted, Dilling et al. 1988; Sabljic & Güsten 1990); rate constant of (2.86 ± 0.40)$x10^{-12}$ cm$^3$ molecule$^{-1}$ sec$^{-1}$ with reference to *n*-butane for the gas-phase reaction with OH radicals in air at (23.2 ± 1.1)°C with atmospheric lifetime of 4.0 days (Edney et al. 1986); rate constant of (1.5 ± 0.7)$x10^{-16}$ cm$^3$ molecule$^{-1}$ sec$^{-1}$ for the gas-phase reaction with $NO_3$ radicals in air at 298 ± 2 K (Atkinson et al. 1987); rate constant of (2.81 ± 0.17)$x10^{-16}$ cm$^3$ molecule$^{-1}$ sec$^{-1}$ relative to ethene for the gas-phase reaction with $NO_3$ radicals in air at 298 ± 2 K (Atkinson et al. 1987; quoted, Atkinson 1991) while rate constant of $2.93x10^{-16}$ cm$^3$ molecule$^{-1}$ sec$^{-1}$ for the gas-phase reaction with $NO_3$ radicals in air at 298 K (Atkinson et al. 1988; quoted, Sabljic & Güsten 1990); estimated rate constant of $6.92x10^{-13}$ cm$^3$ molecule$^{-1}$ sec$^{-1}$ for the reaction with OH radicals in air (as per Atkinson 1987 and 1988, Müller & Klein 1991).

Hydrolysis: not an important process (Mabey et al. 1982); 0.065 month$^{-1}$ at 25°C with half-life of 10.7 months (Dilling et al. 1975; quoted, Howard et al. 1991); $9.00x10^{-5}$ hour$^{-1}$ (Mackay et al. 1985); abiotic hydrolysis or dehydrohalogenation half-life of 10.7 months (Olsen & Davis 1990).

Biodegradation: can be biodegraded by microorganisms in sea water; aerobic half-life of 6 months to 1 year based on acclimated soil screening test data (Tabak et al. 1981; quoted, Howard et al. 1991); 1.1 year$^{-1}$ with half-life of 230 days (Roberts et al. 1982; quoted, Olsen & Davis 1990); anaerobic half-life of 98 days to 4.5 years, based on sediment grab sample data (Barrio-Lage et al. 1986; quoted, Howard et al. 1991); microcosm constructed with crushed rock and water containing low microbial biomass depleted all TCE in 21 months of incubation (Barrio-Lage et al. 1987); 0.06 $year^{-1}$ with biodegradation half-life of 33 days (Barrio-Lage et al. 1987; quoted, Olsen & Davis 1990); 0.08 $year^{-1}$ with half-life of 43 days (Olsen & Davis 1990).

Biotransformation: estimated rate constant of $1x10^{-10}$ ml $cell^{-1}$ $hour^{-1}$ for bacteria (Mabey et al. 1982).

Bioconcentration, Uptake ($k_1$) and Elimination ($k_2$) Rate Constants or Half-Lives:

$k_1$: 10.5 $hour^{-1}$ (flagfish, calculated-BCFx$k_2$, McCarty et al. 1992)

$k_2$: 0.398 $hour^{-1}$ (flagfish, estimated-one compartment first-order kinetics, McCarty et al. 1992)

Common Name: Tetrachloroethylene
Synonym: ethylene tetrachloride, perchloroethene, perchloroethylene, Perklone, tetrachloroethene, 1,1,2,2-tetrachloroethylene
Chemical Name: tetrachloroethylene
CAS Registry No: 127-18-4
Molecular Formula: $CCl_2=CCl_2$
Molecular Weight: 165.83
Melting Point (°C):

-19.0 (Stull 1947; Mackay & Shiu 1981,1990; Horvath 1982; Weast 1982-83; Stephenson & Malanowski 1987; Suntio et al. 1988; Howard 1990)
-22.35 (McGovern 1943; Riddick et al. 1986)
-22.7 (Verschueren 1977,1983; Callahan et al. 1979; Mabey et al. 1982)
-22.4 (Dean 1985)

Boiling Point (°C):

121.2 (McGovern 1943; McConnell et al. 1975; Pearson & McConnell 1975; Stephenson & Malanowski 1987)
120.97 (Dreisbach 1959)
121.1 (Boublik et al. 1973)
121.0 (Jones et al. 1977/1978; Weast 1977; Callahan et al. 1979; Mackay & Shiu 1981,1990; Horvath 1982; Mabey et al. 1982; Banerjee et al. 1990; Howard 1990; Pavlostathis & Mathavan 1992)
121.4 (Verschueren 1983)
121.1 (Dean 1985; Riddick et al. 1986)

Density ($g/cm^3$ at 20°C):

1.623 (McGovern 1943; Coca & Diaz 1980; Pavlostathis & Mathavan 1992)
1.6227 (Dreisbach 1959; Coca & Diaz 1980; Horvath 1982; Weast 1982-83)
1.6145 (25°C, Dreisbach 1959)
1.626 (Verschueren 1983)
1.6230 (Dean 1985)
1.6228 (Riddick et al. 1986)
1.630 (Gillham & Rao 1990)

Molar Volume ($cm^3/mol$):

1.016 ($V_M/100$, Taft et al. 1985; Leahy 1986)
0.578 (calculated intrinsic volume $V_I/100$, Leahy 1986; Hawker 1990; Kamlet et al. 1988; Thoms & Lion 1992)
128.0 (LeBas method, Abernethy & Mackay 1987; Abernethy et al. 1988)
58.0 (intrinsic volume-van der Waals method, Abernethy et al. 1988)
102.0 (calculated from density, Abernethy et al. 1988)

Molecular Volume ($A^3$):
Total Surface Area, TSA ($A^2$):

129.0 (Valsaraj 1988)
127.8 (Okouchi et al. 1992)

Heat of Fusion, $\Delta H_{fus}$, (kcal/mol):
    2.525 (Riddick et al. 1986)
Entropy of Fusion, $\Delta S_{fus}$ (cal/mol K or e.u.):
Fugacity Ratio at 25°C, F:
    1.0 (Suntio et al. 1988)

Water Solubility (g/m$^3$ or mg/L at 25°C):
    150 (McGovern 1943; quoted, Horvath 1982)
    150 (Gladis 1960; quoted, Horvath 1982)
    150 (Scone 1962; quoted, Horvath 1982)
    150 (O'Connell 1963; quoted, Horvath 1982)
    150 (Irmann 1965; quoted, Hine & Mookerjee 1975; Schwarzenbach et al. 1979)
    200 (Günther et al. 1968; quoted, Kenaga & Goring 1980; Kenaga 1980)
    150 (Miller 1969; quoted, Horvath 1982)
    150 (Riddick & Bunger 1970; quoted, Horvath 1982)
    150 (Hancock 1973; quoted, Horvath 1982)
    400 (Neely et al. 1974; Neely 1976; quoted, Mackay & Shiu 1981; Suntio et al. 1988)
    400 (quoted, Dilling et al. 1975)
    151 (quoted, Hine & Mookerjee 1975)
    400 (Kenaga 1975; quoted, Freed et al. 1977)
    400 (Matthews 1975,1979; quoted, Horvath 1982)
    150 (20°C, shake flask-GC, McConnell et al. 1975; Pearson & McConnell 1975; quoted, Jones et al. 1977/1978; Callahan et al. 1979; Geyer et al. 1980; Mackay & Shiu 1981,1990; Mills et al. 1982; Suntio et al. 1988; Shiu et al. 1990; Thoms & Lion 1992)
    150 (Archer & Stevens 1977; quoted, Horvath 1982)
    400 (Chiou et al. 1977; Chiou 1981; quoted, Horvath 1982)
    140 (quoted average, Dilling 1977; quoted, Smith et al. 1980; Mackay & Shiu 1981; Horvath 1982; Suntio et al. 1988; Shiu et al. 1990)
    150 (Mitchell & Smith 1977; quoted, Horvath 1982)
    200 (20°C, shake flask-GC, Chiou et al. 1977; quoted, Callahan et al. 1979; Mabey et al. 1982; Gillham & Rao 1990; Shiu et al. 1990)
    150 (Verschueren 1977; quoted, Wilson et al. 1981; Warner et al. 1987)
    150 (selected, Nathan 1978; quoted, Horvath 1982)
    150 (Schwarzenbach et al. 1979; quoted, Horvath 1982)
    486 (shake flask-LSC, Banerjee et al. 1980; quoted, Davies & Dobbs 1984; Shiu et al. 1990)
    150 (quoted, Cowen & Baynes 1980)
    400 (Kenaga 1980; quoted, Mackay et al. 1980)
    150 (quoted, Neely 1980; Zoeteman et al. 1981)
    478.3 (shake flask-LSC, Veith et al. 1980)

911.3 (calculated-f const., Veith et al. 1980)
150.3 (recommended, Horvath 1982)
150-200 (selected, Mills et al. 1982)
400 (quoted, Thomas 1982)
150 (quoted, Verschueren 1983)
400 (Dean 1985)
140-150 (quoted, Urano & Murata 1985)
150 (quoted, Riddick et al. 1986)
1503 (quoted from Horvath 1982, Howard 1990)
467.4 (calculated-molar volume & solvatochromic p., Leahy 1986)
489, 117 (quoted, calculated-fragment const., Wakita et al. 1986)
200 (quoted, Isnard & Lambert 1988,1989)
150; 439.5, 392.6 (quoted; predicted- $\chi$ & polarizability, Nirmalakhandan & Speece 1988)
140 (selected, Suntio et al. 1988)
150 (selected, Valsaraj 1988)
150 (quoted from Horvath 1982, Grathwohl 1990)
150 (quoted, Broholm et al. 1992)
242 (23-24°C, shake flask-GC, Broholm et al. 1992)
150 (selected, Pavlostathis & Mathavan 1992)
150 (calculated-AI, Okouchi et al. 1992)
286 (19.5°C, shake flask-GC/TC, Stephenson 1992)
221 (31.1°C, shake flask-GC/TC, Stephenson 1992)

Vapor Pressure (Pa at 25°C):
2666 (McGovern 1943)
2397 (Antoine eqn. regression, Stull 1947)
2462 (calculated-Antoine eqn., Dreisbach 1959)
2394 (calculated-Antoine eqn., Weast 1972-73)
2359 (Weast 1972-73; quoted, Chiou et al. 1988)
2388 (calculated-Antoine eqn., Boublik et al. 1973)
2533 (quoted, Dilling et al. 1975)
2462 (quoted, Hine & Mookerjee 1975)
1866 (20°C, McConnell et al. 1975; Pearson & McConnell 1975; quoted, Callahan et al. 1979; Neely 1980; Mackay & Shiu 1981; Zoeteman et al. 1981; Mabey et al. 1982)
1906 (20°C, Neely 1976; quoted, Thomas 1982)
2480 (Dilling 1977; quoted, Mackay & Shiu 1981; Suntio et al. 1988)
2975 (calculated, Dilling 1977)
2533 (selected, Nathan 1978)
2492 (quoted, Cowen & Baynes 1980)
1866 (20°C, selected, Mills et al. 1982)

1866 (20°C, quoted, Verschueren 1983)
2417, 2456 (calculated-Antoine eqn., Boublik et al. 1984)
2465 (Daubert & Danner 1986; quoted, Howard 1990)
2462 (quoted, Riddick et al. 1986)
2415 (calculated-Antoine eqn., Stephenson & Malanowski 1987)
2614 (interpolated between two data points, Warner et al. 1987)
2100 (selected, Suntio et al. 1988)
2614 (selected, Valsaraj 1988)
2479, 4319 (quoted, calculated-UNIFAC, Banerjee et al. 1990)

Henry's Law Constant (Pa $m^3$/mol):
2718 (calculated $1/K_{AW}$, $C_W/C_A$, reported as exptl., Hine & Mookerjee 1975; quoted, Nirmalakhandan & Speece 1988)
247.9 (calculated-group contribution method, Hine & Mookerjee 1975)
581.1 (calculated-bond contribution method, Hine & Mookerjee 1975)
2033, 2157 (quoted, calculated-P/C, Neely 1976)
1239 (calculated-P/C, Dilling 1977; quoted, Schwarzenbach et al. 1979; Mackay & Shiu 1981)
1998 (20°C, McConnell et al. 1975; quoted, Jones et al. 1977/1978)
1621 (20°C, batch stripping, Mackay et al. 1979; quoted, Yurteri et al. 1987)
1635 (concn. ratio-GC, Leighton & Calo 1981)
2940, 2300 (calculated-P/C, recommended, Mackay & Shiu 1981)
1550 (20°C, calculated-P/C, Mabey et al. 1982)
1528 (20°C, batch stripping, Munz & Roberts 1982; quoted, Roberts & Dändliker 1983; Roberts et al. 1985; Yurteri et al. 1987)
840.9 (calculated-P/C, Thomas 1982)
1316 (20°C, EPICS, Lincoff & Gossett 1983; Gossett 1985; quoted, Yurteri et al. 1987)
1317 (20°C, EPICS, Lincoff & Gossett 1984)
1175 (20°C, batch stripping, Lincoff & Gossett 1984)
2908 (adsorption isotherm, Urano & Murata 1985)
1694 (20°C, EPICS, Ashworth et al. 1986; quoted, Yurteri et al. 1987)
1445 (20°C, multiple equilibration, Munz & Roberts 1986; quoted, Yurteri et al. 1987)
1338 (20°C, EPICS, Gossett 1987; quoted, Yurteri et al. 1987; Grathwohl 1990)
1852 (EPICS, Gossett 1987; quoted, Tancréde & Yanagisawa 1990; Thoms & Lion 1992)
2799 (20°C, calculated-P/C, McKone 1987)
1762 (Munz & Roberts 1987; quoted, Tancréde & Yanagisawa 1990)
2908, 2888 (quoted, calculated-P/C, Warner et al. 1987)
1304 (20°C, EPICS, Yurteri et al. 1987)
1242, 1133 (quoted, calculated-QSAR, Nirmalakhandan & Speece 1988)
1586.4 (quoted, Ryan et al. 1988)

1793 (EPICS, quoted from Gossett 1987, Suntio et al. 1988)
2487 (calculated-P/C, Suntio et al. 1988)
2887 (selected, Valsaraj 1988)
1363 (purge & trap-GC/ECD, Tancréde and Yanagisawa 1990)
262.4 (selected, Pavlostathis & Mathavan 1992)
1686 (quoted from Tancréde & Yanagisawa 1990, Tancréde et al. 1992)

Octanol/Water Partition Coefficient, log $K_{OW}$:

2.88 (Neely et al. 1974; quoted, Callahan et al. 1979; Veith et al. 1979; Kenaga & Goring 1980; Mackay et al. 1980; Schwarzenbach et al. 1979,1983; Mackay & Shiu 1990)
2.60 (Kenaga 1975,1980; quoted, Freed et al. 1977; Mackay et al. 1980)
2.60 (Chiou et al. 1977; quoted, Suntio et al. 1988)
2.60 (Hansch & Leo 1979; quoted, Grathwohl 1990; Ball & Roberts 1991)
2.53 (shake flask-LSC, Banerjee et al. 1980; quoted, Karickhoff 1981; Davies & Dobbs 1984; Abernethy & Mackay 1987; Abernethy et al. 1988; Suntio et al. 1988; Olsen & Davis 1990)
2.88 (calculated, Mackay et al. 1980; quoted, Schwarzenbach et al. 1983)
2.53 (shake flask-LSC, Veith et al. 1980,1983)
2.67 (estimated-RPHPLC, Veith et al. 1980)
3.03 (calculated, Veith et al. 1980)
2.88 (calculated, Mabey et al. 1982)
2.88 (Mackay 1982; quoted, Schüürmann & Klein 1988)
2.88 (selected, Mills et al. 1982)
2.53 (Veith & Kosian 1982; quoted, Saito et al. 1992)
2.88 (Veith & Kosian 1982; quoted, Saito et al. 1992)
2.53 (quoted actual exptl. value from Banerjee et al. 1980, Arbuckle 1983)
2.60, 2.76 (quoted, calculated-molar refraction, Yoshida et al. 1983)
3.40 (calculated, Hansch & Leo 1985; quoted, Howard 1990; Thoms & Lion 1992)
2.88, 2.80 (quoted, calculated-molar volume & solvatochromic p., Taft et al. 1985)
2.60 (quoted, Hawker & Connell 1985,1988)
2.88, 2.96 (quoted, calculated-molar volume & solvatochromic p., Leahy 1986)
2.64 (calculated, Leo 1986; quoted, Schüürmann & Klein 1988)
2.39 (THOR 1986; quoted, Schüürmann & Klein 1988)
2.88 (quoted, Isnard & Lambert 1988,1989)
2.88, 3.08 (quoted, calculated-molar volume & solvatochromic p., Kamlet et al. 1988)
2.57 (selected, Mailhot & Peters 1988)
2.65 (selected, Suntio et al. 1988)
2.61 (selected, Valsaraj 1988)
2.85 (quoted, Dobbs et al. 1989)
3.38 (quoted, Lee et al. 1989)
2.60 (quoted, Hawker 1990)

2.60 (selected, Ball & Roberts 1991; Pavlostathis & Mathavan 1992)
3.40 (correction of Ball & Roberts 1991 selection, Ball & Roberts 1992)
3.40 (quoted, Thoms & Lion 1992; Van Leeuwen et al. 1992; Verhaar et al. 1992)

Bioconcentration Factor, log BCF:
- 1.59 (trout, Neely et al. 1974; quoted, Veith et al. 1979; Kenaga & Goring 1980; Kenaga 1980; Chiou 1981; Davies & Dobbs 1984; Suntio et al. 1988; Howard 1990)
- 1.598 (trout muscle, calculated-$k_1/k_2$, Neely et al. 1974; quoted, Suntio et al. 1988)
- 1.45-1.88 (calculated-$K_{OW}$, Veith et al. 1979; quoted, Veith et al. 1980)
- -1.70 (male Albino rats, Geyer et al. 1980)
- 1.69 (bluegill sunfish, Barrows et al. 1980; quoted, Howard 1990)
- 1.49 (calculated-S, Kenaga 1980)
- 1.69 (bluegill sunfish, Veith et al. 1980; Veith & Kosian 1982; quoted, Bysshe 1982; Davies & Dobbs 1984; Sabljic 1987; Saito et al. 1992)
- 2.35 (calculated-$K_{OW}$, Lyman et al. 1982; quoted, Howard 1990)
- 2.40 (microorganisms-water, Mabey et al. 1982)
- 1.59 (Mackay 1982; quoted, Schüürmann & Klein 1988)
- 2.06 (rainbow trout, Veith & Kosian 1982; quoted, Saito et al. 1992)
- 1.70 (calculated- $\chi$ , Koch 1983; quoted, Sabljic 1987)
- 1.49 (USEPA 1986; quoted, Yeh & Kastenberg 1991)
- 1.70, 2.30 (quoted, calculated, Sabljic 1987)
- 1.69 (quoted, Isnard & Lambert 1988)
- 1.60 (calculated-solvatochromic p., Hawker 1990)
- 2.40 (calculated-$K_{OW}$, McCarty et al. 1992)
- 1.79 (rainbow trout, Saito et al. 1992)

Sorption Partition Coefficient, log $K_{OC}$:
- 2.38 (calculated, Kenaga & Goring 1978; quoted, Kenaga 1980)
- 2.32 (Chiou et al. 1979; quoted, Karickhoff 1981; Howard 1990)
- 2.14 (calculated-$K_{OW}$, Karickhoff 1981)
- 2.54, 2.89 (calculated-S & M.P., Karickhoff 1981)
- 2.35, 2.57 (calculated-S, Karickhoff 1981)
- 2.04 (Schwarzenbach & Westall 1981; quoted, Howard 1990)
- 3.23 (calculated-$K_{OW}$, Lyman et al. 1982; quoted, Howard 1990)
- 2.38 (peaty soil, calculated-$K_{OM}$, Friesel et al. 1984; quoted, Howard 1990; Pignatello 1991)
- 2.56 (sediment-water, calculated-$K_{OW}$, Mabey et al. 1982)
- 2.56 (USEPA 1986; quoted, Yeh & Kastenberg 1991; Pavlostathis & Mathavan 1992)
- 2.56 (soil, Chiou et al. 1988; quoted, Grathwohl 1990)
- 2.64 (20°C, soil, sand & loess, Grathwohl 1990)

3.29 (20°C, weathered shale & mudrock, Grathwohl 1990)
4.03 (20°C, unweathered shale & mudrock, Grathwohl 1990)
2.39, 2.57 (20°C, calculated, Grathwohl 1990)
2.14 (calculated-molecular conductivity index, Olsen & Davis 1990)
2.45 (2.57% organic carbon in surface soil, Pignatello 1990,1991)
1.81-2.95 (calculated-$K_{OW}$, Ball & Roberts 1991)
3.60 (Borden organic phase with no mineral sorption, Ball & Roberts 1991)
2.90 (Borden organic phase with no mineral sorption but with Curtis et al. 1986 correlation, Ball & Roberts 1992)

Sorption Partition Coefficient, log $K_{OM}$:
1.70 (untreated A horizon Marlette soil, Lee et al. 1989)
2.56, 2.30 (A horizon Marlette soil treated with HDTMA, DDTMA, Lee et al. 1989)
1.64 (untreated $B_t$ horizon Marlette soil, Lee et al. 1989)
2.75, 2.64 (treated $B_t$ horizon Marlette soil with HDTMA, DDTMA, Lee et al. 1989)
2.32, 1.90 (quoted, calculated-molecular conductivity index, Olsen & Davis 1990)

Half-Lives in the Environment:

Air: disappearance half-life larger than 10 days for the reaction with OH radicals in air (EPA 1974; quoted, Darnall et al. 1976); lifetime about 10 days, based on a rate constant of $1.3x10^{-12}$ $cm^3$ $sec^{-1}$ for reaction with hydroxyl radicals (Yung et al. 1975; quoted, Callahan et al. 1979); estimated N. troposphere residence time of 150 days by one compartment nonsteady state model (Singh et al. 1978); 68 days, estimated as toxic chemical residence time with rate constant of $1.7x10^{13}$ $cm^3$ $molecule^{-1}$ $sec^{-1}$ at 300 K for the reaction with OH radicals (Singh et al. 1981); estimated troposphere residence time of 200-390 days (Lyman 1982); 384-3843, based on measured rate data for the vapor phase reaction with hydroxyl radicals in air (Atkinson 1985; quoted, Howard et al. 1991); atmospheric lifetimes: >240 days for the reaction with $NO_3$ for a 12-hour nighttime average concn. of $2.4x10^9$ molecule $cm^{-3}$, 140 days for the reaction with OH radicals for a 12-hour average concn. of $1.0x10^6$ molecule $cm^{-3}$, and >2000 years for the reaction with $O_3$ for a 24-hour average concn. of $7x10^{11}$ molecule $cm^{-3}$ (quoted, Atkinson et al. 1987).

Surface water: 4320-8640 hours, based on aerobic river die-away test data (Mudder 1981; quoted, Howard et al. 1991) and saltwater sample grab data (Jensen & Rosenberg 1975; quoted, Howard et al. 1991); calculated half-lives from concentration reduction between sampling points on the Rhine River and a lake in the Rhine basin were 10 days and 32 days, respectively (Zoeteman et al. 1980; quoted, Howard 1990).

Ground water: 8640-17280 hours, based on estimated aqueous aerobic biodegradation half-life (Howard et al. 1991).

Sediment:

Soil: < 10 days (Ryan et al. 1988); 4320-8640 hours, based on estimated aqueous aerobic biodegradation half-life (Howard et al. 1991).

Biota: < 1 day in tissues of bluegill sunfish (Barrows et al. 1980); 14 hours, clearance from fish in simulated ecosystem (Neely 1980); < 10 days, subject to plant uptake via volatilization (Ryan et al. 1988); half-life of 0.4-0.5 day to eliminate from small fish (McCarty et al. 1992).

Environmental Fate Rate Constant and Half-Lives:

Volatilization: calculated half-life of 0.56 minute (Mackay & Wolkoff 1973; quoted, Dilling et al. 1975); experimental half-life for 1 mg/liter in water to be 27 $\pm$ 3 minutes when stirred at 200 rpm at approximately 25°C in an open container of 65 mm deep (Dilling et al. 1975; quoted, Callahan et al. 1979; Mills et al. 1982) and 20.2-27.1 minutes was obtained under same conditions for 0.92 mg/liter (Dilling 1977; quoted, Callahan et al. 1979); calculated half-lives: 0.2 minute, 26.5 minutes (Dilling 1977); the ratio evaporation rate constant to that of oxygen reaeration: a measured value of 0.44 to that of 0.40 as predicted (Smith et al. 1980); half-lives of 5-12 days from pond, 3 hours-7 days from river, 3.6-14 days from lake were estimated using representative reaeration rates (Lyman et al. 1982; quoted, Howard 1990); estimated half-life of 4.2 hours from water (Thomas 1982); measured half-lives in a mesocosm simulating Narragansett Bay of Rhode Island were 11 days in winter, 25 days in spring, and 14 days in summer (Wakeham 1983; quoted, Howard 1990).

Photolysis: photocatalyzed mineralization by the presence of $TiO_2$ with the rate of 6.8 ppm/min per gram of catalyst (Ollis 1985).

Oxidation: rate constant of < $2x10^{-23}$ $cm^3$ $molecule^{-1}$ $sec^{-1}$ for the gas-phase reaction with ozone at 297 K (Mathias et al. 1974; quoted, Atkinson & Carter 1984; Tuazon et al. 1984; Atkinson et al. 1987); $1.3x10^{-12}$ $cm^3$ $sec^{-1}$ with a lifetime about 10 days for the reaction with hydroxyl radicals (Yung et al. 1975; quoted, Callahan et al. 1979); $1.7x10^{-13}$ $cm^3$ $molecule^{-1}$ $sec^{-1}$ for the reaction with OH radicals in air at 23°C (Howard 1976; quoted, Callahan et al. 1979); estimated photooxidation half-life from 2 months (Howard 1976; Singh et al. 1981; quoted, Howard 1990); calculated rate constant of $1.7x10^{-13}$ $cm^3$ $molecule^{-1}$ $sec^{-1}$ for reaction with OH radicals at 298 K with lifetime of 1.0 year in troposphere (Chang & Kaufman 1977; quoted, Altshuller 1980); $1.70x10^{-13}$ $cm^3$ $molecule^{-1}$ $sec^{-1}$ for room temperature reaction with OH radicals (Atkinson et al. 1979,1982; quoted, Tuazon et al. 1984; Atkinson et al. 1987); $1.00x10^{11}$ $cm^3$ $mol^{-1}$ $sec^{-1}$ estimated rate constant for the reaction with OH radicals and $1.0x10^3$ $cm^3$ $mol^{-1}$ $sec^{-1}$ for ozone both at 300 K (Lyman 1982); calculated rate constant at 25°C of << 100 $M^{-1}$ $hour^{-1}$ for singlet oxygen (Mabey et al. 1982); completely degraded in an hour (Dimitriades et al. 1983; quoted, Howard 1990); photooxidation half-life of 383-3843 hours, based on measured rate constant of $1.70x10^{-13}$ $cm^3$ $molecule^{-1}$ $sec^{-1}$ for the vapor phase reaction with hydroxyl radicals in air (Atkinson 1985; quoted, Dilling et al. 1988; Howard et al. 1991); rate constant of < $4.0x10^{-17}$ $cm^3$ $molecule^{-1}$ $sec^{-1}$ for the gas-phase reaction with $NO_3$ radicals at 298 ± 2 K (Atkinson et al. 1987); rate constant of < $5.20x10^{-17}$ $cm^3$ $molecule^{-1}$ $sec^{-1}$ relative to ethene for the gas-phase reaction with $NO_3$ radicals at 298 ± 2 K (Atkinson et al. 1987; quoted, Atkinson 1991) while a room temperature rate constant of > $6.19x10^{-17}$ $cm^3$ $molecule^{-1}$ $sec^{-1}$ for reaction with $NO_3$ radicals in the gas-phase (Atkinson et al. 1988; quoted, Sabljic & Güsten 1990) and $2.86x10^{-16}$ $cm^3$ $molecule^{-1}$ $sec^{-1}$ as the predicted rate constant for the same reaction (Sabljic & Güsten 1990).

Hydrolysis: 0.079 $month^{-1}$ at 25°C with a half-life of 8.8 months (Dilling et al. 1975; quoted, Callahan et al. 1979); abiotic hydrolysis or dehydrohalogenation half-life of 8.8 months (Olsen & Davis 1990).

Biodegradation: aqueous aerobic half-life of 4320-8640 hours, based on aerobic river die-away test data (Mudder 1981; quoted, Howard et al. 1991) and saltwater grab sample test data (Jensen & Rosenberg 1975; quoted, Howard et al. 1991); aqueous anaerobic half-life of 2352-39672 hours, based on anaerobic screening test data (Bouwer et al. 1981; quoted, Howard et al. 1991); first-order rate constant of 1.1 $year^{-1}$ with a half-life of 230 days (Roberts et al. 1982; quoted, Olsen & Davis 1990); half-life of 34 days (Wood et al. 1985; quoted, Olsen & Davis 1990).

Biotransformation: estimated rate constant for bacteria of $1x10^{-10}$ ml $cell^{-1}$ $hour^{-1}$ (Mabey et al. 1982).

Bioconcentration, Uptake ($k_1$) and Elimination ($k_2$) Rate Constants or Half-Lives:

$k_1$: 3.323 $hour^{-1}$ (trout muscle, Neely et al. 1974)
$k_2$: 0.0823 $hour^{-1}$ (trout muscle, Neely et al. 1974)
$k_1$: 11.4 $hour^{-1}$ (flagfish, calculated-BCFx$k_1$, McCarty et al. 1992)
$k_2$: 0.0454 $hour^{-1}$ (flagfish, estimated-one compartment first-order kinetic, McCarty et al. 1992)
$k_1$: 3.30 $hour^{-1}$ (trout, Hawker & Connell 1985)
$k_2$: 0.0833 $hour^{-1}$ (trout, Hawker & Connell 1985,1988)

Common Name: 1,3-Dichloropropene
Synonym: 1,3-dichloropropylene, 1,3-dichloro-1-propene, Telone II
Chemical Name: 1,3-dichloropropene
CAS Registry No: 542-75-6
Molecular Formula: $CH_2ClCH=CHCl$
Molecular Weight: 110.97
Melting Point (°C):
Boiling Point (°C):

112.0 (stereoisomer I, Horvath 1982; McNally & Grob 1984; Dean 1985)
104.3 (stereoisomer II, Horvath 1982; McNally & Grob 1984; Dean 1985)

Density ($g/cm^3$ at 20°C):

1.224 (stereoisomer I, Horvath 1982; McNally & Grob 1984)
1.217 (stereoisomer II, Horvath 1982; McNally & Grob 1984)

Molar Volume ($cm^3/mol$):

108.40 (calculated-LeBas method)
90.96 (calculated from density)

Molecular Volume ($A^3$):
Total Surface Area, TSA ($A^2$):
Heat of Fusion, $\Delta H_{fus}$, (kcal/mol):
Entropy of Fusion, $\Delta S_{fus}$ (cal/mol K or e.u.):
Fugacity Ratio at 25°C, F: 1.0

Water Solubility ($g/m^3$ or mg/L at 25°C):

2800 (stereoisomer I, Dilling 1977; quoted, Horvath 1982)
2700 (stereoisomer II, Dilling 1977; quoted, Horvath 1982)
2723 (recommended, Horvath 1982)
2700 (selected, Mills et al. 1982)
1071.0 (stereoisomer I, headspace-GC, McNally & Grob 1983)
1088.1 (stereoisomer II, headspace-GC, McNally & Grob 1983)
1019.9 (30°C, stereoisomer I, headspace-GC, McNally & Grob 1984)
911.2 (30°C, stereoisomer II, headspace-GC, McNally & Grob 1984)
2700 (quoted, Warner et al. 1987)
1000 (20°C, quoted, Wauchope et al. 1992)
2250 (quoted, Wauchope et al. 1992)
4531, 4581 (20°C, calculated-activity coeff., Wright et al. 1992)

Vapor Pressure (Pa at 25°C):

435, 569 (quoted, calculated-solvatochromic p. & UNIFAC, Banerjee et al. 1990)
3333 (20°C, selected, Mills et al. 1982)
3334 (20°C, quoted, Warner et al. 1987)
3866 (20°C, quoted, Wauchope et al. 1992)

Henry's Law Constant (Pa $m^3$/mol):

360, 137 (quoted, calculated-P/C, Warner et al. 1987)

429 (20°C, calculated-P/C using Wauchope et al. 1992 data)

Octanol/Water Partition Coefficient, log $K_{OW}$:

1.98 (selected, Mills et al. 1982)

1.41 (quoted, Verhaar et al. 1992)

Bioconcentration Factor, log BCF:

Sorption Partition Coefficient, log $K_{OC}$:

1.505 (soil, selected, Wauchope et al. 1992)

Half-Lives in the Environment:

Air: disappearance half-life of 0.24-2.4 hours from air for the reaction with OH radicals (EPA 1974; quoted, Darnall et al. 1976); 4.66-80.3 hours, based on measured rate constants for reaction with hydroxyl radicals (Atkinson & Carter 1984; quoted, Howard et al. 1991) and ozone (Pitts Jr. et al. 1984 for cis- and trans-1,3-dichloropropene; quoted, Howard et al. 1991).

Surface water: 133-271 hours, based on measured rate of hydrolysis at pH 7 and 25°C (Mill et al. 1985; quoted, Howard et al. 1991) and at pH 5 and 20°C (McCall 1987; quoted, Howard et al. 1991).

Ground water: 133-271 hours, based on measured rate of hydrolysis at pH 7 and 25°C (Mill et al. 1985; quoted, Howard et al. 1991) and at pH 5 and 20°C (McCall 1987; quoted, Howard et al. 1991).

Sediment:

Soil: 133-271 hours, based on measured rate of hydrolysis at pH 7 and 25°C (Mill et al. 1985; quoted, Howard et al. 1991) and at pH 5 and 20°C (McCall 1987; quoted, Howard et al. 1991).

Biota:

Environmental Fate Rate Constant and Half-Lives:

Volatilization: half-life of 31 minutes in water from stirring in an open container of depth 65 mm at 200 rpm (Dilling et al. 1975; quoted, Mills et al. 1982)

Photolysis:

Oxidation: photooxidation half-life of 4.66-80.3 hours, based on measured rate constants for reaction with hydroxyl radicals (Atkinson & Carter 1984; quoted, Howard et al. 1991) and ozone (Pitts Jr. et al. 1984 for both cis- and trans-1,3-

dichloropropene; quoted, Howard et al. 1991); $(0.774 \pm 0.02)\times10^{11}$ $cm^3$ $molecule^{-1}$ $sec^{-1}$ (cis-1,3-dichloropropene) and $(1.31 \pm 0.05)\times10^{11}$ $cm^3$ $molecule^{-1}$ $sec^{-1}$ (trans-1,3-dichloropropene) both for the reaction with OH radicals in air at 22 ± 2°C (Tuazon et al. 1984).

Hydrolysis: half-life of 133-271 hours, based on measured rate constants for hydrolysis at pH 7 and 25°C (Mill et al. 1985; quoted, Howard et al. 1991) and pH 5 and 20°C (McCall 1987; quoted, Howard et al. 1991); rate of hydrolysis is, however, independent of pH over the range of pH 5 to pH 10 (McCall 1987).

Biodegradation: aqueous aerobic half-life of 168-672 hours, based on unacclimated aqueous aerobic biodegradation screening studies (Tabak et al. 1981; Krijgheld & Van der Gen 1986; quoted, Howard et al. 1991); aqueous anaerobic half-life of 672-2688 hours, based on estimated unacclimated aqueous aerobic biodegradation half-life (Howard et al. 1991).

Biotransformation:

Bioconcentration, Uptake ($k_1$) and Elimination ($k_2$) Rate Constants or Half-Lives:

Common Name: Chloroprene
Synonym: 2-chloro-1,3-butadiene
Chemical Name: 2-chloro-1,3-butadiene
CAS Registry No: 126-99-8
Molecular Formula: $CH_2=CClCH=CH_2$
Molecular Weight: 88.54
Melting Point (°C):
-130 (Kirk-Othmer 1985; Verschueren 1983; Howard 1989)
Boiling Point (°C):
59.4 (Weast 1982-83; Verschueren 1983; Dean 1985; Howard 1989)
Density ($g/cm^3$ at 20°C):
0.9583 (Weast 1982-83; Dean 1985)
0.9580 (Verschueren 1983)
Molar Volume ($cm^3/mol$):
102.3 (calculated-LeBas method)
92.39 (calculated from density)
Molecular Volume ($A^3$):
Total Surface Area, TSA ($A^2$):
Heat of Fusion, $\Delta H_{fus}$, (kcal/mol):
Entropy of Fusion, $\Delta S_{fus}$ (cal/mol K or e.u.):
Fugacity Ratio at 25°C, F: 1.0

Water Solubility ($g/m^3$ or mg/L at 25°C):
964000 (20°C, saturation concn., quoted, Verschueren 1983)

Vapor Pressure (Pa at 25°C):
26660 (20°C, quoted, Verschueren 1983)
36658 (30°C, quoted, Verschueren 1983)
23194 (Boublik et al. 1984; quoted, Howard 1989)

Henry's Law Constant (Pa $m^3$/mol):
3242 (estimated by bond contribution method, Howard 1989)

Octanol/Water Partition Coefficient, log $K_{OW}$:
2.03 (GEMS 1982; quoted, Howard 1989)

Bioconcentration Factor, log BCF:
1.342 (estimated-$K_{OW}$, Lyman et al. 1982; quoted, Howard 1989)

Sorption Partition Coefficient, log $K_{OC}$:

2.498 (soil, estimated-$K_{OW}$, Lyman et al. 1982; quoted, Howard 1989)

1.699 (soil, molecular topology & QSAR, Sabljic 1984; quoted, Howard 1989)

Half-Lives in the Environment:

Air: disappearance half-life of 0.24-2.4 hours from air for the reaction with OH radicals (EPA 1974; quoted, Darnall et al. 1976); 1.8 hour for the reaction with photochemically produced hydroxyl radicals and 12 hours for the reaction with ozone in atmosphere (Cupitt 1980; quoted, Howard 1989); 2.9-27.8 hours, based on estimated rate constants for reaction with hydroxyl radicals and ozone in air (Atkinson & Carter 1984; Atkinson 1987; quoted, Howard et al. 1991).

Surface water: 672-4320 hours, based on estimated unacclimated aqueous aerobic biodegradation half-life (Howard et al. 1991).

Ground water: 1344-8640 hours, based on estimated unacclimated aqueous aerobic biodegradation half-life (Howard et al. 1991).

Sediment:

Soil: 672-4320 hours, based on estimated unacclimated aqueous aerobic biodegradation half-life (Howard et al. 1991).

Biota:

Environmental Fate Rate Constant and Half-Lives:

Volatilization: half-life from a model river 1 m deep with a current velocity of 1 m/sec and a wind speed of 3 m/sec is estimated to be approximately 3 hours (Lyman et al. 1982; quoted, Howard 1989).

Photolysis:

Oxidation: photooxidation half-life of 2.9-27.8 hours, based on estimated rate constants for the reaction with hydroxyl radicals and ozone in air (Atkinson & Carter 1984; Atkinson 1987; quoted, Howard et al. 1991).

Hydrolysis:

Biodegradation: aqueous aerobic half-life of 672-4320 hours, based on aqueous aerobic screening test data for vinyl chloride (Helfgott et al. 1977; Freitag et al. 1984; quoted, Howard et al. 1991); aqueous anaerobic half-life of 2688-17280 hours, based on estimated unacclimated aqueous aerobic biodegradation half-life (Howard et al. 1991).

Biotransformation:

Bioconcentration, Uptake ($k_1$) and Elimination ($k_2$) Rate Constants or Half-Lives:

Common Name: Hexachlorobutadiene
Synonym: HCBD
Chemical Name: hexachloro-1,3-butadiene
CAS Registry No: 87-68-3
Molecular Formula: $Cl_2C=CClClC=CCl_2$
Molecular Weight: 260.76
Melting Point (°C):
-21 (Horvath 1982)
Boiling Point (°C):
215 (Horvath 1982)
Density ($g/cm^3$ at 20°C):
1.682 (Horvath 1982)
Molar Volume ($cm^3/mol$):
206.8 (calculated-LeBas method)
155.03 (calculated from density)
Molecular Volume ($A^3$):
Total Surface Area, TSA ($A^2$):
Heat of Fusion, $\Delta H_{fus}$, (kcal/mol):
Entropy of Fusion, $\Delta S_{fus}$ (cal/mol K or e.u.):
Fugacity Ratio at 25°C, F: 1.0

Water Solubility ($g/m^3$ or mg/L at 25°C):
0.20 (Melnikov 1971)
2.0 (20°C, Pearson & McConnell 1975; quoted, Callahan et al. 1979, Könemann 1981, Warner et al. 1987)
3.23 (shake flask-HPLC, Banerjee et al. 1980; quoted, Horvath 1982)

Vapor Pressure (Pa at 25°C):
20.00 (20°C, Pearson & McConnell 1975; quoted, Callahan et al. 1979)
19.96 (quoted, Dobbs et al. 1980; Warner et al. 1987)

Henry's Law Constant (Pa $m^3/mol$):
2604 (calculated-P/C, Dobbs et al. 1980, Warner et al. 1987)
1044 (batch stripping, Dobbs et al. 1980, Warner et al. 1987)

Octanol/Water Partition Coefficient, log $K_{OW}$:
3.74 (calculated, Callahan et al. 1979)
4.78 (shake flask-HPLC, Banerjee et al. 1980)
4.63 (calculated-f const., Könemann 1981)

4.78 (quoted, Thomann 1989)

Bioconcentration Factor, log BCF:
4.70 (Thomann 1989)

Sorption Partition Coefficient, log $K_{OC}$:

Half-Lives in the Environment:
Air: 2865-28650 hours, based on estimated photooxidation half-life in air (Howard et al. 1991).
Surface water: 672-4320 hours, based on aqueous aerobic biodegradation half-life (Howard et al. 1991).
Ground water: 1344-8640 hours, based on estimated aqueous aerobic biodegradation half-life (Howard et al. 1991).
Sediment:
Soil: 672-4320 hours, based on estimated aqueous aerobic biodegradation half-life (Howard et al. 1991).
Biota:

Environmental Fate Rate Constant and Half-Lives:
Volatilization:
Photolysis:
Oxidation: photooxidation half-life of 2965-28650 hours, based on an estimated rate constant for vapor phase reaction with OH radicals (Atkinson 1987; quoted, Howard et al. 1991).
Hydrolysis: no hydrolyzable groups, rate constant at pH 7 is zero (Kolling et al. 1987).
Biodegradation: aqueous aerobic half-life of 672-4320 hours, based on monitoring data and acclimated aqueous screen test data (Howard et al. 1991); aqueous anaerobic half-life of 2688-17280 hours, based on estimated aqueous aerobic biodegradation half-life (Howard et al. 1991).
Biotransformation:
Bioconcentration, Uptake ($k_1$) and Elimination ($k_2$) Rate Constants or Half-Lives:

Common Name: Hexachlorocyclopentadiene
Synonym: 1,2,3,4,5,5-hexachlorocyclopentadiene, perchlorocyclopentadiene, HCCPD
Chemical Name: 1,2,3,4,5,5-hexachlorocyclopentadiene
CAS Registry No: 77-47-4
Molecular Formula: $C_5Cl_6$
Molecular Weight: 272.77
Melting Point (°C): -9.90, 11.34
Boiling Point (°C): 239
Density (g/cm$^3$ at 20°C): 1.702
Molar Volume (cm$^3$/mol):

210.1 (calculated-LeBas method)
160.3 (calculated from density, Stephenson & Malanowski 1987)

Molecular Volume (A$^3$):
Total Surface Area, TSA (A$^2$):
Heat of Fusion, $\Delta H_{fus}$, (kcal/mol):
Entropy of Fusion, $\Delta S_{fus}$ (cal/mol K or e.u.):
Fugacity Ratio at 25°C, F: 1.0

Water Solubility (g/m$^3$ or mg/L at 25°C):

0.80 (Lu et al. 1974)
1.80 (Zepp et al. 1979)
0.805 (quoted, Callahan et al. 1979; Dobbs et al. 1980; Geyer et al. 1981; Warner et al. 1987; Isnard & Lambert 1988)
1.80 (28°C, vapor saturation or shake flask-GC, Wolfe et al. 1982)

Vapor Pressure (Pa at 25°C):

10.84 (quoted, Verschueren 1977; Dobbs et al. 1980; Warner et al. 1987)
10.67 (Wolfe et al. 1982)
11.90 (extrapolated-Antoine eqn., Stephenson & Malanowski 1987)

Henry's Law Constant (Pa m$^3$/mol):

2736 (concentration ratio-GC, Wolfe et al. 1982)
1621 (calculated-P/C, Wolfe et al. 1982)
3668 (calculated-P/C, Dobbs et al. 1980; Warner et al. 1987)
1662 (batch stripping, Dobbs et al. 1980; Warner et al. 1987)

Octanol/Water Partition Coefficient, log $K_{OW}$:

5.51 (calculated-BCF, Veith et al. 1979; quoted, Isnard & Lambert 1988)
3.99 (Zepp et al. 1979; quoted, Callahan et al. 1979)

5.00 (HPLC-RT, McDuffie 1981)
5.04 (28°C, concentration ratio-GC, Wolfe et al. 1982)
5.51 (quoted, Mackay 1982)

Bioconcentration Factor, log BCF:
2.53, 2.97, 3.21, 2.65 (algae, snail, mosquito, fish, Lu et al. 1975; quoted, Callahan et al. 1979)
1.47 (fathead minnnows, Veith et al. 1979; quoted, Isnard & Lambert 1988)
3.04 (*chlorella*, Geyer et al. 1981)
3.21 (calculated-S, Geyer et al. 1981)
1.47, 4.19 (quoted, calculated-$K_{OW}$, Mackay 1982)
3.18, 3.89 (estimated-S & $K_{OW}$, Isnard & Lambert 1988)

Sorption Partition Coefficient, log $K_{OC}$:
4.08 (sediment organic carbon, calculated-$K_{OW}$, Wolfe et al. 1982)

Half-Lives in the Environment:
Air: 1.0 to 8.9 hours, based on estimated photooxidation half-life in air (Howard et al. 1991).
Surface water: 1.0 minute to 173 hours, based on photolysis and hydrolysis half-lives (Howard et al. 1991).
Ground water: 173 to 1344 hours, based on aerobic aqueous biodegradation and hydrolysis half-lives (Howard et al. 1991).
Sediment: hydrolysis rate constant in the sediment was assumed to be the same as that of water and decay rate constant of 1.5 to $5.4 \times 10^{-5}$ $sec^{-1}$ for natural pond sediment-water system (Wolfe et al. 1982).
Soil: 168 to 672 hours, based on aerobic aqueous biodegradation half-life (Howard et al. 1991).
Biota:

Environmental Fate Rate Constant and Half-Lives:
Volatilization: appears to be important in flowing waters (Callahan et al. 1979).
Photolysis: near-surface photolysis is an important process with a rate constant of 4.9 $day^{-1}$ with a half-life of 11 minutes (Callahan et al. 1979); direct photolysis in natural waters in midday sunlight of Athens, Georgia at latitude 34°N and 83°W with half-life of less than 10 minutes, and near-surface photolysis rate constant on cloudless days averaged over both light and dark periods for a year was computed to be 3.9 $hour^{-1}$ (Wolfe et al. 1982); photolysis half-life of 1.0 minute

to 10.7 minutes, based on photolysis studies in aqueous solutions (Butz et al. 1982; Wolfe et al. 1982; quoted, Howard et al. 1991).

Oxidation: photooxidation half-life of 1.0 to 8.9 hours, based on calculated rate constants for the vapor phase reactions with OH radicals and $O_3$ in air (Cupitt 1980; quoted, Howard et al. 1991).

Hydrolysis: appears to be an important fate process with a rate constant of $5.6 \times 10^{-7}$ $sec^{-1}$ at 25°C with a half-life of 14 days (Callahan et al. 1979); reaction rate constant was independent of pH range of about 3 to 10 under most environmental conditions and an extrapolated rate constant of $4 \times 10^{-3}$ $sec^{-1}$ was found at 25°C with a half-life of 173 hours (Wolfe et al. 1982; quoted, Howard et al. 1991).

Biodegradation: aqueous aerobic half-life of 168 to 672 hours, based on aerobic aqueous screening test data (Tabak et al. 1981; Freitag et al. 1982; quoted, Howard et al. 1991); aqueous anaerobic half-life of 672 to 2688 hours, based on aerobic aqueous biodegradation half-life (Howard et al. 1991).

Biotransformation:

Bioconcentration, Uptake ($k_1$) and Elimination ($k_2$) Rate Constants or Half-Lives:

Common Name: Methyl bromide
Synonym: bromomethane, monobromomethane, Embafume, Terabol
Chemical Name: bromomethane, methyl bromide
CAS Registry No: 74-83-9
Molecular Formula: $CH_3Br$
Molecular Weight: 94.94
Melting Point (°C):
-93.0 (Stull 1947; Verschueren 1983)
-94.07 (Dreisbach 1959; Riddick et al. 1986)
-93.6 (Weast 1977; Callahan et al. 1979; Mackay & Shiu 1981,1990; Mabey et al. 1982)
-93.66 (Howard 1989)
-93.7 (Stephenson & Malanowski 1987)
Boiling Point (°C):
3.56 (Dreisbach 1959; Weast 1972-73; Mackay & Shiu 1981,1990; Mabey et al. 1982; Stephenson & Malanowski 1987)
3.55 (Kudchadker et al. 1979; Riddick et al. 1986)
4.60 (Verschueren 1977,1983; Callahan et al. 1979)
3.60 (Howard 1989)
Density ($g/cm^3$ at 20°C):
1.6755
Molar Volume ($cm^3/mol$):
52.90 (calculated-LeBas method)
55.76 (calculated from density)
Molecular Volume ($A^3$):
Total Surface Area, TSA ($A^2$):
164.7 (Iwase et al. 1985)
70.50 (Okouchi et al. 1992)
Heat of Fusion, $\Delta H_{fus}$, (kcal/mol):
Entropy of Fusion, $\Delta S_{fus}$ (cal/mol K or e.u.):
Fugacity Ratio at 25°C, F: 1.0

Water Solubility ($g/m^3$ or mg/L at 25°C):
13410 (Haight 1951; quoted, Horvath 1982)
20700 (20°C, gravitational method, Glew & Moelwyn-Hughes 1953; quoted, Horvath 1982)
24137 (25°C, gravitational method, Glew & Moelwyn-Hughes 1953; quoted, Horvath 1982)
900 (Patty 1962; quoted, Horvath 1982)
15539 (Glew & Moelwyn-Hughes 1964; quoted, Horvath 1982)
14400 (quoted, Irmann 1965)
13400 (Jolles 1966; quoted, Horvath 1982)

13400 (Günther et al. 1968; quoted, Horvath 1982)
12933 (Korenman et al. 1971; quoted, Horvath 19982)
14704 (quoted, Hine & Mookerjee 1975)
900 (20°C, Verschueren 1977,1983; quoted, Callahan et al. 1979; Mabey et al. 1982; Mills et al. 1982; Mackay & Shiu 1990)
13000 (quoted, Thomas 1982)
17500 (20°C, Windholz et al. 1983; quoted, Howard 1989)
15210, 3119 (quoted, predicted- $\chi$ & polarizability, Nirmalakahanden & Speece 1988)

Vapor Pressure (Pa at 25°C):
256554 (Antoine eqn. regression, Stull 1947)
217678 (calculated-Antoine eqn., Dreisbach 1959; quoted, Riddick et al. 1986)
183846 (calculated-Antoine eqn., Weast 1972-73)
189286 (20°C, USEPA 1976; quoted, Callahan et al. 1979; Mabey et al. 1982; Mills et al. 1982)
183900 (Weast 1972-73; quoted, Mackay & Shiu 1981,1990)
218930 (calculated-Antoine eqn., Kudchadker et al. 1979)
187000 (20°C, quoted, Thomas 1982)
218777, 218967 (calculated-Antoine eqn., Boublik et al. 1984)
217679 (calculated-Antoine eqn., Stephenson & Malanowski 1987)
217700 (quoted from Riddick et al. 1986)

Henry's Law Constant (Pa m$^3$/mol):
631, 621 (exptl., calculated-P/C, Glew & Moelwyn-Hughes 1953)
652 (calculated as $1/K_{AW}$, $C_W/C_A$, reported as exptl., Hine & Mookerjee 1975; quoted, Nirmalakahandan & Speece 1988; Howard 1989)
533 (20°C, calculated-P/C, Mackay & Shiu 1981)
19958 (20°C, calculated-P/C, Mabey et al. 1982)
1317 (calculated-P/C, Thomas 1982)
652, 326.8 (quoted, calculated-QSAR, Nirmalakahanden & Speece 1988)
20262 (20-25°C and low ionic strength, quoted, Pankow & Rosen 1988; Pankow 1990)
10689 (calculated-P/C, Mackay & Shiu 1990)

Octanol/Water Partition Coefficient, log $K_{OW}$:
1.10 (calculated as per Tute 1971, Callahan et al. 1979)
1.19 (shake flask-GC, Leo et al. 1975)
1.19 (Hansch & Leo 1979; quoted, Iwase et al. 1985; Mailhot & Peters 1988; Sangster 1989; Mackay & Shiu 1990)
1.09 (calculated, Mabey et al. 1982)
1.00 (selected, Mills et al. 1982)

1.19 (Hansch & Leo 1985; quoted, Howard 1989)

1.19, 1.13 (quoted, calculated-hydrophobicity const., Iwase et al. 1985)
1.19 (recommended, Sangster 1989)

Bioconcentration Factor, log BCF:
0.623 (microorganisms-water, calculated-$K_{OW}$, Mabey et al. 1982)
0.672 (calculated-$K_{OW}$, Lyman et al. 1982; quoted, Howard 1989)

Sorption Partition Coefficient, log $K_{OC}$:
2.236, 2.241, 2.215 (Naaldwijk loamy sand, Aalsmeer loam, Boskoop peaty clay; Daelmans & Sienbering 1979; quoted, Howard 1989)
0.771 (sediment-water, calculated-$K_{OW}$, Mabey et al. 1982)
2.10 (calculated-S, Lyman et al. 1982; quoted, Howard 1989)

Half-Lives in the Environment:
Air: disappearance half-life of 2.4-24 hours from air for the reaction with OH radicals (EPA 1974; quoted, Darnall et al. 1976); half-life of 0.29 year in the atmosphere for the reaction with $2x10^6$ hydroxy radicals/$cm^3$ at 25°C (Dilling 1982; quoted, Howard 1989); 289 days estimated as toxic chemical residence time with rate constant of $4x10^{-14}$ $cm^3$ $molecule^{-1}$ $sec^{-1}$ for the reaction with OH radicals at 300 K (Singh et al. 1980,1981; quoted, Howard 1989); 1633-16327 hours, based on measured rates for reaction with hydroxyl radicals in air (Atkinson 1985; quoted, Howard et al. 1991).
Surface water: 168-672 hours, based on unacclimated aerobic aqueous screening test data for bromoform from experiments utilizing settled domestic wastewater inoculum (Tabak et al. 1981; quoted, Howard et al. 1991).
Ground water: 336-912 hours, based on unacclimated aerobic aqueous screening test data for bromoform from experiments utilizing settled domestic waste water inoculum and hydrolysis half-life (Tabak et al. 1981; Mabey & Mill 1978; Ehrenberg et al. 1974; quoted, Vogel & Reinhard 1986; Howard et al. 1991).
Sediment:
Soil: 168-672 hours, based on unacclimated aerobic aqueous screening test data for bromoform from experiments utilizing settled domestic wastewater inoculum (Tabak et al. 1981; quoted, Howard et al. 1991).
Biota:

Environmental Fate Rate Constant and Half-Lives:

Volatilization: estimated experimental half-life for 1 mg/liter to be 27 minutes when stirred at 200 rpm in water at approximately 25°C in an open container (Dilling et al. 1975; quoted, Callahan et al. 1979); half-life of approximate 30 minutes (Mills et al. 1982); half-life of 3.0 hours in a model river (estimated, Lyman et al. 1982; quoted, Howard 1989); half-lives for the volatilization from 1 and 10 cm of soil were estimated to be 0.2 and 0.5 day respectively (Jury et al. 1984; quoted, Howard 1989).

Photolysis: photooxidation half-life of 1633-16327 hours in air, based on measured rates for reaction with hydroxyl radicals in air (Atkinson 1985; quoted, Howard et al. 1991).

Oxidation: calculated rate constant of $4.0 \times 10^{-14}$ $cm^3$ $molecule^{-1}$ $sec^{-1}$ for the reaction with OH radicals at 298 K with lifetime of 3.8 years in the troposphere (Davis et al. 1976; quoted, Altshuller 1980); approximately estimated to be $2.4 \times 10^{10}$ $cm^3$ $mol^{-1}$ $sec^{-1}$ for the reaction with OH radicals at 300 K (Lyman 1982); calculated rate constants: $<<360$ $M^{-1}$ $hour^{-1}$ for singlet oxygen and 0.1 $M^{-1}$ $hour^{-1}$ for peroxy radicals both at 25°C (Mabey et al. 1982).

Hydrolysis: first-order rate constant of $4.0 \times 10^{-7}$ $sec^{-1}$ with a maximum half-life of 20 days at pH 7 and 25°C (Radding et al. 1977; Mabey & Mill 1978; quoted, Callahan et al. 1979); half-life of 470-912 hours, based on measured first-order hydrolysis rate constants (Mabey & Mill 1978; Ehrenberg et al. 1974; quoted, Vogel & Reinhard 1986; Howard 1989; Howard et al. 1991); rate constant of $1.44 \times 10^{3}$ $hour^{-1}$ at pH 7 and 25°C (Callahan et al. 1979; quoted, Mabey et al. 1982); $3 \times 10^{-7}$ $sec^{-1}$ at 25°C with half-life of 26.7 days (Castro & Belser 1981; quoted, Howard 1989).

Biodegradation: aerobic aqueous biodegradation half-life of 168-672 hours, based on unacclimated aerobic aqueous screening test data for bromoform from experiments utilizing settled domestic wastewater inoculum (Tabak et al. 1981; quoted, Howard et al. 1991); anaerobic aqueous biodegradation half-life of 672-2688 hours, based on unacclimated aqueous aerobic biodegradation half-life (Tabak et al. 1981; quoted, Howard et al. 1991).

Biotransformation:

Bioconcentration, Uptake ($k_1$) and Elimination ($k_2$) Rate Constants or Half-Lives:

Common Name: Dibromomethane
Synonym: methylene bromide
Chemical Name: dibromomethane, methylene bromide
CAS Registry No: 74-95-3
Molecular Formula: $CH_2Br_2$
Molecular Weight: 173.85
Melting Point (°C):
-52.8 (Stull 1947)
-52.55 (Horvath 1982)
-52.5 (Weast 1982-83)
-52.7 (Dean 1985; Stephenson & Malanowski 1987)
Boiling Point (°C):
97.8 (Rex 1906)
96.95 (Kudchadker et al. 1979)
97.0 (Horvath 1982; Weast 1982-83)
96.5 (Dean 1985)
96.9 (Stephenson & Malanowski 1987)
Density (g/cm$^3$ at 20°C):
2.497 (Horvath 1982; Weast 1982-83)
2.4956 (Dean 1985)
Molar Volume (cm$^3$/mol):
76.20 (calculated-LeBas method)
69.62 (calculated from density)
Molecular Volume (A$^3$):
Total Surface Area, TSA (A$^2$):
95.10 (Okouchi et al. 1992)
Heat of Fusion, $\Delta H_{fus}$, (kcal/mol):
Entropy of Fusion, $\Delta S_{fus}$ (cal/mol K or e.u.):
Fugacity Ratio at 25°C, F: 1.0

Water Solubility (g/m$^3$ or mg/L at 25°C):
11480 (20°C, volumetric, Rex 1906; quoted, Shiu et al. 1990)
11760 (30°C, volumetric, Rex 1906)
11930 (30°C, shake flask-interferometer, Gross & Saylor 1931; quoted, Shiu et al. 1990)
11700 (15°C, shake flask-interferometer, Gross & Saylor 1931; quoted, Shiu et al. 1990)
11486 (Seidell 1940; quoted, Deno & Berkheimer 1960)
17500 (Booth & Everson 1948; quoted, Horvath 1982)
11000 (O'Connell 1963; quoted, Horvath 1982)
11800 (quoted, Irmann 1965)
11000 (Jolles 1966; quoted, Horvath 1982; Shiu et al. 1990)

11800, 8130 (quoted exptl., calculated-group contribution as per Irmann 1965, Horvath 1982)
11486 (quoted, Hine & Mookerjee 1975)
11442 (recommended, Horvath 1982)
11500 (Dean 1985)
11433 (calculated-AI, Okouchi et al. 1992)
12800 (19.3°C, shake flask-GC/TC, Stephenson 1992)
11400 (29.5°C, shake flask-GC/TC, Stephenson 1992)
11107, 11300 (20°C, calculated-activity coefficients, Wright et al. 1992)

Vapor Pressure (Pa at 25°C):
4626 (20°C, Rex 1906)
7518 (30°C, Rex 1906)
5775 (Antoine eqn. regression, Stull 1947)
5767 (calculated-Antoine eqn., Weast 1972-73)
5880 (quoted, Hine & Mookerjee 1975)
5922 (Antoine eqn. regression, Kudchadker et al. 1979)
6034 (calculated-Antoine eqn., Stephenson & Malanowski 1987)
4653 (20°C, quoted from DIPPR, Tse et al. 1992)
7728 (30°C, quoted from DIPPR, Tse et al. 1992)

Henry's Law Constant (Pa $m^3$/mol):
90.0 (calculated-1/$K_{AW}$, $C_W/C_A$, reported as exptl., Hine & Mookerjee 1975; quoted, Nirmalakhandan & Speece 1988)
73.16 (calculated-bond method, Hine & Mookerjee 1975)
86.13 (batch stripping, Nicholson et al. 1984)
90.0, 105.7 (quoted, calculated-QSAR, Nirmalakhandan & Speece 1988)
86.03 (Munz & Roberts 1989)
70.92 (20°C, Tse et al. 1992)
111.4 (30°C, Tse et al. 1992)

Octanol/Water Partition Coefficient, log $K_{OW}$:
2.50 (Hansch et al. 1968)
2.50 (selected, Mailhot & Peters 1988)

Bioconcentration Factor, log BCF:

Sorption Partition Coefficient, log $K_{OC}$:

Half-Lives in the Environment:

Air: disappearance half-life of 2.4-24 hours from air for the reaction with OH radicals (EPA 1974; quoted, Darnall et al. 1976); 46 days estimated as toxic chemical residence time with rate constant of $2.5x10^{-13}$ $cm^3$ $molecule^{-1}$ $sec^{-1}$ for the reaction with OH radicals at 300 K (Singh et al. 1981); 851-8510 hours, based on estimated rate constant for the reaction with hydroxyl radicals in air (Atkinson 1987; quoted, Howard et al. 1991).

Surface water: 168-672 hours, based on unacclimated aqueous aerobic biodegradation half-life (Howard et al. 1991).

Ground water: 336-1344 hours, based on unacclimated aqueous aerobic biodegradation half-life (Howard et al. 1991).

Sediment:

Soil: 168-672 hours, based on unacclimated aqueous aerobic biodegradation half-life (Howard et al. 1991).

Biota:

Environmental Fate Rate Constant and Half-Lives:

Volatilization:

Photolysis: photocatalyzed mineralization by the presence of $TiO_2$ with the rate of 4.1 ppm/min-g of catalyst (Ollis 1985).

Oxidation: rate constant for the reaction with $OH^-$: $(6.1\pm0.61)x10^{-7}$ $M^{-1}$ $sec^{-1}$ was measured in 66.7% dioxane-water at 35.7°C (Hine et al. 1956; quoted, Roberts et al. 1992); $5x10^{11}$ $cm^3$ $mol^{-1}$ $sec^{-1}$, estimated approximately for the reaction with OH radicals at 300 K (Lyman 1982); photooxidation half-life of 851-8510 hours, based on estimated rate constant for the reaction with hydroxyl radicals in air (Atkinson 1987; quoted, Howard et al. 1991); rate constant of $(9.0\pm3.0)x10^7$ $M^{-1}$ $sec^{-1}$ for the reaction with OH radicals in aqueous solution at pH 8.5 (Haag & Yao 1992) with reference to $CH_3CCl_3$ having rate constant of $4.0x10^7$ $M^{-1}$ $sec^{-1}$ for the reaction with OH radicals in aqueous solution (Buxton et al. 1988; quoted, Haag & Yao 1992).

Hydrolysis: half-life of 183 years at pH 7 and 25°C, based on overall hydrolysis rate constant (Mill et al. 1982; quoted, Howard et al. 1991).

Biodegradation: aqueous aerobic half-life of 168-672 hours, based on unacclimated aerobic screening test data for bromoform from experiments utilizing settled domestic wastewater inoculum (Tabak et al. 1981; quoted, Howard et al. 1991); aqueous anaerobic half-life of 672-2688 hours, based on unacclimated aqueous aerobic half-life (Howard et al. 1991).

Biotransformation:

Bioconcentration, Uptake ($k_1$) and Elimination ($k_2$) Rate Constants or Half-Lives:

Common Name: Tribromomethane
Synonym: bromoform, methenyl tribromide
Chemical Name: tribromomethane
CAS Registry No: 75-25-2
Molecular Formula: $CHBr_3$
Molecular Weight: 252.75
Melting Point (°C):

8.5 (Stull 1947)
-8.3 (Weast 1977; Callahan et al. 1979; Mackay & Shiu 1981,1990; Mabey et al. 1982)
8.10 (Dean 1985)
8.05 (Riddick et al. 1986)
7.70 (Stephenson & Malanowski 1987)

Boiling Point (°C):

150.5 (Kahlbaum & Arndt 1898)
149.5 (Weast 1977; Callahan et al. 1979; Mackay & Shiu 1981,1990; Mabey et al. 1982; McNally & Grob 1984; Stephenson & Malanowski 1987)
149.57 (Kudchadker et al. 1979)
149.6 (Dean 1985)
149.21 (Riddick et al. 1986)

Density (g/cm$^3$ at 20°C):

2.891 (Kahlbaum & Arndt 1898)
2.8917, 2.8909 (Kudchadker et al. 1979)
2.890 (McNally & Grob 1984; Gillham & Rao 1990)
2.9031 (15°C, Dean 1985)
2.8909 (Riddick et al. 1986)

Molar Volume (cm$^3$/mol):

99.5 (calculated-LeBas method)
87.46 (calculated from density)

Molecular Volume (A$^3$):
Total Surface Area, TSA (A$^2$):

114.0 (Valsaraj 1988)
122.2 (Okouchi et al. 1992)

Heat of Fusion, $\Delta H_{fus}$, (kcal/mol):
Entropy of Fusion, $\Delta S_{fus}$ (cal/mol K or e.u.):
Fugacity Ratio at 25°C, F: 1.0

Water Solubility (g/m$^3$ or mg/L at 25°C):

3190 (30°C, shake flask-interferometer, Gross & Saylor 1931)
3010 (15°C, shake flask-interferometer, Gross & Saylor 1931)
3110 (Seidell 1940; quoted, Deno & Berkheimer 1960)

3190 (30°C, Seidell 1941; quoted, Verschueren 1977; Callahan et al. 1979; Mackay & Shiu 1981)
3190 (30°C, CRC 1962; quoted, McNally & Grob 1984)
3110 (quoted, Hine & Mookerjee 1975)
3033 (Verschueren 1977; quoted, Mackay & Shiu 1981,1990)
3190 (Cowen & Baynes; quoted, McNally & Grob 1984)
3100, 3470 (quoted exptl., calculated-group contribution as per Irmann 1965, Horvath 1982)
3100 (recommended, Horvath 1982)
3000 (20°C, selected, Mills et al. 1982)
3931.2 (30°C, headspace-GC, McNally & Grob 1984)
3010 (quoted, Neely 1984)
3200 (30°C, Dean 1985)
3180 (30°C, quoted, Riddick et al. 1986)
3130 (interpolated between two data points, Warner et al. 1987)
3097, 1656 (quoted, predicted- $\chi$ & polarizability, Nirmalakhandan & Speece 1988)
3130 (selected, Valsaraj 1988)
3010 (selected, Gillham & Rao 1990)
3513 (calculated-AI, Okouchi et al. 1992)
3974, 4002 (20°C, calculated-activity coefficients, Wright et al. 1992)

Vapor Pressure (Pa at 25°C):
667 (20°C, Jordon 1954; quoted, Mabey et al. 1982; Gillham & Rao 1990)
815 (calculated-Antoine eqn., Weast 1972-73)
715 (calculated-Antoine eqn., Boublik et al. 1973)
833 (quoted, Hine & Mookerjee 1975)
747 (Verschueren 1977,1983; quoted, Mackay & Shiu 1981,1990; Warner et al. 1987)
720 (calculated-Antoine eqn., Kudchadker et al. 1979; quoted, Mackay & Shiu 1981)
798 (quoted, Cowen & Baynes 1980)
1333 (20°C, selected, Mills et al. 1982)
717, 727 (calculated-Antoine eqn., Boublik et al. 1984)
747 (quoted, McNally & Grob 1984)
790 (quoted, Riddick et al. 1986)
727 (calculated-Antoine eqn., Stephenson & Malanowski 1987)
747 (selected, Valsaraj 1988)
720, 1121 (quoted, calculated-solvatochromic p. & UNIFAC, Banerjee et al. 1990)
543 (20°C, quoted from DIPPR, Tse et al. 1992)
981 (30°C, quoted from DIPPR, Tse et al. 1992)

Henry's Law Constant (Pa $m^3$/mol):

68.3 (calculated-$1/K_{AW}$, $C_W/C_A$, reported as exptl., Hine & Mookerjee 1975; quoted, Nirmalakhandan & Speece 1988)
7.66 (calculated-bond method, Hine & Mookerjee 1975)
62.3 (calculated-P/C, Mackay & Shiu 1981,1990)
62.0 (recommended, Mackay & Shiu 1981,1990)
58.8 (20°C, Symons et al. 1981; quoted, Nicholson et al. 1984)
56.7 (20°C, calculated-P/C, Mabey et al. 1982)
43.6 (20°C, batch stripping, Nicholson et al. 1984; quoted, Mackay & Shiu 1990)
43.6 (20°C, calculated-P/C, Nicholson et al. 1984)
61.8 (quoted, Pankow et al. 1984)
54.2 (multiple equilibration, Munz & Roberts 1987,1989)
53.9, 60.3 (quoted, calculated-P/C, Warner et al. 1987)
68.3, 42.1 (quoted, calculated-QSAR, Nirmalakhandan & Speece 1988)
56.7 (20-25°C and low ionic strength, quoted, Pankow & Rosen 1988; Pankow 1990)
60.28 (selected, Valsaraj 1988)
40.52 (20°C, Tse et al. 1992)
70.92 (30°C, Tse et al. 1992)

Octanol/Water Partition Coefficient, log $K_{OW}$:

2.30 (calculated as per Tute 1971, Callahan et al. 1979; quoted, Mackay & Shiu 1990)
2.38 (calculated-f const., Mabey et al. 1982)
2.30 (selected, Mills et al. 1982)
2.38 (selected, Valsaraj 1988)

Bioconcentration Factor, log BCF:

1.80 (microorganisms-water, calculated-$K_{OW}$, Mabey et al. 1982)

Sorption Partition Coefficient, log $K_{OC}$:

2.07 (sediment-water, calculated-$K_{OW}$, Mabey et al. 1982)

Half-Lives in the Environment:

Air: disappearance half-life of 2.4-24 hours from air for the reaction with OH radicals (EPA 1974; quoted, Darnall et al. 1976); 1299-12989 hours, based on estimated rate constant for the reaction with hydroxyl radicals in air (Atkinson 1987; quoted, Howard et al. 1991).

Surface water:

Ground water: 1344-8640 hours, based on unacclimated aqueous aerobic biodegradation half-life (Howard et al. 1991).

Sediment:

Soil: 672-4320 hours, based on unacclimated aerobic aqueous screening test data from experiments utilizing settled domestic wastewater inoculum (Bouwer et al. 1984; quoted, Howard et al. 1991).

Biota:

Environmental Fate Rate Constant and Half-Lives:

Volatilization: calculated half-lives of 63 minutes to 24.2 days (using Langbein & Durum 1967 published $O_2$ reaeration values) and 65.6 days both from rivers and streams (Kaczmar et al. 1984).

Photolysis: photocatalyzed mineralization by the presence of $TiO_2$ with the rate of 6.2 ppm/min per gram of catalyst (Ollis 1985).

Oxidation: calculated rate constants at 25°C: $<<360$ $M^{-1}$ $hour^{-1}$ for singlet oxygen and 0.5 $M^{-1}$ $hour^{-1}$ for peroxy radicals (Mabey et al. 1982); photooxidation half-life of 1299-12989 hours, based on estimated rate constant for the reaction with hydroxyl radicals in air (Atkinson 1987; quoted, Howard et al. 1991); rate constant of $(1.3\pm0.6)x10^{8}$ $M^{-1}$ $sec^{-1}$ for the reaction with OH radicals in aqueous solution at pH 8.5 (Haag & Yao 1992) with reference to $CH_3CCl_3$ having rate constant of $4.0x10^{7}$ $M^{-1}$ $sec^{-1}$ (Buxton et al. 1988; quoted, Haag & Yao 1992).

Hydrolysis: a maximum half-life of 686 years has been estimated at pH 7 and 25°C from experimental data at 100-150°C (Radding et al. 1977) which corresponds to a first-order rate constant of $3.2x10^{-11}$ $sec^{-1}$ (Radding et al. 1977; Mabey & Mill 1978; quoted, Mabey et al. 1982; Howard et al. 1991); rate constant of $2.5x10^{-9}$ $hour^{-1}$ assigned by analogy to trichloromethane (Mabey et al. 1982).

Biodegradation: half-life of 672-4320 hours in soil, based on unacclimated aerobic aqueous screening test data from experiments utilizing settled domestic wastewater inoculum (Bouwer et al. 1984; quoted, Howard et al. 1991); anaerobic half-life of 2688-17280 hours, based on unacclimated aqueous aerobic biodegradation half-life (Howard et al. 1991).

Biotransformation: estimated rate constant of $1x10^{-10}$ ml $cell^{-1}$ $hour^{-1}$ for bacteria (Mabey et al. 1982).

Bioconcentration, Uptake ($k_1$) and Elimination ($k_2$) Rate Constants or Half-Lives:

Common Name: Ethyl bromide
Synonym: bromoethane, monobromoethane
Chemical Name: ethyl bromide, bromoethane
CAS Registry No: 74-96-4
Molecular Formula: $CH_3CH_2Br$
Molecular Weight: 108.97
Melting Point (°C):
-119.33 (Dreisbach 1959)
-118.6 (Horvath 1982; Dean 1985; Riddick et al. 1986)
Boiling Point (°C):
38.40 (Kahlbaum & Arndt 1898; Horvath 1982; Dean 1985)
37.70 (Rex 1906)
38.35 (Dreisbach 1959; Riddick et al. 1986)
Density ($g/cm^3$ at 20°C):
1.4570 (Kahlbaum & Arndt 1898)
1.4594 (Dreisbach 1959)
1.4492 (25°C, Dreisbach 1959)
1.4604 (Horvath 1982)
1.4505 (25°C, Riddick et al. 1986)
Molar Volume ($cm^3/mol$):
75.1 (calculated-LeBas method)
74.64 (calculated from density)
Molecular Volume ($A^3$):
Total Surface Area, TSA ($A^2$):
190.7 (Iwase et al. 1985)
91.4 (Okouchi et al. 1992)
Heat of Fusion, $\Delta H_{fus}$, (kcal/mol):
1.40 (quoted, Riddick et al. 1986)
Entropy of Fusion, $\Delta S_{fus}$ (cal/mol K or e.u.):
Fugacity Ratio at 25°C, F: 1.0

Water Solubility ($g/m^3$ or mg/L at 25°C):
9140 (20°C, Rex 1906)
8960 (30°C, Rex 1906)
9600 (Führer 1924; quoted, Hansch et al. 1968)
9064 (Seidell 1940; quoted, Deno & Berkheimer 1960)
9000 (Kirk-Othmer 1964; quoted, Horvath 1982)
9000 (quoted, Irmann 1965)
6030 (calculated-group contribution as per Irmann 1965, Horvath 1982)
9600, 8817 (quoted, calculated-$K_{OW}$, Hansch et al. 1968)
7890 (calculated-group contribution as per Korenman et al. 1971, Horvath 1982)
9064 (quoted, Hine & Mookerjee 1975)

9020 (selected exptl., Horvath 1982)
9100 (Dean 1985)
9100 (20°C, quoted, Riddick et al. 1986)
9494, 4440 (quoted, calculated-fragment const., Wakita et al. 1986)
10267 (calculated-AI, Okouchi et al. 1992)

Vapor Pressure (Pa at 25°C):
51440 (20°C, Rex 1906)
75141 (30°C, Rex 1906)
62464 (calculated-Antoine eqn., Dreisbach 1959)
63251 (quoted, Hine & Mookerjee 1975)
62470 (quoted, Riddick et al. 1986)

Henry's Law Constant (Pa $m^3$/mol):
766.0 (calculated-1/$K_{AW}$, $C_W/C_A$, reported as exptl., Hine & Mookerjee 1975; quoted, Nirmalakhandan & Speece 1988)
757.4 (calculated-P/C, using Hine & Mookerjee 1975 data)
766.0, 608.5 (quoted, calculated-QSAR, Nirmalakhandan & Speece 1988)

Octanol/Water Partition Coefficient, log $K_{OW}$:
1.60 (calculated-$\pi$, Hansch et al. 1968)
1.61 (shake flask-GC, Hansch et al. 1975)
1.61, 1.62 (quoted, calculated-hydrophobicity const., Iwase et al. 1985)
1.61 (quoted exptl., Sangster 1989)
1.61 (recommended, Sangster 1989)

Bioconcentration Factor, log BCF:

Sorption Partition Coefficient, log $K_{OC}$:

Half-Lives in the Environment:
Air: disappearance half-life of 2.4-24 hours from air for the reaction with OH radicals (USEPA 1974; quoted, Darnall et al. 1976).
Surface water:
Ground water:
Sediment:
Soil:

Biota:

Environmental Fate Rate Constant and Half-Lives:

Volatilization:

Photolysis:

Oxidation:

Hydrolysis: $9.4 \times 10^{-4}$ $hour^{-1}$ at pH 7 and 25°C with estimated half-life of 30 days (Mabey & Mill 1978; quoted, Schwarzenbach et al. 1985; Vogel & Reinhard 1986).

Biodegradation:

Biotransformation:

Bioconcentration, Uptake ($k_1$) and Elimination ($k_2$) Rate Constants or Half-Lives:

Common Name: 1,2-Dibromoethane
Synonym: ethylene bromide, ethylene dibromide, *sym*-dibromoethane, EDB
Chemical Name: 1,2-dibromoethane
CAS Registry No: 106-93-4
Molecular Formula: $CH_2BrCH_2Br$
Molecular Weight: 187.87
Melting Point (°C):
    9.79 (Dreisbach 1959; Horvath 1982; Riddick et al. 1986)
    9.97 (Verschueren 1983)
Boiling Point (°C):
    131.50 (Kahlbaum & Arndt 1898)
    131.36 (Dreisbach 1959; Horvath 1982; Riddick et al. 1986)
    131.0 (Jones et al. 1977/78)
    131.60 (Verschueren 1983)
Density ($g/cm^3$ at 20°C):
    2.1780 (Kahlbaum & Arndt 1898)
    2.1792 (Dreisbach 1959; Horvath 1982)
    2.1688 (25°C, Dreisbach 1959)
    2.1791 (Riddick et al. 1986)
    2.1687 (25°C, Riddick et al. 1986)
Molar Volume ($cm^3$/mol):
    101.40 (calculated-LeBas method)
    86.21 (calculated from density)
Molecular Volume ($A^3$):
Total Surface Area, TSA ($A^2$):
    116.0 (Okouchi et al. 1992)
Heat of Fusion, $\Delta H_{fus}$, (kcal/mol):
    2.397 (calculated, Dreisbach 1959)
    2.616 (quoted, Riddick et al. 1986)
Entropy of Fusion, $\Delta S_{fus}$ (cal/mol K or e.u.):
Fugacity Ratio at 25°C, F: 1.0

Water Solubility ($g/m^3$ or mg/L at 25°C):
    4310 (30°C, shake flask-interferometer, Gross & Saylor 1931)
    3920 (15°C, shake flask-interferometer, Gross & Saylor 1931)
    4017 (Seidell 1940; quoted, Deno & Berkheimer 1960; Hine & Mookerjee 1975)
    4200 (measured by Dow Chemical, Dreisbach 1959; quoted, Irmann 1965)
    4170 (O'Connell 1963; Jolles 1966; quoted, Horvath 1982)
    3370 (Goring 1972; quoted, Kenaga & Goring 1980)
    2910 (shake flask-GC, Jones et al. 1977/78)
    4200 (selected exptl., Horvath 1982)

5130 (calculated-group contribution as per Irmann 1965, Horvath 1982)
4300 (quoted, Thomas 1982)
4310 (30°C, quoted, Verschueren 1983)
4290 (30°C, quoted, Riddick et al. 1986)
4304, 8992 (quoted, calculated-fragment const., Wakita et al. 1986)
4158 (calculated-AI, Okouchi et al. 1992)
4120 (19.5°C, shake flask-GC/TC, Stephenson 1992)
4310 (30.7°C, shake flask-GC/TC, Stephenson 1992)

Vapor Pressure (Pa at 25°C):
1560 (calculated-Antoine eqn., Dreisbach 1959)
1520 (quoted, Hine & Mookerjee 1975)
1546 (20°C, quoted, Thomas 1982)
1466 (20°C, quoted, Verschueren 1983)
2266 (30°C, quoted, Verschueren 1983)
1040 (quoted, Riddick et al. 1986)

Henry's Law Constant (Pa $m^3$/mol):
71.49 (calculated as $1/K_{AW}$, $C_W/C_A$, reported as exptl., Hine & Mookerjee 1975; quoted, Nirmalakhandan & Speece 1988)
15.64 (calculated-group contribution method, Hine & Mookerjee 1975)
110.7 (calculated-bond contribution method, Hine & Mookerjee 1975)
66.86 (calculated-P/C, Thomas 1982)
71.49, 133.12 (quoted, calculated-QSAR, Nirmalakhandan & Speece 1988)
83.07 (20-25°C and low ionic strength, quoted, Pankow & Rosen 1988; Pankow 1990)
86.76 (selected, Jury et al. 1990)

Octanol/Water Partition Coefficient, log $K_{OW}$:

Bioconcentration Factor, log BCF:
0.778 (calculated-S as per Kenaga & Goring 1978, Kenaga 1980)
0.301 (calculated-$K_{OC}$ as per Kenaga & Goring 1978, Kenaga 1980)

Sorption Partition Coefficient, log $K_{OC}$:
1.643 (soil, quoted, Kenaga 1980; Kenaga & Goring 1980)
1.699 (soil, calculated as per Kenaga & Goring 1978, Kenaga 1980)
1.643 (soil, selected, Jury et al. 1990)

Half-Lives in the Environment:

Air: disappearance half-life of 2.4-24 hours for the reaction with OH radicals in air (EPA 1974; quoted, Darnall et al. 1976); 257-2567 hours, based on estimated rate constant for the reaction with hydroxyl radicals in air (Atkinson 1987; quoted, Howard et al. 1991).

Surface water: 672-4320 hours, based on unacclimated aqueous aerobic biodegradation half-life (Howard et al. 1991).

Ground water: 470-2880 hours, based on data from anaerobic ground water ecosystem study (Wilson et al. 1986; quoted, Howard et al. 1991) and data from an aerobic ground water ecosystem study (Swindoll et al. 1987; quoted, Howard et al. 1991).

Sediment: calculated half-life of 1500 days at 25°C and pH 7, based on studies in pure water and in barely saturated subsurface sediment at 25-60°C (Haag & Mill 1988).

Soil: estimated half-life of volatilization loss from soil to be 3650 days (Jury et al. 1990); disappearance half-life of less than 2.0 days was estimated from the volatilization loss of mixtures in soil (Anderson et al. 1991); 672-4320 hours, based on unacclimated aqueous aerobic biodegradation half-life (Howard et al. 1991).

Biota:

Environmental Fate Rate Constant and Half-Lives:

Volatilization: estimated half-life of volatilization from water, 6.1 hours (Thomas 1982)

Photolysis:

Oxidation: photooxidation half-life of 257-2567 hours in air, based on estimated rate constant for the reaction with hydroxyl radicals in air (Atkinson 1987; quoted, Howard et al. 1991).

Hydrolysis: EDB hydrolyzes to ethylene glycol and bromoethanol in water at pH 7 and 25°C with half-life of 5-10 days (Leinster et al. 1978; quoted, Verschueren 1983); $9.9 \times 10^{-6}$ $hour^{-1}$ at pH 7 and 25°C with a calculated half-life of 8.0 years (Jungclaus & Cohen 1986; quoted, Ellington 1989); rate constant of $(8.9 \pm 0.1) \times 10^{-9}$ $sec^{-1}$ in water at 25°C and pH 7.5 with an estimated half-life of 2.5 years (Vogel & Reinhard 1986); half-life of 2.2 years , based on measured neutral hydrolysis rate constant at pH 7 and 25°C (Weintraub et al. 1986; quoted, Howard et al. 1991).

Biodegradation: aqueous aerobic half-life of 672-4320 hours, based on unacclimated aqueous aerobic biodegradation screening test data (Bouwer & McCarty 1983; quoted, Howard et al. 1991); aqueous anaerobic half-life of 48-360 hours, based on anaerobic stream and pond water sediment die-away test data (Jafvert & Wolfe 1987; quoted, Howard et al. 1991).

Biotransformation:

Bioconcentration, Uptake ($k_1$) and Elimination ($k_2$) Rate Constants or Half-Lives:

Common Name: 1-Bromopropane
Synonym: bromopropane, monobromopropane, *n*-propyl bromide, propyl bromide
Chemical Name: *n*-propyl bromide, 1-bromopropane
CAS Registry No: 106-94-5
Molecular Formula: $CH_3CH_2CH_2Br$
Molecular Weight: 123.0
Melting Point (°C):
- -109.85 (Horvath 1982)
- -110.1 (Dean 1985)
- -109.8 (Riddick et al. 1986)

Boiling Point (°C):
- 70.80 (Rex 1906)
- 71.0 (Horvath 1982; Dean 1985)
- 70.97 (Riddick et al. 1986)

Density (g/cm$^3$ at 20°C):
- 1.3537 (Horvath 1982)

Molar Volume (cm$^3$/mol):
- 97.3 (calculated-LeBas method)
- 90.86 (calculated from density)

Molecular Volume (A$^3$):
- 73.7 (Moriguchi et al. 1976)

Total Surface Area, TSA (A$^2$):
- 105.4 (Moriguchi et al. 1976)
- 217.7 (Iwase et al. 1985)
- 112.3 (Okouchi et al. 1992)

Heat of Fusion, $\Delta H_{fus}$, (kcal/mol):
- 2.160 (quoted, Riddick et al. 1986)

Entropy of Fusion, $\Delta S_{fus}$ (cal/mol K or e.u.):
Fugacity Ratio at 25°C, F: 1.0

Water Solubility (g/m$^3$ or mg/L at 25°C):
- 2450 (20°C, volumetric, Rex 1906; quoted, Horvath 1982)
- 2470 (30°C, volumetric, Rex 1906; quoted, Horvath 1982)
- 2275 (Fühner 1924; quoted, Hansch et al. 1968)
- 2310 (30°C, shake flask-interferometer, Gross & Saylor 1931; quoted, Horvath 1982)
- 2312 (30°C, Van Arkel & Vles 1936; quoted, Horvath 1982)
- 2454 (Seidell 1940; quoted, Deno & Berkheimer 1960; Hine & Mookerjee 1975)
- 2500 (20°C, Patty 1962; Lange 1967; quoted, Horvath 1982)
- 2275, 2460 (quoted, calculated-$K_{OW}$, Hansch et al. 1968)
- 2450, 2450 (20°C, exptl., calculated, Korenman et al. 1971; quoted, Horvath 1982)
- 2430 (selected exptl., Horvath 1982)

1910 (calculated-group contribution as per Irmann 1965, Horvath 1982)
2300 (30°C, Dean 1985)
2300 (30°C, quoted, Riddick et al. 1986)
2812, 1511 (quoted, calculated-fragment const., Wakita et al. 1986)
2979 (calculated-AI, Okouchi et al. 1992)

Vapor Pressure (Pa at 25°C):
14770 (20°C, Rex 1906)
22741 (30°C, Rex 1906)
19155 (quoted, Hine & Mookerjee 1975)
18440 (quoted, Riddick et al. 1986)

Henry's Law Constant (Pa $m^3$/mol):
964.4 (calculated-1/$K_{AW}$, $C_W/C_A$, reported as exptl., Hine & Mookerjee 1975; quoted, calculated-QSAR, Nirmalakhandan & Speece 1988)
1133 (calculated-group contribution method, Hine & Mookerjee 1975)
1564 (calculated-bond contribution method, Hine & Mookerjee 1975)
964.4, 766.0 (quoted, calculated-QSAR, Nirmalakhandan & Speece 1988)

Octanol/Water Partition Coefficient, log $K_{OW}$:
2.10 (shake flask-GC, Fujita et al. 1964; quoted, Hansch & Anderson 1967; Hansch et al. 1968; Sangster 1989)
2.10 (Leo et al. 1971; quoted, Moriguchi et al. 1976; Iwase et al. 1985)
2.10, 2.12 (quoted, calculated-hydrophobicity const., Moriguchi et al. 1976)
2.10, 2.14 (quoted, calculated-hydrophobicity const., Iwase et al. 1985)
2.10 (recommended, Sangster 1989)

Bioconcentration Factor, log BCF:

Sorption Partition Coefficient, log $K_{OC}$:

Half-Lives in the Environment:
Air: disappearance half-life of 2.4-24 hours from air for the reaction with OH radicals (USEPA 1974; quoted, Darnall et al. 1976).
Surface water:
Ground water:
Sediment:

Soil:
Biota:

Environmental Fate Rate Constant and Half-Lives:
Volatilization:
Photolysis:
Oxidation:
Hydrolysis: $1.1 \times 10^{-3}$ $hour^{-1}$ at pH 7 and 25°C with estimated half-life of 26 days (Mabey & Mill 1978; quoted, Schwarzenbach et al. 1985; Vogel & Reinhard 1986).
Biodegradation:
Biotransformation:
Bioconcentration, Uptake ($k_1$) and Elimination ($k_2$) Rate Constants or Half-Lives:

Common Name: 2-Bromopropane
Synonym: isopropyl bromide
Chemical Name: 2-bromopropane, isopropyl bromide
CAS Registry No: 75-26-3
Molecular Formula: $CH_3CHBrCH_3$
Molecular Weight: 122.99
Melting Point (°C):
-89.0 (Horvath 1982; Dean 1985; Riddick et al. 1986)
Boiling Point (°C):
59.0 (Rex 1906)
59.38 (Horvath 1982)
59.50 (Dean 1985)
59.41 (Riddick et al. 1986)
Density ($g/cm^3$ at 20°C):
1.3140 (Horvath 1982; Riddick et al. 1986)
1.3060 (Riddick et al. 1986)
Molar Volume ($cm^3/mol$):
97.30 (calculated-LeBas method)
93.90 (calculated from density)
Molecular Volume ($A^3$):
Total Surface Area, TSA ($A^2$):
114.8 (Okouchi et al. 1992)
Heat of Fusion, $\Delta H_{fus}$, (kcal/mol):
Entropy of Fusion, $\Delta S_{fus}$ (cal/mol K or e.u.):
Fugacity Ratio at 25°C, F: 1.0

Water Solubility ($g/m^3$ or mg/L at 25°C):
3180 (20°C, Rex 1906; quoted, Horvath 1982)
3180 (30°C, Rex 1906; quoted, Horvath 1982)
2877 (Fühner 1924; quoted, Hansch et al. 1968)
3198 (30°C, Van Arkel & Vles 1936; quoted, Horvath 1982)
3162 (Seidell 1940; quoted, Deno & Berkheimer 1960; Hine & Mookerjee 1975)
2877, 4304 (quoted, calculated-$K_{OW}$, Hansch et al. 1968)
2880, 2690 (20°C, Korenman et al. 1971; quoted, Horvath 1982)
3000 (selected exptl., Horvath 1982)
2400 (calculated-group contribution as per Irmann 1965, Horvath 1982)
2900 (18°C, Dean 1985)
2860 (18°C, quoted, Riddick et al. 1986)
2883, 1548 (quoted, calculated-fragment const., Wakita et al. 1986)
3089 (calculated-AI, Okouchi et al. 1992)

Vapor Pressure (Pa at 25°C):

23381 (20°C, Rex 1906)
35218 (30°C, Rex 1906)
28473 (quoted, Hine & Mookerjee 1975)
31500 (quoted, Riddick et al. 1986)

Henry's Law Constant (Pa $m^3$/mol):

1107 (calculated-$1/K_{AW}$, $C_W/C_A$, reported as exptl., Hine & Mookerjee 1975; quoted, Nirmalakhandan & Speece 1988)
2479 (calculated-group contribution method, Hine & Mookerjee 1975)
1564 (calculated-bond contribution method, Hine & Mookerjee 1975)
1107, 1082 (quoted, calculated-QSAR, Nirmalakhandan & Speece 1988)

Octanol/Water Partition Coefficient, log $K_{OW}$:

1.90 (calculated-$\pi$, Hansch et al. 1968)

Bioconcentration Factor, log BCF:

Sorption Partition Coefficient, log $K_{OC}$:

Half-Lives in the Environment:

Air: disappearance half-life of 2.4-24 hours from air for the reaction with OH radicals (USEPA 1974; quoted, Darnall et al. 1976).
Surface water:
Ground water:
Sediment: half-life of 2.1 days, based on neutral and base-catalyzed hydrolysis studies at 25°C in pure water and in barely saturated subsurface sediment at 25-60°C (Haag & Mill 1988).
Soil:
Biota:

Environmental Fate Rate Constant and Half-Lives:

Volatilization:
Photolysis:
Oxidation:

Hydrolysis: $1.4 \times 10^{-2}$ hour$^{-1}$ at pH 7 and 25°C with estimated half-life of 2.0 days (Mabey & Mill 1978; quoted, Schwarzenbach et al. 1985; Vogel & Reinhard 1986); rate constants at 25°C: $(379 \pm 41) \times 10^{-8}$ sec$^{-1}$ in distilled water at pH 3-11 for equal or more than 72% conversion (Mill et al. 1980; quoted, Haag & Mill 1988), $(383 \pm 33) \times 10^{-8}$ sec$^{-1}$ in distilled water for 89% conversion, $(372 \pm 64) \times 10^{-8}$ sec$^{-1}$ in sediment-extracted water for 87% conversion, and $(420 \pm 80) \times 10^{-8}$ sec$^{-1}$ in sediment pores at pH 7.3 for 88% conversion (Haag & Mill 1988).

Biodegradation:

Biotransformation:

Bioconcentration, Uptake ($k_1$) and Elimination ($k_2$) Rate Constants or Half-Lives:

Common Name: 1,2-Dibromopropane
Synonym: propylene bromide, propylene dibromide
Chemical Name: 1,2-dibromopropane
CAS Registry No: 78-75-1
Molecular Formula: $CH_3CHBrCH_2Br$
Molecular Weight: 201.90
Melting Point (°C):

-55.5 (Stull 1947; Dreisbach 1959; Dean 1985; Riddick et al. 1986)
-55.35 (Timmermans 1950)
-55.25 (Horvath 1982)
-55.2 (Weast 1982-83)

Boiling Point (°C):

141 (Timmermans 1950)
141.99 (Dreisbach 1959; Riddick et al. 1986)
140 (Horvath 1982; Weast 1982-83)
139.6 (Dean 1985)

Density ($g/cm^3$ at 20°C):

1.9327 (Dreisbach 1959)
1.9234 (25°C, Dreisbach 1959)
1.9324 (Horvath 1982; Weast 1982-83)
1.933 (Dean 1985)

Molar Volume ($cm^3/mol$):

113.2 (calculated-LeBas method)
104.47 (calculated from density)

Molecular Volume ($A^3$):
Total Surface Area, TSA ($A^2$):

139.4 (Okouchi et al. 1992)

Heat of Fusion, $\Delta H_{fus}$, (kcal/mol):

2.136 (calculated, Dreisbach 1959; quoted, Riddick et al. 1986)

Entropy of Fusion, $\Delta S_{fus}$ (cal/mol K or e.u.):
Fugacity Ratio at 25°C, F: 1.0

Water Solubility ($g/m^3$ or mg/L at 25°C):

1430 (measured by Dow Chemical, Dreisbach 1955-1961; quoted, Irmann 1965; Horvath 1982; Riddick et al. 1986)
1463 (quoted, Hine & Mookerjee 1975)
1620 (calculated-group contribution as per Irmann 1965, Horvath 1982)
1428 (recommended, Horvath 1982)
2000 (Dean 1985)
1463, 2986 (quoted, calculated-fragment const., Wakita et al. 1986)
1419, 1919 (quoted, predicted- $\chi$ & polarizability, Nirmalakhandan & Speece 1988)
1420 (calculated-AI, Okouchi et al. 1992)

Vapor Pressure (Pa at 25°C):

- 2000 (37.3°C, Kahlbaum & Arndt 1898; quoted, Timmermans 1950)
- 1036 (Antoine eqn. regression, Stull 1947)
- 1071 (calculated-Antoine eqn., Dreisbach 1959)
- 688 (quoted, Hine & Mookerjee 1975)
- 1072 (quoted from Dreisbach 1955-1961, Riddick et al. 1986)

Henry's Law Constant (Pa $m^3$/mol):

- 94.2 (calculated-1/$K_{AW}$, $C_W/C_A$, reported as exptl., Hine & Mookerjee 1975; quoted, Nirmalakhandan & Speece 1988)
- 21.10 (calculated-group contribution, Hine & Mookerjee 1975)
- 163.8 (calculated-bond contribution, Hine & Mookerjee 1975)
- 94.24, 226.1 (quoted, calculated-QSAR, Nirmalakhandan & Speece 1988)

Octanol/Water Partition Coefficient, log $K_{OW}$:

Bioconcentration Factor, log BCF:

Sorption Partition Coefficient, log $K_{OC}$:

Half-Lives in the Environment:

Air: disappearance half-life of 2.4-24 hours from air for the reaction with OH radicals (EPA 1974; quoted, Darnall et al. 1976).

Surface water:

Ground water:

Sediment:

Soil:

Biota:

Environmental Fate Rate Constant and Half-Lives:

Volatilization:

Photolysis:

Oxidation:

Hydrolysis: rate constant of (2.5 ± 0.5)x$10^{-8}$ $sec^{-1}$ in water at 25°C and pH 7 with an estimated half-life of 320 days (Vogel & Reinhard 1986).

Biodegradation:

Biotransformation:

Bioconcentration, Uptake ($k_1$) and Elimination ($k_2$) Rate Constants or Half-Lives:

Common Name: 1-Bromobutane
Synonym: *n*-butyl bromide, monobromobutane
Chemical Name: 1-bromobutane, n-butyl bromide, monobromobutane
CAS Registry No: 109-65-9
Molecular Formula: $CH_3CH_2CH_2CH_2Br$
Molecular Weight: 137.02
Melting Point (°C):
- -112.65 (Dreisbach 1959)
- -112.4 (Horvath 1982; Dean 1985; Riddick et al. 1986)

Boiling Point (°C):
- 91.44 (Dreisbach 1959)
- 101.6 (Horvath 1982; Dean 1985; Riddick et al. 1986)

Density (g/cm$^3$ at 20°C):
- 1.2609 (Dreisbach 1959)
- 1.2535 (25°C, Dreisbach 1959)
- 1.2758 (Horvath 1982; Riddick et al. 1986)
- 1.2686 (25°C, Dean 1985)
- 1.2687 (Riddick et al. 1986)

Molar Volume (cm$^3$/mol):
- 95.23 (Miller et al. 1985)
- 119.5 (calculated-LeBas method)
- 107.9 (calculated from density)

Molecular Volume (A$^3$):
Total Surface Area, TSA (A$^2$):
- 133.2 (Okouchi et al. 1992)

Heat of Fusion, $\Delta H_{fus}$, (kcal/mol):
- 1.646 (calculated, Dreisbach 1959)
- 1.600 (quoted, Riddick et al. 1986)

Entropy of Fusion, $\Delta S_{fus}$ (cal/mol K or e.u.):
Fugacity Ratio at 25°C, F: 1.0

Water Solubility (g/m$^3$ or mg/L at 25°C):
- 590 (Fühner 1924; quoted, Hansch et al. 1968)
- 608 (30°C, shake flask-interferometer, Gross & Saylor 1931; quoted, Horvath 1982)
- 598 (Seidell 1940; quoted, Deno & Berkheimer 1960; Hine & Mookerjee 1975)
- 254 (Booth & Everson 1948; quoted, Horvath 1982)
- 617 (Kakovsky 1957; quoted, Horvath 1982)
- 600 (quoted, Irmann 1965)
- 590, 677 (quoted, calculated-$K_{OW}$, Hansch et al. 1968; quoted, Tewari et al. 1982)
- 509, 750 (exptl., calculated-group contribution, Korenman et al. 1971; quoted, Horvath 1982)

600 (selected exptl., Horvath 1982)
600 (calculated-group contribution as per Irmann 1965, Horvath 1982)
869 (gen. col.-GC, Tewari et al. 1982; quoted, Miller et al. 1985)
869, 601 (quoted, calculated-UNIFAC, Arbuckle 1986)
608 (30°C, quoted, Riddick et al. 1986)
584, 336 (quoted, calculated-fragment const., Wakita et al. 1986)
540 (calculated-AI, Okouchi et al. 1992)

Vapor Pressure (Pa at 25°C):
8618 (calculated-Antoine eqn., Dreisbach 1959)
5501 (quoted, Hine & Mookerjee 1975)
5502 (quoted, Riddick et al. 1986)

Henry's Law Constant (Pa m$^3$/mol):
1242 (calculated-1/$K_{AW}$, $C_W/C_A$, reported as exptl., Hine & Mookerjee 1975; quoted, Nirmalakhandan & Speece 1988)
1600 (calculated-group contribution method, Hine & Mookerjee 1975)
2313 (calculated-bond contribution method, Hine & Mookerjee 1975)
1242, 964.4 (quoted, calculated-QSAR, Nirmalakhandan & Speece 1988)

Octanol/Water Partition Coefficient, log $K_{OW}$:
2.60 (calculated-$\pi$, Hansch et al. 1968; quoted, Tewari et al. 1982)
2.75 (gen. col.-GC, DeVoe et al. 1981; Wasik et al. 1981; Tewari et al. 1982; quoted, Miller et al. 1985; Sangster 1989)
2.75 (gen. col.-HPLC, DeVoe et al. 1981; quoted, Miller et al. 1985; Schantz & Martire 1987; Sangster 1989)
2.79 (estimated-activity coefficients, Wasik et al. 1981; quoted, Sangster 1989)
2.64 (estimated-activity coefficients, Schantz & Martire 1987; quoted, Sangster 1989)
2.75 (recommended, Sangster 1989)

Bioconcentration Factor, log BCF:

Sorption Partition Coefficient, log $K_{OC}$:

Half-Lives in the Environment:
Air: disappearance half-life of 2.4-24 hours for the reaction with OH radicals in air (USEPA 1974; quoted, Darnall et al. 1976).

Surface water:
Ground water:
Sediment:
Soil:
Biota:

Environmental Fate Rate Constant and Half-Lives:
Volatilization:
Photolysis:
Oxidation:
Hydrolysis:
Biodegradation:
Biotransformation:
Bioconcentration, Uptake ($k_1$) and Elimination ($k_2$) Rate Constants or Half-Lives:

Common Name: 1-Bromopentane
Synonym: *n*-amyl bromide, monobromopentane, pentyl bromide
Chemical Name: *n*-amyl bromide, 1-bromopentane, monobromopentane, pentyl bromide
CAS Registry No: 110-53-2
Molecular Formula: $CH_3CH_2CH_2CH_2CH_2Br$
Molecular Weight: 151.05
Melting Point (°C):
-95.15 (Miller et al. 1985)
-87.9 (Riddick et al. 1986)
Boiling Point (°C):
129.58 (Riddick et al. 1986)
Density ($g/cm^3$ at 20°C):
1.2182 (Riddick et al. 1986)
1.2119 (25°C, Riddick et al. 1986)
Molar Volume ($cm^3/mol$):
104.6 (Miller et al. 1985)
141.7 (calculated-LeBas method)
123.99 (calculated from density)
Molecular Volume ($A^3$):
Total Surface Area, TSA ($A^2$):
112.3 (Okouchi et al. 1992)
Heat of Fusion, $\Delta H_{fus}$, (kcal/mol):
2.740 (Riddick et al. 1986)
Entropy of Fusion, $\Delta S_{fus}$ (cal/mol K or e.u.):
Fugacity Ratio at 25°C, F: 1.0

Water Solubility ($g/m^3$ or mg/L at 25°C):
127 (gen. col.-GC, Tewari et al. 1982; quoted, Miller et al. 1985)
127, 141 (quoted, calculated-UNIFAC, Arbuckle 1986)
127 (quoted, Riddick et al. 1986)

Vapor Pressure (Pa at 25°C):
1680 (quoted, Riddick et al. 1986)

Henry's Law Constant (Pa $m^3$/mol):

Octanol/Water Partition Coefficient, log $K_{OW}$:

3.37 (gen. col.-GC, DeVoe et al. 1981; Tewari et al. 1982; quoted, Miller et al. 1985; Sangster 1989)

3.37 (gen. col.-HPLC, DeVoe et al. 1981; quoted, Schantz & Martire 1987; Sangster 1989)

3.49 (gen. col.-GC, Wasik et al. 1981; quoted, Sangster 1989)

3.32 (estimated-measured activity coefficients, Schantz & Martire 1987; quoted, Sangster 1989)

3.37 (recommended, Sangster 1989)

Bioconcentration Factor, log BCF:

Sorption Partition Coefficient, log $K_{OC}$:

Half-Lives in the Environment:

Air: disappearance half-life of 2.4-24 hours for the reaction with OH radicals in air (EPA 1974; quoted, Darnall et al. 1976).

Surface water:

Ground water:

Sediment:

Soil:

Biota:

Environmental Fate Rate Constant and Half-Lives:

Volatilization:

Photolysis:

Oxidation:

Hydrolysis:

Biodegradation:

Biotransformation:

Bioconcentration, Uptake ($k_1$) and Elimination ($k_2$) Rate Constants or Half-Lives:

Common Name: Vinyl bromide
Synonym: bromoethene, bromoethylene, ethylene bromide
Chemical Name: bromoethene
CAS Registry No: 593-60-2
Molecular Formula: $CH_2=CHBr$
Molecular Weight: 106.96
Melting Point (°C):
-139.54 (Dreisbach 1959; Riddick et al. 1986)
-139.5 (Dean 1985)
Boiling Point (°C):
15.8 (Dreisbach 1959; Dean 1985; Riddick et al. 1986)
Density ($g/cm^3$ at 20°C):
1.4933 (Dreisbach 1959; Riddick et al. 1986)
1.493 (Dean 1985)
1.4738 (25°C, Dreisbach 1959; Riddick et al. 1986)
Molar Volume ($cm^3/mol$):
67.7 (calculated-LeBas method)
71.6 (calculated from density)
Molecular Volume ($A^3$):
Total Surface Area, TSA ($A^2$):
Heat of Fusion, $\Delta H_{fus}$, (kcal/mol):
1.224 (calculated, Dreisbach 1959; quoted, Riddick et al. 1986)
Entropy of Fusion, $\Delta S_{fus}$ (cal/mol K or e.u.):
Fugacity Ratio at 25°C, F: 1.0

Water Solubility ($g/m^3$ or mg/L at 25°C):

Vapor Pressure (Pa at 25°C):
137699 (calculated-Antoine eqn., Dreisbach 1959)
144330, 139840 (calculated-Antoine eqn., Boublik et al. 1984)
137700 (Riddick et al. 1986)
140476 (calculated-Antoine eqn., Stephenson & Malanowski 1987)

Henry's Law Constant (Pa $m^3$/mol):

Octanol/Water Partition Coefficient, log $K_{OW}$:

Bioconcentration Factor, log BCF:

Sorption Partition Coefficient, log $K_{OC}$:

Half-Lives in the Environment:

Air: disappearance half-life of 0.24-2.4 hours from air for the reaction with OH radicals (EPA 1974; quoted, Darnall et al. 1976); 9.4-94 hours, based on measured rate constant for reaction with hydroxyl radicals in air (Atkinson 1985; quoted, Howard et al. 1991).

Surface water: 672-4320 hours, based on estimated unacclimated aqueous aerobic biodegradation half-life (Howard et al. 1991).

Ground water: 1344-69000 hours, based on estimated unacclimated aqueous aerobic biodegradation half-life (Howard et al. 1991) and an estimated half-life for anaerobic biodegradation of vinyl chloride from a ground water field study of chlorinated ethenes (Silka & Wallen 1988; quoted, Howard et al. 1991).

Sediment:

Soil: 672-4320 hours, based on estimated unacclimated aqueous aerobic biodegradation half-life (Howard et al. 1991).

Biota:

Environmental Fate Rate Constant and Half-Lives:

Volatilization:

Photolysis:

Oxidation: $6.81 \times 10^{-12}$ $cm^3$ $molecule^{-1}$ $sec^{-1}$ at 25.4°C for the reaction with OH radicals in air (Perry et al. 1977); photooxidation half-life of 9.4-94 hours in air, based on measured rate constant for the reaction with hydroxyl radicals in air (Atkinson 1985; quoted, Howard et al. 1991).

Hydrolysis:

Biodegradation: aqueous aerobic half-life of 672-4320 hours, based on aqueous screening test data for vinyl chloride (Heffgott et al. 1977; Freitag et al. 1984; quoted, Howard et al. 1991); aqueous anaerobic half-life of 2880-17280 hours, based on estimated unacclimated aqueous aerobic half-life (Howard et al. 1991).

Biotransformation:

Bioconcentration, Uptake ($k_1$) and Elimination ($k_2$) Rate Constants or Half-Lives:

Common Name: Methyl iodide
Synonym: iodomethane, monoiodomethane
Chemical Name: methyl iodide
CAS Registry No: 74-88-4
Molecular Formula: $CH_3I$
Molecular Weight: 141.94
Melting Point (°C):
-64.4 (Stull 1947)
-66.45 (Horvath 1982; Riddick et al. 1986)
-66.1 (Verschueren 1983; Stephenson & Malanowski 1987)
-66.5 (Dean 1985)
Boiling Point (°C):
42.8 (Rex 1906)
42.42 (Kudchadker et al. 1979)
42.4 (Horvath 1982; Dean 1985)
42.5 (Verschueren 1983; Stephenson & Malanowski 1987)
42.43 (Riddick et al. 1986)
Density ($g/cm^3$ at 20°C):
2.279 (Horvath 1982; Verschueren 1983)
2.2789 (Dean 1985)
2.2792 (Riddick et al. 1986)
Molar Volume ($cm^3/mol$):
62.90 (calculated-LeBas method)
62.28 (calculated from density)
Molecular Volume ($A^3$):
51.70 (Moriguchi et al. 1976)
Total Surface Area, TSA ($A^2$):
74.20 (Moriguchi et al. 1976)
179.4 (Iwase et al. 1985)
78.40 (Okouchi et al. 1992)
Heat of Fusion, $\Delta H_{fus}$, (kcal/mol):
Entropy of Fusion, $\Delta S_{fus}$ (cal/mol K or e.u.):
Fugacity Ratio at 25°C, F: 1.0

Water Solubility ($g/m^3$ or mg/L at 25°C):
14190 (20°C, volumetric, Rex 1906; Andelman 1978; quoted, Horvath 1982)
14290 (30°C, volumetric, Rex 1906; Andelman 1978; quoted, Horvath 1982)
13626 (20°C, Merckel 1937; quoted, Horvath 1982)
14195 (Seidell 1940; quoted, Deno & Berkheimer 1960)
14194 (Seidell 1941; quoted, Hansch et al. 1968)
15174 (calculated-$K_{OW}$, Hansch et al. 1968)
14194 (quoted, Hine & Mookerjee 1975)

13871, 8955 (quoted average, calculated-$K_{OW}$, Valvani et al. 1981)
13894 (recommended, Horvath 1982)
14000 (20°C, quoted, Verschueren 1983)
14000 (Dean 1985)
14000 (20°C, quoted, Riddick et al. 1986)
13967 (calculated-AI, Okouchi et al. 1992)

Vapor Pressure (Pa at 25°C):
44176 (20°C, Rex 1906)
64437 (30°C, Rex 1906)
47753 (calculated-Antoine eqn., Weast 1972-73)
54106 (quoted, Hine & Mookerjee 1975)
53975 (calculated-Antoine eqn., Kudchadker et al. 1979)
53320 (quoted, Verschueren 1983)
54120 (quoted, Riddick et al. 1986)
53958 (calculated-Antoine eqn., Stephenson & Malanowski 1987)

Henry's Law Constant (Pa $m^3$/mol):
541, 536 (exptl., calculated-P/C, Glew & Moelwyn-Hughes 1953)
554.9 (calculated-1/$K_{AW}$, $C_W/C_A$, reported as exptl., Hine & Mookerjee 1975; quoted, Nirmalakhandan & Speece 1988)
554.9, 278.1 (quoted, calculated-QSAR, Nirmalakhandan & Speece 1988)

Octanol/Water Partition Coefficient, log $K_{OW}$:
1.69 (19°C, shake flask, Collander 1951; quoted, Sangster 1989)
1.50 (calculated-$\pi$, Hansch et al. 1968)
1.69 (Leo et al. 1971; quoted, Moriguchi et al. 1976; Iwase et al. 1985)
1.51 (shake flask-GC, Hansch et al. 1975)
1.69, 1.52 (quoted, calculated-hydrophobicity const., Moriguchi et al. 1976)
1.51 (Hansch & Leo 1979; quoted, Sangster 1989)
1.69 (quoted, Valvani et al. 1981)
1.69 (quoted, Verschueren 1983)
1.69, 1.61 (quoted, calculated-hydrophobicity const., Iwase et al. 1985)
1.51 (recommended, Sangster 1989)

Bioconcentration Factor, log BCF:

Sorption Partition Coefficient, log $K_{OC}$:

Half-Lives in the Environment:

Air: disappearance half-life of 2.4-24 hours from air for the reaction with OH radicals (USEPA 1974; quoted, Darnall et al. 1976); 535-5348 hours, based on a measured rate constant for the vapor phase reaction with hydroxyl radicals in air (Garraway & Donovan 1979; quoted, Howard et al. 1991).

Surface water: 168-672 hours, based on estimated aqueous aerobic biodegradation half-life (Howard et al. 1991).

Ground water: 336-1344 hours, based on estimated aqueous aerobic biodegradation half-life (Howard et al. 1991).

Sediment:

Soil: 168-672 hours, based on estimated aqueous aerobic biodegradation half-life (Howard et al. 1991).

Biota:

Environmental Fate Rate Constant and Half-Lives:

Volatilization:

Photolysis:

Oxidation: photooxidation half-life of 535-5348 hours, based on a measured rate constant for the vapor phase reaction with hydroxyl radicals in air (Garraway & Donovan 1979; quoted, Howard et al. 1991).

Hydrolysis: $7.28 \times 10^{-8}$ $sec^{-1}$ at pH 7 (extrapolated to 25°C) with half-life of 110 days (Mabey & Mill 1978).

Biodegradation: aqueous aerobic half-life of 1168-672 hours, based on estimated aerobic half-life (Howard et al. 1991); aqueous anaerobic half-life of 672-2688 hours, based on estimated aerobic half-life (Howard et al. 1991).

Biotransformation:

Bioconcentration, Uptake ($k_1$) and Elimination ($k_2$) Rate Constants or Half-Lives:

Common Name: Ethyl iodide
Synonym: iodoethane
Chemical Name: ethyl iodide, iodoethane,
CAS Registry No: 75-03-6
Molecular Formula: $CH_3CH_2I$
Molecular Weight: 155.97
Melting Point (°C):
  -108.0 (Horvath 1982)
  -110.9 (Dean 1985)
  -111.1 (Riddick et al. 1986)
Boiling Point (°C):
  72.40 (Kahlbaum & Arndt 1898; Dean 1985)
  72.30 (Rex 1906; Horvath 1982; Riddick et al. 1986)
Density (g/cm³ at 20°C):
  1.9310 (Kahlbaum & Arndt 1898)
  1.9358 (Horvath 1982; Dean 1985)
  1.9357 (Riddick et al. 1986)
  1.9244 (25°C, Riddick et al. 1986)
Molar Volume (cm³/mol):
  85.1 (calculated-LeBas method)
  80.6 (calculated from density)
Molecular Volume (A³):
  67.10 (Moriguchi et al. 1976)
Total Surface Area, TSA (A²):
  94.20 (Moriguchi et al. 1976)
  203.6 (Iwase et al. 1985)
  99.3 (Okouchi et al. 1992)
Heat of Fusion, $\Delta H_{fus}$, (kcal/mol):
Entropy of Fusion, $\Delta S_{fus}$ (cal/mol K or e.u.):
Fugacity Ratio at 25°C, F: 1.0

Water Solubility (g/m³ or mg/L at 25°C):
  4030 (20°C, volumetric, Rex 1906; quoted, Horvath 1982)
  4150 (30°C, volumetric, Rex 1906; quoted, Horvath 1982)
  4040 (30°C, Gross & Saylor 1931; quoted, Horvath 1982)
  4040 (30°C, Van Arkel & Vles 1936; quoted, Horvath 1982)
  3915 (20°C, Merckel 1937; quoted, Horvath 1982)
  3918 (Seidell 1940,1941; quoted, Deno & Berkheimer 1960; Hansch et al. 1968)
  3915 (22.5°C, Saracco & Spaccamela Marchetti 1958; quoted, Horvath 1982)
  4000 (20°C, Patty 1962; Lange 1967; quoted, Horvath 1982)
  3918, 4121 (quoted, calculated-$K_{OW}$, Hansch et al. 1968)
  3920, 3570 (20°C, exptl., calculated, Korenman et al. 1971; quoted, Horvath 1982)

3918 (quoted, Hine & Mookerjee)
4060 (selected exptl., Horvath 1982)
2820 (calculated-group contribution as per Irmann 1965, Horvath 1982)
3880 (30°C, quoted, Riddick et al. 1986)
3918, 2838 (quoted, calculated-fragment const., Wakita et al. 1986)
3900 (calculated-AI, Okouchi et al. 1992)

Vapor Pressure (Pa at 25°C):
14276 (20°C, Rex 1906)
22154 (30°C, Rex 1906)
18156 (quoted, Hine & Mookerjee 1975)
18160 (quoted, Riddick et al. 1986)

Henry's Law Constant (Pa $m^3$/mol):
731.6 (calculated-1/$K_{AW}$, $C_W/C_A$, reported as exptl., Hine & Mookerjee 1975; quoted, calculated-QSAR, Nirmalakhandan & Speece 1988)
722.8 (calculated-P/C, using Hine & Mookerjee 1975 Data)
731.6, 839.9 (quoted, calculated-QSAR, Nirmalakhandan & Speece 1988)

Octanol/Water Partition Coefficient, log $K_{OW}$:
2.00 (shake flask-GC, Fujita et al. 1964; quoted, Hansch & Anderson 1967; Hansch et al. 1968; Sangster 1989)
2.00 (Leo et al. 1971; quoted, Moriguchi et al. 1976; Iwase et al. 1985)
2.00, 1.94 (quoted, calculated-hydrophobicity const., Moriguchi et al. 1976)
2.00, 2.07 (quoted, calculated-hydrophobicity const., Iwase et al. 1985)
2.00 (recommended, Sangster 1989)

Bioconcentration Factor, log BCF:

Sorption Partition Coefficient, log $K_{OC}$:

Half-Lives in the Environment:
Air: disappearance half-life of 2.4-24 hours from air for the reaction with OH radicals (USEPA 1974; quoted, Darnall et al. 1976).
Surface water:
Ground water:
Sediment:

Soil:
Biota:

Environmental Fate Rate Constant and Half-Lives:
Volatilization:
Photolysis:
Oxidation:
Hydrolysis:
Biodegradation:
Biotransformation:
Bioconcentration, Uptake ($k_1$) and Elimination ($k_2$) Rate Constants or Half-Lives:

Common Name: 1-Iodopropane
Synonym; monoiodopropane, *n*-propyl iodide, propyl iodide
Chemical Name: 1-iodopropane, monoiodopropane, *n*-propyl iodide, propyl iodide
CAS Registry No: 107-08-4
Molecular Formula: $CH_3CH_2CH_2I$
Molecular Weight: 169.99
Melting Point (°C):
-98/-101 (Dean 1985)
-101.3 (Riddick et al. 1986)
Boiling Point (°C):
100.60 (Rex 1906)
102.50 (Dean 1985)
102.45 (Riddick et al. 1986)
Density (g/cm$^3$ at 20°C):
1.7489 (Dean 1985; Riddick et al. 1986)
1.7394 (25°C, Riddick et al. 1986)
Molar Volume (cm$^3$/mol):
107.3 (calculated-LeBas method)
97.20 (calculated from density)
Molecular Volume (A$^3$):
Total Surface Area, TSA (A$^2$):
Heat of Fusion, $\Delta H_{fus}$, (kcal/mol):
Entropy of Fusion, $\Delta S_{fus}$ (cal/mol K or e.u.):
Fugacity Ratio at 25°C, F: 1.0

Water Solubility (g/m$^3$ or mg/L at 25°C):
1070 (20°C, volumetric, Rex 1906)
1030 (30°C, volumetric, Rex 1906)
867 (20°C, Fühner 1924; quoted, Merckel 1937)
1040 (30°C, shake flask-interferometer, Gross & Saylor 1931)
1073 (Seidell 1940; quoted, Deno & Berkheimer 1960)
872 (Seidell 1941; quoted, Hansch et al. 1968)
1110 (calculated-$K_{OW}$, Hansch et al. 1968)
1073 (quoted, Hine & Mookerjee 1975)
1000 (Dean 1985)
1040 (30°C, quoted, Riddick et al. 1986)
872, 759 (quoted, calculated-fragment const., Wakita et al. 1986)

Vapor Pressure (Pa at 25°C):

4679 (20°C, Rex 1906)
7305 (30°C, Rex 1906)
5744 (quoted, Hine & Mookerjee 1975)
5745 (quoted, Riddick et al. 1986)

Henry's Law Constant (Pa m$^3$/mol):

921 (calculated as $1/K_{AW}$, $C_W/C_A$, reported as exptl., Hine & Mookerjee 1975; quoted, Nirmalakhandan & Speece 1988)
1033 (calculated-group contribution method, Hine & Mookerjee 1975)
1082 (calculated-bond contribution method, Hine & Mookerjee 1975)
921, 1057 (quoted, calculated-QSAR, Nirmalakhandan & Speece 1988)

Octanol/Water Partition Coefficient, log $K_{OW}$:

2.50 (calculated-$\pi$, Hansch et al. 1968)

Bioconcentration Factor, log BCF:

Sorption Partition Coefficient, log $K_{OC}$:

Half-Lives in the Environment:

Air: disappearance half-life of 2.4-24 hours from air for the reaction with OH radicals (USEPA 1974; quoted, Darnall et al. 1976).
Surface water:
Ground water:
Sediment:
Soil:
Biota:

Environmental Fate Rate Constant and Half-Lives:

Volatilization:
Photolysis:
Oxidation:
Hydrolysis:
Biodegradation:
Biotransformation:
Bioconcentration, Uptake ($k_1$) and Elimination ($k_2$) Rate Constants or Half-Lives:

Common Name: 2-Iodopropane
Synonym: isopropyl iodide
Chemical Name: 2-iodopropane, isopropyl iodide
CAS Registry No: 75-30-9
Molecular Formula: $CH_3CHICH_3$
Molecular Weight: 169.99
Melting Point (°C):
-90.0 (Riddick et al. 1986)
Boiling Point (°C):
88.20 (Rex 1906)
89.50 (Riddick et al. 1986)
Density ($g/cm^3$ at 20°C):
1.7042 (Riddick et al. 1986)
1.6946 (25°C, Riddick et al. 1986)
Molar Volume ($cm^3/mol$):
107.3 (calculated-LeBas method)
99.70 (20°C, calculated from density)
100.31 (25°C, calculated from density)
Molecular Volume ($A^3$):
Total Surface Area, TSA ($A^2$):
Heat of Fusion, $\Delta H_{fus}$, (kcal/mol):
Entropy of Fusion, $\Delta S_{fus}$ (cal/mol K or e.u.):
Fugacity Ratio at 25°C, F: 1.0

Water Solubility ($g/m^3$ or mg/L at 25°C):
1400 (20°C, Rex 1906)
1340 (30°C, Rex 1906)
1382 (quoted, Hine & Mookerjee 1975)
1400 (20°C, quoted, Riddick et al. 1986)

Vapor Pressure (Pa at 25°C):
7518 (20°C, Rex 1906)
11784 (30°C, Rex 1906)
9199 (quoted, Hine & Mookerjee 1975)
5700 (quoted, Riddick et al. 1986)

Henry's Law Constant (Pa $m^3$/mol):
1133 (calculated-$1/K_{AW}$, $C_W/C_A$, reported as exptl., Hine & Mookerjee 1975; quoted, Nirmalakhandan & Speece 1988)
2479 (calculated-group contribution method, Hine & Mookerjee 1975)

1082 (calculated-bond contribution method, Hine & Mookerjee 1975)
1133, 1837.6 (quoted, calculated-QSAR, Nirmalakhandan & Speece 1988)

Octanol/Water Partition Coefficient, log $K_{OW}$:

Bioconcentration Factor, log BCF:

Sorption Partition Coefficient, log $K_{OC}$:

Half-Lives in the Environment:

Air: disappearance half-life of 2.4-24 hours from air for the reaction with OH radicals (EPA 1974; quoted, Darnall et al. 1976).
Surface water:
Ground water:
Sediment:
Soil:
Biota:

Environmental Fate Rate Constant and Half-Lives:

Volatilization:
Photolysis:
Oxidation:
Hydrolysis:
Biodegradation:
Biotransformation:
Bioconcentration, Uptake ($k_1$) and Elimination ($k_2$) Rate Constants or Half-Lives:

Common Name: 1-Iodobutane
Synonym: *n*-butyl iodide, monoiodobutane
Chemical Name: *n*-butyl iodide, 1-iodobutane, monoiodobutane
CAS Registry No: 542-69-8
Molecular Formula: $CH_3CH_2CH_2CH_2I$
Molecular Weight: 184.02
Melting Point (°C):
-103.5 (Dean 1985)
-103.0 (Riddick et al. 1986)
Boiling Point (°C):
129.50 (Dean 1985)
130.53 (Riddick et al. 1986)
Density (g/cm$^3$ at 20°C):
1.6160 (Dean 1985)
1.6154 (Riddick et al. 1986)
1.6072 (25°C, Riddick et al. 1986)
Molar Volume (cm$^3$/mol):
113.87 (20°C, calculated from density, Dean 1985)
113.92 (20°C, calculated from density, Riddick et al. 1986)
114.50 (25°C, calculated from density, Riddick et al. 1986)
129.50 (calculated-LeBas method)
Molecular Volume (A$^3$):
Total Surface Area, TSA (A$^2$):
141.1 (Okouchi et al. 1992)
Heat of Fusion, $\Delta H_{fus}$, (kcal/mol):
Entropy of Fusion, $\Delta S_{fus}$ (cal/mol K or e.u.):
Fugacity Ratio at 25°C, F: 1.0

Water Solubility (g/m$^3$ or mg/L at 25°C):
202 (20°C, Merckel 1937; quoted, Horvath 1982)
211 (Seidell 1940; quoted, Deno & Berkheimer 1960)
202 (Seidell 1941; quoted, Hansch et al. 1968)
313 (Kakovsky 1957; quoted, Horvath 1982)
280 (18°C, calculated-group contribution, Irmann 1965; quoted, Horvath 1982)
202, 297 (quoted, calculated-$K_{OW}$, Hansch et al. 1968)
211, 320 (20°C, exptl., calculated-group contribution, Korenman et al. 1971; quoted, Horvath 1982)
211 (quoted, Hine & Mookerjee 1975)
182 (selected exptl., Horvath 1982)
202, 202 (quoted, calculated-fragment const., Wakita et al. 1986)
202 (calculated-AI, Okouchi et al. 1992)

Vapor Pressure (Pa at 25°C):

1848 (quoted, Hine & Mookerjee 1975)

1848 (quoted, Riddick et al. 1986)

Henry's Law Constant (Pa $m^3$/mol):

1600 (calculated-1/$K_{AW}$, $C_W/C_A$, reported as exptl., Hine & Mookerjee 1975; quoted, Nirmalakhandan & Speece 1988)

1426 (calculated-group contribution, Hine & Mookerjee 1975)

1600 (calculated-bond contribution, Hine & Mookerjee 1975)

1600, 1331 (quoted, calculated-QSAR, Nirmalakhandan & Speece 1988)

Octanol/Water Partition Coefficient, log $K_{OW}$:

3.00 (calculated-$\pi$, Hansch et al. 1968)

Bioconcentration Factor, log BCF:

Sorption Partition Coefficient, log $K_{OC}$:

Half-Lives in the Environment:

Air: disappearance half-life of 2.4-24 hours from air for the reaction with OH radicals (USEPA 1974; quoted, Darnall et al. 1976).

Surface water:

Ground water:

Sediment:

Soil:

Biota:

Environmental Fate Rate Constant and Half-Lives:

Volatilization:

Photolysis:

Oxidation:

Hydrolysis:

Biodegradation:

Biotransformation:

Bioconcentration, Uptake ($k_1$) and Elimination ($k_2$) Rate Constants or Half-Lives:

Common Name: Bromochloromethane
Synonym: chlorobromomethane, methylene bromochloride
Chemical Name: bromochloromethane, chlorobromomethane
CAS Registry No: 74-97-5
Molecular Formula: $CH_2BrCl$
Molecular Weight: 129.39
Melting Point (°C):
-86.5 (Weast 1982-83)
-88.0 (Dean 1985; Stephenson & Malanowski 1987)
-88.15 (Miller et al. 1985)
-87.95 (Riddick et al. 1986)
Boiling Point (°C):
68.11 (Dreisbach 1959)
68.06 (Kudchadker et al. 1979; Riddick et al. 1986)
68.1 (Weast 1982-83; Stephenson & Malanowski 1987)
67.8 (Dean 1985)
Density ($g/cm^3$ at 20°C):
1.9344 (Dreisbach 1959; Weast 1982-83; Riddick et al. 1986)
1.9229 (25°C, Dreisbach 1959)
Molar Volume ($cm^3/mol$):
73.80 (calculated-LeBas method)
66.89 (calculated from density)
87.25 (Miller et al. 1985)
Molecular Volume ($A^3$):
Total Surface Area, TSA ($A^2$):
87.0 (Okouchi et al. 1992)
Heat of Fusion, $\Delta H_{fus}$, (kcal/mol):
Entropy of Fusion, $\Delta S_{fus}$ (cal/mol K or e.u.):
Fugacity Ratio at 25°C, F: 1.0

Water Solubility ($g/m^3$ or mg/L at 25°C):
15000 (O'Connell 1963; quoted, Horvath 1982)
9000 (quoted, Irmann 1965)
15000 (Jolls 1966; quoted, Horvath 1982)
9000, 10700 (quoted exptl., calculated-group contribution as per Irmann 1965, Horvath 1982)
14778 (recommended, Horvath 1982)
16691 (gen. col.-GC, Tewari et al. 1982; quoted, Miller et al. 1985)
900 (Dean 1985)
16693 (quoted, Miller et al. 1985)
16707, 38896 (quoted, calculated-UNIFAC, Arbuckle 1986)
17000 (quoted, Riddick et al. 1986)

8952, 7978 (quoted, calculated-fragment const., Wakita et al. 1986)
14790, 9977 (quoted, predicted- $\chi$ & polarizability, Nirmalakhandan & Speece 1988)
14788 (calculated-AI, Okouchi et al. 1992)

Vapor Pressure (Pa at 25°C):
19622 (calculated-Antoine eqn., Dreisbach 1959; quoted, Riddick et al. 1986)
19518 (calculated-Antoine eqn., Boublik et al. 1973,84)
19524 (calculated-Antoine eqn., Kudchadker et al. 1979)
18808, 18690 (24°C, quoted exptl., calculated-Antoine eqn., Boublik et al. 1984)
19533.3 (calculated-Antoine eqn., Stephenson & Malanowski 1987)

Henry's Law Constant (Pa $m^3$/mol):

Octanol/Water Partition Coefficient, log $K_{OW}$:
1.41 (gen. col.-GC, Tewari et al. 1982)
1.41 (quoted, Miller et al. 1985)

Bioconcentration Factor, log BCF:

Sorption Partition Coefficient, log $K_{OC}$:

Half-Lives in the Environment:
Air: disappearance half-life of 2.4-24 hours from air for the reaction with OH radicals (EPA 1974; quoted, Darnall et al. 1976).
Surface water:
Ground water:
Sediment:
Soil:
Biota:

Environmental Fate Rate Constant and Half-Lives:
Volatilization:
Photolysis:
Oxidation: rate constant for the reaction with $OH^-$: $(2.5 \pm 0.3) \times 10^{-6}$ $M^{-1}$ $sec^{-1}$ was measured in 66.7% dioxane-water at 35.7°C (Hine et al. 1956; quoted, Roberts et al. 1992).

Hydrolysis:
Biodegradation:
Biotransformation:
Bioconcentration, Uptake ($k_1$) and Elimination ($k_2$) Rate Constants or Half-Lives:

Common Name: Bromodichloromethane
Synonym: dichlorobromomethane
Chemical Name: bromodichloromethane, dichlorobromomethane
CAS Registry No: 75-27-4
Molecular Formula: $CHBrCl_2$
Molecular Weight: 163.83
Melting Point (°C):
-57.1 (Weast 1977; Callahan et al. 1979; Mabey et al. 1982; Howard 1990; Mackay & Shiu 1990)
-55.0 (Verschueren 1983)
Boiling Point (°C):
90.0 (Weast 1977; Callahan et al. 1979; Mabey et al. 1982; Verschueren 1983; McNally & Grob 1984; Howard 1990; Mackay & Shiu 1990)
Density ($g/cm^3$ at 20°C):
1.971 (25°C, Verschueren 1983)
1.980 (McNally & Grob 1984)
1.970 (Gillham & Rao 1990)
Molar Volume ($cm^3/mol$):
94.7 (calculated-LeBas method)
82.95 (calculated from density)
Molecular Volume ($A^3$):
Total Surface Area, TSA ($A^2$):
Heat of Fusion, $\Delta H_{fus}$, (kcal/mol):
Entropy of Fusion, $\Delta S_{fus}$ (cal/mol K or e.u.):
Fugacity Ratio at 25°C, F: 1.0

Water Solubility ($g/m^3$ or mg/L at 25°C):
4700 (22°C, Mabey et al. 1981; quoted, Howard 1990)
4500 (calculated-$K_{OW}$, Mabey et al. 1982; quoted, Gillham & Rao 1990; Mackay & Shiu 1990)
2968 (30°C, headspace-GC, McNally & Grob 1984)

Vapor Pressure (Pa at 25°C):
6665 (20°C, Dreisbach 1952; quoted, Callahan et al. 1979; Mabey et al. 1982; Mills et al. 1982; Gillham & Rao 1990; Howard 1990; Mackay & Shiu 1990)
8555 (20°C, estimated, Tse et al. 1992)
13444 (30°C, estimated, Tse et al. 1992)

Henry's Law Constant (Pa $m^3$/mol):

212.8 (ESE 1980; quoted, Nicholson et al. 1984)
233.0 (20°C, Symons et al. 1981; quoted, Nicholson et al. 1984)
244.2 (20°C, calculated-P/C, Mabey et al. 1982)
162 (20°C, batch stripping, Nicholson et al. 1984; quoted, Howard 1990; Mackay & Shiu 1990)
156 (Munz & Roberts 1987)
214.8 (quoted, Warner et al. 1987)
243.1 (20-25°C and low ionic strength, quoted, Pankow & Rosen 1988; Pankow 1990)
243 (calculated-P/C, Mackay & Shiu 1990)
162.1 (20°C, Tse et al. 1992)
263.4 (30°C, Tse et al. 1992)

Octanol/Water Partition Coefficient, log $K_{OW}$:

1.88 (calculated as per Tute 1971, Callahan et al. 1979)
2.10 (Hansch & Leo 1979; quoted, Mackay & Shiu 1990)
2.10 (Mabey et al. 1981; quoted, Howard 1990)
2.10 (calculated, Mabey et al. 1982)
1.88 (selected, Mills et al. 1982)

Bioconcentration Factor, log BCF:

1.544 (microorganisms-water, calculated-$K_{OW}$, Mabey et al. 1982)
0.72-1.37 (estimated as per Lyman et al. 1982, Howard 1990)

Sorption Partition Coefficient, log $K_{OC}$:

1.785 (sediment-water, calculated-$K_{OW}$, Mabey et al. 1982)
1.724-2.40 (estimated from S & $K_{OW}$, Swann et al. 1983; quoted, Howard 1990)

Half-Lives in the Environment:

Air: disappearance half-life of 2.4-24 hours from air for the reaction with OH radicals (EPA 1974; quoted, Darnall et al. 1976); 3.92 months in the atmosphere, based on an estimated rate constant of $8.522 \times 10^{-14}$ $cm^3$ $molecule^{-1}$ $sec^{-1}$ for the vapor phase reaction with hydroxyl radicals (GEMS 1987; quoted, Howard 1990).
Surface water:
Ground water:
Sediment:
Soil:
Biota:

Environmental Fate Rate Constant and Half-Lives:

Volatilization: typical half-life about 35 hours for a range of 33 minutes to 12 days was estimated from experimentally determined gas transfer rates (Kaczmar et al. 1984; quoted, Howard 1990).

Photolysis:

Oxidation: calculated rate constants at 25°C: $<<360$ $M^{-1}$ $hour^{-1}$ for singlet oxygen and 0.2 $M^{-1}$ $hour^{-1}$ for peroxy radicals (Mabey et al. 1982).

Hydrolysis: first-order rate constant of $1.6 \times 10^{-10}$ $sec^{-1}$ at pH 7 and 25°C with a maximum half-life of 137 years (Mabey & Mill 1978; quoted, Mabey et al. 1982; Howard 1990).

Biodegradation:

Biotransformation: estimated rate constant of $1 \times 10^{-10}$ ml $cell^{-1}$ $hour^{-1}$ for bacteria (Mabey et al. 1982).

Bioconcentration, Uptake ($k_1$) and Elimination ($k_2$) Rate Constants or Half-Lives:

Common Name: Dibromochloromethane
Synonym: chlorodibromomethane
Chemical Name: dibromochloromethane, chlorodibromomethane
CAS Registry No: 124-48-1
Molecular Formula: $CHBr_2Cl$
Molecular Weight: 208.29
Melting Point (°C):
    < -20.0 (Verschueren 1977; Callahan et al. 1979; Mabey et al. 1982)
    -22.0 (Dean 1985)
    -20.0 (Mackay & Shiu 1990)
Boiling Point (°C):
    119-120 (Weast 1977; Callahan et al. 1979; Mabey et al. 1982; Weast 1982-83; Mackay & Shiu 1990)
    120 (Dean 1985)
Density ($g/cm^3$ at 20°C):
    2.451 (Weast 1982-83; Dean 1985)
Molar Volume ($cm^3/mol$):
    97.10 (calculated-LeBas method)
    84.98 (calculated from density)
Molecular Volume ($A^3$):
Total Surface Area, TSA ($A^2$):
Heat of Fusion, $\Delta H_{fus}$, (kcal/mol):
Entropy of Fusion, $\Delta S_{fus}$ (cal/mol K or e.u.):
Fugacity Ratio at 25°C, F: 1.0

Water Solubility ($g/m^3$ or mg/L at 25°C):
    4000 (calculated-$K_{OW}$, Mabey et al. 1982; quoted, Mackay & Shiu 1990)

Vapor Pressure (Pa at 25°C):
    1014 (Weast 1972-73; quoted, Mackay & Shiu 1990)
    10131 (20°C, extrapolated-Antoine eqn., Mabey et al. 1982)
    2000 (20°C, selected, Mills et al. 1982)
    6667 (20°C, quoted, Warner et al. 1987)

Henry's Law Constant (Pa $m^3/mol$):
    88.1 (20°C, ESE 1980; quoted, Nicholson et al. 1984)
    85.1 (20°C, Symsons et al. 1981; quoted, Nicholson et al. 1984)
    100.3 (20°C, calculated-P/C, Mabey et al. 1982)
    88.2 (20°C, batch stripping, Nicholson et al. 1984; quoted, Mackay & Shiu 1990)
    86.03 (Munz & Roberts 1987)

80.35 (quoted, Warner et al. 1987)
100.3 (20-25°C and low ionic strength, quoted, Pankow & Rosen 1988; Pankow 1990)
5276 (calculated-P/C, Mackay & Shiu 1990)

Octanol/Water Partition Coefficient, log $K_{OW}$:
2.09 (calculated as per Tute 1971, Callahan et al. 1979)
2.24 (Hansch & Leo 1979; quoted, Mackay & Shiu 1990)
2.24 (calculated, Mabey et al. 1982)
2.08 (selected, Mills et al. 1982)

Bioconcentration Factor, log BCF:

Sorption Partition Coefficient, log $K_{OC}$:

Half-Lives in the Environment:
Air: disappearance half-life of 2.4-24 hours from air for the reaction with OH radicals (EPA 1974; quoted, Darnall et al. 1976); 1025-10252 hours, based on estimated rate constant for the vapor phase reaction with hydroxyl radicals in air (Atkinson 1987; quoted, Howard et al. 1991).
Surface water: 672-4320 hours, based on aerobic screening test data (Tabak et al. 1981; quoted, Howard et al. 1991).
Ground water: 336-4320 hours, based on estimated aqueous aerobic and anaerobic biodegradation half-life (Howard et al. 1991).
Sediment:
Soil: 672-4320 hours, based on aerobic screening test data (Tabak et al. 1981; quoted, Howard et al. 1991).
Biota:

Environmental Fate Rate Constant and Half-Lives:
Volatilization: calculated half-lives of 43 minute to 16.6 days (using Langbein & Durum 1967 published $O_2$ reaeration values) and 45.9 hours both from rivers and streams (Kaczmar et al. 1984).
Photolysis:
Oxidation: calculated rate constants at 25°C: $<<360\ M^{-1}\ hour^{-1}$ for singlet oxygen and $0.5\ M^{-1}\ hour^{-1}$ for peroxy radicals (Mabey et al. 1982); photooxidation half-life of 1025-10252 hours, based on estimated rate constant for the vapor phase reaction with hydroxyl radicals in air (Atkinson 1987; quoted, Howard et al. 1991).

Hydrolysis: a maximum hydrolytic half-life of 274 years has been reported at pH 7 and 25°C with a corresponding first-order rate constant of $8.0 \times 10^{-11}$ $sec^{-1}$ (Mabey & Mill 1978; quoted, Callahan et al. 1979; Howard et al. 1991); $2.88 \times 10^{-8}$ $hour^{-1}$ at pH 7 and 25°C (analogy to chloroform as per Mabey & Mill 1978, Mabey et al. 1982).

Biodegradation: aqueous aerobic half-life of 672-4320 hours, based on aerobic screening test data (Tabak et al. 1981; quoted, Howard et al. 1991); aqueous anaerobic half-life of 672-4320 hours, based on unacclimated anaerobic screening test data (Bouwer & McCarty 1983; Bouwer et al. 1981; quoted, Howard et al. 1991).

Biotransformation: estimated rate constant of $1.0 \times 10^{-10}$ ml $cell^{-1}$ $hour^{-1}$ for bacteria (Mabey et al. 1982).

Bioconcentration, Uptake ($k_1$) and Elimination ($k_2$) Rate Constants or Half-Lives:

Common Name: Chlorodifluoromethane
Synonym: difluorochloromethane, Freon 22
Chemical Name: chlorodifluoromethane
CAS Registry No: 75-45-6
Molecular Formula: $ClCHF_2$
Molecular Weight: 86.47
Melting Point (°C):

-146.5 (Horvath 1982)
-146 (Weast 1982-83)
-160 (Dean 1985; Stephenson & Malanowski 1987)
-157.42 (Riddick et al. 1986)

Boiling Point (°C):

-40.8 (Horvath 1982; Weast 1982-83; Dean 1985)
-40.83 (Riddick et al. 1986)
-40.75 (Stephenson & Malanowski 1987)

Density ($g/cm^3$ at 20°C):

1.209 (21°C, Dean 1985)
1.194 (25°C, Kirk-Othmer 1985)
1.2136 (Riddick et al. 1986)

Molar Volume ($cm^3/mol$):

60.5 (calculated-LeBas method)
72.06 (calculated from density)

Molecular Volume ($A^3$):
Total Surface Area, TSA ($A^2$):
Heat of Fusion, $\Delta H_{fus}$, (kcal/mol):

0.9855 (Riddick et al. 1986)

Entropy of Fusion, $\Delta S_{fus}$ (cal/mol K or e.u.):
Fugacity Ratio at 25°C, F: 1.0

Water Solubility ($g/m^3$ or mg/L at 25°C):

2800, 2900 (quoted values, Irmann 1965)
3000 (Du Pont 1966,1969; quoted, Horvath 1982; Riddick et al. 1986)
2930 (quoted, Hine & Mookerjee 1975)
3000 (Weast 1976; quoted, Horvath 1982)
2899 (recommended, Horvath 1982)
3000 (Dean 1985)

Vapor Pressure (Pa at 25°C):

1083073 (calculated-Antoine eqn., Weast 1975)
1027940 (calculated-Antoine eqn., Boublik et al. 1984)
1043921 (calculated-Antoine eqn., Stephenson & Malanowski 1987)

Henry's Law Constant (Pa $m^3$/mol):

2980 (calculated as $1/K_{AW}$, $C_W/C_A$, reported as exptl., Hine & Mookerjee 1975; quoted, Nirmalakhandan & Speece 1988)

2980, 2478.8 (quoted, calculated-QSAR, Nirmalakhandan & Speece 1988)

Octanol/Water Partition Coefficient, log $K_{OW}$:

Bioconcentration Factor, log BCF:

Sorption Partition Coefficient, log $K_{OC}$:

Half-Lives in the Environment:

Air: disappearance half-life of 2.4-24 hours from air for the reaction with OH radicals (EPA 1974; quoted, Darnall et al. 1976); estimated tropospheric lifetime of 4.3 years as global average for the reaction with hydroxyl radicals in air (Bunce et al. 1991); estimated tropospheric lifetime of 7.5 years and 15.8 years by rigorous calculation (Nimitz & Skaggs 1992).

Surface water:

Ground water:

Sediment:

Soil:

Biota:

Environmental Fate Rate Constant and Half-Lives:

Volatilization:

Photolysis:

Oxidation: calculated rate constant of $4.7 \times 10^{-15}$ $cm^3$ $molecule^{-1}$ $sec^{-1}$ for the reaction with hydroxyl radicals at 298 K with lifetime of 41 years in troposphere (Atkinson et al. 1975; quoted, Altshuller 1980); calculated rate constant of $4.6 \times 10^{-15}$ $cm^3$ $molecule^{-1}$ $sec^{-1}$ for the reaction with hydroxyl radicals at 298 K with lifetime of 44 years in troposphere (Chang & Kaufman 1977; quoted, Altshuller 1980); calculated rate constant of $4.7 \times 10^{-15}$ $cm^3$ $molecule^{-1}$ $sec^{-1}$ for the reaction with hydroxyl radicals at 298 K with lifetime of 43 years in troposphere (Watson et al. 1977; quoted, Altshuller 1980); $4.7 \times 10^{-15}$ $cm^3$ $molecule^{-1}$ $sec^{-1}$ for the tropospheric reaction with hydroxyl radicals (Atkinson 1985; quoted, Bunce et al. 1991) with an estimated tropospheric lifetime of 4.3 years for a global average concentration of OH radicals (Bunce et al. 1991).

Hydrolysis:
Biodegradation:
Biotransformation:
Bioconcentration, Uptake ($k_1$) and Elimination ($k_2$) Rate Constants or Half-Lives:

Common Name: Dichlorodifluoromethane
Synonym: difluorodichloromethane, Freon 12
Chemical Name: dichlorodifluoromethane
CAS Registry No: 75-71-8
Molecular Formula: $CCl_2F_2$
Molecular Weight: 120.91
Melting Point (°C):
-158 (Horvath 1982; Schmidt-Bleek et al. 1982; Weast 1982-83; Verschueren 1983; Dean 1985)
-158.2 (Riddick et al. 1986)
-155.0 (Stephenson & Malanowski 1987)
Boiling Point (°C):
-29.8 (Kudchadker et al. 1979; Horvath 1982; Weast 1982-83; Verschueren 1983; Dean 1985; Stephenson & Malanowski 1987)
-30 (Schmidt-Bleek et al. 1982)
-29.77 (Riddick et al. 1986)
Density ($g/cm^3$ at 20°C):
1.329 (Timmermans 1950; Schmidt-Bleek et al. 1982; Verschueren 1983)
1.311 (25°C, Kirk-Othmer 1985)
1.3292 (Riddick et al. 1986)
Molar Volume ($cm^3/mol$):
81.4 (calculated-LeBas method)
90.98 (calculated from density)
Molecular Volume ($A^3$):
Total Surface Area, TSA ($A^2$):
Heat of Fusion, $\Delta H_{fus}$, (kcal/mol):
0.99 (Riddick et al. 1986)
Entropy of Fusion, $\Delta S_{fus}$ (cal/mol K or e.u.):
Fugacity Ratio at 25°C, F: 1.0

Water Solubility ($g/m^3$ or mg/L at 25°C):
280 (Du Pont 1966; quoted, Horvath 1982; Riddick et al. 1986)
276 (Gmelin 1974; quoted, Horvath 1982)
280 (shake flask-GC, Pearson & McConnell 1975; quoted, Shiu et al. 1990)
271 (quoted, Hine & Mookerjee 1975)
300 (recommended, Horvath 1982)
280 (selected, Mills et al. 1982)
432 (20°C, quoted, Schmidt-Bleek et al. 1982)
280 (quoted, Verschueren 1983; Dean 1985)
280 (quoted, Kirk-Othmer 1985)

Vapor Pressure (Pa at 25°C):

745.4 (20°C, static method, Gilkey et al. 1931; quoted, Timmermans 1950; Kudchadker et al. 1979)
1031.3 (30°C, static method, Gilkey et al. 1931; quoted, Timmermans 1950; Kudchadker et al. 1979)
214217 (calculated-Antoine eqn., Weast 1975)
573990 (20°C, selected, Mills et al. 1982)
214000 (20°C, quoted, Schmidt-Bleek et al. 1982)
566525 (20°C, quoted, Verschueren 1983)
769941 (30°C, quoted, Verschueren 1983)
566600 (20°C, quoted, Riddick et al. 1986)
651041 (calculated-Antoine eqn., Stephenson & Malanowski 1987)

Henry's Law Constant (Pa $m^3$/mol):

151962 (calculated-P/C, Mackay & Wolkoff 1973; quoted, Roberts & Dändliker 1983)
43077 (calculated-1/$K_{AW}$, $C_W/C_A$, reported as exptl., Hine & Mookerjee 1975; quoted, Nirmalakhandan & Speece 1988)
22798 (estimated, Roberts 1984)
43077, 21098 (quoted, calculated-QSAR, Nirmalakhandan & Speece 1988)
304000 (20-25°C and low ionic strength, quoted, Pankow & Rosen 1988; Pankow 1990)
270000 (selected, Jury et al. 1990)

Octanol/Water Partition Coefficient, log $K_{OW}$:

2.16 (shake flask-GC, Hansch et al. 1975)
2.16, 2.21 (quoted, calculated-f const., Rekker & Kort 1979)
2.16 (selected, Mills et al. 1982)

Bioconcentration Factor, log BCF:

Sorption Partition Coefficient, log $K_{OC}$:

2.05 (soil, selected, Jury et al. 1990)

Half-Lives in the Environment:

Air: disappearance half-life of more than 10 days from air for the reaction with OH radicals (EPA 1974; quoted, Darnall et al. 1976); about 50 years residence in N. troposphere, estimated by one-compartment nonsteady state model (Singh et al. 1978) and 60-70 years by two-compartment nonsteady state model (Singh et al. 1979); estimated residence time in troposphere about one year (Lyman 1982);

2118-21180 hours, based on measured rate data for the vapor phase reaction with hydroxyl radicals in air (Atkinson 1985; quoted, Howard et al. 1991); estimated global average tropospheric lifetime of >50 years for the reaction with hydroxyl radicals in air (Bunce et al. 1991).

Surface water: 672-4320 hours, based on estimated aqueous aerobic biodegradation half-life (Howard et al. 1991).

Ground water: 1344-8640 hours, based on estimated aqueous aerobic biodegradation half-life (Howard et al. 1991).

Sediment:

Soil: estimated half-life of volatilization loss from soil to be 10000 days (Jury et al. 1990); 672-4320 hours, based on estimated aqueous aerobic biodegradation half-life (Howard et al. 1991).

Biota:

Environmental Fate Rate Constant and Half-Lives:

Volatilization: half-life of about few minutes (Mills et al. 1982); loss half-lives in marine mesocosm were estimated to be 20 days in spring at 8-16°C and 13 days in winter at 3-7°C (Wakeham et al. 1983).

Photolysis:

Oxidation: photooxidation half-life of 2118-21180 hours in air, based on measured rate data for the vapor phase reaction with hydroxyl radicals in air (Atkinson 1985; quoted, Howard et al. 1991); rate constant of $< 4 \times 10^{-16}$ $cm^3$ $molecule^{-1}$ $sec^{-1}$ for the reaction with hydroxyl radicals in air (Atkinson 1985; quoted, Bunce et al. 1991) with estimated tropospheric lifetime of > 50 years for a global average concentration of OH radicals (Bunce et al. 1991).

Hydrolysis:

Biodegradation: aqueous aerobic half-life of 672-4320 hours, based on acclimated aerobic screening test data for trichlorofluoromethane (Tabak et al. 1981; quoted, Howard et al. 1991); aqueous anaerobic half-life of 2688-16128 hours, based on estimated aqueous aerobic biodegradation half-life (Howard et al. 1991).

Biotransformation:

Bioconcentration, Uptake ($k_1$) and Elimination ($k_2$) Rate Constants or Half-Lives:

Common Name: Trichlorofluoromethane
Synonym: Arcton 11, fluorotrichloromethane, fluorocarbon-11, Freon-11, Frigen 11
Chemical Name: trichlorofluoromethane, fluorotrichloromethane
CAS Registry No: 75-69-4
Molecular Formula: $CCl_3F$
Molecular Weight: 137.38
Melting Point (°C):
-111.0 (Windholz 1976; Verschueren 1977,1983; Callahan et al. 1979; Horvath 1982; Mabey et al. 1982; Dean 1985; Neely & Blau 1985; Stephenson & Malanowski 1987; Budavari 1989; Howard 1990)
-110.48 (Riddick et al. 1986)
Boiling Point (°C):
23.70 (Windholz 1976; Neely & Blau 1985; Budavari 1989; Howard 1990)
23.8 (Verschueren 1977,1983; Callahan et al. 1979; Mackay & Shiu 1981; Mabey et al. 1982; Dean 1985)
23.82 (Horvath 1982; Stephenson & Malanowski 1987)
23.63 (Riddick et al. 1986)
Density ($g/cm^3$ at 20°C):
1.467 (25°C, Horvath 1982)
1.485 (21°C, Dean 1985)
1.476 (25°C, Kirk-Othmer 1985)
1.494 (17.2°C, Budavari 1989)
Molar Volume ($cm^3/mol$):
97.30 (calculated-LeBas method)
93.65 (calculated from density)
0.820 ($V_M/100$, Leahy 1986)
0.455 (calculated intrinsic volume $V_I/100$, Leahy 1986; Kamlet et al. 1988)
Molecular Volume ($A^3$):
Total Surface Area, TSA ($A^2$):
Heat of Fusion, $\Delta H_{fus}$, (kcal/mol):
1.648 (Riddick et al. 1986)
Entropy of Fusion, $\Delta S_{fus}$ (cal/mol K or e.u.):
Fugacity Ratio at 25°C, F: 1.0

Water Solubility ($g/m^3$ or mg/L at 25°C):
1400 (21°C, quoted, Irmann 1965; Chiou & Freed 1977; quoted, Mackay 1985; Neely & Blau 1985)
1100 (20°C, Du Pont 1966; quoted, Riddick et al. 1986)
1100 (20°C, Pearson & McConnell 1975; quoted, Callahan et al. 1979; Mackay & Shiu 1981; Mabey et al. 1982; Mills et al. 1982)
1100 (Verschueren 1977,1983; quoted, Callahan et al. 1979; Warner et al. 1987)
1080 (recommended, Horvath 1982)

1080 (30°C, Horvath 1982; quoted, Howard 1990)
1100 (quoted, Klöpffer et al. 1982; Kirk-Othmer 1985)
1400 (Dean 1985)
1240 (quoted from Neely & Blau 1985, Karickhoff 1985)
1730 (calculated-molar volume & solvatochromic p., Leahy 1986)

Vapor Pressure (Pa at 25°C):

95668 (calculated-Antoine eqn., Weast 1972-73)
88500 (20°C, Pearson & McConnell 1975; quoted, Callahan et al. 1979; Mackay & Shiu 1981; Mabey et al. 1982)
106391 (Antoine eqn., Reid et al. 1977; quoted, Lyman 1985; Neely & Blau 1985)
95000 (quoted, Klöpffer et al. 1982)
88911 (20°C, selected, Mills et al. 1982)
91582 (20°C, quoted, Verschueren 1983)
107013 (Daubert & Danner 1985; quoted, Howard 1990)
106000 (quoted, Mackay 1985)
102200 (quoted, Riddick et al. 1986)
106343 (calculated-Antoine eqn., Stephenson & Malanowski 1987)
84404 (interpolated between two data points, Warner et al. 1987)

Henry's Law Constant (Pa $m^3$/mol):

81200 (20°C, Pearson & McConnell 1975; quoted, Mackay & Shiu 1981)
12394, 11650 (quoted, calculated-P/C, Neely 1976)
11050 (20°C, calculated-P/C, Mackay & Shiu 1981)
11044 (20°C, calculated-P/C, Mabey et al. 1982)
11754 (estimated-P/C, Lyman 1985)
982.9 (estimated as per Hine & Mookerjee 1975, Lyman 1985)
61.81 (estimated as per Cramer 1980, Lyman 1985)
11800 (calculated-P/C, Mackay 1985)
9827 (Warner & Weiss 1985; quoted, Howard 1990)
5907, 10538 (quoted, calculated-P/C, Warner et al. 1987)
11144 (20-25°C and low ionic strength, quoted, Pankow & Rosen 1988; Pankow 1990)

Octanol/Water Partition Coefficient, log $K_{OW}$:

2.53 (shake flask-GC, Hansch et al. 1975; Hansch & Leo 1979; quoted, Callahan et al. 1979; Karickhoff 1985; Neely & Blau 1985)
2.53, 2.46 (quoted, calculated-f const., Rekker & Kort 1979)
2.52 (calculated, Mabey et al. 1982)
3.53 (selected, Mills et al. 1982)
2.53 (Hansch & Leo 1985; quoted, Howard 1990)

2.53, 2.40 (quoted, calculated-molar volume & solvatochromic p., Leahy 1986)
2.53, 2.29 (quoted, calculated-molar volume & solvatochromic p., Kamlet et al. 1988)

Bioconcentration Factor, log BCF:
1.92 (microorganisms-water, calculated-$K_{OW}$, Mabey et al. 1982)
1.95 (calculated, Klöpffer et al. 1982)

Sorption Partition Coefficient, log $K_{OC}$:
2.20 (sediment-water, calculated-$K_{OW}$, Mabey et al. 1982)
2.31, 2.42, 2.16, 2.21 (estimated-$K_{OW}$, Karickhoff 1985)
2.13, 2.26 (estimated-S, Karickhoff 1985)
2.20 (best estimate, Karickhoff 1985)
2.13 (average from estimates of $K_{OW}$ and S, Karickhoff 1985; quoted, Neely & Blau 1985)

Half-Lives in the Environment:
Air: disappearance half-life of larger than 10 days from air for the reaction with OH radicals (USEPA 1974; quoted, Darnall et al. 1976); estimated residence time about 1000 years in troposphere for the reaction with OH radicals (CEQ 1975); estimated residence time in N. troposphere about 15-20 years by one-compartment nonsteady state model and 40-45 years in troposphere by two-compartment nonsteady state model (Singh et al. 1979); estimated residence time in troposphere about 1-6 years (Lyman 1982); 14.7-147 years, based on an measured rate data for the vapor phase reaction with hydroxyl radicals in air (Atkinson 1985; quoted, Howard et al. 1991).
Surface water: 4032-8640 hours, based on acclimated aerobic screening test data (Tabak et al. 1981; quoted, Howard et al. 1991).
Ground water: 8640-17280 hours, based on estimated aqueous aerobic biodegradation half-life (Howard et al. 1991).
Sediment:
Soil: 4320-8640 hours, based on estimated aqueous aerobic biodegradation half-life (Howard et al. 1991).
Biota:

Environmental Fate Rate Constant and Half-Lives:
Volatilization: estimated half-life of 3.4 hours from Henry's law constant for a river 1.0 m deep with a 3 m/sec wind and 1 m/sec current (Lyman et al. 1982; Cadena et al. 1984; quoted, Howard 1990); half-life of about few minutes (Mills et al. 1982).

Photolysis:

Oxidation: photooxidation half-life of 14.7-147 years, based on measured rate data for the vapor phase reaction with hydroxyl radicals in air (Atkinson 1985; quoted, Howard et al. 1991).

Hydrolysis: estimated exptl. rate constant of $9.5 \times 10^{-7}$ $day^{-1}$ , based on carbon tetrachloride (Neely 1985).

Biodegradation: aqueous aerobic half-life of 4032-8640 hours, based on acclimated aerobic screening test data (Tabak et al. 1981; quoted, Howard et al. 1991); aqueous anaerobic half-life of 16128-34560 hours, based on estimated aqueous aerobic biodegradation half-life (Howard et al. 1991).

Biotransformation:

Bioconcentration, Uptake ($k_1$) and Elimination ($k_2$) Rate Constants or Half-Lives:

Common Name: 1,1,2-Trichloro-1,2,2-trifluoroethane
Synonym: Freon 113, 1,2,2-trifluoro-1,1,2-trichloroethane
Chemical Name: 1,1,2-trichloro-1,2,2-trifluoroethane, Arklone
CAS Registry No: 76-13-1
Molecular Formula: $CClF_2CCl_2F$
Molecular Weight: 187.38
Melting Point (°C):
-36.4 (Dean 1985; Riddick et al. 1986; Howard 1990)
Boiling Point (°C):
48.0 (Jones et al. 1977/1978)
47.6 (Dean 1985)
47.63 (Riddick et al. 1986)
47.7 (Howard 1990)
Density ($g/cm^3$ at 20°C):
1.5635 (25°C, Dean 1985; Riddick et al. 1986)
Molar Volume ($cm^3/mol$):
129.5 (calculated-LeBas method)
119.85 (calculated from density)
Molecular Volume ($A^3$):
Total Surface Area, TSA ($A^2$):
Heat of Fusion, $\Delta H_{fus}$, (kcal/mol):
Entropy of Fusion, $\Delta S_{fus}$ (cal/mol K or e.u.):
Fugacity Ratio at 25°C, F: 1.0

Water Solubility ($g/m^3$ or mg/L at 25°C):
170 (Du Pont 1966; quoted, Riddick et al. 1986; quoted, Howard 1990)
170 (20°C, Jones et al. 1977/1978)
166.4 (recommended, Horvath 1982)
170 (Dean 1985)

Vapor Pressure (Pa at 25°C):
39036 (calculated-Antoine eqn., Weast 1972-73)
48321 (calculated-Antoine eqn., Boublik et al. 1984; quoted, Howard 1990)
48480 (quoted, Riddick et al. 1986)

Henry's Law Constant (Pa $m^3/mol$):
53288 (calculated-P/C, Howard 1990)

Octanol/Water Partition Coefficient, log $K_{OW}$:

3.16 (estimated-HPLC-k', McDuffie 1981; quoted, Howard 1990)

Bioconcentration Factor, log BCF:

1.531 (calculated-S, Lyman et al. 1982; quoted, Howard 1990)

1.041 (calculated-$K_{OW}$, Lyman et al. 1982; quoted, Howard 1990)

Sorption Partition Coefficient, log $K_{OC}$:

2.28 (soil, calculated-$K_{OW}$, Lyman et al. 1982; quoted, Howard 1990)

2.41 (soil, calculated-S, Lyman et al. 1982; quoted, Howard 1990)

Half-Lives in the Environment:

Air: disappearance half-life of larger than 10 days from air for the reaction with OH radicals (EPA 1974; quoted, Darnall et al. 1976); 40-1000 years in troposphere, based on measured rates with singlet oxygen (Davidson et al. 1978; Pitts Jr. et al. 1974; quoted, Howard et al. 1991); half-life of 20 years in troposphere (Dilling 1982; quoted, Howard 1990).:

Surface water: 4320-8640 hours, based on estimated aqueous aerobic biodegradation half-life (Howard et al. 1991).

Ground water: 1440-17280 hours, based on estimated aqueous aerobic biodegradation half-life (Howard et al. 1991).

Sediment:

Soil: 4320-8640 hours, based on estimated aqueous aerobic biodegradation half-life (Howard et al. 1991).

Biota:

Environmental Fate Rate Constant and Half-Lives:

Volatilization: using Henry's law constant, half-life of 4 hours was estimated for a model river 1 m deep flowing 1 m/sec with wind velocity of 3 m/sec (Lyman et al. 1982; quoted, Howard 1990).

Photolysis: both aqueous and atmospheric photolysis half-lives are infinite (Howard et al. 1991).

Oxidation: photooxidation half-life of 40-1000 years in troposphere, based on measured rates with singlet oxygen (Davidson et al. 1978; Pitts Jr. et al. 1974; quoted, Howard et al. 1991).

Hydrolysis:

Biodegradation: aqueous aerobic half-life of 4320-8640 hours, based on a relative resistance of completely halogenated aliphatics to biodegrade (Howard et al. 1991); aqueous anaerobic half-life of 17280-34560 hours, based on estimated aqueous aerobic biodegradation half-life (Howard et al. 1991); degradation rates in microcosms at 20°C: $7.6 \times 10^{-3}$ day$^{-1}$ by control buffer with half-life of 90 days, $9.1 \times 10^{-3}$ day$^{-1}$ by redox buffer with half-life of 75 days, $1.2 \times 10^{-2}$ day$^{-1}$ by 1 mg/L redox-hematin with half-life of 57 days, $1.6 \times 10^{-2}$ day$^{-1}$ by 2 mg/L redox-hematin with half-life of 43 days, $9.9 \times 10^{-2}$ day$^{-1}$ by purged leachate with half-life of 7 days, and 0.14 day$^{-1}$ by leachate with half-life of 5 days (Lesage et al. 1992).

Biotransformation:

Bioconcentration, Uptake ($k_1$) and Elimination ($k_2$) Rate Constants or Half-Lives:

Common Name: 1,1,2,2-Tetrachloro-1,2-difluoroethane
Synonym: Freon-112, F-112
Chemical Name: 1,1,2,2-tetrachloro-1,2-difluoroethane
CAS Registry No: 76-12-0
Molecular Formula: $Cl_2FCCFCl_2$
Molecular Weight: 203.83
Melting Point (°C):
    26.55 (Riddick et al. 1986)
    25.0 (Howard 1990)
Boiling Point (°C):
    91.58 (Boublik et al. 1984)
    92.8 (Riddick et al. 1986)
    93.0 (Howard 1990)
Density (g/cm³ at 20°C):
    1.6447 (25° Riddick et al. 1986)
Molar Volume (cm³/mol):
    145.4 (calculated-LeBas method)
Molecular Volume (A³):
Total Surface Area, TSA (A²):
Heat of Fusion, $\Delta H_{fus}$, (kcal/mol):
Entropy of Fusion, $\Delta S_{fus}$ (cal/mol K or e.u.):
Fugacity Ratio at 25°C, F: 1.0

Water Solubility (g/m³ or mg/L at 25°C):
    120 (Du Pont 1966; quoted, Riddick et al. 1986)

Vapor Pressure (Pa at 25°C):
    6106 (20°C, quoted, Boublik et al. 1984)
    7609 (calculated-Antoine eqn., Boublik et al. 1984)
    8770 (quoted, Riddick et al. 1986; quoted, Howard 1990)

Henry's Law Constant (Pa m³/mol):
    9869 (calculated-1/$K_{AW}$, $C_W/C_A$, reported as exptl., Hine & Mookerjee 1975; quoted, Howard 1990)
    247.88 (calculated-group contribution, Hine & Mookerjee 1975)
    1426 (calculated-bond contribution, Hine & Mookerjee 1975)
    14897 (calculated-P/C using Riddick et al. 1986 data)

Octanol/Water Partition Coefficient, log $K_{OW}$:

3.73 (PCGEMS 1989; quoted, Howard 1990)

Bioconcentration Factor, log BCF:

1.62 (estimated-S, Lyman et al. 1982; quoted, Howard 1990)

Sorption Partition Coefficient, log $K_{OC}$:

2.50 (soil, estimated-linear regression with S, Lyman et al. 1982; quoted, Howard 1990)

Half-Lives in the Environment:

Air: by analogy to other Freon compounds, Freon-112 is predicted to have a stratospheric lifetime on the order of several decades (Chou et al. 1978; quoted, Howard 1990).

Surface water: will volatilize very rapidly (estimated-P/C, Lyman et al. 1982; quoted, Howard 1990).

Ground water: will volatilize very rapidly (estimated-P/C, Lyman et al. 1982; quoted, Howard 1990).

Sediment: will volatilize very rapidly from soil surfaces (estimated-P/C, Lyman et al. 1982; quoted, Howard 1990).

Soil:

Biota:

Environmental Fate Rate Constant and Half-Lives:

Volatilization: volatilization half-life from a model river 1 m deep flowing 1m/sec with a wind velocity of 3 m/sec was estimated to be 4.0 hours, based on estimated Henry's law constant (estimated-P/C, Lyman et al. 1982; quoted, Howard 1990).

Photolysis: will not undergo direct photolysis in the troposphere (Makide et al. 1979; quoted, Howard 1990).

Oxidation: inert to react with photochemically produced radicals and ozone molecules (Dilling 1982; Atkinson 1985,1987; quoted, Howard 1990).

Hydrolysis: not an environmentally significant fate process (Du Pont 1980; quoted, Howard 1990).

Biodegradation:

Biotransformation:

Bioconcentration, Uptake ($k_1$) and Elimination ($k_2$) Rate Constants or Half-Lives:

Common Name: Fluorobenzene
Synonym: phenyl fluoride
Chemical Name: fluorobenzene, phenyl fluoride
CAS Registry No: 462-06-6
Molecular Formula: $C_6H_5F$
Molecular Weight: 96.11
Melting Point (°C):
-41.20 (Horvath 1982)
-42.21 (Riddick et al. 1986)
Boiling Point (°C):
85.10 (Horvath 1982)
84.734 (Riddick et al. 1986)
Density (g/cm$^3$ at 20°C):
1.0225 (Horvath 1982)
1.01314 (30°C, Riddick et al. 1986)
Molar Volume (cm$^3$/mol):
116.0 (calculated-LeBas method)
94.86 (calculated from density)
Molecular Volume (A$^3$):
84.4 (Moriguchi et al. 1976)
Total Surface Area, TSA (A$^2$):
107.8 (Moriguchi et al. 1976)
224.8 (Iwase et al. 1985)
114.6 (Okouchi et al. 1992)
Heat of Fusion, $\Delta H_{fus}$, (kcal/mol):
2.702 (quoted, Riddick et al. 1986)
Entropy of Fusion, $\Delta S_{fus}$ (cal/mol K or e.u.):
Fugacity Ratio at 25°C, F: 1.0

Water Solubility (g/m$^3$ or mg/L at 25°C):
1523 (30°C, Gross & Saylor 1931; quoted, Chiou & Schmedding 1981)
1550 (Andrews & Keefer 1950; quoted, Horvath 1982)
1540 (30°C, Hodgman 1952; quoted, Freed et al. 1977)
1296 (quoted, Deno & Berkheimer 1960)
922.0 (shake flask-GC, Jones et al. 1977/1978)
1559 (Yalkowsky et al. 1979; Yalkowsky & Valvani 1981; quoted, Horvath 1982)
1559, 1238 (quoted, calculated-fragment const., Wakita et al. 1986)
1559 (quoted, Isnard & Lambert 1989)
1559 (calculated-AI, Okouchi et al. 1992)
1700 (19.2°C, shake flask-GC/TC, Stephenson 1992)
1550 (29.7°C, shake flask-GC/TC, Stephenson 1992)

Vapor Pressure (Pa at 25°C):
  10480 (quoted, Riddick et al. 1986)

Henry's Law Constant (Pa $m^3$/mol):
  630 (calculated-P/C from selected data)

Octanol/Water Partition Coefficient, log $K_{OW}$:
  2.27 (shake flask-AS, Fujita et al. 1964; quoted, Freed et al. 1977; Sangster 1989)
  2.27 (Leo et al. 1971; quoted, Moriguchi et al. 1976; Chiou & Schmedding 1981; Iwase et al. 1985)
  2.27, 2.41 (quoted, calculated-hydrophobicity const., Moriguchi et al. 1976)
  2.27, 2.28 (quoted, calculated-f const., Rekker 1977)
  2.27, 2.28 (quoted observed & calculated values, Chou & Jurs 1979)
  2.27, 2.28 (calculated-$\pi$, f consts., Chou & Jurs 1979)
  2.27 (quoted, Kaiser et al. 1983; Isnard & Lambert 1989)
  2.27, 2.29 (quoted, calculated-hydrophobicity const., Iwase et al. 1985)
  2.27 (recommended, Sangster 1989)

Bioconcentration Factor, log BCF:

Sorption Partition Coefficient, log $K_{OC}$:

Half-Lives in the Environment:
  Air:
  Surface water:
  Ground water:
  Sediment:
  Soil:
  Biota:

Environmental Fate Rate Constant and Half-Lives:
  Volatilization:
  Photolysis:
  Oxidation:
  Hydrolysis:
  Biodegradation:
  Biotransformation:
  Bioconcentration, Uptake ($k_1$) and Elimination ($k_2$) Rate Constants or Half-Lives:

Common Name: Bromobenzene
Synonym: phenylbromide
Chemical Name: bromobenzene
CAS Registry No: 108-86-1
Molecular Formula: $C_6H_5Br$
Molecular Weight: 157.02
Melting Point (°C):
-30.82 (Mackay & Shiu 1981,1990; Riddick et al. 1986)
-30.8 (Weast 1982-83)
-31.0 (Verschueren 1983)
-30.72 (Dean 1985)
Boiling Point (°C):
156.0 (Mackay & Shiu 1981,1990; Verschueren 1983)
156.43 (Weast 1982-83)
156.2 (Dean 1985)
155.91 (Riddick et al. 1986)
Density ($g/cm^3$ at 20°C):
1.4950 (Weast 1982-83)
1.4952 (Dean 1985)
Molar Volume ($cm^3/mol$):
105.0 (calculated from density, Chiou 1985)
119.3 (LeBas method, Mackay & Shiu 1990)
Molecular Volume ($A^3$):
Total Surface Area, TSA ($A^2$):
251.1 (Iwase et al. 1985)
133.1 (Valsaraj 1988)
134.3 (Okouchi et al. 1992)
Heat of Fusion, $\Delta H_{fus}$, (kcal/mol):
2.558 (Riddick et al. 1986)
Entropy of Fusion, $\Delta S_{fus}$ (cal/mol K or e.u.):
Fugacity Ratio at 25°C, F: 1.0

Water Solubility ($g/m^3$ or mg/L at 25°C):
442.5 (30°C, Gross & Saylor 1931; quoted, Vesala 1974; Chiou & Schmedding 1981; Chiou et al. 1982; Chiou 1985)
446 (30°C, Hodgman 1952; quoted, Freed et al. 1977; Verschueren 1977,1983)
410 (shake flask-UV, Andrews & Keefer 1950; quoted, Mackay & Shiu 1981,1990)
500 (20°C, Bradly & Cleasby 1953; quoted, Mackay & Shiu 1981; Verschueren 1977,1983)
328 (quoted, Deno & Berkheimer 1960)
446 (shake flask-UV, Vesala 1974)
413 (quoted, Hine & Mookerjee 1975)

360 (Yalkowsky et al. 1979)
378, 410 (quoted values, Wasik et al. 1983)
411 (gen. col.-HPLC, Wasik et al. 1983)
330 (headspace-GC, McNally & Grob 1984)
440 (Dean 1985)
446 (quoted from Gross & Saylor 1931, Riddick et al. 1986)
443, 423 (quoted, calculated-fragment const., Wakita et al. 1986)
409 (selected, Valsaraj 1988)
453 (quoted, Isnard & Lambert 1989)
360 (calculated-AI, Okouchi et al. 1992)

Vapor Pressure (Pa at 25°C):
552 (Boublik et al. 1973; quoted, Mackay & Shiu 1981,1990)
557.2 (quoted, Hine & Mookerjee 1975)
570 (Weast 1972-73; quoted, Mackay & Shiu 1981)
440 (20°C, quoted, Verschueren 1977,1983)
557.6 (quoted, Riddick et al. 1986)
506.5 (selected, Valsaraj 1988)

Henry's Law Constant (Pa $m^3$/mol):
211 (calculated-1/$K_{AW}$, $C_W/C_A$, reported as exptl., Hine & Mookerjee 1975; quoted, Nirmalakhandan & Speece 1988)
221 (calculated-group contribution method, Hine & Mookerjee 1975)
199 (calculated-bond contribution method, Hine & Mookerjee 1975)
247, 211 (exptl., calculated-P/C, Mackay & Shiu 1981,1990)
210 (recommended, Mackay & Shiu 1981)
211, 136.2 (quoted, calculated-QSAR, Nirmalakhandan & Speece 1988)
243.1 (selected, Valsaraj 1988)

Octanol/Water Partition Coefficient, log $K_{OW}$:
2.99 (shake flask-AS, Fujita et al. 1964; quoted, Hansch et al. 1968; Freed et al. 1977; Sangster 1989)
2.99 (Leo et al. 1971; quoted, Freed et al. 1977; Chiou & Schmedding 1981; McDuffie 1981; Chiou et al. 1982; Chiou 1985; Iwase et al. 1985)
2.99, 3.03 (quoted, calculated-f const., Rekker 1977)
2.99 (Hansch & Leo 1979; quoted, Mackay & Shiu 1990)
2.84 (calculated-S, Mackay et al. 1980; quoted, Sangster 1989)
3.01 (calculated-f const. as per Rekker 1977, Hanai et al. 1981)
3.16 (calculated-HPLC-k', McDuffie 1981)
2.98 (gen. col.-HPLC, Wasik et al. 1981; quoted, Sangster 1989)

3.02 (estimated-HPLC, D'Amboise & Hanai 1982; quoted, Sangster 1989)
3.01 (shake flask-GC, Watarai et al. 1982; quoted, Sangster 1989)
3.02 (estimated-HPLC, Eadsforth & Moser 1983; quoted, Sangster 1989)
3.01 (calculated as per Rekker 1977, Hanai et al. 1983)
2.99 (quoted, Kaiser et al. 1983; Verschueren 1983)
2.99, 3.03 (quoted, calculated-molar refraction, Yoshida et al. 1983)
2.99, 3.02 (quoted, calculated-hydrophobicity const., Iwase et al. 1985)
3.15 (estimated-HPLC, Eadsforth 1986; quoted, Sangster 1986)
2.95 (quoted, Hodson & Williams 1988)
2.98 (selected, Valsaraj 1988)
3.02 (quoted, Isnard & Lambert 1989)
2.99 (recommended, Sangster 1989)

Bioconcentration Factor, log BCF:
1.68 (fish, Halfon & Reggiani 1986)
2.28 (algae, Halfon & Reggiani 1986)
3.18 (act. sludge, Halfon & Reggiani 1986)

Sorption Partition Coefficient, log $K_{OC}$:
2.80 (calculated- $\chi$ , Koch 19983)
2.18 (quoted, Hodson & Williams 1988)
2.65 (HPLC, Hodson & Williams 1988)

Half-Lives in the Environment:
Air:
Surface water:
Ground water:
Sediment:
Soil:
Biota:

Environmental Fate Rate Constant and Half-Lives:
Volatilization:
Photolysis:
Oxidation:
Hydrolysis:
Biodegradation:
Biotransformation:
Bioconcentration, Uptake ($k_1$) and Elimination ($k_2$) Rate Constants or Half-Lives:

Common Name: Iodobenzene
Synonym:
Chemical Name: iodobenzene
CAS Registry No: 591-50-4
Molecular Formula: $C_6H_5I$
Molecular Weight: 204.01
Melting Point (°C):
-31.21 (Mackay & Shiu 1981)
-31.15 (Miller et al. 1985)
-31.35 (Riddick et al. 1986)
Boiling Point (°C):
188.30 (Mackay & Shiu 1981)
188.33 (Riddick et al. 1986)
Density ($g/cm^3$ at 20°C):
1.8308 (Riddick et al. 1986)
1.8229 (25°C, Riddick et al. 1986)
Molar Volume ($cm^3/mol$):
126.0 (Miller et al. 1985)
129.3 (calculated-LeBas method)
111.4 (calculated from density)
Molecular Volume ($A^3$):
Total Surface Area, TSA ($A^2$):
264.4 (Iwase et al. 1985)
142.8 (Okouchi et al. 1992)
Heat of Fusion, $\Delta H_{fus}$, (kcal/mol):
2.330 (Riddick et al. 1986)
Entropy of Fusion, $\Delta S_{fus}$ (cal/mol K or e.u.):
Fugacity Ratio at 25°C, F: 1.0

Water Solubility ($g/m^3$ or mg/L at 25°C):
347 (30°C, Gross & Saylor 1933; quoted, Vesala 1974; Chiou & Schmedding 1981)
340 (30°C, Seidell 1941; quoted, Mackay et al. 1980; Mackay & Shiu 1981)
180 (Andrews & Keefer 1950; quoted, Mackay & Shiu 1981)
229 (shake flask-UV, Vesala 1974)
158 (quoted, Deno & Berkheimer 1960)
229 (Yalkowsky et al. 1979; quoted, Mackay & Shiu 1981)
201 (quoted, Miller et al. 1985)
340 (30°C, quoted, Riddick et al. 1986)
346, 331 (quoted, calculated-fragment const., Wakita et al. 1986)
229 (quoted, Isnard & Lambert 1989)
229 (calculated-AI, Okouchi et al. 1992)

Vapor Pressure (Pa at 25°C):

132.0 (Antoine eqn., Boublik et al. 1973; quoted, Mackay & Shiu 1981)
134.5 (quoted, Riddick et al. 1986)

Henry's Law Constant (Pa $m^3$/mol):

134 (calculated-P/C, Mackay & Shiu 1981)
130 (recommended, Mackay & Shiu 1981)

Octanol/Water Partition Coefficient, log $K_{OW}$:

3.25, 3.32 (quoted, calculated-f const., Rekker 1977)
3.25 (Hansch & Leo 1979; quoted, Mackay et al. 1980; Kaiser et al. 1983; Iwase et al. 1985)
3.03 (estimated-S correlation, Mackay et al. 1980; quoted, Sangster 1989)
3.33 (calculated-f const. as per Rekker 1977, Hanai et al. 1981)
3.28 (gen. col.-HPLC, Wasik et al. 1981; quoted, Sangster 1989)
3.28 (gen. col.-HPLC, Tewari et al. 1982; quoted, Miller et al. 1985; Sangster 1989)
3.25, 3.23 (quoted, estimated-HPLC/MS, Burkhard et al. 1985)
3.25, 3.26 (quoted, calculated-f const., Burkhard et al. 1985)
3.25, 3.24 (quoted, calculated-hydrophobicity const., Iwase et al. 1985)
3.37 (estimated-HPLC, Eadsforth 1986; quoted, Sangster 1989)
3.25 (quoted, Isnard & Lambert 1989)
3.25 (recommended, Sangster 1989)

Bioconcentration Factor, log BCF:

Sorption Partition Coefficient, log $K_{OC}$:

Half-Lives in the Environment:

Air:
Surface water:
Ground water:
Sediment:
Soil:
Biota:

Environmental Fate Rate Constant and Half-Lives:

Volatilization:
Photolysis:
Oxidation:
Hydrolysis:
Biodegradation:
Biotransformation:
Bioconcentration, Uptake ($k_1$) and Elimination ($k_2$) Rate Constants or Half-Lives:

## 3.2 Summary Tables and QSPR Plots

Table 3.1 Summary of physical properties of halogenated hydrocarbons at 25 °C

| Compounds | CAS no. | formula | MW g/mol | M.P. °C | Molar volume $V_M$ MW/ρ cm³/mol | Molar volume $V_M$ LeBas cm³/mol | Molar volume $V_I/100$ intrinsic (a) | TSA Å² (b,c) | TSA Å² (d) |
|---|---|---|---|---|---|---|---|---|---|
| *Chloroalkanes*: | | | | | | | | | |
| Chloromethane | 74-87-3 | $CH_3Cl$ | 50.49 | -97.73 | 54.80 | 50.5 | | | 62.5 |
| Dichloromethane | 75-09-2 | $CH_2Cl_2$ | 84.94 | -95 | 64.03 | 71 | 0.338 | 213.49 | 78.9 |
| Chloroform | 67-66-3 | $CHCl_3$ | 119.38 | -63.5 | 80.17 | 92 | 0.427 | 238.68 | 97.9 |
| Carbon tetrachloride | 56-23-5 | $CCl_4$ | 153.82 | -22.9 | 96.50 | 113 | 0.514 | 260.69 | 117 |
| Chloroethane | 75-00-3 | $C_2H_5Cl$ | 64.52 | -136.4 | 72.009 | 72.7 | 0.352 | | 83.3 |
| 1,1-Dichloroethane | 75-34-3 | $C_2H_4CL_2$ | 98.96 | -96.98 | 84.171 | 93.6 | 0.442 | 242.97 | 102.3 |
| 1,2-Dichloroethane | 107-06-2 | $C_2H_4Cl_2$ | 98.96 | -35.36 | 79.035 | 93.6 | 0.442 | 246.27 | 99.8 |
| 1,1,1-Trichloroethane | 71-55-6 | $C_2H_3Cl_3$ | 133.41 | -30.41 | 99.634 | 115 | 0.519 | 263.93 | 121.4 |
| 1,1,2-Trichloroethane | 79-00-5 | $C_2H_3Cl_3$ | 133.41 | -36.3 | 92.654 | 115 | 0.519 | 266.48 | 118.8 |
| 1,1,1,2-Tetrachloroethane | 630-20-6 | $C_2H_2Cl_4$ | 167.85 | -70.2 | 108.951 | 135 | 0.617 | 284.48 | 137.9 |
| 1,1,2,2-Tetrachloroethane | 79-34-5 | $C_2H_2Cl_4$ | 167.85 | -36.0 | 105.215 | 135 | 0.617 | 286.73 | 137.9 |
| Pentachloroethane | 76-01-7 | $C_2HCl_5$ | 202.3 | -29.0 | 120.445 | 156 | 0.7 | 302.47 | 156.9 |
| Hexachloroethane | 67-72-1 | $C_2Cl_6$ | 236.74 | 186.1 | 113.22 | 177 | 0.79 | 318.07 | 176 |
| 1-Chloropropane | 540-54-5 | $C_3H_7Cl$ | 78.54 | -122.8 | 88.35 | 87.5 | | 249.88 | 104.2 |
| 2-Chloropropane | 75-29-6 | $C_3H_7Cl$ | 78.54 | -117.2 | 91.15 | 87.5 | | 246.38 | 106.7 |
| 1,2-Dichloropropane | 78-87-5 | $C_3H_6CL_2$ | 112.99 | -100.4 | 97.74 | 103.3 | 0.54 | 268.64 | 123.2 |
| 1,2,3-Trichloropropane | 96-18-4 | $C_3H_5Cl_3$ | 147.43 | -14.7 | 106.15 | 129.3 | | 287.51 | 139.7 |
| 1-Chlorobutane | 109-69-3 | $C_4H_9Cl$ | 92.57 | -123.1 | 104.52 | 117.1 | 0.548 | 273.22 | 125.1 |
| 2-Chlorobutane | 78-86-4 | $C_4H_9Cl$ | 92.57 | -113.3 | | 117.1 | | | 127.6 |
| 1-Chloropentane | 543-59-6 | $C_5H_{11}Cl$ | 106.6 | -104.4 | 120.86 | 139.3 | 0.645 | 314.4 | 146 |

| Compounds | CAS no. | formula | MW g/mol | M.P. °C | $V_M$ MW/$\rho$ cm$^3$/mol | $V_M$ LeBas cm$^3$/mol | $V_I$/100 intrinsic (a) | TSA Å$^2$ (b,c) | TSA Å$^2$ (d) |
|---|---|---|---|---|---|---|---|---|---|
| *Chloroalkenes*: | | | | | | | | | |
| Chloroethene (vinyl chloride) | 75-01-4 | $C_2H_3Cl$ | 62.5 | -153.8 | 68.64 | 65.3 | | | 77.4 |
| 1,1-Dichloroethene | 75-35-4 | $C_2H_2Cl_2$ | 96.94 | -122.1 | 79.59 | 86.2 | 0.402 | | 94.5 |
| *cis*-1,2-Dichloroethene | 156-59-2 | $C_2H_2Cl_2$ | 96.94 | -80.5 | 75.52 | 86.2 | | | 93.6 |
| *trans*-1,2-Dichloroethene | 156-60-5 | $C_2H_2Cl_2$ | 96.94 | -50 | 77.15 | 86.2 | 0.406 | | |
| Trichloroethylene | 79-01-6 | $C_2HCl_3$ | 131.39 | -73 | 89.74 | 107 | 0.492 | | 110.7 |
| Tetrachloroethylene | 127-18-4 | $C_2Cl_4$ | 165.83 | -19 | 102.17 | 128 | 0.578 | | 127.8 |
| 1,3-Dichloropropene | 542-75-6 | $C_3H_4Cl_2$ | 110.97 | | 91.18 | 86.2 | | | |
| Chloroprene | 129-99-8 | $C_4H_5Cl$ | 88.54 | -130 | 92.39 | 102.3 | | | |
| Hexachloro-1,3-butadiene | 87-68-3 | $C_4Cl_6$ | 260.76 | -21 | 155.03 | 206.8 | | | |
| Hexachlorocyclopentadiene | 77-47-4 | $C_5H_6$ | 272.77 | -11.4 | 160.3 | 210.1 | | | |
| *Bromoalkanes and bromoalkenes*: | | | | | | | | | |
| Bromomethane | 74-83-9 | $CH_3Br$ | 94.94 | -93 | 55.76 | 52.9 | | | 70.5 |
| Dibromomethane | 74-95-3 | $CH_2Br_2$ | 267.84 | 6.1 | 80.54 | 76.2 | | | 95.1 |
| Tribromomethane | 75-25-2 | $CHBr_3$ | 252.75 | -8.3 | 87.43 | 99.5 | | | 122.2 |
| Bromoethane | 74-96-4 | $C_2H_5Br$ | 108.97 | -119 | 72.97 | 75.1 | | | 91.4 |
| 1,2-Dibromoethane | 106-93-4 | $C_2H_4Br_2$ | 187.862 | 9.79 | 86.21 | 101.4 | 0.528 | | 116 |
| 1-Bromopropane | 106-94-5 | $C_3H_7Br$ | 123 | -110 | 90.86 | 89.9 | | | 112.3 |
| 2-Bromopropane | 75-26-3 | $C_3H_7Br$ | 123 | -89 | 93.61 | 89.9 | | | 114.8 |
| 1,2-Dibromopropane | 78-75-1 | $C_3H_6Br_2$ | 201.888 | -55.5 | 104.46 | 113.2 | | | 139.4 |
| 1,3-Dibromopropane | 109-64-8 | $C_3H_6Br_2$ | 201.888 | | | 113.2 | | | 136.9 |

| Compounds | CAS no. | formula | MW g/mol | M.P. °C | $V_M$ MW/ρ cm$^3$/mol | $V_M$ LeBas cm$^3$/mol | $V_I$/100 intrinsic (a) | TSA Å$^2$ (b,c) | TSA Å$^2$ (d) |
|---|---|---|---|---|---|---|---|---|---|
| Bromobutane | 109-65-9 | $C_4H_9Br$ | 137.03 | -112 | | 119.5 | | | 133.2 |
| Isobutyl bromide | 78-77-3 | $C_4H_9Br$ | 137.03 | | | | | | |
| Isoamyl bromide | 107-82-4 | $C_5H_{11}Br$ | 151.05 | -112 | | | | | |
| Bromoethylene | 593-60-2 | $C_2H_3Br$ | 106.95 | -139.54 | 71.62 | 67.7 | | | |
| *Iodoalkanes*: | | | | | | | | | |
| Iodomethane | 74-88-4 | $CH_3I$ | 141.939 | -66.45 | 62.28 | 62.9 | | | |
| Diiodomethane | 75-11-6 | $CH_2I_2$ | 267.836 | 6.1 | 80.64 | 99.9 | | | |
| Iodoethane | 75-03-6 | $C_2H_5I$ | 115.966 | -111.1 | 59.91 | 85.1 | | | |
| Iodopropane | 107-08-4 | $C_3H_7I$ | 169.993 | -101.3 | 97.20 | 107.3 | | | |
| Iodobutane | 542-69-8 | $C_4H_9I$ | 184.02 | -103 | 113.92 | 129.5 | | | |
| *Mixed halides*: | | | | | | | | | |
| Bromochloromethane | 74-97-5 | $CH_2BrCl$ | 129.384 | -87.95 | 66.89 | 73.8 | | | 87 |
| Bromodichloromethane | 75-27-4 | $CHBrCl_2$ | 163.8 | -57.1 | 82.73 | 94.7 | | | |
| Dibromochloromethane | 124-48-1 | $CHBr_2Cl$ | 208.3 | -20 | 84.99 | 97.1 | | | |
| Dibromodichloromethane | 594-18-3 | $CBr_2Cl_2$ | 242.76 | 22 | 100.31 | 118 | | | |
| Chlorotrifluoromethane (freon-13) | 75-72-9 | $CClF_3$ | 104.459 | -189 | 113.25 | | | | |
| Chlorodifluoromethane (freon-22) | 75-45-6 | $CHClF_2$ | 86.469 | -157.42 | 71.25 | 56.8 | | | |
| Chlorofluoromethane (freon-31) | | $CH_2ClF$ | 68.48 | -9.1 | | 40.7 | | | |
| Dichlorofluoromethane (freon-21) | 75-43-4 | $CHCl_2F$ | 102.923 | -135 | 74.67 | 76.4 | | | |
| Dichlorodifluoromethane (freon-12) | 75-71-8 | $CCl_2F_2$ | 120.914 | -158.2 | 90.97 | 81.4 | | | |
| Trichlorofluoromethane | 75-69-4 | $CCl_3F$ | 137.368 | -110.48 | 92.32 | 97.3 | | | |

| Compounds | CAS no. | formula | MW g/mol | M.P. °C | $V_M$ MW/$\rho$ $cm^3/mol$ | $V_M$ LeBas $cm^3/mol$ | $V_I/100$ intrinsic (a) | TSA $Å^2$ (b,c) | TSA $Å^2$ (d) |
|---|---|---|---|---|---|---|---|---|---|
| 1,2-Dichlorotetrafluoro-ethane (freon-114) | 76-14-2 | $C_2Cl_2F_4$ | 170.922 | -92.53 | | 113.6 | | | |
| 1,1,2-Trichloro-1,2,2-trifluoroethane (freon-113) | 76-13-1 | $C_2F_3Cl_3$ | 187.376 | -36.4 | | 123.5 | | | |
| Chloropentafluoroethane | 76-15-3 | $C_2ClF_5$ | 154.467 | -106 | | 97.9 | | | |
| *Aromatics*: | | | | | | | | | 141.1 |
| Fluorobenzene | 462-06-6 | $C_6H_5F$ | 96.104 | -42.21 | 93.22 | 101 | 0.59 | | 114.6 |
| Bromobenzene | 108-86-1 | $C_6H_5Br$ | 157.02 | -30.8 | 104.97 | 119.3 | 0.624 | | 134.3 |
| Iodobenzene | 591-50-4 | $C_6H_5I$ | 204.01 | -31.35 | 111.43 | 129.3 | 0.671 | | 142.8 |
| Chlorobenzene | 108-90-7 | $C_6H_5Cl$ | 112.56 | -45.6 | 132.2 | 117 | | | |

(a) Kamlet et al. 1987, 1988
(b) Amidon et al. 1975
(c) Horvath 1982
(d) Okouchi et al. 1992

Table 3.2 Summary of selected physical-chemical properties of halogenated hydrocarbons at 25 °C

| Compounds | Selected properties | | | | | | | |
|---|---|---|---|---|---|---|---|---|
| | Vapor pressure | | Solubility | | | | Henry's law const. | |
| | $P^S$ Pa | $P_L$ Pa | S $g/m^3$ | $C^S$ $mol/m^3$ | $C_L$ $mol/m^3$ | log $K_{OW}$ | H, calc, P/C $Pa \cdot m^3/mol$ | exptl |
| *Chloroalkanes*: | | | | | | | | |
| Chloromethane | 570000 | 570000 | 5235 | 103.68 | 103.68 | 0.91 | 977.25* | |
| Dichloromethane | 26222 | 26222 | 13200 | 155.40 | 155.40 | 1.25 | 168.73 | 300 |
| Chloroform | 26244 | 26244 | 8200 | 68.69 | 68.69 | 1.97 | 382.07 | 427 |
| Carbon tetrachloride | 15250 | 15250 | 800 | 5.20 | 5.20 | 2.64 | 2932.19 | 2989 |
| Chloroethane | 16000 | 16000 | 5700 | 88.34 | 88.34 | 1.43 | 181.11 | 1023 |
| 1,1-Dichloroethane | 30260 | 30260 | 4767 | 48.17 | 48.17 | 1.79 | 628.18 | 633 |
| 1,2-Dichloroethane | 10540 | 10540 | 8606 | 86.96 | 86.96 | 1.48 | 121.20 | 143 |
| 1,1,1-Trichloroethane | 16500 | 16500 | 1495 | 11.21 | 11.21 | 2.49 | 1472.42 | 1763 |
| 1,1,2-Trichloroethane | 3220 | 3220 | 4394 | 32.94 | 32.94 | 2.38 | 97.77 | 92.2 |
| 1,1,1,2-Tetrachloroethane | 1580 | 1580 | 1100 | 6.55 | 6.55 | | 241.09 | |
| 1,1,2,2-Tetrachloroethane | 793 | 793 | 2962 | 17.65 | 17.65 | 2.39 | 44.94 | 25.7 |
| Pentachloroethane | 625 | 625 | 500 | 2.47 | 2.47 | 2.89 | 252.9 | |
| Hexachloroethane | 50 | 1960 | 50 | 0.2112 | 8.28 | 3.93 | | 846 |
| 1-Chloropropane | 45920 | 45920 | 2561 | 32.61 | 32.61 | | 1408.26 | |
| 2-Chloropropane | 68700 | 68700 | 2945 | 37.50 | 37.50 | 1.69 | 1832.16 | |
| 1,2-Dichloropropane | 6620 | 6620 | 2800 | 24.78 | 24.78 | 2.0 | 267.14 | 287 |
| 1,2,3-Trichloropropane | 492 | 492 | 1896 | 12.86 | 12.86 | 2.63 | 38.26 | |
| 1-Chlorobutane | 13700 | 13700 | 615 | 6.64 | 6.64 | 2.64 | 2062.13 | 1537 |
| 2-Chlorobutane | 20210 | 20210 | 1000 | 10.80 | 10.80 | | 1870.84 | 2267 |
| 1-Chloropentane | 4142 | 4142 | 198 | 1.86 | 1.86 | | 2229.99 | 2375 |

| Compounds | Vapor pressure | | Solubility | | | | Henry's law const. | |
|---|---|---|---|---|---|---|---|---|
| | $P^S$ | $P_L$ | S | $C^S$ | $C_L$ | log $K_{OW}$ | H, calc, P/C | exptl |
| | Pa | Pa | $g/m^3$ | $mol/m^3$ | $mol/m^3$ | | $Pa \cdot m^3/mol$ | |
| *Chloroalkenes*: | | | | | | | | |
| Chloroethene | 354600 | 354600 | 2763 | 44.21 | 44.21 | 1.38 | 8021.17 | 2685 |
| 1,1-Dichloroethene | 80500 | 80500 | 3344 | 34.50 | 34.50 | 2.13 | 2333.63 | 2624 |
| *cis*-1,2-Dichloroethene | 27000 | 27000 | 3500 | 36.10 | 36.10 | 1.86 | 747.82 | 460 |
| *trans*-1,2-Dichloroethene | 44400 | 44400 | 6260 | 64.58 | 64.58 | 1.93 | 687.56 | 958 |
| Trichloroethylene | 9900 | 9900 | 1100 | 8.37 | 8.37 | 2.53 | 1183.70 | 1034 |
| Tetrachloroethylene | 2415 | 2415 | 150 | 0.90 | 0.90 | 2.88 | 2669.86 | 1733 |
| 3-Chloroprene | 148900 | 148900 | 3600 | 47.04 | 47.04 | | 2154* | |
| 1,3-Dichloropropene | | | | | | 2.28 | | |
| Chloroprene | 23194 | 23194 | 96400 | 1089 | 1089 | 2.03 | 21.30 | |
| Hexachloro-1,3-butadiene | 20.0 | 20.0 | 3.2 | 0.013 | 0.013 | 4.70 | 1540 | |
| Hexachlorocyclopentadiene | 10.9 | 10.9 | 1.80 | 0.0066 | 0.0066 | 5.04 | 1670 | |
| *Bromoalkanes and bromoalkenes*: | | | | | | | | |
| Bromomethane | 217700 | 217700 | 15223 | 160.34 | 160.34 | 1.19 | 632* | |
| Dibromomethane | 6034 | 6034 | 11442 | 42.72 | 42.72 | 2.3 | 141.25 | 86.13 |
| Tribromomethane | 727 | 727 | 3100 | 12.27 | 12.27 | 2.38 | 59.27 | 46.61 |
| Bromoethane | 140000 | 140000 | 8939 | 82.03 | 82.03 | 1.6 | 1235* | |
| 1,2-Dibromoethane | 10400 | 10400 | 4152 | 22.10 | 22.10 | | 470 | 65.86 |
| 1-Bromopropane | 18440 | 18440 | 598 | 4.86 | 4.86 | 2.1 | 3792 | |
| 2-Bromopropane | 31940 | 31940 | 3086 | 25.09 | 25.09 | 1.9 | 1273 | |
| 1,2-Dibromopropane | 1040 | 1040 | 1428 | 7.07 | 7.07 | | | |
| 1,3-Dibromopropane | 2433 | 2433 | 548 | 2.71 | 2.71 | 2.7 | 896.34 | |

| Compounds | Vapor pressure | | Solubility | | | | Henry's law const. | |
|---|---|---|---|---|---|---|---|---|
| | $P^S$ Pa | $P_L$ Pa | S g/m³ | $C^S$ mol/m³ | $C_L$ mol/m³ | log $K_{OW}$ | H, calc, P/C Pa·m³/mol | exptl |
| Bromobutane | 5500 | 5500 | | | 0.00 | 2.6 | | |
| Isobutyl bromide | 9190 | 9190 | | | 0.00 | 2.4 | | |
| Isoamyl bromide | 2716 | 2716 | 199 | 1.32 | 1.32 | 2.9 | 2061 | |
| Bromoethylene | 140476 | 140476 | | | | | | |
| *Iodoalkanes*: | | | | | | | | |
| Iodomethane | 54000 | 54000 | 13894 | 97.89 | 97.89 | 1.5 | 551.66 | 541 |
| Diiodomethane | 160 | 160 | 1350 | 5.04 | 5.04 | 2.5 | 31.74 | |
| Iodoethane | 18160 | 18160 | 4041 | 34.85 | 34.85 | 2 | 521.14 | |
| Iodopropane | 5745 | 5745 | 1051 | 6.18 | 6.18 | 2.5 | 929.22 | |
| Iodobutane | 1848 | 1848 | 182 | 0.99 | 0.99 | 3 | 1868.51 | |
| *Mixed halides*: | | | | | | | | |
| Bromochloromethane | 19600 | 19600 | 14778 | 114.22 | 114.22 | 1.41 | 171.60 | |
| Bromodichloromethane | 6670 | 6670 | 4500 | 27.47 | 27.47 | 2.1 | 242.79 | 162 |
| Dibromochloromethane | | | 4000 | 19.20 | 19.20 | 2.24 | | 86.13 |
| Dibromodichloromethane | | | | | 0.00 | | | |
| Chlorotrifluoromethane | 340000 | 340000 | 1540 | 14.74 | 14.74 | | 6874* | |
| Chlorodifluoromethane | 104400 | 104400 | 2899 | 33.53 | 33.53 | | 3022* | |
| Chlorofluoromethane | | | 10522 | 153.65 | 153.65 | | | |
| Dichlorofluoromethane | 225000 | 225000 | 18800 | 182.66 | 182.66 | | 555* | 26639 |
| Dichlorodifluoromethane | 651000 | 651000 | 300 | 2.48 | 2.48 | 2.16 | 40860* | |

| Compounds | Vapor pressure | | Solubility | | | | Henry's law const. | |
|---|---|---|---|---|---|---|---|---|
| | $P^S$ Pa | $P_L$ Pa | S g/m³ | $C^S$ mol/m³ | $C_L$ mol/m³ | log $K_{OW}$ | H, calc, P/C Pa·m³/mol | exptl |
| Trichlorofluoromethane | 102200 | 102200 | 1080 | 7.86 | 7.86 | 2.53 | 12900* | 10243 |
| 1,2-Dichlorotetrafluoro-ethane (freon-114) | 268000 | 268000 | 137 | 0.80 | 0.80 | | 126660* | |
| 1,1,2-Trichloro-1,2,2-trifluoroethane (freon-113) | 483200 | 483200 | 166 | 0.89 | 0.89 | 3.16 | 113840* | 32323 |
| Chloropentafluoroethane | 149000 | 149000 | 60 | 0.39 | 0.39 | | 259810* | |
| *Aromatics*: | | | | | | | | |
| Fluorobenzene | 10480 | 10480 | 1430 | 14.88 | 14.88 | 2.27 | 704 | |
| Bromobenzene | 552 | 552 | 410 | 2.61 | 2.61 | 2.99 | 211.40 | |
| Iodobenzene | 130 | 130 | 340 | 1.67 | 1.67 | 3.28 | 78.00 | |
| Chlorobenzene | 1380 | 1380 | 484 | 4.30 | 4.30 | 2.80 | 368 | 382 |

* vapor pressure exceeds atmosphere pressure, Henry's law constant H (Pa·m³/mol) = 101325 (Pa)/$C^S$ (mol/m³)

Table 3.3 Suggested half-life classes of halogenated hydrocarbons in various environmental compartments

| Compounds | Air class | Water class | Soil class | Sediment class |
|---|---|---|---|---|
| Dichloromethane | 6 | 6 | 7 | 8 |
| Chloroform | 6 | 6 | 7 | 8 |
| Carbon tetrachloride | 8 | 6 | 7 | 8 |
| Chloroethane | 6 | 6 | 7 | 8 |
| 1,2-Dichloroethane | 6 | 6 | 7 | 8 |
| 1,1,2,2-Tetrachloroethane | 8 | 6 | 7 | 8 |
| Pentachloroethane | 8 | 6 | 7 | 8 |
| Hexachloroethane | 8 | 6 | 7 | 8 |
| Vinyl chloride | 3 | 5 | 6 | 7 |
| Trichloroethylene | 4 | 5 | 6 | 7 |
| Tetrachloroethylene | 5 | 5 | 6 | 7 |
| 1,2-Dichloropropane | 5 | 6 | 7 | 8 |
| 1,2,3-Trichloropropane | 5 | 6 | 7 | 8 |
| Chloroprene | 3 | 5 | 6 | 7 |
| Tribromomethane | 6 | 6 | 7 | 8 |
| Bromochloromethane | 5 | 5 | 6 | 7 |
| Bromodichloromethane | 5 | 5 | 6 | 7 |
| Trichlorofluoromethane | 8 | 8 | 9 | 9 |
| Fluorobenzene | 2 | 4 | 5 | 6 |
| Bromobenzene | 4 | 6 | 7 | 8 |
| Iodobenzene | 4 | 6 | 7 | 8 |

where,

| Class | Mean half-life (hours) | Range (hours) |
|---|---|---|
| 1 | 5 | < 10 |
| 2 | 17 (~ 1 day) | 10-30 |
| 3 | 55 (~ 2 days) | 30-100 |
| 4 | 170 (~ 1 week) | 100-300 |
| 5 | 550 (~ 3 weeks) | 300-1,000 |
| 6 | 1700 (~ 2 months) | 1,000-3,000 |
| 7 | 5500 (~ 8 months) | 3,000-10,000 |
| 8 | 17000 (~ 2 years) | 10,000-30,000 |
| 9 | 55000 (~ 6 years) | > 30,000 |

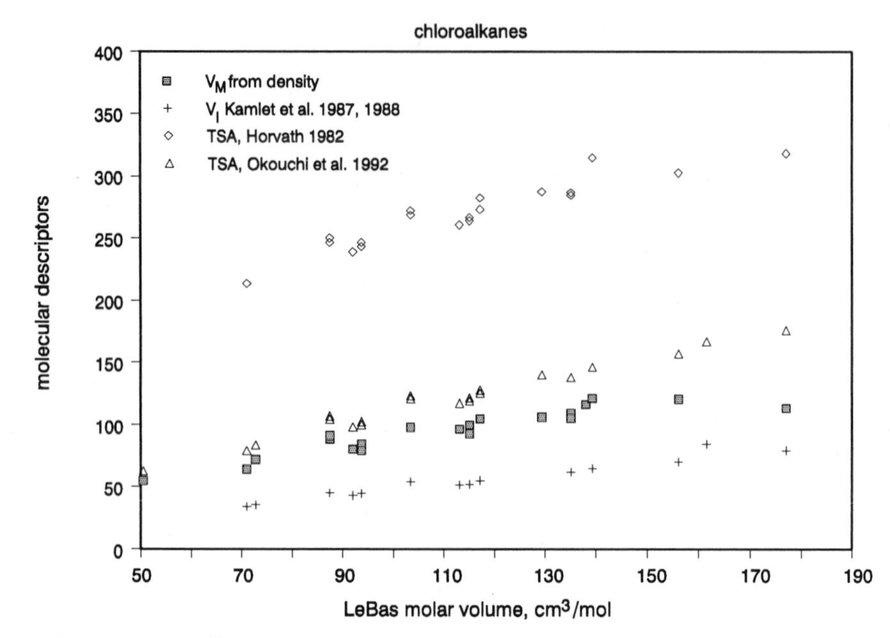


Figure 3.1a Plot of molecular descriptors versus LeBas molar volume.

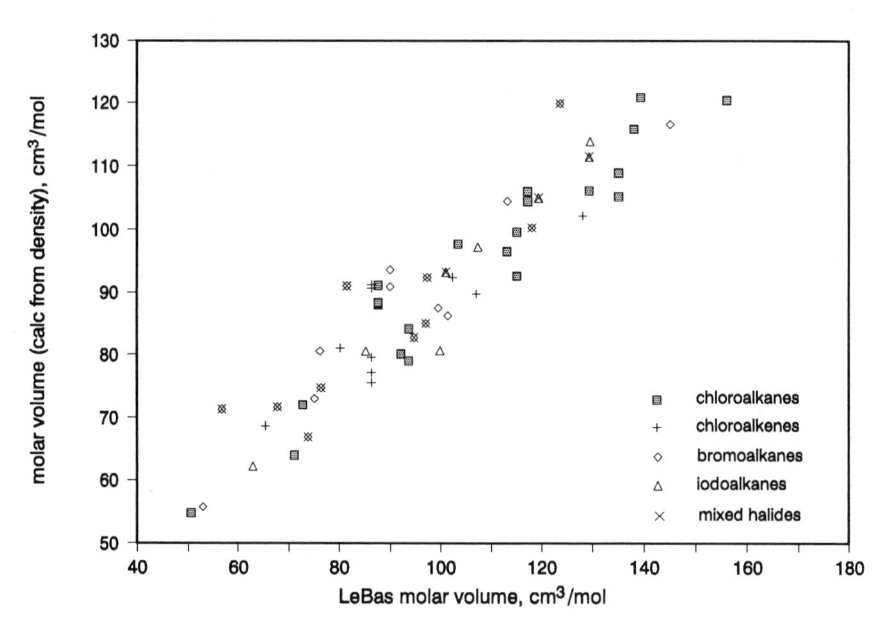


Figure 3.1b Plot of liquid molar volume versus LeBas molar volume.

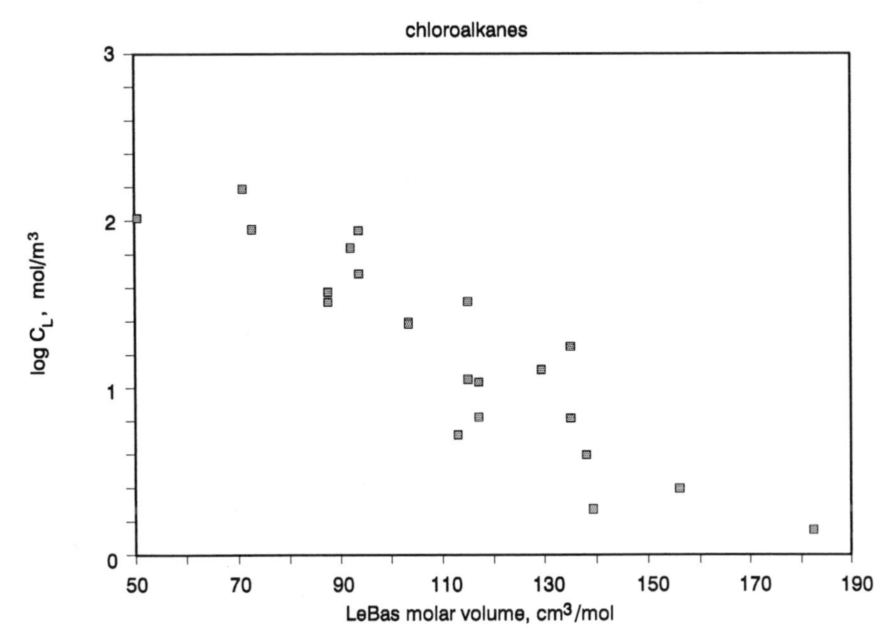


Figure 3.2a Plot of log solubility versus LeBas molar volume.

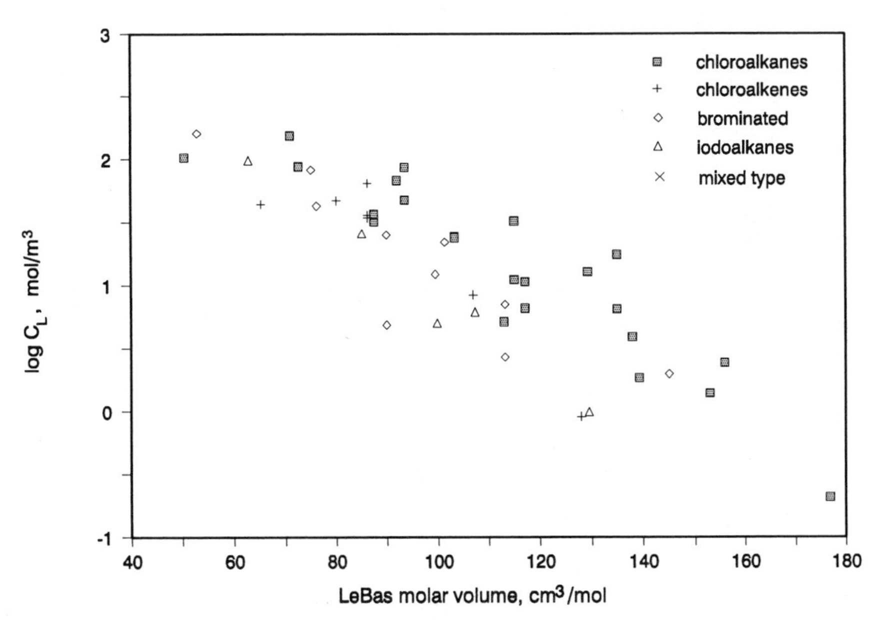


Figure 3.2b Plot of log solubility versus LeBas molar volume.

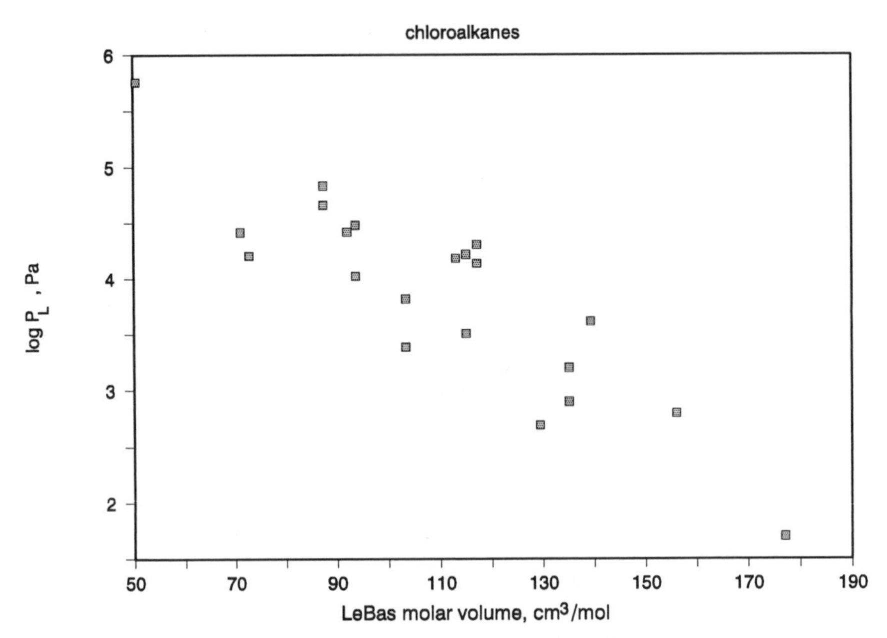


Figure 3.3a Plot of log vapor pressure versus LeBas molar volume.

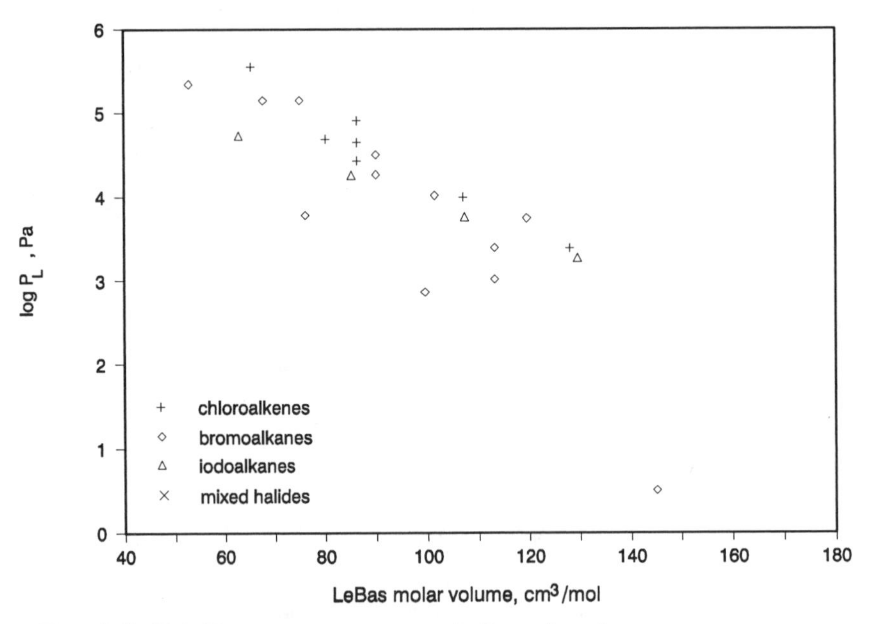


Figure 3.3b Plot of log vapor pressure versus LeBas molar volume.

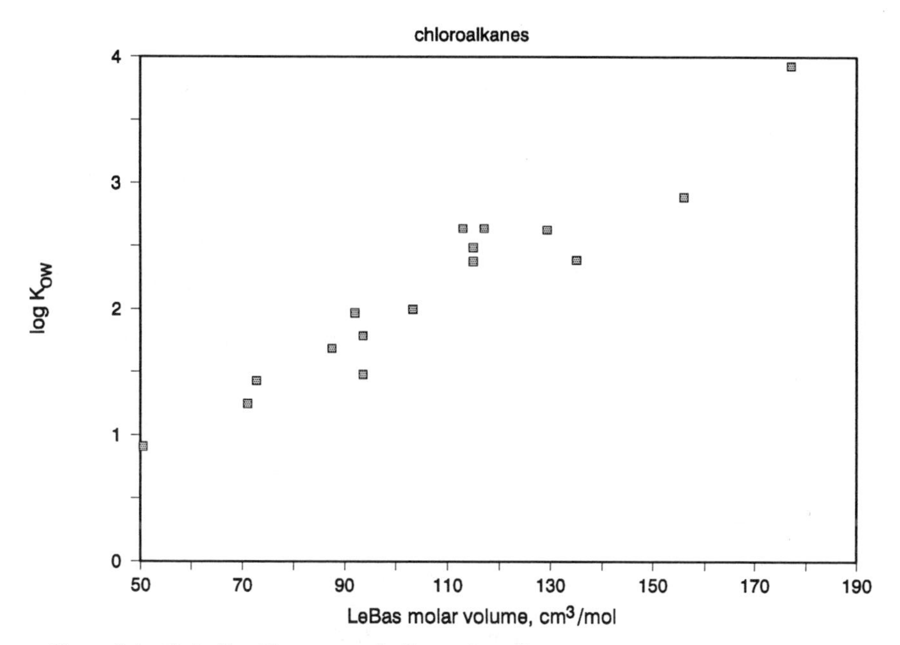


Figure 3.4a Plot of log $K_{OW}$ versus LeBas molar volume.

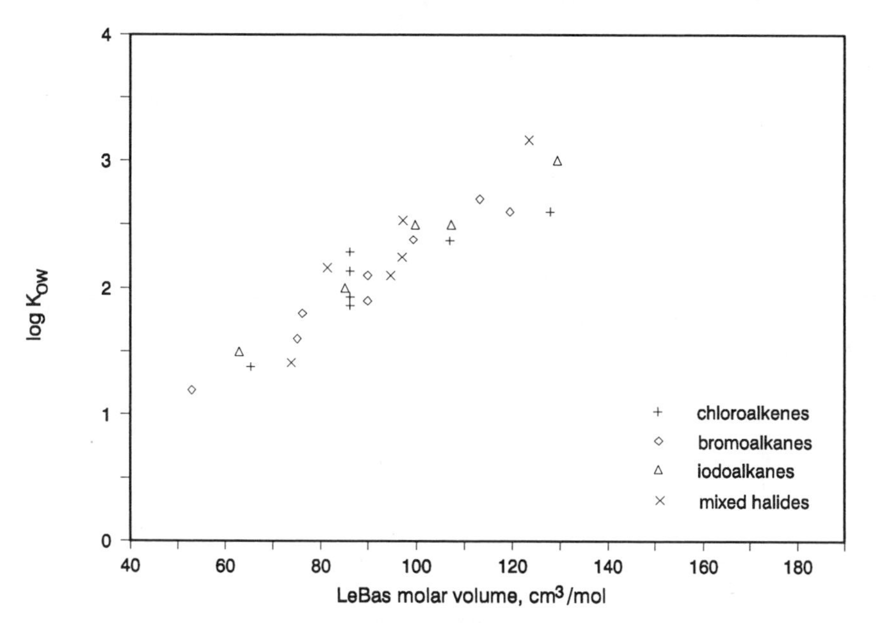


Figure 3.4b Plot of log $K_{OW}$ versus LeBas molar volume.

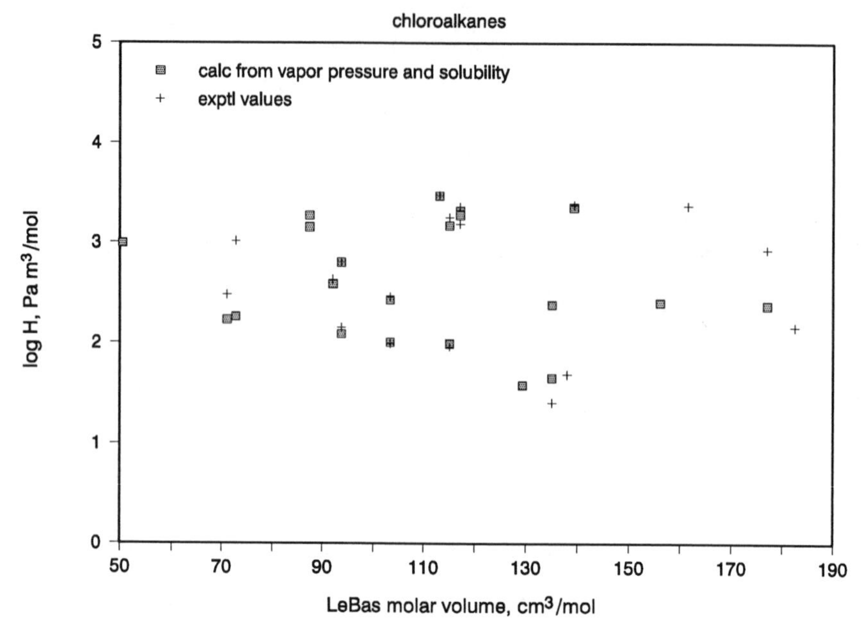


Figure 3.5a Plot of log Henry's law constant versus LeBas molar volume.

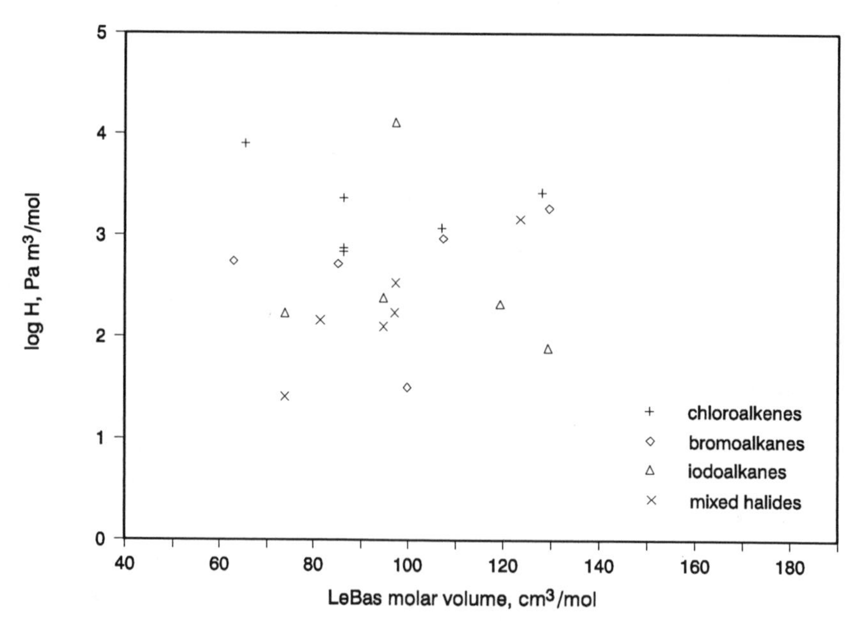


Figure 3.5b Plot of log Henry's law constant versus LeBas molar volume.

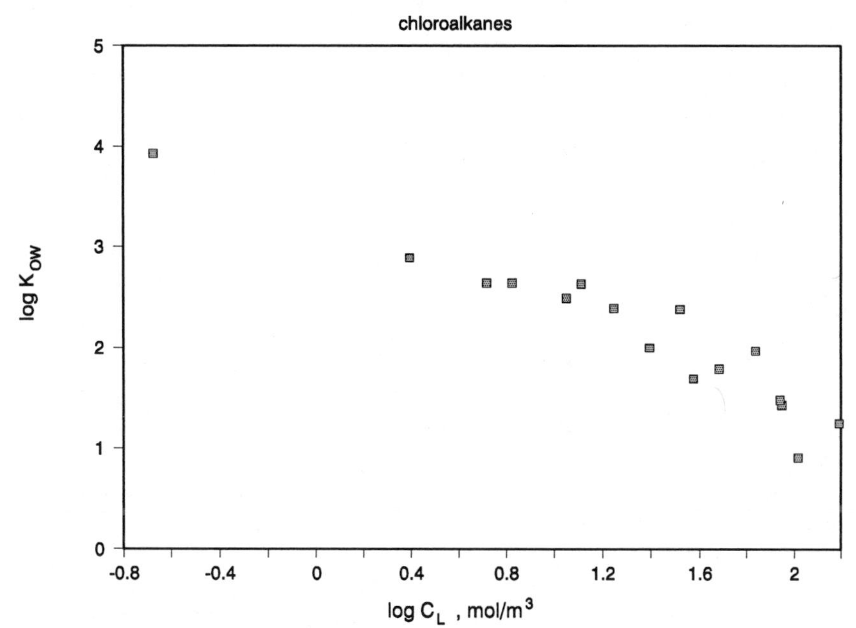


Figure 3.6a Plot of log $K_{OW}$ versus log $C_L$.

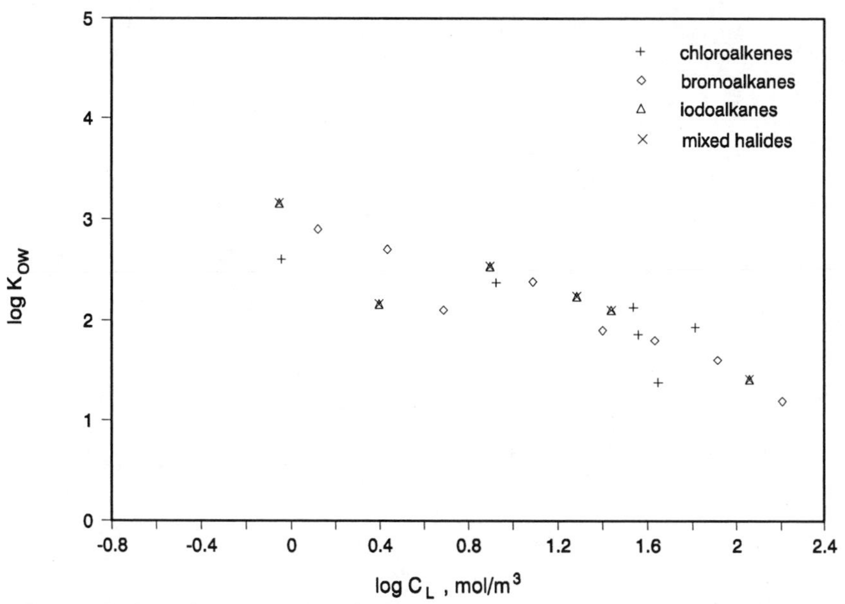


Figure 3.6b Plot of log $K_{OW}$ versus log $C_L$.

## 3.3 Illustrative Fugacity Calculations: Levels I, II and III

Chemical name: Dichloromethane

Level I calculation: (six-compartment model)

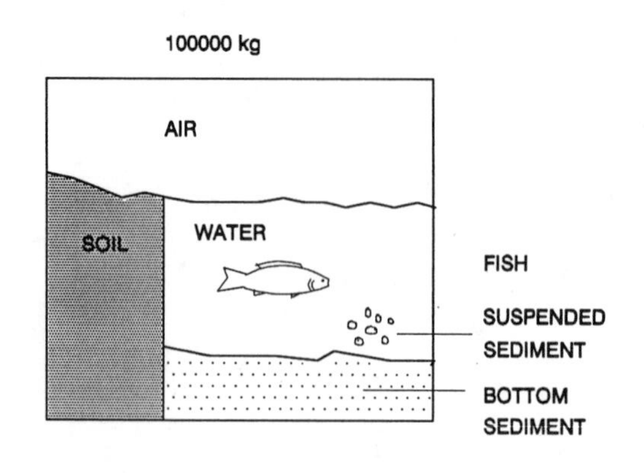


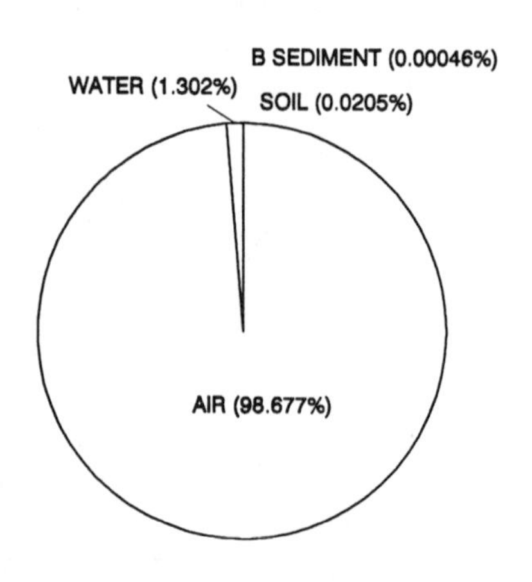


Distribution of mass

physical-chemical properties:

MW: 84.94 g/mol

M.P.: - 95.0 °C

Fugacity ratio: 1.0

vapor pressure: 58400 Pa

solubility: 13200 g/m$^3$

log $K_{OW}$: 1.25

| Compartment | Z | Concentration | | | Amount | Amount |
|---|---|---|---|---|---|---|
| | mol/m3 Pa | mol/m3 | mg/L (or g/m3) | ug/g | kg | % |
| Air | 4.034E-04 | 1.162E-08 | 9.868E-07 | 8.324E-04 | 98677 | 98.677 |
| Water | 2.661E-03 | 7.663E-08 | 6.509E-06 | 6.509E-06 | 1301.79 | 1.3018 |
| Soil | 9.313E-04 | 2.682E-08 | 2.278E-06 | 9.491E-07 | 20.501 | 0.0205 |
| Biota (fish) | 2.366E-03 | 6.813E-08 | 5.787E-06 | 5.787E-06 | 1.16E-03 | 1.16E-06 |
| Suspended sediment | 5.820E-03 | 1.676E-07 | 1.424E-05 | 9.491E-06 | 0.0142 | 1.42E-05 |
| Bottom sediment | 1.863E-03 | 5.364E-08 | 4.556E-06 | 1.898E-06 | 0.4556 | 4.56E-04 |
| | Total | | | | 100000 | 100 |

f = 2.880E-05 Pa

Chemical name: Dichloromethane

Level II calculation: (six-compartment model)

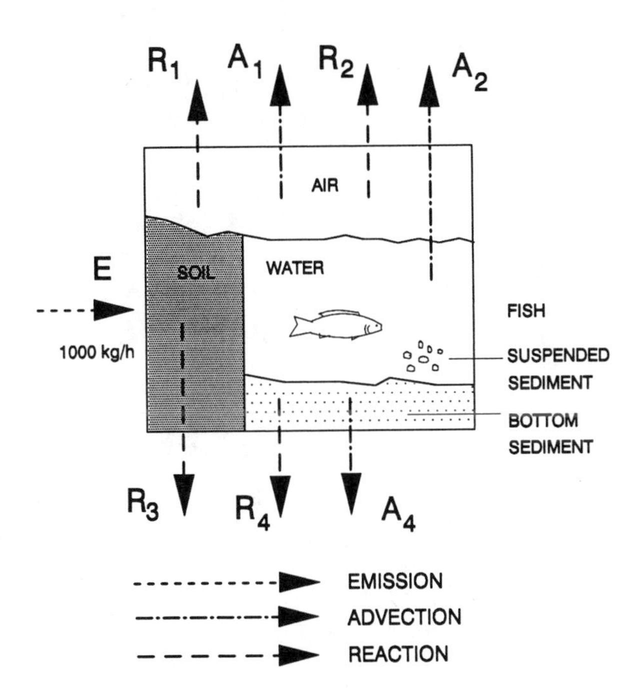


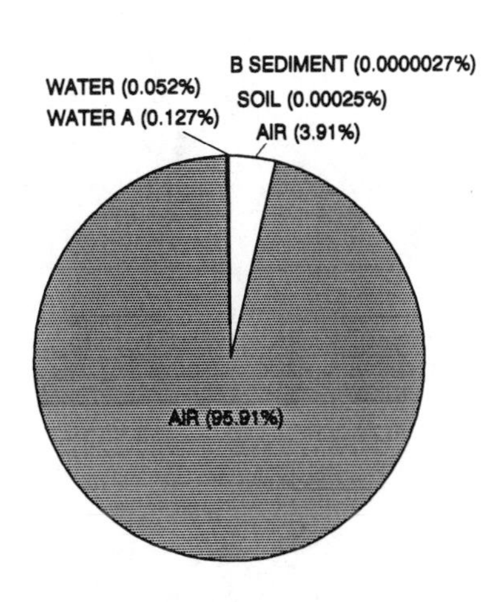


Distribution of removal rates

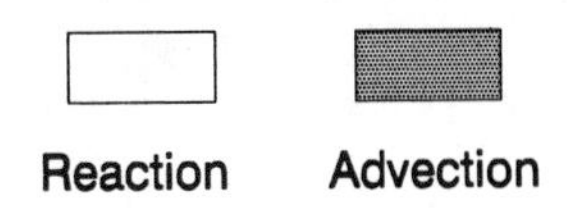


| Compartment | Half-life h | D Values Reaction mol/Pa h | D Values Advection mol/Pa h | Conc'n mol/m3 | Loss Reaction kg/h | Loss Advection kg/h | Removal % |
|---|---|---|---|---|---|---|---|
| Air | 1700 | 1.64E+07 | 4.03E+08 | 1.13E-08 | 39.098 | 959.118 | 99.8216 |
| Water | 1700 | 2.17E+05 | 5.32E+05 | 7.45E-08 | 0.5158 | 1.2653 | 0.1781 |
| Soil | 5500 | 1.06E+03 | | 2.61E-08 | 2.51E-03 | | 2.51E-04 |
| Biota (fish) | | | | 6.62E-08 | | | |
| Suspended sediment | | | | 1.63E-07 | | | |
| Bottom sediment | 17000 | 7.59E+00 | 3.73E+00 | 5.21E-08 | 1.81E-05 | 8.86E-06 | 2.69E-06 |
| | Total | 1.67E+07 | 4.04E+08 | | 39.62 | 960.38 | 100 |
| | R + A | | 4.21E+08 | | | 1000 | |

f = 2.799E-05 Pa
Total Amt= 97198 kg

Overall residence time = 97.20 h
Reaction time = 2453.46 h
Advection time = 101.21 h

## Fugacity Level III calculations: (four-compartment model)

Chemical name: Dichloromethane

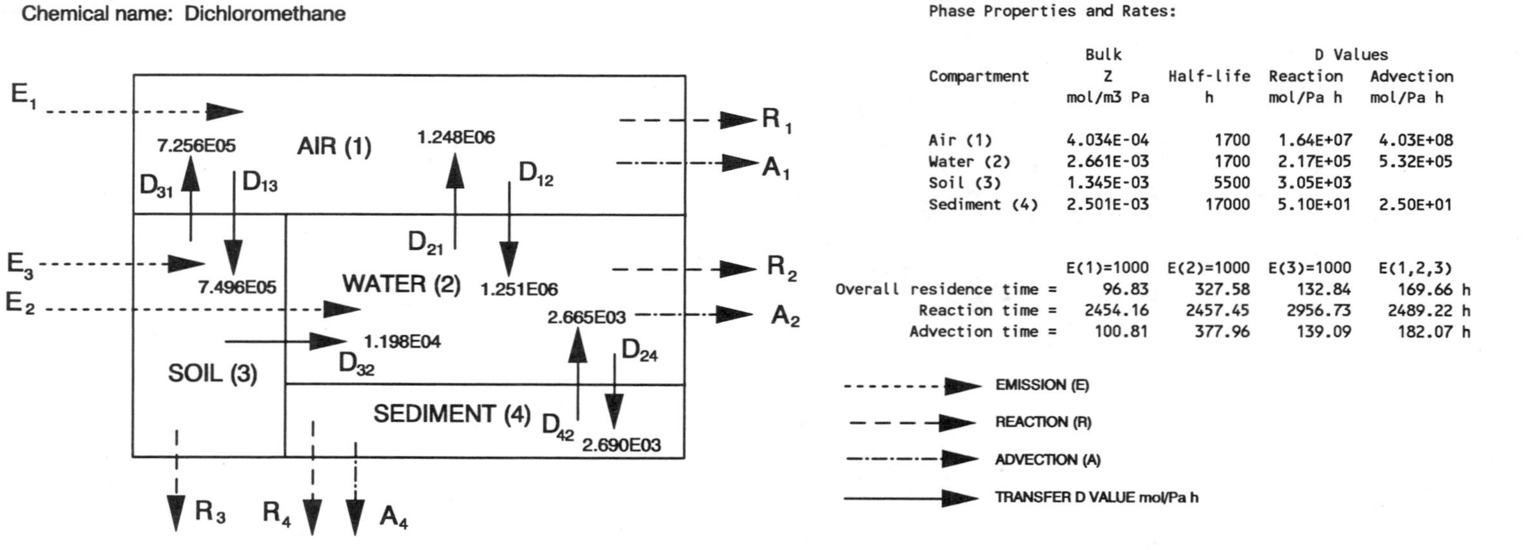


Phase Properties and Rates:

| Compartment | Bulk Z mol/m3 Pa | Half-life h | D Values Reaction mol/Pa h | D Values Advection mol/Pa h |
|---|---|---|---|---|
| Air (1) | 4.034E-04 | 1700 | 1.64E+07 | 4.03E+08 |
| Water (2) | 2.661E-03 | 1700 | 2.17E+05 | 5.32E+05 |
| Soil (3) | 1.345E-03 | 5500 | 3.05E+03 | |
| Sediment (4) | 2.501E-03 | 17000 | 5.10E+01 | 2.50E+01 |

| | E(1)=1000 | E(2)=1000 | E(3)=1000 | E(1,2,3) |
|---|---|---|---|---|
| Overall residence time = | 96.83 | 327.58 | 132.84 | 169.66 h |
| Reaction time = | 2454.16 | 2457.45 | 2956.73 | 2489.22 h |
| Advection time = | 100.81 | 377.96 | 139.09 | 182.07 h |

Phase Properties, Compositions, Transport and Transformation Rates:

| Emission, kg/h | | | Fugacity, Pa | | | | Concentration, g/m3 | | | | Amounts, kg | | | | Total Amount, kg |
|---|---|---|---|---|---|---|---|---|---|---|---|---|---|---|---|
| E(1) | E(2) | E(3) | f(1) | f(2) | f(3) | f(4) | C(1) | C(2) | C(3) | C(4) | m(1) | m(2) | m(3) | m(4) | |
| 1000 | 0 | 0 | 2.801E-05 | 1.771E-05 | 2.835E-05 | 1.738E-05 | 9.597E-07 | 4.003E-06 | 3.237E-06 | 3.693E-06 | 9.597E+04 | 8.006E+02 | 5.827E+01 | 1.847E+00 | 9.683E+04 |
| 0 | 1000 | 0 | 1.750E-05 | 5.905E-03 | 1.771E-05 | 5.796E-03 | 5.997E-07 | 1.335E-03 | 2.023E-06 | 1.231E-03 | 5.997E+04 | 2.669E+05 | 3.642E+01 | 6.157E+02 | 3.276E+05 |
| 0 | 0 | 1000 | 2.772E-05 | 1.128E-04 | 1.592E-02 | 1.107E-04 | 9.500E-07 | 2.550E-05 | 1.819E-03 | 2.353E-05 | 9.500E+04 | 5.101E+03 | 3.274E+04 | 1.176E+01 | 1.328E+05 |
| 600 | 300 | 100 | 2.483E-05 | 1.793E-03 | 1.615E-03 | 1.760E-03 | 8.508E-07 | 4.054E-04 | 1.844E-04 | 3.740E-04 | 8.508E+04 | 8.108E+04 | 3.319E+03 | 1.870E+02 | 1.697E+05 |

| Emission, kg/h | | | Loss, Reaction, kg/h | | | | Loss, Advection, kg/h | | | Intermedia Rate of Transport, kg/h T12 | T21 | T13 | T31 | T32 | T24 | T42 |
|---|---|---|---|---|---|---|---|---|---|---|---|---|---|---|---|---|
| E(1) | E(2) | E(3) | R(1) | R(2) | R(3) | R(4) | A(1) | A(2) | A(4) | air-water | water-air | air-soil | soil-air | soil-water | water-sed | sed-water |
| 1000 | 0 | 0 | 3.912E+01 | 3.264E-01 | 7.34E-03 | 7.527E-05 | 9.597E+02 | 8.006E-01 | 3.693E-05 | 2.976E+00 | 1.878E+00 | 1.783E+00 | 1.747E+00 | 2.883E-02 | 4.047E-03 | 3.934E-03 |
| 0 | 1000 | 0 | 2.445E+01 | 1.088E+02 | 4.59E-03 | 2.510E-02 | 5.997E+02 | 2.669E+02 | 1.231E-02 | 1.860E+00 | 6.261E+02 | 1.114E+00 | 1.092E+00 | 1.802E-02 | 1.349E+00 | 1.312E+00 |
| 0 | 0 | 1000 | 3.873E+01 | 2.079E+00 | 4.12E+00 | 4.796E-04 | 9.500E+02 | 5.101E+00 | 2.353E-04 | 2.945E+00 | 1.196E+01 | 1.765E+00 | 9.814E+02 | 1.620E+01 | 2.578E-02 | 2.507E-02 |
| 600 | 300 | 100 | 3.468E+01 | 3.305E+01 | 4.18E-01 | 7.623E-03 | 8.508E+02 | 8.108E+01 | 3.740E-03 | 2.638E+00 | 1.901E+02 | 1.581E+00 | 9.952E+01 | 1.642E+00 | 4.098E-01 | 3.984E-01 |

# Level III Distribution

Chemical name: Dichloromethane

Distribution of mass

WATER (0.83%)
SEDIMENT (0.002%)
SOIL (0.06%)
AIR (99.11%)

SOIL (0.0111%)
SEDIMENT (0.188%)
AIR (18.31%)
WATER (81.49%)

SEDIMENT (0.0089%)
SOIL (24.64%)
WATER (3.84%)
AIR (71.51%)

SOIL (1.96%)
SEDIMENT (0.11%)
WATER (47.79%)
AIR (50.15%)

Distribution of removal rates

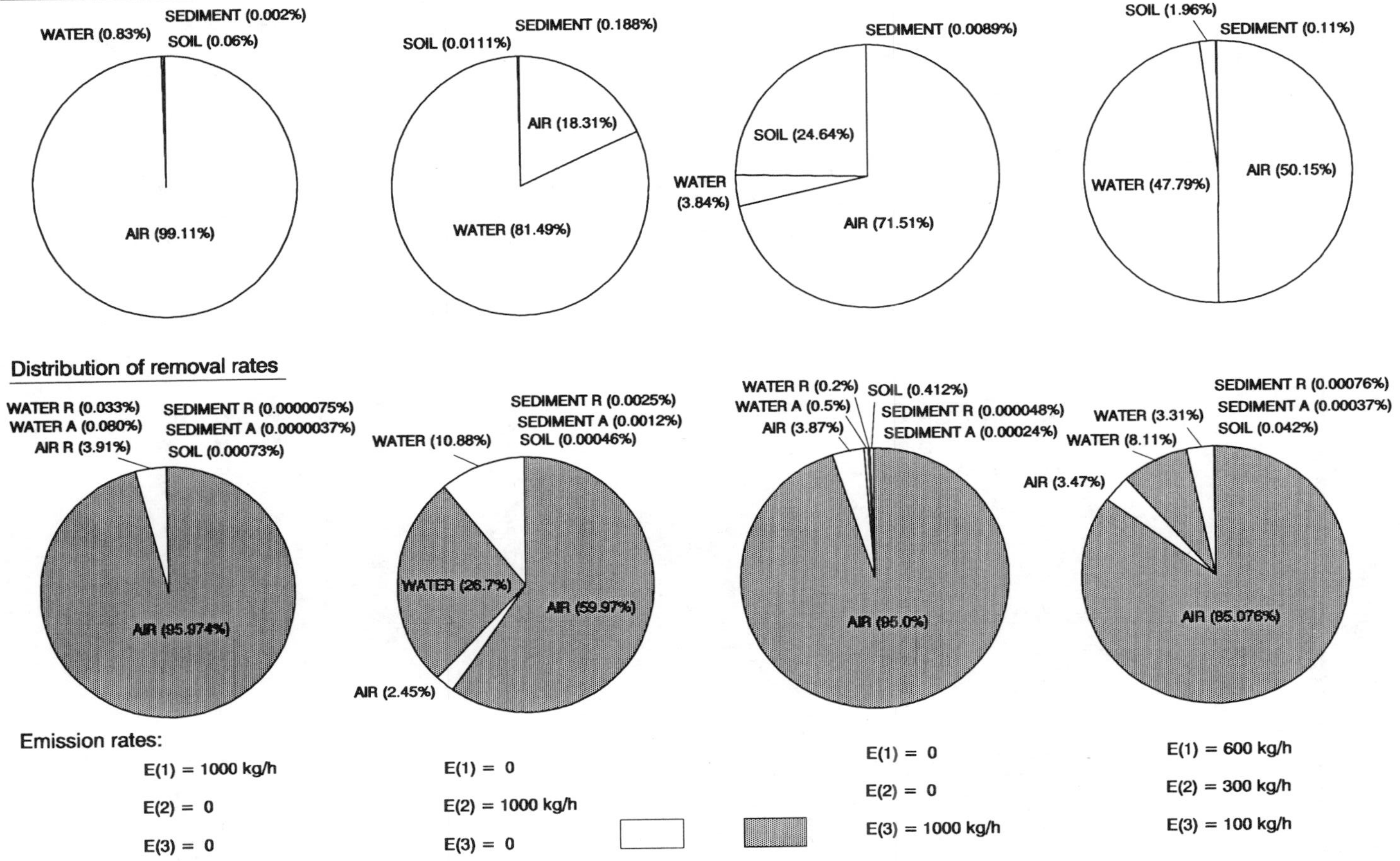

WATER R (0.033%)
WATER A (0.080%)
AIR R (3.91%)
SEDIMENT R (0.0000075%)
SEDIMENT A (0.0000037%)
SOIL (0.00073%)
AIR (95.974%)

SEDIMENT R (0.0025%)
SEDIMENT A (0.0012%)
SOIL (0.00046%)
WATER (10.88%)
WATER (26.7%)
AIR (59.97%)
AIR (2.45%)

WATER R (0.2%)
WATER A (0.5%)
AIR (3.87%)
SOIL (0.412%)
SEDIMENT R (0.000048%)
SEDIMENT A (0.00024%)
AIR (95.0%)

SEDIMENT R (0.00076%)
SEDIMENT A (0.00037%)
SOIL (0.042%)
WATER (3.31%)
WATER (8.11%)
AIR (3.47%)
AIR (85.076%)

Reaction
Advection

Emission rates:

| | | | |
|---|---|---|---|
| E(1) = 1000 kg/h | E(1) = 0 | E(1) = 0 | E(1) = 600 kg/h |
| E(2) = 0 | E(2) = 1000 kg/h | E(2) = 0 | E(2) = 300 kg/h |
| E(3) = 0 | E(3) = 0 | E(3) = 1000 kg/h | E(3) = 100 kg/h |

Residence time:

| | | | |
|---|---|---|---|
| t = 96.83 h | t = 328 h | t = 132.8 h | t = 169.7 h |

Chemical name: Chloroform

Level I calculation: (six-compartment model)

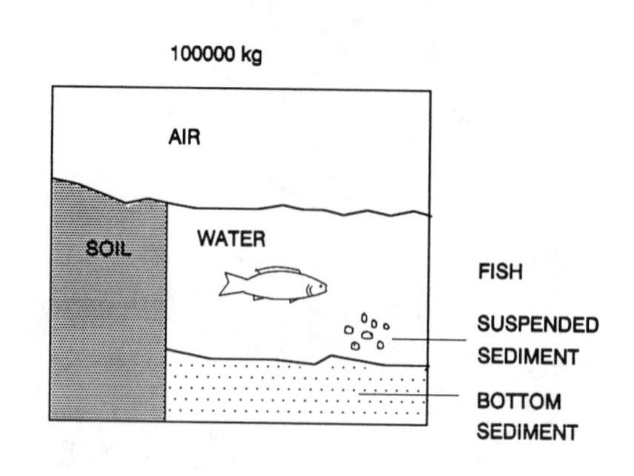


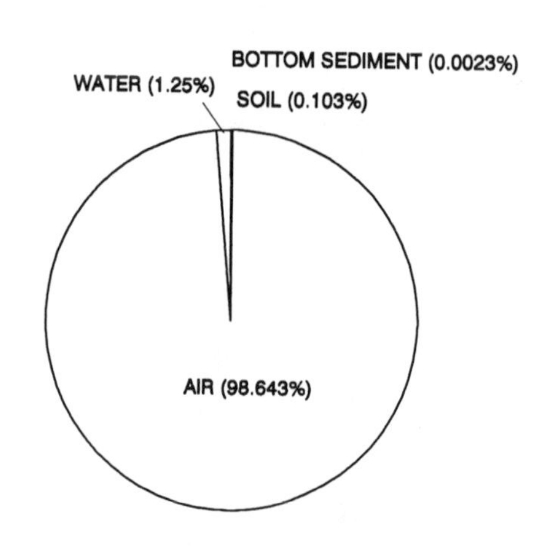


Distribution of mass

physical-chemical properties:

MW: 119.38 g/mol

M.P.: - 63.5 °C

Fugacity ratio: 1.0

vapor pressure: 26200 Pa

solubility: 8000 g/m$^3$

log $K_{OW}$: 1.97

| Compartment | Z | Concentration | | | Amount | Amount |
|---|---|---|---|---|---|---|
| | mol/m3 Pa | mol/m3 | mg/L (or g/m3) | ug/g | kg | % |
| Air | 4.034E-04 | 8.263E-09 | 9.864E-07 | 8.321E-04 | 98643 | 98.643 |
| Water | 2.558E-03 | 5.239E-08 | 6.254E-06 | 6.254E-06 | 1250.84 | 1.2508 |
| Soil | 4.698E-03 | 9.622E-08 | 1.149E-05 | 4.786E-06 | 103.380 | 0.1034 |
| Biota (fish) | 1.194E-02 | 2.445E-07 | 2.918E-05 | 2.918E-05 | 5.84E-03 | 5.84E-06 |
| Suspended sediment | 2.936E-02 | 6.014E-07 | 7.179E-05 | 4.786E-05 | 0.0718 | 7.18E-05 |
| Bottom sediment | 9.395E-03 | 1.924E-07 | 2.297E-05 | 9.572E-06 | 2.2973 | 2.30E-03 |
| | Total | | | | 100000 | 100 |

f = 2.048E-05 Pa

Chemical name: Chloroform

Level II calculation: (six-compartment model)

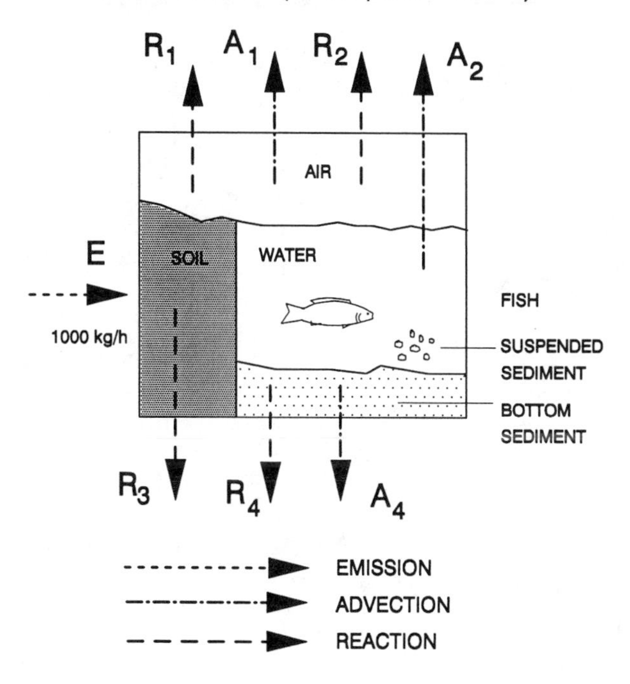


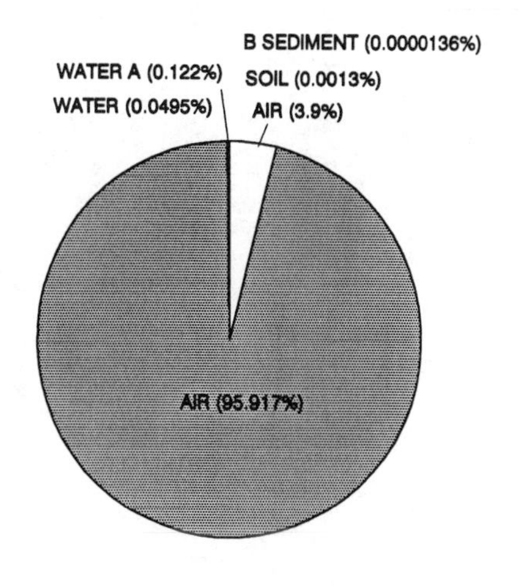


Distribution of removal rates

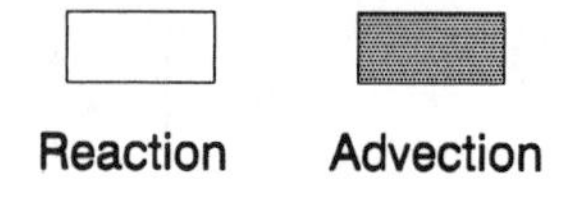


| Compartment | Half-life h | D Values Reaction mol/Pa h | D Values Advection mol/Pa h | Conc'n mol/m3 | Loss Reaction kg/h | Loss Advection kg/h | Removal % |
|---|---|---|---|---|---|---|---|
| Air | 1700 | 1.64E+07 | 4.03E+08 | 8.03E-09 | 39.100 | 959.175 | 99.8275 |
| Water | 1700 | 2.09E+05 | 5.12E+05 | 5.09E-08 | 0.4958 | 1.2163 | 0.1712 |
| Soil | 5500 | 5.33E+03 | | 9.36E-08 | 1.27E-02 | | 1.27E-03 |
| Biota (fish) | | | | 2.38E-07 | | | |
| Suspended sediment | | | | 5.85E-07 | | | |
| Bottom sediment | 17000 | 3.83E+01 | 1.88E+01 | 1.87E-07 | 9.11E-05 | 4.47E-05 | 1.36E-05 |
| | Total | 1.67E+07 | 4.04E+08 | | 39.61 | 960.39 | 100 |
| | R + A | | 4.21E+08 | | | 1000 | |

f = 1.992E-05 Pa
Total Amt= 97237 kg

Overall residence time = 97.24 h
Reaction time = 2454.91 h
Advection time = 101.25 h

## Fugacity Level III calculations: (four-compartment model)

Chemical name: Chloroform

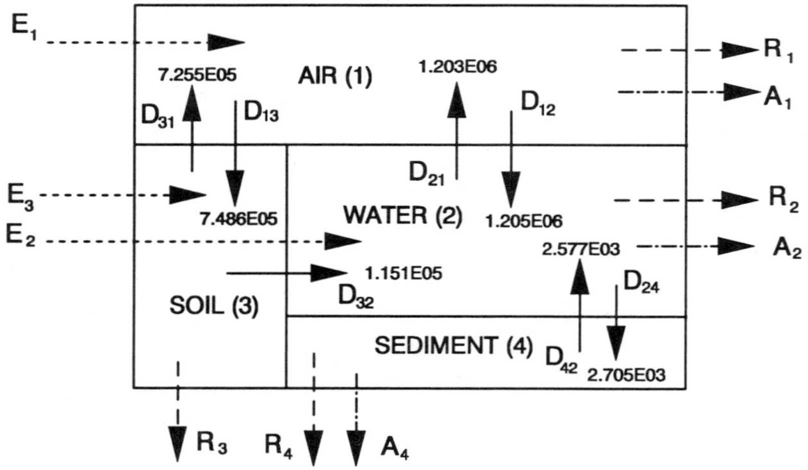

E1 E3 E2
AIR (1) WATER (2) SOIL (3) SEDIMENT (4)
$D_{31}$ 7.255E05 $D_{13}$ 7.486E05
$D_{21}$ 1.203E06 $D_{12}$ 1.205E06
$D_{32}$ 1.151E05
$D_{42}$ 2.577E03 $D_{24}$ 2.705E03
R1 A1 R2 A2 R3 R4 A4

EMISSION (E)
REACTION (R)
ADVECTION (A)
TRANSFER D VALUE mol/Pa h

Phase Properties and Rates:

| Compartment | Bulk Z mol/m3 Pa | Half-life h | D Values Reaction mol/Pa h | D Values Advection mol/Pa h |
|---|---|---|---|---|
| Air (1) | 4.034E-04 | 1700 | 1.64E+07 | 4.03E+08 |
| Water (2) | 2.558E-03 | 1700 | 2.09E+05 | 5.12E+05 |
| Soil (3) | 3.197E-03 | 5500 | 7.25E+03 | |
| Sediment (4) | 3.925E-03 | 17000 | 8.00E+01 | 3.93E+01 |

| | E(1)=1000 | E(2)=1000 | E(3)=1000 | E(1,2,3) | |
|---|---|---|---|---|---|
| Overall residence time = | 96.89 | 327.67 | 176.83 | 174.12 | h |
| Reaction time = | 2455.58 | 2460.48 | 3518.01 | 2536.22 | h |
| Advection time = | 100.87 | 378.01 | 186.18 | 186.95 | h |

Phase Properties, Compositions, Transport and Transformation Rates:

| Emission, kg/h | | | Fugacity, Pa | | | | Concentration, g/m3 | | | | Amounts, kg | | | | Total Amount, kg |
|---|---|---|---|---|---|---|---|---|---|---|---|---|---|---|---|
| E(1) | E(2) | E(3) | f(1) | f(2) | f(3) | f(4) | C(1) | C(2) | C(3) | C(4) | m(1) | m(2) | m(3) | m(4) | |
| 1000 | 0 | 0 | 1.993E-05 | 1.261E-05 | 2.004E-05 | 1.265E-05 | 9.598E-07 | 3.851E-06 | 7.649E-06 | 5.929E-06 | 9.598E+04 | 7.702E+02 | 1.377E+02 | 2.964E+00 | 9.689E+04 |
| 0 | 1000 | 0 | 1.246E-05 | 4.364E-03 | 1.254E-05 | 4.378E-03 | 6.003E-07 | 1.333E-03 | 4.784E-06 | 2.052E-03 | 6.003E+04 | 2.665E+05 | 8.611E+01 | 1.026E+03 | 3.277E+05 |
| 0 | 0 | 1000 | 1.962E-05 | 7.980E-05 | 1.127E-02 | 8.006E-05 | 9.449E-07 | 2.437E-05 | 4.303E-03 | 3.752E-05 | 9.449E+04 | 4.874E+03 | 7.745E+04 | 1.876E+01 | 1.768E+05 |
| 600 | 300 | 100 | 1.766E-05 | 1.325E-03 | 1.143E-03 | 1.329E-03 | 8.504E-07 | 4.045E-04 | 4.363E-04 | 6.228E-04 | 8.504E+04 | 8.091E+04 | 7.853E+03 | 3.114E+02 | 1.741E+05 |

| Emission, kg/h | | | Loss, Reaction, kg/h | | | | Loss, Advection, kg/h | | | Intermedia Rate of Transport, kg/h T12 | T21 | T13 | T31 | T32 | T24 | T42 |
|---|---|---|---|---|---|---|---|---|---|---|---|---|---|---|---|---|
| E(1) | E(2) | E(3) | R(1) | R(2) | R(3) | R(4) | A(1) | A(2) | A(4) | air-water | water-air | air-soil | soil-air | soil-water | water-sed | sed-water |
| 1000 | 0 | 0 | 3.912E+01 | 3.140E-01 | 1.73E-02 | 1.208E-04 | 9.598E+02 | 7.702E-01 | 5.929E-05 | 2.867E+00 | 1.811E+00 | 1.781E+00 | 1.736E+00 | 2.755E-02 | 4.072E-03 | 3.891E-03 |
| 0 | 1000 | 0 | 2.447E+01 | 1.087E+02 | 1.09E-02 | 4.182E-02 | 6.003E+02 | 2.665E+02 | 2.052E-02 | 1.793E+00 | 6.266E+02 | 1.114E+00 | 1.086E+00 | 1.723E-02 | 1.409E+00 | 1.347E+00 |
| 0 | 0 | 1000 | 3.852E+01 | 1.987E+00 | 9.76E+00 | 7.647E-04 | 9.449E+02 | 4.874E+00 | 3.752E-04 | 2.823E+00 | 1.146E+01 | 1.753E+00 | 9.765E+02 | 1.550E+01 | 2.577E-02 | 2.463E-02 |
| 600 | 300 | 100 | 3.467E+01 | 3.298E+01 | 9.89E-01 | 1.269E-02 | 8.504E+02 | 8.091E+01 | 6.228E-03 | 2.541E+00 | 1.902E+02 | 1.578E+00 | 9.902E+01 | 1.571E+00 | 4.277E-01 | 4.088E-01 |

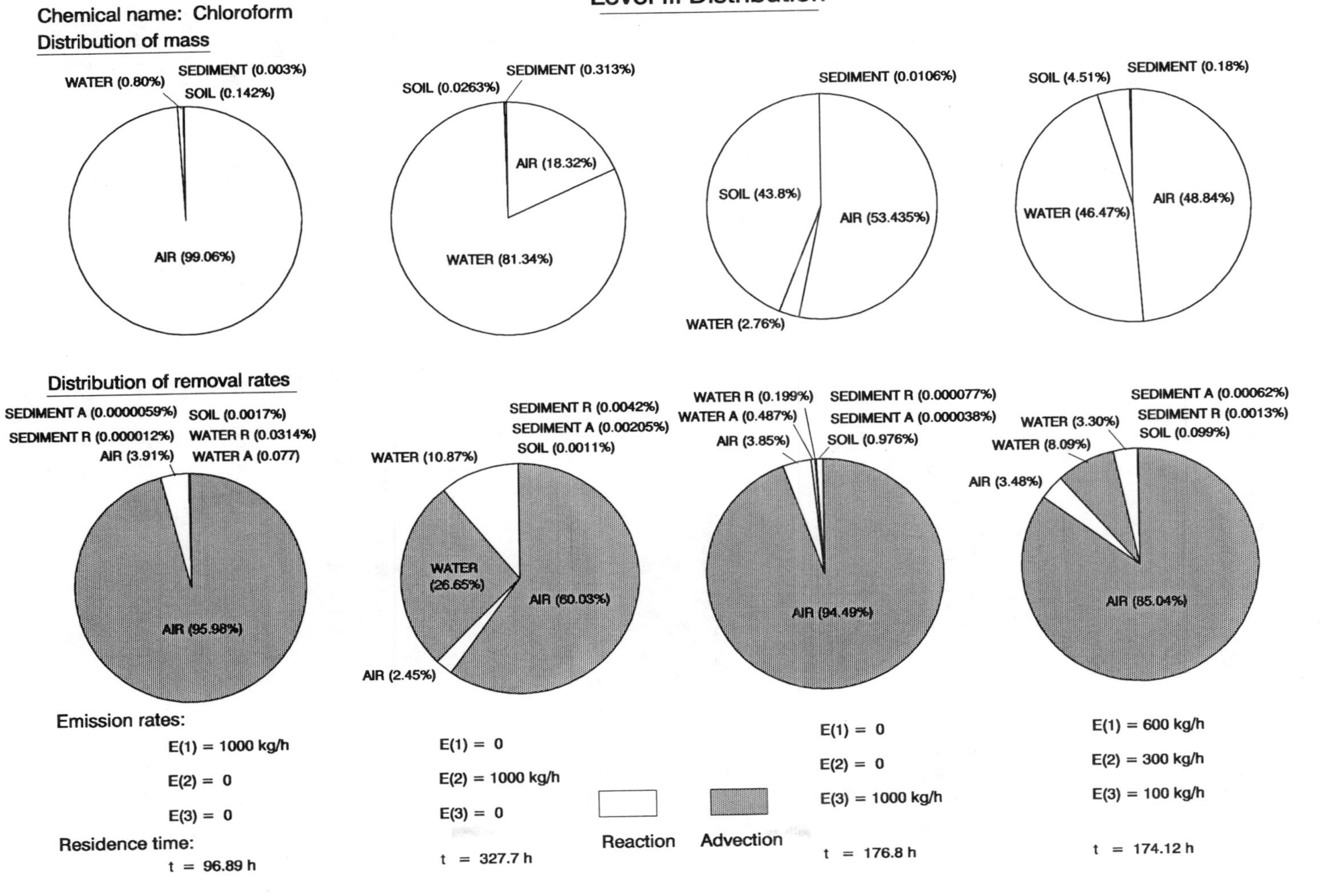

Level III Distribution
Chemical name: Chloroform
Distribution of mass
WATER (0.80%)
SEDIMENT (0.003%)
SOIL (0.142%)
AIR (99.06%)
SOIL (0.0263%)
SEDIMENT (0.313%)
AIR (18.32%)
WATER (81.34%)
SEDIMENT (0.0106%)
SOIL (43.8%)
AIR (53.435%)
WATER (2.76%)
SOIL (4.51%)
SEDIMENT (0.18%)
WATER (46.47%)
AIR (48.84%)
Distribution of removal rates
SEDIMENT A (0.0000059%)
SOIL (0.0017%)
SEDIMENT R (0.000012%)
WATER R (0.0314%)
AIR (3.91%)
WATER A (0.077)
AIR (95.98%)
SEDIMENT R (0.0042%)
SEDIMENT A (0.00205%)
SOIL (0.0011%)
WATER (10.87%)
WATER (26.65%)
AIR (60.03%)
AIR (2.45%)
WATER R (0.199%)
WATER A (0.487%)
AIR (3.85%)
SEDIMENT R (0.000077%)
SEDIMENT A (0.000038%)
SOIL (0.976%)
AIR (94.49%)
SEDIMENT A (0.00062%)
SEDIMENT R (0.0013%)
SOIL (0.099%)
WATER (3.30%)
WATER (8.09%)
AIR (3.48%)
AIR (85.04%)
Emission rates:
E(1) = 1000 kg/h
E(2) = 0
E(3) = 0
Residence time:
t = 96.89 h
E(1) = 0
E(2) = 1000 kg/h
E(3) = 0
t = 327.7 h
Reaction
Advection
E(1) = 0
E(2) = 0
E(3) = 1000 kg/h
t = 176.8 h
E(1) = 600 kg/h
E(2) = 300 kg/h
E(3) = 100 kg/h
t = 174.12 h

Chemical name: Carbon tetrachloride

Level I calculation: (six-compartment model)

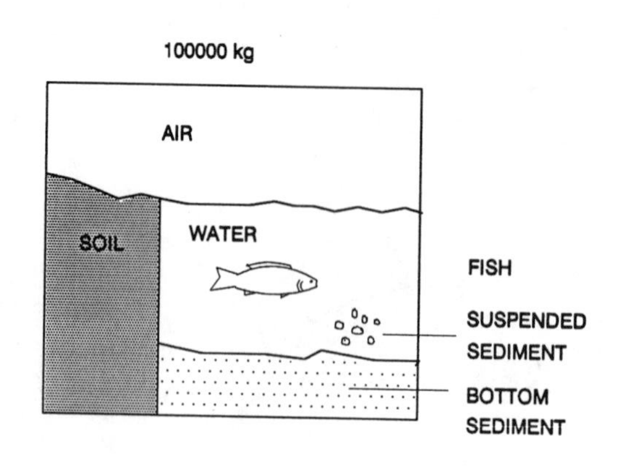


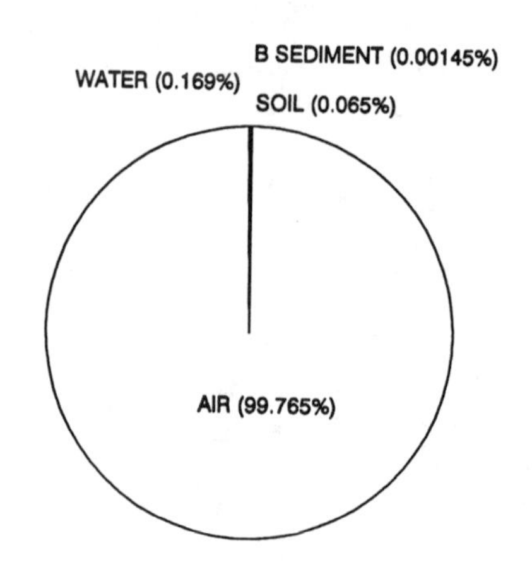


Distribution of mass

physical-chemical properties:

MW: 153.82 g/mol

M.P.: - 22.9 °C

Fugacity ratio: 1.0

vapor pressure: 15250 Pa

solubility: 800 g/m³

log $K_{OW}$: 2.64

| Compartment | Z | Concentration | | | Amount | Amount |
|---|---|---|---|---|---|---|
| | mol/m3 Pa | mol/m3 | mg/L (or g/m3) | ug/g | kg | % |
| Air | 4.034E-04 | 6.486E-09 | 9.976E-07 | 8.416E-04 | 99765 | 99.765 |
| Water | 3.410E-04 | 5.483E-09 | 8.434E-07 | 8.434E-07 | 168.68 | 0.1687 |
| Soil | 2.930E-03 | 4.710E-08 | 7.245E-06 | 3.019E-06 | 65.207 | 0.0652 |
| Biota (fish) | 7.444E-03 | 1.197E-07 | 1.841E-05 | 1.841E-05 | 3.68E-03 | 3.68E-06 |
| Suspended sediment | 1.831E-02 | 2.944E-07 | 4.528E-05 | 3.019E-05 | 0.0453 | 4.53E-05 |
| Bottom sediment | 5.860E-03 | 9.420E-08 | 1.449E-05 | 6.038E-06 | 1.4491 | 1.45E-03 |
| | Total | | | | 100000 | 100 |

f = 1.608E-05 Pa

Chemical name: Carbon tetrachloride

Level II calculation: (six-compartment model)

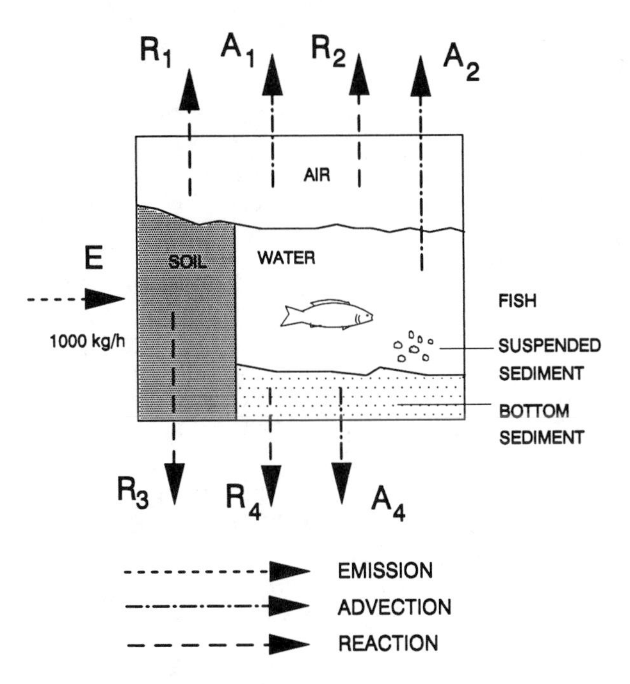


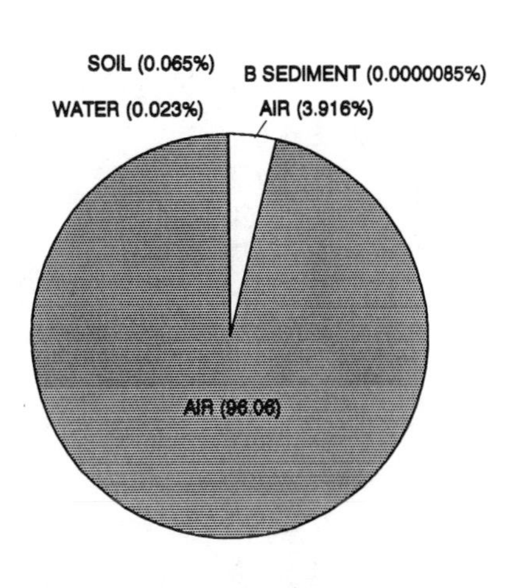


Distribution of removal rates

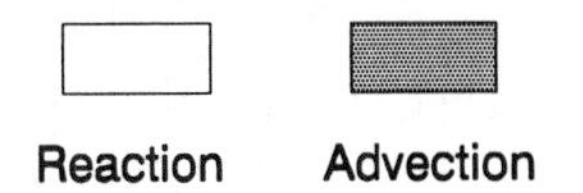


| Compartment | Half-life h | D Values Reaction mol/Pa h | D Values Advection mol/Pa h | Conc'n mol/m3 | Loss Reaction kg/h | Loss Advection kg/h | Removal % |
|---|---|---|---|---|---|---|---|
| Air | 1700 | 1.64E+07 | 4.03E+08 | 6.24E-09 | 39.159 | 960.605 | 99.9763 |
| Water | 1700 | 2.78E+04 | 6.82E+04 | 5.28E-09 | 0.0662 | 0.1624 | 0.0229 |
| Soil | 5500 | 3.32E+03 | | 4.54E-08 | 7.91E-03 | | 7.91E-04 |
| Biota (fish) | | | | 1.15E-07 | | | |
| Suspended sediment | | | | 2.83E-07 | | | |
| Bottom sediment | 17000 | 2.39E+01 | 1.17E+01 | 9.07E-08 | 5.69E-05 | 2.79E-05 | 8.48E-06 |
| | Total | 1.65E+07 | 4.03E+08 | | 39.23 | 960.77 | 100 |
| | R + A | | 4.20E+08 | | | 1000 | |

f = 1.548E-05 Pa

Total Amt= 96287 kg

Overall residence time = 96.29 h

Reaction time = 2454.24 h

Advection time = 100.22 h

## Fugacity Level III calculations: (four-compartment model)

Chemical name: Carbon tetrachloride

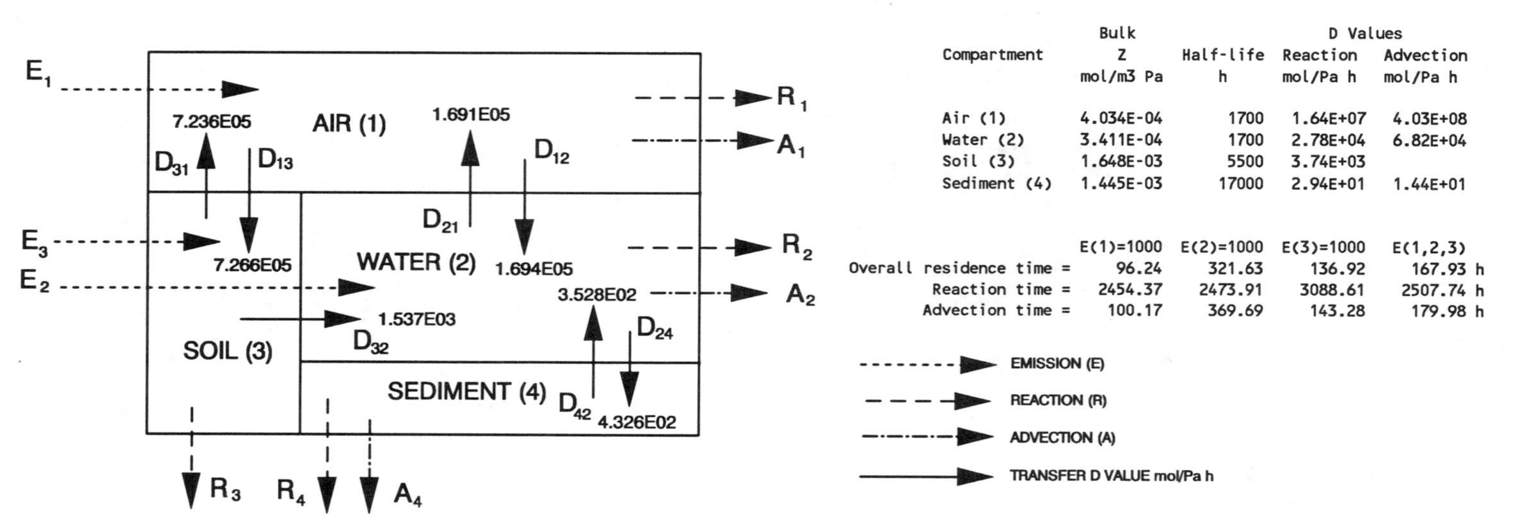


Phase Properties and Rates:

| Compartment | Bulk Z mol/m3 Pa | Half-life h | D Values Reaction mol/Pa h | D Values Advection mol/Pa h |
|---|---|---|---|---|
| Air (1) | 4.034E-04 | 1700 | 1.64E+07 | 4.03E+08 |
| Water (2) | 3.411E-04 | 1700 | 2.78E+04 | 6.82E+04 |
| Soil (3) | 1.648E-03 | 5500 | 3.74E+03 | |
| Sediment (4) | 1.445E-03 | 17000 | 2.94E+01 | 1.44E+01 |

| | E(1)=1000 | E(2)=1000 | E(3)=1000 | E(1,2,3) | |
|---|---|---|---|---|---|
| Overall residence time = | 96.24 | 321.63 | 136.92 | 167.93 | h |
| Reaction time = | 2454.37 | 2473.91 | 3088.61 | 2507.74 | h |
| Advection time = | 100.17 | 369.69 | 143.28 | 179.98 | h |

EMISSION (E)
REACTION (R)
ADVECTION (A)
TRANSFER D VALUE mol/Pa h

Phase Properties, Compositions, Transport and Transformation Rates:

| Emission, kg/h | | | Fugacity, Pa | | | | Concentration, g/m3 | | | | Amounts, kg | | | | Total Amount, kg |
|---|---|---|---|---|---|---|---|---|---|---|---|---|---|---|---|
| E(1) | E(2) | E(3) | f(1) | f(2) | f(3) | f(4) | C(1) | C(2) | C(3) | C(4) | m(1) | m(2) | m(3) | m(4) | |
| 1000 | 0 | 0 | 1.548E-05 | 9.981E-06 | 1.543E-05 | 1.089E-05 | 9.607E-07 | 5.238E-07 | 3.912E-06 | 2.419E-06 | 9.607E+04 | 1.048E+02 | 7.042E+01 | 1.210E+00 | 9.624E+04 |
| 0 | 1000 | 0 | 9.872E-06 | 2.452E-02 | 9.842E-06 | 2.674E-02 | 6.126E-07 | 1.287E-03 | 2.495E-06 | 5.943E-03 | 6.126E+04 | 2.574E+05 | 4.490E+01 | 2.972E+03 | 3.216E+05 |
| 0 | 0 | 1000 | 1.539E-05 | 6.163E-05 | 8.935E-03 | 6.722E-05 | 9.550E-07 | 3.234E-06 | 2.265E-03 | 1.494E-05 | 9.550E+04 | 6.468E+02 | 4.077E+04 | 7.469E+00 | 1.369E+05 |
| 600 | 300 | 100 | 1.379E-05 | 7.369E-03 | 9.057E-04 | 8.036E-03 | 8.557E-07 | 3.867E-04 | 2.296E-04 | 1.786E-03 | 8.557E+04 | 7.733E+04 | 4.132E+03 | 8.930E+02 | 1.679E+05 |

| Emission, kg/h | | | Loss, Reaction, kg/h | | | | Loss, Advection, kg/h | | | Intermedia Rate of Transport, kg/h T12 | T21 | T13 | T31 | T32 | T24 | T42 |
|---|---|---|---|---|---|---|---|---|---|---|---|---|---|---|---|---|
| E(1) | E(2) | E(3) | R(1) | R(2) | R(3) | R(4) | A(1) | A(2) | A(4) | air-water | water-air | air-soil | soil-air | soil-water | water-sed | sed-water |
| 1000 | 0 | 0 | 3.916E+01 | 4.270E-02 | 8.87E-03 | 4.931E-05 | 9.607E+02 | 1.048E-01 | 2.419E-05 | 4.035E-01 | 2.596E-01 | 1.730E+00 | 1.718E+00 | 3.650E-03 | 6.642E-04 | 5.907E-04 |
| 0 | 1000 | 0 | 2.497E+01 | 1.049E+02 | 5.66E-03 | 1.211E-01 | 6.126E+02 | 2.574E+02 | 5.943E-02 | 2.573E-01 | 6.378E+02 | 1.103E+00 | 1.095E+00 | 2.327E-03 | 1.632E+00 | 1.451E+00 |
| 0 | 0 | 1000 | 3.893E+01 | 2.637E-01 | 5.14E+00 | 3.045E-04 | 9.550E+02 | 6.468E-01 | 1.494E-04 | 4.011E-01 | 1.603E+00 | 1.720E+00 | 9.945E+02 | 2.113E+00 | 4.101E-03 | 3.647E-03 |
| 600 | 300 | 100 | 3.488E+01 | 3.153E+01 | 5.21E-01 | 3.640E-02 | 8.557E+02 | 7.733E+01 | 1.786E-02 | 3.594E-01 | 1.917E+02 | 1.541E+00 | 1.008E+02 | 2.142E-01 | 4.903E-01 | 4.361E-01 |

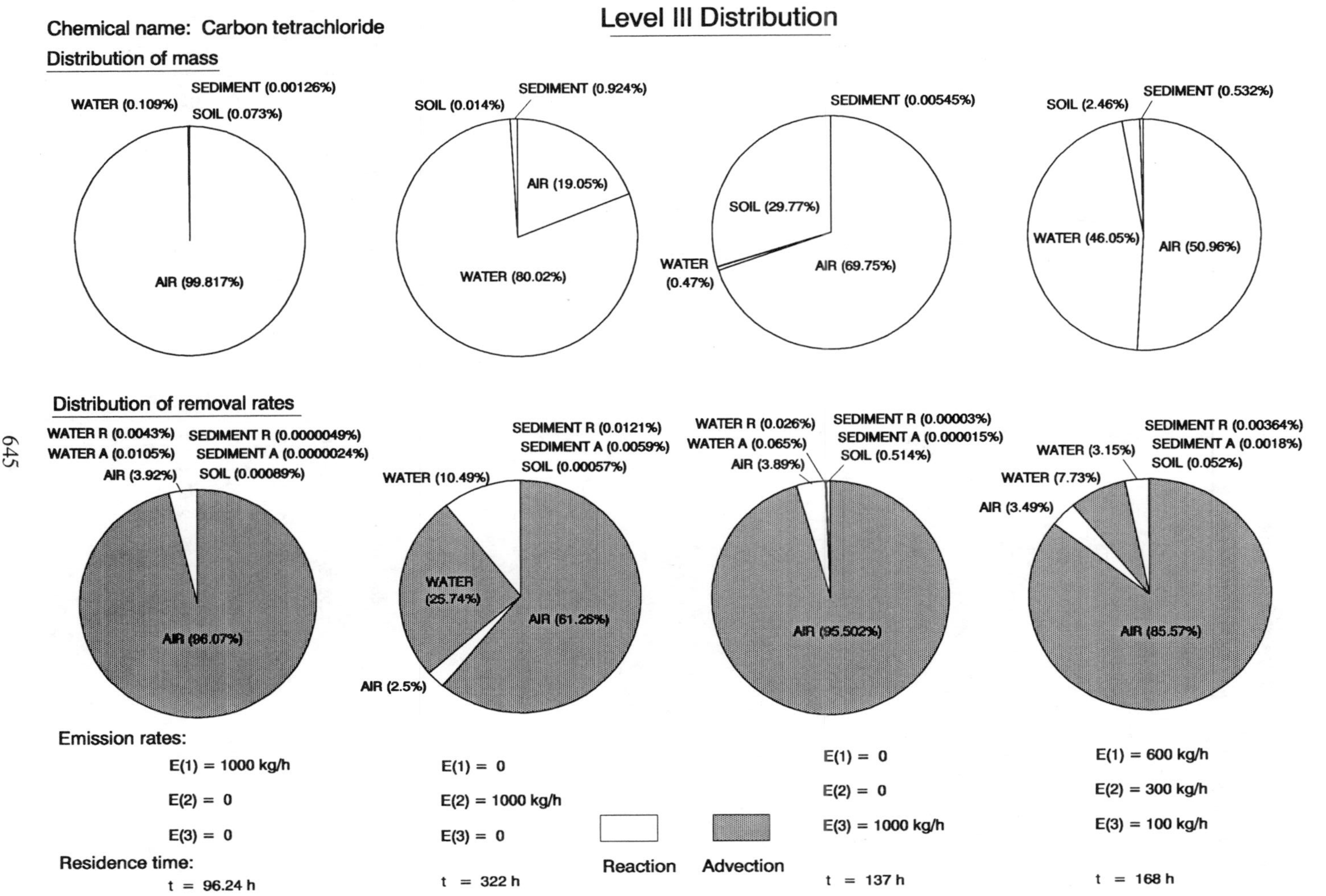

Chemical name: Carbon tetrachloride
Level III Distribution
Distribution of mass
WATER (0.109%)
SEDIMENT (0.00126%)
SOIL (0.073%)
AIR (99.817%)
SOIL (0.014%)
SEDIMENT (0.924%)
AIR (19.05%)
WATER (80.02%)
SEDIMENT (0.00545%)
SOIL (29.77%)
WATER (0.47%)
AIR (69.75%)
SOIL (2.46%)
SEDIMENT (0.532%)
WATER (46.05%)
AIR (50.96%)
Distribution of removal rates
WATER R (0.0043%)
WATER A (0.0105%)
AIR (3.92%)
SEDIMENT R (0.0000049%)
SEDIMENT A (0.0000024%)
SOIL (0.00089%)
AIR (96.07%)
SEDIMENT R (0.0121%)
SEDIMENT A (0.0059%)
SOIL (0.00057%)
WATER (10.49%)
WATER (25.74%)
AIR (61.26%)
AIR (2.5%)
WATER R (0.026%)
WATER A (0.065%)
AIR (3.89%)
SEDIMENT R (0.00003%)
SEDIMENT A (0.000015%)
SOIL (0.514%)
AIR (95.502%)
SEDIMENT R (0.00364%)
SEDIMENT A (0.0018%)
SOIL (0.052%)
WATER (3.15%)
WATER (7.73%)
AIR (3.49%)
AIR (85.57%)
Emission rates:
E(1) = 1000 kg/h
E(2) = 0
E(3) = 0
Residence time:
t = 96.24 h
E(1) = 0
E(2) = 1000 kg/h
E(3) = 0
t = 322 h
Reaction
Advection
E(1) = 0
E(2) = 0
E(3) = 1000 kg/h
t = 137 h
E(1) = 600 kg/h
E(2) = 300 kg/h
E(3) = 100 kg/h
t = 168 h

Chemical name: 1,2-Dichloroethane

Level I calculation: (six-compartment model)

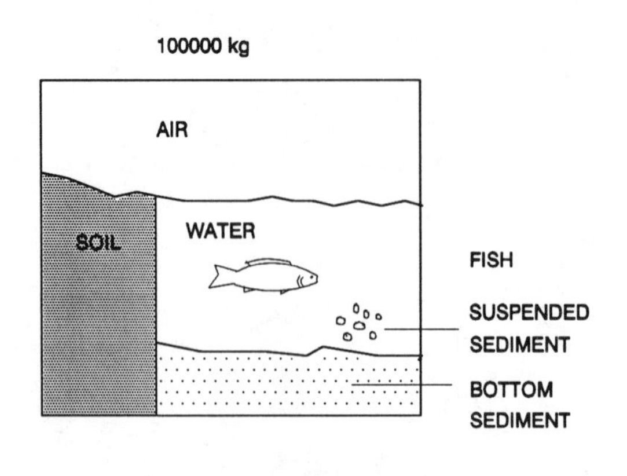


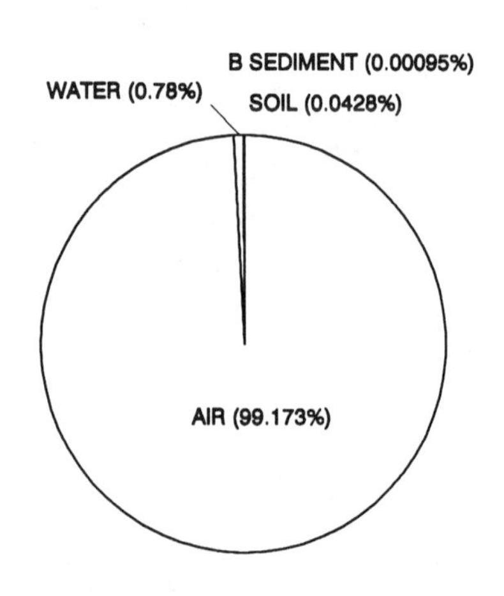


Distribution of mass

physical-chemical properties:

MW: 98.96 g/mol

M.P.: - 35.36°C

Fugacity ratio: 1.0

vapor pressure: 30260 Pa

solubility: 4770 g/$m^3$

log $K_{OW}$: 1.79

| Compartment | Z | Concentration | | | Amount | Amount |
|---|---|---|---|---|---|---|
| | mol/m3 Pa | mol/m3 | mg/L (or g/m3) | ug/g | kg | % |
| Air | 4.034E-04 | 1.002E-08 | 9.917E-07 | 8.366E-04 | 99173 | 99.173 |
| Water | 1.593E-03 | 3.957E-08 | 3.916E-06 | 3.916E-06 | 783.17 | 0.7832 |
| Soil | 1.933E-03 | 4.802E-08 | 4.752E-06 | 1.980E-06 | 42.766 | 0.0428 |
| Biota (fish) | 4.911E-03 | 1.220E-07 | 1.207E-05 | 1.207E-05 | 2.41E-03 | 2.41E-06 |
| Suspended sediment | 1.208E-02 | 3.001E-07 | 2.970E-05 | 1.980E-05 | 0.0297 | 2.97E-05 |
| Bottom sediment | 3.866E-03 | 9.603E-08 | 9.503E-06 | 3.960E-06 | 0.9503 | 9.50E-04 |
| | Total | | | | 100000 | 100 |

f = 2.484E-05 Pa

Chemical name: 1,2-Dichloroethane

Level II calculation: (six-compartment model)

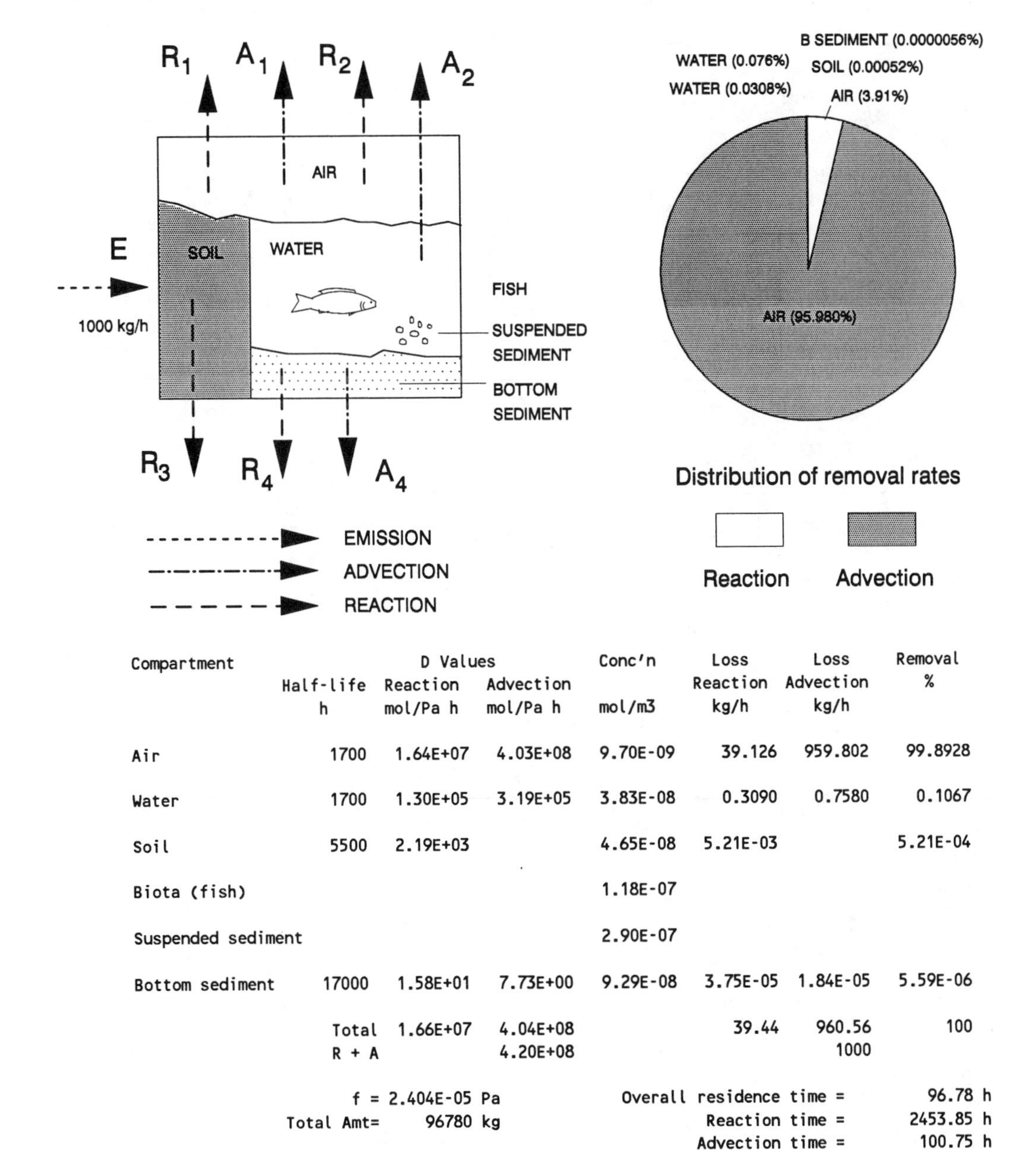


| Compartment | Half-life h | D Values Reaction mol/Pa h | D Values Advection mol/Pa h | Conc'n mol/m3 | Loss Reaction kg/h | Loss Advection kg/h | Removal % |
|---|---|---|---|---|---|---|---|
| Air | 1700 | 1.64E+07 | 4.03E+08 | 9.70E-09 | 39.126 | 959.802 | 99.8928 |
| Water | 1700 | 1.30E+05 | 3.19E+05 | 3.83E-08 | 0.3090 | 0.7580 | 0.1067 |
| Soil | 5500 | 2.19E+03 | | 4.65E-08 | 5.21E-03 | | 5.21E-04 |
| Biota (fish) | | | | 1.18E-07 | | | |
| Suspended sediment | | | | 2.90E-07 | | | |
| Bottom sediment | 17000 | 1.58E+01 | 7.73E+00 | 9.29E-08 | 3.75E-05 | 1.84E-05 | 5.59E-06 |
| | Total | 1.66E+07 | 4.04E+08 | | 39.44 | 960.56 | 100 |
| | R + A | | 4.20E+08 | | | 1000 | |

f = 2.404E-05 Pa

Total Amt= 96780 kg

Overall residence time = 96.78 h

Reaction time = 2453.85 h

Advection time = 100.75 h

## Fugacity Level III calculations: (four-compartment model)

Chemical name: 1,2-Dichloroethane

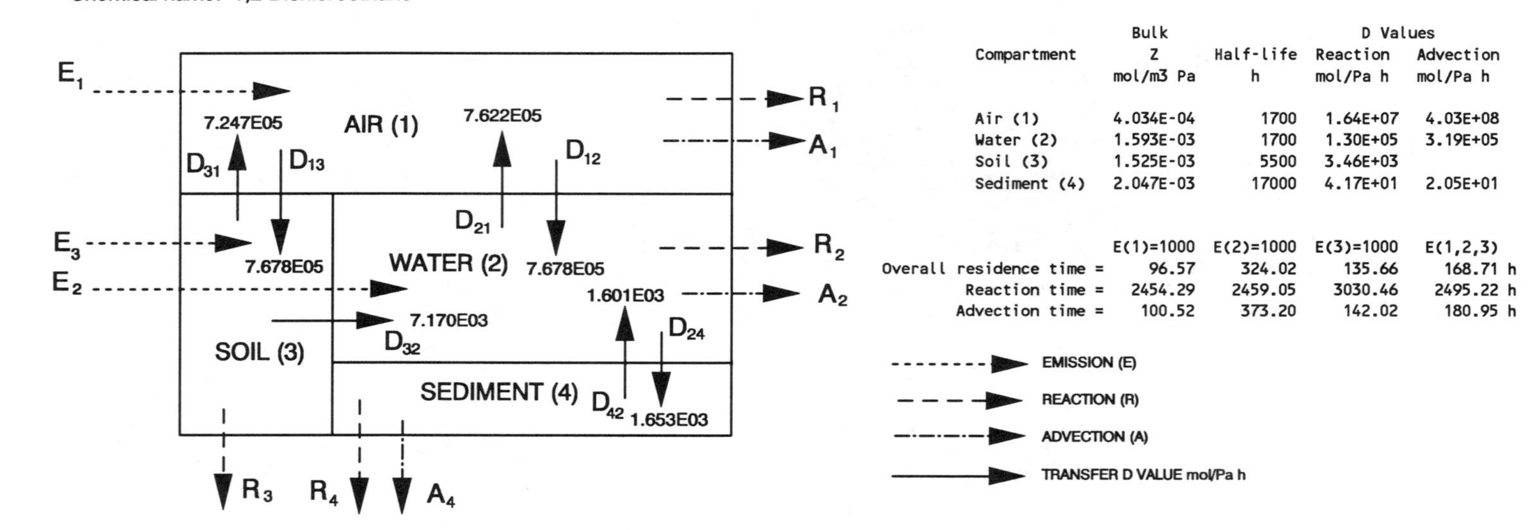


Phase Properties and Rates:

| Compartment | Bulk Z mol/m3 Pa | Half-life h | D Values Reaction mol/Pa h | D Values Advection mol/Pa h |
|---|---|---|---|---|
| Air (1) | 4.034E-04 | 1700 | 1.64E+07 | 4.03E+08 |
| Water (2) | 1.593E-03 | 1700 | 1.30E+05 | 3.19E+05 |
| Soil (3) | 1.525E-03 | 5500 | 3.46E+03 | |
| Sediment (4) | 2.047E-03 | 17000 | 4.17E+01 | 2.05E+01 |

| | E(1)=1000 | E(2)=1000 | E(3)=1000 | E(1,2,3) |
|---|---|---|---|---|
| Overall residence time = | 96.57 | 324.02 | 135.66 | 168.71 h |
| Reaction time = | 2454.29 | 2459.05 | 3030.46 | 2495.22 h |
| Advection time = | 100.52 | 373.20 | 142.02 | 180.95 h |

EMISSION (E)

REACTION (R)

ADVECTION (A)

TRANSFER D VALUE mol/Pa h

Phase Properties, Compositions, Transport and Transformation Rates:

| Emission, kg/h E(1) | E(2) | E(3) | Fugacity, Pa f(1) | f(2) | f(3) | f(4) | Concentration, g/m3 C(1) | C(2) | C(3) | C(4) | Amounts, kg m(1) | m(2) | m(3) | m(4) | Total Amount, kg |
|---|---|---|---|---|---|---|---|---|---|---|---|---|---|---|---|
| 1000 | 0 | 0 | 2.405E-05 | 1.534E-05 | 2.417E-05 | 1.526E-05 | 9.602E-07 | 2.419E-06 | 3.648E-06 | 3.091E-06 | 9.602E+04 | 4.838E+02 | 6.566E+01 | 1.546E+00 | 9.657E+04 |
| 0 | 1000 | 0 | 1.517E-05 | 8.328E-03 | 1.525E-05 | 8.281E-03 | 6.056E-07 | 1.313E-03 | 2.301E-06 | 1.678E-03 | 6.056E+04 | 2.626E+05 | 4.142E+01 | 8.389E+02 | 3.240E+05 |
| 0 | 0 | 1000 | 2.385E-05 | 9.633E-05 | 1.377E-02 | 9.578E-05 | 9.522E-07 | 1.519E-05 | 2.078E-03 | 1.941E-05 | 9.522E+04 | 3.037E+03 | 3.740E+04 | 9.703E+00 | 1.357E+05 |
| 600 | 300 | 100 | 2.137E-05 | 2.517E-03 | 1.396E-03 | 2.503E-03 | 8.530E-07 | 3.968E-04 | 2.106E-04 | 5.071E-04 | 8.530E+04 | 7.937E+04 | 3.791E+03 | 2.536E+02 | 1.687E+05 |

| Emission, kg/h E(1) | E(2) | E(3) | Loss, Reaction, kg/h R(1) | R(2) | R(3) | R(4) | Loss, Advection, kg/h A(1) | A(2) | A(4) | Intermedia Rate of Transport, kg/h T12 air-water | T21 water-air | T13 air-soil | T31 soil-air | T32 soil-water | T24 water-sed | T42 sed-water |
|---|---|---|---|---|---|---|---|---|---|---|---|---|---|---|---|---|
| 1000 | 0 | 0 | 3.914E+01 | 1.972E-01 | 8.27E-03 | 6.301E-05 | 9.602E+02 | 4.838E-01 | 3.091E-05 | 1.827E+00 | 1.163E+00 | 1.759E+00 | 1.734E+00 | 1.715E-02 | 2.511E-03 | 2.417E-03 |
| 0 | 1000 | 0 | 2.469E+01 | 1.070E+02 | 5.22E-03 | 3.420E-02 | 6.056E+02 | 2.626E+02 | 1.678E-02 | 1.153E+00 | 6.315E+02 | 1.109E+00 | 1.093E+00 | 1.082E-02 | 1.363E+00 | 1.312E+00 |
| 0 | 0 | 1000 | 3.882E+01 | 1.238E+00 | 4.71E+00 | 3.956E-04 | 9.522E+02 | 3.037E+00 | 1.941E-04 | 1.812E+00 | 7.304E+00 | 1.744E+00 | 9.873E+02 | 9.768E+00 | 1.576E-02 | 1.517E-02 |
| 600 | 300 | 100 | 3.477E+01 | 3.235E+01 | 4.78E-01 | 1.034E-02 | 8.530E+02 | 7.937E+01 | 5.071E-03 | 1.623E+00 | 1.909E+02 | 1.563E+00 | 1.001E+02 | 9.903E-01 | 4.119E-01 | 3.965E-01 |

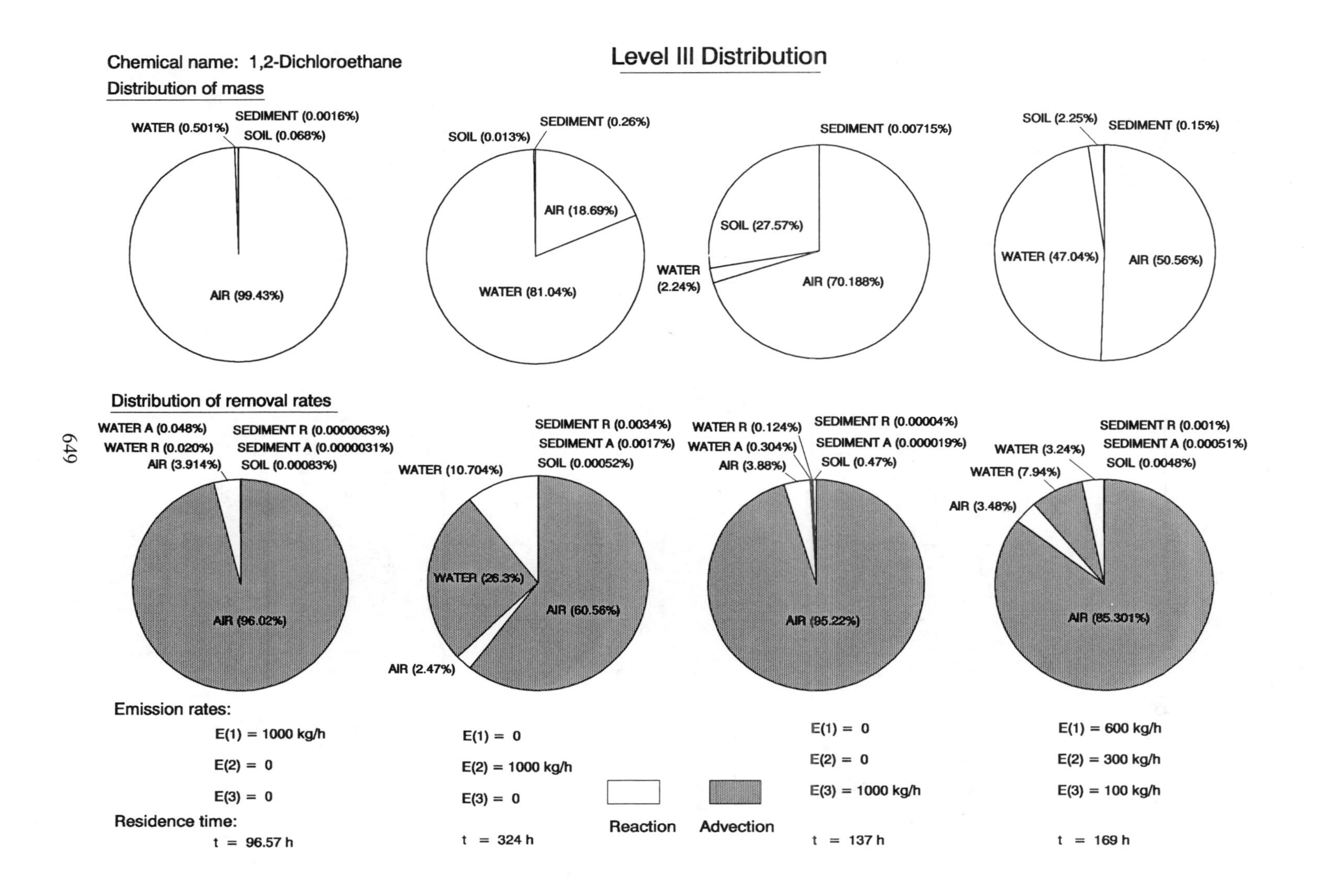
Level III Distribution
Chemical name: 1,2-Dichloroethane
Distribution of mass
WATER (0.501%)
SEDIMENT (0.0016%)
SOIL (0.068%)
AIR (99.43%)
SOIL (0.013%)
SEDIMENT (0.26%)
AIR (18.69%)
WATER (81.04%)
SEDIMENT (0.00715%)
SOIL (27.57%)
WATER
(2.24%)
AIR (70.188%)
SOIL (2.25%)
SEDIMENT (0.15%)
WATER (47.04%)
AIR (50.56%)
Distribution of removal rates
WATER A (0.048%)
WATER R (0.020%)
AIR (3.914%)
SEDIMENT R (0.0000063%)
SEDIMENT A (0.0000031%)
SOIL (0.00083%)
AIR (96.02%)
WATER (10.704%)
SEDIMENT R (0.0034%)
SEDIMENT A (0.0017%)
SOIL (0.00052%)
WATER (26.3%)
AIR (60.56%)
AIR (2.47%)
WATER R (0.124%)
WATER A (0.304%)
AIR (3.88%)
SEDIMENT R (0.00004%)
SEDIMENT A (0.000019%)
SOIL (0.47%)
AIR (85.22%)
WATER (3.24%)
WATER (7.94%)
AIR (3.48%)
SEDIMENT R (0.001%)
SEDIMENT A (0.00051%)
SOIL (0.0048%)
AIR (85.301%)
Emission rates:
E(1) = 1000 kg/h
E(2) = 0
E(3) = 0
Residence time:
t = 96.57 h
E(1) = 0
E(2) = 1000 kg/h
E(3) = 0
t = 324 h
Reaction
Advection
E(1) = 0
E(2) = 0
E(3) = 1000 kg/h
t = 137 h
E(1) = 600 kg/h
E(2) = 300 kg/h
E(3) = 100 kg/h
t = 169 h

## Chemical name: 1,1,2,2-Tetrachloroethane

Level I calculation: (six-compartment model)

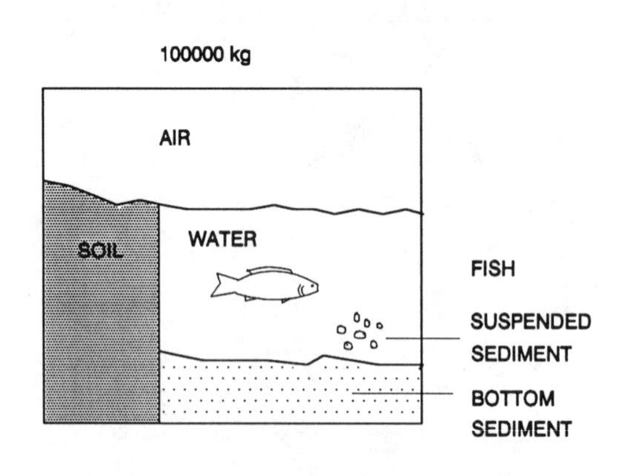


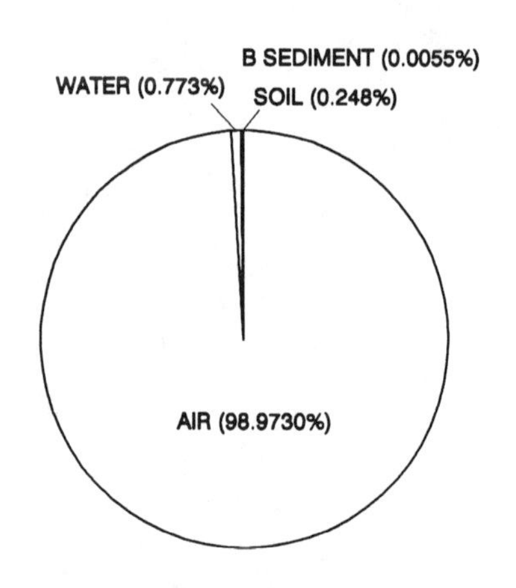


Distribution of mass

physical-chemical properties:

MW: 167.85 g/mol

M.P.: - 36.0 °C

Fugacity ratio: 1.0

vapor pressure: 3000 Pa

solubility: 793 g/m$^3$

log $K_{OW}$: 2.56

| Compartment | Z | Concentration | | | Amount | Amount |
|---|---|---|---|---|---|---|
| | mol/m3 Pa | mol/m3 | mg/L (or g/m3) | ug/g | kg | % |
| Air | 4.034E-04 | 5.897E-09 | 9.897E-07 | 8.349E-04 | 98973 | 98.973 |
| Water | 1.575E-03 | 2.302E-08 | 3.864E-06 | 3.864E-06 | 772.72 | 0.7727 |
| Soil | 1.125E-02 | 1.645E-07 | 2.761E-05 | 1.150E-05 | 248.462 | 0.2485 |
| Biota (fish) | 2.859E-02 | 4.179E-07 | 7.014E-05 | 7.014E-05 | 1.40E-02 | 1.40E-05 |
| Suspended sediment | 7.033E-02 | 1.028E-06 | 1.725E-04 | 1.150E-04 | 0.1725 | 1.73E-04 |
| Bottom sediment | 2.251E-02 | 3.289E-07 | 5.521E-05 | 2.301E-05 | 5.5214 | 5.52E-03 |
| | Total | | | | 100000 | 100 |

f = 1.462E-05 Pa

Chemical name: 1,1,2,2-Tetrachloroethane

Level II calculation: (six-compartment model)

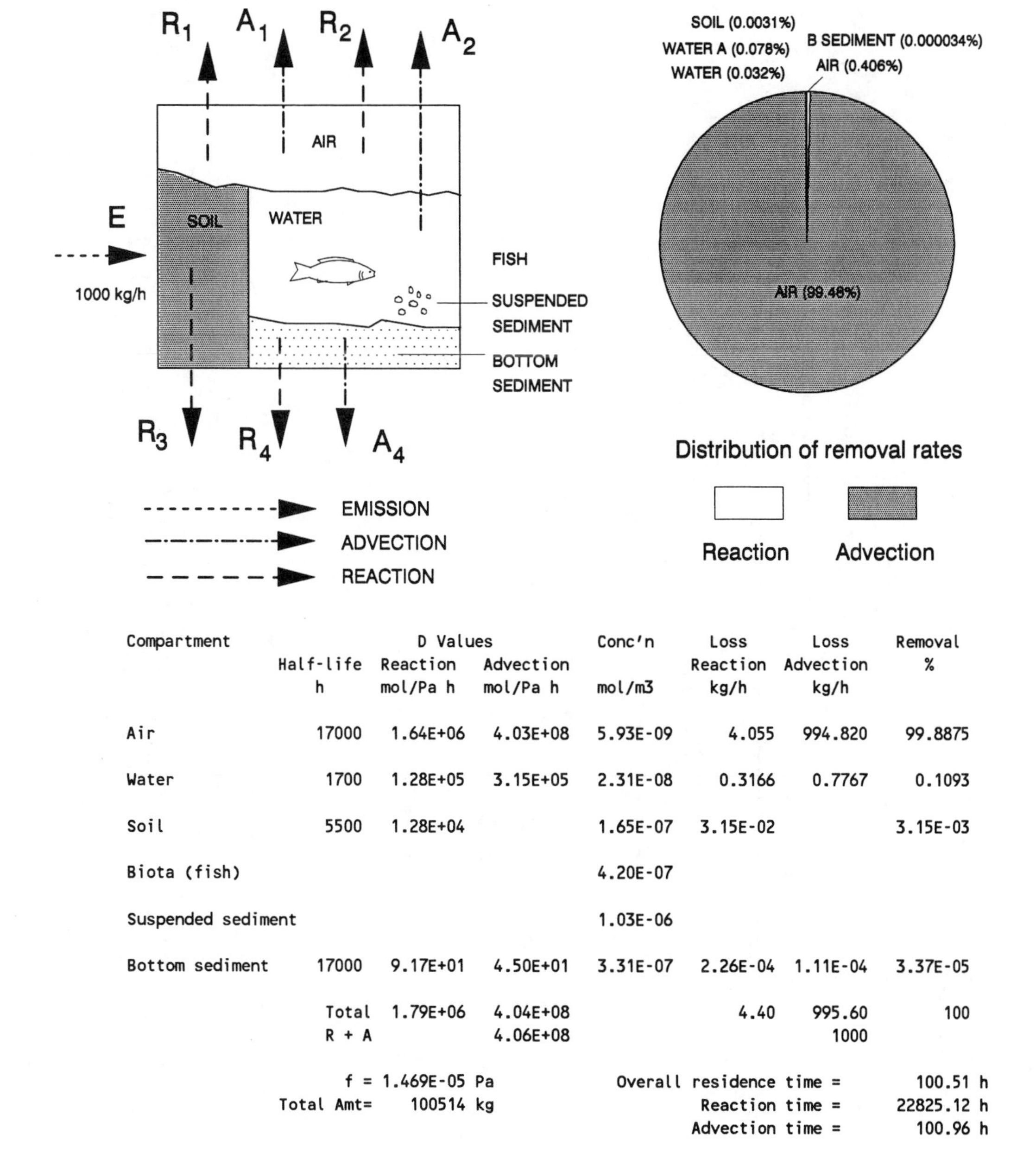


| Compartment | Half-life h | D Values Reaction mol/Pa h | D Values Advection mol/Pa h | Conc'n mol/m3 | Loss Reaction kg/h | Loss Advection kg/h | Removal % |
|---|---|---|---|---|---|---|---|
| Air | 17000 | 1.64E+06 | 4.03E+08 | 5.93E-09 | 4.055 | 994.820 | 99.8875 |
| Water | 1700 | 1.28E+05 | 3.15E+05 | 2.31E-08 | 0.3166 | 0.7767 | 0.1093 |
| Soil | 5500 | 1.28E+04 | | 1.65E-07 | 3.15E-02 | | 3.15E-03 |
| Biota (fish) | | | | 4.20E-07 | | | |
| Suspended sediment | | | | 1.03E-06 | | | |
| Bottom sediment | 17000 | 9.17E+01 | 4.50E+01 | 3.31E-07 | 2.26E-04 | 1.11E-04 | 3.37E-05 |
| | Total | 1.79E+06 | 4.04E+08 | | 4.40 | 995.60 | 100 |
| | R + A | | 4.06E+08 | | | 1000 | |

f = 1.469E-05 Pa
Total Amt= 100514 kg

Overall residence time = 100.51 h
Reaction time = 22825.12 h
Advection time = 100.96 h

Fugacity Level III calculations: (four-compartment model)

Chemical name: 1,1,2,2-Tetrachloroethane

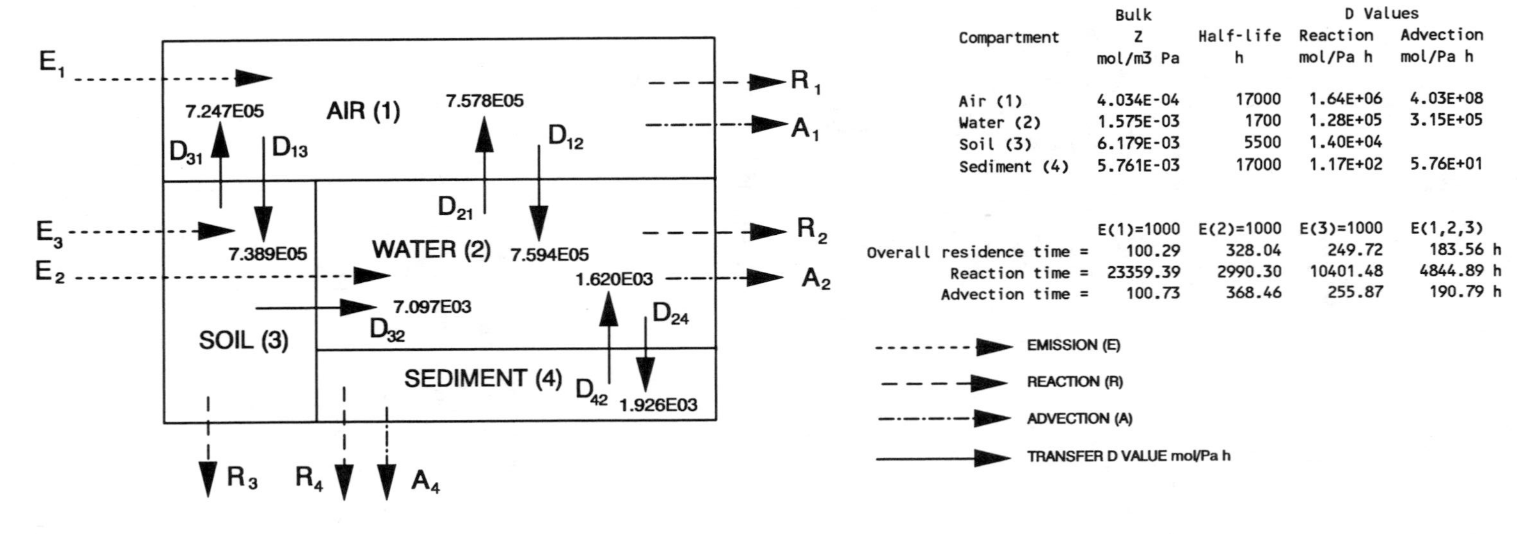


Phase Properties and Rates:

| Compartment | Bulk Z mol/m3 Pa | Half-life h | D Values Reaction mol/Pa h | Advection mol/Pa h |
|---|---|---|---|---|
| Air (1) | 4.034E-04 | 17000 | 1.64E+06 | 4.03E+08 |
| Water (2) | 1.575E-03 | 1700 | 1.28E+05 | 3.15E+05 |
| Soil (3) | 6.179E-03 | 5500 | 1.40E+04 | |
| Sediment (4) | 5.761E-03 | 17000 | 1.17E+02 | 5.76E+01 |

| | E(1)=1000 | E(2)=1000 | E(3)=1000 | E(1,2,3) |
|---|---|---|---|---|
| Overall residence time = | 100.29 | 328.04 | 249.72 | 183.56 h |
| Reaction time = | 23359.39 | 2990.30 | 10401.48 | 4844.89 h |
| Advection time = | 100.73 | 368.46 | 255.87 | 190.79 h |

Phase Properties, Compositions, Transport and Transformation Rates:

| Emission, kg/h | | | Fugacity, Pa | | | | Concentration, g/m3 | | | | Amounts, kg | | | | Total Amount, kg |
|---|---|---|---|---|---|---|---|---|---|---|---|---|---|---|---|
| E(1) | E(2) | E(3) | f(1) | f(2) | f(3) | f(4) | C(1) | C(2) | C(3) | C(4) | m(1) | m(2) | m(3) | m(4) | |
| 1000 | 0 | 0 | 1.470E-05 | 9.376E-06 | 1.456E-05 | 1.006E-05 | 9.952E-07 | 2.479E-06 | 1.510E-05 | 9.731E-06 | 9.952E+04 | 4.958E+02 | 2.719E+02 | 4.865E+00 | 1.003E+05 |
| 0 | 1000 | 0 | 9.270E-06 | 4.965E-03 | 9.185E-06 | 5.329E-03 | 6.277E-07 | 1.313E-03 | 9.526E-06 | 5.153E-03 | 6.277E+04 | 2.625E+05 | 1.715E+02 | 2.576E+03 | 3.280E+05 |
| 0 | 0 | 1000 | 1.437E-05 | 5.635E-05 | 8.003E-03 | 6.048E-05 | 9.730E-07 | 1.490E-05 | 8.301E-03 | 5.849E-05 | 9.730E+04 | 2.980E+03 | 1.494E+05 | 2.924E+01 | 2.497E+05 |
| 600 | 300 | 100 | 1.304E-05 | 1.501E-03 | 8.118E-04 | 1.611E-03 | 8.827E-07 | 3.968E-04 | 8.420E-04 | 1.557E-03 | 8.827E+04 | 7.935E+04 | 1.516E+04 | 7.787E+02 | 1.836E+05 |

| Emission, kg/h | | | Loss, Reaction, kg/h | | | | Loss, Advection, kg/h | | | Intermedia Rate of Transport, kg/h T12 | T21 | T13 | T31 | T32 | T24 | T42 |
|---|---|---|---|---|---|---|---|---|---|---|---|---|---|---|---|---|
| E(1) | E(2) | E(3) | R(1) | R(2) | R(3) | R(4) | A(1) | A(2) | A(4) | air-water | water-air | air-soil | soil-air | soil-water | water-sed | sed-water |
| 1000 | 0 | 0 | 4.057E+00 | 2.021E-01 | 3.43E-02 | 1.983E-04 | 9.952E+02 | 4.958E-01 | 9.731E-05 | 1.873E+00 | 1.193E+00 | 1.823E+00 | 1.771E+00 | 1.735E-02 | 3.032E-03 | 2.736E-03 |
| 0 | 1000 | 0 | 2.559E+00 | 1.070E+02 | 2.16E-02 | 1.050E-01 | 6.277E+02 | 2.625E+02 | 5.153E-02 | 1.182E+00 | 6.315E+02 | 1.150E+00 | 1.117E+00 | 1.094E-02 | 1.605E+00 | 1.449E+00 |
| 0 | 0 | 1000 | 3.966E+00 | 1.215E+00 | 1.88E+01 | 1.192E-03 | 9.730E+02 | 2.980E+00 | 5.849E-04 | 1.832E+00 | 7.168E+00 | 1.782E+00 | 9.734E+02 | 9.533E+00 | 1.822E-02 | 1.645E-02 |
| 600 | 300 | 100 | 3.598E+00 | 3.235E+01 | 1.91E+00 | 3.174E-02 | 8.827E+02 | 7.935E+01 | 1.557E-02 | 1.662E+00 | 1.909E+02 | 1.617E+00 | 9.874E+01 | 9.670E-01 | 4.852E-01 | 4.379E-01 |

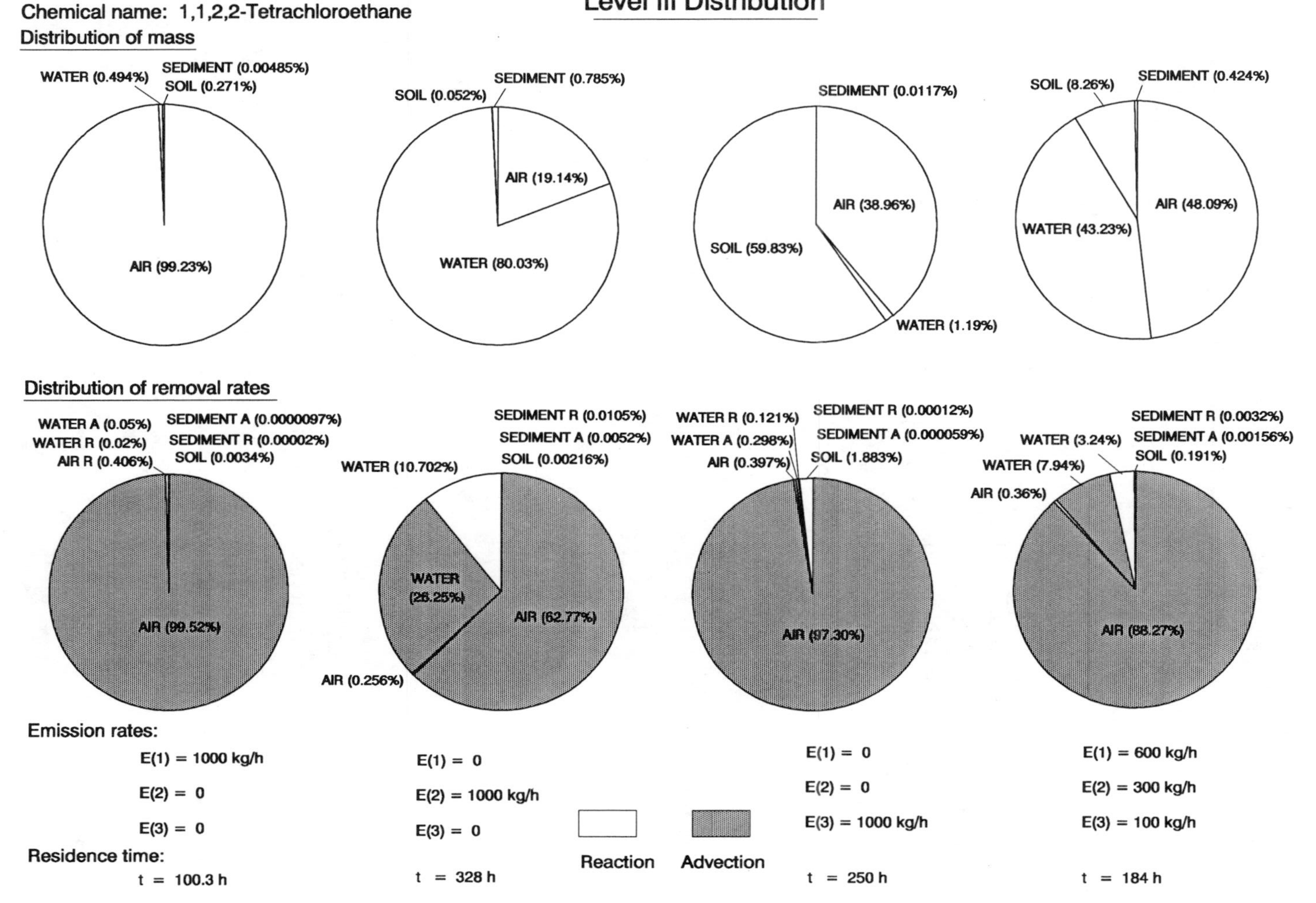

Level III Distribution
Chemical name: 1,1,2,2-Tetrachloroethane
Distribution of mass
WATER (0.494%)
SEDIMENT (0.00485%)
SOIL (0.271%)
AIR (99.23%)
SOIL (0.052%)
SEDIMENT (0.785%)
AIR (19.14%)
WATER (80.03%)
SEDIMENT (0.0117%)
AIR (38.96%)
SOIL (59.83%)
WATER (1.19%)
SOIL (8.26%)
SEDIMENT (0.424%)
AIR (48.09%)
WATER (43.23%)
Distribution of removal rates
WATER A (0.05%)
WATER R (0.02%)
AIR R (0.406%)
SEDIMENT A (0.0000097%)
SEDIMENT R (0.00002%)
SOIL (0.0034%)
AIR (99.52%)
SEDIMENT R (0.0105%)
SEDIMENT A (0.0052%)
SOIL (0.00216%)
WATER (10.702%)
WATER (26.25%)
AIR (62.77%)
AIR (0.256%)
WATER R (0.121%)
WATER A (0.298%)
AIR (0.397%)
SEDIMENT R (0.00012%)
SEDIMENT A (0.000059%)
SOIL (1.883%)
AIR (97.30%)
SEDIMENT R (0.0032%)
SEDIMENT A (0.00156%)
SOIL (0.191%)
WATER (3.24%)
WATER (7.94%)
AIR (0.36%)
AIR (88.27%)
Emission rates:
E(1) = 1000 kg/h
E(2) = 0
E(3) = 0
Residence time:
t = 100.3 h
E(1) = 0
E(2) = 1000 kg/h
E(3) = 0
t = 328 h
Reaction
Advection
E(1) = 0
E(2) = 0
E(3) = 1000 kg/h
t = 250 h
E(1) = 600 kg/h
E(2) = 300 kg/h
E(3) = 100 kg/h
t = 184 h

Chemical name: Pentachloroethane

Level I calculation: (six-compartment model)

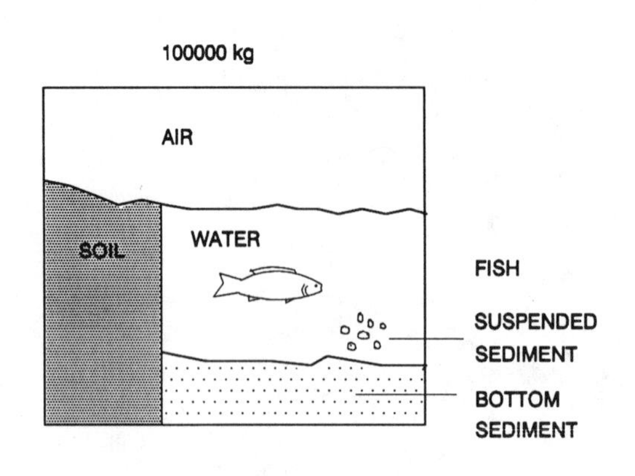


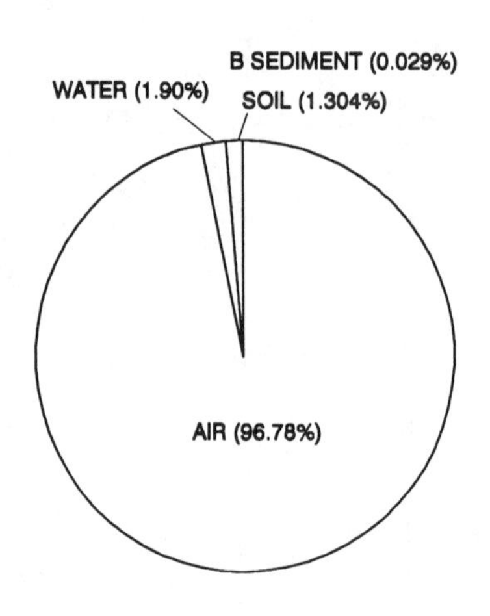


Distribution of mass

physical-chemical properties:

MW: 202.3 g/mol

M.P.: - 29.0 °C

Fugacity ratio: 1.0

vapor pressure: 625 Pa

solubility: 500 g/m$^3$

log $K_{OW}$: 2.89

| Compartment | Z | Concentration | | | Amount | Amount |
|---|---|---|---|---|---|---|
| | mol/m3 Pa | mol/m3 | mg/L (or g/m3) | ug/g | kg | % |
| Air | 4.034E-04 | 4.783E-09 | 9.677E-07 | 8.163E-04 | 96769 | 96.769 |
| Water | 3.955E-03 | 4.689E-08 | 9.486E-06 | 9.486E-06 | 1897.16 | 1.8972 |
| Soil | 6.041E-02 | 7.163E-07 | 1.449E-04 | 6.038E-05 | 1304.19 | 1.3042 |
| Biota (fish) | 1.535E-01 | 1.820E-06 | 3.682E-04 | 3.682E-04 | 7.36E-02 | 7.36E-05 |
| Suspended sediment | 3.776E-01 | 4.477E-06 | 9.057E-04 | 6.038E-04 | 0.9057 | 9.06E-04 |
| Bottom sediment | 1.208E-01 | 1.433E-06 | 2.898E-04 | 1.208E-04 | 28.9820 | 2.90E-02 |
| | Total | | | | 100000 | 100 |

f = 1.186E-05 Pa

Chemical name: Pentachloroethane

Level II calculation: (six-compartment model)

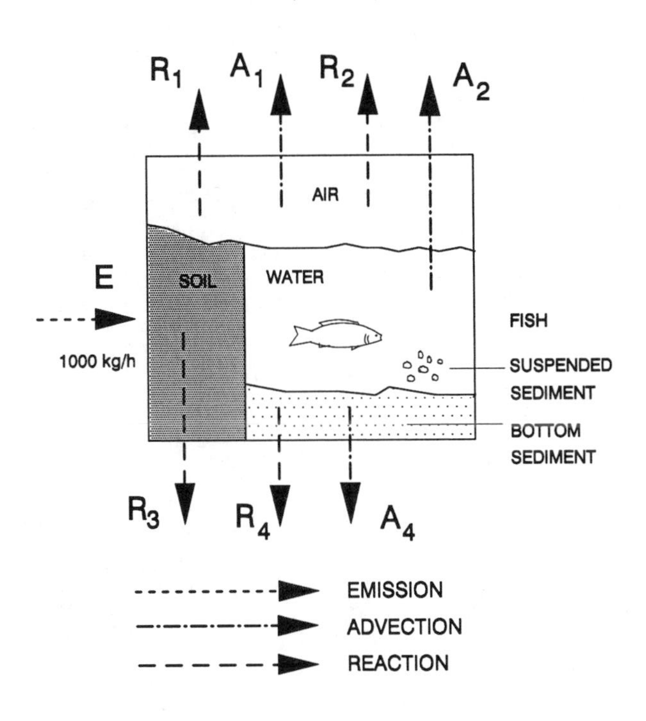


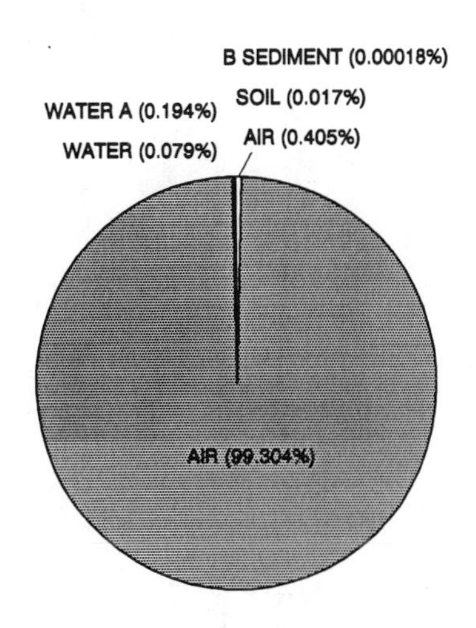


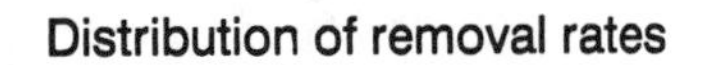


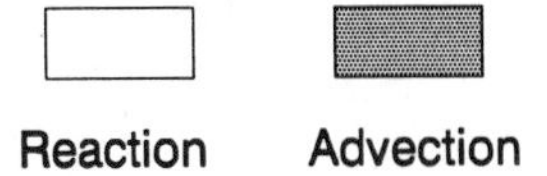


| Compartment | Half-life h | D Values Reaction mol/Pa h | D Values Advection mol/Pa h | Conc'n mol/m3 | Loss Reaction kg/h | Loss Advection kg/h | Removal % |
|---|---|---|---|---|---|---|---|
| Air | 17000 | 1.64E+06 | 4.03E+08 | 4.91E-09 | 4.048 | 993.041 | 99.7089 |
| Water | 1700 | 3.22E+05 | 7.91E+05 | 4.81E-08 | 0.7936 | 1.9469 | 0.2741 |
| Soil | 5500 | 6.85E+04 | | 7.35E-07 | 0.1686 | | 1.69E-02 |
| Biota (fish) | | | | 1.87E-06 | | | |
| Suspended sediment | | | | 4.59E-06 | | | |
| Bottom sediment | 17000 | 4.93E+02 | 2.42E+02 | 1.47E-06 | 1.21E-03 | 5.95E-04 | 1.81E-04 |
| | Total | 2.04E+06 | 4.04E+08 | | 5.01 | 994.99 | 100 |
| | R + A | | 4.06E+08 | | | 1000 | |

f = 1.217E-05 Pa
Total Amt= 102620 kg

Overall residence time = 102.62 h
Reaction time = 20476.58 h
Advection time = 103.14 h

Fugacity Level III calculations: (four-compartment model)

Chemical name: Pentachloroethane

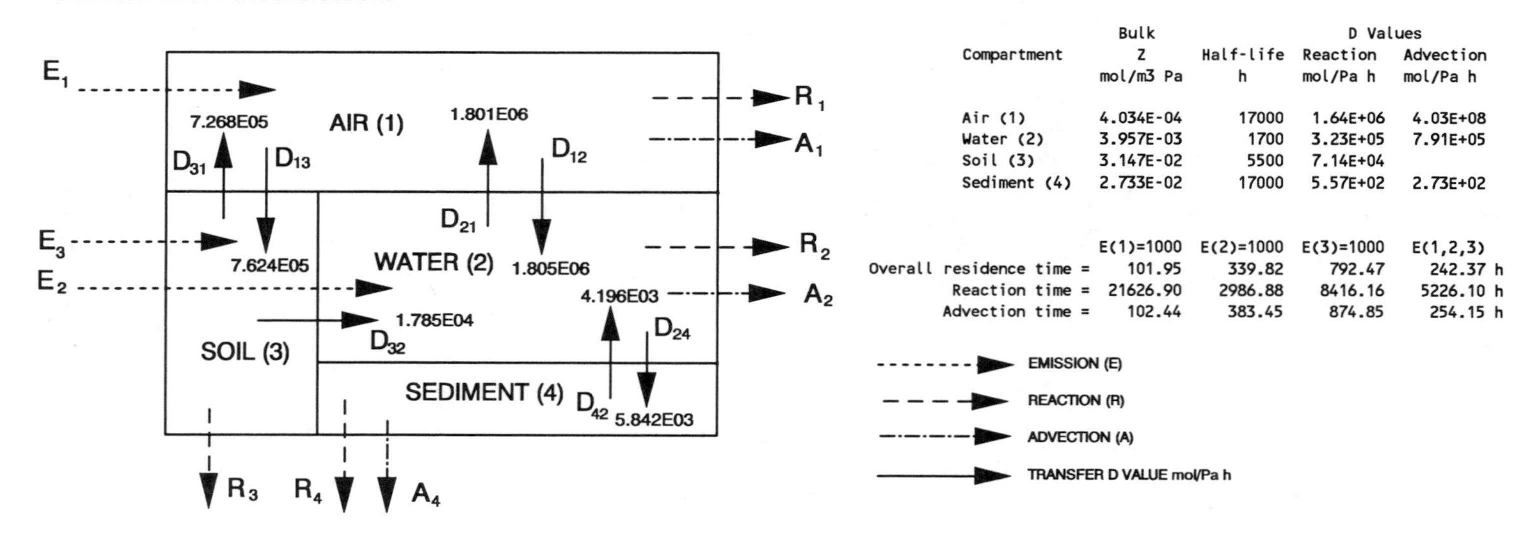


Phase Properties and Rates:

| Compartment | Bulk Z mol/m3 Pa | Half-life h | D Values Reaction mol/Pa h | D Values Advection mol/Pa h |
|---|---|---|---|---|
| Air (1) | 4.034E-04 | 17000 | 1.64E+06 | 4.03E+08 |
| Water (2) | 3.957E-03 | 1700 | 3.23E+05 | 7.91E+05 |
| Soil (3) | 3.147E-02 | 5500 | 7.14E+04 | |
| Sediment (4) | 2.733E-02 | 17000 | 5.57E+02 | 2.73E+02 |

| | E(1)=1000 | E(2)=1000 | E(3)=1000 | E(1,2,3) |
|---|---|---|---|---|
| Overall residence time = | 101.95 | 339.82 | 792.47 | 242.37 h |
| Reaction time = | 21626.90 | 2986.88 | 8416.16 | 5226.10 h |
| Advection time = | 102.44 | 383.45 | 874.85 | 254.15 h |

Phase Properties, Compositions, Transport and Transformation Rates:

| Emission, kg/h | | | Fugacity, Pa | | | | Concentration, g/m3 | | | | Amounts, kg | | | | Total Amount, kg |
|---|---|---|---|---|---|---|---|---|---|---|---|---|---|---|---|
| E(1) | E(2) | E(3) | f(1) | f(2) | f(3) | f(4) | C(1) | C(2) | C(3) | C(4) | m(1) | m(2) | m(3) | m(4) | |
| 1000 | 0 | 0 | 1.218E-05 | 7.609E-06 | 1.138E-05 | 8.844E-06 | 9.941E-07 | 6.091E-06 | 7.247E-05 | 4.890E-05 | 9.941E+04 | 1.218E+03 | 1.305E+03 | 2.445E+01 | 1.020E+05 |
| 0 | 1000 | 0 | 7.523E-06 | 1.700E-03 | 7.030E-06 | 1.976E-03 | 6.140E-07 | 1.361E-03 | 4.476E-05 | 1.092E-02 | 6.140E+04 | 2.722E+05 | 8.057E+02 | 5.462E+03 | 3.398E+05 |
| 0 | 0 | 1000 | 1.101E-05 | 4.397E-05 | 6.068E-03 | 5.110E-05 | 8.988E-07 | 3.519E-05 | 3.863E-02 | 2.825E-04 | 8.988E+04 | 7.038E+03 | 6.954E+05 | 1.413E+02 | 7.925E+05 |
| 600 | 300 | 100 | 1.067E-05 | 5.190E-04 | 6.157E-04 | 6.032E-04 | 8.705E-07 | 4.154E-04 | 3.920E-03 | 3.335E-03 | 8.705E+04 | 8.308E+04 | 7.057E+04 | 1.668E+03 | 2.424E+05 |

| Emission, kg/h | | | Loss, Reaction, kg/h | | | | Loss, Advection, kg/h | | | Intermedia Rate of Transport, kg/h T12 | T21 | T13 | T31 | T32 | T24 | T42 |
|---|---|---|---|---|---|---|---|---|---|---|---|---|---|---|---|---|
| E(1) | E(2) | E(3) | R(1) | R(2) | R(3) | R(4) | A(1) | A(2) | A(4) | air-water | water-air | air-soil | soil-air | soil-water | water-sed | sed-water |
| 1000 | 0 | 0 | 4.052E+00 | 4.966E-01 | 1.64E-01 | 9.966E-04 | 9.941E+02 | 1.218E+00 | 4.890E-04 | 4.447E+00 | 2.772E+00 | 1.879E+00 | 1.674E+00 | 4.110E-02 | 8.993E-03 | 7.508E-03 |
| 0 | 1000 | 0 | 2.503E+00 | 1.109E+02 | 1.02E-01 | 2.227E-01 | 6.140E+02 | 2.722E+02 | 1.092E-01 | 2.747E+00 | 6.193E+02 | 1.161E+00 | 1.034E+00 | 2.539E-02 | 2.009E+00 | 1.677E+00 |
| 0 | 0 | 1000 | 3.664E+00 | 2.869E+00 | 8.76E+01 | 5.758E-03 | 8.988E+02 | 7.038E+00 | 2.825E-03 | 4.021E+00 | 1.602E+01 | 1.699E+00 | 8.922E+02 | 2.191E+01 | 5.196E-02 | 4.338E-02 |
| 600 | 300 | 100 | 3.549E+00 | 3.387E+01 | 8.89E+00 | 6.798E-02 | 8.705E+02 | 8.308E+01 | 3.335E-02 | 3.894E+00 | 1.891E+02 | 1.646E+00 | 9.053E+01 | 2.223E+00 | 6.134E-01 | 5.121E-01 |

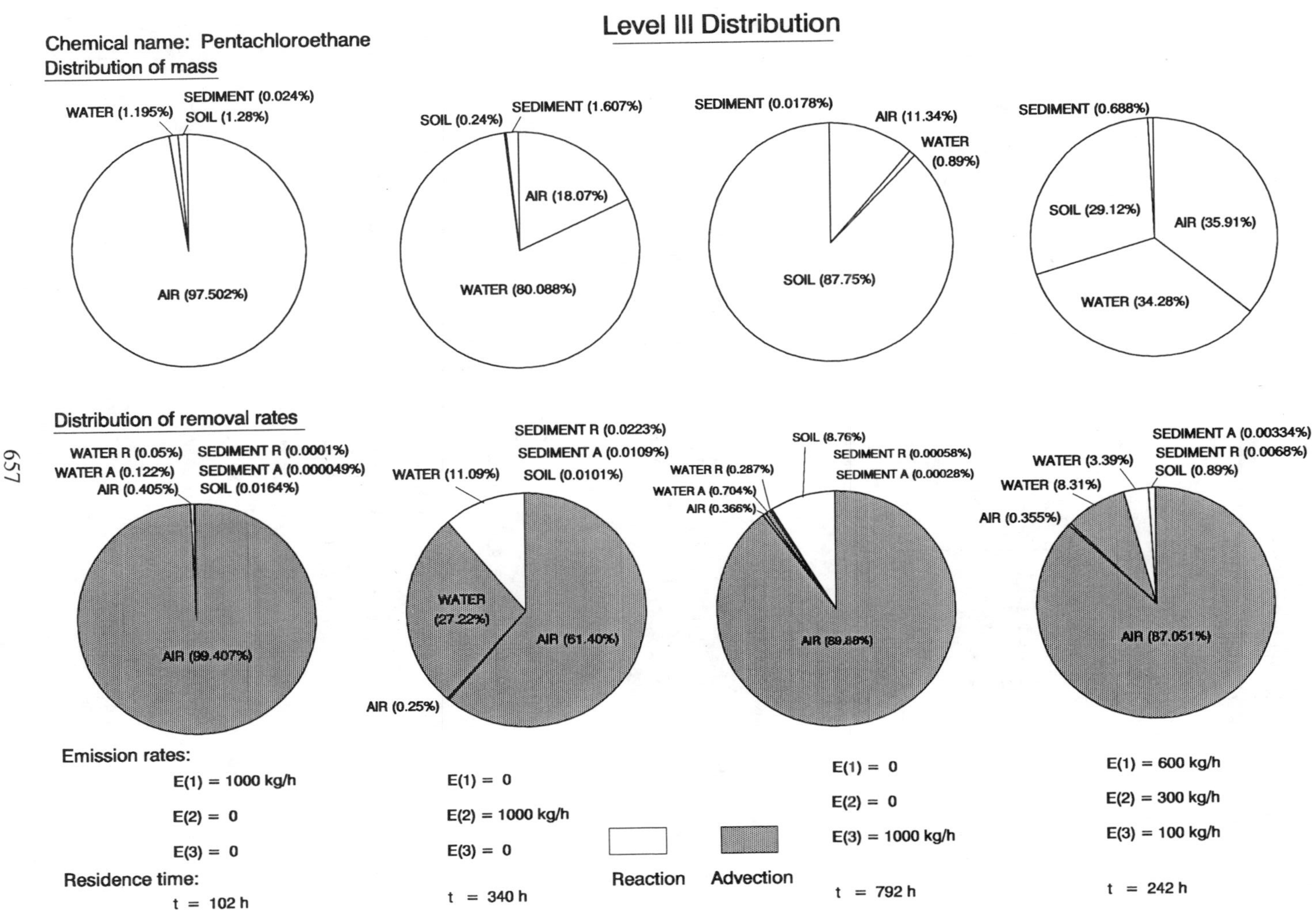

Level III Distribution
Chemical name: Pentachloroethane
Distribution of mass
WATER (1.195%)
SEDIMENT (0.024%)
SOIL (1.28%)
AIR (97.502%)
SOIL (0.24%)
SEDIMENT (1.607%)
AIR (18.07%)
WATER (80.088%)
SEDIMENT (0.0178%)
AIR (11.34%)
WATER (0.89%)
SOIL (87.75%)
SEDIMENT (0.688%)
SOIL (29.12%)
AIR (35.91%)
WATER (34.28%)
Distribution of removal rates
WATER R (0.05%)
WATER A (0.122%)
AIR (0.405%)
SEDIMENT R (0.0001%)
SEDIMENT A (0.000049%)
SOIL (0.0164%)
AIR (99.407%)
WATER (11.09%)
SEDIMENT R (0.0223%)
SEDIMENT A (0.0109%)
SOIL (0.0101%)
WATER (27.22%)
AIR (61.40%)
AIR (0.25%)
WATER R (0.287%)
WATER A (0.704%)
AIR (0.366%)
SOIL (8.76%)
SEDIMENT R (0.00058%)
SEDIMENT A (0.00028%)
AIR (89.88%)
WATER (3.39%)
WATER (8.31%)
AIR (0.355%)
SEDIMENT A (0.00334%)
SEDIMENT R (0.0068%)
SOIL (0.89%)
AIR (87.051%)
Emission rates:
E(1) = 1000 kg/h
E(2) = 0
E(3) = 0
Residence time:
t = 102 h
E(1) = 0
E(2) = 1000 kg/h
E(3) = 0
t = 340 h
Reaction
Advection
E(1) = 0
E(2) = 0
E(3) = 1000 kg/h
t = 792 h
E(1) = 600 kg/h
E(2) = 300 kg/h
E(3) = 100 kg/h
t = 242 h

Chemical name: Hexachloroethane

Level I calculation: (six-compartment model)

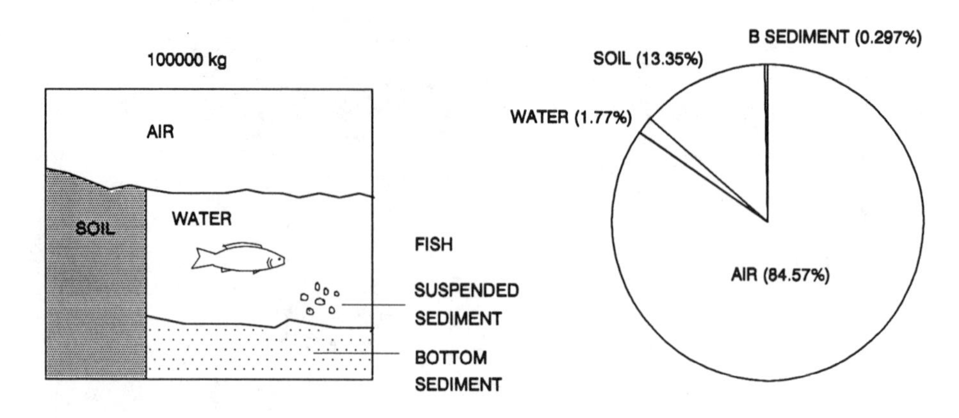


Distribution of mass

physical-chemical properties:

MW: 236.74 g/mol

M.P.: 186.1°C

Fugacity ratio: 0.0255

vapor pressure: 50 Pa

solubility: 50 g/m$^3$

log $K_{OW}$: 3.93

| Compartment | Z | Concentration | | | Amount | Amount |
|---|---|---|---|---|---|---|
| | mol/m3 Pa | mol/m3 | mg/L (or g/m3) | ug/g | kg | % |
| Air | 4.034E-04 | 3.572E-09 | 8.457E-07 | 7.134E-04 | 84573 | 84.573 |
| Water | 4.224E-03 | 3.741E-08 | 8.855E-06 | 8.855E-06 | 1771.06 | 1.7711 |
| Soil | 7.075E-01 | 6.266E-06 | 1.483E-03 | 6.180E-04 | 13349.67 | 13.350 |
| Biota (fish) | 1.798E+00 | 1.592E-05 | 3.769E-03 | 3.769E-03 | 0.7537 | 7.54E-04 |
| Suspended sediment | 4.422E+00 | 3.916E-05 | 9.271E-03 | 6.180E-03 | 9.2706 | 9.27E-03 |
| Bottom sediment | 1.415E+00 | 1.253E-05 | 2.967E-03 | 1.236E-03 | 296.66 | 0.2967 |
| | Total | | | | 100000 | 100 |

f = 8.855E-06 Pa

Chemical name: Hexachloroethane

Level II calculation: (six-compartment model)

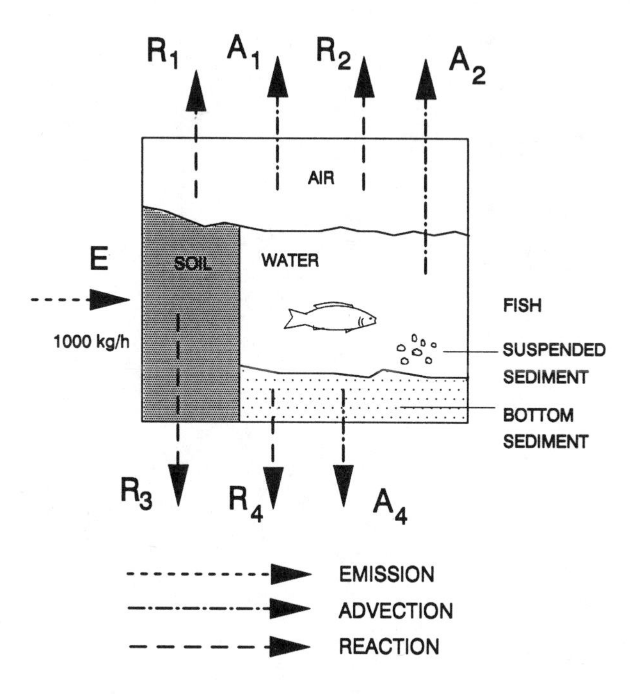


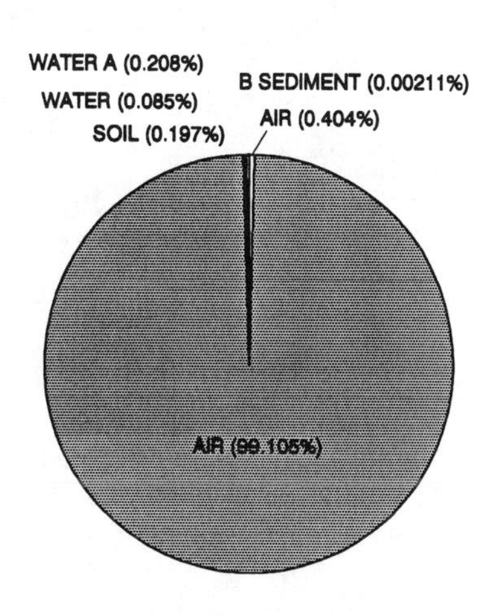


Distribution of removal rates

| Compartment | Half-life h | D Values Reaction mol/Pa h | D Values Advection mol/Pa h | Conc'n mol/m3 | Loss Reaction kg/h | Loss Advection kg/h | Removal % |
|---|---|---|---|---|---|---|---|
| Air | 17000 | 1.64E+06 | 4.03E+08 | 4.19E-09 | 4.040 | 991.046 | 99.5086 |
| Water | 1700 | 3.44E+05 | 8.45E+05 | 4.38E-08 | 0.8460 | 2.0754 | 0.2921 |
| Soil | 5500 | 8.02E+05 | | 7.34E-06 | 1.9711 | | 0.1971 |
| Biota (fish) | | | | 1.87E-05 | | | |
| Suspended sediment | | | | 4.59E-05 | | | |
| Bottom sediment | 17000 | 5.77E+03 | 2.83E+03 | 1.47E-05 | 1.42E-02 | 6.95E-03 | 2.11E-03 |
| | Total | 2.79E+06 | 4.04E+08 | | 6.87 | 993.13 | 100 |
| | R + A | | 4.07E+08 | | | 1000 | |

f = 1.038E-05 Pa

Total Amt= 117183 kg

Overall residence time = 117.18 h

Reaction time = 17054.09 h

Advection time = 117.99 h

Fugacity Level III calculations: (four-compartment model)

Chemical name: Hexachloroethane

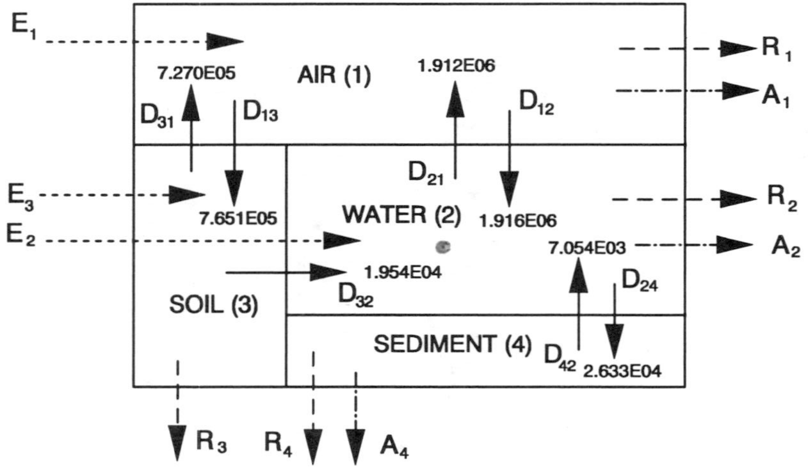

Phase Properties and Rates:

| Compartment | Bulk Z mol/m3 Pa | Half-life h | D Values Reaction mol/Pa h | D Values Advection mol/Pa h |
|---|---|---|---|---|
| Air (1) | 4.034E-04 | 17000 | 1.64E+06 | 4.03E+08 |
| Water (2) | 4.248E-03 | 1700 | 3.46E+05 | 8.50E+05 |
| Soil (3) | 3.551E-01 | 5500 | 8.05E+05 | |
| Sediment (4) | 2.864E-01 | 17000 | 5.84E+03 | 2.86E+03 |

| | E(1)=1000 | E(2)=1000 | E(3)=1000 | E(1,2,3) |
|---|---|---|---|---|
| Overall residence time = | 108.72 | 415.33 | 4174.61 | 607.30 h |
| Reaction time = | 19531.59 | 3536.10 | 7981.81 | 6682.53 h |
| Advection time = | 109.33 | 470.61 | 8752.10 | 668.00 h |

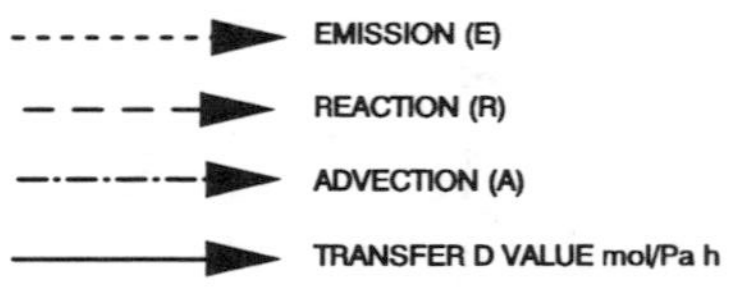

Phase Properties, Compositions, Transport and Transformation Rates:

| Emission, kg/h | | | Fugacity, Pa | | | | Concentration, g/m3 | | | | Amounts, kg | | | | Total Amount, kg |
|---|---|---|---|---|---|---|---|---|---|---|---|---|---|---|---|
| E(1) | E(2) | E(3) | f(1) | f(2) | f(3) | f(4) | C(1) | C(2) | C(3) | C(4) | m(1) | m(2) | m(3) | m(4) | |
| 1000 | 0 | 0 | 1.040E-05 | 6.414E-06 | 5.126E-06 | 1.072E-05 | 9.931E-07 | 6.450E-06 | 4.310E-04 | 7.268E-04 | 9.931E+04 | 1.290E+03 | 7.757E+03 | 3.634E+02 | 1.087E+05 |
| 0 | 1000 | 0 | 6.367E-06 | 1.357E-03 | 3.139E-06 | 2.268E-03 | 6.081E-07 | 1.364E-03 | 2.639E-04 | 1.538E-01 | 6.081E+04 | 2.729E+05 | 4.750E+03 | 7.688E+04 | 4.153E+05 |
| 0 | 0 | 1000 | 4.952E-06 | 2.018E-05 | 2.724E-03 | 3.373E-05 | 4.729E-07 | 2.029E-05 | 2.290E-01 | 2.287E-03 | 4.729E+04 | 4.058E+03 | 4.122E+06 | 1.143E+03 | 4.175E+06 |
| 600 | 300 | 100 | 8.645E-06 | 4.129E-04 | 2.764E-04 | 6.901E-04 | 8.256E-07 | 4.152E-04 | 2.324E-02 | 4.679E-02 | 8.256E+04 | 8.305E+04 | 4.183E+05 | 2.340E+04 | 6.073E+05 |

| Emission, kg/h | | | Loss, Reaction, kg/h | | | | Loss, Advection, kg/h | | | Intermedia Rate of Transport, kg/h T12 | T21 | T13 | T31 | T32 | T24 | T42 |
|---|---|---|---|---|---|---|---|---|---|---|---|---|---|---|---|---|
| E(1) | E(2) | E(3) | R(1) | R(2) | R(3) | R(4) | A(1) | A(2) | A(4) | air-water | water-air | air-soil | soil-air | soil-water | water-sed | sed-water |
| 1000 | 0 | 0 | 4.048E+00 | 5.259E-01 | 9.77E-01 | 1.481E-02 | 9.931E+02 | 1.290E+00 | 7.268E-03 | 4.717E+00 | 2.903E+00 | 1.884E+00 | 8.823E-01 | 2.384E-02 | 3.999E-02 | 1.790E-02 |
| 0 | 1000 | 0 | 2.479E+00 | 1.112E+02 | 5.98E-01 | 3.134E+00 | 6.081E+02 | 2.729E+02 | 1.538E+00 | 2.888E+00 | 6.141E+02 | 1.153E+00 | 5.402E-01 | 1.460E-02 | 8.459E+00 | 3.787E+00 |
| 0 | 0 | 1000 | 1.928E+00 | 1.654E+00 | 5.19E+02 | 4.661E-02 | 4.729E+02 | 4.058E+00 | 2.287E-02 | 2.246E+00 | 9.132E+00 | 8.969E-01 | 4.688E+02 | 1.267E+01 | 1.258E-01 | 5.632E-02 |
| 600 | 300 | 100 | 3.366E+00 | 3.385E+01 | 5.27E+01 | 9.538E-01 | 8.256E+02 | 8.305E+01 | 4.679E-01 | 3.921E+00 | 1.869E+02 | 1.566E+00 | 4.758E+01 | 1.286E+00 | 2.574E+00 | 1.153E+00 |

## Level III Distribution

Chemical name: Hexachloroethane

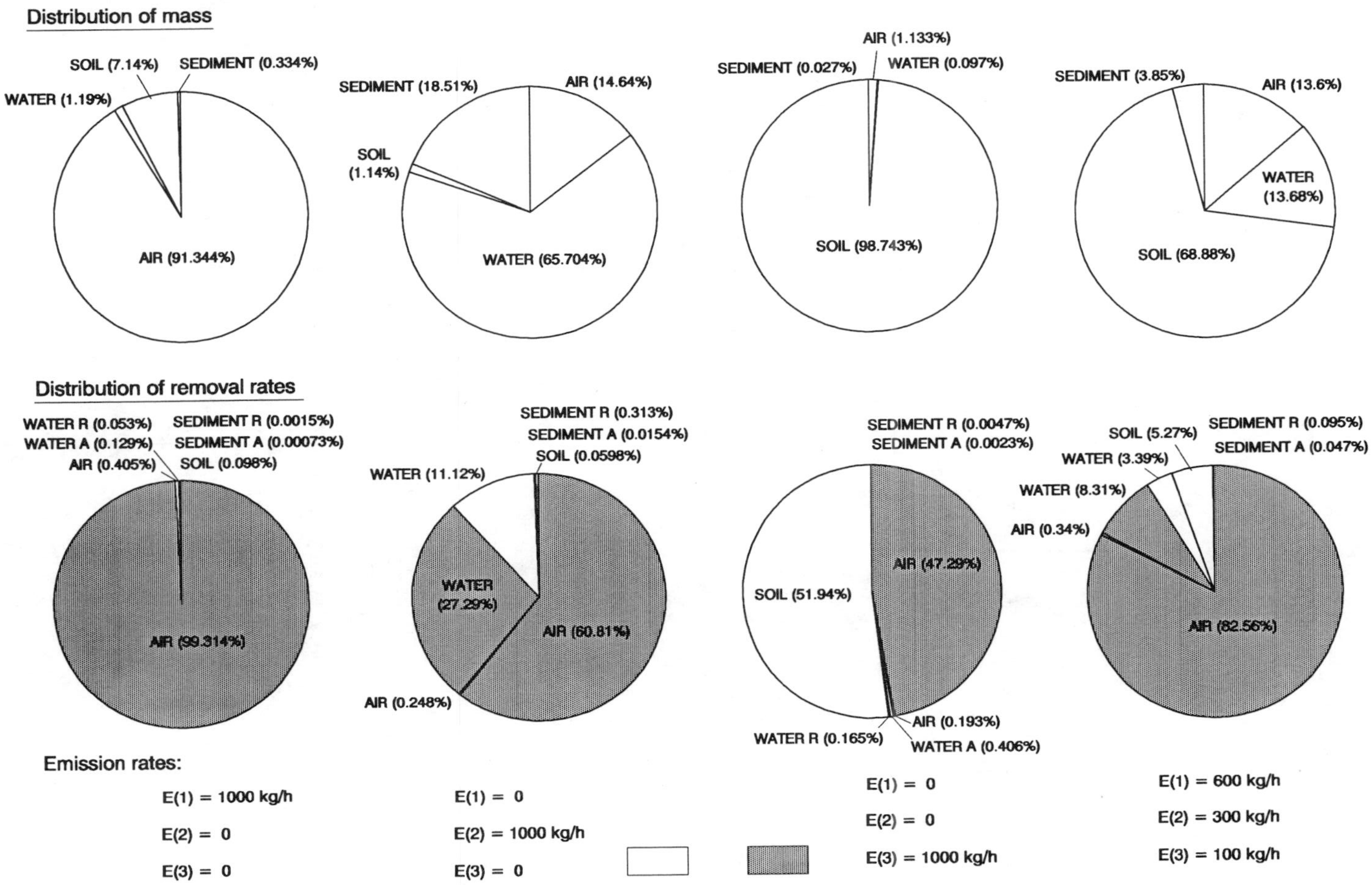

Distribution of mass



Distribution of removal rates



Emission rates:

| | | | |
|---|---|---|---|
| E(1) = 1000 kg/h | E(1) = 0 | E(1) = 0 | E(1) = 600 kg/h |
| E(2) = 0 | E(2) = 1000 kg/h | E(2) = 0 | E(2) = 300 kg/h |
| E(3) = 0 | E(3) = 0 | E(3) = 1000 kg/h | E(3) = 100 kg/h |

Residence time:

| | | | |
|---|---|---|---|
| t = 108.7 h | t = 415 h | t = 4175 h | t = 607 h |

Chemical name: 1,2-Dichloropropane

Level I calculation: (six-compartment model)

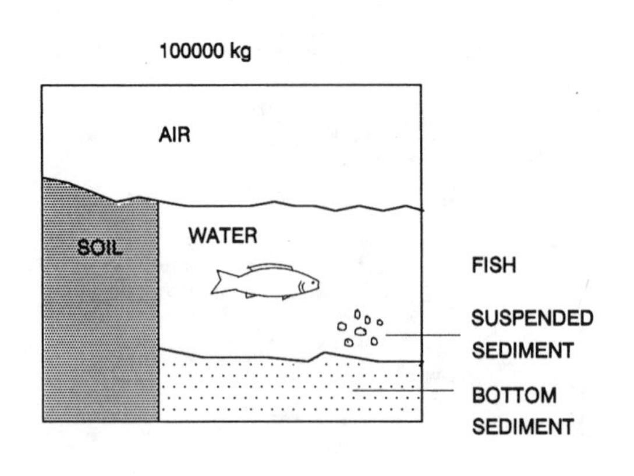


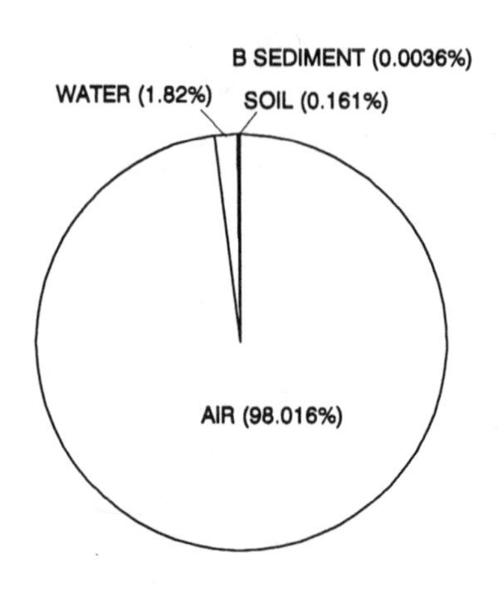


Distribution of mass

physical-chemical properties:

MW: 112.99 g/mol

M.P.: -100.4°C

Fugacity ratio: 1.0

vapor pressure: 6620 Pa

solubility: 2800 g/m³

log $K_{OW}$: 2.0

| Compartment | Z | Concentration | | | Amount | Amount |
|---|---|---|---|---|---|---|
| | mol/m3 Pa | mol/m3 | mg/L (or g/m3) | ug/g | kg | % |
| Air | 4.034E-04 | 8.675E-09 | 9.802E-07 | 8.269E-04 | 98016 | 98.016 |
| Water | 3.743E-03 | 8.049E-08 | 9.095E-06 | 9.095E-06 | 1819.00 | 1.8190 |
| Soil | 7.367E-03 | 1.584E-07 | 1.790E-05 | 7.458E-06 | 161.09 | 0.1611 |
| Biota (fish) | 1.872E-02 | 4.025E-07 | 4.548E-05 | 4.548E-05 | 9.10E-03 | 9.10E-06 |
| Suspended sediment | 4.604E-02 | 9.901E-07 | 1.119E-04 | 7.458E-05 | 0.1119 | 1.12E-04 |
| Bottom sediment | 1.473E-02 | 3.168E-07 | 3.580E-05 | 1.492E-05 | 3.580 | 3.58E-03 |
| | Total | | | | 100000 | 100 |

f = 2.150E-05 Pa

Chemical name: 1,2-Dichloropropane

Level II calculation: (six-compartment model)

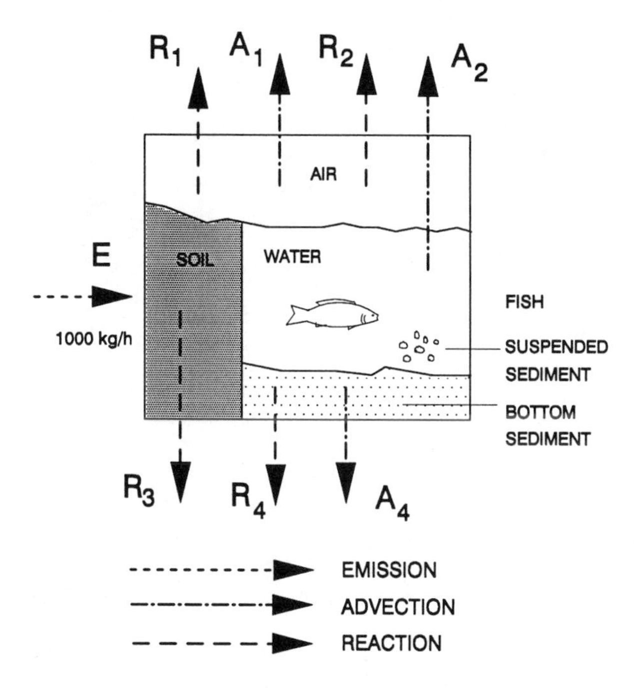


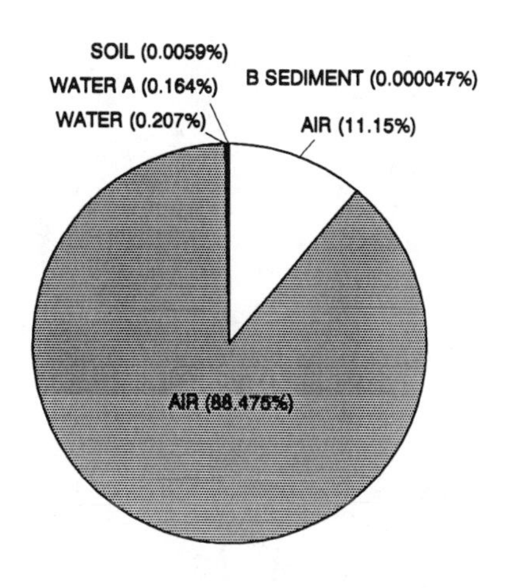


Distribution of removal rates

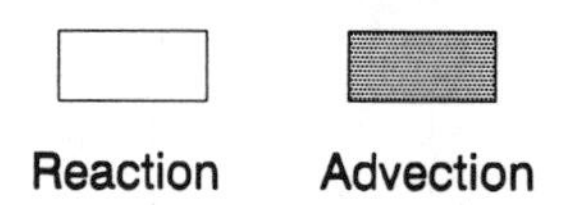


| Compartment | Half-life h | D Values Reaction mol/Pa h | D Values Advection mol/Pa h | Conc'n mol/m3 | Loss Reaction kg/h | Loss Advection kg/h | Removal % |
|---|---|---|---|---|---|---|---|
| Air | 550 | 5.08E+07 | 4.03E+08 | 7.83E-09 | 111.479 | 884.751 | 99.6229 |
| Water | 550 | 9.43E+05 | 7.49E+05 | 7.27E-08 | 2.0688 | 1.6419 | 0.3711 |
| Soil | 1700 | 2.70E+04 | | 1.43E-07 | 0.0593 | | 5.93E-03 |
| Biota (fish) | | | | 3.63E-07 | | | |
| Suspended sediment | | | | 8.94E-07 | | | |
| Bottom sediment | 5500 | 1.86E+02 | 2.95E+01 | 2.86E-07 | 4.07E-04 | 6.46E-05 | 4.72E-05 |
| | Total | 5.18E+07 | 4.04E+08 | | 113.61 | 886.39 | 100 |
| | R + A | | 4.56E+08 | | | 1000 | |

f = 1.941E-05 Pa
Total Amt= 90266 kg

Overall residence time = 90.27 h
Reaction time = 794.54 h
Advection time = 101.83 h

## Fugacity Level III calculations: (four-compartment model)

Chemical name: 1,2-Dichloropropane

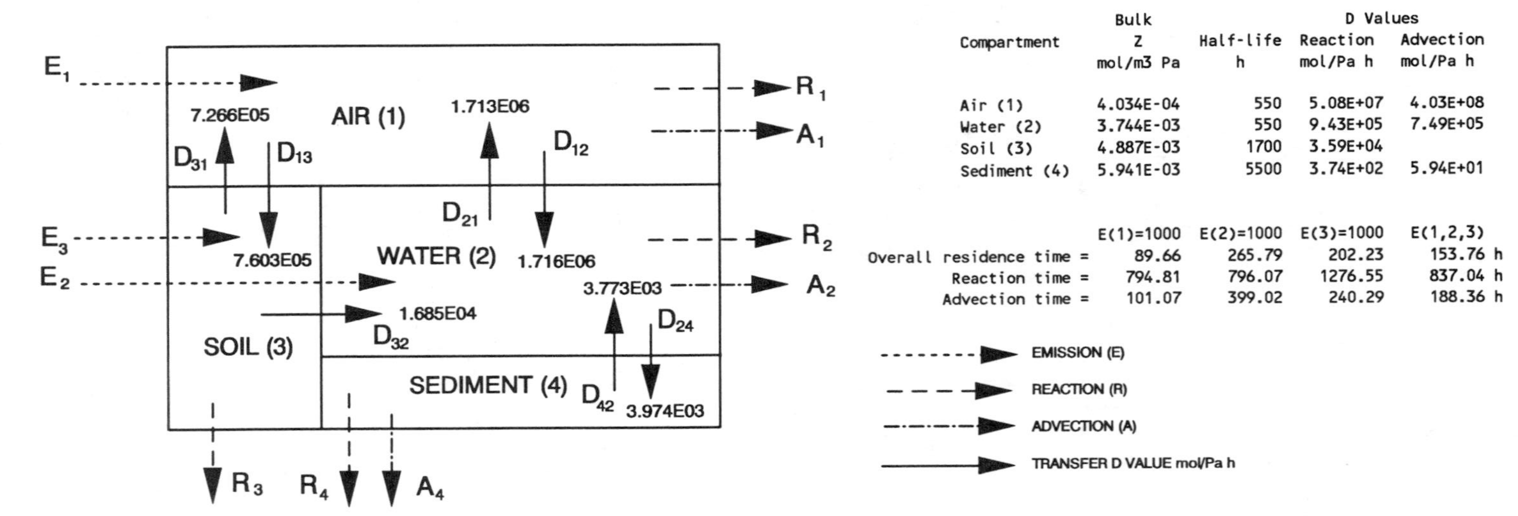


Phase Properties and Rates:

| Compartment | Bulk Z mol/m3 Pa | Half-life h | D Values Reaction mol/Pa h | D Values Advection mol/Pa h |
|---|---|---|---|---|
| Air (1) | 4.034E-04 | 550 | 5.08E+07 | 4.03E+08 |
| Water (2) | 3.744E-03 | 550 | 9.43E+05 | 7.49E+05 |
| Soil (3) | 4.887E-03 | 1700 | 3.59E+04 | |
| Sediment (4) | 5.941E-03 | 5500 | 3.74E+02 | 5.94E+01 |

| | E(1)=1000 | E(2)=1000 | E(3)=1000 | E(1,2,3) |
|---|---|---|---|---|
| Overall residence time = | 89.66 | 265.79 | 202.23 | 153.76 h |
| Reaction time = | 794.81 | 796.07 | 1276.55 | 837.04 h |
| Advection time = | 101.07 | 399.02 | 240.29 | 188.36 h |

Phase Properties, Compositions, Transport and Transformation Rates:

| Emission, kg/h | | | Fugacity, Pa | | | | Concentration, g/m3 | | | | Amounts, kg | | | | Total |
|---|---|---|---|---|---|---|---|---|---|---|---|---|---|---|---|
| E(1) | E(2) | E(3) | f(1) | f(2) | f(3) | f(4) | C(1) | C(2) | C(3) | C(4) | m(1) | m(2) | m(3) | m(4) | Amount, kg |
| 1000 | 0 | 0 | 1.945E-05 | 9.896E-06 | 1.897E-05 | 9.348E-06 | 8.864E-07 | 4.186E-06 | 1.048E-05 | 6.275E-06 | 8.864E+04 | 8.371E+02 | 1.886E+02 | 3.138E+00 | 8.966E+04 |
| 0 | 1000 | 0 | 9.780E-06 | 2.604E-03 | 9.542E-06 | 2.460E-03 | 4.458E-07 | 1.101E-03 | 5.269E-06 | 1.651E-03 | 4.458E+04 | 2.203E+05 | 9.484E+01 | 8.257E+02 | 2.658E+05 |
| 0 | 0 | 1000 | 1.834E-05 | 6.553E-05 | 1.137E-02 | 6.190E-05 | 8.360E-07 | 2.772E-05 | 6.281E-03 | 4.156E-05 | 8.360E+04 | 5.544E+03 | 1.131E+05 | 2.078E+01 | 2.022E+05 |
| 600 | 300 | 100 | 1.644E-05 | 7.937E-04 | 1.152E-03 | 7.497E-04 | 7.492E-07 | 3.357E-04 | 6.360E-04 | 5.033E-04 | 7.492E+04 | 6.714E+04 | 1.145E+04 | 2.517E+02 | 1.538E+05 |

| Emission, kg/h | | | Loss, Reaction, kg/h | | | | Loss, Advection, kg/h | | | Intermedia Rate of Transport, kg/h T12 | T21 | T13 | T31 | T32 | T24 | T42 |
|---|---|---|---|---|---|---|---|---|---|---|---|---|---|---|---|---|
| E(1) | E(2) | E(3) | R(1) | R(2) | R(3) | R(4) | A(1) | A(2) | A(4) | air-water | water-air | air-soil | soil-air | soil-water | water-sed | sed-water |
| 1000 | 0 | 0 | 1.117E+02 | 1.055E+00 | 7.69E-02 | 3.953E-04 | 8.864E+02 | 8.371E-01 | 6.275E-05 | 3.771E+00 | 1.915E+00 | 1.670E+00 | 1.557E+00 | 3.612E-02 | 4.443E-03 | 3.985E-03 |
| 0 | 1000 | 0 | 5.617E+01 | 2.776E+02 | 3.87E-02 | 1.040E-01 | 4.458E+02 | 2.203E+02 | 1.651E-02 | 1.897E+00 | 5.039E+02 | 8.402E-01 | 7.834E-01 | 1.817E-02 | 1.169E+00 | 1.049E+00 |
| 0 | 0 | 1000 | 1.053E+02 | 6.986E+00 | 4.61E+01 | 2.618E-03 | 8.360E+02 | 5.544E+00 | 4.156E-04 | 3.557E+00 | 1.268E+01 | 1.576E+00 | 9.338E+02 | 2.166E+01 | 2.942E-02 | 2.639E-02 |
| 600 | 300 | 100 | 9.439E+01 | 8.460E+01 | 4.67E+00 | 3.171E-02 | 7.492E+02 | 6.714E+01 | 5.033E-03 | 3.188E+00 | 1.536E+02 | 1.412E+00 | 9.455E+01 | 2.193E+00 | 3.563E-01 | 3.196E-01 |

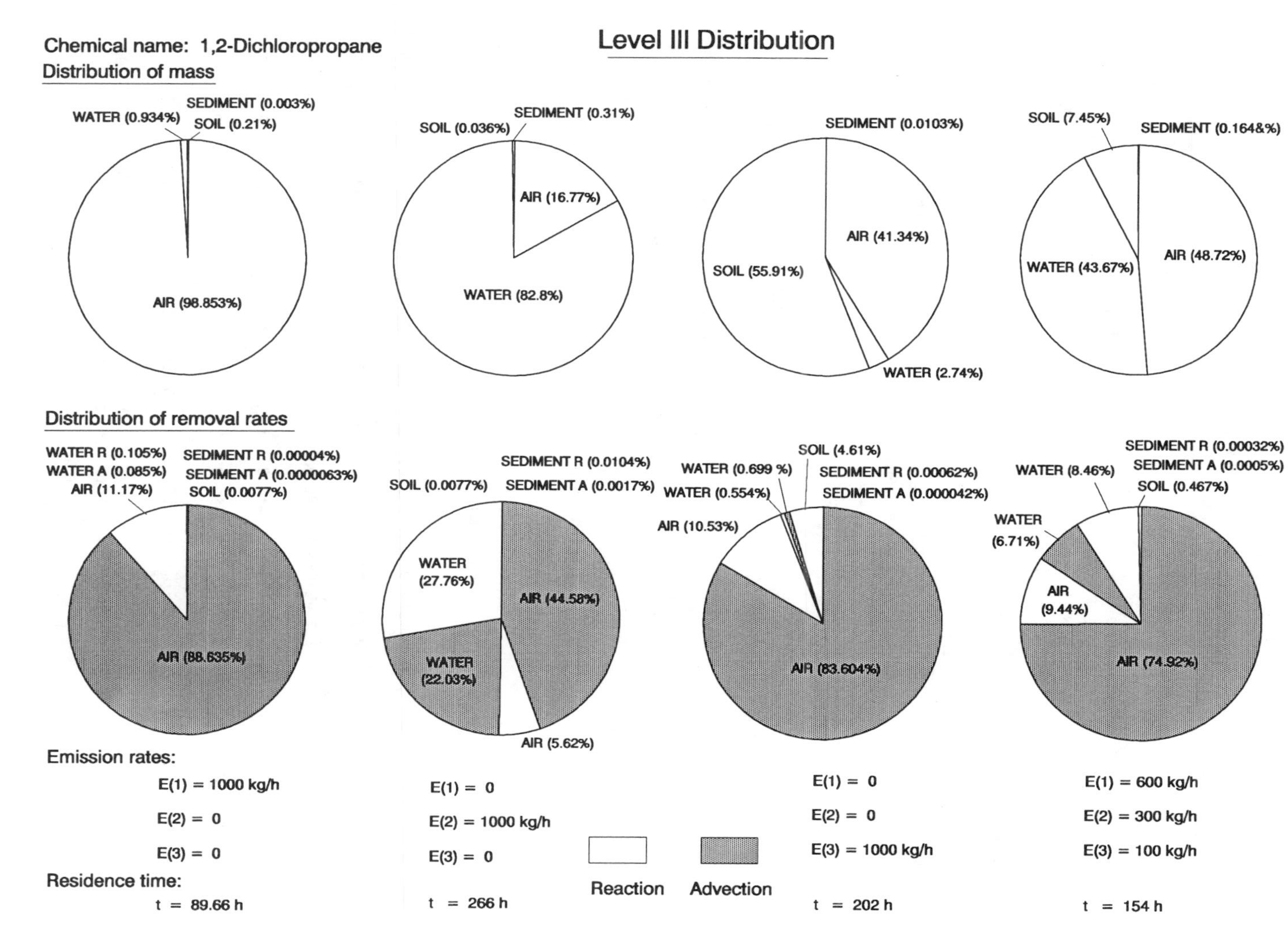
Chemical name: 1,2-Dichloropropane
Level III Distribution
Distribution of mass
WATER (0.934%)
SEDIMENT (0.003%)
SOIL (0.21%)
AIR (98.853%)
SOIL (0.036%)
SEDIMENT (0.31%)
AIR (16.77%)
WATER (82.8%)
SEDIMENT (0.0103%)
AIR (41.34%)
SOIL (55.91%)
WATER (2.74%)
SOIL (7.45%)
SEDIMENT (0.164&%)
WATER (43.67%)
AIR (48.72%)
Distribution of removal rates
WATER R (0.105%)
WATER A (0.085%)
AIR (11.17%)
SEDIMENT R (0.00004%)
SEDIMENT A (0.0000063%)
SOIL (0.0077%)
AIR (88.635%)
SOIL (0.0077%)
SEDIMENT R (0.0104%)
SEDIMENT A (0.0017%)
WATER (27.76%)
AIR (44.58%)
WATER (22.03%)
AIR (5.62%)
SOIL (4.61%)
WATER (0.699 %)
WATER (0.554%)
SEDIMENT R (0.00062%)
SEDIMENT A (0.000042%)
AIR (10.53%)
AIR (83.604%)
SEDIMENT R (0.00032%)
SEDIMENT A (0.0005%)
SOIL (0.467%)
WATER (8.46%)
WATER (6.71%)
AIR (9.44%)
AIR (74.92%)
Emission rates:
E(1) = 1000 kg/h
E(2) = 0
E(3) = 0
E(1) = 0
E(2) = 1000 kg/h
E(3) = 0
E(1) = 0
E(2) = 0
E(3) = 1000 kg/h
E(1) = 600 kg/h
E(2) = 300 kg/h
E(3) = 100 kg/h
Reaction
Advection
Residence time:
t = 89.66 h
t = 266 h
t = 202 h
t = 154 h

Chemical name: 1,2,3-Trichloropropane

Level I calculation: (six-compartment model)

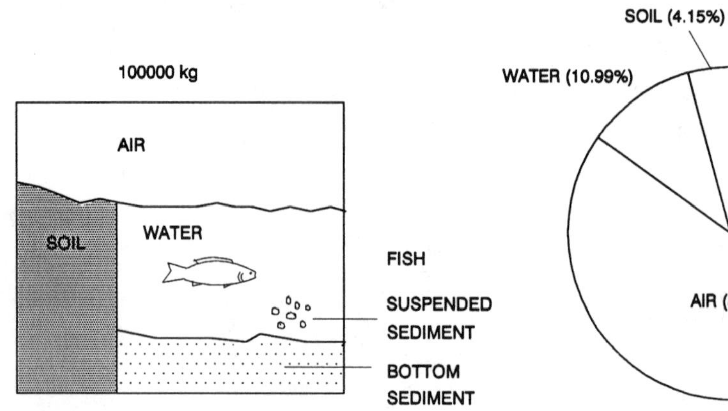


Distribution of mass

physical-chemical properties:

MW: 147.43 g/mol

M.P.: - 14.7°C

Fugacity ratio: 1.0

vapor pressure: 492 Pa

solubility: 1896 g/m$^3$

log $K_{OW}$: 2.63

| Compartment | Z | Concentration | | | Amount | Amount |
|---|---|---|---|---|---|---|
| | mol/m3 Pa | mol/m3 | mg/L (or g/m3) | ug/g | kg | % |
| Air | 4.034E-04 | 5.750E-09 | 8.477E-07 | 7.151E-04 | 84770 | 84.770 |
| Water | 2.614E-02 | 3.726E-07 | 5.493E-05 | 5.493E-05 | 10985 | 10.985 |
| Soil | 2.194E-01 | 3.128E-06 | 4.611E-04 | 1.921E-04 | 4149.93 | 4.1499 |
| Biota (fish) | 5.575E-01 | 7.946E-06 | 1.172E-03 | 1.172E-03 | 0.2343 | 2.34E-04 |
| Suspended sediment | 1.371E+00 | 1.955E-05 | 2.882E-03 | 1.921E-03 | 2.8819 | 2.88E-03 |
| Bottom sediment | 4.389E-01 | 6.255E-06 | 9.222E-04 | 3.843E-04 | 92.221 | 9.22E-02 |
| | Total | | | | 100000 | 100 |

f = 1.425E-05 Pa

Chemical name: 1,2,3-Trichloropropane

Level II calculation: (six-compartment model)

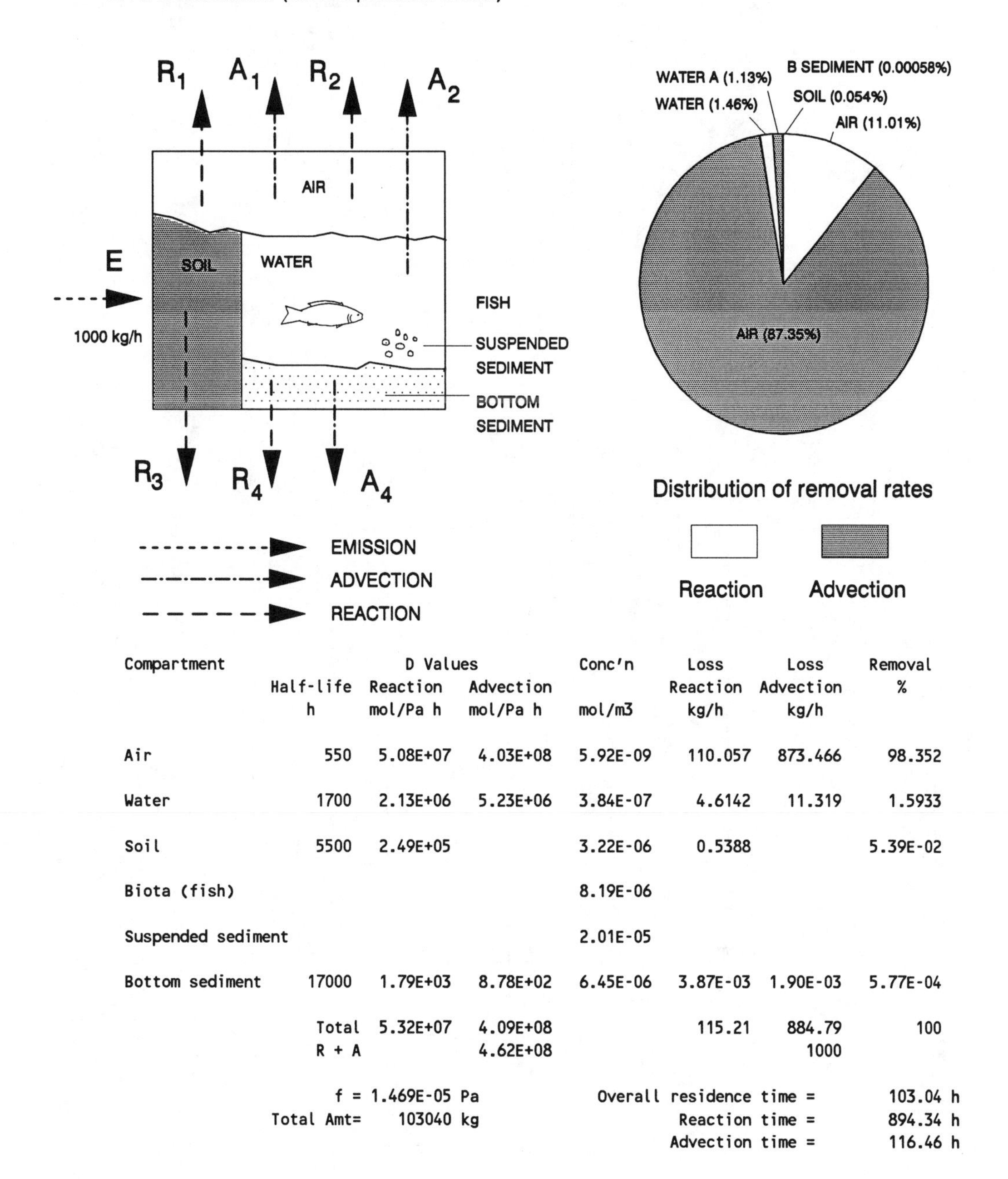


| Compartment | Half-life h | D Values Reaction mol/Pa h | D Values Advection mol/Pa h | Conc'n mol/m3 | Loss Reaction kg/h | Loss Advection kg/h | Removal % |
|---|---|---|---|---|---|---|---|
| Air | 550 | 5.08E+07 | 4.03E+08 | 5.92E-09 | 110.057 | 873.466 | 98.352 |
| Water | 1700 | 2.13E+06 | 5.23E+06 | 3.84E-07 | 4.6142 | 11.319 | 1.5933 |
| Soil | 5500 | 2.49E+05 | | 3.22E-06 | 0.5388 | | 5.39E-02 |
| Biota (fish) | | | | 8.19E-06 | | | |
| Suspended sediment | | | | 2.01E-05 | | | |
| Bottom sediment | 17000 | 1.79E+03 | 8.78E+02 | 6.45E-06 | 3.87E-03 | 1.90E-03 | 5.77E-04 |
| | Total | 5.32E+07 | 4.09E+08 | | 115.21 | 884.79 | 100 |
| | R + A | | 4.62E+08 | | | 1000 | |

f = 1.469E-05 Pa
Total Amt= 103040 kg

Overall residence time = 103.04 h
Reaction time = 894.34 h
Advection time = 116.46 h

## Fugacity Level III calculations: (four-compartment model)

Chemical name: 1,2,3-Trichloropropane

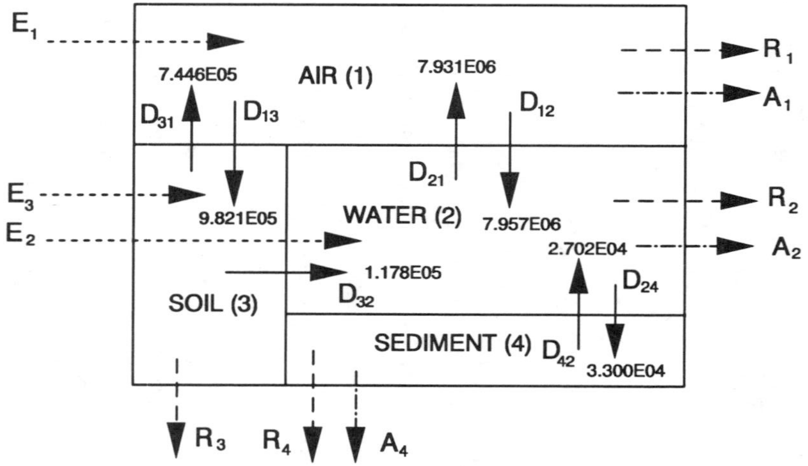

Phase Properties and Rates:

| Compartment | Bulk Z mol/m3 Pa | Half-life h | D Values Reaction mol/Pa h | D Values Advection mol/Pa h |
|---|---|---|---|---|
| Air (1) | 4.034E-04 | 550 | 5.08E+07 | 4.03E+08 |
| Water (2) | 2.615E-02 | 1700 | 2.13E+06 | 5.23E+06 |
| Soil (3) | 1.176E-01 | 5500 | 2.67E+05 | |
| Sediment (4) | 1.087E-01 | 17000 | 2.22E+03 | 1.09E+03 |

| | E(1)=1000 | E(2)=1000 | E(3)=1000 | E(1,2,3) |
|---|---|---|---|---|
| Overall residence time = | 98.10 | 396.62 | 1977.96 | 375.64 h |
| Reaction time = | 861.65 | 1997.47 | 5963.39 | 2332.47 h |
| Advection time = | 110.71 | 494.89 | 2959.61 | 447.75 h |

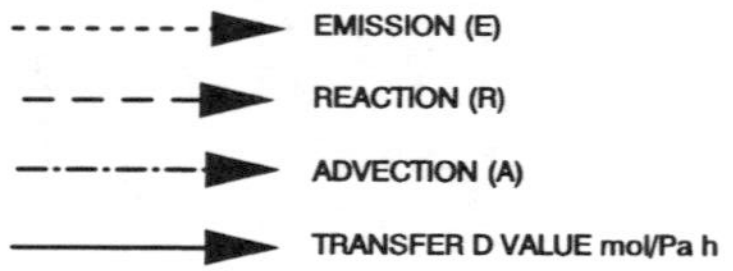

Phase Properties, Compositions, Transport and Transformation Rates:

| Emission, kg/h | | | Fugacity, Pa | | | | Concentration, g/m3 | | | | Amounts, kg | | | | Total |
|---|---|---|---|---|---|---|---|---|---|---|---|---|---|---|---|
| E(1) | E(2) | E(3) | f(1) | f(2) | f(3) | f(4) | C(1) | C(2) | C(3) | C(4) | m(1) | m(2) | m(3) | m(4) | Amount, kg |
| 1000 | 0 | 0 | 1.480E-05 | 7.797E-06 | 1.285E-05 | 8.486E-06 | 8.801E-07 | 3.006E-05 | 2.228E-04 | 1.360E-04 | 8.801E+04 | 6.011E+03 | 4.011E+03 | 6.799E+01 | 9.810E+04 |
| 0 | 1000 | 0 | 7.673E-06 | 4.475E-04 | 6.662E-06 | 4.870E-04 | 4.564E-07 | 1.725E-03 | 1.155E-04 | 7.804E-03 | 4.564E+04 | 3.450E+05 | 2.080E+03 | 3.902E+03 | 3.966E+05 |
| 0 | 0 | 1000 | 1.057E-05 | 5.176E-05 | 6.005E-03 | 5.633E-05 | 6.284E-07 | 1.995E-04 | 1.042E-01 | 9.026E-04 | 6.284E+04 | 3.990E+04 | 1.875E+06 | 4.513E+02 | 1.978E+06 |
| 600 | 300 | 100 | 1.224E-05 | 1.441E-04 | 6.102E-04 | 1.568E-04 | 7.278E-07 | 5.555E-04 | 1.058E-02 | 2.513E-03 | 7.278E+04 | 1.111E+05 | 1.905E+05 | 1.257E+03 | 3.756E+05 |

| Emission, kg/h | | | Loss, Reaction, kg/h | | | | Loss, Advection, kg/h | | | Intermedia Rate of Transport, kg/h T12 | T21 | T13 | T31 | T32 | T24 | T42 |
|---|---|---|---|---|---|---|---|---|---|---|---|---|---|---|---|---|
| E(1) | E(2) | E(3) | R(1) | R(2) | R(3) | R(4) | A(1) | A(2) | A(4) | air-water | water-air | air-soil | soil-air | soil-water | water-sed | sed-water |
| 1000 | 0 | 0 | 1.109E+02 | 2.450E+00 | 5.05E-01 | 2.771E-03 | 8.801E+02 | 6.011E+00 | 1.360E-03 | 1.736E+01 | 9.117E+00 | 2.143E+00 | 1.414E+00 | 2.232E-01 | 3.793E-02 | 3.380E-02 |
| 0 | 1000 | 0 | 5.750E+01 | 1.406E+02 | 2.62E-01 | 1.591E-01 | 4.564E+02 | 3.450E+02 | 7.804E-02 | 9.001E+00 | 5.232E+02 | 1.111E+00 | 7.332E-01 | 1.157E-01 | 2.177E+00 | 1.940E+00 |
| 0 | 0 | 1000 | 7.918E+01 | 1.627E+01 | 2.36E+02 | 1.840E-02 | 6.284E+02 | 3.990E+01 | 9.026E-03 | 1.239E+01 | 6.051E+01 | 1.530E+00 | 6.610E+02 | 1.043E+02 | 2.518E-01 | 2.244E-01 |
| 600 | 300 | 100 | 9.171E+01 | 4.529E+01 | 2.40E+01 | 5.122E-02 | 7.278E+02 | 1.111E+02 | 2.513E-02 | 1.436E+01 | 1.685E+02 | 1.772E+00 | 6.717E+01 | 1.060E+01 | 7.010E-01 | 6.247E-01 |

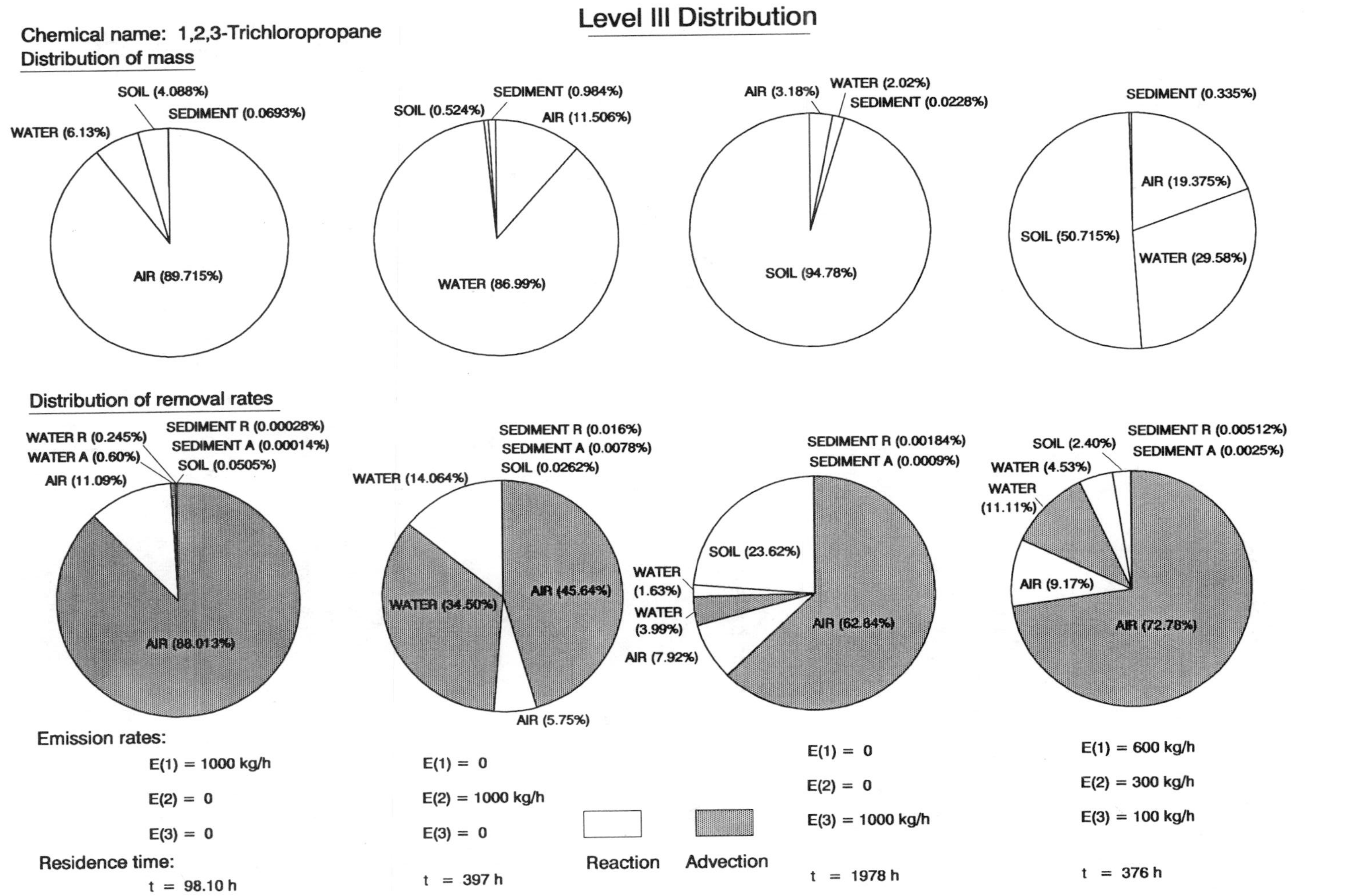
Level III Distribution
Chemical name: 1,2,3-Trichloropropane
Distribution of mass
SOIL (4.088%)
SEDIMENT (0.0693%)
WATER (6.13%)
AIR (89.715%)
SEDIMENT (0.984%)
SOIL (0.524%)
AIR (11.506%)
WATER (86.99%)
AIR (3.18%)
WATER (2.02%)
SEDIMENT (0.0228%)
SOIL (94.78%)
SEDIMENT (0.335%)
AIR (19.375%)
SOIL (50.715%)
WATER (29.58%)
Distribution of removal rates
SEDIMENT R (0.00028%)
WATER R (0.245%)
SEDIMENT A (0.00014%)
WATER A (0.60%)
SOIL (0.0505%)
AIR (11.09%)
AIR (88.013%)
SEDIMENT R (0.016%)
SEDIMENT A (0.0078%)
SOIL (0.0262%)
WATER (14.064%)
AIR (45.64%)
WATER (34.50%)
AIR (5.75%)
SEDIMENT R (0.00184%)
SEDIMENT A (0.0009%)
SOIL (23.62%)
WATER (1.63%)
WATER (3.99%)
AIR (7.92%)
AIR (62.84%)
SOIL (2.40%)
SEDIMENT R (0.00512%)
SEDIMENT A (0.0025%)
WATER (4.53%)
WATER (11.11%)
AIR (9.17%)
AIR (72.78%)
Emission rates:
E(1) = 1000 kg/h
E(2) = 0
E(3) = 0
Residence time:
t = 98.10 h
E(1) = 0
E(2) = 1000 kg/h
E(3) = 0
t = 397 h
Reaction
Advection
E(1) = 0
E(2) = 0
E(3) = 1000 kg/h
t = 1978 h
E(1) = 600 kg/h
E(2) = 300 kg/h
E(3) = 100 kg/h
t = 376 h

Chemical name: 1,1-Dichloroethene

Level I calculation: (six-compartment model)

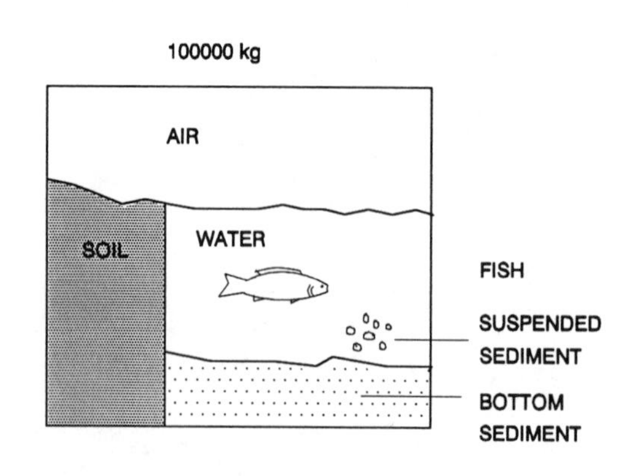


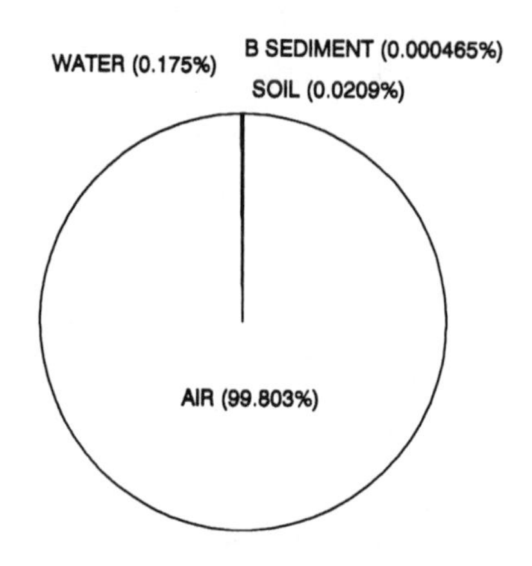


Distribution of mass

physical-chemical properties:

MW: 96.94 g/mol

M.P.: - 122.1 °C

Fugacity ratio: 1.0

vapor pressure: 80500 Pa

solubility: 2763 g/m $^3$

log $K_{OW}$: 2.13

| Compartment | Z | Concentration | | | Amount | Amount |
|---|---|---|---|---|---|---|
| | mol/m3 Pa | mol/m3 | mg/L (or g/m3) | ug/g | kg | % |
| Air | 4.034E-04 | 1.030E-08 | 9.980E-07 | 8.419E-04 | 99803 | 99.803 |
| Water | 3.541E-04 | 9.036E-09 | 8.759E-07 | 8.759E-07 | 175.19 | 0.1752 |
| Soil | 9.400E-04 | 2.399E-08 | 2.325E-06 | 9.689E-07 | 20.929 | 0.0209 |
| Biota (fish) | 2.388E-03 | 6.095E-08 | 5.908E-06 | 5.908E-06 | 1.18E-03 | 1.18E-06 |
| Suspended sediment | 5.875E-03 | 1.499E-07 | 1.453E-05 | 9.689E-06 | 1.45E-02 | 1.45E-05 |
| Bottom sediment | 1.880E-03 | 4.798E-08 | 4.651E-06 | 1.938E-06 | 0.4651 | 4.65E-04 |
| | Total | | | | 100000 | 100 |

f = 2.552E-05 Pa

Chemical name: 1,1-Dichloroethene

Level II calculation: (six-compartment model)

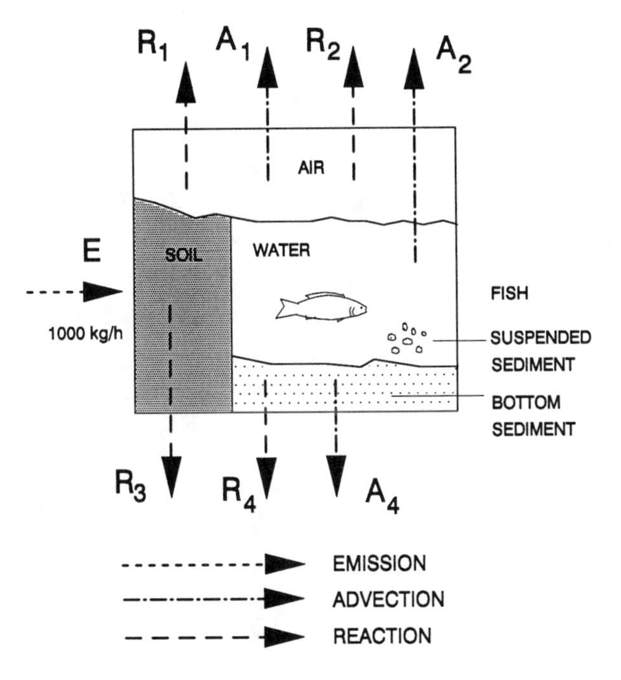


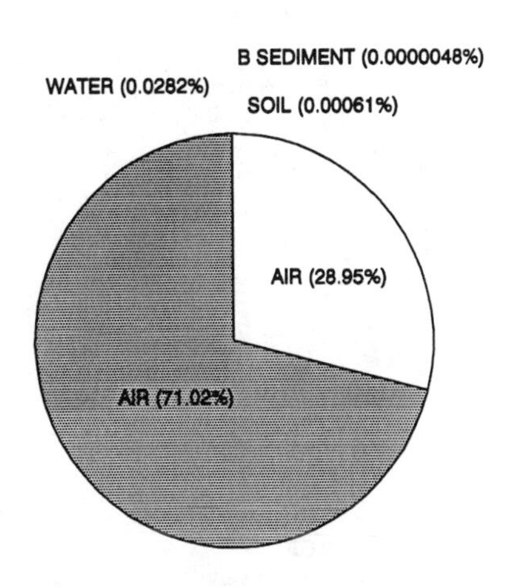


Distribution of removal rates

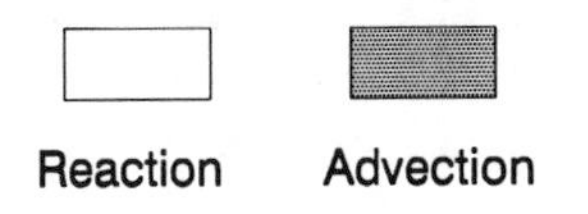


| Compartment | Half-life h | D Values Reaction mol/Pa h | D Values Advection mol/Pa h | Conc'n mol/m3 | Loss Reaction kg/h | Loss Advection kg/h | Removal % |
|---|---|---|---|---|---|---|---|
| Air | 170 | 1.64E+08 | 4.03E+08 | 7.33E-09 | 289.511 | 710.201 | 99.971 |
| Water | 550 | 8.92E+04 | 7.08E+04 | 6.43E-09 | 0.1571 | 0.1247 | 0.0282 |
| Soil | 1700 | 3.45E+03 | | 1.71E-08 | 6.07E-03 | | 6.07E-04 |
| Biota (fish) | | | | 4.34E-08 | | | |
| Suspended sediment | | | | 1.07E-07 | | | |
| Bottom sediment | 5500 | 2.37E+01 | 3.76E+00 | 3.41E-08 | 4.17E-05 | 6.62E-06 | 4.83E-06 |
| | Total | 1.65E+08 | 4.03E+08 | | 289.67 | 710.33 | 100 |
| | R + A | | 5.68E+08 | | | 1000 | |

f = 1.816E-05 Pa

Total Amt= 71160 kg

Overall residence time = 71.16 h

Reaction time = 245.65 h

Advection time = 100.18 h

Fugacity Level III calculations: (four-compartment model)

Chemical name: 1,1-Dichloroethene

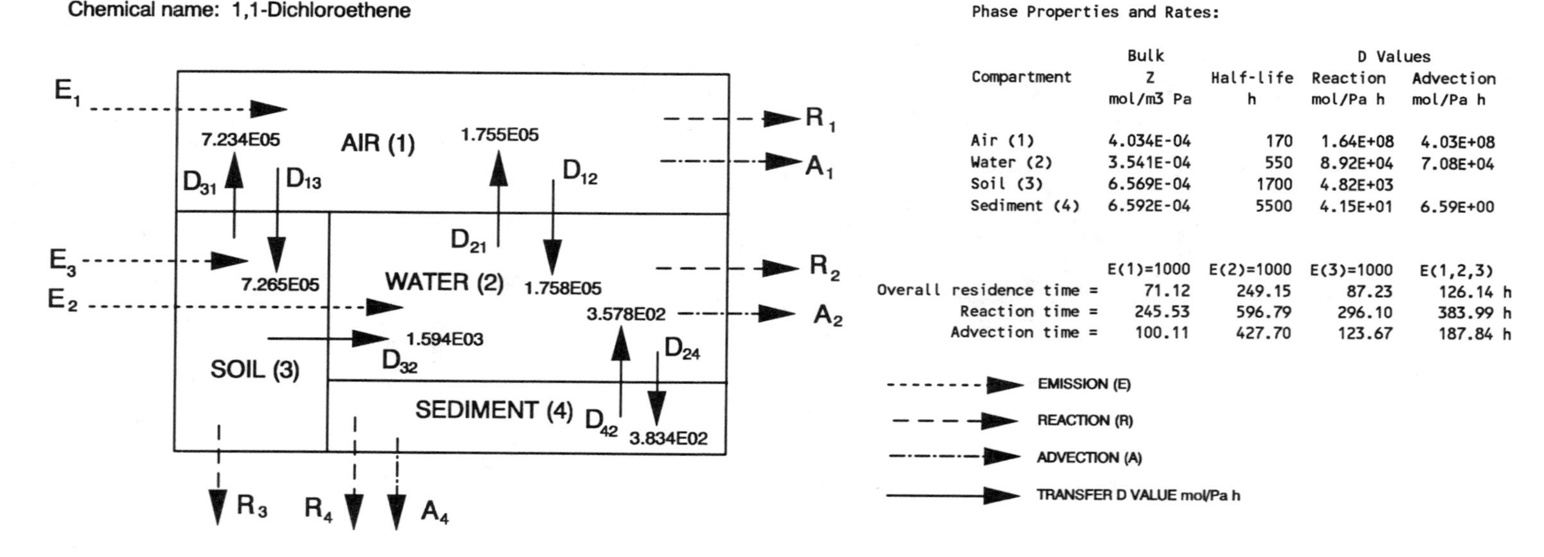


Phase Properties and Rates:

| Compartment | Bulk Z mol/m3 Pa | Half-life h | D Values Reaction mol/Pa h | D Values Advection mol/Pa h |
|---|---|---|---|---|
| Air (1) | 4.034E-04 | 170 | 1.64E+08 | 4.03E+08 |
| Water (2) | 3.541E-04 | 550 | 8.92E+04 | 7.08E+04 |
| Soil (3) | 6.569E-04 | 1700 | 4.82E+03 | |
| Sediment (4) | 6.592E-04 | 5500 | 4.15E+01 | 6.59E+00 |

| | E(1)=1000 | E(2)=1000 | E(3)=1000 | E(1,2,3) |
|---|---|---|---|---|
| Overall residence time = | 71.12 | 249.15 | 87.23 | 126.14 h |
| Reaction time = | 245.53 | 596.79 | 296.10 | 383.99 h |
| Advection time = | 100.11 | 427.70 | 123.67 | 187.84 h |

EMISSION (E)

REACTION (R)

ADVECTION (A)

TRANSFER D VALUE mol/Pa h

Phase Properties, Compositions, Transport and Transformation Rates:

| Emission, kg/h | | | Fugacity, Pa | | | | Concentration, g/m3 | | | | Amounts, kg | | | | Total |
|---|---|---|---|---|---|---|---|---|---|---|---|---|---|---|---|
| E(1) | E(2) | E(3) | f(1) | f(2) | f(3) | f(4) | C(1) | C(2) | C(3) | C(4) | m(1) | m(2) | m(3) | m(4) | Amount, kg |
| 1000 | 0 | 0 | 1.816E-05 | 9.603E-06 | 1.808E-05 | 9.071E-06 | 7.103E-07 | 3.296E-07 | 1.151E-06 | 5.797E-07 | 7.103E+04 | 6.593E+01 | 2.073E+01 | 2.898E-01 | 7.112E+04 |
| 0 | 1000 | 0 | 9.498E-06 | 3.074E-02 | 9.456E-06 | 2.904E-02 | 3.714E-07 | 1.055E-03 | 6.021E-07 | 1.856E-03 | 3.714E+04 | 2.111E+05 | 1.084E+01 | 9.279E+02 | 2.491E+05 |
| 0 | 0 | 1000 | 1.802E-05 | 7.666E-05 | 1.415E-02 | 7.241E-05 | 7.049E-07 | 2.631E-06 | 9.010E-04 | 4.627E-06 | 7.049E+04 | 5.263E+02 | 1.622E+04 | 2.314E+00 | 8.723E+04 |
| 600 | 300 | 100 | 1.555E-05 | 9.237E-03 | 1.429E-03 | 8.724E-03 | 6.081E-07 | 3.171E-04 | 9.097E-05 | 5.575E-04 | 6.081E+04 | 6.341E+04 | 1.637E+03 | 2.788E+02 | 1.261E+05 |

| Emission, kg/h | | | Loss, Reaction, kg/h | | | | Loss, Advection, kg/h | | | Intermedia Rate of Transport, kg/h | | | | | | |
|---|---|---|---|---|---|---|---|---|---|---|---|---|---|---|---|---|
| | | | | | | | | | | T12 | T21 | T13 | T31 | T32 | T24 | T42 |
| E(1) | E(2) | E(3) | R(1) | R(2) | R(3) | R(4) | A(1) | A(2) | A(4) | air-water | water-air | air-soil | soil-air | soil-water | water-sed | sed-water |
| 1000 | 0 | 0 | 2.895E+02 | 8.307E-02 | 8.45E-03 | 3.652E-05 | 7.103E+02 | 6.593E-02 | 5.797E-06 | 3.096E-01 | 1.634E-01 | 1.280E+00 | 1.268E+00 | 2.794E-03 | 3.569E-04 | 3.146E-04 |
| 0 | 1000 | 0 | 1.514E+02 | 2.659E+02 | 4.42E-03 | 1.169E-01 | 3.714E+02 | 2.111E+02 | 1.856E-02 | 1.619E-01 | 5.230E+02 | 6.692E-01 | 6.633E-01 | 1.461E-03 | 1.143E+00 | 1.007E+00 |
| 0 | 0 | 1000 | 2.873E+02 | 6.631E-01 | 6.61E+00 | 2.915E-04 | 7.049E+02 | 5.263E-01 | 4.627E-05 | 3.072E-01 | 1.304E+00 | 1.270E+00 | 9.925E+02 | 2.187E+00 | 2.849E-03 | 2.512E-03 |
| 600 | 300 | 100 | 2.479E+02 | 7.990E+01 | 6.68E-01 | 3.513E-02 | 6.081E+02 | 6.341E+01 | 5.575E-03 | 2.651E-01 | 1.571E+02 | 1.095E+00 | 1.002E+02 | 2.208E-01 | 3.433E-01 | 3.026E-01 |

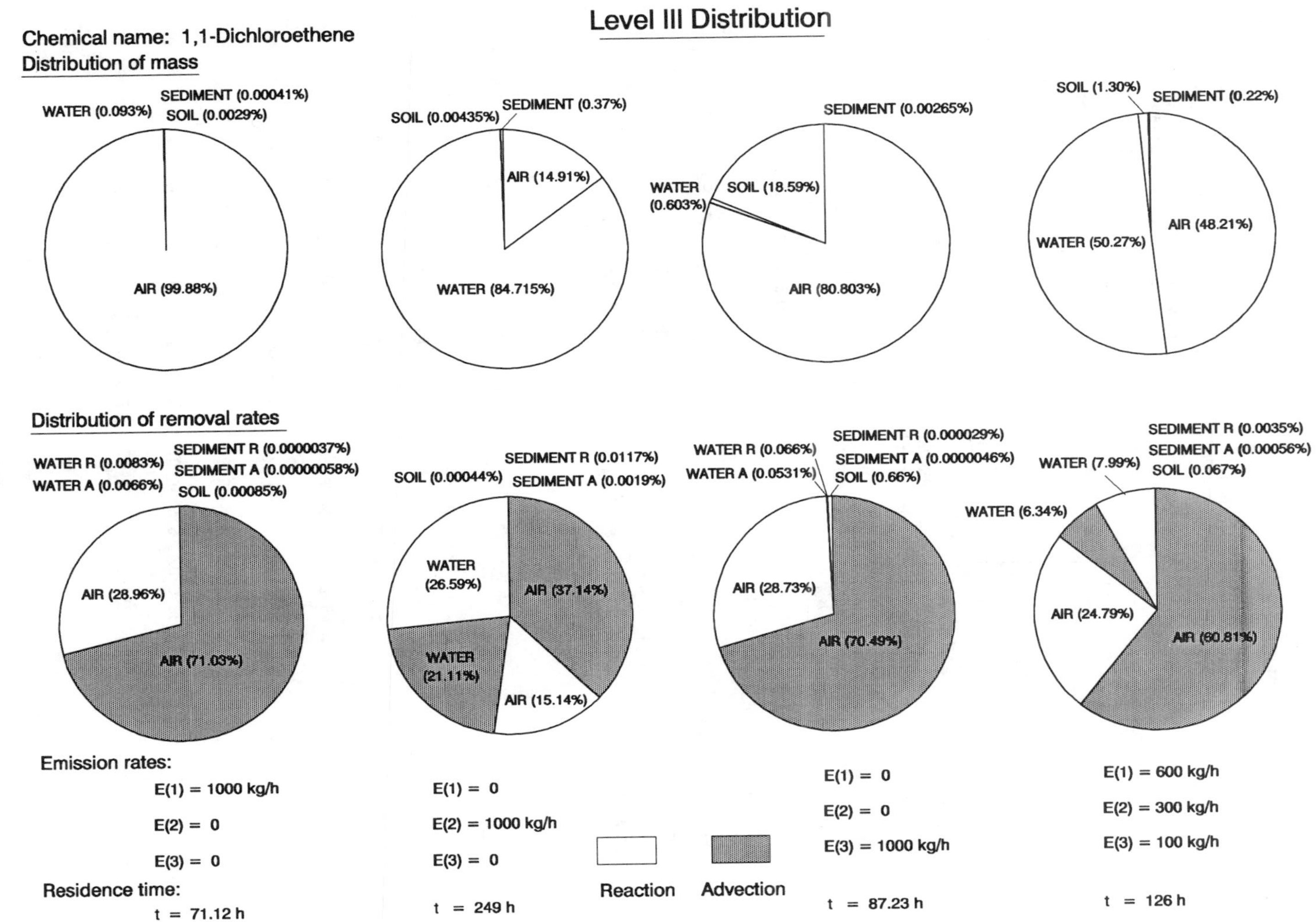
Level III Distribution
Chemical name: 1,1-Dichloroethene
Distribution of mass
WATER (0.093%)
SEDIMENT (0.00041%)
SOIL (0.0029%)
AIR (99.88%)
SOIL (0.00435%)
SEDIMENT (0.37%)
AIR (14.91%)
WATER (84.715%)
SEDIMENT (0.00265%)
WATER (0.603%)
SOIL (18.59%)
AIR (80.803%)
SOIL (1.30%)
SEDIMENT (0.22%)
WATER (50.27%)
AIR (48.21%)
Distribution of removal rates
WATER R (0.0083%)
WATER A (0.0066%)
SEDIMENT R (0.0000037%)
SEDIMENT A (0.00000058%)
SOIL (0.00085%)
AIR (28.96%)
AIR (71.03%)
SOIL (0.00044%)
SEDIMENT R (0.0117%)
SEDIMENT A (0.0019%)
WATER (26.59%)
AIR (37.14%)
WATER (21.11%)
AIR (15.14%)
WATER R (0.066%)
WATER A (0.0531%)
SEDIMENT R (0.000029%)
SEDIMENT A (0.0000046%)
SOIL (0.66%)
AIR (28.73%)
AIR (70.49%)
SEDIMENT R (0.0035%)
SEDIMENT A (0.00056%)
SOIL (0.067%)
WATER (7.99%)
WATER (6.34%)
AIR (24.79%)
AIR (60.81%)
Emission rates:
E(1) = 1000 kg/h
E(2) = 0
E(3) = 0
Residence time:
t = 71.12 h
E(1) = 0
E(2) = 1000 kg/h
E(3) = 0
t = 249 h
Reaction
Advection
E(1) = 0
E(2) = 0
E(3) = 1000 kg/h
t = 87.23 h
E(1) = 600 kg/h
E(2) = 300 kg/h
E(3) = 100 kg/h
t = 126 h

Chemical name: Trichloroethylene

Level I calculation: (six-compartment model)

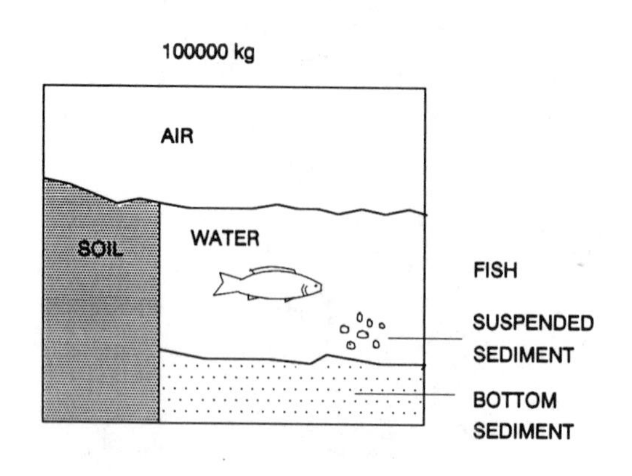


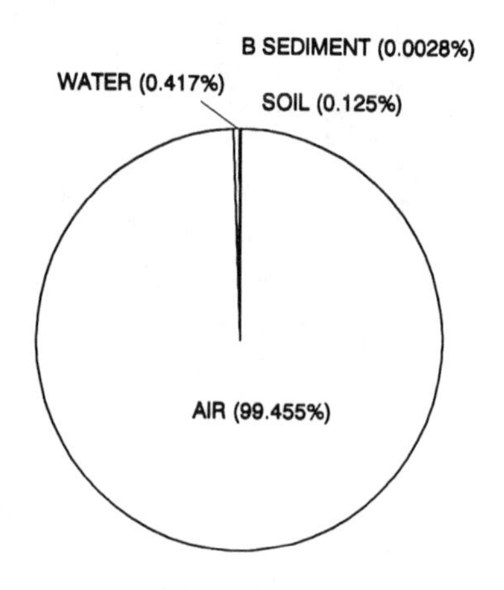


Distribution of mass

physical-chemical properties:

MW: 131.39 g/mol

M.P.: - 73 °C

Fugacity ratio: 1.0

vapor pressure: 9900 Pa

solubility: 1100 g/m³

log $K_{OW}$: 2.53

| Compartment | Z | Concentration | | | Amount | Amount |
|---|---|---|---|---|---|---|
| | mol/m3 Pa | mol/m3 | mg/L (or g/m3) | ug/g | kg | % |
| Air | 4.034E-04 | 7.569E-09 | 9.946E-07 | 8.390E-04 | 99455 | 99.455 |
| Water | 8.457E-04 | 1.587E-08 | 2.085E-06 | 2.085E-06 | 416.96 | 0.4170 |
| Soil | 5.639E-03 | 1.058E-07 | 1.390E-05 | 5.793E-06 | 125.12 | 0.1251 |
| Biota (fish) | 1.433E-02 | 2.688E-07 | 3.532E-05 | 3.532E-05 | 7.06E-03 | 7.06E-06 |
| Suspended sediment | 3.525E-02 | 6.613E-07 | 8.689E-05 | 5.793E-05 | 8.69E-02 | 8.69E-05 |
| Bottom sediment | 1.128E-02 | 2.116E-07 | 2.780E-05 | 1.159E-05 | 2.7805 | 2.78E-03 |
| | Total | | | | 100000 | 100 |

f = 1.876E-05 Pa

Chemical name: Trichloroethylene

Level II calculation: (six-compartment model)

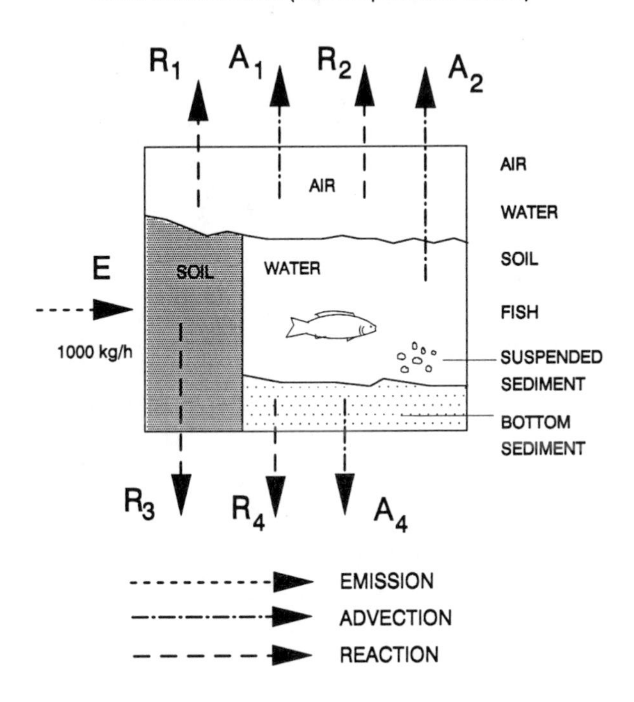


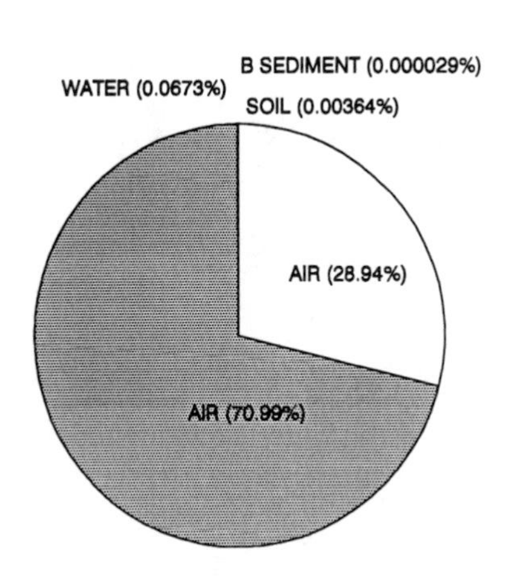


Distribution of removal rates

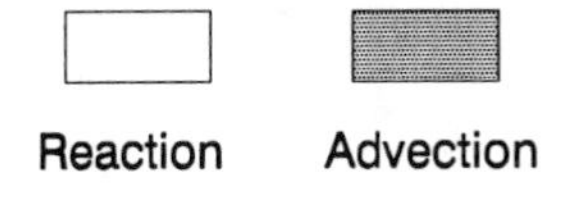


| Compartment | Half-life h | D Values Reaction mol/Pa h | D Values Advection mol/Pa h | Conc'n mol/m3 | Loss Reaction kg/h | Loss Advection kg/h | Removal % |
|---|---|---|---|---|---|---|---|
| Air | 170 | 1.64E+08 | 4.03E+08 | 5.40E-09 | 289.389 | 709.901 | 99.929 |
| Water | 550 | 2.13E+05 | 1.69E+05 | 1.13E-08 | 0.3750 | 0.2976 | 0.0673 |
| Soil | 1700 | 2.07E+04 | | 7.55E-08 | 0.0364 | | 3.64E-03 |
| Biota (fish) | | | | 1.92E-07 | | | |
| Suspended sediment | | | | 4.72E-07 | | | |
| Bottom sediment | 5500 | 1.42E+02 | 2.26E+01 | 1.51E-07 | 2.50E-04 | 3.97E-05 | 2.90E-05 |
| | Total | 1.65E+08 | 4.04E+08 | | 289.80 | 710.20 | 100 |
| | R + A | | 5.68E+08 | | | 1000 | |

f = 1.339E-05 Pa
Total Amt= 71379 kg

Overall residence time = 71.38 h
Reaction time = 246.30 h
Advection time = 100.51 h

**Fugacity Level III calculations:** (four-compartment model)

Chemical name: Trichloroethylene

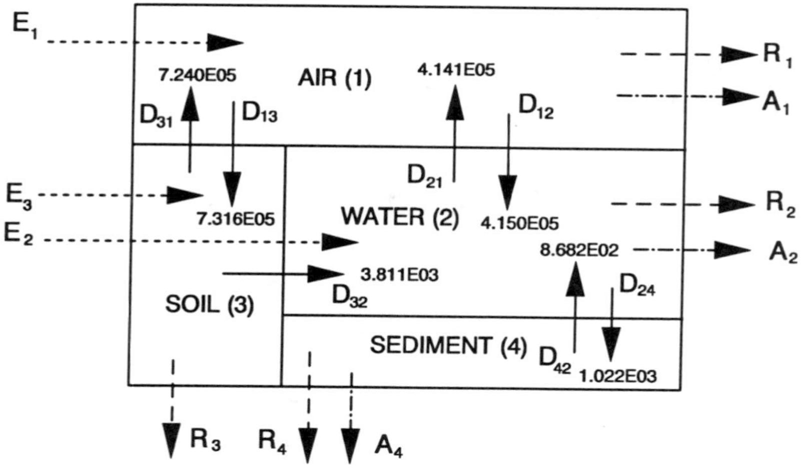

E1, E2, E3; AIR (1); WATER (2); SOIL (3); SEDIMENT (4)

$D_{31}$ 7.240E05; $D_{13}$ 7.316E05; $D_{21}$ 4.141E05; $D_{12}$ 4.150E05; $D_{32}$ 3.811E03; $D_{42}$ 8.682E02; $D_{24}$ 1.022E03

R1, A1, R2, A2, R3, R4, A4

EMISSION (E)

REACTION (R)

ADVECTION (A)

TRANSFER D VALUE mol/Pa h

Phase Properties and Rates:

| Compartment | Bulk Z mol/m3 Pa | Half-life h | D Values Reaction mol/Pa h | D Values Advection mol/Pa h |
|---|---|---|---|---|
| Air (1) | 4.034E-04 | 170 | 1.64E+08 | 4.03E+08 |
| Water (2) | 8.458E-04 | 550 | 2.13E+05 | 1.69E+05 |
| Soil (3) | 3.154E-03 | 1700 | 2.31E+04 | |
| Sediment (4) | 2.932E-03 | 5500 | 1.85E+02 | 2.93E+01 |

| | E(1)=1000 | E(2)=1000 | E(3)=1000 | E(1,2,3) |
|---|---|---|---|---|
| Overall residence time = | 71.27 | 251.13 | 145.58 | 132.66 h |
| Reaction time = | 245.99 | 600.24 | 466.22 | 401.30 h |
| Advection time = | 100.34 | 431.78 | 211.68 | 198.17 h |

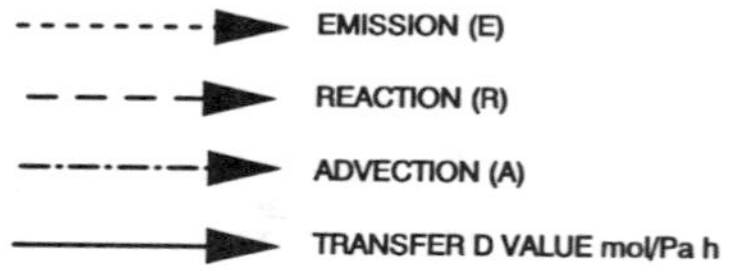

Phase Properties, Compositions, Transport and Transformation Rates:

| Emission, kg/h | | | Fugacity, Pa | | | | Concentration, g/m3 | | | | Amounts, kg | | | | Total |
|---|---|---|---|---|---|---|---|---|---|---|---|---|---|---|---|
| E(1) | E(2) | E(3) | f(1) | f(2) | f(3) | f(4) | C(1) | C(2) | C(3) | C(4) | m(1) | m(2) | m(3) | m(4) | Amount, kg |
| 1000 | 0 | 0 | 1.340E-05 | 7.041E-06 | 1.305E-05 | 6.648E-06 | 7.101E-07 | 7.825E-07 | 5.409E-06 | 2.561E-06 | 7.101E+04 | 1.565E+02 | 9.736E+01 | 1.281E+00 | 7.127E+04 |
| 0 | 1000 | 0 | 6.965E-06 | 9.557E-03 | 6.785E-06 | 9.024E-03 | 3.692E-07 | 1.062E-03 | 2.812E-06 | 3.477E-03 | 3.692E+04 | 2.124E+05 | 5.061E+01 | 1.738E+03 | 2.511E+05 |
| 0 | 0 | 1000 | 1.295E-05 | 5.528E-05 | 1.015E-02 | 5.220E-05 | 6.865E-07 | 6.144E-06 | 4.205E-03 | 2.011E-05 | 6.865E+04 | 1.229E+03 | 7.569E+04 | 1.006E+01 | 1.456E+05 |
| 600 | 300 | 100 | 1.142E-05 | 2.877E-03 | 1.025E-03 | 2.716E-03 | 6.055E-07 | 3.197E-04 | 4.246E-04 | 1.047E-03 | 6.055E+04 | 6.394E+04 | 7.643E+03 | 5.233E+02 | 1.327E+05 |

| Emission, kg/h | | | Loss, Reaction, kg/h | | | | Loss, Advection, kg/h | | | Intermedia Rate of Transport, kg/h | | | | | | |
|---|---|---|---|---|---|---|---|---|---|---|---|---|---|---|---|---|
| | | | | | | | | | | T12 | T21 | T13 | T31 | T32 | T24 | T42 |
| E(1) | E(2) | E(3) | R(1) | R(2) | R(3) | R(4) | A(1) | A(2) | A(4) | air-water | water-air | air-soil | soil-air | soil-water | water-sed | sed-water |
| 1000 | 0 | 0 | 2.895E+02 | 1.972E-01 | 3.97E-02 | 1.614E-04 | 7.101E+02 | 1.565E-01 | 2.561E-05 | 7.305E-01 | 3.831E-01 | 1.288E+00 | 1.242E+00 | 6.535E-03 | 9.454E-04 | 7.584E-04 |
| 0 | 1000 | 0 | 1.505E+02 | 2.677E+02 | 2.06E-02 | 2.190E-01 | 3.692E+02 | 2.124E+02 | 3.477E-02 | 3.798E-01 | 5.200E+02 | 6.695E-01 | 6.455E-01 | 3.397E-03 | 1.283E+00 | 1.029E+00 |
| 0 | 0 | 1000 | 2.799E+02 | 1.548E+00 | 3.09E+01 | 1.267E-03 | 6.865E+02 | 1.229E+00 | 2.011E-04 | 7.062E-01 | 3.008E+00 | 1.245E+00 | 9.653E+02 | 5.080E+00 | 7.423E-03 | 5.955E-03 |
| 600 | 300 | 100 | 2.468E+02 | 8.057E+01 | 3.12E+00 | 6.593E-02 | 6.055E+02 | 6.394E+01 | 1.047E-02 | 6.228E-01 | 1.565E+02 | 1.098E+00 | 9.747E+01 | 5.130E-01 | 3.863E-01 | 3.099E-01 |

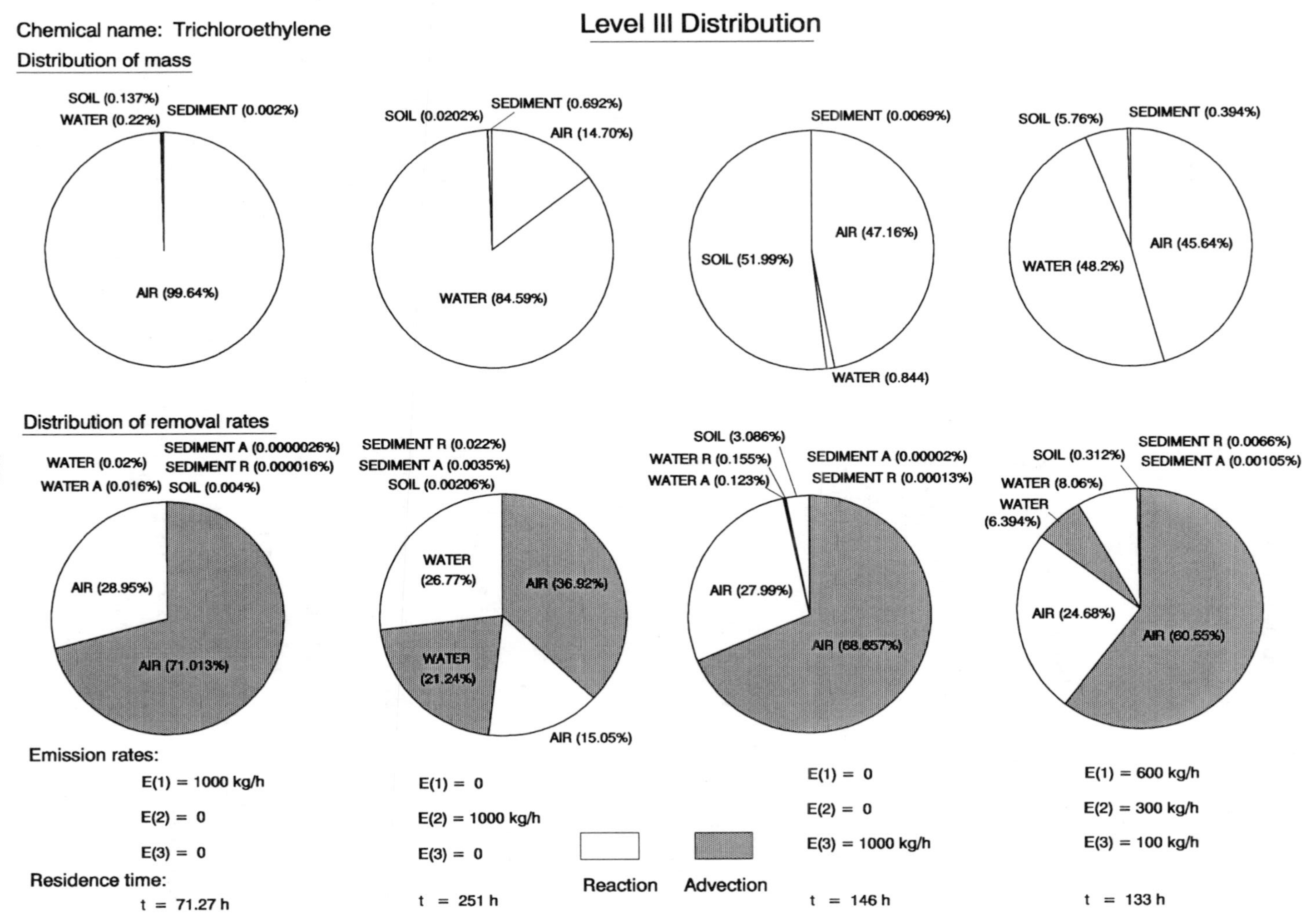
Level III Distribution
Chemical name: Trichloroethylene
Distribution of mass
SOIL (0.137%)
WATER (0.22%)
SEDIMENT (0.002%)
AIR (99.64%)
SOIL (0.0202%)
SEDIMENT (0.692%)
AIR (14.70%)
WATER (84.59%)
SEDIMENT (0.0069%)
AIR (47.16%)
SOIL (51.99%)
WATER (0.844)
SOIL (5.76%)
SEDIMENT (0.394%)
AIR (45.64%)
WATER (48.2%)
Distribution of removal rates
SEDIMENT A (0.0000026%)
WATER (0.02%)
SEDIMENT R (0.000016%)
WATER A (0.016%)
SOIL (0.004%)
AIR (28.95%)
AIR (71.013%)
SEDIMENT R (0.022%)
SEDIMENT A (0.0035%)
SOIL (0.00206%)
WATER (26.77%)
AIR (36.92%)
WATER (21.24%)
AIR (15.05%)
SOIL (3.086%)
WATER R (0.155%)
WATER A (0.123%)
SEDIMENT A (0.00002%)
SEDIMENT R (0.00013%)
AIR (27.99%)
AIR (68.657%)
SOIL (0.312%)
SEDIMENT R (0.0066%)
SEDIMENT A (0.00105%)
WATER (8.06%)
WATER (6.394%)
AIR (24.68%)
AIR (60.55%)
Emission rates:
E(1) = 1000 kg/h
E(2) = 0
E(3) = 0
Residence time:
t = 71.27 h
E(1) = 0
E(2) = 1000 kg/h
E(3) = 0
t = 251 h
Reaction
Advection
E(1) = 0
E(2) = 0
E(3) = 1000 kg/h
t = 146 h
E(1) = 600 kg/h
E(2) = 300 kg/h
E(3) = 100 kg/h
t = 133 h

Chemical name: Tetrachloroethylene

Level I calculation: (six-compartment model)

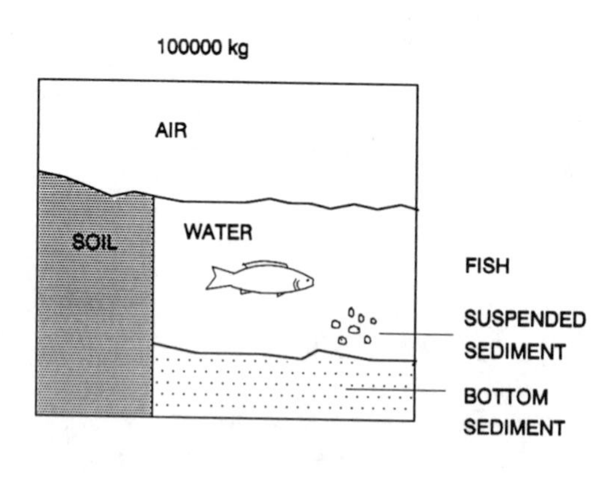


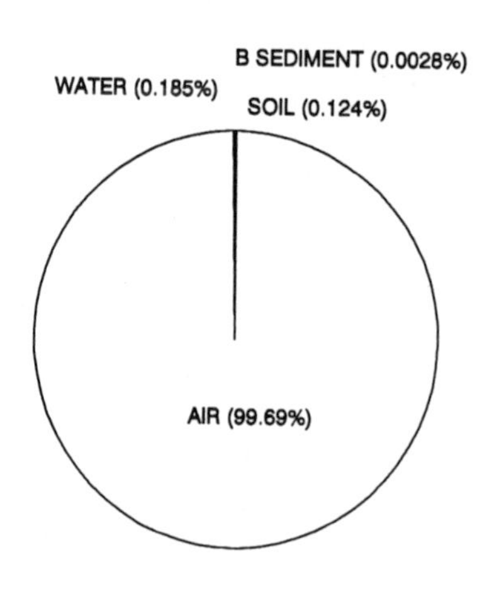


Distribution of mass

physical-chemical properties:

MW: 165.83 g/mol

M.P.: - 19℃

Fugacity ratio: 1.0

vapor pressure: 2415 Pa

solubility: 150 g/m³

log $K_{OW}$: 2.88

| Compartment | Z | Concentration | | | Amount | Amount |
|---|---|---|---|---|---|---|
| | mol/m3 Pa | mol/m3 | mg/L (or g/m3) | ug/g | kg | % |
| Air | 4.034E-04 | 6.011E-09 | 9.969E-07 | 8.410E-04 | 99688 | 99.688 |
| Water | 3.746E-04 | 5.581E-09 | 9.255E-07 | 9.255E-07 | 185.11 | 0.1851 |
| Soil | 5.592E-03 | 8.332E-08 | 1.382E-05 | 5.757E-06 | 124.355 | 0.1244 |
| Biota (fish) | 1.421E-02 | 2.117E-07 | 3.510E-05 | 3.510E-05 | 7.02E-03 | 7.02E-06 |
| Suspended sediment | 3.495E-02 | 5.208E-07 | 8.636E-05 | 5.757E-05 | 0.0864 | 8.64E-05 |
| Bottom sediment | 1.118E-02 | 1.666E-07 | 2.763E-05 | 1.151E-05 | 2.7635 | 2.76E-03 |
| | Total | | | | 100000 | 100 |

f = 1.490E-05 Pa

Chemical name: Tetrachloroethylene

Level II calculation: (six-compartment model)

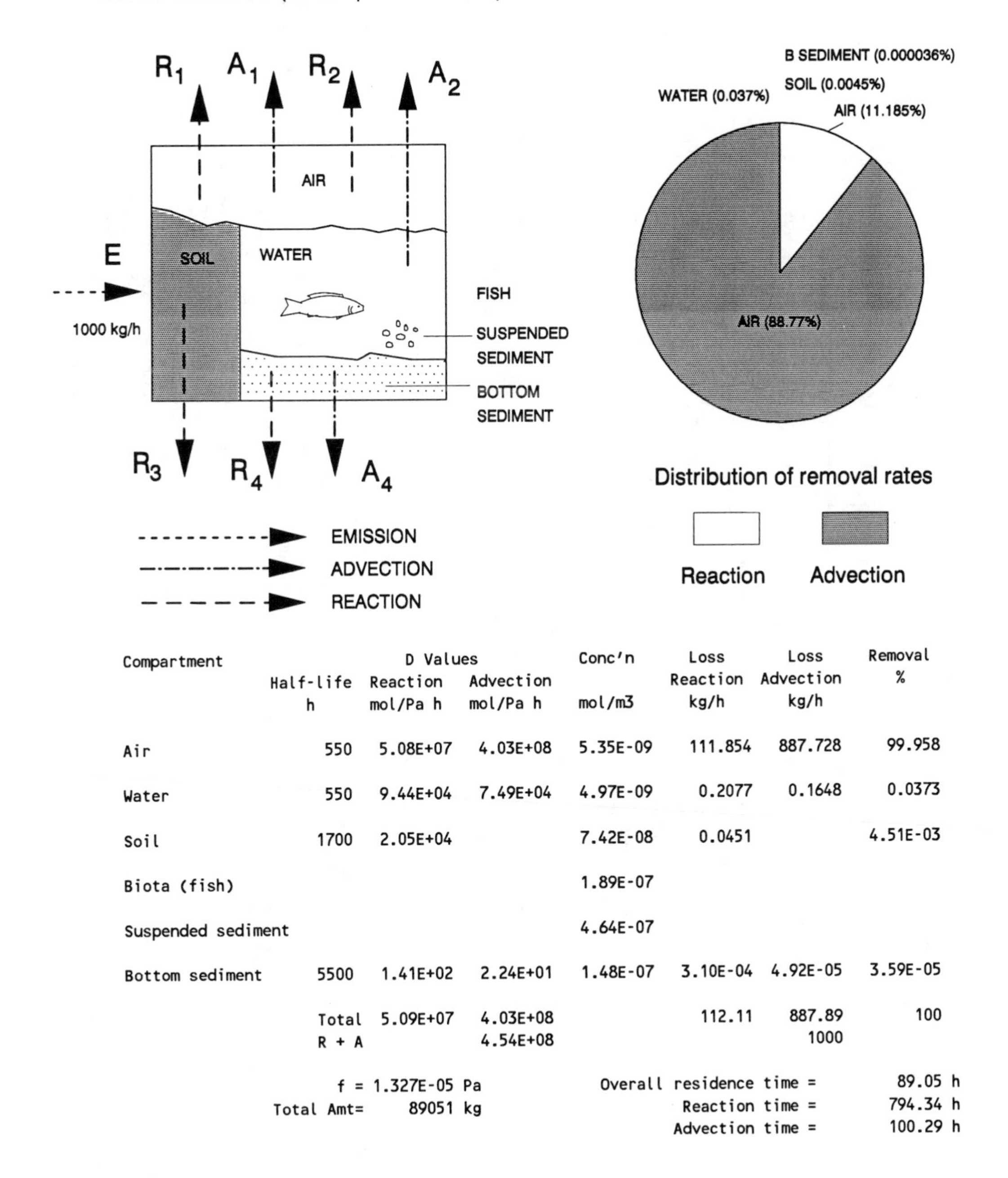


| Compartment | Half-life h | D Values Reaction mol/Pa h | D Values Advection mol/Pa h | Conc'n mol/m3 | Loss Reaction kg/h | Loss Advection kg/h | Removal % |
|---|---|---|---|---|---|---|---|
| Air | 550 | 5.08E+07 | 4.03E+08 | 5.35E-09 | 111.854 | 887.728 | 99.958 |
| Water | 550 | 9.44E+04 | 7.49E+04 | 4.97E-09 | 0.2077 | 0.1648 | 0.0373 |
| Soil | 1700 | 2.05E+04 | | 7.42E-08 | 0.0451 | | 4.51E-03 |
| Biota (fish) | | | | 1.89E-07 | | | |
| Suspended sediment | | | | 4.64E-07 | | | |
| Bottom sediment | 5500 | 1.41E+02 | 2.24E+01 | 1.48E-07 | 3.10E-04 | 4.92E-05 | 3.59E-05 |
| | Total | 5.09E+07 | 4.03E+08 | | 112.11 | 887.89 | 100 |
| | R + A | | 4.54E+08 | | | 1000 | |

f = 1.327E-05 Pa

Total Amt= 89051 kg

Overall residence time = 89.05 h

Reaction time = 794.34 h

Advection time = 100.29 h

## Fugacity Level III calculations: (four-compartment model)

Chemical name: Tetrachloroethylene

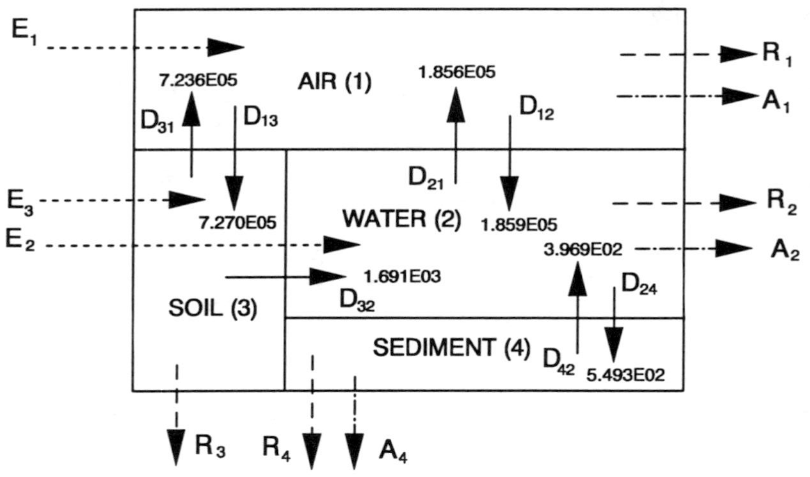

E1 E3 E2
AIR (1) WATER (2) SOIL (3) SEDIMENT (4)
$D_{31}$ 7.236E05 $D_{13}$ 7.270E05
$D_{21}$ 1.856E05 $D_{12}$ 1.859E05
$D_{32}$ 1.691E03
$D_{42}$ 3.969E02 $D_{24}$ 5.493E02
$R_1$ $A_1$ $R_2$ $A_2$ $R_3$ $R_4$ $A_4$

EMISSION (E)
REACTION (R)
ADVECTION (A)
TRANSFER D VALUE mol/Pa h

Phase Properties and Rates:

| Compartment | Bulk Z mol/m3 Pa | Half-life h | D Values Reaction mol/Pa h | D Values Advection mol/Pa h |
|---|---|---|---|---|
| Air (1) | 4.034E-04 | 550 | 5.08E+07 | 4.03E+08 |
| Water (2) | 3.747E-04 | 550 | 9.44E+04 | 7.49E+04 |
| Soil (3) | 2.989E-03 | 1700 | 2.19E+04 | |
| Sediment (4) | 2.536E-03 | 5500 | 1.60E+02 | 2.54E+01 |

| | E(1)=1000 | E(2)=1000 | E(3)=1000 | E(1,2,3) |
|---|---|---|---|---|
| Overall residence time = | 88.99 | 260.93 | 158.77 | 147.55 h |
| Reaction time = | 794.36 | 803.11 | 1145.73 | 826.41 h |
| Advection time = | 100.22 | 386.50 | 184.31 | 179.62 h |

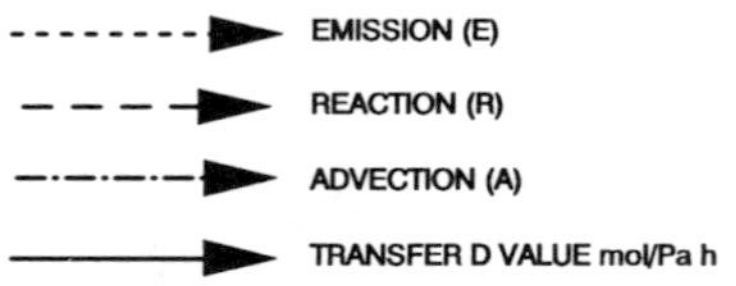

Phase Properties, Compositions, Transport and Transformation Rates:

| Emission, kg/h | | | Fugacity, Pa | | | | Concentration, g/m3 | | | | Amounts, kg | | | | Total Amount, kg |
|---|---|---|---|---|---|---|---|---|---|---|---|---|---|---|---|
| E(1) | E(2) | E(3) | f(1) | f(2) | f(3) | f(4) | C(1) | C(2) | C(3) | C(4) | m(1) | m(2) | m(3) | m(4) | |
| 1000 | 0 | 0 | 1.327E-05 | 7.011E-06 | 1.291E-05 | 6.616E-06 | 8.879E-07 | 4.357E-07 | 6.400E-06 | 2.783E-06 | 8.879E+04 | 8.713E+01 | 1.152E+02 | 1.391E+00 | 8.899E+04 |
| 0 | 1000 | 0 | 6.935E-06 | 1.699E-02 | 6.747E-06 | 1.603E-02 | 4.639E-07 | 1.056E-03 | 3.344E-06 | 6.742E-03 | 4.639E+04 | 2.111E+05 | 6.020E+01 | 3.371E+03 | 2.609E+05 |
| 0 | 0 | 1000 | 1.287E-05 | 4.522E-05 | 8.083E-03 | 4.267E-05 | 8.609E-07 | 2.810E-06 | 4.006E-03 | 1.795E-05 | 8.609E+04 | 5.620E+02 | 7.211E+04 | 8.973E+00 | 1.588E+05 |
| 600 | 300 | 100 | 1.133E-05 | 5.104E-03 | 8.181E-04 | 4.817E-03 | 7.580E-07 | 3.172E-04 | 4.055E-04 | 2.026E-03 | 7.580E+04 | 6.344E+04 | 7.298E+03 | 1.013E+03 | 1.475E+05 |

| Emission, kg/h | | | Loss, Reaction, kg/h | | | | Loss, Advection, kg/h | | | Intermedia Rate of Transport, kg/h T12 | T21 | T13 | T31 | T32 | T24 | T42 |
|---|---|---|---|---|---|---|---|---|---|---|---|---|---|---|---|---|
| E(1) | E(2) | E(3) | R(1) | R(2) | R(3) | R(4) | A(1) | A(2) | A(4) | air-water | water-air | air-soil | soil-air | soil-water | water-sed | sed-water |
| 1000 | 0 | 0 | 1.119E+02 | 1.098E-01 | 4.70E-02 | 1.753E-04 | 8.879E+02 | 8.713E-02 | 2.783E-05 | 4.092E-01 | 2.157E-01 | 1.600E+00 | 1.550E+00 | 3.620E-03 | 6.386E-04 | 4.355E-04 |
| 0 | 1000 | 0 | 5.846E+01 | 2.660E+02 | 2.45E-02 | 4.247E-01 | 4.639E+02 | 2.111E+02 | 6.742E-02 | 2.138E-01 | 5.226E+02 | 8.361E-01 | 8.097E-01 | 1.892E-03 | 1.547E+00 | 1.055E+00 |
| 0 | 0 | 1000 | 1.085E+02 | 7.081E-01 | 2.94E+01 | 1.131E-03 | 8.609E+02 | 5.620E-01 | 1.795E-04 | 3.968E-01 | 1.391E+00 | 1.551E+00 | 9.699E+02 | 2.266E+00 | 4.119E-03 | 2.809E-03 |
| 600 | 300 | 100 | 9.551E+01 | 7.993E+01 | 2.98E+00 | 1.276E-01 | 7.580E+02 | 6.344E+01 | 2.026E-02 | 3.494E-01 | 1.571E+02 | 1.366E+00 | 9.816E+01 | 2.293E-01 | 4.649E-01 | 3.170E-01 |

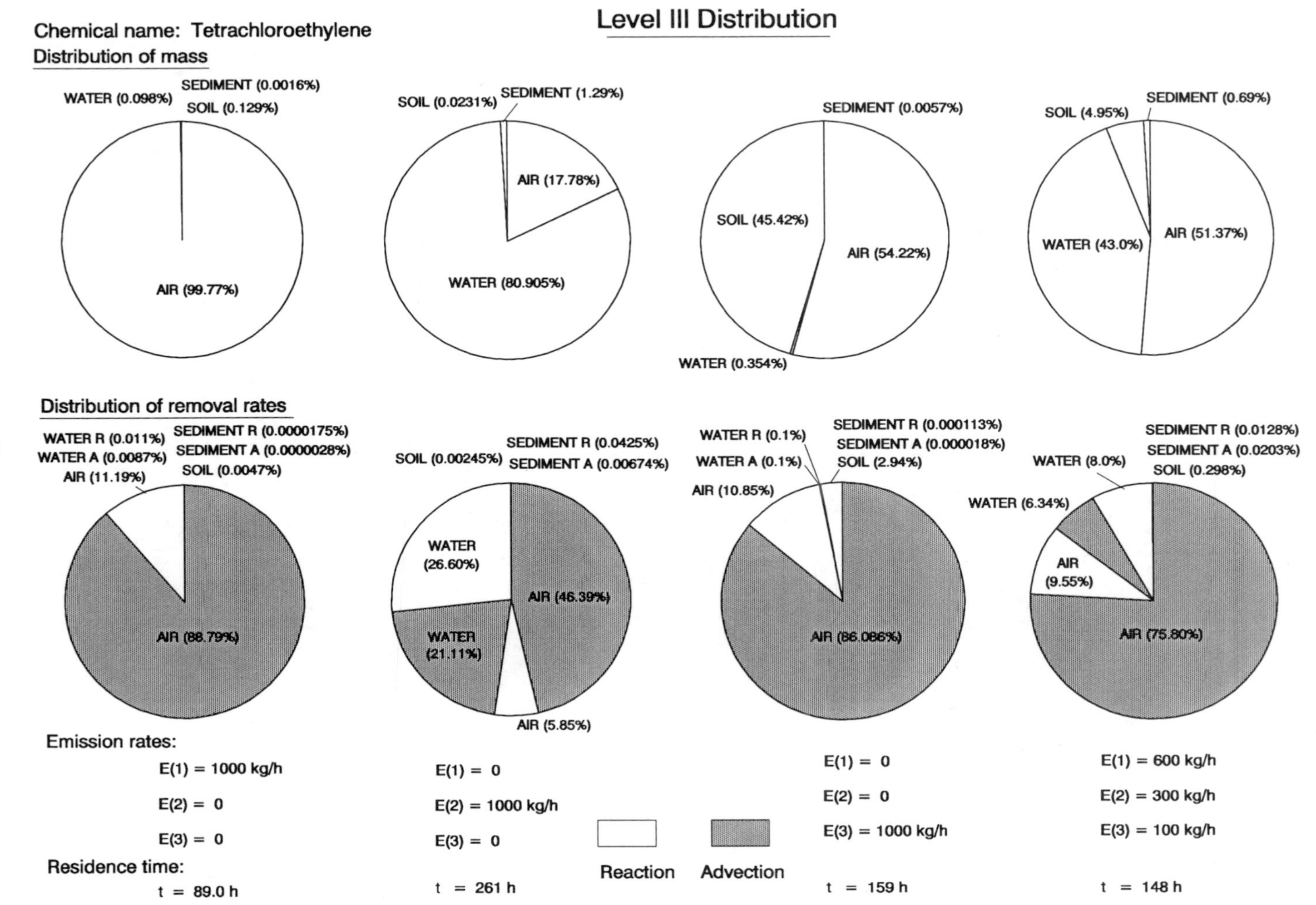
Level III Distribution
Chemical name: Tetrachloroethylene
Distribution of mass
WATER (0.098%)
SEDIMENT (0.0016%)
SOIL (0.129%)
AIR (99.77%)
SOIL (0.0231%)
SEDIMENT (1.29%)
AIR (17.78%)
WATER (80.905%)
SEDIMENT (0.0057%)
SOIL (45.42%)
AIR (54.22%)
WATER (0.354%)
SOIL (4.95%)
SEDIMENT (0.69%)
WATER (43.0%)
AIR (51.37%)
Distribution of removal rates
WATER R (0.011%)
WATER A (0.0087%)
AIR (11.19%)
SEDIMENT R (0.0000175%)
SEDIMENT A (0.0000028%)
SOIL (0.0047%)
AIR (88.79%)
SOIL (0.00245%)
SEDIMENT R (0.0425%)
SEDIMENT A (0.00674%)
WATER (26.60%)
AIR (46.39%)
WATER (21.11%)
AIR (5.85%)
WATER R (0.1%)
WATER A (0.1%)
AIR (10.85%)
SEDIMENT R (0.000113%)
SEDIMENT A (0.000018%)
SOIL (2.94%)
AIR (86.086%)
WATER (8.0%)
WATER (6.34%)
SEDIMENT R (0.0128%)
SEDIMENT A (0.0203%)
SOIL (0.298%)
AIR (9.55%)
AIR (75.80%)
Emission rates:
E(1) = 1000 kg/h
E(2) = 0
E(3) = 0
Residence time:
t = 89.0 h
E(1) = 0
E(2) = 1000 kg/h
E(3) = 0
t = 261 h
Reaction
Advection
E(1) = 0
E(2) = 0
E(3) = 1000 kg/h
t = 159 h
E(1) = 600 kg/h
E(2) = 300 kg/h
E(3) = 100 kg/h
t = 148 h

Chemical name: Chloroprene

Level I calculation: (six-compartment model)

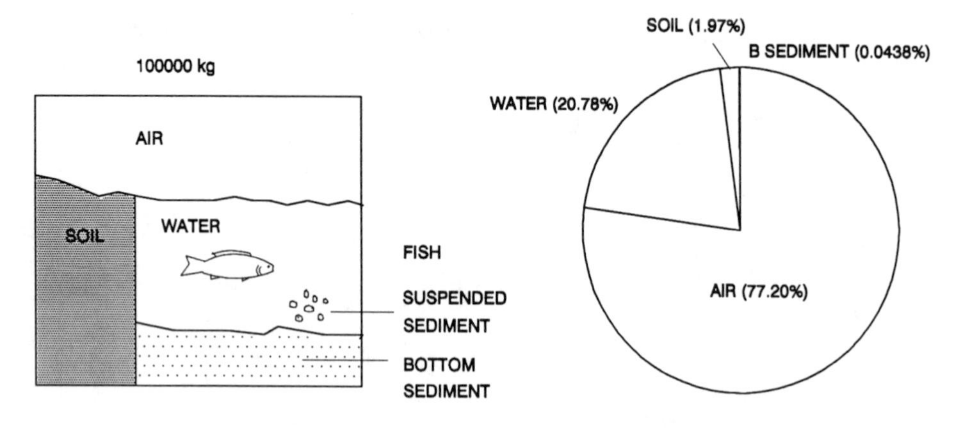


Distribution of mass

physical-chemical properties:

MW: 76.53 g/mol

M.P.: - 135°C

Fugacity ratio: 1.0

vapor pressure: 23194 Pa

solubility: 96400 g/m $^3$

log $K_{OW}$: 2.03

| Compartment | Z | Concentration | | | Amount | Amount |
|---|---|---|---|---|---|---|
| | mol/m3 Pa | mol/m3 | mg/L (or g/m3) | ug/g | kg | % |
| Air | 4.034E-04 | 1.009E-08 | 7.720E-07 | 6.512E-04 | 77197 | 77.197 |
| Water | 5.431E-02 | 1.358E-06 | 1.039E-04 | 1.039E-04 | 20785 | 20.7849 |
| Soil | 1.145E-01 | 2.864E-06 | 2.192E-04 | 9.131E-05 | 1972.36 | 1.9724 |
| Biota (fish) | 2.910E-01 | 7.275E-06 | 5.568E-04 | 5.568E-04 | 0.1114 | 1.11E-04 |
| Suspended sediment | 7.158E-01 | 1.790E-05 | 1.370E-03 | 9.131E-04 | 1.3697 | 1.37E-03 |
| Bottom sediment | 2.290E-01 | 5.727E-06 | 4.383E-04 | 1.826E-04 | 43.8301 | 4.38E-02 |
| | Total | | | | 100000 | 100 |

f = 2.500E-05 Pa

Chemical name: Chloroprene

Level II calculation: (six-compartment model)

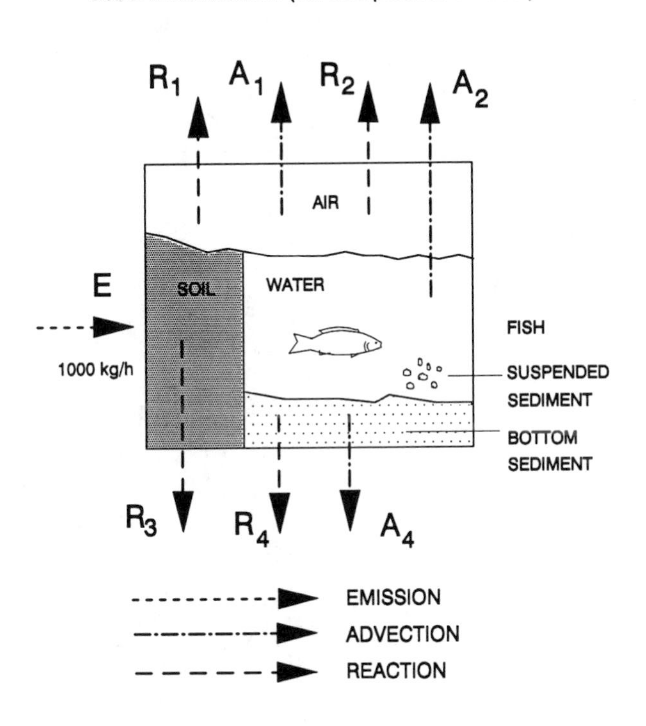


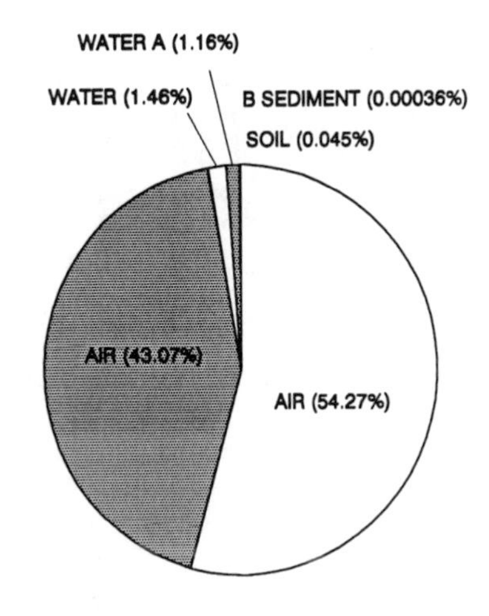


Distribution of removal rates

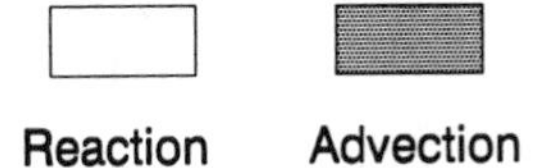


| Compartment | Half-life h | D Values Reaction mol/Pa h | D Values Advection mol/Pa h | Conc'n mol/m3 | Loss Reaction kg/h | Loss Advection kg/h | Removal % |
|---|---|---|---|---|---|---|---|
| Air | 55 | 5.08E+08 | 4.03E+08 | 5.63E-09 | 542.659 | 430.682 | 97.334 |
| Water | 550 | 1.37E+07 | 1.09E+07 | 7.58E-07 | 14.6107 | 11.5958 | 2.6207 |
| Soil | 1700 | 4.20E+05 | | 1.60E-06 | 0.4486 | | 4.49E-02 |
| Biota (fish) | | | | 4.06E-06 | | | |
| Suspended sediment | | | | 9.98E-06 | | | |
| Bottom sediment | 5500 | 2.89E+03 | 4.58E+02 | 3.20E-06 | 3.08E-03 | 4.89E-04 | 3.57E-04 |
| | Total | 5.22E+08 | 4.14E+08 | | 557.72 | 442.28 | 100 |
| | R + A | | 9.37E+08 | | | 1000 | |

f = 1.395E-05 Pa
Total Amt= 55790 kg

Overall residence time = 55.79 h
Reaction time = 100.03 h
Advection time = 126.14 h

Fugacity Level III calculations: (four-compartment model)

Chemical name: Chloroprene

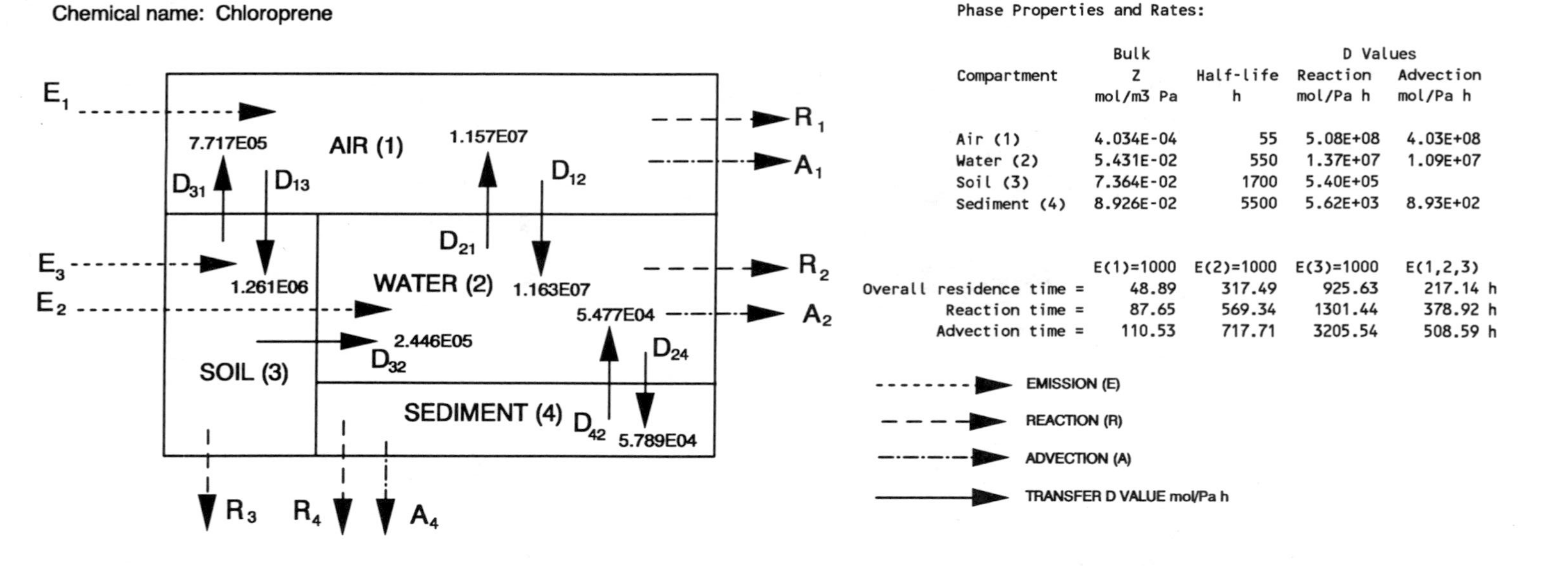


Phase Properties and Rates:

| Compartment | Bulk Z mol/m3 Pa | Half-life h | D Values Reaction mol/Pa h | D Values Advection mol/Pa h |
|---|---|---|---|---|
| Air (1) | 4.034E-04 | 55 | 5.08E+08 | 4.03E+08 |
| Water (2) | 5.431E-02 | 550 | 1.37E+07 | 1.09E+07 |
| Soil (3) | 7.364E-02 | 1700 | 5.40E+05 | |
| Sediment (4) | 8.926E-02 | 5500 | 5.62E+03 | 8.93E+02 |

| | E(1)=1000 | E(2)=1000 | E(3)=1000 | E(1,2,3) |
|---|---|---|---|---|
| Overall residence time = | 48.89 | 317.49 | 925.63 | 217.14 h |
| Reaction time = | 87.65 | 569.34 | 1301.44 | 378.92 h |
| Advection time = | 110.53 | 717.71 | 3205.54 | 508.59 h |

EMISSION (E)

REACTION (R)

ADVECTION (A)

TRANSFER D VALUE mol/Pa h

Phase Properties, Compositions, Transport and Transformation Rates:

| Emission, kg/h | | | Fugacity, Pa | | | | Concentration, g/m3 | | | | Amounts, kg | | | | Total Amount, kg |
|---|---|---|---|---|---|---|---|---|---|---|---|---|---|---|---|
| E(1) | E(2) | E(3) | f(1) | f(2) | f(3) | f(4) | C(1) | C(2) | C(3) | C(4) | m(1) | m(2) | m(3) | m(4) | |
| 1000 | 0 | 0 | 1.420E-05 | 4.648E-06 | 1.150E-05 | 4.391E-06 | 4.384E-07 | 1.932E-05 | 6.480E-05 | 2.999E-05 | 4.384E+04 | 3.864E+03 | 1.166E+03 | 1.500E+01 | 4.889E+04 |
| 0 | 1000 | 0 | 4.549E-06 | 3.632E-04 | 3.684E-06 | 3.430E-04 | 1.404E-07 | 1.509E-03 | 2.076E-05 | 2.343E-03 | 1.404E+04 | 3.019E+05 | 3.737E+02 | 1.172E+03 | 3.175E+05 |
| 0 | 0 | 1000 | 7.755E-06 | 5.935E-05 | 8.401E-03 | 5.606E-05 | 2.394E-07 | 2.467E-04 | 4.734E-02 | 3.829E-04 | 2.394E+04 | 4.934E+04 | 8.522E+05 | 1.915E+02 | 9.256E+05 |
| 600 | 300 | 100 | 1.066E-05 | 1.177E-04 | 8.481E-04 | 1.112E-04 | 3.291E-07 | 4.891E-04 | 4.779E-03 | 7.593E-04 | 3.291E+04 | 9.782E+04 | 8.603E+04 | 3.796E+02 | 2.171E+05 |

| Emission, kg/h | | | Loss, Reaction, kg/h | | | | Loss, Advection, kg/h | | | Intermedia Rate of Transport, kg/h T12 | T21 | T13 | T31 | T32 | T24 | T42 |
|---|---|---|---|---|---|---|---|---|---|---|---|---|---|---|---|---|
| E(1) | E(2) | E(3) | R(1) | R(2) | R(3) | R(4) | A(1) | A(2) | A(4) | air-water | water-air | air-soil | soil-air | soil-water | water-sed | sed-water |
| 1000 | 0 | 0 | 5.524E+02 | 4.869E+00 | 4.75E-01 | 1.889E-03 | 4.384E+02 | 3.864E+00 | 2.999E-04 | 1.264E+01 | 4.117E+00 | 1.370E+00 | 6.792E-01 | 2.152E-01 | 2.059E-02 | 1.840E-02 |
| 0 | 1000 | 0 | 1.770E+02 | 3.804E+02 | 1.52E-01 | 1.476E-01 | 1.404E+02 | 3.019E+02 | 2.343E-02 | 4.048E+00 | 3.217E+02 | 4.388E-01 | 2.176E-01 | 6.893E-02 | 1.609E+00 | 1.438E+00 |
| 0 | 0 | 1000 | 3.017E+02 | 6.216E+01 | 3.47E+02 | 2.412E-02 | 2.394E+02 | 4.934E+01 | 3.829E-03 | 6.901E+00 | 5.257E+01 | 7.481E-01 | 4.962E+02 | 1.572E+02 | 2.629E-01 | 2.350E-01 |
| 600 | 300 | 100 | 4.147E+02 | 1.233E+02 | 3.51E+01 | 4.783E-02 | 3.291E+02 | 9.782E+01 | 7.593E-03 | 9.486E+00 | 1.042E+02 | 1.028E+00 | 5.009E+01 | 1.587E+01 | 5.213E-01 | 4.659E-01 |

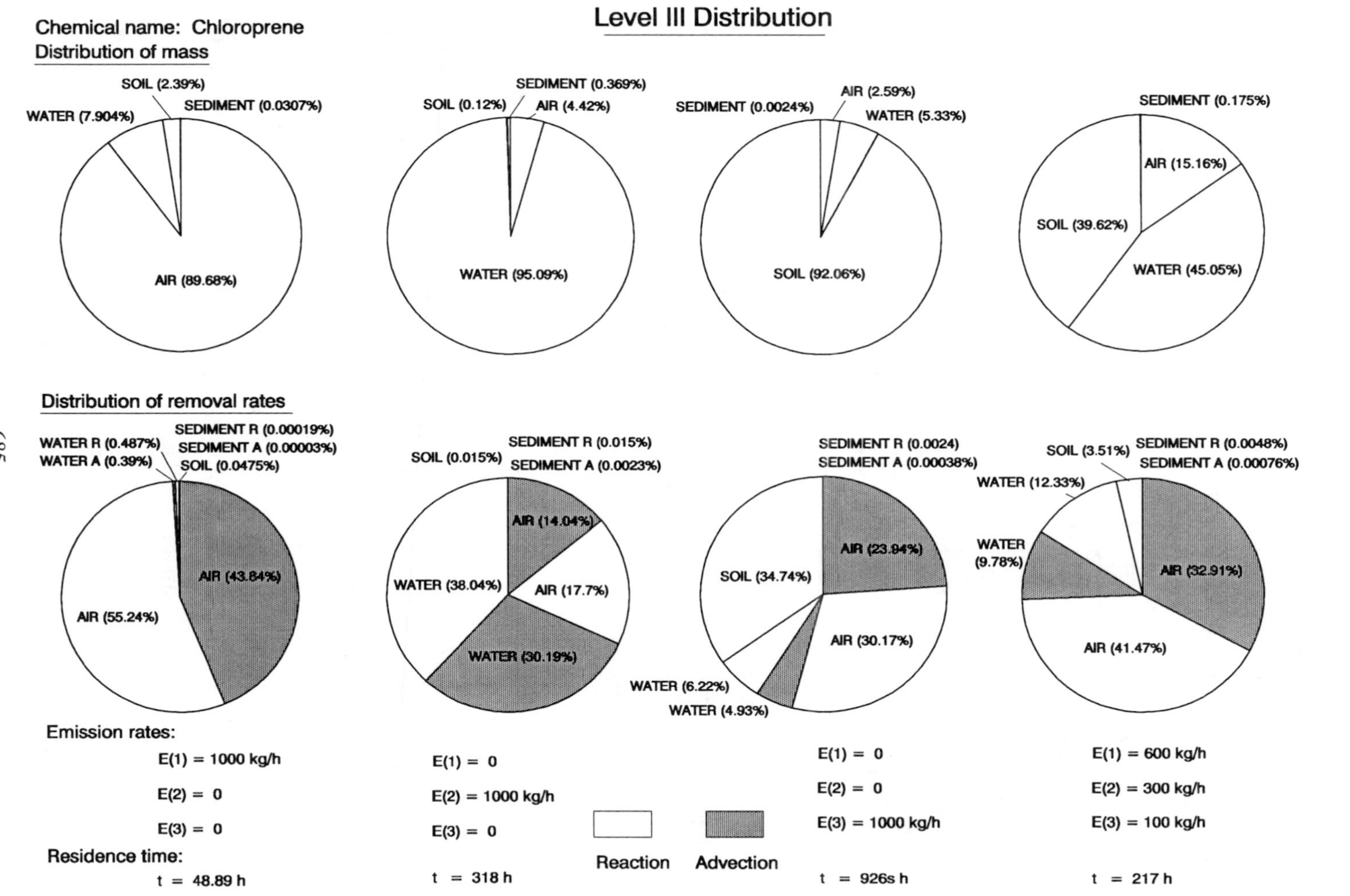
Level III Distribution
Chemical name: Chloroprene
Distribution of mass
SOIL (2.39%)
WATER (7.904%)
SEDIMENT (0.0307%)
AIR (89.68%)
SEDIMENT (0.369%)
SOIL (0.12%)
AIR (4.42%)
WATER (95.09%)
AIR (2.59%)
SEDIMENT (0.0024%)
WATER (5.33%)
SOIL (92.06%)
SEDIMENT (0.175%)
AIR (15.16%)
SOIL (39.62%)
WATER (45.05%)
Distribution of removal rates
SEDIMENT R (0.00019%)
WATER R (0.487%)
SEDIMENT A (0.00003%)
WATER A (0.39%)
SOIL (0.0475%)
AIR (43.64%)
AIR (55.24%)
SEDIMENT R (0.015%)
SOIL (0.015%)
SEDIMENT A (0.0023%)
AIR (14.04%)
WATER (38.04%)
AIR (17.7%)
WATER (30.19%)
SEDIMENT R (0.0024)
SEDIMENT A (0.00038%)
AIR (23.94%)
SOIL (34.74%)
AIR (30.17%)
WATER (6.22%)
WATER (4.93%)
SOIL (3.51%)
SEDIMENT R (0.0048%)
SEDIMENT A (0.00076%)
WATER (12.33%)
WATER
(9.78%)
AIR (32.91%)
AIR (41.47%)
Emission rates:
E(1) = 1000 kg/h
E(2) = 0
E(3) = 0
Residence time:
t = 48.89 h
E(1) = 0
E(2) = 1000 kg/h
E(3) = 0
t = 318 h
Reaction
Advection
E(1) = 0
E(2) = 0
E(3) = 1000 kg/h
t = 926s h
E(1) = 600 kg/h
E(2) = 300 kg/h
E(3) = 100 kg/h
t = 217 h

Chemical name: Hexachlorocyclopentadiene

Level I calculation: (six-compartment model)

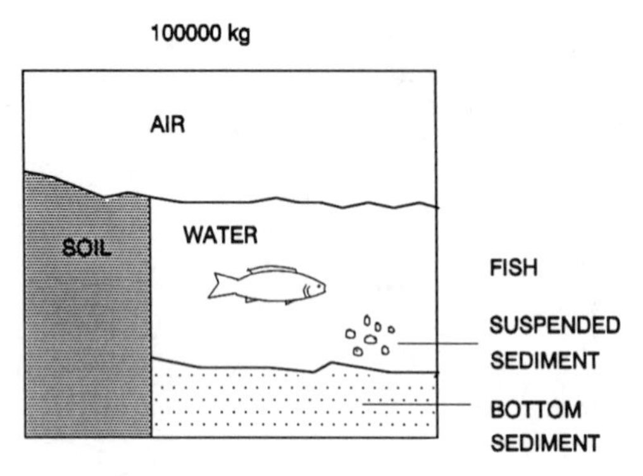


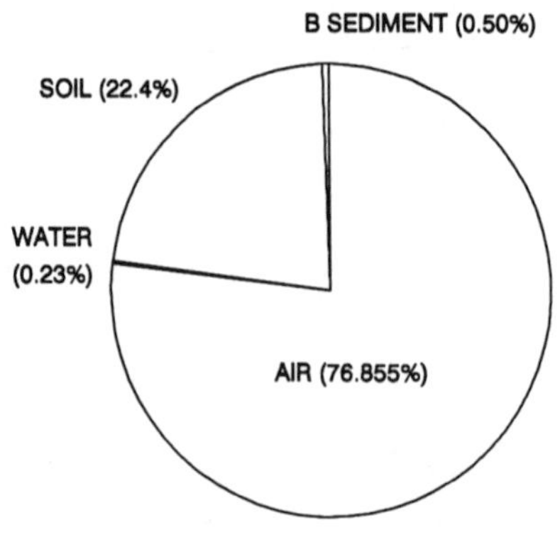


Distribution of mass

physical-chemical properties:

MW: 272.77 g/mol

M.P.: - 10°C

Fugacity ratio: 1.0

vapor pressure: 10.9 Pa

solubility: 1.80 g/m³

log $K_{OW}$: 5.04

| Compartment | Z | Concentration | | | Amount | Amount |
|---|---|---|---|---|---|---|
| | mol/m3 Pa | mol/m3 | mg/L (or g/m3) | ug/g | kg | % |
| Air | 4.034E-04 | 2.818E-09 | 7.686E-07 | 6.483E-04 | 76855 | 76.855 |
| Water | 6.054E-04 | 4.228E-09 | 1.153E-06 | 1.153E-06 | 230.67 | 0.2307 |
| Soil | 1.306E+00 | 9.124E-06 | 2.489E-03 | 1.037E-03 | 22399 | 22.399 |
| Biota (fish) | 3.319E+00 | 2.318E-05 | 6.323E-03 | 6.323E-03 | 1.2646 | 1.26E-03 |
| Suspended sediment | 8.165E+00 | 5.703E-05 | 1.556E-02 | 1.037E-02 | 15.555 | 1.56E-02 |
| Bottom sediment | 2.613E+00 | 1.825E-05 | 4.978E-03 | 2.074E-03 | 497.76 | 0.4978 |
| | Total | | | | 100000 | 100 |

f = 6.984E-06 Pa

Chemical name: Hexachlorocyclopentadiene

Level II calculation: (six-compartment model)

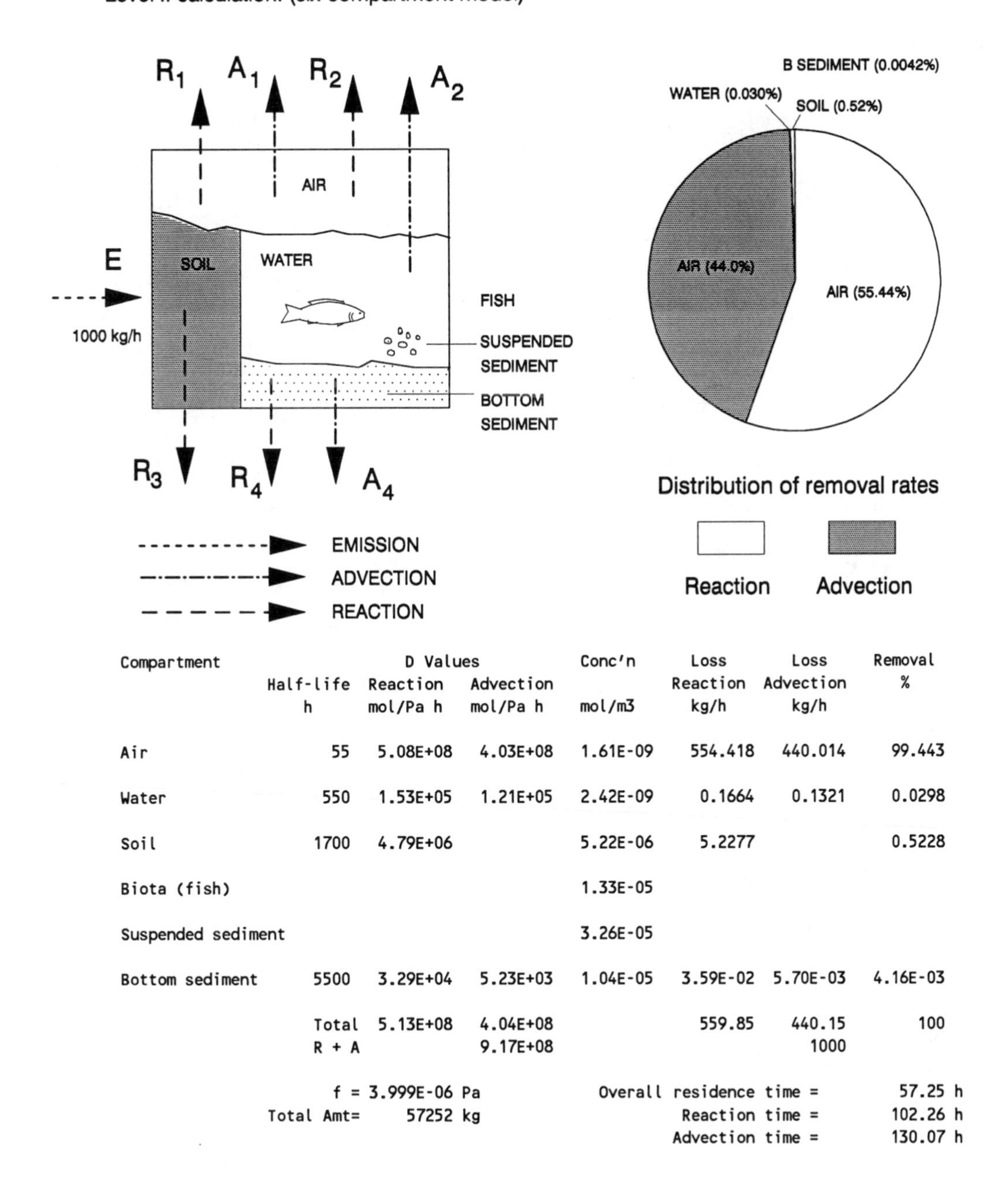


| Compartment | Half-life h | D Values Reaction mol/Pa h | D Values Advection mol/Pa h | Conc'n mol/m3 | Loss Reaction kg/h | Loss Advection kg/h | Removal % |
|---|---|---|---|---|---|---|---|
| Air | 55 | 5.08E+08 | 4.03E+08 | 1.61E-09 | 554.418 | 440.014 | 99.443 |
| Water | 550 | 1.53E+05 | 1.21E+05 | 2.42E-09 | 0.1664 | 0.1321 | 0.0298 |
| Soil | 1700 | 4.79E+06 | | 5.22E-06 | 5.2277 | | 0.5228 |
| Biota (fish) | | | | 1.33E-05 | | | |
| Suspended sediment | | | | 3.26E-05 | | | |
| Bottom sediment | 5500 | 3.29E+04 | 5.23E+03 | 1.04E-05 | 3.59E-02 | 5.70E-03 | 4.16E-03 |
| | Total | 5.13E+08 | 4.04E+08 | | 559.85 | 440.15 | 100 |
| | R + A | | 9.17E+08 | | | 1000 | |

f = 3.999E-06 Pa

Total Amt= 57252 kg

Overall residence time = 57.25 h

Reaction time = 102.26 h

Advection time = 130.07 h

Fugacity Level III calculations: (four-compartment model)

Chemical name: Hexachlorocyclopentadiene

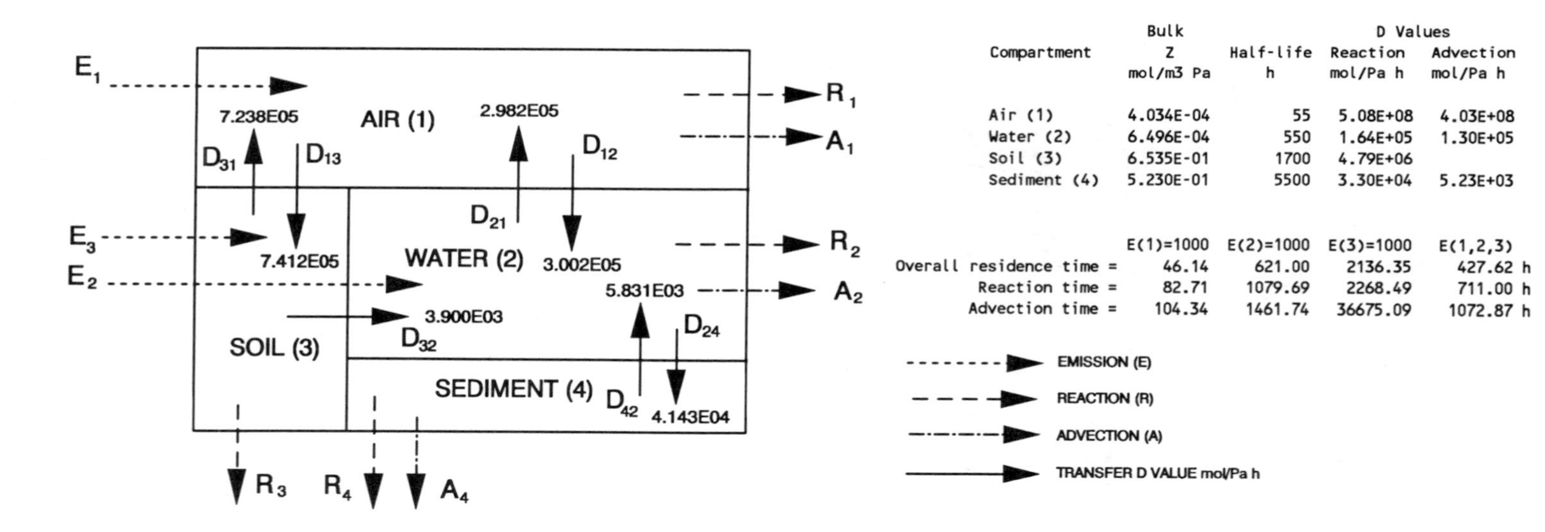


Phase Properties and Rates:

| Compartment | Bulk Z mol/m3 Pa | Half-life h | D Values Reaction mol/Pa h | D Values Advection mol/Pa h |
|---|---|---|---|---|
| Air (1) | 4.034E-04 | 55 | 5.08E+08 | 4.03E+08 |
| Water (2) | 6.496E-04 | 550 | 1.64E+05 | 1.30E+05 |
| Soil (3) | 6.535E-01 | 1700 | 4.79E+06 | |
| Sediment (4) | 5.230E-01 | 5500 | 3.30E+04 | 5.23E+03 |

| | E(1)=1000 | E(2)=1000 | E(3)=1000 | E(1,2,3) |
|---|---|---|---|---|
| Overall residence time = | 46.14 | 621.00 | 2136.35 | 427.62 h |
| Reaction time = | 82.71 | 1079.69 | 2268.49 | 711.00 h |
| Advection time = | 104.34 | 1461.74 | 36675.09 | 1072.87 h |

Phase Properties, Compositions, Transport and Transformation Rates:

| Emission, kg/h | | | Fugacity, Pa | | | | Concentration, g/m3 | | | | Amounts, kg | | | | Total Amount, kg |
|---|---|---|---|---|---|---|---|---|---|---|---|---|---|---|---|
| E(1) | E(2) | E(3) | f(1) | f(2) | f(3) | f(4) | C(1) | C(2) | C(3) | C(4) | m(1) | m(2) | m(3) | m(4) | |
| 1000 | 0 | 0 | 4.017E-06 | 1.924E-06 | 5.392E-07 | 1.811E-06 | 4.421E-07 | 3.409E-07 | 9.611E-05 | 2.584E-04 | 4.421E+04 | 6.819E+01 | 1.730E+03 | 1.292E+02 | 4.614E+04 |
| 0 | 1000 | 0 | 1.909E-06 | 5.841E-03 | 2.562E-07 | 5.498E-03 | 2.100E-07 | 1.035E-03 | 4.566E-05 | 7.844E-01 | 2.100E+04 | 2.070E+05 | 8.219E+02 | 3.922E+05 | 6.210E+05 |
| 0 | 0 | 1000 | 5.279E-07 | 4.377E-06 | 6.639E-04 | 4.120E-06 | 5.809E-08 | 7.755E-07 | 1.183E-01 | 5.878E-04 | 5.809E+03 | 1.551E+02 | 2.130E+06 | 2.939E+02 | 2.136E+06 |
| 600 | 300 | 100 | 3.036E-06 | 1.754E-03 | 6.679E-05 | 1.651E-03 | 3.341E-07 | 3.107E-04 | 1.191E-02 | 2.355E-01 | 3.341E+04 | 6.215E+04 | 2.143E+05 | 1.178E+05 | 4.276E+05 |

| Emission, kg/h | | | Loss, Reaction, kg/h | | | | Loss, Advection, kg/h | | | Intermedia Rate of Transport, kg/h T12 | T21 | T13 | T31 | T32 | T24 | T42 |
|---|---|---|---|---|---|---|---|---|---|---|---|---|---|---|---|---|
| E(1) | E(2) | E(3) | R(1) | R(2) | R(3) | R(4) | A(1) | A(2) | A(4) | air-water | water-air | air-soil | soil-air | soil-water | water-sed | sed-water |
| 1000 | 0 | 0 | 5.570E+02 | 8.592E-02 | 7.05E-01 | 1.628E-02 | 4.421E+02 | 6.819E-02 | 2.584E-03 | 3.289E-01 | 1.565E-01 | 8.123E-01 | 1.065E-01 | 5.736E-04 | 2.175E-02 | 2.881E-03 |
| 0 | 1000 | 0 | 2.646E+02 | 2.608E+02 | 3.35E-01 | 4.942E+01 | 2.100E+02 | 2.070E+02 | 7.844E+00 | 1.563E-01 | 4.751E+02 | 3.859E-01 | 5.058E-02 | 2.725E-04 | 6.601E+01 | 8.745E+00 |
| 0 | 0 | 1000 | 7.319E+01 | 1.954E-01 | 8.68E+02 | 3.703E-02 | 5.809E+01 | 1.551E-01 | 5.878E-03 | 4.322E-02 | 3.561E-01 | 1.067E-01 | 1.311E+02 | 7.063E-01 | 4.946E-02 | 6.553E-03 |
| 600 | 300 | 100 | 4.209E+02 | 7.831E+01 | 8.74E+01 | 1.484E+01 | 3.341E+02 | 6.215E+01 | 2.355E+00 | 2.486E-01 | 1.427E+02 | 6.138E-01 | 1.319E+01 | 7.105E-02 | 1.982E+01 | 2.626E+00 |

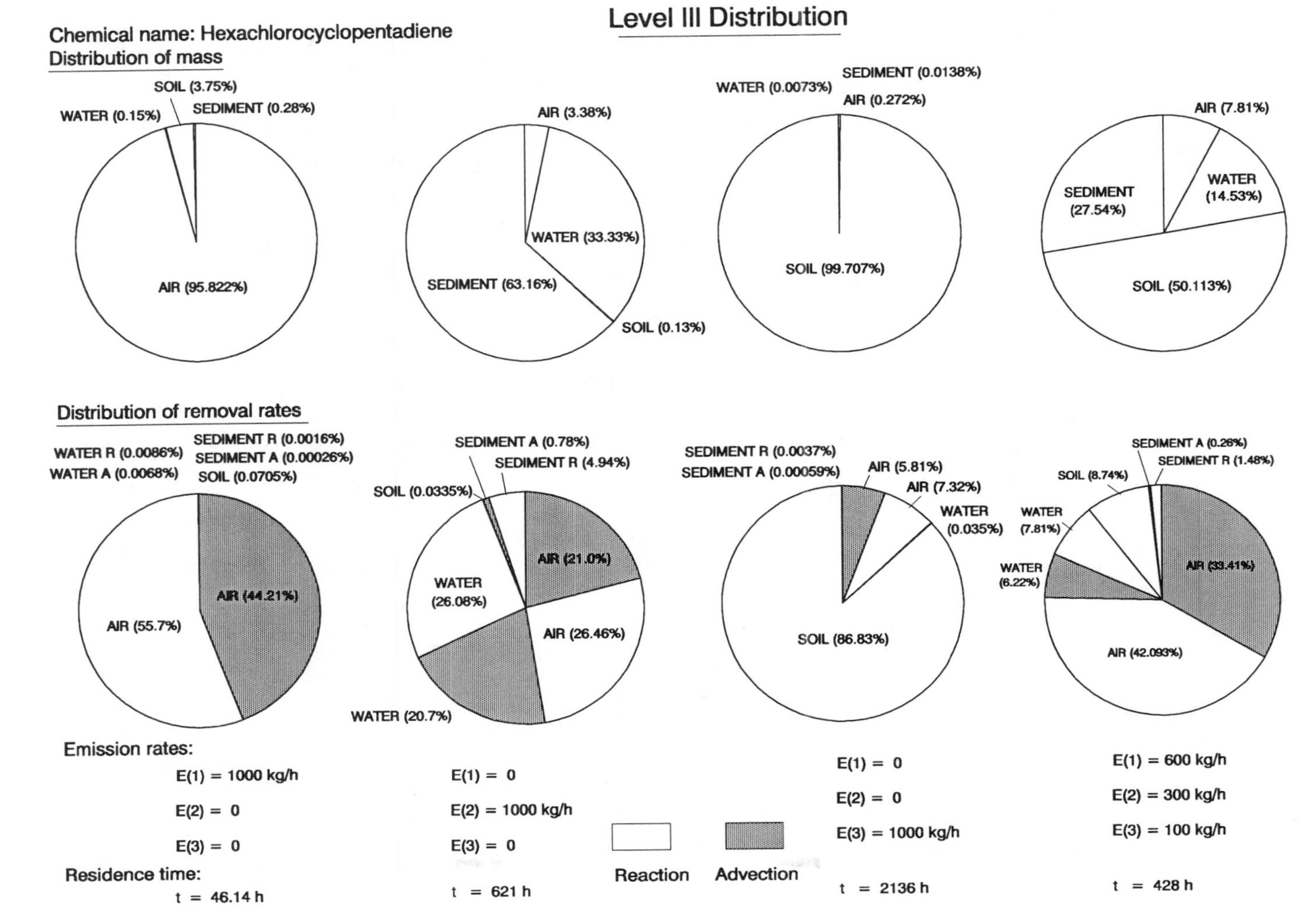
Level III Distribution
Chemical name: Hexachlorocyclopentadiene
Distribution of mass
WATER (0.15%)
SOIL (3.75%)
SEDIMENT (0.28%)
AIR (95.822%)
AIR (3.38%)
WATER (33.33%)
SEDIMENT (63.16%)
SOIL (0.13%)
WATER (0.0073%)
SEDIMENT (0.0138%)
AIR (0.272%)
SOIL (99.707%)
AIR (7.81%)
WATER (14.53%)
SEDIMENT (27.54%)
SOIL (50.113%)
Distribution of removal rates
WATER R (0.0086%)
WATER A (0.0068%)
SEDIMENT R (0.0016%)
SEDIMENT A (0.00026%)
SOIL (0.0705%)
AIR (44.21%)
AIR (55.7%)
SEDIMENT A (0.78%)
SEDIMENT R (4.94%)
SOIL (0.0335%)
AIR (21.0%)
WATER (26.08%)
AIR (26.46%)
WATER (20.7%)
SEDIMENT R (0.0037%)
SEDIMENT A (0.00059%)
AIR (5.81%)
AIR (7.32%)
WATER (0.035%)
SOIL (86.83%)
SEDIMENT A (0.26%)
SEDIMENT R (1.48%)
SOIL (8.74%)
WATER (7.81%)
WATER (6.22%)
AIR (33.41%)
AIR (42.093%)
Emission rates:
E(1) = 1000 kg/h
E(2) = 0
E(3) = 0
Residence time:
t = 46.14 h
E(1) = 0
E(2) = 1000 kg/h
E(3) = 0
t = 621 h
Reaction
Advection
E(1) = 0
E(2) = 0
E(3) = 1000 kg/h
t = 2136 h
E(1) = 600 kg/h
E(2) = 300 kg/h
E(3) = 100 kg/h
t = 428 h

Chemical name: Tribromomethane

Level I calculation: (six-compartment model)

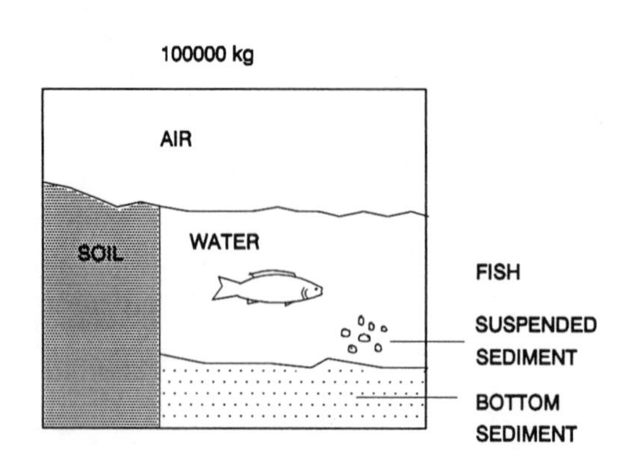


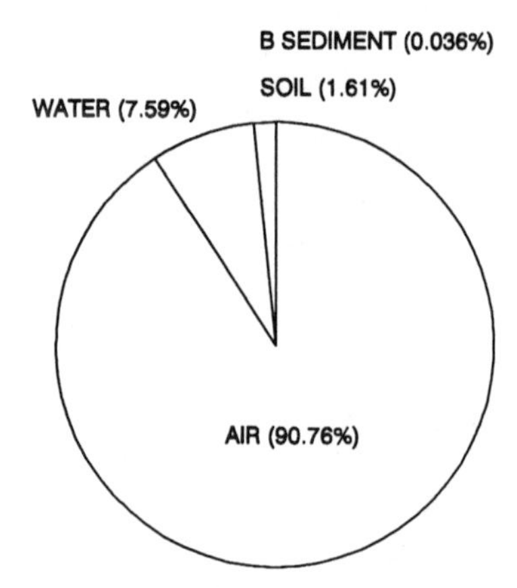


Distribution of mass

physical-chemical properties:

MW: 252.75 g/mol

M.P.: - 8.3 °C

Fugacity ratio: 1.0

vapor pressure: 727 Pa

solubility: 3100 g/m$^3$

log $K_{OW}$: 2.38

| Compartment | Z | Concentration | | | Amount | Amount |
|---|---|---|---|---|---|---|
| | mol/m3 Pa | mol/m3 | mg/L (or g/m3) | ug/g | kg | % |
| Air | 4.034E-04 | 3.591E-09 | 9.076E-07 | 7.656E-04 | 90759 | 90.759 |
| Water | 1.687E-02 | 1.502E-07 | 3.796E-05 | 3.796E-05 | 7591.05 | 7.5911 |
| Soil | 7.965E-02 | 7.089E-07 | 1.792E-04 | 7.466E-05 | 1612.65 | 1.613 |
| Biota (fish) | 2.024E-01 | 1.801E-06 | 4.552E-04 | 4.552E-04 | 0.0910 | 9.10E-05 |
| Suspended sediment | 4.978E-01 | 4.431E-06 | 1.120E-03 | 7.466E-04 | 1.1199 | 1.12E-03 |
| Bottom sediment | 1.593E-01 | 1.418E-06 | 3.584E-04 | 1.493E-04 | 35.837 | 0.0358 |
| | Total | | | | 100000 | 100 |

f = 8.901E-06 Pa

Chemical name: Tribromomethane

Level II calculation: (six-compartment model)

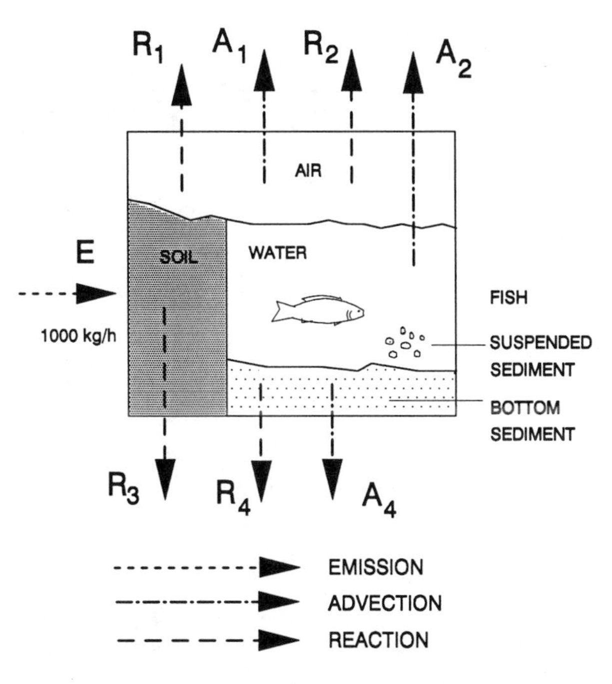


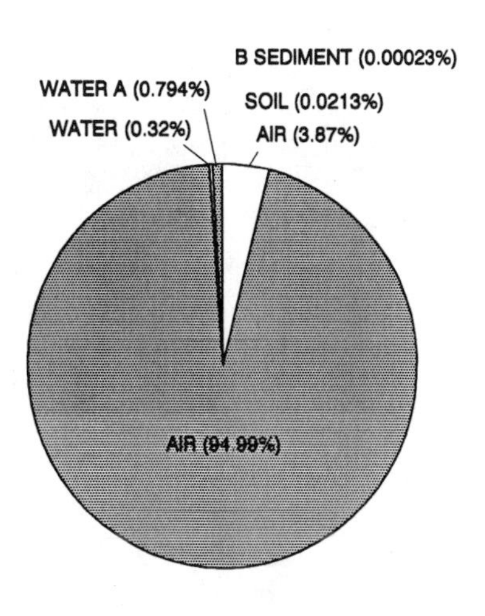


Distribution of removal rates

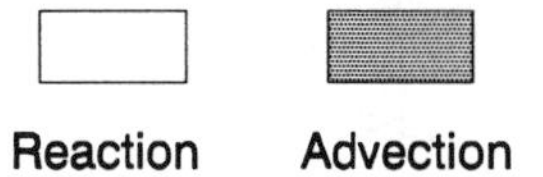


| Compartment | Half-life h | D Values Reaction mol/Pa h | Advection mol/Pa h | Conc'n mol/m3 | Loss Reaction kg/h | Loss Advection kg/h | Removal % |
|---|---|---|---|---|---|---|---|
| Air | 1700 | 1.64E+07 | 4.03E+08 | 3.76E-09 | 38.722 | 949.880 | 98.860 |
| Water | 1700 | 1.38E+06 | 3.37E+06 | 1.57E-07 | 3.2387 | 7.9447 | 1.1183 |
| Soil | 5500 | 9.03E+04 | | 7.42E-07 | 0.2127 | | 0.0213 |
| Biota (fish) | | | | 1.89E-06 | | | |
| Suspended sediment | | | | 4.64E-06 | | | |
| Bottom sediment | 17000 | 6.49E+02 | 3.19E+02 | 1.48E-06 | 1.53E-03 | 7.50E-04 | 2.28E-04 |
| | Total | 1.79E+07 | 4.07E+08 | | 42.17 | 957.83 | 100 |
| | R + A | | 4.25E+08 | | | 1000 | |

f = 9.316E-06 Pa
Total Amt= 104659 kg

Overall residence time = 104.66 h
Reaction time = 2481.58 h
Advection time = 109.27 h

## Fugacity Level III calculations: (four-compartment model)

Chemical name: Tribromomethane

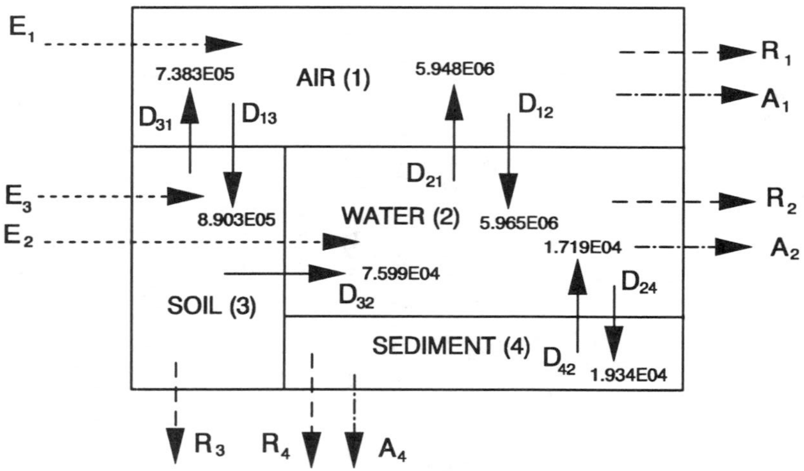

E1, E2, E3; AIR (1); WATER (2); SOIL (3); SEDIMENT (4); R1, A1, R2, A2, R3, R4, A4

D31 7.383E05; D13 8.903E05; D21 5.948E06; D12 5.965E06; D32 7.599E04; D42 1.719E04; D24 1.934E04

EMISSION (E)

REACTION (R)

ADVECTION (A)

TRANSFER D VALUE mol/Pa h

Phase Properties and Rates:

| Compartment | Bulk Z mol/m3 Pa | Half-life h | D Values Reaction mol/Pa h | D Values Advection mol/Pa h |
|---|---|---|---|---|
| Air (1) | 4.034E-04 | 1700 | 1.64E+07 | 4.03E+08 |
| Water (2) | 1.687E-02 | 1700 | 1.38E+06 | 3.37E+06 |
| Soil (3) | 4.496E-02 | 5500 | 1.02E+05 | |
| Sediment (4) | 4.535E-02 | 17000 | 9.24E+02 | 4.54E+02 |

| | E(1)=1000 | E(2)=1000 | E(3)=1000 | E(1,2,3) | |
|---|---|---|---|---|---|
| Overall residence time = | 101.85 | 374.23 | 996.41 | 273.02 | h |
| Reaction time = | 2485.17 | 2471.06 | 6351.44 | 3185.34 | h |
| Advection time = | 106.20 | 441.02 | 1181.81 | 298.61 | h |

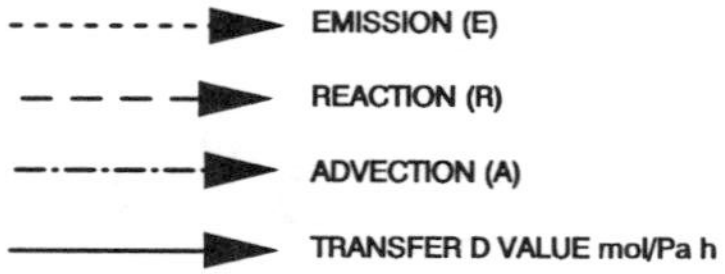

Phase Properties, Compositions, Transport and Transformation Rates:

| Emission, kg/h | | | Fugacity, Pa | | | | Concentration, g/m3 | | | | Amounts, kg | | | | Total Amount, kg |
|---|---|---|---|---|---|---|---|---|---|---|---|---|---|---|---|
| E(1) | E(2) | E(3) | f(1) | f(2) | f(3) | f(4) | C(1) | C(2) | C(3) | C(4) | m(1) | m(2) | m(3) | m(4) | |
| 1000 | 0 | 0 | 9.361E-06 | 5.283E-06 | 9.096E-06 | 5.509E-06 | 9.545E-07 | 2.253E-05 | 1.034E-04 | 6.315E-05 | 9.545E+04 | 4.506E+03 | 1.861E+03 | 3.157E+01 | 1.018E+05 |
| 0 | 1000 | 0 | 5.204E-06 | 3.727E-04 | 5.057E-06 | 3.886E-04 | 5.306E-07 | 1.590E-03 | 5.747E-05 | 4.455E-03 | 5.306E+04 | 3.179E+05 | 1.034E+03 | 2.227E+03 | 3.742E+05 |
| 0 | 0 | 1000 | 7.975E-06 | 3.517E-05 | 4.326E-03 | 3.667E-05 | 8.131E-07 | 1.500E-04 | 4.916E-02 | 4.203E-04 | 8.131E+04 | 3.000E+04 | 8.849E+05 | 2.102E+02 | 9.964E+05 |
| 600 | 300 | 100 | 7.975E-06 | 1.185E-04 | 4.395E-04 | 1.236E-04 | 8.132E-07 | 5.054E-04 | 4.995E-03 | 1.416E-03 | 8.132E+04 | 1.011E+05 | 8.992E+04 | 7.082E+02 | 2.730E+05 |

| Emission, kg/h | | | Loss, Reaction, kg/h | | | | Loss, Advection, kg/h | | | Intermedia Rate of Transport, kg/h T12 | T21 | T13 | T31 | T32 | T24 | T42 |
|---|---|---|---|---|---|---|---|---|---|---|---|---|---|---|---|---|
| E(1) | E(2) | E(3) | R(1) | R(2) | R(3) | R(4) | A(1) | A(2) | A(4) | air-water | water-air | air-soil | soil-air | soil-water | water-sed | sed-water |
| 1000 | 0 | 0 | 3.891E+01 | 1.837E+00 | 2.34E-01 | 1.287E-03 | 9.545E+02 | 4.506E+00 | 6.315E-04 | 1.411E+01 | 7.943E+00 | 2.107E+00 | 1.697E+00 | 1.747E-01 | 2.585E-02 | 2.393E-02 |
| 0 | 1000 | 0 | 2.163E+01 | 1.296E+02 | 1.30E-01 | 9.080E-02 | 5.306E+02 | 3.179E+02 | 4.455E-02 | 7.846E+00 | 5.603E+02 | 1.171E+00 | 9.436E-01 | 9.712E-02 | 1.824E+00 | 1.688E+00 |
| 0 | 0 | 1000 | 3.315E+01 | 1.223E+01 | 1.11E+02 | 8.567E-03 | 8.131E+02 | 3.000E+01 | 4.203E-03 | 1.202E+01 | 5.287E+01 | 1.795E+00 | 8.072E+02 | 8.308E+01 | 1.721E-01 | 1.593E-01 |
| 600 | 300 | 100 | 3.315E+01 | 4.120E+01 | 1.13E+01 | 2.887E-02 | 8.132E+02 | 1.011E+02 | 1.416E-02 | 1.202E+01 | 1.781E+02 | 1.795E+00 | 8.202E+01 | 8.442E+00 | 5.798E-01 | 5.368E-01 |

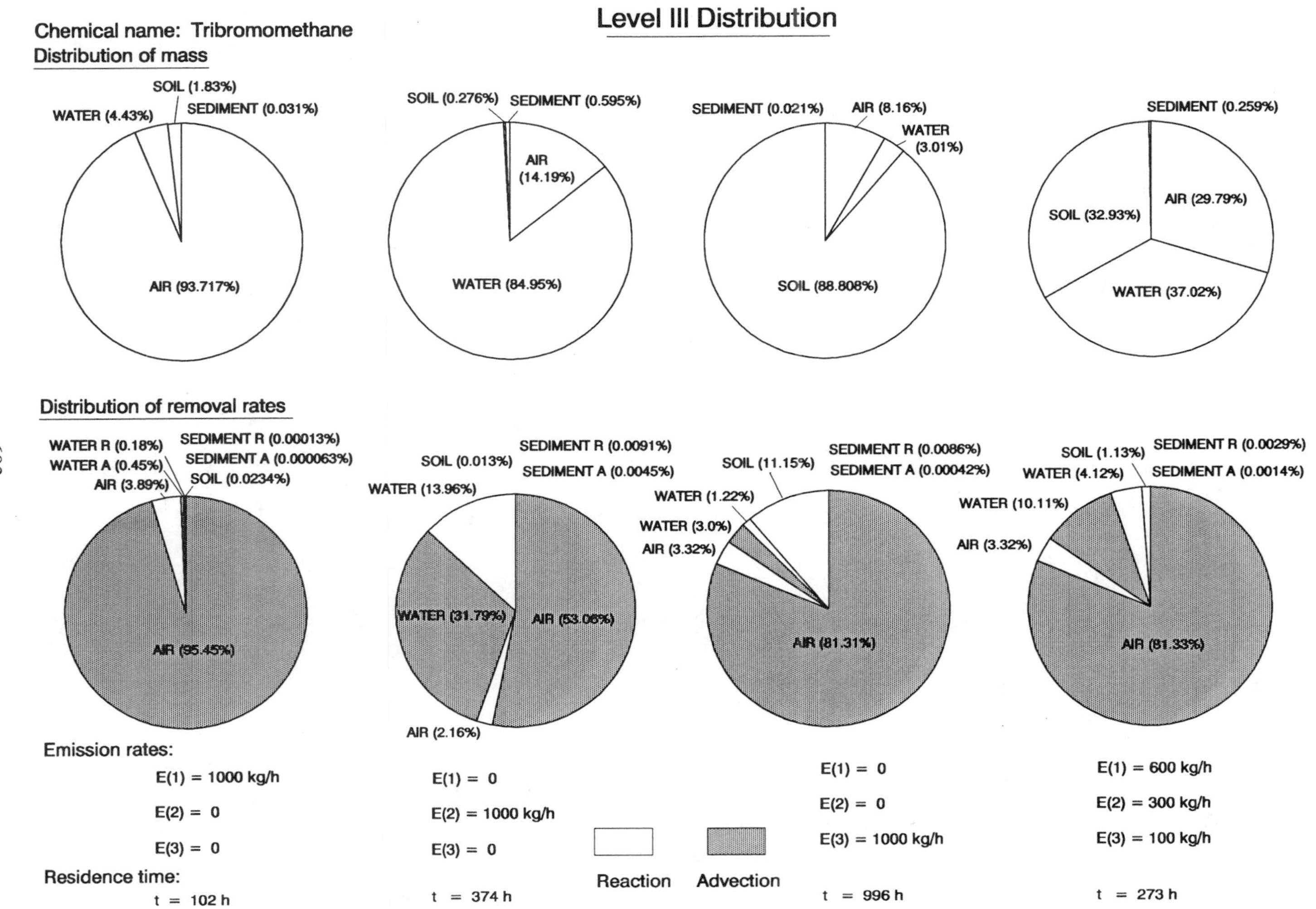
Level III Distribution
Chemical name: Tribromomethane
Distribution of mass
SOIL (1.83%)
WATER (4.43%)
SEDIMENT (0.031%)
AIR (93.717%)
SOIL (0.276%)
SEDIMENT (0.595%)
AIR (14.19%)
WATER (84.95%)
SEDIMENT (0.021%)
AIR (8.16%)
WATER (3.01%)
SOIL (88.808%)
SEDIMENT (0.259%)
SOIL (32.93%)
AIR (29.79%)
WATER (37.02%)
Distribution of removal rates
WATER R (0.18%)
WATER A (0.45%)
AIR (3.89%)
SEDIMENT R (0.00013%)
SEDIMENT A (0.000063%)
SOIL (0.0234%)
AIR (95.45%)
SOIL (0.013%)
SEDIMENT R (0.0091%)
SEDIMENT A (0.0045%)
WATER (13.96%)
WATER (31.79%)
AIR (53.06%)
AIR (2.16%)
SOIL (11.15%)
SEDIMENT R (0.0086%)
SEDIMENT A (0.00042%)
WATER (1.22%)
WATER (3.0%)
AIR (3.32%)
AIR (81.31%)
SOIL (1.13%)
WATER (4.12%)
SEDIMENT R (0.0029%)
SEDIMENT A (0.0014%)
WATER (10.11%)
AIR (3.32%)
AIR (81.33%)
Emission rates:
E(1) = 1000 kg/h
E(2) = 0
E(3) = 0
Residence time:
t = 102 h
E(1) = 0
E(2) = 1000 kg/h
E(3) = 0
t = 374 h
Reaction
Advection
E(1) = 0
E(2) = 0
E(3) = 1000 kg/h
t = 996 h
E(1) = 600 kg/h
E(2) = 300 kg/h
E(3) = 100 kg/h
t = 273 h

Chemical name: Dibromochloromethane

Level I calculation: (six-compartment model)

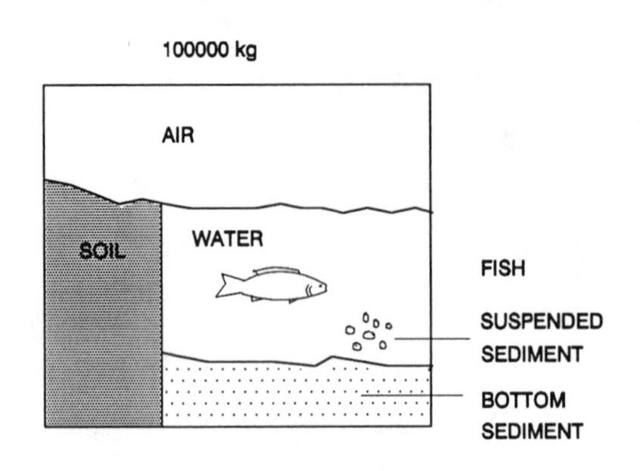


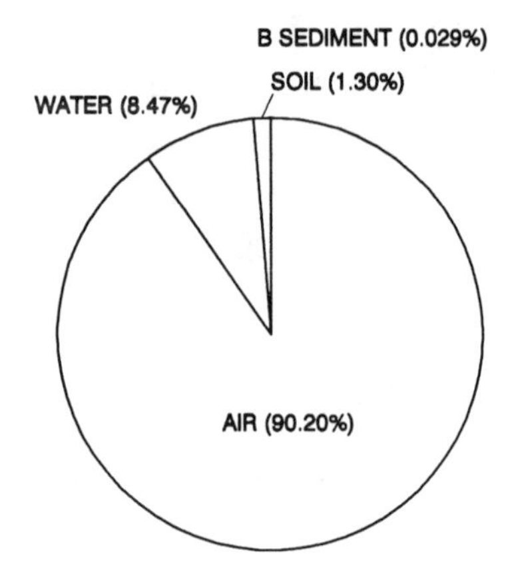


Distribution of mass

physical-chemical properties:

MW: 208.3 g/mol

M.P.: - 20 °C

Fugacity ratio: 1.0

vapor pressure: 1014 Pa

solubility: 4000 g/m$^3$

log $K_{OW}$: 2.24

| Compartment | Z | Concentration | | | Amount | Amount |
|---|---|---|---|---|---|---|
| | mol/m3 Pa | mol/m3 | mg/L (or g/m3) | ug/g | kg | % |
| Air | 4.034E-04 | 4.330E-09 | 9.020E-07 | 7.609E-04 | 90198 | 90.198 |
| Water | 1.894E-02 | 2.033E-07 | 4.234E-05 | 4.234E-05 | 8468.49 | 8.4685 |
| Soil | 6.477E-02 | 6.952E-07 | 1.448E-04 | 6.034E-05 | 1303.30 | 1.3033 |
| Biota (fish) | 1.646E-01 | 1.766E-06 | 3.679E-04 | 3.679E-04 | 0.0736 | 7.36E-05 |
| Suspended sediment | 4.048E-01 | 4.345E-06 | 9.051E-04 | 6.034E-04 | 0.9051 | 9.05E-04 |
| Bottom sediment | 1.295E-01 | 1.390E-06 | 2.896E-04 | 1.207E-04 | 28.962 | 0.0290 |
| | Total | | | | 100000 | 100 |

f = 1.073E-05 Pa

Chemical name: Dibromochloromethane

Level II calculation: (six-compartment model)

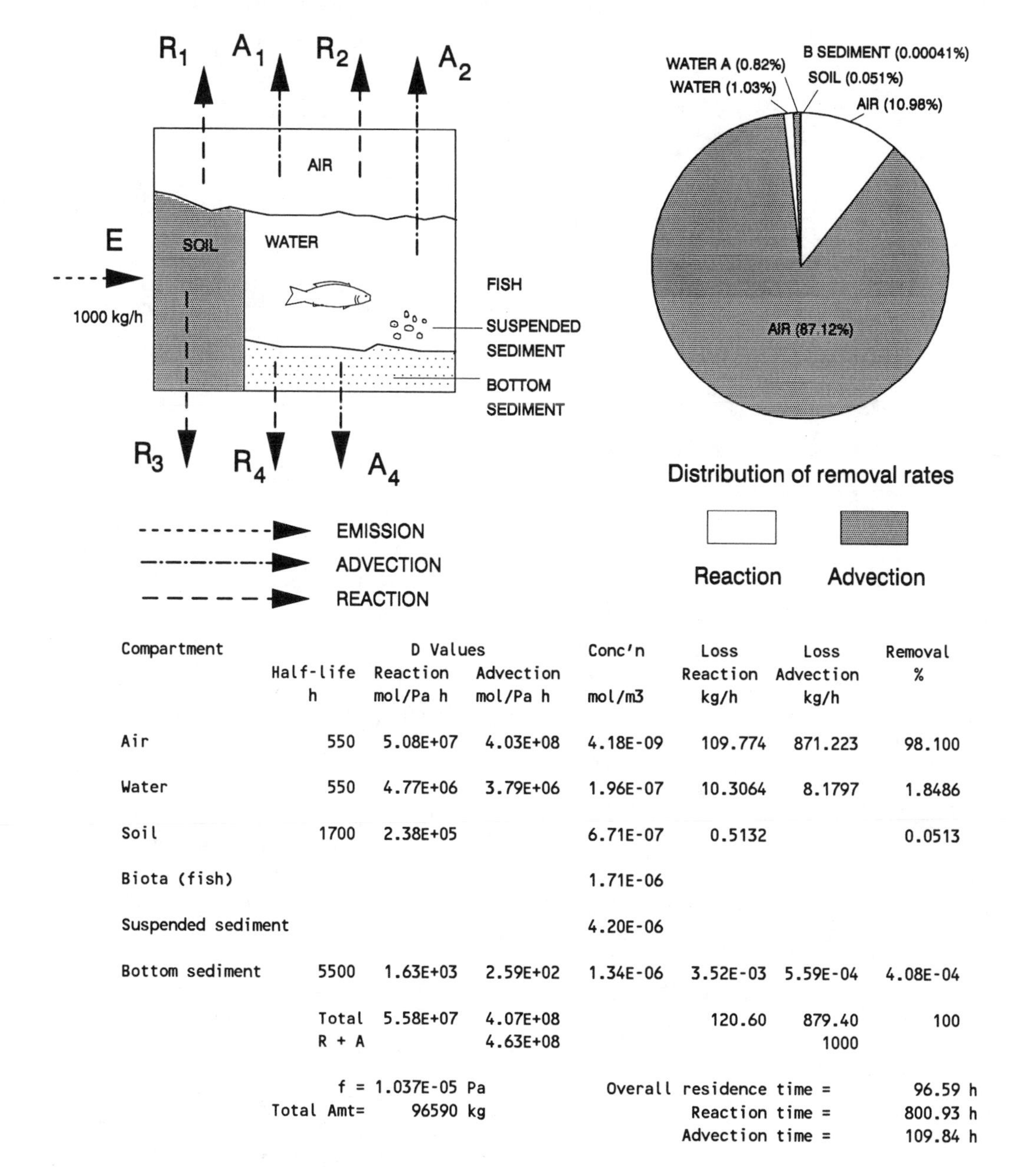


| Compartment | Half-life h | D Values Reaction mol/Pa h | D Values Advection mol/Pa h | Conc'n mol/m3 | Loss Reaction kg/h | Loss Advection kg/h | Removal % |
|---|---|---|---|---|---|---|---|
| Air | 550 | 5.08E+07 | 4.03E+08 | 4.18E-09 | 109.774 | 871.223 | 98.100 |
| Water | 550 | 4.77E+06 | 3.79E+06 | 1.96E-07 | 10.3064 | 8.1797 | 1.8486 |
| Soil | 1700 | 2.38E+05 | | 6.71E-07 | 0.5132 | | 0.0513 |
| Biota (fish) | | | | 1.71E-06 | | | |
| Suspended sediment | | | | 4.20E-06 | | | |
| Bottom sediment | 5500 | 1.63E+03 | 2.59E+02 | 1.34E-06 | 3.52E-03 | 5.59E-04 | 4.08E-04 |
| | Total | 5.58E+07 | 4.07E+08 | | 120.60 | 879.40 | 100 |
| | R + A | | 4.63E+08 | | | 1000 | |

f = 1.037E-05 Pa
Total Amt= 96590 kg

Overall residence time = 96.59 h
Reaction time = 800.93 h
Advection time = 109.84 h

Fugacity Level III calculations: (four-compartment model)

Chemical name: Dibromochloromethane

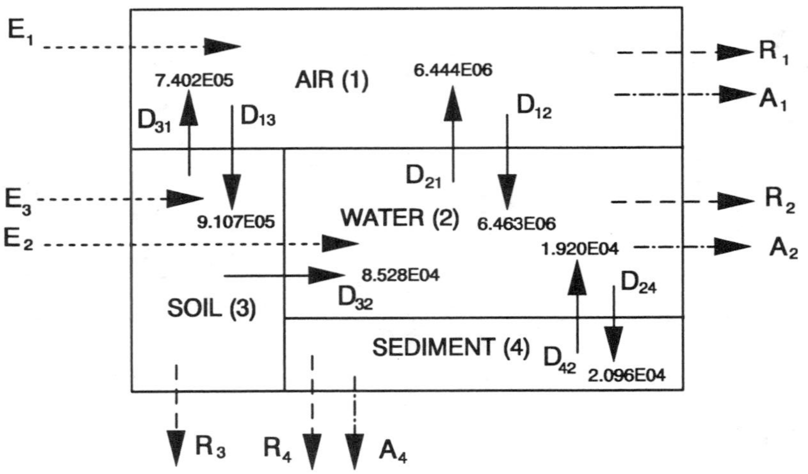

Phase Properties and Rates:

| Compartment | Bulk Z mol/m3 Pa | Half-life h | D Values Reaction mol/Pa h | D Values Advection mol/Pa h |
|---|---|---|---|---|
| Air (1) | 4.034E-04 | 550 | 5.08E+07 | 4.03E+08 |
| Water (2) | 1.894E-02 | 550 | 4.77E+06 | 3.79E+06 |
| Soil (3) | 3.815E-02 | 1700 | 2.80E+05 | |
| Sediment (4) | 4.106E-02 | 5500 | 2.59E+03 | 4.11E+02 |

| | E(1)=1000 | E(2)=1000 | E(3)=1000 | E(1,2,3) | |
|---|---|---|---|---|---|
| Overall residence time = | 92.89 | 293.58 | 706.04 | 214.42 | h |
| Reaction time = | 801.00 | 797.80 | 1965.19 | 993.09 | h |
| Advection time = | 105.08 | 464.52 | 1101.94 | 273.46 | h |

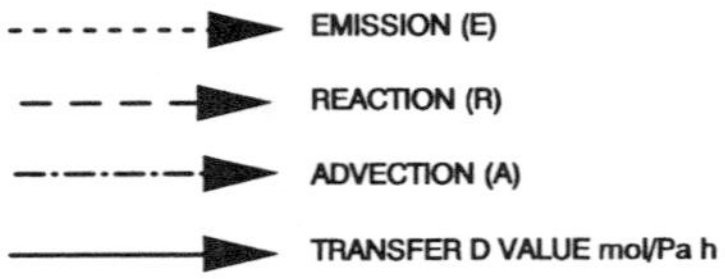

Phase Properties, Compositions, Transport and Transformation Rates:

| Emission, kg/h | | | Fugacity, Pa | | | | Concentration, g/m3 | | | |
|---|---|---|---|---|---|---|---|---|---|---|
| E(1) | E(2) | E(3) | f(1) | f(2) | f(3) | f(4) | C(1) | C(2) | C(3) | C(4) |
| 1000 | 0 | 0 | 1.048E-05 | 4.561E-06 | 8.633E-06 | 4.308E-06 | 8.804E-07 | 1.799E-05 | 6.859E-05 | 3.684E-05 |
| 0 | 1000 | 0 | 4.499E-06 | 3.219E-04 | 3.707E-06 | 3.040E-04 | 3.780E-07 | 1.270E-03 | 2.945E-05 | 2.600E-03 |
| 0 | 0 | 1000 | 7.363E-06 | 2.788E-05 | 4.349E-03 | 2.634E-05 | 6.187E-07 | 1.100E-04 | 3.456E-02 | 2.252E-04 |
| 600 | 300 | 100 | 8.372E-06 | 1.021E-04 | 4.412E-04 | 9.641E-05 | 7.035E-07 | 4.027E-04 | 3.506E-03 | 8.245E-04 |

| Emission, kg/h | | | Amounts, kg | | | | Total Amount, kg |
|---|---|---|---|---|---|---|---|
| E(1) | E(2) | E(3) | m(1) | m(2) | m(3) | m(4) | |
| 1000 | 0 | 0 | 8.804E+04 | 3.599E+03 | 1.235E+03 | 1.842E+01 | 9.289E+04 |
| 0 | 1000 | 0 | 3.780E+04 | 2.540E+05 | 5.301E+02 | 1.300E+03 | 2.936E+05 |
| 0 | 0 | 1000 | 6.187E+04 | 2.200E+04 | 6.221E+05 | 1.126E+02 | 7.060E+05 |
| 600 | 300 | 100 | 7.035E+04 | 8.054E+04 | 6.311E+04 | 4.123E+02 | 2.144E+05 |

| Emission, kg/h | | | Loss, Reaction, kg/h | | | | Loss, Advection, kg/h | | |
|---|---|---|---|---|---|---|---|---|---|
| E(1) | E(2) | E(3) | R(1) | R(2) | R(3) | R(4) | A(1) | A(2) | A(4) |
| 1000 | 0 | 0 | 1.109E+02 | 4.535E+00 | 5.03E-01 | 2.321E-03 | 8.804E+02 | 3.599E+00 | 3.684E-04 |
| 0 | 1000 | 0 | 4.763E+01 | 3.200E+02 | 2.16E-01 | 1.638E-01 | 3.780E+02 | 2.540E+02 | 2.600E-02 |
| 0 | 0 | 1000 | 7.796E+01 | 2.772E+01 | 2.54E+02 | 1.419E-02 | 6.187E+02 | 2.200E+01 | 2.252E-03 |
| 600 | 300 | 100 | 8.865E+01 | 1.015E+02 | 2.57E+01 | 5.195E-02 | 7.035E+02 | 8.054E+01 | 8.245E-03 |

| Emission, kg/h | | | Intermedia Rate of Transport, kg/h | | | | | | |
|---|---|---|---|---|---|---|---|---|---|
| | | | T12 | T21 | T13 | T31 | T32 | T24 | T42 |
| E(1) | E(2) | E(3) | air-water | water-air | air-soil | soil-air | soil-water | water-sed | sed-water |
| 1000 | 0 | 0 | 1.410E+01 | 6.122E+00 | 1.988E+00 | 1.331E+00 | 1.533E-01 | 1.991E-02 | 1.723E-02 |
| 0 | 1000 | 0 | 6.056E+00 | 4.320E+02 | 8.534E-01 | 5.715E-01 | 6.584E-02 | 1.405E+00 | 1.216E+00 |
| 0 | 0 | 1000 | 9.912E+00 | 3.743E+01 | 1.397E+00 | 6.706E+02 | 7.726E+01 | 1.218E-01 | 1.053E-01 |
| 600 | 300 | 100 | 1.127E+01 | 1.370E+02 | 1.588E+00 | 6.803E+01 | 7.838E+00 | 4.457E-01 | 3.855E-01 |

# Level III Distribution

Chemical name: Dibromochloromethane

## Distribution of mass

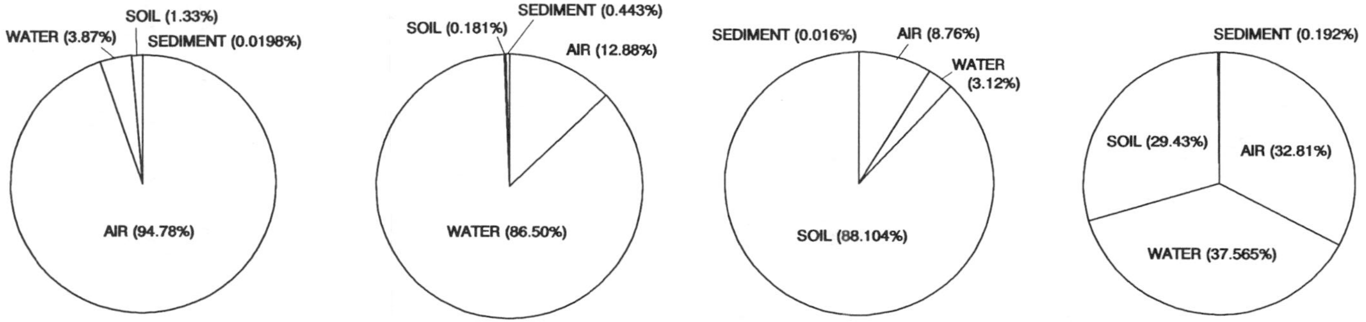

WATER (3.87%)
SOIL (1.33%)
SEDIMENT (0.0198%)
AIR (94.78%)

SOIL (0.181%)
SEDIMENT (0.443%)
AIR (12.88%)
WATER (86.50%)

SEDIMENT (0.016%)
AIR (8.76%)
WATER (3.12%)
SOIL (88.104%)

SEDIMENT (0.192%)
SOIL (29.43%)
AIR (32.81%)
WATER (37.565%)

## Distribution of removal rates

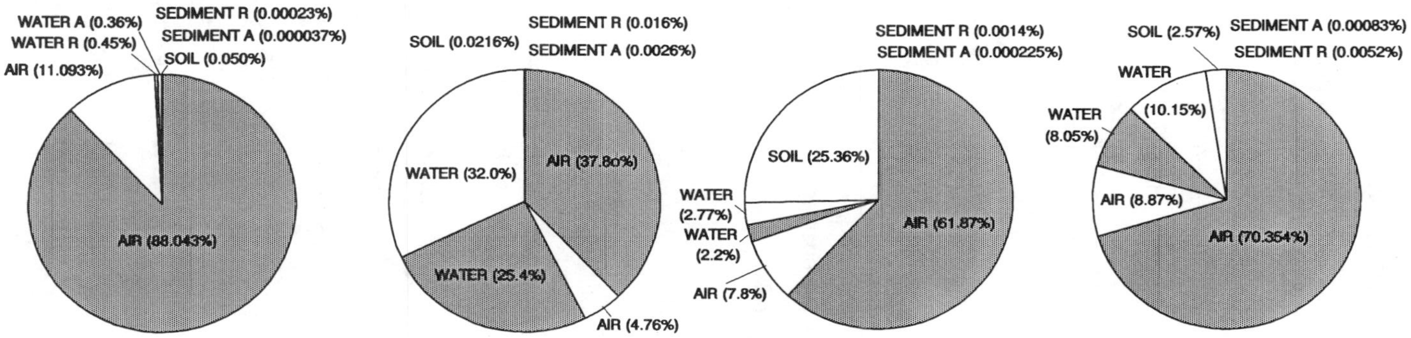

WATER A (0.36%)
WATER R (0.45%)
AIR (11.093%)
SEDIMENT R (0.00023%)
SEDIMENT A (0.000037%)
SOIL (0.050%)
AIR (88.043%)

SOIL (0.0216%)
SEDIMENT R (0.016%)
SEDIMENT A (0.0026%)
WATER (32.0%)
AIR (37.80%)
WATER (25.4%)
AIR (4.76%)

SEDIMENT R (0.0014%)
SEDIMENT A (0.000225%)
SOIL (25.36%)
WATER (2.77%)
WATER (2.2%)
AIR (7.8%)
AIR (61.87%)

SOIL (2.57%)
SEDIMENT A (0.00083%)
SEDIMENT R (0.0052%)
WATER (10.15%)
WATER (8.05%)
AIR (8.87%)
AIR (70.354%)

Reaction
Advection

Emission rates:

| E(1) = 1000 kg/h | E(1) = 0 | E(1) = 0 | E(1) = 600 kg/h |
|---|---|---|---|
| E(2) = 0 | E(2) = 1000 kg/h | E(2) = 0 | E(2) = 300 kg/h |
| E(3) = 0 | E(3) = 0 | E(3) = 1000 kg/h | E(3) = 100 kg/h |

Residence time:

| t = 92.89 h | t = 294 h | t = 706 h | t = 214 h |
|---|---|---|---|

Chemical name: Trichlorofluoromethane

Level I calculation: (six-compartment model)

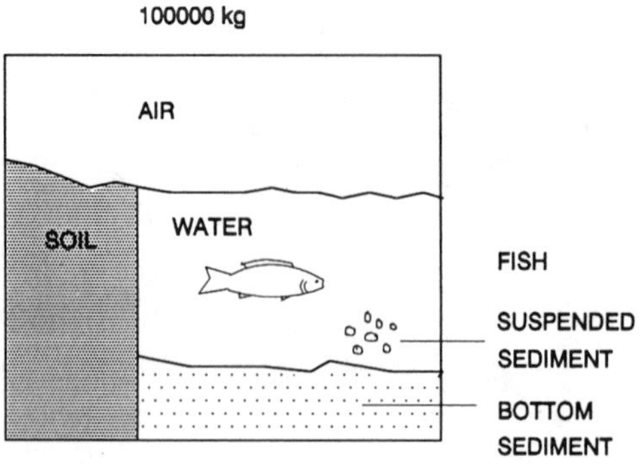

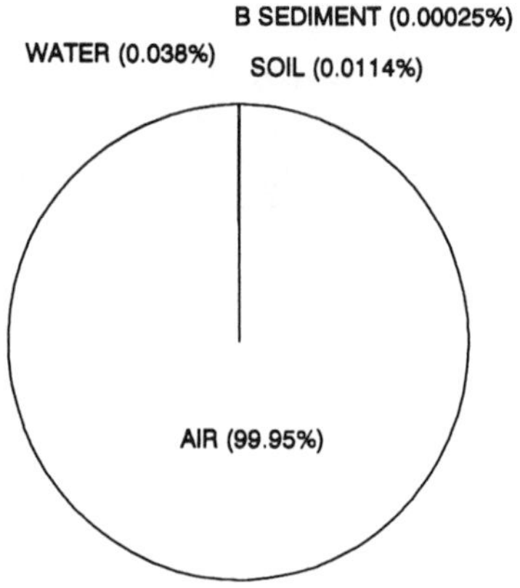

Distribution of mass

physical-chemical properties:

MW: 137.37 g/mol

M.P.: - 110.5 °C

Fugacity ratio: 1.0

vapor pressure: 102200 Pa

solubility: 1080 g/m $^3$

log $K_{OW}$: 2.53

| Compartment | Z | Concentration | | | Amount | Amount |
|---|---|---|---|---|---|---|
| | mol/m3 Pa | mol/m3 | mg/L (or g/m3) | ug/g | kg | % |
| Air | 4.034E-04 | 7.276E-09 | 9.995E-07 | 8.432E-04 | 99950 | 99.950 |
| Water | 7.693E-05 | 1.387E-09 | 1.906E-07 | 1.906E-07 | 38.119 | 0.0381 |
| Soil | 5.130E-04 | 9.252E-09 | 1.271E-06 | 5.296E-07 | 11.439 | 0.0114 |
| Biota (fish) | 1.303E-03 | 2.351E-08 | 3.229E-06 | 3.229E-06 | 6.46E-04 | 6.46E-07 |
| Suspended sediment | 3.206E-03 | 5.783E-08 | 7.944E-06 | 5.296E-06 | 7.94E-03 | 7.94E-06 |
| Bottom sediment | 1.026E-03 | 1.850E-08 | 2.542E-06 | 1.059E-06 | 0.2542 | 2.54E-04 |
| | Total | | | | 100000 | 100 |

f = 1.804E-05 Pa

Chemical name: Trichlorofluoromethane

Level II calculation: (six-compartment model)

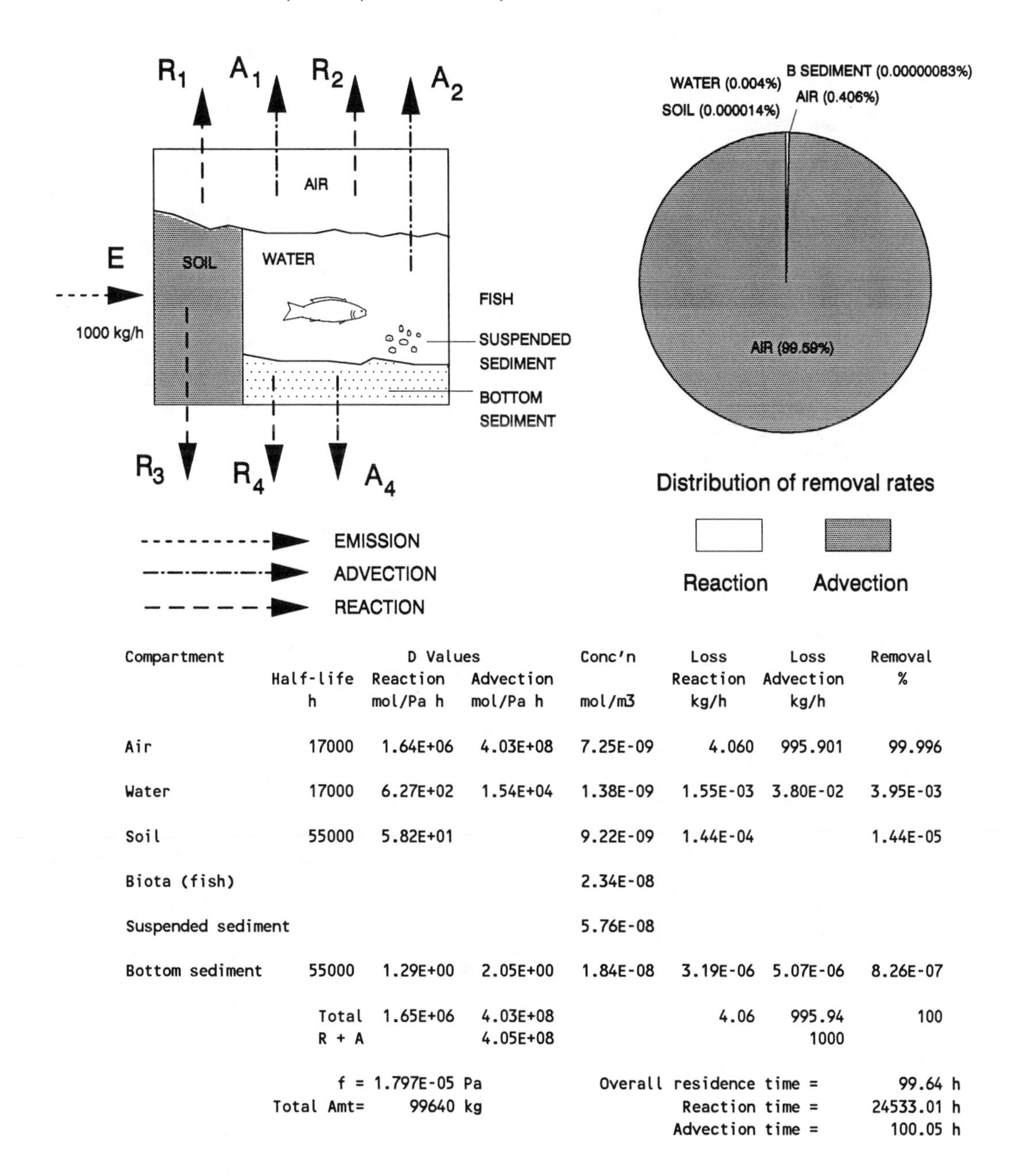


| Compartment | Half-life h | D Values Reaction mol/Pa h | D Values Advection mol/Pa h | Conc'n mol/m3 | Loss Reaction kg/h | Loss Advection kg/h | Removal % |
|---|---|---|---|---|---|---|---|
| Air | 17000 | 1.64E+06 | 4.03E+08 | 7.25E-09 | 4.060 | 995.901 | 99.996 |
| Water | 17000 | 6.27E+02 | 1.54E+04 | 1.38E-09 | 1.55E-03 | 3.80E-02 | 3.95E-03 |
| Soil | 55000 | 5.82E+01 | | 9.22E-09 | 1.44E-04 | | 1.44E-05 |
| Biota (fish) | | | | 2.34E-08 | | | |
| Suspended sediment | | | | 5.76E-08 | | | |
| Bottom sediment | 55000 | 1.29E+00 | 2.05E+00 | 1.84E-08 | 3.19E-06 | 5.07E-06 | 8.26E-07 |
| | Total | 1.65E+06 | 4.03E+08 | | 4.06 | 995.94 | 100 |
| | R + A | | 4.05E+08 | | | 1000 | |

f = 1.797E-05 Pa

Total Amt= 99640 kg

Overall residence time = 99.64 h

Reaction time = 24533.01 h

Advection time = 100.05 h

## Fugacity Level III calculations: (four-compartment model)

Chemical name: Trichlorofluoromethane

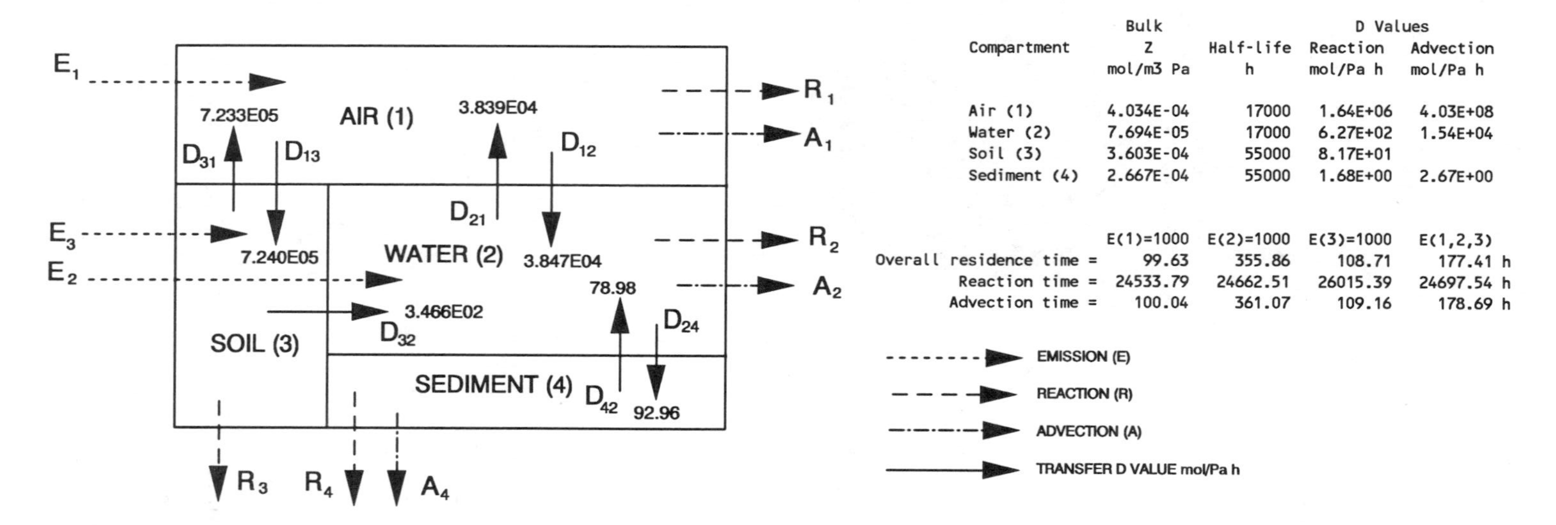


Phase Properties and Rates:

| Compartment | Bulk Z mol/m3 Pa | Half-life h | D Values Reaction mol/Pa h | D Values Advection mol/Pa h |
|---|---|---|---|---|
| Air (1) | 4.034E-04 | 17000 | 1.64E+06 | 4.03E+08 |
| Water (2) | 7.694E-05 | 17000 | 6.27E+02 | 1.54E+04 |
| Soil (3) | 3.603E-04 | 55000 | 8.17E+01 | |
| Sediment (4) | 2.667E-04 | 55000 | 1.68E+00 | 2.67E+00 |

| | E(1)=1000 | E(2)=1000 | E(3)=1000 | E(1,2,3) | |
|---|---|---|---|---|---|
| Overall residence time = | 99.63 | 355.86 | 108.71 | 177.41 | h |
| Reaction time = | 24533.79 | 24662.51 | 26015.39 | 24697.54 | h |
| Advection time = | 100.04 | 361.07 | 109.16 | 178.69 | h |

Phase Properties, Compositions, Transport and Transformation Rates:

| Emission, kg/h | | | Fugacity, Pa | | | | Concentration, g/m3 | | | | Amounts, kg | | | | Total |
|---|---|---|---|---|---|---|---|---|---|---|---|---|---|---|---|
| E(1) | E(2) | E(3) | f(1) | f(2) | f(3) | f(4) | C(1) | C(2) | C(3) | C(4) | m(1) | m(2) | m(3) | m(4) | Amount, kg |
| 1000 | 0 | 0 | 1.797E-05 | 1.282E-05 | 1.798E-05 | 1.430E-05 | 9.959E-07 | 1.355E-07 | 8.897E-07 | 5.240E-07 | 9.959E+04 | 2.710E+01 | 1.601E+01 | 2.620E-01 | 9.963E+04 |
| 0 | 1000 | 0 | 1.268E-05 | 1.338E-01 | 1.268E-05 | 1.493E-01 | 7.027E-07 | 1.414E-03 | 6.277E-07 | 5.469E-03 | 7.027E+04 | 2.828E+05 | 1.130E+01 | 2.735E+03 | 3.559E+05 |
| 0 | 0 | 1000 | 1.797E-05 | 7.689E-05 | 1.008E-02 | 8.578E-05 | 9.957E-07 | 8.127E-07 | 4.986E-04 | 3.143E-06 | 9.957E+04 | 1.625E+02 | 8.976E+03 | 1.572E+00 | 1.087E+05 |
| 600 | 300 | 100 | 1.638E-05 | 4.015E-02 | 1.022E-03 | 4.480E-02 | 9.079E-07 | 4.244E-04 | 5.059E-05 | 1.641E-03 | 9.079E+04 | 8.489E+04 | 9.106E+02 | 8.207E+02 | 1.774E+05 |

| Emission, kg/h | | | Loss, Reaction, kg/h | | | | Loss, Advection, kg/h | | | Intermedia Rate of Transport, kg/h | | | | | | |
|---|---|---|---|---|---|---|---|---|---|---|---|---|---|---|---|---|
| | | | | | | | | | | T12 | T21 | T13 | T31 | T32 | T24 | T42 |
| E(1) | E(2) | E(3) | R(1) | R(2) | R(3) | R(4) | A(1) | A(2) | A(4) | air-water | water-air | air-soil | soil-air | soil-water | water-sed | sed-water |
| 1000 | 0 | 0 | 4.060E+00 | 1.105E-03 | 2.02E-04 | 3.301E-06 | 9.959E+02 | 2.710E-02 | 5.240E-06 | 9.496E-02 | 6.761E-02 | 1.787E+00 | 1.786E+00 | 8.560E-04 | 1.637E-04 | 1.552E-04 |
| 0 | 1000 | 0 | 2.864E+00 | 1.153E+01 | 1.42E-04 | 3.446E-02 | 7.027E+02 | 2.828E+02 | 5.469E-02 | 6.700E-02 | 7.056E+02 | 1.261E+00 | 1.260E+00 | 6.040E-04 | 1.709E+00 | 1.619E+00 |
| 0 | 0 | 1000 | 4.059E+00 | 6.626E-03 | 1.13E-01 | 1.980E-05 | 9.957E+02 | 1.625E-01 | 3.143E-05 | 9.494E-02 | 4.055E-01 | 1.787E+00 | 1.001E+03 | 4.798E-01 | 9.819E-04 | 9.307E-04 |
| 600 | 300 | 100 | 3.701E+00 | 3.460E+00 | 1.15E-02 | 1.034E-02 | 9.079E+02 | 8.489E+01 | 1.641E-02 | 8.657E-02 | 2.118E+02 | 1.629E+00 | 1.016E+02 | 4.867E-02 | 5.128E-01 | 4.860E-01 |

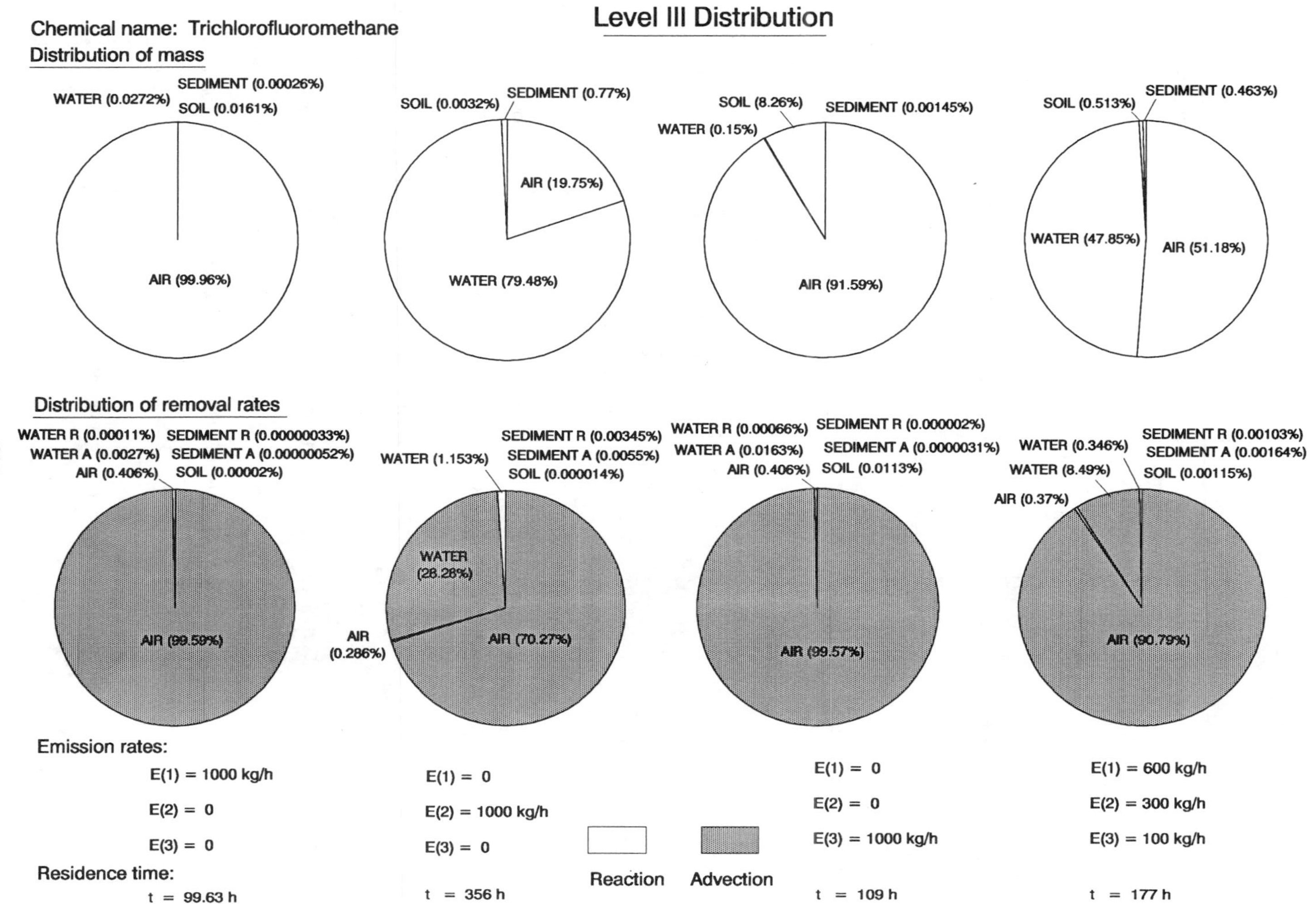

Level III Distribution
Chemical name: Trichlorofluoromethane
Distribution of mass
WATER (0.0272%)
SEDIMENT (0.00026%)
SOIL (0.0161%)
AIR (99.96%)
SOIL (0.0032%)
SEDIMENT (0.77%)
AIR (19.75%)
WATER (79.48%)
SOIL (8.26%)
SEDIMENT (0.00145%)
WATER (0.15%)
AIR (91.59%)
SOIL (0.513%)
SEDIMENT (0.463%)
WATER (47.85%)
AIR (51.18%)
Distribution of removal rates
WATER R (0.00011%)
SEDIMENT R (0.00000033%)
WATER A (0.0027%)
SEDIMENT A (0.00000052%)
AIR (0.406%)
SOIL (0.00002%)
AIR (99.59%)
SEDIMENT R (0.00345%)
SEDIMENT A (0.0055%)
WATER (1.153%)
SOIL (0.000014%)
WATER (28.28%)
AIR (0.286%)
AIR (70.27%)
WATER R (0.00066%)
SEDIMENT R (0.000002%)
WATER A (0.0163%)
SEDIMENT A (0.0000031%)
AIR (0.406%)
SOIL (0.0113%)
AIR (99.57%)
WATER (0.346%)
SEDIMENT R (0.00103%)
SEDIMENT A (0.00164%)
WATER (8.49%)
SOIL (0.00115%)
AIR (0.37%)
AIR (90.79%)
Emission rates:
E(1) = 1000 kg/h
E(2) = 0
E(3) = 0
Residence time:
t = 99.63 h
E(1) = 0
E(2) = 1000 kg/h
E(3) = 0
t = 356 h
Reaction
Advection
E(1) = 0
E(2) = 0
E(3) = 1000 kg/h
t = 109 h
E(1) = 600 kg/h
E(2) = 300 kg/h
E(3) = 100 kg/h
t = 177 h

Chemical name: Fluorobenzene

Level I calculation: (six-compartment model)

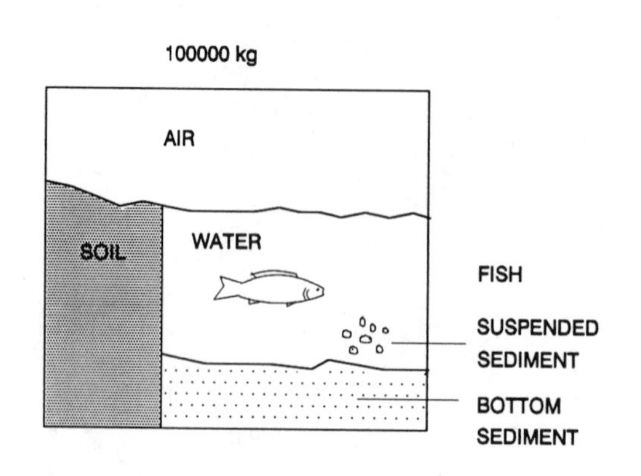


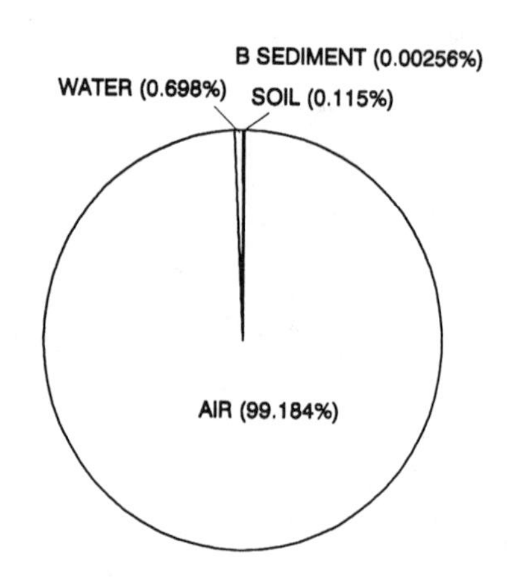


Distribution of mass

physical-chemical properties:

MW: 96.104 g/mol

M.P.: -42.21°C

Fugacity ratio: 1.0

vapor pressure: 10480 Pa

solubility: 1430 g/m$^3$

log $K_{OW}$: 2.27

| Compartment | Z | Concentration | | | Amount | Amount |
|---|---|---|---|---|---|---|
| | mol/m3 Pa | mol/m3 | mg/L (or g/m3) | ug/g | kg | % |
| Air | 4.034E-04 | 1.032E-08 | 9.918E-07 | 8.367E-04 | 99184 | 99.184 |
| Water | 1.420E-03 | 3.632E-08 | 3.491E-06 | 3.491E-06 | 698.152 | 0.6982 |
| Soil | 5.203E-03 | 1.331E-07 | 1.279E-05 | 5.330E-06 | 115.130 | 0.1151 |
| Biota (fish) | 1.322E-02 | 3.382E-07 | 3.250E-05 | 3.250E-05 | 6.50E-03 | 6.50E-06 |
| Suspended sediment | 3.252E-02 | 8.319E-07 | 7.995E-05 | 5.330E-05 | 8.00E-02 | 8.00E-05 |
| Bottom sediment | 1.041E-02 | 2.662E-07 | 2.558E-05 | 1.066E-05 | 2.5584 | 2.56E-03 |
| | Total | | | | 100000 | 100 |

f = 2.558E-05 Pa

Chemical name: Fluorobenzene

Level II calculation: (six-compartment model)

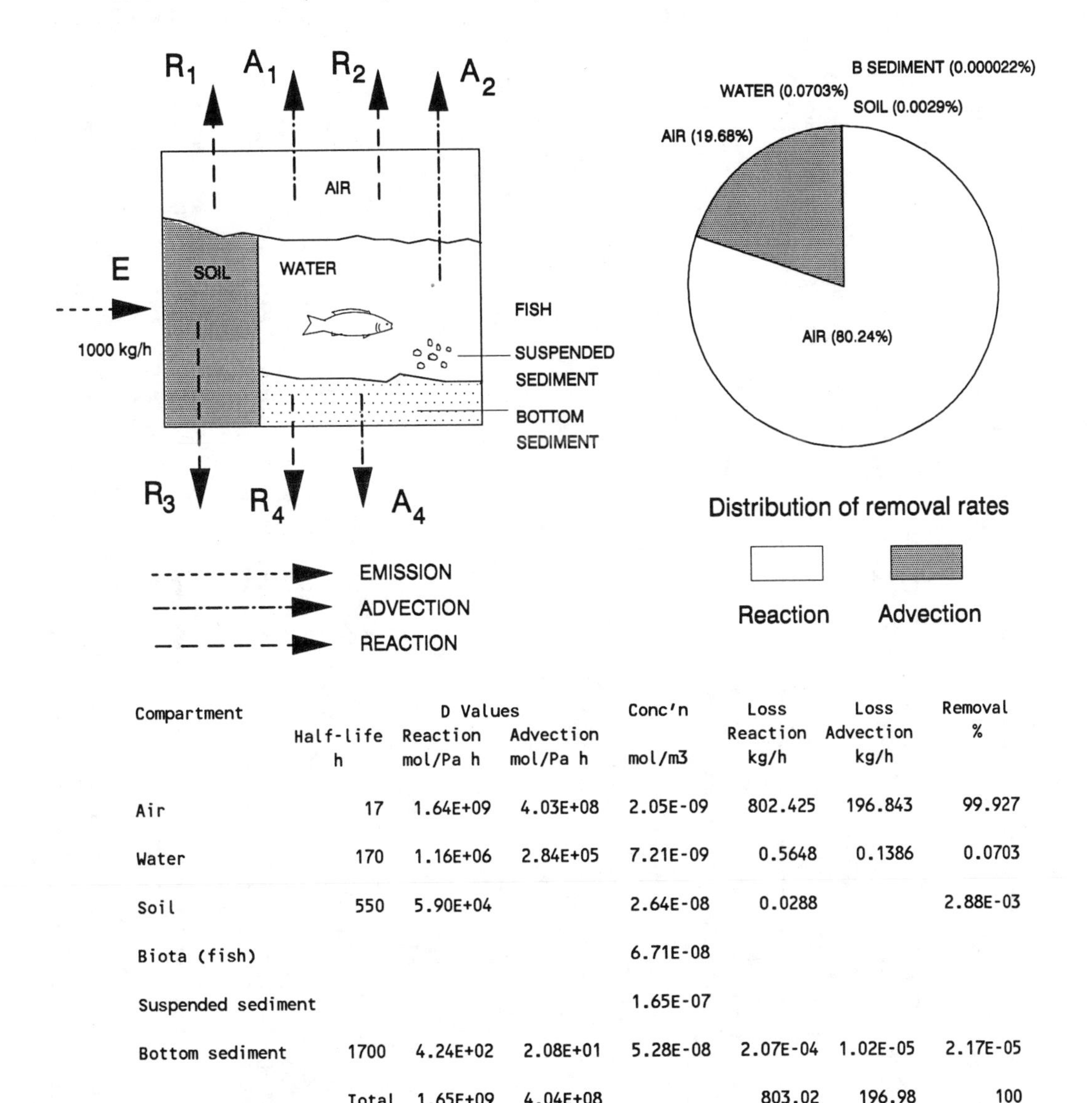


| Compartment | Half-life h | D Values Reaction mol/Pa h | D Values Advection mol/Pa h | Conc'n mol/m3 | Loss Reaction kg/h | Loss Advection kg/h | Removal % |
|---|---|---|---|---|---|---|---|
| Air | 17 | 1.64E+09 | 4.03E+08 | 2.05E-09 | 802.425 | 196.843 | 99.927 |
| Water | 170 | 1.16E+06 | 2.84E+05 | 7.21E-09 | 0.5648 | 0.1386 | 0.0703 |
| Soil | 550 | 5.90E+04 | | 2.64E-08 | 0.0288 | | 2.88E-03 |
| Biota (fish) | | | | 6.71E-08 | | | |
| Suspended sediment | | | | 1.65E-07 | | | |
| Bottom sediment | 1700 | 4.24E+02 | 2.08E+01 | 5.28E-08 | 2.07E-04 | 1.02E-05 | 2.17E-05 |
| | Total | 1.65E+09 | 4.04E+08 | | 803.02 | 196.98 | 100 |
| | R + A | | 2.05E+09 | | | 1000 | |

f = 5.077E-06 Pa

Total Amt= 19846 kg

Overall residence time = 19.85 h

Reaction time = 24.71 h

Advection time = 100.75 h

## Fugacity Level III calculations: (four-compartment model)

Chemical name: Fluorobenzene

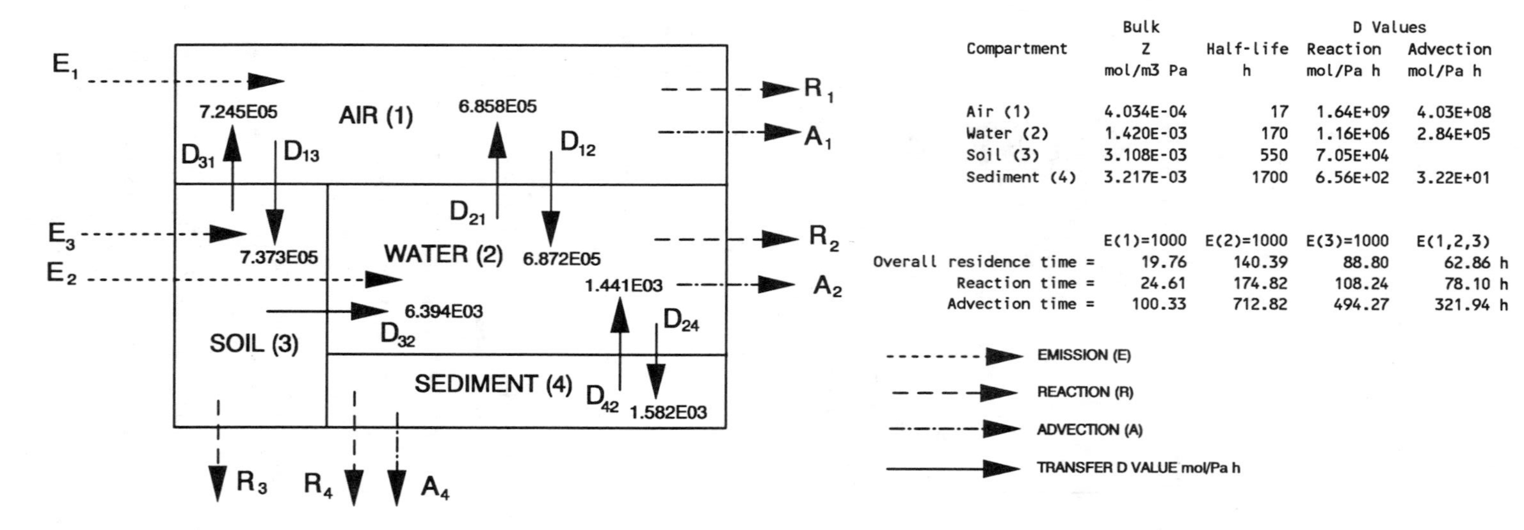


Phase Properties and Rates:

| Compartment | Bulk Z mol/m3 Pa | Half-life h | D Values Reaction mol/Pa h | D Values Advection mol/Pa h |
|---|---|---|---|---|
| Air (1) | 4.034E-04 | 17 | 1.64E+09 | 4.03E+08 |
| Water (2) | 1.420E-03 | 170 | 1.16E+06 | 2.84E+05 |
| Soil (3) | 3.108E-03 | 550 | 7.05E+04 | |
| Sediment (4) | 3.217E-03 | 1700 | 6.56E+02 | 3.22E+01 |

| | E(1)=1000 | E(2)=1000 | E(3)=1000 | E(1,2,3) |
|---|---|---|---|---|
| Overall residence time = | 19.76 | 140.39 | 88.80 | 62.86 h |
| Reaction time = | 24.61 | 174.82 | 108.24 | 78.10 h |
| Advection time = | 100.33 | 712.82 | 494.27 | 321.94 h |

Phase Properties, Compositions, Transport and Transformation Rates:

| Emission, kg/h | | | Fugacity, Pa | | | | Concentration, g/m3 | | | | Amounts, kg | | | | Total Amount, kg |
|---|---|---|---|---|---|---|---|---|---|---|---|---|---|---|---|
| E(1) | E(2) | E(3) | f(1) | f(2) | f(3) | f(4) | C(1) | C(2) | C(3) | C(4) | m(1) | m(2) | m(3) | m(4) | |
| 1000 | 0 | 0 | 5.080E-06 | 1.654E-06 | 4.673E-06 | 1.230E-06 | 1.969E-07 | 2.258E-07 | 1.396E-06 | 3.803E-07 | 1.969E+04 | 4.515E+01 | 2.513E+01 | 1.901E-01 | 1.976E+04 |
| 0 | 1000 | 0 | 1.637E-06 | 4.890E-03 | 1.506E-06 | 3.636E-03 | 6.347E-08 | 6.674E-04 | 4.499E-07 | 1.124E-03 | 6.347E+03 | 1.335E+05 | 8.098E+00 | 5.620E+02 | 1.404E+05 |
| 0 | 0 | 1000 | 4.605E-06 | 4.051E-05 | 1.299E-02 | 3.012E-05 | 1.785E-07 | 5.529E-06 | 3.880E-03 | 9.312E-06 | 1.785E+04 | 1.106E+03 | 6.983E+04 | 4.656E+00 | 8.880E+04 |
| 600 | 300 | 100 | 3.999E-06 | 1.472E-03 | 1.302E-03 | 1.094E-03 | 1.551E-07 | 2.009E-04 | 3.889E-04 | 3.384E-04 | 1.551E+04 | 4.018E+04 | 7.001E+03 | 1.692E+02 | 6.286E+04 |

| Emission, kg/h | | | Loss, Reaction, kg/h | | | | Loss, Advection, kg/h | | | Intermedia Rate of Transport, kg/h T12 | T21 | T13 | T31 | T32 | T24 | T42 |
|---|---|---|---|---|---|---|---|---|---|---|---|---|---|---|---|---|
| E(1) | E(2) | E(3) | R(1) | R(2) | R(3) | R(4) | A(1) | A(2) | A(4) | air-water | water-air | air-soil | soil-air | soil-water | water-sed | sed-water |
| 1000 | 0 | 0 | 8.028E+02 | 1.841E-01 | 3.17E-02 | 7.751E-05 | 1.969E+02 | 4.515E-02 | 3.803E-06 | 3.355E-01 | 1.090E-01 | 3.599E-01 | 3.254E-01 | 2.872E-03 | 2.516E-04 | 1.703E-04 |
| 0 | 1000 | 0 | 2.587E+02 | 5.441E+02 | 1.02E-02 | 2.291E-01 | 6.347E+01 | 1.335E+02 | 1.124E-02 | 1.081E-01 | 3.223E+02 | 1.160E-01 | 1.049E-01 | 9.254E-04 | 7.437E-01 | 5.034E-01 |
| 0 | 0 | 1000 | 7.278E+02 | 4.507E+00 | 8.80E+01 | 1.898E-03 | 1.785E+02 | 1.106E+00 | 9.312E-05 | 3.041E-01 | 2.670E+00 | 3.263E-01 | 9.044E+02 | 7.981E+00 | 6.161E-03 | 4.170E-03 |
| 600 | 300 | 100 | 6.321E+02 | 1.638E+02 | 8.82E+00 | 6.897E-02 | 1.551E+02 | 4.018E+01 | 3.384E-03 | 2.641E-01 | 9.702E+01 | 2.834E-01 | 9.066E+01 | 8.001E-01 | 2.239E-01 | 1.515E-01 |

# Level III Distribution

Chemical name: Fluorobenzene

Distribution of mass

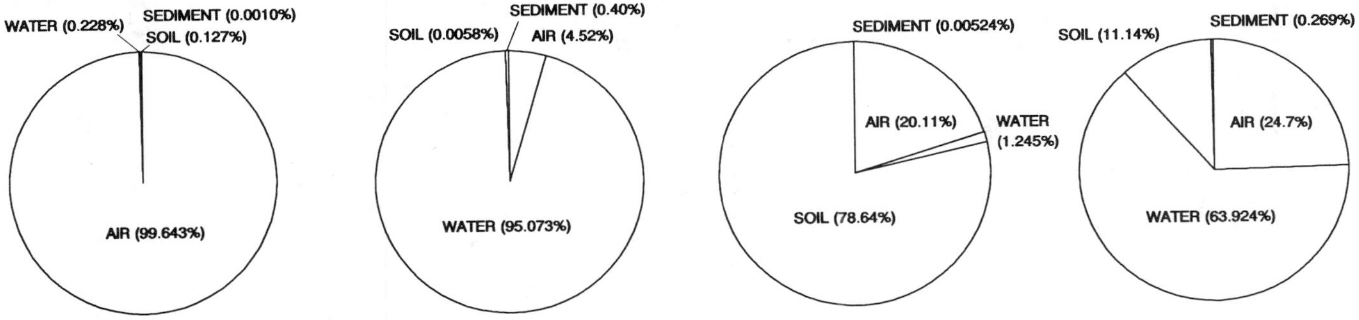

WATER (0.228%) SEDIMENT (0.0010%) SOIL (0.127%) AIR (99.643%)

SOIL (0.0058%) SEDIMENT (0.40%) AIR (4.52%) WATER (95.073%)

SEDIMENT (0.00524%) AIR (20.11%) WATER (1.245%) SOIL (78.64%)

SOIL (11.14%) SEDIMENT (0.269%) AIR (24.7%) WATER (63.924%)

Distribution of removal rates

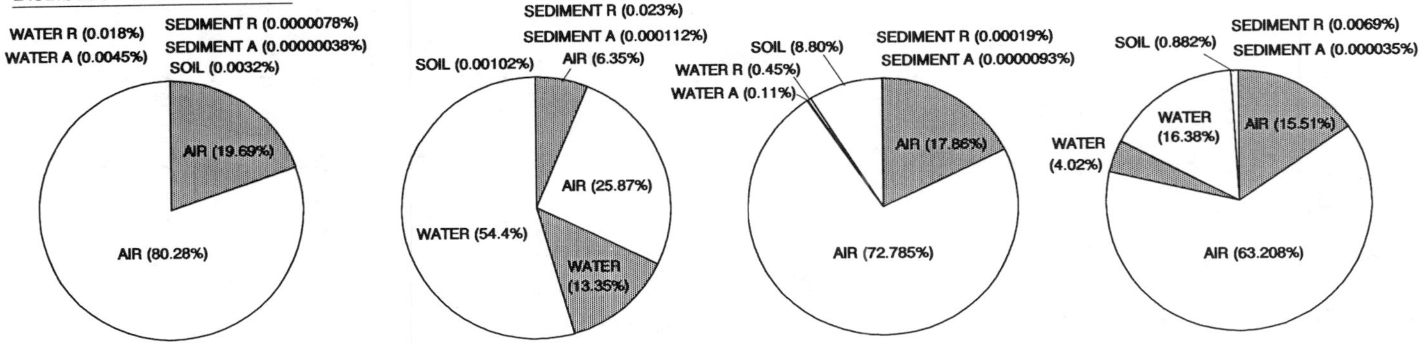

WATER R (0.018%) WATER A (0.0045%) SEDIMENT R (0.0000078%) SEDIMENT A (0.00000038%) SOIL (0.0032%) AIR (19.69%) AIR (80.28%)

SEDIMENT R (0.023%) SEDIMENT A (0.000112%) SOIL (0.00102%) AIR (6.35%) AIR (25.87%) WATER (54.4%) WATER (13.35%)

SOIL (8.80%) WATER R (0.45%) WATER A (0.11%) SEDIMENT R (0.00019%) SEDIMENT A (0.0000093%) AIR (17.86%) AIR (72.785%)

SOIL (0.882%) SEDIMENT R (0.0069%) SEDIMENT A (0.000035%) WATER (16.38%) AIR (15.51%) WATER (4.02%) AIR (63.208%)

Reaction Advection

Emission rates:

| | | | |
|---|---|---|---|
| E(1) = 1000 kg/h | E(1) = 0 | E(1) = 0 | E(1) = 600 kg/h |
| E(2) = 0 | E(2) = 1000 kg/h | E(2) = 0 | E(2) = 300 kg/h |
| E(3) = 0 | E(3) = 0 | E(3) = 1000 kg/h | E(3) = 100 kg/h |

Residence time:

| | | | |
|---|---|---|---|
| t = 19.76 h | t = 140.4 h | t = 88.8 h | t = 62.86 h |

Chemical name: Bromobenzene

Level I calculation: (six-compartment model)

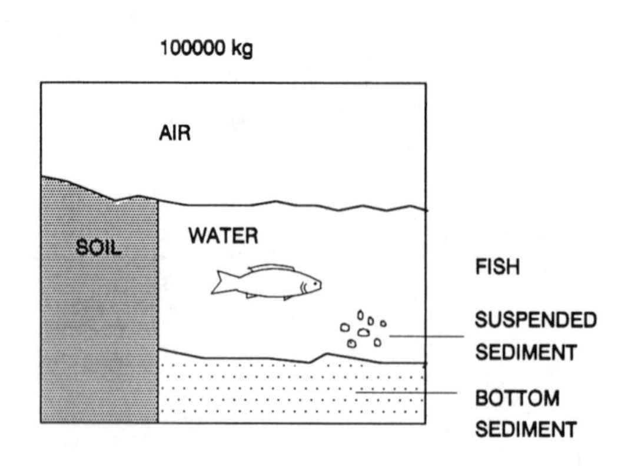


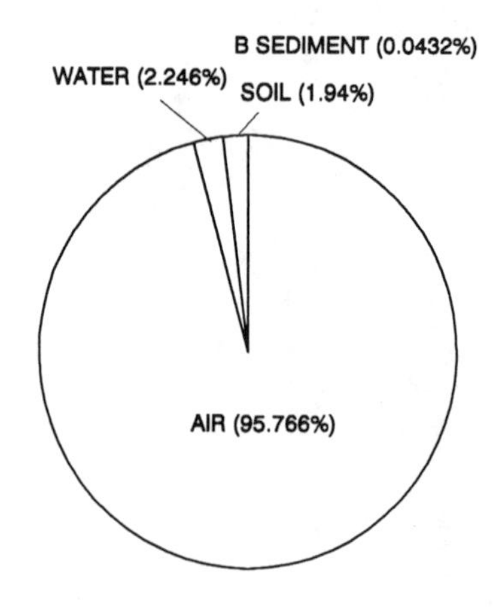


Distribution of mass

physical-chemical properties:

MW: 157.02 g/mol

M.P.: - 30.8 °C

Fugacity ratio: 1.0

vapor pressure: 552 Pa

solubility: 410 g/m $^3$

log $K_{OW}$: 2.99

| Compartment | Z | Concentration | | | Amount | Amount |
|---|---|---|---|---|---|---|
| | mol/m3 Pa | mol/m3 | mg/L (or g/m3) | ug/g | kg | % |
| Air | 4.034E-04 | 6.099E-09 | 9.577E-07 | 8.079E-04 | 95766 | 95.766 |
| Water | 4.730E-03 | 7.151E-08 | 1.123E-05 | 1.123E-05 | 2245.82 | 2.2458 |
| Soil | 9.097E-02 | 1.375E-06 | 2.160E-04 | 8.998E-05 | 1943.63 | 1.9436 |
| Biota (fish) | 2.311E-01 | 3.494E-06 | 5.487E-04 | 5.487E-04 | 0.1097 | 1.10E-04 |
| Suspended sediment | 5.686E-01 | 8.596E-06 | 1.350E-03 | 8.998E-04 | 1.350 | 1.35E-03 |
| Bottom sediment | 1.819E-01 | 2.751E-06 | 4.319E-04 | 1.800E-04 | 43.192 | 4.32E-02 |
| | Total | | | | 100000 | 100 |

f = 1.512E-05 Pa

Chemical name: Bromobenzene

Level II calculation: (six-compartment model)

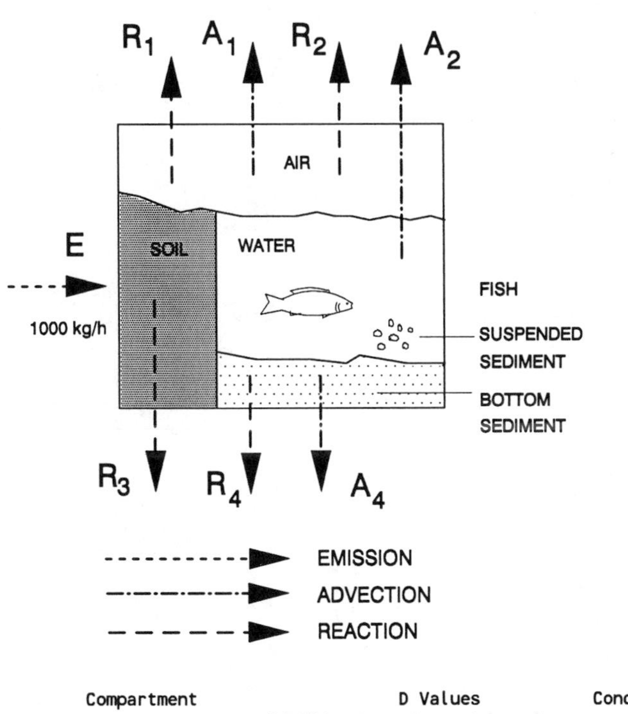


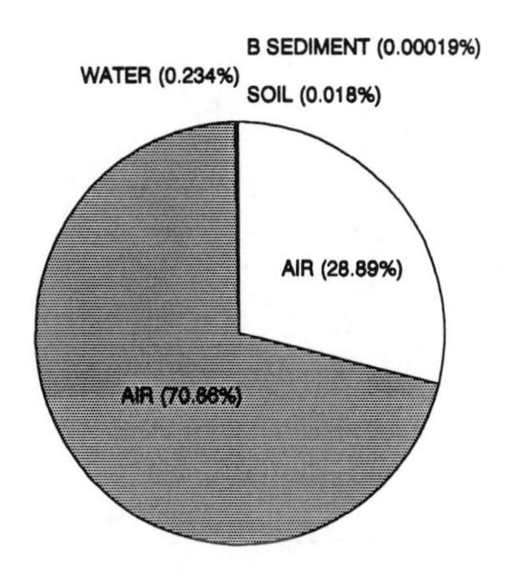


Distribution of removal rates

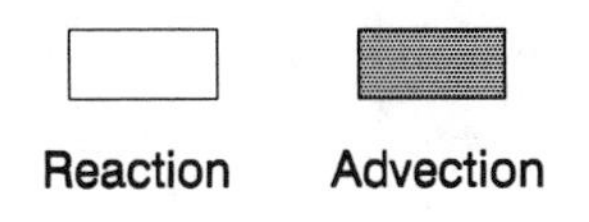


| Compartment | Half-life h | D Values Reaction mol/Pa h | D Values Advection mol/Pa h | Conc'n mol/m3 | Loss Reaction kg/h | Loss Advection kg/h | Removal % |
|---|---|---|---|---|---|---|---|
| Air | 170 | 1.64E+08 | 4.03E+08 | 4.51E-09 | 288.864 | 708.613 | 99.748 |
| Water | 1700 | 3.86E+05 | 9.46E+05 | 5.29E-08 | 0.6774 | 1.6618 | 0.2339 |
| Soil | 5500 | 1.03E+05 | | 1.02E-06 | 0.1812 | | 0.0181 |
| Biota (fish) | | | | 2.59E-06 | | | |
| Suspended sediment | | | | 6.36E-06 | | | |
| Bottom sediment | 17000 | 7.42E+02 | 3.64E+02 | 2.04E-06 | 1.30E-03 | 6.39E-04 | 1.94E-04 |
| | Total | 1.65E+08 | 4.04E+08 | | 289.72 | 710.28 | 100 |
| | R + A | | 5.69E+08 | | | 1000 | |

f = 1.119E-05 Pa

Total Amt= 73994 kg

Overall residence time = 73.99 h

Reaction time = 255.40 h

Advection time = 104.18 h

## Fugacity Level III calculations: (four-compartment model)

Chemical name: Bromobenzene

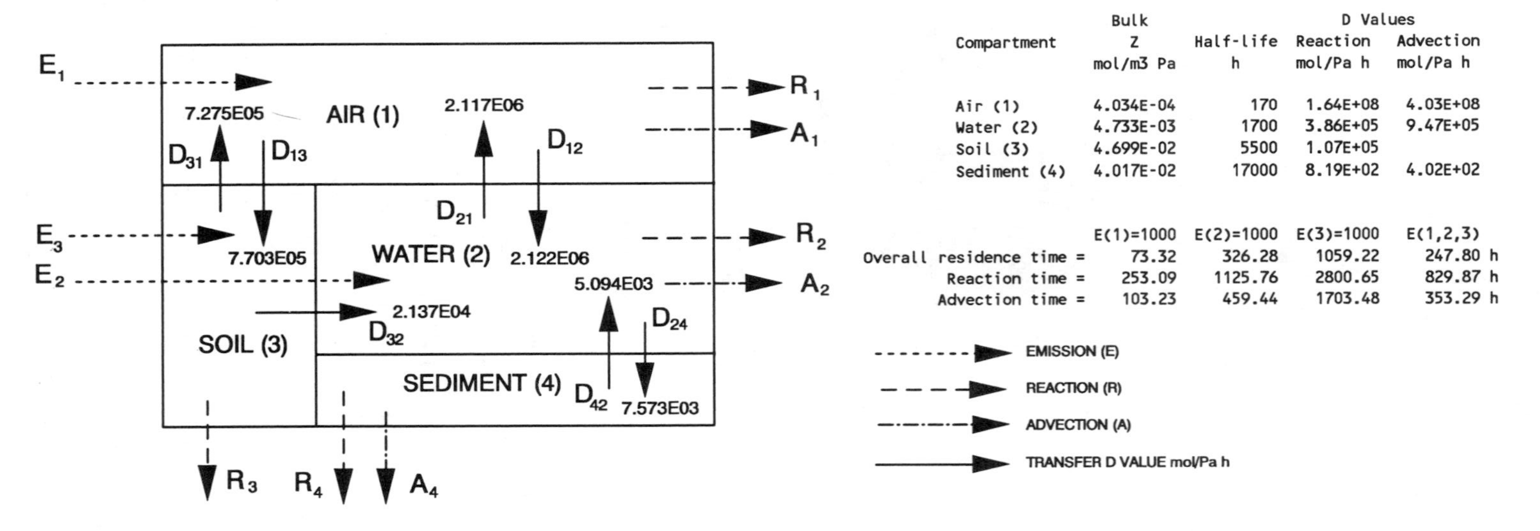


Phase Properties and Rates:

| Compartment | Bulk Z mol/m3 Pa | Half-life h | D Values Reaction mol/Pa h | D Values Advection mol/Pa h |
|---|---|---|---|---|
| Air (1) | 4.034E-04 | 170 | 1.64E+08 | 4.03E+08 |
| Water (2) | 4.733E-03 | 1700 | 3.86E+05 | 9.47E+05 |
| Soil (3) | 4.699E-02 | 5500 | 1.07E+05 | |
| Sediment (4) | 4.017E-02 | 17000 | 8.19E+02 | 4.02E+02 |

| | E(1)=1000 | E(2)=1000 | E(3)=1000 | E(1,2,3) |
|---|---|---|---|---|
| Overall residence time = | 73.32 | 326.28 | 1059.22 | 247.80 h |
| Reaction time = | 253.09 | 1125.76 | 2800.65 | 829.87 h |
| Advection time = | 103.23 | 459.44 | 1703.48 | 353.29 h |

EMISSION (E)
REACTION (R)
ADVECTION (A)
TRANSFER D VALUE mol/Pa h

Phase Properties, Compositions, Transport and Transformation Rates:

| Emission, kg/h | | | Fugacity, Pa | | | | Concentration, g/m3 | | | | Amounts, kg | | | | Total Amount, kg |
|---|---|---|---|---|---|---|---|---|---|---|---|---|---|---|---|
| E(1) | E(2) | E(3) | f(1) | f(2) | f(3) | f(4) | C(1) | C(2) | C(3) | C(4) | m(1) | m(2) | m(3) | m(4) | |
| 1000 | 0 | 0 | 1.120E-05 | 6.946E-06 | 1.008E-05 | 8.331E-06 | 7.093E-07 | 5.163E-06 | 7.439E-05 | 5.255E-05 | 7.093E+04 | 1.033E+03 | 1.339E+03 | 2.627E+01 | 7.332E+04 |
| 0 | 1000 | 0 | 6.868E-06 | 1.850E-03 | 6.185E-06 | 2.218E-03 | 4.351E-07 | 1.375E-03 | 4.563E-05 | 1.399E-02 | 4.351E+04 | 2.750E+05 | 8.214E+02 | 6.997E+03 | 3.263E+05 |
| 0 | 0 | 1000 | 9.694E-06 | 5.211E-05 | 7.454E-03 | 6.250E-05 | 6.140E-07 | 3.873E-05 | 5.499E-02 | 3.942E-04 | 6.140E+04 | 7.746E+03 | 9.899E+05 | 1.971E+02 | 1.059E+06 |
| 600 | 300 | 100 | 9.748E-06 | 5.643E-04 | 7.533E-04 | 6.767E-04 | 6.175E-07 | 4.194E-04 | 5.558E-03 | 4.269E-03 | 6.175E+04 | 8.388E+04 | 1.000E+05 | 2.134E+03 | 2.478E+05 |

| Emission, kg/h | | | Loss, Reaction, kg/h | | | | Loss, Advection, kg/h | | | Intermedia Rate of Transport, kg/h T12 | T21 | T13 | T31 | T32 | T24 | T42 |
|---|---|---|---|---|---|---|---|---|---|---|---|---|---|---|---|---|
| E(1) | E(2) | E(3) | R(1) | R(2) | R(3) | R(4) | A(1) | A(2) | A(4) | air-water | water-air | air-soil | soil-air | soil-water | water-sed | sed-water |
| 1000 | 0 | 0 | 2.891E+02 | 4.209E-01 | 1.69E-01 | 1.071E-03 | 7.093E+02 | 1.033E+00 | 5.255E-04 | 3.730E+00 | 2.309E+00 | 1.354E+00 | 1.152E+00 | 3.383E-02 | 8.260E-03 | 6.664E-03 |
| 0 | 1000 | 0 | 1.774E+02 | 1.121E+02 | 1.03E-01 | 2.852E-01 | 4.351E+02 | 2.750E+02 | 1.399E-01 | 2.288E+00 | 6.148E+02 | 8.307E-01 | 7.065E-01 | 2.075E-02 | 2.200E+00 | 1.774E+00 |
| 0 | 0 | 1000 | 2.503E+02 | 3.158E+00 | 1.25E+02 | 8.036E-03 | 6.140E+02 | 7.746E+00 | 3.942E-03 | 3.229E+00 | 1.732E+01 | 1.172E+00 | 8.514E+02 | 2.501E+01 | 6.197E-02 | 4.999E-02 |
| 600 | 300 | 100 | 2.517E+02 | 3.419E+01 | 1.26E+01 | 8.701E-02 | 6.175E+02 | 8.388E+01 | 4.269E-02 | 3.247E+00 | 1.876E+02 | 1.179E+00 | 8.605E+01 | 2.527E+00 | 6.710E-01 | 5.413E-01 |

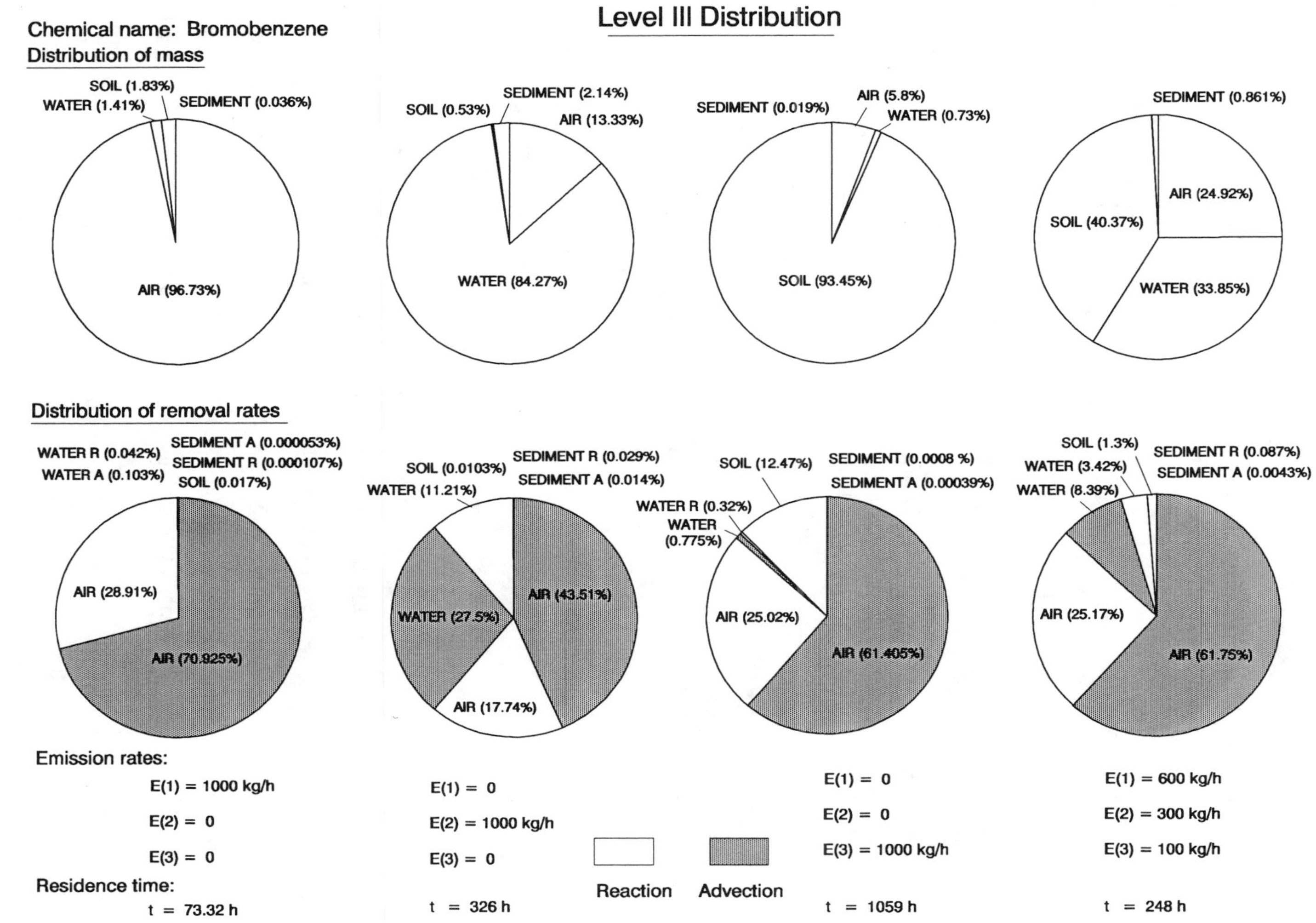
Level III Distribution
Chemical name: Bromobenzene
Distribution of mass
SOIL (1.83%)
WATER (1.41%)
SEDIMENT (0.036%)
AIR (96.73%)
SEDIMENT (2.14%)
SOIL (0.53%)
AIR (13.33%)
WATER (84.27%)
SEDIMENT (0.019%)
AIR (5.8%)
WATER (0.73%)
SOIL (93.45%)
SEDIMENT (0.861%)
AIR (24.92%)
SOIL (40.37%)
WATER (33.85%)
Distribution of removal rates
WATER R (0.042%)
WATER A (0.103%)
SEDIMENT A (0.000053%)
SEDIMENT R (0.000107%)
SOIL (0.017%)
AIR (28.91%)
AIR (70.925%)
SOIL (0.0103%)
WATER (11.21%)
SEDIMENT R (0.029%)
SEDIMENT A (0.014%)
AIR (43.51%)
WATER (27.5%)
AIR (17.74%)
SOIL (12.47%)
SEDIMENT (0.0008 %)
SEDIMENT A (0.00039%)
WATER R (0.32%)
WATER
(0.775%)
AIR (25.02%)
AIR (61.405%)
SOIL (1.3%)
WATER (3.42%)
WATER (8.39%)
SEDIMENT R (0.087%)
SEDIMENT A (0.0043%)
AIR (25.17%)
AIR (61.75%)
Emission rates:
E(1) = 1000 kg/h
E(2) = 0
E(3) = 0
Residence time:
t = 73.32 h
E(1) = 0
E(2) = 1000 kg/h
E(3) = 0
t = 326 h
Reaction
Advection
E(1) = 0
E(2) = 0
E(3) = 1000 kg/h
t = 1059 h
E(1) = 600 kg/h
E(2) = 300 kg/h
E(3) = 100 kg/h
t = 248 h

Chemical name: Iodobenzene

Level I calculation: (six-compartment model)

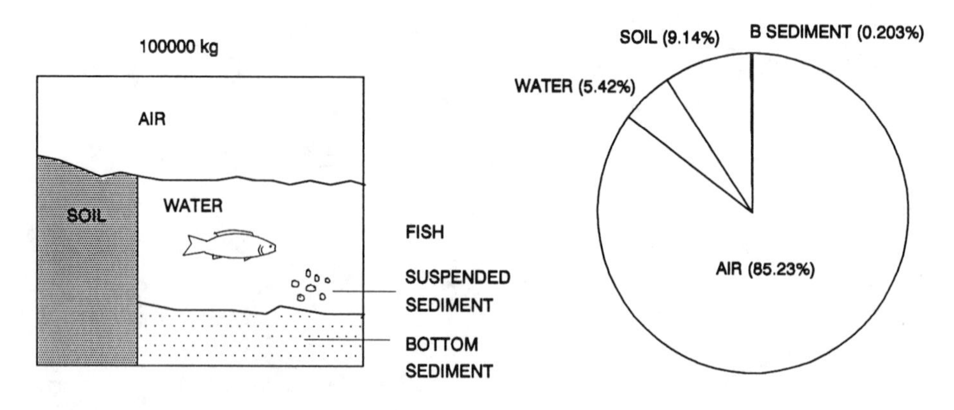


Distribution of mass

physical-chemical properties:

MW: 204.01 g/mol

M.P.: - 31.35 °C

Fugacity ratio: 1.0

vapor pressure: 130 Pa

solubility: 340 g/m $^3$

log $K_{OW}$: 3.28

| Compartment | Z | Concentration | | | Amount | Amount |
|---|---|---|---|---|---|---|
| | mol/m3 Pa | mol/m3 | mg/L (or g/m3) | ug/g | kg | % |
| Air | 4.034E-04 | 4.178E-09 | 8.523E-07 | 7.190E-04 | 85232 | 85.232 |
| Water | 1.282E-02 | 1.328E-07 | 2.709E-05 | 2.709E-05 | 5417.03 | 5.4170 |
| Soil | 4.807E-01 | 4.979E-06 | 1.016E-03 | 4.232E-04 | 9141.10 | 9.1411 |
| Biota (fish) | 1.221E+00 | 1.265E-05 | 2.580E-03 | 2.580E-03 | 0.5161 | 5.16E-04 |
| Suspended sediment | 3.005E+00 | 3.112E-05 | 6.348E-03 | 4.232E-03 | 6.3480 | 6.35E-03 |
| Bottom sediment | 9.615E-01 | 9.957E-06 | 2.031E-03 | 8.464E-04 | 203.136 | 0.2031 |
| | Total | | | | 100000 | 100 |

f = 1.036E-05 Pa

Chemical name: Iodobenzene

Level II calculation: (six-compartment model)

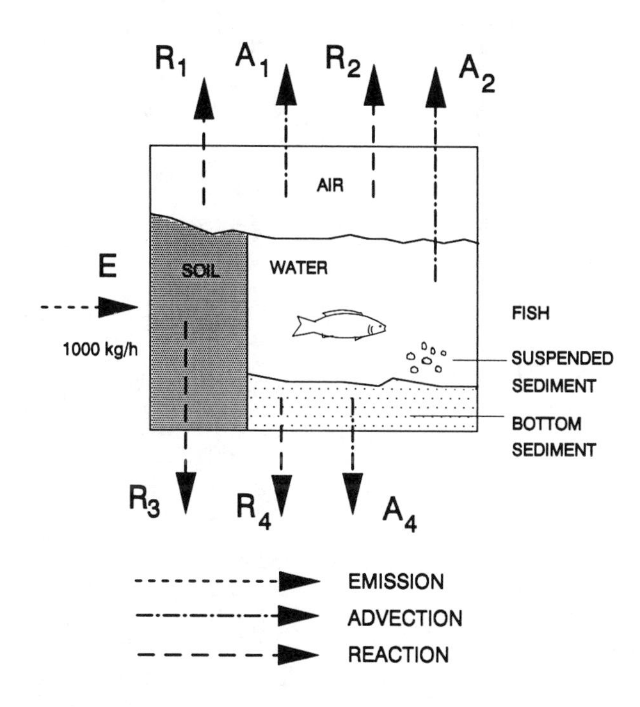


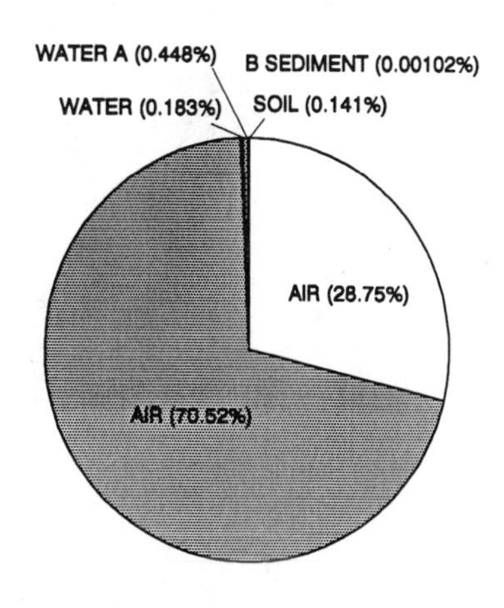


Distribution of removal rates

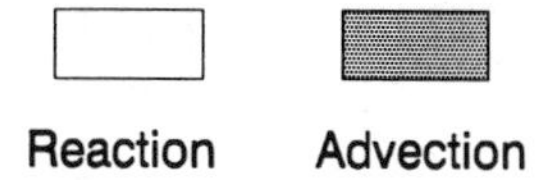


| Compartment | Half-life h | D Values Reaction mol/Pa h | D Values Advection mol/Pa h | Conc'n mol/m3 | Loss Reaction kg/h | Loss Advection kg/h | Removal % |
|---|---|---|---|---|---|---|---|
| Air | 170 | 1.64E+08 | 4.03E+08 | 3.46E-09 | 287.489 | 705.239 | 99.273 |
| Water | 1700 | 1.05E+06 | 2.56E+06 | 1.10E-07 | 1.8272 | 4.4822 | 0.6309 |
| Soil | 5500 | 5.45E+05 | | 4.12E-06 | 0.9530 | | 0.0953 |
| Biota (fish) | | | | 1.05E-05 | | | |
| Suspended sediment | | | | 2.57E-05 | | | |
| Bottom sediment | 17000 | 3.92E+03 | 1.92E+03 | 8.24E-06 | 6.85E-03 | 3.36E-03 | 1.02E-03 |
| | Total | 1.66E+08 | 4.06E+08 | | 290.28 | 709.72 | 100 |
| | R + A | | 5.72E+08 | | | 1000 | |

f = 8.569E-06 Pa
Total Amt= 82744 kg

Overall residence time = 82.74 h
Reaction time = 285.05 h
Advection time = 116.59 h

## Fugacity Level III calculations: (four-compartment model)

Chemical name: Iodobenzene

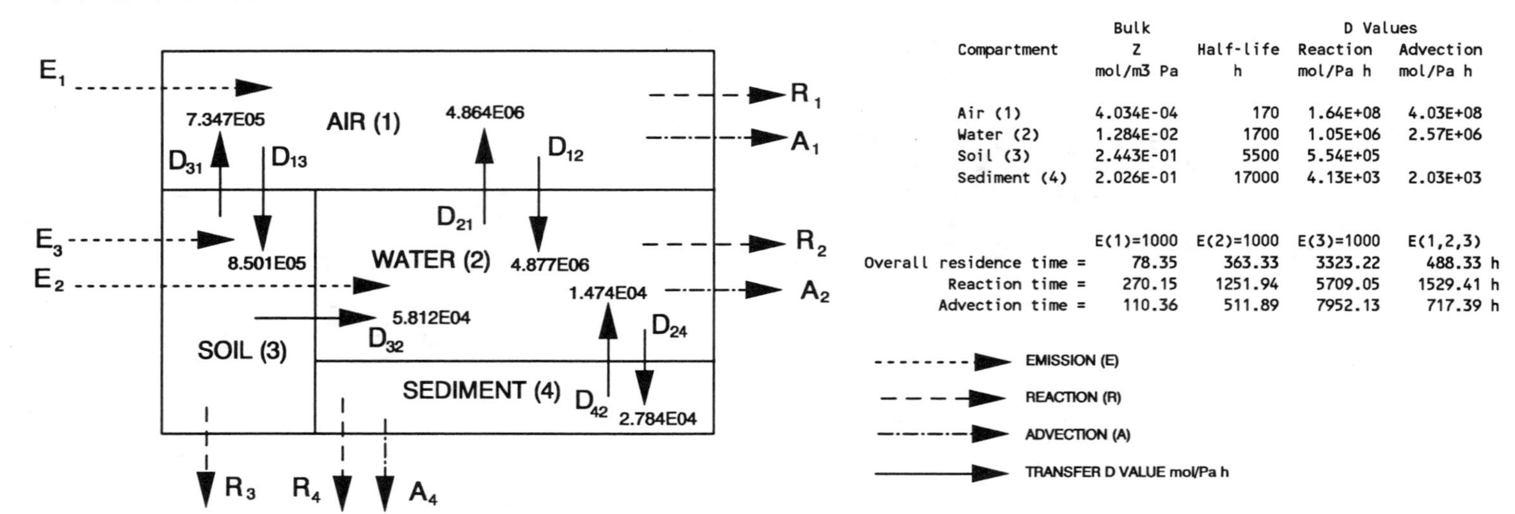


Phase Properties and Rates:

| Compartment | Bulk Z mol/m3 Pa | Half-life h | D Values Reaction mol/Pa h | Advection mol/Pa h |
|---|---|---|---|---|
| Air (1) | 4.034E-04 | 170 | 1.64E+08 | 4.03E+08 |
| Water (2) | 1.284E-02 | 1700 | 1.05E+06 | 2.57E+06 |
| Soil (3) | 2.443E-01 | 5500 | 5.54E+05 | |
| Sediment (4) | 2.026E-01 | 17000 | 4.13E+03 | 2.03E+03 |

| | E(1)=1000 | E(2)=1000 | E(3)=1000 | E(1,2,3) |
|---|---|---|---|---|
| Overall residence time = | 78.35 | 363.33 | 3323.22 | 488.33 h |
| Reaction time = | 270.15 | 1251.94 | 5709.05 | 1529.41 h |
| Advection time = | 110.36 | 511.89 | 7952.13 | 717.39 h |

Phase Properties, Compositions, Transport and Transformation Rates:

| Emission, kg/h | | | Fugacity, Pa | | | | Concentration, g/m3 | | | | Amounts, kg | | | | Total |
|---|---|---|---|---|---|---|---|---|---|---|---|---|---|---|---|
| E(1) | E(2) | E(3) | f(1) | f(2) | f(3) | f(4) | C(1) | C(2) | C(3) | C(4) | m(1) | m(2) | m(3) | m(4) | Amount, kg |
| 1000 | 0 | 0 | 8.595E-06 | 4.977E-06 | 5.431E-06 | 6.631E-06 | 7.074E-07 | 1.303E-05 | 2.707E-04 | 2.740E-04 | 7.074E+04 | 2.607E+03 | 4.872E+03 | 1.370E+02 | 7.835E+04 |
| 0 | 1000 | 0 | 4.926E-06 | 5.805E-04 | 3.113E-06 | 7.734E-04 | 4.055E-07 | 1.520E-03 | 1.551E-04 | 3.196E-02 | 4.055E+04 | 3.040E+05 | 2.793E+03 | 1.598E+04 | 3.633E+05 |
| 0 | 0 | 1000 | 4.901E-06 | 2.776E-05 | 3.642E-03 | 3.699E-05 | 4.033E-07 | 7.270E-05 | 1.815E-01 | 1.529E-03 | 4.033E+04 | 1.454E+04 | 3.268E+06 | 7.643E+02 | 3.323E+06 |
| 600 | 300 | 100 | 7.125E-06 | 1.799E-04 | 3.684E-04 | 2.397E-04 | 5.864E-07 | 4.711E-04 | 1.836E-02 | 9.905E-03 | 5.864E+04 | 9.422E+04 | 3.305E+05 | 4.953E+03 | 4.883E+05 |

| Emission, kg/h | | | Loss, Reaction, kg/h | | | | Loss, Advection, kg/h | | | Intermedia Rate of Transport, kg/h T12 | T21 | T13 | T31 | T32 | T24 | T42 |
|---|---|---|---|---|---|---|---|---|---|---|---|---|---|---|---|---|
| E(1) | E(2) | E(3) | R(1) | R(2) | R(3) | R(4) | A(1) | A(2) | A(4) | air-water | water-air | air-soil | soil-air | soil-water | water-sed | sed-water |
| 1000 | 0 | 0 | 2.884E+02 | 1.063E+00 | 6.14E-01 | 5.585E-03 | 7.074E+02 | 2.607E+00 | 2.740E-03 | 8.552E+00 | 4.939E+00 | 1.492E+00 | 8.140E-01 | 6.440E-02 | 2.827E-02 | 1.994E-02 |
| 0 | 1000 | 0 | 1.653E+02 | 1.239E+02 | 3.52E-01 | 6.514E-01 | 4.055E+02 | 3.040E+02 | 3.196E-01 | 4.902E+00 | 5.760E+02 | 8.554E-01 | 4.666E-01 | 3.691E-02 | 3.297E+00 | 2.326E+00 |
| 0 | 0 | 1000 | 1.644E+02 | 5.928E+00 | 4.12E+02 | 3.116E-02 | 4.033E+02 | 1.454E+01 | 1.529E-02 | 4.876E+00 | 2.755E+01 | 8.509E-01 | 5.459E+02 | 4.319E+01 | 1.577E-01 | 1.113E-01 |
| 600 | 300 | 100 | 2.390E+02 | 3.841E+01 | 4.16E+01 | 2.019E-01 | 5.864E+02 | 9.422E+01 | 9.905E-02 | 7.089E+00 | 1.785E+02 | 1.237E+00 | 5.522E+01 | 4.369E+00 | 1.022E+00 | 7.210E-01 |

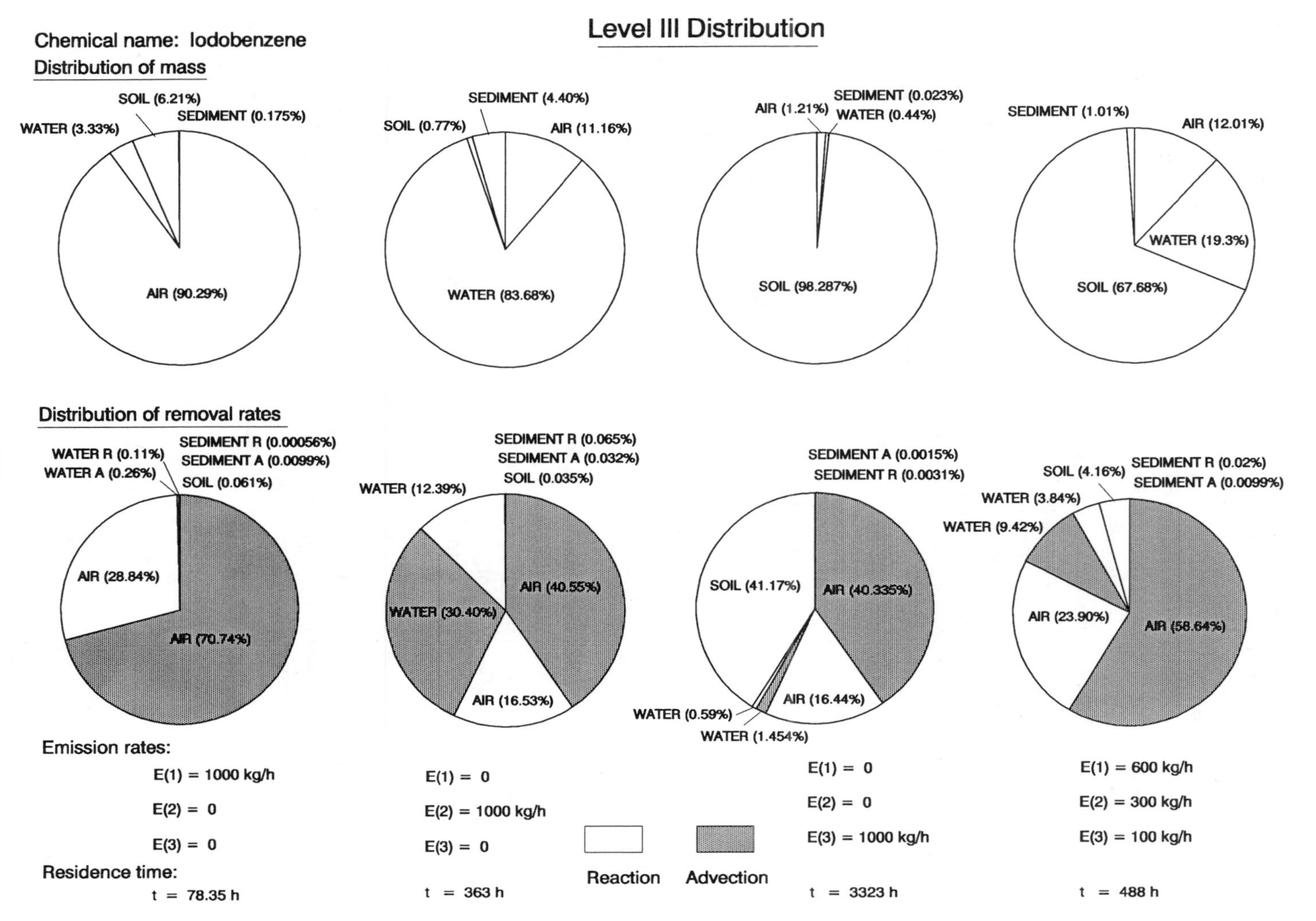
Level III Distribution
Chemical name: Iodobenzene
Distribution of mass
WATER (3.33%)
SOIL (6.21%)
SEDIMENT (0.175%)
AIR (90.29%)
SOIL (0.77%)
SEDIMENT (4.40%)
AIR (11.16%)
WATER (83.68%)
AIR (1.21%)
SEDIMENT (0.023%)
WATER (0.44%)
SOIL (98.287%)
SEDIMENT (1.01%)
AIR (12.01%)
WATER (19.3%)
SOIL (67.68%)
Distribution of removal rates
WATER R (0.11%)
WATER A (0.26%)
SEDIMENT R (0.00056%)
SEDIMENT A (0.0099%)
SOIL (0.061%)
AIR (28.84%)
AIR (70.74%)
WATER (12.39%)
SEDIMENT R (0.065%)
SEDIMENT A (0.032%)
SOIL (0.035%)
AIR (40.55%)
WATER (30.40%)
AIR (16.53%)
SEDIMENT A (0.0015%)
SEDIMENT R (0.0031%)
SOIL (41.17%)
AIR (40.335%)
AIR (16.44%)
WATER (0.59%)
WATER (1.454%)
SOIL (4.16%)
SEDIMENT R (0.02%)
SEDIMENT A (0.0099%)
WATER (3.84%)
WATER (9.42%)
AIR (23.90%)
AIR (58.64%)
Emission rates:
E(1) = 1000 kg/h
E(2) = 0
E(3) = 0
Residence time:
t = 78.35 h
E(1) = 0
E(2) = 1000 kg/h
E(3) = 0
t = 363 h
Reaction
Advection
E(1) = 0
E(2) = 0
E(3) = 1000 kg/h
t = 3323 h
E(1) = 600 kg/h
E(2) = 300 kg/h
E(3) = 100 kg/h
t = 488 h

## 3.4 COMMENTARY ON THE PHYSICAL-CHEMICAL PROPERTIES AND ENVIRONMENTAL FATE

**Reactivity**

The half-life classes in Table 3.3 reflect the generally slow reaction rates of halogenated hydrocarbons in the atmosphere. Reaction is probably mainly with hydroxyl (OH) radicals. Increasing halogenation seems to prolong the half-life. The unsaturated compounds are more reactive. These rates are so slow that advection in air dominates the chemicals' fate when viewed in a multimedia context. When discharged to soil or water the rate of evaporation usually controls the overall residence time, because once evaporated these substances are rapidly removed by advection. The water class was selected to be 6 (usually) for the saturated compounds, 5 for the unsaturated compounds, with longer half-lives for some freons. The soil is a further one class slower than water, and with bottom sediment 2 classes slower than water. This class of compounds is thus among the least reactive and longest-lived contaminants. As is discussed later, it must be appreciated that advective removal, which is usually the dominant mechanism of loss, does not destroy the compound, it merely relocates it. This series thus has the potential for long distance transport in the atmosphere and especially to the stratosphere where the chlorine released may reduce ozone concentrations. The obvious remedy is to substitute the long-lived compounds with others which decay in the troposphere to relatively harmless chloride and fluoride ions.

**QSPR Plots**

The QSPR plots show the usual trends of property changes within this group of chemicals. There are obvious differences between halogenated alkanes and alkenes.

Figures 3.1a and 3.1b show that the LeBas molar volume is well correlated with other molecular volumes and molecular areas and is thus an adequate descriptor for most purposes.

The solubility data for chloroalkanes (Figure 3.2a) show the steady drop in solubility with molecular size which corresponds to a factor of 3.0 (0.47 log units) per 25 $cm^3/mol$ added. This is approximately the volume of a chlorine or methyl group. The other halogens show a similar trend, but there is more scatter. There appears to be a trend for the bromo- and especially the iodo-compounds to have lower solubilities at the same molar volume. This is not surprising because the iodo-group is significantly larger in volume than the chlorine (37 versus 24.6 $cm^3/mol$). The more sophisticated descriptors such as TSA may be more successful in correlating the range of halogens. There is an obvious need to correlate solubility separately for each halogen when using only molar volume as a single descriptor.

The vapor pressure data in Figures 3.3a and 3.3b show a similar trend with each 25 $cm^3/mol$ addition causing a drop in vapor pressure by a factor of about 5.9 or 0.77 log units. Again, the nature of the halogen substituent in the molecule influences vapor pressure, there being a factor of about 10 spread at any given molar volume. The iodo compounds display lower vapor pressures.

The log $K_{OW}$ data in Figures 3.4a and 3.4b show the expected increase in hydrophobicity which is largely attributable to the drop in solubility in water, or equivalently the increase in activity coefficient in the water phase. The slope corresponds to a factor of about 3.9 or 0.59 log units per 25 $cm^3/mol$ added. The similarity in slope to that of water solubility suggests that these compounds have approximately equal solubilities, or affinities for the octanol phase and the variation in $K_{OW}$ is primarily due to changes in water solubility.

The Henry's law constants (H) in Figures 3.5a and 3.5b show a fairly wide scatter with most compounds having values of 2.0 to 3.0 in log H, or 100 to 1000 $Pa \cdot m^3/mol$ in H. This corresponds to an air-water partition coefficient, $K_{AW}$ ranging from 0.04 to 0.40; thus these compounds have a fairly strong partitioning tendency into air. There is a slight tendency for H to decrease with molar volume because the solubility drops less than the vapor pressure.

The log $K_{OW}$ - solubility plots in Figures 3.6a and 3.6b show slopes of about 0.90 confirming the almost reciprocal linear relationship between $K_{OW}$ and solubility.

The general picture which emerges from these plots is that each chlorine substitution for hydrogen produces a drop in solubility and an increase in vapor pressure and in $K_{OW}$ of about a factor of about 5, thus H is relatively unchanged. Substitution with bromine with its 10% larger molar volume produces a factor of about 6 and iodine with its much larger volume (by about 50%) produces about a factor of 11 change. Fluorine, which has only about one third the molar volume of chlorine tends to be anomalous. Although it is interesting to explore these QSPRs, especially to probe the fundamental causes of the phenomena, there is little to gain from a data acquisition viewpoint because experimental measurement is relatively straightforward and the number of candidate chemicals is limited. There is thus no substitute for accurate experimentally determined values.

These plots can be used, with appropriate caution, to deduce the properties of other volatile hydrocarbons. The preferred procedure is to plot the property as a function of molar volume for a group of structurally similar compounds and interpolate. Extrapolation brings a high degree of uncertainty.

**Evaluative Calculations**

Rather than give diagrams for all the chemicals in this class, only a selection is presented. The fate of the others can be inferred from those of the closest homolog for which a diagram is given.

The Level I calculations show almost total partitioning into the atmosphere with only minor amounts into the other compartments. The chemicals with higher $K_{OW}$, notably the aromatics, tend to partition slightly into soil.

The Level II diagrams show the dominance of atmospheric advection and in some cases, reaction. Even if soil, water, or sediment rate constants were increased by a factor of 100 (which is difficult to conceive) the rates of loss from these media would remain negligible. Obviously, these chemicals tend to become widely distributed.

The Level III diagrams show that the chemical discharged into air experiences very little tendency to enter soil or water because the rates of loss by advection and reaction in air greatly exceed the rates of absorption or deposition. The overall residence times are generally about 3 days as controlled by the assumed advection rate.

When discharged to water there is usually competition between evaporation, advection from water and reaction in water. The more hydrophobic chemicals also tend to partition into sediments. The overall residence times are much longer (e.g. 1 to 2 weeks) as a result of slower reaction and advection in water and the constraint imposed by evaporation. Essentially, it is diffusion through the boundary layer in the water and the air at the air-water interface which limits the evaporation rate. Substances of low H tend to be particularly slow to evaporate.

When discharged to soil, most chemical evaporates and again the rate of loss is limited by diffusion processes, but in this case, at the soil-air interface.

The Level III pie charts show these trends pictorially.

The aromatics are somewhat anomalous in this context and they are more appropriately assessed along with the aromatic hydrocarbons which were reviewed in Volume I of this series.

From the perspective of the environmental impact and toxicology of this series, it is apparent that regardless of how the volatile organic chemicals (VOC) are discharged, much will eventually enter the atmosphere. In the atmosphere they tend to be long-lived and become subject to long-distance transport, including transport to the stratosphere. The aquatic and soil environments are usually susceptible only when the VOC, are discharged directly to these media. Of particular concern are situations in which evaporation is impeded, as occurs to VOCs deep in soils, often resulting from solvent discharges. Apparently, the direct ecotoxicological effects of these compounds in the soil or water ecosystems result from direct discharges to these ecosystems, and not the result of air emissions. The evaluative model used here does not do justice to the fate of chemicals in the subsurface (groundwater) environments. There is an extensive and growing literature on the transport and transformation of chlorinated solvents in this region which is a highly specialized "single media" issue and is beyond the scope of this volume. The primary contribution of multimedia modelling to VOCs is to place them in a multimedia perspective and highlight the importance of the more detailed assessments needed to elucidate their behavior in the subsurface environment.

## 3.5 REFERENCES

Abernethy, S., Mackay, D. (1987) A discussion of correlations for narcosis in aquatic species. In: *QSAR in Experimental Toxicology II.* Kaiser, K.L.E., Editor, pp.1-16, Reidel Publication Co., Dordrecht, Holland.

Abernethy, S., Mackay, D., McCarty, L.S. (1988) "Volume fraction" correlation for narcosis in aquatic organisms: the key role of partitioning. *Environ. Toxicol & Chem.* 7, 469-481.

Afghan, B.K., Mackay, D., Editors (1980) *Hydrocarbons and Halogenated Hydrocarbons in the Aquatic Environment.* Environmental Science Research, Volume 16, Plenum Press, New York.

Altshuller, A.P. (1980) Lifetimes of organic molecules in the troposphere and lower stratosphere. *Adv. Environ. Sci. Technol.* 10, 181-217.

Amidon, G.L., Williams, N.A. (1982) A solubility equation for non-electrolytes in water. *Int'l J. Pharm.* 11, 249-256.

Amidon, G.L., Yalkowsky, S.H., Anik, S.T., Valvani, S.C. (1975) Solubilities of nonelectrolytes in polar solvents. V. Estimation of the solubility of aliphatic monofunctional compounds in water using a molecular area approach. *J. Phys. Chem.* 79, 2239-2245.

Andelman, J.B. (1978) *Chloroform, Carbon Tetrachloride, and Other Halomethanes: An Environmental Assessment.* National Academy of Sciences, Washington.

Anderson, T.A., Beauchamp, J.J., Walton, B.T. (1991) Organic chemicals in the environment. *J. Environ. Qual.* 20, 420-424.

Anderson, Y., Ljungström, E. (1989) Gas phase reaction of the $NO_3$ radical with organic compound in the dark. *Atmos. Environ.* 23, 1153-1155.

Andrews, L.J., Keefer, R.M. (1950) Cation complexes of compounds containing of carbon-carbon double bonds. VI. The argentation of substituted benzenes. *J. Am. Chem. Soc.* 72, 3110-3116.

Arbuckle, W.B. (1983) Estimating activity coefficients for use in calculating environmental parameters. *Environ. Sci. Technol.* 17, 537-542.

Arbuckle, W.B. (1986) Using UNIFAC to calculate aqueous solubilities. *Environ. Sci. Technol.* 20(10) 1060-1064.

Archer, W.L., Stevens, V.L. (1977) Comparison of chlorinated aliphatic, aromatic, and oxygenated hydrocarbons as solvents. *Ind. Eng. Chem. Prod. Res. Dev.* 16(4), 319.

Aref'eva, R.P., Korenman, I.M., Gorokhov, A.A. (1979) Determination of the solubility of liquid and solid organic substances in water, using diphenylthiocarbazone. *USSR Patent* 672548(July 5).

Ashworth, R.A., et al. (1986) Air-water partitioning coefficients of organics in dilute aqueous solutions. Paper presented at the American Institute of Chemical Engineers 1986 National Meeting in Boston, Mass.

Ashworth, R.A., Howe, G.B., Mullins, M.E., Rogers, T.N. (1988) Air-water partitioning coefficients of organics in dilute aqueous solutions. *J. Hazard. Materials* 18, 25-36.

Atkinson, R. (1985) Kinetics and mechanisms of the gas-phase reactions of hydroxyl radical with organic compounds under atmospheric conditions. *Chem. Rev.* 85, 69-201.

Atkinson, R. (1987) Structure-activity relationship for the estimation rate constants for the gas phase reaction of OH radicals with organic compounds. *Int'l. J. Chem. Kinetics* 19, 799-828.

Atkinson, R. (1988) Estimation of gas-phase hydroxyl radical rate constants for organic chemicals. *Environ. Toxicol. & Chem.* 7, 435-442.

Atkinson, R. (1991) Kinetics and mechanisms of the gas-phase reactions of the $NO_3$ radical with organic compounds. *J. Phys. Chem. Ref. Data* 20, 459-507.

Atkinson, R., Ashman, S.M., Carter, W.P.L., Winer, A.M., Pitts Jr., J.N. (1982) Kinetics of the reactions of OH radicals with n-alkanes at 299 $\pm$ 2K. *Int'l. Chem. Kinetics* 14, 781-788.

Atkinson, R., Ashman, S.M., Fitz, D.R., Winer, A.M., Pitts Jr., J.N. (1982) Rate constants for the gas-phase reactions of $O_3$ with selected organics at 296 K. *Int'l. Chem. Kinetics* 14, 13-18.

Atkinson, R., Ashman, S.M., Goodman, M.A. (1987) Kinetics of the gas-phase reactions of $NO_3$ radicals with a series of alkynes, haloalkenes, and $\alpha,\beta$-unsaturated aldehydes. *Int'l. Chem. Kinetics* 19, 299-307.

Atkinson, R., Ashman, S.M., Pitts Jr., J.N. (1988) Rate constants for the gas-phase reactions of the $NO_3$ radicals with a series of organic compounds at 296 $\pm$ 2 K. *J. Phys. Chem.* 92, 3454-3457.

Atkinson, R., Ashman, S.M., Winer, A.M., Carter, W.P.L. (1985) Rate constants for the gas-phase reactions of $NO_3$ radicals with furan, thiophene, and pyrrole at 295 $\pm$ 1 K and atmosphere pressure. *Environ. Sci. Technol.* 19, 87-90.

Atkinson, R. Darnall, K.R., Lloyd, A.C., Winer, A.M., Pitts Jr., J.N. (1979) Kinetics and mechanisms of the reaction of the hydroxyl radicals with organic compounds in the gas phase. In: *Advances in Photochemistry.* Volume II, Pitts Jr., J.N., Hammond, G.S., Golnick, K., Editors, John Wiley & Sons, New York. pp.375-488.

Atkinson, R., Carter, W.P.L. (1984) Kinetics and mechanisms of gas-phase reactions of ozone with organic compounds under atmospheric conditions. *Chem. Rev.* 84, 437-470.

Atkinson, R., Hansen, D.A., Pitts Jr., J.N. (1975) Rate constants for the reaction of hydroxyl radicals with chlorodifluoromethane, dichlorodifluoromethane, trichlorofluoromethane and molecular hydrogen over temperature range 297-434 K. *J. Chem. Phys.* 63, 1703-1706.

Aviado, D.M., Zakhari, S., Siman, J.A., Ulsamer, A.G. (1976) *Methyl chloroform and Trichloroethylene in the Environment.* Chemical Rubber Co., Cleveland, Ohio.

Bahnick, D.A., Doucette, W.J. (1988) Use of molecular connectivity indices to estimate soil sorption coefficients for organic chemicals. *Chemosphere* 17(9), 1703-1715.

Ball, W.P., Roberts, P.V. (1991) Long term sorption of halogenated organic chemicals by aquifer materials. 1. Equilibrium. *Environ. Sci. Technol.* 25(7), 1223-1237.

Ball, W.P., Roberts, P.V. (1992) Comment on "Long term sorption of halogenated organic chemicals by aquifer materials. 1. Equilibrium". *Environ. Sci. Technol.* 26(11), 2301-2302.

Banerjee, S. (1985) Calculation of water solubility of organic compounds with UNIFAC-derived parameters. *Environ. Sci. Technol.* 19,369-370.

Banerjee, S., Baughman, G.L. (1991) Bioconcentration factors and lipid solubility. *Environ. Sci. Technol.* 25(3), 536-539.

Banerjee, S., Howard, P.H. (1988) Improved estimation of solubility and partitioning through correction of UNIFAC-derived activity coefficients. *Environ. Sci. Technol.* 22(7), 839-841.

Banerjee, S., Howard, P.H., Lande, S.S. (1990) General structure-vapor pressure relationships for organics. *Chemosphere* 21(10-11), 1173-1180.

Banerjee, S., Yalkowsky, S.H., Valvani, S.C. (1980) Water solubility and octanol/water partition coefficient of organics. Limitations of solubility-partition coefficient correlation. *Environ. Sci. Technol.* 14, 1227-1229.

Barrio-Lage, G., Parsons, F.Z., Nassar, R.S. (1987) Kinetics of the depletion of trichloroethene. *Environ. Sci. Technol.* 21, 366-370.

Barrio-Lage, G., Parsons, F.Z., Nassar, R.S., Lorenzo, P.A. (1986) Sequential dehalogenation of chlorinated ethenes. *Environ. Sci. Technol.* 20(1), 96-99.

Barrows, M.E., Petrocelli, S.R., Macek, K.J. (1980) Bioconcentration and elimination of selected water pollutants by bluegill sunfish (*lepomis machrochirus*). In: *Dynamic, Exposure, Hazard Assesstment of Toxic Chemicals.* Haque, R., Editor, pp.379-392, Ann Arbor Science Publishers Inc., Ann Arbor, Michigan.

Bissell, T.G., Williamson, A.G. (1975) Vapour pressures and excess Gibbs energies of *n*-hexane and of *n*-heptane + carbon tetrachloride and + chloroform at 298.15 K. *J. Chem. Thermodynamics* 7, 131-136.

Blume, C.W., Hisatsune, I.C., Heicklen, J. (1976) Gas phase ozonolysis of *cis*- and *trans*-dichloroethylene. Int'l J. Chem. Kinet. 8, 235-258.

Bobra, A.M., Shiu, W.Y., Mackay, D. (1984) Structure-activity relationships for toxicity of hydrocarbons, chlorinated hydrocarbons and oils to *daphnia magna*. In: *QSAR in Experimental Toxicology.* Kaiser, K.L.E., Editor, pp.3-16, Reidel Publishing Company, Dordrecht, Holland.

Bodor, N., Garbanyi, Z., Wong, C.K. (1989) A new method for the estimation of partition coefficient. *J. Am. Chem. Soc.* 111, 3783-3786.

Booth, H.S., Everson, H.E. (1948) Hydrocarbon solubilities: solubilities in 40 per cent sodium xylenesulfonate. *Ind. Eng. Chem.* 40(8), 1491-1493.

Boublik, T., Fried, V., Hala, E. (1973) *The Vapor Pressures of Pure Substances.* Elsevier, Amsterdam.

Boublik, T., Fried, V., Hala, E. (1984) *The Vapor Pressures of Pure Substances.* 2nd Edition, Elsevier, Amsterdam.

Bouwer, E.J., McCarty, P.L. (1983) Transformations of 1- and 2-carbon halogenated aliphatic organic compounds under methanogenic conditions. *Appl. Environ. Microbiol.* 45, 1286-1294.

Bouwer, E.J., McCarty, P.L., Bouwer, H., Rice, R.C. (1984) Organic contaminant behavior during rapid infiltration of secondary wastewater at Phoenix 23rd Avenue Project. *Water Res.* 18, 463-472.

Bouwer, E.J., Rittman, B., McCarty, P.L. (1981) Anaerobic degradation of 1- and 2-carbon halogenated aliphatic organic compounds. *Environ. Sci. Technol.* 15, 596-599.

Bradley, R.S., Cleasby, T.G. (1953) The vapor pressure and lattice energy of some aromatic ring compounds. *J. Chem. Soc.* 1953, 1690-1692.

Broholm, K., Cherry, J.A., Feenstra, S. (1992) Dissolution of heterogeneously distributed solvents residuals. In: *Proceedings of Subsurface Restoration Conference, 3rd Int'l Conf. on Groundwater Quality Research*, June 21-24, 1992, Dallas, Texas.

Brüggemann, R., Trapp, S., Matthies, M. (1991) Behavior assessment of a volatile chemical in the Rhine River. *Environ. Toxicol. & Chem.* 10, 1097-1103.

Bunce, N.J., Nakai, J.S., Yawching, M. (1991) Estimates of the tropospheric lifetimes of short- and long-lived atmospheric pollutants. *J. Photochem. Photobiol. A: Chem.* 57, 429-439.

Burkhard, L.P., Kuehl, D.W., Veith, G.D. (1985) Evaluation of reverse phase liquid chromatography/mass spectrometry for estimating of *n*-octanol/water partition coefficients. *Chemosphere* 14(10), 1551-1560.

Burns, L.H. (1981) *Exposure Analysis Modeling System.* Environmental Research Laboratory, U.S. Environmental Protection Agency.

Butz, R.G., Yu, C.C., Atallah, Y.H. (1982) Photolysis of hexachlorocyclopentadiene in water. *Ecotox. Environ. Safety* 6, 347-357.

Buxton, G.V., Greenstock, C.L., Helman, W.P., Ross, A.B. (1988) Critical review of rate constants for reactions of hydrated electrons, hydrogen atoms and hydroxyl radicals in aqueous solution. *J. Phys. Chem. Ref. Data* 17, 513-886.

Bysshe, S.E. (1982) Bioconcentration factor in aquatic organisms. Chapter 5. In: *Handbook of Chemical Property Estimation Methods.* Lyman, W.J., Reehl, W.F., Rosenblatt, D.H., Editors, McGraw-Hill Book Company, New York.

Cadena, F., Eiceman, G.A., Vandiver, V.J. (1984) Removal of volatile organic pollutants from rapid streams. *J. Water Pollut. Control Fd.* 56, 460-463.

Callahan, M.A., Slimak, M.W., Gabel, N.W., May, I.P., Fowler, C.F., Freed, J.R., Jennings, P., Durfee, R.L., Whitemore, F.C., Maestri, B., Mabey, W.R., Holt, B.R., Gould, C. (1979) *Water-Related Environmental Fate of 129 Priority Pollutants.* Vol.II, Halogenated aliphatic hydrocarbons, halogenated ethers, monocyclic aromatics, phthalate esters, polycyclic aromatic hydrocarbons, nitrosoamines and miscellaneous compounds. EPA-440/4-79-029b.

Castro, C.E., Belser, N.O. (1968) Biodehalogenation. Reductive dehalogenation of biocides ethylene dibromide, 1,2-dibromo-3-chloropropane, and 2,3-dibromobutane in soil. *Environ. Sci. Technol.* 2, 779-783.

Castro, C.E., Belser, N.O. (1981) Photohydrolysis of methyl bromide and chloropicrin. *J. Agric. Food & Chem.* 29, 1005-1008.

CEQ (1975) *Council on Environment Quality. "Fluorocarbons and the Environment"* Report of Federal Task Force on Inadvertent Modification of the Stratosphere (IMOS). U. S. Government Printing Office, Washington DC.

Chang, J.S., Kaufman, F. (1977) Kinetics of the reactions of hydroxyl radicals with some halocarbons: $CHFCl_2$, $CHF_2Cl$, $CH_3CCl_3$, $C_2HCl_3$ and $C_2Cl_4$. *J. Chem. Phys.* 66, 4989.

Chiou, C.T. (1981) Partition coefficient and water solubility in environmental chemistry. In: *Hazard Assessment of Chemicals, Current Developments.* Volume 1, Academic Press, New York. pp.117-153.

Chiou, C.T. (1985) Partition coefficients of organic compounds in lipid-water systems and correlations with fish bioconcentration factors. *Environ. Sci. Technol.* 19, 57-62.

Chiou, C.T., Freed, V.H. (1977) *Chemodynamics Studies on Bench Mark Industrial Chemicals.* NSF/RA-770286. National Science Foundation, Washington D.C.

Chiou, C.T., Freed, V.H., Schmedding, D.W., Kohnert, R. (1977) Partition coefficient and bioaccumulation of selected organic chemicals. *Environ. Sci. Technol.* 11(5), 475-478.

Chiou, C.T., Kile, D.E., Malcolm, R.L. (1988) Sorption of vapors of some organic liquids on soil humic acid and its relation to partitioning of organic compounds in soil organic matter. *Environ. Sci. Technol.* 22(3), 298-303.

Chiou, C.T., Peters, L.J., Freed, A.H. (1979) A physical concept of soil-water equilibria for nonionic organic compounds. *Science* 206, 831-832.

Chiou, C.T., Schmedding, D.W. (1981) Measurement and determination of octanol-water partition coefficient and water solubility of organic chemicals. In: *Test Protocols for Environmental Fate and Movement of Toxicants. J. Assoc. Anal. Chem.* , Arlington, Virginia.

Chiou, C.T., Schmedding, D.W., Manes, M. (1982) Partitioning of organic compounds in octanol-water systems. *Environ. Sci. Technol.* 16, 4-10.

Chiou, C.T., Freed, V.H., Peters, L.J., Kohnert, R.L. (1980) Evaporation of solutes from water. *Environ. Inter.* 3, 231-236.

Chou, C.C., Milstein, R.J., Smith, W.S., Vera Ruiz, H., Molina, M.J., Rowland, F.S. (1978) Stratospheric photodissociation of several saturated perhalochlorofluorocarbon compounds in current technological use (fluorocarbons-13, -113, -114, and -115). *J. Phys. Chem.* 82, 1-7.

Chou, J.T., Jurs, P.C. (1979) Computation of partition coefficients from molecular structures by a fragment addition method. In: *Physical Chemical Properties of Drugs. Medical Research Series, Vol.10.* Yalkowsky, S.H., Sindula, A.A., Valvani, S.C., Editors, Marcel Dekker, Inc., New York. pp.163-199.

Cline, P.V., Delfino, J.J. (1987) Am. Chem. Soc., Division of Environmental Chemistry, Preprint 27, 577-579. New Orleans, La.

Coca, J., Diaz, R. (1980) Extraction of furfural from aqueous solutions with chlorinated hydrocarbons. *J. Chem. Eng. Data* 25, 80-83.

Cohen, Y., Ryan, P.A. (1985) Multimedia modeling of environmental transport: Trichloroethylene test case. *Environ. Sci. Technol.* 19, 412-417.

Collander, R. (1951) Partition of organic compounds between higher alcohols and water. *Acta. Chem. Scand.* 5, 774-780.

Connell, D.W., Hawker, D.W. (1988) Use of polynomial expressions to describe the bioconcentration of hydrophobic chemicals by fish. *Ecotoxicol. Environ. Safety* 16, 242-257.

Cooper, W.J., Mehran, M., Riusech, D.J., Joens, J.A. (1987) Abiotic transformation of halogenated organics. 1. Elimination reaction of 1,1,2,2-tetrachloroethane and formation of 1,1,2-trichloroethene. *Environ. Sci. Technol.* 21(11), 1112-1114.

Cowen, W.F., Baynes, R.K. (1980) Estimated application of gas chromatographic headspace analysis for priority pollutants. *J. Environ. Sci. Health* A15, 413-427.

Cox, R.A., Derwent, R.G., Eggleton, A.E.J., Lovelock, J.E. (1976) Photochemical oxidation of halocarbons in the troposphere. *Atmos. Environ.* 10, 305-308.

Cramer, R.D. (1980) BC(DCF) parameters. I. The intrinsic dimensionality of intermolecular interactions in the liquid state. *J. Am. Chem. Soc.* 102, 1837.

Cramer, R.D. (1980) BC(DCF) parameters. II. An empirical structure-based scheme for the prediction of some physical properties. *J. Am. Chem. Soc.* 102, 1849.

CRC (1962) *Handbook of Chemistry and Physics*. 44th Ed., CRC Press, Boca Raton, Florida.

Cupitt, L.T. (1980) *Fate of Toxic and Hazardous Materials in the Environment.* EPA-600/3-80-084. U.S. Environmental Protection Agency, Research Triangle Park, North Carolina.

Curtis, G.P., Reinhard, M., Roberts, P.V. (1986) In: *Geochemical Processes at Mineral Surfaces*. pp.191-216. Davis, J.A., Hayes, K.F., Editors, ACS Symposium Series 323, American Chemical Society, Washington DC.

D'Amboise, M., Hanai, T. (1982) Hydrophobicity and retention in reversed phase liquid chromatography. *J. Liquid Chromatogr.* 5, 229-244.

Daniels, S.L., Hoerger, F.D., Moolenaar, R.J. (1985) Environmental exposure assessment: Experience under the toxic substances control act. *Environ. Toxicol. & Chem.* 4, 107-117.

Darnell, K.R., Lloyd, A.C., Winer, A.M., Pitts, J.N. (1976) Reactivity scale for atmospheric hydrocarbons based on reaction with hydroxyl radical. *Environ. Sci. Technol.* 10, 692-696.

Daubert, T.E., Danner, R.P. (1985) *Data Compilation Tables of Properties of Pure Compounds.* American Institute of Chemical Engineering.

Davidson, J.A., Schiff, H.I., Brown, T.J., Howard, C.J. (1978) Temperature dependence of the rate constant for reactions of oxygen (1D) atoms with a number of hydrocarbons. *J. Chem. Phys.* 69, 4277-4279.

Davies, R.P., Dobbs, A.J. (1984) The prediction of bioconcentration in fish. *Water Res.* 18, 1253-1262.

Davis, D.D., Machado, G., Conaway, B., Oh, Y., Watson, R. (1976) A temperature dependent kinetic study of reaction of hydroxyl radicals with chloromethane, dichloromethane, trichloromethane and bromomethane. *J. Chem. Phys.* 65, 1268-1274.

Davis, J.W., Carpenter, C.L. (1990) Aerobic biodegradation of vinyl chloride in groundwater samples. *Appl. & Environ. Microbiol.* 56(12), 3878-3880.

Davis, J.W., Madsen, S.S. (1991) The biodegradation of methylene chloride in soils. *Environ. Toxicol. & Chem.* 10, 465-474.

Davis, E.M., Murray, H.E., Liehr, J.G., Powers, E.L. (1981) Basic microbial degradation rates and chemical by products of selected organic compounds. *Water Res.* 15, 1125-1127.

Dean, J.A., Editor (1985) *Lange's Handbook of Chemistry.* 13th Edition, McGraw-Hill Book Company, New York.

DeLassus, P.T., Schmidt, D.D. (1981) Solubilities of vinyl chloride and vinylidene chloride in water. *J. Chem. Eng. Data* 26, 274-276.

Deno, N.C., Berkheimer, H.E. (1960) Part I, Phase equilibria molecular transport thermodynamics. Activity coefficients as a function of structure and media. *J. Chem. Eng. Data* 5(1), 1-5.

DeVoe, H., Miller, M.M., Wasik, S.P. (1981) Generator columns and high pressure liquid chromatography for determining of solubilities and octanol-water partition coefficients of hydrophobic substances. *J. Res. Nat. Bur. Stand.* 86, 361.

Dilling, W.L. (1977) Interphase transfer processes. II. Evaporation rates of chloromethanes, ethanes, ethylenes, propanes, and propylenes from dilute aqueous solutions. Comparisons with theoretical predictions. *Environ. Sci. Technol.* 11(4), 405-409.

Dilling, W.L. (1982) In: *Environmental Risk Analysis for Chemicals.* pp.154-197. Conway, R.A., Editor, Van Nostrand Reinhold Company, New York.

Dilling, W.L., Bredeweg, C.J., Terfertiller, N.B. (1976) Organic photochemistry. Simulated atmospheric rates of methylene chloride, 1,1,1-trichloroethane, trichloroethylene, tetrachloroethylene, and other compounds. *Environ. Sci. Technol.* 10(4), 351-356.

Dilling, W.L., Goersch, H.K. (1979) *Organic Photochemistry-XVI: Tropospheric photodecomposition of Methylene Chloride.* Preprints of papers presented at the 177th National Meeting, American Chemical Society, Division of Environmental Chemistry.

Dilling, W.L., Terfertiller, N.B., Kallos, G.J. (1975) Evaporation rates of methylene chloride, chloroform, 1,1,1-trichloroethane, trichloroethylene, tetrachloroethylene, and other chlorinated compounds in dilute aqueous solutions. *Environ. Sci. Technol.* 9(9), 833-838.

Dilling, W.L., Gonsior, S.J., Boggs, G.U., Mendoza, C.G. (1988) Organic photochemistry. 20. A method for estimating gas-phase rate constants for reactions of hydroxyl radicals with organic compounds from their relative rates of reaction with hydrogen peroxide under photolysis in 1,1,2-trichlorotrifluoroethane solution. *Environ. Sci. Technol.* 22, 1447-1453.

DIPPR Design Institute for Physical Property Data compilation on the Scientific and Technical International Network (STN).

Dimitriades, B., Gay, Jr., B.W., Arnts, R.R., Seila, R.L. (1983) Photochemical reactivity of perchloroethylene: A new appraisal. *J. Air. Pollut. Control Assoc.* 33, 575-587.

Dobbs, A.J., Williams, N. (1983) Fat solubility-A property of environmental relevance? *Chemosphere* 12(1), 97-104.

Dobbs, R.A., Wang, L., Govind, R. (1989) Sorption of toxic compounds on wastewater solids: Correlation with fundamental properties. *Environ. Sci. Technol.* 23(9), 1092-1097.

Dorfman, L.M., Adams, G.E. (1973) Reactivity of the hydroxyl radical in aqueous solutions. *NSRD-NDB*-46. *NTIS COM*-73-50623. 51 pp. National Bureau of Standards, Washington DC.

Dow Chemical Company (1972) *Chlorothene VG Cleaning Solvent.* Bulletin No. 100-5343-72. Midland, Michigan.

Dreisbach, R.R. (1952) *Pressure-Volume-Temperature Relationships of Organic Compounds.* Handbook Publishers, Inc., Sandusky, Ohio.

Dreisbach, R.R. (1955-1961) *Physical Properties of Chemical Compounds. Am. Chem. Soc. Adv. Chem. Series* 15 (1955), 22 (1959) and 29 (1961). Washington DC.

Du Pont (1966) *Solubility Relationships of the Freon Fluorocarbon Compounds.* Technical Bulletin B-7, Du Pont de Nemours & Company, Wilmington, Delaware.

Du Pont (1969) *Freon Fluorocarbons Properties and Applications.* Technical Bulletin B-2. Du Pont de Nemours & Company, Wilmington, Delaware.

Du Pont (1980) *Freon Products Information B-2.* A98825 12/80.

Eadsforth, C.V. (1986) Application of reverse-phase HPLC for the determination of partition coefficients. *Pest. Sci.* 17, 311-325.

Eadsforth, C.V., Moser, P. (1983) Assessment of reversed phase chromatographic methods for determining partition coefficients. *Chemosphere* 12, 1459-1475.

Edney, E., et al. (1983) *Atmospheric Chemistry of Several Toxic Compounds.* U.S. EPA-600/53-82-092.

Edney, E.O., Kleindienst, T.E., Corse, E.W. (1986) Room temperature rate constants for the reaction of OH with selected chlorinated and oxygenated hydrocarbons. *Int'l. J. Chem. Kinet.* 18, 1355-1371.

Ehrenberg, L., Osterman-Golkar, S., Singh, D., Lundqvist, U. (1974) On the reaction kinetics and mutagenic activity of methylating and beta-halogenoethylating gasoline additives. *Radiation Bot.* 15, 185-194.

Ellington, J.J. (1989) *Hydrolysis Rate Constants for Enhancing Property-Reactivity Relationships.* EPA/600/3-89/063. NTIS PB89-220479. U.S. Environmental Protection Agency, Environmental Research Laboratory, Athens, Georgia.

Ellington, J.J., Stancil, F.E., Payne, W.D. (1987) *Measurements of Hydrolysis rate constants for Evaluation of Hazardous Waste Land Disposal.* Volume 3, Data on 70 chemicals (preprint). EPA/600/3-86-043. NTIS PB87-140-349/GAR.

Ellington, J.J., Stancil, F.E., Payne, W.D. Trusty, C. (1987) *Measurement of Hydrolysis Rate Constants for Evaluation of Hazardous Waste Land Disposal.* Volume 2, Data on 54 chemicals. USEPA 600/53-87/019. U.S. Environmental Protection Agency, Washinton DC.

Environment Canada (1984) Enviro technical information for problem spills. *Ethylene Dichloride.* Environment Canada, Environmental Protection Service, Ottawa, Ontario.

ESE (1980) *Trace Organics Removal by Air Stripping.* ESE No. 79-337-001, May 1980, prepared by Environmental Science and Engineering, Inc., for American Water Works Association Research Foundation.

Farrington, J.W. (1991) Biogeochemical processes governing exposure and uptake of organic pollutant compounds in aquatic organisms. *Environ. Health Perspectives* 90, 75-84.

Flathman, P.E., Dahlgran, J.R. (1982) Correspondence on: Anaerobic degradation of halogenated 1- and 2-carbon organic compounds. *Environ. Sci. Technol.* 16, 130.

Foco, G.M., Bottini, S.B., Brignole, E.A. (1992) Isothermal vapor-liquid equilibria for 1,2-dichloroethane-anisole and trichloroethylene-anisole systems. *J. Chem. Eng. Data* 37, 17-19.

Freed, V.H., Chiou, C.T., Haque, R. (1977) Chemodynamics: Transport and behavior of chemicals in the environment-A problem in environmental health. *Environ. Health Perspectives* 20, 55-70.

Freitag, D., Balhorn, L. Geyer, H., Korte, F. (1985) Environmental hazard profile of organic chemicals. An experimental method for the assessment of the behavior of chemicals in the ecosphere by simple laboratory tests with C-14 labelled chemicals. *Chemosphere* 14, 1589-1616.

Freitag, D., Geyer, H., Kraus, A., Viswanathan, R., Kotzias, D., Attar, A., Klein, W., Korte, F. (1982) Ecotoxicological profile analysis. VII. Screening chemicals for their environmental behavior by comparative evaluation. *Ecotox. Environ. Safety* 6, 60-81.

Freitag, D., Lay, J.P., Korte, F. (1984) Environmental hazard profile-test rules as related to structures and translation into the environment. In: *QSAR Environ. Toxicol. Proc. Workshop Quant. Struct-Act. Relat.* Kaiser, K.L.E., Editor, pp.111-136. Reidell Publ., Dordrecht.

Friesel, P., Milde, G., Steiner, B. (1984) Interactions of halogenated hydrocarbons with soils. *Fresenius Z. Anal. Chem.* 319, 160-164.

Führer, H. (1924) Die wasserlöslichkeit in homologen reihen. *Chem. Ber.* 57, 510-515.

Fujita, T., Iwasa, J., Hansch, C. (1964) A new substituent constant, "pi" derived from partition coefficients. *J. Am. Chem. Soc.* 86, 5175-5180.

Gallant, R.W. (1966) Physical properties of hydrocarbons. V. Chlorinated methanes. *Hydrocarbon Process* 45, 161-169.

Garbarini, D.R., Lion, L.W. (1985) Evaluation of sorptive partitioning of nonionic pollutants in closed systems by headspace analysis. *Environ. Sci. Technol.* 19(11), 1122-1128.

Garbarini, D.R., Lion, L.W. (1986) Influence of the nature of soil organics on the sorption of toluene and trichloroethylene. *Environ. Sci. Technol.* 20(12), 1263-1269.

Garraway, J., Donnovan, J. (1979) Gas phase reaction of hydroxyl radical with alkyl iodides. *J. Chem. Soc. Chem. Comm.* 23, 1108.

Gay, B.W., Jr., Hanst, P.L., Bulfalini, J.J., Noonen, R.C. (1976) Atmospheric oxidation of chlorinated ethylenes. *Environ. Sci. Technol.* 10(1), 58-67.

GEMS (1982) *Graphical Exposure Modeling Systems.* CLOG3.

GEMS (1987) *Graphical Exposure Modeling Systems.* FAP. Fate of Atmospheric Pollutants.

Gerhold, R.M., Malaney, G.W. (1966) Structural determinants in the oxidation of aliphatic compounds by activated sludge. *J. Water Pollut. Control Fed.* 38, 562-579.

Geyer, H., Kraus, A.G., Klein, W., Richter, E., Korte, F. (1980) Relationship between water solubility and bioaccumulation potential of organic chemicals in rats. *Chemosphere* 9, 277-291.

Geyer, H., Politzki, G., Freitag, D. (1984) Prediction of ecotoxiological behavior of chemicals: Relationship between n-octanol/water partition coefficient and bioaccumulation of organic chemicals by *alga chlorella*. *Chemosphere* 13(2), 269-284.

Geyer, H., Viswanathan, R., Freitag, D., Korte, F. (1981) Relationship between water solubility of organic chemicals and their bioaccumulation by the *alga chlorella*. *Chemosphere* 10(11/12), 1307-1313.

Gilkey, W.K., Gerard, F.W., Bixler, M.E. (1931) Thermodynamic properties of dichlorodifluoromethane, a new refrigerant. II. Vapor pressure. *Ind. Eng. Chem.* 23, 364.

Gillham, R.W., Rao, P.S.C. (1990) Transport, distribution, and fate of volatile organic compounds in groundwater. In: *Significance and Treatment of Volatile Organic Compounds in Water Supplies*. Ram, A.M., Christman, R.F., Cantor, K.P., Editors, Chapter 9, pp.141-181, Lewis Publishers, Inc., Chelsea, Michigan.

Gladis, G.P. (1960) Effects of moisture on corrosion in petrochemical environments. *Chem. Eng. Prog.* 56(10), 43-51.

Glew, D.N., Moelwyn-Hughes, E.A. (1953) Chemical statics of the methyl halides in water. *Disc. Farad. Soc.* 15, 150-161.

Gmelins (1974) *Gmelins Handbuch der Anorganischen Chemie*. Kohnlenstoff, Kohlenstoff-halogen-verbindungen. Syst. No.14, Part D2. Springer-Verlag, Berlin. 386pp.

Goodman, M.A., Tuazon, E.C., Atkinson, R., Winer, A.M. (1986) A study of atmospheric reactions of chloroethenes with OH radicals. In: *Am. Chem. Soc. Environ. Chem. 192nd Nat'l Meeting* 26, 169-171.

Goring, C.A.I. (1972) In: *Organic Chemicals in the Soil Environment*. Goring, C.A.I., Hamaker, J.W., Editors, Volume 2, Marcel Dekker, New York. Chapter 9, pp.569-632.

Gossett, J.M. (1985) Anaerobic degradation of $C_1$ and $C_2$ chlorinated hydrocarbons. Report #ESL-TR-8538, U.S. A.F. Engineering and Services Center, Tyndall Air Force Base, Florida.

Gossett, J.M. (1987) Measurement of Henry's law constants for $C_1$ and $C_2$ chlorinated hydrocarbons. *Environ. Sci. Technol.* 21, 202-208.

Grathwohl, P. (1990) Influence of organic matter from soils and sediments from various origins on the sorption of some chlorinated aliphatic hydrocarbons: Implications on $K_{OC}$ correlations. *Environ. Sci. Technol.* 24, 1687-1693.

Green, W.J., Lee, G.F., Jones, R.A., Palit, T. (1983) Interaction of clay soils with water and organic solvents: Implications for the disposal of hazardous wastes. *Environ. Sci. Technol.* 17(5), 278-282.

Gross, P.M. (1929a) The determination of the solubility of slightly soluble liquids in water and the solubilities of dichloroethanes and dichloropropanes. *J. Am. Chem. Soc.* 51, 2362-2366.

Gross, P.M. (1929b) *Z. Phys. Chem.* 68, 215-220.

Gross, P.M., Saylor, J.H. (1931) Solubilities of certain slightly soluble organic compounds in water. *J. Am. Chem. Soc.* 53, 1744-1751.

Günther, F.A., Westlake, W.E., Jaglan, P.S. (1968) Reported solubilities of 738 pesticide chemicals in water. *Residue Rev.* 20, 1-148.

Haag, W.R., Mill, T. (1988) Effect of subsurface sediment on hydrolysis of haloalkanes and epoxides. *Environ. Sci. Technol.* 22, 658-663.

Haag, W.R., Yao, D.C.C. (1992) Rate constants for reaction of hydroxyl radicals with several drinking water contaminants. *Environ. Sci. Technol.* 26(5) 1005-1013.

Hafkenscheid, T.L., Tomlinson, E. (1983) Correlations between alkane/water and octan-1-ol/water distribution coefficients and isocratic reversed-phase liquid chromatographic capacity factors of acids, bases and neutrals. *Int'l J. Pharm.* 16, 225-239.

Halfon, E., Reggiani, M.G. (1986) On ranking chemicals for environmental hazard. *Environ. Sci. Technol.* 20, 1173-1179.

Haight, G.P. (1951) Solubility of methyl bromide in water and in some fruit juices. *Ind. Eng. Chem.* 43(8), 1827-1828.

Hallen, R.T., Pyne, J.W., Molton, P.M. (1986) Transformation of chlorinated ethers by anaerobic microorganisms. In: *Am. Chem. Soc. 192nd Nat'l Meeting* 26, 344-346.

Hanai, T., Tran, C., Hubert, J. (1981) An approach to the prediction of retention times in liquid chromatography. *J. HRC & CC* 4, 454-460.

Hancock, E.G., Editor (1973) *Propylene and Its Industrial Derivatives.* Ernest Benn, London. 517pp.

Hansch, C., Anderson, S. (1967) The effect of intramolecular hydrophobic bonding on partition coefficients. *J. Org. Chem.* 32, 2583.

Hansch, C., Leo, A.J. (1979) *Substituents Constants for Correlation Analysis in Chemistry and Biology.* Wiley, New York.

Hansch, C., Leo, A.J. (1985) *Medchem Project Issue No. 26.* Pomona College, Claremont, California.

Hansch, C., Quinlan, J.E., Lawrence, G.L. (1968) The linear free energy relationship between partition coefficients and aqueous solubilities of organic liquids. *J. Org. Chem.* 33, 347.

Hansch, C., Vittoria, A., Silipo, C., Jow, P.Y.C. (1975) Partition coefficient and structure-activity relationship of the anesthetic gases. *J. Med. Chem.* 18(6), 546-548.

Haque, R., Falco, J., Cohen, S., Riordan, C. (1980) Role of transport and fate studies in the exposure, assessment and screening of toxic chemicals. In: *Dynamics, Exposure and Hazard Assessment of Toxic Chemicals.* Haque, R., Editor, pp.47-67, Ann Arbor Science Publisher, Ann Arbor, Michigan.

Hardie, D.W.F. (1964) *Kirk-Othmer Encyclopaedia of Chemical Technology.* Standon, A., Editor, Vol.5, 2nd Edition, Interscience Publisher, New York.

Harnish, M., Möckel, Schulze, G. (1983) Relationship between LOG $P_{OW}$ shake-flask values and capacity factors derived from reversed-phase high-performance liquid chromatography for *n*-alkylbenzenes and some OECD reference substances. *J. Chromatogr.* 282, 315-332.

Hawker, D.W. (1990) Description of fish bioconcentration factors in terms of solvatochromic parameters. *Chemosphere* 20(5), 467-477.

Hawker, D.W., Connell, D.W. (1985) Relationships between partition coefficient, uptake rate constant, clearance rate constant and time to equilibrium for bioaccumulation. *Chemosphere* 14(9), 1205-1219.

Hawker, D.W., Connell, D.W. (1988) Influence of partition coefficient of lipophilic compounds on bioconcentration kinetics with fish. *Water Res.* 23(6), 701-707.

Hayduk, W., Laudie, H. (1974) Vinyl chloride gas compressibility and solubility in water and aqueous potassium laurate solutions. *J. Chem. Eng. Data* 19(3), 253-257.

Helfgott, T.B., Hart, F.L., Bedard, R.G. (1977) *An Index of Refractory Organics.* U.S. EPA-600/2-77-174. Ada, Oklahoma.

Henson, J.M., Yates, M.V., Cochran, J.W. (1989) Metabolisms of methanes, ethanes and ethylenes by a mixed bacterial culture growing on methane. *J. Indust. Microbiol.* 4, 29-35.

Hewitt, A.D., Miyares, P.H., Leggett, D.C., Jenkins, T.F. (1992) Comparison of analytical methods for determination of volatile organic compounds in soils. *Environ. Sci. Technol.* 26(10), 1932-1938.

Hildenbrand, D.L., McDonald, R.A. (1959) The heat of vaporization and vapor pressure of carbon tetrachloride; The entropy from calorimetric data. *J. Phys. Chem.* 63, 1521-22.

Hine, J., Dowell, A.M., Singley, J.E. (1956) Carbon dihalides as intermediates in the basic hydrolysis of haloforms. IV. Relative reactivities of haloforms. *J. Am. Chem. Soc.* 78, 479-482.

Hine, J., Mookerjee, P.K. (1975) The intrinsic hydrophilic character of organic compounds. Correlations in terms of structural contributions. *J. Org. Chem.* 40(3), 292-298.

Hodgman, C.R., Editor (1952) *Handbook of Chemistry and Physics.* 34th Edition. Chemical Rubber Publishing Co., Cleveland, Ohio.

Hodson, J., Williams, N.A. (1988) The estimation of the adsorption coefficient ($K_{OC}$) for soils by high performance liquid chromatography. *Chemosphere* 17(1), 67-77.

Hodson,P.V., Dixon, D.G., Kaiser, K.L.E. (1988) Estimating the acute toxicity of waterborne chemicals in trout from measurements of median lethal dose and the octanol-water partition coefficient. *Environ. Toxicol. & Chem.*7, 443-454.

Horvath, A.L. (1982) *Halogenated Hydrocarbons.* Solubility-Miscibility with Water. Marcel Dekker, Inc., New York and Basel.

Howard, C.J. (1976) Rate constants for the gas-phase reactions of OH radicals with ethylene and halogenated ethylene compounds. *J. Chem. Phys.* 65, 4771-4777.

Howard, C.J., Evenson, K.M. (1976) Rate constants for the reaction of OH with ethane and some halogen substituted ethanes at 296°K. *J. Chem. Phys.* 64(11), 4303-4306.

Howard, P.H., Editor (1989) *Handbook of Environmental Fate and Exposure Data for Organic Chemicals.* Vol.I- Large Production and Priority Pollutants. Lewis Publishers Inc., Chelsea, Michigan.

Howard, P.H., Editor (1990) *Handbook of Environmental Fate and Exposure Data for Organic Chemicals.* Vol.II-Solvents. Lewis Publishers Inc., Chelsea, Michigan.

Howard, P.H., Boethling, R.S., Jarvis, W.F., Meylan, W.M., Michalenko, E.M., Editors (1991) *Handbook of Environmental Degradation Rates.* Lewis Publishers Inc., Chelsea, Michigan.

Hoy, K.L. (1970) New values of the solubility parameters from vapor pressure data. *J. Paint Technol.* 42, 76-118.

Hull, L.A., Hisatsune, I.C., Heicklen, J. (1973) Reaction of ozone with 1,1-dichloroethylene. *Can. J. Chem.* 51, 1504.

Hutchinson, T.C., Hellebust, J.A., Tam, D., Mackay, D., Mascarenhas, R.A., Shiu, W.Y. (1980) The correlation of toxicity to algae of hydrocarbons and halogenated hydrocarbons with their physical chemical properties. In: *Hydrocarbons and Halogenated Hydrocarbons in the Aquatic Environments.* Afghan, B.K., Mackay, D., Editors, pp.577-586. Plenum Press, New York.

IARC (1986) Monograph of some chemicals used in plastics and elastomers. *IARC* 39, 195-226.

Irmann, F. (1965) Eine einfache korrelation zwischen wasserlöslichkeit und struktur von kohlenwasserstoffen und halogenkohlenwasserstoffen. *Chem. Eng. Tech.* 37(8), 789-798.

Isnard, P., Lambert, S. (1988) Estimating bioconcentration factors from octanol-water partition coefficient and aqueous solubility. *Chemosphere* 17, 21-34.

Isnard, P., Lambert, S. (1989) Aqueous solubility/*n*-octanol water partition coefficient correlations. *Chemosphere* 18, 1837-1853.

Iwase, K., Komatsu, K., Hirono, S., Nakagawa, S., Moriguchi, I. (1985) Estimation of hydrophobicity based on the solvent-accessible surface area of molecules. *Che. Pharm. Bull.* 33, 2114-2121.

Jafvert, C.T., Wolfe, N.L. (1987) Degradation of selected halogenated ethanes in anoxic sediment-water systems. *Environ. Toxicol. & Chem.* 6, 827-837.

Jeffers, P.M., Ward, L.M., Woytowitch, L.M., Wolfe, N.L. (1989) Homogeneous hydrolysis rate constants for selected chlorinated methanes, ethanes, and propanes. *Environ. Sci. Technol.* 23, 965-969.

Jensen, S., Rosenberg, R. (1975) Degradability of some chlorinated aliphatic hydrocarbons in seawater and sterilized water. *Water Res.* 9, 659-661.

Jolles, Z.E., Editor (1966) *Bromine and Its Compound.* Ernest Benn, LOndon.

Jones, C.J., Hudson, B.C., McGugan, Smith, A.J. (1977/1978) The leaching of some halogenated organic compounds from domestic waste. *J. Haz. Materials* 2, 227-233.

Jones, D.C., Ottewill, R.H., Chater, A.P.J. (1957) *Proc. 2nd Int'l Congress Surface Activ.*, Vol.1, 188-199. London.

Jorden, T.E. (1954) *Vapor Pressure of Organic Compounds.* Interscience Publishers, Inc., New York.

Jungclaus, G.A., Cohen, S.Z. (1986) Extended Abstracts, 191st National Meeting of the American Chemical Society, Division of Environmental Chemistry, New York, N.Y.; Americam Chemical Society: Washington, DC, 1986: paper 6.

Jury, W.A., Russo, D., Streile, G., El Abd, H. (1990) Evaluation of volatilization by organic chemicals residing below the soil surface. *Water Resources Res.* 26(1), 13-20.

Jury, W.A., Russo, D., Streile, G., El Abd, H. (1992) Correction to "Evaluation of volatilization by organic chemicals residing below the soil surface". *Water Resources Res.* 28(2), 607-608.

Jury, W.A., Spencer, W.F., Farmer, W.J. (1984) Behavior assessment model for trace organics in soil: III. Application of screening model. *J. Environ. Qual.* 13, 573-579.

Kaczmar, S.W., D'Itri, F.M., Zabik, M.J. (1984) Volatilization rates of selected haloforms from aqueous environments. *Environ. Toxicol. & Chem.* 3, 31-35.

Kahlbaum, G.W., Arndt, K. (1898) Studien über dampfspannkraftmessungen. *Z. Phys. Chem.* (Leipzig) 26, 577-658.

Kaiser, K.L.E., Palabrica, V.S., Ribo, J.M. (1987) QSAR of acute toxicity of mono-substituted benzene derivatives to photobacterium phosphoreum. In: *QSAR in Experimental Toxicology II.* Kaiser, K.L.E., Editor, pp.153-168, D. Reidel Publishing Co., Dordrecht, Holland.

Kakovsky, I.A. (1957) *Proc. 2nd Int. Congr. Surf. Activ.* (London) Vol.4, 225-237.

Kamlet, M.J., Doherty, R.M., Abboud, J-L M., Abraham, M.H. (1986) Linear solvation energy relationships: 36. Molecular properties governing solubilities of organic nonelectrolytes in water. *J. Pharm. Sci.* 75(4), 338-349.

Kamlet, M.J., Doherty, R.M., Abraham, M.H., Carr, P.W., Doherty, R.F., Taft, R.W. (1987) Linear solvation energy relationships. 41. Important differences between aqueous solubility relationships for aliphatic and aromatic solutes. *J. Phys. Chem.* 91, 1996-2004.

Kamlet, M.J., Doherty, R.M., Abraham, M.H., Marcus, Y., Taft, R.W. (1988) Linear solvation energy relationships. 46. An improved equation for correlation and prediction of octanol/water partition coefficients of organic nonelectrolytes (including strong hydrogen bond donor solutes). *J. Phys. Chem.* 92, 5244-5255.

Kamlet, M.J., Doherty, R.M., Taft, R.W., Abraham, M.H., Veith, G.D., Abraham, D.J. (1987) Solubility properties in polymers and biological media. 8. An analysis of the factors that influence toxicities of organic nonelectrolytes to the golden orfe fish (*Leuciscus idus melanotus). Environ. Sci. Technol.* 21, 149-155.

Karickhoff, S.W. (1981) Semi-empirical estimation of sorption of hydrophobic pollutants on natural water sediments and soils. *Chemosphere* 10(8), 833-846.

Karickhoff, S.W. (1985) Chapter 3, Pollutant sorption in environmental systems. In: *Environmental Exposure From Chemicals.* Volume I. Neely, W.B., Blau, G.E., Editors, CRC Press, Inc., Boca Raton, Florida. pp.49-64.

Kavanaugh, M.C., Trussell, R.R. (1980) Design of aeration towers to strip volatile contaminants from drinking water. *J. Am. Water Works Assoc.* 72, 684-692.

Kawasaki, M. (1980) Experiences with test scheme under the chemical control law of Japan: An approach to structure-activity correlations. *Ecotox. Environ. Safety* 4, 444-454.

Kay, R.L., Editor (1973) *The Physical Chemistry of Aqueous Systems*. Plenum Press, New York.

Kenaga, E.E. (1975) Partitioning and uptake of pesticides in biological systems. In: *Environmental Dynamics of Pesticides*. Haque, R., Freed, V.H., Editors, Pergamon Press, New York.

Kenaga, E.E. (1980) Predicted bioconcentration factors and soil sorption coefficients of pesticides and other chemicals. *Ecotoxicol. Environ. Safety* 4, 26-38.

Kenaga, E.E., Goring, C.A.I. (1980) Aquatic Toxicology 3rd Annual Symposium on Aquatic Toxicology, American Society for Testing and Materials, Philadelphia, Pa.

Kenaga, E.E., Goring, C.A.I. (1980) Relationship between water solubility, soil sorption, octanol-water partitioning, and concentration of chemicals in biota. In: *Aquatic Toxicology*. ASTM STP 707. Eaton, J.G., Parrish, P.R., Hendricks, A.C., Editors, American Society for Testing and Materials. pp.78-115.

Kier, L.B., Hall, L.H. (1976) Molar properties and molecular connectivity. In: *Molecular Connectivity in Chemistry and Drug Design*. Medicinal Chemistry Vol.14, Academic Press, New York. pp.123-167.

Kirk-Othmer (1964) *Kirk-Othmer Encyclopedia of Chemical Technology. Bromine Compounds, Organic.* Vol. 3, 2nd Edition, Wiley, New York.

Kirk-Othmer (1985) *Kirk-Othmer Concise Encyclopedia of Chemical Technology*. A Wiley-Interscience Publication, John Wiley & Sons, New York.

Klecka, G.M. (1982) Fate and effects of methylene chloride in activated sludge. *Appl. Environ. Microbiol.* 44, 701-707.

Klecka, G.M. (1985) Chapter 6, Biodegradation. In: *Environmental Exposure From Chemicals.* Volume I. Neely, W.B., Blau, G.E., Editors, CRC Press, Inc., Boca Raton, Florida. pp.110-155.

Klöpffer, W., Rippen, G., Frische, R. (1982) Physicochemical properties as useful tools for predicting the environmental fate of organic chemicals. *Ecotoxicol. & Environ. Safety* 6, 294-301.

Klöpffer, W., Kaufmann, G., Rippen, G., Poremski, H-J. (1982) A laboratory method for testing the volatility from aqueous solution: First results and comparison with theory. *Ecotoxicol. Environ. Safety* 6, 545-559.

Koch, R. (1983) Molecular connectivity index for assessing ecotoxicological behavior of organic compounds. *Toxicol. & Environ. Chem.* 6, 87-96.

Kollig, H.P., Ellington, J.J., Hamrick, K.J., Jafverts, C.T., Weber, E.J., Wolfe, N.L. (1987) *Hydrolysis Rate constants, Partition Coefficients, and Water Solubilities for 129 Chemicals.* A summary of fate constants provided for the concentration-based Listing Program, USEPA Environmental Research Lab., Office of Research and Development. Prepublication, Athens, Georgia.

Kollig, H.P., Parrish, R.S., Holm, H.W. (1987) An estimate of variability in biotransformation kinetics of xenobiotics in natural waters by aufwuchs communities. *Chemosphere* 16, 49-60.

Könemann, H., Zelle, R., Busser, F., Hammers, W.E. (1979) Determination of log $P_{oct}$ values of chloro-substituted benzenes, toluenes and anilines by high-performance liquid chromatography on ODS-silica. *J. Chromatography* 178, 559-565.

Korenman, I.M., Gur'ev, I.A., Gur'eva, Z.M. (1971) Solubility of liquid aliphatic compounds in water. *Zh. Fiz. Khim* 45(7), 1866. (VINITI No. 2885-71).

Krijgsheld, K.R., Van der Gen, A. (1986) Assessment of the impact of the emission of certain organochlorine compounds of the aquatic environment. Part II: allylchloride, 1,3- and 2,3-dichloropropene. *Chemosphere* 15, 861-880.

Kudchadker, A.P., Kudchadker, S.A., Shukla, R.P., Patnaik, P.R. (1979) Vapor pressures and boiling points of selected halomethanes. *Phys. Chem. Ref. Data* 8, 449.

Langbein, W.B., Durum, W.H. (1967) The aeration capacity of streams. *Geo. Survey Circular 542.* U.S. Department of Interior, Washington, DC.

Lange, N.A., Editor (1956) *Lange's Handbook of Chemistry.* Handbook Publishers, Sandusky, Ohio.

Leahy, D.E. (1986) Intrinsic molecular volume as a measure of the cavity term in linear solvation energy relationships: Octanol-water partition coefficients and aqueous solubilities. *J. Pharm. Sci.* 75(7), 629-636.

Lee, J.F., Crum, J.R., Boyd, S.A. (1989) Enhanced retention of organic contaminants by soil exchanged with organic cations. *Environ. Sci. Technol.* 23, 1365-1372.

Leighton Jr., D.T., Calo, J.M. (1981) Distribution coefficients of chlorinated hydrocarbons in dilute air-water systems for groundwater contamination applications. *J. Chem. Eng. Data* 26(4), 382-385.

Leinster, P., Perry, R., Young, R.J. (1978) Ethylenebromide in urban air. *Atmos. Envirn.* 12, 2383-2398.

Leo, A. (1986) *CLOGP-3.42 Medchem Software.* Medicinal Chemistry Project, Pomona College, Claremont, California.

Leo, A., Hansch, C., Elkins, D. (1971) Partition coefficients and their uses. *Chem. Rev.* 71(6), 525-612.

Leo, A., Jow, P.Y.C., Silipo, C., Hansch, C. (1975) Calculation of hydrophobic constant (Log *P*) from $\pi$ and $f$ constants. *J. Med. Chem.* 18(9), 865-868.

Lesage, S., Brown, S., Hosler, K.R. (1992) Degradation of chlorofluorocarbon-113 under anaerobic conditions. *Chemosphere* 24(9), 1225-1243.

Lincoff, A.H., Gossett, J.M. (1983) The determination of Henry's law constant for volatile organics by equilibrium partitioning in closed systems. Paper presented at the International Symposium on Gas Transfer at Water Surfaces, Cornell University, Ithaca, New York.

Lincoff, A.H., Gossett, J.M. (1984) The determination of Henry's law constant for volatile organics by equilibrium partitioning in closed systems. In: *Gas Transfer at Water Surfaces.* Brutsaert, W., Jirka, G.H., Editors, Reidel Publishing Company, Dordrecht, Holland. pp.17-25.

Liu, C.C.K., Tamrakar, N.K., Green, R.E. (1987) Biodegradation and adsorption of DBCP and the mathematical simulation of its transport in tropical soils. *Toxic. Asses.* 2, 239-252.

Liu, J.-L., Huang, T.-C. (1961) *Sci. Sin.* (Peking) 10(6), 700-710.

Lo, J.M., Tseng, C.L., Yang, J.Y. (1986) Radiometric method for determining solubility of organic solvents in water. *Anal. Chem.* 58, 1596-1597.

Love, Jr., O.T., Eilers, R.G. (1982) Treatment of drinking water containing trichloroethylene and related industrial solvents. *J. Am. Water Works Assoc.* 413-425.

Lyman, W.J. (1982) Atmospheric residence time. Chapter 10. In: *Handbook of Chemical Property Estimation Methods.* Lyman, W.J., Reehl, W.F., Rosenblatt, D.H., Editors, McGraw-Hill Book Company, New York.

Lyman, W.J. (1985) Chapter 2, Estimation of physical properties. In: *Environmental Exposure From Chemicals.* Volume I. Neely, W.B., Blau, G.E., Editors, CRC Press, Inc., Boca Raton, Florida. pp.13-48.

Lyman, W.J., Reehl, W.F., Rosenblatt, D.H., Editors (1982) *Handbook of Chemical Property Estimation Methods.* McGraw-Hill Book Company, New York.

Ma, K.C., Shiu, W.Y., Mackay, D. (1990) *A Critically Reviewed Compilation of of Physical and Chemical and Persistence Data for 110 Selected EMPPL Substances.* A Report Prepared for the Ontario Ministry of Environment, Water Resources Branch, Toronto, Ontario.

Mabey, W., Mill, T. (1978) Critical review of hydrolysis of organic compounds in water under environmental conditions. *J. Phys. Chem. Ref. Data* 7, 383-415.

Mabey, W., Smith, J.H., Podoll, R.T., Johnson, H.L., Mill, T., Chiou, T.W., Gate, J., Waight-Partridge, I., Jaber, H., Vandenberg, D. (1981/1982) *Aquatic Fate Process for Organic Priority Pollutants.* EPA Report, No.440/4-81-14.

Mabey, W.R., Barich, V., Mill, T. (1983) Hydrolysis of polychlorinated alkanes. In: *Symposium of American Chemical Society* 23, pp. 359-361 of 186th National Meeting, Washington, D.C.

Macy, R. (1948) Partition coefficients of fifty compounds between olive oil and water at 20°C. *J. Ind. Hyg. Toxicol.* 30, 140.

Mackay, D. (1982) Correlation of bioconcentration factors. *Environ. Sci. Technol.* 16, 274-278.

Mackay, D. (1985) Chapter 5, In: *Environmental Exposure From Chemicals.* Volume I. Neely, W.B., Blau, G.E., Editors, CRC Press, Inc., Boca Raton, Florida. pp.91-108.

Mackay, D., Paterson, S. (1991) Evaluating the multimedia fate of organic chemicals: A level III fugacity model. *Environ. Sci. Technol.* 25(3), 427-436.

Mackay, D., Paterson, S., Cheung, B., Neely, W.B. (1985) Evaluating the environmental behavior of chemicals with a level III fugacity model. *Chemosphere* 14(3/4), 335-374.

Mackay, D., Shiu, W.Y. (1981) A critical review of Henry's law constants for chemicals of environmental interest. *J. Phys. Chem. Ref. Data* 10, 1175-1199.

Mackay, D., Shiu, W.Y. (1990) Physical-chemical properties and fate of volatile organic compounds: An application of the fugacity approach. pp.183-204. In: *Significance and Treatment of Volatile Organic Compounds in Water Supplies.* Ram, N.M., Christman, R.F., Cantor, K.P., Editors, Lewis Publishers Inc., Chelsea, Michigan.

Mackay, D., Bobra, A., Shiu, W.Y., Yalkowsky, S.H. (1980) Relationships between aqueous solubility and octanol-water partition coefficient. *Chemosphere* 9, 701-711.

Mackay, D., Leinonen, P.J. (1975) Rate of evaporation of low solubility contaminants from water bodies to atmosphere. *Environ. Sci. Technol.* 9(13), 1178-1180.
Mackay, D., Shiu, W.Y., Bobra, A., Billington, J., Chau, E., Yuen, A., Ng, C., Szeto, F. (1982) *Volatilization of Organic Pollutants from Water.* EPA 600/3-82-019. National Technical Information Service, Springfield, Virginia.
Mackay, D., Shiu, W.Y., Sutherland, R.P. (1979) Determination of air-water Henry's law constants for hydrophobic pollutants. *Environ. Sci. Technol.* 13, 333.
Mackay, D., Wolkoff, A.W. (1973) Rate of evaporation of low solubility contaminants from water bodies to atmosphere. *Environ. Sci. Technol.* 7(7), 611-614.
Mackay, D., Yuen, T.K. (1979) Volatilization rates of organic contaminants from rivers. *Proceedings of the 14th Canadian Water Pollution Symposium.*
Mackay, D., Yuen, T.K. (1983) Mass transfer coefficient correlations for volatilization of organic solutes from water. *Environ. Sci. Technol.* 17, 211-217.
Mailhot, H. (1987) Prediction of algal bioaccumulation and uptake rate of nine organic compounds by ten physicochemical properties. *Environ. Sci. Technol.* 21(10), 1009-1013.
Makide, T. et al. (1979) *Chem. Lett.* 4, 355-358.
Marsden, C. (1963) *Solvent Guides.* 2nd Edition, Cleaver-Humes Press, London, England.
Marsden, C., Mann, S. (1962) *Solvent Guides.* Cleaver-Humes Press, London, England.
Marsh, K.N. (1968) Thermodynamics of octamethylcyclotetrasiloxane mixtures. *Trans. Faraday Soc.* 64, 883-893.
Mathias, E., Sanhueza, E., Hisatsune, I.C., Heicklen, J. (1974) Chlorine atom sensitized oxidation and the ozonolysis of tetrachloroethylene. *Can. J. Chem.* 52, 3852-3862.
Matthews, P.J. (1975) Limits for volatile organic liquids in sewers. Part 1. *Effl. Water Treat. J.* 15(11), 565-567 and 626-627.
Matthews, P.J. (1979) Use of vapor-liquid equilibrium data for estimating trade effluent limits. *Proc. Nat'l Phys. Lab. Conf. on Chemical Thermodynamics Data on Fluids and Fluid Mixtures.* Teddington, Sept.11-12, 1978. IPC Science and Technology Press, Guildford, England. pp.53-61.
McCall, P.J. (1987) Hydrolysis of 1,3-dichloropropene in dilute aqueous solutions. *Pestic. Sci.* 19, 235-242.
McCarty, L.S., Mackay, D., Smith, A.D., Ozburn, G.W., Dixon, D.G. (1992) Residue-based interpretation of toxicity and bioconcentration QSARs from aquatic bioassays: Neutral narcotic organics. *Environ. Toxicol. & Chem.* 11, 917-930.
McCarty, L.S., Ozburn, G.W., Smith, A.D., Dixon, D.G. (1992) Toxicokinetic modeling of mixtures of organic chemicals. *Environ. Toxicol. & Chem.* 11, 1037-1047.
McConnell, G., Ferguson, D.M., Pearson, C.R. (1975) Chlorinated hydrocarbons and the environment. *Endeavor* XXXVI, 13-18.
McDuffie, B. (1981) Estimation of octanol/water partition coefficients for organic pollutants using reversed phase HPLC. *Chemosphere* 10, 73-78.
McGlashan, M.L., Prue, J.E., Sainsbury, I.E. (1954) Equilibrium properties of mixtures of carbon tetrachloride and chloroform. *Trans. Faraday Soc.* 50, 1284-1292.

McGovern, E.W. (1943) Chlorohydrocarbon solvents. *Ind. Eng. Chem.* 35(12), 1230-1239.

McKone, T.E. (1987) Human exposure to volatile organic compounds in household trap water: The indoor inhalation pathway. *Environ. Sci. Technol.* 21(12), 1194-1201.

McNally, M.E., Grob, R.L. (1983) Determination of solubility limits of organic priority pollutants by gas chromatographic headspace analysis. *J. Chromatography* 260, 23-32.

McNally, M.E., Grob, R.L. (1984) Headspace determination of solubility limits of the base neutral and volatile components from environmental protection agency's list of priority pollutants. *J. Chromatography* 284, 105-116.

Merckel, J.H.C. (1937) Die löslichkeit der dicarbonsäuren. *Recl. Trav. Chim.* 56, 810-814.

Merlin, G., Thiebaud, H., Blake, G., Sembiring, S., Alary, J. (1992) Mesocosms' and microcosms' utilization for ecotoxicity evaluation of dichloromethane, a chlorinated solvent. *Chemosphere* 24(1), 37-50.

Metcalf, R.L., Editor (1962) *Advances in Pest Control Research.* Vol.5, 329pp. Interscience, New York.

Mill, T. et al. (1982) *Aquatic Fate Process Data for Organic Priority Pollutants.* U.S. EPA-440/4-80-014.

Mill, T., Winterle, J.S., Fisher, A., Tse, D., Mabey, W.R., Drossman, H., Liu, A., Davenport, J.E. (1985) *Toxic Substances Process Data Generation and Protocol Development.* U.S. EPA Contract No. 68-03-2981. Washington DC.

Miller, M.M., Wasik, S.P., Huang, G.L., Shiu, W.Y., Mackay, D. (1985) Relationships between octanol-water partition coefficient and aqueous solubility. *Environ. Sci. Technol.* 19, 522-529.

Miller, S.A., Editor (1969) *Ethylene and its Industrial Derivatives.* Ernest Benn, London.

Mills, W.B., Dean, J.D., Porcella, D.B., Gherini, S.A., Hudson, R.J.M., Frick, W.E., Rupp, G.L., Bowie, G.L. (1982) *Water Quality Assessment: A Screening Procedure for Toxic and Conventional Pollutants.* EPA-600/6-82-004a, Environmental Research Laboratory, Environmental Protection Agency, Athens, Georgia.

Mitchell, J., Smith, D.M. (1977) *Aquametry. Part 1: A Treatise on Methods for the Determination of Water.* 2nd Edition, Wiley, New York.

Moelwyn-Hughes, E.A., Missen, R.W. (1957) Thermodynamic properties of methyl iodide + chloromethane solutions. *Trans. Faraday Soc.* 53, 607-615.

Molina, M.J., Rowland, F.S. (1974) *Geophys. Res. Lett.* 1, 309-312.

Moriguchi, I., Kanada, Y., Komatsu, K. (1976) Van der Waals volume and the related parameters for hydrophobicity in structure-activity studies. *Chem. Pharm. Bull.* 24(8), 1799-1806.

Mudder, T. (1981) Development of empirical structure-biodegradability relationships and testing protocol for slightly soluble and volatile priority pollutants. *Diss. Abstr. Int.* B. 42, 1804.

Mudder, T.I., Musterman, J.L. (1982) Development of empirical structure biodegradability relationships and biodegradability testing protocol for volatile and slightly soluble priority pollutants. In: *Am. Chem. Soc.* Div. Meeting, pp. 52-53. Kansas City, Mo.

Mueller, C.R., Kearns, E.R. (1958) Thermodynamic studies of system acetone and chloroform. *J. Phys. Chem.* 62, 1441-1445.

Müller, M., Klein, W. (1991) Estimating atmospheric degradation processes by SARs. *Sci. Total Environ.* 109/110, 261-273.

Munz, C., Roberts, P.V. (1982) Technical Report No. 262, Dept. of Civil Engineering, Stanford University, Stanford, California.

Munz, C., Roberts, P.V. (1986) Effects of solute concentration and cosolvents on the aqueous activity coefficient of halogenated hydrocarbons. *Environ. Sci. Technol.* 20, 830-836.

Munz, C., Roberts, P.V. (1987) Air-water phase equilibria of volatile organic solutes. *J. A. W. W. A.*, 62-69.

Murray, W.J., Hall, L.H., Kier, L.B. (1975) Molecular connectivity III: Relationship to partition coefficients. *J. Pharm. Sci.* 64(12), 1978-1981.

Nathan, M.F. (1978) Choosing a process for chloride removal. *Chem. Eng.* 85(3), 93-100.

Neely, W.B., Branson, D.R., Blau, G.E. (1974) Partition coefficient to measure bioconcentration potential of organic chemicals in fish. *Environ. Sci. Technol.* 8, 1113-1115.

Neely, W.B. (1976) Predicting the flux of organics across the air/water interface. In: *National Conference on Control of Hazardous Material Spills*, New Orleans.

Neely, W.B. (1979) Estimating rate constants for the uptake and clearance of chemicals by fish. *Environ. Sci. Technol.* 13, 1506-1510.

Neely, W.B. (1979) A preliminary assessment of the environmental exposure to be expected from the addition of a chemical to a simulated ecosystem. *J. Environ. Stud.* 13, 101.

Neely, W.B. (1980) A method for selecting the most appropriate environmental experiments on a new chemical. In: *Dynamics, Exposure and Hazard Assessment of Toxic Chemicals.* R. Haque, Editor, Ann Arbor Science, Ann Arbor, Michigan.

Neely, W.B. (1982) Organizing data for environmental studies. *Environ. Toxicol. & Chem.* 1, 259-266.

Neely, W.B. (1984) An analysis of aquatic toxicity data: Water quality and acute LC50 fish data. *Chemosphere* 13(7), 813-819.

Neely, W.B. (1985) Chapter 7, Hydrolysis. In: *Environmental Exposure From Chemicals.* Volume I. Neely, W.B., Blau, G.E., Editors, CRC Press, Inc., Boca Raton, Florida. pp.157-174.

Neely, W.B., Blau, G.E. (1985) Chapter 1, Introduction to environmental exposure from chemicals. In: *Environmental Exposure From Chemicals.* Volume I. Neely, W.B., Blau, G.E., Editors, CRC Press, Inc., Boca Raton, Florida. pp.1-11.

Nelson, O.A., Young, H.D. (1933) Vapor pressure of fumigants. V. $\alpha,\beta$-Propylene dichloride. *J. Am. Chem. Soc.* 55, 2429.

Neuhauser, E.F., Loehr, R.C., Malecki, M.R., Milligan, D.L., Durkin, R.P. (1985) The toxicity of selected organic chemicals to earthworm *eisenia fetida*. *J. Environ. Qual.* 14(3), 383-388.

Nicholson, B.C., Maguire, B.P., Bursill, D.B. (1984) Henry's law constants for the trihalomethanes: Effects of water composition and temperature. *Environ. Sci. Technol.* 18, 518-521.

Niki, H., Maker, P.D., Savage, C.M., Breitenbach, L.P. (1983) Atmospheric ozone-olefin reactions. *Environ. Sci. Technol.* 17, 312A.

Nimitz, J.S., Skaggs, S.R. (1992) Estimating tropospheric lifetimes and ozone depletion potentials of one- and two-carbon hydrofluorocarbons and hydrochlorofluorocarbons. *Environ. Sci. Technol.* 26(4), 739-744.

Nirmalakhandan, N.N., Speece, R.E. (1988) QSAR model for predicting Henry's law constants. *Environ. Sci. Technol.* 22, 1349-1357.

O'Connell, W.L. (1963) Properties of heavy liquids. *Trans. Am. Inst. Mech. Eng.* 226(2), 126-132.

Okouchi, S., Saegusa, H., Nojima, O. (1992) Prediction of environmental parameters by adsorbability index: Water solubilities of hydrophobic organic pollutants. *Environ. Int'l* 18, 249-261.

Ollis, D.F. (1985) Contaminant degradation in water. *Environ. Sci. Technol.* 19(6), 480-484.

Olsen, R.L., Davis, A. (1990) Predicting the fate and transport of organic compounds in groundwater. *Haz. Mat. Control* 3, 40-64.

Pankow, J.F. (1990) Minimization of volatilization losses during sampling and analysis of volatile organic compounds in water. In: *Significance and Treatment of Volatile Organic Compounds in Water Supplies.* Ram, N.M., Christman, R.F., Cantor, K.P., Editors, Lewis Publishers Inc., Chelsea, Michigan.

Pankow, J.F., Isabelle, L.M., Asher, W.E. (1984) Trace organic compounds in rain. 1. Sample design and analysis by adsorption/thermal desorption (ATD). *Environ. Sci. Technol.* 18, 310-318.

Pankow, J.F., Rosen, M.E. (1988) The determination of volatile compounds in water by purging directly to a capillary column with whole column cryotrapping. *Environ. Sci. Technol.* 22, 398-405.

Parsons, F., Lage, G.B., Rice, R. (1985) Biotransformation of chlorinated organic solvents in static microcosms. *Environ. Toxicol. & Chem.* 4, 739-742.

Pavlostathis, S.G., Jaglal, K. (1991) Descriptive behavior of trichloroethylene in contaminated soil. *Environ. Sci. Technol.* 25(2), 274-279.

Pavlostathis, S.G., Mathavan, G.N. (1992) Desorption kinetics of selected volatile organic compounds from field contaminated soils. *Environ. Sci. Technol.* 26(3), 532-538.

Pavlou, S.P., Weston, D.P. (1983/1984) *Initial Evaluation of Alternatives for Development of Sediment Related Criteria for Toxic Contaminants in Marine Waters. (Puget Sound) Phase I & II.* EPA Contract No.68-01-6388.

Pearson, C.R., McConnell, G. (1975) Chlorinated $C_1$ and $C_2$ hydrocarbons in the marine environment. *Proc. Roy. Soc. London* B189, 305-322.

Pechiney-Saint-Gobain (1971) *Flugène 113, 113CM, 113M, 113MA. Solvant de Précision.* Seuilly sur Seine, France.

Perry, R.A., Atkinson, R., Pitts Jr., J.N. (1976) Rate constants for the reaction of hydroxyl radicals with dichlorofluoromethane, and chloromethane over temperature range 298-423 K and with dichloromethane at 298 K. *J. Chem. Phys.* 64, 1618-1620.

Perry, R.A., Atkinson, R., Pitts Jr., J.N. (1977) Rate constants for the reaction of OH radicals with $CH_2$=CHF, $CH_2$=CHCl, and $CH_2$=CHBr over temperature range 299-426°K. *J. Chem. Phys.* 67, 458-462.

Perry, R.H., Chilton, C.H. (1973) *Chemical Engineer's Handbook.* 5th Edition, McGraw-Hill, New York.

Peterson, M.S., Lion, L.W., Shoemaker, C.A. (1988) Influence of vapor-phase sorption and diffusion on the fate of trichloroethylene in an unsaturated aquifer system. *Environ. Sci. Technol.* 22(5), 571-578.

Pierotti, R.A. (1976) A scaled particle theory of aqueous and nonaqueous solutions. *Chem. Rev.* 76(6), 717-726.

Pignatello, J.J. (1990) Slow reversible sorption of aliphatic hydrocarbons in soils: I. Formation of residual fractions. *Environ. Toxicol. & Chem.* 9, 1107-1115.

Pignatello, J.J. (1991) Desorption of the tetrachloroethylene and 1,2-dibromo-3-chloropropane from aquifer sediments. *Environ. Toxicol. & Chem.* 10, 1399-1404.

Pinal, R., Rao, P.S.C., Lee, L.S., Cline, P.V. (1990) Cosolvency of partially miscible organic solvents on the solubility of hydrophobic organic chemicals. *Environ. Sci. Technol.* 24(5), 639-647.

Pitts Jr., Atkinson, R., Winer, A.M., Bierman, H.W., Carter, W.P.I., MacLeod, H., Tuazon, E.C. (1984) *Formation and Rate of Toxic Chemicals in California's Atmosphere.* ARB/R-85/239. NTIS PB 85-172-609. Air Resources Board, Sacramento, California.

Pitts Jr., J.N., Sandoval, H.L., Atkinson, R. (1974) Relative rate constants for the reaction of oxygen (1D) atoms with fluorocarbons and nitrous oxide. *Chem. Phys. Lett.* 29, 31-34.

Politzki, G.R., Bieniek, D., Lehaniatis, E.S., Scheunert, I., Klein, W., Korte, F. (1982) Determination of vapour pressures of nine organic chemicals adsorbed on silicagel. *Chemosphere* 11(12), 1217-1229.

Prinn, R., Cunnold, D., Rasmussen, R., Simmonds, P., Alyea, F., Crawford, A., Fraser, P., Rosen, R. (1987) Atmospheric trends in methylchloroform and the global average for the hydroxyl radical. *Science* 238, 945-950.

Radding, S.B., Liu, D.H., Johnson, H.L., Mill, T. (1977) *Review of the Environmental Fate of Selected Chemicals.* U.S. Environmental Protection Agency, Office of Toxic Substances, Washington, D.C., EPA-560/5-77-003.

Rathbun, R.E., Tai, D.Y. (1981) Technique for the determining the volatilization coefficients of priority pollutants in streams. *Water Res.* 15(2), 243-250.

Reid, R.C., Prausnitz, J.M., Sherwood, T.K. (1977) *The Properties of Gases and Liquids.* 3rd. Edition, McGraw-Hill, New York.

Rekker, R.F. (1977) *The Hydrophobic Fragmental Constant.* Elsevier, Amsterdam.

Rekker, R.F., de Kort, H.M. (1979) The hydrophobic fragmental constant; An extension to a 1000 data point set. *Eur. J. Med. Chem.* 14, 479-488.

Rex, A. (1906) Über die löslichkeit der halogenderivate der kohlenwasserstoffe in wasser. *Z. Phys. Chem.*55, 355-370.

Riddick, J.A., Bunger, W.B. (1970) *Organic Solvents: Physical Properties and Methods of Purification.* 3rd Ed., Wiley-Interscience, New York.

Riddick, J.A., Bunger, W.B., Sakano, T.K. (1986) *Organic Solvents: Physical Properties and Methods of Purification*. 4th Edition, John Wiley & Sons, New York.

Rittman, B.E., McCarty, P.L. (1980) Utilization of dichloromethane by suspended and fixed film bacteria. *Appl. Environ. Microbiol.* 39, 1225-1226.

Robbins, D.E. (1976) Photodissociation of methyl chloride and methyl bromide in the atmosphere. *Geophys. Res. Letter* 3(4), 213-216.

Roberts, A.L., Sanborn, P.N., Gschwend, P.M. (1992) Nucleophilic substitution reactions of dihalomethanes with hydrogen sulfide species. *Environ. Sci. Technol.* 26(11), 2263-2274.

Roberts, P.V., Dändliker, P.G. (1983) Mass transfer of volatile organic contaminants from aqueous solution to the atmosphere during surface aeration. *Environ. Sci. Technol.* 17(8), 484-489.

Roberts, P.V. (1984) Comment on "Mass transfer of volatile organic contaminants from aqueous solution to the atmosphere during surface aeration". *Environ. Sci. Technol.* 18, 894.

Roberts, P.V., Hopkins, G.D., Munz, C., Riojas, A.H. (1985) Evaluating two-resistance models for air stripping of volatile organic contaminants in countercurrent, packed column. *Environ. Sci. Technol.* 19(2), 164-173.

Roberts, P.V., Schreiner, J.E., Hopkins, G.D. (1982) Field study of organic water quality changes during groundwater recharge in the Palo Alto Baylands. *Water Res.* 16, 1025-1035.

Roberts, T.R., Stoydin, G. (1976) The degradation of (Z)-and (E)-1,3-dichloropropenes and 1,2-dichloropropane in soil. *Pestic. Sci.* 7, 325-335.

Rutherford, D.W., Chiou, C.T. (1992) Effect of water saturation in soil organic matter on the partition of organic compounds. *Environ. Sci. Technol.* 26, 995-970.

Rutherford, D.W., Chiou, C.T., Kile, D.E. (1992) Influence of soil organic matter composition on the partition of organic compounds. *Environ. Sci. Technol.* 26, 336-340.

Ryan, J.A., Bell, R.M., Davidson, J.M., O'Connor, G.A. (1988) Plant uptake of non-ionic organic chemicals from soil. *Chemosphere* 17, 2299-2323.

Sabljic, A. (1984) Predictions of the nature and strength of soil sorption of organic pollutants from molecular topology. *J. Agric. Food Chem.* 32, 243-246.

Sabljic, A. (1987) Nonempirical modeling of environmental distribution and toxicity of major organic pollutants. In: *QSAR in Environmental Toxicology-II*. Kaiser, K.L.E., Editor, D. Reidel Publishing Co., Dordrecht, Netherlands. pp.309-332.

Sabljic, A., Güsten, H. (1990) Predicting the night-time $NO_3$ radical reactivity in the troposphere. *Atmos. Environ.* 24A(1), 73-78.

Saito, S., Tanoue, A., Matsuo, M. (1992) Applicability of i/o-characters to a quantitative description of bioconcentration of organic chemicals in fish. *Chemosphere* 24(1), 81-87.

Sanhueza, E., Hisatsune, I.C., Heicklen, J. (1976) Oxidation of haloethylenes. *Chem. Rev.* 76, 801-826.

Sangster, J. (1989) Octanol-water partition coefficients of simple organic compounds. *J. Phys. Chem. Ref. Data* 18, 1111-1230.

Saracco, G., Spaccamela Marchetti, E. (1958) *Ann. Chim.* (Rome) 48(12), 1357-1394.

Scatchard, G., Wood, S.E., Mochel, J.M. (1939) Vapor-liquid equilibrium. IV. Carbon tetrachloride-cyclohexene mixtures. *J. Am. Chem. Soc.* 61, 3206.

Schantz, M.M., Martire, D.E. (1987) Determination of hydrocarbon-water partition coefficients from chromatographic data and based on solution thermodynamics and theory. *J. Chromatogr.* 391, 35.

Schmidt-Bleek, F., Haberland, W., Klein, A.W., Caroli, S. (1982) Steps towards environmental hazard assessment of new chemicals. *Chemosphere* 11(4), 383-415.

Schüürmann, G., Klein, W. (1988) Advances in bioconcentration prediction. *Chemosphere* 17(8), 1551-1574.

Schwarz, F.P. (1980) Measurement of the solubilities of slightly soluble organic liquids in water by elution chromatography. *Anal. Chem.* 52, 10-15.

Schwarz, F.P., Miller, J. (1980) Determination of the aqueous solubilities of organic liquids at 10.0, 20.0, and 30.0 °C by elution chromatography. *Anal. Chem.* 52, 2162-2164.

Schwarzenbach, R.P., Giger, W., Hoehn, E., Schneider, J.K. (1983) Behavior of organic compounds during infiltration of river water to groundwater. Field studies. *Environ. Sci. Technol.* 17(8), 472-479.

Schwarzenbach, R.P., Giger, W., Schaffner, C., Wanner, O. (1985) Groundwater contamination by volatile halogenated alkanes: Abiotic formation of volatile sulfur compounds under anaerobic conditions. *Environ. Sci. Technol.* 19(4), 322-327.

Schwarzenbach, R.P., Molnar-Kubica, E., Giger, W., Wakeham, S.G. (1979) Distribution, residence time, and fluxes of tetrachloroethylene and 1,4-dichlorobenzene in Lake Zurich, Switzerland. *Environ. Sci. Technol.* 13(11), 1367-1373.

Schwarzenbach, R.P., Westall, J. (1981) Transport of nonpolar compounds from surface water to groundwater. Laboratory sorption studies. *Environ. Sci. Technol.* 11, 1360-1367.

Sconce, J.S., Editor (1962) *Chlorine: its Manufacture, Properties, and Uses.* Reinhold, New York.

Seidell, A. (1940) *Solubilities.* Van Nostrand, New York.

Seidell, A. (1941) *Solubilities.* 2nd Edition, Van Nostrand, New York.

Seip, H.M., Alstad, J., Carlberg, G.E., Martinsen, K., Skaane, R. (1986) Measurement of mobility of organic compounds in soils. *Sci. Total Environ.* 50, 87-101.

Selenka, F., Bauer, U. (1978) Detection of readily volatile organochloride compounds in water. *Org. Verunreinig. Umwelt: Erkennen, Bewerten.* pp.242-255.

Shiu, W.Y., Ma, K.C., Mackay, D. (1990) Solubilities of pesticides in water. Part 1, Environmental physical chemistry and Part 2, Data compilation. *Reviews Environ. Contam. Toxicol.* 116, 1-187.

Silka, L.R., Wallen, D.A. (1988) Observed rates of biotransformation of chlorinated aliphatics in groundwater. In: *Superfund 88 Proc. 9th Nat'l Conf. Haz. Mat. Control Inst.* pp.138-141.

Singh, H.B. (1977) Atmospheric hydrocarbons: Evidence in favor of reduced average hydroxyl radical concentrations in troposphere. *Geophys. Res. Letter* 4, 101-104.

Singh, H.B., Salas, L.J., Shigeishi, H., Scribner, E. (1979) Atmospheric halocarbons, hydrocarbons and sulfur hexafluoride: Global distribution, sources and sinks. *Science* 203, 899-903.

Singh, H.B., Salas, L.J., Shigeishi, H., Smith, A.H. (1978) *Fate of Halogenated Compounds in Atmosphere.* Interim report, EPA-600/3-78-017. U.S. Environmental Protection Agency.

Singh, H.B., Salas, L.J., Smith, J.A., Shigeishi, H. (1980) *Atmospheric Measurements of Selected Toxic Organic Chemicals.* EPA-600/3-80-072. USEPA, Research Triangle Park, North Carolina.

Singh, H.B., Salas, L.J., Smith, J.A., Shigeishi, H. (1981) Measurements of some potentially hazardous organic chemicals in urban environments. *Atmos. Environ.* 15, 601-612.

Smith, E.L. (1932) Some solvent properties of soap solutions. *J. Phys. Chem.* 36, 1401-1418.

Smith, J.A., Jaffé, P.R. (1991) Comparison of tetrachloromethane sorption to an alkylammonium-clay and an alkyldiammonium-clay. *Environ. Sci. Technol.* 25, 2054-2058.

Smith, J.A., Jaffé, P.R., Chiou, C.T. (1990) Effect of ten quaternary ammonium cations on tetrachloromethane sorption to clay from water. *Environ. Sci. Technol.* 24(8), 1167-1172.

Smith, J.H., Bomberger, D.C., Jr., Haynes, D.L. (1980) Prediction of the volatilization rates of high-volatility chemicals from natural water bodies. *Environ. Sci. Technol.* 14(11), 1332-1337.

Smith, V.L. et al. (1980) Temporal variations in trihalomethane content of drinking water. *Environ. Sci. Technol.* 14(2), 190-196.

Sørensen, J.M., Arit, W. (1979) *Liquid-Liquid Equilibrium Data Collection: Binary Systems.* Dechema Chemistry Data Series, Vol. 1, Part 1. Dechema, Frankfurt.

Stephen, H., Stephen, T. (1963) *Solubilities of Inorganic and Organic Compounds.* Volumes 1 and 2, Pergamon Press, Oxford.

Stephenson, R.M. (1992) Mutual solubilities: Water-ketones, water-ethers, and water-gasoline-alcohols. *J. Chem. Eng. Data* 37, 80-95.

Stephenson, R.M., Malanowski, S. (1987) *Handbook of the Thermodynamics of Organic Compounds.* Elsevier Science Publishing Co., Inc., New York.

Stover, E.L., Kincannon, D.F. (1983) Biological treatability of specific organic compounds found in chemical industry wastewaters. *J. Water Pollt. Control Fed.* 55, 97-109.

Subba-Rao, R.V., Rubin, H.E., Alexander, M. (1982) Kinetics and extent of mineralization of organic chemicals at trace levels in fresh water and sewage. *Appl. Environ. Microbiol.* 43, 1139.

Suntio, L.R., Shiu, W.Y., Mackay, D. (1988) A review of the nature and properties of chemicals present in pulp mill effluents. *Chemosphere* 17, 1249-1290.

Svetlanov, E.B., Velichko, S.M., Levinskii, M.I., Treger, Yu.A., Flid, R.M. (1971) Solubility of chloromethanes and 1,2-dichloroethane in water and hydrochloric acid. *Zh. Fiz. Khim* 45(4), 877-879.

Swann, R.L., Laskowski, D.A., McCall, P.J., Vender Kuy, K., Dishburger, J.J. (1983) A rapid method for the estimation of environmental parameters octanol/water partition coefficient, soil sorption constant, water to air ratio, and water solubility. *Res. Rev.* 85, 17-28.

Swindoll, C.M., Aelion, C.M., Pfaender, F.K. (1987) Inorganic and organic amendment effects of the biodegradation of organic pollutants by groundwater microorganisms. *Am. Soc. Microbiol. Abstr.*, 87th annual meeting, Atlanta, Georgia.

Symons, J.M., Stevens, A.A., Clark, R.M., Geldreich, E.E., Love Jr., O.T., DeMarco, J. (1981) *Treatment Techniques for Controlling Trihalomethanes in Drinking Water. EPA-600/2-81-156.* U.S. Environmental Protection Agency, Drinking Water Research Division, Municipal Environmental Research Laboratory.

Tabak, H.H., Quave, S.A., Mashni, C.I., Barth, E.F. (1981) Biodegradability studies with organic priority pollutant compounds. *J. Water Pollut. Control Fed.* 53, 1503-1518.

Tabak, H.H., Quave, S.A., Mashni, C.I., Barth, E.F. (1981) Biodegradability studies for predicting the environmental fate of organic priority pollutants. In: *Test Protocols for Environmental Fate and Movement of Toxicants.* Proc. Symposium Assoc. Off. Analytical Chemist 94th Annual Meeting, Washington, DC. pp.267-328.

Taft, R.W., Abraham, M.H., Famini, G.R., Doherty, R.M., Abboud, J-L M., Kamlet, M.J. (1985) Solubility properties in polymers and biological media 5: An analysis of the physicochemical properties which influence octanol-water partition coefficients of aliphatic and aromatic solutes. *J. Pharm. Sci.* 74(8), 807-814.

Tancréde, M.V., Yanagisawa, Y. (1990) An analytical method to determine Henry's law constant for selected volatile organic compounds at concentrations and temperatures corresponding to tap water use. *J. Air Waste Manage. Assoc.* 40, 1658-1663.

Tancréde, M.V., Yanagisawa, Y., Wilson, R. (1992) Volatilization of volatile organic compounds from showers-I. Analytical method and quantitative assessment. *Tom. Environ.* 26A(6), 1103-1111.

Tewari,, Y.B., Miller, M.M., Wasik, S.P., Martire, D.E. (1982) Aqueous solubility and octanol/water partition coefficient of organic compounds at 25.0°C. *J. Chem. Eng. Data* 27, 451-454.

Thomann, R.V. (1989) Bioaccumulation model of organic chemical distribution in aquatic food chains. *Environ. Sci. Technol.* 23, 699-707.

Thomas, R.G. (1982) Volatilization from water. Chapter 15. In: *Handbook of Chemical Property Estimation Methods, Environmental Behavior of Organic Compounds.* Lyman, W.J., Reehl, W.F., Rosenblatt, D.H., Editors, McGraw-Hill, New York.

Thoms, S.R., Lion, L.W. (1992) Vapor-phase partitioning of volatile organic compounds: A regression approach. *Environ. Toxicol. & Chem.* 11, 1377-1388.

THOR (1986) *Database of Medicinal Chemistry Project.* Pomona College, Claremont, California.

Timmermans, J. (1950) *Physical-Chemical Constants of Pure Organic Compounds.* Vol.1, Elsevier Publishing Company, Inc., New York.

Timmermans, J. (1965) *Physical-Chemical Constants of Pure Organic Compounds.* Vol.2, Elsevier Publishing Company, Inc., New York.

Tomlinson, E., Hafkenscheid, T.L. (1986) Aqueous solubility and partition coefficient estimation from HPLC data. In: *Partition Coefficient, Determination and Estimation.* Dunn, III, W.J., Block, J.H., Pearlman, R.S., Editors, pp.101-141. Pergamon Press, New York.

Tse, G., Orbey, H., Sandler, S.I. (1992) Infinite dilution activity coefficients and Henry's law coefficients of some priority pollutants determined by a relative gas chromatographic method. *Environ. Sci. Technol.* 26(10), 2017-2022.

Tuazon, E.C., Atkinson, R., Winer, A.M., Pitts Jr., J.N. (1984) A study of the atmospheric reactions of 1,3-dichloropropene and other selected organic compounds. *Arch. Environ. Contam. Toxicol.* 13, 691-700.

Tute, M.S. (1971) Principles and practice of Hansch analysis: A guide to structure-activity correlation for the medicinal chemist. *Adv. Drug Res.* 6, 1-77.

Ullmann (1975) *Ullmanns Encyklopaedie der Technischen Chemie.* Verlag Chemie, Weinheim.

Urano, K., Murata, C. (1985) Adsorption of principal chlorinated organic compounds on soil. *Chemosphere* 14(3/4), 293-299.

USEPA (1976) A literature survey oriented towards adverse environmental effects resultant from the use of azo-compounds, brominated hydrocarbons, EDTA, formaldehyde resins, and o-nitro-chlorobenzene. U.S. Environment Protection Agency, Office of Toxic Substances, Washington, D.C.

USEPA (1980) *Ambient Water Quality Criteria Document for Chlorinated Ethanes.* USEPA-440/5-80-029.

USEPA (1984) *Health Assessment Document for Chloroform.* External review draft. EPA-600/8-84-004A.

USEPA (1986) *Superfund Public Health Evaluation Manual.* EPA-540-1-86-060, PB 87-183125, U.S. Environmental Protection Agency, Office of Emergency and Remedial Response, Edison, New Jersey.

USEPA (1986) *GEMS Graphical Modeling System. CHEMEST.*

USEPA (1987) *GEMS Graphical Modeling System. CLOGP3.*

USEPA (1987) *EXAMS II Computer Simulation.*

Vallaud, A., Raymond, V., Salmon, P. (1957) *Les Solvants Chlores et L'Hygiene Industrielle.* Inst. Nat'l. Securite pour le Prevention des Accidents du Travail et des Maladies Professionelles, Paris.

Valsaraj, K.T. (1988) On the physico-chemical aspects of partition of non-polar hydrophobic organics at the air-water interface. *Chemosphere* 17(5), 875-887.

Valsaraj, K.T., Thibodeaux, L.J. (1989) Relationships between micelle-water and octanol-water partition coefficients for hydrophobic organics of environmental interest. *Water Res.* 23(2), 183-189.

Valvani, S.C., Yalkowsky, S.H., Roseman, T.J. (1981) Solubility and partitioning: IV. Aqueous solubility and octanol-water partition coefficients of liquid nonelectrolytes. *J. Pharm. Sci.* 70, 502-507.

Van Arkel, A.E., Vles, S.E. (1936) Löslichkeit von organischen verbindungen in wasser. *Recl. Trav. Chim. Pays-Bas* 55, 407-411.

Van Leeuwen, C.J., Van Der Zandt, P.T.J., Aldenberg, T., Verhaar, H.J.M., Hermens, J.L.M. (1992) Application of QSARs, extrapolation and equilibrium partitioning in aquatic effects assessment. I. Narcotic industrial pollutants. *Environ. Toxicol. & Chem.* 11(2), 267-282.

Veith, G.D., Defoe, D.L., Bergstedt, B.V. (1979) Measuring and estimating the bioconcentration factor of chemicals in fish. *J. Fish. Res. Board Can.* 26, 1040-1048.

Veith, G.D., Kosian, P. (1983) Estimating bioconcentration potential from octanol/water partition coefficients. In: *Physical Behavior of PCBs in the Great Lakes.* Mackay, D., Paterson, S., Eisenreich, S.J., Simmons, M.S. , Editors, Ann Arbor Science Publishers, Ann Arbor, Michigan. pp.269-282.

Veith, G.D., Macek, K.J., Petrocelli, S.R., Caroll, J. (1980) An evaluation of using partition coefficients and water solubility to estimate bioconcentration factors for organic chemicals in fish. In: *Aquatic Toxicology.* ASTM STP 707. Eaton, J.G., Parrish, P.R., Hendricks, A.C., Editors, American Society for Testing and Materials. pp.117-129.

Veith, G.D., Call, D.J., Brooke, L.T. (1983) Structure-toxicity relationships for the fathead minnow, *pimephales promelas*: Narcotic industrial chemicals. *Can. J. Fish Aquat. Sci.* 40, 743-748.

Verhaar, H.J.M., Van Leeuwen, C.J., Hermens, J.L.M. (1992) Classifying environmental pollutants. 1. Structure-activity relationships for prediction of aquatic toxicity. *Chemosphere* 25(4), 471-491.

Verschueren, K. (1977) *Handbook of Environmental Data on Organic Chemicals.* Van Nostrand Reinhold, New York.

Verschueren, K. (1983) *Handbook of Environmental Data on Organic Chemicals.* Van Nostrand Reinhold, New York.

Vesala, A. (1973) *Thermodynamics of Transfer of Electrolytes from Light to Heavy Water.* Ph.D. Thesis, University of Turku, Turku, Finland.

Vesala, A. (1974) Thermodynamics of transfer of electrolytes from light to heavy water. I. Linear free energy correlations of free energy of transfer with solubility and heat of mixing of a nonelectrolyte. *Acta. Chem. Scad. Ser.A* 28(8), 839-845.

Vogel, T.M., Reinhard, M. (1986) Reaction products and rates of disappearance of simple bromoalkanes, 1,2-dibromopropane, and 1,2-dibromoethane in water. *Environ. Sci. Technol.* 20(10), 992-997.

Wakeham, , S.G., Davis, A.C., Karas, J.L. (1983) Microcosm experiments to determine the fate and persistence of volatile organic compounds in coastal seawater. *Environ. Sci. Technol.* 17, 611-617.

Wakita, K., Yoshimoto, M., Miyamoto, S. (1986) A method for calculation of the aqueous solubility of organic compounds by using fragment solubility constants. *Chem. Pharm. Bull.* 34, 4663-4681.

Walraevens, R., Trouillet, P., Devos, A. (1974) Basic elimination of hydrogen chloride from chlorinated ethanes. *Int'l. J. Chem. Kinet.* 6, 777-786.

Wang, L.S., Zhao, Y.H., Gao, H. (1992) Predicting aqueous solubility and octanol/water partition coefficients of organic chemicals from molar volume. *Chinese Environ. Chem.* 11(1), 55-70.

Warner, H.P., Cohen, J.M., Ireland, J.C. (1987) *Determination of Henry's Law Constants of Selected Priority Pollutants.* EPA/600/D-87/229, U.S. Environmental Protection Agency, Cincinnati, Ohio. PB87-212684, U.S. Department of Commerce, National Technical Information Service.

Warner, M.J., Weiss, R.F. (1985) *Deep-Sea Res. Part A* 32, 1485-1497.

Wasik, S.P., Tewari, Y.B., Miller, M.M., Martire, D.E. (1981) *Octanol/Water Partition Coefficients and Aqueous Solubilities of Organic Compounds.* NBSIR No.81-2406. U.S. Dept. of Commerce, Washington D.C.

Watarai, H., Tanaka, M., Suzuki, N. (1982) Determination of partition coefficients of halobenzenes in heptane/water and 1-octanol/water systems and comparison with the scaled particle calculation. *Anal. Chem.* 54, 702-705.

Watson, R.T., Machado, G., Conway, B., Wagner, S., Davis, D.D. (1977) A temperature dependent kinetics study of the reaction of OH with $CH_2ClF$, $CHCl_2F$, $CHClF_2$, $CH_3CCl_3$, $CH_3CF_2Cl$ and $CF_2ClCFCl_2$. *J. Phys. Chem.* 81, 256-262.

Wauchope, R.D., Buttler, T.M., Hornsby, A.G., Augustijn-Beckers, P.W.M., Burt, J.P. (1992) The SCS/ARS/CES pesticide properties database for environmental decision-making. *Rev. Environ. Contam. & Toxicol.* 123, 1-164.

Weast, R.C., Editor (1972-73) *Handbook of Chemistry and Physics.* 53rd Edition, CRC Press Inc., Cleveland, Ohio.

Weast, R.C., Editor (1973-74) *Handbook of Chemistry and Physics.* 54th Edition, CRC Press Inc., Cleveland, Ohio.

Weast, R.C., Editor (1977) *Handbook of Chemistry and Physics.* 58th Edition, CRC Press Inc., Cleveland, Ohio.

Weast, R.C., Editor (1982-83) *Handbook of Chemistry and Physics.* 63rd Edition, CRC Press Inc., Boca Raton, Florida.

Weintraub, R.A., Jex, G.W., Moye, H.A. (1986) Chemical and microbial degradation of 1,2-dibromomethane (EFB) in Florida groundwater, soil, and sludge. *American Chemical Society Symposium Series 315, Evaluation Pesticides Ground Water.* pp.294-310. Garner, W.Y., et al., Editors. Washington DC.

Wells, M.J.M., Clark, C.R., Patterson, R.M. (1981) Correlation of reversed-phase capacity factors for barbiturates with biological activities, partition coefficients and molecular connectivity indices. *J. Chromatogr. Sci.* 19, 573-582.

Wilson, B.H., Smith, G.B., Rees, J.F. (1986) Biotransformations of selected alkylbenzenes and halogenated aliphatic hydrocarbons in methanogenic aquifer material: A microcosm study. *Environ. Sci. Technol.* 20, 997-1002.

Wilson, J.T., Cosby, R.L., Smith, G.B. (1984) Potential for biodegradation of organo-chlorine compounds in ground water. R.S. Kerr Environmental Research Laboratory, Ada, Oklahoma.

Wilson, J.T., Enfield, C.G., Dunlap, W.J. (1981) Transport and fate of selected organic pollutants in a sandy soil. *J. Environ. Qual.* 10, 501-506.

Wilson, J.T., McNabb, J.F., Blackwill, D.L., Ghiorse, W.C. (1983) Enumeration and characterization of bacteria indigenous to a shallow water-table aquifer. *Ground Water* 21, 134-142.

Wilson, J.T., McNabb, J.F., Wilson, R.H., Noonan, M.J. (1983) Biotransformation of selected organic pollutants in groundwater. *Devel. Indust. Microbiol.* 24, 225-233.

Windholz, M., Budavari, S., Blumetti, R.F., Otterbein, E.S., Editors (1983) *The Merck Index.* 10th Edition, Merck & Co., Inc., Rahway, New Jersey.

Wolfe, N.L. (1980) Determining the role of hydrolysis in the fate of organics in natural waters. In: *Dynamics, Exposure, and Hazard Assessment of Toxic Chemicals.* Haque, R., Editor, pp.163-178. Ann Arbor Publishing Inc., Ann Arbor, Michigan.

Wolfe, N.L., Zepp, R.G., Schlotzhauer, P., Sink, M. (1982) Transformation pathways of hexachlorocyclopentadiene in the aquatic environment. *Chemosphere* 11, 91-101.

Wood, P.R., Lang, R.F., Payan, I.L. (1985) In: *Groundwater Quality.* Ward, C.H., Giger, W., McCarty, P.L., Editors, Wiley, New York.

Wood, P.R., Parsons, F.Z., Demarco, J., Harween, H.J., Lang, R.F., Payan, I.L., Rutz, M.C. (1981) Introductory study of biodegradation of the chlorinated methane, ethane and ethene compounds. presented at American Water Works Annual Conference and Exposition, St. Louis.

Wright, D.A., Sandler, S.I., DeVoll, D. (1992) Infinite dilution activity coefficients and solubilities of halogenated hydrocarbons in water at ambient temperatures. *Environ. Sci. Technol.* 26(9), 1828-1831.

Wright, W.H., Schaffer, J.M. (1932) *Am. J. Hyg.* 16(2), 325-428.

Yalkowsky, S.H., Mishra, D.S. (1991) Vapor pressure estimation for organic compounds. *Sci. Total Environ.* 109/110, 243-250.

Yalkowsky, S.H., Orr, R.J., Valvani, S.C. (1979) Solubility and partitioning. 3. The solubility of halobenzenes in water. *Ind. Eng. Chem. Fundam.* 18, 351-353.

Yeh, H.C., Kastenberg, W.E. (1991) Health risk assessment of biodegradable volatile organic chemicals: A case study of PCE, TCE, DCE and VC. *J. Hazard. Mat.* 27, 111-126.

Yung, Y.L., McElroy, M.B., Wofsy, S.C. (1975) Atmospheric halocarbons: A discussion with emphasis on chloroform. *Geophys. Res Lett.* 2(9), 397-399.

Yurteri, C., Ryan, D.F., Callow, J.J., Gurol, M.D. (1987) The effect of chemical composition of water on Henry's law constant. *J. Water Pollut. Control Fed.* 59, 950-956.

Yoshida, K., Shigeoka, T., Yamauchi, F. (1983) Relationship between molar refraction and n-octanol/water partition coefficient. *Ecotoxicol. & Environ. Safety* 7, 558-565.

Zhang, J., Hatakeyama, S., Akimoto, H. (1983) *Int'l J. Chem. Kinet.* 15, 655.

Zoeteman, B.C.J., De Greef, E., Brinkmann, F.J.J. (1981) Persistency of organic contaminants in groundwater, lessons from soil pollution incidents in the Netherlands. *Sci. Total Environ.* 21, 187-202.

Zoeteman, B.C.J., Harmsen, K.M., Linders, J.B.H.J. (1980) Persistent organic pollutants in river water and groundwater of the Netherlands. *Chemosphere* 9, 231-249.

# 4. Ethers


4.1 List of Chemicals and Data Compilations:

Aliphatic Ethers

Dimethyl ether . . . . . 750
Diethyl ether . . . . . 753
Methyl *t*-butyl ether . . . . . 756
Di-*n*-propyl ether . . . . . 758
Di-isopropyl ether . . . . . 761
Butyl ethyl ether . . . . . 764
Di-*n*-butyl ether . . . . . 766
1,2-Propylene oxide . . . . . 769
Furan . . . . . 772
2-Methylfuran . . . . . 775
Tetrahydrofuran . . . . . 777
Tetrahydropyran . . . . . 780
1,4-Dioxane . . . . . 782

Aromatic Ethers

Anisole . . . . . 785
Phenetole . . . . . 788
Benzyl ethyl ether . . . . . 790
Diphenyl ether . . . . . 792
Styrene oxide . . . . . 795

Halogenated Ethers

Epichlorohydrin . . . . . 797
Chloromethyl methyl ether . . . . . 800
Bis(chloromethyl)ether . . . . . 802
Bis(2-chloroethyl)ether . . . . . 805
Bis(2-chloroisopropyl)ether . . . . . 809
2-Chloroethyl vinyl ether . . . . . 812
4-Chlorophenyl phenyl ether . . . . . 814
4-Bromophenyl phenyl ether . . . . . 816
Bis(2-chloroethoxy)methane . . . . . 818

4.2 Summary Tables and QSPR Plots . . . . . 820

4.3 Illustrative Fugacity Calculations: Levels I, II and III . . . . . 829

Diethyl ether . . . . . 830
Methyl-*t*-butyl ether . . . . . 834
Di-*n*-butyl ether . . . . . 838
1,2-Propylene oxide . . . . . 842
Furan . . . . . 846
Anisole . . . . . 850
Diphenyl ether . . . . . 854

Bis(2-chloroethyl)ether . . . . . . . . . . 858
2-Chloroethyl vinyl ether . . . . . . . . . . 862
4-Chlorophenyl phenyl ether . . . . . . . . . . 866
4-Bromophenyl phenyl ether . . . . . . . . . . 870
4.4 Commentary on the Physical-Chemical Properties and Environmental Fate . . . . . . . 874
4.5 References . . . . . . . . . . 876

## 4.1 List of Chemicals and Data Compilations

Common Name: Dimethyl ether
Synonym: methyl ether, oxapropane, oxybismethane
Chemical Name: dimethyl ether, methyl ether,
CAS Registry No: 115-10-6
Molecular Formula: $CH_3OCH_3$
Molecular Weight: 46.07
Melting Point (°C):
-138.50 (Stull 1947; Stephenson & Malanowski 1987)
-141.49 (Riddick et al. 1986)
Boiling Point (°C):
-24.00 (Seidell 1941; Lange 1971; Urano et al. 1981)
-24.84 (Riddick et al. 1986)
-23.60 (Stephenson & Malanowski 1987)
Density ($g/cm^3$ at 20°C):
0.6689 (Riddick et al. 1986)
0.6612 (25°C, Riddick et al. 1986)
Molar Volume ($cm^3/mol$):
68.87 (calculated from density)
60.90 (LeBas method)
0.706 (intrinsic volume: $V_I/100$, Taft et al. 1985)
0.735 (estimated intrinsic volume: $V_I/100$, Kamlet et al. 1986)
70.0 (calculated from density, Wang et al. 1992)
Molecular Volume ($A^3$):
Total Surface Area, TSA ($A^2$):
Heat of Fusion, $\Delta H_{fus}$, (kcal/mol):
1.181 (Riddick et al. 1986)
Entropy of Fusion, $\Delta S_{fus}$ (cal/mol K or e.u.):
Fugacity Ratio at 25°C, F: 1.0

Water Solubility ($g/m^3$ or mg/L at 25°C):
71000 (Seidell 1941; Lange 1971; quoted, Urano et al. 1981)
270876 (quoted, Kier & Hall 1976)
292479 (calculated- $\chi$ , Kier & Hall 1976)
353000 (24°C, quoted, Riddick et al. 1986)
47483 (calculated-$V_M$, Wang et al. 1992)

Vapor Pressure (Pa at 25°C):
678086 (calculated-Antoine eqn. regression, Stull 1947)
101308 (quoted, Hine & Mookerjee 1975)
593300 (quoted, Riddick et al. 1986)
575530,593338 (calculated-Antoine eqn., Stephenson & Malanowski 1987)

Henry's Law Constant (Pa $m^3$/mol):

101.0 (calculated-1/$K_{AW}$, $C_W/C_A$, reported as exptl., Hine & Mookerjee 1975)
49.50 (calculated-group contribution method, Hine & Mookerjee 1975)
105.7 (calculated-bond contribution method, Hine & Mookerjee 1975)
77.43 (calculated-P/C using Riddick et al. 1986 data)

Octanol/Water Partition Coefficient, log $K_{OW}$:

0.10 (shake flask-GC, Leo et al. 1975)
0.10, 0.12 (quoted, calculated-f const., Chou & Jurs 1979)
0.10, 0.23 (quoted, calculated-molar volume & solvatochromic p., Taft et al. 1985)
0.10, -0.27 (quoted, calculated-MO, Bodor et al. 1989)
0.10, -0.188 (quoted, calculated-$\pi$ substituent consts., Bodor et al. 1989)
0.10 (quoted, Sangster 1989)
0.10 (recommended, Sangster 1989)

Bioconcentration Factor, log BCF:

Sorption Partition Coefficient, log $K_{OC}$:

Half-Lives in the Environment:

Air: disappearance half-life of less than 0.24 hour from air for the reaction with OH radicals (USEPA 1974; quoted, Darnall et al. 1976).
Surface water:
Ground water:
Sediment:
Soil:
Biota:

Environmental Fate Rate Constant and Half-Lives:

Volatilization:
Photolysis:
Oxidation: rate constant of $5.7 \times 10^{-14}$ $cm^3$ $molecule^{-1}$ $sec^{-1}$ for the reaction with $O(_3P)$ at room temperature (Herron & Huie 1973; quoted, Gaffney & Levine 1979); measured rate constant of > $3.0 \times 10^{-15}$ $cm^3$ $molecule^{-1}$ $sec^{-1}$ for the reaction with $NO_3$ radicals at room temperature (Wallington et al. 1986; quoted, Sabljic & Güsten 1990; Atkinson 1991) while predicted rate constant of $2.92 \times 10^{-15}$ $cm^3$ $molecule^{-1}$ $sec^{-1}$ for the same reaction at room temperature (Sabljic & Güsten 1990); experimentally determined rate constant of $2.98 \times 10^{-12}$ $cm^3$ $molecule^{-1}$ $sec^{-1}$

for the reaction with hydroxyl radicals in air at room temperature with an estimated rate constant of $1.8 \times 10^{-12}$ $cm^3$ $molecule^{-1}$ $sec^{-1}$ under same conditions (Atkinson 1987).

Hydrolysis:

Biodegradation:

Biotransformation:

Bioconcentration, Uptake ($k_1$) and Elimination ($k_2$) Rate Constants or Half-Lives:

Common Name: Diethyl ether
Synonym: ether, ethyl ether, ethoxyethane, ethyl oxide, 3-oxapentane, 1,1'-oxybisethane, sulfuric ether
Chemical Name: ether, diethyl ether, ethoxyethane, ethyl oxide, 3-oxapentane, 1,1'-oxybisethane
CAS Registry No: 60-29-7
Molecular Formula: $CH_3CH_2OCH_2CH_3$
Molecular Weight: 74.12
Melting Point (°C):
-116.3 (Stull 1947; Riddick et al. 1986; Stephenson & Malanowski 1987)
-116/-123 (Verschueren 1983)
Boiling Point (°C):
34.9 (Kahlbaum & Arndt 1898)
35.0 (Seidell 1941; Lange 1971; Urano et al. 1981; Verschueren 1983; Michael et al. 1988)
34.43 (Boublik et al. 1984; Riddick et al. 1986)
34.48 (Stephenson & Malanowski 1987)
Density (g/cm³ at 20°C):
0.7160 (Kahlbaum & Arndt 1898)
0.71378 (Kier & Hall 1976)
0.72307 (calculated- $\chi$ , Kier & Hall 1976)
0.7135 (Verschueren 1983)
0.71361 (Riddick et al. 1986)
0.70782 (25°C, Riddick et al. 1986)
Molar Volume (cm³/mol):
1.046 ($V_M$/100, Taft et al. 1985; Leahy 1986; Kamlet et al. 1986,1987)
1.038 ($V_M$/100, Kamlet et al. 1986,1987)
0.505 (intrinsic volume: $V_I$/100, Leahy 1986; Kamlet et al. 1987,1988)
51.0 (intrinsic volume, Abernethy et al. 1988)
92.0 (LeBas method, Abernethy et al. 1988)
106.10 (LeBas method)
103.87 (calculated from density)
Molecular Volume (A³):
Total Surface Area, TSA (A²):
281.1 (Amidon et al. 1975)
Heat of Fusion, $\Delta H_{fus}$, (kcal/mol):
1.630 (quoted, Riddick et al. 1986)
Entropy of Fusion, $\Delta S_{fus}$ (cal/mol K or e.u.):
Fugacity Ratio at 25°C, F: 1.0

Water Solubility (g/m³ or mg/L at 25°C):
60270 (thermostatic volumetric, Hill 1923)
60300 (method of Hill 1923, Bennett & Phillip 1928)

69000 (Seidell 1941; Lange 1971; quoted, Urano et al. 1981)
64111 (quoted, Hansch et al. 1968)
29508 (calculated-$K_{OW}$, Hansch et al. 1968)
42768, 47574 (quoted, calculated- $\chi$ , Kier & Hall 1976)
64556 (quoted average, Valvani et al. 1981; Isnard & Lambert 1989)
35476 (calculated-$K_{OW}$, Valvani et al. 1981)
60500 (quoted, Verschueren 1983)
54946, 54946 (quoted, calculated-molar volume & solvatochromic p., Leahy 1986)
54946, 32355 (quoted, calculated-molar volume & solvatochromic p., Kamlet et al. 1986)
54946, 52473 (quoted, calculated-molar volume & solvatochromic p., Kamlet et al. 1987)
60400 (quoted, Riddick et al. 1986)
64556, 50111 (quoted, calculated-fragment solubility const., Wakita et al. 1986)
64094, 46112 (quoted, calculated-$V_M$, Wang et al. 1992)

Vapor Pressure (Pa at 25°C):
74691 (calculated-Antoine eqn. regression, Stull 1947)
58919 (20°C, quoted, Verschueren 1983)
63343 (21.82°C, quoted, Boublik et al. 1984)
71610, 71238 (calculated-Antoine eqn., Boublik et al. 1984)
71568 (quoted, Kamlet et al. 1986)
71622 (quoted, Riddick et al. 1986)
71623, 71604 (calculated-Antoine eqn., Stephenson & Malanowski 1987)

Henry's Law Constant (Pa $m^3$/mol):
130.1 (calculated-$1/K_{AW}$, $C_W/C_A$, reported as exptl., Hine & Mookerjee 1975)
90.0 (calculated-group contribution method, Hine & Mookerjee 1975)
236.7 (calculated-bond contribution method, Hine & Mookerjee 1975)
87.89 (calculated-P/C using Riddick et al. 1986 data)

Octanol/Water Partition Coefficient, log $K_{OW}$:
0.83 (20°C, shake flask-CR, Collander 1951; quoted, Sangster 1989)
1.03 (Hansch et al. 1968)
1.03, 1.02 (quoted, calculated- $\chi$ , Murray et al. 1975; Kier & Hall 1976)
0.83 (quoted, Valvani et al. 1981; Isnard & Lambert 1989)
0.77, 0.83 (quoted, Verschueren 1983)
0.89, 1.18 (quoted, calculated-molar volume & solvatochromic p., Taft et al. 1985)
0.89, 1.01 (quoted, calculated-molar volume & solvatochromic p., Leahy 1986)
0.88 (selected, Abernethy et al. 1988)

0.89, 0.91 (quoted, calculated-molar volume & solvatochromic p., Kamlet et al. 1988)
0.83, 0.80 (quoted, calculated-MO, Bodor et al. 1989)
0.89, 0.87 (quoted, calculated-$\pi$ consts., Bodor et al. 1989)
0.77, 0.89 (quoted, Sangster 1989)
0.89 (recommended, Sangster 1989)
0.87 (quoted, Van Leeuwen et al. 1992)

Bioconcentration Factor, log BCF:

Sorption Partition Coefficient, log $K_{OC}$:

Half-Lives in the Environment:

Air: disappearance half-life less than 0.24 hour from the air for the reaction with OH radicals (USEPA 1974; quoted, Darnall et al. 1976).
Surface water:
Ground water:
Sediment:
Soil:
Biota:

Environmental Fate Rate Constant and Half-Lives:

Volatilization:
Photolysis:
Oxidation: experimentally determined rate constant of $1.34 \times 10^{-11}$ cm$^3$ molecule$^{-1}$ sec$^{-1}$ for the reaction with hydroxyl radicals in air at room temperature with an estimated rate constant of $1.06 \times 10^{-11}$ cm$^3$ molecule$^{-1}$ sec$^{-1}$ under same conditions (Atkinson 1987).
Hydrolysis:
Biodegradation:
Biotransformation:
Bioconcentration, Uptake ($k_1$) and Elimination ($k_2$) Rate Constants or Half-Lives:

Common Name: Methyl *t*-butyl ether
Synonym: 3-oxa-3,3-dimethylbutane
Chemical Name: methyl *t*-butyl ether
CAS Registry No: 1634-04-4
Molecular Formula: $CH_3$-O-C$(CH_3)_3$
Molecular Weight: 88.15
Melting Point (°C):

-109 (Windholz 1983; Stephenson & Malanowski 1987; Budavari 1989)

Boiling Point (°C):

54.0 (Bennett & Phillip 1928)
55.20 (Kier & Hall 1976; Windholz 1983; Budavari 1989)
53.6 (calculated- $\chi$ , Kier & Hall 1976)
55.10 (Stephenson & Malanowski 1987)

Density (g/cm³ at 20°C):

0.7578 (Bennett & Phillip 1928)
0.7405 (Kier & Hall 1976)
0.73626 (calculated- $\chi$ , Kier & Hall 1976)
0.7404 (Windholz 1983; Budavari 1989)

Molar Volume (cm³/mol):

60.0 (intrinsic volume, Abernethy et al. 1988)
129.0 (LeBas method, Abernethy et al. 1988)
102.0 (calculated-S & $K_{OW}$, Wang et al. 1992)
127.5 (LeBas method)
119.06 (calculated from density)

Molecular Volume (A³):
Total Surface Area, TSA (A²):
Heat of Fusion, $\Delta H_{fus}$, (kcal/mol):
Entropy of Fusion, $\Delta S_{fus}$ (cal/mol K or e.u.):
Fugacity Ratio at 25°C, F: 1.0

Water Solubility (g/m³ or mg/L at 25°C):

51600 (thermostatic volumetric, Bennett & Phillip 1928)
54341 (quoted, Hansch et al. 1968)
32220 (calculated-$K_{OW}$, Hansch et al. 1968)
48000 (quoted, Windholz 1983; Budavari 1989)
54353, 23186 (quoted, calculated-f solubility const., Wakita et al. 1986)
42000 (19.8°C, shake flask-GC/TC, Stephenson 1992)
31000 (29.6°C, shake flask-GC/TC, Stephenson 1992)
54322 (quoted, Wang et al. 1992)

Vapor Pressure (Pa at 25°C):

32659 (quoted, Windholz 1983; Budavari 1989)

33545 (calculated-Antoine eqn., Stephenson & Malanowski 1987)

Henry's Law Constant (Pa $m^3$/mol):

59.46 (calculated as $1/K_{AW}$, $C_W/C_A$, reported as exptl., Hine & Mookerjee 1975)

142.64 (calculated-group contribution method, Hine & Mookerjee 1975)

304.96 (calculated-bond contribution method, Hine & Mookerjee 1975)

59.98 (calculated-P/C using Windholz 1983 data)

Octanol/Water Partition Coefficient, log $K_{OW}$:

1.06 (quoted, Hansch et al. 1968; Kier & Hall 1976)

1.06, 1.16 (quoted, predicted- $\chi$ , Murray et al. 1975)

1.16, 1.01 (calculated- $\chi$ , Kier & Hall 1976)

1.30 (calculated-f. const., Hansch & Leo 1979; quoted, Veith et al. 1983)

0.94 (shake flask-GC, Funasaki et al. 1985; quoted, Sangster 1989)

1.30 (selected, Abernethy et al. 1988)

0.94 (recommended, Sangster 1989)

0.94 (quoted, Van Leeuwen et al. 1992)

Bioconcentration Factor, log BCF:

Sorption Partition Coefficient, log $K_{OC}$:

Half-Lives in the Environment:

Air: disappearance half-life of less than 0.24 hour from air for the reaction with OH radicals (USEPA 1974; quoted, Darnall et al. 1976).

Surface water:

Ground water:

Sediment:

Soil:

Biota:

Environmental Fate Rate Constant and Half-Lives:

Volatilization:

Photolysis:

Oxidation:

Hydrolysis:

Biodegradation:

Biotransformation:

Bioconcentration, Uptake ($k_1$) and Elimination ($k_2$) Rate Constants or Half-Lives:

Common Name: Di-*n*-propyl ether
Synonym: 4-oxaheptane, 1,1'-oxibispropane, 1-propoxypropane, propyl ether
Chemical Name: di-*n*-propyl ether, propyl ether, 4-oxaheptane
CAS Registry No: 111-43-3
Molecular Formula: $(n\text{-}C_3H_7)_2O$
Molecular Weight: 102.18
Melting Point (°C):
- -122.0 (Stull 1947; Stephenson & Malanowski 1987)
- -123.2 (Riddick et al. 1986)

Boiling Point (°C):
- 90.0 (Seidell 1941; Lange 1971; Urano et al. 1981)
- 89.95, 90.10 (Boublik et al. 1984)
- 90.08 (Riddick et al. 1986)
- 90.5 (Stephenson & Malanowski 1987)

Density ($g/cm^3$ at 20°C):
- 0.736 (Kier & Hall 1976)
- 0.7491 (calculated- $\chi$ , Kier & Hall 1976)
- 0.7466 (Riddick et al. 1986)
- 0.7419 (25°C, Riddick et al. 1986)

Molar Volume ($cm^3/mol$):
- 136.86 (calculated from density)
- 151.60 (LeBas method)
- 1.359 ($V_M/100$, Taft et al. 1985; Leahy 1986; Kamlet et al. 1986,1987)
- 0.699 (intrinsic volume: $V_I/100$, Leahy 1986, Kamlet et al. 1987,1988)
- 139.0 (calculated from density, Wang et al. 1992)

Molecular Volume ($A^3$):
Total Surface Area, TSA ($A^2$):
- 344.8 (Amidon et al. 1975)

Heat of Fusion, $\Delta H_{fus}$, (kcal/mol):
- 2.574 (quoted, Riddick et al. 1986)

Entropy of Fusion, $\Delta S_{fus}$ (cal/mol K or e.u.):
Fugacity Ratio at 25°C, F: 1.0

Water Solubility ($g/m^3$ or mg/L at 25°C):
- 2500 (thermostatic volumetric, Bennett & Philip 1928)
- 3000 (Seidell 1941; Lange 1971; quoted, Urano et al. 1981)
- 4925 (quoted, Hansch et al. 1968)
- 2485 (calculated-$K_{OW}$, Hansch et al. 1968)
- 2508 (quoted, Hine & Mookerjee 1975)
- 3535 (quoted, Kier & Hall 1976)
- 3325 (calculated- $\chi$ , Kier & Hall 1976)

3306 (quoted average, Valvani et al. 1981)
2880 (calculated-$K_{OW}$, Valvani et al. 1981)
3710, 5005 (quoted, calculated-molar volume & solvatochromic p., Leahy 1986)
3710, 3383 (quoted, calculated-molar volume & solvatochromic p., Kamlet et al. 1986)
4900 (quoted, Riddick et al. 1986)
4891, 4163 (quoted, calculated-f solubility const., Wakita et al. 1986)
4882, 3378 (quoted, calculated-$V_M$, Wang et al. 1992)

Vapor Pressure (Pa at 25°C):
9072 (calculated-Antoine eqn. regression, Stull 1947)
8666 (quoted, Hine & Mookerjee 1975)

9041 (26.59°C, quoted, Boublik et al. 1984)
8378, 8320 (calculated-Antoine eqn., Boublik et al. 1984)
8331 (quoted, Kamlet et al. 1986)
8334 (quoted, Riddick et al. 1986)
8334 (calculated-Antoine eqn., Stephenson & Malanowski 1987)

Henry's Law Constant (Pa $m^3$/mol):
350.1 (calculated as $1/K_{AW}$, $C_W/C_A$, reported as exptl., Hine & Mookerjee 1975)
175.5 (calculated-group contribution method, Hine & Mookerjee 1975)
594.6 (calculated-bond contribution method, Hine & Mookerjee 1975)
173.8 (calculated-P/C using Riddick et al. 1986 data)

Octanol/Water Partition Coefficient, log $K_{OW}$:
2.03 (Hansch et al. 1968; quoted, Valvani et al. 1981)
2.03 (Leo et al. 1971; quoted, Funasaki et al. 1985)
2.03, 1.99 (quoted, calculated- $\chi$ , Murray et al. 1975; Kier & Hall 1976)
2.03, 2.10 (quoted, calculated-molar volume & solvatochromic p., Taft et al. 1985)
2.03, 2.04 (quoted, calculated-molar volume & solvatochromic p., Leahy 1986)
2.03, 1.99 (quoted, calculated-molar volume & solvatochromic p., Kamlet et al. 1988)
2.03, 1.86 (quoted, calculated-MO, Bodor et al. 1989)
2.03, 1.93 (quoted, calculated-$\pi$ substituent consts., Bodor et al. 1989)
2.03 (quoted, Sangster 1989)
2.03 (recommended, Sangster 1989)

Bioconcentration Factor, log BCF:

Sorption Partition Coefficient, log $K_{OC}$:

Half-Lives in the Environment:

Air: disappearance half-life of less than 0.24 hour from air for the reaction with OH radicals (USEPA 1974; quoted, Darnall et al. 1976).

Surface water:

Ground water:

Sediment:

Soil:

Biota:

Environmental Fate Rate Constant and Half-Lives:

Volatilization:

Photolysis:

Oxidation: experimentally determined rate constant of $1.68 \times 10^{-11}$ $cm^3$ $molecule^{-1}$ $sec^{-1}$ for the reaction with hydroxyl radicals in air at room temperature with an estimated rate constant of $1.57 \times 10^{-11}$ $cm^3$ $molecule^{-1}$ under same conditions (Atkinson 1987).

Hydrolysis:

Biodegradation:

Biotransformation:

Bioconcentration, Uptake ($k_1$) and Elimination ($k_2$) Rate Constants or Half-Lives:

Common Name: Diisopropyl ether

Synonym: diisopropyloxyde, isopropyl ether, 2-isopropoxypropane, 2,2'-oxybispropane, 3-oxa-2,4-dimethylpentane, IPE, DIPE

Chemical Name: diisopropyl ether, isopropyl ether, 2-isopropoxypropane, 2,2'-oxybispropane, 3-oxa-2,4-dimethylpentane

CAS Registry No: 108-20-3

Molecular Formula: $[(CH_3)_2CH]_2O$

Molecular Weight: 102.18

Melting Point (°C):

-60 (Stull 1947)

-86/-60 (Verschueren 1983)

-85.5 (Riddick et al. 1986; Stephenson & Malanowski 1987)

Boiling Point (°C):

91.0 (Bennett & Phillip 1928)

69.0 (Verschueren 1983; Michael et al. 1988)

68.34 (Boublik et al. 1984)

68.51 (Riddick et al. 1986; Stephenson & Malanowski 1987)

68.0 (Banerjee et al. 1990)

Density ($g/cm^3$ at 20°C):

0.7360 (Bennett & Phillip 1928)

0.730 (Verschueren 1983)

0.7239 (Riddick et al. 1986)

0.71854 (25°C, Riddick et al. 1986)

Molar Volume ($cm^3/mol$):

1.411 ($V_M/100$, Kamlet et al. 1986,1987)

0.699 (intrinsic volume: $V_I/100$, Kamlet et al. 1987,1988)

70.0 (intrinsic volume, Abernethy et al. 1988)

152.0 (LeBas method, Abernethy et al. 1988)

141.0 (calculated from density, Wang et al. 1992)

151.6 (LeBas method)

139.97 (calculated from density)

Molecular Volume ($A^3$):

Total Surface Area, TSA ($A^2$):

Heat of Fusion, $\Delta H_{fus}$, (kcal/mol):

2.876 (quoted, Riddick et al. 1986)

Entropy of Fusion, $\Delta S_{fus}$ (cal/mol K or e.u.):

Fugacity Ratio at 25°C, F: 1.0

Water Solubility ($g/m^3$ or mg/L at 25°C):

4900 (thermostatic volumetric, Bennett & Phillip 1929)

2039 (quoted, Hine & Mookerjee 1975)

9000 (20°C, quoted, Verschueren 1983)

2039, 2567 (quoted, calculated-molar volume & solvatochromic p., Kamlet et al. 1986)
2039, 5121 (quoted, calculated-molar volume & solvatochromic p., Kamlet et al. 1987)
12000 (20°C, quoted, Riddick et al. 1986)
2039, 6597 (quoted, calculated-f solubility const., Wakita et al. 1986)
7900 (20°C, shake flask-GC/TC, Stephenson 1992)
5400 (31°C, shake flask-GC/TC, Stephenson 1992)
4996, 2875 (calculated-$K_{OW}$, $V_M$, Wang et al. 1992)

Vapor Pressure (Pa at 25°C):
21408 (calculated-Antoine eqn. regression, Stull 1947)
20132 (quoted, Hine & Mookerjee 1975)
17329 (20°C, Verschueren 1983)
20092 (quoted, Boublik et al. 1984)
19954, 20120 (calculated-Antoine eqn., Boublik et al. 1984)
19940 (quoted, Kamlet et al. 1986)
19880 (quoted, Riddick et al. 1986)
19952, 19887 (calculated-Antoine eqn., Stephenson & Malanowski 1987)
19862, 10851 (quoted, calculated-solvatochromic p. & UNIFAC, Banerjee et al. 1990)

Henry's Law Constant (Pa $m^3$/mol):
1010 (calculated as $1/K_{AW}$, $C_W/C_A$, reported as exptl., Hine & Mookerjee 1975)
483.33 (calculated-group contribution method, Hine & Mookerjee 1975)
594.63 (calculated-bond contribution method, Hine & Mookerjee 1975)

Octanol/Water Partition Coefficient, log $K_{OW}$:
1.52 (shake flask-GC, Funasaki et al. 1985)
1.56 (calculated-f const., Hansch & Leo 1979; quoted, Veith et al. 1983)
1.56 (selected, Abernethy et al. 1988)
2.03, 1.95 (quoted, calculated-molar volume & solvatochromic p., Kamlet et al. 1988)
1.52 (recommended, Sangster 1989)
1.52 (quoted, Van Leeuwen et al. 1992)

Bioconcentration Factor, log BCF:

Sorption Partition Coefficient, log $K_{OC}$:

Half-Lives in the Environment:

Air: disappearance half-life of less than 0.24 hour from air for the reaction with OH radicals (USEPA 1974; quoted, Darnall et al. 1976).

Surface water:

Ground water:

Sediment:

Soil:

Biota:

Environmental Fate Rate Constant and Half-Lives:

Volatilization:

Photolysis:

Oxidation: Rate constant of $9.8 \times 10^{-12}$ $cm^3$ $molecule^{-1}$ $sec^{-1}$ at 298 K for the reaction with hydroxyl radicals in the air was obtained using both relative (at 295 K) and absolute techniques (over 240-440 K) and the products of the simulated atmospheric oxidation were identified as isopropyl acetate and formaldehyde using FT-IR spectroscopy (Wallington et al. 1993).

Hydrolysis:

Biodegradation:

Biotransformation:

Bioconcentration, Uptake ($k_1$) and Elimination ($k_2$) Rate Constants or Half-Lives:

Common Name: Butylethyl ether
Synonym: butylethyl ether, 1-ethoxybutane, *n*-butylethyl ether, 3-oxaheptane
Chemical Name: butylethyl ether, 1-ethoxybutane, *n*-butylethyl ether
CAS Registry No: 628-81-9
Molecular Formula: $C_4H_9OCH_2CH_3$
Molecular Weight: 102.18
Melting Point (°C):
-103 (Riddick et al. 1986)
Boiling Point (°C):
92.27 (Boublik et al. 1984)
92.24 (Riddick et al. 1986)
Density (g/cm³ at 20°C):
0.7490 (Kier & Hall 1976)
0.7491 (calculated- $\chi$ , Kier & Hall 1976)
0.7495 (Riddick et al. 1986)
0.7448 (25°C, Riddick et al. 1986)
Molar Volume (cm³/mol):
150.5 (LeBas method)
136.38 (calculated from density)
Molecular Volume (A³):
Total Surface Area, TSA (A²):
Heat of Fusion, $\Delta H_{fus}$, (kcal/mol):
Entropy of Fusion, $\Delta S_{fus}$ (cal/mol K or e.u.):
Fugacity Ratio at 25°C, F: 1.0

Water Solubility (g/m³ or mg/L at 25°C):
6500 (20°C, shake flask-GC/TC, Stephenson 1992)
5300 (31.2°C, shake flask-GC/TC, Stephenson 1992)
6000 (selected)

Vapor Pressure (Pa at 25°C):
9087 (calculated-Antoine eqn., Boublik et al. 1984)
7461 (quoted, Riddick et al. 1986)
8000 (selected)

Henry's Law Constant (Pa m³/mol):
136 (calculated-P/C from selected data)

Octanol/Water Partition Coefficient, log $K_{OW}$:

2.03 (shake flask-GC, Hansch & Anderson 1967)

2.03, 1.89 (quoted, calculated-MO, Bodor et al. 1989)

2.03, 1.928 (quoted, calculated-$\pi$ substituent consts., Bodor et al. 1989)

2.03 (recommended, Sangster 1989)

Bioconcentration Factor, log BCF:

Sorption Partition Coefficient, log $K_{OC}$:

Half-Lives in the Environment:

Air: disappearance half-life of less than 0.24 hour from air for the reaction with OH radicals (USEPA 1974; quoted, Darnall et al. 1976).

Surface water:

Ground water:

Sediment:

Soil:

Biota:

Environmental Fate Rate Constant and Half-Lives:

Volatilization:

Photolysis:

Oxidation:

Hydrolysis:

Biodegradation:

Biotransformation:

Bioconcentration, Uptake ($k_1$) and Elimination ($k_2$) Rate Constants or Half-Lives:

Common Name: Di-*n*-butyl ether

Synonym: 1-butoxybutane, butyl ether, dibutyl ether, *n*-butyl ether, 5-oxanonane, 1,1'-oxybisbutane

Chemical Name: butyl ether, dibutyl ether, di-*n*-butyl ether, *n*-butyl ether, 5-oxanonane, 1,1'-oxybisbutane

CAS Registry No: 142-96-1

Molecular Formula: $(n\text{-}C_4H_9)_2O$

Molecular Weight: 130.23

Melting Point (°C):

-95.0 (Verschueren 1983)
-95.2 (Riddick et al. 1986)

Boiling Point (°C):

142.0 (Kier & Hall 1976; Boublik et al. 1984; Michael & Pellizzari 1988)
138.2 (calculated- $\chi$ , Kier & Hall 1976)
141.0 (Verschueren 1983)
140.29 (Boublik et al. 1984; Riddick et al. 1986)

Density ($g/cm^3$ at 20°C):

0.769 (Verschueren 1983)
0.7684 (Riddick et al. 1986)
0.7641 (25°C, Riddick et al. 1986)

Molar Volume ($cm^3/mol$):

0.893 (intrinsic volume: $V_I/100$, Leahy 1986)
1.694 ($V_M/100$, Leahy 1986; Kamlet et al. 1986,1987)
1.693 ($V_M/100$, Kamlet et al. 1987)
89.0 (intrinsic volume, Abernethy et al. 1988)
196.0 (LeBas method, Abernethy et al. 1988)
170.0 (calculated from density, Wang et al. 1992)
169.4 (calculated from density)

Molecular Volume ($A^3$):

Total Surface Area, TSA ($A^2$):

Heat of Fusion, $\Delta H_{fus}$, (kcal/mol):

Entropy of Fusion, $\Delta S_{fus}$ (cal/mol K or e.u.):

Fugacity Ratio at 25°C, F: 1.0

Water Solubility ($g/m^3$ or mg/L at 25°C):

249 (quoted, Kier & Hall 1976)
215 (calculated- $\chi$ , Kier & Hall 1976)
300 (20°C, quoted, Verschueren 1983; Riddick et al. 1986)
254, 452 (quoted, calculated-molar volume & solvatochromic P., Leahy 1986)
254, 320 (quoted, calculated-molar volume & solvatochromic P., Kamlet et al. 1986)
221, 320 (quoted, calculated-fragment solubility const., Wakita et al. 1986)

254, 531 (quoted, calculated-molar volume & solvatochromic P., Kamlet et al. 1987)
230 (19.9°C, shake flask-GC/TC, Stephenson 1992)
230 (30.9°C, shake flask-GC/TC, Stephenson 1992)
221, 319 (quoted, calculated-$V_M$, Wang et al. 1992)

Vapor Pressure (Pa at 25°C):
640 (20°C, quoted, Verschueren 1983)
825, 874 (calculated-Antoine eqn., Boublik et al. 1984)
1662 (quoted, Kamlet et al. 1986)
898 (quoted, Riddick et al. 1986)

Henry's Law Constant (Pa $m^3$/mol):
608.5 (calculated-1/$K_{AW}$, $C_W/C_A$, reported as exptl., Hine & Mookerjee 1975)
350.0 (calculated-group contribution method, Hine & Mookerjee 1975)
1362 (calculated-bond contribution method, Hine & Mookerjee 1975)

Octanol/Water Partition Coefficient, log $K_{OW}$:
3.08 (calculated-f const., Hansch & Leo 1979; quoted, Veith et al. 1983)
3.21 (shake flask-GC, Funasaki et al. 1984; quoted, Sangster 1989)
3.08 (calculated-molar volume & solvatochromic P., Leahy 1986)
3.08 (quoted, Abernethy et al. 1988)
3.21 (recommended, Sangster 1989)
3.21 (quoted, Van Leeuwen et al. 1992)

Bioconcentration Factor, log BCF:

Sorption Partition Coefficient, log $K_{OC}$:

Half-Lives in the Environment:
Air: disappearance half-life less than 0.24 hour from air for the reaction with OH radicals (USEPA 1974; quoted, Darnall et al. 1976).
Surface water:
Ground water:
Sediment:
Soil:
Biota:

Environmental Fate Rate Constant and Half-Lives:

Volatilization:

Photolysis:

Oxidation:

Hydrolysis:

Biodegradation:

Biotransformation:

Bioconcentration, Uptake ($k_1$) and Elimination ($k_2$) Rate Constants or Half-Lives:

Common Name: 1,2-Propylene oxide
Synonym: 1,2-epoxypropane, methyloxirane, propylene oxide
Chemical Name: 1,2-propylene oxide, 1,2-epoxypropane, propylene oxide
CAS Registry No: 75-56-9
Molecular Formula: $C_3H_6O$
Molecular Weight: 58.08
Melting Point (°C):
-112.1 (Stull 1947)
-104.4 (Verschueren 1983)
-112.0 (Stephenson & Malanowski 1987)
-112.13 (Howard 1989)
Boiling Point (°C):
34.50 (Verschueren 1983)
34.20 (Stephenson & Malanowski 1987)
34.23 (Howard 1989)
Density ($g/cm^3$ at 20°C):
0.859 (0°C, Verschueren 1983)
Molar Volume ($cm^3/mol$):
69.7 (calculated-LeBas method)
Molecular Volume ($A^3$):
Total Surface Area, TSA ($A^2$):
Heat of Fusion, $\Delta H_{fus}$, (kcal/mol):
Entropy of Fusion, $\Delta S_{fus}$ (cal/mol K or e.u.):
Fugacity Ratio at 25°C, F: 1.0

Water Solubility ($g/m^3$ or mg/L at 25°C):
476000 (USEPA 1981; quoted, Howard 1989)
650000 (30°C, quoted, Verschueren 1983)
405000 (20°C, quoted, Verschueren 1983)

Vapor Pressure (Pa at 25°C):
74521 (calculated-Antoine eqn. regression, Stull 1947)
53320 (18°C, quoted, Verschueren 1983)
70929 (Daubert & Danner 1985; quoted, Howard 1989)
71704 (calculated-Antoine eqn., Stephenson & Malanowski 1987)

Henry's Law Constant (Pa $m^3/mol$):
8.653 (calculated-P/C, Howard 1989)

Octanol/Water Partition Coefficient, log $K_{OW}$:

0.03 (Hansch & Leo 1985; quoted, Howard 1989)
0.23 (Deneer et al. 1988; quoted, Verhaar et al. 1992)
0.03 (recommended, Sangster 1989)

Bioconcentration Factor, log BCF:

-0.20, -0.40 (calculated, Howard 1989)

Sorption Partition Coefficient, log $K_{OC}$:

0.623 (estimated-S, Lyman et al. 1982; quoted, Howard 1989)
1.477 (calculated-QSAR, Sabljic 1984; quoted, Howard 1989)

Half-Lives in the Environment:

Air: disappearance half-life of less than 0.24 hour from air for the reaction with OH radicals (USEPA 1974; quoted, Darnall et al. 1976); 19.3 days, based on estimated photooxidation half-life in air by using Güesten et al. 1981 data (GEMS 1986; quoted, Howard 1989).

Surface water: calculated half-life of 9.15 years in natural water by using Güesten et al. 1981 data (Howard 1989).

Ground water:

Sediment:

Soil:

Biota:

Environmental Fate Rate Constant and Half-Lives:

Volatilization: calculated half-lives from a representative or natural river and oligotrophic lake are 3 and 18 days (USEPA 1986; quoted, Howard 1989).

Photolysis:

Oxidation: experimentally determined rate constant for the gas phase reaction with OH radicals in air was $5.2 \times 10^{-13}$ cm$^3$ molecule$^{-1}$ sec$^{-1}$ at room temperature with an estimated value of $4.5 \times 10^{-13}$ cm$^3$ molecule$^{-1}$ sec$^{-1}$ and rate constant of $2.4 \times 10^8$ molecules sec$^{-1}$ for the reaction with photochemically produced OH radicals in water at room temperature (Güesten et al. 1981; quoted, Howard 1989), hence a half-life of 9.15 years can be calculated assuming an average OH radical concentration of $1 \times 10^{-17}$ M in natural water (Howard 1989); a photooxidation half-life of 19.3 days can be calculated for the gas phase reaction with OH radicals in air by using Güesten 1981 data and assuming an average OH radical concn. of $8 \times 10^5$ molecules/cm$^3$ (GEMS 1986; quoted, Howard 1989); experimentally determined rate constant of $(1.11 \pm 0.75) \times 10^{-12}$ cm$^3$ molecule$^{-1}$

$sec^{-1}$ with reference to *n*-butane for the gas phase reaction with OH radicals in air at (23.1 ± 1.1)°C with an atmospheric lifetime of 10 days for an average concentration of $1x10^6$ molecules/$cm^3$ of OH radicals (Edney et al. 1986); measured rate constant of $5.2x10^{-13}$ $cm^3$ $molecule^{-1}$ $sec^{-1}$ for the gas phase reaction with OH radicals in air at room temperature with an estimated rate constant of $5.4x10^{-13}$ $cm^3$ $molecule^{-1}$ $sec^{-1}$ under same conditions (Atkinson 1987).

Hydrolysis: estimated half-life in fresh water to be 11.6 days at pH 7 to 9 and 6.6 days at pH 5 (Bogyo Da et al. 1980; quoted, Howard 1989).

Biodegradation:

Biotransformation:

Bioconcentration, Uptake ($k_1$) and Elimination ($k_2$) Rate Constants or Half-Lives:

Common Name: Furan
Synonym: 1,4-epoxy-1,3-butadiene, divinylene oxide, furfuran, oxole, tetrole
Chemical Name: 1,4-epoxy-1,3-butadiene, divinylene oxide, furan
CAS Registry No: 110-00-9
Molecular Formula: $C_4H_4O$
Molecular Weight: 68.08
Melting Point (°C):
    -85.5 (Verschueren 1983)
    -85.61 (Riddick et al. 1986)
Boiling Point (°C):
    31.30 (Verschueren 1983)
    31.36 (Riddick et al. 1986)
Density ($g/cm^3$ at 20°C):
    0.9370 (Verschueren 1983)
    0.9378 (Riddick et al. 1986)
Molar Volume ($cm^3/mol$):
    73.5 (calculated-LeBas method)
    72.6 (calculated from density)
Molecular Volume ($A^3$):
Total Surface Area, TSA ($A^2$):
Heat of Fusion, $\Delta H_{fus}$, (kcal/mol):
    0.909 (Riddick et al. 1986)
Entropy of Fusion, $\Delta S_{fus}$ (cal/mol K or e.u.):
Fugacity Ratio at 25°C, F: 1.0

Water Solubility ($g/m^3$ or mg/L at 25°C):
    10070 (quoted average, Valvani et al. 1981; Isnard & Lambert 1989)
    9841 (calculated-$K_{OW}$, Valvani et al. 1981)
    10000 (quoted, Verschueren 1983)
    10000 (quoted, Riddick et al. 1986)

Vapor Pressure (Pa at 25°C):
    70109 (21.61°C, quoted, Boublik et al. 1984)
    79933 (calculated-Antoine eqn., Boublik et al. 1984)
    79934 (calculated-Antoine eqn., Stephenson & Malanowski 1987)
    84526 (quoted, Riddick et al. 1986)

Henry's Law Constant (Pa $m^3/mol$):
    575.5 (calculated-P/C using Riddick et al. 1986 data)

Octanol/Water Partition Coefficient, log $K_{OW}$:

1.34 (Hansch & Leo 1979; quoted, Veith et al. 1983; Bodor et al. 1989)
1.34 (quoted, Valvani et al. 1981; Isnard & Lambert 1989)
1.13 (estimated-HPLC, Garst 1984; quoted, Sangster 1989)
1.14, 1.35 (estimated-MO, $\pi$ substituent consts., Bodor et al. 1989)
1.34 (quoted, Sangster 1989)
1.34 (recommended, Sangster 1989)
1.34 (quoted, Van Leeuwen et al. 1992)

Bioconcentration Factor, log BCF:

Sorption Partition Coefficient, log $K_{OC}$:

Half-Lives in the Environment:

Air: disappearance half-life of less than 0.24 hour from air for the reaction with OH radicals (USEPA 1974; quoted, Darnall et al. 1976).
Surface water: half-life of 1.0 hour, estimated from oxidation rate by singlet oxygen of $1.4 \times 10^{8}$ $M^{-1}$ $sec^{-1}$ (Mill 1980; quoted, Mill & Mabey 1985).
Ground water:
Sediment:
Soil:
Biota:

Environmental Fate Rate Constant and Half-Lives:

Volatilization:
Photolysis:
Oxidation: oxidation rate by singlet oxygen of $1.4 \times 10^{8}$ $M^{-1}$ $sec^{-1}$ (Mill 1980; quoted, Mill & Mabey 1985); rate constant of $(3.98 \pm 0.35) \times 10^{-11}$ $cm^3$ $molecule^{-1}$ $sec^{-1}$ relative to that of isoprene for the reaction with hydroxyl radicals in the gas phase at $22 \pm 2$°C (Tuazon et al. 1984); estimated rate constant of $4.07 \times 10^{-11}$ $cm^3$ $molecule^{-1}$ $sec^{-1}$ for the reaction with hydroxyl radicals in air (Atkinson 1987,1988; quoted, Müller & Klein 1991); rate constant of $4.046 \times 10^{-11}$ $cm^3$ $molecule^{-1}$ $sec^{-1}$ for the reaction with OH radicals in the gas phase at 298 K and $1.439 \times 10^{-12}$ $cm^3$ $molecule^{-1}$ $sec^{-1}$ for the reaction with $NO_3$ radicals in the gas phase at $295 \pm 1$ K (Atkinson et al. 1985 also quoted measured and compiled data from Atkinson 1986 and Atkinson et al. 1988, Sabljic & Güsten 1990; Atkinson 1991).

Hydrolysis:
Biodegradation:
Biotransformation:
Bioconcentration, Uptake ($k_1$) and Elimination ($k_2$) Rate Constants or Half-Lives:

Common Name: 2-Methylfuran
Synonym: silvan, sylvan
Chemical Name: 2-methylfuran
CAS Registry No: 534-22-5
Molecular Formula: $C_5H_6O$
Molecular Weight: 82.10
Melting Point (°C):
-88.7 (Verschueren 1983)
-88.0 (Stephenson & Malanowski 1987)
Boiling Point (°C):
63/65.6 (Verschueren 1983)
63.0 (Stephenson & Malanowski 1987)
Density (g/cm$^3$ at 20°C):
0.913 (Verschueren 1983)
Molar Volume (cm$^3$/mol):
84.20 (calculated-LeBas method)
89.92 (calculated from density)
Molecular Volume (A$^3$):
Total Surface Area, TSA (A$^2$):
Heat of Fusion, $\Delta H_{fus}$, (kcal/mol):
Entropy of Fusion, $\Delta S_{fus}$ (cal/mol K or e.u.):
Fugacity Ratio at 25°C, F: 1.0

Water Solubility (g/m$^3$ or mg/L at 25°C):
3000 (20°C, quoted, Verschueren 1983)

Vapor Pressure (Pa at 25°C):
18929 (20°C, quoted, Verschueren 1983)
29993 (30°C, quoted, Verschueren 1983)
23089 (calculated-Antoine eqn., Boublik et al. 1984)
23248 (calculated-Antoine eqn., Stephenson & Malanowski 1987)

Henry's Law Constant (Pa m$^3$/mol):
518 (20°C, calculated-P/C using Verschueren 1983 data)

Octanol/Water Partition Coefficient, log $K_{OW}$:
1.85 (quoted, Sangster 1989)
1.85 (recommended, Sangster 1989)

Bioconcentration Factor, log BCF:

Sorption Partition Coefficient, log $K_{OC}$:

Half-Lives in the Environment:

Air: disappearance half-life of less than 0.24 hour from air for the reaction with OH radicals (USEPA 1974; quoted, Darnall et al. 1976).

Surface water:

Ground water:

Sediment:

Soil:

Biota:

Environmental Fate Rate Constant and Half-Lives:

Volatilization:

Photolysis:

Oxidation:

Hydrolysis:

Biodegradation:

Biotransformation:

Bioconcentration, Uptake ($k_1$) and Elimination ($k_2$) Rate Constants or Half-Lives:

Common Name: Tetrahydrofuran
Synonym: 1,4-epoxybutane, diethylene oxide, oxacyclopentane, tetramethylene oxide
Chemical Name: 1,4-epoxybutane, diethylene oxide, oxacyclopentane, tetrahydrofuran, tetramethylene oxide
CAS Registry No: 109-99-9
Molecular Formula: $C_4H_8O$
Molecular Weight: 72.11
Melting Point (°C):
-108.5 (Verschueren 1983)
-108.39 (Riddick et al. 1986)
Boiling Point (°C):
65.50 (Verschueren 1983)
65.97 (Boublik et al. 1984; Riddick et al. 1986)
66.0 (Michael et al. 1988)
Density (g/cm$^3$ at 20°C):
0.8880 (Verschueren 1983)
0.8892 (Riddick et al. 1986)
Molar Volume (cm$^3$/mol):
0.911 ($V_M$/100, Taft et al. 1985; Leahy 1986; Kamlet et al. 1987)
0.455 (intrinsic volume: $V_I$/100, Leahy 1986; Kamlet et al. 1987)
81.0 (LeBas method, Abernethy et al. 1988)
41.0 (intrinsic volume, Abernethy et al. 1988)
88.3 (calculated-LeBas method)
81.15 (calculated from density)
Molecular Volume (A$^3$):
Total Surface Area, TSA (A$^2$):
Heat of Fusion, $\Delta H_{fus}$, (kcal/mol):
2.04 (quoted, Riddick et al. 1986)
Entropy of Fusion, $\Delta S_{fus}$ (cal/mol K or e.u.):
Fugacity Ratio at 25°C, F: 1.0

Water Solubility (g/m$^3$ or mg/L at 25°C):
miscible (Verschueren 1983; Riddick et al. 1986)
300600, 212809 (quoted, calculated-fragment solubility const., Wakita et al. 1986)
2177769, 522391 (quoted, calculated-molar volume & solvatochromic p., Leahy 1986)
2177769, 572790 (quoted, calculated-molar volume & solvatochromic p., Kamlet et al. 1987)

Vapor Pressure (Pa at 25°C):
17526 (20°C, quoted, Verschueren 1983)
26340 (30°C, quoted, Verschueren 1983)

21646 (quoted, Boublik et al. 1984)
21610, 21623 (calculated-Antoine eqn., Boublik et al. 1984)
21600 (quoted, Riddick et al. 1986)
21623, 21904 (calculated-Antoine eqn., Stephenson & Malanowski 1987)

Henry's Law Constant (Pa $m^3$/mol):
7.149 (calculated-1/$K_{AW}$, $C_W/C_A$, reported as exptl., Hine & Mookerjee 1975)
10.33 (calculated-group contribution method, Hine & Mookerjee 1975)
142.6 (calculated-bond contribution method, Hine & Mookerjee 1975)

Octanol Water Partition Coefficient, log $K_{OW}$:
0.46 (calculated-f const., Hansch & Leo 1979; quoted, Veith et al. 1983)
0.22 (shake flask-GC, Funasaki et al. 1985; quoted, Sangster 1989)
0.46, 0.23 (quoted, calculated-molar volume & solvatochromic p., Taft et al. 1985)
0.46, 0.06 (quoted, calculated-molar volume & solvatochromic p., Leahy 1986)
0.46 (quoted, Abernethy et al. 1988; Sangster 1989)
0.46 (recommended, Sangster 1989)
0.46 (quoted, Van Leeuwen et al. 1992)

Bioconcentration Factor, log BCF:

Sorption Partition Coefficient, log $K_{OC}$:

Half-Lives in the Environment:
Air: disappearance half-life of less than 0.24 hour from air for the reaction with OH radicals (USEPA 1974; quoted, Darnall et al. 1976).
Surface water:
Ground water:
Sediment:
Soil: disappearance half-life of 5.7 days was calculated from measured first-order rate constant (Anderson et al. 1991).
Biota:

Environmental Fate Rate Constant and Half-Lives:
Volatilization:
Photolysis:

Oxidation: experimentally determined rate constant of $1.50 \times 10^{-11}$ $cm^3$ $molecule^{-1}$ $sec^{-1}$ for the reaction with hydroxyl radicals in air (Atkinson 1986,87; quoted, Sabljic & Güsten 1990); estimated rate constant of $1.28 \times 10^{-11}$ $cm^3$ $molecule^{-1}$ $sec^{-1}$ for the reaction with hydroxyl radicals in air (Atkinson 1987, 1988; quoted, Müller & Klein 1991); $4.875 \times 10^{-15}$ $cm^3$ $molecule^{-1}$ $sec^{-1}$ for the reaction with $NO_3$ radicals in the gas-phase at 296 ± 2 K (Atkinson et al. 1988; quoted, Sabljic & Güsten 1990; Atkinson 1991).

Hydrolysis:

Biodegradation:

Biotransformation:

Bioconcentration, Uptake ($k_1$) and Elimination ($k_2$) Rate Constants or Half-Lives:

Common Name: Tetrahydropyran
Synonym: pentamethylene oxide, oxacyclohexane
Chemical Name: 1,5-epoxypentane, pentamethylene oxide, oxacyclohexane, tetrahydropyran
CAS Registry No: 142-68-7
Molecular Formula: $C_5H_{10}O$
Molecular Weight: 86.13
Melting Point (°C):
-45 (Riddick et al. 1986)
-44.2 (Stephenson & Malanowski 1987)
Boiling Point (°C):
88.0 (Riddick et al. 1986; Stephenson & Malanowski 1987)
Density ($g/cm^3$ at 20°C):
0.8814 (Riddick et al. 1986)
0.8772 (25°C, Riddick et al. 1986)
Molar Volume ($cm^3/mol$):
107.0 (calculated-LeBas method)
97.72 (calculated from density)
1.077 ($V_M/100$, Leahy 1986; Kamlet et al. 1987)
0.553 (intrinsic volume: $V_I/100$, Leahy 1986; Kamlet et al. 1987)
Molecular Volume ($A^3$):
Total Surface Area, TSA ($A^2$):
Heat of Fusion, $\Delta H_{fus}$, (kcal/mol):
Entropy of Fusion, $\Delta S_{fus}$ (cal/mol K or e.u.):
Fugacity Ratio at 25°C, F: 1.0

Water Solubility ($g/m^3$ or mg/L at 25°C):
66858, 127396 (quoted, calculated-molar volume & solvatochromic p., Leahy 1986)
66858, 136507 (quoted, calculated-molar volume & solvatochromic p., Kamlet et al. 1987)
80200 (quoted, Riddick et al. 1986)
85700 (19.9°C, shake flask-GC/TC, Stephenson 1992)
68800 (31°C, shake flask-GC/TC, Stephenson 1992)

Vapor Pressure (Pa at 25°C):
9536 (calculated-Antoine eqn., Stephenson & Malanowski 1987)

Henry's Law Constant (Pa $m^3/mol$):
12.71 (calculated-$1/K_{AW}$, $C_W/C_A$, reported as exptl., Hine & Mookerjee 1975)
13.94 (calculated-group contribution method, Hine & Mookerjee 1975)
215.9 (calculated-bond contribution method, Hine & Mookerjee 1975)

Octanol/Water Partition Coefficient, log $K_{OW}$:

0.64 (shake flask-GC, Funasaki et al. 1985; quoted, Sangster 1989)

0.82 (recommended, Sangster 1989)

Bioconcentration Factor, log BCF:

Sorption Partition Coefficient, log $K_{OC}$:

Half-Lives in the Environment:

Air: disappearance half-life of less than 0.24 hour from air for the reaction with OH radicals (USEPA 1974; quoted, Darnall et al. 1976).

Surface water:

Ground water:

Sediment:

Soil:

Biota:

Environmental Fate Rate Constant and Half-Lives:

Volatilization:

Photolysis:

Oxidation:

Hydrolysis:

Biodegradation:

Biotransformation:

Bioconcentration, Uptake ($k_1$) and Elimination ($k_2$) Rate Constants or Half-Lives:

Common Name: 1,4-Dioxane
Synonym: 1,4-diethylenedioxide, glycolethyleneether, *p*-dioxane
Chemical Name: 1,4-dioxane
CAS Registry No: 123-91-1
Molecular Formula: $C_4H_8O_2$
Molecular Weight: 88.12
Melting Point (°C):
10.0 (Verschueren 1983)
11.8 (Riddick et al. 1986; Stephenson & Malanowski 1987; Howard 1990)
Boiling Point (°C):
101.0 (Verschueren 1983; Stephenson & Malanowski 1987; Michael et al. 1988)
101.32 (Riddick et al. 1986)
101.1 (Howard 1990)
Density ($g/cm^3$ at 20°C):
1.033 (Verschueren 1983)
1.0336 (Riddick et al. 1986)
1.02797 (25°C, Riddick et al. 1986)
Molar Volume ($cm^3/mol$):
91.8 (calculated-LeBas method)
85.38 (calculated from density)
Molecular Volume ($A^3$):
Total Surface Area, TSA ($A^2$):
Heat of Fusion, $\Delta H_{fus}$, (kcal/mol):
2.978 (quoted, Riddick et al. 1986)
Entropy of Fusion, $\Delta S_{fus}$ (cal/mol K or e.u.):
Fugacity Ratio at 25°C, F: 1.0

Water Solubility ($g/m^3$ or mg/L at 25°C):
miscible (Verschueren 1983; Riddick et al. 1986; Howard 1990)

Vapor Pressure (Pa at 25°C):
5065 (Boublik et al. 1984; quoted, Howard 1990)
4932 (quoted, Verschueren 1983)
4950 (quoted, Riddick et al. 1986)
4915 (calculated-Antoine eqn., Stephenson & Malanowski 1987)
5079, 6092 (quoted, calculated-solvatochromic p. & UNIFAC, Banerjee et al. 1990)

Henry's Law Constant (Pa $m^3$/mol):

0.495 (calculated as $1/K_{AW}$, $C_W/C_A$, reported as exptl., Hine & Mookerjee 1975; quoted, Howard 1990)

0.431 (calculated-group contribution method, Hine & Mookerjee 1975)

1.564 (calculated-bond contribution method, Hine & Mookerjee 1975)

Octanol/Water Partition Coefficient, log $K_{OW}$:

-0.42, 0.01 (observed, calculated-f const., Chou & Jurs 1979)

-0.42 (quoted, Verschueren 1983; quoted, Pinal et al. 1990)

-0.27 (Hansch & Leo 1985; quoted, Howard 1990)

Bioconcentration Factor, log BCF:

Sorption Partition Coefficient, log $K_{OC}$:

1.23 (soil, estimated-$K_{OW}$, Lyman et al. 1982)

Half-Lives in the Environment:

Air: disappearance half-life of less than 0.24 hour from air for the reaction with OH radicals (USEPA 1974; quoted, Darnall et al. 1976); 8.1-81 hours, based on estimated photooxidation half-life in air (Atkinson 1987; quoted, Howard et al. 1991); half-life of 6.69-9.6 hours in the atmosphere, based on estimated reaction rate with photochemically produced hydroxyl radicals (Howard 1990).

Surface water: 672-4320 hours, based on estimated unacclimated aqueous aerobic biodegradation half-life (Howard et al. 1991).

Ground water: 1344-8640 hours, based on estimated unacclimated aqueous aerobic biodegradation half-life (Howard et al. 1991).

Sediment:

Soil: 672-4320 hours, based on estimated unacclimated aqueous aerobic biodegradation half-life (Howard et al. 1991).

Biota:

Environmental Fate Rate Constant and Half-Lives:

Volatilization: estimated Henry' law constant suggests that volatilization for 1,4-dioxane from water and moist soil should be slow; however, it has a moderate vapor pressure, so volatilization from dry soil is possible (Howard 1990).

Photolysis:

Oxidation: photooxidation half-life of 8.1-81 hours in air, based on measured rate constant for the reaction of 1,3,5-trioxane with hydroxyl radicals in air

(Atkinson 1987; quoted, Howard et al. 1991); photooxidation half-life of 6.69-9.6 hours in the atmosphere, based on estimated reaction rate with photochemically produced hydroxyl radicals (Howard 1990); photooxidation half-life of 67 days to 9.1 years in water, based on measured rates for the reaction with hydroxyl radicals in water (Anbar & Neta 1967; Dorfman & Adams 1973; quoted, Howard et al. 1991).

Hydrolysis:

Biodegradation: aqueous aerobic half-life of 672-4320 hours, based on unacclimated aerobic aqueous screening test data with confirmed resistance to biodegradation (Sasaki 1978; Kawasaki 1980; quoted, Howard et al. 1991); aqueous anaerobic half-life of 2688-17280 hours, based on estimated aqueous aerobic biodegradation half-life (Howard et al. 1991).

Biotransformation:

Bioconcentration, Uptake ($k_1$) and Elimination ($k_2$) Rate Constants or Half-Lives:

Common Name: Anisole
Synonym: methoxybenzene
Chemical Name: anisole, methoxybenzene, methyl phenyl ether
CAS Registry No: 100-66-3
Molecular Formula: $C_6H_5OCH_3$
Molecular Weight: 108.14
Melting Point (°C):
-37.3 (Stull 1947)
-37.0 (Verschueren 1983)
-37.5 (Riddick et al. 1986)
Boiling Point (°C):
153.8 (Verschueren 1983)
153.60 (Riddick et al. 1986)
156.0 (Michael et al. 1988)
154.0 (Banerjee et al. 1990)
Density (g/cm$^3$ at 20°C):
0.9954 (Verschueren 1983)
0.99402 (Riddick et al. 1986)
0.98932 (25°C, Riddick et al. 1986)
Molar Volume (cm$^3$/mol):
1.186 ($V_M$/100, Kamlet et al. 1987)
0.639 ($V_I$/100, Kamlet et al. 1988)
127.3 (calculated-LeBas method)
108.7 (calculated from density)
Molecular Volume (A$^3$):
Total Surface Area, TSA (A$^2$):
Heat of Fusion, $\Delta H_{fus}$, (kcal/mol):
Entropy of Fusion, $\Delta S_{fus}$ (cal/mol K or e.u.):
Fugacity Ratio at 25°C, F: 1.0

Water Solubility (g/m$^3$ or mg/L at 25°C):
1514 (shake flask-AS, McGowan et al. 1966; quoted, Vesala 1974)
1536 (shake flask-UV, Vesala 1974)
3192 (calculated-solvatochromic parameters & $V_I$, Leahy 1986)
10569 (quoted, Isnard & Lambert 1988,1989)
2030 (20°C, shake flask-GC/TC, Stephenson 1992)
1860 (29.7°C, shake flask-GC/TC, Stephenson 1992)
1600 (selected)

Vapor Pressure (Pa at 25°C):

496 (calculated-Antoine eqn. regression, Stull 1947)
555 (quoted, Hine & Mookerjee 1975)
413 (quoted, Verschueren 1983)
472 (quoted, Riddick et al. 1986)
204, 383 (quoted, calculated-solvatochromic p. & UNIFAC, Banerjee et al. 1990)
460 (selected)

Henry's Law Constant (Pa $m^3$/mol):

430.8 (calculated-1/$K_{AW}$, $C_W/C_A$, reported as exptl., Hine & Mookerjee 1975)
430.8 (calculated-group contribution method, Hine & Mookerjee 1975)
358.3 (calculated-bond contribution method, Hine & Mookerjee 1975)

Octanol/Water Partition Coefficient, log $K_{OW}$:

2.11 (shake flask-AS, Fujita et al. 1964; quoted, Sangster 1989)
2.04 (shake flask-AS, Rogers & Cammarata 1969; quoted, Sangster 1989)
2.10 (estimated-HPLC, Mirrlees et al. 1976; quoted, Sangster 1989)
2.08 (Hansch & Leo 1979; quoted, Haky & Young 1984)
2.11 (quoted, Veith et al. 1979; Isnard & Lambert 1988,1989)
2.04, 2.11 (quoted, Verschueren 1983)
2.24 (estimated-HPLC-k', Haky & Young 1984; quoted, Sangster 1989)
2.11, 2.11 (quoted, calculated-solvatochromic parameters & $V_I$, Leahy 1986)
2.11, 2.13 (quoted, calculated-solvatochromic parameters & $V_I$, Kamlet et al. 1988)
2.08, 1.81 (quoted, calculated-MO, Bodor et al. 1989)
2.11, 2.061 (quoted, calculated-$\pi$ substituent consts., Bodor et al. 1989)
2.16 (estimated-HPLC, Ge et al. 1987; quoted, Sangster 1989)
2.11 (recommended, Sangster 1989)

Bioconcentration Factor, log BCF:

1.34 (Isnard & Lambert 1988)

Sorption Partition Coefficient, log $K_{OC}$:

Half-Lives in the Environment:

Air: disappearance half-life of less than 0.24 hour from air for the reaction with OH radicals (USEPA 1974; quoted, Darnall et al. 1976).
Surface water:
Ground water:

Sediment:
Soil:
Biota:

Environmental Fate Rate Constant and Half-Lives:

Volatilization:

Photolysis:

Oxidation: overall room temperature rate constant of $(1.57 \pm 0.24)\times10^{-11}$ $cm^3$ $molecule^{-1}$ $sec^{-1}$ for the gas-phase reaction with OH radicals was determined by using a flash photolysis-resonance fluorescence technique (Perry et al. 1977); rate constant of $1.57\times10^{-11}$ $cm^3$ $molecule^{-1}$ $sec^{-1}$ for the gas-phase reaction with OH radicals at 298 K (Atkinson 1985; quoted, Sabljic & Güsten 1990); re-evaluated rate constant of $9.0\times10^{-17}$ $cm^3$ $molecule^{-1}$ $sec^{-1}$ for the gas-phase reaction with $NO_3$ radicals at $(298 \pm 2)$ K (Atkinson et al. 1987); rate constant of $2.08\times10^{-16}$ $cm^3$ $molecule^{-1}$ $sec^{-1}$ for the gas-phase reaction with $NO_3$ radicals at 298 K (Atkinson et al. 1988; quoted, Sabljic & Güsten 1990).

Hydrolysis:

Biodegradation:

Biotransformation:

Bioconcentration, Uptake ($k_1$) and Elimination ($k_2$) Rate Constants or Half-Lives:

Common Name: Phenetole
Synonym: ethoxybenzene, ethyl phenyl ether
Chemical Name: ethoxybenzene, ethyl phenyl ether
CAS Registry No: 103-73-1
Molecular Formula: $C_6H_5$-O-$C_2H_5$
Molecular Weight: 122.17
Melting Point (°C):
-29.52 (Riddick et al. 1986)
-33.0 (Stephenson & Malanowski 1987)
Boiling Point (°C):
169.84 (Riddick et al. 1986)
172.0 (Stephenson & Malanowski 1987)
Density (g/cm$^3$ at 20°C):
0.96514 (Riddick et al. 1986)
0.96049 (25°C, Riddick et al. 1986)
Molar Volume (cm$^3$/mol):
151.4 (calculated-LeBas method)
126.6 (calculated from density)
Molecular Volume (A$^3$):
Total Surface Area, TSA (A$^2$):
Heat of Fusion, $\Delta H_{fus}$, (kcal/mol):
Entropy of Fusion, $\Delta S_{fus}$ (cal/mol K or e.u.):
Fugacity Ratio at 25°C, F: 1.0

Water Solubility (g/m$^3$ or mg/L at 25°C):
1160 (residual volume, Booth & Everson 1948)
550 (shake flask-AS, McGowan et al. 1966; quoted, Vesala 1974)
569 (shake flask-UV, Vesala 1974)
1016, 1114 (quoted, calculated-$K_{OW}$, Valvani et al. 1981)
1200 (quoted, Riddick et al. 1986)

Vapor Pressure (Pa at 25°C):
204 (quoted, Riddick et al. 1986)

Henry's Law Constant (Pa m$^3$/mol):
44.5 (calculated-P/C from selected data)

Octanol/Water Partition Coefficient, log $K_{OW}$:

2.51 (Hansch & Leo 1979; quoted, Haky & Young 1984)
2.51 (quoted, Valvani et al. 1981)
2.68 (HPLC-k', Haky & Young 1984)

Bioconcentration Factor, log BCF:

Sorption Partition Coefficient, log $K_{OC}$:

Half-Lives in the Environment:

Air:
Surface water:
Ground water:
Sediment:
Soil:
Biota:

Environmental Fate Rate Constant and Half-Lives:

Volatilization:
Photolysis:
Oxidation:
Hydrolysis:
Biodegradation:
Biotransformation:
Bioconcentration, Uptake ($k_1$) and Elimination ($k_2$) Rate Constants or Half-Lives:

Common Name: Benzyl ethyl ether
Synonym: (ethoxymethyl)benzene, $\alpha$-ethoxytoluene
Chemical Name: benzyl ethyl ether, (ethoxymethyl)benzene, $\alpha$-ethoxytoluene
CAS Registry No: 539-30-0
Molecular Formula: $C_6H_5CH_2$-O-$C_2H_5$
Molecular Weight: 136.19
Melting Point (°C):
Boiling Point (°C):
185.7 (Weast 1982-83)
185.0 (Dean 1985; Riddick et al. 1986)
Density (g/cm³ at 20°C):
0.9490 (Weast 1982-83)
0.9478 (Dean 1985)
Molar Volume (cm³/mol):
173.6 (calculated-LeBas method)
143.6 (calculated from density)
Molecular Volume (A³):
Total Surface Area, TSA (A²):
Heat of Fusion, $\Delta H_{fus}$, (kcal/mol):
Entropy of Fusion, $\Delta S_{fus}$ (cal/mol K or e.u.):
Fugacity Ratio at 25°C, F: 1.0

Water Solubility (g/m³ or mg/L at 25°C):

Vapor Pressure (Pa at 25°C):
135 (calculated-Antoine eqn. regression, Stull 1947)
100 (quoted, Riddick et al. 1986)

Henry's Law Constant (Pa m³/mol):

Octanol/Water Partition Coefficient, log $K_{OW}$:
2.64 (calculated-f const. as per Rekker 1977, Hanai et al. 1981)

Bioconcentration Factor, log BCF:

Sorption Partition Coefficient, log $K_{OC}$:

Half-Lives in the Environment:

Air: disappearance half-life of less than 0.24 hours from air for the reaction with OH radicals (USEPA 1974; quoted, Darnall et al. 1976).

Surface water:

Ground water:

Sediment:

Soil:

Biota:

Environmental Fate Rate Constant and Half-Lives:

Volatilization:

Photolysis:

Oxidation:

Hydrolysis:

Biodegradation:

Biotransformation:

Bioconcentration, Uptake ($k_1$) and Elimination ($k_2$) Rate Constants or Half-Lives:

Common Name: Diphenyl ether
Synonym: diphenyl oxide, phenyl ether, 1,1'-oxybisbenzene, phenoxybenzene
Chemical Name: diphenyl ether, diphenyloxide, phenylether, phenoxybenzene
CAS Registry No: 101-84-8
Molecular Formula: $C_6H_5$-O-$C_6H_5$
Molecular Weight: 170.20
Melting Point (°C):

28.0 (Verschueren 1983; Pearlman et al. 1984)
26.87 (Riddick et al. 1986)

Boiling Point (°C):

258.0 (Verschueren 1983; Boublik et al. 1984)
259.0 (Pearlman et al. 1984; Michael et al. 1988)
258.06 (Riddick et al. 1986)

Density (g/cm$^3$ at 20°C):

1.073 (Verschueren 1983)

Molar Volume (cm$^3$/mol):

195.60 (LeBas method)
158.62 (calculated from density)
160.4 (Hoy 1970; quoted, Amidon & Williams 1982)

Molecular Volume (A$^3$):

162.38 (Pearlman et al. 1984)

Total Surface Area, TSA (A$^2$):

202.05 (Pearlman et al. 1984)

Heat of Fusion, $\Delta H_{fus}$, (kcal/mol):

4.115 (quoted, Riddick et al. 1986)

Entropy of Fusion, $\Delta S_{fus}$ (cal/mol K or e.u.):
Fugacity Ratio at 25°C, F: 1.0

Water Solubility (g/m$^3$ or mg/L at 25°C):

4000 (residue-volume, Booth & Everson 1948)
18.7 (shake flask-UV, Vesala 1974)
18.0 (shake flask-HPLC, Banerjee et al. 1980; Pearlman et al. 1984)
428, 25.2 (quoted, calculated-$K_{OW}$, Valvani et al. 1981; quoted, Amidon & Williams 1982)
428, 56.4 (quoted, calculated-$K_{OW}$, M.P. & $V_M$, Amidon & Williams 1982)
21 (quoted, Verschueren 1983)
18, 3900 (quoted values, Riddick et al. 1986)
20.9 (quoted, Isnard & Lambert 1988,1989)

Vapor Pressure (Pa at 25°C):

- 2.67 (quoted, Verschueren 1983)
- 1.82 (calculated-Antoine eqn., Boublik et al. 1984)
- 2.93 (calculated-Antoine eqn., Stephenson & Malanowski 1987)
- 2.84 (quoted, Riddick et al. 1986)

Henry's Law Constant (Pa $m^3$/mol):

- 25.1 (calculated-P/C using selected data)

Octanol/Water Partition Coefficient, log $K_{OW}$:

- 4.36 (observed, Rekker & Nauta 1961; quoted, Rekker 1977)
- 4.21 (observed, Dearden & Tubby 1974; quoted, Rekker 1977)
- 4.08 (shake flask-HPLC, Banerjee et al. 1980; quoted, Arbuckle 1983; Isnard & Lambert 1988,1989; Sangster 1989)
- 4.20 (shake flask-GC, Chiou et al. 1977; quoted, Sangster 1989)
- 4.21, 4.36; 4.25 (quoted; calculated-f const., Rekker 1977)
- 4.26 (Hansch & Leo 1979; quoted, Veith et al. 1983)
- 4.21 (Veith et al. 1979,1980; quoted, Mackay 1982; Hawker & Connell 1985)
- 4.28 (quoted, Valvani et al. 1981; quoted, Amidon & Williams 1982)
- 4.21 (Mackay 1982; quoted, Schüürmann & Klein 1988)
- 4.20 (quoted, Verschueren 1983)
- 4.26, 3.79 (quoted, estimated-HPLC/MS, Burkhard et al. 1985)
- 4.26, 4.24 (quoted, calculated-f const., Burkhard et al. 1985)
- 4.21 (THOR 1986; quoted, Schüürmann & Klein 1988)
- 4.24 (Leo 1986; quoted, Schüürmann & Klein 1988)
- 4.21, 4.36 (quoted, Sangster 1989)
- 4.21 (recommended, Sangster 1989)
- 4.21 (quoted, Thomann 1989; Van Leeuwen et al. 1992)

Bioconcentration Factor, log BCF:

- 2.29 (rainbow trout, calculated, Veith et al. 1979; quoted, Isnard & Lambert 1988)
- 2.29 (exptl., Mackay 1982; quoted, Schüürmann & Klein 1988)
- 2.89 (calculated-$K_{OW}$, Mackay 1982)

Sorption Partition Coefficient, log $K_{OC}$:

Half-Lives in the Environment:

Air: disappearance half-life of less than 0.24 hours from air for the reaction with OH radicals (USEPA 1974; quoted, Darnall et al. 1976).

Surface water:

Ground water:

Sediment:

Soil:

Biota:

Environmental Fate Rate Constant and Half-Lives:

Volatilization:

Photolysis:

Oxidation:

Hydrolysis:

Biodegradation:

Biotransformation:

Bioconcentration, Uptake ($k_1$) and Elimination ($k_2$) Rate Constants or Half-Lives:

$k_1$: 5.5 $hour^{-1}$ (trout, Hawker & Connell 1985)

$k_2$: 0.0275 $hour^{-1}$ (trout, Hawker & Connell 1985)

$k_2$: 0.676 $hour^{-1}$ (fish, quoted, Thomann 1989)

Common Name: Styrene oxide
Synonym: (1,2-epoxyethyl)benzene, phenylepoxyethane
Chemical Name: (1,2-epoxyethyl)benzene, phenylepoxyethane, styrene oxide
CAS Registry No: 96-09-3
Molecular Formula: $C_8H_8O$
Molecular Weight: 120.15
Melting Point (°C):
-35.6 (Weast 1982-83)
-37.0 (Verschueren 1983; Dean 1985)
Boiling Point (°C):
194.1 (Weast 1982-83)
194.0 (Verschueren 1983; Dean 1985)
Density (g/cm$^3$ at 20°C):
1.0523 (16°C, Weast 1982-83; Dean 1985)
1.050 (Verschueren 1983)
Molar Volume (cm$^3$/mol):
138.0 (calculated-LeBas method)
114.4 (calculated from density)
Molecular Volume (A$^3$):
Total Surface Area, TSA (A$^2$):
Heat of Fusion, $\Delta H_{fus}$, (kcal/mol):
Entropy of Fusion, $\Delta S_{fus}$ (cal/mol K or e.u.):
Fugacity Ratio at 25°C, F: 1.0

Water Solubility (g/m$^3$ or mg/L at 25°C):
2800 (quoted, Verschueren 1983)

Vapor Pressure (Pa at 25°C):
40.0 (quoted, Verschueren 1983)

Henry's Law Constant (Pa m$^3$/mol):

Octanol/Water Partition Coefficient, log $K_{OW}$:
1.84 (shake flask-HPLC, Pratesi et al. 1979; quoted, Sangster 1989)
1.51 (shake flask-GC, Serrentino et al. 1983; quoted, Sangster 1989)
1.43 (Deneer et al. 1988; quoted, Verhaar et al. 1992)
1.61 (quoted & recommended, Sangster 1989)

Bioconcentration Factor, log BCF:

Sorption Partition Coefficient, log $K_{OC}$:

Half-Lives in the Environment:

Air: disappearance half-life of less than 0.24 hours from air for the reaction with OH radicals (USEPA 1974; quoted, Darnall et al. 1976); 12.3-123 hours, based on estimated photooxidation rate constant with hydroxyl radicals (Atkinson 1987; quoted, Howard et al. 1991).

Surface water: 0.00385-27.5 hours, based on an estimation from measured first-order rate constants at 25°C (Haag & Mill 1988; quoted, Howard et al. 1991).

Ground water: 0.00385-27.5 hours, based on an estimation from measured first-order rate constants at 25°C (Haag & Mill 1988; quoted, Howard et al. 1991).

Sediment:

Soil: 0.00385-27.5 hours, based on an estimation from measured first-order rate constants at 25°C (Haag & Mill 1988; quoted, Howard et al. 1991).

Biota:

Environmental Fate Rate Constant and Half-Lives:

Volatilization:

Photolysis:

Oxidation: 12.3-123 hours, based on estimated photooxidation rate constant with hydroxyl radicals (Atkinson 1987; quoted, Howard et al. 1991).

Hydrolysis: rate constants of $(434 \pm 12)\text{x}10^{-8}$ $sec^{-1}$ at pH 7.25 and $(1690 \pm 620)\text{x}10^{-8}$ $sec^{-1}$ at pH 7.3 in sediment pores both at 25°C for water containing 0.1% w/w $CH_2O$ as sterilant with 1-phenyl-1,2-ethanediol as major hydrolyzed product (Haag & Mill 1988); 0.00385-27.5 hours, based on an estimation from measured first-order rate constants at 25°C, the hydrolysis half-lives at pH 5, 7 and 9 are 0.00385, 21.4 and 27.5 hours (Haag & Mill 1988; quoted, Howard et al. 1991).

Biodegradation: aqueous aerobic half-life of 24-168 hours, based on biological screening test data (Schmidt-Bleek et al. 1982; quoted, Howard et al. 1991); aqueous anaerobic half-life of 96-672 hours, based on estimated aqueous aerobic biodegradation half-life (Howard et al. 1991).

Biotransformation:

Bioconcentration, Uptake ($k_1$) and Elimination ($k_2$) Rate Constants or Half-Lives:

Common Name: Epichlorohydrin
Synonym: 1-chloro-2,3-epoxypropane, (chloromethyl)oxirane, $\alpha$-epichlorohydrin, $\gamma$-chloropropylene oxide
Chemical Name: epichlorohydrin, 1-chloro-2,3-epoxypropane, $\alpha$-epichlorohydrin, $\gamma$-chloropropylene oxide
CAS Registry No: 106-89-8
Molecular Formula: $C_3H_5OCl$
Molecular Weight: 92.53
Melting Point (°C):
-57.2 (Riddick et al. 1986; Howard 1989)
Boiling Point (°C):
116.11 (Riddick et al. 1986)
118.0 (Michael et al. 1988)
116.50 (Howard 1989)
Density (g/cm$^3$ at 20°C):
1.18066 (Riddick et al. 1986)
1.17455 (25°C, Riddick et al. 1986)
Molar Volume (cm$^3$/mol):
92.50 (calculated-LeBas method)
78.37 (calculated from density)
Molecular Volume (A$^3$):
Total Surface Area, TSA (A$^2$):
Heat of Fusion, $\Delta H_{fus}$, (kcal/mol):
Entropy of Fusion, $\Delta S_{fus}$ (cal/mol K or e.u.):
Fugacity Ratio at 25°C, F: 1.0

Water Solubility (g/m$^3$ or mg/L at 25°C):
65800 (20°C, quoted, Riddick et al. 1986)
65800 (20°C, Krijgsheld & Vandergen 1986; quoted, Howard 1989)

Vapor Pressure (Pa at 25°C):
2400 (quoted, Riddick et al. 1986)
2192 (Daubert & Danner 1985; quoted, Howard 1989)

Henry's Law Constant (Pa m$^3$/mol):
3.375 (calculated-P/C using Riddick et al. 1986 data)

Octanol/Water Partition Coefficient, log $K_{OW}$:

0.30 (Krijgsheld & Vandergen 1986; quoted, Howard 1989)

0.58 (Deneer et al. 1988; quoted, Verhaar et al. 1992)

Bioconcentration Factor, log BCF:

0.66 (estimated, Santodonato et al. 1980; quoted, Howard 1989)

Sorption Partition Coefficient, log $K_{OC}$:

2.09 (soil, calculated-S, Lyman et al. 1982; quoted, Howard 1989)

Half-Lives in the Environment:

Air: about 4.0 days, based on estimation for the photooxidation with hydroxyl radicals in air (Cupitt 1980; quoted, Howard 1989); 146-1458 hours, based on measured rate constant for the reaction with hydroxyl radicals in air (Atkinson 1985; quoted, Howard et al. 1991).

Surface water: evaporation half-life about 29 hours for a model river 1 m deep with a 1 m/sec current and 3 m/sec wind (Lyman et al. 1982; quoted, Howard 1989); 168-672 hours, based on estimated unacclimated aqueous aerobic biodegradation half-life (Howard et al. 1991).

Ground water: 336-1344 hours, based on estimated unacclimated aqueous aerobic biodegradation half-life (Howard et al. 1991).

Sediment:

Soil: 168-672 hours, based on estimated unacclimated aqueous aerobic biodegradation half-life (Howard et al. 1991).

Biota:

Environmental Fate Rate Constant and Half-Lives:

Volatilization: evaporation half-life about 29 hours for a model river 1 m deep with a 1 m/sec current and 3 m/sec wind (Lyman et al. 1982; quoted, Howard 1989).

Photolysis:

Oxidation: photooxidation half-life about 4.0 days, based on estimation for the photooxidation with hydroxyl radicals in air (Cupitt 1980; quoted, Howard 1989); 146-1458 hours, based on measured rate constant for the reaction with hydroxyl radicals in air (Atkinson 1985; quoted, Howard et al. 1991); experimentally determined rate constant of $5.5 \times 10^{-13}$ $cm^3$ $molecule^{-1}$ $sec^{-1}$ with reference to *n*-butane for the gas phase reaction with OH radicals at $(23.3 \pm 0.9)$°C with an atmospheric lifetime of less than 21 days for an average OH radicals concentration of $1 \times 10^6$ molecules/$cm^3$ (Edney et al. 1986); measured rate constant of $4.4 \times 10^{-13}$ $cm^3$ $molecule^{-1}$ $sec^{-1}$ compared with calculated value of $6.6 \times 10^{-13}$ $cm^3$ $molecule^{-1}$ $sec^{-1}$ for the reaction with OH radicals at room temperature in the gas phase (Atkinson 1987).

Hydrolysis: half-life of 8.2 days in distilled water to hydrolyze to 1-chloropropan-2,3-diol at 20°C and pH 5-9 (Mabey & Mill 1978; quoted, Howard 1989; Howard et al. 1991).

Biodegradation: aqueous aerobic half-life of 168-672 hours, based on estimated unacclimated aqueous aerobic biodegradation screening test data (Bridie et al. 1979; Sasaki 1978; quoted, Howard et al. 1991); aqueous anaerobic half-life of 672-2688 hours, based on estimated unacclimated aqueous aerobic biodegradation half-life (Howard et al. 1991).

Biotransformation:

Bioconcentration, Uptake ($k_1$) and Elimination ($k_2$) Rate Constants or Half-Lives:

Common Name: Chloromethyl methyl ether
Synonym: chloromethyl ether, chloromethoxymethane, CMME, monochlorodimethyl ether
Chemical Name: chloromethyl methyl ether
CAS Registry No: 107-30-2
Molecular Formula: $ClCH_2$-O-$CH_3$
Molecular Weight: 80.51
Melting Point (°C):
    -103.5 (Verschueren 1983; Dean 1985; Stephenson & Malanowski 1987)
Boiling Point (°C):
    57.9 (Dean 1985; Verschueren 1983)
    59.5 (Stephenson & Malanowski 1987)
    59.0 (Budavari 1989)
Density ($g/cm^3$ at 20°C):
    1.0703 (Dean 1985)
    1.0605 (Budavari 1989)
Molar Volume ($cm^3/mol$):
    81.80 (calculated-LeBas method)
    75.60 (calculated from density)
Molecular Volume ($A^3$):
Total Surface Area, TSA ($A^2$):
Heat of Fusion, $\Delta H_{fus}$, (kcal/mol):
Entropy of Fusion, $\Delta S_{fus}$ (cal/mol K or e.u.):
Fugacity Ratio at 25°C, F: 1.0

Water Solubility ($g/m^3$ or mg/L at 25°C):
    decomposes (Verschueren 1983)

Vapor Pressure (Pa at 25°C):
    24903 (calculated-Antoine eqn., Stephenson & Malanowski 1987)

Henry's Law Constant (Pa $m^3/mol$):

Octanol/Water Partition Coefficient, log $K_{OW}$:

Bioconcentration Factor, log BCF:

Sorption Partition Coefficient, log $K_{OC}$:

Half-Lives in the Environment:

Air: 22.7-227 hours, based on estimated rate constant for the reaction with hydroxyl radicals (Atkinson 1987; quoted, Howard et al. 1991).

Surface water: 0.0108-0.033 hour, based on measured hydrolysis rate constant for bis(chloromethyl) ether (Tou et al. 1974; quoted, Howard et al. 1991) and chloromethyl methyl ether (Ellington et al. 1987; quoted, Ellington 1989; Howard et al. 1991).

Ground water: 0.0108-0.033 hour, based on measured hydrolysis rate constant for bis(chloromethyl) ether (Tou et al. 1974; quoted, Howard et al. 1991) and chloromethyl methyl ether (Ellington et al. 1987; quoted, Ellington 1989; Howard et al. 1991).

Sediment:

Soil: 0.0108-0.033 hour, based on measured hydrolysis rate constant for bis(chloromethyl) ether (Tou et al. 1974; quoted, Howard et al. 1991) and chloromethyl methyl ether (Ellington et al. 1987; quoted, Ellington 1989; Howard et al. 1991).

Biota:

Environmental Fate Rate Constant and Half-Lives:

Volatilization:

Photolysis:

Oxidation: photooxidation half-life of 22.7-227 hours, based on estimated rate constant for the reaction with hydroxyl radicals (Atkinson 1987; quoted, Howard et al. 1991).

Hydrolysis: 21 $hour^{-1}$ at pH 7 and 25°C with a calculated half-life of 2.0 minutes (Van Duuren et al. 1972; quoted, Ellington 1989); 0.0108-0.033 hour, based on measured hydrolysis rate constant for bis(chloromethyl) ether (Tou et al. 1974; quoted, Howard et al. 1991) and chloromethyl methyl ether (Ellington et al. 1987; quoted, Ellington 1989; Howard et al. 1991); hydrolyzed very fast in aqueous solutions with half-life less than 1.0 sec (Verschueren 1983).

Biodegradation: aqueous aerobic half-life of 168-672 hours, based on estimated unacclimated aqueous aerobic biodegradation half-life (Howard et al. 1991); aqueous anaerobic half-life of 672-2688 hours, based on estimated unacclimated aqueous aerobic biodegradation half-life (Howard et al. 1991).

Biotransformation:

Bioconcentration, Uptake ($k_1$) and Elimination ($k_2$) Rate Constants or Half-Lives:

Common Name: Bis(chloromethyl)ether
Synonym: BCME, Bis-CME, chloro(chloromethoxy)methane, dichloromethylether, (dichlorodimethyl)ether, sym-dichloromethyl ether, oxybis(chloromethane)
Chemical Name: chloromethyl ether, sym-dichloromethyl ether
CAS Registry No: 542-88-1
Molecular Formula: $ClCH_2$-O-$CH_2Cl$
Molecular Weight: 114.96
Melting Point (°C):
-41.5 (Weast 1977; Callahan et al. 1979; Mabey et al. 1982; Weast 1982-83; Verschueren 1983; Howard 1989)
Boiling Point (°C):
104 (Weast 1977; Callahan et al. 1979; Mabey et al. 1982; Weast 1982-83; Verschueren 1983)
106 (Howard 1989)
Density (g/cm$^3$ at 20°C):
1.328 (15°C, Weast 1982-83)
1.315 (Verschueren 1983)
Molar Volume (cm$^3$/mol):
102.7 (calculated-LeBas method)
87.42 (calculated from density)
Molecular Volume (A$^3$):
Total Surface Area, TSA (A$^2$):
Heat of Fusion, $\Delta H_{fus}$, (kcal/mol):
Entropy of Fusion, $\Delta S_{fus}$ (cal/mol K or e.u.):
Fugacity Ratio at 25°C, F: 1.0

Water Solubility (g/m$^3$ or mg/L at 25°C):
22000 (calculated as per Moriguchi 1975 using Quayle 1953 data, Callahan et al. 1979; quoted, Mabey et al. 1982)

Vapor Pressure (Pa at 25°C):
3999 (22°C, Dreisbach 1952; quoted, Callahan et al. 1979; Mabey et al. 1982; Howard 1989)

Henry's Law Constant (Pa m$^3$/mol):
21.27 (calculated-P/C, Mabey et al. 1982)
21.27 (20-25°C and low ionic strength, quoted, Pankow & Rosen 1988; Pankow 1990)
213.18 (quoted from WERL Treatability Data, Ryan et al. 1988)

Octanol/Water Partition Coefficient, log $K_{OW}$:

-0.38 (calculated, Radding et al. 1977; quoted, Callahan et al. 1979; Ryan et al. 1988)

2.40 (calculated, Mabey et al. 1982)

Bioconcentration Factor, log BCF:

1.041 (bluegill sunfish, Veith et al. 1980; quoted, Howard 1989)

Sorption Partition Coefficient, log $K_{OC}$:

1.20 (sediment-water, calculated-$K_{OW}$, Mabey et al. 1982)

Half-Lives in the Environment:

Air: 0.196-1.96 hours, based on estimated rate constant for reaction with hydroxyl radicals in air (Atkinson 1987; quoted, Howard et al. 1991).

Surface water: 0.0106-0.106 hour, based on estimated hydrolysis half-life in water (Howard et al. 1991).

Ground water: 0.0106-0.106 hour, based on estimated hydrolysis in water (Howard et al. 1991).

Sediment:

Soil: <10 days, via volatilization subject to plant uptake from the soil (Ryan et al. 1988); 0.0106-0.106 hour, based on estimated hydrolysis half-life in water (Howard et al. 1991).

Biota:

Environmental Fate Rate Constant and Half-Lives:

Volatilization:

Photolysis:

Oxidation: << 360 $M^{-1}$ $hour^{-1}$ for singlet oxygen and 3.0 $M^{-1}$ $hour^{-1}$ for peroxy radical (Mabey et al. 1982); photooxidation half-life of 0.196-1.96 hours, based on estimated rate constant for reaction with hydroxyl radicals in air (Atkinson 1987; quoted, Howard et al. 1991).

Hydrolysis: hydrolyzed very fast in aqueous solution with half-life in the order of 10 seconds when extrapolated to pure water (Hammond & Alexander 1972; quoted, Verschueren 1983); 0.018 $sec^{-1}$ with half-life of 38 seconds (Tou et al. 1974; quoted, Callahan et al. 1979; Howard et al. 1991); hydrolysis half-life of 10-38 seconds and will rapidly disappear from any aquatic system (Fishbein 1979; quoted, Howard 1989); 65 $hour^{-1}$ at pH 7.0 at 20°C (quoted, Mabey et al. 1982).

Biodegradation: aqueous aerobic half-life of 168-672 hours, based on scientific judgement (Howard et al. 1991); aqueous anaerobic half-life of 672-2688 hours, based on unacclimated aqueous aerobic biodegradation half-life (Howard et al. 1991).

Biotransformation:

Bioconcentration, Uptake ($k_1$) and Elimination ($k_2$) Rate Constants or Half-Lives:

Common Name: Bis(2-chloroethyl)ether

Synonym: 2-chloroethyl ether, 1,1'-oxybis(2-chloroethane), bis($\beta$-chloroethyl)ether, Chlorex, 1-chloro-2-($\beta$-chloroethoxy)-ethane, $\beta,\beta$'-dichloroethylether, 2,2'-dichloroethylether, di(2-chloroethyl)ether, di(chloroethyl)ether, dichlorodiethyl ether, *sym*-dichlorodiethyl ether

Chemical Name: 2-chloroethyl ether, bis($\beta$-chloroethyl)ether, 1-chloro-2-($\beta$-chloroethoxy)-ethane

CAS Registry No: 111-44-4

Molecular Formula: $ClCH_2CH_2$-O-$CH_2CH_2Cl$

Molecular Weight: 143.02

Melting Point (°C):

- -46.8 (Weast 1977; Callahan et al. 1979; Riddick et al. 1986)
- -50.0 (Verschueren 1983)
- -51.7 (Dean 1985)
- -24.5 (Mabey et al. 1982; Howard 1989)

Boiling Point (°C):

- 178 (Weast 1977; Callahan et al. 1979; Mabey et al. 1982; Howard 1989)
- 178.8 (Dean 1985)
- 178.75 (Riddick et al. 1986)
- 177.50 (Stephenson & Malanowski 1987)

Density ($g/cm^3$ at 20°C):

- 1.220 (Verschueren 1983)
- 1.2192 (Dean 1985; Riddick et al. 1986)

Molar Volume ($cm^3$/mol):

- 147.9 (calculated-LeBas method)
- 117.2 (calculated from density)

Molecular Volume ($A^3$):

Total Surface Area, TSA ($A^2$):

Heat of Fusion, $\Delta H_{fus}$, (kcal/mol):

- 2.070 (Riddick et al. 1986)

Entropy of Fusion, $\Delta S_{fus}$ (cal/mol K or e.u.):

Fugacity Ratio at 25°C, F: 1.0

Water Solubility ($g/m^3$ or mg/L at 25°C):

- 10200 (20°C, Du Pont 1966; quoted, Riddick et al. 1986)
- 10200 (rm. temp., Verschueren 1977,1983; quoted, Callahan et al. 1979; Mabey et al. 1982; Neuhauser et al. 1985)
- 10200 (quoted, Cowen & Baynes 1980)
- 17195 (shake flask-LSC, Veith et al. 1980; quoted, Davies & Dobbs 1984; Isnard & Lambert 1988,1989)
- 58264 (calculated-f const. as per Hansch et al. 1968, Veith et al. 1980)
- 10200 (20°C, Riddick et al. 1986; quoted, Howard 1989)
- 10400 (20°C, shake flask-GC/TC, Stephenson 1992)

10300 (31°C, shake flask-GC/TC, Stephenson 1992)

Vapor Pressure (Pa at 25°C):

156.2 (Antoine eqn. regression, Stull 1947)
94.64 (20°C, Verschueren 1977,1983; quoted, Callahan et al. 1979; Mabey et al. 1982; Neuhauser et al. 1985)
146.9 (quoted, Cowen & Baynes 1980)
186.6 (Verschueren 1983)
206.6 (quoted, Riddick et al. 1986; quoted, Howard 1989; Banerjee et al. 1990)
143.6 (calculated-Antoine eqn., Stephenson & Malanowski 1987)
857.1 (calculated-solvatochromic p. & UNIFAC, Banerjee et al. 1990)

Henry's Law Constant (Pa $m^3$/mol):

28.97 (calculated-P/C, Lyman et al. 1982; quoted, Howard 1989)
1.32 (20°C, calculated-P/C, Mabey et al. 1982)

Octanol/Water Partition Coefficient, log $K_{OW}$:

1.58 (calculated, Leo et al. 1971; quoted, Callahan et al. 1979; Neuhauser et al. 1985)
1.12 (shake flask-LSC, Veith et al. 1980; Veith & Kosian 1982; quoted, Davies & Dobbs 1984; Isnard & Lambert 1988,1989; Saito et al. 1992)
1.46 (calculated, Mabey et al. 1982)
1.29 (Hansch & Leo 1985; quoted, Howard 1989)
1.12 (quoted, Isnard & Lambert 1988,1989)

Bioconcentration Factor, log BCF:

0.25 (calculated-$K_{OW}$, Veith et al. 1979,1980)
1.041 (bluegill sunfish, Barrows et al. 1980; quoted, Howard 1989)
1.040 (bluegill sunfish, LSC-$^{14}$C, Veith et al. 1980; Veith & Kosian 1982; quoted, Davies & Dobbs 1984; Sabljic 1987; Isnard & Lambert 1988; Howard 1989; Saito et al. 1992)
0.964 (microorganisms-water, calculated-$K_{OW}$, Mabey et al. 1982)
1.04, 1.15 (quoted, calculated, Sabljic 1987)
1.04 (quoted, Isnard & Lambert 1988)

Sorption Partition Coefficient, log $K_{OC}$:

1.38 (soil, calculated-S, Lyman et al. 1982; quoted, Howard 1989)
1.14 (sediment-water, calculated-$K_{OW}$, Mabey et al. 1982)

Half-Lives in the Environment:

Air: atmospheric half-life of 13.44 hours was estimated for the reaction with OH radicals (GEMS 1986; quoted, Howard 1989); photooxidation half-life of 9.65-96.5 hours, based on estimated rate constant for the reaction with hydroxyl radicals in air (Atkinson 1987; quoted, Howard et al. 1991).

Surface water: 672-4320 hours, based on estimated unacclimated aqueous aerobic biodegradation half-life (Howard et al. 1991).

Ground water: 1344-8640 hours, based on estimated unacclimated aqueous aerobic biodegradation half-life (Howard et al. 1991).

Sediment:

Soil: 672-4320 hours, based on estimated unacclimated aqueous aerobic biodegradation half-life (Howard et al. 1991).

Biota: half-life of larger than 4.0 days but less than 7.0 days in fish tissues (Barrows et al. 1980).

Environmental Fate Rate Constant and Half-Lives:

Volatilization: calculated half-life of 5.78 days (as per Mackay & Wolkoff 1973 by Durkin et al. 1975; quoted, Callahan et al. 1979); half-lives of 3.5, 4.4 and 180.5 days for the streams, rivers and lakes were estimated using Henry's law constant (Lyman et al. 1982; quoted, Howard 1989).

Photolysis:

Oxidation: photooxidation half-life of 4.0 hours, based on an estimated half-life for ethyl ether in the smog chamber (Altshuller et al. 1962 and Laity et al. 1973; quoted, Callahan et al. 1979); $<<360\ M^{-1}\ hour^{-1}$ for the reaction with singlet oxygen and $24.0\ M^{-1}\ hour^{-1}$ for the reaction with peroxy radical (Mabey et al. 1982); photooxidation half-life of 9.65-96.5 hours, based on estimated rate constant for the reaction with OH radicals in air (Atkinson 1987; quoted, Howard et al. 1991).

Hydrolysis: half-life of 40.0 days was estimated at pH 7 from ethyl chloride data in water at an unspecified temperature (Brown et al. 1975; quoted, Howard 1989); half-life of 0.5-2.0 years, based on data from chlorinated ethanes and propanes (Dilling et al. 1975; quoted, Callahan et al. 1979); first-order hydrolysis half-life of 22 years, based on neutral hydrolysis rate constant at 20°C which was extrapolated from data for hydrolysis of dioxane at 100°C (Mabey et al. 1982; quoted, Howard et al. 1991); $2.6\times10^{-5}\ hour^{-1}$ at pH 7 and 25°C with a calculated half-life of 3.0 years (Ellington et al. 1987; quoted, Ellington 1989).

Biodegradation: aqueous aerobic half-life of 672-4320 hours, based on river die-away test data (Ludzack & Ettinger 1963 and Doljido 1979; quoted, Howard et al. 1991); aqueous anaerobic half-life of 2688-17280 hours, based on estimated unacclimated aqueous aerobic biodegradation half-life (Howard et al. 1991).

Biotransformation: $3x10^{-9}$ ml $cell^{-1}$ $hour^{-1}$ for the bacterial transformation in water (Mabey et al. 1982).

Bioconcentration, Uptake ($k_1$) and Elimination ($k_2$) Rate Constants or Half-Lives:

Common Name: Bis(2-chloroisopropyl)ether
Synonym: bis(2-chloro-1-methylethyl)ether, dichlorodiisopropyl ether, dichloroisopropyl ether, 2,2'-dichloroisopropyl ether, 2,2'-oxybis(1-chloropropane)
Chemical Name: bis(2-chloroisopropyl)ether, dichlorodiisopropyl ether, dichloroisopropyl ether
CAS Registry No: 108-60-1
Molecular Formula: $ClCH_2CH(CH_3)$-O-$CH(CH_3)CH_2Cl$
Molecular Weight: 171.07
Melting Point (°C):
-97 (Weast 1977; Callahan et al. 1979; Mabey et al. 1982; Verschueren 1983)
Boiling Point (°C):
189 (Weast 1977; Callahan et al. 1979; Mabey et al. 1982; Verschueren 1983)
187.3 (Dean 1985)
188.0 (Stephenson & Malanowski 1987)
Density ($g/cm^3$ at 20°C):
1.110 (Verschueren 1983)
1.1122 (Dean 1985)
Molar Volume ($cm^3/mol$):
193.40 (calculated-LeBas method)
153.98 (calculated from density)
Molecular Volume ($A^3$):
Total Surface Area, TSA ($A^2$):
Heat of Fusion, $\Delta H_{fus}$, (kcal/mol):
Entropy of Fusion, $\Delta S_{fus}$ (cal/mol K or e.u.):
Fugacity Ratio at 25°C, F: 1.0

Water Solubility ($g/m^3$ or mg/L at 25°C):
1700 (rm. temp., Verschueren 1977,1983; quoted, Callahan et al. 1979; Mabey et al. 1982)
2450 (19.1°C, shake flask-GC, Stephenson 1992)
2370 (31.0°C, shake flask-GC, Stephenson 1992)

Vapor Pressure (Pa at 25°C):
103.95 (Antoine eqn. regression, Stull 1947)
113.31 (20°C, Verschueren 1977,1983; quoted, Callahan et al. 1979; Mabey et al. 1982)
112.3 (28.85°C, Antoine eqn., Stephenson & Malanowski 1987)

Henry's Law Constant (Pa $m^3/mol$):
11.14 (20°C, calculated-P/C, Mabey et al. 1982)
11.40 (calculated-P/C from Verschueren 1977/83 data)

116.5 (quoted from WERL Treatability Data, Ryan et al. 1988)

Octanol/Water Partition Coefficient, log $K_{OW}$:
2.58 (calculated, Leo et al. 1971; quoted, Callahan et al. 1979; Ryan et al. 1988)
2.10 (calculated, Mabey et al. 1982)

Bioconcentration Factor, log BCF:
1.544 (microorganisms-water, calculated-$K_{OW}$, Mabey et al. 1982)

Sorption Partition Coefficient, log $K_{OC}$:
1.785 (sediment-water, calculated-$K_{OW}$, Mabey et al. 1982)

Half-Lives in the Environment:
Air: 4.61-46.1 hours, based on photooxidation half-life in air (Howard et al. 1991).
Surface water: 432-4320 hours, based on estimated unacclimated aqueous aerobic biodegradation half-life (Howard et al. 1991).
Ground water: 864-8640 hours, based on estimated unacclimated aqueous aerobic biodegradation half-life (Howard et al. 1991).
Sediment:
Soil: 432-4320 hours, based on estimated unacclimated aqueous aerobic biodegradation half-life (Howard et al. 1991).
Biota:

Environmental Fate Rate Constant and Half-Lives:
Volatilization: calculated half-life of 1.37 day (calculated as per Mackay & Wolkoff 1973, Durkin et al. 1975; quoted, Callahan et al. 1979).
Photolysis:
Oxidation: photooxidation half-life of 4.0 hours, based on estimated half-life for ethyl ether in a smog chamber (Altshuller et al. 1962 and Laity et al. 1973; quoted, Callahan et al. 1979); $<<360$ $M^{-1}$ $hour^{-1}$ for the reaction with singlet oxygen and 2.0 $M^{-1}$ $hour^{-1}$ for the reaction with peroxy radical (Mabey et al. 1982).
Hydrolysis: half-life of 0.5-2.0 years, based on data from chlorinated ethanes and propanes (Dilling et al. 1975; quoted, Callahan et al. 1979); $4x10^{-6}$ $hour^{-1}$ at pH 7.0 and 25°C (Mabey et al. 1982).
Biodegradation: aqueous aerobic half-life of 432-4320 hours, based on river die-away test data (Kleopfer & Fairless 1972; quoted, Howard et al. 1991) and aerobic soil column study data (Kincannon & Lin 1985; quoted, Howard et al. 1991);

aqueous anaerobic half-life of 1728-17280 hours, based on estimated aqueous aerobic biodegradation half-life (Howard et al. 1991).

Biotransformation: $1 \times 10^{-10}$ ml $cell^{-1}$ $hour^{-1}$ for bacterial transformation in water (Mabey et al. 1982).

Bioconcentration, Uptake ($k_1$) and Elimination ($k_2$) Rate Constants or Half-Lives:

Common Name: 2-Chloroethyl vinyl ether
Synonym: (2-chloroethoxy)-ethene, $\beta$-chloroethyl vinyl ether, vinyl 2-chloroethyl ether
Chemical Name: $\beta$-chloroethyl vinyl ether, 2-chloroethyl vinyl ether, vinyl 2-chloroethyl ether
CAS Registry No: 110-75-8
Molecular Formula: $ClCH_2CH_2\text{-}O\text{-}CH{=}CH_2$
Molecular Weight: 106.55
Melting Point (°C):
- -69.7 (Dean 1985)

Boiling Point (°C):
- 108 (Weast 1977; Callahan et al. 1979; Mabey et al. 1982; Weast 1982-83)
- 110 (Dean 1985)

Density ($g/cm^3$ at 20°C):
- 1.0475 (Weast 1982-83)
- 1.048 (Dean 1985)

Molar Volume ($cm^3/mol$):
- 72.12 (calculated from density)
- 119.6 (calculated-LeBas method)

Molecular Volume ($A^3$):
Total Surface Area, TSA ($A^2$):
Heat of Fusion, $\Delta H_{fus}$, (kcal/mol):
Entropy of Fusion, $\Delta S_{fus}$ (cal/mol K or e.u.):
Fugacity Ratio at 25°C, F: 1.0

Water Solubility ($g/m^3$ or mg/L at 25°C):
- 15000 (calculated as per Moriguchi 1975, Callahan et al. 1979; quoted, Mabey et al. 1982; Neuhauser et al. 1985)
- 6000 (Dean 1985)

Vapor Pressure (Pa at 25°C):
- 3565.8 (20°C, calculated, Dreisbach 1952; quoted, Callahan et al. 1979; Mabey et al. 1982; Neuhauser et al. 1985)

Henry's Law Constant (Pa $m^3$/mol):
- 0.0253 (20-25°C, calculated-P/C, Mabey et al. 1982)
- 25.33 (20-25°C and low ionic strength, quoted, Pankow & Rosen 1988; Pankow 1990)
- 24.79 (quoted from WERL Treatability Data, Ryan et al. 1988)

Octanol/Water Partition Coefficient, log $K_{OW}$:

1.28 (calculated as per Leo et al. 1971, Callahan et al. 1979; quoted, Neuhauser et al. 1985; Ryan et al. 1988)

1.14 (calculated, Mabey et al. 1982)

Bioconcentration Factor, log BCF:

0.672 (microorganisms-water, calculated-$K_{OW}$, Mabey et al. 1982)

Sorption Partition Coefficient, log $K_{OC}$:

0.820 (sediment-water, calculated-$K_{OW}$, Mabey et al. 1982)

Half-Lives in the Environment:

Air:

Surface water:

Ground water:

Sediment:

Soil: $<10$ days, via volatilization subject to plant uptake from the soil (Ryan et al. 1988).

Biota:

Environmental Fate Rate Constant and Half-Lives:

Volatilization:

Photolysis:

Oxidation: photooxidation half-life of 30 minutes, based on half-life estimated for 2-methyl-2-butene from smog chamber data (Altshuller et al. 1962 and Laity et al. 1973; quoted, Callahan et al. 1979); $1 \times 10^{10}$ $M^{-1}$ $hour^{-1}$ for singlet oxygen and 34 $M^{-1}$ $hour^{-1}$ for peroxy radical (Mabey et al. 1982).

Hydrolysis: $4.4 \times 10^{-10}$ $sec^{-1}$, minimum rate at pH 7 and 25°C in pure water with a maximum half-life of 0.48 year (Jones & Wood 1964; quoted, Callahan et al. 1979); estimated to be $4 \times 10^{-6}$ $hour^{-1}$ at pH 7.0 and 25°C with reference to that of bis(2-chloroethyl)ether (Mabey et al. 1982).

Biodegradation:

Biotransformation: $1 \times 10^{-10}$ ml $cell^{-1}$ $hour^{-1}$ for bacterial transformation to water (Mabey et al. 1982).

Bioconcentration, Uptake ($k_1$) and Elimination ($k_2$) Rate Constants or Half-Lives:

Common Name: 4-Chlorophenyl phenyl ether
Synonym: 1-chloro-4-phenoxybenzene, p-chlorophenyl phenyl ether, 4-chlorodiphenyl ether, monochlorodiphenyl oxide
Chemical Name: 4-chlorophenyl phenyl ether, 4-chlorodiphenyl ether
CAS Registry No: 7005-72-3
Molecular Formula: $C_6H_5$-O-$C_6H_5Cl$
Molecular Weight: 204.66
Melting Point (°C):
-6.0 (Dow Chemical 1979; Callahan et al. 1979)
-8.0 (Mabey et al. 1982)
Boiling Point (°C):
284 (Mailhe & Murat 1912; Callahan et al. 1979)
284.5 (Weast 1982-83)
293 (Mabey et al. 1982)
Density (g/cm$^3$ at 20°C):
1.2026 (15°C, Weast 1982-83)
Molar Volume (cm$^3$/mol):
216.5 (calculated-LeBas method)
Molecular Volume (A$^3$):
Total Surface Area, TSA (A$^2$):
Heat of Fusion, $\Delta H_{fus}$, (kcal/mol):
Entropy of Fusion, $\Delta S_{fus}$ (cal/mol K or e.u.):
Fugacity Ratio at 25°C, F: 1.0

Water Solubility (g/m$^3$ or mg/L at 25°C):
3.30 (Branson 1977; quoted, Callahan et al. 1979; Mabey et al. 1982)
59.04 (Isnard & Lambert 1988,1989)

Vapor Pressure (Pa at 25°C):
0.3599 (calculated, Branson 1977; quoted, Callahan et al. 1979; Mabey et al. 1982)

Henry's Law Constant (Pa m$^3$/mol):
22.19 (calculated-P/C, Mabey et al. 1982)
22.29 (20-25°C and low ionic strength, quoted, Pankow & Rosen 1988; Pankow 1990)
24.79 (quoted from WERL Treatability Data, Ryan et al. 1988)

Octanol/Water Partition Coefficient, log $K_{OW}$:
4.08 (Branson 1977; quoted, Callahan et al. 1979; Ryan et al. 1988)
5.079 (calculated, Mabey et al. 1982)

4.08 (quoted, Isnard & Lambert 1988,1989)

Bioconcentration Factor, log BCF:

2.867 (rainbow trout muscle, Branson 1977; quoted, Callahan et al. 1979)

4.255 (microorganisms-water, calculated-$K_{OW}$, Mabey et al. 1982)

2.87 (quoted, Isnard & Lambert 1988)

Sorption Partition Coefficient, log $K_{OC}$:

4.763 (sediment-water, calculated-$K_{OW}$, Mabey et al. 1982)

Half-Lives in the Environment:

Air:

Surface water:

Ground water:

Sediment:

Soil:

Biota:

Environmental Fate Rate Constant and Half-Lives:

Volatilization:

Photolysis:

Oxidation: $<<360\ M^{-1}\ hour^{-1}$ for singlet oxygen and $<<1.0\ M^{-1}\ hour^{-1}$ for peroxy radical (Mabey et al. 1982)

Hydrolysis:

Biodegradation: half-life of 4.0 hours, measured only in activated sludge (Branson 1978; quoted, Callahan et al. 1979).

Biotransformation: estimated rate constant of $1 \times 10^{-7}$ ml $cell^{-1}\ hour^{-1}$ for bacterial transformation in water (Mabey et al. 1982).

Bioconcentration, Uptake ($k_1$) and Elimination ($k_2$) Rate Constants or Half-Lives:

Common Name: 4-Bromophenyl phenyl ether
Synonym: 1-bromo-4-phenoxybenzene, *p*-bromophenyl phenyl ether, 4-bromodiphenyl ether, 4-bromophenyl ether
Chemical Name: 4-bromodiphenyl ether, bromophenyl ether
CAS Registry No: 101-55-3
Molecular Formula: $C_6H_5$-O-$C_6H_4Br$
Molecular Weight: 249.11
Melting Point (°C):
18.72 (Weast 1977; Callahan et al. 1979; Mabey et al. 1982)
18.0 (Dean 1985)
Boiling Point (°C):
310.14 (Weast 1977; Callahan et al. 1979; Mabey et al. 1982)
305 (Dean 1985)
Density (g/cm$^3$ at 20°C):
1.423 (Dean 1985)
Molar Volume (cm$^3$/mol):
218.90 (calculated-LeBas method)
175.06 (calculated from density)
Molecular Volume (A$^3$):
Total Surface Area, TSA (A$^2$):
Heat of Fusion, $\Delta H_{fus}$, (kcal/mol):
Entropy of Fusion, $\Delta S_{fus}$ (cal/mol K or e.u.):
Fugacity Ratio at 25°C, F: 1.0

Water Solubility (g/m$^3$ or mg/L at 25°C):
4.80 (calculated-$K_{OW}$, Mabey et al. 1982)

Vapor Pressure (Pa at 25°C):
0.20 (20°C, calculated, Dreisbach 1952; quoted, Callahan et al. 1979; Mabey et al. 1982)

Henry's Law Constant (Pa m$^3$/mol):
10.13 (20-25°C, calculated-P/C, Mabey et al. 1982)

Octanol/Water Partition Coefficient, log $K_{OW}$:
4.28 (calculated as per Leo et al. 1971 using data of Branson 1977, Callahan et al. 1979; quoted, Ryan et al. 1988)
4.94 (calculated, Mabey et al. 1982)

5.24 (quoted, Van Leeuwen et al. 1992)

Bioconcentration Factor, log BCF:

4.114 (microorganisms-water, calculated-$K_{OW}$, Mabey et al. 1982)

Sorption Partition Coefficient, log $K_{OC}$:

4.623 (sediment-water, calculated-$K_{OW}$, Mabey et al. 1982)

Half-Lives in the Environment:

Air:

Surface water:

Ground water:

Sediment:

Soil:

Biota:

Environmental Fate Rate Constant and Half-Lives:

Volatilization:

Photolysis:

Oxidation: $<<360$ $M^{-1}$ $hour^{-1}$ for singlet oxygen and $<<1.0$ $M^{-1}$ $hour^{-1}$ for peroxy radical (Mabey et al. 1982).

Hydrolysis:

Biodegradation: estimated half-life of 4.0 hours in activated sludge, based on biodegradation of 4-chlorophenyl phenyl ether in activated sewage sludge (Branson 1978; quoted, Callahan et al. 1979).

Biotransformation: estimated rate constant of $3 \times 10^{-9}$ ml $cell^{-1}$ $hour^{-1}$ for the bacterial transformation in water (Mabey et al. 1982).

Bioconcentration, Uptake ($k_1$) and Elimination ($k_2$) Rate Constants or Half-Lives:

Common Name: Bis(2-chloroethoxy)methane
Synonym: bis($\beta$-chloroethyl)formal, $\beta$,$\beta$-dichlorodiethyl formal, dichlorodiethyl methylal
Chemical Name: bis(2-chloroethoxy)methane
CAS Registry No: 111-91-1
Molecular Formula: $ClCH_2CH_2$-O-$CH_2$-O-$CH_2CH_2Cl$
Molecular Weight: 173.1
Melting Point (°C):
Boiling Point (°C):
218.1 (Webb et al. 1962; Callahan et al. 1979; Mabey et al. 1982)
Density (g/cm$^3$ at 20°C):
Molar Volume (cm$^3$/mol):
180.0 (calculated-LeBas method)
Molecular Volume (A$^3$):
Total Surface Area, TSA (A$^2$):
Heat of Fusion, $\Delta H_{fus}$, (kcal/mol):
Entropy of Fusion, $\Delta S_{fus}$ (cal/mol K or e.u.):
Fugacity Ratio at 25°C, F: 1.0

Water Solubility (g/m$^3$ or mg/L at 25°C):
81000 (calculated as per Moriguchi 1975, Callahan et al. 1979; quoted, Mabey et al. 1982)

Vapor Pressure (Pa at 25°C):
21.58 (Antoine eqn. regression, Stull 1947)
$<$13.3 (calculated as per Dreisbach 1952 using data of Webb et al. 1962, Callahan et al. 1979; quoted, Mabey et al. 1982)

Henry's Law Constant (Pa m$^3$/mol):
0.0284 (20-25°C, calculated-P/C, Mabey et al. 1982)
0.0273 (quoted from WERL Treatability Data, Ryan et al. 1988)

Octanol/Water Partition Coefficient, log $K_{OW}$:
1.26 (calculated as per Leo et al. 1971, Callahan et al. 1979; Ryan et al. 1988)
1.029 (calculated, Mabey et al. 1982)

Bioconcentration Factor, log BCF:
0.568 (microorganisms-water, calculated-$K_{OW}$, Mabey et al. 1982)

Sorption Partition Coefficient, log $K_{OC}$:

0.716 (sediment-water, calculated-$K_{OW}$, Mabey et al. 1982)

Half-Lives in the Environment:

Air:

Surface water:

Ground water:

Sediment:

Soil: >50 days, via volatilization subject to plant uptake from the soil (Ryan et al. 1988).

Biota:

Environmental Fate Rate Constant and Half-Lives:

Volatilization:

Photolysis:

Oxidation: <<360 $M^{-1}$ $hour^{-1}$ for singlet oxygen and 52 $M^{-1}$ $hour^{-1}$ for peroxy radical (Mabey et al. 1982).

Hydrolysis: minimum rate of $2.53 \times 10^{-6}$ liter $mole^{-1}$ $sec^{-1}$ for acid-catalyzed hydrolysis of the acetal linkage at 25°C (Kankaanperä 1969; quoted, Callahan et al. 1979; Mabey et al. 1982); half-life of 0.5-2.0 years, based on data of Dilling et al. 1975 on chlorinated ethanes and propanes (quoted, Callahan et al. 1979); estimated rate constant to be $4 \times 10^{-6}$ $hour^{-1}$ at pH 7.0 and 25° by analogy to bis(2-chloroethyl)ether (Mabey et al. 1982).

Biodegradation:

Biotransformation:

Bioconcentration, Uptake ($k_1$) and Elimination ($k_2$) Rate Constants or Half-Lives:

## 4.2 Summary Tables and QSPR Plots

Table 4.1 Summary of physical properties of ethers and halogenated ethers at 25 °C

| Compounds | CAS no. | formula | MW g/mol | M.P. °C | B.P. °C | density g/cm³ | $V_M$ MW/ρ cm³/mol | $V_M$ LeBas cm³/mol | $V_I$ intrinsic (a,b) | TSA Å² (c) |
|---|---|---|---|---|---|---|---|---|---|---|
| *Aliphatic ethers*: | | | | | | | | | | |
| Methyl ether (dimethyl ether) | 115-10-6 | $(CH_3)_2O$ | 46.07 | -141.49 | -24.84 | 0.6689 | 68.87 | 60.9 | | |
| Ethyl ether (diethyl ether) | 60-29-7 | $(C_2H_5)_2O$ | 74.12 | -116.3 | 34.43 | 0.71361 | 103.87 | 106.1 | 50.5 | 281.1 |
| Methyl *t*-butyl ether | 1634-04-4 | $CH_3OC(CH_3)_3$ | 88.15 | -109 | 55.2 | 0.7404 | 119.06 | 127.5 | | |
| Di-*n*-porpyl ether | 111-43-3 | $(C_3H_7)_2O$ | 102.18 | -123.2 | 90.1 | 0.7466 | 136.86 | 151.6 | 69.9 | 344.8 |
| Di-isopropyl ether | 108-20-3 | $((CH_3)_2CH)_2O$ | 102.18 | -85.5 | 68.51 | 0.7300 | 139.97 | 151.6 | 69.9 | |
| Di-*n*-butyl ether | 142-96-1 | $(C_4H_9)_2O$ | 130.23 | -95.2 | 140.29 | 0.7684 | 169.48 | 196 | 89.3 | |
| Butylethyl ether | 628-81-9 | $(C_4H_9)O(C_2H_5)$ | 102.18 | -103 | 92.24 | 0.7495 | 136.33 | 150.5 | | |
| Ethylene oxide | 75-21-8 | $C_2H_4O$ | 44.05 | -111 | 10.7 | 0.8694 | 50.67 | 60.9 | | |
| Propylene oxide | 75-56-9 | $C_3H_6O$ | 58.08 | -112.13 | 34.23 | 0.8287 | 70.09 | 69.7 | | |
| 1,4-Dioxane | 123-91-1 | $C_4H_8O_2$ | 88.11 | 11.8 | 101.32 | 1.0336 | 85.24 | 91.8 | | |
| Furan | 110-00-9 | $C_4H_4O$ | 68.08 | -85.61 | 31.36 | 0.9378 | 72.59 | 73.5 | | |
| Tetrahydrofuran | 109-99-9 | $C_4H_8O$ | 72.11 | -108.39 | 65.97 | 0.8892 | 81.09 | 88.3 | 45.5 | |
| 2-Methyl tetrahydrofuran | 94-47-9 | $C_5H_{10}O$ | 86.13 | -137.2 | 79.9 | 0.8540 | 100.85 | 106.8 | | |
| Tetrahydropyran | 142-68-7 | $C_5H_{10}O$ | 86.13 | -45 | 88.0 | 0.8814 | 97.72 | 107 | 55.3 | |
| *Aromatic ethers*: | | | | | | | | | | |
| Diphenyl ether | 101-84-8 | $C_6H_5OC_6H_5$ | 170.2 | 28 | 257-258 | | | 194.5 | | |
| Benzyl ethyl ether | 539-30-0 | $C_6H_5CH_2OC_2H_5$ | 136.19 | | 185.7 | 0.9490 | 143.6 | 173.6 | | |
| Anisole (methoxybenzene) | 100-66-3 | $(C_6H_5)O(CH_3)$ | 108.15 | -37.5 | 153.6 | 0.9940 | 108.80 | 127.3 | 63.9 | |
| Phenetole (ethoxybenzene) | 108-95-2 | $(C_6H_5)O(C_2H_5)$ | 112.17 | -30 | 169-170 | 1.065 | 105.32 | 143.5 | 72.7 | |
| Styrene oxide | 96-09-3 | $C_8H_8O$ | 120.15 | -35.6 | 194.1 | 1.0523 | 114.18 | 138 | | |

| Compounds | CAS no. | formula | MW g/mol | M.P. °C | B.P. °C | density g/cm³ | $V_M$ MW/ρ cm³/mol | $V_M$ LeBas cm³/mol | V(I) intrinsic (a,b) | TSA Å² (c) |
|---|---|---|---|---|---|---|---|---|---|---|
| *Halogenated ethers*: | | | | | | | | | | |
| Epichlorohydrin | 106-89-8 | $C_3H_5OCl$ | 92.53 | -57.21 | 116.1 | 1.18066 | 78.37 | 92.15 | | |
| bis(2-chloroethyl)ether | 111-44-4 | $(ClC_2H_4)_2O$ | 143.02 | -46.8 | 178 | 1.2192 | 117.31 | 147.9 | | |
| Bis(2-chloroisopropyl)ether | 108-60-1 | $(ClC_3H_6)_2O$ | 171.07 | -97 | 189 | 1.11 | 153.98 | 193.4 | | |
| Chloromethyl methyl ether | 107-30-2 | $ClCH_2$-O-$CH_3$ | 80.51 | -103.5 | 57.9 | 1.0703 | 75.22 | 81.8 | | |
| 2-Chloroethyl vinyl ether | 110-75-8 | $ClC_2H_4$-O-$C_2H_3$ | 106.55 | -69.7 | 108 | 1.0475 | 101.72 | 119.6 | | |
| 4-Chlorophenyl phenyl ether | 7005-72-3 | $C_6H_5$-O-$C_6H_4Cl$ | 204.66 | -6 | 284 | | | 216.5 | | |
| 4-Bromophenyl phenyl ether | 101-55-3 | $C_6H_5$-O-$C_6H_4Br$ | 249.11 | 18 | 310 | 1.423 | 175.06 | 218.9 | | |
| Bis(2-chloroethoxy)methane | 11-91-1 | $(ClC_2H_4)_2O_2CH_2$ | 173.1 | | 218.1 | | | 178.4 | | |

(a) Kamlet et al. 1987, 1988
(b) Leahy 1986
(c) Amidon et al. 1975

Table 4.2 Summary of selected physical-chemical properties of ethers and halogenated ethers at 25 °C

| Compounds | Selected properties | | | | | | Henry's law const. |
|---|---|---|---|---|---|---|---|
| | Vapor pressure | | Solubility | | | log $K_{OW}$ | H, Pa·m³/mol |
| | $P^S$, Pa | $P_L$, Pa | g/m³ | $C^S$, mol/m³ | $C_L$, mol/m³ | | calc, P/C |
| *Aliphatic ethers*: | | | | | | | |
| Methyl ether | 60000 | 60000 | 35300 | 7662.25 | 7662.25 | 0.1 | 7.83 |
| Ethyl ether (diethyl ether) | 71600 | 71600 | 60500 | 814.89 | 81.89 | 0.89 | 87.86 |
| Methyl *t*-butyl ether | 33500 | 33500 | 42000 | 476.46 | 476.46 | 0.94 | 70.31 |
| Di-*n*-propyl ether | 8334 | 8334 | 3306 | 32.35 | 32.35 | 2.03 | 257.6 |
| Di-isopropyl ether | 20000 | 20000 | 7900 | 77.31 | 77.31 | 1.52 | 258.7 |
| Di-*n*-butyl ether | 850 | 850 | 230 | 1.766 | 1.776 | 3.21 | 481.3 |
| Butylethyl ether | 8200 | 8200 | 6500 | 63.61 | 63.61 | 2.03 | 128.9 |
| Ethylene oxide | 145855 | 145855 | 3833320 | 8700 | 8700 | -0.30 | 11.65* |
| Propylene oxide | 71000 | 71000 | 476000 | 8196 | 8196 | 0.03 | 8.66 |
| 1,4-Dioxane | 5000 | 5000 | miscible | | | -0.27 | |
| Furan | 80000 | 80000 | 10000 | 146.9 | 146.9 | 1.34 | 544.6 |
| Tetrahydrofuran | 21600 | 21600 | miscible | | | | |
| 2-Methyl tetrahydrofuran | 23250 | 23250 | 3000 | 34.83 | 34.83 | 1.85 | 667.5 |
| Tetrahydropyran | 9536 | 9536 | 85700 | 995 | 995 | 0.82 | 9.584 |
| *Aromatic ethers*: | | | | | | | |
| Diphenyl ether | 2.93 | 3.43 | 18.7 | 0.1099 | 0.129 | 4.21 | 26.67 |
| Benzyl ethyl ether | 100 | 100 | | | | 2.64 | |
| Anisole (methoxybenzene) | 472 | 472 | 2030 | 18.77 | 18.77 | 2.11 | 25.15 |
| Phenetole (ethoxybenzene) | 204 | 204 | 569 | 18.10 | 18.10 | | |
| Styrene oxide | 40 | 40 | 2800 | 23.30 | 23.0 | 1.61 | 1.72 |

| Compounds | Vapor pressure | | Solubility | | | log $K_{OW}$ | Henry's law const. |
|---|---|---|---|---|---|---|---|
| | $P^S$, Pa | $P_L$, Pa | g/m$^3$ | $C^S$, mol/m$^3$ | $C_L$, mol/m$^3$ | | H, calc, P/C |
| *Halogenated ethers*: | | | | | | | |
| Epichlorohydrin | 2400 | 2400 | 65800 | 771.1 | 711.1 | 0.30 | 3.38 |
| Bis(chloromethyl)ether | 4000 | 4000 | 22000 | 191.4 | 191.4 | -0.38 | 20.90 |
| Bis(2-chloroethyl)ether | 206 | 206 | 10200 | 71.32 | 71.32 | 1.12 | 2.888 |
| Bis(2-chloroisopropyl)ether | 104 | 104 | 1700 | 9.94 | 9.94 | 2.58 | 10.46 |
| Chloromethyl methyl ether | 24900 | 24900 | decompose | | | | |
| 2-Chloroethyl vinyl ether | 3566 | 3566 | 15000 | 140.8 | 140.8 | 1.28 | 25.330 |
| 4-Chlorophenyl phenyl ether | 0.36 | 0.36 | 3.3 | 0.0161 | 0.0161 | 4.08 | 22.327 |
| 4-Bromophenyl phenyl ether | 0.20 | 0.20 | 4.8 | 0.0193 | 0.0193 | 4.28 | 10.380 |
| Bis(2-chloroethoxy)methane | 21.6 | 21.6 | 8100 | 46.79 | 46.79 | 1.26 | 0.462 |

* vapor pressure exceeds atmospheric pressure, Henry's law constant H (Pa·m$^3$/mol) = 101325 Pa/$C^S$ mol/m$^3$.

Table 4.3 Suggested half-life classes of ethers and halogenated ethers in various environmental compartments

| Compounds | Air class | Water* class | Soil class | Sediment class |
|---|---|---|---|---|
| 1,2-Propylene oxide | 6 | 4 | 5 | 6 |
| 1,4-Dioxane | 2 | 4 | 5 | 6 |
| Styrene oxide | 2 | 4 | 5 | 6 |
| Furan | 2 | 4 | 5 | 6 |
| Tetrahydrofuran | 2 | 4 | 5 | 6 |
| Methyl ether (dimethyl ether) | 2 | 5 | 5 | 6 |
| Ethyl ether (diethyl ether) | 2 | 5 | 5 | 6 |
| Methyl t-butyl ether | 2 | 5 | 5 | 6 |
| Di-n-propyl ether | 2 | 5 | 5 | 6 |
| Diphenyl ether | 2 | 5 | 5 | 6 |
| Anisole (methoxybenzene) | 2 | 5 | 5 | 6 |
| Bis(chloromethyl)ether | 2 | 5 | 5 | 6 |
| Bis(2-chloroethyl)ether | 2 | 5 | 5 | 6 |
| Bis(2-chloroisopropyl)ether | 2 | 5 | 5 | 6 |
| Chloromethyl methyl ether | 2 | 5 | 5 | 6 |
| 2-Chloroethyl vinyl ether | 2 | 5 | 5 | 6 |
| 4-Chlorophenyl phenyl ether | 2 | 5 | 5 | 6 |
| Bis(2-chloroethoxy)methane | 2 | 5 | 5 | 6 |

* certain ethers will have much shorter half-lives because of hydrolysis with singlet oxygen, and biodegradation; this half-life class is conservatively assigned, see Chapter 1 for a discussion.

where,

| Class | Mean half-life (hours) | Range (hours) |
|---|---|---|
| 1 | 5 | < 10 |
| 2 | 17 (~ 1 day) | 10-30 |
| 3 | 55 (~ 2 days) | 30-100 |
| 4 | 170 (~ 1 week) | 100-300 |
| 5 | 550 (~ 3 weeks) | 300-1,000 |
| 6 | 1700 (~ 2 months) | 1,000-3,000 |
| 7 | 5500 (~ 8 months) | 3,000-10,000 |
| 8 | 17000 (~ 2 years) | 10,000-30,000 |
| 9 | 55000 (~ 6 years) | > 30,000 |

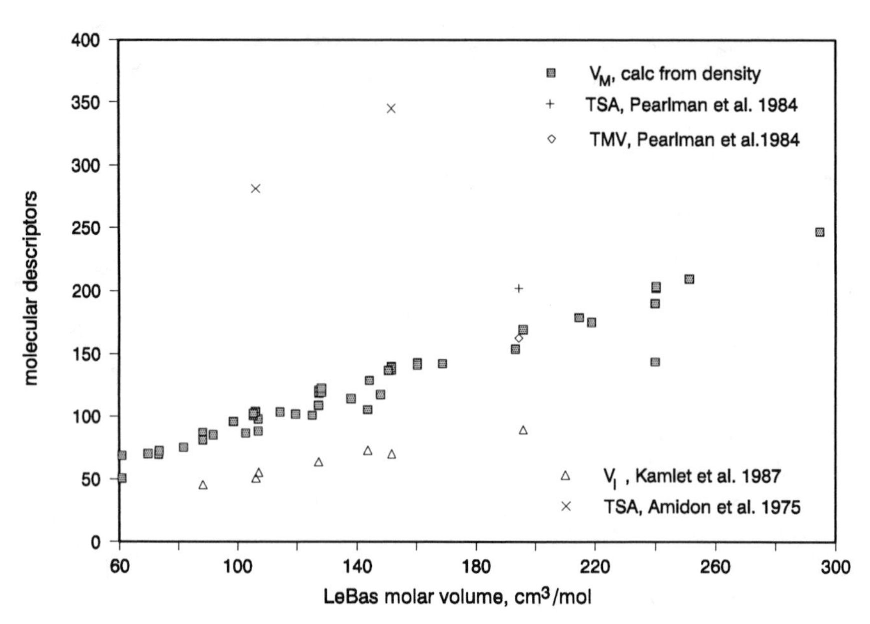


Figure 4.1 Plot of molecular descriptors versus LeBas molar volume.

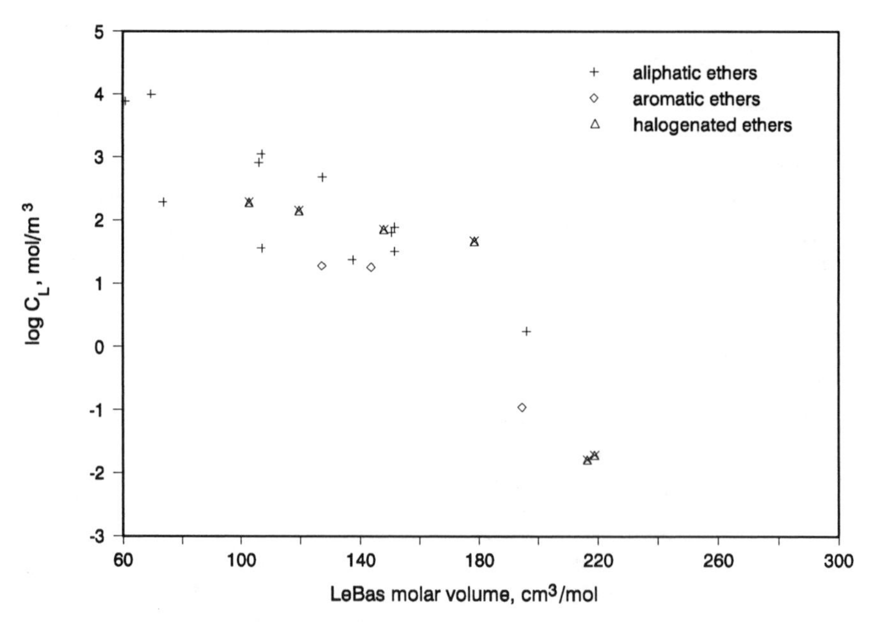


Figure 4.2 Plot of log liquid solubility versus LeBas molar volume.

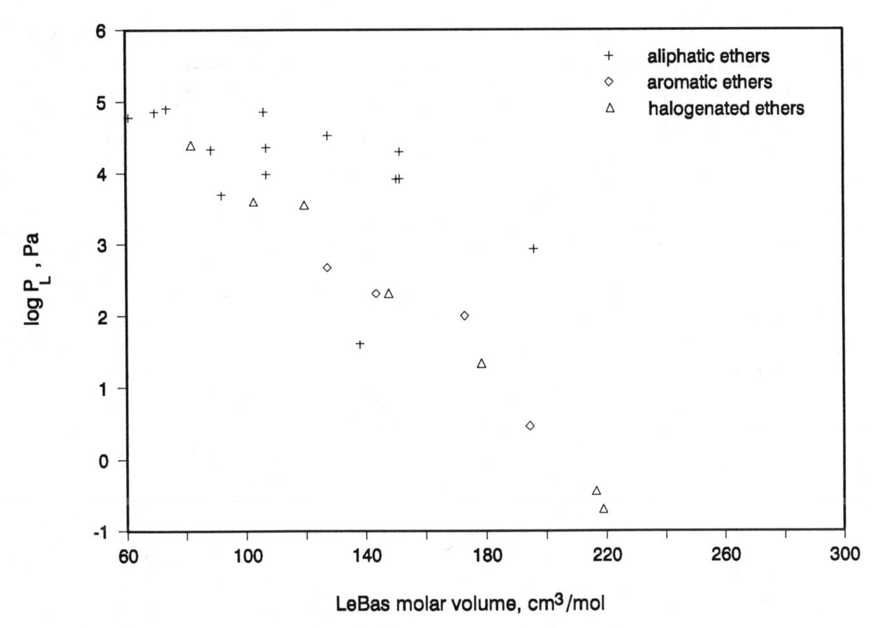


Figure 4.3 Plot of log liquid vapor pressure versus LeBas molar volume.

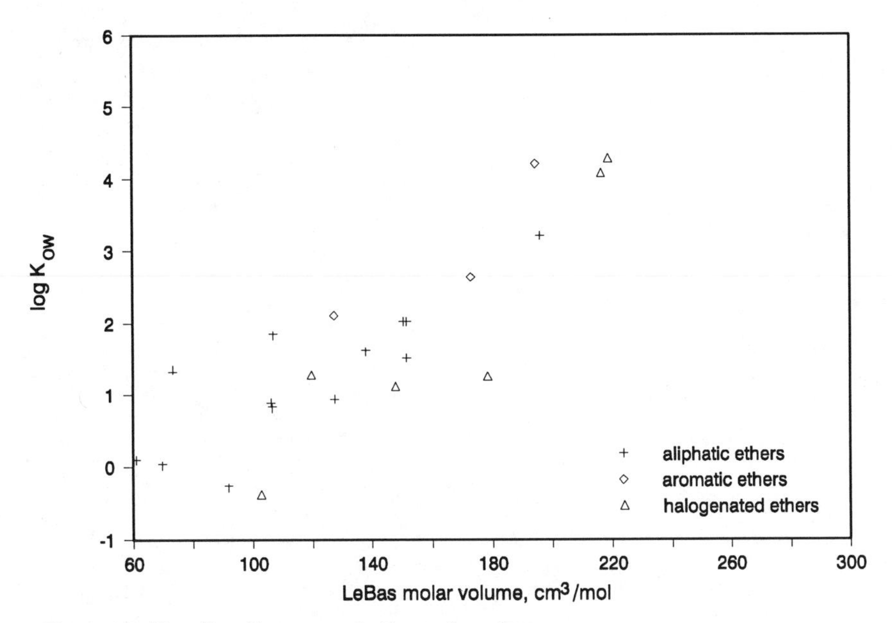


Figure 4.4 Plot of log $K_{OW}$ versus LeBas molar volume.

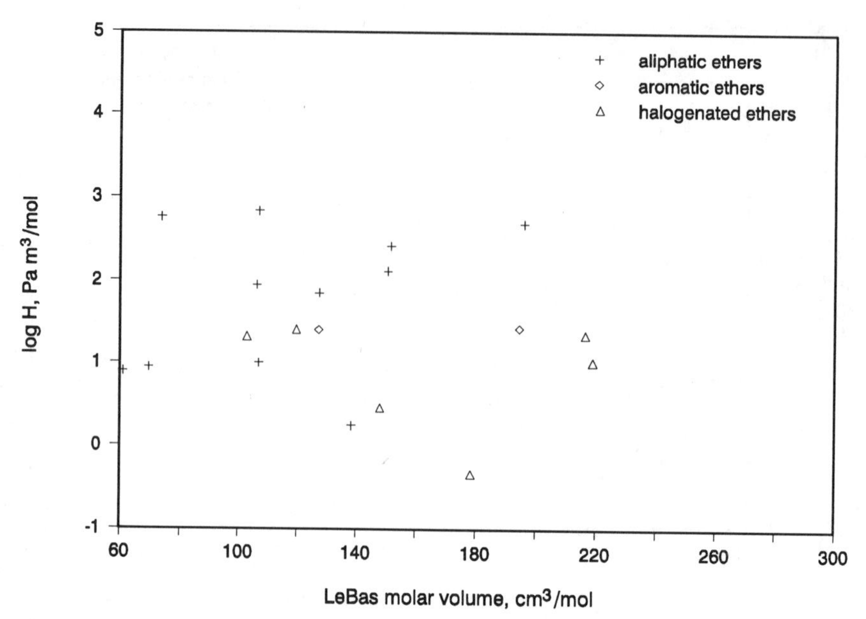


Figure 4.5 Plot of log Henry's law constant versus LeBas molar volume.

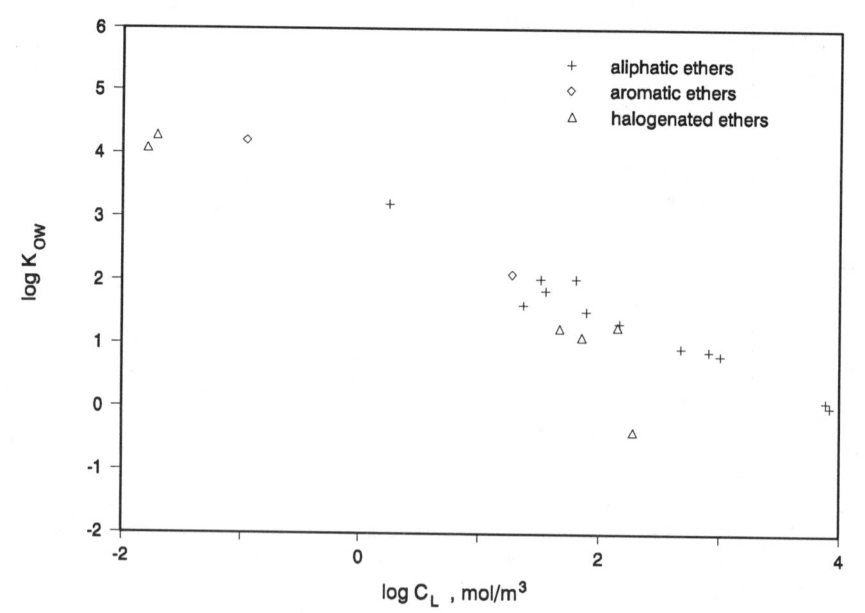


Figure 4.6 Plot of log $K_{OW}$ versus log $C_L$.

## 4.3 Illustrative Fugacity Calculations: Levels I, II and III

Chemical name: Diethyl ether (ethyl ether)

Level I calculation: (six-compartment model)

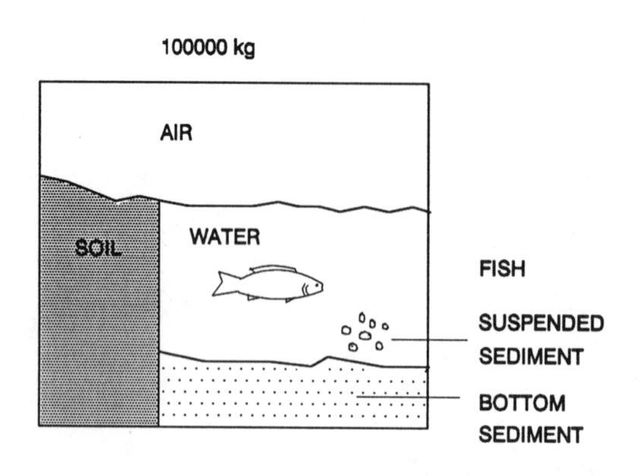


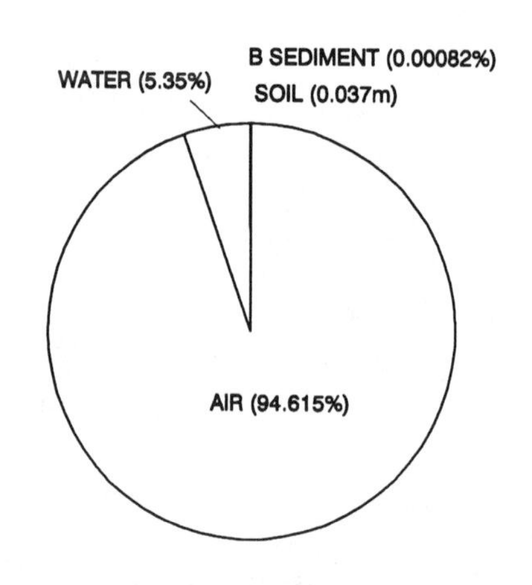


Distribution of mass

physical-chemical properties:

MW: 74.12 g/mol

M.P.: -116.3°C

Fugacity ratio: 1.0

vapor pressure: 71600 Pa

solubility: 60500 g/m³

log $K_{OW}$: 0.89

| Compartment | Z | Concentration | | | Amount | Amount |
|---|---|---|---|---|---|---|
| | mol/m3 Pa | mol/m3 | mg/L (or g/m3) | ug/g | kg | % |
| Air | 4.034E-04 | 1.277E-08 | 9.462E-07 | 7.982E-04 | 94615 | 94.615 |
| Water | 1.140E-02 | 3.607E-07 | 2.674E-05 | 2.674E-05 | 5347.39 | 5.3474 |
| Soil | 1.742E-03 | 5.511E-08 | 4.084E-06 | 1.702E-06 | 36.760 | 0.0368 |
| Biota (fish) | 4.425E-03 | 1.400E-07 | 1.038E-05 | 1.038E-05 | 2.08E-03 | 2.08E-06 |
| Suspended sediment | 1.088E-02 | 3.444E-07 | 2.553E-05 | 1.702E-05 | 0.0255 | 2.55E-05 |
| Bottom sediment | 3.483E-03 | 1.102E-07 | 8.169E-06 | 3.404E-06 | 0.8169 | 8.17E-04 |
| | Total | | | | 100000 | 100 |

f = 3.164E-05 Pa

Chemical name: Diethyl ether (ethyl ether)

Level II calculation: (six-compartment model)

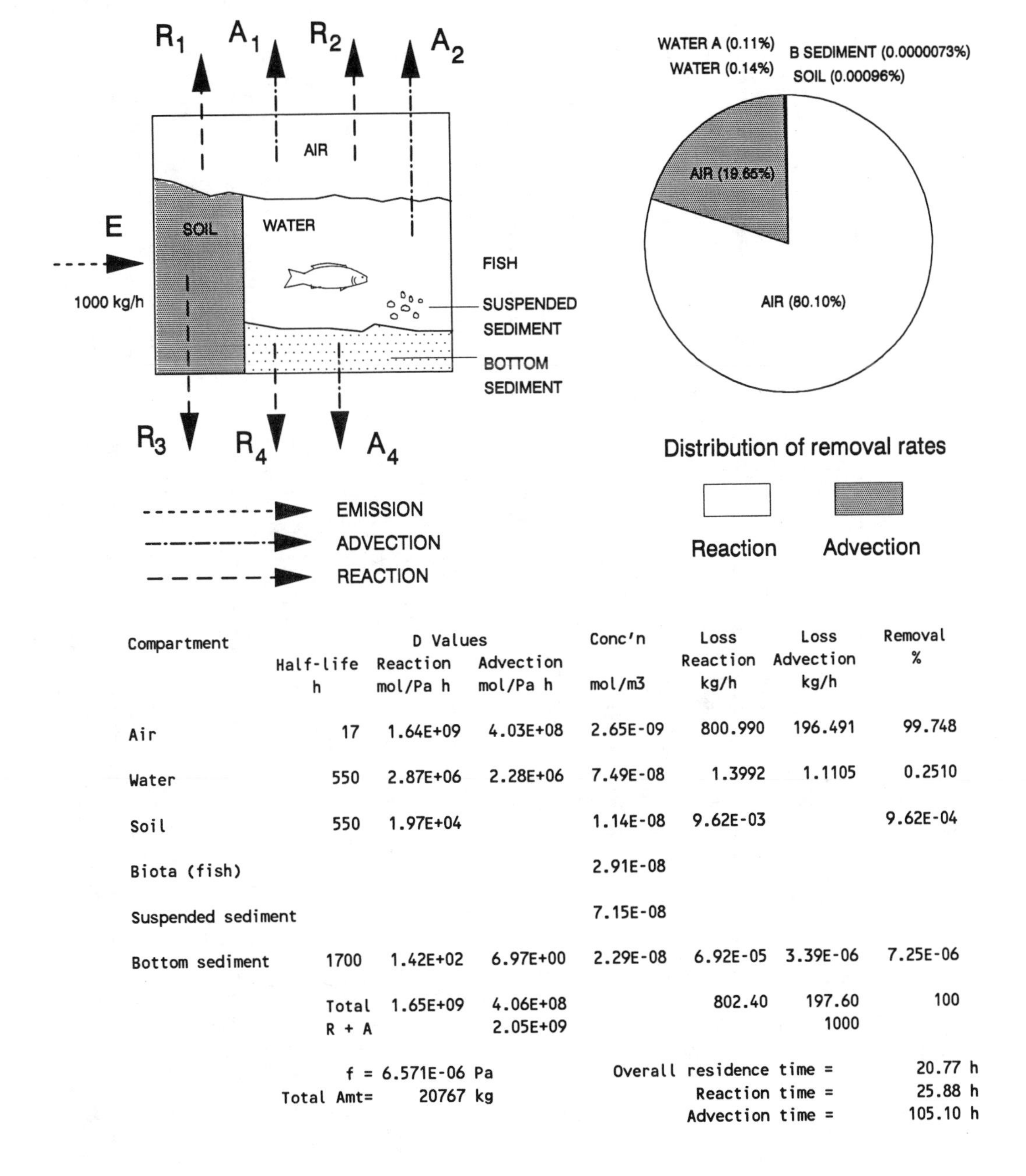


| Compartment | Half-life h | D Values Reaction mol/Pa h | D Values Advection mol/Pa h | Conc'n mol/m3 | Loss Reaction kg/h | Loss Advection kg/h | Removal % |
|---|---|---|---|---|---|---|---|
| Air | 17 | 1.64E+09 | 4.03E+08 | 2.65E-09 | 800.990 | 196.491 | 99.748 |
| Water | 550 | 2.87E+06 | 2.28E+06 | 7.49E-08 | 1.3992 | 1.1105 | 0.2510 |
| Soil | 550 | 1.97E+04 | | 1.14E-08 | 9.62E-03 | | 9.62E-04 |
| Biota (fish) | | | | 2.91E-08 | | | |
| Suspended sediment | | | | 7.15E-08 | | | |
| Bottom sediment | 1700 | 1.42E+02 | 6.97E+00 | 2.29E-08 | 6.92E-05 | 3.39E-06 | 7.25E-06 |
| | Total | 1.65E+09 | 4.06E+08 | | 802.40 | 197.60 | 100 |
| | R + A | | 2.05E+09 | | | 1000 | |

f = 6.571E-06 Pa
Total Amt= 20767 kg

Overall residence time = 20.77 h
Reaction time = 25.88 h
Advection time = 105.10 h

## Fugacity Level III calculations: (four-compartment model)

Chemical name: Diethyl ether (ethyl ether)

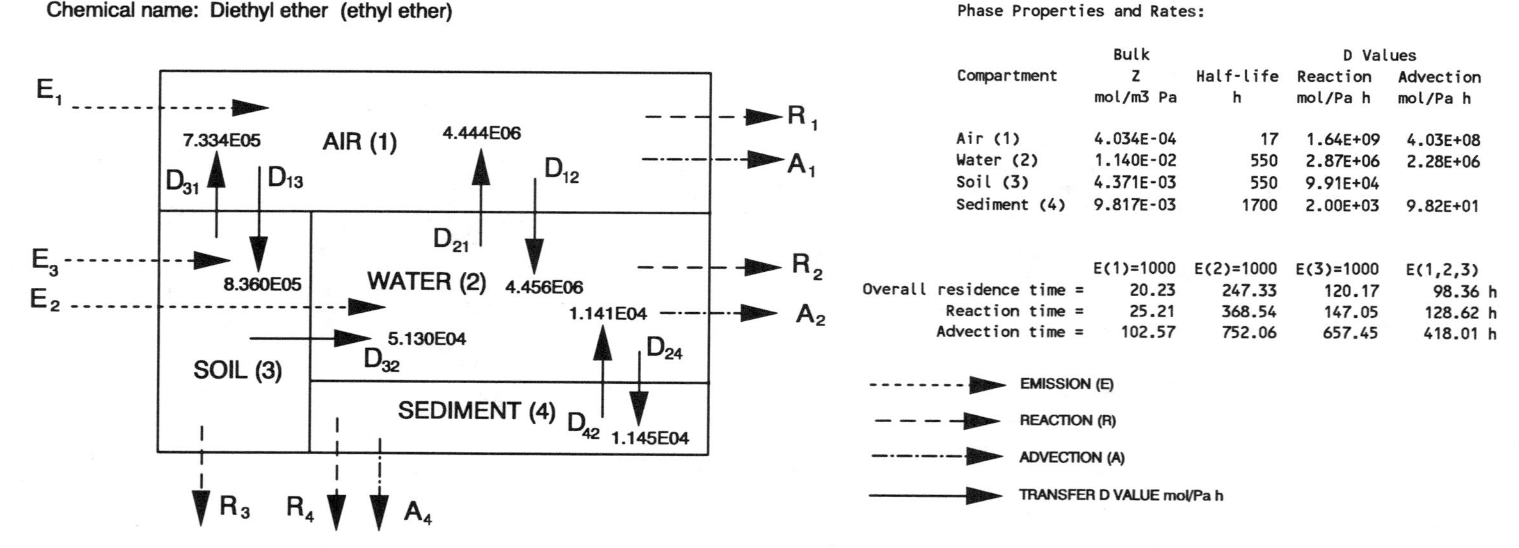


Phase Properties and Rates:

| Compartment | Bulk Z mol/m3 Pa | Half-life h | D Values Reaction mol/Pa h | D Values Advection mol/Pa h |
|---|---|---|---|---|
| Air (1) | 4.034E-04 | 17 | 1.64E+09 | 4.03E+08 |
| Water (2) | 1.140E-02 | 550 | 2.87E+06 | 2.28E+06 |
| Soil (3) | 4.371E-03 | 550 | 9.91E+04 | |
| Sediment (4) | 9.817E-03 | 1700 | 2.00E+03 | 9.82E+01 |

| | E(1)=1000 | E(2)=1000 | E(3)=1000 | E(1,2,3) |
|---|---|---|---|---|
| Overall residence time = | 20.23 | 247.33 | 120.17 | 98.36 h |
| Reaction time = | 25.21 | 368.54 | 147.05 | 128.62 h |
| Advection time = | 102.57 | 752.06 | 657.45 | 418.01 h |

EMISSION (E)
REACTION (R)
ADVECTION (A)
TRANSFER D VALUE mol/Pa h

Phase Properties, Compositions, Transport and Transformation Rates:

| Emission, kg/h | | | Fugacity, Pa | | | | Concentration, g/m3 | | | | Amounts, kg | | | | Total Amount, kg |
|---|---|---|---|---|---|---|---|---|---|---|---|---|---|---|---|
| E(1) | E(2) | E(3) | f(1) | f(2) | f(3) | f(4) | C(1) | C(2) | C(3) | C(4) | m(1) | m(2) | m(3) | m(4) | |
| 1000 | 0 | 0 | 6.580E-06 | 3.087E-06 | 6.224E-06 | 2.618E-06 | 1.967E-07 | 2.609E-06 | 2.017E-06 | 1.905E-06 | 1.967E+04 | 5.218E+02 | 3.630E+01 | 9.526E-01 | 2.023E+04 |
| 0 | 1000 | 0 | 3.046E-06 | 1.407E-03 | 2.882E-06 | 1.193E-03 | 9.109E-08 | 1.189E-03 | 9.337E-07 | 8.682E-04 | 9.109E+03 | 2.378E+05 | 1.681E+01 | 4.341E+02 | 2.473E+05 |
| 0 | 0 | 1000 | 5.637E-06 | 8.423E-05 | 1.527E-02 | 7.143E-05 | 1.685E-07 | 7.117E-05 | 4.947E-03 | 5.197E-05 | 1.685E+04 | 1.423E+04 | 8.905E+04 | 2.599E+01 | 1.202E+05 |
| 600 | 300 | 100 | 5.425E-06 | 4.324E-04 | 1.532E-03 | 3.667E-04 | 1.622E-07 | 3.653E-04 | 4.962E-04 | 2.668E-04 | 1.622E+04 | 7.307E+04 | 8.932E+03 | 1.334E+02 | 9.836E+04 |

| Emission, kg/h | | | Loss, Reaction, kg/h | | | | Loss, Advection, kg/h | | | Intermedia Rate of Transport, kg/h T12 | T21 | T13 | T31 | T32 | T24 | T42 |
|---|---|---|---|---|---|---|---|---|---|---|---|---|---|---|---|---|
| E(1) | E(2) | E(3) | R(1) | R(2) | R(3) | R(4) | A(1) | A(2) | A(4) | air-water | water-air | air-soil | soil-air | soil-water | water-sed | sed-water |
| 1000 | 0 | 0 | 8.020E+02 | 6.574E-01 | 4.57E-02 | 3.883E-04 | 1.967E+02 | 5.218E-01 | 1.905E-05 | 2.173E+00 | 1.017E+00 | 4.077E-01 | 3.383E-01 | 2.367E-02 | 2.621E-03 | 2.214E-03 |
| 0 | 1000 | 0 | 3.713E+02 | 2.996E+02 | 2.12E-02 | 1.770E-01 | 9.109E+01 | 2.378E+02 | 8.682E-03 | 1.006E+00 | 4.635E+02 | 1.888E-01 | 1.566E-01 | 1.096E-02 | 1.195E+00 | 1.009E+00 |
| 0 | 0 | 1000 | 6.871E+02 | 1.793E+01 | 1.12E+02 | 1.059E-02 | 1.685E+02 | 1.423E+01 | 5.197E-04 | 1.861E+00 | 2.774E+01 | 3.493E-01 | 8.301E+02 | 5.806E+01 | 7.151E-02 | 6.039E-02 |
| 600 | 300 | 100 | 6.613E+02 | 9.207E+01 | 1.13E+01 | 5.438E-02 | 1.622E+02 | 7.307E+01 | 2.668E-03 | 1.792E+00 | 1.424E+02 | 3.362E-01 | 8.326E+01 | 5.824E+00 | 3.671E-01 | 3.100E-01 |

# Level III Distribution

Chemical name: Diethyl ether (ethyl ether)

Distribution of mass

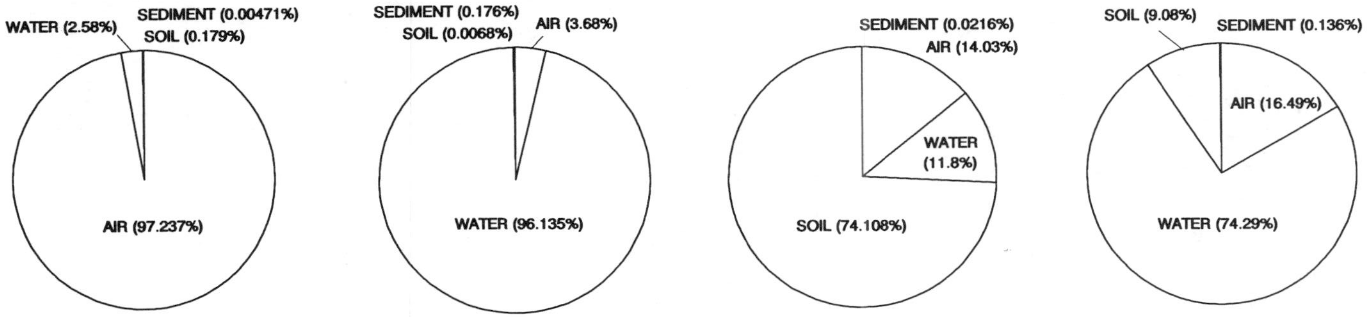


Distribution of removal rates

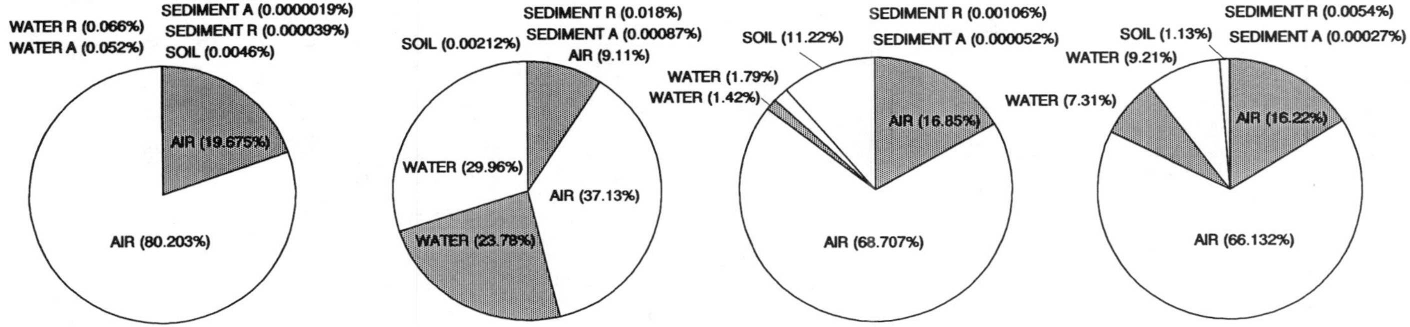


Emission rates:

| | | | |
|---|---|---|---|
| E(1) = 1000 kg/h | E(1) = 0 | E(1) = 0 | E(1) = 600 kg/h |
| E(2) = 0 | E(2) = 1000 kg/h | E(2) = 0 | E(2) = 300 kg/h |
| E(3) = 0 | E(3) = 0 | E(3) = 1000 kg/h | E(3) = 100 kg/h |

Residence time:

| | | | |
|---|---|---|---|
| t = 20.23 h | t = 247 h | t = 120 h | t = 98.36 h |

Reaction   Advection

Chemical name: Methyl t-butyl ether

Level I calculation: (six-compartment model)

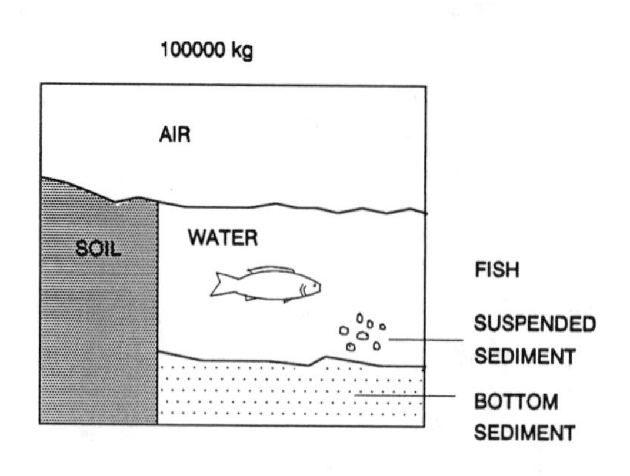


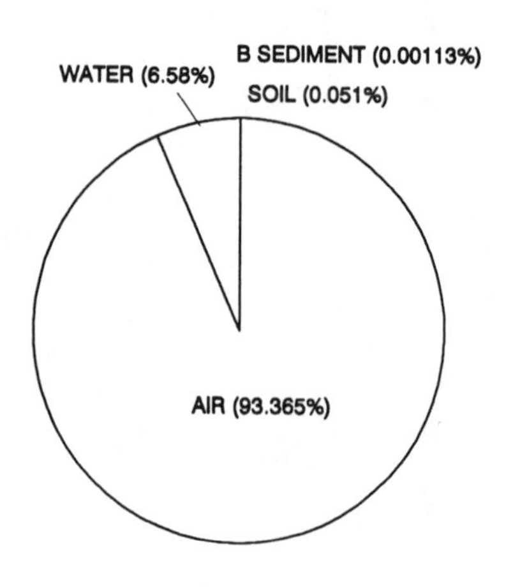


Distribution of mass

physical-chemical properties:

MW: 88.15 g/mol

M.P.: -109 °C

Fugacity ratio: 1.0

vapor pressure: 33500 Pa

solubility: 42000 g/m³

log $K_{OW}$: 0.94

| Compartment | Z | Concentration | | | Amount | Amount |
|---|---|---|---|---|---|---|
| | mol/m3 Pa | mol/m3 | mg/L (or g/m3) | ug/g | kg | % |
| Air | 4.034E-04 | 1.059E-08 | 9.336E-07 | 7.876E-04 | 93365 | 93.365 |
| Water | 1.422E-02 | 3.734E-07 | 3.292E-05 | 3.292E-05 | 6583.25 | 6.5832 |
| Soil | 2.438E-03 | 6.400E-08 | 5.642E-06 | 2.351E-06 | 50.778 | 0.0508 |
| Biota (fish) | 6.194E-03 | 1.626E-07 | 1.433E-05 | 1.433E-05 | 2.87E-03 | 2.87E-06 |
| Suspended sediment | 1.524E-02 | 4.000E-07 | 3.526E-05 | 2.351E-05 | 3.53E-02 | 3.53E-05 |
| Bottom sediment | 4.876E-03 | 1.280E-07 | 1.128E-05 | 4.702E-06 | 1.1284 | 1.13E-03 |
| | Total | | | | 100000 | 100 |

f = 2.625E-05 Pa

Chemical name: Methyl t-butyl ether

Level II calculation: (six-compartment model)

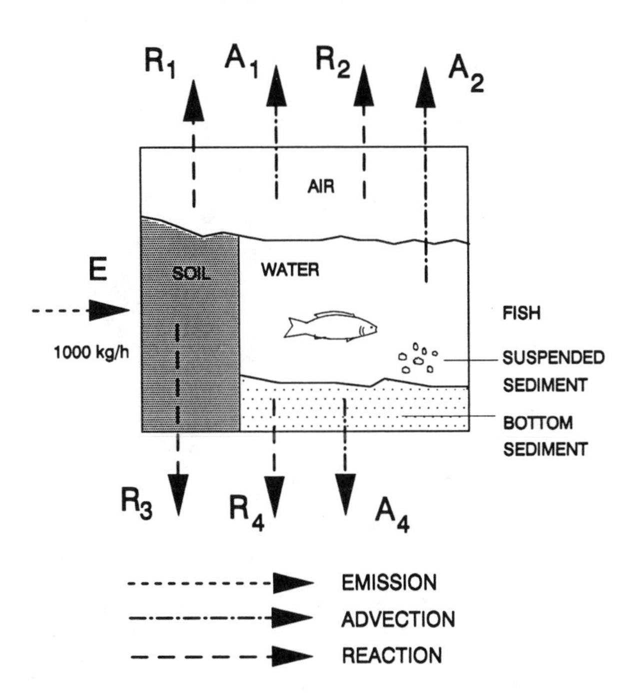


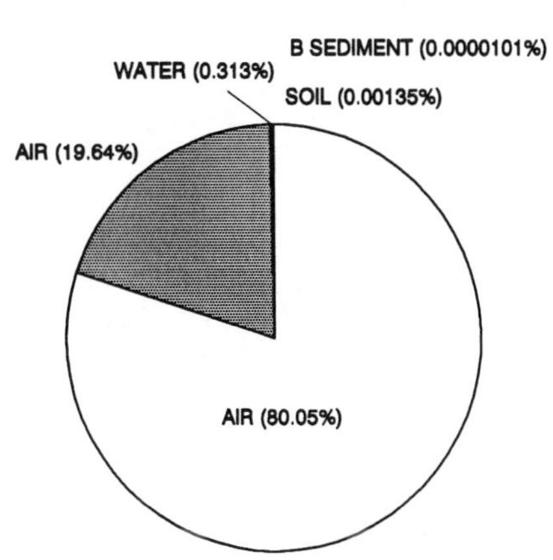


Distribution of removal rates

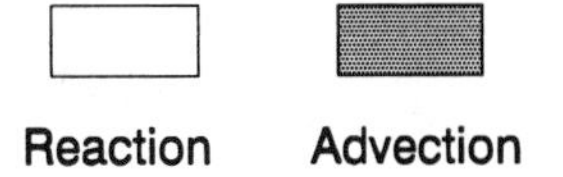


| Compartment | Half-life h | D Values Reaction mol/Pa h | D Values Advection mol/Pa h | Conc'n mol/m3 | Loss Reaction kg/h | Loss Advection kg/h | Removal % |
|---|---|---|---|---|---|---|---|
| Air | 17 | 1.64E+09 | 4.03E+08 | 2.23E-09 | 800.489 | 196.368 | 99.686 |
| Water | 550 | 3.58E+06 | 2.84E+06 | 7.85E-08 | 1.7446 | 1.3846 | 0.3129 |
| Soil | 550 | 2.76E+04 | | 1.35E-08 | 0.0135 | | 1.35E-03 |
| Biota (fish) | | | | 3.42E-08 | | | |
| Suspended sediment | | | | 8.41E-08 | | | |
| Bottom sediment | 1700 | 1.99E+02 | 9.75E+00 | 2.69E-08 | 9.67E-05 | 4.75E-06 | 1.01E-05 |
| | Total | 1.65E+09 | 4.06E+08 | | 802.25 | 197.75 | 100 |
| | R + A | | 2.05E+09 | | | 1000 | |

f = 5.522E-06 Pa

Total Amt= 21032 kg

Overall residence time = 21.03 h

Reaction time = 26.22 h

Advection time = 106.36 h

**Fugacity Level III calculations: (four-compartment model)**

**Chemical name: Methyl t-butyl ether**

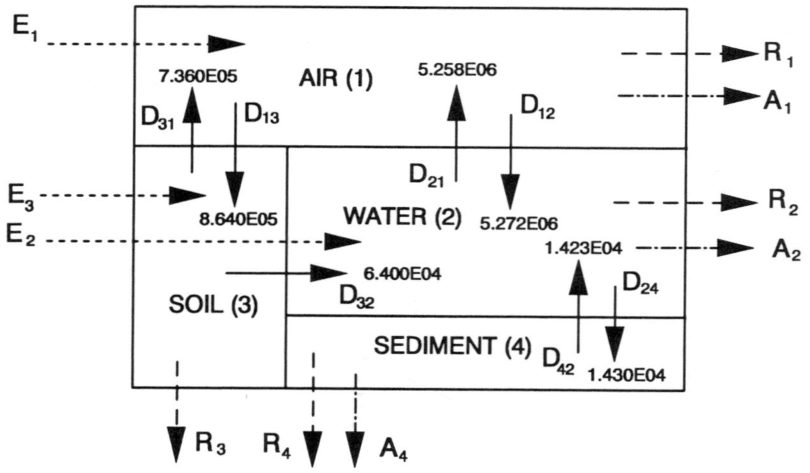

E$_1$ E$_3$ E$_2$

AIR (1) WATER (2) SOIL (3) SEDIMENT (4)

R$_1$ A$_1$ R$_2$ A$_2$ R$_3$ R$_4$ A$_4$

D$_{31}$ 7.360E05 D$_{13}$ 8.640E05 D$_{21}$ 5.258E06 D$_{12}$ 5.272E06 D$_{32}$ 6.400E04 D$_{42}$ 1.423E04 D$_{24}$ 1.430E04

EMISSION (E)
REACTION (R)
ADVECTION (A)
TRANSFER D VALUE mol/Pa h

Phase Properties and Rates:

| Compartment | Bulk Z mol/m3 Pa | Half-life h | D Values Reaction mol/Pa h | D Values Advection mol/Pa h |
|---|---|---|---|---|
| Air (1) | 4.034E-04 | 17 | 1.64E+09 | 4.03E+08 |
| Water (2) | 1.422E-02 | 550 | 3.58E+06 | 2.84E+06 |
| Soil (3) | 5.566E-03 | 550 | 1.26E+05 | |
| Sediment (4) | 1.235E-02 | 1700 | 2.52E+03 | 1.24E+02 |

| | E(1)=1000 | E(2)=1000 | E(3)=1000 | E(1,2,3) |
|---|---|---|---|---|
| Overall residence time = | 20.35 | 252.96 | 141.83 | 102.28 h |
| Reaction time = | 25.35 | 378.76 | 172.91 | 133.88 h |
| Advection time = | 103.12 | 761.63 | 789.05 | 433.37 h |

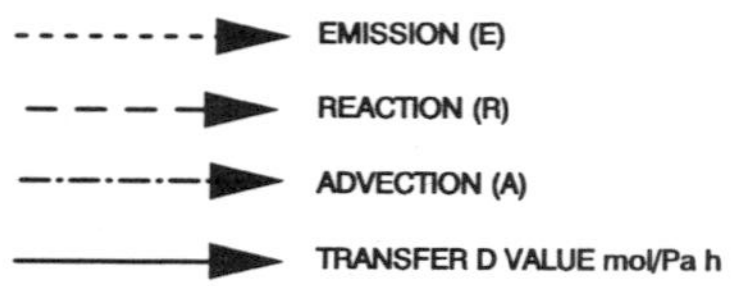

Phase Properties, Compositions, Transport and Transformation Rates:

| Emission, kg/h | | | Fugacity, Pa | | | | Concentration, g/m3 | | | | Amounts, kg | | | | Total Amount, kg |
|---|---|---|---|---|---|---|---|---|---|---|---|---|---|---|---|
| E(1) | E(2) | E(3) | f(1) | f(2) | f(3) | f(4) | C(1) | C(2) | C(3) | C(4) | m(1) | m(2) | m(3) | m(4) | |
| 1000 | 0 | 0 | 5.531E-06 | 2.523E-06 | 5.159E-06 | 2.138E-06 | 1.967E-07 | 3.163E-06 | 2.532E-06 | 2.328E-06 | 1.967E+04 | 6.326E+02 | 4.557E+01 | 1.164E+00 | 2.035E+04 |
| 0 | 1000 | 0 | 2.488E-06 | 9.717E-04 | 2.321E-06 | 8.234E-04 | 8.848E-08 | 1.218E-03 | 1.139E-06 | 8.966E-04 | 8.848E+03 | 2.436E+05 | 2.050E+01 | 4.483E+02 | 2.530E+05 |
| 0 | 0 | 1000 | 4.567E-06 | 6.915E-05 | 1.225E-02 | 5.860E-05 | 1.624E-07 | 8.670E-05 | 6.012E-03 | 6.381E-05 | 1.624E+04 | 1.734E+04 | 1.082E+05 | 3.191E+01 | 1.418E+05 |
| 600 | 300 | 100 | 4.522E-06 | 2.999E-04 | 1.229E-03 | 2.542E-04 | 1.608E-07 | 3.760E-04 | 6.031E-04 | 2.768E-04 | 1.608E+04 | 7.521E+04 | 1.086E+04 | 1.384E+02 | 1.023E+05 |

| Emission, kg/h | | | Loss, Reaction, kg/h | | | | Loss, Advection, kg/h | | | Intermedia Rate of Transport, kg/h T12 | T21 | T13 | T31 | T32 | T24 | T42 |
|---|---|---|---|---|---|---|---|---|---|---|---|---|---|---|---|---|
| E(1) | E(2) | E(3) | R(1) | R(2) | R(3) | R(4) | A(1) | A(2) | A(4) | air-water | water-air | air-soil | soil-air | soil-water | water-sed | sed-water |
| 1000 | 0 | 0 | 8.018E+02 | 7.971E-01 | 5.74E-02 | 4.745E-04 | 1.967E+02 | 6.326E-01 | 2.328E-05 | 2.570E+00 | 1.169E+00 | 4.212E-01 | 3.347E-01 | 2.911E-02 | 3.180E-03 | 2.682E-03 |
| 0 | 1000 | 0 | 3.607E+02 | 3.070E+02 | 2.58E-02 | 1.828E-01 | 8.848E+01 | 2.436E+02 | 8.966E-03 | 1.156E+00 | 4.503E+02 | 1.895E-01 | 1.506E-01 | 1.309E-02 | 1.225E+00 | 1.033E+00 |
| 0 | 0 | 1000 | 6.620E+02 | 2.185E+01 | 1.36E+02 | 1.301E-02 | 1.624E+02 | 1.734E+01 | 6.381E-04 | 2.122E+00 | 3.205E+01 | 3.478E-01 | 7.949E+02 | 6.913E+01 | 8.716E-02 | 7.352E-02 |
| 600 | 300 | 100 | 6.555E+02 | 9.476E+01 | 1.37E+01 | 5.641E-02 | 1.608E+02 | 7.521E+01 | 2.768E-03 | 2.101E+00 | 1.390E+02 | 3.444E-01 | 7.973E+01 | 6.934E+00 | 3.780E-01 | 3.189E-01 |

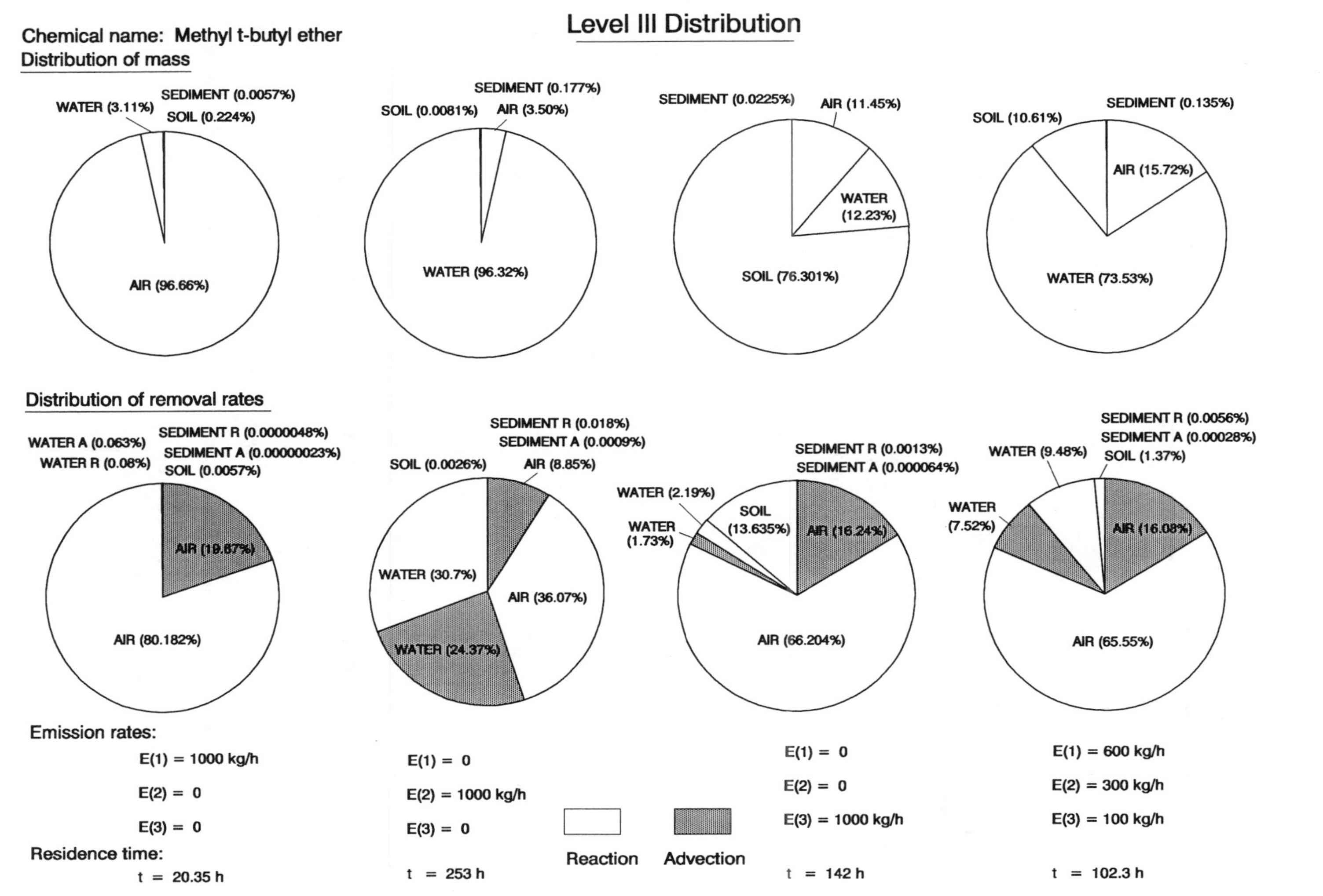

Level III Distribution
Chemical name: Methyl t-butyl ether
Distribution of mass
WATER (3.11%)
SEDIMENT (0.0057%)
SOIL (0.224%)
AIR (96.66%)
SEDIMENT (0.177%)
SOIL (0.0081%)
AIR (3.50%)
WATER (96.32%)
SEDIMENT (0.0225%)
AIR (11.45%)
WATER (12.23%)
SOIL (76.301%)
SOIL (10.61%)
SEDIMENT (0.135%)
AIR (15.72%)
WATER (73.53%)
Distribution of removal rates
WATER A (0.063%)
WATER R (0.08%)
SEDIMENT R (0.0000048%)
SEDIMENT A (0.00000023%)
SOIL (0.0057%)
AIR (19.67%)
AIR (80.182%)
SEDIMENT R (0.018%)
SEDIMENT A (0.0009%)
SOIL (0.0026%)
AIR (8.85%)
WATER (30.7%)
AIR (36.07%)
WATER (24.37%)
WATER (2.19%)
WATER (1.73%)
SOIL (13.635%)
SEDIMENT R (0.0013%)
SEDIMENT A (0.000064%)
AIR (16.24%)
AIR (66.204%)
SEDIMENT R (0.0056%)
SEDIMENT A (0.00028%)
WATER (9.48%)
SOIL (1.37%)
WATER (7.52%)
AIR (16.08%)
AIR (65.55%)
Emission rates:
E(1) = 1000 kg/h
E(2) = 0
E(3) = 0
Residence time:
t = 20.35 h
E(1) = 0
E(2) = 1000 kg/h
E(3) = 0
t = 253 h
Reaction
Advection
E(1) = 0
E(2) = 0
E(3) = 1000 kg/h
t = 142 h
E(1) = 600 kg/h
E(2) = 300 kg/h
E(3) = 100 kg/h
t = 102.3 h

Chemical name: Di-n-butyl ether

Level I calculation: (six-compartment model)

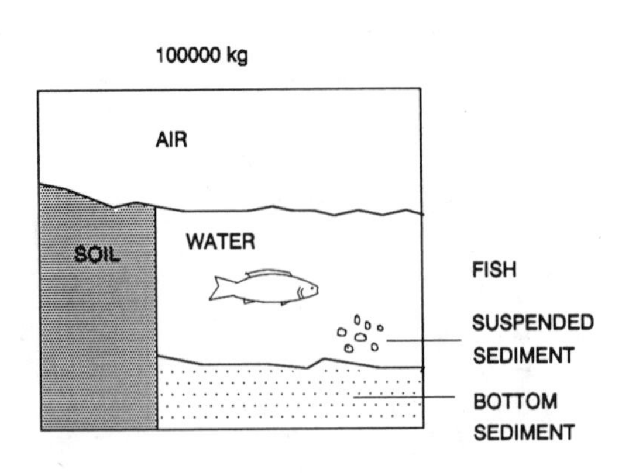


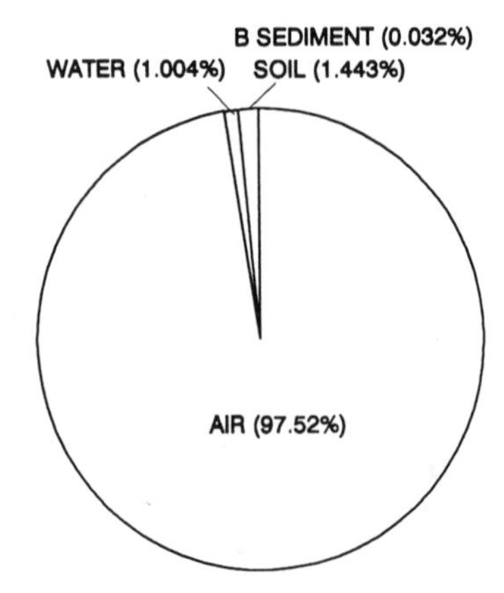


Distribution of mass

physical-chemical properties:

MW: 130.23 g/mol

M.P.: - 95.2 °C

Fugacity ratio: 1.0

vapor pressure: 850 Pa

solubility: 230 g/m$^3$

log $K_{OW}$: 3.21

| Compartment | Z | Concentration | | | Amount | Amount |
|---|---|---|---|---|---|---|
| | mol/m3 Pa | mol/m3 | mg/L (or g/m3) | ug/g | kg | % |
| Air | 4.034E-04 | 7.488E-09 | 9.752E-07 | 8.227E-04 | 97520 | 97.520 |
| Water | 2.078E-03 | 3.857E-08 | 5.023E-06 | 5.023E-06 | 1004.53 | 1.0045 |
| Soil | 6.632E-02 | 1.231E-06 | 1.603E-04 | 6.680E-05 | 1442.79 | 1.4428 |
| Biota (fish) | 1.685E-01 | 3.127E-06 | 4.073E-04 | 4.073E-04 | 8.15E-02 | 8.15E-05 |
| Suspended sediment | 4.145E-01 | 7.694E-06 | 1.002E-03 | 6.680E-04 | 1.0019 | 1.00E-03 |
| Bottom sediment | 1.326E-01 | 2.462E-06 | 3.206E-04 | 1.336E-04 | 32.0619 | 3.21E-02 |
| | Total | | | | 100000 | 100 |

f = 1.856E-05 Pa

Chemical name: Di-n-butyl ether

Level II calculation: (six-compartment model)

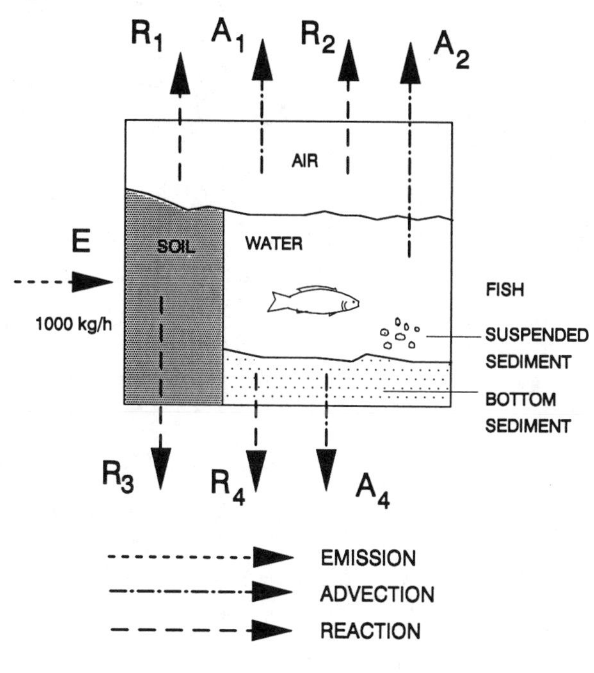


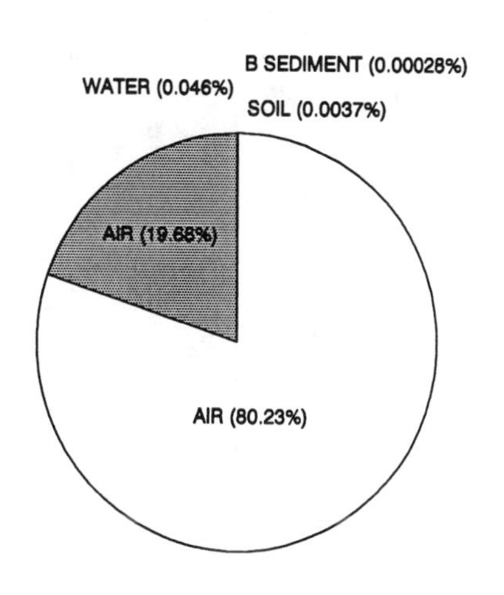


Distribution of removal rates

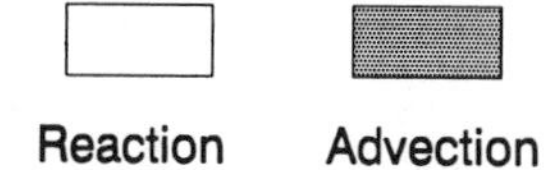


| Compartment | Half-life h | D Values Reaction mol/Pa h | D Values Advection mol/Pa h | Conc'n mol/m3 | Loss Reaction kg/h | Loss Advection kg/h | Removal % |
|---|---|---|---|---|---|---|---|
| Air | 17 | 1.64E+09 | 4.03E+08 | 1.51E-09 | 802.348 | 196.824 | 99.917 |
| Water | 550 | 5.24E+05 | 4.16E+05 | 7.78E-09 | 0.2555 | 0.2027 | 0.0458 |
| Soil | 550 | 7.52E+05 | | 2.48E-07 | 0.3669 | | 3.67E-02 |
| Biota (fish) | | | | 6.31E-07 | | | |
| Suspended sediment | | | | 1.55E-06 | | | |
| Bottom sediment | 1700 | 5.41E+03 | 2.65E+02 | 4.97E-07 | 2.64E-03 | 1.29E-04 | 2.77E-04 |
| | Total | 1.65E+09 | 4.04E+08 | | 802.97 | 197.03 | 100 |
| | R + A | | 2.05E+09 | | | 1000 | |

f = 3.746E-06 Pa
Total Amt= 20183 kg

Overall residence time = 20.18 h
Reaction time = 25.14 h
Advection time = 102.44 h

Fugacity Level III calculations: (four-compartment model)

Chemical name: Di-n-butyl ether

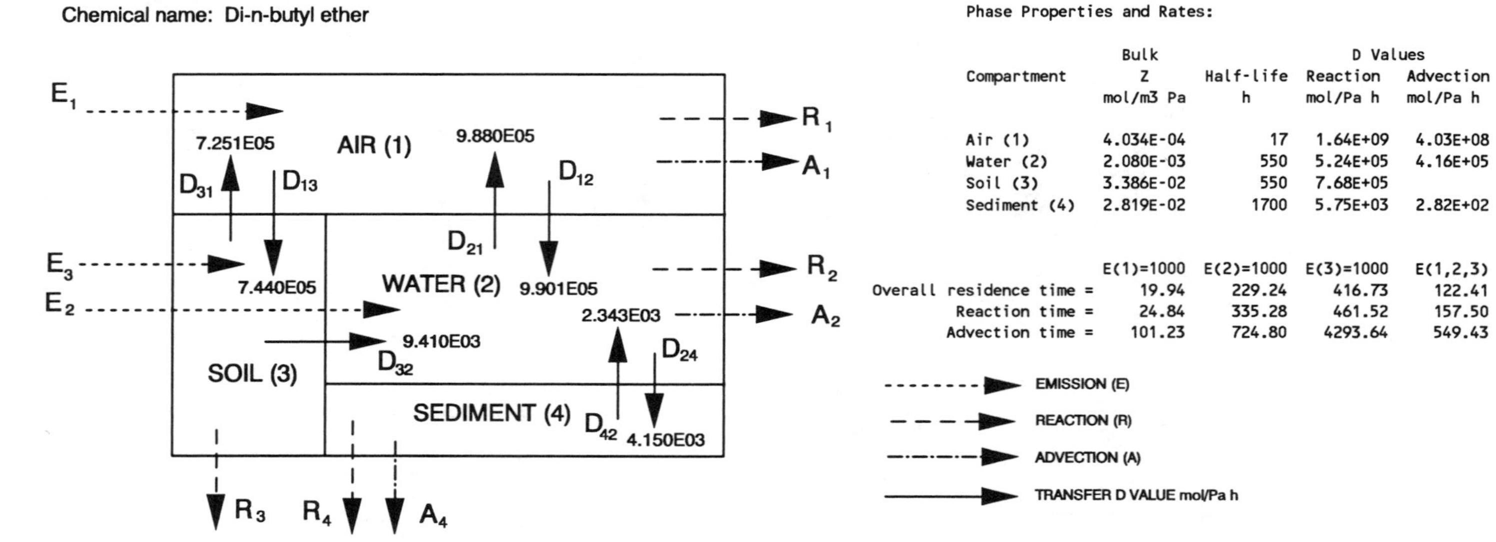


Phase Properties and Rates:

| Compartment | Bulk Z mol/m3 Pa | Half-life h | D Values Reaction mol/Pa h | Advection mol/Pa h |
|---|---|---|---|---|
| Air (1) | 4.034E-04 | 17 | 1.64E+09 | 4.03E+08 |
| Water (2) | 2.080E-03 | 550 | 5.24E+05 | 4.16E+05 |
| Soil (3) | 3.386E-02 | 550 | 7.68E+05 | |
| Sediment (4) | 2.819E-02 | 1700 | 5.75E+03 | 2.82E+02 |

| | E(1)=1000 | E(2)=1000 | E(3)=1000 | E(1,2,3) | |
|---|---|---|---|---|---|
| Overall residence time = | 19.94 | 229.24 | 416.73 | 122.41 | h |
| Reaction time = | 24.84 | 335.28 | 461.52 | 157.50 | h |
| Advection time = | 101.23 | 724.80 | 4293.64 | 549.43 | h |

Phase Properties, Compositions, Transport and Transformation Rates:

| Emission, kg/h | | | Fugacity, Pa | | | | Concentration, g/m3 | | | | Amounts, kg | | | | Total Amount, kg |
|---|---|---|---|---|---|---|---|---|---|---|---|---|---|---|---|
| E(1) | E(2) | E(3) | f(1) | f(2) | f(3) | f(4) | C(1) | C(2) | C(3) | C(4) | m(1) | m(2) | m(3) | m(4) | |
| 1000 | 0 | 0 | 3.748E-06 | 1.931E-06 | 1.856E-06 | 9.572E-07 | 1.969E-07 | 5.230E-07 | 8.184E-06 | 3.514E-06 | 1.969E+04 | 1.046E+02 | 1.473E+02 | 1.757E+00 | 1.994E+04 |
| 0 | 1000 | 0 | 1.917E-06 | 3.977E-03 | 9.494E-07 | 1.972E-03 | 1.007E-07 | 1.077E-03 | 4.187E-06 | 7.239E-03 | 1.007E+04 | 2.155E+05 | 7.536E+01 | 3.620E+03 | 2.292E+05 |
| 0 | 0 | 1000 | 1.821E-06 | 2.584E-05 | 5.111E-03 | 1.281E-05 | 9.566E-08 | 6.999E-06 | 2.254E-02 | 4.703E-05 | 9.566E+03 | 1.400E+03 | 4.057E+05 | 2.352E+01 | 4.167E+05 |
| 600 | 300 | 100 | 3.006E-06 | 1.197E-03 | 5.125E-04 | 5.934E-04 | 1.579E-07 | 3.242E-04 | 2.260E-03 | 2.179E-03 | 1.579E+04 | 6.484E+04 | 4.068E+04 | 1.089E+03 | 1.224E+05 |

| Emission, kg/h | | | Loss, Reaction, kg/h | | | | Loss, Advection, kg/h | | | Intermedia Rate of Transport, kg/h T12 | T21 | T13 | T31 | T32 | T24 | T42 |
|---|---|---|---|---|---|---|---|---|---|---|---|---|---|---|---|---|
| E(1) | E(2) | E(3) | R(1) | R(2) | R(3) | R(4) | A(1) | A(2) | A(4) | air-water | water-air | air-soil | soil-air | soil-water | water-sed | sed-water |
| 1000 | 0 | 0 | 8.027E+02 | 1.318E-01 | 1.86E-01 | 7.162E-04 | 1.969E+02 | 1.046E-01 | 3.514E-05 | 4.833E-01 | 2.484E-01 | 3.631E-01 | 1.752E-01 | 2.274E-03 | 1.043E-03 | 2.921E-04 |
| 0 | 1000 | 0 | 4.107E+02 | 2.715E+02 | 9.50E-02 | 1.475E+00 | 1.007E+02 | 2.155E+02 | 7.239E-02 | 2.472E-01 | 5.117E+02 | 1.858E-01 | 8.966E-02 | 1.163E-03 | 2.150E+00 | 6.017E-01 |
| 0 | 0 | 1000 | 3.899E+02 | 1.764E+00 | 5.11E+02 | 9.586E-03 | 9.566E+01 | 1.400E+00 | 4.703E-04 | 2.348E-01 | 3.325E+00 | 1.764E-01 | 4.827E+02 | 6.264E+00 | 1.397E-02 | 3.909E-03 |
| 600 | 300 | 100 | 6.438E+02 | 8.170E+01 | 5.13E+01 | 4.440E-01 | 1.579E+02 | 6.484E+01 | 2.179E-02 | 3.876E-01 | 1.540E+02 | 2.912E-01 | 4.840E+01 | 6.281E-01 | 6.469E-01 | 1.811E-01 |

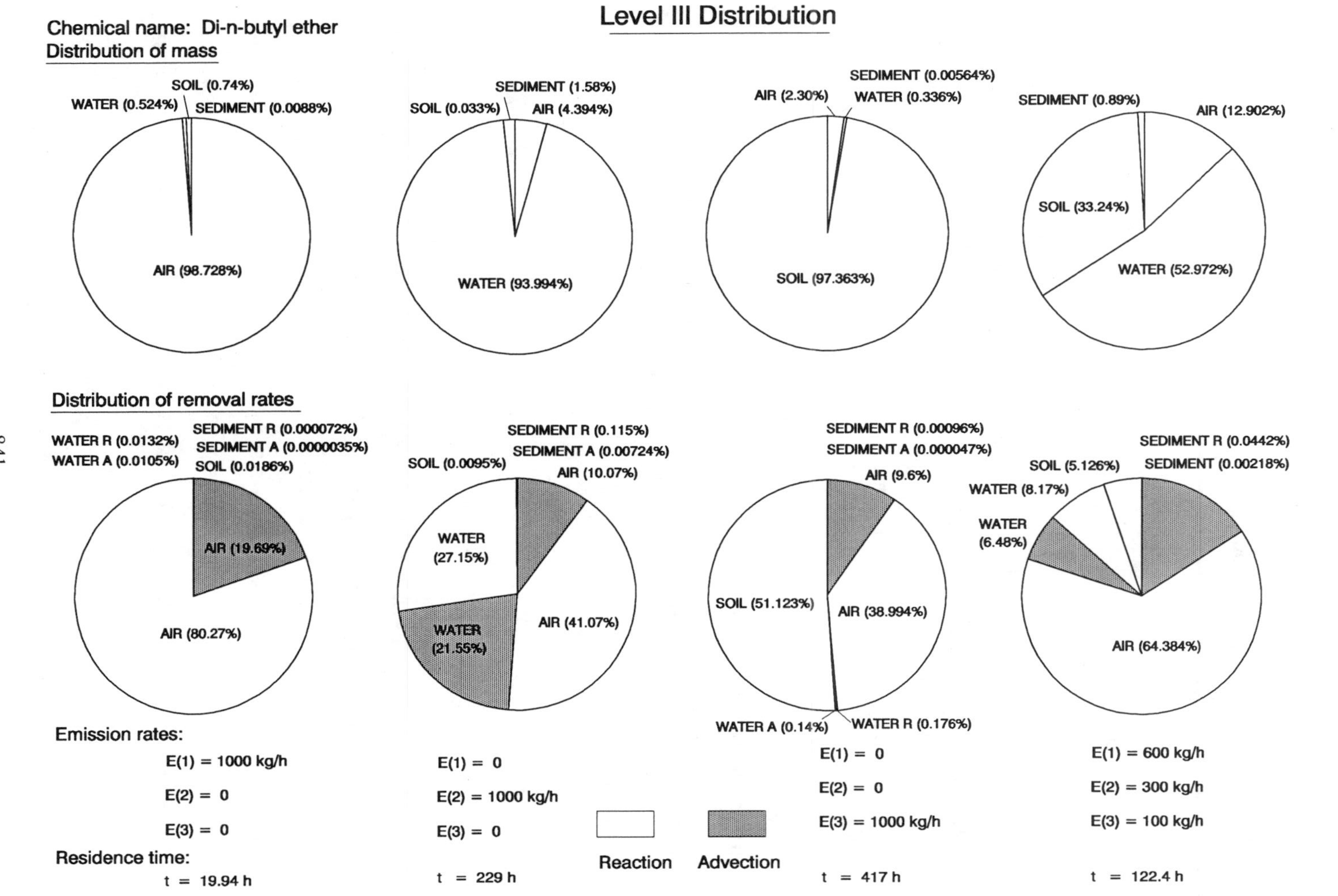
Level III Distribution
Chemical name: Di-n-butyl ether
Distribution of mass
WATER (0.524%)
SOIL (0.74%)
SEDIMENT (0.0088%)
AIR (98.728%)
SOIL (0.033%)
SEDIMENT (1.58%)
AIR (4.394%)
WATER (93.994%)
AIR (2.30%)
SEDIMENT (0.00564%)
WATER (0.336%)
SOIL (97.363%)
SEDIMENT (0.89%)
AIR (12.902%)
SOIL (33.24%)
WATER (52.972%)
Distribution of removal rates
WATER R (0.0132%)
WATER A (0.0105%)
SEDIMENT R (0.000072%)
SEDIMENT A (0.0000035%)
SOIL (0.0186%)
AIR (19.69%)
AIR (80.27%)
SOIL (0.0095%)
SEDIMENT R (0.115%)
SEDIMENT A (0.00724%)
AIR (10.07%)
WATER (27.15%)
WATER (21.55%)
AIR (41.07%)
SEDIMENT R (0.00096%)
SEDIMENT A (0.000047%)
AIR (9.6%)
SOIL (51.123%)
AIR (38.994%)
WATER A (0.14%)
WATER R (0.176%)
SEDIMENT R (0.0442%)
SEDIMENT (0.00218%)
SOIL (5.126%)
WATER (8.17%)
WATER (6.48%)
AIR (64.384%)
Emission rates:
E(1) = 1000 kg/h
E(2) = 0
E(3) = 0
Residence time:
t = 19.94 h
E(1) = 0
E(2) = 1000 kg/h
E(3) = 0
t = 229 h
Reaction
Advection
E(1) = 0
E(2) = 0
E(3) = 1000 kg/h
t = 417 h
E(1) = 600 kg/h
E(2) = 300 kg/h
E(3) = 100 kg/h
t = 122.4 h

Chemical name: 1,2-Propylene oxide

Level I calculation: (six-compartment model)

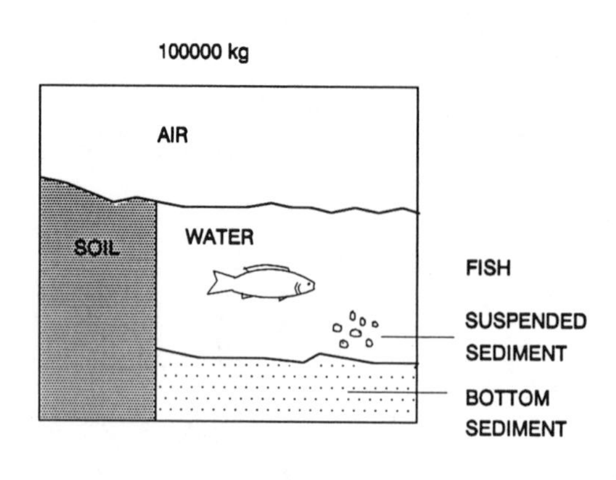


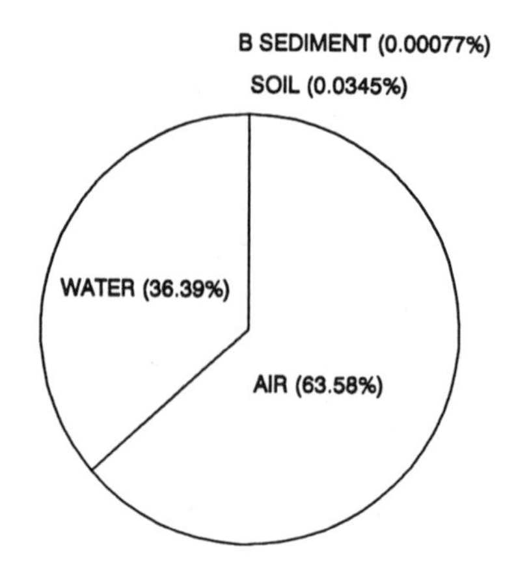


Distribution of mass

physical-chemical properties:

MW: 58.08 g/mol

M.P.: - 112.12 ℃

Fugacity ratio: 1.0

vapor pressure: 71000 Pa

solubility: 476000 g/m $^3$

log $K_{OW}$: 0.03

| Compartment | Z | Concentration | | | Amount | Amount |
|---|---|---|---|---|---|---|
| | mol/m3 Pa | mol/m3 | mg/L (or g/m3) | ug/g | kg | % |
| Air | 4.034E-04 | 7.488E-09 | 9.752E-07 | 8.227E-04 | 97520 | 97.520 |
| Water | 2.078E-03 | 3.857E-08 | 5.023E-06 | 5.023E-06 | 1004.53 | 1.0045 |
| Soil | 6.632E-02 | 1.231E-06 | 1.603E-04 | 6.680E-05 | 1442.79 | 1.4428 |
| Biota (fish) | 1.685E-01 | 3.127E-06 | 4.073E-04 | 4.073E-04 | 8.15E-02 | 8.15E-05 |
| Suspended sediment | 4.145E-01 | 7.694E-06 | 1.002E-03 | 6.680E-04 | 1.0019 | 1.00E-03 |
| Bottom sediment | 1.326E-01 | 2.462E-06 | 3.206E-04 | 1.336E-04 | 32.0619 | 3.21E-02 |
| | Total | | | | 100000 | 100 |

f = 1.856E-05 Pa

Chemical name: 1,2-Propylene oxide

Level II calculation: (six-compartment model)

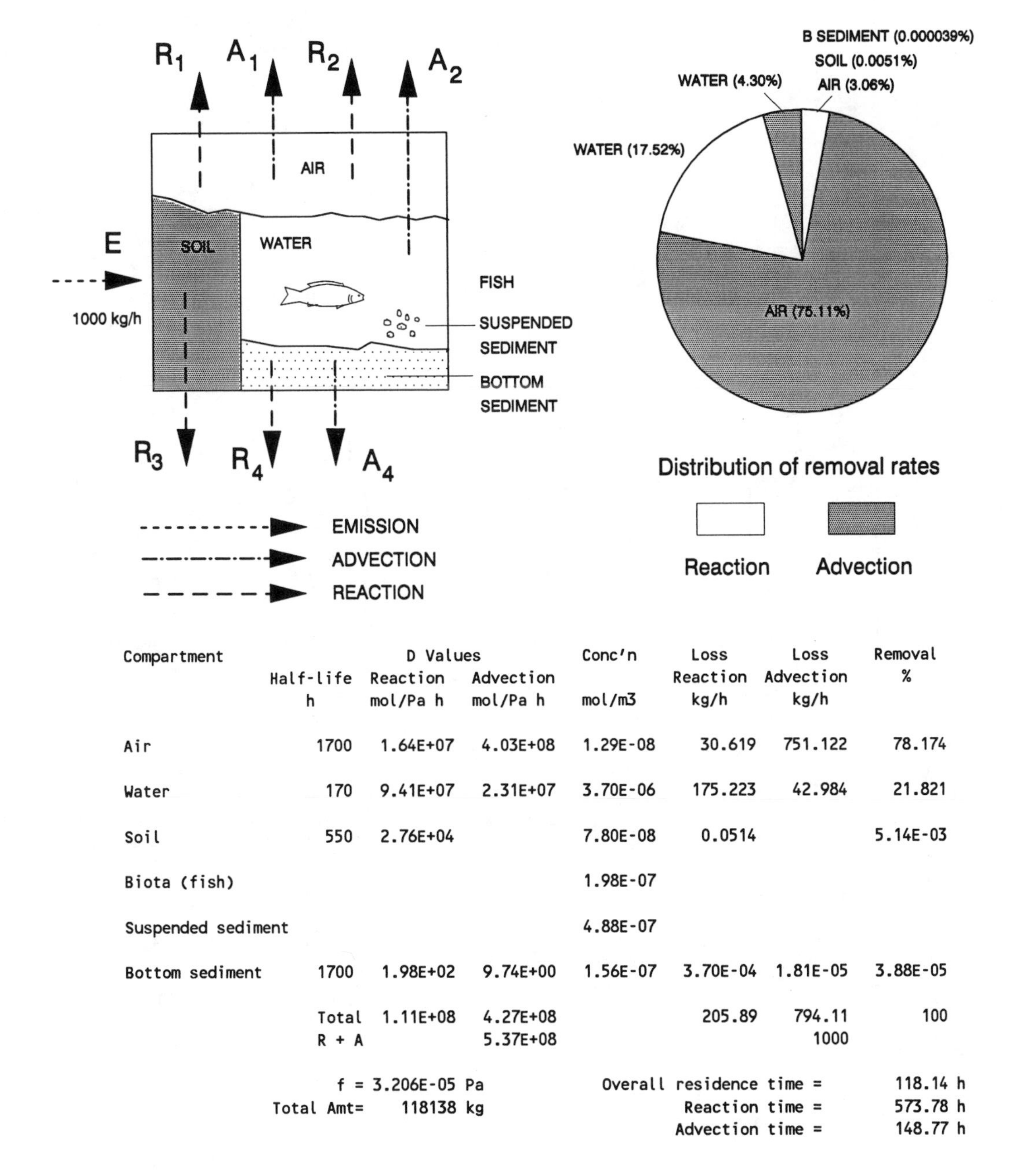


| Compartment | Half-life h | D Values Reaction mol/Pa h | D Values Advection mol/Pa h | Conc'n mol/m3 | Loss Reaction kg/h | Loss Advection kg/h | Removal % |
|---|---|---|---|---|---|---|---|
| Air | 1700 | 1.64E+07 | 4.03E+08 | 1.29E-08 | 30.619 | 751.122 | 78.174 |
| Water | 170 | 9.41E+07 | 2.31E+07 | 3.70E-06 | 175.223 | 42.984 | 21.821 |
| Soil | 550 | 2.76E+04 | | 7.80E-08 | 0.0514 | | 5.14E-03 |
| Biota (fish) | | | | 1.98E-07 | | | |
| Suspended sediment | | | | 4.88E-07 | | | |
| Bottom sediment | 1700 | 1.98E+02 | 9.74E+00 | 1.56E-07 | 3.70E-04 | 1.81E-05 | 3.88E-05 |
| | Total | 1.11E+08 | 4.27E+08 | | 205.89 | 794.11 | 100 |
| | R + A | | 5.37E+08 | | | 1000 | |

f = 3.206E-05 Pa
Total Amt= 118138 kg

Overall residence time = 118.14 h
Reaction time = 573.78 h
Advection time = 148.77 h

## Fugacity Level III calculations: (four-compartment model)

Chemical name: 1,2-Propylene oxide

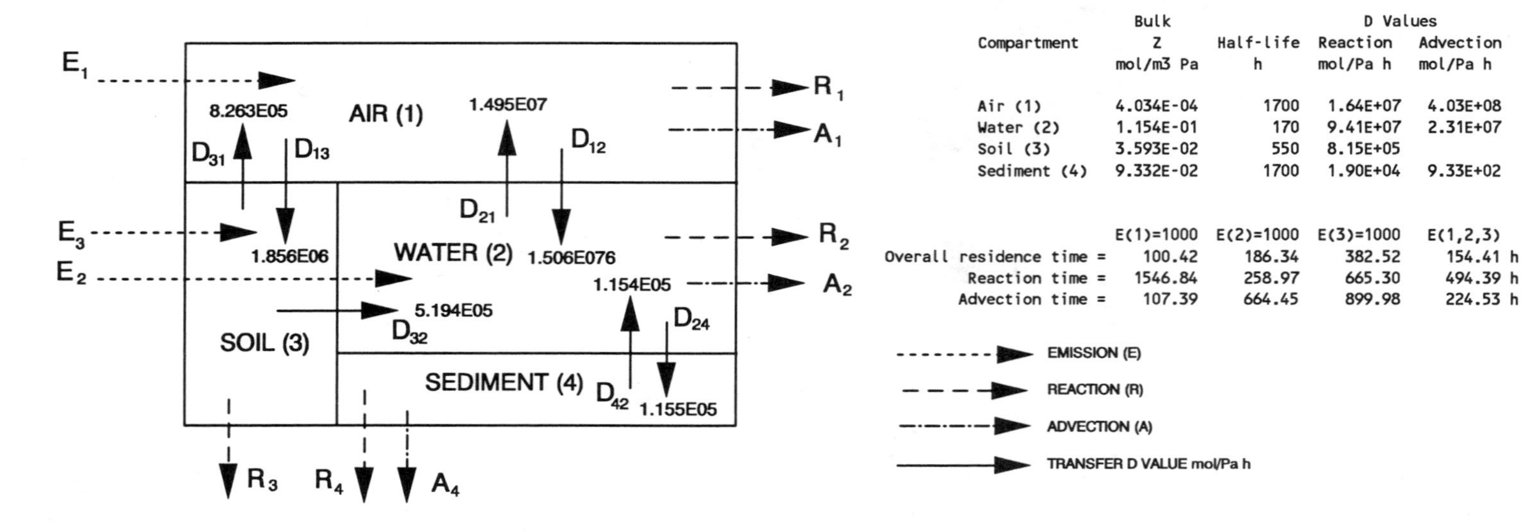


Phase Properties and Rates:

| Compartment | Bulk Z mol/m3 Pa | Half-life h | D Values Reaction mol/Pa h | D Values Advection mol/Pa h |
|---|---|---|---|---|
| Air (1) | 4.034E-04 | 1700 | 1.64E+07 | 4.03E+08 |
| Water (2) | 1.154E-01 | 170 | 9.41E+07 | 2.31E+07 |
| Soil (3) | 3.593E-02 | 550 | 8.15E+05 | |
| Sediment (4) | 9.332E-02 | 1700 | 1.90E+04 | 9.33E+02 |

| | E(1)=1000 | E(2)=1000 | E(3)=1000 | E(1,2,3) |
|---|---|---|---|---|
| Overall residence time = | 100.42 | 186.34 | 382.52 | 154.41 h |
| Reaction time = | 1546.84 | 258.97 | 665.30 | 494.39 h |
| Advection time = | 107.39 | 664.45 | 899.98 | 224.53 h |

Phase Properties, Compositions, Transport and Transformation Rates:

| Emission, kg/h | | | Fugacity, Pa | | | | Concentration, g/m3 | | | | Amounts, kg | | | | Total Amount, kg |
|---|---|---|---|---|---|---|---|---|---|---|---|---|---|---|---|
| E(1) | E(2) | E(3) | f(1) | f(2) | f(3) | f(4) | C(1) | C(2) | C(3) | C(4) | m(1) | m(2) | m(3) | m(4) | |
| 1000 | 0 | 0 | 3.964E-05 | 4.653E-06 | 3.422E-05 | 3.969E-06 | 9.288E-07 | 3.119E-05 | 7.141E-05 | 2.151E-05 | 9.288E+04 | 6.238E+03 | 1.285E+03 | 1.076E+01 | 1.004E+05 |
| 0 | 1000 | 0 | 4.484E-06 | 1.308E-04 | 3.871E-06 | 1.116E-04 | 1.051E-07 | 8.770E-04 | 8.076E-06 | 6.048E-04 | 1.051E+04 | 1.754E+05 | 1.454E+02 | 3.024E+02 | 1.863E+05 |
| 0 | 0 | 1000 | 1.624E-05 | 3.323E-05 | 7.983E-03 | 2.835E-05 | 3.805E-07 | 2.228E-04 | 1.666E-02 | 1.536E-04 | 3.805E+04 | 4.455E+04 | 2.998E+05 | 7.682E+01 | 3.825E+05 |
| 600 | 300 | 100 | 2.675E-05 | 4.536E-05 | 8.200E-04 | 3.869E-05 | 6.269E-07 | 3.041E-04 | 1.711E-03 | 2.097E-04 | 6.269E+04 | 6.081E+04 | 3.080E+04 | 1.049E+02 | 1.544E+05 |

| Emission, kg/h | | | Loss, Reaction, kg/h | | | | Loss, Advection, kg/h | | | Intermedia Rate of Transport, kg/h T12 | T21 | T13 | T31 | T32 | T24 | T42 |
|---|---|---|---|---|---|---|---|---|---|---|---|---|---|---|---|---|
| E(1) | E(2) | E(3) | R(1) | R(2) | R(3) | R(4) | A(1) | A(2) | A(4) | air-water | water-air | air-soil | soil-air | soil-water | water-sed | sed-water |
| 1000 | 0 | 0 | 3.786E+01 | 2.543E+01 | 1.62E+00 | 4.385E-03 | 9.288E+02 | 6.238E+00 | 2.151E-04 | 3.468E+01 | 4.039E+00 | 4.294E+00 | 1.642E+00 | 1.032E+00 | 3.121E-02 | 2.661E-02 |
| 0 | 1000 | 0 | 4.282E+00 | 7.150E+02 | 1.83E-01 | 1.233E-01 | 1.051E+02 | 1.754E+02 | 6.048E-03 | 3.922E+00 | 1.136E+02 | 4.857E-01 | 1.857E-01 | 1.168E-01 | 8.775E-01 | 7.482E-01 |
| 0 | 0 | 1000 | 1.551E+01 | 1.816E+02 | 3.78E+02 | 3.132E-02 | 3.805E+02 | 4.455E+01 | 1.536E-03 | 1.421E+01 | 2.885E+01 | 1.759E+00 | 3.831E+02 | 2.408E+02 | 2.229E-01 | 1.901E-01 |
| 600 | 300 | 100 | 2.555E+01 | 2.479E+02 | 3.88E+01 | 4.275E-02 | 6.269E+02 | 6.081E+01 | 2.097E-03 | 2.341E+01 | 3.937E+01 | 2.898E+00 | 3.935E+01 | 2.474E+01 | 3.043E-01 | 2.594E-01 |

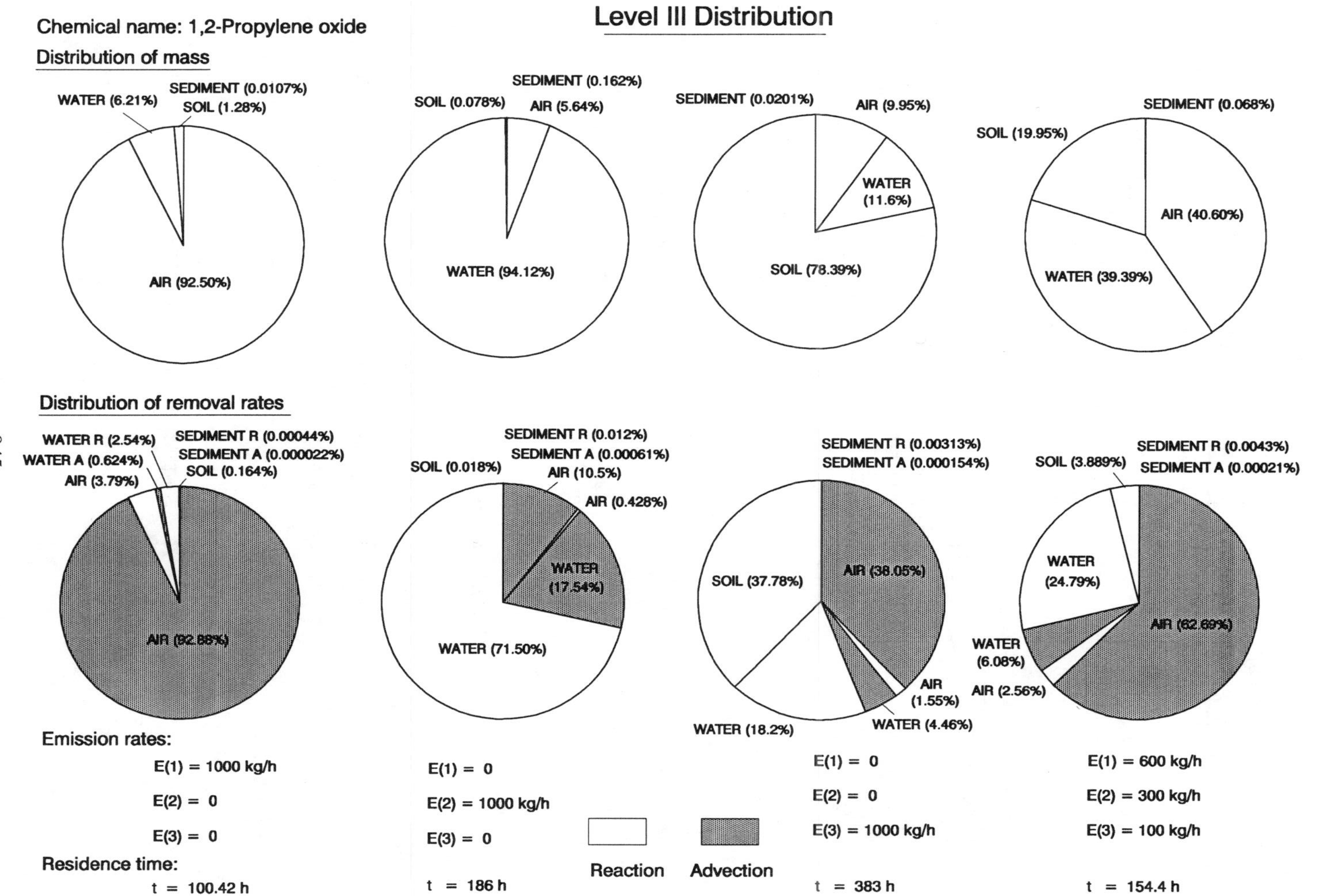
Level III Distribution
Chemical name: 1,2-Propylene oxide
Distribution of mass
WATER (6.21%)
SEDIMENT (0.0107%)
SOIL (1.28%)
AIR (92.50%)
SOIL (0.078%)
SEDIMENT (0.162%)
AIR (5.64%)
WATER (94.12%)
SEDIMENT (0.0201%)
AIR (9.95%)
WATER (11.6%)
SOIL (78.39%)
SOIL (19.95%)
SEDIMENT (0.068%)
AIR (40.60%)
WATER (39.39%)
Distribution of removal rates
WATER R (2.54%)
WATER A (0.624%)
AIR (3.79%)
SEDIMENT R (0.00044%)
SEDIMENT A (0.000022%)
SOIL (0.164%)
AIR (92.88%)
SOIL (0.018%)
SEDIMENT R (0.012%)
SEDIMENT A (0.00061%)
AIR (10.5%)
AIR (0.428%)
WATER (17.54%)
WATER (71.50%)
SEDIMENT R (0.00313%)
SEDIMENT A (0.000154%)
SOIL (37.78%)
AIR (38.05%)
AIR (1.55%)
WATER (18.2%)
WATER (4.46%)
SOIL (3.889%)
SEDIMENT R (0.0043%)
SEDIMENT A (0.00021%)
WATER (24.79%)
AIR (62.69%)
WATER (6.08%)
AIR (2.56%)
Emission rates:
E(1) = 1000 kg/h
E(2) = 0
E(3) = 0
Residence time:
t = 100.42 h
E(1) = 0
E(2) = 1000 kg/h
E(3) = 0
t = 186 h
Reaction
Advection
E(1) = 0
E(2) = 0
E(3) = 1000 kg/h
t = 383 h
E(1) = 600 kg/h
E(2) = 300 kg/h
E(3) = 100 kg/h
t = 154.4 h

## Chemical name: Furan

Level I calculation: (six-compartment model)

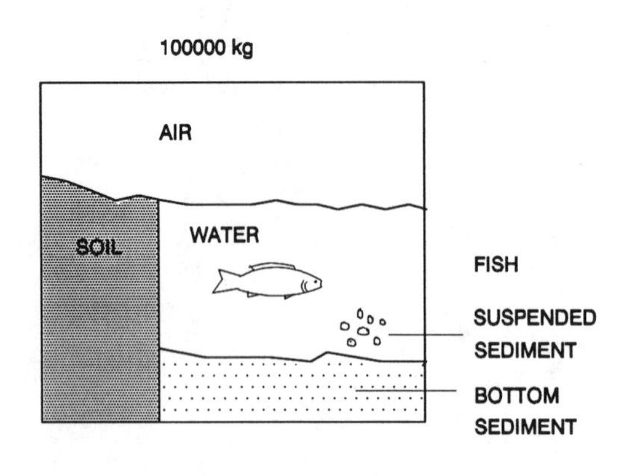


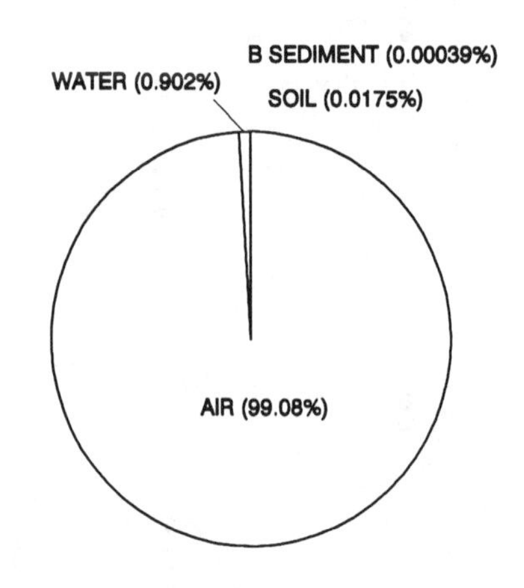


Distribution of mass

physical-chemical properties:

MW: 68.08 g/mol

M.P.: - 85.61 °C

Fugacity ratio: 1.0

vapor pressure: 80000 Pa

solubility: 10000 g/m$^3$

log $K_{OW}$: 1.34

| Compartment | Z | Concentration | | | Amount | Amount |
|---|---|---|---|---|---|---|
| | mol/m3 Pa | mol/m3 | mg/L (or g/m3) | ug/g | kg | % |
| Air | 4.034E-04 | 1.455E-08 | 9.908E-07 | 8.358E-04 | 99080 | 99.080 |
| Water | 1.836E-03 | 6.624E-08 | 4.509E-06 | 4.509E-06 | 901.89 | 0.9019 |
| Soil | 7.905E-04 | 2.852E-08 | 1.942E-06 | 8.090E-07 | 17.474 | 0.0175 |
| Biota (fish) | 2.008E-03 | 7.246E-08 | 4.933E-06 | 4.933E-06 | 9.87E-04 | 9.87E-07 |
| Suspended sediment | 4.941E-03 | 1.782E-07 | 1.213E-05 | 8.090E-06 | 0.0121 | 1.21E-05 |
| Bottom sediment | 1.581E-03 | 5.704E-08 | 3.883E-06 | 1.618E-06 | 0.3883 | 3.88E-04 |
| | Total | | | | 100000 | 100 |

f = 3.608E-05 Pa

Chemical name: Furan

Level II calculation: (six-compartment model)

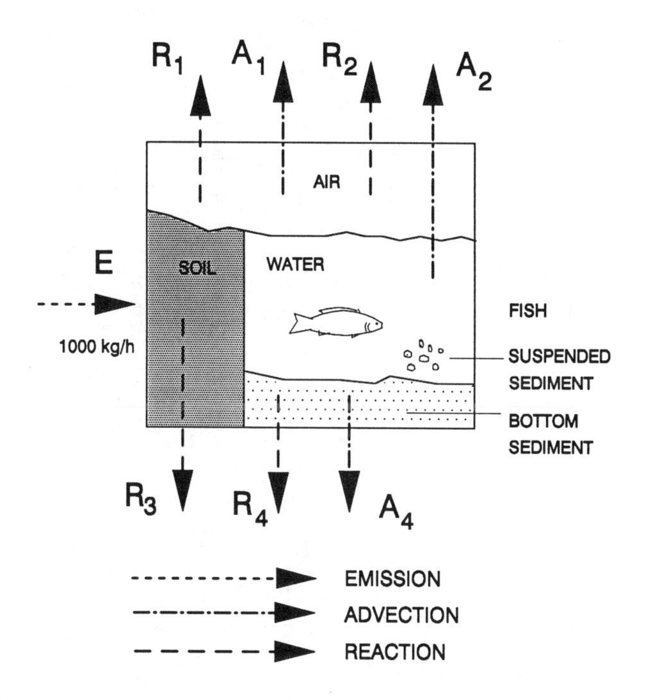


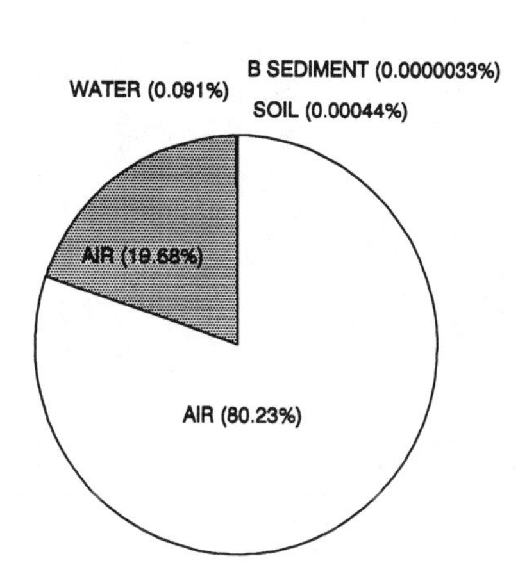


Distribution of removal rates

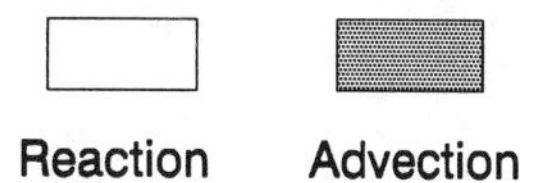


| Compartment | Half-life h | D Values Reaction mol/Pa h | D Values Advection mol/Pa h | Conc'n mol/m3 | Loss Reaction kg/h | Loss Advection kg/h | Removal % |
|---|---|---|---|---|---|---|---|
| Air | 17 | 1.64E+09 | 4.03E+08 | 2.89E-09 | 802.279 | 196.807 | 99.909 |
| Water | 170 | 1.50E+06 | 3.67E+05 | 1.32E-08 | 0.7303 | 0.1791 | 0.0909 |
| Soil | 550 | 8.96E+03 | | 5.66E-09 | 4.37E-03 | | 4.37E-04 |
| Biota (fish) | | | | 1.44E-08 | | | |
| Suspended sediment | | | | 3.54E-08 | | | |
| Bottom sediment | 1700 | 6.45E+01 | 3.16E+00 | 1.13E-08 | 3.14E-05 | 1.54E-06 | 3.30E-06 |
| | Total | 1.65E+09 | 4.04E+08 | | 803.01 | 196.99 | 100 |
| | R + A | | 2.05E+09 | | | 1000 | |

f = 7.166E-06 Pa
Total Amt= 19863 kg

Overall residence time = 19.86 h
Reaction time = 24.74 h
Advection time = 100.84 h

Fugacity Level III calculations: (four-compartment model)

Chemical name: Furan

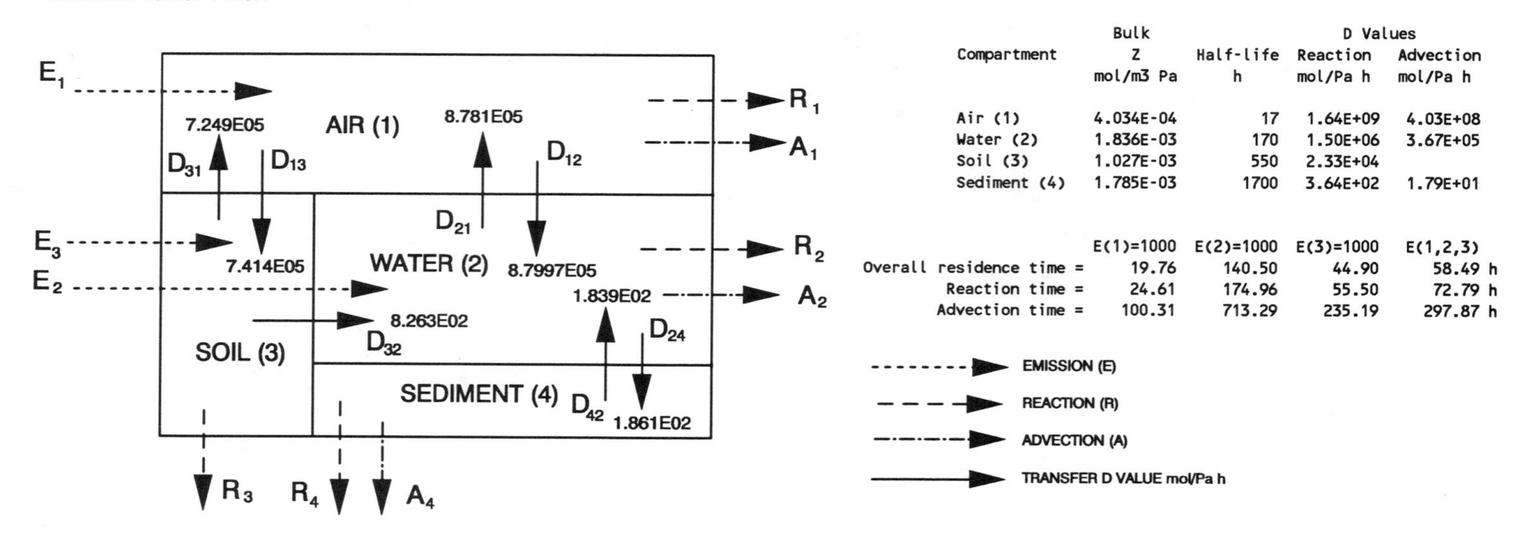


Phase Properties and Rates:

| Compartment | Bulk Z mol/m3 Pa | Half-life h | D Values Reaction mol/Pa h | Advection mol/Pa h |
|---|---|---|---|---|
| Air (1) | 4.034E-04 | 17 | 1.64E+09 | 4.03E+08 |
| Water (2) | 1.836E-03 | 170 | 1.50E+06 | 3.67E+05 |
| Soil (3) | 1.027E-03 | 550 | 2.33E+04 | |
| Sediment (4) | 1.785E-03 | 1700 | 3.64E+02 | 1.79E+01 |

| | E(1)=1000 | E(2)=1000 | E(3)=1000 | E(1,2,3) | |
|---|---|---|---|---|---|
| Overall residence time = | 19.76 | 140.50 | 44.90 | 58.49 | h |
| Reaction time = | 24.61 | 174.96 | 55.50 | 72.79 | h |
| Advection time = | 100.31 | 713.29 | 235.19 | 297.87 | h |

Phase Properties, Compositions, Transport and Transformation Rates:

| Emission, kg/h | | | Fugacity, Pa | | | | Concentration, g/m3 | | | | Amounts, kg | | | | Total Amount, kg |
|---|---|---|---|---|---|---|---|---|---|---|---|---|---|---|---|
| E(1) | E(2) | E(3) | f(1) | f(2) | f(3) | f(4) | C(1) | C(2) | C(3) | C(4) | m(1) | m(2) | m(3) | m(4) | |
| 1000 | 0 | 0 | 7.170E-06 | 2.322E-06 | 7.028E-06 | 1.945E-06 | 1.969E-07 | 2.902E-07 | 4.913E-07 | 2.364E-07 | 1.969E+04 | 5.804E+01 | 8.843E+00 | 1.182E-01 | 1.976E+04 |
| 0 | 1000 | 0 | 2.296E-06 | 5.357E-03 | 2.250E-06 | 4.488E-03 | 6.305E-08 | 6.696E-04 | 1.573E-07 | 5.454E-04 | 6.305E+03 | 1.339E+05 | 2.831E+00 | 2.727E+02 | 1.405E+05 |
| 0 | 0 | 1000 | 6.896E-06 | 6.074E-05 | 1.942E-02 | 5.089E-05 | 1.894E-07 | 7.592E-06 | 1.358E-03 | 6.184E-06 | 1.894E+04 | 1.518E+03 | 2.444E+04 | 3.092E+00 | 4.490E+04 |
| 600 | 300 | 100 | 5.680E-06 | 1.614E-03 | 1.947E-03 | 1.353E-03 | 1.560E-07 | 2.018E-04 | 1.361E-04 | 1.644E-04 | 1.560E+04 | 4.036E+04 | 2.450E+03 | 8.219E+01 | 5.849E+04 |

| Emission, kg/h | | | Loss, Reaction, kg/h | | | | Loss, Advection, kg/h | | | Intermedia Rate of Transport, kg/h T12 | T21 | T13 | T31 | T32 | T24 | T42 |
|---|---|---|---|---|---|---|---|---|---|---|---|---|---|---|---|---|
| E(1) | E(2) | E(3) | R(1) | R(2) | R(3) | R(4) | A(1) | A(2) | A(4) | air-water | water-air | air-soil | soil-air | soil-water | water-sed | sed-water |
| 1000 | 0 | 0 | 8.028E+02 | 2.366E-01 | 1.11E-02 | 4.818E-05 | 1.969E+02 | 5.804E-02 | 2.364E-06 | 4.295E-01 | 1.388E-01 | 3.619E-01 | 3.468E-01 | 3.953E-03 | 2.941E-04 | 2.436E-04 |
| 0 | 1000 | 0 | 2.570E+02 | 5.459E+02 | 3.57E-03 | 1.112E-01 | 6.305E+01 | 1.339E+02 | 5.454E-03 | 1.375E-01 | 3.202E+02 | 1.159E-01 | 1.110E-01 | 1.266E-03 | 6.786E-01 | 5.620E-01 |
| 0 | 0 | 1000 | 7.721E+02 | 6.190E+00 | 3.08E+01 | 1.260E-03 | 1.894E+02 | 1.518E+00 | 6.184E-05 | 4.131E-01 | 3.631E+00 | 3.481E-01 | 9.586E+02 | 1.093E+01 | 7.694E-03 | 6.372E-03 |
| 600 | 300 | 100 | 6.360E+02 | 1.645E+02 | 3.09E+00 | 3.350E-02 | 1.560E+02 | 4.036E+01 | 1.644E-03 | 3.403E-01 | 9.651E+01 | 2.867E-01 | 9.610E+01 | 1.095E+00 | 2.045E-01 | 1.694E-01 |

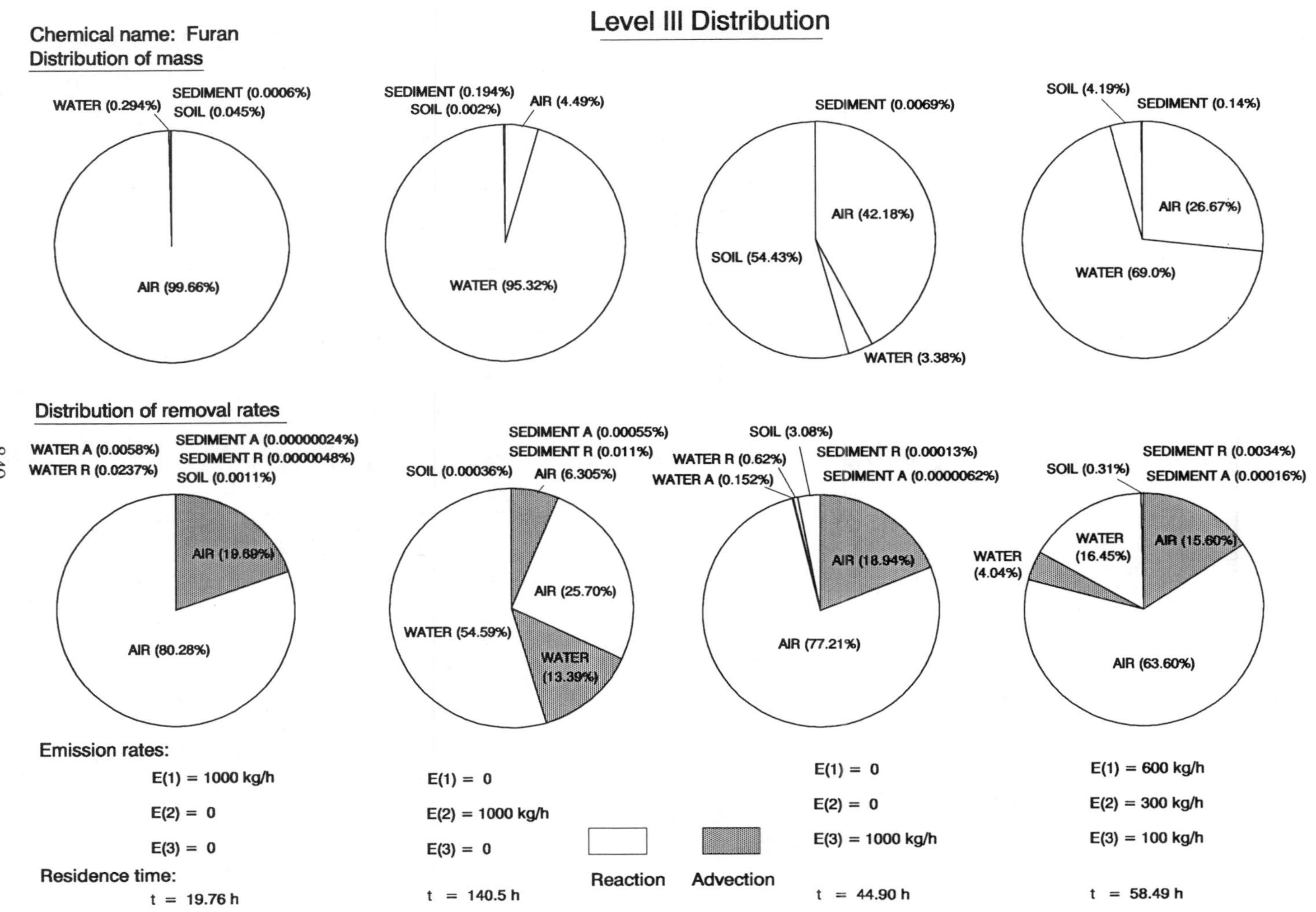
Level III Distribution
Chemical name: Furan
Distribution of mass
WATER (0.294%)
SEDIMENT (0.0006%)
SOIL (0.045%)
AIR (99.66%)
SEDIMENT (0.194%)
SOIL (0.002%)
AIR (4.49%)
WATER (95.32%)
SEDIMENT (0.0069%)
AIR (42.18%)
SOIL (54.43%)
WATER (3.38%)
SOIL (4.19%)
SEDIMENT (0.14%)
AIR (26.67%)
WATER (69.0%)
Distribution of removal rates
WATER A (0.0058%)
WATER R (0.0237%)
SEDIMENT A (0.00000024%)
SEDIMENT R (0.0000048%)
SOIL (0.0011%)
AIR (19.69%)
AIR (80.28%)
SEDIMENT A (0.00055%)
SEDIMENT R (0.011%)
SOIL (0.00036%)
AIR (6.305%)
AIR (25.70%)
WATER (54.59%)
WATER (13.39%)
SOIL (3.08%)
WATER R (0.62%)
WATER A (0.152%)
SEDIMENT R (0.00013%)
SEDIMENT A (0.0000062%)
AIR (18.94%)
AIR (77.21%)
SEDIMENT R (0.0034%)
SOIL (0.31%)
SEDIMENT A (0.00016%)
WATER (16.45%)
AIR (15.60%)
WATER (4.04%)
AIR (63.60%)
Emission rates:
E(1) = 1000 kg/h
E(2) = 0
E(3) = 0
Residence time:
t = 19.76 h
E(1) = 0
E(2) = 1000 kg/h
E(3) = 0
t = 140.5 h
Reaction
Advection
E(1) = 0
E(2) = 0
E(3) = 1000 kg/h
t = 44.90 h
E(1) = 600 kg/h
E(2) = 300 kg/h
E(3) = 100 kg/h
t = 58.49 h

Chemical name: Anisole (methoxybenzene)

Level I calculation: (six-compartment model)

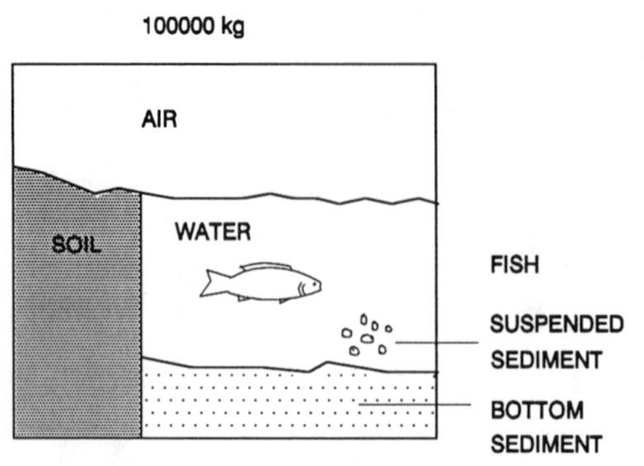


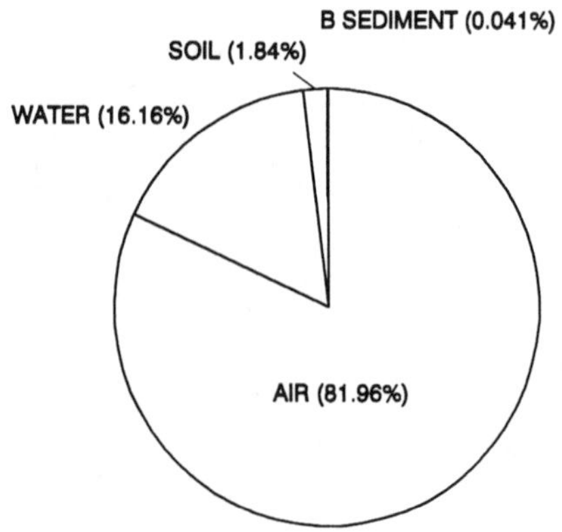


Distribution of mass

physical-chemical properties:

MW: 108.15 g/mol

M.P.: - 37.5 °C

Fugacity ratio: 1.0

vapor pressure: 472 Pa

solubility: 2030 g/m $^3$

log $K_{OW}$: 2.11

| Compartment | Z | Concentration | | | Amount | Amount |
|---|---|---|---|---|---|---|
| | mol/m3 Pa | mol/m3 | mg/L (or g/m3) | ug/g | kg | % |
| Air | 4.034E-04 | 7.578E-09 | 8.196E-07 | 6.914E-04 | 81956 | 81.956 |
| Water | 3.977E-02 | 7.470E-07 | 8.079E-05 | 8.079E-05 | 16158 | 16.1579 |
| Soil | 1.008E-01 | 1.894E-06 | 2.048E-04 | 8.534E-05 | 1843.41 | 1.8434 |
| Biota (fish) | 2.562E-01 | 4.812E-06 | 5.204E-04 | 5.204E-04 | 0.1041 | 1.04E-04 |
| Suspended sediment | 6.301E-01 | 1.184E-05 | 1.280E-03 | 8.534E-04 | 1.2801 | 1.28E-03 |
| Bottom sediment | 2.016E-01 | 3.788E-06 | 4.096E-04 | 1.707E-04 | 40.965 | 4.10E-02 |
| | Total | | | | 100000 | 100 |

f = 1.878E-05 Pa

Chemical name: Anisole (methoxybenzene)

Level II calculation: (six-compartment model)

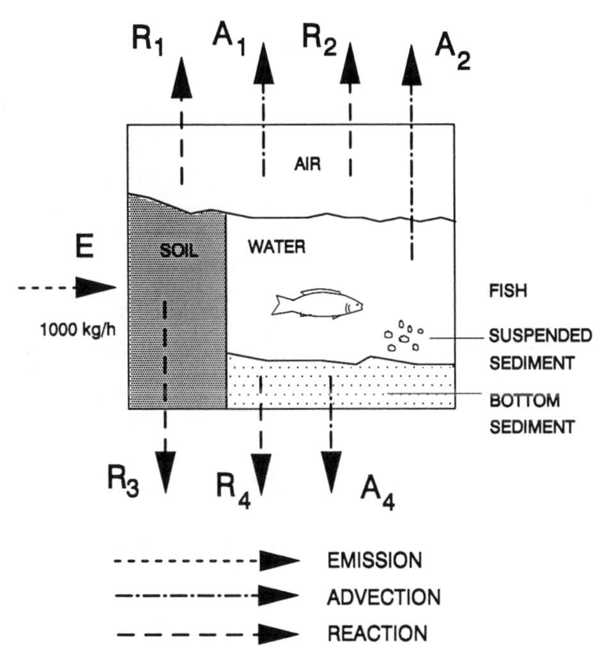


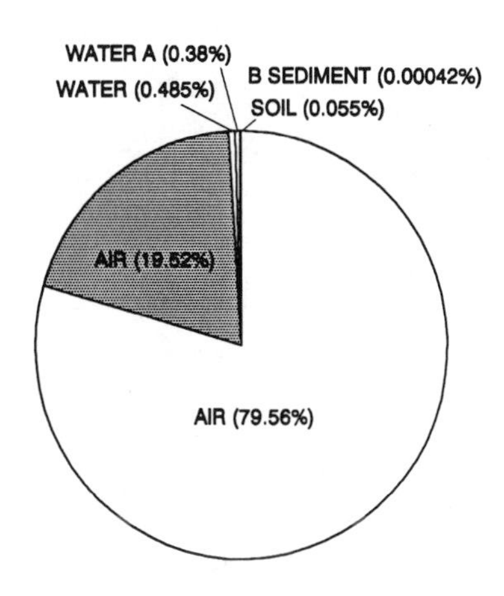


Distribution of removal rates

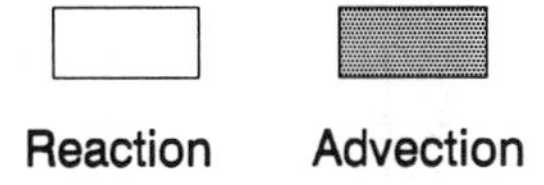


| Compartment | Half-life h | D Values Reaction mol/Pa h | D Values Advection mol/Pa h | Conc'n mol/m3 | Loss Reaction kg/h | Loss Advection kg/h | Removal % |
|---|---|---|---|---|---|---|---|
| Air | 17 | 1.64E+09 | 4.03E+08 | 1.80E-09 | 795.582 | 195.165 | 99.075 |
| Water | 550 | 1.00E+07 | 7.95E+06 | 1.78E-07 | 4.8481 | 3.8477 | 0.8696 |
| Soil | 550 | 1.14E+06 | | 4.51E-07 | 0.5531 | | 5.53E-02 |
| Biota (fish) | | | | 1.15E-06 | | | |
| Suspended sediment | | | | 2.82E-06 | | | |
| Bottom sediment | 1700 | 8.22E+03 | 4.03E+02 | 9.02E-07 | 3.98E-03 | 1.95E-04 | 4.17E-04 |
| | Total | 1.66E+09 | 4.11E+08 | | 800.99 | 199.01 | 100 |
| | R + A | | 2.07E+09 | | | 1000 | |

f = 4.473E-06 Pa
Total Amt= 23813 kg

Overall residence time = 23.81 h
Reaction time = 29.73 h
Advection time = 119.66 h

## Fugacity Level III calculations: (four-compartment model)

Chemical name: Anisole (methoxybenzene)

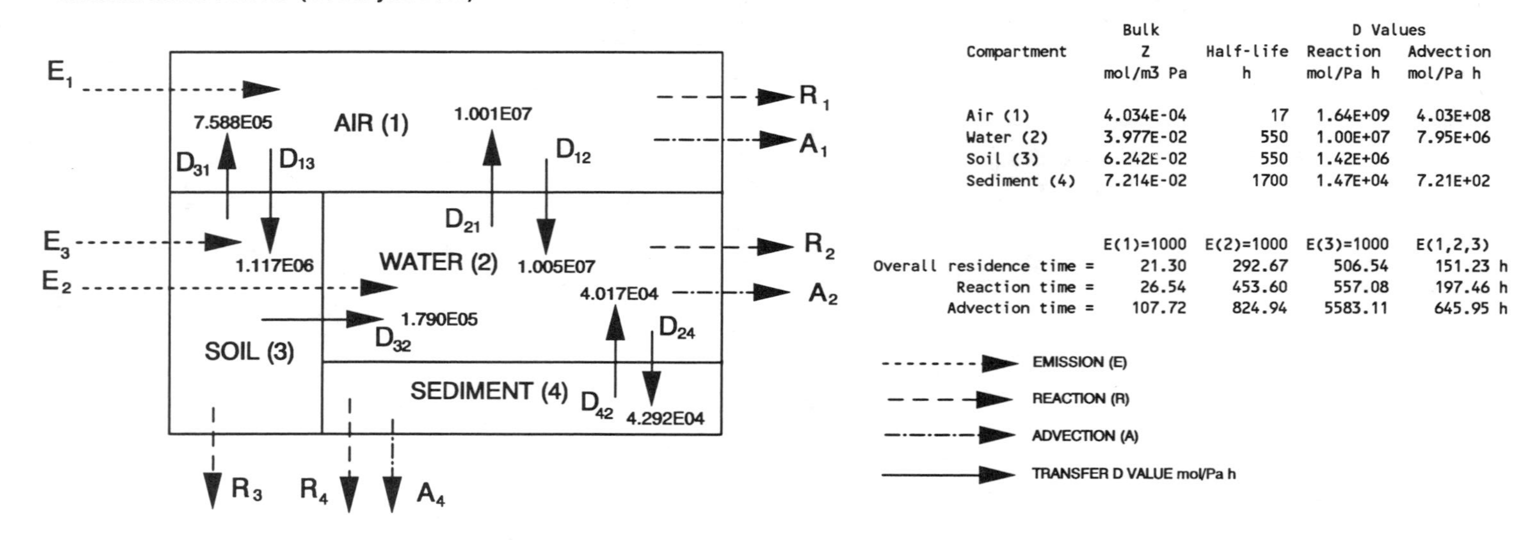


Phase Properties and Rates:

| Compartment | Bulk Z mol/m3 Pa | Half-life h | D Values Reaction mol/Pa h | D Values Advection mol/Pa h |
|---|---|---|---|---|
| Air (1) | 4.034E-04 | 17 | 1.64E+09 | 4.03E+08 |
| Water (2) | 3.977E-02 | 550 | 1.00E+07 | 7.95E+06 |
| Soil (3) | 6.242E-02 | 550 | 1.42E+06 | |
| Sediment (4) | 7.214E-02 | 1700 | 1.47E+04 | 7.21E+02 |

| | E(1)=1000 | E(2)=1000 | E(3)=1000 | E(1,2,3) |
|---|---|---|---|---|
| Overall residence time = | 21.30 | 292.67 | 506.54 | 151.23 h |
| Reaction time = | 26.54 | 453.60 | 557.08 | 197.46 h |
| Advection time = | 107.72 | 824.94 | 5583.11 | 645.95 h |



Phase Properties, Compositions, Transport and Transformation Rates:

| Emission, kg/h | | | Fugacity, Pa | | | | Concentration, g/m3 | | | | Amounts, kg | | | | Total |
|---|---|---|---|---|---|---|---|---|---|---|---|---|---|---|---|
| E(1) | E(2) | E(3) | f(1) | f(2) | f(3) | f(4) | C(1) | C(2) | C(3) | C(4) | m(1) | m(2) | m(3) | m(4) | Amount, kg |
| 1000 | 0 | 0 | 4.499E-06 | 1.629E-06 | 2.135E-06 | 1.257E-06 | 1.963E-07 | 7.006E-06 | 1.441E-05 | 9.811E-06 | 1.963E+04 | 1.401E+03 | 2.595E+02 | 4.906E+00 | 2.130E+04 |
| 0 | 1000 | 0 | 1.609E-06 | 3.308E-04 | 7.636E-07 | 2.554E-04 | 7.020E-08 | 1.423E-03 | 5.155E-06 | 1.992E-03 | 7.020E+03 | 2.846E+05 | 9.278E+01 | 9.962E+02 | 2.927E+05 |
| 0 | 0 | 1000 | 1.573E-06 | 2.569E-05 | 3.930E-03 | 1.983E-05 | 6.863E-08 | 1.105E-04 | 2.653E-02 | 1.547E-04 | 6.863E+03 | 2.210E+04 | 4.775E+05 | 7.737E+01 | 5.065E+05 |
| 600 | 300 | 100 | 3.339E-06 | 1.028E-04 | 3.945E-04 | 7.935E-05 | 1.457E-07 | 4.421E-04 | 2.663E-03 | 6.191E-04 | 1.457E+04 | 8.842E+04 | 4.793E+04 | 3.095E+02 | 1.512E+05 |

| Emission, kg/h | | | Loss, Reaction, kg/h | | | | Loss, Advection, kg/h | | | Intermedia Rate of Transport, kg/h T12 | T21 | T13 | T31 | T32 | T24 | T42 |
|---|---|---|---|---|---|---|---|---|---|---|---|---|---|---|---|---|
| E(1) | E(2) | E(3) | R(1) | R(2) | R(3) | R(4) | A(1) | A(2) | A(4) | air-water | water-air | air-soil | soil-air | soil-water | water-sed | sed-water |
| 1000 | 0 | 0 | 8.002E+02 | 1.766E+00 | 3.27E-01 | 2.000E-03 | 1.963E+02 | 1.401E+00 | 9.811E-05 | 4.892E+00 | 1.764E+00 | 5.435E-01 | 1.752E-01 | 4.135E-02 | 7.561E-03 | 5.463E-03 |
| 0 | 1000 | 0 | 2.862E+02 | 3.586E+02 | 1.17E-01 | 4.061E-01 | 7.020E+01 | 2.846E+02 | 1.992E-02 | 1.749E+00 | 3.582E+02 | 1.944E-01 | 6.266E-02 | 1.479E-02 | 1.535E+00 | 1.109E+00 |
| 0 | 0 | 1000 | 2.797E+02 | 2.785E+01 | 6.02E+02 | 3.154E-02 | 6.863E+01 | 2.210E+01 | 1.547E-03 | 1.710E+00 | 2.782E+01 | 1.900E-01 | 3.225E+02 | 7.609E+01 | 1.192E-01 | 8.616E-02 |
| 600 | 300 | 100 | 5.939E+02 | 1.114E+02 | 6.04E+01 | 1.262E-01 | 1.457E+02 | 8.842E+01 | 6.191E-03 | 3.631E+00 | 1.113E+02 | 4.034E-01 | 3.237E+01 | 7.638E+00 | 4.771E-01 | 3.447E-01 |

# Level III Distribution

Chemical name: Anisole (methyoxybenzene)

Distribution of mass

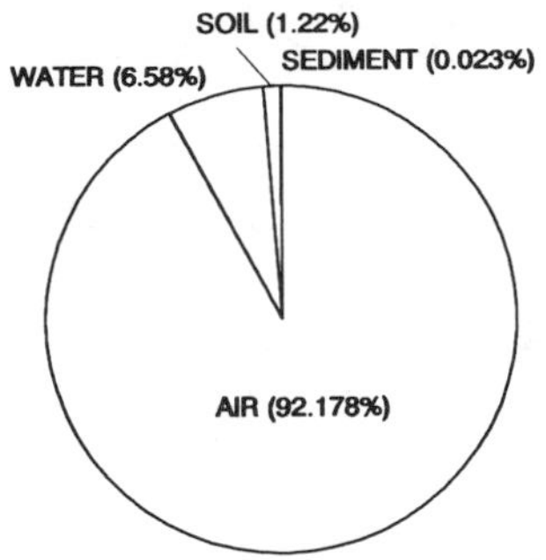

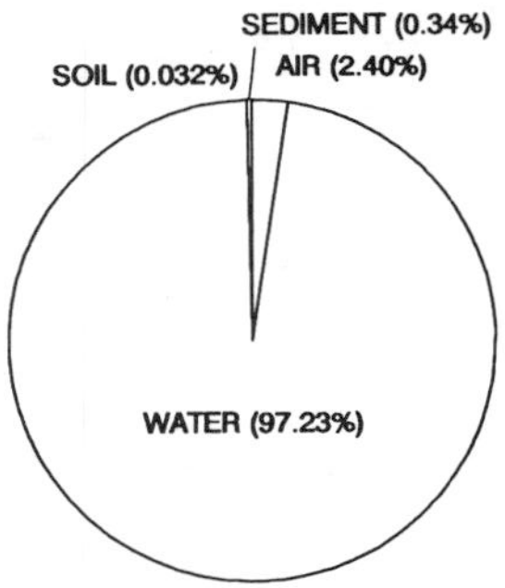

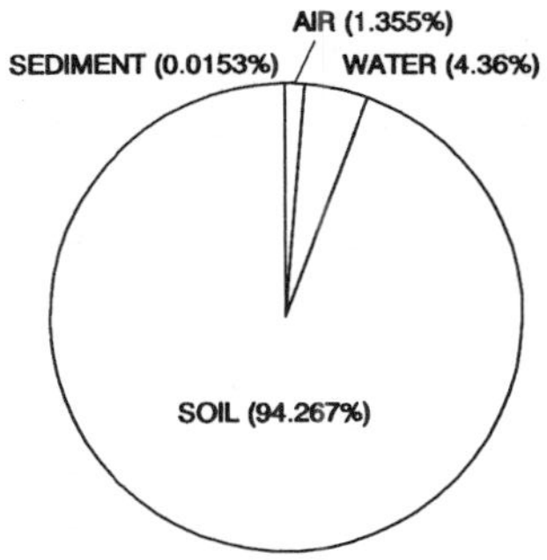

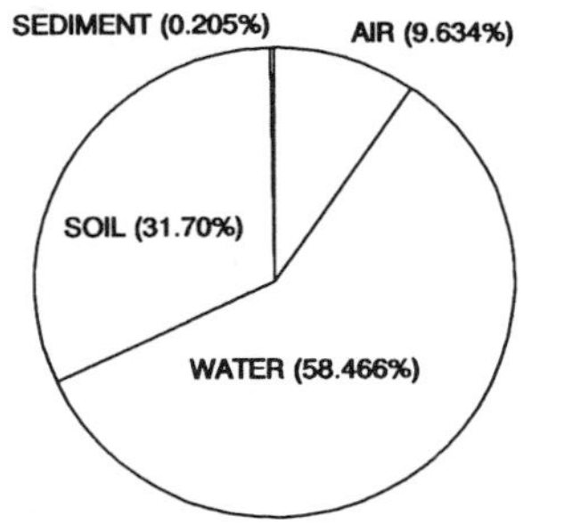

WATER (6.58%)
SOIL (1.22%)
SEDIMENT (0.023%)
AIR (92.178%)

SOIL (0.032%)
SEDIMENT (0.34%)
AIR (2.40%)
WATER (97.23%)

SEDIMENT (0.0153%)
AIR (1.355%)
WATER (4.36%)
SOIL (94.267%)

SEDIMENT (0.205%)
AIR (9.634%)
SOIL (31.70%)
WATER (58.466%)

Distribution of removal rates

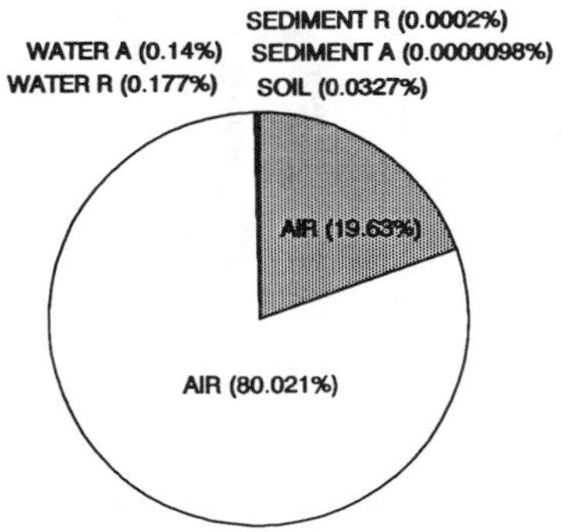

WATER A (0.14%)
WATER R (0.177%)
SEDIMENT R (0.0002%)
SEDIMENT A (0.0000098%)
SOIL (0.0327%)
AIR (19.63%)
AIR (80.021%)

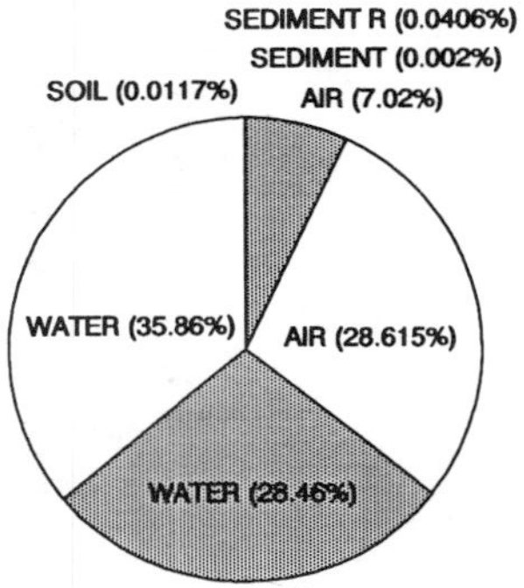

SEDIMENT R (0.0406%)
SEDIMENT (0.002%)
SOIL (0.0117%)
AIR (7.02%)
WATER (35.86%)
AIR (28.615%)
WATER (28.46%)

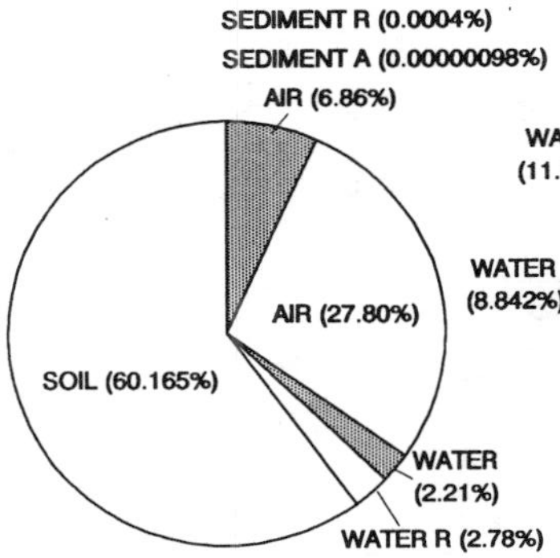

SEDIMENT R (0.0004%)
SEDIMENT A (0.00000098%)
AIR (6.86%)
AIR (27.80%)
SOIL (60.165%)
WATER (2.21%)
WATER R (2.78%)

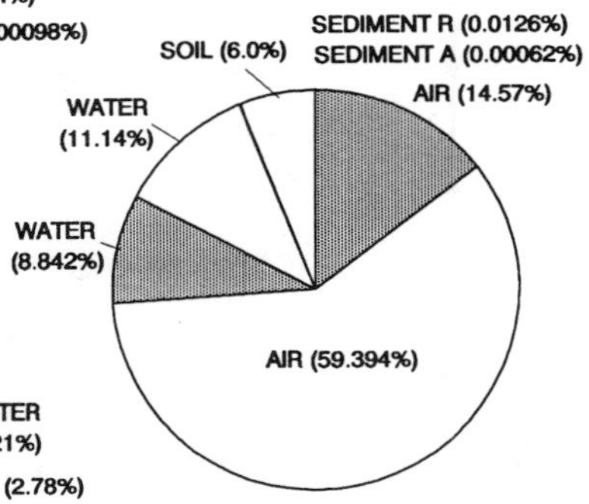

SOIL (6.0%)
SEDIMENT R (0.0126%)
SEDIMENT A (0.00062%)
AIR (14.57%)
WATER (11.14%)
WATER (8.842%)
AIR (59.394%)

Reaction
Advection

Emission rates:

E(1) = 1000 kg/h
E(2) = 0
E(3) = 0

E(1) = 0
E(2) = 1000 kg/h
E(3) = 0

E(1) = 0
E(2) = 0
E(3) = 1000 kg/h

E(1) = 600 kg/h
E(2) = 300 kg/h
E(3) = 100 kg/h

Residence time:

t = 21.30 h

t = 293 h

t = 507 h

t = 151 h

Chemical name: Diphenyl ether

Level I calculation: (six-compartment model)

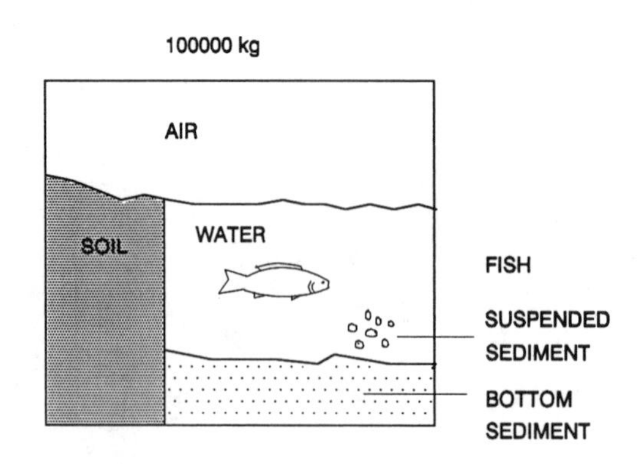


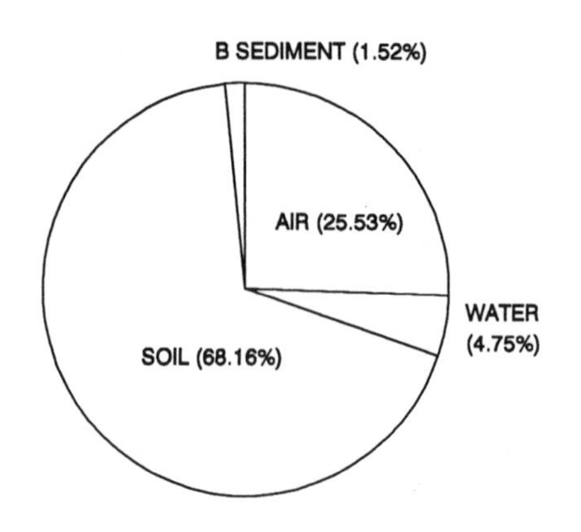


Distribution of mass

physical-chemical properties:

MW: 170.2 g/mol

M.P.: 28 °C

Fugacity ratio: 0.934

vapor pressure: 2.93 Pa

solubility: 18.7 g/m$^3$

log $K_{OW}$: 4.21

| Compartment | Z | Concentration | | | Amount | Amount |
|---|---|---|---|---|---|---|
| | mol/m3 Pa | mol/m3 | mg/L (or g/m3) | ug/g | kg | % |
| Air | 4.034E-04 | 1.500E-09 | 2.553E-07 | 2.153E-04 | 25528 | 25.528 |
| Water | 3.750E-02 | 1.394E-07 | 2.373E-05 | 2.373E-05 | 4745.67 | 4.7457 |
| Soil | 1.197E+01 | 4.450E-05 | 7.573E-03 | 3.156E-03 | 68161 | 68.1609 |
| Biota (fish) | 3.041E+01 | 1.131E-04 | 1.924E-02 | 1.924E-02 | 3.8483 | 3.85E-03 |
| Suspended sediment | 7.480E+01 | 2.781E-04 | 4.733E-02 | 3.156E-02 | 47.334 | 4.73E-02 |
| Bottom sediment | 2.394E+01 | 8.899E-05 | 1.515E-02 | 6.311E-03 | 1514.69 | 1.5147 |
| | Total | | | | 100000 | 100 |

f = 3.718E-06 Pa

Chemical name: Diphenyl ether

Level II calculation: (six-compartment model)

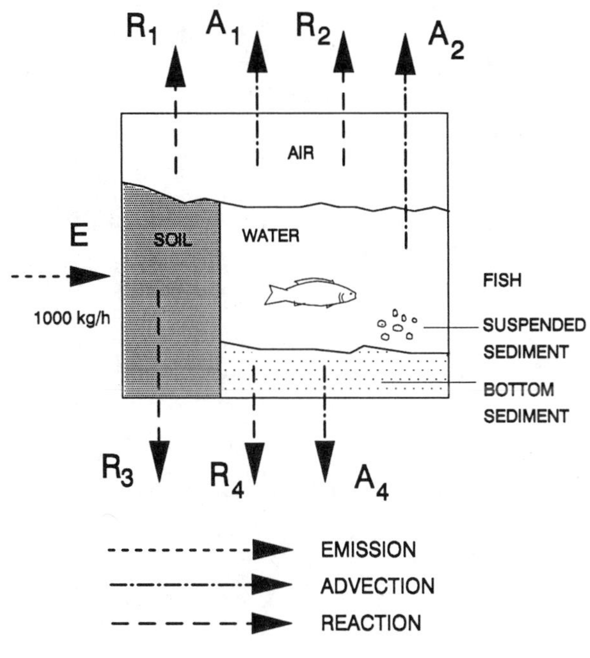


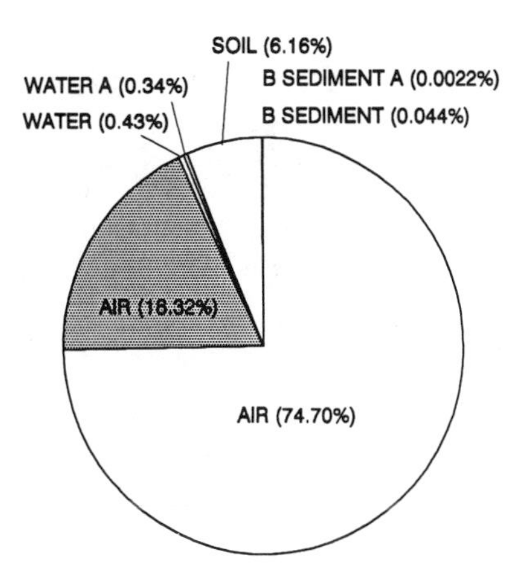


Distribution of removal rates

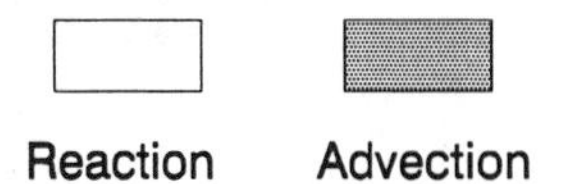

Reaction Advection

| Compartment | Half-life h | D Values Reaction mol/Pa h | D Values Advection mol/Pa h | Conc'n mol/m3 | Loss Reaction kg/h | Loss Advection kg/h | Removal % |
|---|---|---|---|---|---|---|---|
| Air | 17 | 1.64E+09 | 4.03E+08 | 1.08E-09 | 746.955 | 183.236 | 93.019 |
| Water | 550 | 9.45E+06 | 7.50E+06 | 1.00E-07 | 4.2921 | 3.4064 | 0.7699 |
| Soil | 550 | 1.36E+08 | | 3.19E-05 | 61.646 | | 6.1646 |
| Biota (fish) | | | | 8.11E-05 | | | |
| Suspended sediment | | | | 2.00E-04 | | | |
| Bottom sediment | 1700 | 9.76E+05 | 4.79E+04 | 6.39E-05 | 0.4432 | 0.0217 | 0.0465 |
| | Total | 1.79E+09 | 4.11E+08 | | 813.34 | 186.66 | 100 |
| | R + A | | 2.20E+09 | | | 1000 | |

f = 2.669E-06 Pa
Total Amt= 71780 kg

Overall residence time = 71.78 h
Reaction time = 88.25 h
Advection time = 384.54 h

Fugacity Level III calculations: (four-compartment model)

Chemical name: Diphenyl ether

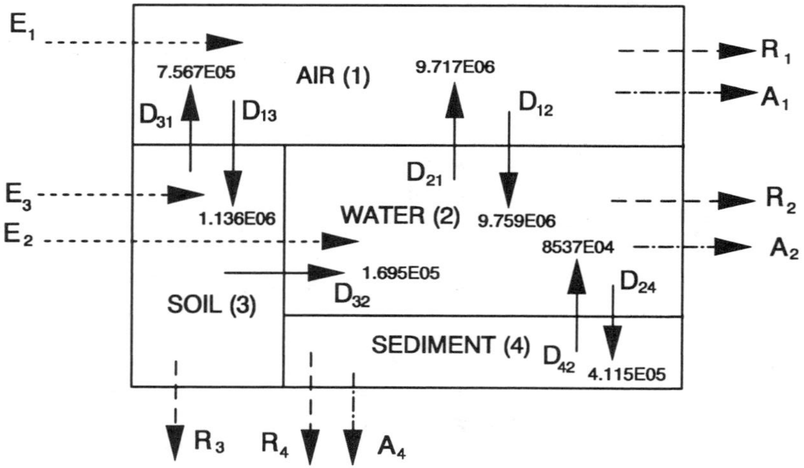

E1 E3 E2 AIR (1) WATER (2) SOIL (3) SEDIMENT (4)
D31 7.567E05 D13 1.136E06 D21 9.717E06 D12 9.759E06 D32 1.695E05 D42 8537E04 D24 4.115E05
R1 A1 R2 A2 R3 R4 A4

EMISSION (E)
REACTION (R)
ADVECTION (A)
TRANSFER D VALUE mol/Pa h

Phase Properties and Rates:

| Compartment | Bulk Z mol/m3 Pa | Half-life h | D Values Reaction mol/Pa h | D Values Advection mol/Pa h |
|---|---|---|---|---|
| Air (1) | 4.034E-04 | 17 | 1.64E+09 | 4.03E+08 |
| Water (2) | 3.790E-02 | 550 | 9.55E+06 | 7.58E+06 |
| Soil (3) | 5.996E+00 | 550 | 1.36E+08 | |
| Sediment (4) | 4.817E+00 | 1700 | 9.82E+05 | 4.82E+04 |

| | E(1)=1000 | E(2)=1000 | E(3)=1000 | E(1,2,3) |
|---|---|---|---|---|
| Overall residence time = | 21.54 | 318.72 | 788.76 | 187.42 h |
| Reaction time = | 26.85 | 490.02 | 789.99 | 241.39 h |
| Advection time = | 109.01 | 911.76 | 508720.44 | 838.23 h |

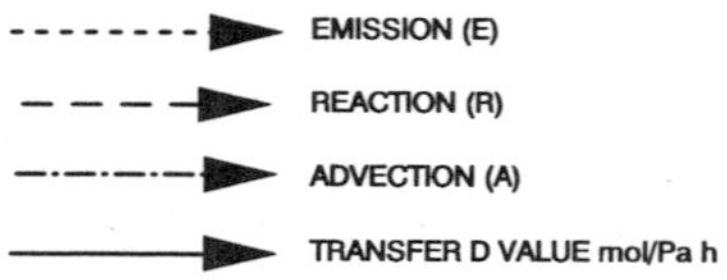

Phase Properties, Compositions, Transport and Transformation Rates:

| Emission, kg/h | | | Fugacity, Pa | | | | Concentration, g/m3 | | | | Amounts, kg | | | | Total |
|---|---|---|---|---|---|---|---|---|---|---|---|---|---|---|---|
| E(1) | E(2) | E(3) | f(1) | f(2) | f(3) | f(4) | C(1) | C(2) | C(3) | C(4) | m(1) | m(2) | m(3) | m(4) | Amount, kg |
| 1000 | 0 | 0 | 2.859E-06 | 1.025E-06 | 2.371E-08 | 3.780E-07 | 1.963E-07 | 6.610E-06 | 2.420E-05 | 3.100E-04 | 1.963E+04 | 1.322E+03 | 4.356E+02 | 1.550E+02 | 2.154E+04 |
| 0 | 1000 | 0 | 1.020E-06 | 2.161E-04 | 8.463E-09 | 7.974E-05 | 7.004E-08 | 1.394E-03 | 8.636E-06 | 6.538E-02 | 7.004E+03 | 2.789E+05 | 1.554E+02 | 3.269E+04 | 3.187E+05 |
| 0 | 0 | 1000 | 1.714E-08 | 2.891E-07 | 4.291E-05 | 1.066E-07 | 1.177E-09 | 1.865E-06 | 4.379E-02 | 8.744E-05 | 1.177E+02 | 3.729E+02 | 7.882E+05 | 4.372E+01 | 7.888E+05 |
| 600 | 300 | 100 | 2.023E-06 | 6.549E-05 | 4.308E-06 | 2.416E-05 | 1.389E-07 | 4.225E-04 | 4.396E-03 | 1.981E-02 | 1.389E+04 | 8.449E+04 | 7.913E+04 | 9.904E+03 | 1.874E+05 |

| Emission, kg/h | | | Loss, Reaction, kg/h | | | | Loss, Advection, kg/h | | | Intermedia Rate of Transport, kg/h T12 | T21 | T13 | T31 | T32 | T24 | T42 |
|---|---|---|---|---|---|---|---|---|---|---|---|---|---|---|---|---|
| E(1) | E(2) | E(3) | R(1) | R(2) | R(3) | R(4) | A(1) | A(2) | A(4) | air-water | water-air | air-soil | soil-air | soil-water | water-sed | sed-water |
| 1000 | 0 | 0 | 8.001E+02 | 1.666E+00 | 5.49E-01 | 6.318E-02 | 1.963E+02 | 1.322E+00 | 3.100E-03 | 4.748E+00 | 1.695E+00 | 5.526E-01 | 3.054E-03 | 7.246E-04 | 7.177E-02 | 5.493E-03 |
| 0 | 1000 | 0 | 2.855E+02 | 3.514E+02 | 1.96E-01 | 1.333E+01 | 7.004E+01 | 2.789E+02 | 6.538E-01 | 1.694E+00 | 3.575E+02 | 1.972E-01 | 1.090E-03 | 2.586E-04 | 1.514E+01 | 1.159E+00 |
| 0 | 0 | 1000 | 4.797E+00 | 4.699E-01 | 9.93E+02 | 1.782E-02 | 1.177E+00 | 3.729E-01 | 8.744E-04 | 2.846E-02 | 4.781E-01 | 3.313E-03 | 5.527E+00 | 1.311E+00 | 2.025E-02 | 1.550E-03 |
| 600 | 300 | 100 | 5.662E+02 | 1.065E+02 | 9.97E+01 | 4.038E+00 | 1.389E+02 | 8.449E+01 | 1.981E-01 | 3.360E+00 | 1.083E+02 | 3.911E-01 | 5.549E-01 | 1.316E-01 | 4.587E+00 | 3.510E-01 |

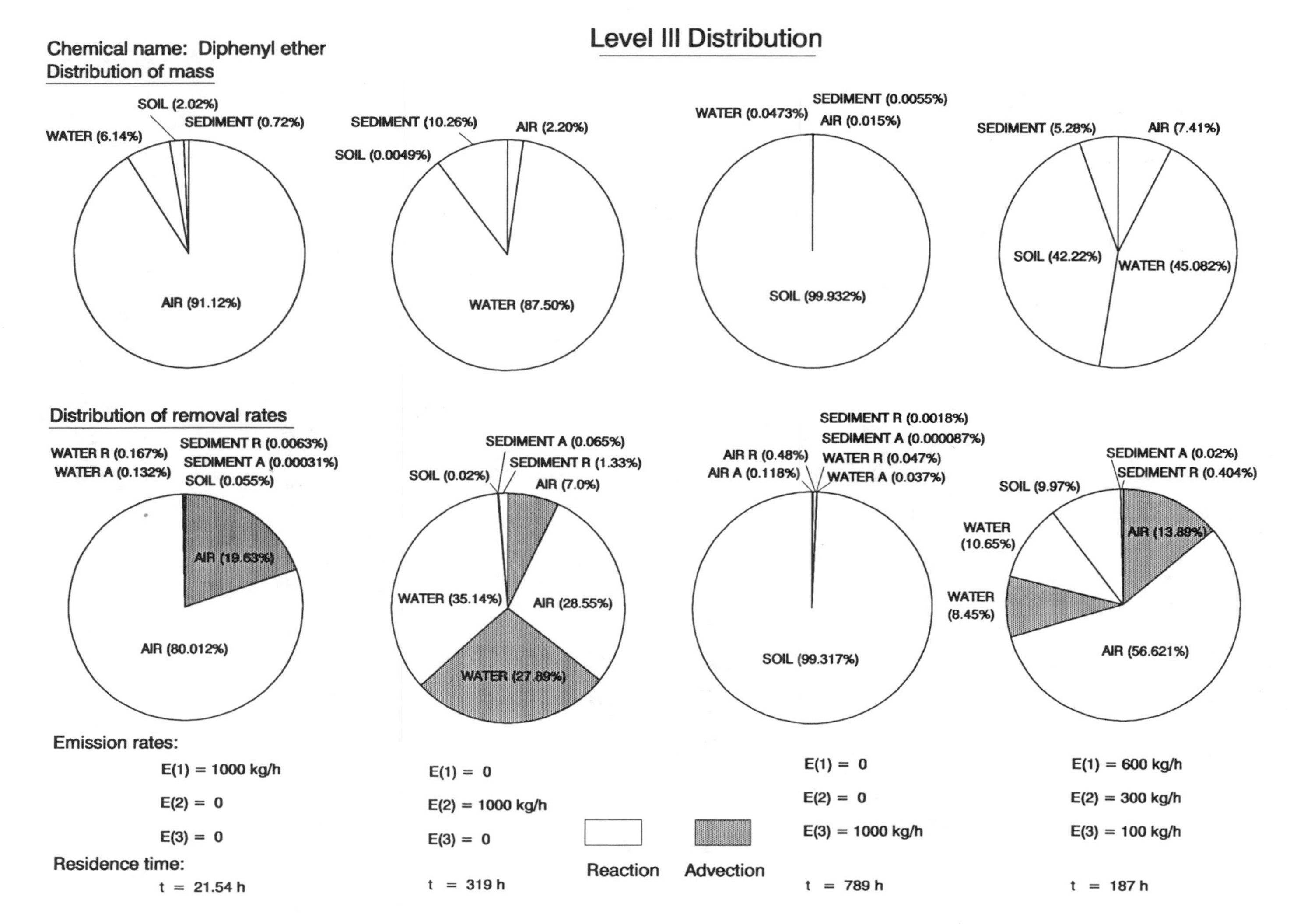
Level III Distribution
Chemical name: Diphenyl ether
Distribution of mass
SOIL (2.02%)
SEDIMENT (0.72%)
WATER (6.14%)
AIR (91.12%)
SEDIMENT (10.26%)
AIR (2.20%)
SOIL (0.0049%)
WATER (87.50%)
SEDIMENT (0.0055%)
WATER (0.0473%)
AIR (0.015%)
SOIL (99.932%)
SEDIMENT (5.28%)
AIR (7.41%)
SOIL (42.22%)
WATER (45.082%)
Distribution of removal rates
WATER R (0.167%)
WATER A (0.132%)
SEDIMENT R (0.0063%)
SEDIMENT A (0.00031%)
SOIL (0.055%)
AIR (19.63%)
AIR (80.012%)
SEDIMENT A (0.065%)
SEDIMENT R (1.33%)
SOIL (0.02%)
AIR (7.0%)
WATER (35.14%)
AIR (28.55%)
WATER (27.89%)
SEDIMENT R (0.0018%)
SEDIMENT A (0.000087%)
AIR R (0.48%)
WATER R (0.047%)
AIR A (0.118%)
WATER A (0.037%)
SOIL (99.317%)
SEDIMENT A (0.02%)
SEDIMENT R (0.404%)
SOIL (9.97%)
WATER (10.65%)
AIR (13.89%)
WATER (8.45%)
AIR (56.621%)
Emission rates:
E(1) = 1000 kg/h
E(2) = 0
E(3) = 0
Residence time:
t = 21.54 h
E(1) = 0
E(2) = 1000 kg/h
E(3) = 0
t = 319 h
Reaction
Advection
E(1) = 0
E(2) = 0
E(3) = 1000 kg/h
t = 789 h
E(1) = 600 kg/h
E(2) = 300 kg/h
E(3) = 100 kg/h
t = 187 h

Chemical name: Bis(2-chloroethyl)ether

Level I calculation: (six-compartment model)

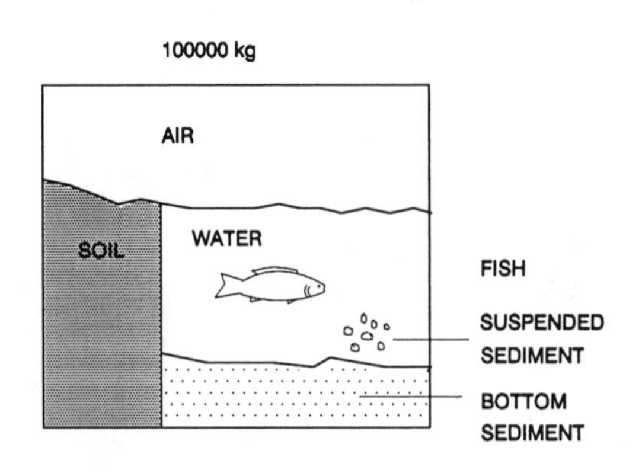


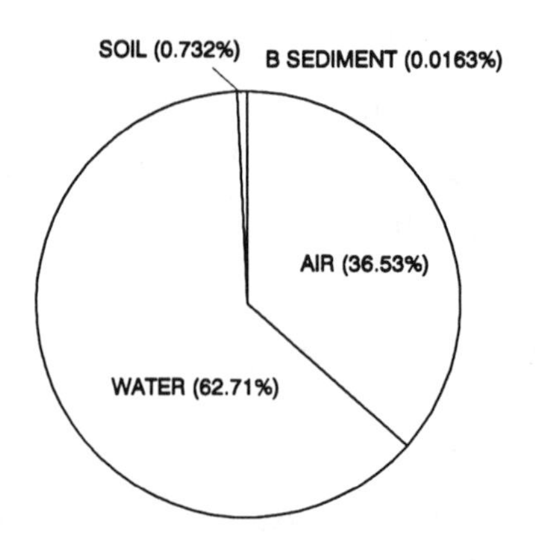


Distribution of mass

physical-chemical properties:

MW: 143.02 g/mol

M.P.: - 46.8 °C

Fugacity ratio: 1.0

vapor pressure: 206 Pa

solubility: 10200 g/m$^3$

log $K_{OW}$: 1.12

| Compartment | Z | Concentration | | | Amount | Amount |
|---|---|---|---|---|---|---|
| | mol/m3 Pa | mol/m3 | mg/L (or g/m3) | ug/g | kg | % |
| Air | 4.034E-04 | 2.555E-09 | 3.654E-07 | 3.082E-04 | 36538 | 36.538 |
| Water | 3.462E-01 | 2.192E-06 | 3.136E-04 | 3.136E-04 | 62713 | 62.7129 |
| Soil | 8.982E-02 | 5.688E-07 | 8.135E-05 | 3.390E-05 | 732.14 | 0.7321 |
| Biota (fish) | 2.282E-01 | 1.445E-06 | 2.067E-04 | 2.067E-04 | 0.0413 | 4.13E-05 |
| Suspended sediment | 5.614E-01 | 3.555E-06 | 5.084E-04 | 3.390E-04 | 0.5084 | 5.08E-04 |
| Bottom sediment | 1.796E-01 | 1.138E-06 | 1.627E-04 | 6.779E-05 | 16.270 | 1.63E-02 |
| | Total | | | | 100000 | 100 |

f = 6.333E-06 Pa

Chemical name: Bis(2-chloroethyl)ether

Level II calculation: (six-compartment model)

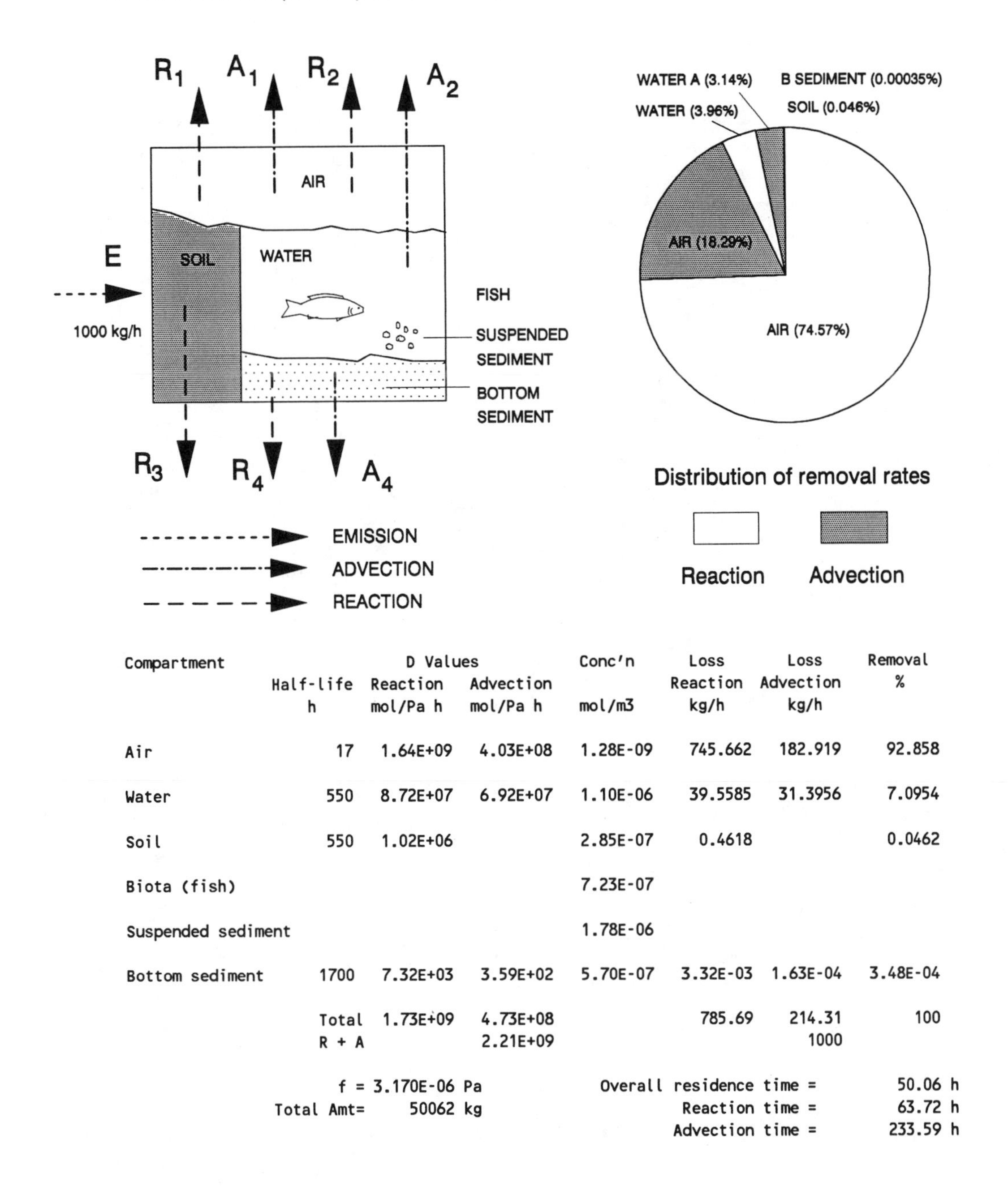


| Compartment | Half-life h | D Values Reaction mol/Pa h | D Values Advection mol/Pa h | Conc'n mol/m3 | Loss Reaction kg/h | Loss Advection kg/h | Removal % |
|---|---|---|---|---|---|---|---|
| Air | 17 | 1.64E+09 | 4.03E+08 | 1.28E-09 | 745.662 | 182.919 | 92.858 |
| Water | 550 | 8.72E+07 | 6.92E+07 | 1.10E-06 | 39.5585 | 31.3956 | 7.0954 |
| Soil | 550 | 1.02E+06 | | 2.85E-07 | 0.4618 | | 0.0462 |
| Biota (fish) | | | | 7.23E-07 | | | |
| Suspended sediment | | | | 1.78E-06 | | | |
| Bottom sediment | 1700 | 7.32E+03 | 3.59E+02 | 5.70E-07 | 3.32E-03 | 1.63E-04 | 3.48E-04 |
| | Total | 1.73E+09 | 4.73E+08 | | 785.69 | 214.31 | 100 |
| | R + A | | 2.21E+09 | | | 1000 | |

f = 3.170E-06 Pa
Total Amt= 50062 kg

Overall residence time = 50.06 h
Reaction time = 63.72 h
Advection time = 233.59 h

## Fugacity Level III calculations: (four-compartment model)

Chemical name: Bis(2-chloroethyl)ether

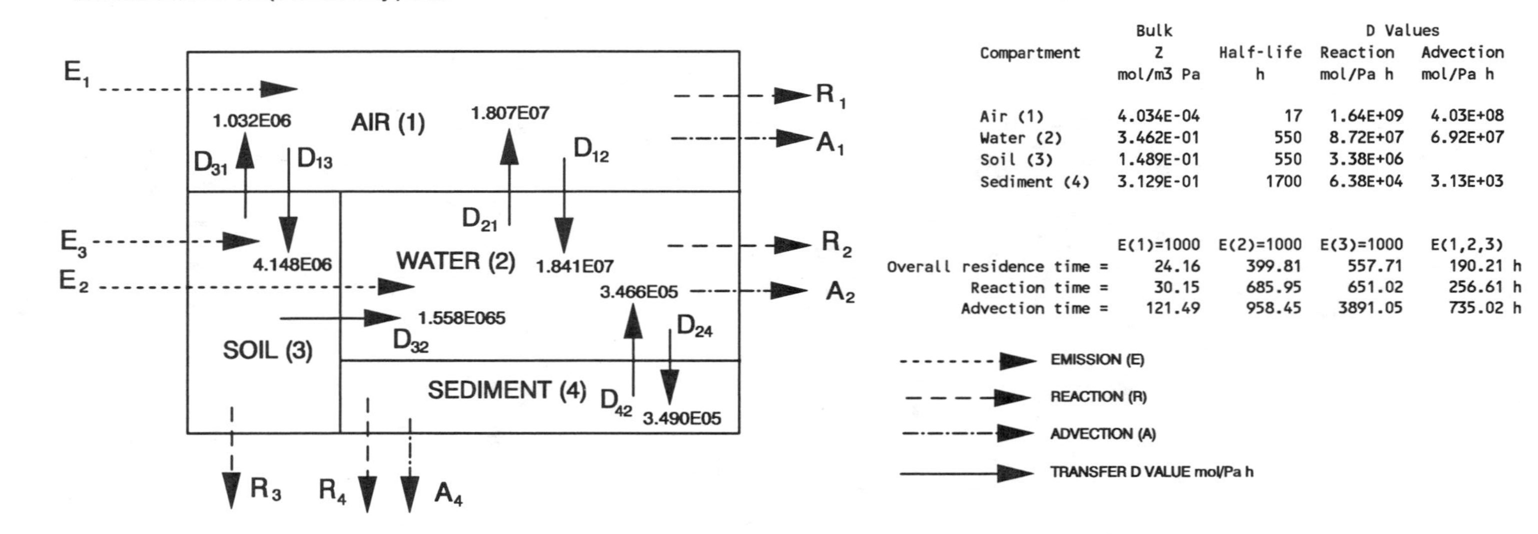


Phase Properties and Rates:

| Compartment | Bulk Z mol/m3 Pa | Half-life h | D Values Reaction mol/Pa h | D Values Advection mol/Pa h |
|---|---|---|---|---|
| Air (1) | 4.034E-04 | 17 | 1.64E+09 | 4.03E+08 |
| Water (2) | 3.462E-01 | 550 | 8.72E+07 | 6.92E+07 |
| Soil (3) | 1.489E-01 | 550 | 3.38E+06 | |
| Sediment (4) | 3.129E-01 | 1700 | 6.38E+04 | 3.13E+03 |

| | E(1)=1000 | E(2)=1000 | E(3)=1000 | E(1,2,3) |
|---|---|---|---|---|
| Overall residence time = | 24.16 | 399.81 | 557.71 | 190.21 h |
| Reaction time = | 30.15 | 685.95 | 651.02 | 256.61 h |
| Advection time = | 121.49 | 958.45 | 3891.05 | 735.02 h |

Phase Properties, Compositions, Transport and Transformation Rates:

| Emission, kg/h | | | Fugacity, Pa | | | | Concentration, g/m3 | | | | Amounts, kg | | | | Total Amount, kg |
|---|---|---|---|---|---|---|---|---|---|---|---|---|---|---|---|
| E(1) | E(2) | E(3) | f(1) | f(2) | f(3) | f(4) | C(1) | C(2) | C(3) | C(4) | m(1) | m(2) | m(3) | m(4) | |
| 1000 | 0 | 0 | 3.381E-06 | 3.775E-07 | 2.351E-06 | 3.187E-07 | 1.951E-07 | 1.869E-05 | 5.006E-05 | 1.426E-05 | 1.951E+04 | 3.739E+03 | 9.010E+02 | 7.131E+00 | 2.416E+04 |
| 0 | 1000 | 0 | 3.499E-07 | 4.008E-05 | 2.433E-07 | 3.384E-05 | 2.019E-08 | 1.985E-03 | 5.179E-06 | 1.514E-03 | 2.019E+03 | 3.970E+05 | 9.323E+01 | 7.571E+02 | 3.998E+05 |
| 0 | 0 | 1000 | 6.762E-07 | 1.053E-05 | 1.172E-03 | 8.891E-06 | 3.902E-08 | 5.216E-04 | 2.496E-02 | 3.979E-04 | 3.902E+03 | 1.043E+05 | 4.493E+05 | 1.989E+02 | 5.577E+05 |
| 600 | 300 | 100 | 2.201E-06 | 1.330E-05 | 1.187E-04 | 1.123E-05 | 1.270E-07 | 6.588E-04 | 2.528E-03 | 5.026E-04 | 1.270E+04 | 1.318E+05 | 4.550E+04 | 2.513E+02 | 1.902E+05 |

| Emission, kg/h | | | Loss, Reaction, kg/h | | | | Loss, Advection, kg/h | | | Intermedia Rate of Transport, kg/h T12 | T21 | T13 | T31 | T32 | T24 | T42 |
|---|---|---|---|---|---|---|---|---|---|---|---|---|---|---|---|---|
| E(1) | E(2) | E(3) | R(1) | R(2) | R(3) | R(4) | A(1) | A(2) | A(4) | air-water | water-air | air-soil | soil-air | soil-water | water-sed | sed-water |
| 1000 | 0 | 0 | 7.953E+02 | 4.711E+00 | 1.14E+00 | 2.907E-03 | 1.951E+02 | 3.739E+00 | 1.426E-04 | 8.904E+00 | 9.755E-01 | 2.006E+00 | 3.470E-01 | 5.239E-01 | 1.885E-02 | 1.580E-02 |
| 0 | 1000 | 0 | 8.229E+01 | 5.002E+02 | 1.17E-01 | 3.086E-01 | 2.019E+01 | 3.970E+02 | 1.514E-02 | 9.213E-01 | 1.036E+02 | 2.076E-01 | 3.590E-02 | 5.421E-02 | 2.001E+00 | 1.677E+00 |
| 0 | 0 | 1000 | 1.590E+02 | 1.314E+02 | 5.66E+02 | 8.110E-02 | 3.902E+01 | 1.043E+02 | 3.979E-03 | 1.781E+00 | 2.722E+01 | 4.012E-01 | 1.730E+02 | 2.613E+02 | 5.258E-01 | 4.407E-01 |
| 600 | 300 | 100 | 5.178E+02 | 1.660E+02 | 5.73E+01 | 1.024E-01 | 1.270E+02 | 1.318E+02 | 5.026E-03 | 5.797E+00 | 3.438E+01 | 1.306E+00 | 1.752E+01 | 2.646E+01 | 6.641E-01 | 5.567E-01 |

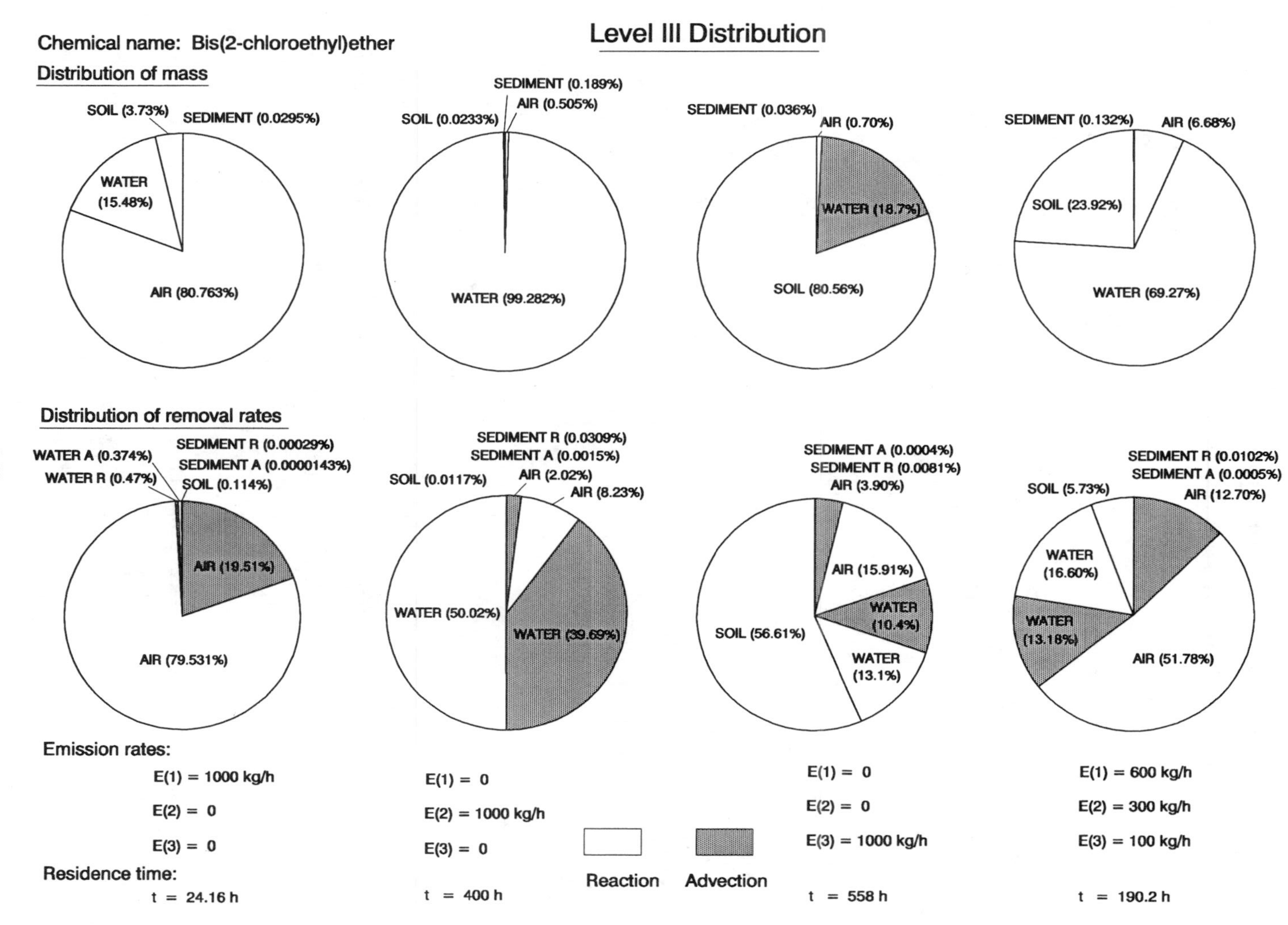

Level III Distribution
Chemical name: Bis(2-chloroethyl)ether
Distribution of mass
SOIL (3.73%)
SEDIMENT (0.0295%)
WATER (15.48%)
AIR (80.763%)
SEDIMENT (0.189%)
SOIL (0.0233%)
AIR (0.505%)
WATER (99.282%)
SEDIMENT (0.036%)
AIR (0.70%)
WATER (18.7%)
SOIL (80.56%)
SEDIMENT (0.132%)
AIR (6.68%)
SOIL (23.92%)
WATER (69.27%)
Distribution of removal rates
WATER A (0.374%)
WATER R (0.47%)
SEDIMENT R (0.00029%)
SEDIMENT A (0.0000143%)
SOIL (0.114%)
AIR (19.51%)
AIR (79.531%)
SEDIMENT R (0.0309%)
SEDIMENT A (0.0015%)
SOIL (0.0117%)
AIR (2.02%)
AIR (8.23%)
WATER (50.02%)
WATER (39.69%)
SEDIMENT A (0.0004%)
SEDIMENT R (0.0081%)
AIR (3.90%)
AIR (15.91%)
WATER (10.4%)
SOIL (56.61%)
WATER (13.1%)
SEDIMENT R (0.0102%)
SEDIMENT A (0.0005%)
SOIL (5.73%)
AIR (12.70%)
WATER (16.60%)
WATER (13.18%)
AIR (51.78%)
Emission rates:
E(1) = 1000 kg/h
E(2) = 0
E(3) = 0
Residence time:
t = 24.16 h
E(1) = 0
E(2) = 1000 kg/h
E(3) = 0
t = 400 h
Reaction
Advection
E(1) = 0
E(2) = 0
E(3) = 1000 kg/h
t = 558 h
E(1) = 600 kg/h
E(2) = 300 kg/h
E(3) = 100 kg/h
t = 190.2 h

Chemical name: 2-Chloroethyl vinyl ether

Level I calculation: (six-compartment model)

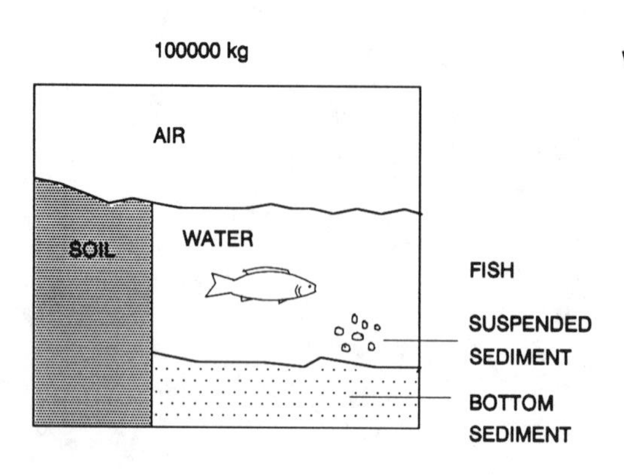


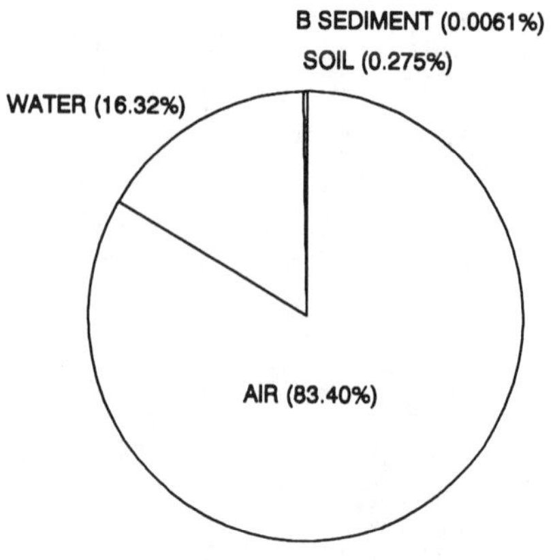


Distribution of mass

physical-chemical properties:

MW: 106.55 g/mol

M.P.: - 69.7 °C

Fugacity ratio: 1.0

vapor pressure: 3566 Pa

solubility: 15000 g/m$^3$

log $K_{OW}$: 1.28

| Compartment | Z | Concentration | | | Amount | Amount |
|---|---|---|---|---|---|---|
| | mol/m3 Pa | mol/m3 | mg/L (or g/m3) | ug/g | kg | % |
| Air | 4.034E-04 | 7.827E-09 | 8.340E-07 | 7.035E-04 | 83396 | 83.396 |
| Water | 3.948E-02 | 7.659E-07 | 8.161E-05 | 8.161E-05 | 16322 | 16.3221 |
| Soil | 1.480E-02 | 2.872E-07 | 3.060E-05 | 1.275E-05 | 275.43 | 0.2754 |
| Biota (fish) | 3.761E-02 | 7.297E-07 | 7.775E-05 | 7.775E-05 | 0.0156 | 1.56E-05 |
| Suspended sediment | 9.253E-02 | 1.795E-06 | 1.913E-04 | 1.275E-04 | 0.1913 | 1.91E-04 |
| Bottom sediment | 2.961E-02 | 5.744E-07 | 6.121E-05 | 2.550E-05 | 6.1207 | 6.12E-03 |
| | Total | | | | 100000 | 100 |

f = 1.940E-05 Pa

Chemical name: 2-Chloroethyl vinyl ether

Level II calculation: (six-compartment model)

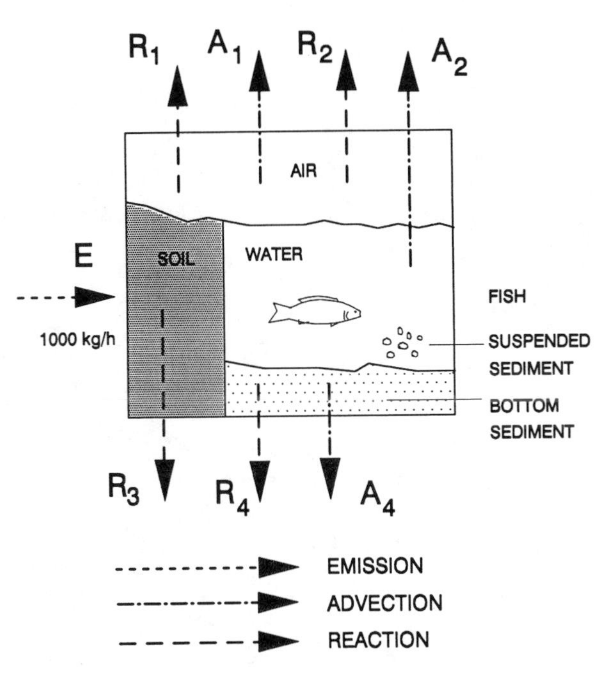


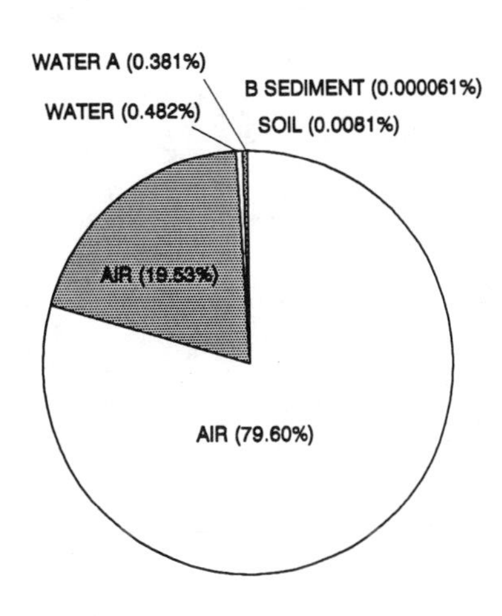


Distribution of removal rates

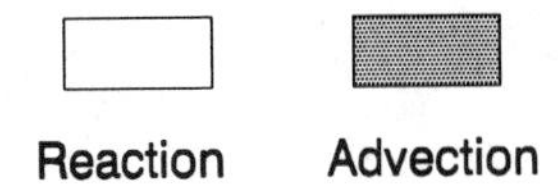


| Compartment | Half-life h | D Values Reaction mol/Pa h | D Values Advection mol/Pa h | Conc'n mol/m3 | Loss Reaction kg/h | Loss Advection kg/h | Removal % |
|---|---|---|---|---|---|---|---|
| Air | 17 | 1.64E+09 | 4.03E+08 | 1.83E-09 | 796.011 | 195.270 | 99.128 |
| Water | 550 | 9.95E+06 | 7.90E+06 | 1.79E-07 | 4.8154 | 3.8218 | 0.8637 |
| Soil | 550 | 1.68E+05 | | 6.73E-08 | 0.0813 | | 8.13E-03 |
| Biota (fish) | | | | 1.71E-07 | | | |
| Suspended sediment | | | | 4.20E-07 | | | |
| Bottom sediment | 1700 | 1.21E+03 | 5.92E+01 | 1.35E-07 | 5.84E-04 | 2.87E-05 | 6.13E-05 |
| | Total | 1.65E+09 | 4.11E+08 | | 800.91 | 199.09 | 100 |
| | R + A | | 2.07E+09 | | | 1000 | |

f = 4.543E-06 Pa
Total Amt= 23415 kg

Overall residence time = 23.41 h
Reaction time = 29.24 h
Advection time = 117.61 h

Fugacity Level III calculations: (four-compartment model)

Chemical name: 2-Chloroethyl vinyl ether

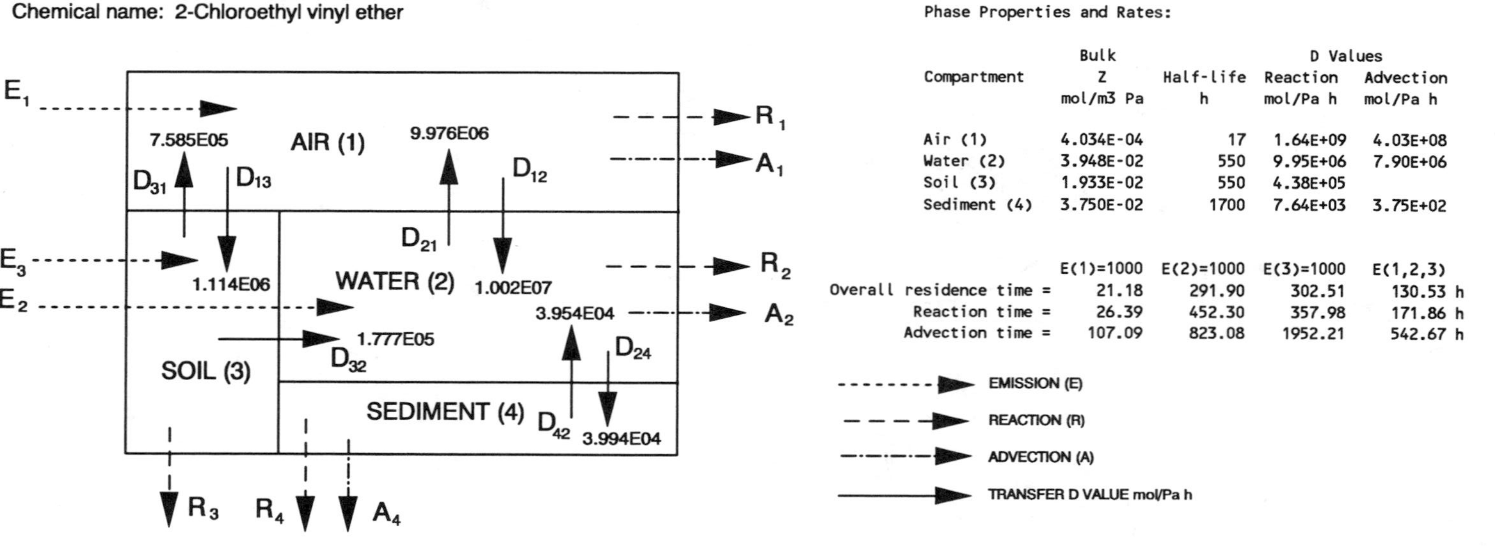


Phase Properties and Rates:

| Compartment | Bulk Z mol/m3 Pa | Half-life h | D Values Reaction mol/Pa h | Advection mol/Pa h |
|---|---|---|---|---|
| Air (1) | 4.034E-04 | 17 | 1.64E+09 | 4.03E+08 |
| Water (2) | 3.948E-02 | 550 | 9.95E+06 | 7.90E+06 |
| Soil (3) | 1.933E-02 | 550 | 4.38E+05 | |
| Sediment (4) | 3.750E-02 | 1700 | 7.64E+03 | 3.75E+02 |

| | E(1)=1000 | E(2)=1000 | E(3)=1000 | E(1,2,3) |
|---|---|---|---|---|
| Overall residence time = | 21.18 | 291.90 | 302.51 | 130.53 h |
| Reaction time = | 26.39 | 452.30 | 357.98 | 171.86 h |
| Advection time = | 107.09 | 823.08 | 1952.21 | 542.67 h |

Phase Properties, Compositions, Transport and Transformation Rates:

| Emission, kg/h | | | Fugacity, Pa | | | | Concentration, g/m3 | | | | Amounts, kg | | | | Total |
|---|---|---|---|---|---|---|---|---|---|---|---|---|---|---|---|
| E(1) | E(2) | E(3) | f(1) | f(2) | f(3) | f(4) | C(1) | C(2) | C(3) | C(4) | m(1) | m(2) | m(3) | m(4) | Amount, kg |
| 1000 | 0 | 0 | 4.567E-06 | 1.668E-06 | 3.701E-06 | 1.401E-06 | 1.963E-07 | 7.015E-06 | 7.622E-06 | 5.597E-06 | 1.963E+04 | 1.403E+03 | 1.372E+02 | 2.798E+00 | 2.118E+04 |
| 0 | 1000 | 0 | 1.637E-06 | 3.379E-04 | 1.327E-06 | 2.838E-04 | 7.039E-08 | 1.421E-03 | 2.733E-06 | 1.134E-03 | 7.039E+03 | 2.842E+05 | 4.919E+01 | 5.670E+02 | 2.919E+05 |
| 0 | 0 | 1000 | 2.732E-06 | 4.459E-05 | 6.830E-03 | 3.745E-05 | 1.174E-07 | 1.876E-04 | 1.407E-02 | 1.497E-04 | 1.174E+04 | 3.752E+04 | 2.532E+05 | 7.483E+01 | 3.025E+05 |
| 600 | 300 | 100 | 3.505E-06 | 1.068E-04 | 6.857E-04 | 8.971E-05 | 1.507E-07 | 4.493E-04 | 1.412E-03 | 3.585E-04 | 1.507E+04 | 8.987E+04 | 2.541E+04 | 1.792E+02 | 1.305E+05 |

| Emission, kg/h | | | Loss, Reaction, kg/h | | | | Loss, Advection, kg/h | | | Intermedia Rate of Transport, kg/h T12 | T21 | T13 | T31 | T32 | T24 | T42 |
|---|---|---|---|---|---|---|---|---|---|---|---|---|---|---|---|---|
| E(1) | E(2) | E(3) | R(1) | R(2) | R(3) | R(4) | A(1) | A(2) | A(4) | air-water | water-air | air-soil | soil-air | soil-water | water-sed | sed-water |
| 1000 | 0 | 0 | 8.003E+02 | 1.768E+00 | 1.73E-01 | 1.141E-03 | 1.963E+02 | 1.403E+00 | 5.597E-05 | 4.874E+00 | 1.773E+00 | 5.421E-01 | 2.991E-01 | 7.007E-02 | 7.097E-03 | 5.900E-03 |
| 0 | 1000 | 0 | 2.869E+02 | 3.582E+02 | 6.20E-02 | 2.311E-01 | 7.039E+01 | 2.842E+02 | 1.134E-02 | 1.748E+00 | 3.591E+02 | 1.943E-01 | 1.072E-01 | 2.512E-02 | 1.438E+00 | 1.195E+00 |
| 0 | 0 | 1000 | 4.787E+02 | 4.727E+01 | 3.19E+02 | 3.050E-02 | 1.174E+02 | 3.752E+01 | 1.497E-03 | 2.916E+00 | 4.740E+01 | 3.243E-01 | 5.520E+02 | 1.293E+02 | 1.898E-01 | 1.578E-01 |
| 600 | 300 | 100 | 6.141E+02 | 1.132E+02 | 3.20E+01 | 7.307E-02 | 1.507E+02 | 8.987E+01 | 3.585E-03 | 3.740E+00 | 1.135E+02 | 4.160E-01 | 5.541E+01 | 1.298E+01 | 4.546E-01 | 3.779E-01 |

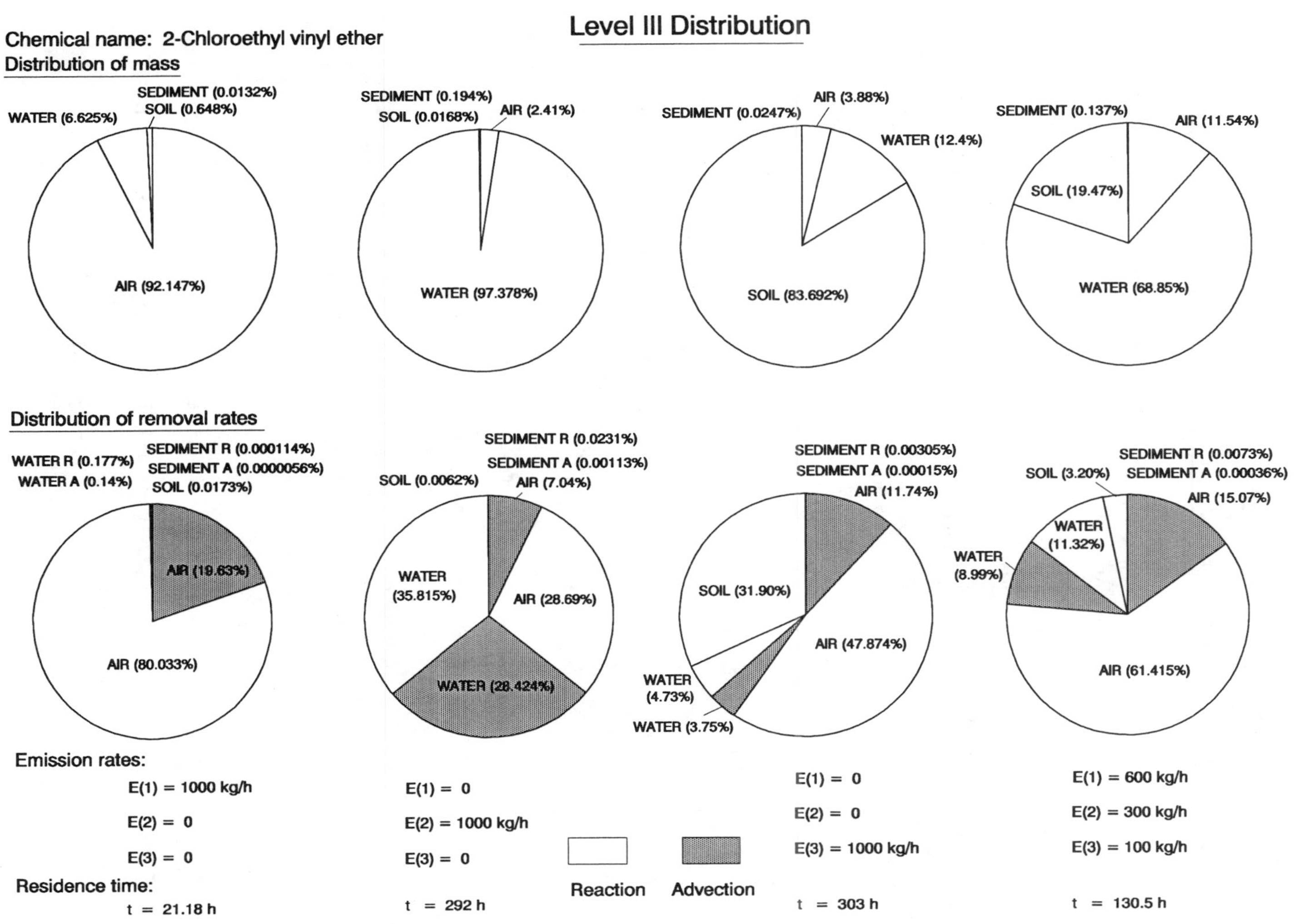

# Level III Distribution

Chemical name: 2-Chloroethyl vinyl ether

## Distribution of mass

WATER (6.625%)
SEDIMENT (0.0132%)
SOIL (0.648%)
AIR (92.147%)

SEDIMENT (0.194%)
SOIL (0.0168%)
AIR (2.41%)
WATER (97.378%)

SEDIMENT (0.0247%)
AIR (3.88%)
WATER (12.4%)
SOIL (83.692%)

SEDIMENT (0.137%)
AIR (11.54%)
SOIL (19.47%)
WATER (68.85%)

## Distribution of removal rates

WATER R (0.177%)
WATER A (0.14%)
SEDIMENT R (0.000114%)
SEDIMENT A (0.0000056%)
SOIL (0.0173%)
AIR (19.63%)
AIR (80.033%)

SEDIMENT R (0.0231%)
SEDIMENT A (0.00113%)
SOIL (0.0062%)
AIR (7.04%)
WATER (35.815%)
AIR (28.69%)
WATER (28.424%)

SEDIMENT R (0.00305%)
SEDIMENT A (0.00015%)
AIR (11.74%)
SOIL (31.90%)
AIR (47.874%)
WATER (4.73%)
WATER (3.75%)

SEDIMENT R (0.0073%)
SOIL (3.20%)
SEDIMENT A (0.00036%)
AIR (15.07%)
WATER (11.32%)
WATER (8.99%)
AIR (61.415%)

Reaction
Advection

Emission rates:

| E(1) = 1000 kg/h | E(1) = 0 | E(1) = 0 | E(1) = 600 kg/h |
|---|---|---|---|
| E(2) = 0 | E(2) = 1000 kg/h | E(2) = 0 | E(2) = 300 kg/h |
| E(3) = 0 | E(3) = 0 | E(3) = 1000 kg/h | E(3) = 100 kg/h |

Residence time:

| t = 21.18 h | t = 292 h | t = 303 h | t = 130.5 h |
|---|---|---|---|

Chemical name: 4-Chlorophenyl phenyl ether

Level I calculation: (six-compartment model)

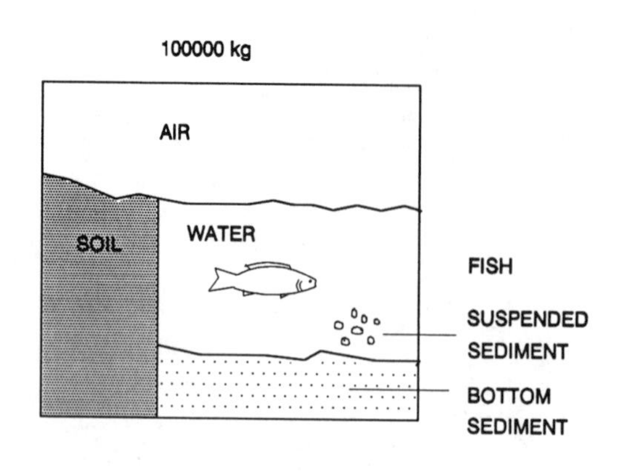


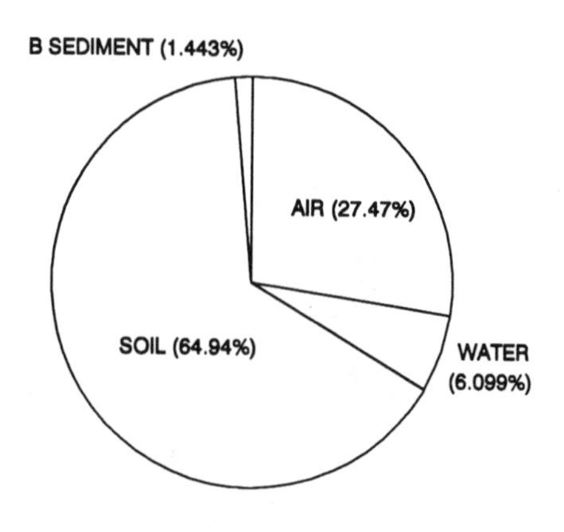


Distribution of mass

physical-chemical properties:

MW: 204.66 g/mol

M.P.: - 6.0 ℃

Fugacity ratio: 1.0

vapor pressure: 0.36 Pa

solubility: 3.30 g/m$^3$

log $K_{OW}$: 4.08

| Compartment | Z | Concentration | | | Amount | Amount |
|---|---|---|---|---|---|---|
| | mol/m3 Pa | mol/m3 | mg/L (or g/m3) | ug/g | kg | % |
| Air | 4.034E-04 | 1.342E-09 | 2.747E-07 | 2.317E-04 | 27468 | 27.468 |
| Water | 4.479E-02 | 1.490E-07 | 3.050E-05 | 3.050E-05 | 6099.30 | 6.0993 |
| Soil | 1.060E+01 | 3.526E-05 | 7.216E-03 | 3.007E-03 | 64941 | 64.941 |
| Biota (fish) | 2.692E+01 | 8.958E-05 | 1.833E-02 | 1.833E-02 | 3.6665 | 3.67E-03 |
| Suspended sediment | 6.623E+01 | 2.204E-04 | 4.510E-02 | 3.007E-02 | 45.098 | 4.51E-02 |
| Bottom sediment | 2.120E+01 | 7.051E-05 | 1.443E-02 | 6.013E-03 | 1443.13 | 1.4431 |
| | Total | | | | 100000 | 100 |

f = 3.327E-06 Pa

Chemical name: 4-Chlorophenyl phenyl ether

Level II calculation: (six-compartment model)

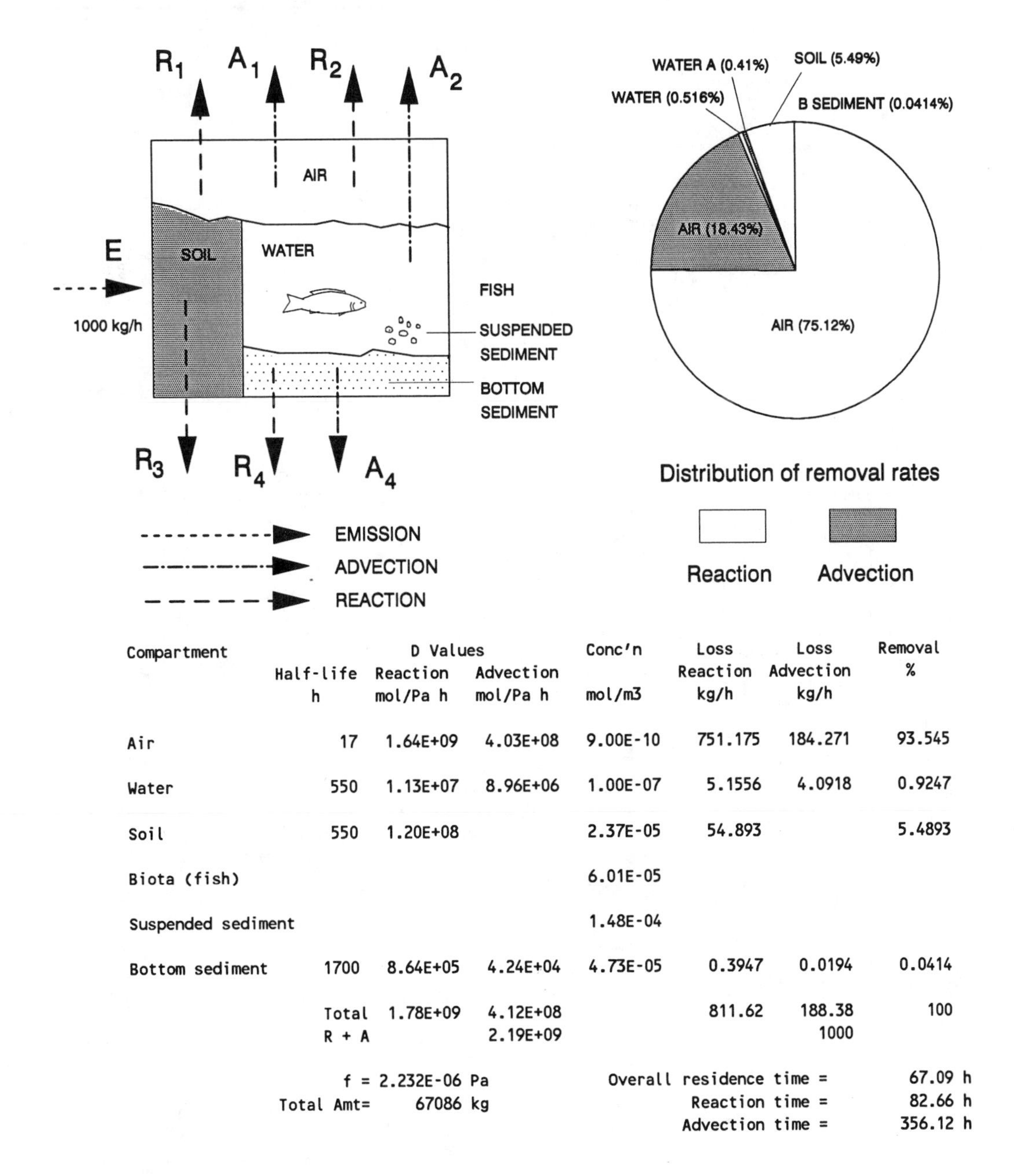


| Compartment | Half-life h | D Values Reaction mol/Pa h | D Values Advection mol/Pa h | Conc'n mol/m3 | Loss Reaction kg/h | Loss Advection kg/h | Removal % |
|---|---|---|---|---|---|---|---|
| Air | 17 | 1.64E+09 | 4.03E+08 | 9.00E-10 | 751.175 | 184.271 | 93.545 |
| Water | 550 | 1.13E+07 | 8.96E+06 | 1.00E-07 | 5.1556 | 4.0918 | 0.9247 |
| Soil | 550 | 1.20E+08 | | 2.37E-05 | 54.893 | | 5.4893 |
| Biota (fish) | | | | 6.01E-05 | | | |
| Suspended sediment | | | | 1.48E-04 | | | |
| Bottom sediment | 1700 | 8.64E+05 | 4.24E+04 | 4.73E-05 | 0.3947 | 0.0194 | 0.0414 |
| | Total | 1.78E+09 | 4.12E+08 | | 811.62 | 188.38 | 100 |
| | R + A | | 2.19E+09 | | | 1000 | |

f = 2.232E-06 Pa

Total Amt= 67086 kg

Overall residence time = 67.09 h

Reaction time = 82.66 h

Advection time = 356.12 h

Fugacity Level III calculations: (four-compartment model)

Chemical name: 4-Chlorophenyl phenyl ether

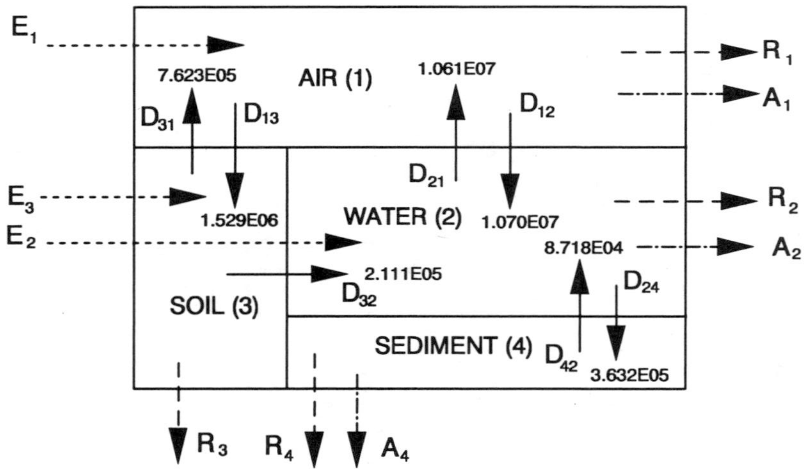

Phase Properties and Rates:

| Compartment | Bulk Z mol/m3 Pa | Half-life h | D Values Reaction mol/Pa h | D Values Advection mol/Pa h |
|---|---|---|---|---|
| Air (1) | 4.036E-04 | 17 | 1.65E+09 | 4.04E+08 |
| Water (2) | 4.515E-02 | 550 | 1.14E+07 | 9.03E+06 |
| Soil (3) | 5.312E+00 | 550 | 1.20E+08 | |
| Sediment (4) | 4.275E+00 | 1700 | 8.71E+05 | 4.27E+04 |

| | E(1)=1000 | E(2)=1000 | E(3)=1000 | E(1,2,3) |
|---|---|---|---|---|
| Overall residence time = | 21.83 | 320.89 | 787.98 | 188.16 h |
| Reaction time = | 27.21 | 497.74 | 789.45 | 242.91 h |
| Advection time = | 110.45 | 903.13 | 423722.44 | 834.89 h |

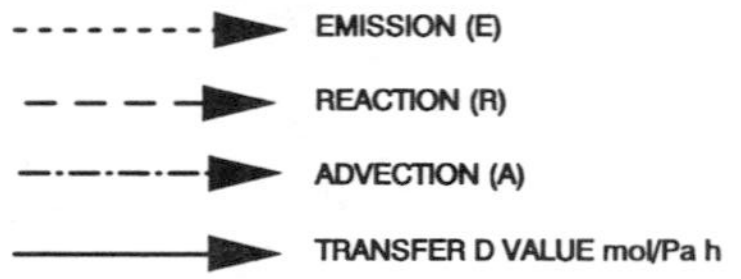

Phase Properties, Compositions, Transport and Transformation Rates:

| Emission, kg/h | | | Fugacity, Pa | | | | Concentration, g/m3 | | | | Amounts, kg | | | | Total Amount, kg |
|---|---|---|---|---|---|---|---|---|---|---|---|---|---|---|---|
| E(1) | E(2) | E(3) | f(1) | f(2) | f(3) | f(4) | C(1) | C(2) | C(3) | C(4) | m(1) | m(2) | m(3) | m(4) | |
| 1000 | 0 | 0 | 2.375E-06 | 8.103E-07 | 2.991E-08 | 3.043E-07 | 1.962E-07 | 7.488E-06 | 3.252E-05 | 2.662E-04 | 1.962E+04 | 1.498E+03 | 5.853E+02 | 1.331E+02 | 2.183E+04 |
| 0 | 1000 | 0 | 8.037E-07 | 1.561E-04 | 1.012E-08 | 5.860E-05 | 6.638E-08 | 1.442E-03 | 1.100E-05 | 5.127E-02 | 6.638E+03 | 2.884E+05 | 1.981E+02 | 2.564E+04 | 3.209E+05 |
| 0 | 0 | 1000 | 1.632E-08 | 2.763E-07 | 4.023E-05 | 1.038E-07 | 1.348E-09 | 2.553E-06 | 4.374E-02 | 9.078E-05 | 1.348E+02 | 5.107E+02 | 7.873E+05 | 4.539E+01 | 7.880E+05 |
| 600 | 300 | 100 | 1.668E-06 | 4.734E-05 | 4.044E-06 | 1.777E-05 | 1.377E-07 | 4.374E-04 | 4.397E-03 | 1.555E-02 | 1.377E+04 | 8.747E+04 | 7.914E+04 | 7.775E+03 | 1.882E+05 |

| Emission, kg/h | | | Loss, Reaction, kg/h | | | | Loss, Advection, kg/h | | | Intermedia Rate of Transport, kg/h T12 | T21 | T13 | T31 | T32 | T24 | T42 |
|---|---|---|---|---|---|---|---|---|---|---|---|---|---|---|---|---|
| E(1) | E(2) | E(3) | R(1) | R(2) | R(3) | R(4) | A(1) | A(2) | A(4) | air-water | water-air | air-soil | soil-air | soil-water | water-sed | sed-water |
| 1000 | 0 | 0 | 7.997E+02 | 1.887E+00 | 7.37E-01 | 5.426E-02 | 1.962E+02 | 1.498E+00 | 2.662E-03 | 5.200E+00 | 1.760E+00 | 7.434E-01 | 4.672E-03 | 1.292E-03 | 6.235E-02 | 5.429E-03 |
| 0 | 1000 | 0 | 2.706E+02 | 3.634E+02 | 2.50E-01 | 1.045E+01 | 6.638E+01 | 2.884E+02 | 5.127E-01 | 1.760E+00 | 3.390E+02 | 2.516E-01 | 1.581E-03 | 4.372E-04 | 1.201E+01 | 1.046E+00 |
| 0 | 0 | 1000 | 5.495E+00 | 6.435E-01 | 9.92E+02 | 1.850E-02 | 1.348E+00 | 5.107E-01 | 9.078E-04 | 3.574E-02 | 6.002E-01 | 5.109E-03 | 6.284E+00 | 1.738E+00 | 2.126E-02 | 1.851E-03 |
| 600 | 300 | 100 | 5.615E+02 | 1.102E+02 | 9.97E+01 | 3.170E+00 | 1.377E+02 | 8.747E+01 | 1.555E-01 | 3.651E+00 | 1.028E+02 | 5.220E-01 | 6.317E-01 | 1.747E-01 | 3.642E+00 | 3.171E-01 |

## Level III Distribution

Chemical name: 4-Chlorophenyl phenyl ether

### Distribution of mass

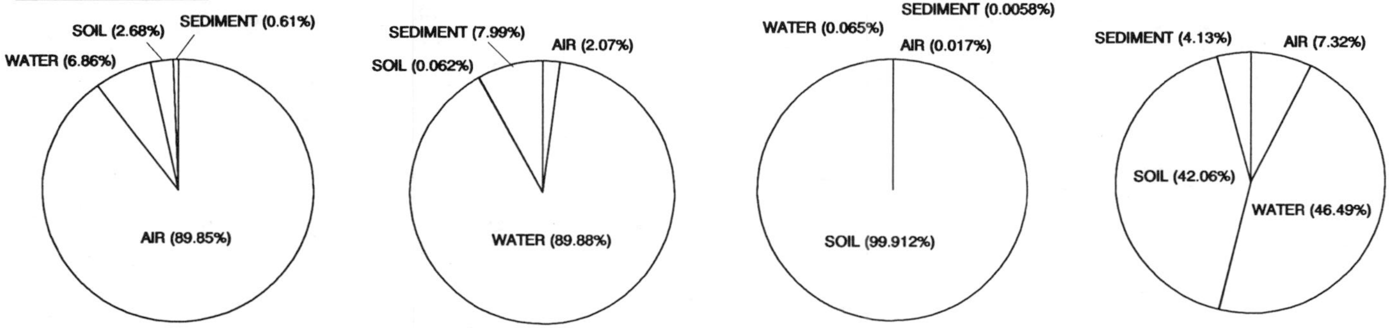

WATER (6.86%)
SOIL (2.68%)
SEDIMENT (0.61%)
AIR (89.85%)

SEDIMENT (7.99%)
SOIL (0.062%)
AIR (2.07%)
WATER (89.88%)

WATER (0.065%)
SEDIMENT (0.0058%)
AIR (0.017%)
SOIL (99.912%)

SEDIMENT (4.13%)
AIR (7.32%)
SOIL (42.06%)
WATER (46.49%)

### Distribution of removal rates

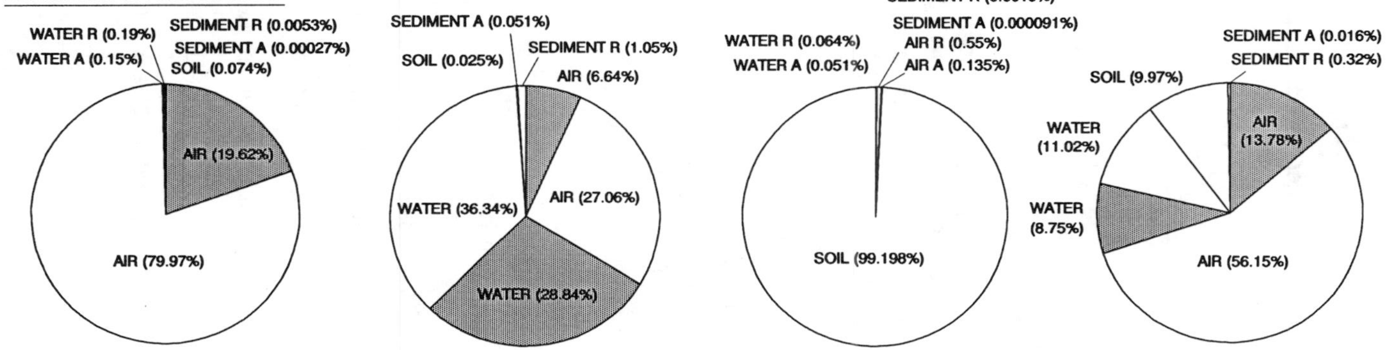

WATER R (0.19%)
WATER A (0.15%)
SEDIMENT R (0.0053%)
SEDIMENT A (0.00027%)
SOIL (0.074%)
AIR (19.62%)
AIR (79.97%)

SEDIMENT A (0.051%)
SOIL (0.025%)
SEDIMENT R (1.05%)
AIR (6.64%)
WATER (36.34%)
AIR (27.06%)
WATER (28.84%)

SEDIMENT R (0.0019%)
SEDIMENT A (0.000091%)
WATER R (0.064%)
WATER A (0.051%)
AIR R (0.55%)
AIR A (0.135%)
SOIL (99.198%)

SEDIMENT A (0.016%)
SEDIMENT R (0.32%)
SOIL (9.97%)
WATER (11.02%)
AIR (13.78%)
WATER (8.75%)
AIR (56.15%)

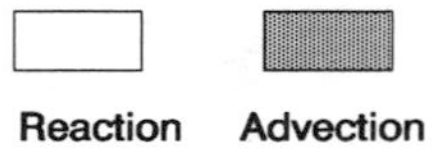

Reaction
Advection

| | | | | |
|---|---|---|---|---|
| Emission rates: | E(1) = 1000 kg/h | E(1) = 0 | E(1) = 0 | E(1) = 600 kg/h |
| | E(2) = 0 | E(2) = 1000 kg/h | E(2) = 0 | E(2) = 300 kg/h |
| | E(3) = 0 | E(3) = 0 | E(3) = 1000 kg/h | E(3) = 100 kg/h |
| Residence time: | t = 21.83 h | t = 321 h | t = 788 h | t = 188 h |

Chemical name: 4-Bromophenyl phenyl ether

Level I calculation: (six-compartment model)

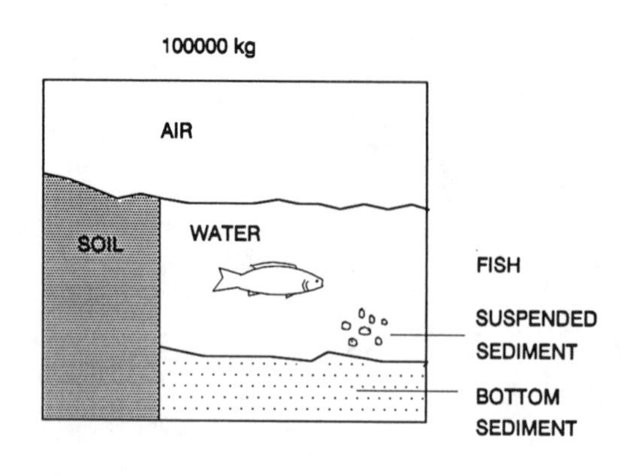


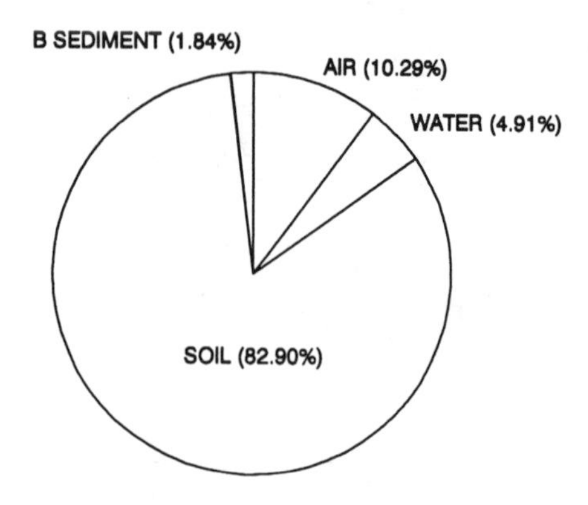


Distribution of mass

physical-chemical properties:

MW: 249.11 g/mol

M.P.: 18.0°C

Fugacity ratio: 1.0

vapor pressure: 0.20 Pa

solubility: 4.80 g/m$^3$

log $K_{OW}$: 4.28

| Compartment | Z | Concentration | | | Amount | Amount |
|---|---|---|---|---|---|---|
| | mol/m3 Pa | mol/m3 | mg/L (or g/m3) | ug/g | kg | % |
| Air | 4.034E-04 | 4.129E-10 | 1.029E-07 | 8.676E-05 | 10285 | 10.285 |
| Water | 9.634E-02 | 9.860E-08 | 2.456E-05 | 2.456E-05 | 4912.54 | 4.913 |
| Soil | 3.613E+01 | 3.698E-05 | 9.211E-03 | 3.838E-03 | 82898 | 82.898 |
| Biota (fish) | 9.179E+01 | 9.394E-05 | 2.340E-02 | 2.340E-02 | 4.6803 | 4.68E-03 |
| Suspended sediment | 2.258E+02 | 2.311E-04 | 5.757E-02 | 3.838E-02 | 57.568 | 5.76E-02 |
| Bottom sediment | 7.226E+01 | 7.395E-05 | 1.842E-02 | 7.676E-03 | 1842.18 | 1.8422 |
| | Total | | | | 100000 | 100 |

f = 1.023E-06 Pa

Chemical name: 4-Bromophenyl phenyl ether

Level II calculation: (six-compartment model)

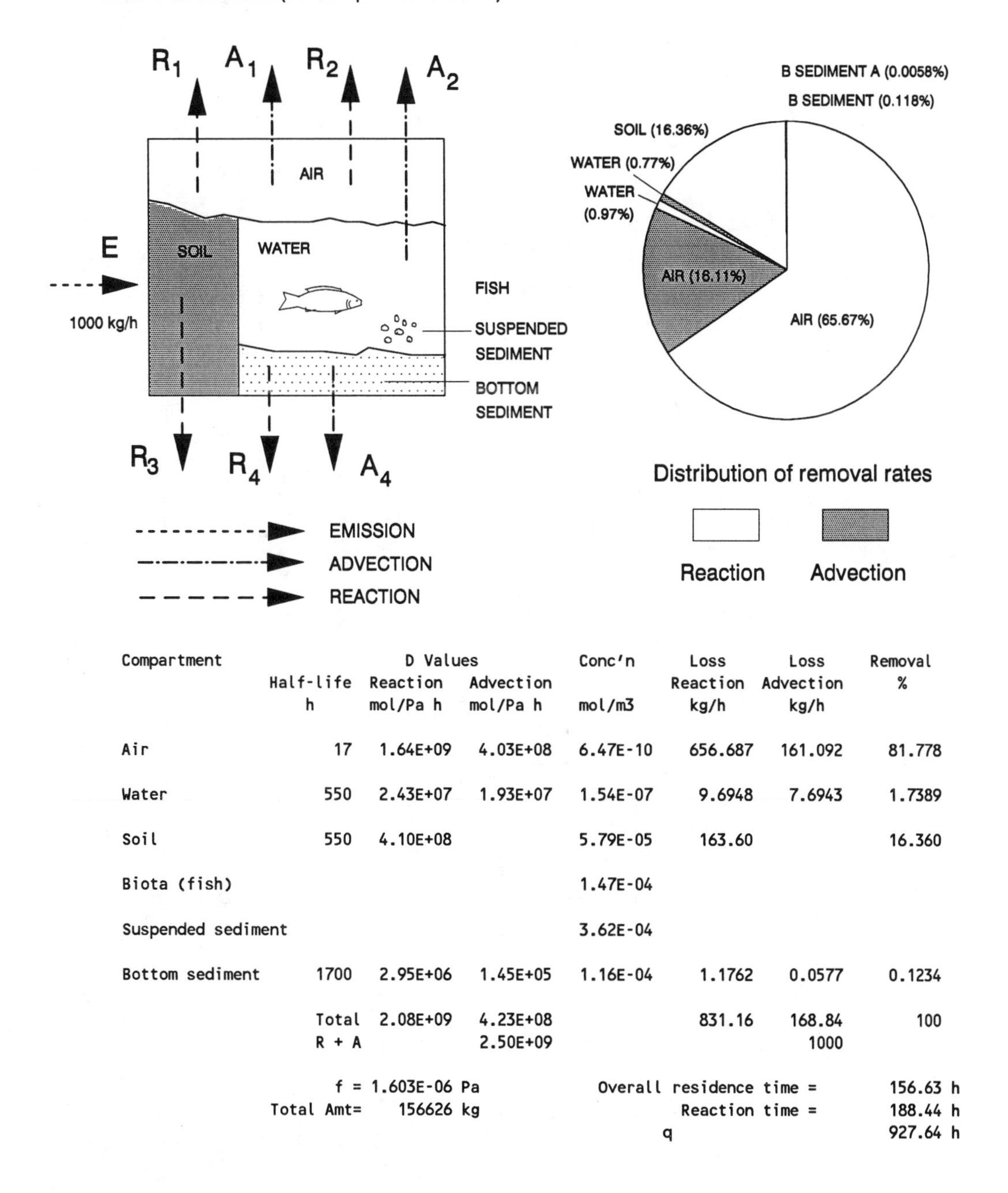


| Compartment | Half-life h | D Values Reaction mol/Pa h | D Values Advection mol/Pa h | Conc'n mol/m3 | Loss Reaction kg/h | Loss Advection kg/h | Removal % |
|---|---|---|---|---|---|---|---|
| Air | 17 | 1.64E+09 | 4.03E+08 | 6.47E-10 | 656.687 | 161.092 | 81.778 |
| Water | 550 | 2.43E+07 | 1.93E+07 | 1.54E-07 | 9.6948 | 7.6943 | 1.7389 |
| Soil | 550 | 4.10E+08 | | 5.79E-05 | 163.60 | | 16.360 |
| Biota (fish) | | | | 1.47E-04 | | | |
| Suspended sediment | | | | 3.62E-04 | | | |
| Bottom sediment | 1700 | 2.95E+06 | 1.45E+05 | 1.16E-04 | 1.1762 | 0.0577 | 0.1234 |
| | Total | 2.08E+09 | 4.23E+08 | | 831.16 | 168.84 | 100 |
| | R + A | | 2.50E+09 | | | 1000 | |

f = 1.603E-06 Pa

Total Amt= 156626 kg

Overall residence time = 156.63 h

Reaction time = 188.44 h

q 927.64 h

## Fugacity Level III calculations: (four-compartment model)

Chemical name: 4-Bromophenyl phenyl ether

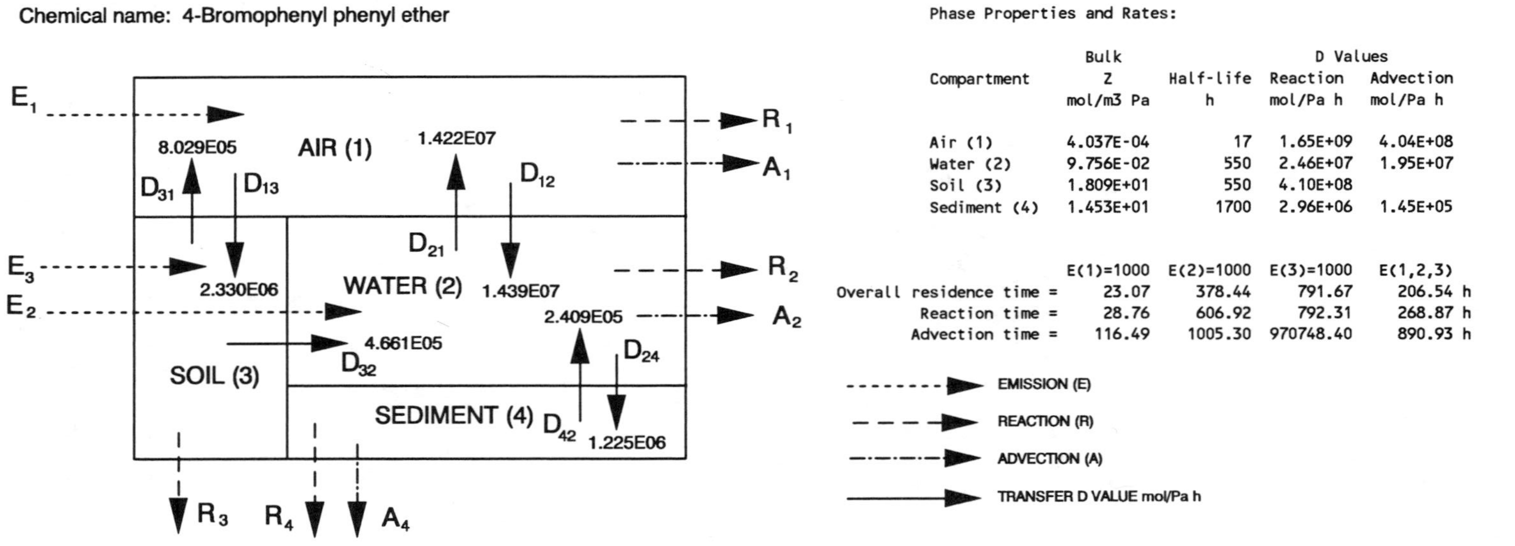


Phase Properties and Rates:

| Compartment | Bulk Z mol/m3 Pa | Half-life h | D Values Reaction mol/Pa h | D Values Advection mol/Pa h |
|---|---|---|---|---|
| Air (1) | 4.037E-04 | 17 | 1.65E+09 | 4.04E+08 |
| Water (2) | 9.756E-02 | 550 | 2.46E+07 | 1.95E+07 |
| Soil (3) | 1.809E+01 | 550 | 4.10E+08 | |
| Sediment (4) | 1.453E+01 | 1700 | 2.96E+06 | 1.45E+05 |

| | E(1)=1000 | E(2)=1000 | E(3)=1000 | E(1,2,3) | |
|---|---|---|---|---|---|
| Overall residence time = | 23.07 | 378.44 | 791.67 | 206.54 | h |
| Reaction time = | 28.76 | 606.92 | 792.31 | 268.87 | h |
| Advection time = | 116.49 | 1005.30 | 970748.40 | 890.93 | h |

Phase Properties, Compositions, Transport and Transformation Rates:

| Emission, kg/h | | | Fugacity, Pa | | | | Concentration, g/m3 | | | | Amounts, kg | | | | Total Amount, kg |
|---|---|---|---|---|---|---|---|---|---|---|---|---|---|---|---|
| E(1) | E(2) | E(3) | f(1) | f(2) | f(3) | f(4) | C(1) | C(2) | C(3) | C(4) | m(1) | m(2) | m(3) | m(4) | |
| 1000 | 0 | 0 | 1.946E-06 | 4.711E-07 | 1.102E-08 | 1.724E-07 | 1.957E-07 | 1.145E-05 | 4.965E-05 | 6.241E-04 | 1.957E+04 | 2.290E+03 | 8.938E+02 | 3.120E+02 | 2.307E+04 |
| 0 | 1000 | 0 | 4.655E-07 | 6.763E-05 | 2.635E-09 | 2.476E-05 | 4.680E-08 | 1.644E-03 | 1.187E-05 | 8.960E-02 | 4.680E+03 | 3.287E+05 | 2.137E+02 | 4.480E+04 | 3.784E+05 |
| 0 | 0 | 1000 | 4.354E-09 | 7.750E-08 | 9.752E-06 | 2.837E-08 | 4.378E-10 | 1.884E-06 | 4.396E-02 | 1.027E-04 | 4.378E+01 | 3.767E+02 | 7.912E+05 | 5.134E+01 | 7.917E+05 |
| 600 | 300 | 100 | 1.308E-06 | 2.058E-05 | 9.826E-07 | 7.534E-06 | 1.315E-07 | 5.002E-04 | 4.429E-03 | 2.727E-02 | 1.315E+04 | 1.000E+05 | 7.972E+04 | 1.363E+04 | 2.065E+05 |

| Emission, kg/h | | | Loss, Reaction, kg/h | | | | Loss, Advection, kg/h | | | Intermedia Rate of Transport, kg/h T12 | T21 | T13 | T31 | T32 | T24 | T42 |
|---|---|---|---|---|---|---|---|---|---|---|---|---|---|---|---|---|
| E(1) | E(2) | E(3) | R(1) | R(2) | R(3) | R(4) | A(1) | A(2) | A(4) | air-water | water-air | air-soil | soil-air | soil-water | water-sed | sed-water |
| 1000 | 0 | 0 | 7.978E+02 | 2.885E+00 | 1.13E+00 | 1.272E-01 | 1.957E+02 | 2.290E+00 | 6.241E-03 | 6.975E+00 | 1.668E+00 | 1.130E+00 | 2.221E-03 | 1.279E-03 | 1.438E-01 | 1.035E-02 |
| 0 | 1000 | 0 | 1.908E+02 | 4.142E+02 | 2.69E-01 | 1.826E+01 | 4.680E+01 | 3.287E+02 | 8.960E-01 | 1.668E+00 | 2.395E+02 | 2.701E-01 | 5.311E-04 | 3.059E-04 | 2.064E+01 | 1.485E+00 |
| 0 | 0 | 1000 | 1.785E+00 | 4.747E-01 | 9.97E+02 | 2.093E-02 | 4.378E-01 | 3.767E-01 | 1.027E-03 | 1.560E-02 | 2.745E-01 | 2.527E-03 | 1.966E+00 | 1.132E+00 | 2.366E-02 | 1.702E-03 |
| 600 | 300 | 100 | 5.361E+02 | 1.260E+02 | 1.00E+02 | 5.557E+00 | 1.315E+02 | 1.000E+02 | 2.727E-01 | 4.687E+00 | 7.289E+01 | 7.591E-01 | 1.981E-01 | 1.141E-01 | 6.282E+00 | 4.520E-01 |

# Level III Distribution

Chemical name: 4-Bromophenyl phenyl ether

## Distribution of mass

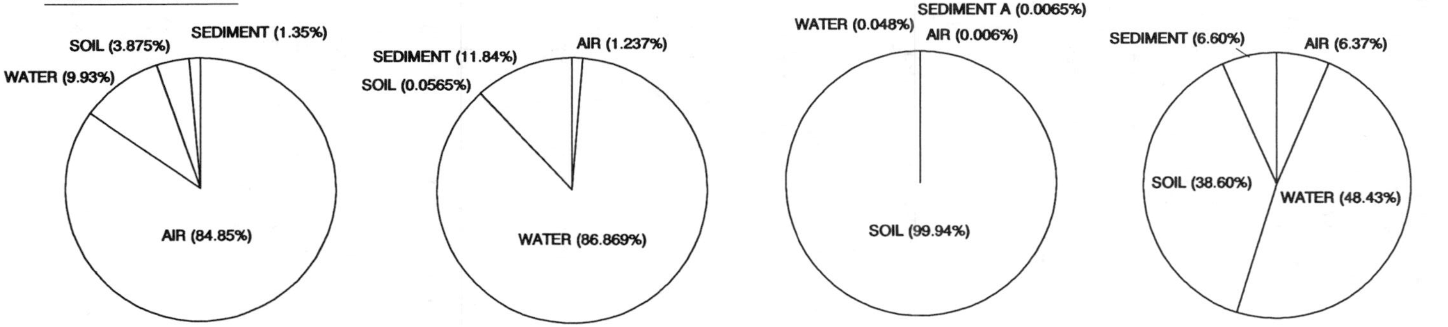

## Distribution of removal rates

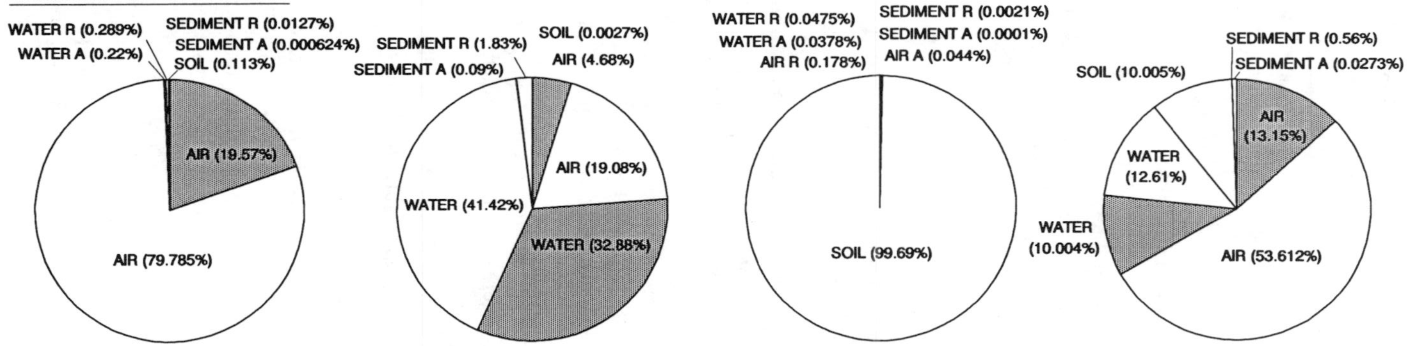

Emission rates:

| | | | |
|---|---|---|---|
| E(1) = 1000 kg/h | E(1) = 0 | E(1) = 0 | E(1) = 600 kg/h |
| E(2) = 0 | E(2) = 1000 kg/h | E(2) = 0 | E(2) = 300 kg/h |
| E(3) = 0 | E(3) = 0 | E(3) = 1000 kg/h | E(3) = 100 kg/h |

Residence time:

| | | | |
|---|---|---|---|
| t = 23.07 h | t = 378 h | t = 792 h | t = 207 h |

## 4.4 COMMENTARY ON THE PHYSICAL-CHEMICAL PROPERTIES AND ENVIRONMENTAL FATE

The ethers treated in this Chapter are a rather diverse group of chemicals which vary considerably in structure and properties. Not all are strictly "volatile" chemicals since some have low vapor pressures and low air-water partition coefficients. As is discussed below, this diversity in structure renders the series less amenable to QSPR analysis and to general statements of environmental fate. To a large extent these chemicals must be treated individually. Comparison of the QSPR plots in this Chapter with those of Chapter 2 for hydrocarbons reveals a much greater scatter. Accordingly, this commentary is less detailed than those of other chapters.

**Reactivity**

The half-life classes in Table 4.3 represent the authors' best judgement about rates of reaction. Of most importance are reactions in the atmosphere, mainly with hydroxyl radicals. These rates are fairly rapid and tend to dominate the chemicals' fate when viewed in a multimedia context. When discharged to soil or water, the rate of evaporation tends to control the overall residence time, because once evaporated, these substances are lost by advection or reaction. The water class was usually selected to be 3 more than that of air (factor of 30 slower), with soil a further one class slower and with bottom sediment one class slower than soil.

**QSPR Plots**

The QSPR plots show greater scatter than those of more structurally similar series such as the hydrocarbons in Chapter 2.

Figure 4.1 shows that the LeBas molar volume is fairly well correlated with the limited information on molecular volumes and molecular areas. It is likely that for this series, other descriptors, notably TSA, will be more useful for correlation purposes.

The solubility data, Figure 4.2, show the trend of decreased solubility with molecular size. No attempt is made to select a slope, although it appears to be similar in magnitude to that of the other QSAR plots discussed in Chapters 2 and 3.

The vapor pressure data in Figure 4.3 shows a similar variability and the expected drop in vapor pressure with increasing molar volume.

The log $K_{OW}$ data in Figure 4.4 shows the expected increase in hydrophobicity which is largely attributable to the drop in solubility, but the scatter extends over two orders of magnitude in $K_{OW}$.

The Henry's law constants, H in Figure 4.5, show that the ethers have relatively high values varying from 0 to 3.0 in log H, or 1 to 1000 $Pa\cdot m^3/mol$ in H. This corresponds to an air-water partition coefficient of 0.0004 to 0.40; thus some ethers have a fair tendency to partition into air but not to the extent of the VOCs discussed in Chapters 2 and 3.

The log $K_{OW}$ - solubility plot in Figure 4.6 has a slope of about 0.75 confirming the generally reciprocal linear relationship between $K_{OW}$ and solubility, but this slope is significantly lower than other VOCs.

It is unwise to use these plots to deduce the properties of other ethers and organic oxides. The preferred procedure is to plot the property as a function of molar volume for a group of structurally similar ethers and to interpolate.

**Evaluative Calculations**

Rather than give diagrams for all the ethers, only a selection is presented. The fate of the others can be inferred from those of the closest homolog for which a diagram is provided.

The Level I calculations show, in most cases, appreciable partitioning into the atmosphere with only minor amounts into the other compartments, The chemicals with higher $K_{OW}$, such as anisole, diphenyl ether and bis(2-chloroethyl)ether partition into water and soil in significant quantities. It is probable that metabolism will reduce the concentrations in fish below those indicated.

The Level II diagrams generally show the dominance of atmospheric reaction and advection, but certain chemicals such as 1,2-propylene oxide react appreciably in water. Even if soil, water, or sediment rate constants were increased by a factor of 1000 (which is inconceivable) the rates of loss from these media would remain negligible.

The Level III diagrams show that the chemical discharged into air experiences very little tendency to enter into soil or water because the rates of reaction and advection in air greatly exceed the rates of absorption or deposition. The overall residence times are generally about 1 day.

When discharged to water there is usually competition between evaporation, advection from water and reaction in water. The more hydrophobic chemicals also tend to partition into sediments. The overall residence times are much longer (e.g., tens of days) as a result of slower reaction and advection in water and the constraint imposed by evaporation which can be fairly slow for compounds with low air-water partition coefficients.

When discharged to soil, most ethers evaporate and again the rate of loss is limited by diffusion processes. It seems likely that some of the less volatile ethers will be subject to appreciable biodegradation, but there is little literature on this topic.

The Level III pie charts show these trends pictorially.

It is difficult, and unwise, to generalize about the environmental fate of this group of chemicals. There are wide differences in volatility and hydrophobicity; thus the fate is quite specific to the individual physical-chemical properties and reactivity.

## 4.5 REFERENCES


Abernethy, S., Mackay, D., McCarty, L.S. (1988) "Volume fraction" correlation for narcosis in aquatic organisms: The key role of partitioning. *Environ. Toxicol. & Chem.* 7, 469-481.

Altshuller, A.P., Cohen, I.R., Sleva,S.F.,Kopczynski,S.L. (1962) Air pollution: Photooxidation of aromatic hydrocarbons. *Science* 138(3538), 442-443.

Amidon, G.L., Yalkowsky, S.H., Anik, S.T., Valvani, S.C. (1975) Solubility of nonelectrolytes in polar solvents. V. Estimation of the solubility of aliphatic monofunctional compounds in water using a molecular surface area approach. *J. Phys. Chem.* 79, 2239-2245.

Anbar, M., Neta, P. (1967) A compilation of specific bimolecular rate and hydroxyl radical with inorganic and organic compounds in aqueous solution. *Int'l J. Appl. Radiation Isotopes* 18, 493-523.

Anderson, T.A., Beauchamp, J.J., Walton, B.T. (1991) Organic chemicals in the environment. *J. Environ. Qual.* 20. 420-424.

Arbuckle, W.B., (1983) Estimating activity coefficients for use in calculating environmental parameters. *Environ. Sci. Technol.* 17, 537-542.

Atkinson, R. (1985) Kinetics and mechanisms of the gas-phase reactions of the hydroxyl radicals with organic compounds under atmospheric conditions. *Chem. Rev.* 85, 69-201.

Atkinson, R. (1987) Structure-activity relationship for the estimation of rate constants for the gas-phase reactions of OH radicals with organic compounds. *Int'l J. Chem. Kinetics* 19, 799-828.

Atkinson, R. (1991) Kinetics and mechanisms of the gas-phase reactions of the $NO_3$ radical with organic compounds. *J. Phys. Chem. Ref. Data* 20(3), 459-507.

Atkinson, R., Aschmann, S.M., Pitts, J.N., Jr. (1988) Rate constants for the gas-phase reactions of the $NO_3$ radicals with a series of organic compounds at 296 $\pm$ 2 K. *J. Phys. Chem.* 92, 3454-3457.

Atkinson, R., Aschmann, S.M., Winer, A.M. (1987) Kinetics of the reactions of $NO_3$ radicals with a series of aromatic compounds. *Environ. Sci. Technol.* 21, 1123-1126.

Atkinson, R., Aschmann, S.M., Winer, A.M., Carter, W.P.L. (1985) Rate constants for the gas-phase reactions of the $NO_3$ radicals with furan, thiophene, and pyrrole at 295 $\pm$ 1 K and atmospheric pressure. *Environ. Sci. Technol.* 19, 87-90.

Banerjee, S., Yalkosky, S.H., Valvani, S.C. (1980) Water solubility and octanol/water partition coefficient of organics. Limitations of solubility-partition coefficient correlation. *Environ. Sci. Technol.* 14, 1227-1229.

Banerjee, S., Howard, P.H., Lande, S.S. (1990) General structure-vapor pressure relationships for organics. *Chemosphere* 21(10-11), 1173-1180.

Barrows, M.E., Petrocelli, S.R., Macek, K.J. (1980) Bioconcentration and elimination of selected water pollutants by bluegill sunfish (*lepomis machrochirus*). In: *Dynamic, Exposure, Hazard Assessment of Toxic Chemicals.* Haque, R., Editor, pp.379-392, Ann Arbor Science Publishers Inc., Ann Arbor, Michigan.

Bennett, G.M., Phillip, W.G. (1928) CCL II.-The influence of structure on the solubilities of ethers. Part I. Aliphatic ethers. *J. Chem. Soc.* 1930-1937.

Bennett, G.M., Phillip, W.G. (1928) CCLIII.-The influence of structure on the solubilities of ethers. Part II. Some cyclic ethers. *J. Chem Soc.* 1937-1943.

Bodor, N., Garbanyi, Z., Wong, C.K. (1989) A new method for the estimation of partition coefficient. *J. Am. Chem. Soc.* 111, 3783-3786.

Bogyo Da et al. (1980) *Investigation of Selected Potential Environmental Contaminants: Epoxies.* pp.201. USEPA-560/11-80-005.

Booth, H.S., Everson, H.E. (1948) Hydrotropic solubilities: Solubilities in 40 per cent sodium xylenesulfonate. *Ind. Eng. Chem.* 40(8), 1491-1493.

Boublik, T., Fried, V. Hala, E. (1973) *The Vapor Pressures of Pure Substances.* Elsevier, Amsterdam.

Boublik, T., Fried, V. Hala, E. (1984) *The Vapor Pressures of Pure Substances. 2nd Edition,* Elsevier, Amsterdam.

Branson, D.R. (1977) A new capacitor fluid - A case study in product stewardship. pp.44-61. In: *Aquatic Toxicology and Hazard Evaluation.* ASTM Special Technical Publication 634. Mayer, F.L., Hamelink, J.L., Editors, American Society for Testing and Materials, Philadelphia, Pennsylvania.

Branson, D.R. (1978) Predicting the fate of chemicals in the aquatic environment from laboratory data. pp.55-70. In: *Estimating the Hazard of Chemical Substances to Aquatic Life.* ASTM Special Technical Publication 657. Cairns, J., Jr., Dickson, K.L., Maki, A.W., Editors, American Society for Testing and Materials, Philadelphia, Pennsylvania.

Bridie, A.L., Wolff, C.J.M., Winter, M. (1979) BOD and COD of some petrochemicals. *Water Res.* 13, 627-630.

Brown S.L., Chan, F.Y., Jones, J.L., Liu, D.H., McCaleb, K.E. (1975) *Research Program on Hazard Priority Ranking of Manufactured Chemicals.* NTIS PB-263163. Stanford Research Institute, Menlo Park, California.

Burkhard, L.P., Kuehl, D.W., Veith, G.D. (1985) Evaluation of reverse phase liquid chromatograph/mass spectrometry for estimating of *n*-octanol/water partition coefficients. *Chemosphere* 14(10) 1551-1560.

Callahan, M.A., Slimak, M.W., Gabel, N.W., May, I.P., Fowler, C.F., Freed, J.R., Jennings, P., Durfee, R.L., Whitemore, F.C., Maestri, B., Mabey, W.R., Holt, B.R., Gould, C. (1979) *Water Related Environmental Fate of 129 Priority Pollutants.* Volume I, EPA Report, No.440/4-79-029a. Versar, Inc., Springfield, Virginia.

Chiou, C.T., Freed, V.H., Schmedding, D.W., Kohnert, R. (1977) Partition coefficient and bioaccumulation of selected, organic compounds. *Environ. Sci. Technol.* 11(5), 475-478.

Chou, J.T., Jurs, P.C. (1979) Computation of partition coefficients from molecular structures by a fragment addition method. In: *Physical Chemical Properties of Drugs.* Medical Research Series Vol. 10, Yalkowsky, S.H., Sinkula, A.A., Valvani, S.C., Editors.

Collander, R. (1951) The partition of organic compounds between higher alcohols and water. *Acta Chem. Scand.* 5, 774-780.

Cowen, W.F., Baynes, R.K. (1980) Estimated application of gas chromatographic headspace analysis for priority pollutants. *J. Environ. Sci. & Health* A15, 413-427.

Cupitt, L.T. (1980) *Fate of Toxic and Hazardous Materials in the Air Environment.* USEPA-600/3-80-084.

Darnall, K.R., Lloyd, A.C., Winer, A.M., Pitts, Jr., J.N. (1976) Reactivity scale for atmospheric hydrocarbons based on reaction with hydroxyl radicals. *Environ. Sci. Technol.* 10(7), 692-696.

Daubert, T.E., Danner, R.P. (1985) *Data Compilation Tables of Properties of Pure Compounds.* American Institute of Chemical Engineering.

Davies, R.P., Dobbs, A.J. (1984) The prediction of bioconcentration in fish. *Water Res.* 18(10), 1253-1262.

Dean, J.D., Editor (1985) *Lange's Handbook of Chemistry.* 13th Edition, McGraw-Hill, Inc., New York.

Dearden, J.C., Tubby, J.H. (1974) Ortho effects in structure-activity studies. Shielding of hydroxyl by alkyl groups. *J. Pharm. Pharmacol.* 26, Suppl. 73p-74p.

Deneer, J.W., Sinnige, T.L., Seinen, W., Hermens, J.L.M. (1988) A quantitative structure-activity relationship for the acute toxicity of some epoxy compounds to the guppy. *Aquat. Toxicol.* 13, 195-204.

Dilling, W.L., Tefertiller, N.B., Kallos, G.J. (1975) Evaporation rates of methylene chloride, chloroform, 1,1,1-trichloroethane, trichloroethylene, tetrachloroethylene, and other chlorinated compounds in dilute aqueous solutions. *Environ. Sci. Technol.* 9(9), 833-838.

Dojlido, J.R. (1979) *Investigations of Biodegradability and Toxicity of Organic Compounds.* Final Report 1975-79. U.S. EPA-600/2-79-163.

Dorfman, L.M., Adams, G.E. (1973) *Reactivity of Hydroxyl Radicals in Aqueous Solution.* NSRD-NDB-46. NTIS COM-73-50623. National Bureau of Standards, Washington, D.C. 51 pp.

Dow Chemical (1979) Personal communication from M. Thomas of Dow Chemical Company to N.W. Gabel of Versar, Inc.

Dreisbach, R.R. (1952) *Pressure-Volume-Temperature Relationships of Organic Compounds.* Handbook Publishers Inc., Sandusky, Ohio.

Du Pont (1966) *Solubility Relationships of the Freon Fluorocarbon Compounds.* Technical Bulletin B-7, Du Pont de Nemours & Company, Wilmington, Delaware.

Durkin, P.R., Howard, P.H., Saxena, J. (1975) *Investigations of Selected Potential Environmental Contaminants: Haloethers.* U.S. EPA-560/2-75-006. U.S. Environmental Protection Agency, Office of Toxic Substances, Washington, D.C.

Edney, E.O., Kleindienst, T.E., Coese, E.W. (1986) Room temperature rate constants for the reaction of OH radicals with selected chlorinated and oxygenated hydrocarbons. *Int'l. J. Chem. Kinet.* 18, 1355-1371.

Ellington, J.J. (1989) *Hydrolysis Rate Constants for Enhancing Property-Reactivity Relationships.* EPA/600/3-89/063. NTIS PB89-220479. U.S. Environmental Protection Agency, Environmental Research Laboratory, Athens, Georgia.

Ellington, J.J., Stancil, F.E., Jr., Payne, W.D., Trusty, C.D. (1987) *Measurements of Hydrolysis Rate Constants for Evaluation of Hazardous Waste Land Disposal: Data on 54 Chemicals.* Volume II. EPA/600/3-87/019. U.S. Environmental Protection Agency, Athens, Georgia.

Fishbein, L. (1978) Potential halogenated industrial carcinogenic and mutagenic chemicals. III. Alkane halides, alkanols and ethers. *Sci. Total Environ.* 11, 223-257.

Fujita, T., Iwasa, J., Hansch, C. (1964) A new substituent constant, "pi" derived from partition coefficients. *J. Am. Chem. Soc.* 86, 5175-5180.

Funasaki, N., Hada, S., Neya, S., Machida, K. (1984) Intramolecular hydrophobic association of two alkyl chains of oligoethylene glycol diethers and diesters in water. *J. Phys. Chem.* 88, 5786-5790.

Funasaki, N., Hada, S., Neya, S. (1985) Partition coefficients of aliphatic ethers-molecular surface area approach. *J. Phys. Chem.* 89, 3046-3049.

Garst, J.E. (1984) Accurate, wide-range, automated, high-performance liquid chromatographic method for the estimation of octanol/water partition coefficients. II: Equilibrium in partition coefficient measurements, additivity of substituent constants, and correlation of biological data. *J. Pharm. Sci.* 73, 1623-1629.

Ge, J., Liu, W., Dong, S. (1987) *Sepu* 5, 182.

GEMS (1986) Graphical Exposure Modeling System. FAP, Fate Atmospheric Pollution. Data Base, Office of Toxic Substances, U.S. Environmental Protection Agency.

Güesten, H., Filby, W.G., Schoop, S. (1981) Prediction of hydroxyl radical reaction rates with organic compounds in the gas phase. *Atom. Environ.* 15, 1763-1765.

Haag, W.R., Mill, T. (1988) Effect of a subsurface sediment on hydrolysis of haloalkanes and epoxides. *Environ. Sci. Technol.* 22, 658-663.

Haky, J.E., Young, A.M. (1984) Evaluation of a simple HPLC correlation method for the estimation of the octanol-water partition coefficients of organic compounds. *J. Liq. Chromatogr.* 7, 675-689.

Hammond, M.W., Alexander, M. (1972) Effects of chemical structure on microbial degradation of methyl substituted aliphatic acids. *Environ. Sci. Technol.* 6(8), 732-735.

Hanai, T., Tran, C., Hubert, J. (1981) An approach to the prediction of retention times in liquid chromatography. *J. HRC & CC* 4, 454-460.

Hansch, C., Anderson, S.M. (1967) The effect of intramolecular hydrophobic bonding on partition coefficients. *J. Org. Chem.* 32, 2583.

Hansch, C., Leo, A. (1979) *Substituents Constants for Correlation Analysis in Chemistry and Biology.* Wiley, New York.

Hansch, C., Leo, A. (1985) *Medchem Project Issue No. 26.* Pomona Colllege, Claremont, California.

Hansch, C., Quinlan, J.E., Lawrence, G.L. (1968) The linear free-energy relationship between partition coefficients and aqueous solubilities of organic liquids. *J. Org. Chem.* 33, 347-350.

Hawker, D.W., Connell, D.W. (1985) Relationship between partition coefficient, uptake rate constant, clearance rate constant and time to equilibrium for bioaccumulation. *Chemosphere* 14(9), 1205-1219.

Hill, A.E. (1923) The mutual solubility of liquids. I. The mutual solubility of ethyl ether and water. II. The solubility of water in benzene. *J. Am. Chem. Soc.* 45, 1143-1152.

Hine, J., Mookerjee, P.K. (1975) The intrinsic hydrophilic character of organic compounds. Correlations in terms of structural contributions. *J. Org. Chem.* 40(3), 292-298.

Howard, P.H., Editor (1989) *Handbook of Fate and Exposure Data for Organic Chemicals. Vol.I-Large Production and Priority Pollutants.* Lewis Publishers, Chelsea, Michigan.

Howard, P.H., Editor (1990) *Handbook of Environmental Fate and Exposure Data for Organic Chemicals. Vol.II-Solvents.* Lewis Publishers Inc., Chelsea, Michigan.

Howard, P.H., Boethling, R.S., Jarvis, W.F., Meylan, W.M., Michalenko, E.M., Editors (1991) *Handbook of Environmental Degradation Rates.* Lewis Publishers, Inc., Chelsea, Michigan.

Isnard, P., Lambert, S. (1988) Estimating bioconcentration factors from octanol-water partition coefficient and aqueous solubility. *Chemosphere* 17, 21-34.

Isnard, P., Lambert, S. (1989) Aqueous solubility/*n*-octanol water partition coefficient correlations. *Chemosphere* 18, 1837-1853.

Jones, D.M., Wood, N.F. (1964) The mechanism of vinyl ether hydrolysis. *J. Chem. Soc.* 5400-5403.

Kamlet, M.J., Abraham, D.J., Doherty, R.M., Taft, R.W., Abraham, M.H. (1986) Solubility properties in polymers and biological media: 6. An equation for correlation and prediction of solubilities of liquid organic nonelectrolytes in blood. *J. Pharmac. Sci.* 75(4), 350-355.

Kamlet, M.J., Doherty, R.M., Abboud, J-L M., Abraham, M.H., Taft, R.W. (1986) Linear solvation energy relationships: 36. Molecular properties governing solubilities of organic nonelectrolytes in water. *J. Pharmac. Sci.* 75(4), 338-349.

Kamlet, M.J., Doherty, R.M., Taft, R.W., Abraham, M.H., Veith, G.D., Abraham, D.J. (1987) Solubility properties in polymers and biological media. 8. An analysis of the factors that influence toxicities of organic nonelectrolytes to the golden orfe fish. (*leuciscus idus melanotus). Environ. Sci. Technol.* 21, 149-155.

Kamlet, M.J., Doherty, R.M., Abraham, M.H., Carr, P.W., Doherty, R.F., Taft, R.W. (1987) Linear solvation energy relationships. 41. Important differences between aqueous solubility relationships for aliphatic and aromatic solutes. *J. Phys. Chem.* 91, 1996-2004.

Kamlet, M.J., Doherty, R.M., Abraham, M.H., Marcus, Y., Taft, R.W. (1988) Linear solvation energy relationships. 46. An improved equation for correlation and prediction of octanol/water partition coefficients of organic nonelectrolytes. (including strong hydrogen bond donor solutes*). J. Phys. Chem.* 92, 5244-5255.

Kankaanperä, A. (1969) Basocoties of the oxygen atoms in symmetrical and unsymmetrical acetals. Part II. The base strengths and their relation to the rate coefficients of the different partial fission reactions of acetal hydrolysis. *Acta. Chem. Scand.* 23(5), 1728-1732.

Kawasaki, M. (1980) Experiences with test scheme under chemical control law of Japan: An approach to structural-activity correlations. *Ecotox. Environ. Safety* 4, 444-454.
Kincannon, D.F., Lin, Y.S. (1985) Microbial degradation of hazardous wastes by land treatment. In: *Proc. Indust. Waste Conf.* 40, 607-619.
Kleopfer, R.D., Fairless, B.J. (1972) Characterization of organic compounds in a municipal water supply. *Environ. Sci. Technol.* 6, 1036-1037.
Krijgsheld, K.R., Van der Gen, A. (1986) Assessment of the impact of the emission of certain organochlorine compounds of the aquatic environment. Part II: Allylchloride, 1,3-- and 2,3-dichloropropene. *Chemosphere* 15, 881-893.
Laity, J.L., Burstain, I.G., Appel, B.R. (1973) Photochemical smog and the atmospheric reactions of solvents. Chapter 7, pp.95-112. In: *Solvents Theory and Practice.* Tess, R.W., Editor, *Advances in Chemistry Series* 124, Am. Chem. Soc., Washington, D.C.
Lange, N.A., Editor (1971) *Lange's Handbook of Chemistry.* McGraw-Hill, New York.
Leahy, D.E. (1986) Intrinsic molecular volume as a measure of the cavity term in linear solvation energy relationships: Octanol-water partition coefficients and aqueous solubilities. *J. Pharmac. Sci.* 75(7), 629-636.
Leo, A. (1986) *CLOGP-3.42 Medchem Software.* Medicinal Chemistry Project, Pomona College, Claremont, California.
Leo, A., Hansch, C., Elkins, D. (1971) Partition coefficients and their uses. *Chem. Rev.* 71, 525-612.
Leo, A., Jow, P.Y.C., Silipo, C., Hansch, C. (1975) Calculation of hydrophobic constant (Log *P*) from $\pi$ and $f$ constants. *J. Med. Chem.* 18(9), 863-868.
Ludzack, F.J., Ettinger, M.B. (1963) Biodegradability of organic chemicals isolated from rivers. *Purdue Univ., Eng. Bull. Ext. Ser.* 115, 278-282.
Lyman, W.J., Reehl, W.F., Rosenblatt, D.H., Editors (1982) *Handbook of Chemical Property Estimation Methods.* McGraw-Hill Book Company, New York.
Mabey, W., Mill, T. (1978) Critical review of hydrolysis of organic compounds in water under environmental conditions. *J. Phys. Chem. Ref. Data* 7, 383-415.
Mabey, W., Smith, J.H., Podoll, R.T., Johnson, H.L., Mill, T., Chiou, T.W., Gate, J., Waight-Partridge, I., Jaber, H., Vandenberg, D. (1982) *Aquatic Fate Process for Organic Priority Pollutants.* EPA Report, No.440/4-81-14.
Mackay, D. (1982) Correlation of bioconcentration factors. *Environ. Sci. Technol.* 16, 274-278.
Mackay, D., Wolkoff, A.W. (1973) Rate of evaporation of low-solubility contaminants from water bodies to atmosphere. *Environ. Sci. Technol.* 7(7), 611-614.
Mailhe, A., Murat, M. (1912) Derivés halogénés de l'oxyde de phényle. *Bull. Soc. Chim. France* 11, 328-332.
McGowan, J.C., Atkinson, P.N., Ruddle, L.H. (1966) The physical toxicity of chemicals. V. Interaction terms for solubilities and partition coefficients. *J. Appl. Chem.* 16, 99-102.
Michael, L.C., Pellizzarl, E.D., Wiseman, R.W. (1988) Development and evaluation of a procedure for determining volatile organics in water. *Environ. Sci. Technol.* 22, 565-570.
Mill, T. (1980) *Handbook of Environmental Chemistry.* Vol.II, Springer-Verlag, Berlin.

Mill, T., Mabey, W. (1985) Chapter 8, Photochemical transformations. In: *Environmental Exposure from Chemicals.* Volume I, Neely, W.B., Blau, G.E., Editors, CRC Press Inc.

Mirrlees, M.S., Moulton, S.J., Murphy, C.T., Taylor, P.J. (1976) Direct measurement of octanol-water coefficients by high-pressure liquid chromatography. *J. Med. Chem.* 19, 615.

Moriguchi, I. (1975) Quantitative structure activity studies. Parameters relating to hydrophobicity. *Chem. Pharm. Bull.* 23(2), 247-257.

Müller, M., Klein, W. (1991) Estimating atmospheric degradation processes by SARs. *Sci. Total Environ.* 109/110, 261-273.

Neuhauser, E.F., Loehr, R.C., Malecki, M.R., Milligan, D.L., Durkin, P.R. (1985) The toxicity of selected organic chemicals to earthworm *eisenia fetida. J.Environ. Qual.* 14(3), 383-388.

Pankow, J.F. (1990) Minimization of volatilization losses during sampling and analysis of volatile organic compounds in water. In: *Significance and Treatment of Volatile Organic Compounds in Water.* Ram, N.M., Christman, R.F., Cantor, K.P., Editors, Lewis Publishers, Inc., Chelsea, Michigan.

Pankow, J.F., Rosen, M.E. (1988) The determination of volatile compounds in water by purging directly to a capillary column with whole column cryotrapping. *Environ. Sci. Technol.* 22, 398-405.

Pearlman, R.S., Yalkowsky, S.H., Banerjee, S. (1984) Water solubilities of polynuclear aromatic and heteroaromatic compounds. *J. Phys. Chem. Ref. Data* 13, 555-562.

Perry, R.A., Atkinson, R., Pitts, Jr., J.N. (1977) Kinetics and mechanisms of the gas phase reaction of OH radicals with methoxybenzene and *o*-cresol over temperature range 299-435 K. *J. Phys. Chem.* 81(17), 1607-1611.

Pinal, R., Rao, P.S.C., Lee, L.S., Cline, P.V. (1990) Cosolvency of partially miscible organic solvents on the solubility of hydrophobic organic chemicals. *Environ. Sci. Technol.* 24(5), 639-647.

Pratesi, P., Villa, L., Ferri, V., de Micheli, C., Grna, E., Grieco, C., Silipo, C., Vittoria, A. (1979) *Il Farmaco Ed. Sci.* 34, 579.

Quayle, O.R. (1953) The parachors of organic compounds. *Chem. Rev.* 53, 439-589.

Radding, S.B., Liu, D.H., Johnson, H.L., Mill, T. (1977) *Review of the Environmental Fate of Selected Chemicals.* EPA Report No.560/5-77-003. U.S. Environmental Protection Agency, Office of Toxic Substances, Washington, D.C.

Rekker, R.F., Nauta, W.T. (1961) Steric effects in the electronic spectra of substituted benzophenones. IV. Unsymmetrically substituted methylbenzophenones. *Rec. Trav. Chim., Pays Bas* 80, 747-763.

Rekker, R.F. (1977) *The Hydrophobic Fragmental Constant.* Elsevier Scientific Publishing Co., Amsterdam.

Riddick, J.A., Bunger, W.B., Sakano, T.K. (1986) *Organic Solvents: Physical Properties and Methods of Purification.* 4th Edition, John-Wiley & Sons, New York.

Rogers, K.S., Cammarata, A. (1969) Superdelocalizability and charge density. A correlation with partition coefficients. *J. Med. Chem.* 12, 692.

Ryan, J.A., Bell, R.M., Davidson, J.M., O'Connor, G.A. (1988) Plant uptake of non-ionic organic chemicals from soils. *Chemosphere* 17(12), 2299-2323.

Sabljic, A. (1984) Predictions of the nature and strength of soil sorption of organic pollutants from molecular topology. *J. Agric. Food & Chem.* 32, 243-246.

Sabljic, A. (1987) Nonempirical modeling of environmental distribution and toxicity of major organic pollutants. In: *QSAR in Environmental Toxicology-II.* Kaiser, K.L.E., Editor, D. Reidel Publishing Co., Dordrecht, Holland. pp.309-332.

Sabljic, A., Güsten, H. (1990) Predicting the night-time $NO_3$ radical reactivity in the troposphere. *Atmos. Environ.* 24A(1), 73-78.

Saito, S., Tanoue, A., Matsuo, M. (1992) Applicability of the i/o-characters to quantitative description of bioconcentration of organic chemicals in fish. *Chemosphere* 24(1), 81-87.

Sangster, J. (1989) Octanol-water partition coefficients of simple organic compounds. *J. Phys. Chem. Ref. Data* 18(3), 1111-1229.

Santodonato, J., et al. (1980) *Investigation of Selected Potential Environmental Contaminants: Epichlorohydrin and Epibromohydrin.* USEPA-560/11-80-006.

Sasaki, S. (1978) The scientific aspects of the chemical substance control law in Japan. In: *Aquatic Pollutants: Transformation and Biological Effects.* Hutzinger, O. et al., Editors, Pergamon, Press, Oxford, U.K. pp.283-298.

Schmidt-Bleek, F., Haberland, W., Klein, A.W., Caroli, S. (1982) Steps toward environmental hazard assessment of new chemicals (including a hazard ranking scheme, based upon directive 79/831/EEC). *Chemosphere* 11, 383-415.

Schüürmann, G., Klein, W. (1988) Advances in bioconcentration prediction. *Chemosphere* 17(8), 1551-1574.

Seidell, A. (1941) *Solubilities.* 2nd. Edition, Van Nostrand, New York.

Serrentino, R., Citti, L., Gervasi, P.G., Turchi, G. (1983) *Ital. J. Biochem.* 32, 111.

Stephenson, R.M. (1992) Mutual solubilities: Water-ketones, water-ethers, and water-gasoline-alcohols. *J. Chem. Eng. Data* 37, 80-95.

Stephenson, R.M., Malanowski, S. (1987) *Handbook of the Thermodynamics of Organic Compounds.* Elsevier Science Publishing Co., Inc., New York.

Stull, D.R. (1947) Vapor pressure of pure substances organic compounds. *Ind. Eng. Chem.* 29, 517-560.

Taft, R.W., Abraham, M.H., Famini, G.R., Doherty, R.M., Abboud, J-L M., Kamlet, M.J. (1985) Solubility properties in polymers and biological media 5: An analysis of the physicochemical properties which influence octanol-water partition coefficients of aliphatic and aromatic solutes. *J. Pharmac. Sci.* 74(8), 807-814.

THOR (1986) *Database of Medicinal Chemistry Project.* Pomona College, Claremont, California.

Tou, J.C., Westover, L.B., Sonnabend, L.F. (1974) Kinetic studies of bis(chloromethyl)ether hydrolysis by mass spectrometry. *J. Phys. Chem.* 78(11), 1096-1098.

Urano, K., Ogura, K., Wada, H. (1981) Direct analytical method for aliphatic compounds in water by steam carrier gas chromatography. *Water Res.* 15, 225-231.

USEPA (1974) *Proceedings of the Solvent Reactivity Conference.* U.S. Environmental Protection Agency, EPA-650/3-74-010. Research Triangle Park, North Carolina.

USEPA (1981) *Treatability Manual I. Treatability Data.* EPA-600/2-82-001a. U.S. Protection Agency, Washington, DC.

USEPA (1986) *EXAMS Computer Database.*

Valvani, S.C., Yalkowsky, S.H., Roseman, T.J. (1981) Solubility and partitioning IV: Aqueous solubility and octanol-water partition coefficients of liquid nonelectrolytes. *J. Pharm. Sci.* 70(5), 502-507.

Van Duuren, B.L., Katz, C., Goldschmidt, B.M., Frenkel, K., Sivak, A. (1972) Carcinogenicity of halo-ethers. II. Structure-activity relationships of analogs of bis(chloromethyl)ether. *J. Nat. Cancer Inst.* 48, 1431-1439.

Van Leeuwen, C.J., Van Der Zandt, P.T.J., Aldenberg, T., Verhaar, H.J.M., Hermens, J.L.M. (1992) Application of QSARs, extrapolation and equilibrium partitioning in aquatic effects assessment. I. Narcotic industrial pollutants. *Environ. Toxicol. & Chem.* 11(2), 267-282.

Veith, G.D., Call, D.J., Brooke, L.T. (1983) Structural-toxicity relationships for the fathead minnow, *pimephales promelas*: Narcotic industrial chemicals. *Can. J. Fish Aquat. Sci.* 40, 743-748.

Veith, G.D., Defoe, D.L., Bergstedt, B.V. (1979) Measuring and estimating the bioconcentration factor of chemicals in fish. *J. Fish Res. Board Can.* 36, 1040-1048.

Veith, G.D., Kosian, P. (1983) Estimating bioconcentration potential from octanol/water partition coefficients. In: *Physical Behavior of PCBs in the Great Lakes.* Mackay, D., Paterson, S., Eisenreich, S.J., Simmons, M.S., Editors, Ann Arbor Science Publishers, Ann Arbor, Michigan. pp.269-282.

Veith, G.D., Macek, K.J., Petrocelli, S.R., Carroll, J. (1980) An evaluation of using partition coefficients and water solubility to estimate bioconcentration factors for organic chemicals in fish. In: *Aquatic Toxicology.* pp.116-129. ASTM STP 707.

Verhaar, H.J.M., Van Leeuwen, C.J., Hermens, J.L.M. (1992) Classifying environmental pollutants. 1. Structure-activity relationships for prediction of aquatic toxicity. *Chemosphere* 25(4), 471-491.

Verschueren, K. (1977) *Handbook of Environmental Data on Organic Chemicals.* Van Nostrand Reinhold, New York.

Verschueren, K. (1983) *Handbook of Environmental Data on Organic Chemicals.* Van Nostrand Reinhold, New York.

Vesala, A. (1974) Thermodynamics of transfer nonelectrolytes from light and heavy water. I. Linear free energy correlations of free energy of transfer with solubility and heat of melting of nonelectrolyte. *Acta Chem. Scand.* 28A(8), 839-845.

Wakita, K., Yoshimoto, M., Miyamoto, S., Watanabe, H. (1986) A method for calculation of aqueous solubility of organic compounds by using new fragment solubility constants. *Chem. Pharm. Bull.* 34, 4663-4681.

Wang, L., Zhao, Y., Hong, G. (1992) Predicting aqueous solubility and octanol/water partition coefficients of organic chemicals from molar volume. *Environ. Chem.* 11, 55-70.

Wallington, T.J., Atkinson, R., Winer, A.M., Pitts, J.N., Jr. (1986) Absolute rate constants for the gas-phase reactions of the $NO_3$ radicals with $CH_3SH$, $CH_3SCH_3$, $CH_3SSCH_3$, $H_2S$, $SO_2$, and $CH_3OCH_3$ over the temperature range 280-350 K. *J. Phys. Chem.* 90, 5393-5396.

Wallington, T.J., Andino, J.M., Potts, A.R., Rudy, S.J., Siegl, W.O., Zhang, Z., Kurylo, M.J., Hule, R.E. (1993) Atmospheric chemistry of automotive fuel additives: Diisopropyl ether. *Environ. Sci. Technol.* 27(1), 98-104.

Weast, R.C. (1977) *CRC Handbook of Chemistry and Physics.* 58th Edition. CRC Press, Inc., Cleveland, Ohio.

Weast, R.C. (1982-83) *CRC Handbook of Chemistry and Physics.* 63rd. Edition, CRC Press, Boca Raton, Florida.

Webb, R.F., Duke, A.J., Smith, L.S.A. (1962) Acetals and oligoacetals. Part I. Preparation and properties of reactive oligoformals. *J. Chem. Soc.* (London), 4307-4319.

**List of Symbols and Abbreviations:**

| | |
|---|---|
| $A_i$ | area of phase i, $m^2$ |
| AI | Adsorbability Index |
| AS | Absorption Spectrometry |
| ALPM | Automated Log-P Measurement |
| BCF | bioconcentration factor |
| C | molar concentration, mol/L or $mmol/m^3$ |
| CR | Chemical Reaction |
| $C^S$ | saturated aqueous concentration, mol/L or $mmol/m^3$ |
| $C_L$ | subcooled liquid concentration, mol/L or $mmol/m^3$ |
| $C_S$ | solid molar concentration, mol/L or $mmol/m^3$ |
| $C_A$ | concentration in air phase, mol/L or $mmol/m^3$ |
| $C_W$ | concentration in water phase, mol/L or $mmol/m^3$ |
| $^{14}C$ | radioactive labelled carbon-14 compound |
| D | D values, mol/Pa·h |
| $D_A$ | D values for advection, mol/Pa·h |
| $D_{Ai}$ | D values for advective loss in phase i, mol/Pa·h |
| $D_R$ | D value for reaction, mol/Pa·h |
| $D_{Ri}$ | D value for reaction loss in phase i, mol/Pa·h |
| $D_{ij}$ | intermedia D values, mol/Pa·h |
| $D_{VW}$ | intermedia D value for air-water diffusion (absorption), mol/Pa·h |
| $D_{RW}$ | intermedia D value for air-water dissolution, mol/Pa·h |
| $D_{QW}$ | D value for total particle transport (dry and wet), mol/Pa·h |
| $D_{RS}$ | D value for rain dissolution (air-soil), mol/Pa·h |
| $D_{QS}$ | D value for wet and dry deposition (air-soil), mol/Pa·h |
| $D_{VS}$ | D value for total soil-air transport, mol/Pa·h |
| $D_S$ | D value for air-soil boundary layer diffusion, mol/Pa·h |
| $D_{SW}$ | D value for water transport in soil, mol/Pa·h |
| $D_{SA}$ | D value for air transport in soil, mol/Pa·h |
| $D_{Ti}$ | total transport D value in bulk phase i, mol/Pa·h |
| DOC | dissolved organic carbon |

| | |
|---|---|
| E | emission rate, mol/h or kg/h |
| EPICS | Equilibrium Partitioning In Closed System |
| F | Fugacity ratio |
| f | fugacity, Pa |
| $f_i$ | fugacity in pure phase i, Pa |
| f-const. | fragmental constants |
| fluo. | fluorescence method |
| G | advective inflow, mol/h |
| $G_B$ | advective inflow to bottom sediment mol/h |
| $\Delta G_v$ | Gibbs's free energy of vaporization kJ/mol or kcal/mol |
| GC | gas chromatography |
| GC/FID | GC analysis with flame ionization detector |
| GC/ECD | GC analysis with electron capture detector |
| GC-RT | GC retention time |
| gen. col. | generator-column |
| H, HLC | Henry's law constant, Pa $m^3$/mol |
| $\Delta H_{fus}$ | enthalpy of fusion, kcal/mol |
| $\Delta H_v$ | enthalpy of vaporization, kJ/mol or kcal/mol |
| HPLC | high pressure liquid chromatography |
| HPLC/UV | HPLC analysis with UV detector |
| HPLC/fluo. | HPLC analysis with fluorescence detector |
| HPLC-k' | HPLC-capacity factor correlation |
| HPLC-RI | HPLC-retention index correlation |
| HPLC-RT | HPLC-retention time correlation |
| HPLC-RV | HPLC-retention volume correlation |
| HPTLC | high pressure thin layer chromatography |
| IP | ionization potential |
| J | intermediate quantities for fugacity calculation |
| k | first-order rate constant, $h^{-1}$ ($hour^{-1}$) |
| $k_i$ | first-order rate constant in phase i, $h^{-1}$ |
| $k_A$ | air-water mass transfer coefficient, air-side, m/h |

| | |
|---|---|
| $k_W$ | air-water mass transfer coefficient, water-side, m/h |
| $K_{AW}$ | dimensionless air/water partition coefficient |
| $K_B$ | bioconcentration factor |
| $K_h$ | association coefficient |
| $K_{OC}$ | organic-carbon sorption partition coefficient |
| $K_{OM}$ | organic-matter sorption partition coefficient |
| $K_{OW}$ | octanol-water partition coefficient |
| $K_p$ | sorption coefficient |
| $k_1$ | uptake/accumulation rate constant, $d^{-1}$ ($day^{-1}$) |
| $k_2$ | elimination/clearance/depuration rate constant, $d^{-1}$ |
| $k_b$ | biodegradation rate constant, $d^{-1}$ |
| $k_h$ | hydrolysis rate constant, $d^{-1}$ |
| $k_p$ | photolysis rate constant, $d^{-1}$ |
| L | lipid content of fish |
| LSC | Liquid Scintillation Counting |
| $m_i$ | amount of chemical in phase i, mol or kg |
| M | total amount of chemical, mol or kg |
| MO | molecular orbital calculation |
| M.P. | melting point, °C |
| MR | molar refraction |
| MS | Mass Spectrometry |
| MW | molecular weight, g/mol |
| $n_C$ | number of carbon atoms |
| $n_{Cl}$ | number of chlorine atoms |
| P | vapor pressure, Pa (Pascal) |
| $P_L$ | liquid or subcooled liquid vapor pressure, Pa |
| $P_S$ | solid vapor pressure, Pa |
| $P^S$ | saturated vapor pressure, Pa |
| Q | scavenge ratio |
| QSPR | Quantitative Structure-Property Relationship |
| QSAR | Quantitative Structural-Activity Relationship |

| | |
|---|---|
| RP-HPLC | Reversed Phase High Pressure Liquid Chromatography |
| RP-TLC | Reversed Phase Thin Layer Chromatography |
| S | water solubility, mg/L or g/$m^3$ |
| $\Delta S_{fus}$ | entropy of fusion, J/mol • K or cal/mol • K (e.u.) |
| $S_{octanol}$ | solubility in octanol |
| t | residence time, h (hour) |
| $t_o$ | overall residence time, h |
| $t_A$ | advection persistence time, h |
| $t_B$ | sediment burial residence time, h |
| $t_R$ | reaction persistence time, h |
| $t_{1/2}$ | half-life, h |
| $T_{ij}$ | intermedia transport rate, mol/h or kg/h |
| T | system temperature, K |
| $T_B$ | boiling point, K |
| $T_M$ | melting point, K |
| TC | Thermal Conductivity |
| TLC | thin-layer chromatography |
| TMV | total molecular volume per molecule, $Å^3$ ($Angstrom^3$) |
| TSA | total surface area per molecule, $Å^2$ |
| $U_1$ | air side, air-water MTC (same as $k_A$), m/h |
| $U_2$ | water side, air-water MTC (same as $k_W$), m/h |
| $U_3$ | rain rate (same as $U_R$), m/h |
| $U_4$ | aerosol deposition rate, m/h |
| $U_5$ | soil-air phase diffusion MTC, m/h |
| $U_6$ | soil-water phase diffusion MTC, m/h |
| $U_7$ | soil-air boundary layer MTC, m/h |
| $U_8$ | sediment-water MTC, m/h |
| $U_9$ | sediment deposition rate, m/h |
| $U^{10}$ | sediment resuspension rate, m/h |
| $U_{11}$ | soil-water run-off rate, m/h |
| $U_{12}$ | soil-solids run-off rate, m/h |

| | |
|---|---|
| $U_R$ | rain rate, m/h |
| $U_Q$ | dry deposition velocity, m/h |
| $U_B$ | sediment burial rate, m/h |
| UV | UV spectrometry |
| UNIFAC | UNIQUAC Functional Group Activity Coefficients |
| $V_i$ | volume of pure phase i, $m^3$ |
| $V_S$ | volume of bottom sediment, $m^3$ |
| $V_{Bi}$ | volume of bulk phase i, $m^3$ |
| $V_I$ | intrinsic molar volume, $cm^3/mol$ |
| $V_M$ | molar volume, $cm^3/mol$ |
| $v_i$ | volume fraction of phase i |
| $v_Q$ | volume fraction of aerosol |
| VOC | volatile organic chemicals |
| W | molecular mass, g/mol |
| $Z_i$ | fugacity capacity of phase i, $mol/m^3$ Pa |
| $Z_{Bi}$ | fugacity capacity of bulk phase i, $mol/m^3$ Pa |

Greek characters:

| | |
|---|---|
| $\pi$-const. | substituent constants |
| $\gamma$ | solute activity coefficient |
| $\gamma_o$ | solute activity coefficient in octanol phase |
| $\gamma_W$ | solute activity coefficient in water phase |
| $\rho_i$ | density of pure phase i, $kg/m^3$ |
| $\rho_{Bi}$ | density of bulk phase i, $kg/m^3$ |
| $\chi$ | molecular connectivity indices |
| $\phi_{OC}$ | organic carbon fraction |
| $\phi_i$ | organic carbon fraction in phase i |

**APPENDICES:**

## A1. BASIC COMPUTER PROGRAM FOR FUGACITY CALCULATIONS

```
10 REM Fugacity Level I,II and III program, 6 compartments,(LEWIS)
20 REM Select condensed print
30 WIDTH "lpt1:",132
40 LPRINT CHR$(15)
50 DIM N$(9),V(9),Z(9),C(9),F(9),M(9),P(9),CG(9),CU(9),
DEN(9),ORG(9),VZ(9),DR(9),DA(9),CB(9),A(9),PA(9),PR(9),RK(9),GA(9)
60 DIM NR(9),NA(9),I(9),GD(9,9),D(9,9),N(9,9),GRA(9),TD(9,9),
HL(9),U(20),UY(20),U$(20)
70 REM N$  = six phases : air, water, soil, sediment, susp sedt and fish
80 REM V   = volume of the six phases  (m3)
90 REM DEN = density of the six phases (kg/m3)
100 REM HT  = depth of air, water, soil and sediment (m)
110 REM AR  = area of air, water, soil and sediment  (m2)
120 REM ORG = the fraction of organic carbons in sediment and susp sedt
130 REM Z   = Z values for each phase (mol/m3.Pa)
140 REM VZ  = VZ values for each phase (mol/Pa)
150 REM F   = fugacity for each phase (Pa)
160 REM C   = concentration of chemical in each phase (mol/m3)
170 REM CG  = concentration of chemical in each phase (g/m3)
180 REM CU  = concentration of chemical in each phase (ug/g)
190 REM M   = the total amount of chemical in each phase (mol)
200 REM MK  = the total amount of chemical in each phase (kg)
210 REM P   = the mole percent of chemical in each phase (%)
212 REM DR  = reaction D values
214 REM DA  = advection D values
216 REM U   = transport velocities
220 N$(1)="Air      "
230 N$(2)="Water    "
240 N$(3)="Soil     "
250 N$(4)="Sediment "
260 N$(5)="Susp sedt"
270 N$(6)="Fish     "
280 ART=100000!*1000000!:FAR(2)=.1:FAR(3)=1-FAR(2)'total area fractions
290 AR(2)=FAR(2)*ART:AR(3)=FAR(3)*ART:AR(1)=ART:AR(4)=AR(2)'areas m2
300 HT(1)=1000:HT(2)=20:HT(3)=.1:HT(4)=.01
310 V(1)=AR(1)*HT(1):V(2)=AR(2)*HT(2):V(3)=AR(3)*HT(3):V(4)=AR(4)*HT(4)
320 V(5)=.000005*V(2):V(6)=.000001*V(2)
330 REM input properties
340 PRINT "Select chemical ,user-spec =1, benzene= 2, HCB= 3, 123TCB= 4, Type 10 to
exit program"
350 INPUT QC
```

```
360 ON QC GOTO 370,510,520,530,4360
370 INPUT "Name of chemical ",CHEM$
380 INPUT "Temperature eg 25 deg C";TC
390 INPUT "Melting point temperature or data temperature if chemical is liquid eg 80 deg
C";TM
400 INPUT "Molecular mass eg 200 g/mol";WM
410 INPUT "Vapor pressure eg 2 Pa ";P
420 INPUT "Water solubility eg 50 g/m3 ";SG
430 INPUT "Log octanol water coefficient eg 4.0 ";LKOW
440 PRINT "Input overall reaction rate half-lives eg 100 h "
450 PRINT "For zero reaction enter a fictitiously long half life eg 1E11 h"
460 INPUT "Half-life in air          ";HL(1)
470 INPUT "Half-life in water        ";HL(2)
480 INPUT "Half-life in soil         ";HL(3)
490 INPUT "Half-life in sediment     ";HL(4)
500 GOTO 590
510 CHEM$="Benzene":TC=25!:WM=78.11:P=12700:SG=1780:LKOW=2.13:TM=5.53:
HL(1)=17:HL(2)=170:HL(3)=550:HL(4)=1700:GOTO590
520 CHEM$="Hexachlorobenzene(HCB)":TC=25!:WM=284.8:P=.0015:SG=.005:
LKOW=5.47:TM=230!:HL(1)=17000:HL(2)=55000!:HL(3)=55000!:HL(4)=55000!:
GOTO 590
530 CHEM$="1,2,3-Triclorobenzene,(123TCB)":TC=25!:WM=181.45:P=28:SG=21:
LKOW=4.10:TM=53:HL(1)=550:HL(2)=1700!:HL(3)=5500!:HL(4)=17000!:GOTO590
550 CHEM$="NAPHTHALENE":TC=25!:WM=128.19:P=10.4:SG=31:LKOW=3.37:
TM=80.5:HL(1)=17:HL(2)170:HL(3)=1700:HL(4)=5500:GOTO590
570 CHEM$="Trichloroethylene,(TCE)":TC=25!: WM=131.39:P=9900: SG=1100:
LKOW=2.54:TM=-73:HL(1)=170:HL(2)=550:HL(3)=1700:HL(4)=5500:GOTO590
590 MTK=100000!
600 MT=MTK*1000/WM
610 REM    Input for Fugacity Level II program
620 EK=1000
630 E=EK*1000/WM
640 PRINT "Input emission rates of chemical for Level III calculation kg/h"
650 INPUT "Emission into air        ";IK(1)
660 INPUT "Emission into water      ";IK(2)
670 INPUT "Emission into soil       ";IK(3)
680 GRA(1)=100
690 GRA(2)=1000
700 GRA(4)=50000!
710 S=SG/WM 'solubility mol/m3
720 H=P/S   'Henry's law constant Pa.m3/mol
730 KOW=10^LKOW   'Octanol-water partition coefficient
740 KOC=.41*KOW   'Organic carbon-water partition coefficient
```

```
750 KFW=.05*KOW  'Fish-water bioconcentration factor
760 TK=TC+273.15 'Temperature K
770 RG=8.314     'Gas constant

780 IF TM>TC GOTO 790 ELSE GOTO 810
790 FR=EXP(6.79*(1-(TM+273.15)/TK))
800 GOTO 820
810 FR=1
820 PL=P/FR
830 ORG(3)=.02:ORG(4)=.04:ORG(5)=.2'Organic carbon contents g/g
840 DEN(1)=.029*101325!/RG/TK:DEN(2)=1000:DEN(3)=2400'Densities kg/m3
850 DEN(4)=2400:DEN(5)=1500:DEN(6)=1000:DEN(7)=2000'Densities kg/m3
860 REM calculate Z values
870 Z(1)=1/RG/TK 'Z values
880 Z(2)=1/H
890 Z(3)=Z(2)*DEN(3)*ORG(3)*KOC/1000
900 Z(4)=Z(2)*DEN(4)*ORG(4)*KOC/1000
910 Z(5)=Z(2)*DEN(5)*ORG(5)*KOC/1000
920 Z(6)=Z(2)*DEN(6)*KFW/1000
930 K71=6000000!/PL
940 Z(7)=Z(1)*K71
950 K12=Z(1)/Z(2) 'Partition coefficients
960 K32=Z(3)/Z(2)
970 K42=Z(4)/Z(2)
980 K52=Z(5)/Z(2)
990 K62=Z(6)/Z(2)
1000 REM calculate distribution
1010 VZT=0
1020 FOR N= 1 TO 6
1030 VZ(N)=V(N)*Z(N)
1040 VZT=VZT+VZ(N)
1050 NEXT N
1060 F1=MT/VZT 'fugacity
1070 FOR N=1 TO 6
1080 F(N)=F1
1090 C(N)=F(N)*Z(N) 'concentration mol/m3
1100 M(N)=C(N)*V(N) 'amount mol
1110 MK(N)=M(N)*WM/1000
1120 P(N)=100*M(N)/MT 'percentages
1130 CG(N)=C(N)*WM  'concentration g/m3
1140 CU(N)=CG(N)*1000/DEN(N)'concentration ug/g
1150 NEXT N
1160 REM print out results
```

```
1170 LPRINT " PROGRAM 'LEWIS':SIX COMPARTMENT FUGACITY LEVEL I
CALCULATION "
1180 LPRINT " "
1190 LPRINT "Properties of "CHEM$
1200 LPRINT " "
1210 LPRINT "Temperature deg C                ";TC
1220 LPRINT "Molecular mass g/mol             ";WM
1230 LPRINT "Melting point deg C              ";TM
1240 LPRINT "Fugacity ratio                   ";FR
1250 LPRINT "Vapor pressure Pa                ";P
1260 LPRINT "Sub-cooled liquid vapor press Pa ";PL
1270 LPRINT "Solubility g/m3                  ";SG
1280 LPRINT "Solubility mol/m3                ";S
1290 LPRINT "Henry's law constant Pa.m3/mol   ";H
1300 LPRINT "Log octanol-water p-coefficient  ";LKOW
1310 LPRINT "Octanol-water partn-coefficient  ";KOW
1320 LPRINT "Organic C-water ptn-coefficient  ";KOC
1330 LPRINT "Fish-water partition coefficient ";KFW
1340 LPRINT "Air-water partition coefficient  ";K12
1350 LPRINT "Soil-water partition coefficient ";K32
1360 LPRINT "Sedt-water partition coefficient ";K42
1370 LPRINT "Susp sedt-water partn coeffnt    ";K52
1380 LPRINT "Aerosol-air partition coeff      ";K71
1390 LPRINT "Aerosol Z value                  ";Z(7)
1400 LPRINT "Aerosol density kg/m3            ";DEN(7)
1410 LPRINT " "
1420 LPRINT "Amount of chemical moles         ";MT
1430 LPRINT "Amount of chemical kilograms     ";MTK
1440 LPRINT "Fugacity Pa                      ";F1
1450 LPRINT "Total of VZ products             ";VZT
1460 LPRINT " "
1470 LPRINT "Phase properties and compositions"
1480 LPRINT " "
1490 LPRINT "Phase       "TAB(15) N$(1) TAB(30) N$(2) TAB(45) N$(3) TAB(60) N$(4)
TAB(75) N$(5) TAB(90) N$(6)
1500 LPRINT "Volume m3   "TAB(15) V(1) TAB(30) V(2) TAB(45) V(3) TAB(60) V(4)
TAB(75) V(5) TAB(90) V(6)
1510 LPRINT "Density kg/m3"TAB(15) DEN(1) TAB(30) DEN(2) TAB(45) DEN(3)
TAB(60) DEN(4) TAB(75) DEN(5) TAB(90) DEN(6)
1520 LPRINT "Depth m     "TAB(15) HT(1) TAB(30) HT(2) TAB(45) HT(3) TAB(60) HT(4)
1530 LPRINT "Area  m2    "TAB(15) AR(1) TAB(30) AR(2) TAB(45) AR(3) TAB(60)AR(4)
1540 LPRINT "Frn org carb " TAB(45) ORG(3) TAB(60) ORG(4) TAB(75) ORG(5)
1550 LPRINT "Z mol/m3.Pa "TAB(15) Z(1) TAB(30) Z(2) TAB(45) Z(3) TAB(60) Z(4)
```

```
TAB(75) Z(5) TAB(90) Z(6)
1560 LPRINT "VZ mol/Pa   "TAB(15) VZ(1) TAB(30) VZ(2) TAB(45) VZ(3) TAB(60) VZ(4
) TAB(75) VZ(5) TAB(90) VZ(6)
1570 LPRINT "Fugacity Pa "TAB(15) F(1) TAB(30) F(2) TAB(45) F(3) TAB(60) F(4)
TAB(75) F(5) TAB(90) F(6)
1580 LPRINT "Conc mol/m3 "TAB(15) C(1) TAB(30) C(2) TAB(45) C(3) TAB(60) C(4)
TAB(75) C(5) TAB(90) C(6)
1590 LPRINT "Conc g/m3   "TAB(15) CG(1) TAB(30) CG(2) TAB(45) CG(3) TAB(60)
CG(4 ) TAB(75) CG(5) TAB(90) CG(6)
1600 LPRINT "Conc ug/g   "TAB(15) CU(1) TAB(30) CU(2) TAB(45) CU(3) TAB(60)
CU(4) TAB(75) CU(5) TAB(90) CU(6)
1610 LPRINT "Amount mol  "TAB(15) M(1) TAB(30) M(2) TAB(45) M(3) TAB(60) M(4)
TAB(75) M(5) TAB(90) M(6)
1620 LPRINT "Amount kg   "TAB(15) MK(1) TAB(30) MK(2) TAB(45) MK(3) TAB(60)
MK(4) TAB(75) MK(5) TAB(90) MK(6)
1630 LPRINT "Amount %    "TAB(15) P(1) TAB(30) P(2) TAB(45) P(3) TAB(60) P(4)
TAB(75) P(5) TAB(90) P(6)
1640 LPRINT CHR$(12)
1650 REM Fugacity Level II program, 6 compartments
1660 REM calculate total inflows
1670 GA(1)=V(1)/GRA(1)
1680 GA(2)=V(2)/GRA(2)
1690 GA(4)=V(4)/GRA(4)
1700 REM calculate D values
1710 NRT=0:NRTK=0:NAT=0:NATK=0:VZT=0:MT=0:DT=0:DTA=0:DTR=0 'set
totals to zero
1720 FOR N= 1 TO 4
1730 RK(N)=.693/HL(N) 'rate constants from half-lives
1740 DR(N)=V(N)*Z(N)*RK(N):DA(N)=GA(N)*Z(N) 'reaction and advection D values
1750 DTR=DTR+DR(N):DTA=DTA+DA(N) 'total D values
1760 NEXT N
1770 DT=DTR+DTA 'total D value
1780 F2=E/DT 'fugacity
1790 FOR N=1 TO 6
1800 F(N)=F2
1810 C(N)=F(N)*Z(N) 'concentration mol/m3
1820 M(N)=C(N)*V(N) 'amount mol
1830 MK(N)=M(N)*WM/1000
1840 MT=MT+M(N) 'total amount
1850 CG(N)=C(N)*WM 'concentration g/m3
1860 CU(N)=CG(N)*1000/DEN(N) 'concentration ug/g
1870 NR(N)=V(N)*C(N)*RK(N):NRK(N)=NR(N)*WM/1000 'reaction rates mol/h and kg/h
1880 NA(N)=GA(N)*C(N):NAK(N)=NA(N)*WM/1000 'advection rates mol/h and kg/h
```

```
1890 NRT=NRT+NR(N):NAT=NAT+NA(N) 'total rates mol/h
1900 NRTK=NRTK+NRK(N):NATK=NATK+NAK(N) 'total rates kg/h
1910 NEXT N
1920 NT=NRT+NAT:NTK=NRTK+NATK
1930 MTK=MT*WM/1000
1940 FOR N=1 TO 6
1950 P(N)=100*M(N)/MT 'percentages of amount
1960 PR(N)=100*NR(N)/NT 'percentages of reaction rate
1970 PA(N)=100*NA(N)/NT 'percentages of advection rate
1980 NEXT N
1990 IF NRT=0 THEN TR=0 ELSE TR=MT/NRT
2000 IF NAT=0 THEN TA=0 ELSE TA=MT/NAT
2010 TOV=MT/NT 'overall residence time h
2020 REM print out results
2030 LPRINT
2040 LPRINT  "SIX COMPARTMENT FUGACITY LEVEL II CALCULATION  ";CHEM$
2050 LPRINT " "
2060 LPRINT "Emission rate of chemical mol/h  ";E
2070 LPRINT "Emission rate of chemical  kg/h  ";EK
2080 LPRINT "Fugacity Pa                      ";F2
2090 LPRINT "Total amount of chemical mol     ";MT
2100 LPRINT "Total amount of chemical kg      ";MTK
2110 LPRINT " "
2120 LPRINT "Phase properties,compositions and rates"
2130 LPRINT " "
2140 LPRINT "Phase       "TAB(15) N$(1) TAB(30) N$(2) TAB(45) N$(3) TAB(60) N$(4)
TAB(75) N$(5) TAB(90) N$(6)
2150 LPRINT "Adv.flow m3/h"TAB(15) GA(1) TAB(30) GA(2) TAB(45) GA(3) TAB(60)
GA(4)
2160 LPRINT "Adv.restime h"TAB(15) GRA(1) TAB(30) GRA(2) TAB(45) GRA(3) TAB(60)
GRA(4)
2170 LPRINT "Rct halflife h"TAB(15) HL(1) TAB(30) HL(2) TAB(45) HL(3) TAB(60)
HL(4)
2180 LPRINT "Rct rate c.h-1"TAB(15) RK(1) TAB(30) RK(2) TAB(45) RK(3) TAB(60)
RK(4)
2190 LPRINT "Fugacity Pa "TAB(15) F(1) TAB(30) F(2) TAB(45) F(3) TAB(60) F(4)
TAB(75) F(5) TAB(90) F(6)
2200 LPRINT "Conc mol/m3 "TAB(15) C(1) TAB(30) C(2) TAB(45) C(3) TAB(60) C(4)
TAB(75) C(5) TAB(90) C(6)
2210 LPRINT "Conc g/m3   "TAB(15) CG(1) TAB(30) CG(2) TAB(45) CG(3) TAB(60)
CG(4 ) TAB(75) CG(5) TAB(90) CG(6)
2220 LPRINT "Conc ug/g   "TAB(15) CU(1) TAB(30) CU(2) TAB(45) CU(3) TAB(60)
CU(4) TAB(75) CU(5) TAB(90) CU(6)
```

```
2230 LPRINT "Amount mol  "TAB(15) M(1) TAB(30) M(2) TAB(45) M(3) TAB(60) M(4)
TAB(75) M(5) TAB(90) M(6)
2240 LPRINT "Amount kg   "TAB(15) MK(1) TAB(30) MK(2) TAB(45) MK(3) TAB(60)
MK(4) TAB(75) MK(5) TAB(90) MK(6)
2250 LPRINT "Amount %    "TAB(15) P(1) TAB(30) P(2) TAB(45) P(3) TAB(60) P(4)
TAB(75) P(5) TAB(90) P(6)
2260 LPRINT "D rct mol/Pa.h"TAB(15) DR(1) TAB(30) DR(2) TAB(45) DR(3) TAB(60)
DR(4)
2270 LPRINT "D adv mol/Pa.h"TAB(15) DA(1) TAB(30) DA(2) TAB(45) DA(3) TAB(60)
DA(4)
2280 LPRINT "Rct rate mol/h"TAB(15) NR(1) TAB(30) NR(2) TAB(45) NR(3) TAB(60)
NR(4)
2290 LPRINT "Adv rate mol/h"TAB(15) NA(1) TAB(30) NA(2) TAB(45) NA(3) TAB(60)
NA(4)
2300 LPRINT "Rct rate kg/h "TAB(15) NRK(1) TAB(30) NRK(2) TAB(45) NRK(3) TAB(60)
NRK(4)
2310 LPRINT "Adv rate kg/h "TAB(15) NAK(1) TAB(30) NAK(2) TAB(45) NAK(3)
TAB(60) NAK(4)
2320 LPRINT "Reaction %    "TAB(15) PR(1) TAB(30) PR(2) TAB(45) PR(3) TAB(60)
PR(4)
2330 LPRINT "Advection %   "TAB(15) PA(1) TAB(30) PA(2) TAB(45) PA(3) TAB(60)
PA(4)
2340 LPRINT " "
2350 LPRINT "Total advection D value     ";DTA
2360 LPRINT "Total reaction D value      ";DTR
2370 LPRINT "Total D value               ";DT
2380 LPRINT " "
2390 LPRINT "Output by reaction   mol/h ";NRT
2400 LPRINT "Output by advection  mol/h ";NAT
2410 LPRINT "Total output by reaction and advection mol/h ";NT
2420 LPRINT" "
2430 LPRINT "Output by reaction   kg/h  ";NRTK
2440 LPRINT "Output by advection  kg/h  ";NATK
2450 LPRINT "Total output by reaction and advection kg/h  ";NTK
2460 LPRINT" "
2470 LPRINT "Overall residence time   h  ";TOV
2480 LPRINT "Reaction residence time  h  ";TR
2490 LPRINT "Advection residence time h  ";TA
2500 LPRINT CHR$(12)
2510 LPRINT
2520 REM Fugacity Level III Program
2530 REM Set bulk phase volumes, densities and Z values
2540 VB(1)=V(1):VB(2)=V(2):VB(3)=1.8E+10:VB(4)=5E+08
```

```
2550 VA(1)=1:VQ(1)=2E-11 'volume fractions
2560 VW(2)=1:VP(2)=.000005:VF(2)=.000001
2570 VA(3)=.2:VW(3)=.3:VE(3)=.5
2580 VW(4)=.8:VS(4)=.2
2590 DENB(1)=VA(1)*DEN(1)+VQ(1)*DEN(7)
2600 DENB(2)=VW(2)*DEN(2)+VP(2)*DEN(5)+VF(2)*DEN(6)
2610 DENB(3)=VW(3)*DEN(2)+VA(3)*DEN(1)+VE(3)*DEN(3)
2620 DENB(4)=VW(4)*DEN(2)+VS(4)*DEN(4)
2630 ZB(1)=VA(1)*Z(1)+VQ(1)*Z(7)
2640 ZB(2)=VW(2)*Z(2)+VP(2)*Z(5)+VF(2)*Z(6)
2650 ZB(3)=VA(3)*Z(1)+VW(3)*Z(2)+VE(3)*Z(3)
2660 ZB(4)=VW(4)*Z(2)+VS(4)*Z(4)
2670 KB12=ZB(1)/ZB(2)
2680 KB32=ZB(3)/ZB(2)
2690 KB42=ZB(4)/ZB(2)
2700 REM Parameters
2710 U(1)=5          :U$(1)="air side air-water MTC               "
2720 U(2)=.05        :U$(2)="water side air-water MTC             "
2730 U(3)=.0001      :U$(3)="rain rate                            "
2740 U(4)=6E-10      :U$(4)="aerosol deposition velocity          "
2750 U(5)=.02        :U$(5)="soil air phase diffusion MTC         "
2760 U(6)=.00001     :U$(6)="soil water phase diffusion MTC       "
2770 U(7)=5          :U$(7)="soil air boundary layer MTC          "
2780 U(8)=.0001      :U$(8)="sediment-water diffusion MTC         "
2790 U(9)=.0000005 :U$(9)="sediment deposition velocity         "
2800 U(10)=.0000002:U$(10)="sediment resuspension velocity      "
2810 U(11)=.00005  :U$(11)="soil water runoff rate              "
2820 U(12)=1E-08   :U$(12)="soil solids runoff rate             "
2830 'Calculate D values
2840 DRW=AR(2)*U(3)*Z(2)
2850 DQW=AR(2)*U(4)*Z(7)
2860 DVWA=AR(2)*U(1)*Z(1)
2870 DVWW=AR(2)*U(2)*Z(2)
2880 DVW=1/(1/DVWA+1/DVWW)
2890 D(2,1)=DVW
2900 D(1,2)=DVW+DQW+DRW
2910 DVSB=AR(3)*U(7)*Z(1)
2920 DVSA=AR(3)*U(5)*Z(1)
2930 DVSW=AR(3)*U(6)*Z(2)
2940 DRS=AR(3)*U(3)*Z(2)
2950 DQS=AR(3)*U(4)*Z(7)
2960 DVS=1/(1/DVSB+1/(DVSW+DVSA))
2970 D(3,1)=DVS
```

```
2980 D(1,3)=DVS+DRS+DQS
2990 DSWD=AR(2)*U(8)*Z(2)
3000 DSD=AR(2)*U(9)*Z(5)
3010 DSR=AR(4)*U(10)*Z(4)
3020 D(2,4)=DSWD+DSD
3030 D(4,2)=DSWD+DSR
3040 DSWW=AR(3)*U(11)*Z(2)
3050 DSWS=AR(3)*U(12)*Z(3)
3060 D(3,2)=DSWW+DSWS
3070 D(2,3)=0
3080 REM calculate total chemical inflows
3090 IN=0:INK=0
3100 FOR N=1 TO 4
3110 I(N)=IK(N)*1000/WM
3120 IN=IN+I(N):INK=INK+IK(N)
3130 NEXT N
3140 REM calculate reaction and advection D values for bulk phases
3150 GAB(1)=VB(1)/GRA(1)
3160 GAB(2)=VB(2)/GRA(2)
3170 GAB(4)=VB(4)/GRA(4)
3180 VZBT=0
3190 FOR N= 1 TO 4
3200 RK(N)=.693/HL(N)
3210 DR(N)=VB(N)*ZB(N)*RK(N):DA(N)=GAB(N)*ZB(N)
3220 VZB(N)=VB(N)*ZB(N)
3230 VZBT=VZBT+VZB(N)
3240 NEXT N
3250 FOR N=1 TO 4
3260 FOR NN=1 TO 4
3270 GD(N,NN)=D(N,NN)/ZB(N)
3280 IF GD(N,NN)=0 GOTO 3300 ELSE GOTO 3290
3290 TD(N,NN)=.693*VB(N)/GD(N,NN)
3300 NEXT NN
3310 NEXT N
3320 DT(1)=DR(1)+DA(1)+D(1,2)+D(1,3)
3330 DT(2)=DR(2)+DA(2)+D(2,1)+D(2,3)+D(2,4)
3340 DT(3)=DR(3)+DA(3)+D(3,1)+D(3,2)
3350 DT(4)=DR(4)+DA(4)+D(4,2)
3360 J1=I(1)/DT(1)+I(3)*D(3,1)/DT(3)/DT(1)
3370 J2=D(2,1)/DT(1)
3380 J3=1-D(3,1)*D(1,3)/DT(1)/DT(3)
3390 J4=D(1,2)+D(3,2)*D(1,3)/DT(3)
3400 F(2)=(I(2)+J1*J4/J3+I(3)*D(3,2)/DT(3)+I(4)*D(4,2)/DT(4))/(DT(2)-J2*J4/J3
```

```
-D(2,4)*D(4,2)/DT(4))
3410 F(1)=(J1+F(2)*J2)/J3
3420 F(3)=(I(3)+F(1)*D(1,3))/DT(3)
3430 F(4)= (I(4)+F(2)*D(2,4))/DT(4)
3440 NRT=0:NAT=0:MT=0
3450 FOR N=1 TO 4
3460 C(N)=F(N)*ZB(N)
3470 M(N)=C(N)*VB(N)
3480 MK(N)=M(N)*WM/1000
3490 MT=MT+M(N)
3500 CG(N)=C(N)*WM
3510 CU(N)=CG(N)*1000/DENB(N)
3520 NR(N)=F(N)*DR(N):NRK(N)=NR(N)*WM/1000
3530 NA(N)=F(N)*DA(N):NAK(N)=NA(N)*WM/1000
3540 NRT=NRT+NR(N):NAT=NAT+NA(N)
3550 NEXT N
3560 MTK=MT*WM/1000
3570 NRTK=NRT*WM/1000:NATK=NAT*WM/1000
3580 NT=NRT+NAT:NTK=NT*WM/1000
3590 FOR N=1 TO 4
3600 P(N)=100*M(N)/MT
3610 PR(N)=100*NR(N)/NT
3620 PA(N)=100*NA(N)/NT
3630 NEXT N
3640 N(1,2)=D(1,2)*F(1):NK(1,2)=N(1,2)*WM/1000
3650 N(1,3)=D(1,3)*F(1):NK(1,3)=N(1,3)*WM/1000
3660 N(2,1)=D(2,1)*F(2):NK(2,1)=N(2,1)*WM/1000
3670 N(2,4)=D(2,4)*F(2):NK(2,4)=N(2,4)*WM/1000
3680 N(3,1)=D(3,1)*F(3):NK(3,1)=N(3,1)*WM/1000
3690 N(3,2)=D(3,2)*F(3):NK(3,2)=N(3,2)*WM/1000
3700 N(4,2)=D(4,2)*F(4):NK(4,2)=N(4,2)*WM/1000
3710 TR=MT/(NRT+.0000001)
3720 TA=MT/(NAT+.0000001)
3730 TOV=MT/NT
3740 TOVD=TOV/24
3750 REM print out results
3760 LPRINT " FOUR COMPARTMENT FUGACITY LEVEL III
CALCULATION",CHEM$
3770 LPRINT
3780 LPRINT "Bulk phase properties,compositions and rates"
3790 LPRINT " "
3800 LPRINT "Phase        "TAB(15) N$(1) TAB(30) N$(2) TAB(45) N$(3) TAB(60) N$(4)
TAB(75) "Total"
```

```
3810 LPRINT "Bulk vol m3 "TAB(15) VB(1) TAB(30) VB(2) TAB(45) VB(3) TAB(60)
VB(4)
3820 LPRINT "Density kg/m3"TAB(15) DENB(1) TAB(30) DENB(2) TAB(45) DENB(3)
TAB(60) DENB(4)
3830 LPRINT "Bulk Z value"TAB(15) ZB(1) TAB(30) ZB(2) TAB(45) ZB(3) TAB(60) ZB(4
)
3840 LPRINT "Bulk VZ     "TAB(15) VZB(1) TAB(30) VZB(2) TAB(45) VZB(3) TAB(60)
VZB(4) TAB(75) VZBT
3850 LPRINT "Emission mol/h"TAB(15) I(1) TAB(30) I(2) TAB(45) I(3) TAB(60) I(4)
TAB(75) IN
3860 LPRINT "Emission kg/h "TAB(15) IK(1) TAB(30) IK(2) TAB(45) IK(3) TAB(60) IK(4)
TAB(75) INK
3870 LPRINT "Fugacity Pa "TAB(15) F(1) TAB(30) F(2) TAB(45) F(3) TAB(60) F(4)
3880 LPRINT "Conc mol/m3 "TAB(15) C(1) TAB(30) C(2) TAB(45) C(3) TAB(60) C(4)
3890 LPRINT "Conc g/m3   "TAB(15) CG(1) TAB(30) CG(2) TAB(45) CG(3) TAB(60)
CG(4 )
3900 LPRINT "Conc ug/g   "TAB(15) CU(1) TAB(30) CU(2) TAB(45) CU(3) TAB(60)
CU(4)
3910 LPRINT "Amount mol  "TAB(15) M(1) TAB(30) M(2) TAB(45) M(3) TAB(60) M(4)
TAB(75) MT
3920 LPRINT "Amount kg   "TAB(15) MK(1) TAB(30) MK(2) TAB(45) MK(3) TAB(60)
MK(4) TAB(75) MTK
3930 LPRINT "Amount %    "TAB(15) P(1) TAB(30) P(2) TAB(45) P(3) TAB(60) P(4)
3940 LPRINT "Adv.flow m3/h"TAB(15) GAB(1) TAB(30) GAB(2) TAB(45) GAB(3)
TAB(60) GAB(4)
3950 LPRINT "D rct mol/Pa.h"TAB(15) DR(1) TAB(30) DR(2) TAB(45) DR(3) TAB(60)
DR(4 )
3960 LPRINT "D adv mol/Pa.h"TAB(15) DA(1) TAB(30) DA(2) TAB(45) DA(3) TAB(60)
DA(4 )
3970 LPRINT "Rct rate mol/h"TAB(15) NR(1) TAB(30) NR(2) TAB(45) NR(3) TAB(60)
NR(4 ) TAB(75) NRT
3980 LPRINT "Rct rate kg/h "TAB(15) NRK(1) TAB(30) NRK(2) TAB(45) NRK(3) TAB(60)
NRK(4 ) TAB(75) NRTK
3990 LPRINT "Adv rate mol/h"TAB(15) NA(1) TAB(30) NA(2) TAB(45) NA(3) TAB(60)
NA(4 ) TAB(75) NAT
4000 LPRINT "Adv rate kg/h "TAB(15) NAK(1) TAB(30) NAK(2) TAB(45) NAK(3)
TAB(60) NAK(4 ) TAB(75) NATK
4010 LPRINT "Reaction %    "TAB(15) PR(1) TAB(30) PR(2) TAB(45) PR(3) TAB(60) PR(4
)
4020 LPRINT "Advection %   "TAB(15) PA(1) TAB(30) PA(2) TAB(45) PA(3) TAB(60)
PA(4 )
4030 LPRINT " "
4040 LPRINT "Overall residence time   h  ";TOV
```

```
4050 LPRINT "Reaction residence time  h  ";TR;
4060 LPRINT "    Advection residence time h  ";TA
4070 LPRINT
4080 LPRINT "Intermedia Data.  Half times   Equiv flows   D values   Rates  of transport "
4090 LPRINT "                              h           m3/h          mol/Pa.h    mol/h       kg/h"
4100 LPRINT "Air to water     ";:LPRINT USING " ##.####^^^^ ";TD(1,2);GD(1,2);D(1,2)
;N(1,2);NK(1,2)
4110 LPRINT "Air to soil      ";:LPRINT USING " ##.####^^^^ ";TD(1,3);GD(1,3);D(1,3)
;N(1,3);NK(1,3)
4120 LPRINT "Water to air     ";:LPRINT USING " ##.####^^^^ ";TD(2,1);GD(2,1);D(2,1)
;N(2,1);NK(2,1)
4130 LPRINT "Water to sediment";:LPRINT USING " ##.####^^^^
";TD(2,4);GD(2,4);D(2,4) ;N(2,4);NK(2,4)
4140 LPRINT "Soil to air      ";:LPRINT USING " ##.####^^^^ ";TD(3,1);GD(3,1),D(3,1)
,N(3,1);NK(3,1)
4150 LPRINT "Soil to water    ";:LPRINT USING " ##.####^^^^ ";TD(3,2);GD(3,2);D(3,2)
;N(3,2);NK(3,2)
4160 LPRINT "Sediment to water";:LPRINT USING " ##.####^^^^
";TD(4,2);GD(4,2),D(4,2) ,N(4,2);NK(4,2)
4170 LPRINT "   Transport velocity parameters              m/h              m/year  "
4180 FOR I=1 TO 12
4190 UY(I)=U(I)*8760
4200 LPRINT TAB(5) I TAB(10) U$(I) TAB(45) U(I) TAB(60) UY(I)
4210 NEXT I
4220 LPRINT "Individual process D values "
4230 LPRINT "Air-water diffusion (air-side)        ";DVWA TAB(50);
4240 LPRINT "Air-water diffusion (water-side)      ";DVWW
4250 LPRINT "Air-water diffusion (overall)         ";DVW
4260 LPRINT "Rain dissolution to water             ";DRW  TAB(50);
4270 LPRINT "Aerosol deposition to water           ";DQW
4280 LPRINT "Rain dissolution to soil              ";DRS  TAB(50);
4290 LPRINT "Aerosol deposition to soil            ";DQS
4300 LPRINT "Soil-air diffusion (air-phase)        ";DVSA TAB(50);
4310 LPRINT "Soil-air diffusion (water-phase)      ";DVSW
4320 LPRINT "Soil-air diffusion (bndry layer)      ";DVSB TAB(50);
4330 LPRINT "Soil-air diffusion (overall)          ";DVS
4340 LPRINT "Water-sediment diffusion              ";DSWD
4350 LPRINT "Water-sediment deposition             ";DSD TAB(50);
4360 LPRINT "Sediment-water resuspension           ";DSR
4370 LPRINT "Soil-water runoff (water)             ";DSWW TAB(50);
4380 LPRINT "Soil-water runoff (solids)            ";DSWS
```

PROGRAM 'GENERIC':SIX COMPARTMENT FUGACITY LEVEL I CALCULATION

Properties of Trichloroethylene

| | |
|---|---|
| Temperature deg C | 25 |
| Molecular mass g/mol | 131.39 |
| Melting point deg C | -73 |
| Fugacity ratio | 1 |
| Vapor pressure Pa | 9900 |
| Sub-cooled liquid vapor press Pa | 9900 |
| Solubility g/m3 | 1100 |
| Solubility mol/m3 | 8.372023 |
| Henry's law constant Pa.m3/mol | 1182.51 |
| Log octanol-water p-coefficient | 2.53 |
| Octanol-water partn-coefficient | 338.8442 |
| Organic C-water ptn-coefficient | 138.9261 |
| Fish-water partition coefficient | 16.94221 |
| Air-water partition coefficient | .4770457 |
| Soil-water partition coefficient | 6.668453 |
| Sedt-water partition coefficient | 13.33691 |
| Susp sedt-water partn coeffnt | 41.67783 |
| Aerosol-air partition coeff | 606.0606 |
| Aerosol Z value | .2444957 |
| Aerosol density kg/m3 | 2000 |

| | |
|---|---|
| Amount of chemical moles | 761093 |
| Amount of chemical kilograms | 100000 |
| Fugacity Pa | 1.876122E-05 |
| Total of VZ products | 4.056736E+10 |

Phase properties and compositions

| Phase | Air | Water | Soil | Sediment | Susp sedt | Fish |
|---|---|---|---|---|---|---|
| Volume m3 | 1E+14 | 2E+11 | 8.999999E+09 | 5E+08 | 999999.9 | 200000 |
| Density kg/m3 | 1.185413 | 1000 | 2400 | 2400 | 1500 | 1000 |
| Depth m | 1000 | 20 | .1 | .05 | | |
| Area m2 | 1E+11 | 1E+10 | 9E+10 | 1E+10 | | |
| Frn org carb | | | .02 | .04 | .2 | |
| Z mol/m3.Pa | 4.034179E-04 | 8.456589E-04 | 5.639236E-03 | 1.127847E-02 | 3.524523E-02 | 1.432733E-02 |
| VZ mol/Pa | 4.034179E+10 | 1.691318E+08 | 5.075312E+07 | 5639236 | 35245.23 | 2865.466 |
| Fugacity Pa | 1.876122E-05 | 1.876122E-05 | 1.876122E-05 | 1.876122E-05 | 1.876122E-05 | 1.876122E-05 |
| Conc mol/m3 | 7.56861E-09 | 1.586559E-08 | 1.057989E-07 | 2.115978E-07 | 6.612433E-07 | 2.687981E-07 |
| Conc g/m3 | 9.944396E-07 | 2.08458E-06 | 1.390092E-05 | 2.780184E-05 | 8.688076E-05 | 3.531738E-05 |
| Conc ug/g | 8.38897E-04 | 2.08458E-06 | 5.79205E-06 | 1.15841E-05 | 5.792051E-05 | 3.531739E-05 |
| Amount mol | 756861 | 3173.118 | 952.1902 | 105.7989 | .6612433 | 5.375962E-02 |
| Amount kg | 99443.97 | 416.9159 | 125.1083 | 13.90092 | 8.688075E-02 | 7.063477E-03 |
| Amount % | 99.44396 | .4169159 | .1251083 | 1.390092E-02 | 8.688076E-05 | 7.063476E-06 |

SIX COMPARTMENT FUGACITY LEVEL II CALCULATION Trichloroethylene

| | |
|---|---|
| Emission rate of chemical mol/h | 7610.929 |
| Emission rate of chemical kg/h | 1000 |
| Fugacity Pa | 1.339307E-05 |
| Total amount of chemical mol | 543321.3 |
| Total amount of chemical kg | 71386.99 |

Phase properties,compositions and rates

| Phase | Air | Water | Soil | Sediment | Susp sedt | Fish |
|---|---|---|---|---|---|---|
| Adv.flow m3/h | 1E+12 | 2E+08 | 0 | 10000 | | |
| Adv.restime h | 100 | 1000 | 0 | 50000 | | |
| Rct halflife h | 170 | 550 | 1700 | 5500 | | |
| Rct rate c.h-1 | 4.076471E-03 | .00126 | 4.076471E-04 | .000126 | | |
| Fugacity Pa | 1.339307E-05 | 1.339307E-05 | 1.339307E-05 | 1.339307E-05 | 1.339307E-05 | 1.339307E-05 |
| Conc mol/m3 | 5.403003E-09 | 1.132597E-08 | 7.552666E-08 | 1.510533E-07 | 4.720417E-07 | 1.918869E-07 |
| Conc g/m3 | 7.099005E-07 | 1.488119E-06 | 9.923449E-06 | 1.98469E-05 | 6.202156E-05 | 2.521202E-05 |
| Conc ug/g | 5.988634E-04 | 1.488119E-06 | 4.13477E-06 | 8.269541E-06 | 4.134771E-05 | 2.521202E-05 |
| Amount mol | 540300.3 | 2265.193 | 679.74 | 75.52666 | .4720417 | 3.837738E-02 |
| Amount kg | 70990.06 | 297.6237 | 89.31102 | 9.923449 | 6.202156E-02 | 5.042403E-03 |
| Amount % | 99.44398 | .4169159 | .1251083 | 1.390092E-02 | 8.688076E-05 | 7.063477E-06 |
| D rct mol/Pa.h | 1.644521E+08 | 213106.1 | 20689.36 | 710.5437 | | |
| D adv mol/Pa.h | 4.034179E+08 | 169131.8 | 0 | 112.7847 | | |
| Rct rate mol/h | 2202.518 | 2.854144 | .277094 | 9.516359E-03 | | |
| Adv rate mol/h | 5403.003 | 2.265193 | 0 | 1.510533E-03 | | |
| Rct rate kg/h | 289.3889 | .3750059 | 3.640738E-02 | 1.250355E-03 | | |
| Adv rate kg/h | 709.9006 | .2976237 | 0 | 1.98469E-04 | | |
| Reaction % | 28.93889 | .0375006 | 3.640738E-03 | 1.250355E-04 | | |
| Advection % | 70.99007 | 2.976238E-02 | 0 | 1.98469E-05 | | |

| | |
|---|---|
| Total advection D value | 4.035871E+08 |
| Total reaction D value | 1.646866E+08 |
| Total D value | 5.682738E+08 |

Output by reaction mol/h 2205.659
Output by advection mol/h 5405.27
Total output by reaction and advection mol/h 7610.929

Output by reaction kg/h 289.8016
Output by advection kg/h 710.1984
Total output by reaction and advection kg/h 999.9999

Overall residence time h 71.387
Reaction residence time h 246.3306
Advection residence time h 100.517

FOUR COMPARTMENT FUGACITY LEVEL III CALCULATION        Trichloroethylene

Bulk phase properties,compositions and rates

| Phase | Air | Water | Soil | Sediment | Total | Fish |
|---|---|---|---|---|---|---|
| Bulk vol m3 | 1E+14 | 2E+11 | 1.8E+10 | 5E+08 | | |
| Density kg/m3 | 1.185413 | 1000.009 | 1500.237 | 1280 | | |
| Bulk Z value | 4.034179E-04 | 8.458495E-04 | 3.153999E-03 | 2.932222E-03 | | 1.432733E-02 |
| Bulk VZ | 4.034179E+10 | 1.691699E+08 | 5.677199E+07 | 1466111 | 4.05692E+10 | |
| Emission mol/h | 7610.929 | 0 | 0 | 0 | 7610.929 | |
| Emission kg/h | 1000 | 0 | 0 | 0 | 1000 | |
| Fugacity Pa | 1.339732E-05 | 7.04122E-06 | 1.305248E-05 | 6.64837E-06 | | 7.04122E-06 |
| Conc mol/m3 | 5.404717E-09 | 5.955813E-09 | 4.116751E-08 | 1.949449E-08 | | 1.008819E-07 |
| Conc g/m3 | 7.101258E-07 | 7.825342E-07 | 5.408999E-06 | 2.561382E-06 | | 1.325487E-05 |
| Conc ug/g | 5.990534E-04 | 7.825276E-07 | 3.60543E-06 | 2.001079E-06 | | 1.325487E-05 |
| Amount mol | 540471.7 | 1191.163 | 741.0152 | 9.747247 | 542413.6 | |
| Amount kg | 71012.58 | 156.5069 | 97.36198 | 1.280691 | 71267.73 | |
| Amount % | 99.64198 | .2196041 | .1366144 | 1.797014E-03 | | |
| Adv.flow m3/h | 1E+12 | 2E+08 | 0 | 10000 | | |
| D rct mol/Pa.h | 1.644521E+08 | 213154.1 | 23142.94 | 184.73 | | |
| D adv mol/Pa.h | 4.034179E+08 | 169169.9 | 0 | 29.32222 | | |
| Rct rate mol/h | 2203.217 | 1.500865 | .3020727 | 1.228153E-03 | 2205.022 | |
| Rct rate kg/h | 289.4807 | .1971986 | 3.968933E-02 | 1.61367E-04 | 289.7178 | |
| Adv rate mol/h | 5404.717 | 1.191163 | 0 | 1.949449E-04 | 5405.908 | |
| Adv rate kg/h | 710.1258 | .1565068 | 0 | 2.561381E-05 | 710.2823 | |
| Reaction % | 28.94807 | 1.971986E-02 | 3.968933E-03 | 1.61367E-05 | | |
| Advection % | 71.01258 | 1.565068E-02 | 0 | 2.561381E-06 | | |

Overall residence time h 71.26773
Reaction residence time h 245.9902    Advection residence time h 100.3372

| Intermedia Data. | Half times h | Equiv flows m3/h | D values mol/Pa.h | Rates of transport mol/h | kg/h |
|---|---|---|---|---|---|
| Air to water | 6.7367E+04 | 1.0287E+09 | 4.1500E+05 | 5.5598E+00 | 7.3050E-01 |
| Air to soil | 3.8211E+04 | 1.8136E+09 | 7.3164E+05 | 9.8020E+00 | 1.2879E+00 |
| Water to air | 2.8307E+02 | 4.8962E+08 | 4.1415E+05 | 2.9161E+00 | 3.8315E-01 |
| Water to sediment | 1.1472E+05 | 1.2081E+06 | 1.0219E+03 | 7.1953E-03 | 9.4539E-04 |
| Soil to air | 5.4340E+01 | 2.2955E+08 | 7.2401E+05 | 9.4502E+00 | 1.2417E+00 |
| Soil to water | 1.0325E+04 | 1.2082E+06 | 3.8105E+03 | 4.9737E-02 | 6.5349E-03 |
| Sediment to water | 1.1702E+03 | 2.9609E+05 | 8.6822E+02 | 5.7722E-03 | 7.5841E-04 |

| | Transport velocity parameters | m/h | m/year |
|---|---|---|---|
| 1 | air side air-water MTC | 5 | 43800 |
| 2 | water side air-water MTC | .05 | 438 |
| 3 | rain rate | .0001 | .876 |
| 4 | aerosol deposition velocity | 6E-10 | 5.256E-06 |
| 5 | soil air phase diffusion MTC | .02 | 175.2 |
| 6 | soil water phase diffusion MTC | .00001 | .0876 |
| 7 | soil air boundary layer MTC | 5 | 43800 |
| 8 | sediment-water diffusion MTC | .0001 | .876 |
| 9 | sediment deposition velocity | .0000005 | .00438 |
| 10 | sediment resuspension velocity | .0000002 | .001752 |
| 11 | soil water runoff rate | .00005 | .438 |
| 12 | soil solids runoff rate | 1E-08 | .0000876 |

Individual process D values

| | | | |
|---|---|---|---|
| Air-water diffusion (air-side) | 2.017089E+07 | Air-water diffusion (water-side) | 422829.5 |
| Air-water diffusion (overall) | 414148 | | |
| Rain dissolution to water | 845.6589 | Aerosol deposition to water | 1.466974 |
| Rain dissolution to soil | 7610.93 | Aerosol deposition to soil | 13.20277 |
| Soil-air diffusion (air-phase) | 726152.2 | Soil-air diffusion (water-phase) | 761.093 |
| Soil-air diffusion (bndry layer) | 1.815381E+08 | Soil-air diffusion (overall) | 724014.2 |
| Water-sediment diffusion | 845.6589 | | |
| Water-sediment deposition | 176.2261 | Sediment-water resuspension | 22.55695 |
| Soil-water runoff (water) | 3805.465 | Soil-water runoff (solids) | 5.075312 |

FOUR COMPARTMENT FUGACITY LEVEL III CALCULATION        Trichloroethylene

Bulk phase properties,compositions and rates

| Phase | Air | Water | Soil | Sediment | Total | Fish |
|---|---|---|---|---|---|---|
| Bulk vol m3 | 1E+14 | 2E+11 | 1.8E+10 | 5E+08 | | |
| Density kg/m3 | 1.185413 | 1000.009 | 1500.237 | 1280 | | |
| Bulk Z value | 4.034179E-04 | 8.458495E-04 | 3.153999E-03 | 2.932222E-03 | | 1.432733E-02 |
| Bulk VZ | 4.034179E+10 | 1.691699E+08 | 5.677199E+07 | 1466111 | 4.05692E+10 | |
| Emission mol/h | 0 | 7610.929 | 0 | 0 | 7610.929 | |
| Emission kg/h | 0 | 1000 | 0 | 0 | 1000 | |
| Fugacity Pa | 6.964544E-06 | 9.557041E-03 | 6.785282E-06 | 9.023825E-03 | | 9.557041E-03 |
| Conc mol/m3 | 2.809622E-09 | 8.083818E-06 | 2.140077E-08 | 2.645985E-05 | | 1.369269E-04 |
| Conc g/m3 | 3.691562E-07 | 1.062133E-03 | 2.811848E-06 | 3.47656E-03 | | 1.799082E-02 |
| Conc ug/g | 3.114156E-04 | 1.062124E-03 | 1.874269E-06 | 2.716063E-03 | | 1.799082E-02 |
| Amount mol | 280962.2 | 1616763 | 385.2139 | 13229.93 | 1911341 | |
| Amount kg | 36915.62 | 212426.6 | 50.61326 | 1738.28 | 251131.1 | |
| Amount % | 14.69974 | 84.58792 | 2.015412E-02 | .6921805 | | |
| Adv.flow m3/h | 1E+12 | 2E+08 | 0 | 10000 | | |
| D rct mol/Pa.h | 1.644521E+08 | 213154.1 | 23142.94 | 184.73 | | |
| D adv mol/Pa.h | 4.034179E+08 | 169169.9 | 0 | 29.32222 | | |
| Rct rate mol/h | 1145.334 | 2037.122 | .1570313 | 1.666971 | 3184.28 | |
| Rct rate kg/h | 150.4855 | 267.6575 | 2.063235E-02 | .2190233 | 418.3826 | |
| Adv rate mol/h | 2809.622 | 1616.763 | 0 | .2645986 | 4426.65 | |
| Adv rate kg/h | 369.1562 | 212.4266 | 0 | .0347656 | 581.6175 | |
| Reaction % | 15.04854 | 26.76574 | 2.063235E-03 | 2.190233E-02 | | |
| Advection % | 36.91562 | 21.24265 | 0 | 3.47656E-03 | | |

Overall residence time  h  251.131
Reaction residence time  h  600.2426    Advection residence time h  431.7805

| Intermedia Data. | Half times h | Equiv flows m3/h | D values mol/Pa.h | Rates of transport mol/h | kg/h |
|---|---|---|---|---|---|
| Air to water | 6.7367E+04 | 1.0287E+09 | 4.1500E+05 | 2.8903E+00 | 3.7975E-01 |
| Air to soil | 3.8211E+04 | 1.8136E+09 | 7.3164E+05 | 5.0955E+00 | 6.6950E-01 |
| Water to air | 2.8307E+02 | 4.8962E+08 | 4.1415E+05 | 3.9580E+03 | 5.2005E+02 |
| Water to sediment | 1.1472E+05 | 1.2081E+06 | 1.0219E+03 | 9.7662E+00 | 1.2832E+00 |
| Soil to air | 5.4340E+01 | 2.2955E+08 | 7.2401E+05 | 4.9126E+00 | 6.4547E-01 |
| Soil to water | 1.0325E+04 | 1.2082E+06 | 3.8105E+03 | 2.5856E-02 | 3.3972E-03 |
| Sediment to water | 1.1702E+03 | 2.9609E+05 | 8.6822E+02 | 7.8346E+00 | 1.0294E+00 |

| | Transport velocity parameters | m/h | m/year |
|---|---|---|---|
| 1 | air side air-water MTC | 5 | 43800 |
| 2 | water side air-water MTC | .05 | 438 |
| 3 | rain rate | .0001 | .876 |
| 4 | aerosol deposition velocity | 6E-10 | 5.256E-06 |
| 5 | soil air phase diffusion MTC | .02 | 175.2 |
| 6 | soil water phase diffusion MTC | .00001 | .0876 |
| 7 | soil air boundary layer MTC | 5 | 43800 |
| 8 | sediment-water diffusion MTC | .0001 | .876 |
| 9 | sediment deposition velocity | .0000005 | .00438 |
| 10 | sediment resuspension velocity | .0000002 | .001752 |
| 11 | soil water runoff rate | .00005 | .438 |
| 12 | soil solids runoff rate | 1E-08 | .0000876 |

Individual process D values

| | | | |
|---|---|---|---|
| Air-water diffusion (air-side) | 2.017089E+07 | Air-water diffusion (water-side) | 422829.5 |
| Air-water diffusion (overall) | 414148 | | |
| Rain dissolution to water | 845.6589 | Aerosol deposition to water | 1.466974 |
| Rain dissolution to soil | 7610.93 | Aerosol deposition to soil | 13.20277 |
| Soil-air diffusion (air-phase) | 726152.2 | Soil-air diffusion (water-phase) | 761.093 |
| Soil-air diffusion (bndry layer) | 1.815381E+08 | Soil-air diffusion (overall) | 724014.2 |
| Water-sediment diffusion | 845.6589 | | |
| Water-sediment deposition | 176.2261 | Sediment-water resuspension | 22.55695 |
| Soil-water runoff (water) | 3805.465 | Soil-water runoff (solids) | 5.075312 |

FOUR COMPARTMENT FUGACITY LEVEL III CALCULATION        Trichloroethylene

Bulk phase properties,compositions and rates

| Phase | Air | Water | Soil | Sediment | Total | Fish |
|---|---|---|---|---|---|---|
| Bulk vol m3 | 1E+14 | 2E+11 | 1.8E+10 | 5E+08 | | |
| Density kg/m3 | 1.185413 | 1000.009 | 1500.237 | 1280 | | |
| Bulk Z value | 4.034179E-04 | 8.458495E-04 | 3.153999E-03 | 2.932222E-03 | | 1.432733E-02 |
| Bulk VZ | 4.034179E+10 | 1.691699E+08 | 5.677199E+07 | 1466111 | 4.05692E+10 | |
| Emission mol/h | 0 | 0 | 7610.929 | 0 | 7610.929 | |
| Emission kg/h | 0 | 0 | 1000 | 0 | 1000 | |
| Fugacity Pa | 1.29518E-05 | 5.528258E-05 | 1.014745E-02 | 5.21982E-05 | | 5.528258E-05 |
| Conc mol/m3 | 5.224989E-09 | 4.676074E-08 | 3.200504E-05 | 1.530567E-07 | | 7.920516E-07 |
| Conc g/m3 | 6.865112E-07 | 6.143893E-06 | 4.205143E-03 | 2.011012E-05 | | 1.040677E-04 |
| Conc ug/g | 5.791324E-04 | 6.143841E-06 | 2.802985E-03 | 1.571103E-05 | | 1.040677E-04 |
| Amount mol | 522498.9 | 9352.146 | 576090.8 | 76.52834 | 1108018 | |
| Amount kg | 68651.12 | 1228.779 | 75692.57 | 10.05506 | 145582.5 | |
| Amount % | 47.15616 | .8440426 | 51.9929 | 6.906776E-03 | | |
| Adv.flow m3/h | 1E+12 | 2E+08 | 0 | 10000 | | |
| D rct mol/Pa.h | 1.644521E+08 | 213154.1 | 23142.94 | 184.73 | | |
| D adv mol/Pa.h | 4.034179E+08 | 169169.9 | 0 | 29.32222 | | |
| Rct rate mol/h | 2129.952 | 11.78371 | 234.8418 | 9.64257E-03 | 2376.587 | |
| Rct rate kg/h | 279.8543 | 1.548261 | 30.85586 | 1.266937E-03 | 312.2597 | |
| Adv rate mol/h | 5224.989 | 9.352147 | 0 | 1.530567E-03 | 5234.343 | |
| Adv rate kg/h | 686.5113 | 1.228779 | 0 | 2.011012E-04 | 687.7403 | |
| Reaction % | 27.98543 | .1548261 | 3.085586 | 1.266937E-04 | | |
| Advection % | 68.65113 | .1228779 | 0 | 2.011012E-05 | | |

Overall residence time h 145.5825

Reaction residence time h 466.2226    Advection residence time h 211.6824

| Intermedia Data. | Half times h | Equiv flows m3/h | D values mol/Pa.h | Rates of transport mol/h | kg/h |
|---|---|---|---|---|---|
| Air to water | 6.7367E+04 | 1.0287E+09 | 4.1500E+05 | 5.3749E+00 | 7.0621E-01 |
| Air to soil | 3.8211E+04 | 1.8136E+09 | 7.3164E+05 | 9.4760E+00 | 1.2451E+00 |
| Water to air | 2.8307E+02 | 4.8962E+08 | 4.1415E+05 | 2.2895E+01 | 3.0082E+00 |
| Water to sediment | 1.1472E+05 | 1.2081E+06 | 1.0219E+03 | 5.6492E-02 | 7.4225E-03 |
| Soil to air | 5.4340E+01 | 2.2955E+08 | 7.2401E+05 | 7.3469E+03 | 9.6531E+02 |
| Soil to water | 1.0325E+04 | 1.2082E+06 | 3.8105E+03 | 3.8667E+01 | 5.0805E+00 |
| Sediment to water | 1.1702E+03 | 2.9609E+05 | 8.6822E+02 | 4.5319E-02 | 5.9545E-03 |

| | Transport velocity parameters | m/h | m/year |
|---|---|---|---|
| 1 | air side air-water MTC | 5 | 43800 |
| 2 | water side air-water MTC | .05 | 438 |
| 3 | rain rate | .0001 | .876 |
| 4 | aerosol deposition velocity | 6E-10 | 5.256E-06 |
| 5 | soil air phase diffusion MTC | .02 | 175.2 |
| 6 | soil water phase diffusion MTC | .00001 | .0876 |
| 7 | soil air boundary layer MTC | 5 | 43800 |
| 8 | sediment-water diffusion MTC | .0001 | .876 |
| 9 | sediment deposition velocity | .0000005 | .00438 |
| 10 | sediment resuspension velocity | .0000002 | .001752 |
| 11 | soil water runoff rate | .00005 | .438 |
| 12 | soil solids runoff rate | 1E-08 | .0000876 |

Individual process D values

| | | | |
|---|---|---|---|
| Air-water diffusion (air-side) | 2.017089E+07 | Air-water diffusion (water-side) | 422829.5 |
| Air-water diffusion (overall) | 414148 | | |
| Rain dissolution to water | 845.6589 | Aerosol deposition to water | 1.466974 |
| Rain dissolution to soil | 7610.93 | Aerosol deposition to soil | 13.20277 |
| Soil-air diffusion (air-phase) | 726152.2 | Soil-air diffusion (water-phase) | 761.093 |
| Soil-air diffusion (bndry layer) | 1.815381E+08 | Soil-air diffusion (overall) | 724014.2 |
| Water-sediment diffusion | 845.6589 | | |
| Water-sediment deposition | 176.2261 | Sediment-water resuspension | 22.55695 |
| Soil-water runoff (water) | 3805.465 | Soil-water runoff (solids) | 5.075312 |

FOUR COMPARTMENT FUGACITY LEVEL III CALCULATION Trichloroethylene

Bulk phase properties,compositions and rates

| Phase | Air | Water | Soil | Sediment | Total | Fish |
|---|---|---|---|---|---|---|
| Bulk vol m3 | 1E+14 | 2E+11 | 1.8E+10 | 5E+08 | | |
| Density kg/m3 | 1.185413 | 1000.009 | 1500.237 | 1280 | | |
| Bulk Z value | 4.034179E-04 | 8.458495E-04 | 3.153999E-03 | 2.932222E-03 | | 1.432733E-02 |
| Bulk VZ | 4.034179E+10 | 1.691699E+08 | 5.677199E+07 | 1466111 | 4.05692E+10 | |
| Emission mol/h | 4566.558 | 2283.279 | 761.093 | 0 | 7610.929 | |
| Emission kg/h | 600 | 300 | 100 | 0 | 1000 | |
| Fugacity Pa | 1.142293E-05 | 2.876865E-03 | 1.024612E-03 | 2.716356E-03 | | 2.876865E-03 |
| Conc mol/m3 | 4.608216E-09 | 2.433395E-06 | 3.231625E-06 | 7.964958E-06 | | 4.121779E-05 |
| Conc g/m3 | 6.054734E-07 | 3.197238E-04 | 4.246033E-04 | 1.046516E-03 | | 5.415606E-03 |
| Conc ug/g | 5.107699E-04 | 3.19721E-04 | 2.830241E-04 | 8.175904E-04 | | 5.415606E-03 |
| Amount mol | 460821.6 | 486679 | 58169.26 | 3982.479 | 1009652 | |
| Amount kg | 60547.35 | 63944.74 | 7642.859 | 523.258 | 132658.2 | |
| Amount % | 45.64162 | 48.20263 | 5.761317 | .3944407 | | |
| Adv.flow m3/h | 1E+12 | 2E+08 | 0 | 10000 | | |
| D rct mol/Pa.h | 1.644521E+08 | 213154.1 | 23142.94 | 184.73 | | |
| D adv mol/Pa.h | 4.034179E+08 | 169169.9 | 0 | 29.32222 | | |
| Rct rate mol/h | 1878.526 | 613.2155 | 23.71253 | .5017923 | 2515.956 | |
| Rct rate kg/h | 246.8195 | 80.57038 | 3.115589 | 6.593049E-02 | 330.5714 | |
| Adv rate mol/h | 4608.216 | 486.6789 | 0 | 7.964958E-02 | 5094.974 | |
| Adv rate kg/h | 605.4735 | 63.94474 | 0 | 1.046516E-02 | 669.4286 | |
| Reaction % | 24.68195 | 8.057038 | .3115589 | 6.593049E-03 | | |
| Advection % | 60.54735 | 6.394474 | 0 | 1.046516E-03 | | |

Overall residence time h 132.6582

Reaction residence time h 401.2997 Advection residence time h 198.1663

| Intermedia Data. | Half times | Equiv flows | D values | Rates of transport | |
|---|---|---|---|---|---|
| | h | m3/h | mol/Pa.h | mol/h | kg/h |
| Air to water | 6.7367E+04 | 1.0287E+09 | 4.1500E+05 | 4.7405E+00 | 6.2285E-01 |
| Air to soil | 3.8211E+04 | 1.8136E+09 | 7.3164E+05 | 8.3575E+00 | 1.0981E+00 |
| Water to air | 2.8307E+02 | 4.8962E+08 | 4.1415E+05 | 1.1914E+03 | 1.5654E+02 |
| Water to sediment | 1.1472E+05 | 1.2081E+06 | 1.0219E+03 | 2.9398E+00 | 3.8626E-01 |
| Soil to air | 5.4340E+01 | 2.2955E+08 | 7.2401E+05 | 7.4183E+02 | 9.7470E+01 |
| Soil to water | 1.0325E+04 | 1.2082E+06 | 3.8105E+03 | 3.9043E+00 | 5.1299E-01 |
| Sediment to water | 1.1702E+03 | 2.9609E+05 | 8.6822E+02 | 2.3584E+00 | 3.0987E-01 |

| Transport velocity parameters | | m/h | m/year |
|---|---|---|---|
| 1 | air side air-water MTC | 5 | 43800 |
| 2 | water side air-water MTC | .05 | 438 |
| 3 | rain rate | .0001 | .876 |
| 4 | aerosol deposition velocity | 6E-10 | 5.256E-06 |
| 5 | soil air phase diffusion MTC | .02 | 175.2 |
| 6 | soil water phase diffusion MTC | .00001 | .0876 |
| 7 | soil air boundary layer MTC | 5 | 43800 |
| 8 | sediment-water diffusion MTC | .0001 | .876 |
| 9 | sediment deposition velocity | .0000005 | .00438 |
| 10 | sediment resuspension velocity | .0000002 | .001752 |
| 11 | soil water runoff rate | .00005 | .438 |
| 12 | soil solids runoff rate | 1E-08 | .0000876 |

Individual process D values

| | | | |
|---|---|---|---|
| Air-water diffusion (air-side) | 2.017089E+07 | Air-water diffusion (water-side) | 422829.5 |
| Air-water diffusion (overall) | 414148 | | |
| Rain dissolution to water | 845.6589 | Aerosol deposition to water | 1.466974 |
| Rain dissolution to soil | 7610.93 | Aerosol deposition to soil | 13.20277 |
| Soil-air diffusion (air-phase) | 726152.2 | Soil-air diffusion (water-phase) | 761.093 |
| Soil-air diffusion (bndry layer) | 1.815381E+08 | Soil-air diffusion (overall) | 724014.2 |
| Water-sediment diffusion | 845.6589 | | |
| Water-sediment deposition | 176.2261 | Sediment-water resuspension | 22.55695 |
| Soil-water runoff (water) | 3805.465 | Soil-water runoff (solids) | 5.075312 |

Appendix 2. Listing of Fugacity calculations for Lotus 123 spreedsheet program.

Fugacity calculations: Trichloroethylene

LEVEL I, II and III

|  |  |  |  | | |
|---|---|---|---|---|---|
| * Amount of chemicals, moles | | 761092.929 | moles | 100000 | kg |
| * Emission rate of chemicals, E = | | 7610.92929 | mol/h | 1000 | kg/h |
| Gas constant, Pa m3/mol K, R= | | 8.314 | | | |
| System temperature, K, T= (t + 273.15) | | 298.15 | | | |
| * Molecular weight, g/mol | MW = | 131.39 | * input data | | |
| * Melting point, t C | M.P. | -73 | * input data | | |
| # If M.P. > 25 C enter Tm (mp+273.15) or else system temp. K | | 298.15 | * input | | |
| Fugacity ratio = exp(6.79(1-Tm/T)) for solid comp'ds | | 1.0000 | | | |
| * Solublitiy, g/m3 or mg/L | S = | 1100 | * input data | | |
| molar solubility, mol/m3, c=S/MW | | 8.37E+00 | | | |
| * Vapor pressure, Pa | P = | 9.90E+03 | * input data | | |
| Vap. pressure, subcooled liquid, Pa | | 9900.000 | | | |
| Henry's law constant, Pa m3/mol, H=p/c | | 1182.51 | | | |
| * Octanol/water partition coefficient, | log Kow | 2.53 | * input data | | |
| Kow = | | 339 | | | |
| Partition coefficient, organic C, Koc = 0.41*Kow*y | | 138.926104 | | | |
| for soil (mole fraction organic C), y(3) = | | 0.02 | | | |
| suspended sediment, y(5) = | | 0.2 | | | |
| bottom sediment, y(6) = | | 0.04 | | | |
| Kp(3) =0.41*Kow*y(3) | | 2.779 | | | |
| Kp(4) =0.41*Kow*y(4) | | 27.785 | | | |
| Kp(5) = 0.41*Kow*y(5) | | 5.557 | | | |
| Bioconcentraion factor, K(6) = 0.050*Kow | | 16.942 | | | |
| Air/water partition coeff., Z(air)/Z(water) | | 0.47704570 | | | |
| Soil/water partition coeff., Z(soil)/Z(water) | | 6.66845299 | | | |
| Sediment/water partition coeff., Z(sediment)/Z(water) | | 13.3369059 | | | |
| Sus. sediment/water partition coeff., Z(ss)/Z(water) | | 41.6778312 | | | |
| Aerosol/water partition coeff., Z(aerosol)/Z(air) | | 606.060606 | | | |
| Densities, g/cm3 or kg/L | | | | | |
| air, d(1) = (0.029*101325/RT) | | 0.00118541 | | | |
| water, d(2) | | 1 | | | |
| soil, d(3) = | | 2.4 | | | |
| bottom sediment, d(4) = | | 2.4 | | | |
| suspended sediment, d(5) = | | 1.5 | | | |
| biota, d(6) = | | 1 | | | |
| aerosol, d(7) | | 2 | | | |
| Fugacity capacities, Z: | | | | | |
| Z(1) or Z(air) = 1/RT | | 4.034E-04 | | | |
| Z(2) or Z(water) = 1/H = c/p | | 8.457E-04 | | | |
| Z(3) or Z(soil) = Kp(s)*Z(water)*d(s) | | 5.639E-03 | | | |
| Z(4) or Z(bottom sediment) = Kp(bs)*Z(water)*d(bs) | | 1.128E-02 | | | |
| Z(5) or Z(suspended sediment) = Kp(ss)*Z(water)*d(ss) | | 3.525E-02 | | | |
| Z(6) or Z(biota) = K *Z(water)*d(B) | | 1.433E-02 | | | |
| Z(7) or Z(aerosol) = 6*E6/p(L)RT | | 2.445E-01 | | | |
| Fugacity (Level I), f = total no. of moles/sum(ViZi) | | 1.876E-05 | | | |
| Fugacity (Level II), f = emission/sum(D values) | | 1.339E-05 | | | |

| | * input | | | | |
|---|---|---|---|---|---|
| | Adv. flow, G(air) | | 1.0000E+12 | mol/hr | |
| | Adv. flow G(water) | | 5.00E+08 | mol/hr | |
| | Adv. flow, G(sed.) | | 10000 | mol/hr | |
| | Residence time: h | | | | |
| | air | 100 | h | | |
| | water | 1000 | h | | |
| | sediment | 50000 | h | | |
| | Half-lives, hours | | | | |
| * | t(air) | | 170 | h | |
| * | t(water) | | 550 | h | |
| * | t(soil) | | 1700 | h | |
| * | t(sediment) | | 5500 | h | |
| | Emission, kg/h : | | | | |
| | | air | water | soil | |
| | E | 600 | 300 | | 100 |
| | E(A) | 1000 | 0 | | 0 |
| | E(B) | 0 | 1000 | | 0 |
| | E(C) | 0 | 0 | | 1000 |

Tansport parameters:

| | m/h<br>k | m/yr<br>k/8760 |
|---|---|---|
| air-water MTC, air side, U(1) | 5 | 43800 |
| air-water MTC, water side, U(2) | 0.05 | 438 |
| rain rate, U(3) = 0.85/8760 | 0.0001 | 0.876 |
| aerosol deposition velocity, U(4) | 6.00E-10 | 0.00000525 |
| soil air phase diffusion MTC, U(5) | 0.02 | 175.2 |
| soil water phase diffusion MTC, U(6) | 1.00E-05 | 0.0876 |
| soil air boundary layer MTC, U(7) | 5 | 43800 |
| sediment-water MTC, U(8) | 0.0001 | 0.876 |
| sediment deposition rate, U(9) | 5.00E-07 | 0.00438 |
| sediment resuspended rate, U(10) | 2.00E-07 | 0.001752 |
| soil water runoff rate, U(11) | 5.00E-05 | 0.438 |
| soil solids runoff rate, U(12) | 1.00E-08 | 0.0000876 |
| sediment burial rate, U(13) | 0 | 0 |
| diffusion to stratosphere, U(14) | 0 | 0 |

Define unit world:

| Compartment | Volume, Vi, m3 | Depth, h, m | Area, A, m2 | Density d, kg/m3 | Fugacity cap., Zi mol/Pa m3 | VZ | *input Advective flow, G mol/h | Residence time, V/G t(R),h | *input Reaction half-life t(1/2),h | *input Rate const. 0.693/t k, 1/hr | *input Emission rate, E mol/h | E(A) mol/h | E(B) mol/h | E(C) mol/h |
|---|---|---|---|---|---|---|---|---|---|---|---|---|---|---|
| Air (1) | 1.00E+14 | 1000 | 1.000E+11 | 1.18541324 | 4.034E-04 | 4.03E+10 | 1.000E+12 | 100 | 170 | 0.004076470 | 4.57E+03 | 7610.93 | 0 | 0 |
| Water (2) | 2.00E+11 | 20 | 1.000E+10 | 1000 | 8.457E-04 | 1.69E+08 | 2.000E+08 | 1000 | 550 | 0.00126 | 2.28E+03 | 0 | 7610.93 | 0 |
| Soil (3) | 9.00E+09 | 0.1 | 9.000E+10 | 2400 | 5.639E-03 | 5.08E+07 | 0 | | 1700 | 0.000407647 | 7.61E+02 | 0 | 0 | 7610.929294 |
| bottom sediment (4) | 1.00E+08 | 0.01 | 1.000E+10 | 2400 | 1.128E-02 | 1.13E+06 | 2000 | 50000 | 5500 | 0.000126 | 0.00 | 0 | 0 | 0 |
| Sus. sediment (5) | 1.00E+06 | | | 1500 | 3.525E-02 | 3.52E+04 | | | 1.0000E+11 | 6.9300E-12 | 7610.93 | | | |
| Biota (fish) (6) | 2.00E+05 | | | 1000 | 1.433E-02 | 2.87E+03 | | | 1.0000E+11 | 6.9300E-12 | | | | |
| Aerosol (7) | 2000 | | | 2000 | 2.445E-01 | 4.06E+10 | | | | | | | | |

Level I calculation:

| Compartment | Volume, Vi, m3 | Fugacity capacity Zi | VZ | Conc'n, c c = f*Z mol/m3 | Amount m= ciVi mol | Amount =m*MW/1E kg | Amount % | Conc'n,S mg/L (or g/m3) | Conc'n (S/d)*1000 ug/g |
|---|---|---|---|---|---|---|---|---|---|
| Air (1) | 1.00E+14 | 4.034E-04 | 4.034E+10 | 7.569E-09 | 7.57E+05 | 9.946E+04 | 9.95E+01 | 9.95E-07 | 8.39E-04 |
| Water (2) | 2.00E+11 | 8.457E-04 | 1.691E+08 | 1.587E-08 | 3.17E+03 | 4.170E+02 | 4.17E-01 | 2.08E-06 | 2.08E-06 |
| Soil (3) | 9.00E+09 | 5.639E-03 | 5.075E+07 | 1.058E-07 | 9.52E+02 | 1.251E+02 | 1.25E-01 | 1.39E-05 | 5.79E-06 |
| Bottom sediment (4) | 1.00E+08 | 1.128E-02 | 1.128E+06 | 2.116E-07 | 2.12E+01 | 2.780E+00 | 2.78E-03 | 2.78E-05 | 1.16E-05 |
| Sus. sediment (5) | 1.00E+06 | 3.525E-02 | 3.525E+04 | 6.613E-07 | 6.61E-01 | 8.689E-02 | 8.69E-05 | 8.69E-05 | 5.79E-05 |
| Biota (fish) (6) | 2.00E+05 | 1.433E-02 | 2.865E+03 | 2.688E-07 | 5.38E-02 | 7.064E-03 | 7.06E-06 | 3.53E-05 | 3.53E-05 |
| | | | 4.056E+10 | | 761092.929 | 100000 | 100 | | |

Level II phase properties and rates:

---

| Compartment | Rate const. k, 1/hr | D(reaction) VZk | D(advec'n) GZ | conc'n c =f*Z mol/m3 | Amount m= ciVi mol | Amount m*MW/1000 kg | conc'n mg/L (or g/m3) | Conc'n ug/g (S/d)*1000 | Loss Reaction mol/h Vck | Loss Advection mol/h Gc | % Loss reaction | % Loss advection | Removal % |
|---|---|---|---|---|---|---|---|---|---|---|---|---|---|
| Air (1) | 0.0040764706 | 1.645E+08 | 4.034E+08 | 5.403E-09 | 5.40E+05 | 7.10E+04 | 7.099E-07 | 5.99E-04 | 2.203E+03 | 5.403E+03 | 2.89E+01 | 70.99 | 99.93 |
| Water (2) | 0.00126 | 2.131E+05 | 1.691E+05 | 1.133E-08 | 2.27E+03 | 2.98E+02 | 1.488E-06 | 1.49E-06 | 2.854E+00 | 2.265E+00 | 3.75E-02 | 0.030 | 0.067 |
| Soil (3) | 0.0004076471 | 2.069E+04 | | 7.553E-08 | 6.80E+02 | 8.93E+01 | 9.923E-06 | 4.13E-06 | 2.771E-01 | | 3.64E-03 | | 3.64E-03 |
| Bottom sediment (4) | 0.000126 | 1.421E+02 | 2.256E+01 | 1.511E-07 | 1.51E+01 | 1.98E+00 | 1.985E-05 | 8.27E-06 | 1.903E-03 | 3.021E-04 | 2.50E-05 | 3.97E-06 | 2.90E-05 |
| Sus. sediment (5) | 6.93000E-12 | 2.442E-07 | | 4.720E-07 | 4.72E-01 | 6.20E-02 | 6.202E-05 | 4.13E-05 | 3.271E-12 | | 4.30E-14 | | 4.30E-14 |
| Biota (fish) (6) | 6.93000E-12 | 1.986E-08 | | 1.919E-07 | 3.84E-02 | 5.04E-03 | 2.521E-05 | 2.52E-05 | 2.660E-13 | | 3.49E-15 | | 3.49E-15 |
| | Total | 1.647E+08 | 4.036E+08 | | 543261.539 | 71379.1336 | | | 2205.65 | 5405.28 | 28.98 | 71.02 | 100 |
| | R + A | | 5.683E+08 | | | | | | | 7610.929596 | | | |

| | | | | |
|---|---|---|---|---|
| Total amount of chemicals, | 543261.539 | moles | 71379 | kg |
| Total reaction D value | 164686058. | | | |
| Total advection D value | 403587033. | | | |
| Total D value | 568273092. | | | |
| Fugacity, E/sum D values | 1.339E-05 | | | |
| Output by reaction, mol/h | 2205.65422 | | 289.800908 | kg/h |
| Output by advection, mol/h | 5405.27536 | | -394.58510 | kg/h |
| Total output, mol/h | 7610.92959 | | 2269.19865 | kg/h |
| Overall resistence time, h | 71.3791336 | | | |
| Reaction resistence time, h | 246.304036 | | | |
| Advection resistence time, h | 100.505802 | | | |

| Compartment | Subcomp't | volume fraction vi | Fugacity capacity Zi | Bulk vol. VB, m3 | Bulk ZB(i) sum(viZi) | partition coeff. (Zi/Zw) | Bulk VZ=VB*Zi | Bulk den. sum(vidi) kg/m3 | Bulk Adv. flow GAB=VB/Gi |
|---|---|---|---|---|---|---|---|---|---|
| Air (1) | Air | 1 | 4.034E-04 | 1.000E+14 | 4.034E-04 | 4.77E-01 | 4.034E+10 | 1.18541 | 1.00E+12 |
| | Aerosol | 2.00000E-11 | 0.24 | | | | | | |
| Water (2) | Water | 1 | 8.457E-04 | 2.000E+11 | 8.458E-04 | 6.6685 | 1.692E+08 | 1000.0085 | 2.00E+08 |
| | Particulate | 0.000005 | 3.525E-02 | | | | | | |
| | Biota(fish) | 0.000001 | 1.433E-02 | | | | | | |
| Soil (3) | Air | 0.2 | 4.034E-04 | 1.800E+10 | 3.15E-03 | 13.3369 | 5.677E+07 | 1500.2371 | |
| | Water | 0.3 | 8.457E-04 | | | | | | |
| | Solids | 0.5 | 5.639E-03 | | | | | | |
| Bottom sediment (4) | Water | 0.8 | 8.457E-04 | 5.00E+08 | 2.932E-03 | 16.9422 | 1.466E+06 | 1280 | 10000 |
| | Solids | 0.2 | 1.128E-02 | | | | | | |
| | | | | | 0.00733548 | | 4.057E+10 | | |

Level III Intermedia Data:

| | D values Dij mol/Pa h | Eq. flows Dij/Z(i) GDij, m3/h | half-life .693Vi/G t(1/2), h | Rate of transport Dij*f(i) N, mol/h | N*MW/1000 Nk, kg/h |
|---|---|---|---|---|---|
| Air to water (D12) | 4.1500E+05 | 1.029E+09 | 6.737E+04 | 4.740E+00 | 6.23E-01 |
| Air to soil (D13) | 7.3164E+05 | 1.814E+09 | 3.821E+04 | 8.357E+00 | 1.10E+00 |
| Water to air (D21) | 4.1415E+05 | 4.896E+08 | 2.831E+02 | 1.191E+03 | 1.57E+02 |
| Water to sed. (D24) | 1.0219E+03 | 1.208E+06 | 1.147E+05 | 2.940E+00 | 3.86E-01 |
| Soil to air (D31) | 7.2401E+05 | 2.296E+08 | 2.717E+01 | 7.418E+02 | 9.75E+01 |
| Soil to water (D32) | 3.8105E+03 | 1.208E+06 | 5.162E+03 | 3.904E+00 | 5.13E-01 |
| Sed. to water (D42) | 8.6822E+02 | 2.961E+05 | 2.340E+02 | 2.358E+00 | 3.10E-01 |
| Water to soil (D23) | 0 | | | | |
| Sed. burial DL(4) | 0.0000E+00 | | | | |
| Stratosphere DL(1) | 0.0000E+00 | | | | |

Equations for Dij values:

D12 = A(2)*(1/(1/U(1)*Z(1)+1/(U(2)*Z(2)+U(4)*Z(7))

D13 = A(3)*(1/(1/(U(5)*Z(1)+U(6)*Z(2)))+U(3)*Z(2)+U(4)*Z(7))

D21 = A(2)*(1/(1/(U(1)*Z(1)+1/(U(2)*Z(2)))

D24 = A(2)*(U(8)*Z(2)+U(9)*Z(5))

D31 = A(3)*(1/(1/(U(5)*Z(1)+U(6)*Z(2)+1/(U(7)*Z(1)))+U(3)*Z(2)+U(4)*Z(7))

D32 = A(3)*(U(11)*Z(2)+U(12)*Z(3))

D42 = A(4)*(U(8)*Z(2)+U(10)*Z(4))

| | | | | E(A) | | E(B) | | E(C) | |
|---|---|---|---|---|---|---|---|---|---|
| | D values Dij mol/Pa h | Eq. flows Dij/Z(i) GDij, m3/h | half-life .693Vi/G t(1/2), h | Rate of transport Dij*f(i) N, mol/h | N*MW/100 Nk, kg/h | Rate of transport Dij*f(i) N, mol/h | N*MW/100 Nk, kg/h | Rate of transport Dij*f(i) N, mol/h | N*MW/1000 Nk, kg/h |
| Air to water (D12) | 4.1500E+05 | 1.029E+09 | 6.737E+04 | 5.560E+00 | 7.31E-01 | 2.890E+00 | 3.80E-01 | 5.375E+00 | 7.062E-01 |
| Air to soil (D13) | 7.3164E+05 | 1.814E+09 | 3.821E+04 | 9.802E+00 | 1.29E+00 | 5.096E+00 | 6.70E-01 | 9.476E+00 | 1.245E+00 |
| Water to air (D21) | 4.1415E+05 | 4.896E+08 | 2.831E+02 | 2.916E+00 | 3.83E-01 | 3.958E+03 | 5.20E+02 | 2.290E+01 | 3.008E+00 |
| Water to sed. (D24) | 1.0219E+03 | 1.208E+06 | 1.147E+05 | 7.195E-03 | 9.45E-04 | 9.766E+00 | 1.28E+00 | 5.649E-02 | 7.423E-03 |
| Soil to air (D31) | 7.2401E+05 | 2.296E+08 | 2.717E+01 | 9.450E+00 | 1.24E+00 | 4.913E+00 | 6.45E-01 | 7.347E+03 | 9.653E+02 |
| Soil to water (D32) | 3.8105E+03 | 1.208E+06 | 5.16E+03 | 4.974E-02 | 6.53E-03 | 2.586E-02 | 3.40E-03 | 3.867E+01 | 5.080E+00 |
| Sed. to water (D42) | 8.6822E+02 | 2.961E+05 | 2.340E+02 | 5.772E-03 | 7.58E-04 | 7.835E+00 | 1.03E+00 | 4.532E-02 | 5.955E-03 |
| Water to soil (D23) | 0 | | | | | | | | |
| Sed. burial DL(4) | 0.0000E+00 | | | | | | | | |
| Stratosphere DL(1) | 0.0000E+00 | | | | | | | | |

Phase properties and rates:

| Compartment | Rate const. k, 1/hr | D Values Reaction VB*ZB*k mol/Pa h | Advection GAB*ZB mol/Pa h | DTs | Js | Js, E(A) | Js, E(B) | Js, E(C) |
|---|---|---|---|---|---|---|---|---|
| Air (1) | 0.0040764706 | 1.645E+08 | 4.034E+08 | 5.690E+08 | 9.315E-06 | 1.338E-05 | 0.000E+00 | 1.290E-05 |
| Water (2) | 0.00126 | 2.132E+05 | 1.692E+05 | 7.975E+05 | 7.278E-04 | 7.278E-04 | 7.278E-04 | 7.278E-04 |
| Soil (3) | 0.0004076471 | 2.314E+04 | 0 | 7.510E+05 | 9.988E-01 | 9.988E-01 | 9.988E-01 | 9.988E-01 |
| Bottom sediment (4) | 0.000126 | 1.847E+02 | 2.932E+01 | 1.082E+03 | 4.187E+05 | 4.187E+05 | 4.187E+05 | 4.187E+05 |
| Total | Total | 1.647E+08 | 4.036E+08 | | | | | |
| | R + A | | 5.683E+08 | | | | | |

Level III calculation for four emission scenarios:

For E(1,2,3), i.e., E(air)=600, E(water)=300 & E(soil)=100 kg/h

| Compartment | Fugacity | Concentration | | | amount | amount | Loss Reaction | Loss Advection | Reaction | Advection | Removal |
|---|---|---|---|---|---|---|---|---|---|---|---|
| | f's | C | S | ug/g | m, mol | % | mol/h | mol/h | % | % | % |
| | Pa | mol/m3 | mg/L | | C*VBi | | (Ci*DRi) | (Ci*DAi) | | | |
| Air (1) | 1.142E-05 | 4.608E-09 | 6.055E-07 | 5.108E-04 | 4.61E+05 | 45.64161 | 1.88E+03 | 4608.2154 | 24.6819 | 60.5473 | 85.23 |
| Water (2) | 2.877E-03 | 2.433E-06 | 3.197E-04 | 3.197E-04 | 4.87E+05 | 48.20263 | 6.13E+02 | 486.6789 | 8.0570 | 6.3945 | 14.45 |
| Soil (3) | 1.025E-03 | 3.232E-06 | 4.246E-04 | 2.830E-04 | 5.82E+04 | 5.76132 | 2.37E+01 | 0.0000 | 0.3116 | 0.0000 | 0.31 |
| Bottom sediment (4) | 2.716E-03 | 7.965E-06 | 1.047E-03 | 8.176E-04 | 3.98E+03 | 0.39444 | 5.02E-01 | 0.0796 | 0.0066 | 0.0010 | 7.64E-03 |
| Total | | | | | 1.01E+06 | 100 | 2.52E+03 | 5094.9740 | | | 100.00 |
| | | | | | | | | 7.61E+03 | | | |

For E(A), i.e., E(1) = 1000 kg/L conditions:

| Compartment | Fugacity | Concentration | | | amount | amount | Loss Reaction | Loss Advection | Reaction | Advection | Removal |
|---|---|---|---|---|---|---|---|---|---|---|---|
| | f, Pa | mol/m3 | mg/L | ug/g | m, mol | % | mol/h | mol/h | % | % | % |
| Air (1) | 1.340E-05 | 5.405E-09 | 7.101E-07 | 5.991E-04 | 5.40E+05 | 99.64198 | 2203.22 | 5404.7168 | 28.9481 | 71.0126 | 99.961 |
| Water (2) | 7.041E-06 | 5.956E-09 | 7.825E-07 | 7.825E-07 | 1.19E+03 | 0.21960 | 1.5009 | 1.1912 | 0.0197 | 0.0157 | 0.035 |
| Soil (3) | 1.305E-05 | 4.117E-08 | 5.409E-06 | 3.605E-06 | 7.41E+02 | 0.13661 | 0.3021 | 0.0000 | 0.0040 | 0.0000 | 3.97E-03 |
| Bottom sediment (4) | 6.648E-06 | 1.949E-08 | 2.561E-06 | 2.001E-06 | 9.75E+00 | 1.797E-03 | 0.0012 | 1.949E-04 | 0.0000 | 2.56E-06 | 1.87E-05 |
| Total | | | | | 5.42E+05 | | 2.21E+03 | 5405.9082 | | | 100.00 |
| | | | | | | | | 7611 | | | |

For E(B), i.e., E(2) = 1000 kg/L conditions:

| Compartment | Fugacity | Concentration | | | amount | amount | Reaction | Advection | Reaction | Advection | Removal |
|---|---|---|---|---|---|---|---|---|---|---|---|
| | f, Pa | mol/m3 | mg/L | ug/g | m, mol | % | mol/h | mol/h | % | % | % |
| Air (1) | 6.965E-06 | 2.810E-09 | 3.692E-07 | 3.114E-04 | 280962 | 14.69974 | 1145.3347 | 2809.6233 | 15.0485 | 36.9156 | 51.96 |
| Water (2) | 9.557E-03 | 8.084E-06 | 1.062E-03 | 1.062E-03 | 1.62E+06 | 84.58792 | 2037.12 | 1616.7645 | 26.7657 | 21.2427 | 48.01 |
| Soil (3) | 6.785E-06 | 2.140E-08 | 2.812E-06 | 1.874E-06 | 385.21 | 0.02015 | 0.1570 | 0.0000 | 0.0021 | 0.0000 | 2.06E-03 |
| Bottom sediment (4) | 9.024E-03 | 2.646E-05 | 3.477E-03 | 2.716E-03 | 13230 | 0.69218 | 1.6670 | 0.2646 | 0.0219 | 3.48E-03 | 2.54E-02 |
| Total | | | | | 1911341.99 | | 3184.28 | 4426.6524 | | | 100 |
| | | | | | | | | 7611 | | | |

For E(C) i.e., E(3) = 1000 kg/L conditions:

| Compartment | Fugacity | Concentration | | | amount | amount | Reaction | Advection | Reaction | Advection | Removal |
|---|---|---|---|---|---|---|---|---|---|---|---|
| | f, Pa | mol/m3 | mg/L | ug/g | m, mol | % | mol/h | mol/h | % | % | % |
| Air (1) | 1.295E-05 | 5.225E-09 | 6.865E-07 | 5.791E-04 | 5.22E+05 | 47.15616 | 2129.95 | 5224.9891 | 27.9854 | 68.6511 | 96.64 |
| Water (2) | 5.528E-05 | 4.676E-08 | 6.144E-06 | 6.144E-06 | 9.35E+03 | 0.84404 | 11.7837 | 9.3521 | 0.1548 | 0.1229 | 0.28 |
| Soil (3) | 1.015E-02 | 3.201E-05 | 4.21E-03 | 2.803E-03 | 5.76E+05 | 51.99289 | 234.84 | 0.0000 | 3.0856 | 0.0000 | 3.09 |
| Bottom sediment (4) | 5.220E-05 | 1.531E-07 | 2.011E-05 | 1.571E-05 | 7.65E+01 | 0.00691 | 0.0096 | 1.531E-03 | 1.27E-04 | 2.01E-05 | 1.47E-04 |
| Total | | | | | 1108018.34 | | 2376.59 | 5234.3428 | | | 100 |
| | | | | | | | | 7611 | | | |

Level III summary:

| | | E(A) | E(B) | E(C) |
|---|---|---|---|---|
| Total emission rate mol/h | 7610.92929 | 7610.92929 | 7610.92929 | 7610.92929 |
| Total VZ products | 4.057E+10 | 4.057E+10 | 4.057E+10 | 4.057E+10 |
| Total amount of chemicals | 1009652.21 | 542413.609 | 1911341.99 | 1108018.34 |
| Total advection D value | 1.647E+08 | 164688604. | 164688604. | 164688604. |
| Total reaction D value | 403587105. | 403587105. | 403587105. | 403587105. |
| Total D value | 5.683E+08 | 568275710. | 568275710. | 568275710. |
| Output by reaction, mol/h | 2515.95525 | 2205.02109 | 3184.28196 | 2376.58649 |
| Output by advection, mol/h | 5094.97403 | 5405.90820 | 4426.65240 | 5234.34279 |
| Overall residence time, h | 132.658204 | 71.2677241 | 251.131057 | 145.582529 |
| Reaction residence time, h | 401.299750 | 245.990213 | 600.242697 | 466.222602 |
| Advection residence time, h | 198.166312 | 100.337184 | 431.780457 | 211.682418 |

Output for Fugacity Levels I, II, and III calculateions:

Trichloroethylene

Level I calculation:

| Compartment | Z | Concentration | | | Amount | Amount |
|---|---|---|---|---|---|---|
| | mol/m3 Pa | mol/m3 | mg/L (or g/m3) | ug/g | kg | % |
| Air | 4.034E-04 | 7.569E-09 | 9.946E-07 | 8.390E-04 | 99455.04 | 99.455 |
| Water | 8.457E-04 | 1.587E-08 | 2.085E-06 | 2.085E-06 | 416.96 | 0.4170 |
| Soil | 5.639E-03 | 1.058E-07 | 1.390E-05 | 5.793E-06 | 125.12 | 0.1251 |
| Biota (fish) | 1.433E-02 | 2.688E-07 | 3.532E-05 | 3.532E-05 | 0.0071 | 7.06E-06 |
| Suspended sediment | 3.525E-02 | 6.613E-07 | 8.689E-05 | 5.793E-05 | 0.0869 | 8.69E-05 |
| Bottom sediment | 1.128E-02 | 2.116E-07 | 2.780E-05 | 1.159E-05 | 2.780 | 2.78E-03 |
| | Total | | | | 100000 | 100 |
| | | f = | 1.876E-05 | Pa | | |

Level II Calculation:

| Compartment | Half-life h | D Values D(reaction) | D(advec'n) | Conc'n mol/m3 | Loss Reaction kg/h | Loss Advection kg/h | Removal % |
|---|---|---|---|---|---|---|---|
| Air | 170 | 1.64E+08 | 4.03E+08 | 5.40E-09 | 289.389 | 709.901 | 99.929 |
| Water | 550 | 2.13E+05 | 1.69E+05 | 1.13E-08 | 0.375 | 0.298 | 0.0673 |
| Soil | 1700 | 2.07E+04 | | 7.55E-08 | 3.64E-02 | | 3.64E-03 |
| Biota (fish) | | | | 1.92E-07 | | | |
| Suspended sediment | | | | 4.72E-07 | | | |
| Bottom sediment | 5500 | 1.42E+02 | 2.26E+01 | 1.51E-07 | 2.50E-04 | 3.97E-05 | 2.90E-05 |
| | Total | 1.65E+08 | 4.04E+08 | | 289.80 | 710.20 | 100 |
| | R + A | | 5.68E+08 | | | 1000 | |

| | f = | 1.339E-05 Pa | Overall residence time = | 71.38 h |
|---|---|---|---|---|
| Total amount = | | 71379 kg | Reaction time = | 246.30 h |
| | | | Advection time = | 100.51 h |

Trichloroethylene

| | |
|---|---|
| MW | 131.39 |
| M.P. | -73 |
| S, mg/L | 1100 |
| P, Pa | 9900 |
| log Kow | 2.53 |
| Fug. ratio | 1.0000 |

Unit world dimensions:

| comp't | volume,m3 | depth, m |
|---|---|---|
| air | 1.00E+14 | 1000 |
| water | 2.00E+11 | 20 |
| soil | 9.00E+09 | 0.1 |
| sediment | 1.00E+08 | 0.01 |

Calc partition coeff.

| | |
|---|---|
| H | 1182.51 |
| Kow | 338.84 |
| Koc | 138.93 |
| Kaw | 0.4770 |
| soil, Ksw | 6.668 |
| sed. Ksw | 13.337 |
| s.sed. Ksw | 41.678 |

| Advective flow: mol/h | | Res. time, h |
|---|---|---|
| G(air) | 1.0000E+12 | 100 |
| G(water) | 5.00E+08 | 1000 |
| G(s.s.) | 10000 | 50000 |

| Emisision, k | air | water | soil |
|---|---|---|---|
| E, kg/h | 600 | 300 | 100 |
| E(A) | 1000 | 0 | 0 |
| E(B) | 0 | 1000 | 0 |
| E(C) | 0 | 0 | 1000 |

Level III Calculation: Trichloroethylene

Phase Properties and Rates:

| Compartment | Bulk Z mol/m3 Pa | Half-life h | D Values Reaction mol/Pa h | D Values Advection mol/Pa h |
|---|---|---|---|---|
| Air (1) | 4.034E-04 | 170 | 1.64E+08 | 4.03E+08 |
| Water (2) | 8.458E-04 | 550 | 2.13E+05 | 1.69E+05 |
| Soil (3) | 3.154E-03 | 1700 | 2.31E+04 | |
| Bottom sediment (4) | 2.932E-03 | 5500 | 1.85E+02 | 2.93E+01 |

Intermedia D values:

| | mol/Pa h |
|---|---|
| diff., DL | 0 |
| air/water D12 | 4.150E+05 |
| air/soil, D13 | 7.316E+05 |
| water/air, D21 | 4.141E+05 |
| water/sed., D24 | 1.022E+03 |
| soil/air, D31 | 7.240E+05 |
| soil/water, D32 | 3.811E+03 |
| sed./water, D42 | 8.682E+02 |
| sed., burial, DL | 0 |

| | E(1)=1000 | E(2)=1000 | E(3)=1000 | E(1,2,3) | |
|---|---|---|---|---|---|
| Overall residence time = | 71.27 | 251.13 | 145.58 | 132.66 | h |
| Reaction time = | 245.99 | 600.24 | 466.22 | 401.30 | h |
| Advection time = | 100.34 | 431.78 | 211.68 | 198.17 | h |

Phase Properties, Compositions, Transport and Transformation Rates:

| Emission, kg/h | | | Fugacity, Pa | | | | Concentration, g/m3 | | | | Amounts, kg | | | | Total |
|---|---|---|---|---|---|---|---|---|---|---|---|---|---|---|---|
| E(1) | E(2) | E(3) | f(1) | f(2) | f(3) | f(4) | C(1) | C(2) | C(3) | C(4) | m(1) | m(2) | m(3) | m(4) | amount, kg |
| 1000 | 0 | 0 | 1.340E-05 | 7.041E-06 | 1.305E-05 | 6.648E-06 | 7.101E-07 | 7.825E-07 | 5.409E-06 | 2.561E-06 | 7.101E+04 | 1.565E+02 | 9.736E+01 | 1.281E+00 | 7.127E+04 |
| 0 | 1000 | 0 | 6.965E-06 | 9.557E-03 | 6.785E-06 | 9.024E-03 | 3.692E-07 | 1.062E-03 | 2.812E-06 | 3.477E-03 | 3.692E+04 | 2.124E+05 | 5.061E+01 | 1.738E+03 | 2.511E+05 |
| 0 | 0 | 1000 | 1.295E-05 | 5.528E-05 | 1.015E-02 | 5.220E-05 | 6.865E-07 | 6.144E-06 | 4.205E-03 | 2.011E-05 | 6.865E+04 | 1.229E+03 | 7.569E+04 | 1.006E+01 | 1.456E+05 |
| 600 | 300 | 100 | 1.142E-05 | 2.877E-03 | 1.025E-03 | 2.716E-03 | 6.055E-07 | 3.197E-04 | 4.246E-04 | 1.047E-03 | 6.055E+04 | 6.394E+04 | 7.643E+03 | 5.233E+02 | 1.327E+05 |

| Emission, kg/h | | | Loss, Reaction, kg/h | | | | Loss, Advection, kg/h | | | Intermedia Rate of Transport, kg/h T12 | T21 | T13 | T31 | T32 | T24 | T42 |
|---|---|---|---|---|---|---|---|---|---|---|---|---|---|---|---|---|
| E(1) | E(2) | E(3) | R(1) | R(2) | R(3) | R(4) | A(1) | A(2) | A(4) | air-water | water-air | air-soil | soil-air | soil-water | water-sed | sed-water |
| 1000 | 0 | 0 | 2.895E+02 | 1.972E-01 | 3.97E-02 | 1.614E-04 | 7.101E+02 | 1.565E-01 | 2.561E-05 | 7.305E-01 | 3.831E-01 | 1.288E+00 | 1.242E+00 | 6.535E-03 | 9.454E-04 | 7.584E-04 |
| 0 | 1000 | 0 | 1.505E+02 | 2.677E+02 | 2.06E-02 | 2.190E-01 | 3.692E+02 | 2.124E+02 | 3.477E-02 | 3.798E-01 | 5.200E+02 | 6.695E-01 | 6.455E-01 | 3.397E-03 | 1.283E+00 | 1.029E+00 |
| 0 | 0 | 1000 | 2.799E+02 | 1.548E+00 | 3.09E+01 | 1.267E-03 | 6.865E+02 | 1.229E+00 | 2.011E-04 | 7.062E-01 | 3.008E+00 | 1.245E+00 | 9.653E+02 | 5.080E+00 | 7.423E-03 | 5.955E-03 |
| 600 | 300 | 100 | 2.468E+02 | 8.057E+01 | 3.12E+00 | 6.593E-02 | 6.055E+02 | 6.394E+01 | 1.047E-02 | 6.228E-01 | 1.565E+02 | 1.098E+00 | 9.747E+01 | 5.130E-01 | 3.863E-01 | 3.099E-01 |